Applied Hydrology

This comprehensive textbook combines the theoretical principles of engineering hydrology together with their practical applications, using modern industry-standard software. The textbook is written by the combination of a practitioner of water resources engineering with over 30 years of professional experience and a highly respected academic and recognized world authority in hydrology. Examples are drawn from global case studies, with exercises available online. The book begins with a review of the necessary mathematics and statistical hydrology. The underlying principles of the geographic information systems are discussed. In addition to topics covering fundamental concepts, separate chapters are devoted to reservoir operations, water resources management, climate change, and various methods of optimizing hydrologic models for calibration and validation. This textbook will prove to be indispensable for advanced students in civil, environmental, and agricultural engineering, preparing them to confidently join the industrial sector. It will also be an indispensable reference textbook for practicing engineers, bringing them up to date with modern techniques in applied hydrology.

DR. BISWAJIT MUKHOPADHYAY is a licensed professional engineer and certified floodplain manager with expertise in hydrology, hydraulics, and water resources engineering and management. He has over three decades of experience in working on solutions to real-world problems in hydrology all over the world, from Alaska to Auckland, and China to Chile. He is one of the first 100 recipients of the Diplomate, Water Resources Engineer (D.WRE) Board Certification conferred by the American Society of Civil Engineers. He is a Fellow of the American Society of Civil Engineers. He has authored over 50 technical articles in international journals, over 100 engineering reports, and presented key findings and critical observations at local, regional, national, and international conferences and seminars. He taught surface water hydrology, groundwater hydrology, and open-channel hydraulics classes at the University of Texas at Arlington and Southern Methodist University. He regularly teaches professional development classes focusing on hydrology and hydraulics at his current place of work.

DR. VIJAY P. SINGH is a Distinguished Professor, a Regents Professor, and the inaugural holder of the Caroline and William N. Lehrer Distinguished Chair in Water Engineering at Texas A&M University. He is a member of the US National Academy of Engineering. He has published extensively in the areas of hydrology, irrigation engineering, hydraulics, groundwater, water quality, water resources, entropy theory, and copula theory with more than 1490 journal articles, 37 books, 85 edited reference books, 122 book chapters, and 330 conference papers. He has received more than 110 national and international awards, including the Norman Medal, the Chow Award, the Torrens Award, the Arid Lands Hydraulic Engineering Award, the Outstanding Projects and Leader (OPAL) Award, the Lifetime Achievement Award, and the Roy C. Tipton Award (all from the American Society of Civil Engineers); the Hancor Award from the American Society of Agricultural and Biological Engineers; the Merriam Improved Irrigation Award from the International Commission on Irrigation and Drainage; the Linsley Award, the Wetzel Award, and the Founders Award from the American Institute of Hydrology; the Crystal Drop Award and the Chow Memorial Award from the International Water Resources Association; the Outstanding Scientist Award from Sigma Xi; the Distinguished Scientist Award from the Chinese Academy of Science; the Distinguished Professor Award from the Mexican Academy of Science; and the Professor Gajendra Singh Gold Medal for Education from the Indian Society of Agricultural Engineers. He has served as President of the American Institute of Hydrology; President of American Academy of Water Resources Engineers; President of International Association for Water, Environment, Energy, and Society; and Chair of the Watershed Council of American Society of Civil Engineers. He has served as editor-in-chief for five journals and two book series and has served on editorial boards of more than 40 journals and three book series. He has given more than 155 keynote lectures and 300 invited seminars all over the world, and has organized more than 25 international conferences.

"The book *Applied Hydrology* by Biswajit Mukhopadhyay and Vijay Singh is a very comprehensive handbook on hydrologic methods. Dr. Mukhopadhyay and Dr. Singh have created a book which is one of the most thorough and in-depth coverages of a wide range of hydrologic methods and topics. Not only is this book a great reference, but it is also an excellent how-to for hydrologists across the US and many other countries."

Jerry L. Cotter, Chief of Water Resources,
US Army Corps of Engineers, Fort Worth District

"Altogether different from many other books which address hydrology: this book addresses all essential theories and methods of applied hydrology in detail, and at the same time seamlessly integrates the principles with practice using the current industry standard software. All examples given in the book are drawn from real-world cases based on the first author's rich experience in working on engineering hydrology for over three decades in various corners of the globe. In addition to the fundamental concepts, the book delves deep into many contemporaneous topics such as climate change, reservoir operations, and model optimization techniques. The book is suitable both for an advanced course in applied hydrology at a university and as a companion for practicing civil, environmental, and agricultural engineers."

Ataul Hannan, Director, Planning Division, Harris County Flood Control District

"*Applied Hydrology* is a well-written textbook that provides an in-depth treatment of the principles of applied hydrology, with excellent illustrations, examples, exercises, and numerous references. This book will serve students – both graduate and undergraduate – as well as practicing professionals, and is a valuable reference for the members of the research community."

Venkataraman Lakshmi, University of Virginia;
President-Elect, Hydrology Section, American Geophysical Union

"*Applied Hydrology* by Mukhopadhyay and Singh provides an advanced quantitative and mathematical treatment of estimation problems that hydrologists have to deal with. The choice of material is thoughtful and comprehensive, emphasizing both conceptual understanding and problem-solving skills. The book will help prepare both undergraduate and graduate students for the problems they will face in practice."

Murugesu Sivapalan, University of Illinois at Urbana–Champaign

"The book *Applied Hydrology* by Mukhopadhyay and Singh provides a clear in-depth presentation of the full range of topics encountered in the professional practice of hydrologic engineering, supported by relevant articulations of foundational principles of mathematics, statistics, and hydrologic science."

Ralph A. Wurbs, Texas A&M University

Applied Hydrology

BISWAJIT MUKHOPADHYAY
DEC
Houston, Texas

VIJAY P. SINGH
Texas A&M University
College Station, Texas

Shaftesbury Road, Cambridge CB2 8EA, United Kingdom

One Liberty Plaza, 20th Floor, New York, NY 10006, USA

477 Williamstown Road, Port Melbourne, VIC 3207, Australia

314–321, 3rd Floor, Plot 3, Splendor Forum, Jasola District Centre, New Delhi – 110025, India

103 Penang Road, #05–06/07, Visioncrest Commercial, Singapore 238467

Cambridge University Press is part of Cambridge University Press & Assessment, a department of the University of Cambridge.

We share the University's mission to contribute to society through the pursuit of education, learning and research at the highest international levels of excellence.

www.cambridge.org
Information on this title: www.cambridge.org/9781009376105

DOI: 10.1017/9781009376082

© Biswajit Mukhopadhyay and Vijay P. Singh 2024

This publication is in copyright. Subject to statutory exception and to the provisions of relevant collective licensing agreements, no reproduction of any part may take place without the written permission of Cambridge University Press & Assessment.

First published 2024

Printed in the United Kingdom by CPI Group Ltd, Croydon CR0 4YY

A catalogue record for this publication is available from the British Library

Library of Congress Cataloging-in-Publication Data
Names: Mukhopadhyay, Biswajit, 1959- author. | Singh, V. P. (Vijay P.), author.
Title: Applied hydrology / Biswajit Mukhopadhyay, Vijay P. Singh.
Description: Cambridge ; New York, NY : Cambridge University Press, 2024. | Includes bibliographical references and index.
Identifiers: LCCN 2023048661 (print) | LCCN 2023048662 (ebook) | ISBN 9781009376105 (hardback) | ISBN 9781009376082 (epub)
Subjects: LCSH: Hydraulic engineering. | Hydrology.
Classification: LCC TC15 .M85 2024 (print) | LCC TC15 (ebook) | DDC 627–dc23/eng/20240415
LC record available at https://lccn.loc.gov/2023048661
LC ebook record available at https://lccn.loc.gov/2023048662

ISBN 978-1-009-37610-5 Hardback

Additional resources for this publication at www.cambridge.org/appliedhydrology.

Cambridge University Press & Assessment has no responsibility for the persistence or accuracy of URLs for external or third-party internet websites referred to in this publication and does not guarantee that any content on such websites is, or will remain, accurate or appropriate.

To the women of my life

Mitsu (Shyamosree Mukhopadhyay)
and
Chini (Anosua Senesac-Mukhopadhyay)

BM

To wife Anita (who is no more), son Vinay, daughter Arti, daughter-in-law Sonali, son-in-law Vamsi, and grandchildren Ronin, Kayden, Devin, and Alivia

VPS

Contents

Preface		*page* xiii
Acknowledgments		xvii
List of Common Abbreviations		xix
1	**Applied Hydrology in the Twenty-First Century**	1
1.1	Concepts and Purpose	1
1.2	Historical Context	1
1.3	Hydrologic Science and Hydrologic Engineering	2
1.4	Hydrologic Systems	3
1.5	Hydrologic Processes	4
1.6	Hydrologic Models	5
1.7	An Industry-Standard Hydrologic Simulation Model	5
1.8	Hydrologic Data Sources	6
1.9	About this Book	6
1.10	Examples	10
2	**Review of Mathematics**	11
2.1	Concepts and Purpose	11
2.2	Derivatives and Integrals	11
2.3	Higher-Order Derivatives	14
2.4	Partial Derivatives	15
2.5	Approximation of Functions by Series Expansion	16
2.6	Taylor Series	16
2.7	Integral Transforms	17
2.8	Laplace Transform	19
2.9	Convolution Integrals	19
2.10	Linear and Time-Invariant Systems	20
2.11	Impulse Functions	21
2.12	Step Functions	22
2.13	Linear First-Order Differential Equations	23
2.14	Growth or Decay Equations	25
2.15	Periodic Functions	26
2.16	Fourier Series	27
2.17	Fourier Transform	28
2.18	Matrix Algebra	29
2.19	Examples	32
3	**Statistical Hydrology**	36
3.1	Concepts and Purpose	36
3.2	Random Variables and Variates	36
3.3	Samples and Populations	36
3.4	Frequency Distribution	36
3.5	Moments of Distribution and Moment Measures	41
3.6	Quantiles	44
3.7	Relationship between the Mean, Median, and Mode	45
3.8	Probability	45

3.9	Hydrologic Return Periods and Exceedance Probability	48
3.10	Probability Distributions	50
3.11	Expected Values	52
3.12	Calculation of Probability Values: Plotting Positions	53
3.13	Quantile Functions	53
3.14	Bivariate Distributions	54
3.15	Discrete Probability Distributions	55
3.16	Probability Density Functions	56
3.17	Probability Plots	62
3.18	Parameter Estimation	62
3.19	Statistical Tests of Hypotheses	66
3.20	Interval Estimation	72
3.21	Covariance and Correlation	73
3.22	Time-Series Analysis	77
3.23	Analysis of Variance (ANOVA)	79
3.24	Examples	80
4	**Rainfall Measurements and Models**	**93**
4.1	Concepts and Purpose	93
4.2	Methods of Rainfall Measurements	93
4.3	Types of Rainfall	96
4.4	Descriptive Statistics of Rainfall	99
4.5	Spatial Distribution of Observed Rainfall	100
4.6	Temporal Distribution of Observed Rainfall	105
4.7	Design Storms	113
4.8	Probable Maximum Precipitation	137
4.9	Gridded Rainfall	147
4.10	Design of Rain Gauge Networks	148
4.11	Examples	151
5	**Streamflow Measurements and Statistics**	**171**
5.1	Concepts and Purpose	171
5.2	Measurement of Streamflow	171
5.3	Stage–Discharge Relationships	175
5.4	Streamflow Statistics	180
5.5	Flow-Duration Analysis	184
5.6	Flood-Frequency Analysis	193
5.7	Regional Flood-Frequency Analysis	203
5.8	Streamflows for Environmental Hydrology	206
5.9	Examples	206
6	**Geomorphologic Concepts and Watershed Characteristics**	**226**
6.1	Concepts and Purpose	226
6.2	Hierarchical Structure of a Drainage Basin	226
6.3	Morphometric Parameters of a Watershed	227
6.4	Hypsometry and Hypsometric Curves	236
6.5	Stream Order	238
6.6	Horton's Laws	238
6.7	Stream Slope and Stream Power	242
6.8	Longitudinal Profile of a Stream	243
6.9	Hydraulic Geometry of Streams	245
6.10	Drainage Density and Stream Frequency	248
6.11	Drainage Pattern	249
6.12	Time of Concentration and Lag Time	250
6.13	Methods for Estimation of Time of Concentration	253
6.14	Examples	258

7	**Abstractions and Effective Rainfall**	**271**
7.1	Concepts and Purpose	271
7.2	Physical Process of Production of Excess Rainfall	271
7.3	Interception	273
7.4	Infiltration	274
7.5	Models of Infiltration	274
7.6	Estimation of Effective Rainfall	286
7.7	Surface Storage	300
7.8	Initial Abstractions	302
7.9	Abstractions during Long-Term Rainfall–Runoff Simulation	302
7.10	Infiltration Measurements	308
7.11	Examples	308
8	**Groundwater and Baseflow**	**324**
8.1	Concepts and Purpose	324
8.2	Groundwater	325
8.3	Common Methods of Baseflow Separation	328
8.4	Baseflow Models	330
8.5	Estimating Baseflow Model Parameters	331
8.6	Examples	332
9	**Unit Hydrograph Models**	**349**
9.1	Concepts and Purpose	349
9.2	Response of a Watershed to Rainfall as a Linear Time-Invariant System	349
9.3	Response Functions and the Convolution	350
9.4	The Unit Hydrograph	353
9.5	Components of a Direct Runoff Hydrograph	354
9.6	Factors Governing Direct Runoff Hydrograph Shape	355
9.7	Derivation of Unit Hydrographs of a Watershed from measured Rainfall–Runoff Data: Nonparametric System Analysis	355
9.8	Synthetic Unit Hydrographs: Parametric System Synthesis	360
9.9	Instantaneous Unit Hydrographs	392
9.10	S-Hydrographs	394
9.11	Applications of Unit Hydrographs	395
9.12	Examples	397
10	**Kinematic Wave Model of Overland Flow**	**424**
10.1	Concepts and Purpose	424
10.2	Kinematic Wave Equation for Channel Flow	424
10.3	Kinematic Wave Equation for Overland Flow	425
10.4	Analytical Solutions of the Kinematic Wave Equation	427
10.5	Numerical Solutions of the Kinematic Wave Equation	432
10.6	Distinguishing Features of the Kinematic Wave Model	435
10.7	Implementation of Kinematic Wave Model in HEC-HMS	435
10.8	Examples	437
11	**The Rational Method**	**442**
11.1	Concepts and Purpose	442
11.2	The Rational Method Equation	442
11.3	The Runoff Coefficient	442
11.4	Drainage Area	445
11.5	Characteristic Time	445
11.6	Implications for the Rational Method	445
11.7	The Modified Rational Method	445
11.8	Implications for the Modified Rational Method	447
11.9	Applications of the Rational Method	447
11.10	Examples	450

12	Channel Routing	454
12.1	Concepts and Purpose	454
12.2	Governing Equations	454
12.3	Characteristics of Flood Wave Movement through a Channel and Flood Hydrographs	456
12.4	Prism Storage and Wedge Storage	458
12.5	Flood Wave Speed or Celerity	460
12.6	Channel Routing Methods	462
12.7	Modified Puls Method	462
12.8	Muskingum Method	468
12.9	Muskingum–Cunge Method	472
12.10	Hydraulic Methods of Channel Routing	480
12.11	Selection of the Routing Method	481
12.12	Comparing Storage from HEC-HMS and HEC-RAS	482
12.13	Conclusions	482
12.14	Examples	482

13	Reservoir Routing	516
13.1	Concepts and Purpose	516
13.2	The Modified Puls Method	516
13.3	Storage–Discharge Relationships	517
13.4	Flood Routing by Numerical Integration of the Continuity Equation	526
13.5	Accuracy of the Routing Method	530
13.6	Detention Basins	530
13.7	Existing and Proposed Reservoirs	532
13.8	Reservoir Routing in HEC-HMS	532
13.9	Examples	534

14	Evaporation and Evapotranspiration	549
14.1	Concepts and Purpose	549
14.2	The Processes of Evaporation and Evapotranspiration	549
14.3	Evaporation Thermodynamics	549
14.4	Controlling Factors of Evaporation	554
14.5	Estimation of Evaporation	554
14.6	Models for Estimation of Evapotranspiration	557
14.7	Measurement of Potential Evaporation	565
14.8	Reference Evapotranspiration	566
14.9	Potential Evapotranspiration	566
14.10	Actual Evapotranspiration	567
14.11	Examples	567

15	Snowmelt	574
15.1	Concepts and Purpose	574
15.2	Energy Flux	574
15.3	Physical Properties of Snow	575
15.4	Metamorphism of Snowpack	576
15.5	Rate of Snowmelt from Energy Flux	577
15.6	Energy Inputs in Snow Melting	578
15.7	Energy Exchange Mechanisms	578
15.8	Runoff Generation from Snowmelt	585
15.9	Snowmelt Modeling	585
15.10	Snow-Covered Areas	586
15.11	Examples	588

16	Erosion and Sedimentation	591
16.1	Concepts and Purpose	591

16.2	Erosion Processes	591
16.3	Types of Land Surface Erosion	591
16.4	Estimation of Erosion from the Land Surface	591
16.5	Sediment Yield	596
16.6	Determination of Sediment Yield	596
16.7	Temporal Distribution of S_y or E_T	597
16.8	Enrichment Ratio	599
16.9	Sediment Loads in Channels	599
16.10	Sediment Transport in the Stream	600
16.11	Sediment Routing	607
16.12	Reservoir Sedimentation	609
16.13	Modeling Erosion and Sedimentation in HEC-HMS	611
16.14	Examples	613

17 Reservoir Operations — 615

17.1	Concepts and Purpose	615
17.2	What is a Reservoir Operation?	615
17.3	Rule Curves	615
17.4	Methods of Mathematical Programming	617
17.5	Basic Principles of Optimization of Reservoir Operations	618
17.6	Linear Programming	619
17.7	Nonlinear Programming	619
17.8	Dynamic Programming	619
17.9	Simulation Models	620
17.10	Reservoir Operations Modeling with HEC-ResSim	620
17.11	Mass Curves	621
17.12	Reservoir Siltation	622
17.13	Examples	622

18 Climate Change — 629

18.1	Concepts and Purpose	629
18.2	Climate and Hydrologic Processes	629
18.3	Evapotranspiration	630
18.4	Precipitation	630
18.5	Snowpack	631
18.6	Glaciers	632
18.7	Streamflow	634
18.8	Urban Climate Change	638
18.9	Examples	638

19 Water Resources Management — 646

19.1	Concepts and Purpose	646
19.2	What is Water Resources Management?	646
19.3	Water Availability	647
19.4	Water Balance	647
19.5	Integrated Water Resource Management	651
19.6	Integrated River Basin Management	652
19.7	Hydrology in a Changing World	653
19.8	Examples	655

20 Geographic Information Systems — 659

20.1	Concepts and Purpose	659
20.2	Database Management Systems	659
20.3	Geodatabases	659
20.4	Data Structure of Geographic Features	661
20.5	Topologic Data Structure	662

20.6	Geographic Data Models	663
20.7	Types of Data Models	664
20.8	The Earth Datum	664
20.9	Map Projections	665
20.10	Map Scales	671
20.11	Geoprocessing and Geovisualization	673
20.12	Delineation of Drainage Areas and Streams	673
20.13	Derivation of Hydrologic Parameters using a GIS	676
20.14	Examples	676

21 Hydrologic Modeling — 683

21.1	Concepts and Purpose	683
21.2	Types of Hydrologic Models	683
21.3	Model Calibration through Optimization	683
21.4	Goodness-of-Fit Indices	684
21.5	Measures of Model Performance	685
21.6	Optimization Methods	686
21.7	Model Validation	691
21.8	Sensitivity Analysis	692
21.9	Optimization of Models in HEC-HMS	692
21.10	Ungauged Watersheds	693
21.11	Applications of Hydrologic Modeling	695
21.12	Hydrologic System Settings	702
21.13	Examples	703

References — 724
Index — 741

Color plates can be found between pages 396 and 397.

Preface

The security of water, food, energy, and environment is essential to sustainable development, as highlighted in the 17 Sustainable Development Goals (SDGs) adopted by the United Nations, for example, ensuring water and sanitation for all (Goal 6), food security and improved nutrition (Goal 2), affordable and reliable energy (Goal 7), and combating climate change and its impacts (Goal 13). Thus, water security, food security, and energy security are amongst the key challenges of the twenty-first century compounded by growing population, rising standard of living, and climate change. Fundamental to meeting these challenges is water security, that is, without water security, food security and energy security cannot possibly be achieved. By definition, water security implies access at all times to sufficient and quality water to satisfy varied needs and is built on three pillars: (1) demand for and use of water, (2) availability and supply of water, and (3) access to water. Essential to achieving water security and to sustainable development is water resources management encompassing water supply, agricultural irrigation, land reclamation, drainage infrastructure design, and flood control measures. *However, water resources cannot be managed without hydrology.* This may partly explain that a course in hydrology is taught in most institutions in the world that offer programs in civil, agricultural, and environmental engineering, watershed management, geography, and earth sciences. Further, there is growing interest in water resources even in socio-economic departments in many universities in the United States. The hydrology course in engineering disciplines is aimed at applications to planning, development, design, and management, and other applied aspects, that is, the course essentially is **applied hydrology**. On the other hand, the hydrology course in science disciplines is more process-oriented and has fringes of engineering applications.

There are several excellent books on hydrology some of which are standard textbooks. However, none of these books focuses on the **contemporaneous standards** of practices for practicing engineers whose work primarily revolves around stormwater management, flood risk management, supporting designs of hydraulic structures related to urban drainage systems, transportation, hydroelectricity, flood control and flood risk management, and water resources management for drinking water supplies. The term "contemporaneous standards" refers to the currently used or most widely adopted methods of computation of hydrologic variables and software applications. There is therefore a need for a standard up-to-date reference for the student preparing to enter the practice of water resources engineering and for the community of practicing hydrologic engineers for quick access to the fundamentals and exploring the areas where there are gaps in knowledge and understanding that can be filled by further research and investigation. This is what constituted the motivation for writing this book.

The subject matter of the book is divided into 21 chapters. Providing a condensed history of hydrology, Chapter 1 introduces hydrology by providing a brief discussion of hydrologic science and engineering, hydrologic system, hydrologic processes, hydrologic modeling, hydrologic models, and hydrologic data sources. The chapter is concluded with the organization and highlights of the book. Hydrologic analysis and synthesis require applications of certain mathematics, which is reviewed in Chapter 2, covering concepts of derivative and integral, differentiation and integration of standard functions, higher-order derivatives, partial derivatives, error propagation, series expansion of functions, integral transforms, convolution integrals, linear systems, linear first-order ordinary differential equations, periodic functions, Fourier series, Fourier transform, and matrix algebra. These aspects are employed in subsequent chapters.

Hydrology involves random variables and stochastic processes. Thus, Chapter 3 discusses statistical methods, covering random variables and variates, sample and population, frequency distributions, moments and moment measures, probability and stochastic processes, discrete and continuous probability distributions, return periods and quantiles, parameter estimation, hypothesis testing, confidence intervals, covariance, regression and correlation, and time-series analysis.

The key input to hydrologic modeling is rainfall which is treated in Chapter 4 dealing with rainfall measurements and models, methods of rainfall measurement, types of rainfall, rainfall statistics, spatial and temporal distributions of rainfall, design storms, frequency analysis of annual maxima and partial duration series, intensity–duration–frequency relationships, depth–area relationships, temporal distribution of design rainfall, probable maximum precipitation, gridded rainfall, and design of rain-gauge networks.

The key output of hydrologic models is streamflow whose measurements and statistics are discussed in Chapter 5, which deals with stage measurement, discharge measurement, stage–discharge relationships, streamflow statistics, flow-duration analysis, flood-frequency analysis, trends and correlation, regional frequency analysis, and environmental flow.

Fundamental to hydrologic modeling is the characterization of watershed wherein hydrologic processes occur. Chapter 6 deals with watershed geomorphology and characteristics, including hierarchical structure of a drainage basin, morphological parameters, hypsometry, stream order, Horton's laws, stream power, longitudinal stream profile, hydraulic geometry, drainage density, drainage pattern, lag time, and time of concentration.

After presenting the essential concepts of mathematics, statistics, rainfall, streamflow, and watershed characterization, which form the foundation of applied hydrology, we proceed to present various hydrologic processes and computational methods that are at the core of hydrologic engineering practice.

All rainfall does not become runoff, for part of it is lost to abstractions, covered in Chapter 7, which discusses effective rainfall, interception, infiltration, infiltration models, effective rainfall estimation, surface storage, initial abstractions, and soil-moisture accounting.

One of the key components of the hydrologic cycle is groundwater which gives rise to baseflow. Groundwater and baseflow are thus discussed in Chapter 8 covering aquifers and their properties, gaining and losing streams, governing equations for groundwater flow, baseflow separation, baseflow models, parameter estimation, and the linear reservoir model.

One of the main methods for surface runoff modeling is the unit hydrograph method. Unit hydrograph models are presented in Chapter 9 dealing with the representaion of a watershed as a linear time-invariant system, response function and convolution, unit hydrograph characteristics and derivation, synthetic unit hydrographs, instantaneous unit hydrographs and their models and parameter estimation, and the application of unit hydrographs and instantaneous unit hydrographs. One of the main physically based methods for overland flow and channel-flow modeling is the kinematic wave method. Chapter 10 discusses kinematic wave models, including kinemtic wave equations for channel flow and overland flow, analytical solutions, numerical solutions, distinguishing features of kinematic wave model, and implementation of kinematic wave model in the Hydrologic Engineering Center Hydrologic Modeling System (HEC-HMS).

The unit hydrograph and kinematic wave methods yield the runoff hydrograph. However, many design problems, such as urban drainage and channel sizing require only peak discharge which is often estimated by the rational method described in Chapter 11, presenting the rational method equation, the rational coefficient, drainage area, characteristic time, implications of the method, the modified rational method and its implications, and applications.

Overland flow is transported by channel flow. Chapter 12 describes channel routing, including governing equations, characteristics of flood wave movement, channel routing methods, a selection of routing methods, comparison of hydrologic and hyraulic methods of routing, and channel routing in HEC-HMS. In real world, there are dams built on rivers which impound water forming reservoirs. Thus, channel flow feeds these reservoirs and flow routing through the reservoirs needs to be modeled. Reservoir routing is discussed in

Chapter 13 dealing with storage–discharge relations, methods of routing, numerical methods of flood routing and accuracy, existing and proposed reservoirs, reservoir routing in HEC-HMS, and storage methods.

A significant part of water from reservoirs and lakes as well as from land evaporates. Likewise, a large quantity of water is evapotranspired from cropland and vegetated surfaces. Evapotranspiration is also a key compoent of the hydrologoc cyle. Chapter 14 deals with evaporation and evapotranspiration. Beginning with a discussion of evaporation thermodynamics, it goes on to discuss factors controlling evaporation, evporation estimation, models of evapotranspiration, selection of a method, estimation of net radiation, measurement of evaporation, potential evapotranspiration, reference and actual evapotranspiration, and estimation of actual evapotranspiration.

In cold regions, a significant part of precipitation occurs in the form of snow, which upon melting either by itself or in combination with rainfall can generate flooding. Hence, snowmelt is an important component of streamflow genration and water supply. Chapter 15 presents snowmelt, discussing energy flux, physical propertirs of snow, metamorphism of snowpack, rate of snowmelt, energy exchange mechanisms, turbulent convection, snowmwlt runoff generation, and snow-covered areas.

Water quality is an integral part of water resource management and it is known that a large pollutant load is carried by sediment that primarily originates in the upland areas. Hence, erosion and sedimentation are discussed in Chapter 16 covering erosion process, types of erosion, estimation of erosion, sediment yield and its determination, temporal distribution of sediment yield, sediment loads in channels, sediment transport, sediment properties, fall velocity, sediment transport functions, sediment routing, reservoir sedimentation, and erosion and sedimentation modeling in HEC-HMS.

For water supply for varied purposes, water is stored in lakes and reservoirs. The reservoirs must be operated considering demand and supply and the purpose of reservoirs. Chapter 17 deals with reservoir operation, including rule curves, methods of mathematical programming, optimization of reservoir operations, simulation models, reservoir operation modeling with the HEC-ResSim software, mass curves, and reservoir siltation.

The hydrologic cycle is being impacted by climate change, which is being manifested by global warming. Therefore, hydraulic structures must be designed, and water resources must be managed considering the impact of climate change, which is the subject of Chapter 18 covering climate and hydrological processes, evapotranspiration, precipitation, snowpack, glaciers, streamflow, and urban climate change.

Chapter 19 deals with water resources management, discussing water availability, water balance, integrated water resources management, integrated river basin management, and hydrology in a changing world. For water resources management and hydrologic modeling, geographic information system (GIS) technology has become essential. This subject is therefore discussed in Chapter 20 encompassing database management, geodatabases, data structure of geographic features, topologic data structure, geographic data models, types of data models, earth datum, map projections, map scales, geoprocessing and geo-visualization, delineation of drainage areas and streams, and derivation of hydrologic parameters using GIS technology.

Finally, Chapter 21 is on hydrologic modeling, with particular focus on model calibration and applications of hydrologic models in water resources engineering. Beginning with a short discussion of hydrologic models, the chapter goes on to discuss model calibration through optimization, goodness-of-fit indices, measures of model performance, optimization methods, model validation, and sensitivity analysis. A part of the chapter is devoted to the discussion of optimization models included in HEC-HMS. The chapter is concluded with several examples of applications of hydrologic models and various physical settings of hydrologic systems.

The subject matter of the book is intended for a broad audience, including students specializing in water resources; instructors offering a course in hydrology for engineers; practitioners of hydrology dealing with real-world problems on a daily basis needing a good grounding of the subject through self-teaching by having one comprehensive text that does not contain too many materials of purely academic interest and methods or concepts that are either obsolete or have not gained wide acceptance or popularity; and for researchers whose curiosity is driven by a desire to contribute towards solutions of problems of practical importance. Although the book focuses on engineering applications, it may be appealing to those working in agricultural, environmental, earth, ecological, and watershed sciences.

Biswajit Mukhopadhyay
Dallas/Houston, Texas

Vijay P. Singh
Texas A&M University
College Station, Texas

Acknowledgments

In addition to the close attention we paid to most of the well-known books on hydrology and published research papers, we have consulted numerous reports prepared by various organizations involved in hydrologic work, particularly the engineering manuals (EM series) of the US Army Corps of Engineers (USACE), several documents available from the Hydrologic Engineering Center (HEC) in Davis, California, and scientific reports of the US Geological Survey (USGS). We are thankful to the Texas A&M University (College Station, Texas) library system for accessing numerous journal articles. Similarly, we have used hydrological and meteorological data that are available from reliable sources via the internet. Examples are streamflow data from the USGS, the Global Runoff Data Centre in Germany, Environment Canada, the UK Centre for Ecology and Hydrology, digital elevation data from the USGS, and rainfall data or hydrologic models from the National Oceanic and Atmospheric Administration of the United States, and various local agencies and governments, such as Harris County Flood Control District in Texas, Clark County Regional Flood Control District in Nevada, Maricopa County Flood Control District in Arizona, City of Dallas, City of Fort Worth, and City of Austin in Texas. Any omission to cite any other agency at this place is unintentional and we offer our apologies in advance.

A concerted effort has been made in this book to use data and examples from the real world. For this reason, some of the models used in various places of this book are modified or scaled-down versions of project models developed by the first author at various organizations with whom he was associated at different stages. Other models presented here he developed himself during the preparation of this book, but still drawing on data and project examples from the real-world experience gained through his association with these outstanding private-sector engineering firms.

In addition to the data and models described above, several workers from different organizations have also kindly provided data and models for use in the book. We sincerely acknowledge their contributions. The individuals to who we owe our gratitude are Dr. Will Asquith of the USGS (Austin/Lubbock, Texas); Jerry Cotter, Landon Erickson, Chris Chiu, John Hunter, and Simeon Benson of the USACE (Fort Worth District office, Texas); Rohit Kumar of Indira Gandhi National Open University (New Delhi, India); Bill Kappel of Applied Weather Associates (Monument, Colorado); Kelly Godsey of the National Weather Service (Tallahassee, Florida); Jonathan Atwell of Southeast River Forecast Center of the National Weather Service (Peachtree City, Georgia); Robbie McKinney of Suwannee River Water Management District (Live Oak, Florida); Craig McDougall and Andrew Trelease of Clark County Regional Flood Control District (Las Vegas, Nevada); Steven Anderson of the City of Dallas; Dr. M. A. Ghorbani of the University of Tabriz (Tabriz, Iran); Dr. M. H. Kashani of the University of Mohaghegh Ardabili (Ardabil, Iran); Rangsarit Vanijjirattikhan of the National Electronics and Computer Technology Center (Pathum Thani, Thailand); Patsorn Rakcheep, Jittiwut Suwatthikul, and Chinoros Thongthamchart of the Electricity Generating Authority of Thailand (Nonthaburi, Thailand); Dr. C. Yoo and Dr. W. Na of Korea University (Seoul, South Korea); Dr. Bo Chen and Dr. Fei Ren of Beijing Normal University (Beijing, China); Dr. Alan Morton of DMap (Aberystwyth, UK), Mr. Jay Kumar Shah of Department of Irrigation, Government of Nepal; Dr. T. Cleveland of Texas Tech University (Lubbock, Texas); Dr. P. K. Bhunya of the Indian Institute of Technology at Bhubaneswar (Bhubaneswar, Odisha, India); Danial Hashmey of Water and Power Development Authority (Lahore, Pakistan); Mr. Ahsan Moeen of 2M Associates (Dallas, Texas); and Xiaohua (Nellie) Yang of DEC (Houston, Texas).

Some figures (Figures 20.4–20.8 and 20.11–20.21) and contents in Chapter 20 are adopted from a lecture presentation given by Dr. D. R. Maidment of the University of Texas at Austin in 2006 and are available in the public domain: "GIS in Water Resources" on SlidePlayer. We acknowledge the contribution of Dr. Maidment, as an educator, for free dissemination of excellent teaching materials outside the regular classroom. We are ever thankful to all listed above. Nevertheless, here again, the authors ask forgiveness from any individual who contributed but was omitted unintentionally.

We are highly indebted to our employers who enabled us to work on the book. We thank and express gratitude to John R. McAdams Company and DE Corp (DEC), the former and present employers of BM, and Texas A&M University where VPS belongs. BM is also thankful to many young engineers who worked under his supervision and other peers he worked with for enlightening him to recognize those areas of applied hydrology needing in-depth exposition in a modern context. All the maps used in this book were created in ArcGIS software by BM either at McAdams (the majority), at 2M Associates, or at DEC.

We acknowledge the stupendous contributions made by the HEC of the USACE for producing HEC-HMS and other software in the field of hydrology and hydraulics, and numerous invaluable technical documents, all available in the public domain. We have drawn heavily from their work.

We are thankful to Dr. Matt Lloyd, Dr. Maya Zakrzewska-Pim, Ms. Jenny van der Meijden, and other staff of the Cambridge University Press for undertaking the project of publishing this book. We extend our special thanks and gratitude to Dr. Zoë Lewin for excellent copy-editing. It has by no means been an easy task to bring a manuscript full of lengthy tables and hundreds of equations and figures into the form of an attractive book. Furthermore, Dr. Lewin's superb editorial work helped us with better clarifications of many concepts and rectify certain errors that were present in the original manuscript.

Nobody can author a book without being indebted to those who developed the concepts, principles, laws, theories, procedures, and analytical tools. We are grateful to all those who have contributed to hydrology and shaped the subject as we know it today. Its importance is growing day by day. The pillars of modern hydrology were erected by those pioneers.

We are grateful to our families who allowed us to work long hours, late in night, and during weekends and holidays, often away from them. Without their support, this book would not have been completed.

Common Abbreviations

ADCP	acoustic Doppler current profiler
AEP	annual exceedance probability
AET	actual evapotranspiration
AMS	annual maximum series
ARF	area reduction factor
ARI	average recurrence interval
CDF	cumulative probability distribution function for a *discrete* random variable
cdf	cumulative probability distribution function for a *continuous* random variable
CWC	Central Water Commission (India)
DAD	depth–area–duration
DDF	depth–duration–frequency
DEM	digital elevation model
DRH	direct runoff hydrograph
EMA	expected moments algorithm
ERH	effective rainfall hyetograph
EV I	extreme value type I
EV III	extreme value type III
FAO	Food and Agricultural Organization of the United Nations
FFA	flood-frequency analysis
FHWA	Federal Highway Administration (under the United States Department of Commerce)
FWS	flood warning system
GEV	generalized extreme value
GIS	geographic information system
HEC	Hydrologic Engineering Center
HEC-HMS	Hydrologic Engineering Center – Hydrologic Modeling System
HEC-RAS	Hydrologic Engineering Center – River Analysis System
HEC-SSP	Hydrologic Engineering Center – Statistical Software Package
HSG	hydrologic soil group
HUC	hydrologic unit code
IDF	intensity–duration–frequency
IDW	inverse-distance weighting
IRF	impulse response function
LP III	log Pearson type III
MAF	moisture adjustment factor
MAP	mean areal precipitation
MRM	modified rational method
MSL	mean sea level
MUSLE	modified universal soil loss equation
NASA	National Aeronautics and Space Administration
NIH	National Institute of Hydrology (Roorkee, India)
NOAA	National Oceanic and Atmospheric Administration of the United States
NRCS	Natural Resources Conservation Commission (under the United States Department of Agriculture)
NWS	National Weather Service (United States)
pdf	probability density function
PDF	probability distribution function
PDS	partial duration series
PET	potential evapotranspiration
PILF	potentially influential low flood
PMF	probable maximum flood
PMP	probable maximum precipitation
PMS	probable maximum storm
POR	period of record
PRF	pulse response function
SCS	Soil Conservation Service (now NRCS)
SRF	step response function
SRTD	synthetic rainfall time distribution
STP	storm total precipitation
TRMM	Tropical Rainfall Measuring Mission
USACE	United States Army Corps of Engineers
USDA	United States Department of Agriculture
USGS	United States Geological Survey
USLE	universal soil loss equation
WMO	World Meteorological Organization

1 Applied Hydrology in the Twenty-First Century

1.1 CONCEPTS AND PURPOSE

The style, contents, and presentation of any subject in a textbook is largely influenced by the philosophy of the author(s) about the subject, developed from experience either in pedagogy or practice and the depth of knowledge gained about the subject. This book is no exception. This chapter presents how the authors view and approach applied hydrology for engineering practice. The concept of a hydrologic system is presented from the view point of thermodynamics, not systems theory. The HEC-HMS software, used throughout the book, is introduced. An abridged account of the history of development of hydrology is presented. This chapter is an orientation of the book.

1.2 HISTORICAL CONTEXT

Hydrology in its broadest definition means science of water. However, today it is used in a much-restricted sense – it is a subject that deals with the amount of water coming to the land surface of the Earth from rainfall, snow, and/or glacial melts and the amount of water available in rivers and streams, lakes, lagoons, wetlands, and groundwater. The development of hydrology began almost concurrently with the flourishment of ancient civilizations in the valleys of the Nile, Tigris-Euphrates, Sindhu (Indus) and Ganga (Ganges), collectively called the Indo-Gangetic Plain, and Huang-Ho (Yellow) rivers. In addition to curiosity-driven developments as a science, hydrology had an early root of development as an engineering discipline because of mankind's invention of irrigation for farming, land development, aqueducts, water supply systems, sanitation systems, and flood control measures from the beginning of human civilizations. Biswas (1970) gave an excellent account of the long history of hydrology tracing back to several millennia. Peters-Lidard (2019) presented a perspective on the development of hydrology during the twentieth century. Giving a historical account of hydrology is not our intention here. However, to put the scope of the present book in the context of development of the subject hydrology, we follow Ven Te Chow, who described the development of hydrology into several distinct periods starting from the ancient civilization to the time when he wrote the account.

Chow (1964) classified the history of development of hydrology into eight periods with general time divisions as: (i) **period of speculation**, from ancient to 1400; (ii) **period of observation**, from 1400 to 1600; (iii) **period of measurements**, from 1600 to 1700; (iv) **period of experimentation**, from 1700 to 1800; (v) **period of modernization**, from 1800 to 1900; (vi) **period of empiricism**, from 1900 to 1930; (vii) **period of rationalization**, from 1930 to 1950; and (viii) **period of theorization**, from 1950 to the time (1964) Chow completed writing the account.

As given above, in the early twentieth century, hydrology was mostly empirical since the physical basis for most quantitative hydrologic determinations was not well known. Rational analysis only started in 1930. LeRoy Kempton Sherman, a civil engineer from Chicago, proposed the unit hydrograph method in 1932 and Robert Elmer Horton, often called the father of American hydrology, propounded the infiltration theory in 1933. French hydraulic engineer Henry Darcy developed the most fundamental law of groundwater flow in 1856. Charles Vernon Theis from the United States Geological Survey (USGS), combined the Darcy law with the continuity equation, which led to a diffusion type equation. By solving this equation, in 1935 Theis derived a relation between the rate of drawdown and the rate and duration of pumping. This revolutionary work laid the foundation of quantitative groundwater hydrology. C. E. Jacob (1943, 1944) correlated infiltration and groundwater which led to the development of techniques for baseflow separation (Barnes, 1940). L. G. Puls (1928) and G. T. McCarthy (1938), both from the United States Army Corps of Engineers (USACE), developed the method of hydrograph routing. The works of Charles Warren Thornthwaite (1948), an American geographer and climatologist and Howard Latimer Penman (1948), an English physicist, paved the path for the quantification of evaporation and evapotranspiration. Emil Julius Gumbel, a German mathematician, introduced the extreme value distribution for frequency analysis of hydrologic data in 1941 and Chow formulated the frequency factor for hydrologic statistics in 1951. These monumental contributions laid the foundation of modern hydrology. Theoretical approaches began about 1950, most notably with the mathematical representation of hydrologic systems as linear and time-invariant systems by stalwarts of hydrology such as James Clement Dooge, James Eamon Nash, both from Ireland, and Chow, who carried out his work at the University of Illinois. An empirical unit hydrograph method developed by Franklin F. Snyder (1938), from the USGS and another unit hydrograph method proposed by

C. O. Clark (1943), from the United States Engineering Office in Virginia, with a theoretical basis, changed the way hydrologic design calculations were made. Stupendous progress has been made since then in theoretical developments with the application of advanced mathematics.

One name that has not been mentioned very often by those who have narrated the history of hydrology in various writings, as a contributor of certain milestone concepts of applied hydrology, is Victor Mockus. This unintentional absence of his recognition is due to the fact that Mockus was an employee of the Soil Conservation Service (SCS) of the United States Department of Agriculture, and his contributions were published in SCS reports in the 1950s and the 1960s, and credits are given to those reports. But the fact is that Mockus was the originator of the curve number method used for the calculation of runoff volumes that contribute to streamflow, the lag method for computation of watershed time of concentration, and the dimensionless unit hydrograph method, all of which are roped together today as the NRCS method. It is not an exaggeration to say that this is the most popular and widely accepted method worldwide; in spite of the fact that it was developed for the watersheds of the United States, it has an inherent simplicity enabling its adaptability to other parts of the world. A noteworthy point about the dimensionless unit hydrograph that was developed by Mockus, essentially from empirical synthesis of rainfall–runoff records from a large number of watersheds, is that its shape mimics the shape of a unit hydrograph derived from a theoretical gamma distribution, which was proposed by theoreticians such as Nash and Dooge.

Following Robert Horton, Luna Leopold, Walter Langbein, Gordon Wolman, and Thomas Maddock, four remarkable theoreticians, armed with meticulously collected large sets of observational data and all from the USGS, formulated certain laws of geomorphology in the 1950s, such as the discovery of hydraulic geometry relationships and the introduction of the concept of entropy in geomorphology, which have had a lasting impact on applied hydrology.

In the mid-1940s, the Federal Government of the United States initiated a major research program as a cooperative effort between the USACE and US Weather Bureau, with the major impetus being to develop procedures to derive spillway design floods for major dams that were being planned for western river basins subject to snowmelt runoff. The Cooperative Snow Investigation Program established three snow laboratories, which were operated until the mid-1950s. Results of the laboratory experiments and other scientific research of the program were documented in numerous technical reports, research notes, and technical bulletins. These were in turn compiled into a summary report, named *Snow Hydrology*, by USACE in 1956. This document was a landmark contribution in the field of hydrology of snowmelt and still remains a valuable resource for hydrologists and engineers working with applications of snow hydrology.

The last but not the least significant theoretical development, which started in the 1950s and shaped the subsequent mathematical modeling of watershed hydrology, was the development of kinematic and diffusion wave theories. Even though shallow water equations were developed by French mathematician Adhémar Jean Claude Barré de Saint-Venant in 1843, their simplified forms were the basis of physically based hydraulic and hydrologic modeling. Iwagaki (1955) from Japan conceived the kinematic wave concept to develop the method of kinematic wave routing for channel flows and also applied diffusion wave approximation to flow routing in streams and canals. Although this approximation has been applied to rainfall–runoff modeling, a further simplification of the Saint Venant equations has been more popular. This was due to a seminal contribution in 1955 by Lighthill and Whitham from the United Kingdom, who developed the kinematic wave theory for describing flood movement in long rivers. The real impetus to the application of kinematic wave theory was due to the development of the kinematic wave number proposed by Woolhiser and Liggett in 1967. A full account of these theories is given by Singh (1995). Diffusion wave approximation to channel flow routing became popular since, in1969, Jean A. Cunge, a Polish–French engineer, improved the Muskingum method developed by McCarthy and others that is now known as the Muskingum–Cunge method.

With the advent of analog and digital computers, a **period of simulation** started in the 1960s. However, computer technology took a quantum jump in the 1980s when microcomputers were commercialized, analog computers became obsolete, and software engineering started to evolve by leaps and bounds, particularly with the invention of object-oriented programming languages and the graphical user interface (GUI). Various software with widespread circulation and application came into existence. Nonetheless, the revolutionary period of hydrologic modeling began in the 1990s with the arrival of internet technology and the adoption of geographic information system (GIS) technology, which started in the 1980s. Thus, the newest period in the development of hydrology, the **period of modeling with information technology**, had begun. Applied or engineering hydrology at the current stage of the twenty-first century constitutes mostly hydrologic computations through modeling, using digital data and advanced software coupled with GIS software, together with mathematical methods that have reached maturity for their robustness and reproducibility to represent the physical principles of rainfall–runoff processes.

1.3 HYDROLOGIC SCIENCE AND HYDROLOGIC ENGINEERING

Hydrology is legitimately regarded as a geophysical science. Dooge strongly advocated the inclusion of hydrology as a distinct branch of geophysical sciences. However, its origin in civil or water resources engineering cannot be overlooked. Indeed, hydrology originated from the need for designing civil infrastructure facilities, such as water supply systems, urban and rural drainage, flood control works, water impoundments, arterial airfields, land reclamation, river training works, and agricultural irrigation. The need for the appropriate design of such facilities, in turn, gave birth to the development of measurement techniques and tools. Nevertheless, hydrologic science and hydrologic engineering are inseparable and have developed in a complementary fashion, with an understanding of the basic scientific principles and appreciation of the practical applications to further basic research. Hydrology can be viewed both as a geophysical science and an engineering discipline, and one can also justifiably consider hydrology as an environmental science as well as a branch of environmental engineering. It is not important what hydrology is called or where it is placed. What is important at the current time is that hydrology is advanced in all aspects, and water-related questions impacting mankind are answered with a scientific basis due to the advancements made in this subject over the past several centuries. The software and information technology available now has enabled scientists and engineers to obtain and process vast amounts of data of hydrologic importance and perform advanced computations with lightning speed and desired accuracy by bridging the theoretical concepts and applied aspects of hydrology. Even though we have treated the subject matter in light of the advancements of the twenty-first century, our presentation of the subject matter is along the direction in which civil engineers, irrigation and agricultural engineers, and applied earth scientists approach hydrology. For this reason, instead of diving deep into the fundamental science that underlies the principles and practice of hydrology in engineering applications, our emphasis in this book is on the purity of the practice, keeping the rigor of the principles but without delving too much into the scientific description.

1.4 HYDROLOGIC SYSTEMS

A **system** is that part of the universe that is set aside for current consideration and the **surroundings** are the rest of the universe. A system has a set of connected parts called components that form the whole. The system approach in hydrology was perhaps first extensively discussed by Dooge, from the Department of Civil Engineering, University College, Dublin, Ireland, in a 1973 publication resulting from lectures given by him in August 1967 at the University of Maryland in the United States, under the sponsorship of the Agricultural Research Service of the US Department of Agriculture. Dooge (1973) approached hydrologic system essentially from an angle of systems theory of operations research. We will introduce this concept in Chapter 2. Here we introduce a hydrologic system, as it is commonly used in the field of thermodynamics, that deals with energy and mass transfer in various kinds of processes, which can be physical, mechanical, chemical, or biological.

Applied hydrology is all about a system that encompasses a component, which is a piece of land surface, subjected to input that is either rainfall or snow and ice melt, or both, producing output that results in water flowing down the streams present on the land surface. The system has sources and sinks. The water that infiltrates into the surface of the land, percolating down to recharge groundwater, and water that returns to the atmosphere above the land surface are examples of sinks. Other examples of sources of water entering the system are water supplied by a pipe or diversion.

The land surface component of the system is called a **drainage basin**, which assumes names like catchment, watershed, and basin but all have the common characteristics of having a single point called an outlet from which all water from within the unit exits the system. The boundary of this land surface is called the drainage divide. The definitions of the synonymous terms given to a **drainage unit** are discussed in Chapter 6. Treating a drainage unit as a *system* with various interconnected parts or components and atmosphere as its *surrounding* is a fundamental concept that has also been called the **watershed concept**.

A pictorial representation of a hydrologic system is given in Figure 1.1. It has a set of connected parts or components, such as the land surface, streams, the overlying atmosphere, and the underlying subsurface, that form a whole. In thermodynamics, three kinds of system are defined. These are the following: (1) An **isolated system** is one that cannot exchange either mass or

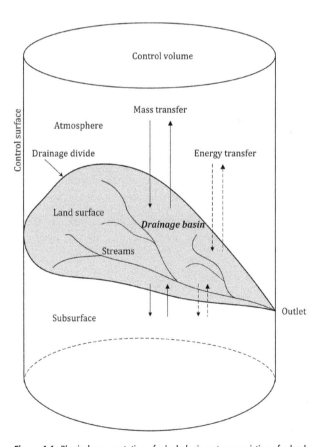

Figure 1.1 Physical representation of a hydrologic system consisting of a land surface component called a drainage basin or drainage unit, the atmosphere over it, and the subsurface. Energy and mass transfer occur in both directions as shown by lines with arrow heads; the dashed lines indicate a process that cannot be "seen."

energy with its surroundings. A perfect isolated system is hard to come by, but an insulated drink cooler with a lid is conceptually similar to a true isolated system. The items inside can exchange energy with each other, which is why the drinks get cold and the ice melts a little, but they exchange very little energy (heat) with the outside environment or its surrounding. (2) A **closed system** can exchange only energy with its surroundings, not mass. (3) An **open system** can exchange both energy and mass with its surroundings. A system of drainage basin is an open system as shown in Figure 1.1. In continuum mechanics, fluid mechanics, as well as in thermodynamics, a **control volume** or **control space** is a mathematical abstraction employed in the representation of mathematical models of the **physical processes** within a system. The closed surface enclosing the region is referred to as the **control surface**. The connected parts are operated by the hydrologic processes driven by heat engines, and the system is a drainage unit or control space in four dimensions (space and time).

1.5 HYDROLOGIC PROCESSES

Within the control volume of a hydrologic system distinct types of physical and biological processes take place by the work done by the force of gravity and the transfer of energy in the form of heat. We state this as the first law of thermodynamics in Chapter 15. The hydrologic processes are overly complex and hardly fully understood even the present day. Nevertheless, this has not deterred scientists and engineers from developing mathematical representations of the processes by sets of equations that have been working quite well for building engineered systems to manage and utilize water resources successfully.

The hydrologic processes that are of immediate concern for hydrologic engineering depends on the purpose of the engineering project where this engineering has a role. If a water resources project is designed and built to manage water from rainfall–runoff events of a certain magnitude and fixed duration, all of the major hydrologic processes need not be considered. On the other hand, if the project entails a fuller understanding of the hydrology of a drainage basin for the long term, then considerations must be given to a broader set of hydrologic processes. This is illustrated in Figure 1.2 in which the principal hydrologic processes that are typically considered for two distinct types of simulations are shown.

As stated above, for the purpose of engineering, the hydrologic processes are described by sets of equations. These equations contain a set of **variables** and a set of **parameters**. For this reason, engineering hydrology can be called **parametric hydrology**, a term coined by Dooge (1973). Modern parametric hydrology is essentially a simulation of hydrologic processes to attempt an accurate

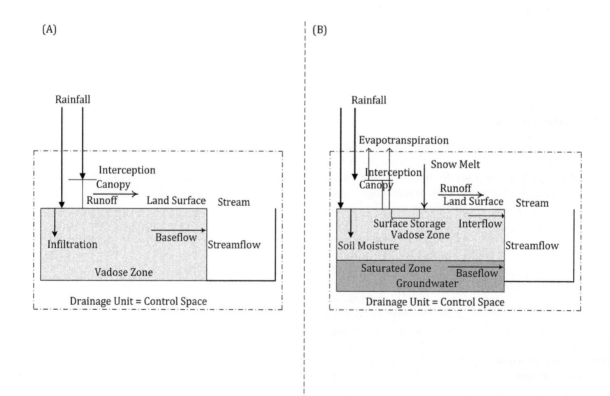

Figure 1.2 Hydrologic processes usually considered in modeling two types of simulations. (A) Processes considered for simulation of rainfall events of finite duration. (B) Processes considered for long-term simulation, also called continuous simulation of rainy and dry periods spanning several months, years, or decades.

representation of the real-world hydrologic phenomena. This is the field of hydrologic modeling.

1.6 HYDROLOGIC MODELS

The development of hydrologic modeling also has a long history that will not be described here. Singh and Woolhiser (2002) and Singh (2018) provided a perspective account of the development of such models. Many hydrologic models have been developed in the world and most of these are described by Singh (1995) and Singh and Frevert (2002a, b, 2006). The current hardware and software technologies have provided the prowess to model hydrologic processes at varied spatial and temporal scales. Several models have been developed by various governmental and commercial organizations as well as researchers from academia. However, a selected number of those models are currently in widespread use and have become industry standards, particularly for hydrologic engineering practice. There are several reasons behind their widespread adoption world over. First, the models that are in the public domain and can be easily downloaded have a natural appeal to their intended users. In choosing those models, the second consideration is given to the model originators. If a model is developed by a reliable and well-respected organization, particularly if the model is approved by governmental and regulatory agencies, its acceptability is backed by engineering communities involved in developing models for real-world problems and, in particular, projects that entail design for the construction of hydraulic structures. The third consideration given in the adoption of a model is to the methods and concepts that are imbedded in the models. If a model implements well-established methods and principles that have reached their full maturity and are recognized as a standard practice, the model is also considered a standard model. Sometimes consideration is also given to choosing a hydrologic model for its capability to be integrated with GIS software. After all, watershed hydrology is a spatial science and GIS technology offers a tremendous power of combining analytical or computational components of modelling with the spatially distributed nature of the science, which is at the very heart of hydrology of large areas. In addition to these primary considerations, other factors that also come into play for the selection and acceptance of a hydrologic model are the level of training required to learn how to use the model, the continuity of its development, and maintenance or upgrades by the developers, together with the availability of technical support.

1.7 AN INDUSTRY-STANDARD HYDROLOGIC SIMULATION MODEL

1.7.1 Hydrologic Simulation

Applied hydrology revolves around the calculations of volumetric flow rates of water, which result from complex interactions of various hydrologic processes in a matrix of human-made and natural settings. In the modern era of highly advanced software and hardware engineering, virtually all hydrologic calculations are routinely carried out by using a simulation model. Many hydrologic simulation models, either proprietary or in the public domain, are now available. A discussion covering aspects of all of the commonly known hydrologic simulation models is not within the scope of this chapter. Here we only introduce the model that is used in this book, the industry-standard hydrologic simulation model.

1.7.2 Introduction to the HEC-HMS Software

1.7.2.1 Highlights of HEC-HMS

The USACE has led the developments of many methods of applied hydrology for over a century. The Hydrologic Engineering Center (HEC) of USACE developed the numerical model (computer program) HEC-1 in the later part of twentieth century as the age of digital computers emerged. It was a first-generation hydrologic simulation model. With the invention of GUI, HEC-1 transitioned from a computer program to the HEC-HMS software at the advent of twenty-first century. HEC-HMS, which stands for Hydrologic Engineering Center – Hydrologic Modeling System, includes a large set of methods to simulate drainage basin, channel, and water-control structure behavior, thus predicting flow, stage, and timing. HEC-HMS is used throughout all chapters of this book for the following reasons.

1. In the United States as well as in many other countries, HEC-HMS is the current industry standard for simulation and calculations pertinent to hydrologic engineering practice.
2. In the United States it has been used in many studies for achieving goals of flood-damage reduction, reservoir and system operation, floodplain regulation, environmental restoration, water-supply planning, among others.
3. In many countries, including the United States, all governmental agencies have officially adopted HEC-HMS as the hydrologic simulation model to use for studies pertinent to both planning and design.
4. HEC-HMS is maintained and continues to be updated or improved by a well-respected and reliable agency of the Federal Government of the United States.
5. The software is very adaptable because it includes a variety of model choices for each segment of the hydrologic cycle.
6. While almost all well-established, traditional, or conventional methods of engineering hydrology have been implemented in HEC-HMS, it also continues to incorporate recent advances in methodologies that rely on other technologies or disciplines such as GIS, advanced meteorological data input and processing, and statistical methods.
7. The HEC-HMS software mostly utilizes simulation components built from conceptual models. These

models typically rely on empirical data to make predictions about water movement. Nevertheless, many of these models contain parameters with a physical basis and may be estimated from measurable properties of the watershed. These models can function very effectively when calibration data are available. In the ungagged case, it is generally accepted that physically based models are a better choice. HEC-HMS includes many physically based simulation components such as Green–Ampt and Smith–Parlange infiltration components, kinematic wave surface runoff components, Priestley–Taylor potential evapotranspiration components, etc.

8. With the very first release, HEC-HMS was in the public domain. So, it is available to anyone, at any corner of the globe, free of charge.
9. The GUI in HEC-HMS is very well designed. It is relatively easy for the first-time student or practitioner to learn how to use it as opposed to many other models that have very steep learning curves due to cumbersome user interfaces.
10. HEC-HMS is an excellent educational tool for any first-time student in the field of hydrologic models.

1.7.2.2 Basic Structure of HEC-HMS

At the very core of any hydrologic modeling lies the concept of a drainage unit, which is the system under consideration. In HEC-HMS this unit is called a **basin**. This is the same concept discussed in Section 1.4. In HEC-HMS, there are seven elements within a basin, the entire system. The foundation of an HEC-HMS model then rests on the concept of spatial connectivity of these elements. One of the seven elements within a basin is the **subbasin**, the first element of the system. Simply put, either a basin or a subbasin is a land surface area defined by a boundary on which a **sink**, possessing the property of being a singular point, to which all surface water from within that boundary flows into. This is same as the outlet shown in Figure 1.1. So, the sink is the second element of the system in HEC-HMS. The surface water is carried by a channel and its tributaries. These are **reaches**, the third element. On the course of a reach there can be one or more open water bodies like lakes, detention ponds, or storage or flood control reservoirs. Any of these make the fourth element called a **reservoir**, of the system. The fifth element of a HEC-HMS model is a **junction**. A junction is a point where water from one or more elements joins to form the flow to the element immediately downstream of the junction. The sixth element of the system represented by a HEC-HMS model is the **source**. It is a virtual point representing a physical entity that contributes water flow to any of the other six elements of the system. The seventh element that can be present within a basin or the system is a **diversion**. This is also a virtual representation of an entity such as a pump station that changes the direction and rate of water flow. At each computational point, all flows contributing from all the upstream elements are joined to form the combined flow that moves to the next downstream element to which it is connected.

A basin can be variously subdivided into multiple subbasins. However, the governing principle here is that the subbasins are spatially connected from higher to successively lower elevations for the surface water to respect gravity as well as to follow the flow patterns of streams that ultimately drain the basin to that singular point called outlet. Figure 1.3 A shows a small drainage basin in Utah, USA. The area is drained by several streams, all of which go to an outlet located on a major stream named Clear Creek. The HEC-HMS basin model of this tributary system is shown in Figure 1.3B.

Once a basin model is created in HEC-HMS, the next step involves selecting the hydrologic processes that must be considered for a specific model. Each of the processes is modeled according to a method. HEC-HMS offers the option to choose from several common methods or models implemented in the program to represent a process. Each of the methods in turn requires one or more variables and parameters. Inputs of rainfall, and in some cases of energy, are provided and the time window for the simulation also needs to be specified.

1.8 HYDROLOGIC DATA SOURCES

As noted above, the current age of information technology has completely changed the way hydrology is practiced today from the way it was practiced 30 years ago. One of the greatest impacts that the internet era has had on hydrologic studies is the availability of hydrologic data in digital formats from numerous sources. We recognize that it is not possible to list or discuss all the sources of hydrologic data that are presently available. Instead, in each chapter, we refer to sources that are reliable, robust, and predictable in relation to the data required for the simulation of the hydrologic processes and methods discussed in that chapter. For example, we use NOAA, FAO, USGS, and GRDC sources for climatic and streamflow data. References to various modern hydrologic databases such as CLIMWAT and ResOpsUS are made in appropriate places. In the same vein, in many parts of the world such data are still not available, either because they do not exist in the very first place or there is no system implemented for the dissemination of information. In addition, there are examples, like in India, where flow data for transboundary rivers are not made available to the public due to restrictions posed by the government. On the other hand, there are many transboundary rivers; the Danube is an example where intergovernmental agreements have made data freely available.

1.9 ABOUT THIS BOOK

1.9.1 Audience

This book has been prepared for three groups of audiences. The first group consists of senior undergraduate and beginning graduate students in civil and environmental

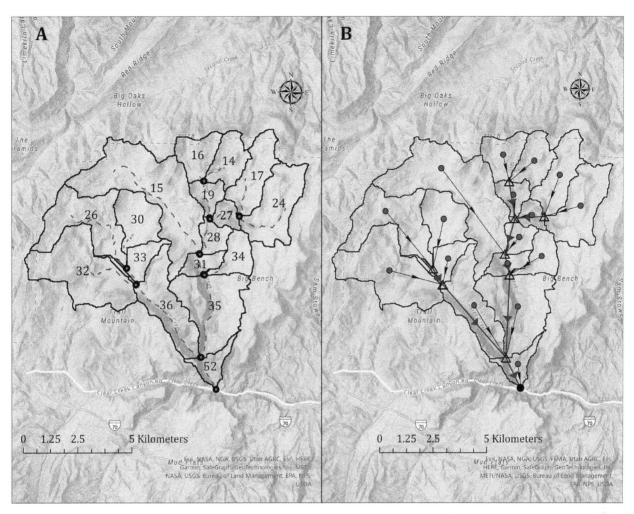

Figure 1.3 (A) A small drainage basin drained by a few tributary channels of Clear Creek in Utah, USA. The drainage basin is subdivided into several subbasins (numbered) and confluences of two streams forming a junction (solid dots). Streams are shown by dashed lines. (B) Abstraction of the physical basin in an HEC-HMS model. Each subbasin is represented by a solid circle. Flow of water from each subbasin to a junction is represented by a thin solid line with a thin arrow pointing in the direction of flow. This is overland flow. Each junction is represented by an open triangle. Flow along a reach from one junction to the next junction downstream is represented by a thick solid line with a thick arrow pointing to the direction of flow.

engineering, agricultural engineering, biological or biosystems engineering, natural resources management, and applied earth sciences and their instructors for a serious course in hydrology offered and taken with the intention that the knowledge gained from this course will equip them adequately to enter the profession where they will be practicing hydrologic engineering of the present time. After absorbing the material presented in this book, they should be able to perform their job with confidence and credibility. The second group includes those who are already in the profession of practicing hydrology for engineering applications but, for whatever reason, did not attain the background necessary for the modern standard practice of hydrology. This book can be a companion to them to gain proficiency in the work they do and develop a better understanding and level of confidence in building hydrologic models and perform other calculations, meeting quality requirements that will eventually be used for either design and construction of hydraulic structures or for planning watershed and water resources management. The third group who could also benefit from this book are researchers in hydrology for whom it may be a handy reference, which they can consult either to refresh their knowledge or to see which areas of hydrology are currently important in the practical world.

Each chapter concludes with some worked examples of questions that will help students to test their understanding of the topics covered. Sample exercises are provided as supplementary material and are available at www.cambridge.org/appliedhydrology.

1.9.2 Organization

We have not adopted a traditional approach of writing a book on hydrology, starting with some basics such as the hydrologic cycle, elementary principles of fluid and energy flow;

we assume that those should be common knowledge to those dealing with water science and engineering. We go straight to the point after giving in Chapter 2 an account of the mathematical methods that are necessary to understand various concepts presented in subsequent chapters. A chapter such as this has been conspicuously absent in the currently available textbooks on engineering hydrology. Since most students who need to take an applied hydrology course (with some exceptions), complete college level calculus courses with only an introductory knowledge of differential equations, this chapter will be useful for them as well as to the practitioners who left mathematics classes years ago. An instructor may wish to skip this chapter and ask the students to keep this in mind if they encounter difficulties in following the mathematics in various other chapters.

Then in Chapter 3 we present the principles of probability distribution and certain statistical methods that are also essential to fully comprehend various concepts and methods covered in various chapters. However, the most important contents of this chapter represent hydrologic frequency analysis, which is at the core of hydrologic design, using probabilistic models of rainfall and streamflow.

Throughout the book, we have kept all concepts with a common thread or lineage in one place and have not repeated materials, to aid convenience in learning, consulting, and teaching. For example, in Chapter 3, where we present probability density functions, we only discuss basic types of function, which are referred to in multiple places, but omit those that are specifically used for certain areas, discussing those when we come to the discussions of those areas. For example, we discuss three types of extreme value distributions in Chapter 4 and log Pearson distributions in Chapter 5 in relation to frequency analysis of rainfall and streamflow records, respectively. We have also made a concerted effort to include only those methods and principles that are most often and commonly used in practice. Chapter 3 is another good example of this. There is a multitude of probability distribution functions and, for this reason, most authors tended to list many distributions in a table. We have not attempted to do so since in practice only a few of those are really used and the rest are purely of academic interest, at least for now.

Some topics, which have almost invariably been covered in most existing texts on hydrology, have not been included deliberately in this book, because those subjects are too vast and too important to cover in a single chapter of a textbook primarily focusing on surface water hydrology. It would be an injustice to these very important subjects to include them here. For example, subjects such as groundwater and water quality, in their own right, deserve their own dedicated texts. Fortunately there are some good books that have been written exclusively on subjects like groundwater hydrology and hydraulics and water quality, which are specialized branches of water science and engineering. Anyone interested in any of these subjects should take courses on these subjects that are also routinely offered in most civil engineering and earth sciences programs. In this book, the essential concepts related to groundwater are covered in Chapter 8 in relation to baseflow. After all, groundwater is what makes baseflow.

After presenting methods of measurements and analysis of rainfall and streamflow data in Chapters 4 and 5, respectively, Chapters 6 through 13 cover the topics that are of utmost importance for event-based rainfall–runoff modeling. These models are typically used for hydrologic design or planning purposes.

Chapters 14 and 15 cover evapotranspiration and snowmelt, respectively. Evapotranspiration is especially important for continuous simulation or long-term models. An instructor may choose to cover the evapotranspiration chapter after Chapter 7, dealing with abstraction, since both abstraction and evapotranspiration have the subtraction operation or can be viewed as sinks in accounting of rainfall to runoff.

An important topic that also has been ignored in most texts on hydrology, except for the book by Singh (1992), is erosion and sedimentation. The estimation of erosion potential and sediment yield from a drainage basin has importance in many areas of conservation and resource management. Chapter 16 covers this subject in depth.

Another topic that has also not been discussed in the most well-known texts on hydrology is reservoir operations, which is extremely important for water resources management. This is covered in Chapter 17. The instructors are encouraged to include this topic in their course curriculum to prepare their students for future applications of this knowledge.

Climate change is an extremely important topic of contemporaneous hydrology and is considered in Chapter18. The effects of climate change on rainfall, evapotranspiration, snow covers, and glaciers are discussed in this chapter.

Practicing hydrologists should be cognizant about contemporaneous thoughts and trends in the management of water resources. For example, the Integrated Water Resources Management (IWRM) approach started to emerge during the last couple of decades and is considered by multilateral agencies such as the World Bank and the Asian Development Bank when funding and approving water resources projects in developing and emerging economies. However, the scope of water resources management is not very well established. Davie (2002) delved into this subject, but not in very much depth. We have elaborated on this this topic in Chapter 19, which also covers the water balance of a watershed because of its relevance to the assessment of long-term water availability in these systems.

GIS technology is now fully integrated with modern hydrology. In Chapter 20, we discuss a few specific GIS methods that are commonly used in the development of hydrologic models. This chapter is not a philosophical discussion of GIS technology in hydrology but is written with due recognition that it is not possible to cover all conceivable GIS applications in hydrology in one chapter. Again, this is a topic that merits a book of its own, and indeed some books have already been written on that line. On the other hand, this is a subject that cannot be omitted altogether in a book on applied hydrology prepared in the twenty-first century. So, we have chosen to present certain principles, such as map projection, which are important for an engineer to understand to apply GIS technology in hydrologic studies. We present a few specific applications with hands-on examples to illustrate how GIS software is used to develop and analyze certain data and models for hydrologic modeling. Emphasis has been given to the principles, such as projection systems, geodatabases, processing of vector and raster data etc., so that the concepts are clear to take advantage of the real analytical power of GIS software and not to be viewed simply as a platform to make pretty maps.

Chapter 21 contains the important topic of the optimization of model parameters through the calibration of models. This is a critically important topic for developing reliable models that can be used for the design of hydraulic structures but have been noticeably ignored by most authors of published textbooks on engineering or applied hydrology. Jain and Singh (2019) attempted to give an overview of this topic. HEC-HMS has the capability to use sophisticated optimization techniques for model calibration and these are covered in this chapter. A modern hydrologic design project typically involves the development of hydrologic models of an entire drainage unit and the models integrate some or all of the individual components presented in Chapters 6 through 16 and use GIS software at various stages for varied purposes and by obtaining data from diverse sources. For this reason, we present a holistic discussion of hydrologic modeling as a conclusion to this book. Uncertainty and errors in computations or modeling are also discussed in this chapter.

The organization of the materials discussed above is due to the fact that, for engineering projects where hydrology is involved, only certain processes of the hydrologic cycle are considered at a time, and then only a certain drainage unit of limited spatial extent covering the land surface relative to the land area at the continental scale. Such projects can be major structural flood control engineering measures. Examples include the design of high-volume pump stations on the dry side of levees (in some countries like New Zealand this is called stop banks) and in-line or off-line detention basins with inflow and outflow controls, such as dams, weirs and spillways. Structural flood control measures may also include diversion channels or channel improvements for which event-based hydrologic models of a watershed are necessary. Non-structural measures like flood warning systems require real-time hydrologic simulations. Other examples of projects where hydrologic models are indispensable include urban stormwater management systems involving detention basins and intricate storm drainage networks comprising open and closed conduits; land development projects for the design of new subdivisions or tracts; culverts and bridge crossings along major highways, city arterials, and minor streets; erosion control measures along a reach of a stream; conveyance systems for irrigation and water supply, etc.

1.9.3 Highlights

This book is written within the context of hydrologic modeling with the most up-to-date software packages that are in the public domain but developed by respected and reliable organizations and recognized and adopted by numerous regulatory or governmental agencies worldwide. We have taken advantage, in every aspect of the presentations, of being able to use GIS software. However, fundamentals have not been ignored at any expense. As a matter of fact, the first author finds in his supervisory role in the private sector that one problem being posed by the current era of using sophisticated software as a standard practice is that while the generation of freshly educated engineers entering the field of hydrologic engineering profession gain proficiency in using the software quite quickly they may lack the required depth of knowledge and understanding of the principles. The problem can be attributed to the fact that people can get the results of lengthy calculations without working through the mathematics or calculation steps involved, but by simply entering a set of inputs without realizing that correct inputs in fact do require proper knowledge of the underlying theories or principles of the methods being used. This book is written in order to make a seamless mosaic of the principles and practice of hydrology in engineering applications.

In addition to HEC-HMS, other programs also developed by the HEC are used for specialized applications such as HEC-SSP (Statistical Software Package) for hydrologic frequency analysis, HEC-ResSim (Reservoir Simulation) for reservoir operations, and HEC-MetVue (for visualization and processing of meteorological data). In addition to the hydrologic software developed by the HEC, some other programs or software such as PeakFQ developed by the United States Geologic Survey for flood-frequency analysis are also used in relevant places.

1.9.4 Unit Systems

When an author of a book on the subject of hydrology aimed for engineers and applied scientists is based in the United States and anticipates that practicing engineers in the United States will also use the book, the author faces a dilemma for the choice of the unit system.

The metric system or International System of Units (SI) is the world standard with the exception that in the United States the customary unit is still the so-called English System or foot–pound–second (FPS) system. The availability and accessibility of hydrologic and meteorological data is most ubiquitous in the United States. Those data are also in FPS system, and we use those data extensively throughout the book.

For the reasons given above, we use both the FPS system and SI. We have mostly used FPS system where the original data are obtained in the United States; if the data are obtained in SI from some other sources such as in Canada, UK, Europe, and Australia, we used SI. If we have created any new set of data, then we convert those into SI. For most of the exercises, we used SI. For the same reasons, Chow et al. (1988) and Singh (1992) dexterously used both systems in their examples and exercises.

1.10 EXAMPLES

Example 1.1: Represent the hydrologic cycle in the form of a process flow diagram and discuss how a system of applied hydrology can be considered an open system when the hydrologic cycle at a global scale is a closed system since water circulating within the system always remains within the system.

Solution: There are numerous ways by which the hydrologic cycle in the form of a process flow diagram can be represented. One such diagram, shown in Figure 1.4, is modified from Kulandaiswamy (1964). However, in this diagram the circulation of the runoff from the basin that ultimately flows to the sea from where it returns back into the atmosphere is not shown. In applied hydrology, this process is largely ignored and thereby the system becomes an open system. Thus, at the basin scale, the hydrologic cycle is not a closed system.

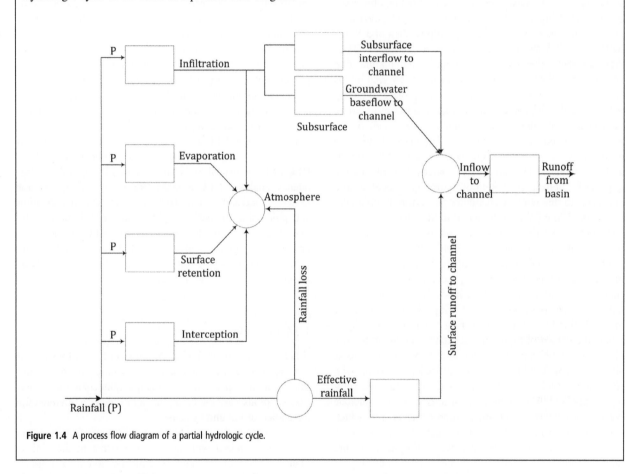

Figure 1.4 A process flow diagram of a partial hydrologic cycle.

Exercises: A selection of exercises on this topic is available at www.cambridge.org/appliedhydrology.

2 Review of Mathematics

2.1 CONCEPTS AND PURPOSE

In applied hydrology, hydrologic systems and processes are represented by a set of mathematical equations. This is done so because, after all, the ultimate objective of engineering hydrology is to obtain quantitative estimations of variables of hydrologic interest, such as streamflow at a point of interest resulting from a rainfall event, the amount of snowmelt that can be generated from a snowpack, the amount of evaporation that can result from a reservoir, etc. These equations that are presented in subsequent chapters invariably contain one or more parameters. Dooge (1973) termed this **parametric hydrology**. In parametric hydrology, there are frequent appearances of many mathematical concepts and methods that may not be adequately known to those whose primary interest lies in the practice of hydrology. The purpose of this chapter is to serve as a ready reference for the mathematical concepts and formulas that are often used in parametric hydrology. Those interested in further exploration of the beauty of applied mathematics should consult many excellent texts written on each of these topics individually or collectively. Those who find this of little interest should not worry unduly about it. They can concentrate on the quantitative representation of the hydrologic systems and simply accept the mathematical aspects with a belief that these are based on well-founded and powerful mathematical tools. What is important for both groups is to have a familiarity with the concepts covered in this chapter. This should aid in an appreciation of the developments that follow in the rest of the book.

2.2 DERIVATIVES AND INTEGRALS

This section is primarily aimed as a refresher in the context of the relatively more advanced concepts that are presented in subsequent sections.

2.2.1 Fundamental Theorem of Calculus

If

$$\frac{d}{dt}[f(t)] = g(t) \tag{2.1}$$

then,

$$\int g(t)\,dt = f(t) \tag{2.2}$$

Using the differential notation,

$$d[f(t)] = g(t)\,dt \tag{2.3}$$

If Eq. 2.1 holds true, the following also holds true:

$$\frac{d}{dt}[f(t) + c] = g(t) \tag{2.4}$$

in which c is a constant. Therefore,

$$\int g(t)\,dt = f(t) + c \tag{2.5}$$

Here, the constant c is called the **constant of integration** and the integral is called the **indefinite integral**.

If Eq. 2.1 holds true, for all values of t between a and b,

$$\int_a^b g(t)\,dt = f(b) - f(a) \tag{2.6}$$

Here, b is the **upper limit** and a is the **lower limit** of integration. When a and b are finite or zero. it is a **definite integral**. It they are at infinity, it is called an **improper integral**. The quantity on the right-hand side of Eq. 2.6 gives the area underneath the curve $g(t)$ from a to b.

If the upper limit of integration is generalized as a variable, we use a **dummy variable of integration** to avoid confusion; for example, x is used instead of t as

$$\int_0^t g(x)\,dx \tag{2.7}$$

The reason we use x rather than t is because t represents the end of the interval over which we are evaluating the integral, so it does not make sense to use t to represent an arbitrary point inside the interval.

The fundamental theorem of calculus has two parts.

Part I: If f is a continuous function on $[a, b]$, $F(x) = \int_a^x f(t)\,dt$ is also continuous on $[a, b]$ and differentiable on $[a, b]$, with derivative $f(x)$. One can also write $F(x) - F(a)$:

$$F'(x) = \frac{d}{dx}\int_a^x f(t)\,dt = f(x) \tag{2.8}$$

where $x \leq b$.

11

Part II: If f is continuous on $[a, b]$ and $F(x)$ is an antiderivative of f on $[a, b]$,

$$\int_a^b f(x)dx = F(b) - F(a) \qquad (2.9)$$

It must be borne in mind that the concept and derivation of the definite integral are completely different from those for the indefinite integral. These by definitions are *two different types* of operations. However, the formal operation $\int$, as it turns out, treats the integrand in the same way for both. This results from the fundamental theorem of calculus. The indefinite integral is an operator, and the process is called **antidifferentiation**. The amazing link between antidifferentiation and finding the area underneath the curve is that in order to find the area underneath a function f over some interval $[a, b]$, we simply look at the difference of the values of the antiderivative at the endpoints of the interval $F(b) - F(a)$.

2.2.2 Rules of Differentiation and Integration

A **function** is an expression, rule, or law that defines a relationship between one variable, called the **independent variable**, and another variable, called the **dependent variable**. In the following, t is the independent variable, y is the dependent variable $[y = f(t)]$; and a, b, c, m, etc., where used, denote constants.

1. The derivative of the product of a constant and a function is the product of the constant and the derivative of the function where the derivative of a constant is zero.

$$\frac{d}{dt}[cf(t)] = c\frac{df(t)}{dt} \qquad (2.10)$$

$$\frac{d}{dt}(c) = 0 \qquad (2.11)$$

In Eq. 2.11, the function is a constant function, that is, $f(t) = c$.

2. The operation of integration is commutative with regard to a constant.

$$\int cf(t)\,dt = c\int f(t)\,dt \qquad (2.12)$$

3. The derivative of the sum or the difference of two or more functions is the sum or difference of their derivatives.

$$\frac{d}{dt}[f_1(t) \pm f_2(t) \pm \ldots \pm f_n(t)]$$
$$= \frac{df_1(t)}{dt} \pm \frac{df_2(t)}{dt} \pm \ldots \pm \frac{df_n(t)}{dt} \qquad (2.13)$$

Combining the results of Eqs. 2.10 and 2.13 we can also write

$$\frac{d}{dt}[a_1 f_1(t) \pm a_2 f_2(t) \pm \ldots \pm a_n f_n(t)]$$
$$= a_1 \frac{df_1(t)}{dt} \pm a_2 \frac{df_2(t)}{dt} \pm \ldots \pm a_n \frac{df_n(t)}{dt} \qquad (2.14)$$

where a_i is a constant ($i = 1, 2, \ldots, n$).

4. The integral of the sum or difference of any finite number of functions is equal to the sum or difference of the integrals of the functions taken separately.

$$\int [f_1(t) \pm f_2(t) \pm \ldots \pm f_n(t)]dt$$
$$= \int f_1(t)dt \pm \int f_2(t)dt \pm \ldots \pm \int f_n(t)dt \qquad (2.15)$$

Combining the results of Eqs. 2.12 and 2.15 we can also write.

$$\int [a_1 f_1(t) \pm a_2 f_2(t) \pm \ldots \pm a_n f_n(t)]dt$$
$$= a_1 \int f_1(t)dt \pm a_2 \int f_2(t)dt \pm \ldots \pm a_n \int f_n(t)dt \qquad (2.16)$$

5. The derivative of the product of two functions is equal to the first function times the derivative of the second function plus the second function times the derivative of the first function. This is called the **product rule** of derivatives.

$$\frac{d}{dt}[f_1(t)f_2(t)] = f_1(t)\frac{d[f_2(t)]}{dt} + f_2(t)\frac{d[f_1(t)]}{dt} \qquad (2.17)$$

This holds true for any number of functions.

$$\frac{d}{dt}[f_1(t)f_2(t)\ldots f_n(t)] = \frac{d[f_1(t)]}{dt}\{f_2(t)f_3(t)\ldots f_n(t)\}$$
$$+ \frac{d[f_2(t)]}{dt}\{f_1(t)f_3(t)\ldots f_n(t)\} + \ldots$$
$$+ \frac{d[f_n(t)]}{dt}\{f_1(t)f_2(t)\ldots f_{n-1}(t)\} \qquad (2.18)$$

6. The integral of the product of two functions is equal to the first function times the integral of the second function minus the integral of the derivative of the first function times integral of the second function. This is also known as **integration by parts**.

$$\int f_1(t)f_2(t)dt = f_1(t)\int f_2(t)dt$$
$$- \int \left[\frac{d}{dt}\{f_1(t)\} \times \int f_2(t)dt\right]dt \qquad (2.19)$$

7. The derivative of the quotient of two functions equals:

$$\frac{\text{Derivative of Numerator} \times \text{Denominator} - \text{Derivative of Denominator} \times \text{Numerator}}{\text{Square of Denominator}}$$

$$\frac{d}{dt}\left[\frac{f_1(t)}{f_2(t)}\right] = \frac{f_2(t)\frac{d}{dt}[f_1(t)] - f_1(t)\frac{d}{dt}[f_2(t)]}{[f_2(t)]^2} \qquad (2.20)$$

Note that for integrals of the quotient of two functions there is no such rule as given for derivatives. Integrals of products or the quotient of two or more functions must be evaluated either using the method of integration by parts or using integration by substitution.

8. The derivative of a function of a function or a composite function is called the **chain rule** of differentiation.

If $y = f(u)$ and $u = g(t)$, y is a function of a function of t and is called a **composite function** denoted by $f \circ g$ (read f circle g), which is simply

$$\frac{dy}{dt} = \frac{d}{dt}[f\{g(t)\}] = \frac{d}{du}[f(u)]\frac{d}{dt}[g(t)] = \frac{dy}{du} \times \frac{du}{dt} \qquad (2.21)$$

The chain rule applied to some specific functions are given below:

$$\frac{d}{dt}([f(t)]^n) = n[f(t)]^{n-1}f'(t) \qquad (2.22)$$

$$\frac{d}{dt}\left(e^{f(t)}\right) = f'(t)e^{f(t)} \qquad (2.23)$$

$$\frac{d}{dt}(\ln[f(t)]) = \frac{f'(t)}{f(t)} \qquad (2.24)$$

9. If y is a function of x but the relationship is given as an **implicit function**, $f(x, y) = 0$, each term in the equation is differentiated with respect to x, regarding y as an unknown function of x having a derivative $\frac{dy}{dx}$, and then the resulting equation is solved for $\frac{dy}{dx}$.

10. If x and y are expressed in terms of a third variable, t, called a parameter: $x = \phi(t)$ and $y = \varphi(t)$, and we need to find $\frac{dy}{dx}$, then

$$\frac{dy}{dx} = \frac{dy}{dt} \cdot \frac{dt}{dx} = \frac{dy}{dt} / \frac{dx}{dt} \text{ provided, } \frac{dx}{dt} \neq 0 \qquad (2.25)$$

2.2.3 Standard Derivatives and Integrals

Derivatives and integrals of standard functions are listed below. Note that, for the indefinite integrals below, the constant of integration is omitted for writing so that the essence of the formulas can be remembered (it is implicit that it is present on the right-hand side).

1. The derivative of a variable with respect to itself is unity:

$$\frac{d}{dt}[t] = 1 \qquad (2.26)$$

2. The integral of one is the variable itself:

$$\int dt = t \qquad (2.27)$$

3. The integral of a constant with respect to a variable is the constant times the variable of integration:

$$\int c\, dt = ct \qquad (2.28)$$

4. Derivatives and integrals of a **power function:**

A power function is a function with a single term that is the product of a real number, a coefficient, and a variable (at its base) raised to a fixed real number. A power function is a function that can be represented in this form: $f(x) = ax^m$ in which both the coefficient, a, and the exponent, m, are **real numbers**, positive, negative, integer, or fraction. In many areas of hydrology, power functions play central roles. Box 2.1 lists the laws of exponents.

The **power rule** of differentiation is

$$\frac{d}{dt}[t^n] = nt^{n-1} \qquad (2.29)$$

$$\int t^n dt = \frac{1}{n+1}t^{(n+1)} \text{ for } n \neq -1 \qquad (2.30)$$

If $n = -1$,

$$\int t^{-1} dt = \int \frac{1}{t} dt = \ln|t| \qquad (2.31)$$

This sometimes is also given by the common logarithm.

$$\int \frac{1}{t} = \log|t| \qquad (2.32)$$

5. Derivatives and integrals of **exponential functions:**

An exponential function is given as: $y = f(t) = a^t$. Note that it is not a power function since it has a constant base raised to a variable power:

Box 2.1 Laws of exponents

If x and y are real numbers and $a, b > 0$:

(1) $a^{x+y} = a^x a^y$; (2) $a^{x-y} = \dfrac{a^x}{a^y}$; (3) $(a^x)^y = a^{xy}$; (4) $(ab)^x = a^x b^x$

---

2 Review of Mathematics

Box 2.2 Laws of logarithms

If there is an exponential function $a^x = N$, the exponent x is called the logarithm of N to the base a. This is written as $x = \log_a N$ (both $a^x = N$ and $x = \log_a N$ mean the same thing). If $a = 10$, it is called a **common logarithm** and the base is usually dropped and is written as $\log(N)$. If $a = e$, it is called a **natural logarithm** or **Napierian logarithm** after Napier, the inventor of logarithms, and written as $\ln(N)$. In all cases, $N > 0$.

If x and y are **positive numbers** and a is a **rational number,**

(1) $\log(xy) = \log x + \log y$; (2) $\log\left(\dfrac{x}{y}\right) = \log x - \log y$; (3) $\log(x)^a = a \log x$

These same rules can be written using ln, for example, $\ln(M \times N \times P \times \ldots) = \ln M + \ln N + \ln P + \ldots$

If, say, logarithms of all numbers to base a are known and it is required to find the logarithms to base b, $\log_b(N) = \dfrac{\log_a(N)}{\log_a(b)}$.

The number e is the limit of the infinite series, $\left(1 + \dfrac{1}{n}\right)^n$. Its value is approximately 2.71828 and is known as **Euler's number** (not to be confused with Euler's constant, γ).

$$\frac{d}{dt}[a^t] = a^t \ln(a) \tag{2.33}$$

This when written in terms of common logarithm,

$$\frac{d}{dt}[a^t] = a^t \log_e(a) \tag{2.34}$$

It follows from the two above, when, $f(t) = e^t$,

$$\frac{d}{dt}[e^t] = e^t \tag{2.35}$$

If $y = e^u$, where $u = f(t)$, from the chain rule of differentiation,

$$\frac{d}{dt}[e^u] = e^u \frac{du}{dt} \tag{2.36}$$

In particular, if $f(t) = e^{mt}$

$$\frac{d}{dt}[e^{mt}] = me^{mt} \tag{2.37}$$

For integrals,

$$\int a^t dt = \frac{a^t}{\ln(a)} \quad a > 0 \tag{2.38}$$

$$\int e^t dt = e^t \tag{2.39}$$

$$\int e^{mt} dt = \frac{1}{m} e^{mt} \tag{2.40}$$

6. Derivatives and integrals of **logarithmic functions:**

A logarithmic function is $f(t) = \log(t)$ or $f(t) = \ln(t)$. Box 2.2 summarizes the laws of logarithms.

$$\frac{d}{dt}[\log(t)] = \frac{1}{t}; t > 0 \tag{2.41}$$

$$\frac{d}{dt}[\log_a t] = \frac{1}{t} \log_a(e) = \frac{1}{t \ln(a)} \tag{2.42}$$

$$\frac{d}{dt}[\ln(t)] = \frac{1}{t} \quad \text{for } t > 0 \,(t \neq 0) \tag{2.43}$$

If $y = \ln(u)$ and $u = f(t)$,

$$\frac{dy}{dt} = \frac{1}{u}\frac{du}{dt}$$

or,

$$\frac{d}{dt}[\ln f(t)] = \frac{1}{f(t)}\frac{d}{dt}[f(t)] \tag{2.44}$$

For integrals,

$$\int \ln(t)\,dt = t\ln(t) - t \tag{2.45}$$

$$\int \frac{1}{ax+b}\,dx = \frac{1}{a}\ln|ax+b| \tag{2.46}$$

2.3 HIGHER-ORDER DERIVATIVES

The derivative of a function $f(t)$, denoted by f' or D, itself is also a function of t. If this is differentiable, it can be differentiated, and the result is called the second derivative. If $y = f(t)$, its second derivative is denoted by $\frac{d^2y}{dt^2}$ or f'' or $f^{(2)}$, implying the derivative of the derivative. This is the order of the derivative. The nth derivative of a function is usually denoted by D^n or $f^{(n)}$. Higher-order derivatives can be obtained by successive differentiation. The derivative of the $(n-1)$th derivative, $f^{(n-1)}(t)$, is

$$f^{(n)}(t) = \left(f^{(n-1)}(t)\right)' \tag{2.47}$$

Leibnitz[1] Theorem: If u and v are two functions of x, the nth derivative of their product is given as

---

[1] Gottfried Wilhelm (von) Leibniz (1646–1716) was a German polymath active as a mathematician, philosopher, scientist, and diplomat. He is one of the most prominent figures in both the history of philosophy and the history of mathematics.

> **Box 2.3 Combinatorial formula or binomial coefficient**
>
> Each of the groups or selections that can be made by taking some of a number of objects from a set is called a **combination**. For example, the combinations which can be made by taking the letters a, b, c, d, two at a time are six, namely, ab, ac, ad, bc, bd, and cd; each of these presenting a different selection of two letters. The number of combinations that can be made from a set of n dissimilar objects by taking r at a time is given by the combinatorial formula noted as
>
> $$\binom{n}{r} = \frac{n!}{r!(n-r)!}$$
>
> where $n! = n(n-1)(n-2)\ldots 1$ or **factorial** n. In older texts of mathematics, notation ^nC_r can be found.

$$(uv)_n = u_n v + \binom{n}{1} u_{n-1} v_1 + \binom{n}{2} u_{n-2} v_2 + \ldots$$
$$+ \binom{n}{r} u_{n-r} v_r + \ldots + u v_n \quad (2.48)$$

where subscripts of u and v denote the order of differentiations of u and v with respect to x. The notation $\binom{n}{r}$ is the binomial coefficient derived from Newton's binomial formula. This is given by the combinatorial formula, shown in Box 2.3, which often arises in many equations and formulas such as the binomial probability distribution discussed in Chapter 3.

Two of the common formulas of successive derivatives include:

$$f^{(n)}(x^n) = n! \quad (2.49)$$

$$f^n(e^{ax}) = a^n e^{ax} \Rightarrow f^n(e^x) = e^x \quad (2.50)$$

2.4 PARTIAL DERIVATIVES

Partial derivatives appear in many equations presented in subsequent chapters. If $z = f(x, y)$, differentiation of z with respect to x by holding y constant gives the partial derivative of z with respect to x, and differentiation of z with respect to y by holding x constant gives the partial derivative of z with respect to x. These are denoted by $\frac{\partial z}{\partial x}$ or f_x and $\frac{\partial z}{\partial y}$ or f_y, respectively. Such an operation can be done on a function of multiple variables. Similarly, partial derivatives can be higher orders also, just like ordinary derivatives. An important rule, which generally holds true, about successive partial derivatives is as follows:

If $u = f(x, y)$, then

$$\frac{\partial^2 u}{\partial x \partial y} = \frac{\partial^2 u}{\partial y \partial x} \quad (2.51)$$

Equation 2.51 implies that the partial derivative has the same value whether we differentiate partially first with respect to x and then with respect to y or the reverse. In general, for the second-order differentiation, the order of differentiation is immaterial. However, the equality in Eq. 2.51 implies that the two limiting operations involved therein should be commutative, which may not always be true. If $\frac{\partial^2 u}{\partial x \partial y}$ and $\frac{\partial^2 u}{\partial y \partial x}$ both exist for a set of values of x and y and one of them is continuous, Eq. 2.51 holds true.

An important theorem related to partial derivatives is Euler's theorem for homogeneous functions. In most mathematical texts, the definition of a homogeneous function is given as: if $f(\lambda x, \lambda y) = \lambda^n f(x, y)$ for some real number n, then $f(x, y)$ is said to be a homogeneous function of degree n. Stated slightly differently, a homogeneous function is a function of several variables such that if all its arguments are multiplied by a scalar, its value is multiplied by some power of this scalar, called the **degree of homogeneity**, or simply the **degree**. A more simplified definition can be: a homogenous function of degree n of the variables x, y is a function in which all terms are of degree n, for example, $ax^2 + 2hxy + by^2$ is a homogeneous function of degree 2.

Euler's[2] theorem: If $f(x, y)$ is a homogeneous function of x and y of degree n, then

$$x \frac{\partial f}{\partial x} + y \frac{\partial f}{\partial y} = n f(x, y) \quad (2.52)$$

This theorem can be generalized for functions of multiple variables.

An important application of partial derivative is the evaluation of the **total differential** of a function of multiple independent variables. We define it for the case of two independent variables.

Let $z = f(x, y)$ be continuous on open $[a, b]$, and dx and dy represent changes in x and y, respectively. Where the partial derivatives, f_x and f_y, exist, the total differential of z is

$$dz = f_x dx + f_y dy \quad (2.53)$$

The total differential is important for the approximation of quantities and estimation of errors in measurements or calculations where the estimated quantity is a function of several variables and, in general, the functional relationships amongst the variables are known. The well-known error

---

[2] Leonhard Euler (1707–1783) was a Swiss mathematician, physicist, astronomer, geographer, logician, and engineer who founded the studies of graph theory and topology and made pioneering and influential discoveries in numerous branches of mathematics, such as analytic number theory, complex analysis, and infinitesimal calculus. He introduced much of modern mathematical terminology and notation, including the notion of a mathematical function. He is regarded as the all-time absolute mathematical genius.

propagation formula, used in almost all branches of science and engineering, is derived from the concept of the total differential.

Let some quantity u be a function of the quantities $x, y, z, \ldots, t$: $u = f(x, y, z, \ldots, t)$ and let there be errors $\Delta x, \Delta y, \Delta z, \ldots, \Delta t$ associated with the estimation or measurement of the values of $x, y, z, \ldots, t$. If the magnitudes of these errors associated with each of the independent variables are known or can be estimated otherwise, the errors present in the estimation of z using the functional relationships existing with them is given as

$$\Delta u = f_x \Delta x + f_y \Delta y + f_z \Delta z + \ldots + f_t \Delta t \qquad (2.54)$$

The values of the partial derivatives and the errors of the arguments may be either positive or negative. Replacing them by the absolute values, we get the inequality:

$$|\Delta u| \leq |f_x||\Delta x| + |f_y||\Delta y| + |f_z||\Delta z| + \ldots + |f_t||\Delta t| \qquad (2.55)$$

The inequality 2.55 arises because errors are always additive.

The ratio of error Δx of some quantity to the approximate value of x of this quantity is called the **relative error** of the quantity. Let us designate it by δx,

$$\delta x = \frac{\Delta x}{x} \qquad (2.56)$$

The **maximum relative error** of a quantity x is the ratio of the maximum absolute error to the absolute value of x and is denoted by $|\delta * x|$

$$|\delta * x| = \frac{|\Delta * x|}{|x|} \qquad (2.57)$$

For the multivariate function u, described above, it can be shown from the theory of the total differential, that,

$$|\delta * u| = |\Delta * \ln|f|| \qquad (2.58)$$

Thus, the maximum relative error of the function is equal to the maximum absolute error of the logarithm of the function.

In terms of $|\Delta * x|$, $|\Delta * y|$, ..., $|\Delta * u|$ we denote the maximum absolute errors of the corresponding quantities (the boundaries for the absolute values of the errors), and it is obviously possible to take:

$$|\Delta * u| = |f_x||\Delta * x| + |f_y||\Delta * y| + |f_z||\Delta * z| + \ldots + |f_t||\Delta * t| \qquad (2.59)$$

Equation 2.59 can be used to estimate measurement errors.

Consider, for example, the hypotenuse c and leg or perpendicular a of a right-angled triangle ABC, determined with maximum absolute errors $|\Delta * c| = 0.2$, $|\Delta * a| = 0.1$, respectively, are $c = 75$ m and $a = 32$ m and we want to estimate the maximum absolute error in the calculation of angle A.

From elementary trigonometry, $\sin A = \frac{a}{c} \therefore A = \arcsin\left(\frac{a}{c}\right)$. Using differentiation formula from any calculus text, $\frac{\partial A}{\partial a} = \frac{1}{\sqrt{c^2 - a^2}}$, $\frac{\partial A}{\partial c} = \frac{a}{c\sqrt{c^2 - a^2}}$

From Eq. 2.59 we get,

$$|\Delta A| = \frac{1}{\sqrt{(75)^2 - (32)^2}}(0.1) + \frac{32}{75\sqrt{(75)^2 - (32)^2}}(0.2)$$
$$= 0.00273 \text{ radian} = 9'24''$$

Thus, $A = \arcsin \frac{32}{75} \pm 9'24''$

2.5 APPROXIMATION OF FUNCTIONS BY SERIES EXPANSION

In many problems in engineering, it is often necessary to approximate a function. For example, a situation may arise where there is a finite set of data points and it is necessary to determine the underlying functional form. Let us suppose one knows that the next period's value of a variable, r_{t+1}, is some function of its current value, but the functional form is not known. In this case, there is an empirical relationship: $r_{t+1} = f(r_t) + \varepsilon_{t+1}$, and the question is then whether, with enough data, the functional form of f can be discovered.

Approximation of functions is done by series expansion. A series is a sum of an infinite, ordered list of numbers or functions, that is, the sum of the first term, the second term, the third term, and so forth, for an infinite list of terms. The limit of the series can be thought of as the sum of all terms, even though there are infinitely many. In some cases, there are ways to figure out the limit, and in other cases summing some number of terms starting with the first gives a reasonable approximation of the limit. Thus, approximation of functions is an expansion of a function into a series, or infinite sum. It is a method of calculating a function that cannot be expressed by just elementary operators or can be expressed by some direct formula, that is to say that no equation has been solved for the function. The resulting series often can be limited to a finite number of terms, thus yielding an approximation of the function. The fewer terms of the sequence are used, the simpler this approximation will be. Often, the resulting inaccuracy (i.e., the partial sum of the omitted terms) can be described by an equation involving the Big O notation.

There are several kinds of series expansions. We only present the Taylor series and, later, the Fourier series, after discussing the elements of periodic functions. This is mainly because in certain hydrologic models these two series find applications.

2.6 TAYLOR SERIES

The Taylor series is often applied in numerical or finite difference approximation of derivatives appearing in various types of differential equations. An example of the application of a Taylor series in the optimization of a hydrologic model can be seen in Chapter 21. Exercise 2.15 is an example where a Taylor series can be used to represent a hydrologic system, discussed in Chapter 1.

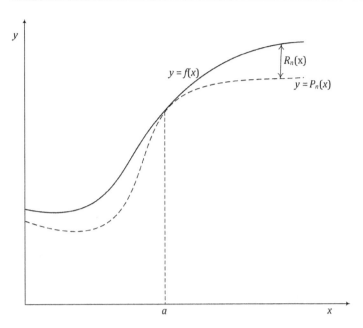

Figure 2.1 The Taylor polynomial approximation of a function.

The Taylor series provides an approximation or series expansion for a function. This is useful to evaluate numerically certain functions that do not have a simple formula. The Taylor series of a function f at a (or about a or centered at a) is given as

$$f(x) = f(a) + \frac{f'(a)}{1!}(x-a) + \frac{f''(a)}{2!}(x-a)^2 \\ + \frac{f'''(a)}{3!}(x-a)^3 + \ldots \quad (2.60)$$

which when written in compact notation is

$$f(x) = \sum_0^\infty \frac{f^{(n)}(a)}{n!}(x-n)^n \quad (2.61)$$

Equation 2.60 or 2.61 gives a polynomial approximation of a function $f(x)$ if n is finite. The term $f^n(a)$ means the nth order derivative evaluated at $x = a$. This is also denoted as $P_n(x)$ or the Taylor polynomial. Since $P_n(x) \sim f(x)$, the difference between these is called $R_n(x)$: $R_n(x) = f_n(x) - P_n(x)$. Figure 2.1 shows the concept. Probably the most important application of the Taylor series is to use partial sums to approximate functions. These partial sums are (finite) polynomials and are easy to compute. We call them Taylor polynomials. The Taylor series can be used to calculate the value of an entire function at every point if the value of the function and of all its derivatives are known at a single point. This is illustrated with a simple function discussed below.

Let us construct the Taylor polynomial of degrees 5 and 10 about $x = 0$ for the function $f(x) = e^x$. We have $f(0) = 1$. Since the derivative of e^x is equal to e^x, all the higher-order derivatives are equal to e^x. Consequently, for any $n = 1, 2, 3, \ldots, 10$, $f^{(n)}(x) = e^x$ and $f^{(n)}(0) = e^0 = 1$. Therefore, the Taylor polynomial approximations of degree 5 and 10 are given by

$$P_5(x) = 1 + x + \frac{x^2}{2!} + \frac{x^3}{3!} + \frac{x^4}{4!} + \frac{x^5}{5!}$$

$$P_{10}(x) = 1 + x + \frac{x^2}{2!} + \frac{x^3}{3!} + \frac{x^4}{4!} + \frac{x^5}{5!} + \frac{x^6}{6!} + \frac{x^7}{7!} + \frac{x^8}{8!} \\ + \frac{x^9}{9!} + \frac{x^{10}}{10!}$$

Figure 2.2 shows comparisons of the actual function $f(x) = e^x$ and the Taylor series polynomial for $n = 5$ and $n = 10$. As can be seen from these two graphs, for $n = 5$ the polynomial approximation diverges as we move further from the point of evaluation $x = 0$. But as n increases to 10, the polynomial almost coincides with the true function. This demonstration is to elucidate two key points: (i) the Taylor series can be used to calculate the value of an entire function at every point, if the value of function, and of all its derivatives, are known at a single point; and (ii) the quality of approximation increases as the number of terms included in the approximation increases. This example also shows that in certain cases the infinite series approximation converges rapidly or, in other words, with the inclusion of just a few terms in the series approximation of functions, as stated in Section 2.5.

2.7 INTEGRAL TRANSFORMS

Integral equations are equations in which an unknown function appears under an integral sign. The following is an integral equation:

$$f(x) = g(x) + \int_a^b k(x,y)f(y)dy \quad (2.62)$$

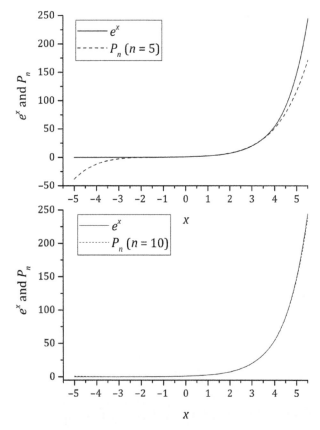

Figure 2.2 Taylor polynomial approximation of the function $f(x) = e^x$ with number of terms in the Taylor series, $n = 5$ and 10, respectively.

In Eq. 2.62, functions $k(x, y)$ and $g(x)$ are given, $f(y)$ is unknown and $f(x)$ is the function sought. A **kernel** is a known function that appears in the integrand of an integral equation. Thus, in Eq. 2.62, $k(x, y)$ is the kernel.

An **integral transform** is a mathematical operator that produces a new function $f(y)$ by integrating the **product of an existing function** $F(x)$ and a **kernel function** $K(x, y)$ between suitable limits. The process, which is called **transformation**, is symbolized by the equation $f(y) = \int K(x, y) F(x) dx$. Several transforms are commonly named after the mathematicians who introduced them: in the **Laplace transform**, introduced by Pierre-Simon, Marquis de Laplace (1749–1827), a French scholar and polymath, the kernel is e^{-xy} and the limits of integration are zero and plus infinity; in the **Fourier transform**, introduced by Jean-Baptiste Joseph Fourier (1768–1830), another French mathematician and physicist, the kernel is $(2\pi)^{-\frac{1}{2} e^{-ixy}}$ and the limits are minus and plus infinity.

If there is a function $f(x)$ and a function $k(x, s)$ then as long as the product of $f(x)$ times $k(x, s)$ is integrable on the set X, another function can always be formed of a new variable s by an operation

$$F(s) = \int_X k(x, s) f(x) dx \qquad (2.63)$$

In the operation given as Eq. 2.63, the function $f(x)$ is *transformed* into the function $F(s)$ via an integral transform.

The question may arise in the mind of someone not in the business of mathematics as to why we want to do this, when performing an integration operation often does not give much enjoyment because, in general, it is a tedious process, and integral transformation involves the integration of the product of two functions, which can become quite complicated if the functions in question are not simple functions. The main reason such a transformation is done is because the function $F(s)$ may be easier to work with than $f(x)$ itself or provides some interesting information about $f(x)$ that it would be hard to figure out in other ways.

A particularly interesting class of functions $k(x, s)$ are ones that produce invertible transformations, which implies that the transform destroys no information contained in the original function. This occurs when there exists a function $K(x, s)$ [the inverse of $k(x, s)$] and a set S such that

$$f(x) = \int_S K(x, s) F(s) ds \qquad (2.64)$$

which undoes the original transformation or, at least, undoes it for some large class of functions $f(x)$.

Whenever this is the case, *the operation can be viewed as changing the domain from x space to s space*. Each function f of x becomes a function F of s that we can convert back to f later if we so choose to. This is a new way of looking at the original function!

Integral transforms are valuable for the simplification that they bring about, most often in dealing with differential equations subject to particular boundary conditions. A proper choice of the class of transformation usually makes it possible to convert not only the derivatives in an intractable differential equation but also the boundary values into terms of an algebraic equation that can easily be solved. The solution obtained is, of course, the transform of the solution of the original differential equation, and it is necessary to invert this transform to complete the operation. For the common transformations, tables are available that list many functions and their transforms.

An important type of integral transform gives rise to **gamma function**, which was first introduced by none other than Euler in his goal to generalize the factorial to non-integer values. In many places in subsequent chapters of this book, the gamma function appears.

The gamma function denoted by $\Gamma(n)$ is defined by the **convergent improper** or **singular** integral:

$$\Gamma(n) = \int_0^\infty e^{-t} t^{n-1} dt \qquad (2.65)$$

Note that n is positive, but it needs not to be integer. For any positive integer, n,

$$\Gamma(n) = (n - 1)! \qquad (2.66)$$

2.8 LAPLACE TRANSFORM

Consider a quantity which is a function of time; we can talk of this quantity being in the **time domain**. We can represent such a function as $f(t)$. In most problems, we are concerned with $t > 0$. To obtain the **Laplace**[3] **transform** of this function we multiply it by e^{-st}, and then integrate it with respect to time from zero to infinity. Here, s is a constant with unit of 1/time. The result is what we now call the Laplace transform and the equation is then said to be in the **s-domain**. Thus, the Laplace transform of the function $f(t)$ is

$$\mathcal{L}\{f(t)\} = \int_0^\infty e^{-st} f(t) dt \qquad (2.67)$$

The transform is *one-sided* in that values are only considered between 0 and $+\infty$, and not over the full range of time from $-\infty$ to $+\infty$. In the above notation, $\mathcal{L}$ is called the **Laplace operator**. Thus, $\mathcal{L}\{f(t)\}$ indicates that the Laplace transform is taken of $f(t)$ to give the function in the *s*-domain. When in the *s*-domain, the function is usually written as $F(s)$. Thus, $\mathcal{L}\{f(t)\} = F(s)$. We can then carry out algebraic manipulations on the quantity in the *s*-domain, that is to say we can add, subtract, divide, and multiply in the normal way we can with any algebraic quantities.

For the inverse operation, when the function of time is obtained from the Laplace transform, we write: $f(t) = \mathcal{L}^{-1}\{F(s)\}$, which is read as: $f(t)$ is the inverse transform of the Laplace transform $F(s)$. There are tables available in mathematical texts dealing with the Laplace transform, giving Laplace transforms of many standard functions.

The Laplace transform is extensively used in the solutions of differential equations. For this reason, Laplace transforms of derivatives and integrals are done routinely. These procedures are not discussed here, since those are the subjects of detailed studies of differential equations.

2.9 CONVOLUTION INTEGRALS

Convolution is a mathematical operation on two functions (f and g) that produces a third function ($f * g$), which expresses how the shape of one is modified by the other. The term convolution refers to both the resulting function and the process of computing it. It is defined as the integral of the *product of the two functions* after one is *reversed* and *shifted*. The integral is evaluated for all values of shift, producing the **convolution function**. As such, it is a particular kind of integral transform.

Figure 2.3 Conceptual representation of the transfer (black box) function taking an input signal and outputting a time-delayed signal.

The convolution of f and g is written $(f * g)$, denoting the operator with the symbol $*$.

$$(f * g)(t) = \int_{-\infty}^{+\infty} f(\tau) g(t - \tau) d\tau \qquad (2.68)$$

In hydrology, we often use $\{f(t) * g(t)\}$.

An equivalent definition resulting from commutativity is

$$(g * f)(t) = \int_{-\infty}^{+\infty} f(t - \tau) g(\tau) d\tau \qquad (2.69)$$

This is a nice property of convolution integrals, because during the integration process it gives us the freedom of choosing the function that can facilitate the operation of *integration by parts*. For functions f, g supported only on $[0, \infty)$, (i.e., zero for negative arguments), the integration limits can be truncated. Thus, we can also write

$$\int_0^t f(\tau) g(t - \tau) d\tau = \int_0^t f(t - \tau) g(t) d\tau \qquad (2.70)$$

Or

$$(f * g)(t) = (g * f)(t) \qquad (2.71)$$

While the symbol t is used above, it need not represent the time domain. But in that context, the convolution formula can be described as the area under the function $f(\tau)$ weighted by the function $g(-\tau)$ shifted by the amount t. As t changes, the weighting function $g(t - \tau)$ emphasizes different parts of the input function $f(\tau)$.

For the theory of the unit hydrograph presented in Chapter 9, it is important to understand what convolution does mathematically, and what this mathematical operation represents conceptually.

Consider a dynamic system in which an input signal, say $x(t)$, enters a "black box," H, and its output is $y(t)$, as shown in Figure 2.3. Mathematically, we represent the black-box system H as

$$y(t) = H[x(0 \text{ to } t)] = \int_0^t x(\tau) h_\tau(t - \tau) d\tau \qquad (2.72)$$

---

[3] Pierre-Simon Marquis de Laplace, sometimes referred to as "Newton of France," is regarded as one of the greatest scientists of all time. He formulated Laplace equation (discussed in Chapter 8) and pioneered the Laplace transform both of which appear in many branches of mathematics, physics, and engineering. His extensive work shaped the developments of engineering, mathematics, statistics, physics, astronomy, and philosophy.

In Eq. 2.72 $h_\tau(t-\tau)$ is the **transfer function**, which is also referred to as the **impulse response function**. The $t-\tau$ term is referred to as the **lag**, as it represents the time lag between the input and the output time. The transfer function is given a subscript τ to indicate that the function itself may take on different forms at different points in time. That is, the response to an input impulse at $t=\tau_1$ could have a different form than the response to an impulse at $t=\tau_2$.

Convolution elements require that the input signal, $x(t)$, and the transfer function, $h(t-\tau)$, are supplied. The element then computes the output signal, $y(t)$. Note that the convolution integral is a **linear operation**. That is, for any two functions $x_1(t)$ and $x_2(t)$, and any constant a, the following holds:

$$H[x_1(t)+x_2(t)] = H[x_1(t)] + H[x_2(t)] \tag{2.73}$$

$$H[ax_1(t)] = aH[x_1(t)] \tag{2.74}$$

Having defined mathematically what a convolution integral does, let us now try to understand what it represents conceptually. Perhaps the simplest way to think about the convolution integral is that it is simply a linear superposition of response functions, $h(t-\tau_i)$, each of which is multiplied by the impulse $x(\tau_i)\delta\tau$.

A more common way to interpret the convolution integral is that the output represents a weighted sum of the present and past input values. We can see this if we write the integral in terms of a sum by assuming that the system is discretized by a single unit of time:

$$y(t) = x(0)h(t) + x(1)h(t-1) + x(2)h(t-2) + \ldots \tag{2.75}$$

The convolution operation may also be thought of as a *filtering* operation on the signal $x(t)$, where the transfer function is acting as the filter. The shape of the transfer function determines which properties of the original signal $x(t)$ are "filtered out." Equation 2.75 shows that convolution is the time-slicing technique. It is used in signal processing to implement digital filters. Convolution performs the same function in the time domain as multiplication in the frequency domain and vice versa.

Note that if the transfer function was a constant, the integral would collapse, and the output signal would just be a scaled version of the input signal. Integration is necessary when the transfer function is not constant. Conceptually, this indicates that the system has memory and does not respond instantaneously to a signal. Rather, the response is delayed and spread out over time.

2.10 LINEAR AND TIME-INVARIANT SYSTEMS

Most natural systems are nonlinear, meaning the change in output is not proportional to the change in input. Mathematical representation of such systems is often quite complex. However, a special class of systems that have the properties of linearity and time invariance is of particular importance to describe a hydrologic system because with just these two properties it is possible to develop most of the practical tools or methods for analysis and design of a hydrologic system. Linear and time-invariant systems, often known as LTI systems, constitute a very important class of systems, considered in many branches of engineering. The book by Hallauer (2022) gives an excellent comprehensive account of the subject of LTI systems.

A **linear system** has the property that the response to a linear combination of inputs is the same linear combination of the individual responses. A **linear function** or a **linear mapping** is one that follows both of the following two principles.

1. Principle of additivity or superposition

$$f(x+y) = f(x) + f(y) \tag{2.76}$$

2. Principle of homogeneity or proportionality

$$f(cx) = cf(x) \tag{2.77}$$

where c is a constant.

The conditions of additivity (superposition) and homogeneity (proportionality) can be combined as

$$f(ax+by) = af(x) + bf(x) \tag{2.78}$$

where a and b are constant.

A time-invariant system is not sensitive to the time of origin. More specifically, if the input is shifted in time by some amount, the output is simply shifted by the same amount. In other words, a system is said to be time-invariant when its parameters do not change with time. For such a system, the form of the output depends only on the form of the input and not on the time at which the input is applied. Thus, if $x(t) \to y(t)$, for a time-invariant system: $x(t+\tau) \to y(t+\tau)$, where τ is a time shift, which may be either positive or negative.

The importance of linearity derives from the basic notion that for a linear system, if the system inputs can be decomposed as a linear combination of some basic inputs and the system response is known for each of the basic inputs, the response can be constructed as the same linear combination of the responses to each of the basic inputs.

In Chapter 9 we develop in detail the representation of both continuous-time and discrete-time signals (rainfall) as a linear combination of delayed impulses and the consequences (runoff hydrographs) for representing drainage basins as LTI systems. The resulting representation is convolution, which as noted in Eqs. 2.73 and 2.74 is a linear operator. The principles used in the unit hydrograph theory presented in Chapter 9 are derived from the fact that superposition of ideal impulse responses can be represented by convolution integrals. This is explained below.

Suppose that an LTI system has an input that is an arbitrary physically realistic function, $u(t)$ for $t>0$. Let us apply directly the principle of superposition to derive an equation for response $x(t)$ at some arbitrary instant of time $t>0$. At any instant τ less than t, $0<\tau<t$, the input $u(\tau)$ imposes onto the system a differentially small impulse of

2.11 Impulse Functions

it appears in the solutions of the second-order ordinary differential equation representing the response of a linear, undamped system to a time-varying mechanical excitation.

2.11 IMPULSE FUNCTIONS

One of the more useful functions in the study of linear systems is the **unit impulse function**. The impulse function is a very short pulse (in theory, infinitely short) used to evaluate system dynamics. The impulse function is often written as $\delta(t)\delta t$. In the real world, an impulse function is a pulse that is much shorter than the time response of the system. The system's response to an impulse can be used to determine the output of a system to any input using convolution.

The unit impulse function, also known as the **Dirac delta function**,[4] denoted by $\delta(t)$, is usually taken to mean a *rectangular pulse* of unit area, and in the limit the width of the pulse tends to zero whilst its magnitude tends to infinity. Thus, the special property of the unit impulse function is

$$\int_{-\infty}^{+\infty} \delta(t-t_0)\,dt = 1 \qquad (2.80)$$

An ideal impulse function is a function that is zero everywhere but at the origin t_0, where it is infinitely high. However, the *area* of the impulse is finite. This is, at first, hard to visualize. In a formal way, we can visualize $\delta(t)$ as a concentrated spike of unit area at $t = 0$.

The integral of the impulse is one. Consider the integral (with $b > a$):

$$\int_a^b \delta(t)\,dt = \begin{cases} 1 & a < 0 < b \\ 0 & \text{otherwise} \end{cases} \qquad (2.81)$$

In other words, if the integral includes the origin (where the impulse lies), the integral is one. If it doesn't include the origin, the integral is zero.

Likewise, by similar reasoning, if the impulse is not at the origin (and $b > a$),

$$\int_a^b \delta(t-\lambda)\,dt = \begin{cases} 1 & a < \lambda < b \\ 0 & \text{otherwise} \end{cases} \qquad (2.82)$$

Informally, this function is one that is infinitesimally narrow, infinitely tall, yet integrates to one. Perhaps the simplest way to visualize this is as a rectangular pulse from $a - \frac{\varepsilon}{2}$ to $a + \frac{\varepsilon}{2}$ with a height of $\frac{1}{\varepsilon}$ (Figure 2.5A). As we take the limit of this setup as ε approaches 0, we see that the width tends to zero and the height tends to infinity as the total area remains

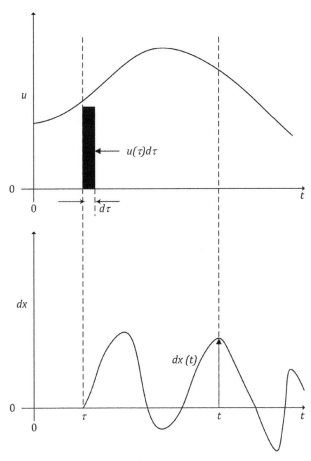

Figure 2.4 Conceptual representation of convolution as a superposition of ideal impulse responses.

magnitude $dI_U = u(\tau)d\tau$, as is indicated conceptually in Figure 2.4. Then, the differentially small response at time $t > \tau$ due to this impulse at τ is $dx(t) = dI_U \times h(t-\tau) = u(\tau)h(t-\tau)d\tau$, in which $h(t-\tau)$ is called the **impulse response function** due to a unit impulse acting at instant τ. Because this system is linear, we can express the total response as the superposition of responses due to all separate inputs, as is stated in Eq. 2.76. In this case, there are an infinite number of instants between time zero and time t, so the superposition, or summation, becomes a definite integral:

$$x(t)|_{x(0)=0} = \int_{\tau=0}^{\tau=t} dx(t) = \int_{\tau=0}^{\tau=t} u(\tau)h(t-\tau)d\tau \qquad (2.79)$$

Equations such as Eq. 2.79 and its versions for specific systems [with explicit functions for $h(t-\tau)$] are often called **Duhamel integrals** (after Jean-Marie Duhamel [1797–1872], a French mathematician and physicist), especially in the literature of structural dynamics. For example,

---

[4] Named after the English physicist Paul Adrien Maurice Dirac (1902–1984), who introduced this function in theoretical physics.

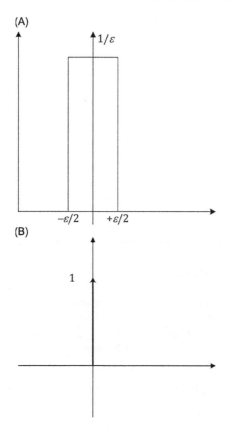

Figure 2.5 (A) One way to visualize the Dirac delta function. (B) Unit impulse (no scaling).

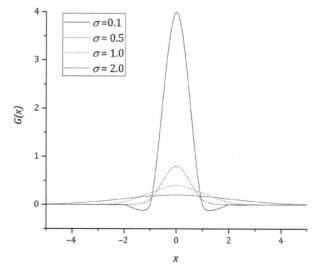

Figure 2.6 Gaussian function for different values of σ.

constant at one. Since it is quite difficult to draw something that is infinitely tall, we represent the Dirac delta function with an arrow centered at the point it is applied. If we wish to scale it, we may write the value which it is scaled by next to the point of the arrow (Figure 2.5B).

There are other equally valid definitions of an impulse function. The only important result is that the function has width approaching zero, height approaching infinity, and the area under the curve is unity. As an illustration, consider a Gaussian curve about mean zero, defined by the function:

$$f(x) = \frac{1}{\sigma\sqrt{2\pi}} e^{-\frac{x^2}{2\sigma^2}}$$

The area under the curve is one. As shown in Figure 2.6, as $\sigma \to 0, f(x) \to \infty$, but the area under the curve always remains one.

Below is a brief list of a few important properties of the unit impulse without going into detail of their proofs:

$$\delta(at) = \frac{1}{|a|}\delta(t) \qquad (2.83)$$

$$\delta(t) = \delta(-t) \qquad (2.84)$$

$$\delta(t) = \frac{d}{dt}u(t) \qquad (2.85)$$

where $u(t)$ is the unit step, and

$$f(t)\delta(t) = f(0)\delta(t) \qquad (2.86)$$

The last of these is especially important as it gives rise to the shifting property of the Dirac delta function, which selects the value of a function at a specific time and is especially important in studying the relationship of the convolution operation in the time-domain analysis of LTI systems. The shifting property is shown and derived below:

$$\int_{-\infty}^{+\infty} f(t)\delta(t)\,dt = \int_{-\infty}^{+\infty} f(0)\delta(t)\,dt = f(0)\int_{-\infty}^{+\infty} \delta(t)\,dt \qquad (2.87)$$

Equation 2.87 indicates that the action of $\delta(t)$ on $f(t)$ is to pick out its value at $t=0$.

2.12 STEP FUNCTIONS

The **Heaviside step function**[5] is a mathematical function denoted as $H(x)$, or sometimes $\theta(x)$ or $u(x)$ and is known as the unit step function. The function was originally developed in operational calculus for the solution of differential equations, where it represents a signal that switches on at a specified time and stays switched on indefinitely. Oliver

---

[5] Introduced by English mathematician and physicist Oliver Heaviside (1850–1925), who changed the face of telecommunications, invented a new technique equivalent to the Laplace transform for solving differential equations, developed vector calculus, and rewrote the modern version of Maxwell's equations of thermodynamics.

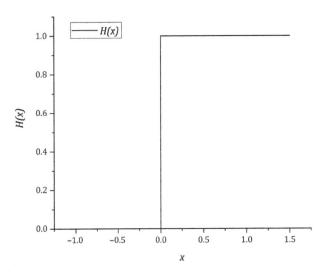

Figure 2.7 Heaviside step function.

Heaviside, who developed operational calculus as a tool in the analysis of telegraphic communications, represented the function as 1. The value of the step function is zero for negative arguments and one for positive arguments (Figure 2.7). It is an example of the general class of step functions, all of which can be represented as linear combinations of translations of this one. The term Heaviside step function and its symbol can represent either a piecewise constant function or a generalized function:

$$H(t) = \begin{cases} 1 & \text{for } t > 1 \\ 0 & \text{for } t \leq 1 \end{cases} \quad (2.88)$$

It is implied that the origin is 1, otherwise it should 0.

The Dirac delta function is the derivative of the Heaviside function:

$$\delta(t) = \frac{d}{dt}[H(t)] \quad (2.89)$$

Hence, the Heaviside function can be the integral of the Dirac delta function. This is sometimes written as

$$H(t) = \int_{-\infty}^{x} \delta(t)dt \quad (2.90)$$

However, the expansion Eq. 2.90 may not hold or even make sense for $t = 0$, depending on which formalism one uses to give meaning to integrals involving δ.

A **general step function**, also called as **staircase function**, is defined as a piecewise constant function, which has only a finite number of pieces.

2.13 LINEAR FIRST-ORDER DIFFERENTIAL EQUATIONS

In engineering and science there are many relationships that give a relation between the rates of change of quantities, rather than between the quantities themselves. The derivative of a function represents the rate of change and the equation involving derivatives is called a **differential equation**. Thus, an equation containing the derivatives or differentials of one or more dependent variables with respect to one or more independent variables is said to be a differential equation. If an equation contains only ordinary derivatives of one or more dependent variables with respect to a single independent variable, it is then said to be an **ordinary differential equation**. An equation involving the partial derivatives of one or more dependent variables of two or more independent variables is called **partial differential equation**.

The **order** of a differential equation is the order of the highest derivative which appears in the equation. The **degree** of a differential equation is the degree (power) of the derivative of the highest order after the equation has been rationalized, that is, after fractional powers of all derivatives have been removed.

Solutions of differential equations of various classes and types are formidable and cannot be covered in a single chapter of a book written for a disciple of a subject other than mathematics. For this reason, we will concentrate here on only one type of ordinary differential equation that is fundamentally important for understanding the concepts presented in relation to the unit hydrograph theory (Chapter 9), which is perhaps the most vital concept in applied hydrology. This is the differential equation of a linear system. A hydrologic system such as a watershed or a stream is said to be a linear system if the relationship between storage, inflow, and outflow is such that it leads to a linear differential equation with a constant coefficient.

The general form of a **linear differential equation** of order n is

$$a_n(x)\frac{d^n y}{dx^n} + a_{n-1}(x)\frac{d^{n-1} y}{dx^{n-1}} + \ldots + a_1(x)\frac{dy}{dx} + a_0(x)y = g(x) \quad (2.91)$$

Linearity means:

1. all coefficients are functions of x only;
2. y and all its derivatives are raised to the first power.

When $n = 1$, the equation is *first order*,

$$a_1(x)\frac{dy}{dx} + a_0(x)y = g(x) \quad (2.92)$$

Dividing Eq. 2.92 by $a_1(x)$, a more useful form of the linear first-order differential equation is obtained. From this point forward, we let t be the independent variable,

$$\frac{dy}{dt} + p(t)y = g(t) \quad (2.93)$$

in which $p(t)$ and $g(t)$ are continuous functions of t alone. Furthermore, $p(t)$ and $g(t)$ can also be constant.

An important example of a first-order linear differential equation in hydrology is $\frac{dQ}{dt} + \frac{1}{k}Q(t) = I(t)$, which is in the form of Eq. 2.93 where $p(t)$ is a constant $\frac{1}{k}$. The procedure for the solution of Eq. 2.93 is as follows.

Let there exist a function $\mu(t)$, which we call an **integrating factor**, without worrying at this point about not having any knowledge of this magical function. Multiplying Eq. 2.93 by this integrating factor, we obtain

$$\mu(t)\frac{dy}{dt} + \mu(t)p(t)y = \mu(t)g(t) \tag{2.94}$$

Again, without worrying at this point about $\mu(t)$, let us further assume that $\mu(t)$ is such that the following relationship holds true:

$$\mu(t)p(t) = \mu'(t) \tag{2.95}$$

Note that the left side of Eq. 2.95 is the second term on the left-hand side of Eq. 2.94. So, substituting Eq. 2.95 into Eq. 2.94 we arrive at

$$\mu(t)\frac{dy}{dt} + \mu'(t)y = \mu(t)g(t) \tag{2.96}$$

Now we see that it follows the product rule of the derivative given in Section 2.2.2 (No. 5). Thus,

$$\frac{d}{dt}[\mu(t)y(t)] = \mu'(t)y(t) + \frac{dy(t)}{dt}\mu(t) \tag{2.97}$$

So, Eq. 2.96 can be written as

$$[\mu(t)y(t)]' = \mu(t)g(t) \tag{2.98}$$

Now, integrating both sides of Eq. 2.98 we arrive at

$$\int [\mu(t)y(t)]' dt = \int \mu(t)g(t) dt \tag{2.99}$$

$$\mu(t)y(t) + C_1 = \int \mu(t)g(t) dt \tag{2.100}$$

It is very important to include the constant of integration, C_1, in the above. From Eq. 2.100 we get $y(t)$ as

$$y(t) = \frac{\int \mu(t)g(t)dt - C_1}{\mu(t)} \tag{2.101}$$

Since C_1 is a constant, we can write Eq. 2.101 as well:

$$y(t) = \frac{\int \mu(t)g(t)dt + C_1}{\mu(t)} \tag{2.102}$$

Now let us turn our attention to finding the integrating factor, $\mu(t)$, from where we started, and which will solve for Eq. 2.102.

Recall that we assumed Eq. 2.95 holds true to arrive at Eq. 2.102. Dividing both sides of Eq. 2.95 by $\mu(t)$,

$$\frac{\mu'(t)}{\mu(t)} = p(t) \tag{2.103}$$

From Eq. 2.43 in Section 2.2.3,

$$\frac{d}{dt}[\ln\{\mu(t)\}] = \frac{1}{\mu(t)}$$

Therefore, we can write Eq. 2.103 as

$$[\ln\{\mu(t)\}]' = p(t) \tag{2.104}$$

Now integrating Eq. 2.104 we get

$$\ln \mu(t) + C_2 = \int p(t) dt \tag{2.105}$$

$$\ln \mu(t) = \int p(t) dt + C_2 \tag{2.106}$$

Again, C_2 is the constant of integration, and the minus sign is absorbed when it is moved to the right-hand side.

Now, we can exponentiate both sides of Eq. 2.106 to get

$$\mu(t) = e^{\int p(t) dt + C_2} \tag{2.107}$$

Or

$$\mu(t) = e^{C_2} \cdot e^{\int p(t) dt}. \tag{2.108}$$

Writing $C_3 = e^{C_2}$, we rewrite Eq. 2.108 as

$$\mu(t) = C_3 e^{\int p(t) dt} \tag{2.109}$$

Substituting Eq. 2.109 into Eq. 2.101 we get what we sought, $y(t)$ as

$$y(t) = \frac{\int \left[C_3 e^{\int p(t) dt} g(t)\right] dt + C_1}{C_3 e^{\int p(t) dt}} \tag{2.110}$$

$$y(t) = \frac{C_3 \int \left[e^{\int p(t) dt} g(t)\right] dt + C_1}{C_3 e^{\int p(t) dt}} \tag{2.111}$$

$$y(t) = \frac{\int \left[e^{\int p(t) dt} g(t)\right] dt + \frac{C_1}{C_3}}{e^{\int p(t) dt}} \tag{2.112}$$

Now we can deal with only one constant. Thus, the solution of Eq. 2.93 is given as

$$y(t) = \frac{\int \mu(t)g(t) dt + C}{\mu(t)} \tag{2.113}$$

in which

$$\mu(t) = e^{\int p(t) dt} \tag{2.114}$$

Equation 2.113 is the general solution of Eqs. 2.93 and 2.114 gives the integrating factor.

Once the constant of integration is evaluated, a particular solution can be obtained. It should be borne in mind that Eq. 2.113, in many cases, may not be applicable directly. Instead, the process that has been described above to arrive at the solution should be followed to solve a particular problem.

The solution process of a first-order linear differential equation is as follows:

1. The given equation must be put in the form of Eq. 2.93.
2. Find the integrating factor using Eq. 2.114.
3. Multiply all terms in the differential equation by the integrating factor. Verify that the left side becomes the product rule $[\mu(t)y(t)]'$ and write the equation in the form $\frac{d}{dt}\left[e^{\int p(t)\,dt} \cdot y(t)\right] = e^{\int p(t)\,dt} g(t)$. This is the happy property of linear differential equations and therein lies the magic of the integrating factor.
4. Integrate both sides of the equation obtained in step 3 and make sure to include the constant of integration on both sides. This process results in $e^{\int p(t)\,dt} y(t) + C_1 = \int e^{\int p(t)\,dt} g(t)\,dt$
5. Solve the equation obtained in step 4 to get $y(t)$. First, evaluate the integral and bring C_1 to the right and add to C_2, which is the second constant after evaluating the integral. Then, either divide both sides of the equation by $e^{\int p(t)\,dt}$ or multiply by $e^{-\int p(t)\,dt}$ (multiplication often looks easier).

Two useful corollaries can be made from the presentation given above.

Corollary 1: $g(t) = 0$

In this case it can be seen from Eq. 2.102

$$y(t) = \frac{C_1}{\mu(t)} \qquad (2.115)$$

which can be stated as

$$y(t) = Ce^{-\int p(t)\,dt} \qquad (2.116)$$

Corollary 2: $p(t) = \text{constant}$

Let $p(t) = m$

It follows from Eq. 2.114 that

$$\mu(t) = e^{-\int m\,dt} = e^{-mt} \qquad (2.117)$$

From Eq. 2.102

$$y(t) = \frac{\int e^{-mt} g(t)\,dt + C}{e^{-mt}} \qquad (2.118)$$

$$y(t) = e^{mt}\left[\int e^{-mt} g(t)\,dt + C\right] \qquad (2.119)$$

Further, if $g(t)$ is also a constant, it can be easily simplified by letting $g(t) = I$:

$$y(t) = -\frac{I}{m} + Ce^{mt} \qquad (2.120)$$

2.14 GROWTH OR DECAY EQUATIONS

A second type of differential equation that is quite often found in science and engineering is of the form:

$$\frac{dy}{dt} = ky(t) \qquad (2.121)$$

in which k is a constant. Equation 2.121 implies that the rate of change of quantity y is proportional to the amount of quantity present at that time. This equation gives either growth or decay of quantity y over time, depending on whether k is positive (for growth) or negative (or decay). In Chapter 8, certain baseflow equations are modeled after this equation, which also finds applications elsewhere.

Equation 2.121 is a linear first-order differential equation. So, it can be solved using an integrating factor. However, in addition to being linear, it is a type of differential equation that is called **separable differential equation**. A differential equation that can be written as

$$\frac{dy}{dx} = \frac{g(x)}{h(y)} \qquad (2.122)$$

is called an equation with separable or separated variables. This type of equation can be solved by integrating both sides, since from Eq. 2.122 we deliberately write

$$h(y)\,dy = g(x)\,dx \qquad (2.123)$$

We obviously cannot separate the derivative like that to be mathematically correct, but if we pretend that we can do it, we arrive at the answer with less work by integrating both sides of

$$\int h(y)\,dy = \int g(x)\,dx + c \qquad (2.124)$$

Let us do this for Eq. 2.121. We write,

$$\frac{1}{y(t)}\,dy = k\,dt \qquad (2.125)$$

Now we integrate Eq. 2.125:

$$\int \frac{1}{y(t)}\,dy = \int k\,dt \qquad (2.126)$$

Now using the integration Eq. 2.32 given in Section 2.2.3,

$$\ln[y(t)] = kt + \ln c \qquad (2.127)$$

Note that we use one constant of integration that is absorbed from both sides into one side. From Eq. 2.127,

$$\ln\left[\frac{y(t)}{c}\right] = kt \qquad (2.128)$$

$$y(t) = ce^{kt} \qquad (2.129)$$

Equation 2.129 is the general solution of Eq. 2.121.

To find the constant, c, in Eq. 2.129, we have a problem that is known as the **initial value problem**. In these problems, the value of the quantity y is either given or known at

$t = 0$. This is called the **initial condition**. Let the initial condition be given as

$$y(t) = y_0 @ t = 0 \tag{2.130}$$

Substituting, Eq. 2.130 into Eq. 2.129 we can write

$$y_0 = ce^{(k \times 0)} \tag{2.131}$$

This gives

$$c = y_0 \tag{2.132}$$

Thus, the solution of Eq. 2.121 with the initial condition given as Eq. 2.130 is

$$y(t) = y_0 e^{kt} \tag{2.133}$$

2.15 PERIODIC FUNCTIONS

A function is said to be **periodic** with a period τ if

$$f(t + \tau) = f(t) \tag{2.134}$$

for all t.

Both sine and cosine functions are periodic functions. They are sinusoidal functions. Sin (x) is periodic with period 2π since $\sin(x + 2\pi) = \sin x$. Actually, $\sin(x)$ is periodic with period, 4π, 6π, ..., but when we refer to the period, we generally mean the smallest possible one. In other words, the period is the smallest value of τ in a function f for which there exists some constant τ such that $f(t) = f(t + \tau)$ for every number t in the domain of f. Sin $t = 0$ when $t = n\pi$ where n is an integer (positive or negative, even, or odd) and cos $t = 0$ when $t = m\frac{\pi}{2}$, where m is an odd integer (positive or negative but only odd).

Due to the periodic nature of sine and cosine functions, the following properties hold true.

$$\sin(t \pm 2\pi) = \sin(t) \tag{2.135}$$

$$\cos(t \pm 2\pi) = \cos(t) \tag{2.136}$$

$$\frac{d}{dt}[\sin(t)] = \cos(t) \tag{2.137}$$

$$\frac{d}{dt}[\cos(t)] = -\sin(t) \tag{2.138}$$

$$\int \sin(mt)\,dt = -\frac{\cos(mt)}{m} \tag{2.139}$$

$$\int \cos(mt)\,dt = \frac{\sin(mt)}{m} \tag{2.140}$$

Several terms that are used to describe periodic functions are described below (see Figure 2.8)

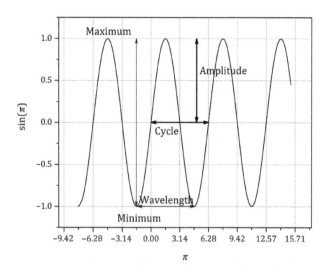

Figure 2.8 Elements of a periodic function illustrated with the sine function. In this graphic, the maximum of $y = \sin x$ is 1 and the minimum is -1. So the amplitude is ½[1 − (−1)] = 1.

1. **Amplitude**: The amplitude A is the distance from the midline or centerline (generally the x-axis if there is no vertical shift) to the highest or lowest point on the function. The amplitude, A, will always be $|A|$ since distance (a magnitude) is always positive. It is also the half of the distance from the crest or peak (maximum) to the trough (minimum) or the height of the oscillation (wave) of a sinusoidal function: $A = \frac{1}{2}|y_{max} - y_{min}|$.

2. **Period**: As noted above, a periodic function is the one which repeats a pattern of y-values at regular intervals. One complete repetition of the pattern is called a **cycle**. The period, T, of a periodic function is the horizontal distance of one cycle. It is also the distance between two successive peaks (maximum) or two successive troughs (they are same). The term period is equivalent to the term **wavelength**, λ. In general, the period refers to time (if the x-axis is representing time) whereas the term wavelength is used for distance (if the x-axis is representing distance or space). Thus, $T \equiv \lambda$. A period can be the distance from any point on one wave form to the equivalent point on the next wave form, that is, the distance between two repeating points on the function. The sin function $y = \sin x$, shown in Figure 2.8 has a period of 2π radians (360°), which is the horizontal length of one complete cycle. Note that we really should not say radian after 2π. It is a number only.

3. **Frequency**: This is the number of cycles a sinusoidal function completes in a given interval. This interval is generally 2π for sine and cosine functions. The relationship between frequency and period is given as

$$f = \frac{1}{T} \equiv \frac{1}{\lambda} \tag{2.141}$$

So, frequency is per period [e.g., per second (time)]. An example is Hertz: 10 Hertz means 10 times per second. For example, $y = \sin x$ completes one cycle in the interval 0 to 2π. So, its frequency is 1 in the interval $[0, 2\pi]$. Figure 2.8 illustrates this. The **natural frequency** or the **fundamental frequency** is defined as

$$\omega = \frac{2\pi}{T} \qquad (2.142)$$

4. **Phase shift**: The phase shift, ϕ, of a sinusoidal function is the horizontal shift of the curve from the y-axis.
5. **Vertical shift**: The vertical shift, D, of a sinusoidal curve is the vertical shift of the curve from the x-axis.

The general forms of a sine or cosine function are given as:

$$y = A\sin[B(t-\phi)] + D \qquad (2.143)$$

$$y = A\cos[B(t-\phi)] + D \qquad (2.144)$$

Note that we have already defined A, ϕ, and D above but have not yet defined B. In Eqs. 2.143 and 2.144, B is a parameter that controls the period T (hence frequency, f, via Eq. 2.141). It is defined as

$$B = \frac{2\pi}{T} \qquad (2.145)$$

Once B is defined this way, we can write Eqs. 2.143 and 2.144 in two other ways if we simply want to specify either period or frequency.

$$y = A\sin\left[\frac{2\pi}{T}(t-\phi)\right] + D \qquad (2.146)$$

$$y = A\cos\left[\frac{2\pi}{T}(t-\phi)\right] + D \qquad (2.147)$$

Alternatively,

$$y = A\sin[2\pi f(t-\phi)] + D \qquad (2.148)$$

$$y = A\cos[2\pi f(t-\phi)] + D \qquad (2.149)$$

If we want to control the period or frequency, we use the parameter B, which affects or changes the period (wavelength) and hence the frequency in various ways. For example, for a cosine function, if B is equal to 1, it takes 2π to complete a period. If B is equal to 2, it takes only π to complete a period. If B is equal to 3, we can see that there will be nine complete waves in a distance along the x-axis of 6π, thus the period is $6\pi/9 = 2\pi/3$. If B is equal to $1/2$, it takes only 4π to complete a period, which is twice as long as the period for $y = \cos x$. Similarly, for a sine function, if $B = 1/2$, it takes 4 π units to complete a full period, which is twice as long as the period for $y = \sin x$, which is equal to 2π.

Finally, a note about the sign of ϕ (phase shift) that controls the horizontal shift of the curves. If ϕ is negative in Eqs. 2.143 and 2.144, the curve is shifted to the left. If it is positive, the curve is shifted to the right. Also, in these equations, if we used + instead of − before ϕ, the opposite would hold true. These forms of sine and cosine functions can model some functions in hydrology.

The tangent and cotangent functions have a period of π:

$$\tan(\theta \pm n\pi) = \tan(\theta) \qquad (2.150)$$

$$\cot(\theta \pm n\pi) = \cot(\theta) \qquad (2.151)$$

$$\frac{d}{dx}[\tan(x)] = \sec^2 x \qquad (2.152)$$

$$\frac{d}{dx}[\cot(x)] = -\csc^2 x \qquad (2.153)$$

$$\int \tan x = \ln|\sec x| \qquad (2.154)$$

2.16 FOURIER SERIES

Like a Taylor series, which provides a polynomial approximation of a function, a **Fourier**[6] **series** provides an approximation of a function by an infinite trigonometric series. It should be noted here that all functions cannot be represented by a Fourier series. There are certain conditions for a function to be met for it to be represented by a Fourier series. We will not go into the mathematical details here.

If a Fourier series for a function exists, then the Fourier series of the function is given by

$$f(t) = \frac{a_0}{2} + \sum_{n=1}^{\infty}(a_n \cos nt + b_n \sin nt) \qquad (2.155)$$

where $n = 1, 2, 3, \ldots$; and a_0, a_n, and b_n are called **Fourier coefficients**. When these coefficients are determined, the function can be represented by a Fourier series.

The Fourier coefficients are defined by the integrals

$$a_0 = \frac{1}{\pi}\int_{-\pi}^{\pi} f(t)\,dt \qquad (2.156)$$

---

[6] Jean-Baptiste Joseph Fourier (1768–1830) was a French mathematician and physicist best known for initiating the investigation of Fourier series, which eventually developed into Fourier analysis and harmonic analysis (an example given in Chapter 5), and their applications to problems of heat transfer, vibrations, or wave propagation. The Fourier transform and Fourier's law of conduction are also named in his honor. Fourier is also generally credited with the discovery of the greenhouse effect. Interestingly, Laplace was critical of his work.

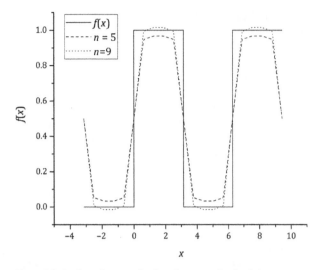

Figure 2.9 Fourier series approximation of a rectangular signal given by: $f(x) = \begin{cases} 1 & 0 < x \leq \pi \\ 0 & \pi < x < 2\pi \\ 0 & -\pi < x < 0 \end{cases}$

$$a_n = \frac{1}{\pi} \int_{-\pi}^{\pi} f(t) \cos(nt)\, dt \qquad (2.157)$$

$$b_n = \frac{1}{\pi} \int_{-\pi}^{\pi} f(t) \sin(nt)\, dt \qquad (2.158)$$

The process of finding the values of a_0, a_n, and b_n is called **Fourier analysis**. Equations 2.156–2.158 require that $f(t)$ is continuous over an interval $[-\pi, \pi]$. If the interval is $[0, 2\pi]$, only the limits of integration change. Note that a_0, is the average value of the function over the limits of integration.

As an example, consider a rectangular signal that can be described by a function

$$f(x) = \begin{cases} 1 & 0 < x \leq \pi \\ 0 & \pi < x < 2\pi \\ 0 & -\pi < x < 0 \end{cases}$$

Such a function cannot be represented by an analytical form that can be manipulated by mathematical operations. Here comes the utility of a series expansion of a function. If we want to represent it or approximate it by a Fourier series, we find the Fourier coefficients as (derive these by doing the integration given in Eqs. 2.156–158):

$$a_0 = 1$$

$$a_n = \begin{cases} 1 & \text{for } n = 0 \\ 0 & \text{for } n > 0 \end{cases}$$

$$b_n = \begin{cases} 0 & \text{for } n = \text{Even} \\ \dfrac{2}{\pi n} & \text{for } n = \text{Odd} \end{cases}$$

So, the Fourier series approximation is given as

$$f(x) = \frac{1}{2} + \frac{2}{\pi} \sum_{n=1,3,5,7,\ldots}^{n} \frac{\sin(nx)}{n}$$

or

$$f(x) = \frac{1}{2} + \frac{2}{\pi} \sum_{n=1}^{n} \frac{\sin[(2n-1)x]}{(2n-1)}$$

Figure 2.9 shows the approximation for two values of n. Obviously, the approximation for $n = 9$ is better than that for $n = 5$.

2.17 FOURIER TRANSFORM

The Fourier series expresses any periodic function into a sum of sinusoids. The **Fourier transform** is the extension of this idea to non-periodic functions by taking the limiting form of the Fourier series when the fundamental period is made very large (infinite). A Fourier transform is another integral transform that finds its applications in time-series analysis, signal processing, LTI systems etc. It transforms a function from the time domain (t) to the frequency (ω) domain and it is also invertible.

Given a function $f(t)$, its Fourier transform $\mathcal{F}(\omega)$, written as $\mathcal{F}\{f(t)\}$, is given by

$$\mathcal{F}(\omega) = \mathcal{F}\{f(t)\} = \int_{-\infty}^{+\infty} f(t) e^{-i\omega t}\, dt \qquad (2.159)$$

The inverse transform that brings the function back from the frequency domain to its original time domain, written as $\mathcal{F}^{-1}\{F(\omega)\}$, is given as

$$f(t) = \mathcal{F}^{-1}\{\mathcal{F}(\omega)\} = \frac{1}{2\pi} \int_{-\infty}^{+\infty} \mathcal{F}(\omega) e^{i\omega t}\, d\omega \qquad (2.160)$$

Like any integral transform, a Fourier transform requires the evaluation of the integral of the product of two functions. However, in this case a knowledge of the algebra and calculus involving complex numbers is required. In the equations above, $i = \sqrt{-1}$. For example, if we define a function

$$f(t) = \begin{cases} 0 & t < 0 \\ e^{-1} & t \geq 0 \end{cases}$$

which is called a one-sided decaying exponential function, then its Fourier transform is given as

$$\int_0^\infty e^{-t} e^{-i\omega t}\, dt = \frac{1}{i\omega + 1} = \mathcal{F}(i\omega)$$

2.18 MATRIX ALGEBRA

Applications of matrix algebra and certain theories of linear algebra are used in various places in this book, and especially extensively in Chapter 9, which is an extremely important chapter.

2.18.1 Definition of a Matrix

If m and n denote positive integers, a rectangular array of numbers (entries) arranged in m rows and n columns is called an m **by** n **matrix**. The entries of a matrix are called **elements**. The following is a symbolic representation of a matrix.

$$A = \begin{bmatrix} a_{11} & a_{12} & \ldots & a_{1n} \\ a_{21} & a_{22} & \ldots & a_{2n} \\ . & . & & . \\ & & \ldots & \\ . & . & & . \\ a_{m1} & a_{m2} & \ldots & a_{mn} \end{bmatrix} \qquad (2.161)$$

The element of a matrix is denoted by a_{ij} where i denotes the row number and j denotes the column number. In general, entries of a matrix are **scalars**, which include real and complex numbers or can even be a functional value thereof.

One use of matrix notation is to enable us to manipulate large rectangular arrays of numbers as single entities. Algebraic operations can be done on the entire array at a time.

A common notation for a matrix is by a capital letter and is written as, for example, $A = [a_{ij}]_{(m,n)}$, where $i = 1, 2, \ldots, m$ and $j = 1, 2, \ldots, n$. The double subscript can be called the address of the entry.

The dimensions of the array, that is, the number of rows and columns, determine the **order** of the matrix. It is designated as m by n. If $m = n$, it is a called a **square matrix**.

A matrix with a single row is called **row matrix** and a matrix with a single column is called a **column matrix**.

Two matrices, A and B, are said to be equal if $a_{ij} = b_{ij}$.

The **transpose** A^T of an m by n matrix $A = [a_{ij}]_{(m,n)}$ is the n by m matrix whose entry in row i and column j is a_{ji}, the entry in row j and column i of A. Thus, the rows of A are the columns of A^T and the columns of A are the rows of A^T.

A **vector** of dimension n is a list (in specific order) of n scalars, $a_1, a_2, a_3, \ldots, a_n$. From this definition, a vector is either a row matrix or a column matrix. It has its own properties.

2.18.2 Matrix Operations

Matrix addition and subtraction can only be performed if the orders of two matrices are equal.

If $A = [a_{ij}]_{(m,n)}$ and $B = [b_{ij}]_{(m,n)}$, then

$$A + B = [a_{ij} + b_{ij}]_{(m,n)} \qquad (2.162)$$

Obviously, matrix addition is commutative: $A + B = B + A$.

Given a scalar c and a matrix $A = [a_{ij}]_{(m,n)}$,

$$cA = [ca_{ij}]_{(m,n)} \qquad (2.163)$$

In order for two matrices to be multipliable, the number of columns of the first matrix has to be equal to the number of rows of the second matrix. Thus, let A be an m by p matrix and B be a p by n matrix. The product $C = AB$ is an m by n matrix where each entry c_{ij} of C is obtained by multiplying the corresponding entries of the ith row of A by those of the jth column of B and then adding the result. Thus,

$$c_{1j} = a_{i1}b_{1j} + a_{i2}b_{2j} + \ldots + a_{ip}b_{pj} \qquad (2.164)$$

In a compact summation notation,

$$c_{ij} = \sum_{r=1}^{p} a_{ip}b_{pj} \qquad (2.165)$$

for $i = 1, 2, \ldots, m$ and $j = 1, 2, \ldots, p$.

Many rules of real and complex number algebra are applicable to matrices. However, there are certain exceptions. For example, it must be remembered that matrix multiplication, in general, is noncommutative, i.e.,

$$A \times B \neq B \times A \qquad (2.166)$$

Some important rules of algebraic operations of matrices are summarized below.

Let A, B, and C be matrices of appropriate sizes so that the indicted operations can be performed, and r and s are any scalars. Then, the following holds true.

$$(AB)C = A(BC) \qquad (2.167)$$

$$A(B + C) = AB + AC \qquad (2.168)$$

$$(A + B)C = AC + BC \qquad (2.169)$$

$$r(AB) = (rA)B = A(rB) \qquad (2.170)$$

$$-1(A) = -A \qquad (2.171)$$

$$-(-A) = A \qquad (2.172)$$

$$A - B = -(B - A) \qquad (2.173)$$

$$A(B - C) = AB - AC \qquad (2.174)$$

$$(A - B)C = AC - BC \qquad (2.175)$$

$$(r - s)A = rA - sA \qquad (2.176)$$

$$r(A - B) = rA - rB \qquad (2.177)$$

$$0A = A0 = 0 \qquad (2.178)$$

2.18.3 Inverse of a Matrix

A matrix $A = [a_{ij}]_{(n,n)}$ is called a square matrix because here the number of rows equals the number of columns. The entries a_{ij}, whose row and column indices are the same, are said to lie on the main diagonal of A.

For each positive integer n, the n by n matrix I_n, whose main diagonal entries are all equal to 1 and all the rest of the entries are zero, is called the n by n **identity matrix**. This is because $I_n M = M I_n = M$ for any n by n matrix M.

If for a given n by n matrix A, there is an n by n matrix, designated A^{-1} such that

$$AA^{-1} = A^{-1}A = I_n \qquad (2.179)$$

A^{-1} is an *inverse* of matrix A with respect to matrix multiplication. An n by n matrix A is said to be invertible if A^{-1} exists and noninvertible if A does not have an inverse.

There are a couple of different ways to calculate the inverse of a matrix. We present one equation by which the inverse of a matrix can be calculated after introducing the determinant of a matrix.

If A and B are n by n matrices and $AB = I$ or $BA = I$, then A has an inverse B.

The transpose of the inverse of a matrix is referred to as the **reciprocal matrix**. A matrix is said to be **orthogonal** if its inverse is equal to its transpose, that is,

$$A^T = A^{-1} \qquad (2.180)$$

In addition, the following holds true.

$$(A^T)^T = A \qquad (2.181)$$

$$(rA)^T = rA^T \qquad (2.182)$$

$$(A+B)^T = A^T + B^T \qquad (2.183)$$

$$(AB)^T = B^T A^T \qquad (2.184)$$

A square matrix A is said to be **symmetric** if $A = A^T$.

2.18.4 Matrix Representation of a System of Linear Equations

A **linear equation** in n variables is an equation of the form:

$$a_1 x_1 + a_2 x_2 + a_3 x_3 + \ldots + a_n x_n = b \qquad (2.185)$$

where b and a_i, $i = 1, 2, \ldots, n$ are all real numbers and constants and x_i, $i = 1, 2, 3, \ldots, n$ are variables. A **system of linear equations** is a collection of such equations. A **solution** to a system of linear equations is any ordered n-tuple $(c_1, c_2, \ldots, c_n)$ of real numbers such that each of the equations of the system is satisfied when the values $c_1, c_2, \ldots, c_n$ are substituted for $x_1, x_2, \ldots, x_n$, respectively.

Consider, therefore, a system of m equations in n variables:

$$\begin{aligned} a_{11}x_1 + a_{12}x_2 + \ldots\ldots + a_{1n} &= b_1 \\ a_{21}x_1 + a_{22}x_2 + \ldots\ldots + a_{2n} &= b_2 \\ \ldots\ldots\ldots\ldots\ldots\ldots\ldots\ldots\ldots\ldots\ldots& \\ a_{m1}x_1 + a_{m2}x_2 + \ldots\ldots + a_{mn} &= b_m \end{aligned} \qquad (2.186)$$

Arthur Cayley, an English mathematician, first used a matrix notation in 1858 to represent Eq. 2.186 in an abbreviated form. Equation 2.186 can be expressed in matrix notation as

$$AX = B \qquad (2.187)$$

where, from the rule of matrix multiplication, we can write

$$A = \begin{bmatrix} a_{11} & a_{12} & \ldots & a_{1n} \\ a_{21} & a_{22} & \ldots & a_{2n} \\ . & . & & . \\ & & \ldots & \\ . & . & & . \\ a_{m1} & a_{m2} & \ldots & a_{mn} \end{bmatrix}, X = \begin{bmatrix} x_1 \\ x_2 \\ . \\ . \\ . \\ x_n \end{bmatrix}, B = \begin{bmatrix} b_1 \\ b_2 \\ . \\ . \\ . \\ b_m \end{bmatrix} \qquad (2.188)$$

In this notation, the matrix A is called the **coefficient matrix**. The system of linear equations represented by Eq. 2.187 has a **unique solution**, if there exists a vector $(c_1, c_2, \ldots, c_n)$ such that n identities result when its n components are substituted for the respective unknowns of the system.

If a unique solution to the system represented in Eq. 2.187 exists, it can be found by the matrix method. The procedure is to *pre-multiply* both sides of A^{-1} and proceed as follows:

$$\begin{aligned} A^{-1}(AX) &= A^{-1}B, \\ (A^{-1}A)X &= A^{-1}B, \\ I_n X &= A^{-1}B \\ X &= A^{-1}B \end{aligned} \qquad (2.189)$$

Thus, $A^{-1}B$ is a column vector that gives the values of all the x_i and the following fundamental theorem of linear algebra is proved:

If the system $AX = B$, where A is invertible, has a unique solution, the solution is $X = A^{-1}B$.

This method of solving a system of linear equations is valid only when the number of equations equals the number of unknowns and when A is invertible.

A lemma of linear algebra is this: Every system of linear equations has either no solutions, or exactly one solution called the unique solution, or infinitely many solutions.

2.18.5 Determinant of a Matrix

It should have been observed by now that the definition of a matrix A does not assign a numerical value to A. We now define a function of the entries of a square matrix in such a way that to every square matrix of scalars there corresponds a unique scalar. This function is called the **determinant** or the **determinant function**. Its value is denoted by det A or $|A|$ and is called the determinant of A.

The idea of the determinant was started by Seki Kowa, the great Japanese mathematician, in about 1683 in Japan, and Gottfried Leibniz in Germany, in about 1693. However, it was not until 1812 when Baron Augustin-Louis Cauchy

(1789–1857), a French mathematician, engineer, and physicist who made pioneering contributions to several branches of mathematics, including mathematical analysis and continuum mechanics, gave the function its present name: determinant.

The determinant of $A = [a_{i,j}]_{(m,n)}$ is the sum of all terms of the form $(-1)^t a_{1j_1} a_{2j_2} \ldots a_{nj_n}$, where the column subscripts assume all possible arrangements in which each column is represented exactly once in each term of the sum, and the exponent t is the number of interchanges necessary to bring the column subscript into natural order, that is, $1, 2, 3, \ldots, n$. Let us illustrate this apparently confusing definition, with two examples below:

$$\det \begin{bmatrix} a_{11} & a_{12} \\ a_{21} & a_{22} \end{bmatrix} = |A| = (-1)^0 a_{11} a_{22} + (-1)^1 a_{12} a_{21}$$
$$= a_{11} a_{22} - a_{12} a_{21}$$

$$\det \begin{bmatrix} a_{11} & a_{12} & a_{13} \\ a_{21} & a_{22} & a_{23} \\ a_{31} & a_{32} & a_{33} \end{bmatrix} = |A|$$
$$= (-1)^0 a_{11} a_{22} a_{33} + (-1)^1 a_{11} a_{23} a_{32}$$
$$+ (-1)^1 a_{12} a_{21} a_{33} + (-1)^2 a_{12} a_{23} a_{31}$$
$$+ (-1)^2 a_{13} a_{21} a_{32} + (-1)^1 a_{13} a_{22} a_{31}$$
$$= a_{11} a_{22} a_{33} - a_{11} a_{23} a_{32} - a_{12} a_{21} a_{33} + a_{12} a_{23} a_{31}$$
$$+ a_{13} a_{21} a_{32} - a_{13} a_{22} a_{31}$$

The expansion of $[a_{ij}]_{(3,3)}$, given above, can be also written using summation or sigma notation as

$$\det A = \sum_{(j)} (-1)^t a_{1j_1} a_{2j_2} a_{3j_3}$$

When A is of order n by n, we say that $\det A$ is of order n.

The **minor** of the entry a_{ij} in row i and column j in matrix A in Eq. 2.161 is the determinant of the submatrix obtained by deleting row i and column j from A. The **cofactor** of the entry a_{ij} in matrix A is the product of the minor of a_{ij} and $(-1)^{i+j}$. If the cofactor of a_{ij} is noted by A_{ij}, then

$$A_{ij} = \text{Cofactor of } a_{ij} = (-1)^{i+j} \cdot (\text{Minor of } a_{ij}) \quad (2.190)$$

Having defined the cofactor of a matrix, we are now able to give a better formula for the expansion of determinants, known as the **Laplace expansion of the determinant**. Once again, we encounter one of the numerous fundamental contributions made by Laplace.

$$\det A = a_{i1} A_{i1} + a_{i2} A_{i2} + a_{i3} A_{i3} + \ldots + a_{in} A_{in} = \sum_{j=1}^{n} a_{ij} A_{ij}$$
$$(2.191)$$

Equation 2.191 indicates that this operation, namely the expansion, can be performed on any row $i = 1, 2, \ldots, n$ to arrive at the same result.

2.18.6 Cramer's Rule

One way, and probably not the best way to solve certain systems of n linear equation in n unknowns is called Cramer's rule, named after Swiss mathematician Gabriel Cramer (1704–1752).

Consider the system of equations given by Eq. 2.186. Let (^jA) denote the matrix obtained from A by replacing jth column of A by vector B. Then, Cramer's rule is as follows.

If $\det A \neq 0$, the system $AX = B$ has exactly one solution, which is given by

$$x_j = \frac{\det(^jA)}{\det A} \quad (2.192)$$

Consider for example the system of equations

$$2x_1 - 3x_2 = 1$$
$$5x_1 + x_2 = 11$$

$$\Delta = \det \begin{bmatrix} 2 & -3 \\ 5 & 1 \end{bmatrix} = (2)(1) - (-3)(5) = 17 \neq 0$$

Thus, we may apply Cramer's rule to obtain

$$x_1 = \frac{\det \begin{bmatrix} 1 & -3 \\ 11 & 1 \end{bmatrix}}{\Delta} = \frac{(1)(1) - (-3)(11)}{17} = 2$$

$$x_2 = \frac{\det \begin{bmatrix} 2 & 1 \\ 5 & 11 \end{bmatrix}}{\Delta} = \frac{(2)(11) - (1)(5)}{17} = 1$$

2.18.7 Invertibility of a Matrix: Singular and Nonsingular Matrices

A fact of real and complex number algebra is that for any nonzero number a, the equation $ax = 1$ has a unique solution $x = a^{-1} = \frac{1}{a}$. The number a^{-1} is called the multiplicative **inverse** of a. In contrast, it is not always true that a given nonzero matrix A has a multiplicative inverse. In those cases where A is an n by n matrix for which there does exist an n by n matrix B,

$$AB = BA = I_n \quad (2.193)$$

If Eq. 2.193 holds true, we call A an **invertible** or **nonsingular matrix** and say that the matrix B is the inverse. In this case we write $B = A^{-1}$. When such a matrix, B, does not exist, we say that A is a **singular matrix**.

We defined the cofactor of an element of a matrix in Eq. 2.190. Now we define the **cofactor matrix**. The cofactor matrix of $A = (a_{ij})$, denoted by $\text{Cof } A$, is defined as the n by n matrix whose entry in row i and column j is A_{ij}; that is,

$$\text{Cof} A = (A_{ij}) \quad (2.194)$$

Another useful definition is the **adjoint** or **adjugate** of A, denoted by adj A, and defined as the transpose of the cofactor matrix A. That is,

$$\text{adj} A = (\text{Cof} A)^T \quad (2.195)$$

Now we state an important theorem:

> If A is an n by n matrix, the inverse of A, A^{-1} exists if and only if $\det A \neq 0$.

Thus, for a singular matrix, $\det A = 0$. In other words, if the determinant of a matrix is zero, it is not invertible. For a nonsingular matrix,

$$A^{-1} = \frac{1}{\det A} \cdot (\text{adj} A) \quad (2.196)$$

Equation 2.196 provides a means by which to calculate the inverse of a matrix.

2.18.8 Rank of a Matrix

A submatrix of a matrix A is the rectangular array that remains if certain rows or columns or both of A are deleted. The **rank** of a matrix A is the greatest integer r for which A has an r by r submatrix whose determinant is zero.

2.18.9 Norm of a Matrix

A matrix **norm** is a number defined in terms of the entries of the matrix. The norm is a useful quantity that can give important information about a matrix. The norm of a matrix is a measure of how large its elements are. It is a way of determining the size of a matrix, which is not necessarily related to how many rows or columns the matrix has. The norm of a square matrix A is a non-negative real number denoted $\|A\|$. There are several different ways of defining a matrix norm. One way to define the norm of a matrix is

$$\|A\|_1 = \max_{1 \leq j \leq n} \left(\sum_{i=1}^{n} |a_{ij}| \right) \quad (2.197)$$

Equation 2.197 is called the 1-norm of a square matrix, which is the maximum of the absolute column sums. Put simply, we sum the absolute values down each column and then take the biggest answer.

2.18.10 Condition Number of a Matrix

If in a system of linear equations, represented in the matrix form such as Eq. 2.187, the coefficient matrix is such that a small change in the constant coefficients results in a large change in the solution, the coefficient matrix is said to be **ill-conditioned**. A **condition number**, defined in more advanced courses, is used to measure the degree of ill-conditioning of a matrix. A condition number of a matrix measures how sensitive the answer is to perturbations in the input data and to roundoff errors made during the solution process. The concept of condition numbers appears in Chapter 9 in relation to the determination of unit hydrographs by certain matrix operations.

2.19 EXAMPLES

Example 2.1: Determine from first principles the Laplace transform of the function $f(t) = t^3$.

Solution:

$$F(s) = \int_0^\infty e^{-st} t^3 \, dt$$

This can be solved using integration by parts or the integration rule of products of two functions given in Section 2.2.2:

$$\int u \, dv = uv - \int v \, du$$

with $dv = e^{-st}$ and $u = t^3$. Thus,

$$F(s) = \left[\frac{t^3 e^{-st}}{-s} \right]_0^\infty - \int_0^\infty \frac{3t^2 e^{-st}}{-s} \, dt$$

This integral can then have integration by parts applied to it, with $dv = e^{-st}$ and $u = t^2$

$$F(s) = \left[\frac{t^3 e^{-st}}{-s} \right]_0^\infty - \left[\frac{3t^2 e^{-st}}{s^2} \right]_0^\infty - \int_0^\infty \frac{3 \times 2t e^{-st}}{s^2} \, dt$$

Again, integration by parts with $dv = e^{-st}$ and $u = t$ gives

$$F(s) = \left[\frac{t^3 e^{-st}}{-s} - \frac{3t^2 e^{-st}}{s^2} - \frac{3 \times 2t e^{-st}}{s^3} - \frac{3 \times 2t e^{-st}}{s^4} \right]_0^\infty$$

$$= \frac{3 \times 2}{s^4} = \frac{3!}{s^4}$$

> **Notes on Infinite Series**
>
> The study of infinite series is an important topic in applied mathematics. Situations like this arise often for evaluation of definite integrals when the upper limit is ∞. The summation can be evaluated when the series converges, either rapidly, or slowly. A few examples of infinite series are given below.
>
> A. Exponential series
>
> $$e^x = 1 + x + \frac{x^2}{2!} + \frac{x^3}{3!} + \frac{x^4}{4!} + \ldots$$
>
> $$e^{cx} = 1 + cx + \frac{c^2 x^2}{2!} + \frac{c^3 x^3}{3!} + \frac{c^4 x^4}{4!} + \ldots$$
>
> $$a^x = 1 + x \ln a + \frac{1}{2!} x^2 (\ln a)^2 + \frac{1}{3!} x^3 (\ln a)^3 + \ldots$$
>
> B. Logarithmic series
>
> $$\ln(1+x) = x - \frac{x^2}{2} + \frac{x^3}{3} - \frac{x^4}{4} + \ldots$$

Example 2.2: Find the Taylor series for $f(x) = 7x^2 - 6x + 1$ at about $x = 2$.

Solution: We start by taking successive derivatives (order n) of the function and evaluating them at $x = 2$

$n = 0 : f(x) = 7x^2 - 6x + 1 \quad f(2) = 17$
$n = 1 : f'(x) = 14x - 6 \quad f'(2) = 22$
$n = 2 : f''(x) = 14 \quad f''(2) = 14$
$n \geq 3 : f^{(n)}(x) = 0 \quad f^{(n)}(2) = 0$

For all polynomial functions, after some point all the derivatives will be zero. Because all the derivatives are zero after some point, we do not need a formula for the general term. All we need are the values of the non-zero derivative terms. Thus, Taylor's polynomial is written as

$$P_n(x) = f(2) + f'(2)(x-2) + \frac{1}{2} f''(2)(x-2)^2$$
$$= 17 + 22(x-2) + 7(x-2)^2$$

No further simplification is done. This is the answer, albeit it may appear strange.

Example 2.3: A time-dependent input to a system is given by the function $g(t) = \sin(t)$. The system processes the input according to the function $f(t) = e^{-t}$. What is the function for the output signal? Illustrate graphically for a conceptual representation of what happens in the process.

Solution: The problem at hand is a convolution problem where the transfer function is e^{-t}.

From the definition of convolution operation, we write $(f * g)(t) = \int_0^t e^{-\tau} \sin(t - \tau) d\tau$
Now we integrate it in parts twice:

$$\int_0^t e^{-\tau} \sin(t-\tau) d\tau$$
$$= [e^{-\tau} \cos(t-\tau)]_0^t - [e^{-\tau} \sin(t-\tau)]_0^t - \int_0^t e^{-\tau} \sin(t-\tau) d\tau$$

$$2 \int_0^t e^{-\tau} \sin(t-\tau) d\tau = [e^{-\tau} \cos(t-\tau)]_0^t - [e^{-\tau} \sin(t-\tau)]_0^t$$

$$2(f * g)(t) = e^{-t} - \cos(t) - 0 + \sin(t)$$

$$(f * g)(t) = \frac{1}{2} [e^{-t} - \cos(t) + \sin(t)]$$

Figure 2.10 shows the graphs of the three functions. From the plots of the original function and convoluted response function it can be said that the input signal is dampened and delayed as the response.

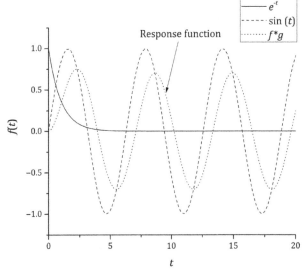

Figure 2.10 Convolution of e^{-t} and $\sin t$.

Application of first-order linear differential equation to an electrical circuit

Figure 2.11 shows a simple resistor–inductor (RL) circuit where the resistance by the resistor is R ohms (Ω) and the inductance of the inductor is L Henries (H). Ohm's law gives the drop in voltage due to the resistor is iR. The drop in voltage due to the inductor is $\frac{di}{dt}$. Kirchhoff's second law states that the sum of the voltage drops across the inductor, L, $\frac{di}{dt}$ and the voltage drops across a resistor, R (iR) is the same as the impressed voltage $E(t)$, by an electromotive force, usually a battery or generator. The battery or the generator produces a voltage of $E(t)$ volts (V) and a current of $i(t)$ amperes (A) at the time t. Thus, Kirchhoff's second law giving the sum of the voltage drop equaling the supplied voltage $E(t)$ can be written as a first-order linear differential equation:

$$L\frac{di}{dt} + Ri = E(t)$$

The solution gives current i at time t.

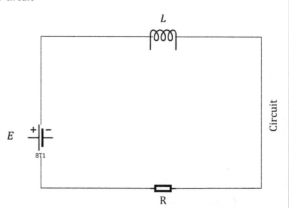

Figure 2.11 A simple RL circuit.

Example 2.4: In an RL circuit, instead of a battery, a generator is used that produces a variable voltage of $E(t) = 60 \sin 30t$ volts. $L = 4$ H and $R = 12$ Ω. Determine $i(t)$.

Solution: The differential equation for this circuit is written as

$$4\frac{di}{dt} + 12i = 60 \sin 30t$$

Divide the equation by 4 to get

$$\frac{di}{dt} + 3i = 15 \sin 30t$$

IF $= e^{\int 3dt} = e^{3t}$

Now multiplying the differential equation by the integrating factor (IF)

$$e^{3t}\frac{di}{dt} + e^{3t}3i = e^{3t}15 \sin 30t$$

$$\frac{d}{dt}\left(e^{3t}i\right) = e^{3t}15 \sin 30t$$

Integrating both sides and, as usual, using the constant of integration only on the right-hand side:

$$e^{3t}i = \int 15 e^{3t} \sin 30t\, dt$$

Using a table of integrals given in most college-level texts on calculus:

$$e^{3t}i = \frac{15e^{3t}}{909}(3\sin 30t - 30\cos 30t) + C$$

$$\therefore i(t) = \frac{5}{101}(\sin 30t - 10\cos 30t) + Ce^{-3t}$$

Setting $i(0) = 0$ for $t = 0$ we get $-\frac{50}{101} + C = 0$. Therefore,

$$i(t) = \frac{5}{101}(\sin 30t - 10\cos 30t) + \frac{50}{101}e^{-3t}$$

Example 2.5: Find the inverse of the matrix A given below.

$$A = \begin{bmatrix} 2 & 1 & 0 \\ 0 & 2 & 1 \\ 3 & 0 & 2 \end{bmatrix}$$

Solution: We first find the cofactor matrix of A (C of A).

$$\text{Cof } A = \begin{bmatrix} +\begin{vmatrix} 2 & 1 \\ 0 & 2 \end{vmatrix} & -\begin{vmatrix} 0 & 1 \\ 3 & 2 \end{vmatrix} & +\begin{vmatrix} 0 & 2 \\ 3 & 0 \end{vmatrix} \\ -\begin{vmatrix} 1 & 0 \\ 0 & 2 \end{vmatrix} & +\begin{vmatrix} 2 & 0 \\ 3 & 2 \end{vmatrix} & -\begin{vmatrix} 2 & 1 \\ 3 & 0 \end{vmatrix} \\ +\begin{vmatrix} 1 & 0 \\ 2 & 1 \end{vmatrix} & -\begin{vmatrix} 2 & 0 \\ 0 & 1 \end{vmatrix} & +\begin{vmatrix} 2 & 1 \\ 0 & 2 \end{vmatrix} \end{bmatrix}$$

$$= \begin{bmatrix} 4 & 3 & -6 \\ -2 & 4 & 3 \\ 1 & -2 & 4 \end{bmatrix}$$

Now we find the adjoint of matrix A. By definition, it is the transpose of the cofactor matrix.

$$\text{adj} A = (\text{Cof} A)^T = \begin{bmatrix} 4 & 3 & -6 \\ -2 & 4 & 3 \\ 1 & -2 & 4 \end{bmatrix}^T = \begin{bmatrix} 4 & -2 & 1 \\ 3 & 4 & -2 \\ -6 & 3 & 4 \end{bmatrix}$$

Now we find the determinant of A:

$$\det A = \begin{bmatrix} 2 & 1 & 0 \\ 0 & 2 & 1 \\ 3 & 0 & 2 \end{bmatrix}$$

$$= (-1)^0 2.2.2 + (-1)^1 2.1.0 + (-1)^1 1.0.2 \\ + (-1)^2 1.1.3 + (-1)^2 0.0.0 + (-1)^1 0.2.3$$

$$= 8 + 0 + 0 + 3 = 11$$

$$A^{-1} = \frac{\text{adj} A}{|A|} = \begin{bmatrix} \frac{4}{11} & -\frac{2}{11} & \frac{1}{11} \\ \frac{3}{11} & \frac{4}{11} & -\frac{2}{11} \\ -\frac{6}{11} & \frac{3}{11} & \frac{4}{11} \end{bmatrix}$$

Now it can be verified by matrix multiplication:

$$AA^{-1} = \begin{bmatrix} 1 & 0 & 0 \\ 0 & 1 & 0 \\ 0 & 0 & 1 \end{bmatrix} = I_3 = A^{-1}A$$

Exercises: A selection of exercises on this topic is available at www.cambridge.org/appliedhydrology.

3 Statistical Hydrology

3.1 CONCEPTS AND PURPOSE

The utility of statistics lies in its power of representing a large set of numbers by a few key numbers or a comprehensive set of mathematical equations, which succinctly convey the major characteristics of the dataset. In hydrology, important inferences are drawn from sets of large numbers, such as daily discharge of a river, recorded at one or more gauging stations, for many years. This is the main reason why hydrologists must have a sound grounding in statistics. This chapter covers the principal statistical concepts useful in applied hydrology. The concepts necessary to extract meaningful information from the apparently random behavior of hydrologic quantities are emphasized. This is because hydrologic problems involve variables that are stochastic in their very nature, yet hydrologic problems are visualized and analyzed using quantitative models, which can be deterministic, parametric, or probabilistic. The probabilistic models capture the stochastic behavior of hydrologic quantities. The principles of statistics and probability theory required for building stochastic models used in applied hydrology are the major topics of this chapter. Haan (1977) presented a detailed account of statistical hydrology.

3.2 RANDOM VARIABLES AND VARIATES

Hydrologic observations through either direct measurements or through calculations are made largely with **random variables**, meaning variables whose values are subject to chance fluctuations and cannot be known exactly beforehand. In statistics, the term **variate** is used to distinguish between the variables and the numbers the variables can assume. Variate means the numbers that represent a variable with units such as $m^3 s^{-1}$ for river flows and $mm\, h^{-1}$ for rainfall. River flow is the variable and a specific value of this variable such as $25\,000\ m^3 s^{-1}$ is the variate. The variable is a quantity, which may take on any value from a given set of values of variates, called the **domain**.

Variates are of two kinds: **discrete** and **continuous**. Discrete variates have domains restricted to isolated or discrete real numbers, such as a set of finite or countably infinite numbers. For example, the number of rainy days in Kolkata for a 10-year period of 2000–2010. Discrete variates can be *counted*. The domain of a continuous variate, on the other hand, is an interval, or set of intervals, on the real number axis, such as an infinite set of numbers. Continuous variates correspond to variables which are *measured*, theoretically to any degree of fineness, but they can assume any value within a domain of possibility. For example, the values of 24-hour rainfall depths measured at Kolkata airport for the same 10-year period, 2000–2010. These rainfall depths lie within a range (domain) and can plausibly take on any value lying within that range.

Both discrete and continuous variates correspond to random variables. For this reason, for convenience, the term random variable is used as a synonym of random variate. However, random variables are denoted by capital letters, such as X, Y, and Z and their corresponding variates will be represented by lower case letters, such as x, y, and z. A function of a random variable is also a random variable. For example, if $Z = h(X)$, where X is a random variable, Z is also a random variable. As a result, the probability, P, of a random variable, Y, is a random variable.

3.3 SAMPLES AND POPULATIONS

A **sample** is a collection of objects selected from a larger collection of these same objects. The larger collection of objects, if it contains all of the objects possible, is called the **population**. For example, a dataset comprising records of instantaneous maximum discharge of every year, for a certain period, from a certain river is a sample of all the possible peak flows of the river. All possible peak flows of that set constitute the population. The task in hand is to decipher the characteristics of the population from the sample.

A **random sample** is one that is selected in such a fashion that any other sample could have resulted with equal likelihood. If the 50 years of peak flow data from a given river for a specific period are considered a random sample, it is assumed that these 50 years of data are just as likely as any other possible 50 years of data and vice versa. The concept of a random sample is used throughout subsequent sections.

3.4 FREQUENCY DISTRIBUTION

3.4.1 Tabular Form of Frequency Distribution

The raw numbers subject to statistical analysis, are derived from measurements or counting. These are collectively referred to as **data**. Raw data do not convey any clear idea of the subject matter unless they are organized and condensed in a systematic way. As a first task in organizing raw data, they are partitioned into **classes** of appropriate size, showing the corresponding **frequency**, that is, the number of occurrences of variates in each class. This arrangement is called **frequency distribution** in **tabular form**.

The width of a class is called **class interval** or **bin** and the midvalue of a class or bin center is variously called a class mark, midvalue, or central value. The upper and lower bounds of a class are called **class limits**. In general, the class intervals are of equal width. But sometimes the upper and lower classes can be set as greater than or less than a limit. Grouping variates into the most appropriate number of classes is a matter of judgment. The number should be as small as possible but should contain sufficient observations to be meaningful. The number of classes can be determined from one of the following equations:

$$k = 1 + 3.3 \log_{10}(N) \quad (3.1)$$

$$k = 1 + \log_2(N) \quad (3.2)$$

where k denotes the number of classes, and N is the total number of observations. Equation 3.1 was recommended by Sturges (1926), whereas Zhang and Singh (2019) suggested Eq. 3.2 to determine an appropriate number of classes. Helsel et al. (2020) suggested using the condition $2^k \geq N$, satisfied by the smallest integer value of k.

The number of occurrences of an observation, $n(x_i)$ in a class interval i is its **absolute frequency**. The total of absolute frequencies must be equal to the total number of observations, N. The ratio of the absolute frequency and the total number of observations of a class is its **relative frequency**, $f(x_i)$. Thus,

$$\sum_{i=1}^{k} n(x_i) = N \quad (3.3)$$

$$f(x_i) = \frac{n(x_i)}{N} \quad (3.4)$$

The **cumulative absolute frequency** and **cumulative relative frequency** for a particular class interval are the sum of absolute frequencies and relative frequencies, respectively, for that interval and all intervals lying below it. Thus, if $F(x_i)$ denotes the cumulative relative frequency at the end of the ith class interval, then

$$F(x_i) = \sum_{j=1}^{i} f(x_j) \quad (3.5)$$

$$F(x_i) = F(x_{i-1}) + f(x_i) \quad (3.6)$$

Similarly, the cumulative absolute frequency at the ith class interval is

$$N(x_i) = \sum_{j=1}^{i} n(x_j) \quad (3.7)$$

Sometimes, the cumulative frequency is expressed as a percentage, instead of fractional numbers. It is worth recapping the notation convention used. The variate is denoted by x, its subscript i denotes the class to which it belongs, f and F denote relative and cumulative relative frequencies, respectively, and n and N respectively denote absolute and cumulative absolute frequencies. It is also implicit in this notation that all these are *functional values* and are based on *sample data*.

Tabular representations of data and the accompanying frequency analysis are illustrated in Tables 3.1 and 3.2. Table 3.1 gives the monthly precipitation (inches) in Dallas, Texas, for 123 years. Table 3.2 shows the frequency analysis of the data given in Table 3.1. In this analysis, the data listed in the last column of Table 3.1 are excluded, since they are the annual totals, but they will be used in another context later.

Concerns in hydrology are often with the number or percentage of observations that are less than or greater than a given value. This is why accumulated or cumulative frequencies of several class intervals up to a certain class boundary are quite important. The cumulative frequency corresponding to the upper boundary of any class interval is the total absolute frequency of all values less than that boundary. Sometimes, this is denoted by $N_<$ (read as "N less than"). Thus, $N_<$ is the **absolute cumulative frequency**. Relative cumulative frequencies, which are denoted by F_i above, are obtained by dividing $N_<$ by N, which gives n/N values. For example, in the frequency distribution shown in Table 3.2, the value of $N_<$ (9) is 97.54%, meaning that 97.54% of the monthly precipitation total in Dallas is less than 9 inches.

3.4.2 Graphical Forms of Frequency Distribution

3.4.2.1 Significance of Graphical Representation

The results of a frequency distribution can be presented more effectively in graphical forms that not only convey important insights about the data under scrutiny but also relay the overall data picture that is easier to grasp than tabular results. Four common types of graphical plots can be used to represent the frequency distribution of a set of data. Figure 3.1 shows such plots for the data of Table 3.1.

3.4.2.2 Frequency Polygons

A statistical distribution, more simply called a distribution of values (x) of a **discrete variate** may be represented graphically by plotting the points $(x_1, f_1), (x_2, f_2), \ldots, (x_k, f_k)$ or, alternatively, $(x_1, n_1), (x_2, n_2), \ldots, (x_k n_k)$, and drawing a dashed line through them. Such a graph is called a frequency polygon because it is a polygon formed by connecting the tops of a series of ordinates whose lengths are proportional to the various frequencies and whose abscissas correspond to variate values of the distribution.

Table 3.1 *Monthly precipitation (inches) in Dallas, Texas, for 123 consecutive years*

Year	Jan	Feb	Mar	Apr	May	Jun	Jul	Aug	Sep	Oct	Nov	Dec	Total
2021	0.85	2.22	3.03	4.5	7.77	2.15	1.50	4.82	0.25	2.96	3.11	0.43	33.59
2020	5.00	3.88	6.75	1.9	7.54	5.35	2.31	1.28	3.87	1.74	1.08	3.00	43.70
2019	1.58	1.29	2.01	6.75	8.15	4.13	0.78	2.44		4.42	1.8	1.17	34.52
2018	0.85	11.31	2.9	0.77	1.87	1.27	0.25	2.99	12.69	15.66	0.86	4.55	55.97
2017	4.39	2.33	1.06	3.38	0.7	8.44	4.12	4.24	0.47	2.12	0.81	4.56	36.62
2016	1.04	2.2	2.67	4.6	6.25	3.60	3.89	4.42	0.98	2.01	3.22	0.60	35.48
2015	3.62	2.96	2.53	5.56	16.96	3.95	0.92	0.46	2.14	9.82	9.86	3.83	62.61
2014	0.33	0.41	1.45	1.74	3.4	3.26	0.98	4.34	0.06	2.09	2.13	1.13	21.32
2013	4.06	1.68	2.27	1.98	3.17	2.14	2.05	1.32	2.72	3.13	2.12	2.76	29.40
2012	6.18	1.88	5.74	4.24	1.66	2.82	0.78	3.19	1.75	1.02	0.05	1.95	31.26
2011	1.60	0.92	0.07	2.46	7.95	2.84	0.09	0.96	0.66	3.12	0.86	4.35	25.88
2010	2.76	2.83	3.57	2.03	1.09	2.08	3.13	0.41	9.09	1.16	1.5	2.05	31.70
2009	0.82	0.72	5.56	3.54	4.36	3.98	2.09	1.64	6.52	8.05	1.76	1.85	40.89
2008	0.27	2.3	6.07	3.85	2.21	0.84	0.81	2.82	0.84	2.29	4.53	0.27	27.10
2007	5.58	0.43	3.81	2.82	8.34	11.10	5.54	0.35	4.99	3.53	1.22	2.34	50.05
2006	2.25	3.85	4.4	1.86	1.9	0.34	1.78	0.52	2.60	4.34	2.58	3.33	29.75
2005	4.33	1.62	2.17	0.56	3.35	1.14	0.74	2.46	1.36	0.89	0.02	0.33	18.97
2004	3.04	3.84	1.71	2.96	4.73	10.49	4.16	4.24	1.02	5.72	5.01	0.65	47.57
2003	0.22	3.07	0.85	1.9	2.53	5.17	0.08	1.85	3.99	0.78	3.15	0.96	24.55
2002	4.90	0.94	7.39	5.68	5.4	3.10	3.07	1.47	1.38	6.44	0.52	4.13	44.42
2001	2.44	6.17	5.27	0.89	5.58	1.28	3.85	2.72	3.72	1.87	1.11	3.24	38.14
2000	1.59	3.3	2.91	4.28	3.17	5.93		0.00	0.17	4.38	6.95	3.57	36.26
1999	1.44	0.48	2.84	2.74	6.91	0.99	0.77		2.30	2.26	0.31	2.55	23.59
1998	5.07	3.22	4.45	1.25	2.38	1.75	0.11	0.35	0.68	5.64	4.91	4.43	34.24
1997	0.33	7.4	2.21	6.73	3.92	3.99	1.68	3.13	2.01	5.66	1.01	6.93	45.00
1996	0.97	0.35	2.36	2.14	0.95	3.42	3.85	5.02	1.51	6.56	5.54	0.47	33.14
1995	2.11	0.44	6.69	6.83	7.5	2.41	3.45	0.86	1.54	0.75	0.74	2.07	35.39
1994	1.43	2.01	1.69	3.62	5.8	2.05	4.58	4.89	1.39	8.19	6.03	2.42	44.10
1993	1.74	5.78	3.03	3.49	1.75	3.75	0.00	0.75	3.28	5.10	1.62	2.54	32.83
1992	3.25	2.4	3.24	2.46	6.93	5.23	2.48	2.08	3.25	3.05	3.56	4.26	42.19
1991	2.72	2.6	1.35	3.63	6.97	4.26	3.99	4.30	4.61	9.32	1.04	8.75	53.54
1990	4.54	4.72	5.89	6.9	7.16	1.89	2.60	2.37	1.12	2.81	3.81	1.46	45.27
1989	2.56	3.7	3.72	1.86	9.62	8.75	2.61	1.89	2.40	2.02	0.49	0.33	39.95
1988	0.88	1.23	2.03	2.21	2.11	3.23	2.47	0.44	4.04	1.64	2.28	2.48	25.04
1987	1.22	3.67	1.7	0.11	5.95	3.45	1.77	0.81	1.38	0.12	4.17	2.90	27.25
1986		2.49	1.08	5.3	5.52	3.92	0.41	1.63	4.60	1.81	3.25	2.44	32.45
1985	0.81	2.62	3.7	3.75	2.13	3.78	2.40	0.53	3.35	3.91	3.11	0.61	30.70
1984	1.07	3.11	4.92	1.41	3.04	2.79	0.43	1.47	0.09	6.50	2.97	6.09	33.89
1983	2.55	1.25	4.36	0.59	5.83	2.07	1.56	5.55	0.22	4.04	2.22	0.83	31.07
1982	2.33	1.89	1.71	2.71	13.66	4.28	2.73	0.52	0.58	3.36	4.22	2.76	40.75
1981	0.58	1.44	3.39	2.69	6.24	7.85	1.81	2.32	2.40	14.18	1.53	0.17	44.60
1980	2.52	0.84	1.24	2.23	3.01	1.25	0.71		6.54	1.08	1.23	1.43	22.08
1979	3.35	1.52	6.33	2.03	5.9	1.36	1.94	2.47	0.99	3.38	0.43	2.72	32.42
1978	1.41	3.33	2.66	1.34	8.01	0.77	0.33	1.53	0.93	0.55	2.73	0.78	24.37
1977	2.39	1.68	5.88	4.31	0.99	0.69	2.20	2.33	1.72	2.96	1.79	0.25	27.19
1976	0.13	0.52	2.29	5.71	6.03	1.40	3.83	4.75	5.02	3.46	0.5	1.99	35.63
1975	3.34	3.72	1.67	3.4	6.88	1.95	5.06	0.30	0.87		0.42	1.49	29.10
1974	1.79	1.01	0.8	2.51	6	5.44	0.67	4.19	6.04	5.93	3.32	1.93	39.63
1973	3.26	1.92	2.28	6.06	3.18	5.88	11.13	0.01	7.16	6.85	2.06	0.83	50.62
1972	1.09	0.26	0.1	3.25	2.35	1.50	0.59	0.81	2.42	6.89	2.36	0.61	22.23
1971	0.19	1.32	0.34	2.76	1.88	0.83	3.60	5.70	3.24	7.64	1.77	6.99	36.26
1970	0.72	4.78	3.49	4.68	3.62	0.61	0.94	6.85	6.25	2.95	0.2	1.01	36.10
1969	1.26	1.99	3.62	3.4	7.12	0.63	0.77	2.56	4.55	5.82	1.22	2.75	35.69
1968	3.60	1.48	6.39	2.41	6.02	3.50	1.88	2.71	2.53	2.18	4.58	1.20	38.48
1967	0.28	0.32	2.09	3.84	4.02	0.72	2.20	0.48	5.94	4.19	0.92	2.30	27.30
1966	1.68	2.84	1.38	10.74	3.13	5.47	3.26	3.38	4.23	1.48	0.53	1.17	39.29

Table 3.1 (cont.)

Year	Jan	Feb	Mar	Apr	May	Jun	Jul	Aug	Sep	Oct	Nov	Dec	Total
1965	2.77	6.2	1.45	2.15	8.97	1.50	0.09	2.26	5.04	1.97	2.43	1.73	36.56
1964	3.53	1.17	3.35	2.71	2.85	0.40	0.25	2.43	9.52	0.62	6.23	1.25	34.31
1963	0.86	0.15	0.48	6.2	2.52	0.57	2.28	2.73	1.70	0.23	1.29	1.45	20.46
1962	1.00	2.01	1.8	5.66	1.58	6.94	6.36	3.22	3.79	4.15	3.93	0.99	41.43
1961	3.29	2.2	2.95	2.23	1.06	5.93	2.32	0.02	2.92	2.82	2.72	2.12	30.58
1960	2.29	2.16	0.74	1.67	1.89	1.72	3.96	2.76	1.25	1.82	0.49	4.22	24.97
1959	0.36	1.61	2.31	0.92	3.27	5.27	3.27	0.93	2.40	9.22	1.74	2.84	34.14
1958	1.70	0.84	5.49	8.63	1.5	0.67	3.69	3.64	5.10	1.07	2.26	1.09	35.68
1957	1.72	1.77	4.18	12.19	12.64	3.96	0.65	0.12	3.23	3.53	4.72	1.78	50.49
1956	1.34	2.54	0.11	3.12	3.83	0.88	0.38	0.23	0.23	1.20	2.61	2.08	18.55
1955	1.17	2.01	2.15	1.94	6.58	4.99	0.64	1.00	3.68	0.20	0.59	0.21	25.16
1954	2.08	0.73	0.66	3.62	4.38	1.20	0.24	0.81	1.46	2.35	1.24	0.78	19.55
1953	0.54	1.34	2.52	4.82	3.55	0.55	0.97	1.09	1.68	4.27	2.09	1.32	24.74
1952	0.58	1.12	1.39	6.51	3.21		0.56	0.44	0.54	0.01	5.84	2.49	22.69
1951	1.39	2.42	1.33	2.27	4.6	4.12	2.22	0.47	1.84	1.62	1	0.09	23.37
1950	5.01	2.47	1.58	4.73	6.16	3.16	4.53	3.05	3.21	0.30	0.02		34.22
1949	5.45	4.75	3.69	2.47	10.64	3.52	0.10	2.27	3.13	6.50	0.09	1.04	43.65
1948	0.96	4.12	1.07	1.11	4.34	2.46	1.93	0.90	0.19	2.09	0.5	0.44	20.11
1947	1.21	0.55	2.92	2.98	2.5	4.08	0.10	4.18	2.81	2.14	2.23	4.50	30.20
1946	2.79	2.93	2.8	2.49	12.09	0.65	0.90	6.84	2.69	1.31	6.5	3.40	45.39
1945	1.92	6.96	6.19	2.87	1.81	4.12	3.07	0.62	2.17	2.31	1.13	0.55	33.72
1944	2.58	4.81	1.3	2.7	6.42	0.76	2.52	2.65	0.80	2.53	3.82	3.60	34.49
1943	0.20	0.51	4.05	1.63	7.83	3.93	0.73		7.31	0.73	0.51	3.32	30.75
1942	0.39	0.64	1.37	16.97	2.85	3.23	0.62	4.69	3.82	6.18	0.92	1.59	43.27
1941	1.45	3.42	1.52	3.52	2.02	7.12	1.49	2.71	1.28	3.68	1.08	1.88	31.17
1940	0.59	2	0.4	5.97	7.15	7.30	2.86	2.16	0.68	1.47	6.35	4.72	41.65
1939	2.66	2.42	1.64	1.48	2.54	4.04	2.02	1.44	0.12	0.55	2.72	0.68	22.31
1938	2.74	4.57	3.89	3.03	2.8	1.61	2.16	0.11	0.78	0.11	1.17	1.26	24.23
1937	1.71	0.3	3.88	0.58	1	5.74	1.93	1.02	0.32	3.55	4.39	5.31	29.73
1936	0.67	0.45	0.63	0.99	9.48	0.03	2.35	0.23	7.30	3.72	0.46	1.84	28.15
1935	3.70	3.29	1.4	3.06	9.15	7.22	0.89	0.70	3.61	4.01	1.65	2.26	40.94
1934	1.86	1.67	4.26	2.39	0.82		0.08	0.13	4.90	0.12	2.3	0.56	19.09
1933	1.96	2.47	2.18	1.57	4.67	0.03	5.70	2.25	4.94	1.24	0.66	2.13	29.80
1932	9.07	4.92	0.63	3.43	6.03	3.04	2.07	2.92	10.80	1.66	1.56	4.90	51.03
1931	1.79	2.84	4.2	1.97	2.42	2.43	0.44	3.38	1.25	3.39	2.78	2.73	29.62
1930	0.84	1.08	2.86	2.37	10.37	1.87	0.37	3.12	1.19	7.96	1.71	2.08	35.82
1929	2.08	2.78	1.39	2.06	5.83	0.20	0.43		2.29	2.12	1.5	0.41	21.09
1928	0.46	3.53	1.1	5.7	3.77	11.58	4.24	2.13	0.45	4.15	1.97	5.50	44.58
1927	1.45	1.77	2.19	3.66	0.44	3.33	1.53	0.80	4.00	4.47	0.58	2.59	26.81
1926	4.04	0.08	3.6	3.73	3.79	3.32	4.13	4.39	1.41	3.16	0.73	3.03	35.41
1925	1.44	0.74	0.02	3.59	8.11	0.29	0.98	0.40	1.79	3.77	2.05	0.04	23.22
1924	0.89	1.97	4.66	2.33	4	1.25	0.96	3.77	3.78		1.6	1.23	26.44
1923	4.60	2.05	1.52	5.3	0.54	6.74	0.99	1.68	2.06	6.05	1.63	4.68	37.84
1922	1.63	2	1.57	17.64	4.58	1.76	1.35	0.52	0.41	2.33	2.57	0.06	36.42
1921	2.87	2.62	2.67	1.99	1.04	2.63	1.14	0.95	0.11	0.31	1.24	0.34	17.91
1920	3.48	0.76	4.42	0.51	8.66	2.33	3.49	4.22	2.76	6.52	1.7	1.31	40.16
1919	3.03	2.03	3.34	2.06	3.99	3.72	5.25	5.00	4.12	9.44	3.32	0.44	45.74
1918	1.36	0.01	0.93	6.21	1.99	5.16	1.10	0.29	2.09	3.31	7.94	4.08	34.47
1917	1.43	1.47	2.42	4.11	3.92	1.97	2.65	1.92	2.41	0.17	1.35	0.05	23.87
1916	4.01	0.01	3.68	6.99	3.7	3.30	1.38	3.84	0.73	1.89	1.82	0.11	31.46
1915	1.32	2.18	1.4	4.98	2.49	6.88	0.30	10.33	1.62	2.58	0.29	1.99	36.36
1914	0.43	1.17	2.89	5.99	10.71	2.97	0.73	9.02	1.61	0.28	6.44	4.40	46.64
1913	2.30	0.87	1.04	2.47	2.74	3.03	4.36		7.29	2.28	5.9	5.42	37.70
1912	0.17	1.22	3.34	3.2	2.71	4.26	0.27	6.56	0.83	1.51	0.33	1.95	26.35
1911	0.21	3.84	1.87	3.33	0.22	0.43	6.26	2.39	1.38	0.99	1.05	5.06	27.03
1910	1.36	1.14	1.02	2.65	5.76	1.38	0.14	0.26	2.21	0.68	0.14	1.23	17.97

Table 3.1 (cont.)

Year	Jan	Feb	Mar	Apr	May	Jun	Jul	Aug	Sep	Oct	Nov	Dec	Total
1909	0.09	0.11	0.41	1.66	1.09	2.97	0.02	2.38	2.08	2.20	5.11	2.81	20.93
1908	0.96	2.45	2.95	9.63	10.69	2.90	2.66	2.74	3.52	4.49	2.05	0.03	45.07
1907	0.51	1.9	0.7	1.31	6.53	2.22	4.15	0.29	1.92	3.01	5.81	2.18	30.53
1906	0.93	2.08	1.99	2.56	8.24	4.13	2.56	4.98	4.16	0.91	2.19	1.22	35.95
1905	1.52	1.93	3.39	7.73	5.45	2.69	8.35	0.56	0.83	4.21	3.19	3.60	43.45
1904	1.30	1.79	4.01	2.21	3.86	5.42	2.15	3.26	2.63	5.29	0.02	0.36	32.30
1903	1.83	5.07	2.03	0.59	1.84	4.84	1.84	1.57	2.70	4.53	0	0.30	27.14
1902	0.42	0.36	3.8	1.81	4.31	0.58	6.29		2.40	1.40	6.89	1.05	29.31
1901	0.08	1.59	1.57	2.04	4.5	0.33	1.99	1.29	1.67	1.90	2.1	0.59	19.65
1900	0.70	0.12	0.73	7.34	6.58	0.84	5.90	1.43	9.12	3.22	0.56	0.35	36.89
1899	1.24	0.21	0.32	2.32	2.01	2.78	1.66	0.02	1.02	1.49	3.06	1.98	18.11

Table 3.2 *Frequency analysis of the precipitation data given in Table 3.1*

Class interval (inches)	Absolute frequency (n_i); $N = 1462$	Cumulative absolute frequency (N_i)	Relative frequency (f_i)	Cumulative relative frequency (F_i)
0.0–1.5	476	476	32.56	32.56
1.5–3.0	453	929	30.98	63.54
3.0–4.5	272	1201	18.60	82.15
4.5–6.0	122	1323	8.34	90.49
6.0–7.5	79	1402	5.40	95.90
7.5–9.0	24	1426	1.64	97.54
9.0–10.5	17	1443	1.16	98.70
10.5–12.0	9	1452	0.62	99.32
12.0–13.5	4	1456	0.27	99.59
13.5–15.0	2	1458	0.14	99.73
15.0–16.5	1	1459	0.07	99.79
16.5–18.0	3	1462	0.21	100.00

3.4.2.3 Histograms

A histogram is a set of rectangles with bases along the intervals between the class boundaries and with **areas** proportional to the frequencies in the corresponding classes. If the class intervals are equal, the heights of the rectangles are also proportional to the frequencies. Thus, one property of the histogram is

$$\sum_{i=1}^{k} f_i = 1 \tag{3.8}$$

Histograms can be constructed for both discrete and continuous variates as opposed to frequency polygons, which are used only for discrete variates. For data measured on a continuous scale, such as streamflow, the sensitivity is lost if too few or too many classes are used. Actually, this is a primary deficiency of histograms. The visual impression about the data, given by histograms, depends on the number of classes selected. On the other hand, they are excellent when displaying data that have natural groupings. Furthermore, histograms can nicely depict the symmetry or asymmetry of the data.

3.4.2.4 Frequency Curves

In a histogram, it is assumed that, within any one class, the values of the variates are uniformly spread out between the class boundaries, which is unlikely to be the case. However, if we suppose that the size of the sample is increased indefinitely so that, even with very small class intervals, there are many observations in each class throughout the domain of the variate, the outline of the histogram will approximate to a smooth curve. This curve can be regarded as the **frequency curve** of the parent population from which the sample taken. The sample is considered to be a random sample. In practice, frequency curves are fitted to histograms either by eye or more usually by calculation, using the known mathematical properties of certain curves that seem to be of about the right shape. This will become apparent in Section 3.10, covering probability distribution functions, which are mathematical functions giving curves of different shapes as a function of the variate and few parameters.

3.4.2.5 Cumulative Frequency Polygons or Curves

If the cumulative frequency $F_<$ is plotted against the upper-class boundary (x_e) and the points are joined by

3.5 Moments of Distribution and Moment Measures

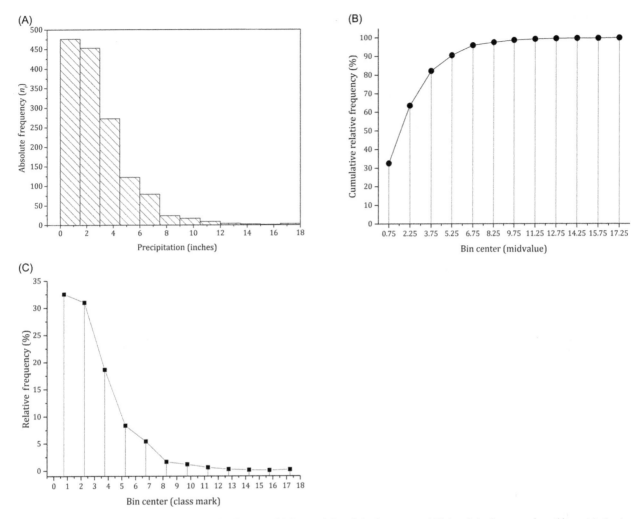

Figure 3.1 Graphical representations of (A) the absolute frequency, (B) the cumulative relative frequency, and (C) the relative frequency of monthly precipitation in Dallas based on records of 123 consecutive years (Table 3.1). The histogram (A) is based on 12 classes (according to both Eqs. 3.1 and 3.2) with a class interval of 1.5.

straight lines, a **cumulative frequency polygon** is generated. The polygon should start from zero at the lower boundary of the first interval. Strictly speaking, $F_<$ for a continuous variate is defined for the end values x_e only, but if the assumption is made that the observations in any one class are uniformly spread out over the whole interval, the intermediate points on the polygon also represent the cumulative frequencies at the corresponding values of x. This means that we can interpolate linearly between the class boundaries.

If relative frequencies, instead of absolute frequencies, are used, the cumulative polygon rises from the value 0 at the left to the value 1 at the right. These numbers are sometimes multiplied by 100 to give percentage cumulative frequency polygons. The points are often joined such that a smooth curve is generated. These are also called cumulative frequency curves. Cumulative frequency distributions show the frequency of variates less than (or greater than) some given value.

3.5 MOMENTS OF DISTRIBUTION AND MOMENT MEASURES

3.5.1 Classical Moments

Moments are statistical descriptors of a distribution and reflect its quantitative properties. Furthermore, the moments are used to compare distributions and derive their properties. Several types of moment occur in the statistical literature. Here we will discuss two types of classical moment: (1) general or noncentral moments and (2) central moments. These are powerful statistics that reduce data down to a few values.

To rephrase the definitions in Section 3.3, a **population** represents the set of all conceivable observations in their entirety that are plausible about a subject matter. A **sample** is just a subset of this set comprising the actual observations made. There are three kinds of population. The first includes a large population that is **finite** but can be enumerated if necessary. For example, all streams including major, minor, ephemeral, perennial etc. in the Amazon River basin constitute a large stream population. For any study at the scale of such a vast river basin, only a limited number of streams can reasonably be sampled. The second kind of population is indefinitely large or **infinite**, as, for example, all possible values of discharge in those streams of the Amazon River basin. A third kind is a purely **hypothetical** population, which can be completely described mathematically. Here, the distribution of values in the population is given by a mathematical model. The quality of the match between the observed or actually occurring distribution of the data and the equation describing the distribution, which seems to have similar distribution characteristics, is judged by one or more goodness-of-fit tests. If the fit between the observed and model distributions is satisfactory, we can regard the observed sample as coming from a population that has the characteristics of the mathematical distribution. Moments play an important part in the developments of various mathematical equations representing all kinds of population, as we will see in subsequent sections.

3.5.2 Moments about the Origin: General Moments

General moments or **noncentral moments** are the **moments about the origin** but can be about any arbitrarily chosen point. The term moment was introduced in analogy with the moment of a force in solid mechanics. This is illustrated in Figure 3.2. A hypothetical frequency distribution of a discrete random variable X in the form of a histogram is shown. The values, $x_1, x_2, x_3 \ldots$ have corresponding relative frequencies $f(x_1), f(x_2), f(x_3), \ldots$ on either side of the origin O. Now, if we assign a hypothetical weight (force) to each of the values proportional to its relative frequency, the total moment that will tend to rotate the abscissa about its origin is $\sum_i x_i f_i = \sum_i x_i (n_i/N) = (1/N) \sum_i x_i n_i$, which we will define shortly as the mean of the sample. If the variable is a continuous random variable, its distribution is shown as a smooth curve, $f(x)$, defined by the class marks. Now we formalize the definitions.

For a *sample of discrete random variables*, X, with a total frequency, N, the r th moment about the origin is defined as

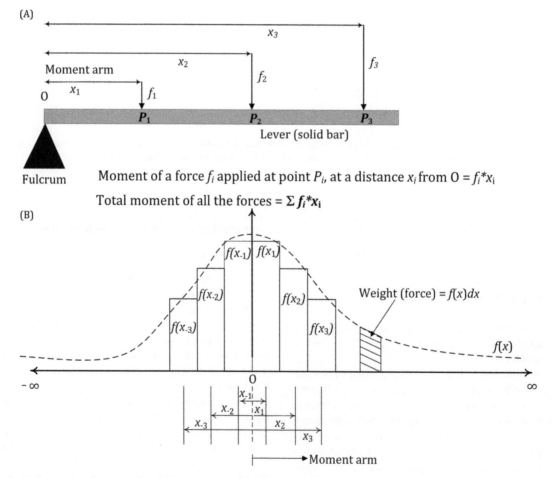

Figure 3.2 (A) The moments of a system of parallel forces acting on a solid lever resting on a fulcrum. (B) The concept of moment of a discrete function about an arbitrary point.

$$M_r^0 = \frac{1}{N}\sum_{i=1}^{N} f(x_i)(x_i)^r \tag{3.9}$$

For a *sample of continuous random variables*, the rth moment about the origin is defined as

$$M_r^0 = \frac{\int_{-\infty}^{\infty} x^r f(x)\,dx}{\int_{-\infty}^{\infty} f(x)\,dx} \tag{3.10}$$

If we consider a finite population with a total population frequency, M, so that $\sum f_i = M$, the proportion of values p_i lying in the ith class is $p_i = f_i/M$ and the variate, x_i, represents the class mark, Eq. 3.9 can be written as

$$M_r^0 = \frac{1}{M}\sum_{i=1}^{M} f(x_i)(x_i)^r \tag{3.11}$$

Now, if $r = 1$, we write Eq. 3.9 as

$$M_1^0 = \bar{x} = \sum_{i=1}^{N} \frac{n(x_i)}{N} = \frac{1}{N}\sum_{i=1}^{N} n_i x_i \tag{3.12}$$

The first sample moment about the origin, $\bar{x}$, given in Eq. 3.12, is the **mean** of the sample. Note that $\bar{x}$ is a point on the abscissa and r is called the **order** of the moment.

3.5.3 Moments about the Mean: Central Moments

Moments can be defined about any point located either on the abscissa or the ordinate at a finite distance away from the origin. If we only consider that the point is located at a distance, a, from the origin on the abscissa, the rth moment about a of a continuous random variable is given by

$$M_r^a = \frac{\int_{-\infty}^{\infty} (x-a)^r f(x)\,dx}{\int_{-\infty}^{\infty} f(x)\,dx} \tag{3.13}$$

This is called a **central moment**. If the point is shifted to the point representing the mean, three central moments of order 2, 3, and 4 about the mean of the sample of a discrete random variable are important in statistics, and further higher-order moments are of less significance.

For a discrete random variable, the rth moment about mean is given by

$$M_r^{\bar{x}} = \sum_{i=1}^{N} f(x_i)[x_i - \bar{x}]^r \tag{3.14}$$

When $r = 2$,

$$M_2^{\bar{x}} = \frac{1}{N}\sum_{i=1}^{N} (x_i - \bar{x})^2 \tag{3.15}$$

When $r = 3$,

$$M_3^{\bar{x}} = \frac{N}{(N-1)(N-2)}\sum_{i=1}^{N}(x_i - \bar{x})^3 \tag{3.16}$$

When $r = 4$

$$M_4^{\bar{x}} = \frac{N^2}{(N-1)(N-2)(N-3)}\sum_{i=1}^{N}(x_i - \bar{x})^4 \tag{3.17}$$

These three moments are called sample **variance, skewness,** and **kurtosis,** respectively.

3.5.4 Moment Measures

3.5.4.1 Mean: A Measure of Central Tendency

The mean or **average** is the first moment of a distribution, measured about the origin, and it is perhaps the most important moment measure of a distribution. An average is a value which is intended to be in some sense typical of a whole distribution. It is more or less the *central value* and may be regarded as a *measure of the location* of the distribution on the axis of the variate x. The arithmetic mean of a sample is given as

$$\bar{x} = \frac{1}{N}\sum_{i=1}^{N} x_i = \frac{1}{N}\sum_{i=1}^{k} n_i x_i = \sum_{i=1}^{k} f(x_i)x_i \tag{3.18}$$

The expansion of the expression of the mean in alternative forms in Eq. 3.18 is due to the fact that the sum of the frequencies is the total number of observations. The numerical values of $\sum_1^k f_i x_i$ and $\sum_1^N x_i$ are equal. N refers to the total number of values in the set (some of which may be alike), whereas k refers to the number of *different* values of x in the set and hence to the number of products of the form $x_i f_i$, where f_i is the number of times x_i occurs. Here, the value x_i is said to be weighted, the weight being the corresponding frequency f_i, and the arithmetic mean so obtained is called a weighted arithmetic mean.

Equation 3.18 gives a sample mean that is denoted by, $\bar{x}$. The population mean is denoted by μ. In Section 3.19.5.2, we use the t-test to determine the confidence level (Section 3.20.2) at which the sample mean approximates the population mean.

Root Mean Square

Another average that is sometimes used is the root mean square (RMS) defined as the positive square root of the mean of the squares of the values x_i, that is

$$\text{RMS} = \sqrt{\bar{x^2}} = \sqrt{\frac{1}{N}\sum_{i=1}^{N} x_i^2} \tag{3.19}$$

The RMS. can be used when the x_i values are both positive and negative. However, its chief application is the measurement of deviations of the values x_i of a variable from the mean, $\bar{x}$. The RMS of these deviations is called the standard deviation indicating the spread of the values of x, and is described in the following section

3.5.4.2 Variance and Standard Deviation: Measures of Dispersion

The dispersion of a distribution is indicated by the extent to which observed values of the variate tend to spread over an interval rather than to cluster closely around a central average. The most important measure of dispersion is the **standard deviation**, which is the square root of the second-order moment of the distribution, measured about the mean.

The common measure of dispersion, preferred in most circumstances, is the RMS average of the deviations from the mean. The name standard deviation for this quantity was proposed by Karl Pearson, a distinguished English statistician (1857–1936). We denote it by s, with a subscript indicating the variable.

$$s_x = \sqrt{\frac{1}{N} \sum_{i=1}^{N} n_i (x - \bar{x})^2} \qquad (3.20)$$

where $N = \sum_i n_i$.

The square root has the effect of making the unit of s_x the same as the unit of x_i. If x_i is in m, $(x_i - \bar{x})^2$ is in m^2, and s_x again in m. The quantity s_x^2 is called the **variance**, and in many ways it is more fundamental than the standard deviation because it is a moment measure of the distribution.

The letter s is used to denote the **sample standard deviation** and the Greek letter σ is used to denote the **population standard deviation**. It has been proved that if s_x^2 is the variance of a sample of N, the best estimate of the population variance σ_x^2 is not s_x^2 but

$$\sigma_x^2 = \frac{N s_x^2}{N - 1} \qquad (3.21)$$

In Section 3.19.5.3, we use the F-test to determine the confidence of the estimation of population variance from sample variance.

Coefficient of Variation

The ratio σ/μ is known as the **coefficient of variation** (C_V) and is useful as it is an expression of the relative variability. A large C_V indicates the data are spread widely whereas a small C_V indicates that the data are clustered around the mean.

The **range**, defined by the difference between the maximum and minimum values, seems to be an obvious statistic for dispersion but it lacks the powerful properties of variance.

3.5.4.3 Skewness: Measure of Symmetry

Skewness, also called the **coefficient of skewness** is defined as the ratio obtained by dividing the third-order central moment (Eq. 3.16) by the cube of the standard deviation:

$$g_x = \frac{N}{(N-1)(N-2)s_x^3} \left[\sum_{i=1}^{N} (x_i - \bar{x})^3 \right] \qquad (3.22)$$

This is a measure of asymmetry of the distribution about the mean. For a symmetrical distribution, $g_x = 0$. If there is a long tail to the right, $g_x > 0$, the distribution is positively skewed but if the distribution has a long tail to the left, it has negative skewness, $g_x < 0$.

As a general rule of thumb:

- If the skewness is less than -1 or greater than 1, the distribution is highly skewed.
- If the skewness is between -1 and -0.5 or between 0.5 and 1, the distribution is moderately skewed.
- If skewness is between -0.5 and 0.5, the distribution is slightly skewed and can be considered approximately symmetrical.

The population skewness is denoted by γ.

3.5.4.4 Kurtosis: Measure of Thickness of Distribution Tails

Kurtosis, also called the **coefficient of kurtosis**, is defined by

$$k_x = \frac{N^2}{(N-1)(N-2)(N-3)s_x^4} \sum_{i=1}^{N} (x_i - \bar{x})^4 \qquad (3.23)$$

The population kurtosis is denoted by κ. In most engineering or science disciplines this is considered as a measure of peakedness of a distribution and a high value of kurtosis is thought to represent a sharply humped or peaked distribution and a kurtosis of low value represents a relatively flat-topped distribution. However, in statistical literature, studies have shown that kurtosis has not necessarily anything to do with peakedness. Rather, it describes the thickness of the tails of a distribution.

3.6 QUANTILES

In addition to the moment measures, another set of quantitative measures are also important in characterizing the frequency distribution of a dataset. These are called **quantiles**. Quantiles are cut points dividing the variates into intervals or groups in different proportions. There is one fewer quantile than the number of groups created. Common quantiles have special names, such as quartiles (four groups), deciles (10 groups), and percentiles (100 groups). In general,

q-quantiles are values that partition a finite set of values into q subsets of nearly equal sizes

An important quartile is the *median*, which is defined as the central value of a distribution, a value such that greater and smaller values occur with equal frequency. If N values of x are arranged in order from the least to the greatest, so that $x_1 \leq x_2 \leq x_3 \ldots \leq x_n$, the median is the value x_k if N is odd ($N = 2k - 1$), and is not uniquely defined if N is even ($N = 2k$), unless $x_k = x_{k+1}$.

For a frequency distribution of a continuous variate, grouped in classes, the median is that value of x at which the ordinate divides the histogram into two parts of equal area and, on a cumulative frequency diagram, the median, denoted by $\tilde{x}$, is that value of x for which the relative cumulative frequency, $F(\tilde{x})$, is exactly 0.5.

Just as the ordinate at the median of a grouped distribution divides the histogram into two parts of equal area, the ordinates at quartiles q_1 and q_3 cut off one-quarter of the area at each end. Thus, the first quartile, denoted by q_1, is that value of x for which $N_< = N/4$. The second quartile, q_2, is that value of x for which $N_<$ is $N/2$ and *is therefore the median*. The third quartile, denoted by q_3, is that value of x for which $N_< = 3N/4$. Hence, 50% of the total frequency is included between q_1 and q_3. On the relative cumulative frequency polygon for a continuous variate, the ordinates at q_1, q_2, q_3. Are 0.25, 0.50, and 0.75, respectively. Quartiles are calculated by interpolation using the cumulative frequency table or quantile plots.

Many other quantities like quartiles can be calculated but corresponding to different fractions of the total frequency. These statistics are collectively called quantiles or sometimes **fractiles**, which are the inverse of quantiles.

With a large sample it may be useful to calculate **percentiles**. The kth percentile, P_k, is that value of x, say x_k, which corresponds to a cumulative frequency of $(Nk)/100$. Thus, the 25th percentile is q_1 and 50th percentile is q_2, that is, the median.

The **percentile rank** of P_k is k. Thus, instead of saying that the 20th percentile is 57 (just an arbitrary example), we could say that the percentile rank of 57 is 20. Both statements mean that, in the particular sample studied, 20% of the observations had a value of the variate x less than 57. If $F_<$ is the cumulative frequency corresponding to the kth percentile, then

$$\frac{F_<}{N} = \frac{k}{100}$$

Quantiles can be used for an approximate characterization of the distribution of a sample. If, in a large sample, we know the median, the quartiles, and the two extreme values, we can form a fairly good idea of the whole distribution by plotting these five points and sketching in a percentage cumulative frequency curve joining them. This is seen in Figure 3.8, constructed from real data and discussed in relation to quantile functions in Section 3.13.

3.7 RELATIONSHIP BETWEEN THE MEAN, MEDIAN, AND MODE

The value of the variable which occurs most frequently in a distribution is called the **mode**. In a frequency distribution of continuous variables, it is more convenient to talk about the modal class, instead of the mode. For a distribution with equal class intervals, the modal class is the class having the greatest frequency. For unequal intervals, it is the class with the greatest frequency per unit of x.

Having defined the mean, median, and mode, it is now worth noting the relationship between them since all these three statistics in separate ways signify the central tendency of a set of data and their distribution. If the distribution is presented in the form of a histogram, an ordinate through the median divides the area into two equal parts. An ordinate through the mean passes through the centroid of the area. An ordinate through the mode passes through the highest point of the frequency curve if there is only one mode. However, in the case of a bimodal or multimodal distribution, there can be multiple lines representing the mode.

Figure 3.3 shows the relationship of mean, median, and mode in a frequency distribution of 44 years (1977–2020) of daily discharge of the Rio Negro, the major tributary of the River Amazon. The moment measures and other statistics, all usually referred to as the descriptive statistics of a dataset, are also shown in a table in this figure.

The median is considered to be a more robust statistic than mean.

3.8 PROBABILITY

3.8.1 Stochastic Processes

Hydrologic processes are stochastic, meaning processes that involve a variate at each instant of time where the variate may take on any of the values of a specified set with a certain probability. The annual maximum daily rainfall in an area or the instantaneous maximum peak flow of a river in a year at a gauge are examples of a stochastic hydrologic process. If we have a set of observations and pick a value from it, we want to estimate the probability of occurrence of this magnitude in a year or any instant of time.

The stochastic nature of the process means that one can never estimate with certainty the exact value of the process such as the peak discharge of a river, based solely on past observations. We will see that the probability of getting an exact value of a variate from its population is actually zero. This poses little discouragement for hydrologists, since they are really not interested in the probability of the flow rate in a stream assuming a certain exact value, instead, indeed they want to estimate the probability that the flow will exceed a certain value or be less than that value or between a specific range.

In statistical hydrology, the main effort is geared toward determining the probability that should be assigned to

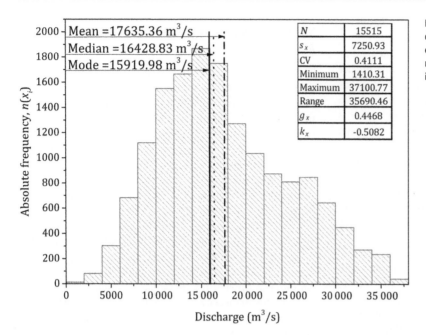

Figure 3.3 Frequency distribution of the daily discharge of the Rio Negro at Serrinha, upstream of its confluence with Amazon River at Manaus, Brazil. The relationship between the mean, median, and mode is illustrated.

events. In mathematical statistics, the concern is not so much determining the probability but what to do with that probability. In that respect, the appeal of probability to a hydrologist is somewhat different to that of a mathematician or statistician.

3.8.2 Definition of Probability

The simplest interpretation of probability is that of frequency, which we have already discussed. If a random event can occur in n equally likely and mutually exclusive ways, and if n_a of these ways have an attribute A, the classical definition of probability of the occurrence of the event having attribute A is

$$P(A) = \frac{n_a}{n} \qquad (3.24)$$

In this equation, there is nothing different to what is already given in Eq. 3.4 presented above except that instead of $f(x_i)$, we introduce the term $P(A)$, denoting the probability of an event A. However, this is an *a priori definition* because it assumes that one can determine, before the fact, all of the equally likely and mutually exclusive ways that an event can occur (n) and all of the ways (n_a) that an event with attribute A can occur.

Equation 3.24 takes on a greater value in hydrology in terms of relative frequencies and limits. We will see later that the relative frequency concept of probability is the source of the relationship between the return period T of an event and the probability that it will occur. Some forms of Eq. 3.24, known as plotting-position formula, will be used in which a number will be assigned to n_a by ranking the data in either ascending or descending order of its magnitude. Greater details on this will come later but here we are focused on mathematical formalisms.

Mathematically speaking, if a random event occurs a large number of times, n, and the event has the attribute A in n_a of these occurrences, then the probability of the occurrence of the event having attribute A is

$$P(A) = \lim_{n \to \infty} \frac{n_a}{n} \qquad (3.25)$$

Equation 3.25 allows us to estimate probabilities based on observations and does not require that any outcome is equally likely or that the outcomes will be enumerated. However, the estimates of probability based on observations are empirical and will only **stochastically converge**[1] to the true probability as the number of observations becomes large.

From the two equations given above, it can be seen that probability ranges from 0 to 1; where 0 means *nearly impossible* and 1 means *almost certain*. Probability is often expressed as a percentage. However, percentage chance values and true probability values are not the same thing. For example, 1 percent chance is not the same as having a

---

[1] Stochastic convergence is a mathematical concept intended to formalize the idea that a sequence of random or unpredictable events sometimes is expected to settle into a pattern. The pattern may for instance be: (1) convergence in the classical sense to a fixed value, itself coming from a random event; (2) an increasing similarity of outcomes to what a purely deterministic function would produce; (3) an increasing preference toward a certain outcome; or (4) an increasing aversion against straying far away from a certain outcome.

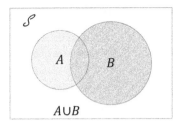
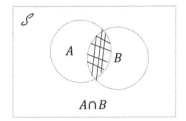
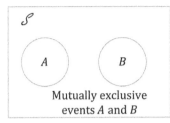
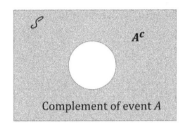
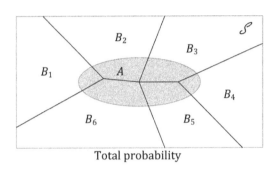

Figure 3.4 A representation of Venn diagrams. The interior of a rectangle represents the sample space, S. Any event, A or B, is represented by a circle, contained within the sample space.

probability of 1% or 0.01. The former means there is a 1% chance of occurrence, whereas the latter means it will probably happen one time in 100.

3.8.3 Rules of Probability

Having introduced a few key terms in relation to probability, certain concepts of probability are presented below with the aid of Venn diagrams, which are pictorials of overlapping circles or other shapes to illustrate the logical relationships between two or more sets (Figure 3.4).

An **experiment** is any process that generates values of random variables. All possible outcomes of an experiment constitute a **sample space**, denoted by S. Any particular point in the sample space is a **sample point** or **element** denoted by E. Thus, E_i represents an element of a sample space S where $i = 1, 2, 3, \ldots$ An **event** is a collection of elements or sample outcomes known as a **set**, s, contained in the sample space, written as $s \in S$.

The daily discharge data of Rio Negro for 44 years, shown as histograms in Figure 3.3, can be viewed as an experiment. The sample space representing all the possible outcomes of the experiment would have positive numbers. The possible daily maximum discharge during this period of 44 years could be considered as an element or sample point in the sample space. Any daily discharge above 20 000 m³/s could represent an event.

As illustrated in Figure 3.4, the notation $\cup$ represents a **union** so that $A \cup B$ represents all elements in A or B. Note that in probability the word "or" means "either or both." The notation $\cap$ represents an intersection so that $A \cap B$ represents all elements in both A and B. If an event A is defined in the sample space S, the complement of A, designated by A^c or A', is the event consisting of all the outcomes in S other than those contained in A. When A and B have no outcomes in common, they are said to be **mutually exclusive**, or **disjoint** events.

For any event A,

$$P(A) \geq 0$$
$$P(S) = 1 \tag{3.26}$$

$$P(A^C) = 1 - P(A) \tag{3.27}$$

If A and B are mutually exclusive,

$$P(A \cap B) = 0 \tag{3.28}$$

3.8.4 Joint Probability

Joint probability specifies the probability of a combination of events. If A and B are two events in S that are *not* mutually exclusive, the probability of A or B (either or both) is given by

$$P(A \cup B) = P(A) + P(B) - P(A \cap B) \tag{3.29}$$

This is so because $P(A)$ and $P(B)$ both include $P(A \cap B)$, so it must be subtracted.

If A and B are **mutually exclusive**, both cannot occur and, in that case, $P(A \cap B) = 0$. Therefore,

$$P(A \cup B) = P(A) + P(B) \tag{3.30}$$

Thus, for a finite collection of mutually exclusive events, $A_1, A_2, \ldots, A_n$

$$P(A_1 \cup A_2 \cup \ldots \cup A_n) = \sum_{i=1}^{n} P(A_i) \tag{3.31}$$

So, the probability of at least one of a number of events occurring, but of no two of the events occurring simultaneously, is the sum of the probabilities of the individual events.

If the events A and B are independent but not mutually exclusive, their joint probability is given by

$$P(A \cap B) = P(A)P(B) \tag{3.32}$$

Equation 3.32, which gives the products of probabilities of independent events can be extended to any number of events and the events may or may not occur at the same time.

3.8.5 Conditional Probability and Probabilistic Independence

If the probability of an event B depends on the occurrence of an event A, then we write $P(B|A)$ which is called the **conditional probability**, which denotes probability of B given that A has occurred. It is given by

$$P(B|A) = \frac{P(A \cap B)}{P(A)} \tag{3.33}$$

provided $P(A) \neq 0$, i.e., $P(A) > 0$.

A and B are said to be independent events, if the occurrence or nonoccurrence of one event has no bearing on the chance that the other will occur. If $P(B|A) = P(B)$, we say that B is independent of A. Thus, for independent events,

$$P(A \cap B) = P(A)P(B) \tag{3.34}$$

and $P(A|B) = P(A)$.

Thus, the probability that both events will occur is given by

$$P(A \text{ and } B) = P(A)P(B) \tag{3.35}$$

This is the same as the relationship given in Eq. 3.32 for two independent but not mutually exclusive events.

3.8.6 Total Probability Theorem

This theorem concerns the probability of a compound event A in a random experiment. Given a set of mutually exclusive, collectively exhaustive events, $B_1, B_2, \ldots, B_n$, the probability of any event A can be expressed as

$$P(A) = P(A \cap B_1) + P(A \cap B_2) + \ldots + P(A \cap B_n) \tag{3.36}$$

Every term of Eq. 3.36 can be expressed in the form of a conditional probability to yield

$$P(A) = \sum_{i=1}^{n} P(A|B_i)P(B_i) \tag{3.37}$$

which is the most common version of the total probability theorem. Equation 3.36 is also illustrated in Figure 3.4.

3.8.7 Bayes' Theorem

Bayes' theorem is very important in hydrology when considering the conditional probability of an event, B_j, given another event, A. Rewriting the conditional probability Eq. 3.33 as

$$P(A)P(B_j|A) = P(B_j)P(A|B_j) \tag{3.38}$$

and then substituting from Eq. 3.37, the total probability theorem, we get

$$P(B_j|A) = \frac{P(A|B_j)P(B_j)}{\sum_{i=1}^{n} P(A|B_i)P(B_i)} \tag{3.39}$$

Equation 3.39 is called **Bayes' theorem** or **Bayes' rule**.

3.9 HYDROLOGIC RETURN PERIODS AND EXCEEDANCE PROBABILITY

The concept of the **return period** of hydrologic events needs to be clearly understood. The return period is defined as the *average elapsed time* between occurrences of *an event with certain magnitude or greater* based on observations of a large number of occurrences.

Consider, for example, the maximum daily discharge of the Amazon River downstream of Manaus (Brazil) observed in a year for a period of 37 consecutive years from 1978 to 2014, shown in Figure 3.5. There were six instances where daily flow exceeded 200 000 m³/s (>7 million ft³/s). The interval between these occurrences varied from 1 to 10 years. But from this observation we can say that, on average, the return period for flow to equal or exceed 200 000 m³/s is $\frac{37}{6} = 6.2$ year. We can also restate this arithmetically, as follows. If, in an interval of 6.2 years, there is one occurrence, in an interval of one year there is an occurrence of $\frac{1}{6.2} = 0.1622$. This value of the reciprocal of return period is called the **exceedance probability**. Sometimes, the actual

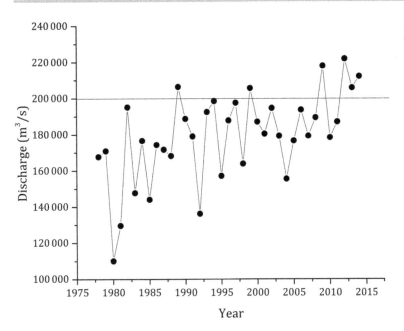

Figure 3.5 Maximum daily discharge of the Amazon River downstream of its confluence with the Rio Negro River at Manaus, Brazil. The daily maximum is for a certain year.

time between exceedances is called the **recurrence interval**. Thus, the average recurrence interval for a certain event is equal to the return period of that event. The recurrence interval and return period generally denote the same concept and are often used interchangeably.

The exceedance probability, $P_{E(X)}(x)$, of a variable X is defined as

$$P_{E(X)}(x) = P(X \geq x_T) = 1 - F_X(x_T) \qquad (3.40)$$

where, $P(X \geq x_T)$ means the probability of the variate being either equal to or greater than a specific value of x_T, and $F_X(x)$ denotes all probabilities of all variates accumulated up to this value of x_T. When applied to time-series values, $P_{E(X)}(x)$ is the probability that $X > x$ within any time interval of a specified length, Δt. Since, in most cases of applying the probability concept to hydrologic time series, Δt is a year, the term **annual exceedance probability** (AEP) is used. If expressed in terms of return period, then the corresponding term is **annual recurrence interval** (ARI).

From the example of the Amazon River discharge cited above, the return period, T_R, or recurrence interval is related to the probability of an exceedance or simply the exceedance probability, P_E, as

$$P_E = \frac{1}{T_R} \qquad (3.41)$$

This is a fundamental definition in statistical hydrology.

It is a common practice to denote the magnitude of a variable with a certain recurrence interval by simply referring it by the return period it represents. For example, a 100-year peak discharge is a discharge that is equalled or exceeded *on average* once every 100 years over a long period of time. It does not mean that an exceedance occurs every 100 years, but that the average time between exceedances is 100 years. An **exceedance** is an event with a magnitude equal to or greater than a certain value. The AEP is the probability of occurrence of the event with the specified magnitude in any given year. Again, turning to our example of the Amazon River flow, we estimated that the number of occurrences of a discharge of magnitude 200 000 m³/s in one year is 0.1622. Stated differently, the probability of 200 000 m³/s discharge occurring in any year is 0.1622 or 16.22%, and this one year can be any year in an interval of average length 6.2 years.

Thus, if an exceedance occurs on the average once every 50 years, the probability or chance that the event occurs in any given year is 1/50 = 0.02 or 2%. Since this probability is for any given year, a more appropriate term for this is *annual exceedance probability* or AEP and in the same vein, the more appropriate term for the return period is *annual recurrence interval* ARI as noted above.

We can define the return period as a function of the magnitude of the event (x_T) by expressing the definitions given above in Eqs. 3.40 and 3.41:

$$T_{R(X)}(x_T) = \frac{1}{P_{E(X)}(x_T)} = \frac{1}{[1 - F_X(x_T)]} \qquad (3.42)$$

If an exceedance occurs on average once every T years, the probability or chance that the event occurs in any given year is $1/T$. For example, 12 000 m³/s is the estimated discharge value of the River Rhine at Lobith, on the German–Dutch border, which has an AEP of 0.01 or return period of 100 years. This implies that there is 1% chance that 12 000 m³/s will be equaled or exceeded in any given year at Lobith but not that 12,000 m³/s flow occurs at Lobith once in every 100 years. The AEPs for events with various ARIs are summarized in the Table. 3.3.

Table 3.3 *Correspondence between AEP and ARI*

Return period or annual recurrence interval (ARI), T	Annual chance or annual exceedance probability (AEP), P
10 years	$\frac{1}{10} = 0.1 = 10\%$
25 years	$\frac{1}{25} = 0.04 = 4\%$
50 years	$\frac{1}{50} = 0.02 = 2\%$
100 years	$\frac{1}{100} = 0.01 = 1\%$
500 years	$\frac{1}{500} = 0.002 = 0.2\%$
1000 years	$\frac{1}{1000} = 0.001 = 0.1\%$

Since a T-year flood simply implies a probability of its occurrence in any given year, a flood with a T-year return period may occur any number of times in any given year or any number of times within a given period, or may not even occur at all within any given period of time such as $T, 2T, 3T \ldots$ years.

The concept of the return period can also be applied to low flows and droughts. In this case, the return period would be the average time between events with a certain magnitude or less. Such an event might still be called an exceedance in the sense that the severity of a drought exceeds some preset level.

3.10 PROBABILITY DISTRIBUTIONS

3.10.1 Probability Distribution of Discrete Random Variables

When probabilities are assigned to various outcomes in S, these in turn determine probabilities associated with the values of any particular random variable X. The **probability distribution** of X shows how the total probability of 1 is distributed among, or allocated to, the various possible variates of X. This distribution for a discrete random variable results in a function that is called a **probability distribution function** (PDF) or a **probability mass function** (pmf).

Thus, the function, designated by $P_X(x_i)$, called the PDF or pmf, gives the values of the probability of the discrete random variable, X, taking on any possible values x_i. It is a nonnegative function in accordance with the definition of probability and

$$\sum_{\text{all } i} P_X(x_i) = 1.0 \tag{3.43}$$

For $x_j < X < x_k$,

$$P(x_j \leq X \leq x_k) = \sum_{i=j}^{k} P_X(x_i) \tag{3.44}$$

The **cumulative distribution function** (CDF) of any variable gives the probability of the event that the random variable takes on a value equal to or less than the argument. Therefore, the cumulative probability of any variable is

$$F_X(x) = P(X \leq x) \tag{3.45}$$

For discrete random variables we have

$$F_X(x) = \sum_{\text{all } i} P_X(x_i) \tag{3.46}$$

It should be borne in mind that a discrete random variable X can assume values $x_1, x_2 \ldots, x_n$ with discrete probabilities $P(x_1), P(x_2), \ldots, P(x_n)$ and $\sum_i P(x_i) = 1$. From Eq. 3.46, we can see that, for a *discrete* random variable, the CDF is obtained by discrete summation and for this reason we use all capital letters to distinguish it from the cumulative distribution function for *continuous* random variables, cdf, which is obtained by integration of another function, as discussed in the next section.

With a discrete random variable, there are spikes of probability associated with the values that the random variable can assume but the function $F_X(x_k)$ represents the probability that X is less than or equal to x_k. Therefore, Eq. 3.46 can be rewritten as

$$F_X(x) = \sum_{x_i \leq x} P_X(x_i) \tag{3.47}$$

The cumulative distribution has jumps in it at each x_i equal in magnitude to $P(x_i)$, the probability that $X = x_i$, can be determined from

$$P_X(x_i) = F_X(x_i) - F_X(x_{i-1}) \tag{3.48}$$

The notation $P_X(x)$ and $F_X(x)$ denote the **probability distribution** and **cumulative probability distribution** of the discrete random variable X evaluated at $X = x$. Since

$$P(A) = \frac{n_a}{n} = \frac{n_i}{n} = P_x \tag{3.49}$$

the relative frequency, $f(x_i)$, can be interpreted as a **probability estimate**, the frequency histogram can be interpreted as an approximation for a probability distribution, and the cumulative frequency can be interpreted as an approximation for a cumulative probability distribution.

3.10.2 Probability Distribution of Continuous Random Variables

Since a continuous random variable can assume any value on the real line, the probability that a continuous random variable X falls between x and $x + dx$ is given by $p_X(x)dx$, where the function $p_X(x)$ is called the **probability density function** (pdf) of X. In this case, we cannot define a probability function like Eq. 3.44. Probabilities can be defined only over intervals on the real number line.

The occurrence of x in different intervals dx constitutes mutually exclusive events and thus, according to the theorem of total probability, we can compute the probability that X takes a value in an interval of finite length:

$$\int_a^b p_X(x)dx = P(a \leq X \leq b) \qquad (3.50)$$

The value of $p_X(x)$ is *not* itself a probability. It is a measure of *probability density*. That is, the probability that X takes on a value in the interval $[a,b]$ is the area under the graph of the density function as shown in Figure 3.6. Furthermore, from the axioms of probability theory, any function $p_X(x)$ defined on the real line can be a pdf if and only if

$$p_X(x) \geq 0 \qquad (3.51)$$

$$\int_{-\infty}^{+\infty} p_X(x)dx = 1 \qquad (3.52)$$

From Eqs. 3.51 and 3.52 we can see that $p_X(-\infty) = 0$ and $p_X(+\infty) = 1$. It is also apparent that the probability that X takes on a value between a and b is given by

$$P(a \leq X \leq)b = \int_a^b p_X(x)dx = p_X(b) - p_X(a) \qquad (3.53)$$

Equation 3.53 is another way of stating that $P(a \leq X \leq b)$ is the area under the *pdf* between a and b (Figure 3.6). The probability of *any specific value from a continuous distribution is zero*. This can be proved from

$$P(X = c) = \int_c^c p_X(x)dx = P(c) - P(c) = 0 \qquad (3.54)$$

where, c is any number in the interval $[a, b]$. Since the probability that a continuous random variable takes on a specified value is zero, the expressions $P(a \leq X \leq b)$, $P(a < X \leq b), P(a \leq X < b)$, and $P(a < X < b)$ are all equivalent. Since $P(X=0)$, $p_X(x)$ can be interpreted as the probability that X is strictly less than x. This is why we will see later that return periods are defined in terms of events that equal or exceed a given value or in terms of events that strictly exceed the given value since the probability of equaling the value in the case of a continuous random variable is zero.

The **cumulative probability distribution function (cdf)** of a *continuous* random variable is given by

$$F_X(x) = P(X \leq x) = \int_{-\infty}^x p_X(u)du \qquad (3.55)$$

where, u is a dummy variable of integration. From Eq. 3.55, the pdf and the cdf are related by

$$\frac{dF_X(x)}{dx} = p_X(x) \qquad (3.56)$$

It should be remembered that the pdf denoted by $p_X(x)$ is not a probability but a probability density and can have values exceeding one, but the capitalized abbreviation PDF ($\equiv$ pmf) denoted by $P_X(x)$ is a probability $[P(X \leq x) \leq P_x(x)]$ and must have values ranging from 0 to 1. The cdf for a continuous random variable is obtained by the integration of the mathematical function pdf and for this reason we have used small letters for its abbreviation to distinguish it from the CDF of a discrete random variable. The notation $p_X(x)$ and $F_X(x)$ denote the pdf and cdf of a continuous random variable, X evaluated at $X = x$.

Figure 3.7 shows the histogram of the annual rainfall in Dallas based on records of 123 years (the last column of Table 3.1). A pdf called a **normal distribution**, discussed below, is calculated based on the mean and standard deviation of the sample. This is a density function and is purely a model predicting the mathematical nature of the distribution. How well it really models reality will be determined later when we discuss the goodness-of-fit test in Section 3.19.6. A pdf can have several types of shapes, such as symmetrical bell-shaped curve, as shown in Figure 3.7, or other types resembling an exponentially decreasing function, an asymmetrical bell curve, etc. Various types of formulations have been developed to describe a pdf that can represent the observed CDF of a random variable. Also plotted on the histogram in Figure 3.7A is the CDF, which is based on the absolute frequency distribution. This is given by cumulative percent and shown by the stepped

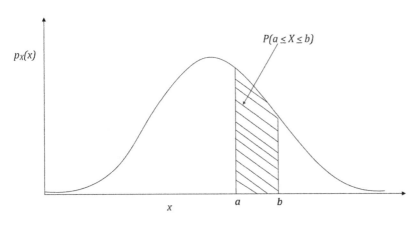

Figure 3.6 A probability density function (pdf) and the probability of a random variable X taking values between a and b on the real line.

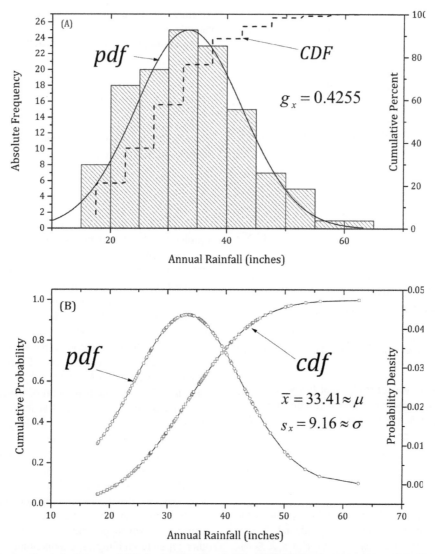

Figure 3.7 Frequency functions for sample data and probability functions from the population. (A) The absolute frequency represented as a histogram is computed from sample data of annual rainfall in Dallas for a period of record of 123 years. The cumulative frequency distribution (CDF) shown as a stepped curve is computed from this after converting the absolute frequency to relative frequency. The theoretical probability density function (pdf) computed from sample statistics (mean and standard deviation) is shown in this plot for a visual inspection of the match between a hypothesized distribution and observed distribution. Assuming the sample statistics adequately represent the population parameters (μ and σ) the pdf and cumulative distribution function (cdf) are computed and plotted in (B) to represent population characteristics. The pdf represents the variation of the derivative of the cdf in the domain of the variable (annual rainfall).

curve. Many times, continuous data are treated as though they are discrete, as in in Figure 3.7. It is also desirable to treat continuous random variable directly. Probability distributions of continuous random variables are smooth curves that are probability *density* functions $[p_X(x)]$ but for the random variable it is the probability *distribution* function $[P_x(x)]$. To reiterate, the CDF $[F_X(x)]$ represents the probability that X is less than or equal to x, i.e., $P_X(x) = P(X \leq x)$.

3.11 EXPECTED VALUES

The **expected value** of a random variable gives a measure of the center of the distribution of the variable. Essentially, it is the long-term average value of the variable. Because of the law of large numbers, the average value of the variable converges to the expected value as the number of repetitions approaches infinity. The expected value is also known as the expectation, the mean, or the first moment, and can be calculated for single discrete and continuous variables as well as multiple discrete and continuous variables.

If X is a random variable characterized with either a pdf (continuous X) or PDF (discrete X), $f(x)$, and $g(x)$, another function of x defining a given system, the expected value of the function $g(x)$, denoted by $E[g(x)]$, is defined as

$$E[g(x)] = \int_{-\infty}^{\infty} g(x)f(x)dx \tag{3.57}$$

when X is a continuous random variable, or

$$E[g(x)] = \sum_{\text{all } x} g(x) f(x) \tag{3.58}$$

when X is a discrete random variable.

In other words, the expectation of a function $g(x)$ is a weighted average of its possible values determined at various X values that it can take on. Each $g(x)$ value is weighted by its corresponding probability.

If $g(x) = x$, Eqs. 3.57 and 3.58 can be respectively rewritten as

$$E[X] = \int_{-\infty}^{\infty} xf(x)dx \tag{3.59}$$

$$E[X] = \sum_{\text{all } x} xf(x) \tag{3.60}$$

The expectation of a constant c is the constant itself, that is, $E[X] = c$, if $X = c$. The expectation of a product of a constant c and X is equal to the constant multiplied by the expectation of X, that is, $E[cX] = cE[X]$.

A simple statistical parameter of a population is its mean, μ. It is the expected value of the random variable X and can be evaluated from the pdf of the variable as

$$\mu_X = E(X) = \int_{-\infty}^{\infty} x p_X(x) dx \tag{3.61}$$

3.12 CALCULATION OF PROBABILITY VALUES: PLOTTING POSITIONS

As noted above, a major task in statistical hydrology is to assign probability values to observation. Grouping of values into classes and assigning frequencies to the classes does not work when each value in a dataset requires unique probability values. This stems from the fact that one type of data that plays a vital role in statistical hydrology is extreme events, such as the annual peak discharge in a river or the maximum depth of rainfall for a certain duration in a year. These types of datasets constitute what is called an **annual series** and mostly pertain to rainfall and streamflows.

To assign probability values to each of the datapoints in a series, a simple method is used, in which the data are arranged either in ascending or descending order following their magnitudes. Then a **rank** starting from 1 is assigned sequentially to these ordered data and a formula is used to calculate the probability based on the assertion that the rank of a value in the set signifies its frequency. If the order is in descending magnitude, the rank signifies the exceedance probability and for the reverse case it is known as the non-exceedance probability. For example, if the data are arranged in descending order of magnitude, a value getting a rank of 2 (i.e., the second highest value in the set) means that its value

Table 3.4 *Values of plotting-position coefficients for common plotting-position formulas*

a	b	Formula name*
0	0	Weibull
0.3	0.3	Median plotting position or Chegodayev formula
0.44	0.44	Gringorten
0.40	0.40	Cunnane
0.375	0.375	Blom
0.5	0.5	Hazen

* See Singh (1992) and Helsel et al. (2020).

is either equaled or exceeded only twice out of the given number of observations.

If the variates of a random variable X are such that $x_n \leq x_{n-1} \leq x_{n-2} \leq \ldots \leq x_1$, where x_1 is the largest value in the set,

$$U_i = 1 - F_X(x_i) \tag{3.62}$$

where U_i is the random variable giving the true exceedance probability of x_i. If the observations are independent, U_i can assume a *beta distribution* with the mean

$$E(U_i) = \frac{i}{n+1} \tag{3.63}$$

and variance

$$V(U_i) = \frac{i(n+1-i)}{(n+1)^3(n+2)} \tag{3.64}$$

Equation 3.63 is known as the **Weibull** plotting-position formula. There are several plotting-position formulas to calculate the empirical probability associated with the discharge or rainfall values in a *ranked* series. Most plotting-position formulas can be represented by the following form:

$$p_m = \frac{m-a}{N+1-a-b} \tag{3.65}$$

where p_m refers to the probability value (exceedance probability) assigned to the data point (variate of a variable such as discharge) whose rank or order number in the series (dataset) is m, N is the number of data points, that is, events in the series being analyzed, and a and b are parameters called **plotting-position coefficients**, which have different values in various formulations. It should be also noted that $1 - p_m \equiv P(X \geq x_m)$, which denotes the cumulative probability. Table 3.4. gives the values of a and b for the most common plotting-position formulas.

3.13 QUANTILE FUNCTIONS

The **quantile function**, associated with a probability distribution of a random variable, specifies the value of the random variable such that the probability of the variable being less than or equal to that value equals the given probability. Intuitively, the quantile function associates with a range, at and below a probability input, the likelihood that a random variable is realized in that range for some probability distribution. It is also called the **percentile function**, **percent-point function**, or **inverse cumulative distribution function**. The latter name means that, if a value of the probability is specified, the function gives the value of the random variable that can assume that probability. Thus, a quantile function can be given as

$$x = x(F_x) = F_X^{-1}(x) \tag{3.66}$$

Quantile functions offer one of the simplest ways of describing the distribution of any random variable where several quantiles of the distribution are plotted. As noted in Section 3.6, the qth quantile of the variable X is the value x_q that is larger than $(100 - q)$ percent of all values; for example, the median is x_{50}.

Quantile functions can be also calculated using frequency factors after ranking the data as noted above. Figure 3.8 shows a comparison of the quantile plots generated from detailed calculations of the cumulative probabilities of 15 467 data points compared to five key points read from the graphical frequency plots.

3.14 BIVARIATE DISTRIBUTIONS

In certain situations, it may become necessary to understand the simultaneous behavior of two or more random variables. An example would be the flow rates on two streams near their confluence. One might like to know the probability of both streams having a peak flow exceeding a given value. Another example is in a hydrologic simulation, where often say a 100-year rainfall (P) of certain duration is used to compute the peak discharge (Q) resulting from this rainfall with the assumption that this will also produce 100-year peak discharge from direct runoff. To verify this assumption, it is necessary to study the *joint probabilities* of the two random variables P and Q.

If X and Y are two continuous random variables, their joint pdf, denoted by $p_{X,Y}(x, y)$, and their corresponding cdf, denoted by $F_{X,Y}(x, y)$, are related by

$$p_{X,Y}(x, y) = \frac{\partial^2}{\partial x \partial y} \left[P_{x,y}(x, y) \right] \tag{3.67}$$

$$F_{X,Y}(x, y) = P(X \leq x \text{ and } Y \leq y) = \int_{-\infty}^{x} \int_{-\infty}^{y} p_{X,Y}(s, t) ds dt \tag{3.68}$$

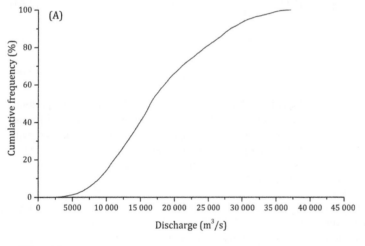
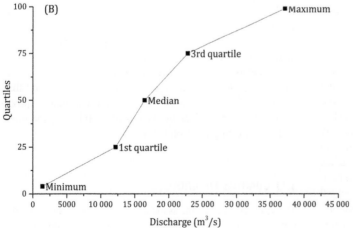

Figure 3.8 (A) Quantile plots calculated using a plotting-position formula and (B) plots of three quartile values along with minimum and maximum values read from the frequency plots of daily discharge values of Rio Negro for a 44-year period (see Figure 3.3).

Where s and t are dummy variables of integration. The corresponding relationships for X and Y being discrete random variables are

$$f_{X,Y}(x_i, y_j) = P(X = x_i \text{ and } Y = y_j) \tag{3.69}$$

$$F_{X,Y}(x, y) = P(X \leq x \text{ and } Y \leq y) = \sum_{x_i \leq x} \sum_{y_j \leq y} f_{X,Y}(x_i, y_j) \tag{3.70}$$

$$f_{X,Y}(x_i, y_j) = F_{X,Y}(x_i, y_j) - F_{X,Y}(x_i, y_{j-1}) - F_{X,Y}(x_{i-1}, y_j) + F_{X,Y}(x_{i-1}, y_{j-1}) \tag{3.71}$$

3.15 DISCRETE PROBABILITY DISTRIBUTIONS

3.15.1 The Bernoulli Process

If there is a probability of either an occurrence or nonoccurrence of a specified random event at an ith point on a discrete time scale with $i = 1, 2, \ldots, n$ and the occurrence of the event at any point during the period is independent of the history of any prior occurrence or nonoccurrence on the time scale, then such a stochastic process is called a **Bernoulli process**. Thus, in a Bernoulli process, at each instant of time, an event may either occur with probability p or not occur with probability $q = 1 - p$. The probability of the event occurring is independent of the time and independent of the past history of occurrences.

Any random variable whose only possible values are 0 and 1 is called a **Bernoulli random variable**.

An example of a Bernoulli process would be, during any year, the probability of the maximum discharge in a river exceeding Q^*, where the annual discharge in the river is independent from year to year. We can define a discrete random variable, X, that takes on a value of 1 if a discharge greater than or equal to Q^* occurs in a year. The variable assumes a value of 0 if the discharge is less than Q^*. This variable X is then a Bernoulli variable. The PDF is given by

$$P_X(k) = \begin{cases} p \text{ if } k = 1 \\ 1 - p \text{ if } k = 0 \end{cases}$$

Since the expectation is defined as the operation of taking the mean of a function, the mean or expected value of X in this case is

$$\mu_X = E(X) = 1p + 0(1 - p)$$

and the variance is

$$E\left[(X - \mu_X)^2\right] = (1 - p)p$$

3.15.2 Binomial Distribution

From the rule of probability of intersection of two events (say A for occurrence and B for nonoccurrence of the specified event), of a Bernoulli process where the probability p of the variable X assuming a value of k (i.e., $X = k$) during a period n (n is discrete, $i = 1, 2, \ldots, n$) is designated by: $f_X(k : n, p)$. Note that here we have chosen k to denote the variate, instead of x, to remember that it is a discrete variable. The probability distribution, $f_X(k : n, p)$, is known as **binomial distribution**.

The probability of $X = k$ occurrences of an event in n independent trials with p as the probability of an occurrence in a single trial and q as the probability of nonoccurrence is given as

$$f_X(k : n, p) = \binom{n}{k} p^k (1 - p)^{n-k} = \binom{n}{k} p^k q^{n-k} \tag{3.72}$$

$k = 0, 1, 2, \ldots, n$

Even though Eq. 3.72 is most well known in hydrology to estimate the probability of k floods in n years, the binomial distribution and Bernoulli process are not limited to a time scale. Any process that may occur with probability p at discrete points in time or space or in individual trials may be a Bernoulli process and may follow the binomial distribution (Haan, 1977).

The cumulative binomial distribution is

$$F_X(k : n, p) = \sum_{i=0}^{k} \binom{n}{i} p^i q^{n-i} \tag{3.73}$$

Equation 3.73 gives the probability of k or fewer occurrences of an event in n independent trials if the probability of an occurrence in any trial is p and the probability of nonoccurrence is q.

The mean and variance of binomial distribution are

$$E(X) = np \tag{3.74}$$

$$V(X) = npq \tag{3.75}$$

The coefficient of skewness is $(q - p)/\sqrt{npq}$ so that the distribution is symmetrical for $p = q$, skewed to the right for $q > p$, and skewed to the left for $q < p$. Note that calculation of mean and variance of a binomial random variable does not necessitate evaluating a summation.

3.15.3 Geometric Distribution

The probability that the first occurrence of a Bernoulli trial occurs on the nth trial is found by noting that for the first occurrence to be on the nth trial, there must be $(n - 1)$ preceding trials without occurrence of the event. Thus, the probability of this first occurrence of the event is pq^{n-1}. This is known as the **geometric distribution** and is given as

$$f_X(k : p) = pq^{n-1} \tag{3.76}$$

$k = 1, 2, \ldots, n$

The mean and variance of the geometric distributions are

$$E(X) = \frac{1}{p} \qquad (3.77)$$

$$V(X) = \frac{q}{p^2} = \frac{1-p}{p^2} \qquad (3.78)$$

It can be noted from Eq. 3.77, that the mean of the geometric distribution is the recurrence interval or the average number of years that will pass before the event (e.g., $x > x_T$) in question occurs.

Thus, while the number of occurrences of a Bernoulli process in a given time interval follows binomial distribution, the geometric distribution describes the probability that the first occurrence of an event in a Bernoulli process is at the nth time. Figure 3.9 shows the various forms of these distributions as functions of p.

3.15.4 Applications of Binomial and Geometric Distributions

We have already stated that an event with a return period T_R has the probability of occurrence in any year is $1/T_R$. We designate this by the small letter p to imply that it has a value ranging from 0 to 1 for the function $P_{E(X)}$.

The question of what the probability is of the T-year event occurring in n successive years can be answered with binomial distribution. This is also known as the **risk** since it gives the risk of having at least one event of recurrence interval T in an n-year period. This probability or risk can be calculated as

$$\begin{aligned} P_{E(X)}(X \geq x_T \text{ at least once in } n \text{ years}) \\ = 1 - P_X(X < x_T \text{ each year during } n \text{ years}) \end{aligned} \qquad (3.79)$$

where the probability of nonoccurrence is given by (from Eq. 3.72)

$$\begin{aligned} P_X(X < x_T \text{ no event in } n \text{ years}) &= \binom{n}{0} p^0 (1-p)^{n-0} \\ &= (1-p)^n = \left(1 - \tfrac{1}{T}\right)^n \end{aligned} \qquad (3.80)$$

Substitution of Eq. 3.80 into Eq. 3.79 gives

$$\begin{aligned} P_{E(X)}(X \geq x_T \text{ 1 or more in } n \text{ years}) &= 1 - (1-p)^n \\ &= 1 - \left(1 - \tfrac{1}{T}\right)^n \end{aligned} \qquad (3.81)$$

Note that Eq. 3.81 can also be derived from the rule of probability of intersections of independent events as given in Eq. 3.33 which states that, if A and B are independent events, $P(A \cap B) = P(A) P(B)$. Thus, the probability of nonoccurrence of the event $X \geq x_T$ in n successive years is the product of n probability values, each of which is given by $(1-p)$. In other words,

$$\begin{aligned} P_X(X < x_T \text{ for each year in } n \text{ consecutive years}) \\ = (1-p) \times (1-p) \times \ldots n \text{ times} = (1-p)^n \end{aligned} \qquad (3.82)$$

Equation 3.80 is an important equation in flood risk management. It has found numerous applications in the National Flood Insurance Program of the United States. More importantly, it defines the return period for hydraulic designs at specified levels of risk.

3.16 PROBABILITY DENSITY FUNCTIONS

3.16.1 Preliminaries

Numerous mathematical models have been developed to represent various kinds of frequency distributions of continuous random variables. Singh and Zhang (2017) presented an exhaustive catalogue of the pdfs and corresponding cdfs that have found applications in hydrology. Here we only present the ones that are widely accepted and used for certain common tasks of applied hydrology. Further details of some of these are covered in Chapters 4 and 5.

3.16.2 Normal Distribution

3.16.2.1 Probability Density and Cumulative Density Functions

The most widely used and one of the most important continuous probability distributions is the **Gaussian** or **normal distribution**. We will see from the **central limit theorem** the conditions under which a random variable can be expected to be normally distributed. Theoretically, most hydrological variables cannot be normally distributed because the range of any random variable that is normally distributed is the entire real line ($-\infty$ to $+\infty$). Thus, nonnegative variables such as rainfall, streamflow, reservoir storage, etc. cannot be strictly normally distributed.

The normal distribution is a two-parameter distribution whose density function is given as

$$p_X(x) = \frac{1}{\sigma_X \sqrt{2\pi}} \exp\left[-\frac{1}{2}\left(\frac{x - \mu_X}{\sigma_X}\right)^2\right], \ -\infty \leq x \leq \infty \qquad (3.83)$$

where $p_X(x)$ is the pdf of X, μ_X is the average value of X and σ_X is the standard deviation of X. The normal distribution is denoted as $N(\mu, \sigma^2)$ which means that X is normally distributed with a mean μ and a variance σ^2, respectively, known as **scale** and **shape** parameters of this continuous distribution. It can be shown that $p_X(x)$ is symmetrical about the mean and that it decreases on either side of the mean without ever reaching zero. Therefore, a normal distribution is a bell-shaped, continuous curve symmetrical about μ with a coefficient of skewness of zero.

The range of a normally distributed variable is from $-\infty$ to $+\infty$ but most hydrologic variables vary from zero to some high positive value. Strictly speaking, such variables

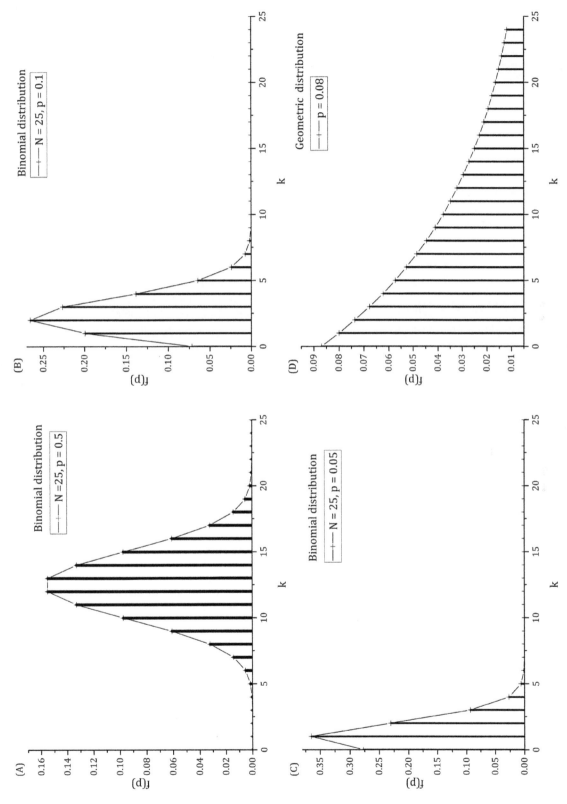

Figure 3.9 (A)–(C) Binomial and (D): geometric distributions for $n = 25$ and different values of p. In the case of binomial distributions, n is also a parameter but, in this figure, it is kept constant.

cannot follow a normal distribution. However, if μ_X of a random variable is more than three times σ_X, the probability of the variable acquiring negative values is very small and hence the normal distribution can be applied without incurring unacceptable error. Furthermore, hydrologic data such as extreme values of rainfall and river flows tend to be skewed and a normal distribution does not approximate this well.

The cdf, $F_X(x)$, can be expressed as

$$F_X(x) = \int_{-\infty}^{x} \frac{1}{\sigma_X \sqrt{2\pi}} \exp\left[-\frac{1}{2}\left(\frac{x - \mu_X}{\sigma_X}\right)^2\right] dx, \quad -\infty \leq X \leq \infty \tag{3.84}$$

Equation 3.84 cannot be evaluated in a closed form but by the expression in an infinite series and integrating term by term.

Equation 3.83 is simplified by defining a **standard normal variable**, Z, given by

$$Z = \frac{x - \mu_X}{\sigma_X} \tag{3.85}$$

The quantity $Z = \dfrac{x - \mu_X}{\sigma_X}$ can be interpreted as the number of deviations by which x differs from the mean. For the sample, the **standard normal variate** is

$$z = \frac{x - \bar{x}}{s_x} \tag{3.86}$$

The standardized variable Z has mean 0 and standard deviation 1 and is normally distributed, $N(0, 1)$. Thus, the pdf of Z is

$$p_Z(z) = \frac{1}{\sqrt{2\pi}} \exp\left(-\frac{z^2}{2}\right) \quad -\infty \leq z \leq \infty \tag{3.87}$$

The cdf can be expressed as

$$F_X(x) = P(X \leq x) = P\left[Z \leq \frac{X - \mu_X}{\sigma_X}\right]$$
$$= F_Z\left(\frac{x - \mu_X}{\sigma_X}\right) = F_Z(z) \tag{3.88}$$

$$F_Z(z) = \frac{1}{\sqrt{2\pi}} \int_{-\infty}^{z} \exp\left[-\frac{u^2}{2}\right] du \quad -\infty \leq u \leq \infty \tag{3.89}$$

The mean and variance of Z can be shown to be

$$E[Z] = \frac{1}{\sigma_X}[E(X) - \mu_X] = \frac{1}{\sigma_X}[\mu_X - \mu_X] = 0 \tag{3.90}$$

$$V[Z] = \frac{1}{\sigma_X^2}[Var(x)] = \frac{1}{\sigma_X^2}\sigma_X^2 = 1 \tag{3.91}$$

The integration of Eq. 3.89 cannot be performed analytically, and it can be only evaluated by series expansion.

Therefore, the values of $F_Z(z)$ for various values of z are listed in tables for $z \geq 0$, found in any statistical text. Thus, by expressing Eq. 3.83 in the standardized form, the values of the cumulative distribution are obtained from the standard normal distribution tables. A numerical approximation of $F(z)$ was given by Abramowitz and Stegun (1970) as

$$F(z) = (2.5052367 + 1.2831204 z^2 + 0.2264718 z^4 \\ + 0.1306469 z^6 - 0.0202490 z^8 + 0.003932 z^{10})^{-1}$$

The error of this approximation is $< 2.3 \times 10^{-4}$. An approximate formula to calculate $F(z)$ is

$$F(z) = 0.5 + [0.5 + 0.064 \times \{\exp(-0.4 z^2)\}] \\ \times \sqrt{\left[1 - \exp\left\{-\left(\frac{z^2}{2}\right)\right\}\right]} \tag{3.92}$$

However, $p_X(x)$, $F_X(x)$, $F_Z(z)$, and z for A given value of $F_Z(z)$ can be easily obtained from Excel without recourse to looking at tables or using the approximations given above. Excel offers two functions that are especially useful in applications of the normal distribution to statistical hydrology. These two functions with inputs in the arguments are the following two functions.

1. NORM.DIST (x, mean, standard deviation, TRUE/FALSE) (TRUE for cumulative and FALSE for density functions).
2. NORM.S.INV(z) – returns the value of z for a given value of $F_Z(z)$.

Figure 3.10A shows the frequency distribution of mean annual discharge of Potomac River to the northwest of Washington, DC. These data are based on 126 years (1896–2021) of mean annual flow at the river gauging station at the Round of Rocks, Maryland. The skewness of the distribution is 0.49. Since this is less than 1, the distribution is considered nearly symmetrical and a pdf assuming normal distribution has been fitted to the frequency histogram. The pdfs and cdfs (Eqs 3.83 and 3.84), calculated using the NORM.DIST function, are shown in Figure 3.10B.

3.16.2.2 Central Limit Theorem

Let us consider $X_1, X_2, \ldots, X_n$ to be a sequence of n independent identically distributed random variables, each having mean μ and standard deviation σ. Then the distribution of the sum, $S_n = X_1 + X_2 + \ldots + X_n$, tends to a normal distribution with mean $n\mu$ and standard deviation $\sigma\sqrt{n}$, if n tends to infinity. This is true no matter what the probability distribution function of X is. Hydrologic variables, such as annual precipitation, calculated as the sum of the effects of many independent events, tend to follow the normal distribution. A case has already been demonstrated with the aid of

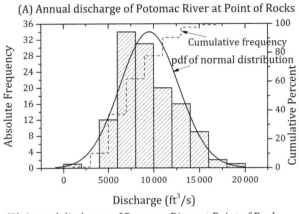

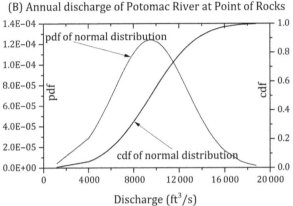

Figure 3.10 (A) Frequency distribution of 126 years (1896–2021) of mean annual discharge of Potomac River to the northwest of Washington, DC shown by the histogram and its approximation by normal or Gaussian distribution. The skewness of the distribution is 0.49. The CDF is shown as the stepped function. (B) The calculated pdf and cdf (Gaussian distribution) for the data presented as histogram in A.

123 years of annual precipitation data from Dallas, as shown in Figure 3.7. The frequency distribution can be well represented by a normal distribution.

The mean and variance of the random variable S_n can be derived as:

$$E[S_n] = E[X_1 + X_2 + X_3 + \ldots + X_n]$$
$$= E[X_1] + E[X_2] + \ldots + E[X_n] = n\mu \qquad (3.93)$$

$$V[S_n] = Var[X_1 + X_2 + \ldots + X_n] = nV[X] = n\sigma^2 \qquad (3.94)$$

Thus, the standard deviation of S_n is $\sigma\sqrt{n}$.

Mathematically, the central limit theorem indicates that

$$\lim_{n->\infty} P[S_n \leq s] = \lim_{n->\infty} P\left[\frac{S_n - n\mu}{\sigma\sqrt{n}} \leq \frac{y - n\mu}{\sigma\sqrt{n}}\right] = \frac{1}{\sqrt{2\pi}} \int_{-\infty}^{z} e^{-y^2/2} dy$$
$$(3.95)$$

where y is a particular value of S_n, and $z = \frac{y - n\mu}{\sigma\sqrt{n}}$. Further, if there exists a constant A, such that $|X_n| \leq A$ for all n, for $a < b$,

$$\lim_{n->\infty} P\left[a \leq \frac{S_n - n\mu}{\sigma\sqrt{n}} \leq b\right] = \frac{1}{\sqrt{2\pi}} \int_a^b e^{-y^2/2} dy \qquad (3.96)$$

Similarly, one can obtain that the average $\overline{X} = (X_1 + X_2 + \ldots\ldots + X_n)/n$ tends toward a normal distribution function with mean μ and standard deviation $\sigma/\sqrt{n}$.

The central limit theorem helps approximate the sampling distribution of the sum by an appropriate normal curve regardless of the form of the parent pdf from which individual observations were derived. To explain it further, let us consider measurements of discharge of a river being made at a gauging station. The technician takes note of the computed discharge after rounding it off to the nearest integer. These data are subsequently used in hydrologic analysis and design. For example, monthly flow at the station is obtained by adding daily values. What is the rounding error in the sum of n measurements? The rounding error in a single measurement display is a random variable, called X. It is assumed that this variable is uniformly distributed. Suppose now that n rounding errors are summed. The sum is a random variable, denoted as S_n. Quite likely, the values of X will be clustered around zero since positive and negative values in the individual measurements tend to cancel each other out to some extent. Values near the extremes will have a very low probability density since outcomes near extremes will occur if all or most individual measurements have large rounding errors of the same sign.

3.16.2.3 Frequency Factor for Normal Distribution

Calculation of magnitudes of extreme events such as rainfall and flood flow for desired probability levels (AEP or ARI), from frequency analysis requires that the cdf of the probability distribution is invertible. The problem can be posed as

Given exceedance probability:

$$P_{E(X)}(X > x_T) = 1 - F_X(x_T) : 1 - \frac{1}{T_R} = 1 - P_{E(X)}$$

Solve for : x_T

The problem statement is simply determining quantile values from a quantile plot, which is the inverse of the cdf.

The cdfs of certain probability distributions are not readily invertible. The cdf of normal distribution is one such example. Another is the cdf of log Pearson type III distribution that is discussed in Chapter 5 and is extensively used in flood frequency analysis.

Chow (1951) developed a method to derive quantile functions using a factor called the **frequency factor**. A random variate x can be expressed as the sum of its mean value plus the departure of the variate from the mean given as

$$x = \overline{x} + \Delta x \qquad (3.97)$$

The departure Δx depends on the statistical characteristics of the distribution and can be assumed to be some multiple of either its standard deviation s_x or coefficient of variation ($C_v = s_x/\overline{x}$) and a factor K_T that is called the frequency factor. Thus,

$$x = \bar{x} + s_x K_T \tag{3.98}$$

$$x = \bar{x} + C_v K_T \tag{3.99}$$

The frequency factor depends upon the statistical characteristics of the frequency or probability distribution and the recurrence interval or return period, T_R. For a particular distribution, the relationship between K_T and the recurrence interval can be presented through tables or graphs.

For normal distribution, the frequency factor is the standardized normal variate z, as can be seen from Eq. 3.98:

$$K_T = \frac{x_T - \bar{x}}{s_x} \tag{3.100}$$

As noted above, z can be found for a desired value of $F(z)$, either from tables in any statistical text or using the NORM.S.INV(z) function offered in Excel.

From Eq. 3.89, the exceedance probability can be computed by

$$P(X \geq x) = 1.0 - \frac{1}{\sigma_X \sqrt{2\pi}} \int_{-\infty}^{K_T} \exp\left(-\frac{K_T^2}{2}\right) dK \tag{3.101}$$

Thus, this is just a simple matter of finding the value of z for a nonexceedance probability. Once z is found which is K_T for normal distribution, various quantiles of the variable x_T can be calculated from Eq. 3.100.

For example, the gauging station of Potomac River, where the mean annual flows were obtained and presented in relation to Figure 3.10, also recorded the instantaneous maximum discharge in a year. The data for 128 years (1889–2021) are given in Table 5.9 (Chapter 5, Example 5.3). Figure 3.11 shows the pdf and cdf *assuming* that the frequency distribution can be adequately modeled as a normal distribution. The mean and standard deviation of the data are: $\bar{x} = 121{,}839.1 \text{ ft}^3/\text{s}$ and $s_x = 76{,}956.1 \text{ ft}^3/\text{s}$. We desire estimates of 100-year, 50-year, and 10-year discharges through Potomac River at this location. In other words, we need to determine x_T for AEP $(P) = 0.01, 0.02,$ and 0.10. The z value for the corresponding $F(z) = 1 - P = 0.99, 0.98,$ and 0.90, respectively. The corresponding z values are: $2.3263, 2.0537,$ and 1.2815. These sample statistics and the frequency factors give the quantile estimates: $Q_{100} = 300{,}865.8$, $Q_{50} = 279{,}887.6,$ and $Q_{10} = 220{,}462.3 \text{ ft}^3/\text{s}$.

3.16.3 Lognormal Distribution

3.16.3.1 Probability Density and Cumulative Distribution Functions

Many probability distributions encountered in hydrology are skewed. The distribution of annual flood peaks is an example. Hazen (1914) found that the frequency curves of flood peaks generally showed a marked upward curvature, but curves could be straightened in most cases if a logarithmic scale, instead of a linear scale, was used along the horizontal axis, that is, plotting the logarithms of the data instead of the regular data. If this results in a straight line on a probability graph, the logarithms of X are normally distributed. In many situations,

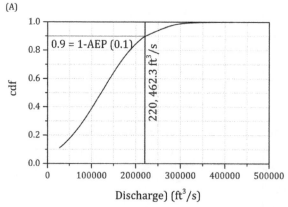

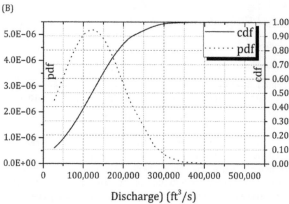

Figure 3.11 (A) Normal distribution approximation of annual maxima of peak discharge of 126 years (1896–2021) of Potomac River at the Round of Rocks, Maryland (B) The calculated pdf and cdf (Gaussian distribution) for the data presented as histogram in Figure 3.12.

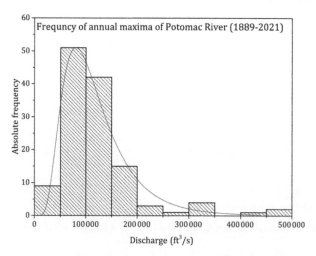

Figure 3.12 Lognormal distribution of annual maxima of peak discharge of 126 years (1896–2021) of Potomac River at the Round of Rocks, Maryland.

the hydrologic variables, such as daily streamflows and annual flood peaks, follow a lognormal distribution. Figure 3.12 shows the annual maximum series of Potomac River, discussed above, with a pdf of lognormal distribution.

A random variable, X, is said to be lognormally distributed if its logarithm, $Y = \ln(X)$, can be characterized by a normal distribution with parameters μ_Y and σ_Y. The pdf of Y can be written as

$$p_Y(y) = \frac{1}{\sigma_Y \sqrt{2\pi}} \exp\left[-\frac{1}{2}\left(\frac{y-\mu_Y}{\sigma_Y}\right)^2\right], \quad -\infty \leq Y \leq \infty \tag{3.102}$$

where $\mu_Y = \mu_{\ln(Y)}$ and $\sigma_Y = \sigma_{\ln(Y)}$. Parameters μ_Y and σ_Y can be estimated by the transformed sample data using a logarithmic transformation such that $y_i = \ln(x_i)$. The sample mean and standard deviation are used as the estimates of μ_Y and σ_Y. Thus,

$$\mu_Y \approx \bar{y} = \frac{\sum y_i}{n} \text{ and } \sigma_Y \approx s_Y = \sqrt{\frac{(y_i - \bar{y})^2}{n-1}}.$$

One can determine the distribution of X by the technique of variable transformation as given below:

$$p_X(x) = p_Y(y)\left|\frac{dy}{dx}\right|$$

Differentiating $Y = \ln(X)$ with respect to X, one gets $\frac{dY}{dX} = \frac{1}{X}$. Substituting this relationship and Eq. 3.102 into the above relation, the pdf of X is given as

$$p_X(x) = \frac{1}{x\sigma_Y\sqrt{2\pi}} \exp\left[-\frac{1}{2}\left(\frac{\ln x - \mu_Y}{\sigma_Y}\right)^2\right], \quad x \geq 0 \tag{3.103}$$

Equation 3.103 represents the lognormal distribution. Note that the range of Y is $-\infty$ to $+\infty$ while that of X is 0 to ∞.

It should be noted that the mean of X, μ_X, should not be interpreted as a 50% probable value. Instead, the median value is the 50% probable value of a lognormally distributed variable X on either side of which half of the distribution lies. Let M_X be the median value and the geometric mean of X. Thus, we can write $P(X \leq M_X) = 0.5$. Further, based on the definition of the normal distribution, this relationship can be rewritten as $F_u\left(\frac{\ln M_X - \mu_Y}{\sigma_Y}\right) = 0.5$. Thus, $\frac{\ln M_X - \mu_Y}{\sigma_Y} = F_u^{-1}(0.5) = 0$. Therefore, one can write

$$\mu_Y = \ln(M_X) \tag{3.104}$$

Equation 3.103 can be rewritten in terms of the median value of X

$$p_X(x) = \frac{1}{x\sigma_Y\sqrt{2\pi}} \exp\left[-\frac{1}{2}\left\{\frac{1}{\sigma_Y}\ln\left(\frac{x}{M_X}\right)\right\}^2\right], \quad x \geq 0 \tag{3.105}$$

denoted as $\ln(M_X, \sigma_{\ln X})$, M_X and $\sigma_{\ln X}$ are parameters.

One can use normal tables as follows. If U is a standardized $N(0,1)$ variable, then

$$p_X(x) = \frac{1}{x\sigma_{\ln X}} f_U\left(\frac{\ln(x/M_X)}{\sigma_{\ln X}}\right) \tag{3.106}$$

In other words,

$$p_X(x) = \frac{1}{x\sigma_{\ln X}} f_U(u) \tag{3.107}$$

where

$$u = \frac{1}{\sigma_{\ln X}} \ln \frac{x}{M_X} \tag{3.108}$$

Tables of the function $f_U(u)$ are widely available.

The cdf of X is easily calculated from the tables of the normal distribution:

$$F_X(x) = P[X \leq x] = P[\ln X \leq \ln x] = P[Y \leq \ln x] = F_Y[\ln x] \tag{3.109}$$

Y is $N(\mu_Y, \sigma_Y^2)$ or $N[\ln(M_X, \sigma_Y^2)]$. Hence,

$$F_X(x) = F_U\left(\frac{\ln x - \ln M_X}{\sigma_{\ln X}}\right) = F_U\left(\frac{\ln(y/M_X)}{\sigma_{\ln X}}\right) = F_U(u) \tag{3.110}$$

where u is defined as above.

3.16.3.2 Frequency Factor for Lognormal Distribution

Here X is normally distributed while $Y(=\exp[X])$ is lognormally distributed. Chow (1964) presented the frequency factor as:

$$K_T = \frac{\exp(s_X K_X - s_X^2/2) - 1}{\sqrt{[\exp(s_X^2 - 1)]}} \tag{3.111}$$

where $K_X = (X_T - \bar{x}_X)/s_X$. For a given value of X_T, K_X and thereby K_T can be calculated. The coefficient of variation and the coefficient of skewness are computed as

$$C_v = \left[\exp(\sigma_x^2) - 1\right]^{0.5} \tag{3.112}$$

$$C_s = 3C_v + C_v^3 \tag{3.113}$$

Using the computed value of C_s and the exceedance probability, K_T can be read from tables of frequency factors for the lognormal distribution.

3.16.4 Gamma Distribution

3.16.4.1 Probability Density and Cumulative Distribution Functions

This density function is also important in hydrology. Here we present the well-known two-parameter gamma distribution. An extension of this distribution with three parameters with logarithmic transformation of the variable, known as log Pearson type III distribution, is discussed in detail in Chapter 5 for its wide application in flood frequency analysis. The gamma distribution is also discussed in Chapter 9, where it finds important usage in synthetic unit hydrograph development.

The pdf of the gamma distribution is

$$p_X(x;\alpha,\beta) = \frac{1}{\beta^\alpha \Gamma(\alpha)} x^{\alpha-1} \exp\left(-\frac{x}{\beta}\right) \quad (3.114)$$

where α and β are the *shape* and *scale* parameters, respectively, and $\Gamma(\alpha)$ is the gamma function given by

$$\Gamma(\alpha) = \int_0^\infty x^{\alpha-1} \exp(-x) dx \quad (3.115)$$

The standard gamma distribution ($\beta = 1$) is

$$p_X(x;\alpha) = \frac{x^{\alpha-1}}{\Gamma(\alpha)} \exp(-x) \quad (3.116)$$

When $\alpha = 1$, the distribution with $\lambda = \frac{1}{\beta}$ is known as the **exponential distribution**.

The cdf is known as incomplete gamma function.

3.16.5 Extreme Value Distributions

Frequently in the design and management of major flood control and water management systems, the largest or the smallest value of a rainfall or flood flow is of concern. For example, the largest flood resulting from extreme rainfall is critical to the design of major dams, spillways, and levees. If Y is the maximum of n random variables, $X_1, X_2, \ldots, X_n$, the probability

$$F_Y(y) = P[Y \leq y] = P \text{ [all } n \text{ random variables } X_i \leq y]$$

If random variables X_i are independent,

$$\begin{aligned} F_Y(y) &= P[X_1 \leq y]P[X_2 \leq y]\ldots P[X_n \leq y] \\ &= F_{X_1}(y) \cdot F_{X_2}(y) \cdot \ldots \cdot F_{X_n}(y) \end{aligned} \quad (3.117)$$

If all X_i are identically distributed with the cdf, $F_X(x)$, then

$$F_Y(y) = \{F_X(y)\}^n \quad (3.118)$$

Assuming X_i to be continuous random variables with the pdf, $f_X(x)$, one obtains

$$f_Y(y) = \frac{d}{dy}[F_Y(y)]^n = n[F_X(y)]^{n-1} f_X(y) \quad (3.119)$$

Equations 3.118 and 3.119 can be used to determine the distribution of Y if X_i are mutually independent, and identically distributed. Three limiting forms of $f_Y(y)$ for large values of n are found, depending on (1) the interest in the largest or smallest value, and (2) the behavior of the appropriate table of X_i.

Three extreme value distributions known as extreme value type I (EV I) or Gumbel distributions, extreme value type III (EV III) or Weibull distributions, and generalized extreme value (GEV) distributions are presented in greater detail in Chapter 4. These distributions are widely used in frequency analysis of rainfall data. They are also used in flood frequency analysis, discussed in Chapter 5.

3.17 PROBABILITY PLOTS

Probability plots are graphical measures to assess if a set of data plausibly came from some theoretical distribution, such as a normal or lognormal distribution. Even though it is just a visual check, it allows us to inspect at a glance if an assumption about the distributional characteristics of the population from which the sample is obtained is valid and, if not, how the assumption is violated and what data points contribute to the violation. Determining how well data fit a theoretical distribution can be done by visually comparing histograms of sample data to the pdf of the theoretical distributions as shown in Figures 3.10A and 3.12. However, a kind of plot that is meant for such purpose is called a **quantile–quantile plot** or **Q–Q plot**. A Q–Q plot is generally more diagnostic than comparing the histogram of a sample and a theoretical distribution. The term probability plot generally refers specifically to a Q–Q plot.

A Q–Q plot is a scatterplot created by plotting two sets of quantiles against one another. If both sets of quantiles come from the same distribution, then the points define a line that is nearly straight. Figure 3.13 shows the Q–Q plots of mean annual discharge of Potomac River at the Point of Rocks in which a Q–Q plot is generated by plotting the quantiles of two distributions against each other in two alternative ways. In the first alternative (Figure 3.13A), the inverse of the cumulative distribution derived from the Blom plotting-position formula is plotted against the observed discharge values. In the second alternative (Figure 3.13B, the inverse of the plotting positions of the standard normal variates of the observed discharge are plotted. Because it is unbiased for the normal distribution, it can be argued that the Blom plotting position is preferable for this kind of plot. Nevertheless, the plots indicate a nearly normal distribution of the data. However, in the Q–Q plots, particular attention must be paid toward the high and low tails of the data, which can deviate from the straight-line definition of the plots.

3.18 PARAMETER ESTIMATION

3.18.1 Methods of Parameter Estimation

Quantities that are descriptive of a population are called **parameters**. These are usually estimated from samples of data. Sample statistics are estimates for population parameters; they are functions of values that are random variables and therefore they themselves are also random variables. The samples must be representative of the population so that the sample statistics are representatives of population parameters.

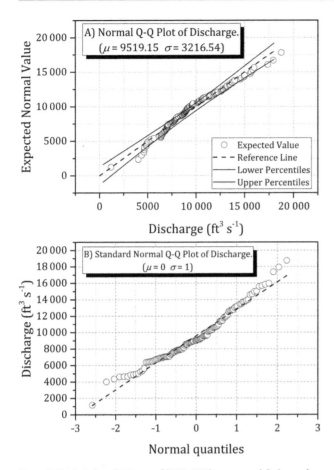

Figure 3.13 Q–Q plots of 126 years of (1896–2021) mean annual discharge of Potomac River at the Point of Rocks. (A) Observed discharge plotted on the x-axis and expected value assuming normal distribution plotted on the y-axis. (B) Quantiles of standard normal variates on the x-axis and quantiles of the observed values, which are simply the observation values themselves, are plotted on the y-axis. (A)

A number of methods are available to estimate parameters of probability distributions. The methods commonly used in hydrology include (1) the method of product moments, (2) the method of probability-weighted moments, (3) the method of mixed moments, (4) the method of L-moments, (5) maximum-likelihood estimation, and (6) the least-squares method. A convenient way of quantifying the location and some measures of the shape of a probability distribution is by computing the moments of the distribution. Possibly for this reason, the method of product moments was the most commonly used method for the estimation of parameters. However, in recent year, the method of L-moments has become the standard for estimating the parameters of probability distributions. These two methods are presented here. However, before discussing L-moments, it is necessary to introduce probability-weighted moments.

3.18.2 Method of Product Moments

This method was developed by Karl Pearson in 1902 based on the premise that when the parameters of a probability distribution are estimated correctly, the moments of the pdf are equal to the corresponding moments of the sample data.

In the method of moments, distribution moments are equated to the sample moments to estimate distribution parameters. The advantage of the method of moments is that moments are simple to understand and interpret, and are easy to calculate. But the disadvantage is that they are often not available for all the probability distribution functions and they do not have the desirable optimality properties as other methods, such as the maximum-likelihood and least-squares estimators, have.

The main steps for parameter estimation for a distribution having k parameters using the method of moments are:

1. Compute the algebraic expressions for the first k moments from the assumed distribution.
2. Compute the numerical values for the first k sample moments from the sample data.
3. Equate each algebraic expression of the assumed distribution moment to the corresponding numerical value of the sample moment. This will form k algebraic equations involving k sample moments and k distribution parameters, $\alpha_1, \alpha_2, \ldots, \alpha_k$.
4. Solve these equations to determine k unknown parameters.

The rth moment of $f(x)$ about an arbitrary point is denoted as $\mu_r^a(f)$. The algebraic expression for this rth moment of the function $f(x)$ can be defined as

$$\mu_r^a(f) = \int_{-\infty}^{\infty} (x-a)^r f(x) dx \tag{3.120}$$

In Eq. 3.120, μ denotes the moment, the subscript ($r \geq 0$) denotes the order of the moment, the superscript a denotes the point about which the moment is taken, and the quantity within the parentheses denotes the function, in normalized form, whose moment is to be taken. Thus, the area under the curve $f(x)$ is unity. For example, let us consider the gamma distribution given by Eq. 3.114. This is rewritten as

$$f(x) = \frac{1}{\beta^\alpha \Gamma(\alpha)} x^{\alpha-1} \exp\left(-\frac{x}{\beta}\right)$$

Let the two parameters be designated as $\alpha \equiv \alpha_1$ and $\beta \equiv \alpha_2$.

The first moment of any distribution about the origin is the mean of the distribution. For the gamma distribution, this is given as (from Eq. 3.120 by changing the lower limit of the integration since the function is only defined for $x > 0$, and dropping the superscript 0 that is used to denote the moment about the origin (in Eq. 3.120, $a = 0$))

$$\mu_1 = \int_0^\infty x f(x) dx = \int_0^\infty \frac{\alpha_2^{\alpha_1} x^{\alpha_1}}{\Gamma(\alpha_1)} \exp(-\alpha_2 x) dx$$

$$= \frac{\alpha_2^{\alpha_1}}{\Gamma(\alpha_1)} \int_0^\infty x^{\alpha_1} \exp(-\alpha_2 x) dx = \frac{\alpha_2^{\alpha_1}}{\Gamma(\alpha_1)} \frac{\Gamma(\alpha_1+1)}{\alpha_2^{\alpha_1+1}} = \frac{\alpha_1}{\alpha_2}$$

The second moment of any distribution is the variance, which is the second moment about the mean. Therefore, for the gamma distribution we can write (dropping the superscript $a = \mu_1$)

$$\mu_2 = \int_0^\infty (x-\mu_1)^2 f(x)\,dx = \int_0^\infty \left(x - \frac{\alpha_1}{\alpha_2}\right)^2 \frac{\alpha_2^{\alpha_1}}{\Gamma(\alpha_1)} \exp(-\alpha_2 x) = \frac{\alpha_1}{\alpha_2^2}$$

Since the method of moments assumes that the sample moments are equal to the population moments,

$$\mu_1 = \bar{x} = \frac{\alpha_1}{\alpha_2} \text{ and } \mu_2 = (s_x)^2 = \frac{\alpha_1}{\alpha_2^2}$$

These two equations can be solved to obtain two distributional parameters from the sample statistics, namely the mean and variance.

3.18.3 Method of Probability-Weighted Moments

Greenwood et al. (1979) introduced the method of probability-weighted moments (PWMs) and showed its usefulness in deriving explicit expressions for parameters of distributions whose inverse forms $X = X(F)$ can be explicitly defined. They derived relations between parameters and PWMs for several distributions. Hosking (1986) developed the theory of PWMs and applied it to estimate parameters of several distributions.

Let a PDF be denoted as $F = F(X) = P[X \leq x]$. The PWMs of this function can be defined as

$$M_{i,j,k} = E\left[x^i F^j (1-F)^k\right] = \int_0^1 [x(F)]^i F^j (1-F)^k\,dF \quad (3.121)$$

where $M_{i,j,k}$ is the PWM of order (i,j,k), E is the expectation operator, and i, j, and k are real numbers. If $j = k = 0$ and i is a nonnegative integer, $M_{i,0,0}$ represents the conventional moment of order i about origin. If $M_{i,0,0}$ exists and X is a continuous function of F, $M_{i,j,k}$ exists for all nonnegative real numbers j and k.

For nonnegative integers j, k, we can express

$$M_{i,0,k} = \sum_{j=0}^{k} \binom{k}{j} (-1)^j M_{i,j,0} \quad (3.122)$$

$$M_{i,j,0} = \sum_{k=0}^{j} \binom{j}{k} (-1)^k M_{i,0,k} \quad (3.123)$$

If $M_{i,0,k}$ exists and X is a continuous function of F then $M_{i,j,0}$ exists. When the inverse $X = X(F)$ of the distribution $F = F(X)$ cannot be analytically defined, it may, in general, be difficult to derive $M_{i,j,k}$ analytically.

We normally work with the moments $M_{i,j,k}$ into which x enters linearly. In particular, if we consider an ordered sample in which $x_1 \leq x_2 \leq \ldots \leq x_n$, the PWMs for hydrologic applications are defined as

$$M_{1,0,s} = a_s = \frac{1}{n} \sum_{i=1}^{n} \binom{n-i}{s} x_i \Big/ \binom{n-1}{s} \quad (3.124)$$

$$M_{1,0,r} = b_r = \frac{1}{n} \sum_{i=1}^{n} \binom{i-1}{r} x_i \Big/ \binom{n-1}{r} \quad (3.125)$$

where $n > r$ and $n > s$.

In general, a_s and b_r are functions of each other as

$$a_s = \sum_{k=0}^{s} (-1)^k \binom{s}{k} b_k \quad (3.126)$$

$$b_r = \sum_{k=0}^{r} (-1)^k \binom{r}{k} a_k \quad (3.127)$$

Therefore,

$$\begin{aligned}
a_0 &= b_0 & b_0 &= a_0 \\
a_1 &= b_0 - b_1 & b_1 &= a_0 - a_1 \\
a_2 &= b_0 - 2b_1 + b_2 & b_2 &= a_0 - 2a_1 + a_2 \\
a_3 &= b_0 - 3b_1 + 3b_2 - b_3 & b_3 &= a_0 - 3a_1 + 3a_2 - a_3
\end{aligned}$$
$$(3.128)$$

A complete set of these a or b PWMs characterizes a distribution. The theory of PWMs parallels the theory of conventional moments. The main advantage of PWMs over conventional moments is that PWMs, being linear functions of the data, suffer less from the effects of sampling variability: PWMs are more robust than conventional moments to outliers in the data, enable more secure inferences to be made from small samples about an underlying probability distribution, and frequently yield more efficient parameter estimates than the conventional moment estimates.

3.18.4 Method of L-Moments

The PWMs characterize a distribution but are not meaningful by themselves. L-moments were developed by Hosking (1986) as functions of PWMs that provide a descriptive summary of the location, scale, and shape of the probability distribution. These moments are analogous to ordinary moments and are expressed as *linear* combinations of **order statistics**, hence the name. They can also be expressed by linear combinations of PWMs. Thus, ordinary moments, PWMs, and L-moments are related to each other. L-moments are known to have several important advantages over ordinary moments. They have less bias than ordinary moments because they are linear combinations of ranked observations. As an example, the variance (second moment) and skewness (third moment) involve squaring and cubing of observations, respectively, which compel them to give greater weight to the observations far from the mean. As a consequence, they result in substantial bias and variance.

If X is a real-value ordered random variate of a sample of size n, such that $x_{1:n} \leq x_{2:n} \leq \ldots \leq x_{n:n}$ with the cumulative distribution $F(x)$ and quantile function $x(F)$, the rth

L-moment of X (Hosking 1990) can be defined as a linear function of the expected order statistics as

$$\lambda_r = \frac{1}{r}\sum_{k=0}^{r-1}(-1)^k\binom{r-1}{k}E\{X_{r-k:r}\}, \quad r=1,2,\ldots \tag{3.129}$$

where r is the order of the L-moment and $E\{X_{r-k:r}\}$ is the expectation of the $r-k$ order statistic of a sample of size r and is equal to

$$E\{X_{j:r}\} = \frac{r!}{(r-j)!(j-1)!}\int_0^1 x\{F(x)\} \times \{F(x)\}^{j-1}\{1-F(x)\}^{r-j}dF(x) \tag{3.130}$$

As noted by Hosking (1990), the natural estimator of λ_r, based on an observed sample of data, is a linear combination of the ordered data values, i.e., an L-statistic. Substituting Eq. 3.130 in Eq. 3.129, expanding the binomials of $F(x)$ and summing the coefficients of each power of $F(x)$, one can write

$$\lambda_r = E[xP^*_{r-1}\{F(x)\}] = \int_0^1 x(F)P^*_{r-1}(F)dF, \quad r=1,2,\ldots \tag{3.131}$$

where $P^*_r(F)$ is the rth shifted Legendre polynomial expressed as

$$P^*_r(F) = \sum_r^k (-1)^{r-k}\binom{r}{k}\binom{r+k}{k}F^k \tag{3.132}$$

Equation 3.132 can simply be written as

$$P^*_r(F) = \sum_{k=0}^r P_{r,k}F^k \tag{3.133}$$

and

$$P_{r,k} = (-1)^{r-k}\binom{r}{k}\binom{r+k}{k} \tag{3.134}$$

The shifted Legendre polynomials are related to the ordinary Legendre polynomials $P_r(u)$ as $P^*{}_r(u) = P_r(2u-1)$ and are orthogonal on the interval (0,1) with a constant weight function.

The first four L-moments are:

$$\lambda_1 = EX = \int_0^1 x(F)\,dF \tag{3.135}$$

$$\lambda_2 = \frac{1}{2}E(x_{2:2}-x_{1:2}) = \int_0^1 x(F)\times(2F-1)\,dF \tag{3.136}$$

$$\lambda_3 = \frac{1}{3}E(x_{3:3}-2x_{2:3}+x_{1:3}) = \int_0^1 x(F)\times(6F^2-6F+1)\,dF \tag{3.137}$$

$$\lambda_4 = \frac{1}{4}E(x_{4:4}-3x_{3:4}+3x_{2:4}-x_{1:4})$$
$$= \int_0^1 x(F)\times(20F^3-30F^2+12F-1)\,dF \tag{3.138}$$

The L-moments can also be defined in terms of PWMs α and β as

$$\lambda_{r+1} = (-1)^r\sum_{k=0}^r P_{r,k}\alpha_k = \sum_{k=0}^r P_{r,k}\beta_k \tag{3.139}$$

Parallel to conventional moment ratios, L-moment ratios are the dimensionless quantities defined by

$$\tau_2 = \frac{\lambda_2}{\lambda_2} = \text{L-}C_V \text{ or coefficient of L-variation} \tag{3.140}$$

$$\tau_3 = \frac{\lambda_3}{\lambda_4} = \text{L-}C_S \text{ or L-skewness} \tag{3.141}$$

$$\tau_4 = \frac{\lambda_4}{\lambda_2} = \text{L-}C_K \text{ or L-kurtosis} \tag{3.142}$$

The sample L-moments are computed from the sample order statistics $x_{1:n} \leq x_{2:n} \leq \cdots \leq x_{n:n}$. These are:

$$\lambda_r = \frac{1}{r}\sum_{i=1}^n\left[\frac{\sum_{j=0}^{r-1}(-1)^j\binom{r-1}{j}\binom{i-1}{r-1-j}\binom{n-i}{j}}{\binom{n}{r}}\right]x_{i:n} \tag{3.143}$$

In practice, L-moments are calculated as follows. The data of a sample size N are first ranked in ascending order, $x_1 \leq x_n \ldots \leq x_N$. Then the PWMs are estimated using the general formula (Hosking and Wallis, 1997)

$$b_r = \frac{1}{N(N-1)(N-2)\cdots(N-r)}\sum_{i=r+1}^N (i-1)(i-2)\ldots(i-r)x_i \tag{3.144}$$

Thus,

$$b_0 = \frac{1}{N}\sum_{i=1}^N x_i \tag{3.145}$$

$$b_1 = \frac{1}{N(N-1)}\sum_{i=2}^N (i-1)x_i \tag{3.146}$$

$$b_2 = \frac{1}{N(N-1)(N-2)}\sum_{i=3}^N (i-1)(i-2)x_i \tag{3.147}$$

$$b_3 = \frac{1}{N(N-1)(N-2)(N-3)}\sum_{i=4}^N (i-1)(i-2)(i-3)x_i \tag{3.148}$$

From these, the sample L-moments are calculated as

$$\lambda_1 = b_0 \tag{3.149}$$

$$\lambda_2 = 2b_1 - b_0 \tag{3.150}$$

$$\lambda_3 = 6b_2 - 6b_1 + b_0 \tag{3.151}$$

$$\lambda_4 = 20b_3 - 30b_2 + 12b_1 - b_0 \tag{3.152}$$

Once the L-moments of a sample are calculated, parameters of a distribution can be calculated. Hosking (1990) provided the equations for the estimation of parameters of most common probability distributions using L-moments.

3.19 STATISTICAL TESTS OF HYPOTHESES

3.19.1 Hypotheses and Test Procedures

A **statistical hypothesis** or simply a **hypothesis**, is a claim either about the value of a single population characteristic or about the values of several population characteristics. It can be best illustrated by an analogy with the American judicial system where, in a courtroom trial, a defendant is assumed to be not guilty and the burden of proof is on the plaintiff to disprove the claim. In testing a statistical hypothesis, the problem is formulated so that one of the claims is initially favored. This initially favored claim will not be rejected in favor of the alternative claim unless sample evidence contradicts it and provides strong support for the alternative assertion.

The assertion, initially believed to be true, is called the **null hypothesis**. This is the *hypothesis of no difference* and is denoted by H_0. The other claim in this hypothesis testing problem is called the **alternative hypothesis** and is denoted by H_a. An example is that we claim that the frequency distribution of a sample follows normal distribution. In this case, the null hypothesis is that there is no difference between the distribution factually exhibited by the sample and a normal distribution. The alternative hypothesis is that there is overwhelming evidence that the data do not follow normal distribution.

The procedure for statistical hypothesis testing has two constituents:

I. A **test statistic**, which is a function of the sample data on which the decision on either to reject H_0 or not to reject H_0 is based.
II. A **rejection region**, which is the set of all test statistic values for which H_0 will be rejected.

The null hypothesis will be rejected if and only if the observed or computed test statistic value falls in the rejection region.

3.19.2 Errors in Hypothesis Testing and Level of Significance

Once a hypothesis is expressed, we can make a decision to either accept or reject it on the basis of a statistical test. There are also two possible states of the hypothesis: it may be **true** or **false**. This combination produces four possible outcomes, of which two are correct and two are incorrect. The possibilities can be categorized as presented in Table 3.5.

Table 3.5 *Four possible outcomes of a statistical hypothesis test*

	Hypothesis (H_0) is correct	Hypothesis (H_0) is incorrect
Hypothesis is accepted	Correct decision (no error committed); probability $= 1 - \alpha$.	Type II error; probability $= \beta$
Hypothesis is rejected	Type I error; probability $= \alpha$	Correct decision (no error committed); probability $= 1 - \beta$

Thus, in a statistical test two types of errors can occur:

1. A **type I error** involves rejecting H_0 when it is true
I2. A **type II error** involves not rejecting H_0 when it is false.

The probability of committing a type I error is called **significance level** and is designated by α. This probability is set before running a test. In order to minimize the possibility of a committing type II error, we write the null hypothesis with the intention that it will be rejected. If the null hypothesis is rejected, there is no chance of a type II error, and the probability of a type I error is known because it is specified. If, however, the test fails and does not reject the null hypothesis, the probability of a mistake in the form of a type II error remains. This probability, called β, is not known.

The complement of the significance level is called the **confidence level**. Thus, the confidence level $= 1 - \alpha$. An $\alpha = 0.05$ (5% significance level) is equivalent to a confidence level of 0.95 (95%).

3.19.3 p-Values

The null hypothesis is either accepted or rejected at a specified significance level, α. However, this criterion alone may introduce subjectivity since someone may choose $\alpha = 0.05$ while for the same test another person may choose $\alpha = 0.01$. To avoid this type of situation, a *p*-value must be specified for a test.

The *p*-value is the smallest level of significance at which H_0 would be rejected when a specified test procedure is used on a given dataset. Once the *p*-value has been determined, the conclusion at any particular level α results from comparing the *p*-value to α:

1. *p*-value $\leq \alpha \Rightarrow$ reject H_0 at level α
2. *p*-value $> \alpha \Rightarrow$ do not reject H_0 at level α.

3.19.4 Structure of Hypothesis Tests

The structure of a standard statistical hypothesis testing involves the following steps:

1. Select the appropriate test.
2. Establish the null hypothesis.

3. Establish an appropriate significance level, α.
4. Compute the test statistics from the data.
5. Compute the *p*-value.
6. Reject the null hypothesis if $p \leq \alpha$.

Test procedures are usually selected on the basis of the characteristics of data and the objectives of study. The characteristics of the data lead to the choice between **parametric** and **nonparametric** tests.

Parametric tests assume that the data have a particular type of frequency distribution, such as a normal distribution, where the information contained in the data can be summarized by parameters, such as sample mean and standard deviation, and the test statistic is computed using these parameters. The strength of parametric tests not to reject H_0, when it is actually false, can be quite low when the data do not follow a well-defined distribution, particularly a normal distribution. This results in type II errors and this loss of strength in parametric tests is a major concern if they are applied to an inappropriate dataset.

Nonparametric tests do not require an assumption that the data follow a particular type of frequency distribution and therefore the tests are not based on parameters. In this case, information is extracted from the data by comparing each value with all others, and ranking the data instead of computing parameters.

There can be **two-sided** or **one-sided** tests to reject H_0, which is what is assumed to be true about the system under study. Two-sided tests occur when the evidence in either direction from the null hypothesis, such as larger or smaller, positive or negative, would cause null hypothesis to be rejected. One-sided tests occur when there is a departure in only one direction from the null hypothesis that would cause the null hypothesis to be rejected in favor of the alternative hypothesis. With one-sided tests, it is considered supporting evidence for H_0, if the data indicate differences opposite in direction to the alternative hypothesis. If it cannot be stated prior to looking at any data that departures for H_0 are only in one direction of interest, a two-sided test should be performed. For example, in a one-sided test: $H_0 \to$ the 100-year flood is less than or equal to the design discharge; $H_a \to$ the 100-year flood exceeds the design discharge. On the other hand, in the two-sided test: $H_0 \to$ the 100-year flood is either smaller or greater than design discharge; $H_a \to$ the 100-year flood equals the design discharge.

There is a clear distinction between the α-value and *p*-value. The α-value, or the significance level, is the probability of incorrectly rejecting H_0. It is not based on data but merely gives the risk of a type I error deemed acceptable. In typical cases of statistical tests, α is set to 5% (0.05) but any other value can be chosen. The *p*-value, on the other hand, is calculated based on data. It is the probability of obtaining the computed test statistic. The smaller the *p*-value, the less likely is the observed test statistic when H_0 is true, and the stronger the evidence for rejection of the H_0.

An example of setting up a hypothesis test is given below where we test if the population mean is equal to the sample mean.

1. Set up the hypothesis:

$H_0 : \bar{x} = \mu_0$
$H_a : \bar{x} \neq \mu_0$

2. Set up the significance level:

$\alpha = 0.05$

3. Calculate the value of the test statistic:

$$z = \frac{\bar{x} - \mu_0}{\sigma/\sqrt{n}}$$

4. Set up the rection region or critical region for the value of z at α significance level.
   a. For a one-sided upper-tailed test
   b. For a one-sided lower-tailed test
   c. For a two-sided test: $z_{c(-)} \leq z \leq z_{c(+)}$

In this example, the null hypothesis states that the mean μ has a particular numerical value, denoted by μ_0. Let $X_1, \ldots, X_n$ represent a random sample of size n drawn from the normal population. Then the sample mean, $\bar{X}$, has a normal distribution with expected value $\mu_{\bar{X}} = \mu$ and standard deviation $\sigma_{\bar{X}} = \sigma/\sqrt{n}$. When H_0 is true, $\mu_{\bar{X}} = \mu_0$. Consider now the statistic Z obtained by standardizing $\bar{X}$ under the assumption that H_0 is true:

$$Z = \frac{\bar{X} - \mu_0}{\sigma/\sqrt{n}}$$

In the above test statistics, if we substitute the computed sample mean, $\bar{x}$, we get the test statistic, z, the distance between $\bar{x}$ and μ_0, expressed in standard deviation units. For example, if $\mu_0 = 100$, $\sigma/\sqrt{n} = 2.0$, and $\bar{x} = 103$, then $z = 1.5$. This implies that the observed mean, $\bar{x}$, is 1.5 standard deviations of $\bar{X}$ above what we expect it to be when H_0 is true. The statistic Z is a natural measure of the distance between $\bar{X}$, the estimator of μ, and the expected value when H_0 is true. If this distance is too great in the direction consistent with H_a, the null hypothesis should be rejected.

Tables of standard normal distribution in which the probability density values for a range of values of Z can be found in any textbook of elementary statistics. From these tables, the critical value of z, designated by z_{crit}, can be obtained for a probability level α. However, Excel has a function that can be used to get the critical value z_{crit}, at a desired significance level, α. Box 3.1 lists the Excel functions related to standard normal distribution.

Box 3.1 Excel functions for standard normal distribution

- NORM.S.DIST (z, TRUE/FALSE): Returns the standard normal distribution of argument z; returns the probability density if FALSE or cumulative probability if TRUE).
- NORM.S.INV (probability): Returns the inverse (z) of the standard normal cumulative distribution.

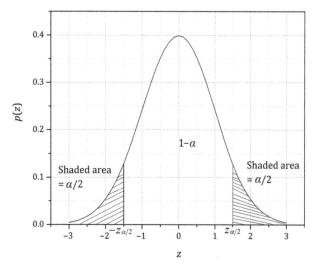

Figure 3.14 A pdf of standard normal distribution or *z*-curve. Two shaded regions are shown where the probability of a value of *z* either greater than the value $z_{\alpha/2}$ or less than the value $-z_{\alpha/2}$ is α. These are called rejection regions and the values $z_{\alpha/2}$ and $-z_{\alpha/2}$ are called z_{crit}, the critical value of *z*. The unshaded region is the sum of all probabilities of all *z* values lying between these two critical values of *z* is $1 - \alpha$. In other words, $P(-z_{\alpha/2} \leq z \leq z_{\alpha/2}) = 1 - \alpha$.

The function NORM.S.INV (probability) is most useful for our purpose, since we need to specify the probability level and get the value of *z* at that specified probability level. In order to get the two-sided value (e.g., 0.025 on each side), the following form should be used: NORM.S.INV $(1 - \alpha/2)$.

Figure 3.14 shows the probability density curve of standard normal distribution and the concept of critical regions.

3.19.5 Important Distributions of Sample Statistics

3.19.5.1 Concept of Test Statistics

Sample statistics as functions of random variables are themselves random variables. Statistical tests depend on the probability distribution of test statistics, which are merely sample statistics. In the example given above, the test statistic, *z*, is the standard normal variate and its probability distribution is given by the normal standard curve. However, tests involving the *z*-statistic or the Z-test are not so useful for dealing with natural variables since we don't know the population mean and variance a priori. Three other distributions of sample statistics or test statistics that are frequently useful in statistical tests are discussed in this section.

A concept that is very important in this relation is the **degrees of freedom**. In tests based on samples, we must estimate a number of population parameters in order to calculate test statistic. It is unwise to both estimate the parameters and perform the test from the same set of data without somehow compensating for the double use of observations. This is done by considering the degrees of freedom, which may be defined as the number of observations in a sample minus the number of parameters estimated from the sample. In other words, the degrees of freedom are the number of observations in excess of those necessary to estimate the parameters of the distribution. Degrees of freedom are symbolically indicated by *v* and are *always a positive integer*.

The general idea underlying the degrees of freedom is that the N deviations $(x_1 - \bar{x})$ that are used in calculating the sample variance s_x^2 are not all independent, since there is a linear relation connecting them, namely

$$\sum_{i=1}^{N}(x_i - \bar{x}) = 0$$

Because of this, only $N - 1$ of the quantities are actually independent, and we say that there are $N - 1$ degrees of freedom in the calculation of s_x. An unbiased estimate of the population variance is

$$\frac{Ns_x^2}{N-1} = \frac{\sum_{i=1}^{N}(x_i - \bar{x})^2}{N-1} \qquad (3.153)$$

and is given by dividing the sum of squares of deviations from the mean by the number of degrees of freedom.

3.19.5.2 Student's *t*-Distribution

We have seen that for samples from a normal parent population, the quantity $x = \dfrac{\bar{x} - \mu}{\sigma/\sqrt{N}}$ is a normal standard variate. In most practical situations, σ is unknown. If, instead of σ, we substitute the estimate $\hat{\sigma} = s_x\sqrt{N/(N-1)}$, as was first shown by W. S. Glosset, writing under the pen name of "Student" in a paper that later became a classic, the variable or the sample statistic, $t = \dfrac{\bar{x} - \mu}{s_x\sqrt{N/(N-1)}}$ has a distribution that can be represented mathematically, and for which tables like those of the normal distribution can be calculated. The numerator of *t* is normally distributed with mean zero, and the denominator is an estimate, from the sample, of the standard deviation of the numerator.

The *t*-distribution with a degree of freedom, *v*, is given as

$$p_T(t|v) = \Gamma\left(\frac{v+1}{2}\right)\left[\frac{1+t^2}{v}\right]^{-(v+1)/2}\left\{\frac{1}{\sqrt{\pi v}\left[\Gamma\left(\frac{v}{2}\right)\right]}\right\}$$
$$-\infty < t < \infty; v > 0$$
$$(3.154)$$

The mean and variance of the *t*-distribution are

$$E(T) = 0 \qquad (3.155)$$

$$V(t) = \frac{v}{v-2} \quad \text{for } v > 2 \qquad (3.156)$$

The cumulative *t*-distribution, which is given in tables, is

$$F_T(t) = \int_{-\infty}^{t} p_T(x)dx \qquad (3.157)$$

$$P(T \geq t) = \int_x^\infty \frac{\Gamma[(v+1)/2]}{\sqrt{\pi v}\,\Gamma(v/2)} \left(1 + \frac{t^2}{v}\right)^{-(v+1)/2} dt \quad t \geq 0 \quad v \geq 1 \tag{3.158}$$

The important point about the distribution of t is that it depends only on the sampling size n and not on the variance, σ^2, of the population. The curve of $p(t)$ plotted against t is a symmetrical, hump-backed curve, not unlike the normal curve in shape but with higher tails, that is, wider spread.

A theorem can be stated as follows. When $\overline{X}$ is the mean of a random sample of size N from a normal distribution with mean μ, the random variable,

$$T = \frac{\overline{X} - \mu}{s_x/\sqrt{N}} \tag{3.159}$$

has a probability distribution called a t-distribution with $N - 1$ degrees of freedom. We denote the variable by T to emphasize that it does not have a standard normal distribution when N is small. A normal distribution is governed by two parameters, the mean, μ, and standard deviation, σ. A t-distribution is governed by only one parameter, the number of degrees of freedom, v. For a fixed value of v, the density function that specifies the associated t-curve has the following properties that are illustrated in Figure 3.15:

1. Each t-curve is bell-shaped and centered at 0.
2. Each t-curve is more spread out than the standard normal curve or z curve.
3. As n increases, the spread of the corresponding t-curve decreases.
4. As $v \to \infty$, the sequence of t-curves approaches the standard normal curve.

These tests, called t-tests, and based on the t-distribution are useful for establishing the likelihood that a given sample could be a member of a population with specified characteristics, or for testing the hypothesis about the equivalency of two samples, such as whether the mean derived from a sample of a parent population with mean μ_1 is equivalent to a population with known mean μ_0. Then, the test statistic t is established as

$$t = \frac{\overline{x} - \mu_0}{s_x/\sqrt{n}} \tag{3.160}$$

with the hypothesis

$H_0 : \mu_1 = \mu_0$
$H_a : \mu_1 \neq \mu_0$

The null hypothesis is rejected if $t \geq t_{\alpha,v}$, where $t_{\alpha,v}$ is called the *t* **critical value**. If $t_{\alpha,v}$ is the number on the measurement axis for which the area under the t-curve with v degrees of freedom to the right of $t_{\alpha,v}$ is α, $t_{\alpha,v}$ is called a t critical value. The critical value of t with a known degree of freedom, v, for a given significance level, α, can be obtained from published tables given in statistical texts or using the Excel function T.INV(α, df).

The discussion above is concerned with a t-test involving one set of samples with size N. However, in certain situations two sets of observations are made with the same sample size and t-test involves testing the equivalency of the means of two samples. This is called *paired t-test*. In some other situations, a t-test may involve two samples assumed to be drawn from the same population but with either equal or unequal variances. In each of these situations, the test statistic is formulated differently. An example of two sample cases with unequal variances is provided in Example 3.3.

Box 3.2 lists the Excel functions that can be used in relation to t-distribution and t-tests.

Box 3.2 Excel functions related to t-distribution and *t*-tests

- T.DIST(x, v,TRUE/FALSE): Returns the *left-tailed t-distribution*; cumulative probability (for TRUE) or density value (for FALSE)
- T.DIST.RT(x, v): Returns the *right-tailed t-distribution*
- T.DIST.2T(x, v): Returns the *two-tailed t-distribution*
- T.INV(α, v): Returns the *left-tailed* inverse of t-distribution at a the probability level of α
- T.INV.2T(α, v): Returns the *two-tailed* inverse of the t-distribution. The significance level at each tail is $\alpha/2$
- T.TEST(Arry1, Array2, Tails, Type): Returns the probability (*p*-value) associated with a t-test, that the two samples in Array 1 come from the same two underlying populations that have the same mean. Tails (1/2) specifies the number of distribution tails; 1 uses a one-tailed distribution and 2 uses two-tailed distribution. Type (1/2/3) specifies the kind of t-test; 1 performs a *paired* test, 2 performs test for two-sample equal variance (homoscedastic), and 3 performs two-sample unequal variance (heteroscedastic).

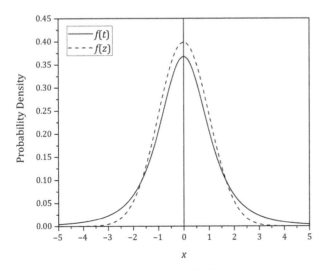

Figure 3.15 Student's t-distribution (degrees of freedom = 3) compared to the standard normal curve $N(0,1)$.

3.19.5.3 The F-Distribution

Methods for comparing two population variances or standard deviations are occasionally needed, although such problems arise less frequently than those involving means. For the case in which the populations under investigation are normal, the procedures are based on a family of probability distribution known as the **F-distribution**.

The F-distribution has two parameters, denoted by v_1 and v_2 both of which are positive integers. The parameter v_1 is called the **number of numerator degrees of freedom** and v_2 is the **number of denominator degrees of freedom**. The form of the density function is quite complicated. The cdf is given as

$$F(X \geq x) = \int_x^\infty \frac{\Gamma[(v_1+v_2)/2](v_1/v_2)^{v_1/2}}{\Gamma(v_1/2)\Gamma(v_2/2)} (f)^{(v_1/2)-1}$$
$$\times \left(1 + \frac{v_1 f}{v_2}\right)^{-(v_1+v_2)/2} df; \quad x \geq 0 \quad v_1, v_2 \geq 1 \quad (3.161)$$

The inverse function is given as

$$p = P(X \geq X_p) = \int_{X_p}^\infty \frac{\Gamma[(v_1+v_2)/2](v_1/v_2)^{v_1/2}}{\Gamma(v_1/2)\Gamma(v_2/2)} (f)^{(v_1/2)-1}$$
$$\times \left(1 + \frac{v_1 f}{v_2}\right)^{-(v_1+v_2)/2} df; \quad X_p \geq 0 \quad v_1, v_2 \geq 1 \quad (3.162)$$

The essence of the F-distribution is as follows. Suppose we have two independent random samples of sizes N_1 and N_2 with variances s_1^2 and s_2^2. We need to test whether the null hypothesis, that the two samples come from populations with the same variance σ^2, is justified. We make the assumption, for instance, in testing the significance of the difference of the means, and it would be satisfactory, before going ahead with the t-test, to be assured that the assumption is a reasonable one.

It turns out that instead of using the difference of the two variances, it is more convenient to use their ratio. We find the quantity

$$F = \frac{N_1 s_1^2}{v_1} \bigg/ \frac{N_2 s_2^2}{v_2} \quad (3.163)$$

which is the ratio of the estimate of σ_1^2 from the first sample to the estimate of σ_2^2 from the second, has a distribution that, on the hypothesis that the two samples are from the same population with variance σ^2, is independent of σ^2 and depends only on the two numbers v_1 and v_2. As would be expected, F fluctuates around the value of 1.

We now arrive at the theorem of F-test. Let $X_1, \ldots, X_m$ be a random sample from a normal distribution with variance σ_1^2, let $Y_1, \ldots, Y_n$ be another independent random sample from a normal distribution with σ_2^2 and let s_1^2 and s_2^2 denote two sample variances. Then, the random variable

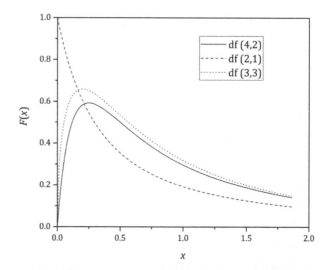

Figure 3.16 Density functions of F-distribution. Three curves are calculated for three different degrees of freedom (df) with three sets of v_1 and v_2 as given in parenthesis.

$$F = \frac{s_1^2 \sigma_1^2}{s_2^2 \sigma_2^2} \quad (3.164)$$

has an F-distribution with $v_1 = m - 1$ and $v_2 = n - 1$.

Figure 3.16 illustrates the graph of a typical F density function. Analogous to the notion $t_{\alpha,v}$, we use F_{α,v_1,v_2} for the point on the axis that captures part of the area under the density function with v_1 and v_2 in the upper tail. The density function is not symmetrical, so it would seem that both upper- and lower-tail critical values are required or need to be tabulated separately. This is not necessary because of the following property

$$F_{1-\alpha,v_1,v_2} = \frac{1}{F_{\alpha,v_2,v_1}} \quad (3.165)$$

A typical F-test is set up as follows by calculating the test statistic simply as the ratio of the sample variances:

Null hypothesis: $H_0 : \sigma_1^2 = \sigma_2^2$
Test statistic value: $f = \frac{s_1^2}{s_2^2}$
Alternative hypothesis:

$H_a : \sigma_1^2 > \sigma_2^2$

$H_a : \sigma_1^2 < \sigma_2^2$

$H_a : \sigma_1^2 \neq \sigma_2^2$

Rejection region of a level of α-test:

$f \geq F_{\frac{\alpha}{2},m-1,n-1}$ or $f \leq F_{1-\alpha/2,m-1,n-1}$
$f \geq F_{\alpha,m-1,n-1}$
$f \leq F_{1-\alpha,m-1,n-1}$

In standard texts on statistics, tables giving values of F_{α,v_1,v_2} for various values of v_1 and v_2 and for certain α values (usually 0.01 and 0.05) are available. However, Excel offers the necessary inverse functions to calculate either F_{α,v_1,v_2} or

$F_{1-\alpha,v_2,v_1}$ Box 3.3 lists the Excel functions related to F distribution.

3.19.5.4 Chi-Squared Distribution

A random variable X is said to have a chi-squared (χ^2) distribution with parameter v if the pdf of X is the two-parameter gamma density function (Eq. 3.114) with $\alpha = \frac{v}{2}$ and $\beta = 2$. Thus, the pdf of a χ^2-distribution is given by

$$p_X(x|v) = \frac{1}{2^{v/2}\Gamma\left(\frac{v}{2}\right)} x^{(v/2-1)} \exp\left[-\frac{x}{2}\right] \text{ for } x \geq 0 \quad (3.166)$$

Figure 3.17 shows the pdf of the χ^2-distribution for $v = 3$ and $v = 9$. Tables of the pdf for different v and range of x are available in statistical texts.

If a sample of size n is taken from a normal population having a mean, μ, and standard deviation, σ, each observation within the sample can be standardized to the standard normal form:

$$z_i = \frac{x_i - \mu}{\sigma} \quad (3.167)$$

If the standardized values are squared and summed, they form a new statistic

$$\sum z^2 = \sum_{i=1}^{n}\left(\frac{x_i - \mu}{\sigma}\right)^2 \quad (3.168)$$

If we draw all possible samples of size n from a normal population and plot the values of $\sum z^2$, they will form a χ^2-distribution. In other words, the random variable χ^2 with n degrees of freedom is defined as

$$\chi^2_v = \sum_{i=1}^{v} z_i^2 \quad (3.169)$$

Box 3.3 Excel functions related to the F-distribution

- F.DIST(x, v_1, v_2, TRUE/FALSE): Returns the F-distribution; cumulative probability (for TRUE) or density value (for FALSE)
- F.DIST.RT(x, v_1, v_2): Returns the right-tailed F-distribution
- F.INV($1 - \alpha$, v_1, v_2): Returns the value of F at a probability level of $1 - \alpha$, e.g., for $\alpha = 0.05$, F at 0.99

  If p = F.DIST(x, v_1, v_2) then F.INV(p, v_1, v_2) = x

- F.INV.RT(α, v_1, v_2): Returns the value of F at a probability level of α, e.g., for $\alpha = 0.05$, F at 0.0.05

$$\text{F.INV}(\alpha, v_1, v_2) = \frac{1}{\text{F.INV.RT}(\alpha, v_2, v_1)}$$

- F.TEST(Arry1, Array2): Returns the result of an F-test, the two-tailed probability that the variances in Array 1 and Array 2 are not significantly different. The function returns the p-value

---

3.19 Statistical Tests of Hypotheses

A common problem of statistical hydrology is that of comparing a distribution of sample observations to some predefined standard distribution. By the use of the χ^2-distribution we can attach probabilistic significance of correspondence between the form of two distributions, one derived from a sample and the other assumed or specified.

Box 3.4 lists the Excel functions related to χ^2 distribution and related tests.

3.19.6 Statistical Tests

3.19.6.1 Statistical Tests in Applied Hydrology

A plethora of statistical tests used in water resources have been presented and discussed by Helsel et al (2020).

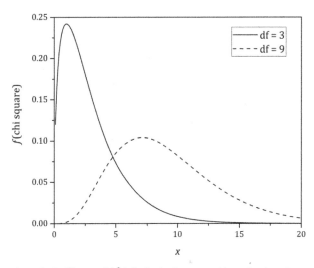

Figure 3.17 Chi-squared (χ^2-) distribution for curves with $v = 3$ and $v = 9$.

Box 3.4 Excel functions related to χ^2-distribution and the χ^2-test.

- CHISQ.DIST(x, df, $logical$): Returns the pdf value, $p(x)$, for the χ^2-distribution with df degrees of freedom when $logical$ = FALSE and the corresponding cdf, $F(x)$, when $logical$ = TRUE.
- CHISQ.DIST.RT(x, df) returns the right tail at x of the χ^2-distribution with df. The function calculates the right-tailed probability of a χ^2-distribution, calculates a level of significance using the χ^2-value and df.
- CHISQ.INV (p, df) returns the left-tailed probability of a χ^2-distribution, where p refers to the probability or the level of significance.
- CHISQ.INV.RT (p, df) returns the inverse of the right-tailed probability of a χ^2-square distribution, where p (probability) equals the level of significance (α). The function returns the critical value.
- CHISQ.TEST(actual range, expected range) returns the p-value, or probability, indicating the percentage chance exists that the differences between the observed and expected outcomes stem from a sampling error.

A particular test is selected to address a specific problem. It is not the intent of this chapter to provide all plausible statistical tests that can have applications in hydrology. Only two tests are presented below because these are used to test the goodness of fit of a theoretical probability distribution with the observed distribution. In this regard too, several other tests are also available to do the same tasks. The two tests presented below are quite customary and easy to implement.

3.19.6.2 Chi-Squared Test

The χ^2-distribution can be used to compare the distribution of a sample with a theoretical distribution if the observed data can be analyzed by the frequency distribution method described in Section 3.4. The test statistic is set up as

$$\chi^2 = \sum_{j=1}^{k} \frac{(O_j - E_j)^2}{E_j} \quad (3.170)$$

where O_j is the number of observations in the jth class, E_j is the number of observations *expected* in the jth class, and k is the number of intervals. From Eq. 3.4

$$O_j = f_s(x_j) = \frac{n_j}{N} \quad (3.171)$$

and from Eq. 3.6

$$E_j = p_X(x_i) = F_X(x_i) - F_X(x_{i-1}) \quad (3.172)$$

The cumulative probabilities in Eq. 3.172 are obtained from the assumed theoretical distribution.

When comparing two distributions, the degree of freedom is given by

$$v = k - m - 1 \quad (3.173)$$

where m is the number of parameters used in fitting the proposed distribution.

The χ^2-test statistic is given by

$$\chi^2 = \sum_{i=1}^{k} \frac{N[f_s(x_i) - p_X(x_i)]^2}{p_X(x_i)} \quad (3.174)$$

In the statistical test, the null hypothesis gets rejected if the calculated value of χ^2 (Eq. 3.174) is larger than the limiting value, with $\chi^2_{v,a}$ determined from chi-squared distribution tables or using Excel function CHISQ.INV.RT(p, df).

3.19.6.3 Kolmogorov–Smirnov Test

The Kolmogorov–Smirnov test (K–S test) is used to decide if a sample comes from a population with a specific distribution. The K–S test is based on the **empirical distribution function**. Given N ordered data points $x_1, x_2, \ldots, x_n$, the empirical distribution function is defined as

$$S_n(x) = \frac{n(i)}{N} \quad (3.175)$$

where $n(i)$ is the number of points less than x_i, and x_i are ordered from smallest to largest value. In other words, $n(i)$ is the cumulative number of sample events in the class limit n. This is a step function that increases by $1/N$ at the value of each ordered data point.

Let $x_1, \ldots, x_n$ be an ordered sample with $x_1 \leq \ldots \leq x_n$ and define $S_n(x)$ as follows:

$$S_n(x) = \begin{cases} 0, & x < x_1 \\ k/n, & x_k \leq x \leq x_{k+1} \\ 1, & x \geq x_n \end{cases} \quad (3.176)$$

Now suppose that the sample comes from a population with the cdf, $F(x)$, and define D_n as follows:

$$D_n = \max_x |F(x) - S_n(x)| \quad (3.177)$$

The distribution of D_n can be calculated (known as the Kolmogorov distribution) but, for our purposes, the important aspect of this distribution is the table of critical values. These can be found in the Kolmogorov–Smirnov table. If $D_{n,\alpha}$ is the critical value from the table, $P(D_n \leq D_{n,\alpha}) = 1 - \alpha$. D_n can be used to test the hypothesis that a random sample came from a population with a specific distribution function $F(x)$. If

$$\max_x |F(x) - S_n(x)| \leq D_{n,\alpha} \quad (3.178)$$

the sample data is a good fit with $F(x)$.

Also, from the definition of D_n given above, it follows that

$$\begin{aligned} 1 - \alpha &= P(D_n \leq D_{n,\alpha}) P\left(\max_x |F(x) - S_n(x)| \leq D_{n,x}\right) \\ &= P\{S_n(x) - D_{n,\alpha} \leq F(x) \leq S_n(x) + D_{n,\alpha} \text{ for all } x\} \\ &= P\{|F(x) - S_n(x)| \leq D_{n,\alpha} \text{ for all } x\} \end{aligned}$$
$$(3.179)$$

Thus, $S_n(x) \pm D_{n,\alpha}$ provides a confidence interval for $F(x)$.

The general steps involved in a K–S test are:

1. An empirical distribution function of the sample data is created.
2. A parent or theoretical distribution against which the test will be performed is specified.
3. The two distributions are plotted together.
4. The greatest vertical distance between the two plots is measured.
5. The test statistic D_n is calculated.
6. The critical value of $D_{n,\alpha}$ is obtained from the Kolmogorov–Smirnov table, which is available in advanced statistical texts.
7. Calculated test statistic are compared to the critical value to reject or accept the null hypothesis.

3.20 INTERVAL ESTIMATION

3.20.1 Concept of Intervals

Any parameter of a distribution or sample we estimate has its set of possible values, an interval of numbers. The single value that is estimated is called the **point estimate**. Point estimates do not portray the reliability, or lack of reliability or variability, of these estimates. As covered in the previous section, the probability that this interval will not cover the true value is called the significance level, α. The probability that the interval will contain the point estimate is called confidence level which is $1 - \alpha$.

The single point value obtained from the sample is peculiar to the particular sample used. Indeed, any other sample would yield a different point estimate. Even the addition of a single observation might measurably change the point estimate. For this reason, it is of interest to know something about the range of values a sample statistic may assume by the very reason of its randomness. In other words, we want to know something about the probability distribution of the sample statistic. Given the sample, one can determine the limits within which one can expect the distribution of parameters to lie with a predetermined probability of, say, 80%, 90%, or 95%. These limits are called **confidence limits** and the interval of uncertainty, the **confidence interval**. This statement can be written as

$$P(L < \theta < U) = 1 - \alpha \qquad (3.180)$$

where L and U are the lower and upper confidence limits so that the interval from L to U is the confidence interval.

Interval estimation does not improve the accuracy of estimation but allows quantification of the uncertainty involved. An appreciation of this uncertainty is needed to avoid making unwarranted assertions.

3.20.2 Confidence Limits for the Sample Mean and Standard Deviation

A sample mean $\bar{x}$ is normally distributed if the variable is normally distributed or if the sample is large. The normal distribution of a sample mean can be written as

$$p_X(\bar{x}) = N\left(\mu, \frac{\sigma}{\sqrt{n}}\right) \qquad (3.181)$$

From Eq. 3.159, we know that the quantity $T = [\sqrt{N}(\bar{x} - \mu)]/s_x$ has a symmetrical t distribution. Thus, symmetrical confidence intervals for the mean are computed from

$$\bar{x} + t_{(\frac{\alpha}{2}, N-1)} \cdot \sqrt{\frac{s_x^2}{N}} \leq \mu \leq \bar{x} + t_{(1-\frac{\alpha}{2}, N-1)} \cdot \sqrt{\frac{s_x^2}{N}} \qquad (3.182)$$

If $1 - \alpha$ were the desired confidence level and the sample size was N, the critical t-values would be $t_{\alpha/2, N-1}$ and $t_{1-\alpha/2, N-1}$.

If the data do not follow normal distribution, such as in a skewed distribution, asymmetrical confidence intervals for the means may also be computed for the case where the logarithms $y = \ln(x)$ approximate a normal distribution. If the logarithmically transformed data are approximately normal, this approach will give a lower variance of the estimate of the mean than will the computation of the usual sample mean without transformation and would thereby provide more reliable confidence limits.

To obtain the confidence interval of sample standard deviation, Eq. 3.169 is written in a slightly different form

$$\chi^2 = \frac{(N-1)s_x^2}{\sigma_x^2} \qquad (3.183)$$

from which

$$P\left(\chi^2_{N-1, \alpha/2} \leq \frac{(N-1)s_x^2}{\sigma_X^2} \leq \chi^2_{N-1, 1-\alpha/2}\right) = 1 - \alpha \qquad (3.184)$$

Manipulating the quantities inside the bracket, we obtain

$$\frac{(N-1)s_x^2}{\chi^2_{N-1, 1-\alpha/2}} \leq \sigma_X^2 \leq \frac{(N-1)s_x^2}{\chi^2_{N-1, \alpha/2}} = 1 - \alpha \qquad (3.185)$$

3.21 COVARIANCE AND CORRELATION

3.21.1 Relationship between Covariance and Correlation

Covariance and correlation are two related and important concepts in probability and statistics. Both covariance and correlation measure the relationship and the dependency between two random variables. Covariance indicates the direction of the linear relationship between variables, while correlation measures both the strength and direction of the linear relationship between two variables. Correlation is a function of covariance. What sets these two concepts apart is the fact that correlation values are standardized, whereas covariance values are not. Covariance can assume any value from $+\infty$ to $-\infty$ and the higher this value, the more dependent is the relationship between the two variables. The correlation value on the other hand ranges from $+1$ to -1 and the higher the value the stronger is the relationship between the two. In both cases, negative values indicate an inverse relationship, whereas positive values indicate direct proportionality. And in both cases, values closer to zero indicate poor or no direct relationship between the two variables.

3.21.2 Covariance

Just like the variance of the N values of x_i is given by $s_x^2 = \sum (x - \bar{x})^2 / N$ we define a quantity s_{xy}, called the **covariance** of the N pairs of values (x_i, y_i) by the relationship

$$s_{xy} = \frac{1}{N} \sum_{i=1}^{N} (x_i - \bar{x})(y_i - \bar{y}) \qquad (3.186)$$

Unlike the variance, this quantity may be positive or negative. The covariance of independent variables is zero.

The covariance is a second moment about the centroidal axes and can also be defined as

$$\text{Cov}(x,y) = \sigma_{xy} = E[(x - \mu_X)(Y - \mu_Y)]$$
$$= \int_{-\infty}^{+\infty} \int_{-\infty}^{+\infty} (x - \mu_x)(y - \mu_y) f(x,y) \, dx \, dy \qquad (3.187)$$

or

$$\text{Cov}(x,y) = \sigma_{xy} = \sum_{i=1}^{n} \sum_{j=1}^{m} (x_i - \mu_x)(y_j - \mu_y) p(x_i, y_j) \qquad (3.188)$$

3.21.3 Correlation Coefficient

The covariance has a dimension equal to the product of the dimensions of the random variables. For that reason, the magnitude of the parameter does not say much about the degree of dependence between the random variables. It is, therefore, customary to divide the covariance by the product of the standard deviations of the respective random variables, which results in a dimensionless parameter, called the **correlation coefficient**:

$$r = \frac{1}{N} \sum_{i=1}^{N} \left(\frac{x_i - \bar{x}}{s_x} \right) \left(\frac{y_i - \bar{y}}{s_y} \right) \qquad (3.189)$$

From Eq. 3.186 it can be seen that

$$\text{Correlation}(x,y) = \frac{\text{Cov}(x,y)}{\sqrt{\text{Var}(x)} \cdot \sqrt{\text{Var}(y)}}$$

This is known as **Pearson's correlation coefficient** or **Pearson's** *r* or the product moment coefficient of correlation.

For computation, it is convenient to write

$$r = \frac{n \sum x_i y_i - (\sum x_i)(\sum y_i)}{\sqrt{n \sum x_i^2 - (\sum x_i)^2} \cdot \sqrt{n \sum y_i^2 - (\sum y_i)^2}} \qquad (3.190)$$

The correlation coefficient lies between -1 and $+1$. Variables having either of the two extreme values of the correlation coefficient are said to be highly correlated. However, a high correlation does not mean that the variables have a cause-and-effect relationship. A value of zero is obtained when the variables are independent but uncorrelated variables are not necessarily independent.

3.21.4 Hypothesis Test for Pearson's *r*

Pearson's *r* is not resistant to outliers because it is computed by using nonresistant measured means and standard deviations. Pearson's *r* also assumes that the variability in *y* cannot increase or decrease with increasing *x*. In linear regression, variables with the property of having constant variability in *y* with increasing *x* are said to be **homoscedastic**. Skewed variables often demonstrate outliers and increasing variance.

For this reason, *r* is often not useful for describing the correlation between skewed hydrologic variables. Transforming the data to reduce skewness and linearize the relation between *x* and *y* in order to compute Pearson's *r* is a common practice that is often used in hydrologic data analysis. If these assumptions are met, the statistical significance of *r* can be tested under the null hypothesis that *r* is not significantly different from zero. that is, there is no correlation or, in terms of the null and alternative hypotheses, $H_0 : r = 0$ or $H_a : r \neq 0$. The test statistic t_r is computed as

$$t_r = \frac{r\sqrt{N-2}}{\sqrt{1-r^2}} \qquad (3.191)$$

and compared to the table of the *t*-distribution with $n - 2$ degrees of freedom.

3.21.5 Regression

Regression is a method of developing a relation between a **criterion variable** (*y*) and one or more **predictor** or **explanatory variables** (*x*), with the objective of predicting the criterion variable for given values of the predictor variables. The term regression was first used by the British polymath Francis Galton (1822–1911) in studying the inheritance of stature. He found that offspring of abnormally tall or short parents tended to step back or regress to the ordinary population height. However, as now widely used, regression has no reference to biometry, but is merely a convenient term.

Correlation analysis is quite different from regression analysis, although they are frequently used together. Regression is a predictive technique that distinguishes between the predictor and criterion variables. A regression equation that is developed to predict *y* should not be transformed to predict the *x* variable for a given value of *y*. Regression is based on the assumption that no error exists in the independent variable; errors occur only in the dependent variable. Thus, regression is directional. Correlation is not directional in that the correlation between *Y* and *X* is the same as that between *x* and *y*. Also, correlation is different from regression in that correlation is only a standardized index of the degree of a linear relation.

Forms of regression analysis include linear bivariate, linear multiple, and curvilinear. The linear bivariate regression relates a criterion variable (*Y*) and a single predictor variable (*X*) by using:

$$Y = a + bX \qquad (3.192)$$

It can be shown that

$$\left. \begin{array}{l} b = \dfrac{s_{xy}}{s_x^2} \\ a = \bar{y} - b\bar{x} \end{array} \right\} \qquad (3.193)$$

The line therefore passes through the point whose coordinates are $(\bar{x}, \bar{y})$. The **residual** or the **error** is defined as

$$e = Y - (a + bX) \tag{3.194}$$

Similarly, equations have been developed for linear multiple regression and curvilinear or nonlinear regression to express various kinds of relationships between more than one criterion variables and a single predictor variable.

In a multiple regression, that is, one in which there is more than one predictor variable, the linear-regression equation is written as:

$$Y = a + b_1 X_1 + b_2 X_2 + \ldots + b_p X_p \tag{3.195}$$

If there are n observations of the dependent variables and there are p number of predictor variables, in matrix notation

$$[Y]_{(n,1)} = [X]_{(n,p)} \times [b]_{(p,1)} \tag{3.196}$$

The values of regression parameters are obtained by minimizing the residual sum of squares $e^T e$:

$$e^T e = (Y - Xb)^T (Y - Xb) \tag{3.197}$$

Differentiation with respect to b and equating the resulting equation to zero gives what is known as the **normal equation**:

$$X^T X b = X^T Y \tag{3.198}$$

from which b can be given explicitly by

$$b = \left(X^T X\right)^{-1} \times X^T \times Y \tag{3.199}$$

Multiple linear or nonlinear equations find extensive applications in ungauged basins (Chapter 5) where discharge is expressed as a function of multiple measurable variables.

3.21.6 Coefficient of Determination

The **coefficient of determination** is a measure of the degree to which the variance in the dependent variable is explained by the linear-regression relation between the two variables. In other words, it measures the strength of a regression model. This coefficient is computed as $(1 - \Delta)$ where Δ is the difference of the variance of the observed values of the dependent variable and the variance of the values of the dependent variable that have been computed using the regression relation. Clearly, as Δ becomes smaller, the coefficient of determination becomes larger or the regression improves. This quantity is calculated as

$$R^2 = \frac{\sum_{i=1}^{N} \left(\widehat{Y}_i - \overline{Y}\right)^2}{\sum_{i=1}^{N} \left(Y_i - \overline{Y}\right)^2} \tag{3.200}$$

The value of R^2 ranges from zero to one. It should be noted that the Pearson correlation coefficient (r) is used to identify patterns in things, whereas the coefficient of determination $\left(R^2\right)$ is used to identify the strength of a model. Usually, the coefficient of determination is taken as the square of the correlation coefficient and equals the amount of explained variance. However, the squared Pearson correlation coefficient is not always equal to the coefficient of determination: $r^2 \neq R^2$. In Eq. 3.200, $\widehat{Y}_i$ represents the predicted value corresponding to the observed value, Y_i.

3.21.7 Evaluation of Regression Models

A **residual** is the difference between the value predicted with the regression equation and the criterion variable. A residual measures the amount of criterion variation left unexplained by the regression equation. Even though the individual residuals are of interest, the moments of the residuals are also worth examining. While the mean of the residuals is zero, the standard deviation of the residuals is called the **standard error of estimate**, which is denoted by S_e, and is computed as

$$S_e = \sqrt{\frac{1}{N-2} \sum_{i=1}^{n} \left(Y_i - \widehat{Y}_i\right)^2} = \sqrt{\frac{S_r}{N-2}} \tag{3.201}$$

where S_r is the sum of the squares of the errors. The standard error of regression, S_e, provides a measure of goodness of a regression model. It quantifies the spread of data around the regression line of fit and can be referred to as the unexplained sum of squares. It is worth mentioning that if any of the regression assumptions (independence, zero mean error, common variance) concerning the residual error $\left(e_i = Y_i - \widehat{Y}\right)$ is incorrect, S_e may not be a useful estimate of the scale or the dispersion for residual errors.

Another measure, also called the coefficient of determination, is a useful measure of the goodness of fit for evaluating simple regression models. It is defined as

$$C_d = 1 - \frac{s_\varepsilon^2}{\sigma_o^2} = \frac{\sigma_o^2 - s_\varepsilon^2}{\sigma_o^2} = \frac{\text{Fraction of } \sigma_o^2 \text{ explained by regression}}{\text{Variance of the observed data}} \tag{3.202}$$

where σ_o^2 is the variance of the observed data, which are a measure of the variability associated with dependent variable before regression. As an important note about the application of Eq. 3.202, its useful characteristics are dependent on partitioning the variance of the observed data into error and regression components using the minimization criteria of ordinary least squares. If either of these criteria is not used, the interpretations associated with Eq. 3.202 are no longer valid.

The significance of predictor variables and the total equation are evaluated by using F-tests. Two F-tests are used. The partial F-test (F_P) checks the significance of predictor variables that are added or deleted from a regression equation. The total F-test (F_t) checks the significance of the entire regression equation.

3.21.8 Evaluation of Nonregression Models

While evaluating the efficacy of nonregression models such as mechanistic hydrologic models, characteristics of the goodness of fit are determined by analyzing the differences between observed and predicted values. Although the linear-regression concepts are still useful, much of the interpretative power is lost because the minimization criteria are no longer valid. The most commonly used error statistic is the **root-mean-square error**, designated as **RMSE,** which may be an adequate statistic for summarizing the predictive accuracy of models for the same dataset. The *RMSE* describes the magnitude of the direct error and hence is used in decision making when one needs to know the implication of a model's uncertainty. It is given as

$$RMSE = \sqrt{\frac{\sum_{i=1}^{n}(O_i - P_i)^2}{n}} \qquad (3.203)$$

where O_i and P_i represent the ith observed and predicted values of y, respectively, and n is the number of observations. The *RMSE* should be as small as possible. The *RSME* is affected by the units used for expressing the parameter of concern. For this reason it cannot be used to compare a model's efficacy across parameters having different units of measurement. To overcome this shortcoming, **normalized error statistics** are used. One such statistic is the **normalized mean-square error**, designated as *NMSE*, which is defined as

$$NMSE = \frac{\sum_{i=1}^{n}(O_i - P_i)^2}{\sum_{i=1}^{n}(O_i - \overline{O})^2} \qquad (3.204)$$

where $\overline{O}$ is the mean of the observed data. The *NMSE* ranges from 0 to $+\infty$. When it is zero, the model is perfect; when it is 1, the model is as good as the observed mean value; when the *NMSE* is greater than 1, the model is poor. The other measure of goodness of fit that has been widely used to evaluate the performance of hydrologic models is the **coefficient of efficiency** developed by Nash and Sutcliffe (1970). Mathematically, the coefficient of efficiency, η_1, is defined as

$$\eta_1 = 1 - \frac{\sum_{i=1}^{n}(O_i - P_i)^2}{\sum_{i=1}^{n}(O_i - \overline{O})^2} = 1 - NMSE \qquad (3.205)$$

The **Nash–Sutcliff coefficient** ranges from $-\infty$ to 1. The model performance is measured as follows:

1. If $\eta_1 < 0$, the model is poor, the observed mean is better than the model predictions.
2. If $\eta_1 = 0$, the model is as good as the observed mean value.
3. If $\eta_1 = 1$, the model is perfect.

The decision is subjective when η_1 ranges between 0 and 1 depending upon what is considered acceptable. Both the NMSE and η_1 contain square terms, which make them sensitive to outliers. Willmott et al. (1985) suggested a modified coefficient of efficiency, η_2, also known as the **index of agreement**. Mathematically, η_2 is defined as

$$\eta_2 = 1 - \frac{\sum_{i=1}^{n}|O_i - P_i|}{\sum_{i=1}^{n}|O_i - \overline{O}|} \qquad (3.206)$$

A similar interpretation can be made for η_2.

3.21.9 Spearman's Rho

Spearman's rho (ρ) is a nonparametric, rank-based correlation coefficient, developed by noted British psychologist Charles Edward Spearman (1863–1945), that depends only on the ranks of the data and not the observations themselves. Therefore, ρ is resistant to outliers and can be implemented even in cases where some of the data are censored. With ρ, differences between data ranked further apart are given more weight. ρ is perhaps easiest to understand as the linear correlation coefficient computed on the ranks of the data rather than the data themselves.

To compute ρ, the data for the two variables are separately ranked from smallest to largest. If two or more values of x or y are same, they are initially assigned a unique rank. The average rank is then computed from these unique ranks and assigned to each of the tied observations, replacing the unique ranks. For example, in a sequence of $x \equiv \{4, 5, 7, 8, 8, 12\}$, we can assign the number 4 the lowest rank :1 and the number 12 the highest rank 6. Since 8 appears in both the fourth and fifth positions in our ordered list, we will assign each instance of 8 a rank equivalent to the average of their positions, or a rank of $(4+5)/2 = 4.5$. Using the ranks of x and ranks of y, ρ is computed as

$$\rho = \frac{\sum_{i=1}^{N}(R_{x_i} R_{y_i}) - N\left(\frac{N+1}{2}\right)^2}{N\frac{(N^2-1)}{12}} \qquad (3.207)$$

where R_{x_i} is the rank of x_i, R_{y_i} is the rank of y_i, and $(N+1)/2$ is the mean rank of both x and y. Eq. 3.207 can be derived by substituting R_{x_i} and R_{y_i} for x_i and y_i in the equation for Pearson's r, and simplifying. If there is a positive correlation, the higher ranks of x will be paired with the higher ranks of y, and their product will be large. For a negative correlation, the higher ranks of x will be paired with lower ranks of y, and their product will be smaller. When there is no correlation there will be nothing other than a random pattern in the association between x and y ranks, and their product will be similar to the product of their average rank, $(N^2 - 1)$. Thus, ρ will be close to zero.

3.21.10 Kendall's Tau

Kendall's tau (τ) (Kendall, 1938, 1975), much like Spearman's ρ, measures the strength of the monotonic relation between x and y and is a rank-based procedure. It is used to determine trends of time-series data. Just as with ρ, τ is resistant to the effect of outliers and, because τ also depends only on the ranks of data and not the observations themselves, it can be implemented even in cases where some of the data are categorical, such as censored observations like observations stated as less than a threshold of flood discharge that is perceived as a warning value or threat.

Despite having similar properties, ρ and τ use different scales to measure the same correlation. Though τ is generally lower than ρ in magnitude, their p-values for significance should be quite similar when computed on the same data. In general, τ will be lower than values of r for linear associations for any given linearly related data. Strong linear correlations of $r \geq 0.9$ typically correspond to $\tau \geq 0.7$. These lower values do not mean that τ is less sensitive than r, but simply that a different scale of correlation is being used. As it is a rank correlation method, τ is unaffected by monotonic power transformations of one or both variables. For example, τ for the correlation of $\log(y)$ versus $\log(x)$ will be identical to that of y versus $\log(x)$, and of y versus x.

Kendall's τ examines every possible pair of data points, (x_i, y_i) and (x_j, y_j), to determine if the pairs have the same relation to one another, for example, if x_i is greater than y_i and x_j is greater than y_j, or if x_i is less than y_i and x_j is less than y_j. Each pair is assessed in this way, keeping track of the number of pairs that have the same relation to one another versus the number of pairs that do not. Kendall's τ is computed by ordering all data pairs by increasing x. If a positive correlation exists, the y observations will increase more often than decrease as x increases. For a negative correlation, the y observations will decrease more often than increase as x increases. If no correlation exists, the y observations will increase and decrease about the same number of times. Kendall's τ is related to the sign test in that positive differences between data pairs are assigned $+1$ without regard to the magnitude of those differences and negative differences are assigned -1.

The calculation of τ begins with the calculation of Kendall's S, the test statistic given as

$$S = P - M \tag{3.208}$$

Kendall's S measures the monotonic dependence of y on x. The S statistic is simply the number of concordant pairs, denoted by P, when the slope between the two points is a positive minus the number of discordant pairs, denoted by M, when the slope between the two points is negative.

A concordant pair is a pair of observations where the difference between the y observations is of the same sign as the difference between the x observations. A discordant pair is a pair of observations where the difference in the y observations and the difference in the x observations is of the opposite sign. The computation can be simplified by rearranging the data pairs, placing the N observations in order based on the x observations with x_1 being the smallest x to x_n being the largest x. After this rearrangement, all pairwise comparisons of the y observations is considered, where the pairs are sorted by their x rank. Then, (x_i, y_i) is compared to (x_j, y_j) where $i < j$. There are $N(N-1)/2$ possible comparisons to be made among the N data pairs. If all y observations increase along with the x observations, $S = N(N-1)/2$. In this situation, the Kendall's τ correlation coefficient should equal $+1$. When all y observations decrease with increasing x, $S = -N(N-1)/2$ and Kendall's τ should equal -1. Therefore, dividing S by $N(N-1)/2$ will give a value always falling between -1 and $+1$. This is the definition of Kendall's τ, which measures the strength of the monotonic association between two variables,

$$\tau = \frac{S}{N\frac{(N-1)}{2}} \tag{3.209}$$

To test for the significance of τ, S is compared to what would be expected when the null hypothesis is true for a given N. For a two-sided test, $H_0 : \tau = 0$, or $H_a : \tau \neq 0$. If τ is further from 0 than expected, H_0 is rejected. When $N \leq 10$, the table of exact p-values using the S and N values should be used. Such a table can be found in Hollander and Wolfe (1999).

When $N > 10$, a large-sample approximation can be used. The test statistic, Z_s, given by

$$Z_s = \begin{cases} \frac{S-1}{\sigma_s} & \text{if } S > 0 \\ 0 & \text{if } S = 0 \\ \frac{S+1}{\sigma_s} & \text{if } S < 0 \end{cases} \tag{3.210}$$

closely approximates a normal distribution with the mean, μ_s, equal to zero and the variance, $\sigma_s = \sqrt{(N/18)(N-1)(2N+5)}$.

The null hypothesis is rejected at the significance level α if $|Z_s| > Z_{crit}$, where Z_{crit} is the value of the standard normal distribution with a probability of exceedance of $\alpha/2$ (for a two-sided test). A continuity correction must be applied as reflected in Eq. 3.210 by the -1 or $+1$.

3.22 TIME-SERIES ANALYSIS

3.22.1 Hydrologic Time Series

Analysis of hydrologic time series is an extensive subject that merits an exclusive treatment of its own. A hydrologic time series can be completely stochastic or can be composed of a combination of deterministic and stochastic components. The major characteristics of a hydrologic time series that are fundamentally important include homogeneity, stationarity, trend, periodicity, and persistence. Some of the basic concepts pertinent to these characteristics are presented here.

3.22.2 Autocorrelation

Autocorrelation is a mathematical representation of the degree of similarity between a given time series and a lagged version of this time series over successive time intervals. It is conceptually similar to the correlation between two different time series, but autocorrelation uses the same time series twice: once in its original form and once lagged one or more time periods.

Autocorrelation is also called **serial correlation** and is used to describe the relationship between observations of the same variable over specific periods. Therefore, it is an important statistical method in time-series analysis. If the serial correlation of a variable is measured as zero, there is no correlation, and each of the observations is independent of one another. Conversely, if the serial correlation of a variable skews toward one, the observations are serially correlated, and future observations are affected by past values. Essentially, a variable that is serially correlated has a pattern and is not random.

The **autocorrelation function** (**ACF**) at lag k, denoted by ρ_k, of a stationary stochastic process, is defined as

$$\rho_k = \frac{\gamma_k}{\gamma_0} = \frac{\mathrm{Cov}(y_i, y_{i+k})}{\mathrm{Var}(y_i)} \quad \text{for any } i. \tag{3.211}$$

where γ_0 is the variance of the stochastic process. The mean of a time series $y(t) = y_1, \ldots, y_n$ is

$$\bar{y} = \frac{1}{n} \sum_{i=1}^{n} y_i \tag{3.212}$$

The **autocovariance function** at lag k, for $k \geq 0$, of the time series is defined by

$$s_k = \frac{1}{n} \sum_{i=1}^{n-k} (y_i - \bar{y})(y_{i+k} - \bar{y}) = \frac{1}{n} \sum_{i=k+1}^{n} (y_i - \bar{y})(y_{i-k} - \bar{y}) \tag{3.213}$$

The ACF at lag k, for $k \geq 0$, of the time series is defined by

$$r_k = \frac{s_k}{s_0} \tag{3.214}$$

The **variance** of the time series is s_0. The definition of autocovariance given above is a little different from the usual definition of covariance between $\{y_1, \ldots, y_{n-k}\}$ and $\{y_{k+1}, \ldots, y_n\}$ in two respects: (1) We divide by n instead of $n-k$ and we subtract the overall mean instead of the means of $\{y_1, \ldots, y_{n-k}\}$ and $\{y_{k+1}, \ldots, y_n\}$, respectively. For large values of n with respect to k, the difference is small. (2) Theoretical advantages of using division by n instead of $n-k$ in the definition of s_k, namely that the covariance and correlation matrices will always be definite nonnegative. Even though the definition of autocorrelation is slightly different from that of correlation, r_k (or ρ_k) still takes a value between -1 and 1.

A plot of r_k against k is known as a **correlogram**, provided autocorrelation is calculated at every lag position. Correlograms help reveal the characteristics of a time series. A typical correlogram, in the case of random data, will fall from a value of $+1.0$ at zero lag, to possibly negative values.

On the other hand, if the successive data are not independent, at lags of near coincidence of the elements, the correlogram will show a peak of high autocorrelation. Examination of the correlogram will disclose intervals of time at which the time series has a repetitive or periodic nature.

3.22.3 Persistence

Persistence is the tendency for the high values of a variable in a time series to follow high values, and low values to follow low values. Here, the high and low values considered in this regard are relative to the average values. The presence of the most common type of persistence is reflected in the autocorrelation coefficient.

Variables with seasonal variation tend to have significant autocorrelation at a lag equal to the period of the seasonality. For example, monthly average streamflow values in many regions have significant correlation at lag $k = 12$. The presence of persistence makes it more likely that a time-series sample was taken from a period when values tended to be higher or lower than average, and when the variability was not representative of the population. Thus, persistence reduces the confidence with which one can estimate the quantile and moment statistics of the time series.

3.22.4 Spectral Analysis

The ACF is a time-domain representation of signals. However, analysis of random processes in the frequency domain allows one to observe several characteristics of the signal that are not easily discernable in the time domain. For instance, frequency-domain analysis becomes useful to reveal the cyclical behavior or periodicities of a signal or random processes.

Let there be a random process $X(t)$ with an ACF, $R_X(\tau)$. The **spectral density function** of $X(t)$, designated by $S_X(f)$, is the Fourier transform of $R_X(\tau)$. We can write

$$S_X(f) = \mathcal{F}\{R_X(\tau)\} = \int_{-\infty}^{\infty} R_X(\tau) \exp(-2j\pi f \tau) d\tau \tag{3.215}$$

where $j = \sqrt{-1}$, and f is frequency in cycles per time.

The spectral density function shows the amount of variance per interval of frequency. A variance spectrum partitions the variance into a number of intervals or bands of frequency. For a completely random series of uncorrelated numbers, $S_X(f)$ is a constant and is termed white noise. This indicates that no frequency interval contains any more variance than any other frequency interval. In the case of an ACF, this independence would be shown by $\rho_k = 0$ for $k \neq 0$.

A **periodogram** is used to identify the dominant periods or frequencies of a time series. This is a helpful plot for identifying the dominant cyclical behavior in a series, particularly when the cycles are not related to the commonly encountered seasonality.

3.23 ANALYSIS OF VARIANCE (ANOVA)

3.23.1 One-Factor ANOVA

Analysis of variance (ANOVA) determines whether the mean of at least one group of samples of a variable differs from the means for other groups. In one-factor or one-way ANOVA, the samples of the variable of interest are grouped on the basis of only one factor, such as rainfall totals of a regional storm event in different watersheds of a river basin. If the group means are dissimilar, some of them will differ from the overall mean. If the group means are similar, they will also be similar to the overall mean.

Why should a test of differences between means be named as analysis of variance? It is because the tests are actually based on variance of the variable, not the means. The group means are viewed as true signals and the variances are viewed as noise or random errors. The procedure essentially attempts to separate the noise, also called random errors, from the signals to identify the differences.

In order to determine if the differences between group means or signals can be seen above the variation within the noise present in the groups, the total noise or total variance, denoted by SS_T in the data as measured by the total sum of squares is split into two parts,

$$\sum_{j=1}^{k}\sum_{i=1}^{n_i}(x_{ij}-\bar{x})^2 = \sum_{j=1}^{k}N_j(\bar{x}_j-\bar{x})^2 + \sum_{j=1}^{k}\sum_{i=1}^{n_i}(x_{ij}-\bar{x}_i)^2 \quad (3.216)$$

where x_{ij} is the ith observation in the jth group, there are k groups, and N_j designates that the sample size within the jth group may or may not be equal to those in other groups. In Eq. 3.216, the first factor on the right-hand side of the equation gives the difference between group means and the overall mean of all samples pooled together. This is called **sum of squares among samples** or **factor sum of squares** denoted by SS_A or SS_F. The second term on the right-hand side of Eq. 3.216 measures variation within groups or **residual sum of squares** designated by SS_W or SS_E.

If the total sum of squares is divided by $N - 1$, where N is the total number of observations, it equals the variance of the x_{ij}s. Thus, ANOVA partitions the variance of the data into two parts: one measuring the signal or factor mean square, representing differences between groups; and the other measuring the noise or residual mean square, representing differences within groups. If the signal or factor mean square is large compared to the noise or residual mean square, the means are found to be significantly different.

If x_{ij} is the ith replicate or observation of the jth sample or group and the total number of observations, N, is equal to the number of replicates per sample times the number of samples, or $N = n \times m$, the source of variation has the following components.

The first source of variance among the samples with degree of freedom $(m - 1)$ is

$$SS_A = \sum_{j=1}^{m}\left[\frac{\left(\sum_{i=1}^{n}x_{ij}\right)^2}{n}\right] - \frac{\left(\sum_{j=1}^{m}\sum_{i=1}^{n}x_{ij}\right)^2}{N} \quad (3.217)$$

The second source of variance, that within samples or replications with $(N - m)$ degrees of freedom, is

$$SS_W = \sum_{j=1}^{m}\sum_{i=1}^{n}x_{ij}^2 - \sum_{j=1}^{m}\frac{\left(\sum_{i=1}^{n}x_{ij}\right)^2}{n} \quad (3.218)$$

The total variance with $(N - 1)$ degrees of freedom, of all observations is given by

$$SS_T = \sum_{j=1}^{m}\sum_{i=1}^{n}x_{ij}^2 - \frac{\left(\sum_{j=1}^{m}\sum_{i=1}^{n}x_{ij}\right)^2}{N} \quad (3.219)$$

Therefore,

$$SS_W = SS_T - SS_A \quad (3.220)$$

To detect a difference between means, the variation within a group around its mean must be sufficiently small in comparison to the difference between group means so that the group means may be seen as different. The noise within groups is estimated by the residual or error mean square (MS_E or MS_W), and the signal between group means is estimated by the factor or treatment mean square (MS_F or MS_A). These two statistics are computed as follows.

The residual or error sum of squares SS_E, calculated as

$$SS_E = \sum_{j=1}^{k}\sum_{i=1}^{n_i}(x_{ij}-\bar{x}_j)^2 \quad (3.221)$$

estimates the total within-group noise using departures from the sample group mean, $\bar{x}_j$. Error in this context refers not to a mistake, but to the inherent noise or random variation within a group. The factor sum of squares SS_F, calculated as

$$SS_F = \sum_{j=1}^{k}N_j(\bar{x}_j-\bar{x})^2 \quad (3.222)$$

estimates the factor effect using differences between group means, $\bar{x}_j$, and the overall mean, $\bar{x}$, weighted by sample size.

Each sum of squares has an associated number of degrees of freedom. For the factor sum of squares (SS_F or SS_A) this equals $k - 1$, as when $k - 1$ of the group means are known, the kth group mean can be calculated. The total sum of squares (SS_T) has $N - 1$ degrees of freedom. The residual sum of squares (SS_E or SS_W) has degrees of freedom equal to the difference between the above two, or $N - k$.

The test to compare the two estimates of variance, MS_F and MS_E, is whether their ratio equals 1, the test statistic is computed as

$$F = \frac{MS_F}{MS_E} \quad (3.223)$$

The null and alternative hypotheses for the analysis of variance are

$$H_0: \mu_1 = \mu_2 = \cdots = \mu_k$$

H_a: At least one mean is different

This is always a two-sided test. The test statistic F is compared to quantiles of an F-distribution and H_0 is rejected for large F. Equivalently, H_0 is rejected if the p-value for the test $< \alpha$.

3.23.2 Two-Factor ANOVA

In certain situations, more than one factor may simultaneously influence the magnitudes of observations. In such cases, groups of the variables are based on more than one factor. Multifactor tests can evaluate the influence of all factors simultaneously, in a way similar to multiple regression. The influence of one factor can be determined while compensating for the others. This is the objective of a factorial analysis of variance and its nonparametric alternatives. A factorial ANOVA occurs when none of the factors is a subset of the others. If subset factors do occur, the design includes nested factors and the equations for computing the F-test will differ from those presented here. Two-way ANOVA can further be classified as two-factor ANOVA with or without replication.

Two-factor ANOVA is only used to address certain specialized cases in statistical hydrology and for this reason it is not presented here. A detailed discussion about this can be found in Helsel et al. (2020).

3.24 EXAMPLES

Example 3.1: The design discharge frequency of freeway culverts in Texas is 25 years as recommended by the Texas Department of Transportation. On average, how many times can the culvert be overtopped by a 25-year flood in a 50-year period? What is the probability that exactly this number of 25-year floods will occur in a 50-year period?

Solution: A 25-year flood has, $p = \frac{1}{25} = 0.04$
The average number of occurrences is given by the mean of the binomial distribution (Eq. 3.74),

$E(np) = 50(0.04) = 2$

Probability of two occurrences is given as

$$P(k = 4) = \binom{50}{2}(0.04)^2(1 - 0.04)^{50-2}$$
$$= 1225 \times 0.0016 \times 0.140935 = 0.2762$$
$$= 27.62\% \text{ chance.}$$

Example 3.2: Define the limiting probability for occurrence of an n-year event in an n-year period.

Solution: The probability of an event with an n-year recurrence interval during an n-year period is given as

The limiting probability is when n becomes large.
Figure 3.18 shows the plot of p against n. As n increases (>25) $p \to \left(1 - \frac{1}{e}\right) = 0.6321$.

Figure 3.18 Plot of probability p against year n.

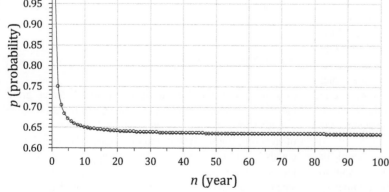

Example 3.3: The effect of climate change on river flows has been an important topic in contemporaneous hydrologic research. Both increasing and decreasing trends have been documented for major rivers of the world. Dai et al. (2009) presented the statistical significance of such trends drawing from a carefully developed dataset for the 925 largest ocean-reaching rivers. The Colorado River of the United States was not specifically mentioned in their discussion of these large rivers. Nonetheless, it is a vitally important river for the arid southwestern United States and electronic media is replete with reports that discharge in the Colorado River is declining due to climate change. Scientific evidence indicates that global warming has been pronounced since middle of twentieth century and is becoming more so in the twenty-first century. Discharge data from the Colorado River near Peach Springs, Arizona, for 32 years are given in Table 3.6. For the annual discharge of rivers covering a long recording period, a normal distribution often fits the observations. However, the period of available record from the USGS at this gauging station begins near the end of twentieth century. Divide this period into two and conduct statistical tests to infer whether within the period of changing climate significant variations have occurred. In this test case, assume that the parent population follows a normal distribution. Test the validity of this assumption.

Solution:

Step 1: The data are plotted to divide the annual flows into two groups (Figure 3.19).
The data are divided into two groups. The first group is the period 1990–2002 (end of the twentieth century) and the second group is the period 2003–2021 (beginning of the twenty-first century).

Step 2: *t*-test.
Are the means of the two samples the same? This is a somewhat different problem than typical *t*-tests where the mean of the parent population is either known or assumed. Here we are comparing statistics of two samples against one another, rather than against a proposed population parameter. The appropriate test is a *t*-test but the test statistic is calculated in a different manner.

The hypothesis we are testing is

$$H_0: \mu_1 = \mu_2$$

which states that the mean of the population from which the first group is drawn is the same as the mean of the parent population of the second group. The hypothesis is posed against the alternative

$$H_a: \mu_1 \neq \mu_2$$

that the means of the two populations are not equal.

Table 3.6 *Annual discharge (ft^3/s = cfs) of the Colorado River near Peach Springs, Arizona (USGS Gauge 09404200)*

Year	Flow (cfs)	Year	Flow (cfs)
1990	12 260	2006	12 450
1991	12 310	2007	12 710
1992	12 350	2008	13 850
1993	12 890	2009	12 840
1994	12 190	2010	13 100
1995	13 910	2011	18 600
1996	16 500	2012	14 180
1997	20 010	2013	12 700
1998	19 480	2014	11 560
1999	16 720	2015	13 620
2000	14 110	2016	13 550
2001	12 530	2017	13 860
2002	12 270	2018	13 570
2003	12 370	2019	13 900
2004	12 360	2020	12 540
2005	13 000	2021	12 400

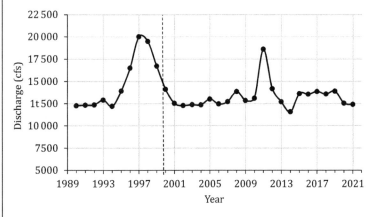

Figure 3.19 Colorado River annual discharge near Peach Springs, AZ.

The test statistic has the form

$$t = \frac{\bar{x}_1 - \bar{x}_2}{s_p \sqrt{\frac{1}{N_1} + \frac{1}{N_2}}}$$

where the quantity s_p is the pooled estimate of the population standard deviation, based on two samples. In order to estimate the variance of the parent population from which both samples are assumed to have been drawn, we must estimate the variance from each sample and then combine or pool them. The estimated variance is given by

$$s_p^2 = \frac{(N_1 - 1)s_1^2 + (N_2 - 1)s_2^2}{N_1 + N_2 - 2}$$

where subscripts refer respectively to the samples from two groups or sub-periods.
This process requires the estimation of two parameters: σ_1^2 and σ_2^2. Therefore, the degrees of freedom

$$v = N_1 + N_2 - 2$$

The significance level is $\alpha = 0.1$. This is for a two-tailed test, with each tail having a significance level of 0.05 (95% confidence limit).
From our data:

$\bar{x}_1 = 14425.4$ and $\bar{x}_2 = 13324.2$; $N_1 = 13$ and $N_2 = 19$
$s_1 = 2824.5$ and $s_2 = 1453.0$

$$s_p^2 = \frac{(13 - 1)2824.5^2 + (18 - 1)1453^2}{13 + 19 - 2} = 4457933$$

$$T = \frac{14425.4 - 13324.2}{\sqrt{4457933} \times \sqrt{\left(\frac{1}{13}\right) + \left(\frac{1}{19}\right)}} = 1.449$$

$$v = 13 + 19 - 2 = 30$$

Using Excel function T.INV.2T(0.1,30) we obtain

$$t_{crit} = -1.697 < t_{0.05,30} < +1.697$$

Since the computed value of the test statistic T does not fall into either critical region H_0 cannot be rejected.
The p-value is obtained from Excel function T.TEST (Array1, Array2, 2,3) where Array 1 and Array 2 are given as:

Year	Array 1	Year	Array 2
1990	12 260	2003	12 370
1991	12 310	2004	12 360
1992	12 350	2005	13 000
1993	12 890	2006	12 450
1994	12 190	2007	12 710
1995	13 910	2008	13 850
1996	16 500	2009	12 840
1997	20 010	2010	13 100
1998	19 480	2011	18 600
1999	16 720	2012	14 180
2000	14 110	2013	12 700
2001	12 530	2014	11 560
2002	12 270	2015	13 620

(cont.)

Year	Array 2
2016	13 550
2017	13 860
2018	13 570
2019	13 900
2020	12 540
2021	12 400

p-value $= 0.2138 > \alpha$

H_0 cannot be rejected.

Step 3: F-test.
The t-test presented above involves two samples with unequal variances. So an F-test is performed to ensure that the null hypothesis

$$H_0: \sigma_1^2 = \sigma_2^2$$

can be rejected to accept

$$H_a: \sigma_1^2 \neq \sigma_2^2$$

In this case the test statistic

$$f = \frac{13.2824.5^2}{12} \bigg/ \frac{19.1453.02^2}{18} = 3.878$$

Significance level, $\alpha = 0.05 \Rightarrow$ Confidence level $= 1 - \alpha = 0.95$

Excel function F.INV.RT(0.05,12,19) $= F_{\alpha,v_1,v_2} = 1/F_{\alpha,v_2,v_1} = 2.3421 < f$

Excel function F.TEST (Array1, Array2,12,19) $= 0.0111 = p$ value $< \alpha$

Therefore, H_0 is rejected. The variability in the first group of samples is greater than the second group.

Step 4: χ^2-test to check the validity of the assumption of normal distribution of the population.
The calculations necessary to perform the χ^2-test are presented in Table 3.7.

a. The range of discharge is divided into nine bins (intervals) given in column 1.
b. The observed count or frequency within each bin is given in column 2. Note that the numbers in column 1 represent the upper limits of the bins.
c. Column 3 lists the cumulative frequency at the upper limit of each class or interval.
d. The mean and standard deviation of all discharge values from 1990–2021 (Table 3.6) are calculated ($\bar{x} = 13771.6$; $s_x = 2148.5$) and are used to obtain the standard normal variate, z, corresponding to the upper limit of each of the class intervals. These are given in column 4 (note: Excel has a function called STANDARDIZE [x, mean, standard deviation] which can accomplish this step).
e. Column 5 gives the cumulative probability values of standard normal distribution of the z values of column 4.

Table 3.7 Computations of the χ^2-test

Bin	Frequency	Cumulative frequency	z	NORM.S.DIST (Cumulative)	Relative frequency (observed frequency)	Expected probability	χ^2
Column 1	Column 2	Column 3	Column 4	Column 5	Column 6	Column 7	Column 8
11 000	0	0.0000					
12 200	2	0.0625	−0.7315	0.2322	0.0625	0.2322	3.9701
13 400	16	0.5625	−0.1729	0.4313	0.5000	0.1991	14.5515
14 600	9	0.8438	0.3856	0.6501	0.2813	0.2188	0.5714
15 800	0	0.8438	0.9441	0.8274	0.0000	0.1773	5.6751
17 000	2	0.9063	1.5026	0.9335	0.0625	0.1061	0.5731
18 200	0	0.9063	2.0612	0.9804	0.0000	0.0468	1.4983
19 400	1	0.9375	2.6197	0.9956	0.0313	0.0152	0.5379
20 600	2	1.0000	3.1782	0.9993	0.0625	0.0037	30.2756

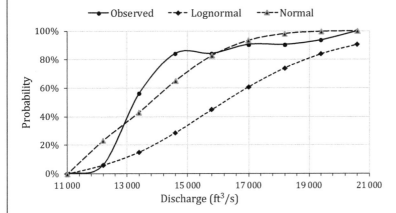

Figure 3.20 Observed and theoretical distributions.

The Excel function NORM.S.DIST(z, TRUE) is used to obtain these cumulative probability values. These values represent the function $F_X(x_i)$.

f. The observed relative frequency for each class is listed in column 6. These are calculated as $f_s(x_i) = \frac{n_i}{N}$. The values of n_i are those given in column 2 (absolute frequency) and, in this example, $N = 32$.

g. The expected probability if the data follow a normal distribution, given in column 7, is calculated as the incremental probability that can be obtained from the cumulative probability (column 5) using: $p_X(x_i) = F_X(x_i) - F_X(x_{i-1})$.

h. The χ^2 values given in column 8 are obtained from

$$\chi_i^2 = \frac{N[f_s(x_i) - p_X(x_i)]^2}{p_X(x_i)}$$

The test statistic χ^2 is obtained as:

$$\chi^2 = \sum_{i=1}^{k} \chi_i^2 = 57.65$$

In this case the number of class intervals is $k = 8$ and for normal distribution the number of parameters $m = 2$. Thus, the degrees of freedom is

$$df = 8 - 2 - 1 = 5$$

Significance level, $\alpha = 0.5 \Rightarrow$ confidence limit $= 1 - \alpha = 0.95$ (95%)

Critical value of χ^2 = CHISQ.INV.RT(0.05,5) = $\chi^2_{(1-\alpha),df} = 11.07$

Since the computed test statistic value of χ^2 is greater than the critical value of χ^2, the null hypothesis is rejected.

It should be pointed out here that the F-test may yield a significant value of F not only when H_0 is false but also when the underlying populations are not normal. It is sensitive to departures from normality as well as from H_0. On the other hand, the t-test is a good test as long as the underlying populations are not too nonnormal. Figure 3.20 shows the observed cdf compared with the computed cdfs for the normal and lognormal distribution.

Example 3.4: In Chapter 5 we discuss a method called the drainage-area ratio method which is based on a premise that a correlation exists between drainage area and discharge. Emerson et al. (2005) in the Scientific Investigations Report 2005-5017 of the USGS, provided the drainage area and mean annual discharge at 27 gauging stations located within Red River basin of North Dakota and Minnesota. The data are given in Table 3.8. Determine the correlation by calculating Pearson's r. Also, determine the coefficient of determination R^2 and make observation.

Solution:

Step 1: A table is set up to compute the factors of the equation that is used to calculate Pearson's r, shown as Table 3.9.

The factors are computed at the bottom of each column of Table 3.9.

Step 2: Calculate r.

$$r = \frac{N\sum XY - \sum X \sum Y}{\sqrt{\left[N\sum X^2 - (\sum X)^2\right] \cdot \left[N\sum Y^2 - (\sum Y)^2\right]}}$$

$$= \frac{27 \times 1.2 \times 10^7 - (26430 \times 5037)}{\sqrt{(27 \times 5.7 \times 5.7^7 - 26430^2) \times (27 \times 2899148 - 5037^2)}}$$

$$= 0.90298$$

Step 3: Determine the regression coefficients.

Since Pearson's r is reasonably high, the data can be represented by a linear-regression model:

$$Y = a + bX$$

The slope coefficient is calculated from

$$b = \frac{N\sum XY = \sum X \sum Y}{N\sum X^2 - (\sum X)^2}$$

$$= \frac{27 \times 1.2 \times 10^7 - 26430 \times 5036.9}{N \times 5.7 \times 10^7 - 26430^2} = 0.2264$$

Table 3.8 *Drainage area and annual flow in Red River basin of North Dakota and Minnesota*

River	Gauging Station	Drainage area (mi^2)	Annual streamflow (ft^3/s)
Wild Rice River	5053000	1490	129.43
Sheyenne River	5054500	154	14.72
Sheyenne River	5056000	760	79.82
Sheyenne River	5057000	1270	155.27
Baldhill Creek	5057200	351	30.48
Maple River	5059700	796	65.69
Rush River	5060500	116	15.71
Buffalo River	5061000	325	92.78
Buffalo River	5062000	975	188.01
Wild Rice River	5062500	934	245.55
Wild Rice River	5064000	1560	382.6
Goose River	5066500	1093	128.2
Sand Hill River	5069000	420	95.82
Thief River	5076000	985	226.78
Clearwater River	5078000	555	177.8
Lost River	5078230	254	73.71
Clearwater River	5078500	1380	370.52
Red Lake River	5079000	5270	1446.46
Forest River	5084000	336	43.96
Forest River	5085000	620	55.38
Middle River	5087500	255	48.57
Park River	5090000	695	68.02
South Branch River	5094000	422	100.8
Pembina River	5100000	3410	320.6
Tongue River	5101000	160	20.64
Roseau River	5104500	424	127.35
Roseau River	5112000	1420	332.23

Table 3.9 *Computation of the factors in the equation for Pearson's r*

X	Y	XY	X^2	Y^2
1490	129.43	192850.7	2220100	16752.1249
154	14.72	2266.88	23716	216.6784
760	79.82	60663.2	577600	6371.2324
1270	155.27	197192.9	1612900	24108.7729
351	30.48	10698.48	123201	929.0304
796	65.69	52289.24	633616	4315.1761
116	15.71	1822.36	13456	246.8041
325	92.78	30153.5	105625	8608.1284
975	188.01	183309.8	950625	35347.7601
934	245.55	229343.7	872356	60294.8025
1560	382.6	596856	2433600	146382.76
1093	128.2	140122.6	1194649	16435.24
420	95.82	40244.4	176400	9181.4724
985	226.78	223378.3	970225	51429.1684
555	177.8	98679	308025	31612.84
254	73.71	18722.34	64516	5433.1641
1380	370.52	511317.6	1904400	137285.07
5270	1446.46	7622844	27772900	2092246.53
336	43.96	14770.56	112896	1932.4816
620	55.38	34335.6	384400	3066.9444
255	48.57	12385.35	65025	2359.0449
695	68.02	47273.9	483025	4626.7204
422	100.8	42537.6	178084	10160.64
3410	320.6	1093246	11628100	102784.36
160	20.64	3302.4	25600	426.0096
424	127.35	53996.4	179776	16218.0225
1420	332.23	471766.6	2016400	110376.773
26430	5036.9	1.2E+07	5.7E+07	2899147.8
$\sum X$	$\sum Y$	$\sum XY$	$\sum X^2$	$\sum Y^2$

$$a = \frac{\sum Y - b\sum X}{N} = \frac{5037 - 0.2264 \times 26430}{N} = -35.1115$$

Step 4: Compute the factors for the equation giving coefficient of determination.

Table 3.10 shows the computation of the factors for the calculation of the coefficient of determination. The observed Y is denoted by Y_i and the predicted Y calculated from the regression model is denoted by Y_p.

Step 5: Calculate R^2.

$$R^2 = \frac{\left(\sum Y_p - Y_{av}\right)^2}{\left(\sum Y_i - Y_{av}\right)^2} = \frac{1597743}{1959505} = 0.8154$$

Step 6: Observation.

In this particular example,

$$r^2 = 0.9023^2 = 0.81534 = R^2$$

Table 3.10 *Computation of the factors for the coefficient of determination, R^2*

X	Y_i	Y_p	$(Y_p - Y_{av})^2$	$(Y_i - Y_{av})^2$
1490	129.43	302.2898	13395.27	3262.906
154	14.72	−0.23913	34890.87	29526.19
760	79.82	136.9858	2456.792	11391.69
1270	155.27	252.4722	4345.487	978.5543

Table 3.10 (*cont.*)

X	Y_i	Y_p	$(Y_p - Y_{av})^2$	$(Y_i - Y_{av})^2$
351	30.48	44.3703	20215.59	24358.42
796	65.69	145.1378	1715.124	14607.59
116	15.71	−8.844	38179.54	29186.94
325	92.78	38.48276	21924.46	8793.16
975	188.01	185.6712	0.775482	2.126196
934	245.55	176.387	103.3234	3480.781
1560	382.6	318.1409	17315.67	38434.88
1093	128.2	212.3916	667.6929	3404.939
420	95.82	59.99492	16016.66	8232.269
985	226.78	187.9357	1.914967	1618.304
555	177.8	90.56484	9213.507	76.59491
254	73.71	22.40525	26944.11	12733.28
1380	370.52	277.381	8249.931	33844.28
5270	1446.46	1158.247	944192.3	1587369
336	43.96	40.97364	21193.02	20332.44
620	55.38	105.2837	6604.515	17206.05
255	48.57	22.63169	26869.82	19038.99
695	68.02	122.267	4132.546	14049.8
422	100.8	60.44781	15902.23	7353.38
3410	320.6	737.0619	303061.3	17968.91
160	20.64	1.11953	34385.15	27526.74
424	127.35	60.9007	15788.21	3504.859
1420	332.23	286.4387	9977.389	21222.12
	186.552		**1597743**	**1959505**
	Y_{av}		$-(Y_p - Y_{av})^2$	$-(Y_i - Y_{av})^2$

Example 3.5: In Example 5.3 of Chapter 5, annual peak flows of Potomac River at the Point of Rocks for 128 years are given in Table 5.9. Calculate correlogram and spectral density functions of this time series to prove that the annual flood flows at this station is a purely random stochastic process.

Step 1: Determine the autocorrelation for various lags and create a correlogram.

The ACF, r_k, for lag $k = 1, \ldots, 10$ is calculated using

$$r_k = \frac{\text{Cov}\{Q(t), Q(t+k)\}}{\text{Var}\{Q(t)\}}$$

The correlogram is shown in Figure 3.21. The value of r_k for $k = 0$ is 1 and is not plotted in this figure, which shows that the autocorrelation is very low and remains fairly constant for various lags. This proves that the successive annual flood flows are independent of each other.

Step 2: Calculate the spectral density function and plot amplitude against frequency.

The magnitude or the amplitude in the frequency domain is calculated by a Fourier transform of the time-domain-data:

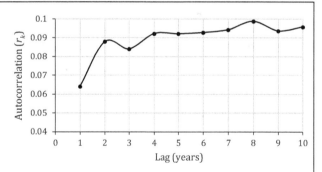

Figure 3.21 Correlogram of annual peak flows.

$$S(f) = \mathcal{F}[Q(t)]$$

The Fourier transform is done using Excel's Fourier analysis program. However, Excel outputs N complex amplitudes:

$$A_n = \sum_{k=0}^{N-1} a_k e^{jk\frac{2\pi n}{N}}$$

Excel provides a function called IMABS (x)x that gives the absolute value of a complex number. Thus, the complex amplitudes are transformed to real amplitudes.

The frequency is calculated as

$$f = \frac{1}{T}$$

where T denotes the period, which is the length of the time series. In our case it is 128 years.

Figure 3.22 shows the spectral density function.

The spectral density function shown in Figure 3.22 oscillates about the mean value of the function. This is a typical white noise spectrum, indicating that the variance is equally distributed among all frequencies. This proves that the annual peak flood flow at this station of Potomac River is a purely random stochastic process.

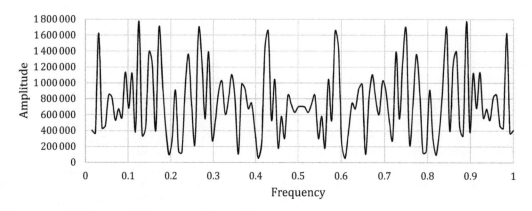

Figure 3.22 Spectral density of annual peak flows.

Example 3.6: Table 3.11 gives the rainfall recorded in 5-minute intervals at three adjacent watersheds in Houston on the first day of tropical storm Allison. Chapter 4 has much more detail about this severe rainfall event that caused havoc in Houston in early June of 2021. Each column in Table 3.11 is the mean value of the rainfall recorded at multiple rain gauges within each watershed. Was there any significant difference in the mean rainfall in these three watersheds?

Table 3.11 *Rainfall records (inches) in three adjacent watersheds of Houston on the first day of Tropical Storm Allision*

Date/Time	Buffalo Bayou Watershed	Brays Bayou Watershed	White Oak Bayou Watershed
6/5/2001 0:00	0	0	0
6/5/2001 0:05	0	0	0
6/5/2001 0:10	0	0	0
6/5/2001 0:15	0	0	0
6/5/2001 0:20	0	0	0
6/5/2001 0:25	0	0	0
6/5/2001 0:30	0	0	0
6/5/2001 0:35	0	0	0
6/5/2001 0:40	0	0	0
6/5/2001 0:45	0	0	0
6/5/2001 0:50	0	0	0
6/5/2001 0:55	0	0	0
6/5/2001 1:00	0	0	0
6/5/2001 1:05	0	0	0
6/5/2001 1:10	0	0	0
6/5/2001 1:15	0	0	0
6/5/2001 1:20	0	0	0
6/5/2001 1:25	0	0	0
6/5/2001 1:30	0	0	0

Table 3.11 (cont.)

Date/Time	Buffalo Bayou Watershed	Brays Bayou Watershed	White Oak Bayou Watershed
6/5/2001 1:35	0	0	0
6/5/2001 1:40	0	0	0
6/5/2001 1:45	0	0	0
6/5/2001 1:50	0	0	0
6/5/2001 1:55	0	0.0066667	0
6/5/2001 2:00	0.0285714	0	0
6/5/2001 2:05	0.0114286	0.01	0.0057143
6/5/2001 2:10	0.0285714	0	0.0142857
6/5/2001 2:15	0.0228571	0	0.0171429
6/5/2001 2:20	0	0	0.04
6/5/2001 2:25	0	0	0.0342857
6/5/2001 2:30	0	0	0.0371429
6/5/2001 2:35	0	0	0.0171429
6/5/2001 2:40	0	0	0
6/5/2001 2:45	0	0	0
6/5/2001 2:50	0	0	0.0028571
6/5/2001 2:55	0	0	0
6/5/2001 3:00	0	0	0
6/5/2001 3:05	0	0	0.0028571
6/5/2001 3:10	0	0	0
6/5/2001 3:15	0	0	0
6/5/2001 3:20	0	0	0
6/5/2001 3:25	0	0	0
6/5/2001 3:30	0	0	0
6/5/2001 3:35	0	0.0166667	0
6/5/2001 3:40	0	0.0066667	0
6/5/2001 3:45	0.0057143	0.0033333	0
6/5/2001 3:50	0.0057143	0.0033333	0
6/5/2001 3:55	0.0057143	0.0033333	0.0028571
6/5/2001 4:00	0	0	0
6/5/2001 4:05	0	0	0.0028571
6/5/2001 4:10	0.0057143	0	0
6/5/2001 4:15	0.0057143	0	0
6/5/2001 4:20	0	0	0
6/5/2001 4:25	0	0	0
6/5/2001 4:30	0	0	0
6/5/2001 4:35	0	0	0
6/5/2001 4:40	0	0	0
6/5/2001 4:45	0	0	0
6/5/2001 4:50	0	0	0
6/5/2001 4:55	0	0	0
6/5/2001 5:00	0	0	0
6/5/2001 5:05	0	0	0
6/5/2001 5:10	0	0	0
6/5/2001 5:15	0	0	0
6/5/2001 5:20	0	0	0
6/5/2001 5:25	0	0	0
6/5/2001 5:30	0	0	0
6/5/2001 5:35	0	0	0
6/5/2001 5:40	0	0	0
6/5/2001 5:45	0	0	0
6/5/2001 5:50	0	0	0
6/5/2001 5:55	0	0	0
6/5/2001 6:00	0	0	0
6/5/2001 6:05	0.0057143	0	0
6/5/2001 6:10	0	0	0

Table 3.11 (cont.)

Date/Time	Buffalo Bayou Watershed	Brays Bayou Watershed	White Oak Bayou Watershed
6/5/2001 6:15	0	0	0
6/5/2001 6:20	0	0.0033333	0
6/5/2001 6:25	0	0.0033333	0
6/5/2001 6:30	0	0.01	0
6/5/2001 6:35	0	0.0033333	0
6/5/2001 6:40	0.0057143	0.0066667	0
6/5/2001 6:45	0	0.0066667	0
6/5/2001 6:50	0.0057143	0.0133333	0
6/5/2001 6:55	0.0171429	0.0033333	0
6/5/2001 7:00	0	0	0
6/5/2001 7:05	0.0114286	0.0066667	0
6/5/2001 7:10	0	0	0
6/5/2001 7:15	0	0	0
6/5/2001 7:20	0	0	0
6/5/2001 7:25	0	0	0
6/5/2001 7:30	0	0.0033333	0
6/5/2001 7:35	0	0	0
6/5/2001 7:40	0	0	0
6/5/2001 7:45	0	0	0
6/5/2001 7:50	0	0	0
6/5/2001 7:55	0	0	0
6/5/2001 8:00	0	0	0
6/5/2001 8:05	0	0	0
6/5/2001 8:10	0	0	0
6/5/2001 8:15	0	0	0
6/5/2001 8:20	0	0	0
6/5/2001 8:25	0	0	0
6/5/2001 8:30	0	0	0
6/5/2001 8:35	0	0	0
6/5/2001 8:40	0	0	0
6/5/2001 8:45	0	0	0
6/5/2001 8:50	0	0	0
6/5/2001 8:55	0	0	0
6/5/2001 9:00	0	0	0
6/5/2001 9:05	0	0	0
6/5/2001 9:10	0	0	0
6/5/2001 9:15	0	0	0
6/5/2001 9:20	0	0	0
6/5/2001 9:25	0	0	0.0028571
6/5/2001 9:30	0	0	0
6/5/2001 9:35	0	0	0
6/5/2001 9:40	0	0	0
6/5/2001 9:45	0	0	0
6/5/2001 9:50	0	0	0.0057143
6/5/2001 9:55	0	0	0
6/5/2001 10:00	0	0	0
6/5/2001 10:05	0	0	0
6/5/2001 10:10	0	0	0
6/5/2001 10:15	0	0	0
6/5/2001 10:20	0	0.0033333	0
6/5/2001 10:25	0	0	0
6/5/2001 10:30	0	0	0
6/5/2001 10:35	0	0	0
6/5/2001 10:40	0	0.0033333	0.0028571
6/5/2001 10:45	0	0	0.0114286
6/5/2001 10:50	0	0	0.0114286

Table 3.11 (*cont.*)

Date/Time	Buffalo Bayou Watershed	Brays Bayou Watershed	White Oak Bayou Watershed
6/5/2001 10:55	0	0	0
6/5/2001 11:00	0	0	0.0057143
6/5/2001 11:05	0	0	0
6/5/2001 11:10	0	0	0
6/5/2001 11:15	0.0114286	0.0033333	0
6/5/2001 11:20	0.0057143	0	0
6/5/2001 11:25	0	0.0133333	0.0057143
6/5/2001 11:30	0.0171429	0.0033333	0.0057143
6/5/2001 11:35	0	0.0066667	0
6/5/2001 11:40	0.0114286	0.01	0
6/5/2001 11:45	0.0114286	0.0233333	0.0057143
6/5/2001 11:50	0.0171429	0.0033333	0.0028571
6/5/2001 11:55	0	0.0033333	0.0028571
6/5/2001 12:00	0	0	0.0028571
6/5/2001 12:05	0.0114286	0	0.0028571
6/5/2001 12:10	0	0	0.0114286
6/5/2001 12:15	0.0057143	0	0.0085714
6/5/2001 12:20	0	0	0.0057143
6/5/2001 12:25	0	0	0
6/5/2001 12:30	0	0	0.0028571
6/5/2001 12:35	0	0	0
6/5/2001 12:40	0	0	0
6/5/2001 12:45	0	0	0
6/5/2001 12:50	0	0	0
6/5/2001 12:55	0	0	0
6/5/2001 13:00	0	0	0
6/5/2001 13:05	0	0	0
6/5/2001 13:10	0	0.0066667	0.0028571
6/5/2001 13:15	0	0	0
6/5/2001 13:20	0	0	0
6/5/2001 13:25	0	0.0066667	0.0028571
6/5/2001 13:30	0	0	0.0085714
6/5/2001 13:35	0.0114286	0	0
6/5/2001 13:40	0	0.0033333	0
6/5/2001 13:45	0	0	0
6/5/2001 13:50	0	0	0
6/5/2001 13:55	0	0	0
6/5/2001 14:00	0	0	0
6/5/2001 14:05	0	0	0
6/5/2001 14:10	0	0	0.0028571
6/5/2001 14:15	0	0.0033333	0
6/5/2001 14:20	0	0	0
6/5/2001 14:25	0	0	0.0028571
6/5/2001 14:30	0	0	0.0057143
6/5/2001 14:35	0.0114286	0	0
6/5/2001 14:40	0.0057143	0.0033333	0
6/5/2001 14:45	0	0.0066667	0
6/5/2001 14:50	0.0057143	0	0.0028571
6/5/2001 14:55	0.0057143	0	0.0028571
6/5/2001 15:00	0.0057143	0	0
6/5/2001 15:05	0	0.0033333	0
6/5/2001 15:10	0.0057143	0	0.0028571
6/5/2001 15:15	0	0	0
6/5/2001 15:20	0	0	0
6/5/2001 15:25	0	0	0
6/5/2001 15:30	0	0	0

Table 3.11 (cont.)

Date/Time	Buffalo Bayou Watershed	Brays Bayou Watershed	White Oak Bayou Watershed
6/5/2001 15:35	0.0114286	0.0033333	0
6/5/2001 15:40	0.0057143	0	0
6/5/2001 15:45	0.0114286	0.0033333	0
6/5/2001 15:50	0.0114286	0.01	0
6/5/2001 15:55	0.0228571	0.01	0
6/5/2001 16:00	0.0171429	0.0066667	0.0028571
6/5/2001 16:05	0.04	0.0166667	0
6/5/2001 16:10	0.0228571	0.03	0.0085714
6/5/2001 16:15	0.0514286	0.0533333	0.0057143
6/5/2001 16:20	0.0285714	0.0133333	0.0114286
6/5/2001 16:25	0.0571429	0.0333333	0.02
6/5/2001 16:30	0.0171429	0.03	0.0142857
6/5/2001 16:35	0.0457143	0.03	0.0057143
6/5/2001 16:40	0.0628571	0.04	0.0057143
6/5/2001 16:45	0.04	0.0766667	0.0114286
6/5/2001 16:50	0.0342857	0.0333333	0.0257143
6/5/2001 16:55	0.0285714	0.0366667	0.0085714
6/5/2001 17:00	0.0342857	0.0433333	0.0057143
6/5/2001 17:05	0.0171429	0.06	0.0171429
6/5/2001 17:10	0.0342857	0.0433333	0.0114286
6/5/2001 17:15	0.0114286	0.0333333	0.0057143
6/5/2001 17:20	0.0571429	0.0333333	0.0142857
6/5/2001 17:25	0.0114286	0.0791667	0.0114286
6/5/2001 17:30	0.0057143	0.0233333	0.0028571
6/5/2001 17:35	0.0342857	0.02	0.02
6/5/2001 17:40	0.0114286	0.0133333	0.0142857
6/5/2001 17:45	0.0342857	0.0133333	0.0228571
6/5/2001 17:50	0	0.0033333	0.0228571
6/5/2001 17:55	0.0457143	0.0266667	0.0171429
6/5/2001 18:00	0.0057143	0.0066667	0.0028571
6/5/2001 18:05	0.0228571	0.0066667	0.0171429
6/5/2001 18:10	0.0114286	0.0133333	0.0028571
6/5/2001 18:15	0.0171429	0.01	0.0085714
6/5/2001 18:20	0.0457143	0.04	0.0171429
6/5/2001 18:25	0.0228571	0.0366667	0.0028571
6/5/2001 18:30	0.0685714	0.0825	0.0028571
6/5/2001 18:35	0.04	0.0533333	0.0371429
6/5/2001 18:40	0.08	0.0991667	0.0028571
6/5/2001 18:45	0.0514286	0.09	0.0371429
6/5/2001 18:50	0.0685714	0.0866667	0.0342857
6/5/2001 18:55	0.0857143	0.1125	0.0478571
6/5/2001 19:00	0.1014286	0.1508333	0.0142857
6/5/2001 19:05	0.0685714	0.17	0.0571429
6/5/2001 19:10	0.1257143	0.1316667	0.0542857
6/5/2001 19:15	0.0857143	0.08	0.0342857
6/5/2001 19:20	0.1628571	0.0233333	0.0735714
6/5/2001 19:25	0.0557143	0.0325	0.0278571
6/5/2001 19:30	0.0557143	0.02	0.0557143
6/5/2001 19:35	0.1114286	0.0325	0.0171429
6/5/2001 19:40	0.1114286	0.0325	0.0557143
6/5/2001 19:45	0	0.0325	0
6/5/2001 19:50	0.1028571	0.0783333	0.1264286
6/5/2001 19:55	0.0285714	0.0266667	0.1628571
6/5/2001 20:00	0.0914286	0.0858333	0.1271429
6/5/2001 20:05	0.0857143	0.0791667	0.1228571
6/5/2001 20:10	0.0228571	0.0591667	0.1585714

Table 3.11 (cont.)

Date/Time	Buffalo Bayou Watershed	Brays Bayou Watershed	White Oak Bayou Watershed
6/5/2001 20:15	0.0457143	0.0266667	0.1142857
6/5/2001 20:20	0.1071429	0.0133333	0.145
6/5/2001 20:25	0.0857143	0.0466667	0.0771429
6/5/2001 20:30	0.0571429	0.0133333	0.1
6/5/2001 20:35	0.0571429	0.0333333	0.0257143
6/5/2001 20:40	0.0557143	0.0133333	0.0628571
6/5/2001 20:45	0.0628571	0.0558333	0.0314286
6/5/2001 20:50	0.0957143	0.0266667	0.04
6/5/2001 20:55	0.0914286	0.02	0.0428571
6/5/2001 21:00	0.0171429	0.0133333	0.0257143
6/5/2001 21:05	0.0342857	0.0266667	0.02
6/5/2001 21:10	0	0	0.0057143
6/5/2001 21:15	0.0057143	0.0433333	0.1085714
6/5/2001 21:20	0.0285714	0.0333333	0.0542857
6/5/2001 21:25	0.0628571	0.03	0.0542857
6/5/2001 21:30	0.0342857	0.0466667	0.0571429
6/5/2001 21:35	0.0342857	0.0133333	0.0371429
6/5/2001 21:40	0.0171429	0.01	0.0228571
6/5/2001 21:45	0.0685714	0.0066667	0.0457143
6/5/2001 21:50	0.0514286	0.0066667	0.04
6/5/2001 21:55	0.0285714	0.0133333	0.0342857
6/5/2001 22:00	0.0342857	0.0033333	0.0428571
6/5/2001 22:05	0.0571429	0.0066667	0.0114286
6/5/2001 22:10	0.0571429	0	0.0114286
6/5/2001 22:15	0.0228571	0.0166667	0.0628571
6/5/2001 22:20	0	0	0.0428571
6/5/2001 22:25	0	0.0033333	0.0428571
6/5/2001 22:30	0	0.01	0.02
6/5/2001 22:35	0.0114286	0.0033333	0.0257143
6/5/2001 22:40	0.0114286	0	0.0171429
6/5/2001 22:45	0.0114286	0	0
6/5/2001 22:50	0.0057143	0	0
6/5/2001 22:55	0	0.0033333	0.0028571
6/5/2001 23:00	0	0	0.0085714
6/5/2001 23:05	0.0114286	0	0
6/5/2001 23:10	0.0057143	0.0066667	0.0028571
6/5/2001 23:15	0	0	0
6/5/2001 23:20	0	0	0
6/5/2001 23:25	0	0	0.0057143
6/5/2001 23:30	0	0	0.0057143
6/5/2001 23:35	0	0	0.0028571
6/5/2001 23:40	0	0	0
6/5/2001 23:45	0	0	0
6/5/2001 23:50	0	0	0
6/5/2001 23:55	0	0	0.0085714

Solution:

Here, a one-factor or one-way ANOVA is conducted where the factor is the watershed boundary. The results of ANOVA conducted using Excel's ANOVA program are presented below.

Since the calculated F-statistic (1.143) is less than the critical value of F (3.01) at $\alpha = 0.5$. the null hypothesis cannot be rejected. In other words, there was no statistically significant difference in rainfall in these three watersheds during that day of the event.

SUMMARY

Groups	Count	Sum	Average	Variance
Buffalo Bayou Watershed	288	4.114286	0.014286	0.000728
Brays Bayou Watershed	288	3.254167	0.011299	0.000602
White Oak Bayou Watershed	288	3.355	0.011649	0.000685

ANOVA

Source of variation	SS	df	MS	F	p-value	F_{crit}
Between groups	0.001535	2	0.000768	1.142775	0.319416	3.00618
Within groups	0.578365	861	0.000672			
Total	0.5799	863				

Exercises: A selection of exercises on this topic is available at www.cambridge.org/appliedhydrology.

4 Rainfall Measurements and Models

4.1 CONCEPTS AND PURPOSE

Precipitation, which includes rainfall, snowfall, hail, and dew, together with net radiation are two inputs for a hydrologic system. Rainfall is the most important input to produce immediate surface runoff contributing to stream flow. Measured rainfall is one of the most important types of observational data used in hydrology. Observed rainfall data are used for (1) calibration and verification of hydrologic models; (2) real-time forecasting; (3) statistical analyses to develop models of rainstorms used in hydrologic designs; (4) evaluating the performance of both proposed designs and constructed hydraulic structures; and (5) assessing the efficacy of regulations stipulated for hydrologic designs. Accurate and adequate measurement of rainfall is therefore of fundamental importance in applied hydrology.

Rainfall is a stochastic natural process and is a scale-independent, spatially and temporally distributed quantity. Therefore, there are certain means of quantitative analysis and specifications for its space–time variabilities and probabilistic models that are used in hydrologic design calculations and modeling. This chapter deals with these aspects of rainfall.

4.2 METHODS OF RAINFALL MEASUREMENTS

Recording rainfall is one of the oldest scientific measurement practices, which started 400 BCE in India and 500 BCE in Greece due to ancient civilization's dependence on farming. The earliest records of measuring rainfall appear to be in Korea, in the 1400s. The development of instrumentation for accurate measurements of rainfall has progressed since then. Details of the requirements of various types of devices for precipitation measurements are described by the World Meteorological Organization (WMO) (WMO, 2020). Rainfall is measured by (1) rain gauges; (2) weather radars; and (3) satellites.

4.2.1 Rain Gauges

Rain gauges are broadly classified into two types: (1) **nonrecording** and (2) **recording**. The non-recording types give only total rainfall amount during a particular time period, such as daily or even monthly. Rainfall amount is always expressed in **depth** (inches or mm). The recording-type rain gauges provide rainfall amounts continuously in time, which can be transformed into amounts at specific intervals such as every 5 minutes or every hour. In other words, a recording gauge provides information about the temporal distribution of rainfall at the location of rain gauge. The basic features of the principal types of rain gauges are described below.

4.2.1.1 Nonrecording Rain Gauges

There are two types of nonrecording rain gauges.

Standard Gauges

A **standard rain gauge** is the simplest rain gauge device that consists of a collector above a funnel leading to a receiver, which is like either a measuring cylinder or a graduated flask, or one with a graduated dip stick. The volume of rain collected in the receiver is measured in terms of depth, typically at 24-hour intervals. These are sometimes called Symons gauges, named after the British meteorologist George James Symons, who published the first annual volumes of British Rainfall in 1860 from measurements at 168 stations in England and Wales using this type of gauges.

Storage Gauges

A **storage rain gauge** consists of a collector above a funnel leading to a receiver which, compared to a standard gauge, is large enough to store rainfall for a relatively long period, such as a month or an entire season. In this case also, the depth of rain water collected in the receiver is either measured in a measuring cylinder or by a dip stick. These are usually used to measure total seasonal precipitation in remote, sparsely inhabited areas.

4.2.1.2 Recording Rain Gauges

Three types of recording rain gauges are most common. These are **weighing**, **float** or **siphon**, and **tipping** or **tilting bucket** gauges. In addition, there are also **optical** and **acoustic** gauges.

4 Rainfall Measurements and Models

Weighing Gauges

In a weighing rain gauge, the weight of precipitation accumulating in a vessel, mounted on a spring balance or with a system of balance weights, is recorded continuously. The weight, converted to depth, provides the amount of accumulated precipitation at regular intervals.

Float or Siphon Gauges

In a float rain gauge, rainfall is fed into a float chamber (container) containing a light float whose vertical movement with rise in water level in the container is recorded. Measurements of the rise of the float at regular times give rainfall depths at those intervals. These gauges respond continuously to water depth in the float container. When the float reaches a preset maximum level, a siphon arrangement empties the float chamber.

Tipping Bucket Gauges

In a tipping or tilting bucket gauge, rain is fed through a conventional collecting funnel into a container divided into two vertical compartments, which can contain a predetermined amount of rain and are balanced on a fulcrum in an unstable equilibrium about a horizontal axis in its normal position. After rain is collected in one compartment to its capacity, it tips over, discharging its contents into a tank. Meanwhile the other compartment captures the rain, and the movement momentarily activates a magnetic switch that sends a signal to a data logger, which records the timing of the tipping; the timing between tipping or rotation and spillage volume provides the incremental rainfall depth.

Figure 4.1 schematically shows the basic principles of most common types of rain gauges. The parts and the design of a particular device vary according to manufacturers. All modern rain gauges record rainfall depths electronically and transmit the data through telemetry to a central station. Therefore, in addition to the mechanical parts of a device there are also electrical components in any type of modern rain gauge. Each type has advantages and disadvantages. Problems associated with rainfall measurements with gauges include effects of topography and nearby trees, buildings, and other structures or obstructions.

There are certain special types of recording gauges also, such as distrometers and acoustic gauges.

Distrometers

A **distrometer** measures the spectrum of precipitation particles either through momentum transferred to a transducer or through the reflectivity of the hydrometers.

Acoustic Gauges

An **acoustic gauge** is used for measuring rainfall over a water body. It records the noise spectrum generated by raindrops hitting water surface.

4.2.2 Measurements of Rainfall using Weather Radars

Ground-based weather radar (radio detection and ranging) devices provide estimates of rainfall that are continuous in time and space, covering large areal extents. The evolution of radars has a long history starting from the work of Heinrich Hertz on radio waves in the 1880s and Christian Hülsmeyer in the early 1900s, both in Germany. The technology developed considerably during World War II in Britain and the United States, in an effort to detect advancing hostile aircraft. During that time, it was noticed that radars were also capable of detecting precipitation. Sir Robert Watson-Watt, in Britain, who led Britain's wartime effort to develop radars, had already developed methods to detect thunderstorms based on various types of waves emitted by the radar prior to the wartime developments. After the end of World War II, the US Weather Bureau (now the National Weather Service) received 25 radars that had been used by the US Navy aircraft during the war. These weather surveillance radars (WSRs) were modified for meteorological use and were deployed across the United States.

A weather radar emits a focused pulse of microwave energy, an electromagnetic wave with long wavelengths (wavelengths typically range between 1 cm and 10 cm and have a frequency of $\sim 10^8$–10^{12} Hz). Part of this energy is reflected back or backscattered when it hits falling raindrops. The reflected energy or echo is used to calculate reflectivity, Z, which depends on the number of raindrops and their diameters. The reflectivity increases to the sixth power of raindrop diameter. The rate of rainfall, R, on the other hand increases to the third power of the drop diameter. Thus, if the drop diameter doubles, the reflectivity increases by 64 and the rate increases by a factor of 8. In nature, the characteristics of the drop size distribution are not random but rather well defined. The radar return from the rain and the rainfall rate both depend upon the moments of the drop size distribution, and the number density distribution $N(D)$. That is the reason why the radar return from precipitation is quantitatively linked to the precipitation rate. The relationship between reflectivity and rate of rainfall, known as the Z–R relationship, is given by a power function (Sauvageot, 1994):

$$Z = aR^b \qquad (4.1)$$

where a and b, respectively, are the coefficient and the exponent of the power function and depend on the $N(D)$ parameters. Simply speaking, $N(D)$ is the number of drops with diameters between D and $D + dD$ in a unit volume of air. The rainfall rate, R, is given in mm/h and the reflectivity, Z, is given in decibels of reflectivity (dBZ). The unit of reflectivity is mm^6/m^3. It is converted to echo strength of Z or decibels of Z using the relationship $dBZ = 10 \log_{10} Z$.

The drop size distribution is rarely known, and it varies in time and space. Thus, the Z–R relationship is not unique and has empirical forms. It has been measured for many types of rain and characteristic Z–R relationships have been reported. A widely used value of a is 200 and that of b is 1.6, given by Marshall and Palmer (1948). Nevertheless, the values of parameters a and b vary widely with location, storm type,

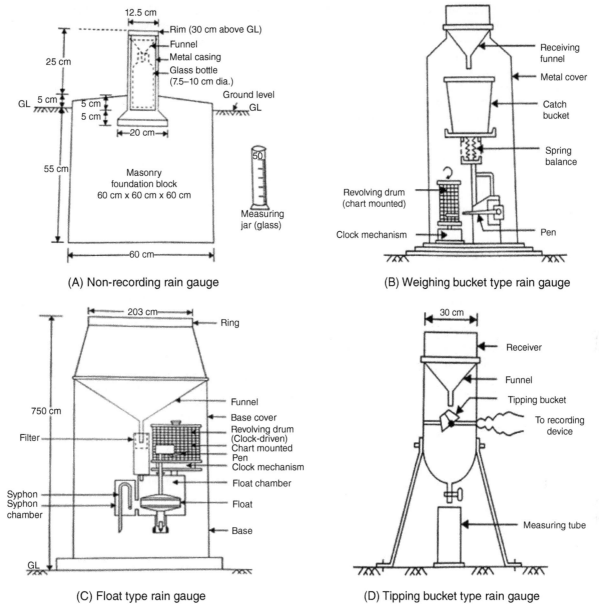

Figure 4.1 Schematics of common types of rain gauges to illustrate the principles of their operations: (A) Nonrecording rain gauge; (B) weighing gauge; (C) float gauge; and (D) tipping bucket gauge (taken from Hydrology Archives – Cement Concrete, courtesy Jay Kumar Shah).

drop size, and various other factors. These relationships have been developed for various regions based on these factors and empirical determinations.

In addition to measuring reflectivity, the modern weather radars also have the ability to detect the *shift in phase* of the pulse of energy, known as the **Doppler effect**, named after its discoverer Christian Doppler, an Austrian physicist. The present Next-Generation Radar (NEXRAD) system WSR-88D radars of the US Weather Service are Doppler radars, having the capabilities to determine rainstorms advancing or moving further away.

Numerous factors can cause errors in radar rainfall measurements, such as errors in estimating the radar reflectivity factor and variations in the Z–R relationship, backscatter, attenuation, absorption, and reflection of signals, particularly in areas of varying relief and signal calibration. Some studies in the past have shown that radars generally underestimate the rainfall amount when compared to the gauge data (e.g., Smith et al., 1996; Johnson et al., 1999; Mukhopadhyay et al., 2003). The cause of the discrepancy between radar rain gauge observations has been attributed to several factors, such as range-dependence sampling by radars, bias in reflectivity, bright bands, anomalous propagation, and the season. However, more recent developments for the calibration of Z–R relationships, such as use of multi-radar, multi-sensor datasets has substantially improved the accuracy of rainfall estimates made from radar observation (e.g., Rivera-Giboyeaux, 2020). Nonetheless, the decisive advantage of using radar over a single point for precipitation measurements is the real-time monitoring of a wide area with high spatial and temporal continuity and resolution. Typically, the maximum range of radars to receive signals is 300 km. For hydrologic purposes, the effective range is usually 40–200 km. The **hydrologic range** is defined as the maximum range over which the relationship between the radar echo intensity and rainfall intensity remains reasonably valid. The WSR-88D units of the National Weather Service in the United States provide a coverage of a 230-km-radius circular area.

Presently, the density of weather radar networks is not uniform in the world. It is very good in Europe and North America. However, although the networks operated in North American and European countries by the national meteorological agencies adequately service the primary meteorological requirement of daily weather reporting and forecasting, the quantitative estimates of rainfall to support applications in hydrology and water resources, especially flood forecasting, are even here still limited, to various extents. The high temporal and spatial resolution of radar rainfall data provide the data necessary for gridded rainfall models that can be used in spatially distributed event-based hydrologic models. But they are less immediately suited for applications to design rainfall estimation.

Radar-based rainfall data are often extremely useful, particularly when there are no gauge data. For example, Mukhopadhyay et al. (2003) investigated a severe flash flood event that occurred on June 17, 1999, in the East Mesa area of Las Cruces, New Mexico, for the purpose of evaluating the performance of a constructed channel that was blamed for flooding several properties and causing considerable damage by erosion and mass movement of sediments. There was no rain gauge within the watershed that was most hard hit by that unusual rainstorm. So Mukhopadhyay et al. (2003) used NEXRAD-derived rainfall data for hydrologic modeling. All of the rain gauges that were present outside the watershed under investigation, but in surrounding areas, recorded a significantly lesser amount of rainfall to that calculated from the NEXRAD data covering the subject watershed. So, the radar data provided a more realistic picture of what actually happened in the most hard-hit watershed.

4.2.3 Rainfall Measurements by Satellites

With the advent of geostationary weather satellites in the 1960s and the 1970s, positioned above the equator at five to six locations around the globe to provide complete coverage, various techniques have been developed to estimate rainfall from visible and infrared (IR) radiation upwelling from the Earth into space. The higher the cloud albedo, the more droplets and/or ice crystals the cloud contains and the deeper it tends to be, so the more likely rainfall is on the ground. And the lower the IR brightness temperature, the higher the cloud top, and the more likely there is rainfall. A combination of both channels works best (Arkin and Ardanuy, 1989). The visible/IR rain-retrieval algorithms work best at low latitudes, because at higher latitudes the view is more slanted, confusion arises with high-albedo surfaces of snow or ice, and deep-convective precipitation is less common.

The rainfall rate, $R(\text{mm/h})$, depends on the cloud-top temperature, T (degrees kelvin), and a relationship is given as (Vicente et al, 1998)

$$R = 1.1183 \times 10^{11} \times \exp\left(-3.6382 \times 10^{-2} \times \sqrt{T}\right) \quad (4.2)$$

Adjustments are then made according to the precipitable water and surface relative humidity. The rate of change of cloud-top temperature can be used as well. It indicates the speed of cloud growth, and hence the areas of heavy rainfall. The major problem with measurements of rainfall from a satellite is the indirect relation between the rainfall on the ground and the measured signals at the satellite.

In November 1997, NASA launched the Tropical Rainfall Measuring Mission (TRMM) satellite. This satellite was in operation from 1997 to 2015 and carried passive microwave, infrared, and visible sensors, plus an active radar. The TRMM product (version 3B42) covers the tropics between 50° N and 50° S, with grid cells of spatial resolution of 0.25° by 0.25°. The NASA TRMM daily and 3-hourly rainfall products are available from 1998 to 2014. The Global Precipitation Measurement (GPM) is an international satellite mission to provide next-generation observations of rain and snow worldwide every 3 hours. NASA and the Japanese Aerospace Exploration Agency launched the GPM Core Observatory satellite on February 27, 2014, carrying advanced instruments that set a new standard for precipitation measurements from space.

In addition, a number of satellite-based daily 0.25° precipitation products are currently under development by several research groups. However, hydrologic applications typically require higher spatial and temporal resolutions. For example, sub-hourly to hourly data are typically required for flash-flood forecasting and control, daily data are needed for reservoir management, and weekly to monthly data are required for water resources management. The spatial scales required are determined by the sizes of watersheds and sub-watersheds, typically on the order of a few square kilometers to hundreds of square kilometers in size.

Satellite rainfall measurements can be used under the following circumstances: (1) When neither gauge nor radar data are available; (2) to get supporting evidence for model predictions from hydrologic models; and (3) in typical situations of large-scale events such as tropical or mid-latitude convection and storms moving on shore from sea, especially tropical cyclones.

4.3 TYPES OF RAINFALL

An **air mass** is a large body of air that is fairly homogeneous horizontally in temperature and moisture content. Vertical lifting of an air mass containing water vapor followed by **adiabatic cooling** and **condensation** in the upper atmosphere (cloud formation) is the primary cause of rainfall. According to the factor responsible for the lifting of the air mass, the majority of principal rainfall types are as follows. The majority of principal rainfall types can be categorized according to the factor responsible for the lifting of the air mass, and these are described in the sections below.

4.3.1 Convective Rainfall

The essence of convective rainfall is typified as a vertical column of air or convective cell consisting of: an inflow region near the Earth's surface, where warm moist air is drawn into the cell and expands by near-surface heating; an uplift region in the middle, where the warm or heated moisture-laden air mass rises by a density differential but condenses by dynamic cooling with altitude at lower temperature and thereby becomes unstable, producing vertical currents and rainfall; and an outflow region at the higher level of the atmosphere where cooler and drier air flows out from the column. The outflowing air can again descend, pick up moisture, gets heated at the lower level, and can again reenter the cell. Thus, convective rainfalls result from the *uplift of moisture-laden air mass due to convection.*

Convective rainfall can be light showers or very intense thunderstorms. The areal extent of such rainstorms is usually small, limited to a diameter of 5–10 km. Nonetheless, convective thunderstorms often are fairly violent and produce rainfall with very high intensities that is highly significant for the design of stormwater management measures, such as storm drainage systems and detention basins at local levels, particularly in urban settings. Examples of flooding caused by convective rainstorms abound all over the world.

4.3.2 Frontal Rainfall

Unequal heating of air masses by conductive and radiative heat transfer from the Earth's surface produces cold and warm air masses in different parts of a region. When a warm air mass meets a cold air mass, typically there is no mixing but a definite surface of discontinuity, known as a **front** or **frontal surface** occurs between them. A frontal surface is actually a layer or zone of transition between two air masses of different temperatures and moisture contents. Its horizontal width, however, is small with respect to dimensions of air masses. If the cold air mass advances toward the warm air mass, the leading edge of the cold air mass is called a **cold front** and is nearly vertical in slope. On the other hand, if the warm air mass advances toward a cold air mass, the leading edge is a **warm front**, which has a very flat slope, the warm air flowing up and over the cold air. A front can also be **stationary** if it is not in motion.

Frontal rainfall results from the lifting of warm air on one side of a frontal surface over colder and denser air on the other side of the contact zone. This is called **frontal lifting**. Heavy rainfall occurs at or near the front. **Warm-front rainfall** is formed in the warm air advancing upward over a colder air mass. The rate of ascent is relatively slow, since the average slope of the frontal surface is gentle, usually between 1/100 and 1/300 (Linsley et al., 1982). Rainfall can extend up to 300–500 km ahead of the surface front and is generally light to moderate and nearly continuous until the passage of the front. **Cold-front rainfall**, on the other hand, is showery in nature and is formed in the warm air forced upward by an advancing cold front with steep frontal surface slopes averaging 1/50 to 1/150. Cold fronts move faster than warm fronts. The warm air mass is forced upward more rapidly than the warm front. The rainfall rates are generally much higher, with the heaviest amounts and intensities occurring near the front.

An important *seasonal* rainfall that is of frontal type is **monsoonal** rainfall. This type of rainfall is characterized by seasonal reversal of winds that carry oceanic moisture with them and cause extensive rainfall. The summer monsoon is a very important type of rainfall in south and southeast Asia that includes the Indian subcontinent, Indochina peninsula, and southeast China. This is also known as the southwest monsoon, because winds blow from the southwest direction from the cooler Indian Ocean toward the warmer landmass of south and southeast Asia. Monsoons due to seasonal reversals of wind flow in other parts of the globe includes the North American monsoon in northwest Mexico to southwest United States, the South American monsoon that forms on the Atlantic side of Brazil to the Pacific side of Columbia and Peru, the West African monsoon in the northern hemisphere summer, and the East African monsoon during the southern hemisphere summer. Nevertheless, it should be borne in mind that a monsoon is not a single storm event, rather it is a seasonal wind shift over a region. However, a **monsoonal rain event** indeed often causes torrential rains that can last from a few hours to several days or even several weeks. Monsoonal rains are a major cause of flooding in many countries, such as the devastating flood of August 2010 in 30 provinces of Thailand, including the capital city of Bangkok, and Malaysia, and the 2017 flood in many states in India, including the overwhelming Mumbai flood of August.

4.3.3 Cyclonic Rainfall

A cyclone is a region of low pressure around which air flows in a counterclockwise direction in the northern hemisphere but clockwise in the southern hemisphere. The low-pressure region is surrounded by a region of larger pressure. Cyclonic rainfall is associated with the lifting of air masses converging from high-pressure regions to low-pressure regions. This is known as **uplift due to convergence**. The pressure differential is created by unequal heating of the Earth's surface. The low-pressure areas are loci of convergence. The air converging from several directions is forced to rise, and the rising produces adiabatic cooling, which leads to precipitation.

Cyclonic rainfall can be further classified as **frontal** or **nonfrontal**, depending on whether the convergence is due to frontal activity or not. Nonfrontal rainfall occurs due to production of a barometric low. In this case the air mass is lifted through horizontal convergence of the inflow into a low-pressure area. This is mainly convective rainfall and occurs on a large scale, such as a tropical cyclone, which, as the name implies, is mostly a tropical phenomenon that occurs within a mass of warm moist air (warm core). Cyclones due to frontal convergence, on the other hand, are characteristic of the mid-latitudes or a temperate climate and occur at the boundaries between air masses of contrasting temperature and humidity. These are also known as extratropical cyclones or winter storms.

Tropical cyclones form at low latitudes. There can be several variants with variable severity.

1. **Tropical depression**: this is the first stage of development of a tropical cyclone with winds that gust at 33 knots or less. This is also known as a **tropical wave** or **disturbance**. This can bring significant amounts of rain and thunderstorms.
2. **Tropical storm**: the next stage of development of a tropical cyclone is a tropical storm when the cyclone's circulation is more organized and has sustained wind speeds of 34–63 knots. Tropical storms are further categorized according to Saffir–Simpson scale. These storms produce large amounts of rain.
3. **Hurricane**: when a storm system has a sustained wind speed over 64 knots and a spiral arrangement of thunderstorms with low-pressure centers, it is called a **hurricane** or **typhoon**. It has a central region with relatively little cloud and a light wind called the eye. Usually, the storms over the Atlantic Ocean are called hurricanes and the ones over the Pacific Ocean are called typhoons. These are the most devastating types of tropical weather. The Earth's rotation causes hurricanes to accelerate toward the poles if a current does not steer them, a force known as Coriolis effect. According to the Saffir–Simpson scale, hurricanes are classified from category 1 to category 5. Hurricanes or typhoons form over the ocean, often beginning as a tropical wave, which is a low-pressure area that moves through the moisture-rich tropics. Because these systems are fed by the energy from a warm-water environment, they dissipate when they move over land or cold water (Singh, 1992). Tropical storms and hurricanes or typhoons often cause major flooding inland near the coast such as Category 4 Hurricane Harvey of August 2017, which caused unprecedented flooding and damage in the Houston metropolitan area and southeast Texas and Category 5 Hurricane Katrina of August 2005, which caused destruction and havoc of historical magnitude in the city of New Orleans and the surrounding areas of Louisiana.

Extratropical cyclones are formed when warm and cold air masses, initially flowing horizontally in opposite directions adjacent to one another, begin to interact and whirl together in a circular motion, creating both a warm front and a cold front centered on a low-pressure zone (cold core), that is, forming an **occluded front**. A front is said to be occluded where the cold front overtakes the worm front, so that there is colder air everywhere at the surface with warm air above. Extratropical cyclones can cause heavy rainstorms and snowstorms, which in some situations are known as **blizzards**.

4.3.4 Orographic Rainfall

Orographic rainfall occurs when the warm and most air mass near the Earth's surface gets mechanically uplifted due to the presence of topographic barriers, principally mountains. This is known as **uplift due to orography**. As the warm and moisture-laden air mass rises, it undergoes cooling and condensation at higher altitude, causing rainfall. Rainfall is heavier on the windward slopes and lighter on leeward slopes.

Unusual orographic rainfall incidents have also been causes of flooding in many instances. For example, the June 1999 flood event of Las Cruces, New Mexico, mentioned above was primarily an orographic rainfall due to the effect of the Organ Mountains of the Las Cruces area.

Any of the storms noted above can also be classified in accordance with its spatial and temporal characteristics. A **microscale** event is defined by convective cells of short duration and small spatial extent such as 5 km in diameter. A **mesoscale** event consists of a unit of developing cells, each at various stages of development but all in unison motion in a preferred direction. Small mesoscale units typically extend over 150–400 km^2 and large mesoscale events can cover areas as large as 2300–4700 km^2 (Bras, 1990). A thunderstorm, consisting of a conglomerate of active cells, can occur at the mesoscale. A hurricane, with rotating arms extending with a large areal extent, also corresponds to a mesoscale event. The agglomeration of mesoscale units forms an event at the **synoptic scale** that can cover several thousand square kilometers. A typical hurricane with several hundred kilometers, such as 500 km, in diameter is an event at a synoptic scale.

A well-documented case of extensive flooding over a large area due to orographic rainfall originating from a synoptic system is the flood in southern California during February 27 to March 3, 1938 (Troxell, 1942). A storm centered along the western slopes of the San Bernardino and San Gabriel Mountain areas, ranging in altitude from 1000 feet to 11 500 feet, caused a series of heavy rainstorms in the coastal area of southern California encompassing the City of Los Angeles and the City of San Diego, producing an average of 22.5 inches (571.5 mm) of rain. Figure 4.2 shows the locations of the standard gauges from where rainfall data were collected (this figure does not include the area and the gauges to the further south near the San Diego area, which also experienced this storm event). As per the analysis of Pierce (1938), the storm originated in Siberia, circled southward over the Pacific Ocean, swinging eastward near Hawaii, and thence to the California coast. In its long, encircling course over the

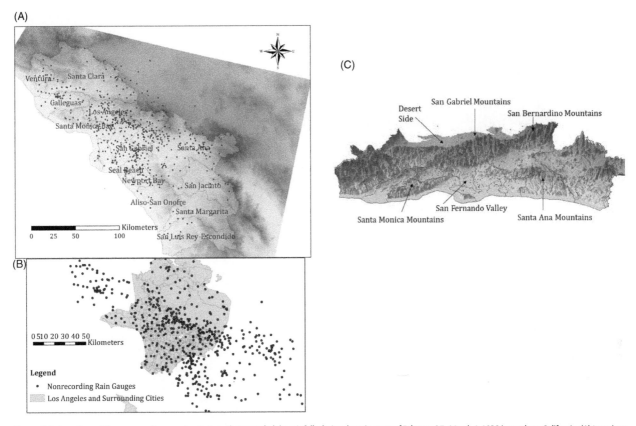

Figure 4.2 Locations of the nonrecording gauging stations that recorded the rainfalls during the rain event of February 27–March 4, 1938 in southern California. (A) Locations of the rain gauges in relation to the watersheds on the western side of the mountains that separate the coastal area from Mojave Desert. (B) Locations of the rain gauges in relation to the City of Los Angeles and surrounding areas. (C) Locations of the rain gauges in relation to the topographic features that controlled the rainfall. The digital elevation model shown in C is created from the National Elevation Dataset provided by USGS. The watershed boundaries shown in A are from the National Hydrography Dataset provided by USGS. (A black and white version of this figure will appear in some formats. For the color version, please refer to the plate section.)

Pacific Ocean, the storm absorbed a great amount of moisture and, when it reached southern California, the air masses swept at almost right angles to the mountain ranges. As the moist air was thrust upward by the high mountains that stood as barriers, it encountered the colder upper air. The resulting rapid condensation caused excessively heavy rainfall on the west side of the mountains, but rainfall decreased sharply toward the Mojave Desert area on the east side of the mountain ranges. Figure 4.3 shows the isopluvial (line of equal rainfall or rainfall contours) map of the area, constructed from four to five days (it varied from location to location) of rainfall totals given by Troxell (1942). The close match between rainfall profiles and elevation profiles shown in Figure 4.3B clearly demonstrate the orographic effects. On the coastal side of the mountains, rainfall increased with elevation but much of the moisture was pulled out of the air mass by the time it reached the desert side of the ranges, thereby pouring a significantly lesser amount of rainfall on the desert side of San Gabriel and San Bernardino mountains.

The above provides a simplified summary of the meteorological processes that cause rainfall. More in-depth knowledge about the physics of each of these processes as well as various subprocesses within those can be gained from a good textbook on meteorology (e.g., Holton and Hakim, 2013).

A basic understanding of types of rainfall is important for hydrologic engineering, because different types of rainstorms produce different types of rainfall records, as shown by the schematic illustration in Figure 4.4. This theme constitutes the basis of much of what will be covered in subsequent sections of this chapter.

4.4 DESCRIPTIVE STATISTICS OF RAINFALL

The meteorological characterization of a rainstorm is not of much use for hydrologic engineering. For engineering purposes, rain events are described by their **exterior** and **interior** statistics. The exterior of a rainstorm is defined by the **depth** of rainfall (P) and the **duration** (T_D) of the rainfall. It also includes the time between storms if there are multiple pulses of rainfall during a storm event.

The **intensity** (I) of a rainstorm is the time rate of rainfall, that is, the depth of rainfall per unit time (e.g., mm/h or in/h). It can be either instantaneous intensity (I) or the average intensity (I_{ave}) over the duration of rainfall. The average intensity is commonly used and can be expressed as

$$I_{ave} = \frac{P}{T_D} \qquad (4.3)$$

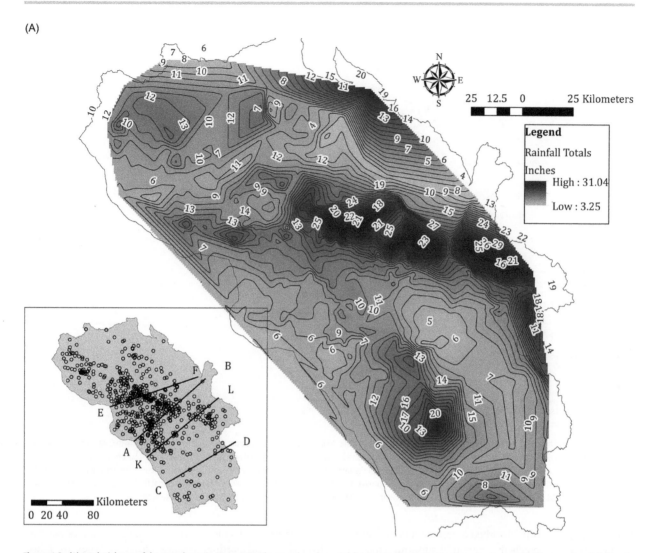

Figure 4.3 (A) Isopluvial map of the area shown in Figure 4.2B. These are based on rainfall totals for the four to five days of rain event. In this map, the rainfall totals are given in inches taken from the original source. The inset shows the rainfall and elevation profiles along four lines. (A black and white version of this figure will appear in some formats. For the color version, please refer to the plate section.) (B) Rainfall profiles are shown by lines with filled circles (dots) and elevation profiles are shown by lines without any symbol.

The interior of a storm refers to the **spatial** and **temporal** distribution of depths throughout the duration of a rain event. These characteristics of rainfall are probabilistic in nature but are not statistically independent. For example, a spatial correlation exists between point rainfall depths, which vary with duration. The most significant aspect of these characteristics, for hydrologic engineering, is that they depend on the climatic characteristics of a location or region.

4.5 SPATIAL DISTRIBUTION OF OBSERVED RAINFALL

4.5.1 Mean Areal Precipitation

At any spatial scale, there is always great variation of rainfall. However, in certain studies, an estimation of the mean depth of rainfall over an area for a given duration of a rain event is required. This is called the **mean areal precipitation** (MAP). For example, in some hydrologic modeling, it is sufficient to specify the MAP over a watershed. The required watershed rainfall depth is inferred from the measured rainfall depths at multiple gauges using an averaging scheme that uses the general formula

$$P_{MAP} = \frac{\sum_i \left[w_i \sum_t p_i(t) \right]}{\sum_i w_i} \quad (4.4)$$

where P_{MAP} is the total storm MAP over the watershed; $p_i(t)$ is the rainfall depth measured at time t at gauge i; and w_i represents a weighting factor assigned to gauge (observation point) I $\left(\sum_i w_i = 1 \right)$. If gauge i is not a recording gauge, only the quantity $\sum p_i(t)$, which represents the storm total precipitation (STP), is available and is used in Eq. 4.4. The time scale for MAP can be any duration such as a day, a month, or a year.

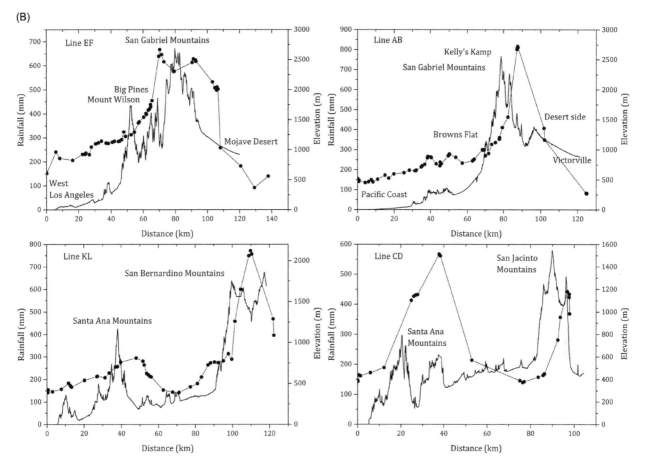

Figure 4.3 (cont.)

There are a variety of ways by which MAP can be estimated from multiple gauge data. Four common methods for the estimation of MAP according to the way the weighting factor of Eq. 4.4 is determined, are detailed below.

4.5.2 Arithmetic Mean Method

This is the simplest method and assigns a weight to each gauge equal to the reciprocal of the total number of gauges used for the computation of MAP. Gauges in or adjacent to the watershed can be selected. The method is applicable if the gauges are uniformly distributed in a watershed and there is no great variation in individual gauge measurements.

4.5.3 Thiessen Polygon Method

This is an area-based weighting scheme that was proposed by the American meteorologist Alfred Thiessen (1911). The concept was based on the Voronoi diagram developed by the eighteenth-century Russian mathematician Gregory Voronoy. This is an essential method for the analysis of proximity and neighborhood. In the case of MAP estimation, it assumes that the rainfall depth at any point within a watershed is the same as the rainfall depth at the nearest gauge in or near the watershed and thereby in this method a weight is assigned to each gauge in proportion to the area of the watershed that is closest to that gauge.

The method entails a graphical construction of Thiessen polygons or the areas that surround each gauge. First, the gauges are connected by straight lines and then perpendicular bisectors of each of these lines are drawn. These, along with the watershed boundary, form the polygons surrounding each gauge. The area within each polygon is nearest to the enclosed gauge. Thereby, the weight assigned to each gauge is the fraction of the total area of the watershed that the polygon represents. The MAP is then calculated from

$$P_{MAP} = \frac{1}{A}\sum_{i}^{NG} A_i \left[\sum_{j}^{NT} p(t)_j\right] \quad (4.5)$$

where A is the area of the watershed and A_i is the area of the polygon that encloses gauge i, NG is the total number of gauges, which is equal to the number of Thiessen polygons formed, and NT is the final time step of rain recording. Note that

$$P_i = \sum_{j}^{NT} p(t)_j \quad (4.6)$$

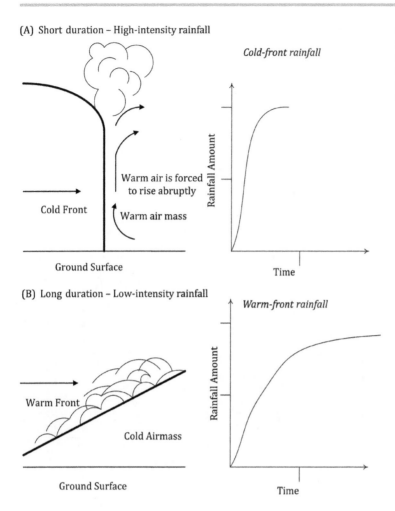

Figure 4.4 Schematic illustration of conditions for two types of storms and the resulting rainfall records. In the case shown in (A) a thunderstorm is common whereas in that shown in (B) day-long drizzle is common. The rainfall records show the same amount of rainfall for these two different types of rain events occurring in different time windows.

4.5.4 Isohyetal Method

This is also an area-based weighting scheme. First, contour lines of equal rainfall are either estimated or calculated by the interpolation of point rainfall measurements at gauges. The contours of equal rainfall depth are called **isohyets.** Manual construction of the isohyets requires proper judgment and knowledge of the watershed but is a preferred way when there are a limited number of rain gauges present in and around a watershed. On the other hand, if there is a fairly dense network of rain gauges, **isohyetal maps** can also be constructed using computer programs or GIS technology for automated contouring. Automated contouring requires choosing one of the various mathematical models that can be used for interpolation. The mathematical model can be as simple as linear interpolation between two points or assigning weights to the interpolation points according to the inverse of the distance, raised to a power (typically it is 2), between the point of interpolation and the point with a known value. More complex models, ranging from spline interpolation to advanced geostatistical methods, chiefly kriging, or trend surface analysis, can also be applied, depending on the density of the gauging points. The selection of correct geostatistical models requires a proper understanding of the mathematical principles involved in a method whose aptness depends on the geospatial characteristics of rainfall distribution. Details of common geostatistical models can be found in Isaaks and Srivastava (1989). In any event, the choice of a model for automated contouring requires proper judgment and understanding of the methods involved in the models and the nature or degree of complexity of the spatial variation of the rain event.

Once the isohyetal map of a watershed for a rainfall event is constructed, the MAP is estimated by finding the average rainfall depth between each pair of isohyets instead of rainfall at individual gauges and weighting these depths by the fraction of the total area enclosed by the pair of isohyets. Equation 4.5 is used in this case also, but here A_i represents the area between each pair of isohyets and P_i (Eq. 4.6) is the average rainfall depth between the bounding isohyets.

4.5.5 Inverse-Distance-Squared Method

In this method, the influence of rainfall at a gauged point on the computation of rainfall at an ungauged point is inversely proportional to the square of the distance between these two points. Since this method estimates rainfall at a point from observations made at several other points, it does not truly provide an estimate of the MAP for a watershed (it can be used as an interpolation technique as noted above). However, the technique is often used for the estimation of the MAP, by taking the *centroid* of a watershed as the ungauged point with the assumption that the estimate made at this point adequately represents the spatially lumped rainfall over the entire watershed. This is an acceptable approach since in many hydrologic models, point estimates of rainfall are assumed to be uniformly distributed over the entire watershed. In this method, the weight for a gauge i is calculated as

$$w_i = \frac{\frac{1}{d_i^2}}{\sum_i \left(\frac{1}{d_i^2}\right)} \quad (4.7)$$

where d_i is the distance between gauge i and the computation point.

Selected rainfall data from Tropical Storm Allison, one of the most devastating rain events that occurred over a large economic and populous urban hub in the United States, are used to illustrate the methods described above. Tropical Storm Allison had a complex meteorological history. Simply put, originally it started as a tropical wave off the west coast of Africa on May 21, 2001, and travelled westward. It was still a mere disturbance off the Gulf of Mexico on June 4 but moved inland and significant rainfall over much of Houston and surrounding parts of Harris County occurred on June 5. The storm weakened by becoming a tropical depression and moved further inland. But the storm looped back over Harris County on June 8, drew heavy moisture from the Gulf of Mexico and poured torrential rain from the afternoon of June 8 to the morning of June 9. The highest recorded rainfall total in Harris County during this 5-day event was 38 inches (965.2 mm). Tropical Storm Allison left Harris County, Texas, with 22 fatalities, 95 000 damaged automobiles, 73 000 damaged residents, 30 000 stranded residents in shelters, and over $5 billion (2001 estimate) in property damage.

Figure 4.5 shows the 24-hour STP for June 5, 2001, rainfall at the recording gauges within Greens Bayou Watershed, located in the northeast and central parts of Harris County that covers a significant portion of the City of Houston. Greens Bayou Watershed experienced the most severe rainfall. Up to 28 inches of STP was recorded in a 12-hour period from June 8 to June 9.

The MAP, calculated as the arithmetic mean of the gauge STP, is 5.95 inches (Figure 4.5A). The Thiessen polygon method shown in Figure 4.5B gives a MAP of 5.90 inches. The close match in the estimates of the MAP by these two methods is due to the nature of spatial coverage and density of the rain gauge network within the watershed.

For the construction of isohyets, advantages are taken to include the data from a few additional gauges that were present around the Greens Bayou watershed. Based on these additional rain gauges, the isohyets for the June 5, 2001 STP, using inverse-distance weighting (IDW) and by spline methods, are shown in Figure 4.6A and B. The MAP obtained by these two methods is 5.93 and 5.88 inches, respectively. Figures 4.7A and B show the isohyets constructed by using ordinary kriging method and by fitting a trend surface. A spherical semivariogram model is used in the kriging method. For the trend surface model, a third-degree surface is used. The MAP obtained by these two methods is 5.81 and 5.87 inches, respectively. Thus, all the contouring methods, when done carefully, yield similar results as far as MAP is concerned. However, the isohyetal patterns vary, depending on the method employed in the construction of the isohyets, particularly if a trend surface model is used without the presence of a real spatial trend. For example, the first three methods, in the example presented here, show that there are two regions of high rainfall within the watershed. The spline method defines these areas more closely relative to the inverse-distance-squared and ordinary kriging methods. On the other hand, the trend surface model indicates that the rainfall increased from west to east and the area with the highest rainfall was to the extreme southeast of the watershed. If a first-degree surface would be chosen, then it would be equivalent to a simple linear interpolation method.

Another factor that has a strong influence on the isohyetal pattern is the density of the gauge network, that is, the number of gauges per unit area. Huff and Neill (1957) demonstrated this fact by constructing a series of isohyetal maps for the same storm event on a watershed in central Illinois by selectively choosing various gauge densities from an existing network of rain gauges.

Singh and Chowdhury (1986) evaluated the various methods for the computation of MAP and concluded that all the methods provided comparable results, especially when the time period was long. The results from different methods varied more when applied to short-duration rainfall events.

The accuracy of areal average rainfall estimates from a system of point rain gauge measurements, i.e., a rain gauge network, depends on the distance between a gauge and the point of application of the estimate, and the arrangement of the gauges in the network. Huff and Neill (1957) conducted a pioneering investigation, based on seven dense rain gauge networks in central Illinois, to determine various aspects of the spatial variability of rainfall. One of their findings showed that areal estimates of rainfall linearly decrease with watershed area. However, they noted that such linear relationships may not hold true for very large basins. From an extensive analysis, they also developed quantitative measures of rainfall estimates at arbitrary points at certain distances from gauged points within areas ranging from 19 acres to 400 square miles and having a dense network of rain gauges. Their procedures can be used to develop empirical equations to calculate rainfall depth at an ungauged point within a network from the measured rainfall depth at a gauging point located at a known distance from the ungauged point. However, great caution must be

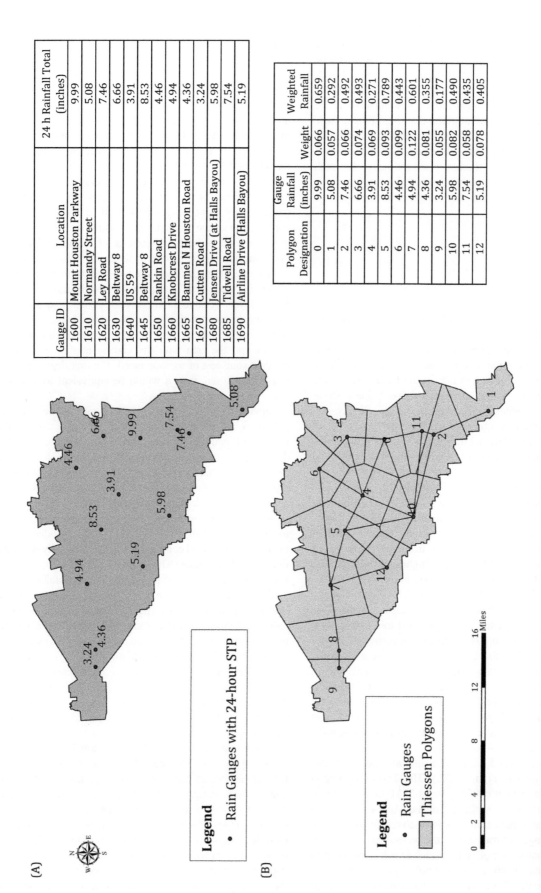

Figure 4.5 (A) June 5, 2001, 24-hour storm total precipitation (STP) at the recording rain gauges within Greens Bayou Watershed, Harris County, Texas. MAP estimated by the arithmetic mean (5.95 inches). (B) Thiessen polygons (with polygon designations) constructed from the rain gauges shown by solid dots within each polygon. Calculation of MAP by Thiessen polygon method is shown in the table to the right of the map. MAP = 5.90 inches.

4.6 Temporal Distribution of Observed Rainfall

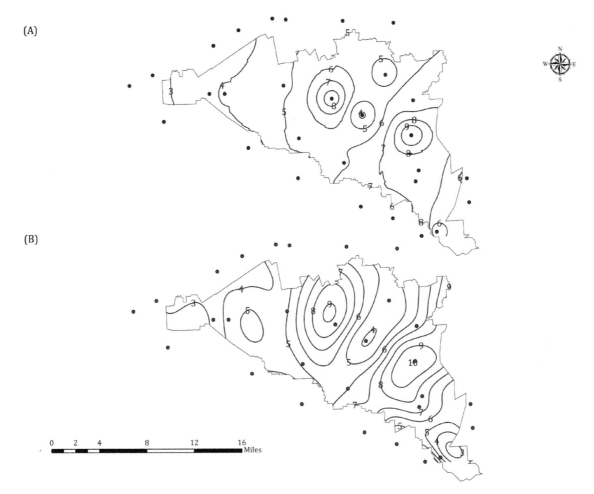

Figure 4.6 (A) Isohyets constructed from the data shown in Figure 4.5 with additional rain gauges surrounding Greens Bayou Watershed (solid dots) using the inverse-distance weighting (IDW) method (in this case the square of the inverse distance is used). (B) Isohyets constructed by the spline method using the same data shown in A.

exercised in areas where there are pronounced orographic effects on rainfall.

4.6 TEMPORAL DISTRIBUTION OF OBSERVED RAINFALL

4.6.1 Representation of Temporal Variation of Rainfall

When observed rainfall data come from recording gauges, the time distribution of rainfall is represented by two types of graphical procedures, called **hyetographs** and **mass curves** (or **rainfall curves**). Leaving aside the theoretical aspects of storm structure, quantification of the time distribution of rainfall is extremely important for two practical purposes: (1) it is necessary to have this information for the computation of hydrographs (Chapter 9) and (2) it has a significant effect on the shape of the hydrographs, particularly the peak discharge.

The recording gauges provide rainfall depths at specific intervals, such as 5-minutes, 15-minutes, 1-hour, etc. This is called incremental rainfall. Table 4.1 gives the incremental rainfall values at Gauge 1600 of Harris County from 4:30 P.M. on June 8 to 6:00 A.M. June 9 (2001) during which

Tropical Storm Allison poured 28.51 inches of rain in 13.5 hours at this gauge location.

4.6.2 Hyetographs

Recording rain gauges provide a set of rainfall depths recorded at successive increments of time, as shown in Table 4.1. A rainfall hyetograph is simply a plot of either depth or intensity of rainfall as a function of time, typically presented as a bar diagram as shown in Figure 4.8 for the data given in Table 4.1.

4.6.3 Mass Curves, Percentage Mass Curves, and Huff Curves

By summing the incremental rainfall amounts through successive time intervals, a cumulative distribution curve giving total rainfall depth up to the end of a time period is obtained. This is called a **mass curve** or rainfall curve (Figure 4.9). The derivative (slope) of this curve gives the rainfall

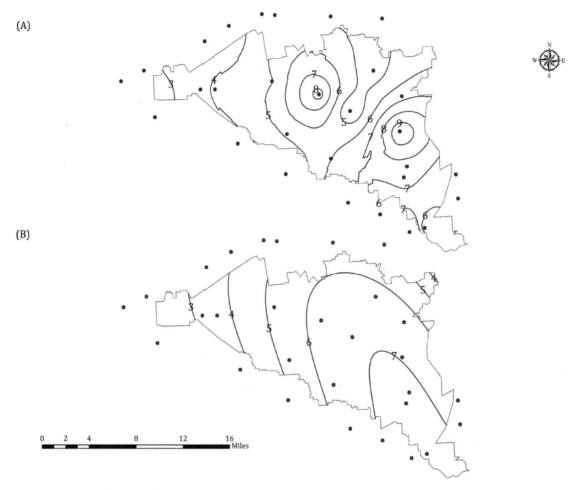

Figure 4.7 (A) Isohyets constructed from the data shown in Figure 4.5 with additional rain gauges surrounding Greens Bayou Watershed (solid dots) using the ordinary kriging method (the semivariogram model is spherical). (B) Isohyets constructed by a third-degree trend surface model using the same data shown in A.

intensity. Since the rainfall distribution represents the cumulative rainfall from the beginning to the end of the storm, the resulting variation of the intensities is the hyetograph. A mass curve of a recorded rain event readily provides information on storm duration, total depth of rainfall, and the average rainfall, as well as the maximum rainfall intensity for any certain time interval.

The maximum depth recorded in a specified time interval gives the maximum intensity for that duration. That provides an index of how severe a particular rain event is, compared to other events recorded at the same location. Such data are useful for the design of flow control structures. The maximum rainfall intensity recorded in a chosen time interval is found by computing a series of running totals for the rainfall depth for that time interval starting at various points in the storm, then selecting the maximum value of this series. Figure 4.9 shows that the maximum depths for durations of 30 minutes, 1 hour, 2 hours, and 3 hours of the storm recorded at Gauge 1600 (Table 4.1) are 2.66, 4.94, 9.31, and 13.44 inches, respectively (these are maximum precipitation values for the periods corresponding to absolute times). Thus, the maximum intensities for these durations are 5.32, 4.94, 4.65, and 4.48 in/h, respectively. On the other hand, the average intensity of an arbitrary 30-minute interval, as shown in Figure 4.9, is 5.18 in/h. The average intensity for any specified duration varies throughout the storm, as shown in the inset of Figure 4.9. Thus, unless it is necessary to know the average intensity for a specific time window of a storm, the maximum intensities for various durations provide more meaningful information about the exterior characteristics of a rainstorm event.

A mass curve can also be expressed as a **percentage mass curve** where each cumulative rainfall amount is expressed as percent of the total rainfall. Thus, the percentage mass curve is a plot of normalized cumulative depth against normalized duration. Percentage mass curves offer the opportunity to compare the nature of variation of temporal distributions of various types of rainfall. This is because rainfall resulting from different types of storms, such as stratiform precipitation, convective storms, cyclones, etc., each has distinct characteristics of duration, intensity, and spatial extent, producing different types of temporal patterns of rainfall, which can be observed from

4.6 Temporal Distribution of Observed Rainfall

Table 4.1 *Rainfall depths (inches) recorded at Gauge 1600 within Greens Bayou Watershed in Houston during Tropical Storm Allison*

Date Time	Incremental rainfall	Cumulative rainfall
6/8/01 16:30	0	0
6/8/01 16:35	0	0
6/8/01 16:40	0	0
6/8/01 16:45	0.04	0.04
6/8/01 16:50	0	0.04
6/8/01 16:55	0	0.04
6/8/01 17:00	0	0.04
6/8/01 17:05	0	0.04
6/8/01 17:10	0.08	0.12
6/8/01 17:15	0	0.12
6/8/01 17:20	0.08	0.2
6/8/01 17:25	0.08	0.28
6/8/01 17:30	0.04	0.32
6/8/01 17:35	0.28	0.6
6/8/01 17:40	0.28	0.88
6/8/01 17:45	0.2	1.08
6/8/01 17:50	0.12	1.2
6/8/01 17:55	0.04	1.24
6/8/01 18:00	0.04	1.28
6/8/01 18:05	0.04	1.32
6/8/01 18:10	0.04	1.36
6/8/01 18:15	0.08	1.44
6/8/01 18:20	0	1.44
6/8/01 18:25	0.08	1.52
6/8/01 18:30	0.24	1.76
6/8/01 18:35	0	1.76
6/8/01 18:40	0.2	1.96
6/8/01 18:45	0.12	2.08
6/8/01 18:50	0.12	2.2
6/8/01 18:55	0.04	2.24
6/8/01 19:00	0.08	2.32
6/8/01 19:05	0.16	2.48
6/8/01 19:10	0.28	2.76
6/8/01 19:15	0.12	2.88
6/8/01 19:20	0.47	3.35
6/8/01 19:25	0.2	3.55
6/8/01 19:30	0.39	3.94
6/8/01 19:35	0.12	4.06
6/8/01 19:40	0.28	4.34
6/8/01 19:45	0.16	4.5
6/8/01 19:50	0.16	4.66
6/8/01 19:55	0.08	4.74
6/8/01 20:00	0	4.74
6/8/01 20:05	0.12	4.86
6/8/01 20:10	0.08	4.94
6/8/01 20:15	0	4.94
6/8/01 20:20	0	4.94
6/8/01 20:25	0.24	5.18
6/8/01 20:30	0.39	5.57
6/8/01 20:35	0.39	5.96
6/8/01 20:40	0.28	6.24
6/8/01 20:45	0.39	6.63
6/8/01 20:50	0.35	6.98
6/8/01 20:55	0.16	7.14
6/8/01 21:00	0.08	7.22
6/8/01 21:05	0.31	7.53

Table 4.1 *(cont.)*

Date Time	Incremental rainfall	Cumulative rainfall
6/8/01 21:10	0.12	7.65
6/8/01 21:15	0.28	7.93
6/8/01 21:20	0.32	8.25
6/8/01 21:25	0.08	8.33
6/8/01 21:30	0.16	8.49
6/8/01 21:35	0.04	8.53
6/8/01 21:40	0	8.53
6/8/01 21:45	0	8.53
6/8/01 21:50	0	8.53
6/8/01 21:55	0.08	8.61
6/8/01 22:00	0.08	8.69
6/8/01 22:05	0.08	8.77
6/8/01 22:10	0	8.77
6/8/01 22:15	0	8.77
6/8/01 22:20	0.12	8.89
6/8/01 22:25	0	8.89
6/8/01 22:30	0	8.89
6/8/01 22:35	0.28	9.17
6/8/01 22:40	0.24	9.41
6/8/01 22:45	0.2	9.61
6/8/01 22:50	0.35	9.96
6/8/01 22:55	0.35	10.31
6/8/01 23:00	0.47	10.78
6/8/01 23:05	0.39	11.17
6/8/01 23:10	0.43	11.6
6/8/01 23:15	0.55	12.15
6/8/01 23:20	0.47	12.62
6/8/01 23:25	0.35	12.97
6/8/01 23:30	0.28	13.25
6/8/01 23:35	0.43	13.68
6/8/01 23:40	0.51	14.19
6/8/01 23:45	0.28	14.47
6/8/01 23:50	0.24	14.71
6/8/01 23:55	0.47	15.18
6/9/01 0:00	0.12	15.3
6/9/01 0:05	0.24	15.54
6/9/01 0:10	0.63	16.17
6/9/01 0:15	0.12	16.29
6/9/01 0:20	0.39	16.68
6/9/01 0:25	0.35	17.03
6/9/01 0:30	0.39	17.42
6/9/01 0:35	0.35	17.77
6/9/01 0:40	0.2	17.97
6/9/01 0:45	0.39	18.36
6/9/01 0:50	0.39	18.75
6/9/01 0:55	0.59	19.34
6/9/01 1:00	0.16	19.5
6/9/01 1:05	0.75	20.25
6/9/01 1:10	0.35	20.6
6/9/01 1:15	0.32	20.92
6/9/01 1:20	0.35	21.27
6/9/01 1:25	0.2	21.47
6/9/01 1:30	0.35	21.82
6/9/01 1:35	0.28	22.1
6/9/01 1:40	0.28	22.38
6/9/01 1:45	0.47	22.85
6/9/01 1:50	0.55	23.4
6/9/01 1:55	0.35	23.75

Table 4.1 (cont.)

Date Time	Incremental rainfall	Cumulative rainfall
6/9/01 2:00	0.35	24.1
6/9/01 2:05	0.35	24.45
6/9/01 2:10	0.59	25.04
6/9/01 2:15	0.28	25.32
6/9/01 2:20	0.63	25.95
6/9/01 2:25	0.39	26.34
6/9/01 2:30	0.31	26.65
6/9/01 2:35	0.39	27.04
6/9/01 2:40	0.24	27.28
6/9/01 2:45	0	27.28
6/9/01 2:50	0.28	27.56
6/9/01 2:55	0	27.56
6/9/01 3:00	0	27.56
6/9/01 3:05	0	27.56
6/9/01 3:10	0.39	27.95
6/9/01 3:15	0	27.95
6/9/01 3:20	0	27.95
6/9/01 3:25	0	27.95
6/9/01 3:30	0	27.95
6/9/01 3:35	0	27.95
6/9/01 3:40	0	27.95
6/9/01 3:45	0	27.95
6/9/01 3:50	0.24	28.19
6/9/01 3:55	0	28.19
6/9/01 4:00	0.04	28.23
6/9/01 4:05	0.04	28.27
6/9/01 4:10	0	28.27
6/9/01 4:15	0.04	28.31
6/9/01 4:20	0	28.31
6/9/01 4:25	0	28.31
6/9/01 4:30	0	28.31
6/9/01 4:35	0	28.31
6/9/01 4:40	0	28.31
6/9/01 4:45	0	28.31
6/9/01 4:50	0.08	28.39
6/9/01 4:55	0	28.39
6/9/01 5:00	0	28.39
6/9/01 5:05	0	28.39
6/9/01 5:10	0.04	28.43
6/9/01 5:15	0	28.43
6/9/01 5:20	0	28.43
6/9/01 5:25	0	28.43
6/9/01 5:30	0.04	28.47
6/9/01 5:35	0	28.47
6/9/01 5:40	0	28.47
6/9/01 5:45	0	28.47
6/9/01 5:50	0	28.47
6/9/01 5:55	0	28.47
6/9/01 6:00	0.04	28.51

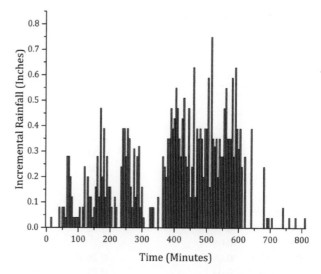

Figure 4.8 Hyetograph showing incremental rainfall at Gauge 1600 (Greens Bayou at Mount Houston Parkway) for the period of most intense rainfall on the final day of Tropical Storm Allison over Harris County, Texas.

the percentage mass curves. For example, convective thunderstorms typically exhibit convex upward curves, whereas tropical cyclones give rise to S-shaped curves. Figure 4.10 shows the contrasting types of percentage mass curves resulting from three distinct types of rain events recorded in a geographical area (Raleigh-Durham region of North Carolina). The summer thunderstorms are mostly due to developments of convective cells, whereas the rain during fall can be warm-front rains when warmer air from the Atlantic (ocean water does not cool rapidly from summer to fall) rides over the cold air over the land.

Patterns of temporal distributions of rainfall are often necessary for hydrologic design. For example, during the design of developments of Downtown South area of Raleigh, North Carolina, it was necessary to develop temporal rainfall distribution patterns resulting from typical short-duration summer thunderstorms for the purpose of stormwater management. In this case, mass curves were produced for recent rainfall records for five summer thunderstorm events from four recording gauges within the watershed of concern shown in Figure 4.10. From these, a best-fit curve (Figure 4.11) was obtained as the representative mass curve for summer convective thunderstorms in the Raleigh area of North Carolina.

The example cited above is for the design where the local site-specific rainfall pattern needs to be considered in the absence of better information. However, a rainfall distribution curve derived as a best-fit curve from a limited set of observations may not be very reliable in conveying the intricate distribution characteristics of heavy rainstorms as it does not provide the measures of the varying types of storm profiles that occur in nature or of the inter-storm variability of a particular type of profile, both of which are important factors in the application of temporal distributions. Huff (1967, 1990) and Huff and Angel (1989) presented a comprehensive statistical analysis of time distribution characteristics of heavy rainfalls, based on rainfall data from 261 individual storms with durations ranging from 1 hour

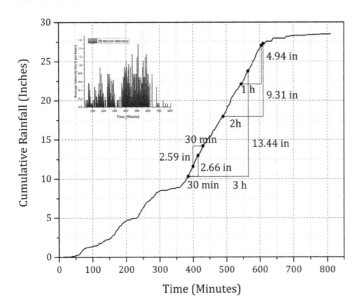

Figure 4.9 Mass curve or cumulative rainfall for the data shown in Table 4.1 and the hyetograph shown in Figure 4.8. Examples of maximum intensities for 30 minutes, 1 hour, 2 hours, and 3 hours are shown. On the other hand, the average intensity for a given time slice and for a selected time window can be computed using the incremental depth for that time slice, as shown by another example of the 30-minute window with an incremental depth of 2.59 inches. The inset figure shows variations of intensities for a 30-minute duration.

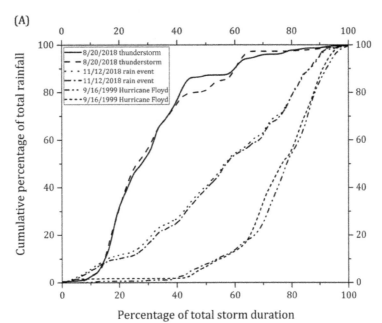

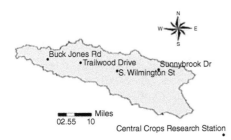

Event Date	Event Type	Recording Station	Duration (hours)	Depth (inches)	Depth (mm)
August 20 2018	Thunderstorm	Buck Jones Road	4	2.28	57.9
August 20 2019	Thunderstorm	Trailwood Drive	4	3.74	95.0
November 12 2018	Long Rain	Buck Jones Road	15.1	3.52	89.4
November 12 2018	Long Rain	Trailwood Drive	15.3	3.37	85.6
September 14-16 1999	Hurricane	RDU International Airport	50	6.49	164.8
September 14-16 2000	Hurricane	Central Crops Research Station	50	7.33	186.2

Figure 4.10 (A) Contrasting types of percentage mass curves for three types of rainfall events recorded at the same location. The location selected for this illustration is Raleigh-Durham area of North Carolina, United States. (B) The locations of the gauging stations from where records are derived. The rain gauges are located within Walnut Creek Watershed as shown in this figure. Rainfall duration and depth for each of the events are given in the table insert.

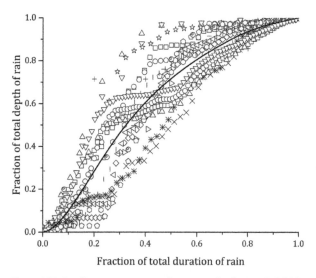

Figure 4.11 Best-fit mass curve representing summer thunderstorm in Raleigh, North Carolina, United States, constructed from the rainfall records from four rain gauges shown in Figure 4.10 and from five summer thunderstorm events (possibly all convective systems). The best-fit curve is shown as a solid line whereas the data for the events from the individual gauges are shown by different symbols.

to 48 hours, that occurred during a 12-year period (1955–1966) on a 400-square mile area with a network of 49 recording rain gauges in east-central Illinois. The rainstorms were mostly typical mid-continental thunderstorms, which were categorized as first-, second-, third-, and fourth-quartile storms depending on whether the greatest percentage of STP occurred in the first-, second, third-, or fourth quarter of the storm period. Within each category, storms were further classified according to their frequency or probability of occurrence from 10% to 90%. The statistical models were smooth curves, called **storm profiles**, reflecting the average temporal distribution of rainfall without exhibiting the burst characteristics of observed storms (Figure 4.12). This type of mass curve, derived empirically, is referred to as a **Huff curve**, honoring Floyd A. Huff who first produced curves like these, which convey varying characteristics of temporal patterns of rainfall. These curves provide the quantitative measures of the variabilities that occur in the storm profiles belonging to different quartile distributions as well as the variabilities that can occur in the storm profiles for a certain quartile classification but having a different probability of occurrence. For example, Figure 4.12A indicates that there is a 50% chance for all first-quartile storms to have 50% of its rainfall total within the first 20% of total storm duration, whereas for the fourth-quartile storms that much rainfall would occur within 75% of total storm duration. Similarly, Figure 4.12B indicates that, for all first-quartile storms, there is a 10% chance that 89% of rainfall total will occur within the first quarter of the total storm duration, a 50% chance that 62% of rainfall total will occur within the first quarter of the total storm duration, and a 90% chance that 39% of rainfall total will occur within the first quarter of the storm duration

Huff (1990) calculated that the percentages of first-, second-, third-, and fourth-quartile storms were 33%, 33%, 23%, and 11%, respectively, of all storms considered. Huff also concluded that first-quartile storms often had durations of less than 6 hours, second-quartile storms often had durations of 6–12 hours, third-quartile storms often had durations of 12–24 hours, and the fourth-quartile storms often had durations of more than 24 hours. Even though Huff noted that of the 261 storms, 110 (42%) had durations less than or equal to 12 hours, 86 (33%) lasted from 12.1 to 24 hours, and 65 (25%) had durations exceeding 24 hours, he did not classify the temporal patterns of rainfall distribution based on the duration of storms.

Subsequent development of Huff curves in Texas based on a limited set of data by Pani and Haragan (1981), Asquith (2003), and Williams-Sether et al. (2004) showed that the relative frequencies of quartile distributions of rainfall can vary among areas with different precipitation climate regions. For example, in the case of Illinois, the first- and second-quartile storms were the most common types but that may not be the case in every region. From a study conducted based on rainfall data from Ohio, aimed to identify the factors that affect Huff curves, Bonta and Rao (1987) concluded that Huff curves have potential for more widespread practical use as design storms.

Even though temporal patterns of rainfall vary widely from one geographical location to another that differs in regional climatology, it is observed that for given locations and climatic conditions some type of events exhibit similar histories of rainfall accumulation. Such observations have led to developments of regional synthetic mass curves for design storms discussed in Section 4.7. We note that, starting from 2003, NOAA has produced sets of data to generate Huff curves for certain durations and for different climatic regions of the United States. These curves bear the potential for use in characterizing the temporal patterns of design storms.

4.6.4 Hyetographs of MAP

The procedures described in Section 4.5 yield a time-integrated static value of MAP or STP for a watershed. However, the computation of a runoff hydrograph, which represents stream flow variation with time (Chapter 9), resulting from an observed rain event, requires a specification of the temporal distribution of the MAP. The variation of MAP with time from multiple gauge observations can be modeled in several different ways. For example, if there is a dense network of rain gauges, which is common in most important urban centers in the United States, a watershed can be subdivided into smaller catchments such that it is reasonable to assign only one rain gauge for each catchment. Nonetheless, three alternative ways for the derivation of time-dependent MAP from rainfall records from multiple gauges are presented below.

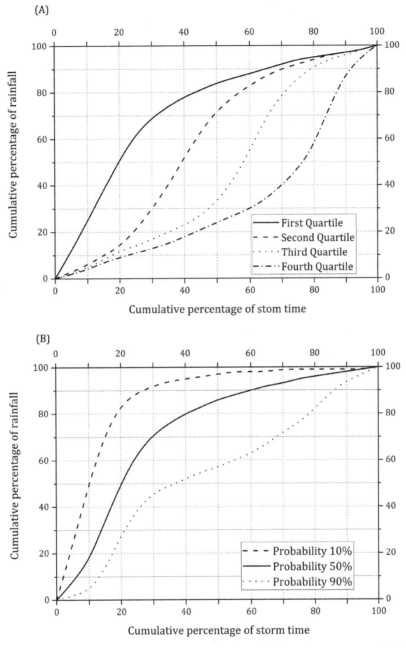

Figure 4.12 (A) Median (50th-percentile) time distribution of rainfall of first-, second-, third-, and fourth-quartile storms on areas of 10–50 square miles in Illinois (Huff, 1990). The median distribution is statistically most useful since it is the 50% probability level of occurrence of a particular category of storm. (B) Time distributions of areal mean rainfall (areas of 50–400 square miles) for first-quartile storms at the 10%, 50%, and 90% probability levels. The probability shown is the chance that the observed storm pattern will lie to the left of the curve. The 50th-percentile curve represents a cumulative rainfall pattern that should be exceeded in about half of the storms. The 90th-percentile curve can be interpreted as a storm distribution that is equaled or exceeded in 10% or less of the storms. The 10th-percentile curve is typical of storms in which the rainfall is concentrated in an unusually short portion of a storm. It shows that at least about 88% of its rainfall in the first quarter and about 95% of it will occur in the first half of the storm. Only one out of ten first-quartile storms will exhibit such a pattern; 90% or more of the storms will not show such an unusual distribution pattern.

If the pattern of temporal variation of rainfall is defined for a watershed, the temporal variation of MAP is computed from

$$p_{MAP}(t) = \left[\frac{p_p(t)}{\sum_t p_p(t)}\right] P_{MAP} \qquad (4.8)$$

where $p_p(t)$ is the ordinate of the watershed-wide rainfall pattern, which usually is the percentage mass curve, and $p_{MAP}(t)$ is the watershed MAP at time t.

The watershed precipitation pattern, $p_p(t)$ in Eq. 4.8, can be inferred from gauge observations with a second weighting scheme

$$p_p(t) = \frac{\sum_i w_i(t) p_i(t)}{\sum_i w_i(t)} \qquad (4.9)$$

where $p_i(t)$ denotes rainfall measured at gauge i at time t, and $w_i(t)$ is the weighting factor assigned to gauge i at time t (applicable only for recording gauges).

Equation 4.8 implies that, if only one recording gauge is used, the resulting MAP hyetograph will have the same relative distribution as the observed hyetograph. If there are two or more gauges, the resulting pattern will depend upon the weights given to the gauges and the relative patterns of rainfall time distributions at those gauges. In many cases, the weights are all equal by being the reciprocal of the number of gauges. However, such a temporal distribution, represented as the average of the patterns observed at multiple gauges, may lead to significant errors if the rainfall amounts recorded at those gauges clearly differ in their time distributions. This can occur when a storm moves over a watershed. For example, if a storm moves from the upstream to the downstream direction of a watershed, the gauges located in the upstream section of the watershed will record rainfalls at earlier times than the downstream gauges and when the downstream gauges record higher values of rainfall there will be lesser amounts of rainfall recorded in the upstream gauges. For this reason, a weighting scheme here should be applied to give different weights to the observations at various gauges. A common practice is to give different weights to rainfall depths but using one of the gauges as a pattern for the watershed average, i.e., giving a time weight of 1 to one of the gauges that best represents the watershed-wide rainfall pattern.

As an alternative to separately calculating the total static MAP of the watershed and combining this with a temporal distribution pattern for the watershed, a weighting scheme can also be applied dynamically to gauge rainfall values, to compute $P(t)$, the watershed rainfall at time t, which yields the MAP hyetograph. A common weighting scheme is the IDW method described above, in which the weight to rainfall depths measured at a rain gauge is proportional to the inverse of the square of the distance of the gauge from certain nodes within the watershed (Eq. 4.7). With the weights thus computed, the node hyetograph ordinate at time t is computed as

$$p_n(t) = \sum_i^{NG} w_i p_i(t) \qquad (4.10)$$

In general, the centroid of a watershed or sub-watershed (catchment) is taken as the node where the hyetograph is computed. The choice of centroid of a watershed as the node for watershed-wide hyetograph calculation also minimizes the errors introduced by lumping gauge hyetograph data that show distinct temporal patterns and/or accumulated rainfall depths due to storm movements or intrinsic spatial variability of rainfall.

Even though the temporal variation of MAP can be calculated in different ways, great care must be exercised in the selection of a method, because the results can vary significantly from one method to another, and those results have a strong influence on the characteristics of flood hydrographs calculated from using those results (Chapter 9). The following example is used as an illustration.

During Tropical Storm Allison, Harris County collected rainfall data from seven recording gauges within Buffalo Bayou watershed, which straddles the important uptown business district as well as downtown Houston (Figure 4.13). As mentioned above, when the storm moved inland, severe rainfall occurred over much of Harris County on June 5, 2001, for a 24-hour period. From June 6 to June 7 there was a lull in rainfall. But the storm intensified again and looped back over the Houston area from west to east producing havoc-causing rainfall starting in the late hours of June 8 and ending in the early hours of June 9, 2001. The table insert in Figure 4.13, indicates that, on June 5, the most severe intensity of rainfall in this watershed occurred within a short time window of one hour from 7:20 to 8:20 P.M. But during June 8 and 9, the period of maximum intensity rainfall varied noticeably from west to east in accordance with the loop-back movement of Tropical Storm Allison. Heavy flooding occurred over much of Buffalo Bayou watershed, particularly in downtown Houston (Mukhopadhyay et al., 2009b) located near the outlet of the watershed.

The station hyetographs shown in Figure 4.14 are converted to MAP mass curves using the approaches presented above. Since, during the rainfall on June 5, the $p_i(t)$ patterns were not strikingly different, an averaging scheme where w_i in Eq. 4.9 is the inverse of the number of gauges can be used to derive the temporal variation of MAP (Figure 4.15A). The averaging can be done either by separately calculating a static MAP and then applying the average pattern (Eqs. 4.8 and 4.9) or by dynamically averaging the gauge patterns. In this case, the former method, designated as static averaging and the latter method, called dynamic averaging, yield similar results as shown in Figure 4.15A. On the other hand, due to noticeable differences in the timing of the high-intensity rainfalls recorded at different gauging stations during the rain event on June 8–9, a weighting scheme should be applied in the calculation of the time variance of

4.7 Design Storms

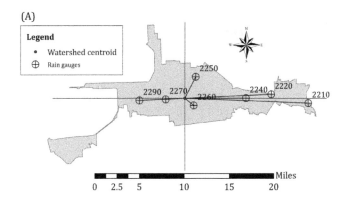

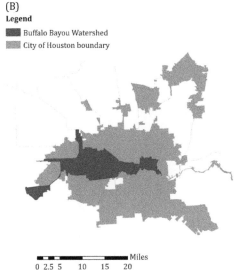

Gauge No.	Location	Distance from centroid of the watershed (miles)	6/5/2001 24 h STP (inches)	Time of maximum intensity rainfall	6/8-6/9/2001 16 h STP (inches)	Time of maximum intensity rainfall
2210	Turning Basin	13.87	7.61	6-5-01 8:20 PM	14.46	6-9-01 3:35 AM
2220	Milam Street	9.65	5.82	6-5-01 7:20 PM	10.02	6-9-01 12:55 AM
2240	Shepherd Drive	6.79	4.82	6-5-01 7:30 PM	12.51	6-9-01 12:10 AM
2250	Bingle Road	2.63	4.09	6-5-01 7:20 PM	7.95	6-9-01 4:15 AM
2260	San Felipe Drive	1.25	3.11	6-5-01 7:40 PM	3.87	6-8-01 11:00 PM
2270	West Beltway 8	2.15	1.91	6-5-01 7:35 PM	3.83	6-9-01 12:25 AM
2290	Dairy Ashford Road	5.07	1.44	6-5-01 8:00 PM	4.32	6-9-01 12:20 AM

Figure 4.13 (A) Map showing the locations of the recording rain gauges within Buffalo Bayou Watershed during Tropical Storm Allison. The table insert lists key features of rainfall data during June 5, 8, and 9, 2001. (B) Location of Buffalo Bayou Watershed in relation to the limits of City of Houston, Texas.

MAP. Figure 4.15B shows that the three different weighting schemes produce markedly different temporal variations of MAP. The dynamic averaging scheme produces a significantly different pattern from the ones calculated by applying an IDW scheme. Here also, due to differential weights given to the gauges, the rainfall depths obtained from dynamic weighting and static weighting are different. However, the overall pattern for the IDW method is the same for both approaches.

4.7 DESIGN STORMS

4.7.1 Definitions

A **design storm** is a hypothetical precipitation pattern defined for use in the engineering or design of a system that operates to respond to rainfall. It can be defined by a value of rainfall depth at a point. However, there must be a duration of rainfall specified for that depth. Furthermore, water resource projects involving the design of various types of hydraulic structures at various scales also require an association of probability of occurrence of that depth of rainfall for that duration. Large water resource projects, such as large dams, spillways, levees, major river bridges, major irrigation canals, etc., require the specification of rainfall depth that has a very low probability or frequency of occurrence. Small projects, such as local urban storm drainage systems, small detention basins etc., on the other hand, can be designed for rainfall events that are more frequent or have a higher probability of occurrence. Table 4.2 gives a few examples of storm frequencies used in the design of certain hydraulic structures.

In many other situations, the most extreme event from historical records is used as the design value. In such cases, the probability, Pr,[1] of the most extreme event of the last N years will be exceeded once in the next n years can be given by a plotting position formula, such as Gringorten formula:

---

[1] In this chapter we denote probability by Pr because we use P to denote precipitation. However, in Chapters 3 and 5 we use P to denote probability following common convention found in statistical literature.

114 4 Rainfall Measurements and Models

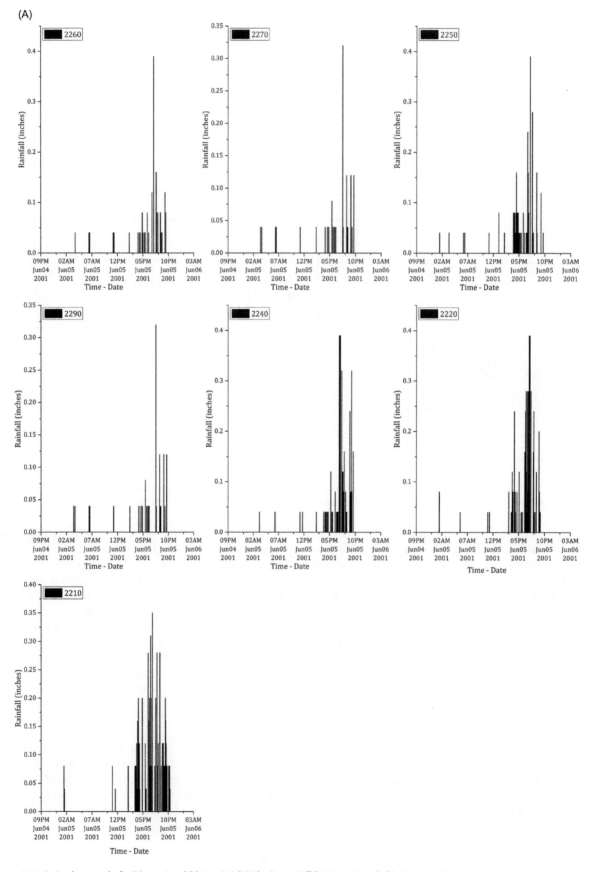

Figure 4.14 Station hyetographs for (A) June 5 and (B) June 8–9 (2001) rains on Buffalo Bayou Watershed in Houston, Texas.

4.7 Design Storms

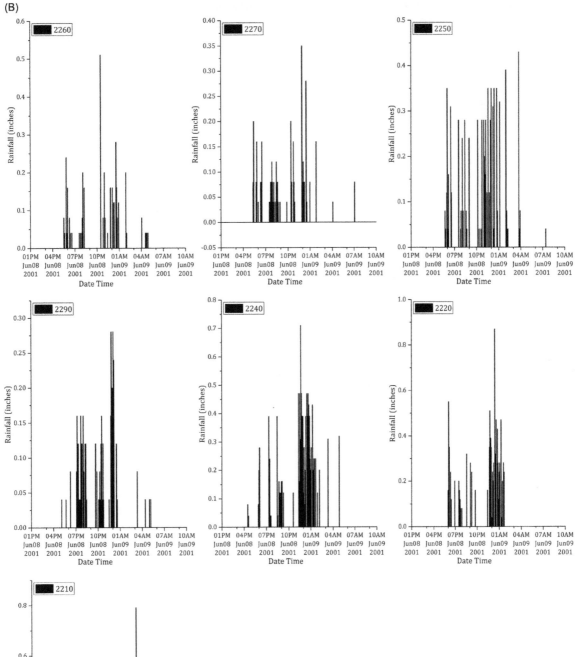

Figure 4.14 (cont.)

4 Rainfall Measurements and Models

Table 4.2 *Examples of design frequencies (years) for certain types of hydraulic structures*

Structure type	Design frequency (return period in year)	Duration (hour)	Examples of agencies adopting the design standards
Highway (freeway) culverts	50	24	Texas Department of Transportation
Highway (freeway) bridges	50	24	Texas Department of Transportation
Urban storm drainage system	100	24	City of Dallas
Urban storm drainage system	2	24	City of Houston
Emergency spillways (drainage area 20 acres or less; dam height 20 ft or less	10	24	NRCS Dams
Emergency spillways (drainage area greater than 20 acres; dam height 20 ft or more	50	24	NRCS Dams

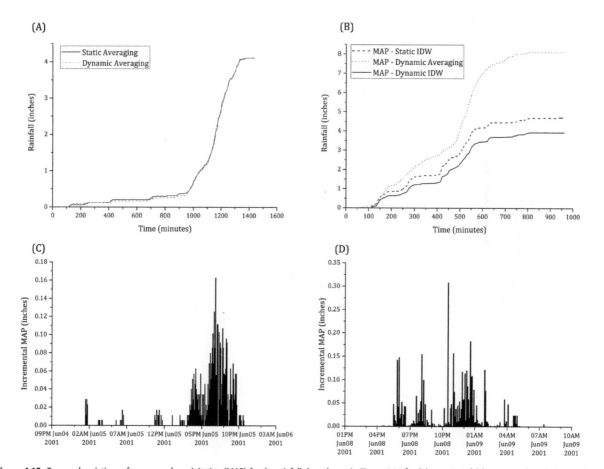

Figure 4.15 Temporal variations of mean areal precipitation (MAP) for the rainfall data shown in Figure 4.14 for (A) June 5 and (B) June 8–9 (2001). (C) and (D) Hyetographs are constructed from the mass curves derived by dynamic averaging for June 5 and June 8–9, respectively.

$$Pr(N, n) = \frac{n}{N + n} \quad (4.11)$$

In addition to the specification of depth, duration, and frequency, a design storm may also require the specification of the time distribution of rainfall since this information is needed for the computation of design hydrographs (Chapter 9).

Rainstorms with high intensity generally have short durations and small areal extents. Rainstorms occurring over large areal extents are usually of low intensity with longer durations. The infrequent combination of relatively high-intensity and long-duration rainfall events produces large total amounts of rainfall. These rainfall events are generally associated with a slow-moving warm front or the development of a stationary front and can cause significant floods and damage.

There are several possible methods of arriving at a design storm. The two most common methods are either to adopt an appropriate historical rain event that occurred in the vicinity of the project area and caused considerable damage or to develop a **synthetic** or **hypothetical** design rainstorm. The historical rainfall approach has the advantage of being readily identified and explained to engineers and the public (Haan et al., 1994). However, it has the distinct disadvantage of an unknown frequency of occurrence. It is very difficult to assign a return period (frequency) with a particular rainfall hyetograph. For this reason, design storms are typically based on synthetic or hypothetical rainfall patterns, which have been constructed using the general characteristics of rainfall derived from time-series data of historical rainfall in the region where the project site is located.

Rainfall depth–duration–frequency (DDF) or intensity–duration–frequency (IDF) data in the form of tables, graphs, or isohyetal maps for a region have been developed in many countries. Such data have been derived using frequency analysis of historical rainfall data of the region. Johnson and Sharma (2017) have provided some details about the derivation of design storms from historical rainfall records.

The probability of occurrence is usually measured in terms of either **annual exceedance probability (AEP)** or **average recurrence interval (ARI)** or **return period**. The AEP or "1-in-N events" refers to the probability associated with exceeding a given amount of precipitation for a specified duration at least once in any given year. For example, an AEP of 1-in-100 or AEP = 0.01 equates to a 1% chance of the amount being exceeded at least once in any given year. The ARI or "N-year event" refers to the average time between precipitation events exceeding the particular magnitude for a specified duration. For example, a 100-year amount *on average* occurs once every 100 years. From the definitions, given in Chapter 3 it is apparent that AEP = 1/ARI. Since the probability of occurrence is the probability in percent of an event equaling or exceeding a given event will occur in a given year, the relationship between the return period, T, in years and the probability of occurrence, Pr, is also expressed as $T = 100/Pr$.

The above definitions assume that precipitation events in the time series are independent, i.e., the event magnitudes are not correlated with each other in time, and that the process is stationary, i.e., the probability distribution of events is not changing through time.

4.7.2 Frequency Analysis

Rainfall is a process of random occurrence that cannot be predicted with absolute certainty. For this reason, the analysis of rainfall data involves well-established statistical procedures, which give one of the most common types of design storms that have a certain probability of occurrence with given depth and duration.

Design rainfalls are based on fitting a probability distribution to a suitable set of extreme rainfalls chosen from the full time series of recorded rainfall. Four major elements are entailed in this process: (1) construction of databases of rainfall to derive design rainfall; (2) selection between an annual maximum series (AMS) and a partial duration series (PDS); (3) selection of one or more appropriate probability density functions (pdfs) or cumulative distribution functions (cdfs) and fitting the assumed distribution function to the data, conducting a goodness-of-fit test to determine which distribution model best fits the observed data; and (4) regionalization.

4.7.2.1 Construction of Rainfall Databases

A database of large or extreme rainfall totals in a region, recorded in a network of rain gauges with long periods of records, is required to develop the IDF relationships of the region. The period of record (POR) should be at least 30 years or more for a statistically significant sample (Singh, 1992). If the data come from recording gauges, the largest totals of desired durations are selected and separated into different time-series data. Several durations are used. If the fine resolution of the time window is not present in certain records, where data are collected at longer intervals such as 1 hour or 1 day, a further step is required to ensure that the largest totals are found. Nevertheless, such a database contains rainfalls of large or extreme rainfall totals and therefore does not represent the usual rainfall values typically observed in that region. For this reason, hydrologic and meteorological expertise must be exercised so as not to include values that are really unusual or outliers. Other potential problems can arise from gauge locations, wrong aggregation of rainfall values (this arises mainly from standard gauges where observed rainfall totals are not recorded at correct intervals), gauge undercatch, and mechanical errors. In summary, a rigorous quality control process must be implemented to develop such a database.

4.7.2.2 AMS versus PDS

An AMS is a time series comprising the largest rainfall totals from each year for a particular duration. A PDS is a time series comprising all rainfall totals that are larger than some predefined threshold depth or intensity (usually the smallest maximum annual event in a series). This means that in some years there can be more than one event selected, whereas in other years there are totals that are not larger than the threshold value and those values are excluded from the series. In a similar way, an AMS can contain values that are too low if there are records from years that are particularly dry. An argument put forth in favor of PDS over AMS is that an AMS ignores large values of rainfall depth if there are several major rainfall events in one or more years within the POR. On the other hand, for PDS, it is also difficult to set the threshold value with a real basis. However, it has been shown that for records greater than 10 years, both series yield comparable results.

It has been found that the pdf that fits the selected series depends on whether an AMS or PDS is used. AMS data lead to an estimate of the AEP whilst PDS data give estimates of the ARI. The relationship between the return period obtained from AMS, denoted by T_A, and the return period obtained from PDS, denoted by T_P, are related by the following equation first derived by Langbein (1949) and later confirmed by Rosbjerg (1977):

$$\frac{1}{T_A} = 1 - \exp\left(\frac{-1}{T_P}\right) \quad (4.12)$$

In most studies, for each duration selected, the AMS depths are extracted from historical rainfall records and subsequently frequency analysis is applied to the annual data and the AMS is converted to a PDS. Frederick et al. (1977) developed empirical factors to convert AMS values to equivalent PDS values. Hershfield (1961a), on the other hand, also used empirical factors to convert PDS values to AMS values.

When only a few years of data are available, an annual exceedance series for each duration can be determined by ranking the depths and choosing the N largest values for a record of n years. If only a short record of rainfall is available, it may be desirable to use a PDS rather than an AMS so that more than one data point per year can be included in the series. Subsequently, a PDS can be converted to an AMS.

4.7.2.3 Selection of Appropriate Probability Density Function

The probability associated with a certain rainfall depth occurring in the time series (e.g., an AMS) of a certain duration is obtained by fitting a cdf to the selected time series. Thus, the values of the rainfall depths for various AEPs are obtained from a fitted pdf.

A multitude of probability distributions are used in hydrologic frequency analysis. For rainfall DDF analysis, some of those that have been applied in many studies in various regions of the world in recent years include generalized extreme value (GEV), generalized logistics, generalized normal, generalized log-normal, Burr distribution, generalized Pareto, and log Pearson type III (LP III) distributions. Of these, most often the GEV distribution fits the observed data quite well. This distribution has been adopted in the United States and United Kingdom. Furthermore, there are two special cases of GEV distribution, known as the extreme value type I (EV I), or *Gumbel distribution*, and extreme value type III (EV III) or *Weibull distribution*, which have historically been used quite extensively in different parts of the world for rainfall frequency analysis. These are called extreme value distributions because they are applicable to very high or low magnitudes, not the commonly or regularly observed values.

Earlier, parameters of the distributions were mostly estimated by the method of conventional or product moments or in some cases by maximum likelihood estimation. For example, Hershfield (1961a) and Frederick et al. (1977) used the fitting procedure of Gumbel (1958), which is also called the Fisher–Tippett type I distribution in some of the literature, and involves the method of moments (the distribution parameters as a function of the first two moments). The 2-year value is a measure of the first moment, or central tendency of the distribution. The relation of the 2-year to the 100-year value is a measure of the second moment, or dispersion of the distribution. Values for other return periods were derived mathematically from the 2-year and 100-year rainfall values. However, with the inception of the Atlas 14 project in the United States to update the DDF estimates of rainfall from extensive networks of rain gauges with longer periods of data, NOAA adopted the method of L-moments, developed by Hosking (1989, 1990) for the estimation of the parameters of various distributions for the variety of advantages that L-moments provide (Bonnin, 2003).

Gumbel or Extreme Value Type I Distribution

The EV I or Gumbel distribution is given by the two-parameter cdf

$$Pr(X \leq x) = F(x) = \exp\left[-\exp\left(-\frac{x-\mu}{\alpha}\right)\right] \quad (4.13)$$

where α and μ are the two parameters of the distribution, estimated from the sample mean ($\bar{x}$) and sample standard deviation (s) as

$$\alpha = \frac{\sqrt{6}s}{\pi} \quad (4.14)$$

$$\mu = \bar{x} - 0.5772\alpha \quad (4.15)$$

For a given value of x, $F(x)$ is the cumulative probability $Pr(X \leq x)$. Here, the random variable x is the rainfall depth. Thus, parameters of the distributions are estimated by the method of moments. The distribution has a skewness of 0.5772.

Equation 4.13 can be rewritten as

$$F(x) = \exp[-\exp(-y)] \quad (4.16)$$

where y is a **reduced variate** defined as

$$y = \frac{x-\mu}{\alpha} \quad (4.17)$$

Solving for y in terms of $F(x)$,

$$y = -\ln\left[\ln\left(\frac{1}{F(x)}\right)\right] \quad (4.18)$$

A plot of x against y shows the type of distribution (EV I, II, and II). A straight-line distribution indicates that it is an EV I (Gumbel) distribution.

If T is the return period,

$$Pr(X \geq x_T) = \frac{1}{T} \quad (4.19)$$

or

$$1 - F(x_T) = \frac{1}{T} \tag{4.20}$$

which gives $F(x_T)$ in terms of return period as

$$F(x_T) = \frac{T-1}{T} \tag{4.21}$$

Substituting Eq. 4.21 into Eq. 4.18, we obtain

$$y_T = -\ln\left[\ln\left(\frac{T}{T-1}\right)\right] \tag{4.22}$$

Once y_T is calculated for a return period T, x_T is calculated from Eq. 4.17 as

$$y_T = \frac{x_T - \mu}{\alpha} \tag{4.23}$$

or

$$x_T = \mu + \alpha y_T \tag{4.24}$$

The parameter μ is the mode (most probable) of the distribution, that is, the point of maximum probability density.

The procedure is illustrated with rainfall data from the Mandakini catchment, which is a small drainage area drained by the Mandakini River, within the Ganga basin in the middle Himalayas of India. A calamity claiming thousands of lives happened in this catchment due to an unprecedented downpour from June 10 to 18 and devastating flooding on June 16 and 17, 2013. Rainfall analysis as part of later flood protection work involved frequency analysis of rainfall data provided by the Indian Meteorological Department (IMD). The IMD has produced gridded rainfall data with 0.25° × 0.25° latitude–longitude grid spatial resolution for various durations for the years 1901–2015 based on gauged rainfall records from all over India (e.g., see Pai et al., 2014). Figure 4.16 shows the point estimations of 100-year and 500-year, 24-hour rainfall employing the Gumbel distribution of a 24-hour AMS. Due to the catastrophic nature of the event, no gauge-based rainfall records could be obtained. In this case, satellite data were used to estimate 24-hour rainfalls for the nine rainy days as given in the caption of Figure 4.16.

Extreme Value Type III or Weibull Distribution

The EV III or Weibull distribution can be used to approximate exponential, normal, or skewed distributions and hence is quite versatile (Weibull, 1951). The Weibull cdf is given as

$$Pr(X = x) = \exp\left[-\left(\frac{x}{\alpha}\right)^\beta\right] \tag{4.25}$$

where α and β are the parameters of the distributions known as the characteristic depth (L) and shape factor, respectively. The cdf, $Pr(X < x)$, is given as

$$F(x) = 1 - Pr(x) \tag{4.26}$$

Equation 4.25 can be algebraically manipulated to

$$\ln\left[\ln\left(\frac{1}{Pr(x)}\right)\right] = \beta \ln x - \beta \ln \alpha \tag{4.27}$$

Equation 4.27 is in the slope–intercept form ($y = Mx + C$) of a straight-line equation in the x–y space, where y represents the left-hand side of Eq. 4.27. Thus, in this case parameters of the distribution, α and β, are determined by the method of least-squares. However, the first step of analysis requires a rearrangement of the data in decreasing order of magnitude. Subsequently, the data are assigned a rank, m, from 1 to N where N is the number of observations. From ranking, a plotting position (frequency) is determined from

$$Pr_m = \frac{m-a}{N+1-2a} \tag{4.28}$$

where a is a parameter that depends on the distribution and varies from 0 for the original Weibull formula to 0.375 for normal or lognormal distributions, 0.3 for median rank, and 0.44 for the Gumbel distribution. When the distribution is unknown, a can be set as 0.4.

Once the plotting positions Pr_m have been determined, the left-hand side of Eq. 4.27 (y) can be computed by setting $Pr(x) = Pr_m$, which is the plotting position of observation x. A linear regression of y values and natural logarithms of rainfall values produces the slope $M = \beta$. The characteristic depth, α is calculated from the slope (β) and intercept (c) as

$$\alpha = \exp\left(-\frac{c}{\beta}\right) \tag{4.29}$$

For a given $P_T (= 1/T)$, the value of x can be calculated from Eq. 4.25. Solving for x_T, the result is

$$x_T = \alpha[-\ln(Pr_T)]^{\frac{1}{\beta}} \tag{4.30}$$

The same dataset used in Figure 4.16 is employed to estimate the Weibull distribution parameters (Figure 4.17). Using these parameter values, the 100-year and 500-year, 24-hour point estimates of rainfall at the same grid point of the Mandakini catchment in Figure 4.16, are shown in Figure 4.17.

Generalized Extreme Value Distribution

The GEV distribution was introduced by Jenkinson (1955) to show that the three limiting forms of extreme value distribution are special cases of this distribution. The GEV distribution is a three-parameter distribution. For a random variable X, the pdf of the GEV is expressed as (Kotz and Nadarajah, 2000)

$$p_X(x) = \frac{1}{\sigma}\left[1 + \kappa\left(\frac{x-\mu}{\sigma}\right)\right]^{\frac{-1-\kappa}{\kappa}} \exp\left[-\left\{1 + \kappa\left(\frac{x-\mu}{\sigma}\right)\right\}^{-\frac{1}{\kappa}}\right] \tag{4.31}$$

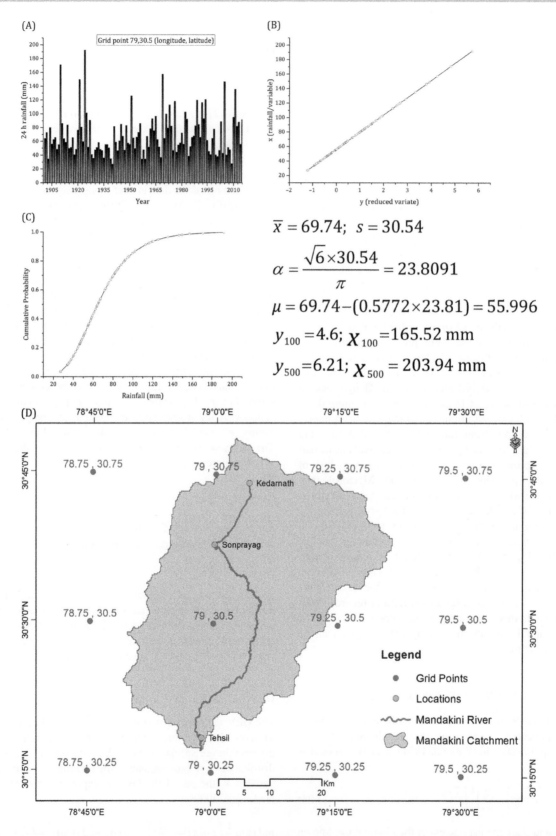

Figure 4.16 (A) Frequency analysis employing a Gumbel distribution to a 24-hour AMS at grid point 79°E, 30.5°N (longitude, latitude), located near the centroid of the Mandakini catchment, shown in D. (B) Plots of the variable, x (rainfall), against the reduced variate, y, defining a perfect straight line, shows the validity of the Gumbel distribution. (C) Plot of the cumulative pdf. The calculations presented in the lower right show that the 100-year, 24-hour rainfall is 165.52 mm and the 500-year, 24-hour rainfall is 203.94 mm. The rainfalls estimated from satellite data (TRMM data) for 24-hour periods from June 10 to June 18, 2013 (nine days) are 181, 341, 217, 114, 94, 71, 332, 347, and 233 mm respectively.

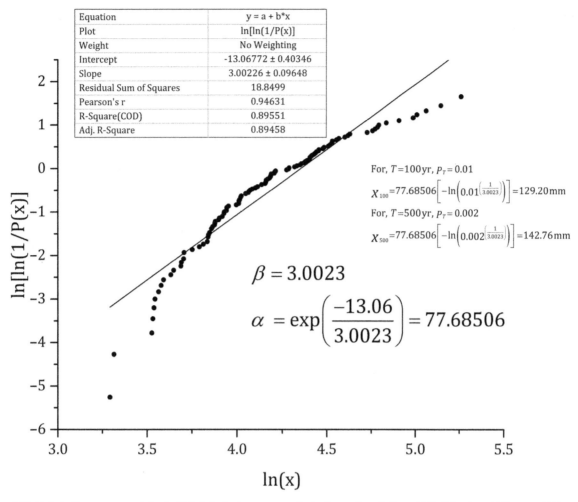

Figure 4.17 Method of least squares to derive the Weibull distribution parameters for the dataset shown in Figure 4.16A. Estimations of 100-year and 500-year, 24-hour rainfalls at the grid point shown in Figure 4.16A, using Weibull distribution.

where $p_X(x)$ is the pdf of X, $\sigma(>0)$ and μ, respectively, are the *scale* and *location* parameters, and κ is the shape parameter. The range of X depends on the value of κ; it is bounded by $\mu + (\sigma/\kappa)$ from above for $\kappa > 0$, that is, $-\infty < x < \mu + (\sigma/\kappa)$; it is bounded from below for $\kappa < 0$, that is, $\mu + (\sigma/\kappa) < x < \infty$. Depending on the value of κ, different extreme value distributions are represented by Eq. 4.31. For example, the GEV distribution corresponds to the EV I distribution for $\kappa = 0$, the EV II distribution for $\kappa < 0$, and the EV III distribution for $\kappa > 0$.

The cdf of the GEV distribution can be expressed as

$$F_X(x) = \exp\left[-\left\{1+\kappa\left(\frac{x-\mu}{\sigma}\right)\right\}^{-\frac{1}{\kappa}}\right] \quad (4.32)$$

The quantile function, given as the inverse of the cdf, is

$$X_p(F) = x_T = \mu + \frac{\sigma}{\kappa}[1 - \{-\ln(F)\}^\kappa] \quad (4.33)$$

The parameters of the distribution can be best estimated by the method of L-moments as given by Hosking (1990):

$$Z = \frac{2}{(3+\tau_3)} - \frac{\ln(2)}{\ln(3)} \quad (4.34)$$

$$\kappa = 7.8590\,Z + 2.9554\,Z^2 \quad (4.35)$$

$$\sigma = \frac{\lambda_2 \kappa}{(1-2^{-\kappa})\Gamma(1+\kappa)} \quad (4.36)$$

$$\mu = \lambda_1 + \frac{\sigma}{\kappa}[\Gamma(1+\kappa) - 1] \quad (4.37)$$

where, λ_1 and λ_2 are the L-moments and τ_3 is the L-moment ratio as discussed in Chapter 3.

The GEV distribution is illustrated in Figure 4.18. The data shown in this figure are from the Nigerian Meteorological Agency and include monthly precipitation data for 40 years from 1974 to 2013. The data are sorted to obtain annual maximum precipitation records. L-moments are calculated from the parameters of GEV distribution shown in Figure 4.18A. The pdf and cdf, as calculated from Eqs. 4.31 and 4.32, are shown in Figure 4.18B and C, respectively. From the ranking of data, the observed frequency is calculated using the plotting position (Eq. 4.28);

the probabilities calculated using this formula are used in Eq. 4.33 to obtain the predicted values. The correspondence between observed and predicted values of precipitation is shown in Figure 4.18D.

Test of Goodness of Fit of a pdf

Once a theoretical model of probability distribution is applied to an AMS or a PDS, it is necessary to evaluate the goodness of fit of the theoretical model to the observed frequency distribution. There are several methods to test the goodness of fit as discussed by Zhang and Singh (2019). A common and well-known method is χ^2-test.

Figure 4.19 shows the probability distributions of observed data shown in Figure 4.16A and those calculated using Gumbel and Weibull distributions. The results of the χ^2-test are also presented in this figure. This particular example shows that the goodness of fit of the Gumbel distribution to the observed data is acceptable but that of Weibull distribution is not statistically significant. It should be noted that the Gumbel distribution is mostly applied for maxima series, whereas the Weibull distribution is mostly applied to minima series.

Isopluvial Maps

Typically, several point estimates of DDF are made to produce isopluvial maps of a region, which are contours of rainfall depths of certain frequency and duration. These maps are used to extract the DDF values for a project site by interpolation. In the United States, such maps produced by Hershfield (1961a) and and Frederick et al. (1977) were used for many years (discussed in further detail below).

As an example, Figure 4.20 shows the isopluvial maps of 100-year and 500-year, 24-hour rainfall, constructed from AMS data for the period 1900–2015 at the 12 grid points shown in Figure 4.16. Frequency analysis of the AMS at each of the grid points was conducted using the Gumbel distribution.

4.7.2.4 Regionalization

One of the primary problems in precipitation-frequency analysis is the need to provide estimates for the ARI that are significantly longer than available records. In regional frequency analysis, data from multiple stations are used concurrently. Thus, the reliability of estimates of design rainfall is increased by creating regionalized relationships that are developed by grouping together multiple stations from climatically homogeneous regions or regions with geographical proximity to have a common frequency distribution (pdf).

Several regionalization approaches have been proposed by various workers in this field. However, regional frequency analysis using L-moments, developed by Hosking and Wallis (1993, 1997), is considered perhaps the most efficient and robust method compared to any other approach (Bonnin et al., 2003, 2004). In the L-moment approach three statistics are used in regional frequency analysis. These include a discordancy measure for identifying unusual sites in a region, a heterogeneity measure for assessing whether a proposed region is indeed homogeneous, and a goodness-of-fit measure, for assessing whether a candidate pdf provides an adequate fit to the data. Five basic steps involved in regional frequency analysis with the L-moment approach are briefly enumerated below.

Identification of Homogeneous Regions

The most practical method of forming homogeneous regions from groups of sites is based upon a multivariate statistical method called **cluster analysis**. Cluster analysis is a method of grouping sites based on several hydrologic variables in such a way that sites in the same group or clusters are hydrologically more alike to each other than to those in other groups. The similarity of sites in the group can be defined by using geographical and physical site characteristics or hydrologic features, such as latitude, longitude, elevation, and mean daily, monthly, or annual precipitation, etc. The minimum variance hierarchical clustering algorithm, developed by Ward (1963), is a widely used method to form homogeneous regions. This method employs analysis of variance to determine the distance between sites and clusters.

Test for Discordancy

The discordancy measure, D_i, is based on L-moment ratios and can be used to determine an unusual site. It is defined as

$$D_i = \frac{1}{3}\left[(u_i - \overline{u})^T (u_i - \overline{u}) S^{-1}\right] \tag{4.38}$$

where u_i is the vector containing three L-moment ratios (L-Cv, L-skewness, and L-kurtosis) for site i, and $\overline{u}$ is the vector containing the simple average L-moment ratios given by

$$\overline{u} = N_S - 1 \sum_{i=1}^{N_S} u_i \tag{4.39}$$

where N_s is the total number of sites. S is the sample covariance matrix of L-moments of all sites given by

$$S = (N_S - 1)^{-1} \sum_{i=1}^{N_S} (u_i - \overline{u})(u_i - \overline{u})^T \tag{4.40}$$

Generally, any site with $D_i \geq 3$ is considered discordant (Hosking and Wallis, 1993).

Test for Homogeneity of Regions

For validating the homogeneity of a cluster, a heterogeneity measure called the H-statistic is calculated using

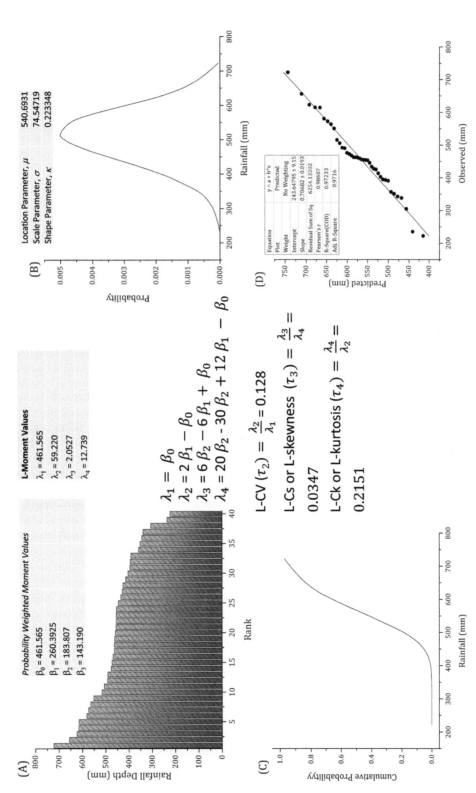

Figure 4.18 Frequency analysis of 40 years annual rainfall maxima employing generalized extreme value (GEV) distribution of rainfall data from Benin City, Nigeria (5°35′E; 6°17′N). (A) Ranked annual rainfall maxima (1974–2013); (B) GEV pdf; (C) GEV cdf; and (D) plot of observed versus predicted values using GEV rainfall.

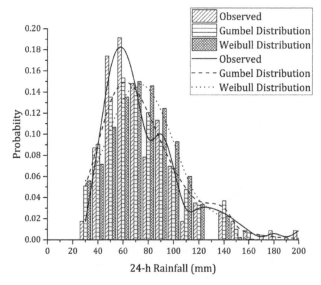

Figure 4.19 Comparisons of observed frequency distributions with Gumbel and Weibull distributions for the data used in Figure 4.16.

$$H_i = \frac{V_i - \mu_v}{\sigma_v} \qquad (4.41)$$

where μ_v and σ_v respectively is the mean and standard deviation of the N_{sim} values of V-statistics. N_{sim} represents the number of simulation data. V_i is calculated from the regional data, based on the corresponding V-statistic defined as:

$$V_1 = \left\{ \sum_{i=1}^{N} n_i \left[t^{(i)} - t^R \right]^2 \bigg/ \sum_{i=1}^{N} n_i \right\}^{\frac{1}{2}} \qquad (4.42)$$

$$V_2 = \sum_{i=1}^{N} n_i \left\{ \left[t^{(i)} - t^R \right]^2 + \left[t_3^{(i)} - t_3^R \right]^2 \right\}^{1/2} \bigg/ \sum_{i=1}^{N} n_i \qquad (4.43)$$

$$V_3 = \sum_{i=1}^{N} n_i \left\{ \left[t_3^{(i)} - t_3^R \right]^2 + \left[t_4^{(i)} - t_4^R \right]^2 \right\}^{1/2} \bigg/ \sum_{i=1}^{N} n_i \qquad (4.44)$$

where $t^{(i)}$, $t_3^{(i)}$, $t_4^{(i)}$ are the sample L-moment ratios for site I and t^R, t_3^R, t_4^R are the L-moment ratios for a given region. According to Hosking and Wallis (1993), if $H_i < 1$, the region is acceptably homogeneous; if $1 \leq H_i < 2$ it is possibly heterogeneous; and the region is definitely heterogeneous when $H_i > 2$.

Goodness-of-fit Measure

For each homogeneous cluster, a distribution is fitted to the data from all sites. The aim is to choose a distribution that not only fits the data approximately but also gives reasonable estimates of the quantiles at each site. Hosking and Wallis (1993) introduced a goodness-of-fit measure (Z^{Dist}) that measures how well the L-kurtosis of the fitted distribution matches the regional averaged L-kurtosis of the sample data. The goodness-of-fit measure is defined as

$$Z^{Dist} = \frac{\tau_4^{Dist} - t_4^R + B_4}{\sigma_4} \qquad (4.45)$$

where τ_4^{Dist} is the theoretical L-kurtosis of the fitted distribution, t_4^R is the regional averaged L-kurtosis calculated for a given region using observed data (estimated by weighting site-specific L-kurtosis coefficients by their sample size), σ_4 is the estimate of the standard deviation of t_4^R, and B_4 represents the bias value for t_4^R. A given distribution is considered a good fit if $|Z^{Dist}| \leq 1.64$. When more than one distribution qualifies for the goodness-of-fit measurement criteria, the preferred distribution will be the one that has the minimum $|Z^{Dist}|$ value at a confidence level of 90%.

Quantile Estimation

Quantile estimation is the next step after selecting the regional frequency distribution. The regional quantile, $q(F)$, is estimated for various nonexceedance probability (F) levels or return periods for each homogeneous region. At station rainfall quantiles, $Q(F)$ values are estimated using the **index-flood** method as discussed in Chapter 5. In this method it is assumed that rainfall values of different stations in a homogeneous region have the same frequency distribution except for a site-specific scale factor (Dalrymple, 1960). This scale factor is known as the index flood. The quantile estimates, $\hat{Q}(F)$, at the ith rain gauge station with nonexceedance probability F are calculated as

$$\hat{Q}_i(F) = \hat{l}_j^i \hat{q}(F) \qquad (4.46)$$

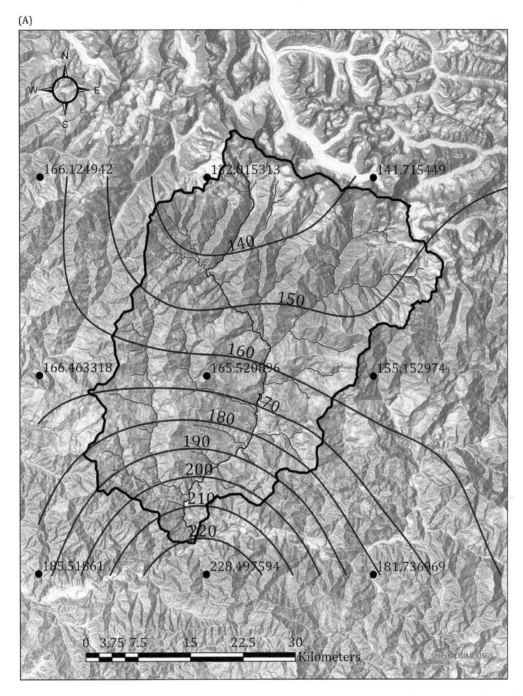

Figure 4.20 Isopluvial maps of (A) 100-year and (B) 500-year, 24-hour rainfall (mm) constructed from AMS data at the 12 grid points shown in Figure 4.16. The contours are drawn using the spline method. (A black and white version of this figure will appear in some formats. For the color version, please refer to the plate section.)

where $\hat{q}(F)$ is the regional quantile estimate or regional growth curve for the region and $\hat{l}_j^i$ is the index rainfall value for the site i which is usually *average maximum monthly* rainfall.

4.7.3 Intensity–Duration–Frequency Relationships

4.7.3.1 Historical DDF/IDF Data in the United States

For a long time, design engineers in the United States used the isopluvial maps, produced by Hershfield (1961a) from the US Weather Bureau and Frederick et al. (1977) from NOAA, to read rainfall depths at a point of interest for certain durations and return periods (frequencies). These two sources of design rainfall data in the United States, popularly known as TP-40 and Hydro-35, respectively, were developed using rainfall data from rain gauges that existed at that time. Another source of design rainfall for the western United States was the NOAA Atlas 2 (Miller et al., 1973). The design rainfall data for durations of 2 to 10 days were given by Miller (1964).

126 4 Rainfall Measurements and Models

(B)

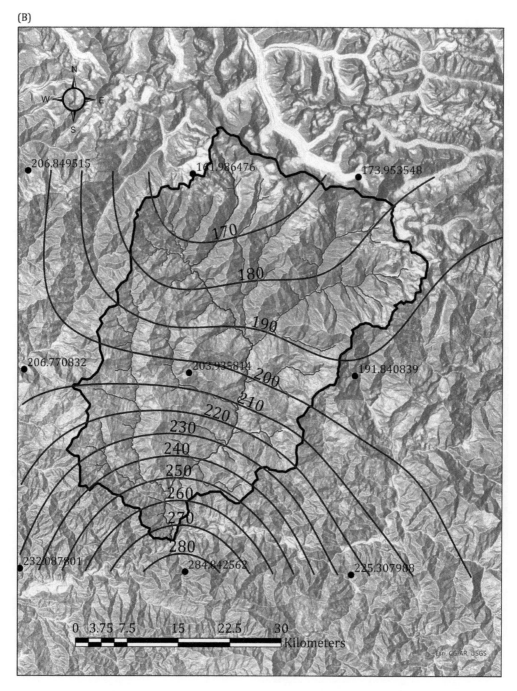

Figure 4.20 (cont.)

These sources of data were widely used and accepted. Rainfall depths were converted to rainfall intensities (Eq. 4.1), and the information was presented as a graph, with duration plotted on the horizontal (x) axis and intensity on the vertical (y) axis. Typically, the graph contained a series of curves, one for each return period. The graph was typically prepared as a log–log plot showing the decrease in rainfall intensity with increasing rainfall duration and increasing rainfall intensity with decreasing frequency. Figure 4.21 is an example. Almost all municipalities or authorities having jurisdiction over drainage design in an area developed such IDF curves.

Various algebraic forms have been used to represent IDF curves to fit the IDF values of a location and thus produce wholly smooth definitions of rainfall IDF. It has been found that relations between intensity, duration, and frequency in

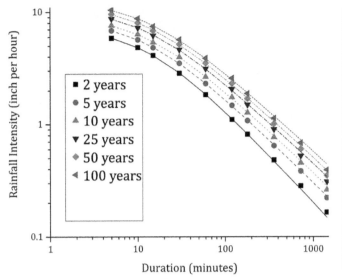

Duration (minutes)	Frequency (Return Period in Year)					
	2	5	10	25	50	100
	Rainfall Depth (inches)					
5	0.490	0.572	0.634	0.726	0.798	0.870
10	0.809	0.948	1.060	1.210	1.340	1.460
15	1.030	1.210	1.350	1.550	1.710	1.870
30	1.430	1.750	1.990	2.320	2.590	2.850
60	1.840	2.310	2.650	3.130	3.510	3.880
120	2.210	2.960	3.490	4.130	4.630	5.150
180	2.450	3.230	3.820	4.500	5.090	5.650
360	2.890	3.870	4.600	5.410	6.060	6.800
720	3.400	4.600	5.450	6.290	7.230	8.200
1440	3.950	5.320	6.260	7.420	8.420	9.420
	Rainfall Intensity (inch per hour)					
5	5.88	6.864	7.608	8.712	9.576	10.44
10	4.854	5.688	6.36	7.26	8.04	8.76
15	4.12	4.84	5.4	6.2	6.84	7.48
30	2.86	3.5	3.98	4.64	5.18	5.7
60	1.84	2.31	2.65	3.13	3.51	3.88
120	1.105	1.48	1.745	2.065	2.315	2.575
180	0.8167	1.0767	1.2733	1.5000	1.6967	1.8833
360	0.4817	0.6450	0.7667	0.9017	1.0100	1.1333
720	0.2833	0.3833	0.4542	0.5242	0.6025	0.6833
1440	0.1646	0.2217	0.2608	0.3092	0.3508	0.3925

$$i = \frac{KF^x}{(T_D+b)^n}$$

	2 year	5 year	10 year	25 year	50 year	100 year
K	6.5831	3.5995	2.8354	2.0070	1.7141	1.7103
x	3.4232	1.7895	1.3691	1.1218	0.9903	0.8544
b	13.7653	12.9084	12.9593	13.2532	13.7393	13.7640
n	0.8468	0.7745	0.7498	0.7389	0.7359	0.7264

Figure 4.21 Rainfall depth duration for certain frequencies at Lewisville (north of Dallas), Texas, as derived from the isopluvial maps in TP-40 and Hydro-35. The intensity–duration–frequency (IDF) values are modeled using Eq. 4.47. The values of the parameters are derived by nonlinear regression optimization.

most locations can be represented by equations such as those given by Bernard (1932) and Chow et al. (1988). These two forms are

$$I = \frac{KF^x}{(T_D+b)^n} \quad (4.47)$$

$$I = \frac{KF^x}{T_D^e + f} \quad (4.48)$$

where F denotes the frequency or return period (years) and K, x, b, and n (in Eq. 4.47) and K, x, e, f (in Eq. 4.48) are coefficients varying with location and return period (other symbols are as defined above). The values of these coefficients can be estimated by nonlinear regression of the IDF data for a particular location. The values of K, x, b, n given in the Table in Figure 4.21 are derived in this manner. Nonetheless, such an equation has no theoretical basis; it is purely empirical and useful for expressing regional or local IDF relations, such as those shown in Figure 4.21. One of the reasons for using an equation like Eq. 4.47 was that the rainfall depths were read from isopluvial maps and, when those data were plotted on the graph described above, they often did not produce smooth curves. However, if the data were fitted with such an equation, smooth curves would result, which would produce reasonably shaped design storm hyetographs. The IDF curves were also expressed as equations to avoid having to read the design rainfall intensity by interpolation from a graph. Furthermore, such an equation is highly useful in the development of synthetic hyetographs as discussed in Section 4.7.5.2.

4.7.3.2 Current Precipitation-Frequency Estimates in the United States: NOAA Atlas 14

Concerns in the United States grew that the design rainfall data obtained from TP-40, Hydro-35, and Atlas 2 were not representative of the current nature of precipitation. The Atlas 14 project was started by the Hydrometeorological Design Studies Center (HDSC) within NOAA in 2000 at the request of, and using funds provided by, a variety of federal, state, and local agencies, to update the precipitation-frequency data for the United States and its affiliated territories. The updates have been made in phases and HDSC started to release the data from 2006. At present, data are available for all 50 states and the territories except for the states of Washington, Oregon, and Nebraska (the data for Montana and Idaho will be released in the future). In the following, the precipitation-frequency estimates at a given location actually indicate precipitation (both liquid and solid) depth of a specific duration that has a certain frequency of occurrence.

The NOAA Atlas 14 data are published through the Precipitation Frequency Data Server (https://hdsc.nws.noaa.gov/pfds). The data can be obtained in the form of point estimates at any point selected on the map, or at a station selected from the station drop-down menu, as cartographic maps, GIS files in both raster and vector formats, metadata, as well as time series.

NOAA Atlas 14 provides precipitation-frequency estimates for durations of 5 minutes to 60 days and return periods of 1, 2, 5, 10, 25, 50, 100, 200, 500, and 1000 years. The estimates, for the first time, are accompanied by assessments of the associated uncertainty, specifically the upper and lower 90% confidence limits. The estimates are based on long periods of records, the average length of record being greater than 60 years, and from a large number of official recording gauges. In Texas alone, for example, data from 11 934 stations have been analyzed (Perica et al., 2018).

The precipitation-frequency estimates are computed using regional frequency analysis based on L-moment statistics calculated from AMS data. NOAA Atlas 14 employs a regionalization approach described by Hosking and Wallis (1997) wherein the L-moment statistics are calculated by grouping stations within a 60-mile radius. The method of L-moments or linear combinations of probability-weighted moments provides great utility in choosing the most appropriate probability distribution function to describe the rainfall frequency distribution. It also provides tools for estimating the shape or higher-order statistical moments of distribution and the uncertainty associated with the estimates. The regionalized approach recognizes that different observing stations can be assembled into groupings of similar climatic regimes or regions. This results in 700–1800 years of data for daily durations and 200–700 years for hourly duration. Several distribution functions were examined and ultimately the GEV distribution was adopted for all stations and durations. The upper and lower 90% confidence intervals are based on a Monte-Carlo simulation approach. The AMS of each station was converted to PDS (Eq. 4.12). Finally, gridded precipitation-frequency estimates at 30-arc-second resolutions were developed. The current approach used in the development of Atlas 14 has the assumption of stationarity, which means that it is assumed that extreme precipitation characteristics do not change in time. NOAA is in the process of developing IDF relationships under nonstationary climate.

Comparisons of results from TP 40 and NOAA Atlas 14 show that, in some instances, the rainfall depth for a given duration and frequency increased, in certain instances it did not change, and in some instances it even decreased. These vary with rainfall frequency, duration, and location. In addition, it has also been observed that in some locations there is a large change in rainfall depth over short distances. For example, from Minneapolis to St. Cloud the 100-year, 24-hour rainfall depth changed from 7.9 inches to 6.1 inches (a difference of 1.8 inches over approximately 65 miles).

4.7.3.3 Precipitation-Frequency Estimates in other Countries

In Australia, a recent revision of design rainfall was undertaken for the whole country in a staged process to revise the previous rainfall estimates by incorporating significant amounts of recent additional data as well as newer analytic techniques (Green et al., 2012). Other large-scale national design rainfall projects include work done to develop DDF rainfall data as part of the Flood Estimation Handbook in the United Kingdom (Centre for Ecology and Hydrology, 2008) to replace the Flood Studies Report (Natural Environment Research Council, 1975), as well as revisions of design rainfalls in New Zealand (Thompson, 2011)

Determination of DDF or IDF curves from local rainfall data requires several years of data. Two of the most widely accepted probability distributions are the Gumbel and GEV. The data consist of the annual maximum rainfall depths for various durations over the POR.

The rainfall intensities corresponding to various return periods can be estimated for the rainfall durations using

$$x_T = \bar{x} + K_T s \tag{4.49}$$

where x_T denotes the rainfall depth with return period T, $\bar{x}$ denotes the mean rainfall depth of the data series (time series), K_T denotes the **frequency factor** for the return period T, and s is the standard deviation of the data series. For example, the frequency factor for the Gumbel distribution is given by

$$K_T = -\frac{\sqrt{6}}{\pi}\left[0.5772 + \ln\left\{\ln\left(\frac{T}{T-1}\right)\right\}\right] \tag{4.50}$$

The data for each duration are analyzed separately and plotted in the form of an IDF curve. Equations 4.49 or 4.50 or a similar equation can then then be used to smooth the data to ensure consistency at a given duration for various frequencies.

4.7.4 Depth–Area Relationships

The precipitation or rainfall estimates made using one of the procedures described above is a measure of **point rainfall**. Point rainfall is rainfall occurring at a single point in space. On the other hand, **areal rainfall** is the average rainfall over a region. An intense point rainfall is unlikely to be distributed uniformly over the entire area of a watershed. In general, for a specified frequency and duration, the average rainfall depth over an area is less than the depth at a point. For this reason, point rainfall depths are usually extended to derive areal rainfall depth using a factor called the **areal reduction factor** (**ARF**), expressed as fraction of point depth. The first attempt to derive an ARF was probably

4.7 Design Storms

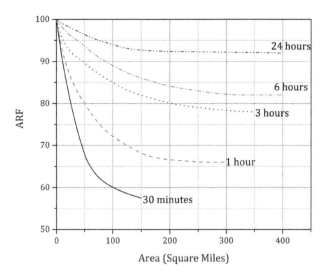

Figure 4.22 Areal reduction factors (ARFs) as a function of area and rainfall duration, as given in Technical Paper No. 29, Part 3 by the US Weather Bureau (1958).

in 1957, given by the in Technical Paper No. 29 Part 1 of the US Weather Bureau (1957). Subsequently, in Technical Paper No. 29 Part 3, the US Weather Bureau (1958) provided a graph that showed ARFs as a function of area and rainfall duration (Figure 4.22), and these curves were developed from averages of annual series of point and areal values for several dense, recording rain gauge networks.

The equation that was used by the US Weather Bureau to derive the ARFs from a dense network of recording rain gauges is

$$ARF = \frac{\sum_{i=1}^{r}\left(\frac{A_i}{r}\right)}{\frac{1}{nr}\sum_{i=1}^{r}\sum_{j=1}^{n}P_{ij}} \quad (4.51)$$

where r is the number of years of rainfall records, n is the number of stations in the network, A_i is the maximum areal rainfall of the ith year for a certain duration, and P_{ij} is the annual maximum rainfall depth for a certain duration at the jth station for the ith year.

Several observations of importance can be made from the studies conducted by the US Weather Bureau: (1) an ARF depends on rainfall duration and area and these relationships can be applied for rainfall for other frequencies as well; in other words, the ARFs do not change with rainfall frequencies; (2) the ARFs are much more significant for short-duration rainfalls than those with longer duration; (3) the ARFs are not significant for watersheds with smaller areas; and (4) the area–depth relationships are independent of geographic location and seasons. Along these lines, according to WMO (1994), ARFs should not be used for areas less than 9.6 square miles and according to USACE (1982) no adjustment to point rainfall depths should be made for durations less than 30 minutes.

It should be noted with due importance that the ARFs developed by the US Weather Bureau and still quite widely used were derived from the data existing at that time. Several studies have suggested that further improvements and better ways to estimate ARFs are necessary (e.g., Wright et al., 2014). Furthermore, Kao et al. (2020), using NOAA Atlas 14 precipitation data across multiple hydrologic regions of the United States and improved analytic methods, concluded that while ARFs decrease with increasing area and increase with increasing rainfall duration, they also decrease with increasing return period. This last conclusion is contradictary to that drawn in the earlier investigation given in TP-29. In addition, Kao et al. (2020) also concluded that ARFs are specific to regional climate patterns and geographical characteristics and should not be applied arbitrarily.

4.7.5 Temporal Distribution of Design Rainfall

4.7.5.1 Synthetic Rainfall Time Distribution

If design of hydrologic or hydraulic systems needs an estimation of peak discharge only, a knowledge of how frequently rainfall of a specific intensity and duration occurs at a location, that is, the IDF data of the location, is sufficient. The rational method, the topic of Chapter 11, utilizes this approach. However, for rainfall-runoff analysis, which requires the time distribution of flows, a **synthetic rainfall time distribution** (SRTD) of the design rainfall depth of a given duration and frequency is required. This is required in the hydrograph methods, discussed in Chapters 9 and 10. For a given rainfall depth, the computed runoff hydrographs can vary dramatically, depending on the **shape** and **duration** of the rainfall curve.

The SRTD is different from the observed time distribution of actual rainstorms that is discussed in Section 4.6. The SRTD includes maximum rainfall intensities for the selected design frequency arranged in a sequence that is critical for producing peak runoff. In other words, for a synthetic or design storm, the temporal rainfall distribution of rainfall intensity smoothly increases and decreases, whereas in actual storms the time distribution of rainfall intensity is irregular. Nevertheless, the intensity of rainfall varies considerably during rain events of various durations as well as geographical region. For this reason, the SRTD for a region is developed from an intensive analysis of historical records of many actual storms that occurred in a region over a long period. Table 4.3 summarizes the differences between design storms and actual storms.

4.7.5.2 Synthetic Rainfall Time Distribution in the United States

Conceptual Patterns or Mathematical Distributions: NRCS-Type Curves

To represent generalized patterns of SRTD in various regions of the United States, the Natural Resources Conservation Service (NRCS) (US Department of

Table 4.3 *Differences between design storms and actual storms*

Storm characteristic	Design storms	Actual storms
Storm duration	Specific duration (e.g., 24 hours, 6 hours, etc.)	Any duration from minutes to days
Temporal rainfall distribution	Smoothly increasing and decreasing rainfall intensity	Irregular rainfall pattern with respect to time, possibly including intervals of no rainfall
Intensity–duration relationship	Based on intensity–duration data for a single return period such as 25 years, 100 years, etc.	Generally, include intensity–duration data for different return periods

Agriculture, Soil Conservation Service, 1973) developed four synthetic 24-hour rainfall distributions known as type I, IA, II, and III distributions or **type curves**. These patterns were developed from DDF data of TP-40, Hydro-35, and local rainfall records. For each rainfall curve, the rainfall values for all durations for a single frequency occur within one 24-hour period. The curves are generated by imbedding a set of ratios of the shorter durations to the 24-hour rainfall depths such that the high-intensity rainfalls are centrally nested and the resulting patterns are broad approximations to observed regional patterns.

The four distributions are shown in Figure 4.23. Types I and IA represent the Pacific maritime climate with wet winters and dry summers. Type III represents the Gulf of Mexico and Atlantic coastal areas where tropical storms bring large 24-hour rainfall amounts. Type II represents the rest of the contiguous United States. The steepest slope of a rainfall distribution curve represents the most intense portion of a rainstorm. Thus, Type IA and Type II represent the least and the most intense rainfalls, respectively. Types II and III curves represent **center-peaking** rainstorms. That is, the beginning and end of the rain events have a relatively low intensity compared to the higher intensity in the middle portions. In other words, these type curves do not really model an intense storm that lasts for 24 hours. They model an intense storm of a few hours that is sandwiched between several hours of mild rains on each side. The four curves mainly differ from one another in the allocation of different proportions of rainfall depths in different time windows of 24 hours.

The NRCS-type curves have been widely used in the United States by design engineers to distribute 24-hour rainfall depths for a desired storm frequency and applicability in the region of interest. This has mainly been due to the unattainability of established local rainfall curves. NRCS (2019) has also developed procedures to transform 24-hour-type curves to rainfall distributions of shorter duration, which have the same distributional characteristics of 24-hour curves but can be used with rainfall depths for shorter durations (Figure 4.24). Since the original NRCS-type curves were mainly based on TP-40 and Hydro-35, NRCS (2019) has prescribed procedures to develop the same standard-type curves with the DDF data from NOAA Atlas 14. Figure 4.25 shows NRCS-type curves, calculated from DDF data given in NOAA Atlas 14, for three stations located in three distinctly different climatic regions of Texas.

Empirical Patterns or Statistical Distributions: Huff Curves

With the Atlas 14 program, NOAA has developed Huff curves for specific durations, based on statistical distributions of rainfall depths in different climatic regions of the United States. These curves capture the regional characteristics of rainfall patterns for specific durations in much greater detail than that given by the NRCS-type curves.

Figure 4.26 shows the subregions (numbered) and regions (same symbols) of the contiguous United States that have temporal patterns of rainfalls of specific duration (6 hours, 12 hours, 24 hours, and 4 days) with distinctive attributes. The temporal patterns of rainfall of these durations are developed from an intensive statistical analyses of precipitation records of storm events of different duration. The basic idea behind this approach is that the longer-duration storms are expected to behave differently than those of shorter duration. For example, the most intense portion of a 24-hour storm is expected to differ from the most intense portion of a 1-hour storm. The temporal distributions for the duration are expressed in probability terms as cumulative percentages of precipitation totals. To provide detailed information on the varying temporal distributions, separate temporal distributions are also derived for four precipitation cases defined by the duration quartile in which the greatest percentage of the total precipitation occurred. Thereby, each of the temporal distributions is further divided into storms where the bulk of precipitation occurred in the first, second, third, or fourth quartile of storm duration, similar to the analyses presented by Huff (1967, 1990) and discussed above.

Figure 4. 27 shows the temporal distributions of 6-hour storms in region 1 of Texas. This is a semi-arid interior highland region. The first- and second-quartile storms of 6-hour duration in this region of Texas bear the characteristics typical of short-duration, high-intensity thunderstorms but the third- and fourth-quartile storms resemble cyclonic rainfall. Figure 4.28 shows the Huff curves of 24-hour storms in region 3 of Texas. This is the wetter Gulf coast plain region where tropical cyclones are common. The temporal distribution patterns of 24 h rainfalls show the range of possibilities that actually exist with such relatively long duration rainfall.

According to the statistical data on 39 550 Texas storms provided by Perica et al. (2018), the first-quartile storms for all durations are most frequent in all three regions of Texas except for the 6-hour-duration storms in region 3 where the third-quartile storms are most frequent. Similarly, the

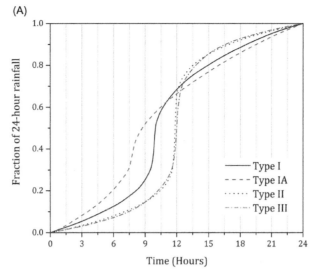

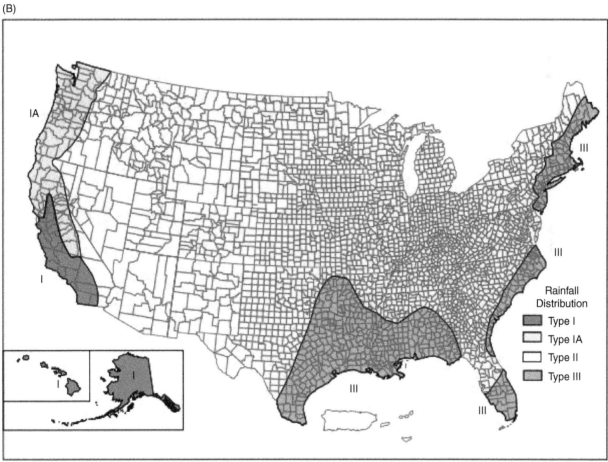

Figure 4.23 (A) Four types of time distributions of rainfall defined by NRCS. These are the mass curves. Each type is for 24-hour duration rain event. Note that each of these four rainfall distributions has an intense rainfall period somewhere near the middle and lesser rainfall intensities at the beginning and end of the storm. (B) Map of the United States showing the regions where these type curves are applicable (from NRCS, 1986). (A black and white version of this figure will appear in some formats. For the color version, please refer to the plate section.)

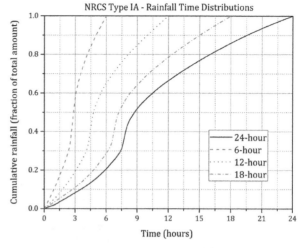
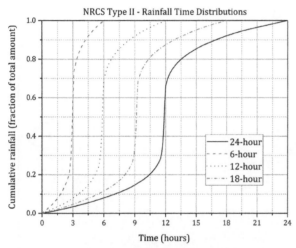
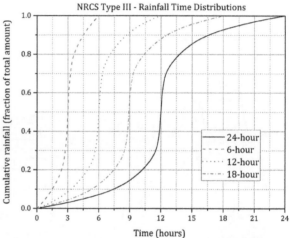

Figure 4.24 Time distribution of rainfall with four different durations for the four types defined by NRCS. Note that these curves of shorter duration exhibit the same temporal distributional patterns that are given by the 24-hour curves shown in Figure 4.23.

fourth-quartile storms are least frequent in all three regions of Texas except for 24-hour-duration storms in region 1 where the third-quartile storms are least frequent.

Huff (1967) showed the percentage of total rainfall as a function of cumulative percent of storm time, for a certain category of quartile storm with various probabilities of occurrence, in the form of bar charts that he called histograms (strictly speaking these are not histograms). Figure 4.29 shows a typical example, from which it can be deduced that for the first-quartile storms with 24-hour duration, in 10% of cases, the bulk of rainfall occurs during the first 10% of the time of the storm duration; in 50% of cases, the bulk of rainfall occurs within the first 30% of the time of the storm duration; and, only in 10% or fewer cases, are the rainfall amounts almost uniformly distributed throughout the storm duration.

4.7.5.3 Synthetic Rainfall Time Distribution in Other Parts of the World

For any location, the best-fit curve can be derived by developing a time distribution of rainfall from the IDF data and using each of the type of curves that best describes that location.

4.7.5.4 Construction of Synthetic Hyetographs and Mass Curves from IDF Data

As noted above, IDF data cannot be used directly for the computation of runoff hydrographs. However, these data can be used to develop hyetographs that are employed for design hydrograph calculations. Several procedures have been developed to construct hyetographs from IDF data. Two of the most common methods are the alternating block or IDF method (Chow et al., 1988; Haan et al., 1994) and the Chicago hyetograph method (Keifer and Chu, 1957).

Alternating Block or IDF Method

The alternating block or IDF method is a simple way of developing a design hyetograph from an IDF curve. The design hyetograph developed by this method specifies the rainfall depth occurring in n successive time intervals, each

4.7 Design Storms

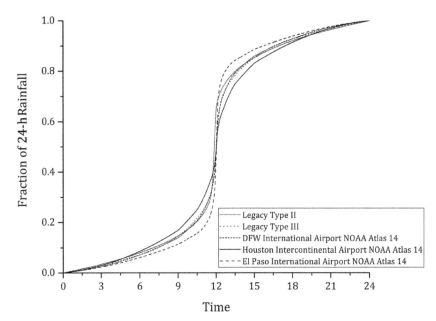

Figure 4.25 NRCS-type curves, calculated from 100-year depth-duration data given in NOAA Atlas 14, for three stations, located in three climatically distinct regions of Texas. The original Type II and Type III curves (legacy curves) are shown for comparison. Note that the curves, updated from rainfall depth-duration data given in NOAA Atlas 14, differ from the legacy curves only in ratios of rainfall depths but not in the overall pattern of temporal distribution.

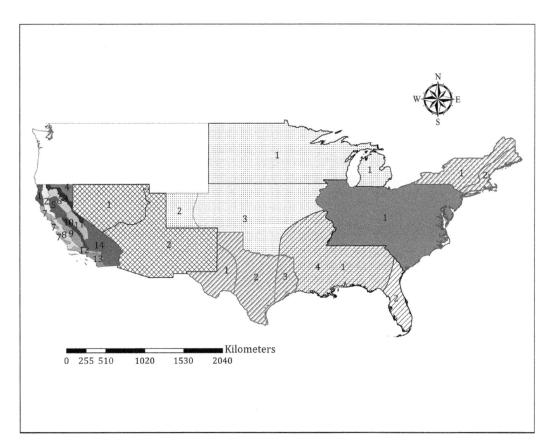

Figure 4.26 Regions and subregions of the contiguous United States that have different characteristics of temporal distribution of rainfall of 6-hour, 12-hour, 24-hour, and 4-day duration according to precipitation-frequency estimates given in NOAA Atlas 14 (source of data: https://hdsc.nws.noaa.gov/pfds). No information has been yet developed for the unshaded region (the northwest). Each region is shown by one type of shading and, within each region, subregions are denoted by numeric labels.

with duration Δt, over a total duration T_D ($T_D = n \times \Delta t$). The steps involved in this method are as follows.

A design return period is selected and the intensity values for each of the durations are computed using an equation such as Eq. 4.47 or 4.48. The corresponding rainfall depths are calculated as the product of intensity and duration. The depths calculated in this step are cumulative depths at each of the time increments. The

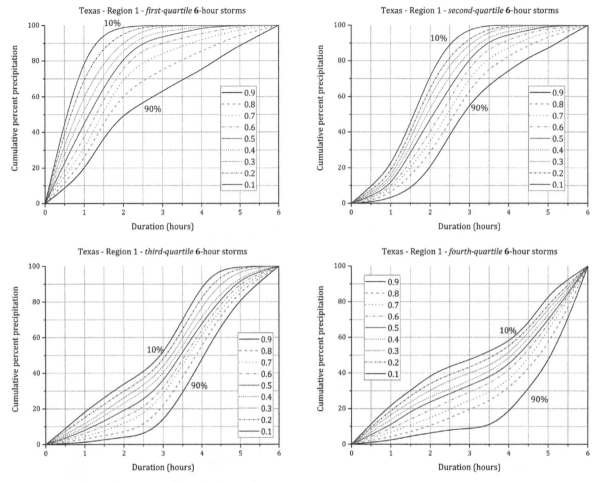

Figure 4.27 Huff curves showing temporal distributional characteristics of 6-hour storms in region 1 of Texas. (A black and white version of this figure will appear in some formats. For the color version, please refer to the plate section.)

incremental depths are then calculated by taking the difference between successive rainfall depth values. These are the amounts of rainfall to be added for each additional unit of time, Δt. The incremental rainfall depths are *rearranged* into a time sequence with maximum intensity occurring at the center of the duration and the remaining blocks arranged in descending order alternatively to the right (time greater than the center) and left (time less than center) of the central block to form a **balanced** storm. The area under this design hyetograph is equal to the design storm depth, P.

Table 4.5 shows the example calculation for a 25-year, 3-hour hyetograph constructed from the IDF data given in Table 4.4, used by the Department of Water Management of the City of Chicago.

Figure 4.30A shows a synthetic hyetograph. The balanced hyetograph constructed from IDF data, using the alternating block method, can be converted to mass curves, as shown in Table 4.5. Note that the mass curve constructed from the IDF data is *center-peaking*, meaning, the most intense portion of the storm is placed in the center of the curve. This is true for storms of all durations because the intensity (slope) of the center portion of each curve created from the IDF data is identical. The beginning and end tails of the same curve are lengthened to attain the desired duration and the corresponding total depth. These types of rainfall curves indicate that the most intense portions of storms are the same, regardless of the storm duration. This concept differs from the Huff curves derived from statistical distributions that are not necessarily center-peaking and might have intensities in different quartiles of storm duration. The center-peaking mass curves derived from the IDF data are similar to the NRCS-type curves.

It should be noted that the NRCS-type curves were developed so that, for a selected frequency, the depth–duration relationship based on the curves would be very close to the depth–duration curve developed from a frequency analysis of actual rainfall data. Thus, the hyetographs developed from the IDF data of a location should be similar to the hyetograph that can also be constructed from an NRCS-type curve with a given rainfall depth. This is shown in Figure 4.30B.

The alternating block method can be used to place the block of maximum incremental depth first at any specified

4.7 Design Storms 135

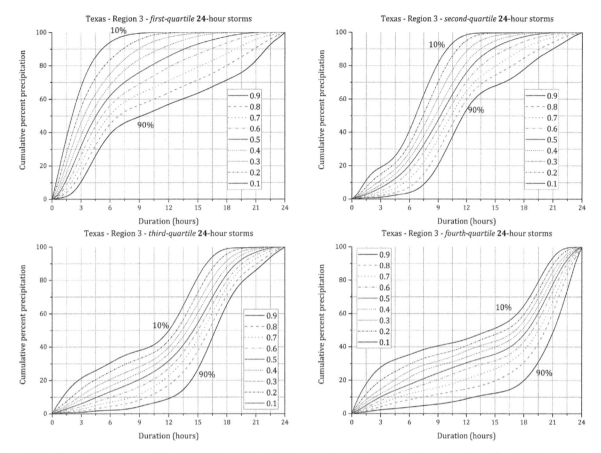

Figure 4.28 Huff curves showing temporal distributional characteristics of 24-hour storms in region 3 of Texas. (A black and white version of this figure will appear in some formats. For the color version, please refer to the plate section.)

location within the storm duration. For example, HEC-HMS gives the options to place the block of maximum incremental rainfall at 25%, 33%, 50%, 67%, or 75% of the time measured from the beginning of the storm to the total storm duration. The remaining blocks are arranged then in descending order, alternating before and after the central block. When the maximum incremental depth is not located at the 50% point, the arranged blocks stop alternating when the first part (25% and 33%) or latter part (67% and 75%) of the remainder of the rainfall is filled.

Chicago Method

Keifer and Chu (1957) partitioned the storm pattern so that, at any intensity, the time from the center of the storm back to the rising limb, t_a, divided by the time from the rising limb to the falling limb of the storm, $t_a + t_b$, that is, the total duration, T_D, was a constant, r (Figure 4.31). The ratio, r, is called the *storm advancement coefficient*. Thus,

$$r = \frac{t_a}{t_a + t_b} = \frac{t_a}{T_D} \tag{4.52}$$

or

$$T_D = \frac{t_a}{r} = \frac{t_b}{1-r} \tag{4.53}$$

For a given duration, T_D, the average intensity of rainfall, is given by Eq. 4.3. The instantaneous intensity, I, is given by

$$\frac{dP}{dT_D} = I = I_{ave} + T_D \frac{dI_{ave}}{dT_D} \tag{4.54}$$

If the average intensity is given by, say Eq. 4.48 where for a given frequency, we use the constant c to represent KF^x,

$$I_{ave} = \frac{c}{T_D^e + f} \tag{4.55}$$

then differentiation of Eq. 4.55 yields for Eq. 4.54

$$I = \frac{c\left[(1-e)T_D^e + f\right]}{\left(T_D^e + f\right)^2} \tag{4.56}$$

Now, substituting t_a and t_b for T_D (Eq. 4.53), the intensities for the rising and falling limbs of the hyetograph are, respectively, obtained as

$$I_a = \frac{c\left[(1-e)\left(\frac{t_a}{r}\right)^e + f\right]}{\left[\left(\frac{t_a}{r}\right)^e + f\right]^2} \tag{4.57}$$

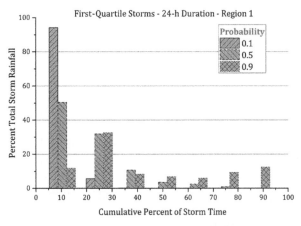
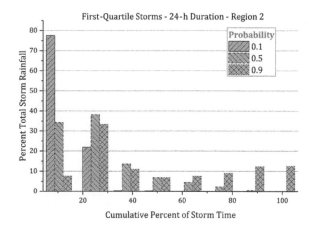
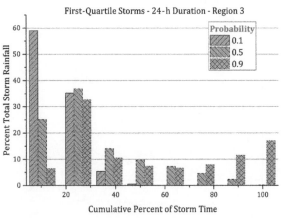

Figure 4.29 Bar charts (Huff histograms) showing time distributions of various percentages of first-quartile storms in all three regions of Texas with 24-hour duration and having a 10%, 50%, and 90% probability of occurrence.

$$I_b = \frac{c\left[(1-e)\left(\frac{t_b}{1-r}\right)^e + f\right]}{\left[\left(\frac{t_b}{1-r}\right)^e + f\right]^2} \quad (4.58)$$

Similarly, if Eq. 4.47 is used to represent the average intensity and a constant a is used to represent KF^x, the corresponding equations are

$$I_a = \frac{a\left[(1-n)\left(\frac{t_a}{r}\right) + b\right]}{\left[\left(\frac{t_a}{r}\right) + b\right]^{(1+n)}} \quad (4.59)$$

$$I_b = \frac{a\left[(1-n)\left(\frac{t_b}{1-r}\right) + b\right]}{\left[\left(\frac{t_b}{1-r}\right) + b\right]^{(1+n)}} \quad (4.60)$$

Note that $r = 0.5$ produces a symmetrical pattern (a balanced storm). A value of r between 0 and 0.5 produces an advanced pattern, and a value between 0.5 and 1.0 produces a delayed pattern.

Figure 4.32A shows the intensity graph (A) and hyetograph with mass curve, calculated using the Chicago method for the same 3 h, 25-year storm for the City of Chicago shown in Figure 4.30.

The intensities calculated from Eqs. 4.57 to 4.60 are instantaneous intensities. From these average intensities, the incremental rainfall depths must be calculated. For this reason, the cumulative rainfall depths thus obtained are not quite accurate. The accuracy increases with decreasing time steps of computation. For example, a time interval of 2 minutes was used to calculate the incremental rainfall depths, as shown in Figure 4.32B. The cumulative depth thus obtained is 3.38 inches, which is close to the value of 3.36 inches obtained in the alternating block method (the value for 3-hour, 25-year rainfall depth is 3.53 inches as shown in Table 4.4)

While the Chicago method offers the flexibility of positioning the peak intensity of a storm event at any desired time, the method suffers from the limitation of significantly inaccurate calculation of the STP when $r \neq 0.5$. This is due to the approximation of average intensities from instantaneous intensity values.

4.7.5.5 Center-Peaking versus Statistical Distribution of Temporal Rainfall Depths

There can be significant variations between center-peaking and statistically derived rainfall distributions. An obvious question then arises as to *which approach is better for hydrologic design work*? It can be argued that a statistically

Table 4.4 *City of Chicago IDF data for design rainfall*

Storm duration (min)	5-Year 20% annual chance		10-Year 10% annual chance		25-Year 4% annual chance		50-Year 2% annual chance		100-Year 1% annual chance	
	Rainfall (in)	Average intensity (in/h)	Rainfall (in)	Average intensity (in/h)	Rainfall (in)	Average intensity (in/h)	Rainfall (in)	Average intensity (in/h)	Rainfall (in)	Average intensity (in/h)
5	0.46	5.520	0.54	6.480	0.66	7.920	0.78	9.360	0.91	10.920
10	0.84	5.040	0.98	5.880	1.21	7.260	1.42	8.520	1.67	10.020
15	1.03	4.120	1.21	4.840	1.49	5.960	1.75	7.000	2.05	8.200
30	1.41	2.820	1.65	3.300	2.04	4.080	2.39	4.780	2.80	5.600
60	1.79	1.790	2.10	2.100	2.59	2.590	3.04	3.040	3.56	3.560
120	2.24	1.120	2.64	1.320	3.25	1.625	3.82	1.910	4.47	2.235
180	2.43	0.810	2.86	0.953	3.53	1.177	4.14	1.380	4.85	1.617
360	2.85	0.475	3.35	0.558	4.13	0.688	4.85	0.808	5.68	0.947
720	3.31	0.276	3.89	0.324	4.79	0.399	5.62	0.468	6.59	0.549
1080	3.50	0.194	4.11	0.228	5.06	0.281	5.95	0.331	6.97	0.387
1440	3.80	0.158	4.47	0.186	5.51	0.230	6.46	0.269	7.58	0.316
2880	4.09	0.085	4.81	0.100	5.88	0.123	6.84	0.143	8.16	0.170
4320	4.44	0.062	5.18	0.072	6.32	0.088	7.41	0.103	8.78	0.122
7200	4.91	0.041	5.70	0.048	6.93	0.058	8.04	0.067	9.96	0.083
14400	6.04	0.025	6.89	0.029	8.18	0.034	9.38	0.039	11.14	0.046

derived rainfall curve better represents the rainfall pattern in the area where the project site in question is located than assuming a uniform distribution of rainfall intensity or a center-peaking pattern. Yet, there is every probability that rainfall events in that area will still occur with significantly different characteristics than that used for a design using a statistically derived rainfall pattern. A site will encounter a wide variety of rain events. Rarely, a site will experience two rain events with identical rainfall depth, duration, and time distribution. An engineer needs to understand the kinds of rain events and the range of possible rainfall patterns that can occur at their project site and ideally should use a range of patterns to ensure that the system that is being designed can adequately respond to various types of events with a high degree of probability.

4.8 PROBABLE MAXIMUM PRECIPITATION

4.8.1 Definition and Scope

Probable maximum precipitation (PMP) is defined as the theoretically greatest depth of precipitation for a given duration that is physically possible over a given size storm area and reasonably characteristic over a given geographic location at a certain time of the year. The PMP is only an analytical estimate representing a theoretical upper limit and therefore cannot be exact. The probability of occurrence of an estimated PMP is not usually given and the impact of climate change is not evaluated.

For the design of certain hydraulic structures, such as reservoirs, dams, and spillways where the loss of life and significant property damage would result from overtopping or failure, it is necessary to obtain the PMP of the project area, because the risks associated with the design of large flood or water control measures are too high to have any room for inadequacy as opposed to the risks involved in the design of structures where frequency-based design storms are used. PMP estimates are also used for dam breach and dam safety analysis.

There are two different approaches or methods for the determination of PMP of a region. The first approach is deterministic and the second is probabilistic. The outlines of the methods are given below. Further details can be found in WMO (1986; 2009) and in Mukhopadhyay and Kappel (2017).

4.8.2 Deterministic Method of Estimation of PMP

The deterministic method of estimation of PMP is based on hydrometeorological methods and is also known as the physical method or synoptic method. This is the traditional approach of PMP determination and involves the following steps. This approach, largely developed by Schreiner and Riedel (1978) for all-season regional PMP estimates, is best suited for areas where there are adequate records of extreme precipitation and no significant orographic effect. However, the approach can be modified when there are inadequate storm samples (e.g., Hansen et al., 1977) and orographic effects (e.g., Miller et al., 1984).

4.8.2.1 Rainfall Records of Outstanding Storms

PMP estimates over a region require compilation of records of severe storm events that produced exceptionally high

Table 4.5 *Calculation of 3-hour, 25-year synthetic hyetograph using the alternative block method from the IDF data given in Table 4.4*

Time (min)	Intensity (in/h)[a]	Depth (in)[b]	Incremental depth at each time interval (in)	Rearranged depth (in)[c]	Cumulative depth (in)[d]
0	9.9705	0.0000	0.0000	0.0000	
5	8.1308	0.6776	0.6776	0.0120	0.0120
10	6.8766	1.1461	0.4685	0.0133	0.0253
15	5.9611	1.4903	0.3442	0.0148	0.0401
20	5.2624	1.7541	0.2639	0.0165	0.0566
25	4.7114	1.9631	0.2090	0.0187	0.0753
30	4.2656	2.1328	0.1697	0.0212	0.0965
35	3.8973	2.2734	0.1406	0.0244	0.1209
40	3.5878	2.3919	0.1185	0.0283	0.1491
45	3.3242	2.4931	0.1013	0.0332	0.1824
50	3.0968	2.5807	0.0876	0.0397	0.2220
55	2.8987	2.6572	0.0765	0.0482	0.2702
60	2.7246	2.7246	0.0674	0.0599	0.3301
65	2.5703	2.7845	0.0599	0.0765	0.4066
70	2.4326	2.8380	0.0536	0.1013	0.5079
75	2.3090	2.8862	0.0482	0.1406	0.6485
80	2.1974	2.9299	0.0436	0.2090	0.8574
85	2.0961	2.9695	0.0397	0.3442	1.2016
90	2.0038	3.0058	0.0362	0.6776	1.8792
95	1.9194	3.0390	0.0332	0.4685	2.3477
100	1.8418	3.0696	0.0306	0.2639	2.6116
105	1.7702	3.0979	0.0283	0.1697	2.7813
110	1.7041	3.1241	0.0262	0.1185	2.8998
115	1.6427	3.1485	0.0244	0.0876	2.9873
120	1.5856	3.1712	0.0227	0.0674	3.0547
125	1.5324	3.1924	0.0212	0.0536	3.1083
130	1.4826	3.2123	0.0199	0.0436	3.1519
135	1.4360	3.2309	0.0187	0.0362	3.1881
140	1.3922	3.2485	0.0176	0.0306	3.2187
145	1.3511	3.2650	0.0165	0.0262	3.2450
150	1.3123	3.2807	0.0156	0.0227	3.2677
155	1.2757	3.2954	0.0148	0.0199	3.2875
160	1.2410	3.3094	0.0140	0.0176	3.3051
165	1.2083	3.3227	0.0133	0.0156	3.3207
170	1.1772	3.3353	0.0126	0.0140	3.3347
175	1.1477	3.3473	0.0120	0.0126	3.3473
180	1.1196	3.3588	0.0114	0.0114	3.3588

[a] Calculated from Eq. 4.48: $K = 5.1409$, $x = 1.1633$, $e = 0.9917$, $f = 21.8029$; [b] From Eq. 4.3; [c] Depth rearranged in alternating blocks by placing the highest incremental depth at 90 minutes; [d] Cumulative depths calculated from rearranged depths in the previous column.

amounts[2] of rainfall in the region from the weather stations having records of rainfall data of various durations, such as 6 hours, 1 day, 2 days, or 3 days. The selected storms are extended to the areas where storms can be transposed (storm transposition is discussed below). These selected storm events that affected the region of interest are noted as **outstanding storms** and are the most influential in setting the level of the PMP of the region in one combination of area, size, and duration. Thus, the first requirement for estimating the PMP of a region is to build a database using precipitation depths recorded for severe storms that brought exceptional rainfall amounts, with variable duration, and affected various areal extents (e.g., 100, 500, 1000, 10 000 km^2).

---

[2] The choice of major storms can be subjective. For this reason, it may be necessary to establish a *threshold* rainfall depth for a region. Storms producing rainfall depths above the threshold value are selected. The threshold value for the rainfall depth depends on the synoptic climatology of the region and size of the area. Thus, a high amount is quantified by the rarity of the event and/or the magnitude compared to other rainfall events at that station location.

4.8 Probable Maximum Precipitation 139

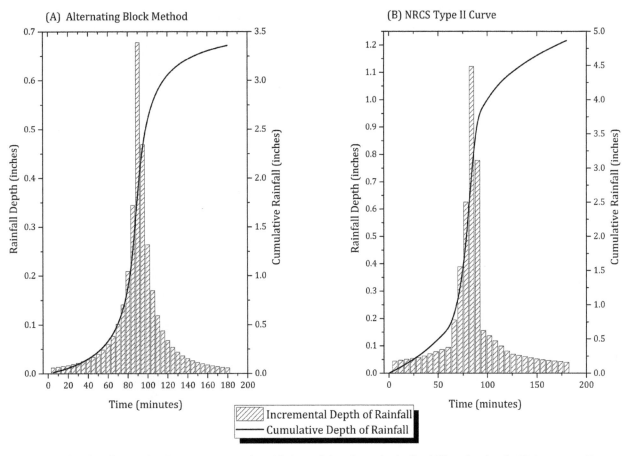

Figure 4.30 (A) Synthetic hyetograph and mass curve representing a 3-h, 25-year balanced storm for the City of Chicago based on the IDF data presented in Table 4.5 using the alternating block or IDF method. (B) Synthetic hyetograph and mass curve representing a 3-hour, 25-year balanced storm for the City of Chicago based on the 24-hour, 25-year rainfall depth presented in Table 4.5 using a Type II NRCS curve.

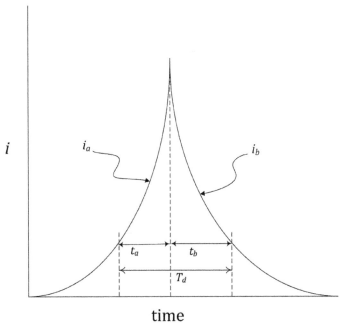

Figure 4.31 Parameters used in the Chicago hyetograph method.

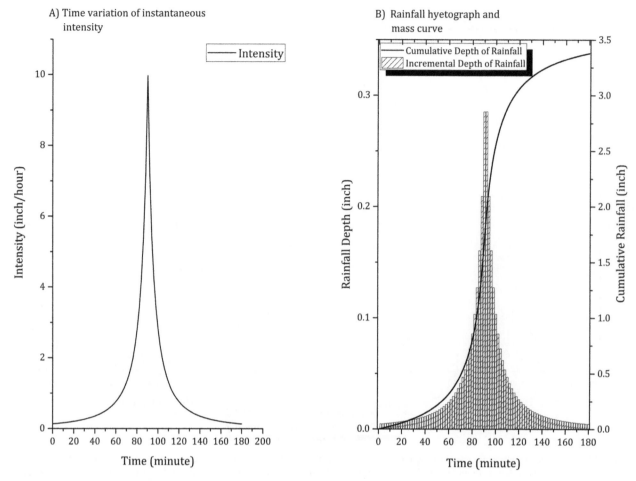

Figure 4.32 (A) Instantaneous intensity and (B) synthetic hyetograph and mass curve representing a 3-hour, 25-year storm for the City of Chicago based on the IDF data presented in Table 4.5 using the Chicago hyetograph method and setting $r = 0.5$ (for comparison with Figure 4.30).

4.8.2.2 Depth–Area–Duration Analysis

The type of rainfall data noted above is used to derive maximum observed areal precipitation depths for various durations. The derivation of area-averaged precipitation depths for various durations from discrete observation points (rain gauge stations) involves a procedure called **depth–area–duration (DAD) analysis**. In the United States, DAD data are available for many major storms that have occurred during certain periods in the past. In many countries, such data are either scant or absent. In such instances, a major effort in relation to the production of PMP charts is directed toward the development of DAD relationships of the outstanding storms.

A DAD analysis requires interpolation and integration of point precipitation estimates for a given storm. Details of the procedure have been described by the WMO (1969) and Singh (1992). Even though with the aid of modern software and GIS technology the procedure can be expedited, it still requires a considerable effort. The essential steps in the DAD analysis of an outstanding storm are as follows.

1. Establish the beginning and ending period of the storm.
2. Compute and tabulate the STP at 1-hour increments at each of the gauging stations.
3. Develop isohyetal maps of the area affected by the storm for each of the 1-hour periods. This step can be accomplished with GIS and by employing a contouring routine. In instances where weather radar data are available, this can provide more accurate spatial representation of the rainfall accumulation between rain gauge locations.
4. The isohyetal map of the lowest time interval (1 hour) is analyzed first. Beginning with the central isohyet (highest value or the storm center), the area enclosed by each isohyet is calculated and then the net area between each pair of isohyets is determined. From this step, the mean precipitation depths (average of two consecutive isohyets) for various areas for the given storm duration are calculated.
5. The procedure outlined in step 4 is repeated for each of the storm durations.
6. In the storm-centered DAD analysis, the last enclosing isohyet is taken as the boundary, whereas the watershed boundary is used in watershed-centered analysis. A large

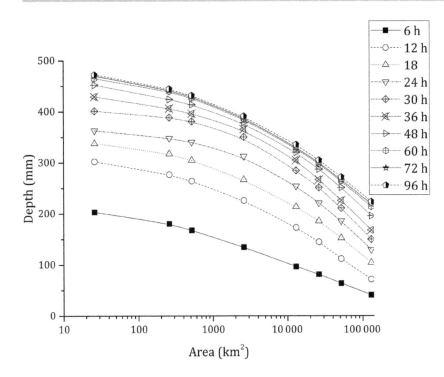

Figure 4.33 Depth–area–duration (DAD) curves of various duration for the storm of June 19–23, 1972, centered at Zerbe, Pennsylvania (data source: Schreiner and Riedel, 1978).

river basin is subdivided into catchments and catchment-centered analysis is performed for the outstanding storm, because a particular storm may not affect the entire basin.

7. The results of tabulations from steps 4 and 5 are plotted on a logarithmic scale and curves are fitted through the data points to obtain depth–area relationships for various durations.

DAD analysis is conducted for the severe rainstorms at the place of their occurrence. Schreiner and Riedel (1978) tabulated the results of DAD analysis of 55 outstanding storms that occurred in the United States to the east of the 105° W meridian between the period 1878 and 1972. Figure 4.33 shows the DAD relationships for a storm that occurred from June 19 to 23, 1972, with the rainfall center at Zerbe, Pennsylvania. It can be seen that the depth of precipitation decreases with increasing area for all durations. With increasing duration, for a given area, the depth of precipitation increases. In addition to making the best use of existing DAD data in a project area, newer storm data, if available, should be used for new sets of DAD analysis.

4.8.2.3 Moisture Maximization of Observed Extreme Precipitation Amounts

Moisture maximization is the process of increasing the observed storm rainfall depths to a value that would result if the moisture available to the storm were the maximum value that is consistent with the maximum moisture in the atmosphere for the storm location and month of occurrence.

The actual precipitation depth, obtained from DAD analysis, is maximized according to

$$P_M = P \frac{W_P^{Max}}{W_P^{Storm}} \quad (4.61)$$

where P_M denotes the maximized precipitation depth, P is the observed precipitation depth, W_P^{Max} signifies the enveloping highest observed precipitable water depth (at maximum dew point at 100 kPa, 12-hour persisting), and W_P^{Storm} denotes the observed precipitable water depth in the storm (at storm dew point 100 kPa, 12-hour persisting).

The ratio W_P^{Max}/W_P^{Storm} is called the **moisture maximization factor** or **moisture adjustment factor** (**MAF**). The precipitable water (W_P) is a measure of the atmospheric moisture. It is the depth of water vapor condensed into liquid in a column of air of unit cross section. For a saturated pseudo-adiabatic atmosphere, tables have been prepared, giving the W_P values based on mean sea level or 1000 mb (100 kPa) dew points (e.g., WMO, 1986). Two dew-point temperatures are required for moisture maximization. One is the dew-point representative of moisture inflow during the storm and the other is the maximum dew point for the same location and time of the year as the occurrence of the storm. Both storm and maximum dew points are reduced pseudo-adiabatically to 1000 mb (100 kPa) in order to normalize differences in recording station elevations and are taken as the highest value persisting for 12 hours or for the average duration representing the storm rainfall accumulation, e.g., 3, 6, 12, or 24 hours.

Dew-point temperatures at station elevations are calculated from recorded data of relative humidity and dry-bulb (air) temperatures. Several relationships exist for these calculations. A simple approximation is given by

$$T_D = T - \frac{(100 - RH)}{5} \quad (4.62)$$

where T_D denotes the dew-point temperature (°C), RH is the relative humidity, and T denotes the air or dry-bulb temperature (°C).

If both dry- and wet-bulb temperatures are available, a closer approximation can be obtained by calculations of saturated water-vapor pressure at dry- and wet-bulb temperatures. As noted above, the dew-point temperatures must be normalized to mean sea-level values (100 kPa). From the calculated values of dew points, the values of precipitable water are obtained from the look-up tables.

As an example of moisture maximization, let us consider a storm that occurred over Warner, Oklahoma, on May 6, 1943. At the place of the storm, the dew-point temperature was 21.1 °C. At this temperature, $W_P^{Storm} = 57.61$ mm. The maximum dew-point temperature was 25 °C. At this temperature, $W_P^{Max} = 81.05$ mm. Thus, at the place of the storm, $W_P^{Max}/W_P^{Storm} = 1.41$. This storm was transposed to Kansas City where the maximum dew-point temperature was 23.9 °C. At this temperature, $W_P^{Max} = 73.56$ mm. Thus, at the place of transposition, $W_P^{Max}/W_P^{Storm} = 0.91$. So, the total adjustment factor was 1.28 at the place of transposition.

4.8.2.4 Storm Transposition

Storm transposition is the procedure whereby an outstanding storm that occurred in one area of a region, which is considered meteorologically and topographically similar to the location being analyzed, is hypothetically moved to this area under consideration. This procedure is often necessary because there may be areas in a project region where there is either an inadequate record of rainfall data or no past occurrence of severe storms. Transposition greatly increases the available data for evaluating the rainfall potential of a region. However, an outstanding storm can only be transposed from one area to another area if the two areas are homogeneous relative to the terrain and meteorological characteristics important to the particular storm rainfall under consideration. Two different areas can be considered meteorologically homogeneous if they are influenced by the same moisture source, experience similar types of storms, have similar topography, and encounter similar orientation to seasonal wind-flow directions. The principal factors that control meteorological homogeneity of two areas are the topography, distance from the sea, direction of the prevailing winds, elevation differences, latitudinal variations, season of occurrence, and mean annual temperature.

After a DAD analysis of the outstanding storms at their place of occurrence within or around the region of interest, it is necessary to establish the transposition limits of these storms. The setting of such limits involves a detailed study of the storms to determine the meteorological controls of the heavy rain producing them.

Storm transposition requires an understanding of the synoptic meteorology of the project region. The orientation of the transposed storm must be such that rainfall over the subbasin or sub-area yields maximum precipitation depths for each of the durations obtained in the previous DAD analysis. Furthermore, when storm transposition is involved, it is necessary to adjust the maximized precipitation depths through moisture maximization process. In this case, the MAF is modified as given by

$$P_M = P \frac{W_P^{Max2}}{W_P^{Storm}} \quad (4.63)$$

where P_M and P are as defined above, W_P^{Max2} is the precipitable water depth corresponding to the maximum dew-point temperature in the new location, and W_P^{Storm} is the precipitable water depth in the storm in its original location.

Once storms are transposed to a new area, several other adjustments are necessary in addition to the adjustment made for the MAF. The most common is the elevation and barrier adjustment.

If a significant mountain barrier exists between the moisture source and the area over which a rainstorm is analyzed or if the area under consideration is at a higher elevation, the MAF is adjusted. In these cases, the mean elevation of the mountain ridge or elevation of the rainfall, rather than mean sea level (100 kPa surface), is used as the base of the column of moisture. If a storm transposition is made across mountains less than 600 meters high, allowance must be made to account for the mountains acting as barriers. This is because when moist air encounters a barrier, it may lose moisture during its ascension to the windward slopes of the mountain. In such cases, the MAF is given as the ratio of the precipitable water at the foothills (W_l) and that at the crest of the barrier (W_c). The value of MAF (W_c/W_l) would be less than one for windward slopes but greater than one for the leeward slopes. Mukhopadhyay and Kappel (2017) give further details on the adjustment of the MAF due to barriers.

When the moist air encounters a barrier, it may lose moisture during its ascension to the windward slopes of the mountain. In that case, the MAF is adjusted as

$$P_M = P \frac{W_P^{Crest}}{W_P^{Foot}} \quad (4.64)$$

where P_M and P are as defined above, W_P^{Crest} is the precipitable water depth at the crest of the barrier, and W_P^{Foot} denotes the precipitable water depth at the foothills. The value of the MAF is less than the one for windward slopes and greater than the one for leeward slopes.

4.8.2.5 Envelopment

The process of maximization of a single storm and its transposition to a basin does not assure that the estimation

4.8 Probable Maximum Precipitation

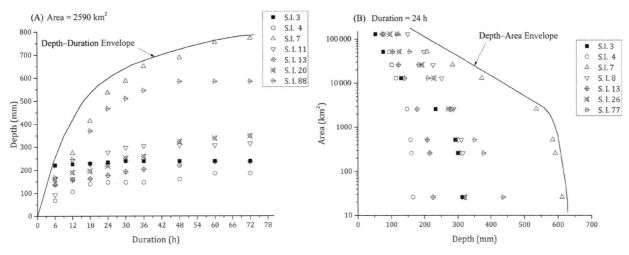

Figure 4.34 (A) Depth–duration envelope of maximized and adjusted storm values for a fixed area (2590 km²). (B) Depth–area envelope of maximized and adjusted storm values for a fixed duration (24 hours). Data are from Schreiner and Riedel (1978). S.I. indicates storm index (given in the data source).

at the level of the PMP has been obtained. Envelopment is a process of selecting the largest value from a set of data. Thus, it is also a rainfall maximization process. This accounts for the likelihood that no storm identified in the storm search process has achieved maximum efficiency and rainfall production at all area sizes and durations being investigated. Thus, an envelope of many rainfall values, maximized and transposed to an area, is quite likely to yield values indicative of the magnitude of the PMP. Three types of envelope curves are produced from analyses of several storms either adjusted at place or adjusted and transposed after the initial DAD analysis.

Constant-Area Depth–Duration Envelope

The depth and duration values for a fixed area of several storms after maximization and other adjustments either at place or transposed from one location to the area or watershed of interest are plotted in Figure 4.34A. A smooth curve (envelope) is drawn through the highest values of the precipitation depth for a fixed area. This is called a **durational envelope**.

Constant-Duration Depth–Area Envelope

The depth and area values for a fixed duration of several storms after maximization and other adjustments either at place or transposed from one location to the area or watershed of interest are shown in Figure 4.34B. A smooth curve (envelope) is drawn through the highest values of the precipitation depth. This is called an **areal envelope**. In this process, the enveloping curve is expected to be as large as or larger than all the maximized data points. For area sizes where the data transitioning between storm types (i.e., from a local storm to a synoptic storm) there may be a significant amount of envelopment to provide a smooth curve.

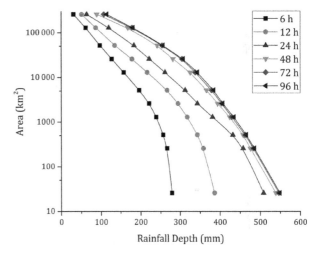

Figure 4.35 Enveloping DAD curves of probable maximum precipitation for English River Basin, Ontario, Canada.

Depth–Area–Duration Curves

The values read from the durational and areal enveloping curves are used to construct a set of DAD curves as shown in Figure 4.35 for English River basin in Ontario, Canada. Such final DAD curves are constructed for a large area or individual river basins or catchments within a basin. From these values, PMP charts or atlases are produced for the area under consideration.

4.8.2.6 Undercutting

The accuracy and reliability of the data used in the construction of an envelope curve are not equal. For this reason, sometimes it is justifiable to place the curves at somewhat lower values than the extremes as shown in Figure 4.34. This is called undercutting. Any undercutting should be done only after a careful review of (1) the meteorological characteristics of storms; (2) the transposition limits; (3) the moisture and other adjustment factors; and (4) any other factors that affect the magnitude of the plotted value.

4.8.2.7 Construction of a Grid System

A grid system, typically consisting of $0.25°–0.5° \times 0.25°–0.5°$ longitude–latitude grid, is placed over the base map of the region under consideration. The intersections of the grid points are the locations where (1) maximized storms are transposed with proper adjustments and (2) the envelope precipitation depths for different standard areas and for different standard durations are plotted. These sets of data are used to construct isohyetal maps of PMP or generalized PMP charts.

4.8.2.8 Generation of Generalized PMP Maps/Charts

For generalized PMP studies (such as HMR 51 produced by Schreiner and Riedel, 1978), the PMP values for specific durations and areas are plotted at various grid points, and smooth isolines are drawn to show the regional variation of areal PMP. The PMP isohyetal maps obtained in this way are further smoothed to remove any discontinuities in the PMP estimates between adjacent areas of a region. This process may sometimes result in undercutting or over-enveloping of the data of a few grid points in order to maintain consistency in the PMP estimates of the region and to produce a smooth PMP isohyetal pattern for the region.

4.8.2.9 Use of PMP Charts

PMP charts for various area sizes and durations can be used to derive the PMP of a specific project with a given area and for all seasons. PMP charts are available for various regions of the United States as hydrometeorological reports prepared by the NWS/NOAA and as technical papers prepared by the US Weather Bureau.

The general procedure for the derivation of project-specific PMP follows the following steps.

1. From the PMP maps, record the average PMP depths for the project location. Several area sizes covering the project area size should be selected and all durations for these area sizes should be selected.
2. Plot PMP depths on semi-log graphs (depth versus area) and draw smooth duration curves through the data points. Mathematical functions can be fitted to the data points to generate smooth curves so that the fitted equations can be used directly to calculate PMP depths for other area sizes for specific projects.
3. From the DAD graph of step 2, determine the PMP depths at the project area size for each duration.
4. Plot these project area PMP values on linear graph paper (depth versus duration for the given area). Draw a smooth curve connecting these points. Interpolate along this curve to obtain PMP depths for other durations, if required.

It should be noted that in probability (frequency)-based point precipitation estimates, there are three variables: depth or intensity, duration, and frequency of occurrence. For PMP estimates, the frequency or probability of occurrence is replaced by storm area as the third variable.

4.8.2.10 Use of Gridded Data to Derive PMP

The gridded PMP datasets are designed to allow for the calculation of basin-average PMP depths for drainage basins within a given project domain for any area size and duration for which the underlying rainfall DAD tables provide data (generally from 1 km^2 to 258 999 km^2 and from 1 hour through 72 hours).

The following steps are followed to obtain basin average PMP:

1. Create or obtain a polygon shapefile of the drainage basin outline and calculate the basin area. The calculated PMP is the average depth for the area of the basin. The areal reduction is inherent within the PMP development process and the DAD tables used from the storms, and *no further areal reduction should be applied*.
2. Using ArcGIS or similar GIS software, for a given duration, import the two PMP grid datasets for standard area sizes that bound the basin area size from step 1.
3. Extract the PMP grid data to the basin shapefile for both of the bounding area sizes. There are numerous methods for extracting data using GIS and the best approach depends on the experience level and needs of the user and the basin itself.
4. Obtain the mean raster value for the extracted area from the grid layers at both bounding area sizes. These values are the basin average PMP depth for each of the bounding standard area sizes.
5. Interpolate the basin-size PMP depth from the basin average values obtained in step 4 for both bounding area sizes. The user can apply a linear interpolation or plot four or more data points and apply a nonlinear curve fitting using a depth–area analysis. The linear interpolation can be done using the following equation:

$$P = \frac{(A - A_1)(P_2 - P_1)}{(A_2 - A_1)} + P_1$$

where A_1 is the smaller bounding area size, P_1 denotes the basin average PMP for the smaller bounding area size, A_2 is the larger bounding area size, P_2 denotes the basin average PMP for the larger bounding area size, A is the target basin area size, and P denotes the interpolated basin average PMP.

4.8.3 Probabilistic Method of Estimation of PMP

4.8.3.1 Frequency-based Equation

Chow (1951) presented a general frequency equation for the analysis of hydrologic time series. This equation was later modified by Hershfield (1961b, 1965) to compute the point PMP from 24-hour annual maxima spanning a certain POR. The Hershfield method is based on

$$P_M = \overline{P}_N + K_P \sigma_N \qquad (4.65)$$

where P_M denotes the maximum rainfall (PMP) of a given duration (D-hour, e.g., 24 hours), $\overline{P}_N$ denotes the mean of the D-hour annual rainfall maxima series covering the POR with N values, K_P is a constant equal to the frequency factor associated with a return period, and σ_N is the standard deviation of the D-hour annual rainfall maxima series. Thus, the basic data requirement of this method is a series of maximum annual D-hour rainfalls at a station. For a region, multiple station records are used, with each station providing a separate annual D-hour AMS. The point PMP estimates are then converted to an areal PMP estimate by using a depth–area relationship.

It follows from Eq. 4.65 that the frequency factor, K_P, is the number of standard deviations to be added to the mean of the annual maxima of D-hour rainfall records. The frequency factor is principally a function of the recurrence interval for a particular probability distribution. The value of K_P increases with increasing return period. Thus, K_P can be used both as an index to compare the relative frequency of rainfalls at different stations or for computing the rainfall probability if a distribution is assumed (Hershfield, 1981). Usually, the value of K_P is selected from the pdf that best fits the time-series data, but it can also be calculated from

$$K_P = \frac{P_N^{Max} - \overline{P}_{N-1}}{\sigma_{N-1}} \qquad (4.66)$$

where P_N^{Max} denotes the highest observed rainfall in a series of N annual maxima; $\overline{P}_{N-1}$ and σ_{N-1}, respectively, denote the mean and standard deviation of the annual maxima of the series excluding the highest value. The ratio of the D-hour PMP at a station and the D-hour highest observed rainfall at that station is an indicator of the magnitude of the PMP at that location (Figure 4.36).

Application of Eq. 4.65 may require adjustments to the values of $\overline{P}_N$ and σ_N for (1) maximum observed events within a short POR, (2) sample size, and (3) fixed observational time intervals (WMO, 1986). Nevertheless, in many places of the world, the lack of adequacy of the hydrometeorological data limits the application of the physical or deterministic method to establish regional PMP estimates. On the other hand, the statistical method has been widely applied in various parts of the world, because it is based on the type of data that are commonly available from rain gauge stations (Figure 4.36).

4.8.3.2 Homogeneity of Data Series

The AMS used in the determination of a PMP should be homogeneous, meaning that it is free from any temporal trends. The presence or absence of trends in an annual maximum rainfall series can be detected by using the

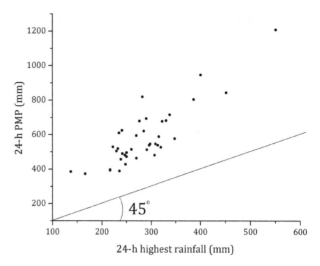

Figure 4.36 Calculated PMP at 39 stations plotted against the highest 24-hour rainfall at the respective stations at Johor, Malaysia (data are from Desa and Rakhecha, 2007). The PMP at a station is calculated based on 24 h AMS. The POR varies from 14 to 55 years with an average POR of 40 years.

Mann–Kendall rank test (WMO, 1966). The test statistic is computed from

$$\tau = \frac{4\sum n_i}{N(N-1)} - 1 \qquad (4.67)$$

where n_i is the number of values larger than the ith value in the series subsequent to its position in the series of N values. Its expected value in a random series, that is, a series without any increasing or decreasing trend, is zero. The variance of τ is given by

$$\sigma_\tau = \sqrt{\frac{4N+10}{9N(N-1)}} \qquad (4.68)$$

The ratio τ/σ_τ provides an indication of trend in the data. For a data series with no trend, τ/σ_τ should be within the limits of ± 1.96 at the 5% level of significance.

4.8.3.3 Areal Reduction

The PMP, calculated from various station records in an area, yields only point PMP estimates. Thus, the second step of the statistical method requires the development of areal reduction curves for adjusting the point values to various sizes of area. For this reason, the statistical method is mostly applicable to regions with area 1000–10 000 km^2.

Depth–area relationships to be used with any PMP studies should be based on the DAD characteristics of the storm types capable of producing the PMP in the region. Based on such relationships, curves for the reduction of point estimates of PMP of a given duration with increasing area have been established in certain parts of the world (e.g., US Weather Bureau, 1960).

Isoline or contour maps of the area are prepared from the point PMP estimates obtained at several stations. The areas between the isolines are calculated and the average PMP value is obtained for consecutive pairs of isolines. The values obtained in this process provide the depth–area relationships for the PMP at certain durations.

4.8.3.4 Depth–Duration Relationships

An AMS can be constructed for storms with various durations, such as 1-day, 2-day, and 3-day storms and, from an analysis of these individual annual series, the depth–duration relationships are established.

4.8.3.5 Generalized PMP Maps

Station PMP values are plotted on a map to generate isopleths at regular, say 100-mm, intervals. However, before isopleths are drawn, the values of the coefficient of variation $\left(C_v = \sigma_N/\overline{P}_N\right)$ at all stations should be plotted to see if there are stations where the C_v values differ significantly from those at the neighboring stations. This is because C_v is considered a more stable statistic than σ_N and $\overline{P}_N$. For such stations where the C_v values can be viewed as aberrant, these values are adjusted to match those of the neighboring stations. Subsequently, using the revised value of C_v of a station, the standard deviation (σ_N) is recalculated using the original value of the mean $(\overline{P}_N)$ of that station. The revised σ_N value thus obtained is then used to recompute the PMP estimates for those stations where σ_N values were readjusted.

4.8.4 Generalized versus Basin-Specific PMP Estimates

Generalized estimates of PMP refer to the PMP values over a large region encompassing several river basins with varying sizes, and they are usually represented by isohyetal maps depicting regional variation of PMP for specific duration and basin size. In typical cases, these maps, commonly known as **generalized PMP charts**, show PMP values for all seasons and thus ignore possible seasonal variations. In contrast to the generalized estimates, PMP estimates can also be made for individual river basins. The individual river basin-wide PMP estimates consider local climatology of the river basin instead of the synoptic climatology of a regional scale.

The conceptual definition of PMP provides the upper limit of the precipitation potential at the center of the precipitation pattern of a storm irrespective of the boundary of a river basin. In other words, the definition is a description of the upper limit of the precipitation potential that is *storm-centered*. The storm-centered PMP cannot be always applied directly to a river basin. It must be modified to develop **basin-averaged** but **spatially varied** PMP estimates. Hansen et al. (1982) defined these estimates as the average PMP depth over the drainage basin after the storm-centered PMP value has been distributed across the drainage basin in accordance with the PMP storm pattern.

4.8.5 Orographic Effects

The hydrometeorological method elaborated in Section 4.8.2 is based on the atmospheric convergence model and is mostly applicable to non-orographic regions. Since topography has a profound effect on rainfall, the PMP estimates for mountainous regions must be based on two rainfall components: (1) orographic precipitation, which results from topographic influences; and (2) convergence precipitation, which results from atmospheric processes independent of orographic influences. Both components must be evaluated in the estimation of the PMP in orographic regions. Two techniques are used: (1) the orographic separation method, using a laminar flow model; and (2) the modification of non-orographic PMP for orography. Details of both these procedures are given in WMO (1986).

4.8.6 Spatial Variation of PMP

The spatial variation of PMP involves how point PMP estimates are distributed over an area. The construction of areal distribution of PMP is actually the development of an idealized pattern of distribution of isohyets. In many cases, the shape and orientation of the storm isohyetal pattern is established empirically. For example, from an analysis of 53 major storms in the eastern United States, Hansen et al. (1982) found that the most representative shape of the isohyets of these storms over the non-orographically influenced regions in these states is an ellipse with a major and minor axes ratio of 2.5:1. They developed such standard ellipses containing areas of 25.9–16 835 km². In addition to elliptical isohyetal patterns, the specific geographic orientation of the isohyets should be known, since it is possible to place the isohyetal pattern of the PMP storm over the drainage basin in any direction without making any meteorological sense. For this reason, Hansen et al. (1982) also established preferred orientations of the major storms in the eastern United States based on the analyses of several hundred large storms. From the DAD data of a design PMP obtained from HMR 51, they provided an algorithm to construct the spatial distribution of the PMP.

In general, an actual storm pattern is extended in one or more directions due to storm movements and an ellipse having a particular ratio of major to minor axis is usually the best fit to the portion of the storm that has the highest density of precipitation (Figure 4.37). Such spatial analysis of precipitation distributions can be done on a basin- or storm-specific basis to determine a basin- or storm-specific shape and orientation of the design PMP. However, this should only be used in non-orographically influenced regions where the terrain does not affect the spatial distribution of rainfall.

4.8.7 Temporal Distribution of PMP – Development of Probable Maximum Storms

The PMP estimates do not provide any temporal pattern of the storm. However, for rainfall-runoff simulation, it is

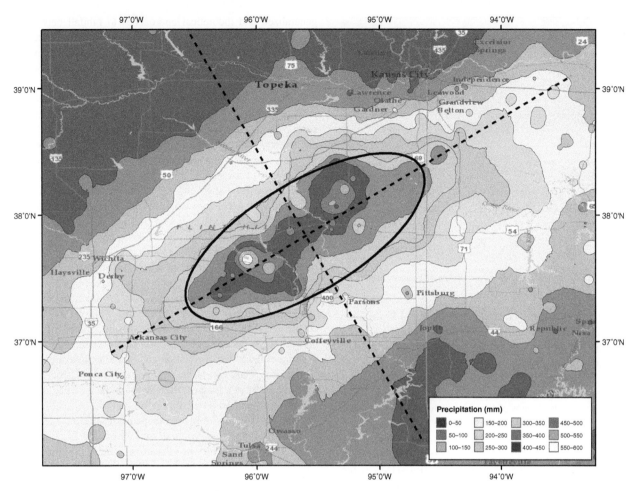

Figure 4.37 Spatial analysis of the distribution of precipitation depths of a specific storm to determine the shape and orientation of the storm (June 28–July 2, 2007 storm over Fall River, eastern Kansas). Such analyses can be done to compare the suggested idealized distributions that can be derived from application of the procedures given in HMR 52 for design PMP in regions without orographic effects. (A black and white version of this figure will appear in some formats. For the color version, please refer to the plate section.)

necessary to have the temporal distribution of the PMP to produce the **probable maximum storm** (**PMS**). The PMS is a hypothetical storm that produces **the probable maximum flood** (**PMF**) in a particular drainage area. The PMS is used as an input to a hydrologic model to produce the flood hydrograph of a PMF. The time distribution of PMP or the hyetograph of a PMS significantly affects the shape and peak of the resulting discharge hydrograph. Thus, when PMP is applied to determine the flood hydrograph, it is necessary to specify in what order various rain increments are arranged with time from the beginning of the storm.

For the areas to the east of 105° W meridian in the United States (USACE, 1987) provided a procedure to develop the temporal distribution of PMP, estimated from HMR 51, in addition to the spatial pattern associated with the PMP as discussed above. Subsequently, in the implementation of the HMR 52 algorithm in the computer program developed by USACE (1984), the temporal distribution of PMP was also modeled. In this algorithm, typically PMP estimates for 6-, 12-, 24-, 48-, and 72-hour durations are used for the development of the mass curve of rainfall. Successive subtraction of the PMP for each of these durations from that 6 hours longer gives 6-hour increments of PMP. In this way, the PMP for all durations (6–72 hours) is used in a single storm. Other studies have produced basin- or region-specific temporal patterns based on actual storm mass curves. These analyses follow the same procedure as that used to derive the PMP depths, that is, they are storm-based and specific to the location being modeled (see Mukhopadhyay and Kappel, 2017 for a specific example).

4.9 GRIDDED RAINFALL

Gridded rainfall is usually used for rainfall data derived from radar observations. It can also be created from rain gauge data by estimating rainfalls at the center of each cell of a grid laid over a network of rain gauges by the inverse-distance-squared method. Here, we focus on gridded rainfall from weather radar.

Radar-derived rainfall data are usually available as grid formats. For example, the radar data, available from the NWS Weather Surveillance Radar Doppler units (WSR-88D) throughout the United States, are in digitized forms

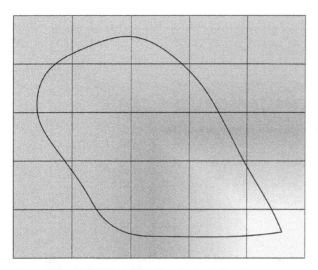

Figure 4.38 Rainfall grid superimposed on a watershed boundary map.

providing an estimate of rainfall in each cell in a Hydrologic Rainfall Analysis Project (HRAP) grid. Cells of the grid are approximately 4 km × 4 km. Figure 4.38 shows an HRAP-like grid system overlaid on a map showing a watershed boundary. The color codes in each grid represent a rainfall depth at a particular time over that entire grid, not at any point. Successive sweeps of the radar provide the time series of rainfall in each cell. From the time series of average rainfall depths, the required MAP series can be computed, accounting explicitly for the spatial variability of rainfall. Another grid system that is available in HEC-HMS is the standard hydrologic grid (SHG). Gridded data explicitly model the temporal and spatial variability of rainfall by providing an estimate for rainfall intensity everywhere.

4.10 DESIGN OF RAIN GAUGE NETWORKS

Since depths of rainfalls during any rain event have tremendous spatial variability, it is extremely important that a rain gauge network in an area is such that it does not contain holes that lead to either an underestimation or overestimation of the rainfall. Thus, among various considerations that must be given to the design of rain gauge networks, the two most important are the number and locations of the rain gauges. The determination of the number and location of the rain gauge stations to provide sufficient information regarding the reasonably accurate estimation of the rainfall falling over a watershed is called **network design**. A rain gauge network serves the general as well as specific purposes such as flood forecasting, flood control, water supply, irrigation, drainage design, and hydropower generation.

4.10.1 Number of Rain Gauges in a Network: Density of Stations

There are various theories on the number of gauges that should be present in a given network. In general, all generalized and theoretical approaches to rainfall network design require some knowledge of the rainfall-process spatial correlation. The spatial correlation measures the level of linear dependence of rainfall at two points separated by a distance v.

Rodriguez-Iturbe and Mejia (1974) expressed the rainfall process in terms of an autocorrelation structure in time and space. The correlation was in the form of e^{-hv}, where the parameter h (km^{-1}) controls the decay in correlation with distance. They developed a general framework to estimate the variance of the sample long-term mean areal rainfall of a storm event. By expressing the variance as a function of correlation in space, correlation in time, length of operation of the network, and the geometry of the gauging array, they developed the trade-off of time versus space.

Bras and Rodriguez-Iturbe (1976) considered rainfall as a multidimensional stochastic process. By using this process and multivariate estimation theory they developed a procedure for designing an optimal network to obtain the mean areal rainfall of an event over a fixed area. This methodology considered three aspects of network design, namely spatial uncertainty and correlation process, error in measurement techniques, and their correlation and nonhomogeneous sampling costs. They found the optimal network (density and location of the rain gauges) together with the resulting costs and mean-square errors of rainfall estimation.

Bras and Colon (1978) developed a procedure for the estimation of mean areal rainfall through a state of augmentation procedure and the use of multivariate linear estimation concepts, in particular the Kalman–Bucy filter. The resulting technique could be used to analyze the existing data network, design new networks, and process data for new networks.

The procedures cited above, and several others suggested along such lines, involve complex statistical methods to achieve better accuracy. However, historically, network design has been strongly influenced by convenience and cost, ignoring the issue of required accuracy. For this reason, for general practical purposes, a simplified approach can be taken to determine the number of rain gauges required in a network. For example, the NWS suggest that the *minimum* number of rain gauges, N_G in a network of an area A should be

$$N_G = A^{0.33} \tag{4.69}$$

However, even if a network contains more than this minimum number of gauges, all the rainfall during a major rain event may not be adequately measured. For example, typical rain gauges are 20–30 cm in diameter. Thus, in a 2.6-km^2 watershed (one square mile), the catch surface of the gauge represents a sample of rainfall on approximately 1/100,000,000th of the total watershed area.

The problem of ascertaining the optimum number of rain gauges in a watershed is of a statistical nature and depends on the spatial variation of rainfall. Thus, if there is already an existing network, the coefficient of spatial variation of rainfall from such a network can be utilized to determine the

4.10 Design of Rain Gauge Networks   149

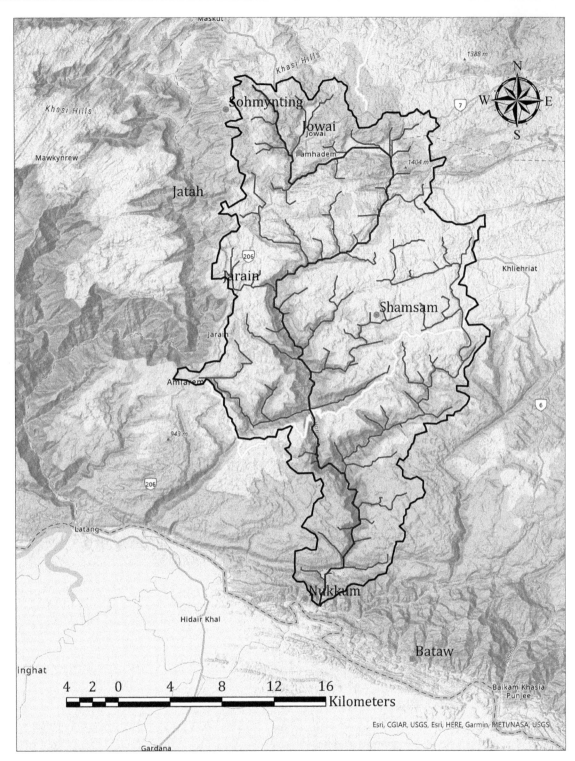

Figure 4.39 Map of Myntdu-Leshka catchment showing the rain gauges (red dots) in and around the catchment that were used to propose rain gauge network design. The main Myntdu-Leshka river course is shown by a bold dark blue line whereas other streams have thinner blue lines. (A black and white version of this figure will appear in some formats. For the color version, please refer to the plate section.)

optimum number of gauges. For such cases, this optimum number, corresponding to an assigned percentage of error in the estimation of the mean areal rainfall, can be obtained as

$$N_G = \left(\frac{C_v}{\varepsilon}\right)^2 \quad (4.70)$$

where C_v is the coefficient of variation of rainfall values at the existing stations and ε is the allowable degree of error (assigned percentage of error) in the estimate of the mean areal rainfall. This approach has been recommended by several governmental agencies, such as the Indian Meteorological Department (1972).

If there are m stations in a watershed and $P_1, P_2, \ldots, P_m$ is the recorded rainfall at a known time at $1, 2, \ldots, m$ stations, the coefficient of variation is given by

$$C_v = \frac{100 \times \sigma_{m-1}}{\overline{P}} \quad (4.71)$$

where

$$\sigma_{m-1} = \sqrt{\frac{\sum_{i=1}^{m} P_i^2 - m\overline{P}^2}{m-1}} \quad (4.72)$$

where P_i is the monthly average rainfall at station i and $\overline{P}$ is the average rainfall of m number of stations given by

$$\overline{P} = \frac{\sum_{i=1}^{m} P_i}{m} \quad (4.73)$$

Usually, ε is set to 10%. Also, if the number of stations, m, in the network is less than 30, Eq. 4.72 is used, otherwise, σ_m can also be used.

4.10.2 Locations of Rain Gauges in a Network

The determination of the locations of the rain gauge stations in a network design is very important. A common method is to construct isohyetal lines of either the mean annual or seasonal rainfall patterns within a watershed and then, based on the spatial pattern of variation of rainfall, divide the watershed into several zones. Each zone should have a number of rain gauges and the number of stations should nearly be equal in each zone to ensure that the distribution of gauges is fairly uniform over the watershed area. Appropriate locations for the rain gauges depend on a variety of factors that can be summarized as follows.

1. Urban versus sparsely populated watersheds: Densely populated urban areas need a very dense rain gauge network for both temporal and spatial resolution of storms and for the design, management, and real-time control of the storm drainage systems and for other engineering applications. On the other hand, sparsely settled zones may not require a high-density network of rain gauges for engineering applications.
2. Location of rain gauges in conjunction with stream gauging stations: To ensure that rainfall data are available for extending streamflow records, flood forecasting, or other hydrologic applications, it is important to coordinate the rain gauge locations relative to the locations of the stream gauges in a watershed. Rain gauges should be located such that the MAP for the drainage area (catchment) of each of the stream gauges can be estimated with reasonable accuracy. These will usually be located at or near the stream gauge and in the upper parts of the catchment. A rain gauge should be located at the site of a stream gauge only if the rainfalls will be representative of the overall catchment area.

The importance and challenges of rain gauge network design is illustrated with an example from Myntdu-Leshka catchment located in the Jaintia Hills of the Meghalaya state in India. Several run-on-the river hydroelectric power projects within this catchment were proposed in the late 1990s and construction of some projects started in the 2010s. Hydrologic studies to assess water availability were initiated by Meghalaya Power Generation Corporation. The catchment had a very limited number of rain gauges. The National Institute of Hydrology (NIH) (NIH, 1998) conducted a study related to rain gauge network design for this catchment. Figure 4.39 shows the existing rain gauges that were used in the proposed network design. The coefficient of variation (C_v) evaluated for the average annual rainfall for the existing rain gauge network was determined as 33. For $C_v = 33$ and $N_G = 7$, from Eq. 4.70, $\varepsilon = 12.5\%$ (error). If a value of 10 (10% error) is assigned to ε, $N_G = 11$. This indicates that four additional gauges are required. However, note that out of the seven gauges considered to estimate C_v, only two of the gauges are inside the catchment boundary, three are close to it but outside, and one (Bataw) is distinctly outside this boundary. The catchment area is 536 km². If Eq. 4.69 is used, $N_G = 8$. This indicates six additional gauges are required inside the catchment boundary. Figure 4.40 shows

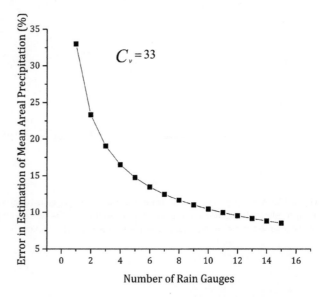

Figure 4.40 Variation of error as a function of the number of rain gauges in a network.

the variation of allowable error as a function of the number of gauges for the given C_v. From this plot it can be also concluded that the error does not substantially decrease if the number of gauges is increased beyond 11. However, the value of C_v would also change if additional gauges within the catchment were considered. This example demonstrates the need for the design of a suitable rain gauge network for proper hydrologic studies of important projects such as hydroelectric power plants.

4.11 EXAMPLES

Example 4.1: The Upper Indus Basin is an interesting river basin. It straddles Greater Himalayas to the south and Karakorum Mountains to the north and is the abode of some of the largest glaciers outside the polar region and high-altitude peaks (> 6000 m), including Godwin Austen, the second-highest peak in the world, and Nanga Parbat (both > 8000 m). The river system in the basin is mostly fed by meltwater from the glaciers and seasonal snows (Mukhopadhyay and Khan, 2015b). The river system of this basin is extremely important to Pakistan for many reasons. Several studies have indicated that the effect of climate change is pronounced in this basin. The Water and Power Development Authority (WPDA) of Pakistan harnesses the water resources of the basin for the production of hydroelectricity. Historically, there are only seven climate stations along the valleys that have precipitation records for a long period. However, due to the importance arising from addressing climate change and hydropower generation, WPDA have started to install more climate stations in the basin since 1999. Figure 4.41 shows the locations of the climate stations within the basin in 2015. Table 4.6 gives the mean annual precipitation at six stations based on the POR noted. The station at Leh also has a long POR but is excluded from the

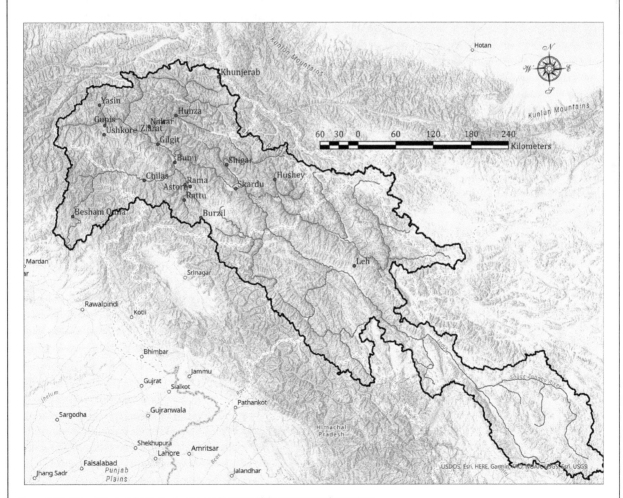

Figure 4.41 Map of Upper Indus Basin showing the locations of the existing weather stations.

Table 4.6 *Mean annual precipitation at six climate stations shown in Figure 4.41*

Station	Mean annual rainfall (mm)	POR used
Gilgit	138.55	1951–2010
Gupis	189.73	1955–2010
Skardu	232.64	1952–2010
Astore	499.48	1954–2010
Bunji	158.04	1953–2010
Chilas	197.03	1953–2010

present analysis. Based on the data given in Table 4.6, determine the number of rain gauges required for a network design in the basin.

Solution: In this case, $m = 6$

$$\sum_{i=1}^{6} P_i = 1415.47$$

$$\overline{P} = 235.91$$

$$\sigma_{m-1} = \sqrt{\frac{422592.68 - 6 \times 235.91^2}{6-1}} = 133.16$$

$$\therefore C_v = \frac{100 \times 133.16}{235.91} = 56.45$$

$$N_G = \left(\frac{56.45}{10}\right)^2 = 31.86$$

Therefore, the network requires 32 stations to allow 10% error in MAP estimates. As of 2015, there are 20 stations as shown in Fig. 4.41. So, 12 more stations are needed.

Example 4.2: The rain gauges of the City of Dallas can be used to construct Theisen polygons which can be employed for any rain event for the calculation of the MAP. Figure 4.42 shows the Theisen polygons of the Turtle Creek watershed, located in the heart of Dallas, and the rain gauges. Determine the weights that should be assigned to each of the gauges to estimate the MAP from any rain event.

Solution: Column 1 of Table 4.7 gives the polygon number assigned to each of the Theisen polygons shown in Figure 4.42. Column 2 lists the areas of each of the Theisen polygons contained within the Turtle Creek watershed. The total area is 21 892.99 acres. The gauges that are associated with each of the polygons are given in column 4, with the weight that should be assigned to each of these gauges shown in column 3. These weights are obtained by dividing each of the values listed in column 2 by 21 892.99. The sum of the values shown in column 3 is 1.

Table 4.7 *Theisen weights to the gauges in and around the Turtle Creek watershed*

Polygon	Area (acres)	Weight	Gauge
1	22.42	0.0010	Cedar Creek at Ewing
2	607.09	0.0277	Sargent Road @ Morrell Avenue
3	1434.13	0.0655	Cedar Spring at Kings Road
4	2803.97	0.1281	Turtle Creek Intake @ Park Bridge
5	2897.62	0.1324	Mill Creek @ Caddo Street
6	2492.87	0.1139	Knights Branch @ Denton Drive
7	1028.31	0.0470	Bachman Dam
8	3264.86	0.1491	Mockingbird @ CRIP RR
9	2757.22	0.1259	Turtle Creek @ Willow Wood Street
10	1794.35	0.0820	NW Hwy @ Edgemere Road
11	574.25	0.0262	Bachman Branch at Midway Rd
12	648.45	0.0296	Skillman @ Southwestern
13	1567.43	0.0716	Inwood Rad at University Blvd

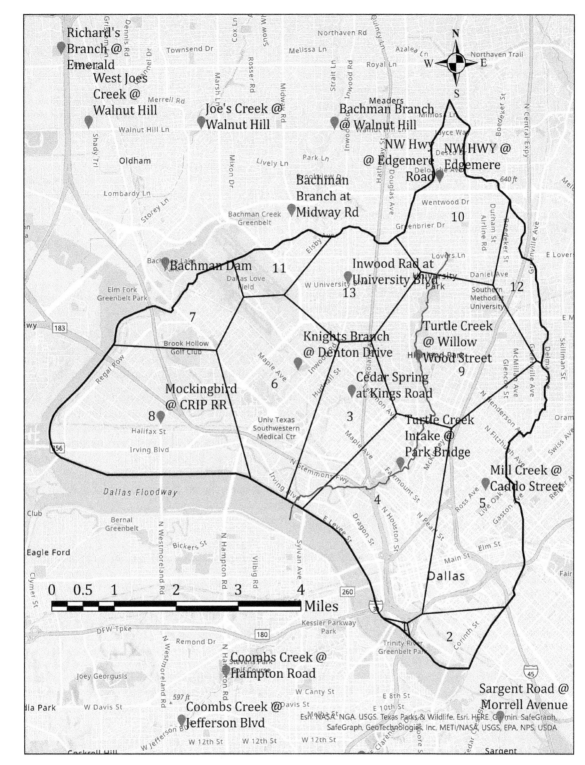

Figure 4.42 Theisen polygons and rain gauges in the Turtle Creek watershed.

Example 4.3: From a study of spatial variation of rainfall totals in two rain gauge networks in Illinois, Huff and Neill (1957) developed two sets of empirical equations to show that such equations can be used to predict the value of rainfall at a point located at a certain distance from a rain gauge where the value of rainfall is known. One such equation is given as:

$$\log E = -0.796 + 0.49 \log P + 0.32 \log D$$

where E (inches) represents the average difference between the total storm rainfall at the gauge and at points located at distance D (miles) from the gauge, when the point rainfall recorded at the gauge is P (inches). Another equation, that gives the standard deviation, s, is

$$\log s = -0.398 + 0.49 \log P + 0.32 \log D$$

During a 24-hour rainfall event on March 19, 2006, the gauge at NW HWWY at Edgemere (Gauge 195) shown in Figure 4.42 recorded a rainfall total of 6.87 inches. The Turtle Creek at Willow Wood Street gauge (Gauge 155) recorded a rainfall total of 6.84 inches during the same event. No data are available for the rainfall value at the Turtle Creek Intake gauge at Park Bridge (Gauge 111). If the two equations developed by Huff and Neill were applicable to Dallas, what would be the value of rainfall expected at Turtle Creek Intake at Park Bridge and what is its probability? Comment on your findings.

Solution: First, the validity of the first equation in the case of Dallas rainfall is checked from the known values of P at gauges 195 and 155. The distance between these two gauges is 2.85 miles.

Using $P = 6.87$ inches at Gauge 195, the value of E is calculated as:

$$E = 10^{[-0.796 + 0.49 \times \log(6.87) + 0.32 \times \log(2.85)]} = 0.57$$

Thus, the value of P at Gauge 155 is expected to be 6.87 ± 0.57 inches. The observed value at Gauge 155 is 6.84 inches, which satisfies this condition.

One standard deviation represents a value that can be either added or subtracted from the mean value of a sample to yield a value with a probability of 0.68 ($\simeq 0.7$) of occurrence. Thus, s is calculated as

$$s = 10^{[-0.398 + 0.49 \times \log(6.87) + 0.32 \log(2.85)]} = 1.44$$

Thus, there is 70% chance that the rainfall value at Gauge 155 will lie within 5.43 inches and 8.31 inches. The observed value lies within this range.

The Gauge 111 is 1.78 miles from Gauge 155. In this case,

$$E = 10^{[-0.796 + 0.49 \times \log(6.84) + 0.32 \times \log(1.78)]} = 0.49$$

Thus, the rainfall at Gauge 111 would be either 6.35 inches or 7.33 inches.

$$s = 10^{[-0.398 + 0.49 \times \log(6.84) + 0.32 \log(1.78)]} = 1.23$$

Thus, there is 70% chance that the rainfall value at Gauge 111 will lie within 5.61 inches and 8.07 inches.

Comment: In many urban centers in the United States the density of rain gauges is very high. However, predictive equations like those developed by Huff and Neill are lacking. As illustrated in this example, such equations when developed for specific areas having a statistically significant POR and dense network of rain gauges can serve as important predictive tools at locations of projects where site-specific rainfall data would be helpful but are absent.

Example 4.4: Tropical moisture associated with Hurricane Norbert generated intense rainfall over large areas of Las Vegas and surrounding areas of Clark County, Nevada on September 8, 2014. The rainfall generated considerable flows in the local washes of this arid desert region and exceeded the capacity of the Interstate Highway-15 drainage system. Large sections of IH-15 were washed away, which resulted in the highway being closed to traffic for several days. Table 4.8 gives the rainfall values measured at Gauge 3064 (Weiser Wash), in the automatic rain gauge network operated and maintained by the Regional Flood Control District, Clark County, Nevada. Determine the period and magnitude of maximum intensity of the rainfall during this event and make qualitative observations on the nature of the mass curve and hyetograph. Determine the quartile storm and compare the mass curve for this event with the Huff curve corresponding to this quartile and duration closest to this event from NOAA Atlas 14. Comment on the differences.

Solution: The period of maximum intensity is the time interval in which the slope of the mass curve is steepest. Figure 4.43 shows the percentage mass curve. From a visual examination the mass curve, this period appears to be from 32% to 40% of the total storm duration. The total duration of this event as recorded at this gauge was 182 minutes. Thus, the period of maximum intensity appears to be from approximately 58 to 73 minutes after the beginning of rainfall. This is a 15-minute interval. Thus, the task in hand is to determine the maximum intensity for a 15-minute period during the entire duration of the event. Table 4.9 gives the incremental rainfall in column 3 and cumulative time in column 5. The rain gauges in this

Table 4.8 *Measured rainfall at Gauge 3064 during the rain event on September 8, 2014, in the Las Vegas area*

Date and time	Cumulative rainfall (in)	Date and time	Cumulative rainfall (in)
9/8/14 13:20	0	9/8/14 14:35	2.67
9/8/14 13:24	0.04	9/8/14 14:36	2.75
9/8/14 13:28	0.08	9/8/14 14:37	2.83
9/8/14 13:32	0.11	9/8/14 14:38	2.91
9/8/14 13:34	0.15	9/8/14 14:39	2.95
9/8/14 13:48	0.23	9/8/14 14:40	2.99
9/8/14 13:50	0.27	9/8/14 14:41	3.03
9/8/14 13:51	0.31	9/8/14 14:42	3.07
9/8/14 13:54	0.35	9/8/14 14:43	3.11
9/8/14 13:57	0.39	9/8/14 14:45	3.15
9/8/14 13:59	0.43	9/8/14 14:46	3.22
9/8/14 14:01	0.47	9/8/14 14:47	3.26
9/8/14 14:02	0.51	9/8/14 14:48	3.3
9/8/14 14:04	0.55	9/8/14 14:49	3.38
9/8/14 14:05	0.59	9/8/14 14:50	3.42
9/8/14 14:06	0.67	9/8/14 14:51	3.5
9/8/14 14:07	0.74	9/8/14 14:52	3.54
9/8/14 14:08	0.78	9/8/14 14:53	3.58
9/8/14 14:10	0.86	9/8/14 14:54	3.62
9/8/14 14:11	0.9	9/8/14 14:55	3.66
9/8/14 14:12	0.94	9/8/14 14:57	3.7
9/8/14 14:13	0.98	9/8/14 14:59	3.74
9/8/14 14:14	1.06	9/8/14 15:00	3.78
9/8/14 14:15	1.14	9/8/14 15:07	3.85
9/8/14 14:16	1.22	9/8/14 15:12	3.93
9/8/14 14:17	1.3	9/8/14 15:14	3.97
9/8/14 14:18	1.37	9/8/14 15:17	4.01
9/8/14 14:19	1.45	9/8/14 15:19	4.05
9/8/14 14:20	1.53	9/8/14 15:22	4.09
9/8/14 14:21	1.61	9/8/14 15:25	4.13
9/8/14 14:22	1.69	9/8/14 15:29	4.17
9/8/14 14:23	1.81	9/8/14 15:35	4.21
9/8/14 14:24	1.85	9/8/14 15:43	4.25
9/8/14 14:25	1.93	9/8/14 15:48	4.33
9/8/14 14:26	2	9/8/14 15:51	4.33
9/8/14 14:27	2.12	9/8/14 15:53	4.37
9/8/14 14:28	2.16	9/8/14 15:55	4.41
9/8/14 14:29	2.28	9/8/14 15:56	4.45
9/8/14 14:30	2.36	9/8/14 15:58	4.48
9/8/14 14:32	2.48	9/8/14 15:59	4.52
9/8/14 14:33	2.56	9/8/14 16:02	4.56
9/8/14 14:34	2.59	9/8/14 16:20	4.64

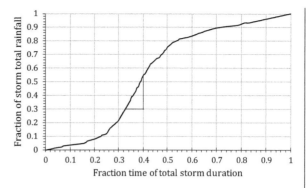

Figure 4.43 Percentage mass curve at Gauge 3064 (September 8, 2014), Las Vegas, Nevada.

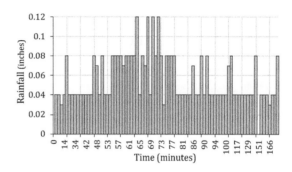

Figure 4.44 Rainfall hyetograph at Gauge 3064 (September 14, 2014), Las Vegas, Nevada.

network are tipping bucket gauges and catch approximately 0.04 inch with each tip of the bucket. For this reason, the time intervals are not uniform throughout the recording. The time interval that is closest to a 15-minute interval is seen to be 14 minutes. Thus, the maximum rainfall depth during every 14-minute interval is calculated in the last column. For example, the first 14-minute interval is at 13:48. The running totals for every 14 minutes in subsequent periods are calculated. The maximum is found to be 1.19 inches during the interval between 14:18 and 14:33, which correspond to 32.42 and 40.11 percent (column 6) of total duration of the rainfall. A rainfall depth of 1.19 inches accumulated for 14 minutes gives an intensity of $1.19\,\text{in}/(14\,\text{min}/60\,\text{min}/\text{h}) = 5.1$ in/h

The rainfall hyetograph is constructed from columns 3 and 4 of Table 4.9 and shown in Figure 4.44. The hyetograph shows that the rainfall had nearly uniform pulses with intermittent bursts.

Table 4.10 gives the quartile distribution of rainfall depths. The greatest accumulation of rainfall was in the second quartile of time (45.5 to 91 minutes).

Since the duration of this storm was 3 hours, the 6-hour second-quartile Huff curves for region 1 (semiarid southwest) are constructed from the data given in Table 4.11 (obtained from Precipitation Frequency Data Server of NOAA Atlas 14) and are shown in Figure 4.45.

Comment: The rainfall event described in this example resulted from the effect of a hurricane. However, it was a relatively short-duration event resembling a thunderstorm, resulting in a convex upward pattern of the mass curves, unlike the long-duration rainfalls following hurricanes and thunderstorms discussed in the text, which produce S-shaped mass curves. On the other hand, it was not a typical short-duration second-quartile storm, characteristic of this region, as shown in Figure 4.44.

Table 4.9 *Calculation*

Date and time	Cumulative Rainfall (in)	Incremental rainfall (in)	Time interval(min)	Cumulative time	Fraction duration	Fraction STP	Running total (14 min)
9/8/14 13:20	0	0	0	0	0	0	
9/8/14 13:24	0.04	0.04	4	4	0.0220	0.0086	
9/8/14 13:28	0.08	0.04	4	8	0.0440	0.0172	
9/8/14 13:32	0.11	0.03	4	12	0.0659	0.0237	
9/8/14 13:34	0.15	0.04	2	14	0.0769	0.0323	0.15
9/8/14 13:48	0.23	0.08	14	28	0.1538	0.0496	0.19
9/8/14 13:50	0.27	0.04	2	30	0.1648	0.0582	
9/8/14 13:51	0.31	0.04	1	31	0.1703	0.0668	
9/8/14 13:54	0.35	0.04	3	34	0.1868	0.0754	
9/8/14 13:57	0.39	0.04	3	37	0.2033	0.0841	
9/8/14 13:59	0.43	0.04	2	39	0.2143	0.0927	
9/8/14 14:01	0.47	0.04	2	41	0.2253	0.1013	0.32
9/8/14 14:02	0.51	0.04	1	42	0.2308	0.1099	0.28
9/8/14 14:04	0.55	0.04	3	45	0.2473	0.1185	0.24
9/8/14 14:05	0.59	0.04	1	46	0.2527	0.1272	
9/8/14 14:06	0.67	0.08	1	47	0.2582	0.1444	
9/8/14 14:07	0.74	0.07	1	48	0.2637	0.1595	0.59
9/8/14 14:08	0.78	0.04	1	49	0.2692	0.1681	
9/8/14 14:10	0.86	0.08	2	51	0.2802	0.1853	0.47
9/8/14 14:11	0.9	0.04	1	52	0.2857	0.1940	
9/8/14 14:12	0.94	0.04	1	53	0.2912	0.2026	0.51
9/8/14 14:13	0.98	0.04	1	54	0.2967	0.2112	
9/8/14 14:14	1.06	0.08	1	55	0.3022	0.2284	0.59
9/8/14 14:15	1.14	0.08	1	56	0.3077	0.2457	0.63
9/8/14 14:16	1.22	0.08	1	57	0.3132	0.2629	
9/8/14 14:17	1.3	0.08	1	58	0.3187	0.2802	
9/8/14 14:18	1.37	0.07	1	59	0.3242	0.2953	0.82
9/8/14 14:19	1.45	0.08	1	60	0.3297	0.3125	0.86
9/8/14 14:20	1.53	0.08	1	61	0.3352	0.3297	0.86
9/8/14 14:21	1.61	0.08	1	62	0.3407	0.3470	0.87
9/8/14 14:22	1.69	0.08	1	63	0.3462	0.3642	0.91
9/8/14 14:23	1.81	0.12	1	64	0.3516	0.3901	
9/8/14 14:24	1.85	0.04	1	65	0.3571	0.3987	0.99
9/8/14 14:25	1.93	0.08	1	66	0.3626	0.4159	1.03
9/8/14 14:26	2	0.07	1	67	0.3681	0.4310	1.06
9/8/14 14:27	2.12	0.12	1	68	0.3736	0.4569	1.14
9/8/14 14:28	2.16	0.04	1	69	0.3791	0.4655	1.1
9/8/14 14:29	2.28	0.12	1	70	0.3846	0.4914	1.14
9/8/14 14:30	2.36	0.08	1	71	0.3901	0.5086	1.14
9/8/14 14:32	2.48	0.12	1	72	0.3956	0.5345	1.18
9/8/14 14:33	2.56	0.08	1	73	0.4011	0.5517	1.19
9/8/14 14:34	2.59	0.03	1	74	0.4066	0.5582	1.14
9/8/14 14:35	2.67	0.08	1	75	0.4121	0.5754	1.14
9/8/14 14:36	2.75	0.08	1	76	0.4176	0.5927	1.14
9/8/14 14:37	2.83	0.08	1	77	0.4231	0.6099	1.14
9/8/14 14:38	2.91	0.08	1	78	0.4286	0.6272	1.1
9/8/14 14:39	2.95	0.04	1	79	0.4341	0.6358	1.1
9/8/14 14:40	2.99	0.04	1	80	0.4396	0.6444	1.06

Table 4.9 (*cont.*)

Date and time	Cumulative Rainfall (in)	Incremental rainfall (in)	Time interval(min)	Cumulative time	Fraction duration	Fraction STP	Running total (14 min)
9/8/14 14:41	3.03	0.04	1	81	0.4451	0.6530	1.03
9/8/14 14:42	3.07	0.04	1	82	0.4505	0.6616	0.95
9/8/14 14:43	3.11	0.04	1	83	0.4560	0.6703	0.95
9/8/14 14:45	3.15	0.04	2	85	0.4670	0.6789	0.79
9/8/14 14:46	3.22	0.07	1	86	0.4725	0.6940	0.74
9/8/14 14:47	3.26	0.04	1	87	0.4780	0.7026	0.7
9/8/14 14:48	3.3	0.04	1	88	0.4835	0.7112	0.71
9/8/14 14:49	3.38	0.08	1	89	0.4890	0.7284	0.71
9/8/14 14:50	3.42	0.04	1	90	0.4945	0.7371	0.67
9/8/14 14:51	3.5	0.08	1	91	0.5000	0.7543	0.67
9/8/14 14:52	3.54	0.04	1	92	0.5055	0.7629	0.63
9/8/14 14:53	3.58	0.04	1	93	0.5110	0.7716	0.63
9/8/14 14:54	3.62	0.04	1	94	0.5165	0.7802	0.63
9/8/14 14:55	3.66	0.04	1	95	0.5220	0.7888	0.63
9/8/14 14:57	3.7	0.04	2	97	0.5330	0.7974	0.59
9/8/14 14:59	3.74	0.04	2	99	0.5440	0.8060	0.59
9/8/14 15:00	3.78	0.04	1	100	0.5495	0.8147	0.56
9/8/14 15:07	3.85	0.07	7	107	0.5879	0.8297	0.27
9/8/14 15:12	3.93	0.08	5	112	0.6154	0.8470	
9/8/14 15:14	3.97	0.04	2	114	0.6264	0.8556	0.19
9/8/14 15:17	4.01	0.04	3	117	0.6429	0.8642	
9/8/14 15:19	4.05	0.04	2	119	0.6538	0.8728	
9/8/14 15:22	4.09	0.04	3	122	0.6703	0.8815	
9/8/14 15:25	4.13	0.04	3	125	0.6868	0.8901	
9/8/14 15:29	4.17	0.04	4	129	0.7088	0.8987	
9/8/14 15:35	4.21	0.04	6	135	0.7418	0.9073	
9/8/14 15:43	4.25	0.04	8	143	0.7857	0.9159	0.08
9/8/14 15:48	4.33	0.08	5	148	0.8132	0.9332	
9/8/14 15:51	4.33	0	3	151	0.8297	0.9332	
9/8/14 15:53	4.37	0.04	4	155	0.8516	0.9418	
9/8/14 15:55	4.41	0.04	4	159	0.8736	0.9504	
9/8/14 15:56	4.45	0.04	4	163	0.8956	0.9591	
9/8/14 15:58	4.48	0.03	3	166	0.9121	0.9655	
9/8/14 15:59	4.52	0.04	4	170	0.9341	0.9741	
9/8/14 16:02	4.56	0.04	4	174	0.9560	0.9828	
9/8/14 16:20	4.64	0.08	8	182	1.0000	1.0000	

Table 4.10 *Quartile distribution of rainfall totals for the event of September 14, 2014, in Las Vegas area*

	Time (min)	Cumulative rainfall (in)	Quartile total (in)
First quartile	45.5	0.55	0.55
Second quartile	91	3.5	2.95
Third quartile	136.5	4.21	0.71
Fourth quartile	182	4.64	0.43

Table 4.11 *Characteristics of second-quartile storms of 6-hour duration in the semiarid southwest of the United States*

Percent of duration	Probability								
	10%	20%	30%	40%	50%	60%	70%	80%	90%
0	0	0	0	0	0	0	0	0	0
8.3	9.4	8.1	7.2	6.8	6	5.5	4.9	4.1	2.5
16.7	20.3	17.5	15.8	14.6	13.4	12.3	11.2	9.5	6.8
25	34.8	29.7	27	24.7	23	21.4	19.8	17.6	14.1
33.3	51.4	43.5	39.7	36.6	34.2	32.1	30.3	27.8	23.8
41.7	67.6	58	53.2	49.9	47.2	44.7	42.6	40.2	36.4
50	81.1	71.4	66.1	62.7	59.8	57	54.5	52	48.8
58.3	88.6	81.2	76	72.5	69.6	66.6	63.8	61.2	58.4
66.7	92.7	88.2	83.1	80.1	77.4	74.5	71.9	69.3	66.5
75	95.9	92.9	89	86.8	84.4	81.7	79.6	77.3	74.6
83.3	97.9	96	94.1	92.4	90.5	88.4	86.8	85.1	82.7
91.7	99.1	98.2	97.7	96.6	95.5	94.4	93.5	92.6	91.2
100	100	100	100	100	100	100	100	100	100

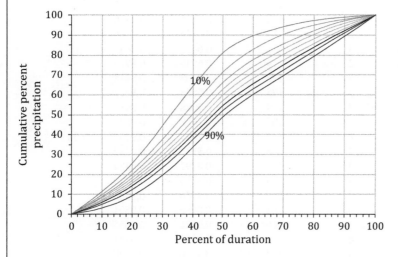

Figure 4.45 Huff curves for 6-hour second-quartile storms in the semiarid southwest United States. (A black and white version of this figure will appear in some formats. For the color version, please refer to the plate section.)

Example 4.5: Rainfall records obtained at the gauge at Sunnybrook Drive crossing over Walnut Creek in Raleigh, North Carolina for a short duration summer thunderstorm are given in Table 4.12. Determine the maximum 15-minute and 30-minute intensities. What quartile storm was this one? Plot a percentage mass curve and hyetograph and make observations.

Solution: Table 4.13 gives the cumulative rainfall in column 2 and running totals for 15-minute and 30-minute durations starting from the first 15 minute and 30-minute intervals in columns 3 and 4, respectively.

The maximum 15-minute and 30-minute duration rainfall totals are found to be 0.96 inches and 1.56 inches respectively. These correspond to 3.84 in/h and 3.12 in/h respectively.

From Table 4.14 it can be seen that the maximum accumulation of rainfall was during the second quartile. Thus, this was a second-quartile storm.

Figures 4.46 and 4.47 show the mass curve and hyetograph, respectively. The mass curve is convex upward.

Table 4.12 *Rainfall measurements at the gauge at Sunnybrook Drive over Walnut Creek, North Carolina*

Time	Incremental precipitation (in)	Time	Incremental precipitation (in)
8/31/20 21:25	0.03	8/31/20 23:10	0.07
8/31/20 21:30	0.02	8/31/20 23:15	0.08
8/31/20 21:35	0	8/31/20 23:20	0.07
8/31/20 21:40	0	8/31/20 23:25	0.03
8/31/20 21:45	0.01	8/31/20 23:30	0
8/31/20 21:50	0.05	8/31/20 23:35	0.01
8/31/20 21:55	0.05	8/31/20 23:40	0.01
8/31/20 22:00	0.13	8/31/20 23:45	0.01
8/31/20 22:05	0.17	8/31/20 23:50	0.01
8/31/20 22:10	0.18	8/31/20 23:55	0.07
8/31/20 22:15	0.25	9/1/20 0:00	0.09
8/31/20 22:20	0.17	9/1/20 0:05	0.08
8/31/20 22:25	0.33	9/1/20 0:10	0.09
8/31/20 22:30	0.34	9/1/20 0:15	0.05
8/31/20 22:35	0.29	9/1/20 0:20	0.05
8/31/20 22:40	0.17	9/1/20 0:25	0.04
8/31/20 22:45	0.14	9/1/20 0:30	0.03
8/31/20 22:50	0.14	9/1/20 0:35	0.03
8/31/20 22:55	0.13	9/1/20 0:40	0.02
8/31/20 23:00	0.07	9/1/20 0:45	0.01
8/31/20 23:05	0.07	9/1/20 0:50	0
		9/1/20 0:55	0.01

Table 4.13 *Calculations*

Time (min)	Cumulative precipitation (in)	Running total (15 min)	Running total (30 min)
0	0.03		
5	0.05		
10	0.05		
15	0.05	0.05	
20	0.06	0.01	
25	0.11	0.06	
30	0.16	0.11	0.16
35	0.29	0.23	0.24
40	0.46	0.35	0.41
45	0.64	0.48	0.59
50	0.89	0.6	0.83
55	1.06	0.6	0.95
60	1.39	0.75	1.23
65	1.73	0.84	1.44
70	2.02	0.96	1.56
75	2.19	0.8	1.55
80	2.33	0.6	1.44
85	2.47	0.45	1.41
90	2.6	0.41	1.21
95	2.67	0.34	0.94
100	2.74	0.27	0.72
105	2.81	0.21	0.62
110	2.89	0.22	0.56
115	2.96	0.22	0.49

Table 4.13 (*cont.*)

Time (min)	Cumulative precipitation (in)	Running total (15 min)	Running total (30 min)
120	2.99	0.18	0.39
125	2.99	0.1	0.32
130	3	0.04	0.26
135	3.01	0.02	0.2
140	3.02	0.03	0.13
145	3.03	0.03	0.07
150	3.1	0.09	0.11
155	3.19	0.17	0.2
160	3.27	0.24	0.27
165	3.36	0.26	0.35
170	3.41	0.22	0.39
175	3.46	0.19	0.43
180	3.5	0.14	0.4
185	3.53	0.12	0.34
190	3.56	0.1	0.29
195	3.58	0.08	0.22
200	3.59	0.06	0.18
205	3.59	0.03	0.13
210	3.6	0.02	0.1

Table 4.14 *Quartile distribution of rainfall totals*

	Time (min)	Cumulative rainfall (in)	Quartile rainfall (in)
First quartile	52.5	0.975	0.975
Second quartile	105	2.81	1.835
Third quartile	157.5	3.23	0.42
Fourth quartile	210	3.6	0.37

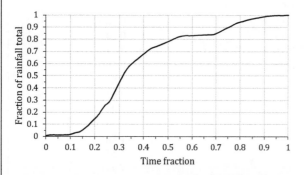

Figure 4.46 Percentage mass curve at Sunnybrook Drive, Raleigh, North Carolina for a short-duration summer thunderstorm on August 31, 2020.

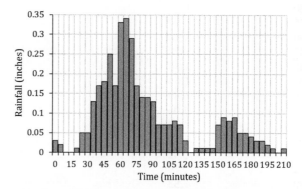

Figure 4.47 Rainfall hyetograph at Sunnybrook Drive, Raleigh, North Carolina for a short duration summer thunderstorm on August 31, 2020.

Example 4.6: For a major flood control project in the Mandakini River catchment in the central Himalayas in India, it was necessary to determine the IDF statistics of rainfall. However, there was no long POR from rain gauges to conduct a frequency analysis. Rainfall data were derived from TRMM shown in Table 4.15. Use frequency factor method to derive the IDF table.

Solution: The IMD prefers the Gumbel distribution. The frequency factor for the Gumbel distribution is given as

$$K_T = -\frac{\sqrt{6}}{\pi}\left\{0.5772 + \ln\left[\ln\left(\frac{T}{T-1}\right)\right]\right\}$$

Using this equation for $T = 2, 5, 10, 25, 50,$ and 100 years, the corresponding values of K_T are obtained as given in Table 4.16.

From the data given in Table 4.15 the mean ($\bar{x}$) and standard deviation (s) are calculated as given in Table 4.17.

Now, using the values of the mean, standard deviation, and frequency factor, the rainfall depths for various frequencies (return periods) are calculated using the equation:

$$x_T = \bar{x} + K_T s$$

These values are shown in Table 4.18.

The corresponding values of the intensities are given in Table 4.19.

Table 4.15 *Maximum rainfall values (mm) for different durations derived from TRMM*

Year	Duration					
	3 h	6 h	12 h	24 h	48 h	72 h
2000	47.74	72.93	88.27	101.30	123.94	127.98
2001	60.00	60.50	60.80	60.80	75.52	75.52
2002	68.99	91.94	102.26	106.78	110.44	121.80
2003	48.72	58.40	74.71	94.15	112.66	117.34
2004	38.54	39.34	67.57	73.29	87.91	100.21
2005	36.49	51.24	66.42	100.63	107.67	119.77
2006	32.64	60.75	66.14	66.14	70.50	80.43
2007	52.04	63.58	110.54	146.40	151.45	151.76
2008	23.23	32.09	38.48	40.72	65.83	87.62
2009	35.64	53.55	58.41	58.41	66.11	100.67
2010	47.17	59.20	78.00	99.54	121.55	137.04
2011	35.71	58.06	109.14	150.29	165.88	168.99
2012	38.57	50.11	60.84	68.06	97.03	129.44
2013	63.81	82.18	101.96	126.34	206.48	218.36
2014	55.56	58.72	87.32	90.48	96.72	124.45
2015	39.88	49.47	49.72	67.23	78.20	92.38

Table 4.16 *Frequency factors for various return periods*

K_2	−0.1641
K_5	0.71881
K_{10}	1.3034
K_{25}	2.04202
K_{50}	2.58997
K_{100}	3.13388

Table 4.17 *Mean and standard deviation of the values given in Table 4.15.*

	3 h	6 h	12 h	24 h	48 h	72 h
Mean	45.2945	58.8789	76.2861	90.6611	108.6172	122.1094
Standard Dev.	12.4416	14.6393	21.7042	31.5038	39.0978	36.1522

Table 4.18 *DDF data from the Mandakini catchment, Uttarakhand, India*

	Rainfall depth (mm) for different duration					
Return period (y)	3 h	6 h	12 h	24 h	48 h	72 h
2	43.25	56.48	72.72	85.49	102.20	116.18
5	54.24	69.40	91.89	113.31	136.72	148.10
10	61.51	77.96	104.58	131.72	159.58	169.23
25	70.70	88.77	120.61	154.99	188.46	195.93
50	77.52	96.79	132.50	172.25	209.88	215.74
100	84.29	104.76	144.30	189.39	231.14	235.41

Table 4.19 *IDF data from the Mandakini catchment, Uttarakhand, India*

	Rainfall intensity (mm/h) for different duration					
Return period (y)	3 h	6 h	12 h	24 h	48 h	72 h
2	14.42	9.41	6.06	3.56	2.13	1.61
5	18.08	11.57	7.66	4.72	2.85	2.06
10	20.50	12.99	8.71	5.49	3.32	2.35
25	23.57	14.80	10.05	6.46	3.93	2.72
50	25.84	16.13	11.04	7.18	4.37	3.00
100	28.10	17.46	12.03	7.89	4.82	3.27

Example 4.7: For the 100-year IDF data given in Table 4.19, determine the values of the coefficients K, b, x, and n.

Solution: The empirical equation used to express the IDF relationship is

$$I = \frac{KF^x}{(T_D + b)^n}$$

As a first step, a set of initial values of the parameters, K, x, b, and n are assumed and, with these values, intensities for various durations are calculated. The error (ε) in this first estimation is calculated as

$$\varepsilon = \left(\frac{I_{observed} - I_{calculated}}{I_{calculated}}\right)^2$$

The errors are summed.

In the second step, the optimization process using Excel's solver function is invoked to minimize the sum of the errors by optimizing the values of K, x, b, and n ($F = 100$). The calculations are shown in Table 4.20A. The values of the parameters shown in Table 4.20B are used to calculate again a set of values of I for given T_D (Table 4.21) with the plots shown in Figure 4.48.

Table 4.20A *Observed and calculated rainfall intensities*

T_D (h)	$I_{observed}$ (mm/h)	$I_{calculated}$ (mm/h)	Error
3	28.09500	27.16889	0.00109
6	17.45942	18.58037	0.00412
12	12.02537	12.01788	0.00000
24	7.89125	7.50801	0.00236
48	4.81552	4.59950	0.00201
72	3.26953	3.43517	0.00257
		Sum of errors	0.01215

Table 4.20B *Calculations of the fitting parameters of the IDF curve*

K	10.9627
b	1.45328
x	0.43637
n	0.73778

Table 4.21 *Values of I calculated for given T_D*

T_D (h)	I (mm/h)
3	27.169
6	18.580
10	13.533
15	10.359
20	8.517
25	7.298
30	6.423
35	5.760
40	5.239
45	4.817
50	4.467
55	4.172
60	3.918
65	3.699
70	3.506
75	3.335

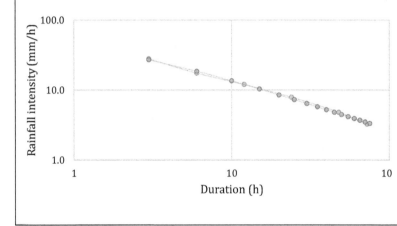

Figure 4.48 IDF curve (F = 100 years) in the Mandakini catchment, central Himalayas (Uttarakhand, India). (A black and white version of this figure will appear in some formats. For the color version, please refer to the plate section.)

Example 4.8: The 24-hour AMS at Grid Point 79° E, 30.5° N of the Mandakini catchment, shown in Figure 4.16, is given in Table 4.22. The L-moments calculated from the AMS are given below. Compute the GEV distribution and compare the observed and predicted values to determine the goodness of fit of the distribution.

Solution: The L-moments of the AMS given in Table 4.22 are obtained as follows:

$\lambda_1 = 69.73826$; $\lambda_2 = 15.99033$; $\lambda_3 = 4.200769$; $\tau_1 = 0.229291$; $\tau_3 = 0.262707$; $\tau_4 = 0.176959$.

From the L-moments, the parameters of the GEV distribution are calculated as:

$$Z = \frac{2}{(3+0.262707)} + \frac{\ln(2)}{\ln(3)} = -0.01794$$

$$\kappa = 7.8590 \times (-0.01794) + 2.9554 \times (-0.01794)^2 = -0.14005$$

$$\sigma = \frac{15.99033 \times (-0.14005)}{(1 - 2^{0.14005}) \times \Gamma(1 - 0.14005)} = 19.91304$$

$$\mu = 69.73826 + \frac{19.91304}{-0.14005}[\Gamma(-0.14005 + 1) - 1]$$
$$= 55.06886$$

Table 4.23 shows the calculations. Figures 4.49 and 4.50 show the pdf and cdf, respectively. Calculated and observed plots are given in Figure 4.51.

Table 4.22 *Twenty-four-hour annual maxima at grid point 79°E, 30.5°N, at the center of the Mandakini catchment*

Year	24-h rainfall (mm)	Year	24-h rainfall (mm)	Year	24-h rainfall (mm)
1901	63.9	1939	34.6	1977	44.3
1902	72.6	1940	26.9	1978	54.4
1903	34.4	1941	81.6	1979	57.3
1904	79.8	1942	58.7	1980	71.8
1905	55.9	1943	46.3	1981	55.7
1906	62.5	1944	60.4	1982	102.5
1907	65.5	1945	84.8	1983	92.1
1908	48.3	1946	67.2	1984	38.4
1909	55.1	1947	46.9	1985	52.1
1910	170.9	1948	82.9	1986	65.8
1911	86	1949	57.3	1987	67.5
1912	63.7	1950	55.4	1988	82.4
1913	58.3	1951	125.7	1989	119.6
1914	83.8	1952	64.6	1990	84.4
1915	49.5	1953	49.3	1991	65.7
1916	50.9	1954	65.9	1992	116.2
1917	65.2	1955	72.8	1993	92.7
1918	40.6	1956	85.8	1994	120.5
1919	48.4	1957	34	1995	61.1
1920	76.1	1958	47.4	1996	45.3
1921	149.3	1959	34.2	1997	40.7
1922	80.6	1960	67	1998	64.1
1923	59.4	1961	51.6	1999	77.5
1924	192.2	1962	78.1	2000	40.1
1925	101.3	1963	92.6	2001	37.8
1926	51.5	1964	75.1	2002	46.6
1927	90.6	1965	96.2	2003	88.5
1928	40.7	1966	62.5	2004	42.6
1929	35.3	1967	51.6	2005	146.3
1930	47.4	1968	36.3	2006	40
1931	50.8	1969	157.2	2007	50.8
1932	58.1	1970	64.1	2008	47.7
1933	48.2	1971	99.6	2009	27.5
1934	46.5	1972	79.3	2010	94.9
1935	35.8	1973	112.8	2011	135.3
1936	55.2	1974	81.9	2012	81.8
1937	55.2	1975	46.5	2013	87.9
1938	50.3	1976	117.9	2014	55.7
				2015	91.5

Table 4.23 *Results of calculations*

Rank	Observed annual maximum 24-h rainfall (mm)	$p = (m - a)/(n + 1 - 2a)$	$p_X(x)$	$F(x)$	Quantile function, X_p	Predicted
1	26.9	0.00521	0.00402	0.02639	26.9	25.6
2	27.5	0.01389	0.00431	0.02888	27.5	28.9
3	34	0.02257	0.00803	0.06841	34	30.9
4	34.2	0.03125	0.00815	0.07003	34.2	32.4
5	34.4	0.03993	0.00828	0.07168	34.4	33.6
6	34.6	0.04861	0.00841	0.07335	34.6	34.7
7	35.3	0.05729	0.00887	0.07939	35.3	35.6
8	35.8	0.06597	0.00920	0.08391	35.8	36.5
9	36.3	0.07465	0.00953	0.08859	36.3	37.3

Table 4.23 (*cont.*)

Rank	Observed annual maximum 24-h rainfall (mm)	$p = (m - a)/(n + 1 - 2a)$	$p_X(x)$	$F(x)$	Quantile function, X_p	Predicted
10	37.8	0.08333	0.01052	0.10362	37.8	38.1
11	38.4	0.09201	0.01092	0.11005	38.4	38.8
12	40	0.10069	0.01197	0.12836	40	39.4
13	40.1	0.10938	0.01203	0.12956	40.1	40.1
14	40.6	0.11806	0.01235	0.13566	40.6	40.7
15	40.7	0.12674	0.01242	0.13689	40.7	41.3
15	40.7	0.12674	0.01242	0.13689	40.7	41.3
17	42.6	0.14410	0.01360	0.16162	42.6	42.5
18	44.3	0.15278	0.01459	0.18559	44.3	43.0
19	45.3	0.16146	0.01514	0.20046	45.3	43.6
20	46.3	0.17014	0.01565	0.21586	46.3	44.1
21	46.5	0.17882	0.01575	0.21900	46.5	44.7
21	46.5	0.17882	0.01575	0.21900	46.5	44.7
23	46.6	0.19618	0.01580	0.22058	46.6	45.7
24	46.9	0.20486	0.01595	0.22534	46.9	46.2
25	47.4	0.21354	0.01618	0.23337	47.4	46.7
25	47.4	0.21354	0.01618	0.23337	47.4	46.7
27	47.7	0.23090	0.01632	0.23825	47.7	47.7
28	48.2	0.23958	0.01654	0.24646	48.2	48.1
29	48.3	0.24826	0.01658	0.24812	48.3	48.6
30	48.4	0.25694	0.01662	0.24978	48.4	49.1
31	49.3	0.26563	0.01698	0.26490	49.3	49.6
32	49.5	0.27431	0.01706	0.26830	49.5	50.0
33	50.3	0.28299	0.01735	0.28207	50.3	50.5
34	50.8	0.29167	0.01751	0.29078	50.8	51.0
34	50.8	0.29167	0.01751	0.29078	50.8	51.0
36	50.9	0.30903	0.01754	0.29253	50.9	51.9
37	51.5	0.31771	0.01772	0.30312	51.5	52.4
38	51.6	0.32639	0.01775	0.30489	51.6	52.8
38	51.6	0.32639	0.01775	0.30489	51.6	52.8
40	52.1	0.34375	0.01789	0.31380	52.1	53.8
41	54.4	0.35243	0.01838	0.35555	54.4	54.2
42	55.1	0.36111	0.01848	0.36845	55.1	54.7
43	55.2	0.36979	0.01849	0.37030	55.2	55.2
43	55.2	0.36979	0.01849	0.37030	55.2	55.2
45	55.4	0.38715	0.01851	0.37400	55.4	56.1
46	55.7	0.39583	0.01855	0.37956	55.7	56.6
46	55.7	0.39583	0.01855	0.37956	55.7	56.6
48	55.9	0.41319	0.01857	0.38327	55.9	57.5
49	57.3	0.42188	0.01865	0.40934	57.3	58.0
49	57.3	0.42188	0.01865	0.40934	57.3	58.0
51	58.1	0.43924	0.01867	0.42427	58.1	59.0
52	58.3	0.44792	0.01866	0.42800	58.3	59.5
53	58.7	0.45660	0.01866	0.43547	58.7	60.0
54	59.4	0.46528	0.01863	0.44852	59.4	60.5
55	60.4	0.47396	0.01855	0.46711	60.4	61.0
56	61.1	0.48264	0.01847	0.48007	61.1	61.5
57	62.5	0.49132	0.01827	0.50580	62.5	62.0
57	62.5	0.49132	0.01827	0.50580	62.5	62.0
59	63.7	0.50868	0.01804	0.52759	63.7	63.1
60	63.9	0.51736	0.01799	0.53119	63.9	63.6
61	64.1	0.52604	0.01795	0.53478	64.1	64.2
61	64.1	0.52604	0.01795	0.53478	64.1	64.2
63	64.6	0.54340	0.01783	0.54373	64.6	65.3
64	65.2	0.55208	0.01768	0.55438	65.2	65.8

Table 4.23 (cont.)

Rank	Observed annual maximum 24-h rainfall (mm)	$p = (m - a)/(n + 1 - 2a)$	$p_X(x)$	$F(x)$	Quantile function, X_p	Predicted
65	65.5	0.56076	0.01760	0.55968	65.5	66.4
66	65.7	0.56944	0.01755	0.56319	65.7	67.0
67	65.8	0.57813	0.01752	0.56495	65.8	67.6
68	65.9	0.58681	0.01750	0.56670	65.9	68.2
69	67	0.59549	0.01717	0.58577	67	68.8
70	67.2	0.60417	0.01711	0.58920	67.2	69.4
71	67.5	0.61285	0.01702	0.59432	67.5	70.0
72	71.8	0.62153	0.01547	0.66428	71.8	70.7
73	72.6	0.63021	0.01514	0.67652	72.6	71.3
74	72.8	0.63889	0.01506	0.67954	72.8	72.0
75	75.1	0.64757	0.01410	0.71308	75.1	72.7
76	76.1	0.65625	0.01366	0.72696	76.1	73.4
77	77.5	0.66493	0.01305	0.74566	77.5	74.1
78	78.1	0.67361	0.01278	0.75341	78.1	74.8
79	79.3	0.68229	0.01225	0.76843	79.3	75.6
80	79.8	0.69097	0.01203	0.77450	79.8	76.3
81	80.6	0.69965	0.01168	0.78399	80.6	77.1
82	81.6	0.70833	0.01124	0.79544	81.6	77.9
83	81.8	0.71701	0.01115	0.79768	81.8	78.8
84	81.9	0.72569	0.01111	0.79880	81.9	79.6
85	82.4	0.73438	0.01089	0.80430	82.4	80.5
86	82.9	0.74306	0.01067	0.80969	82.9	81.4
87	83.8	0.75174	0.01029	0.81912	83.8	82.4
88	84.4	0.76042	0.01003	0.82521	84.4	83.3
89	84.8	0.76910	0.00986	0.82919	84.8	84.4
90	85.8	0.77778	0.00944	0.83884	85.8	85.4
91	86	0.78646	0.00936	0.84072	86	86.5
92	87.9	0.79514	0.00859	0.85778	87.9	87.6
93	88.5	0.80382	0.00836	0.86286	88.5	88.8
94	90.6	0.81250	0.00756	0.87956	90.6	90.1
95	91.5	0.82118	0.00723	0.88621	91.5	91.4
96	92.1	0.82986	0.00701	0.89049	92.1	92.8
97	92.6	0.83854	0.00684	0.89395	92.6	94.2
98	92.7	0.84722	0.00680	0.89463	92.7	95.8
99	94.9	0.85590	0.00606	0.90877	94.9	97.4
100	96.2	0.86458	0.00565	0.91639	96.2	99.1
101	99.6	0.87326	0.00467	0.93390	99.6	101.0
102	101.3	0.88194	0.00423	0.94146	101.3	103.0
103	102.5	0.89063	0.00393	0.94635	102.5	105.2
104	112.8	0.89931	0.00200	0.97605	112.8	107.5
105	116.2	0.90799	0.00156	0.98208	116.2	110.2
106	117.9	0.91667	0.00138	0.98458	117.9	113.0
107	119.6	0.92535	0.00121	0.98677	119.6	116.3
108	120.5	0.93403	0.00113	0.98782	120.5	120.0
109	125.7	0.94271	0.00074	0.99260	125.7	124.2
110	135.3	0.95139	0.00031	0.99735	135.3	129.3
111	146.3	0.96007	0.00009	0.99934	146.3	135.5
112	149.3	0.96875	0.00006	0.99957	149.3	143.4
113	157.2	0.97743	0.00002	0.99988	157.2	154.3
114	170.9	0.98611	0.00000	0.99999	170.9	171.4
115	192.2	0.99479	0.00000	1.00000	192.2	209.7

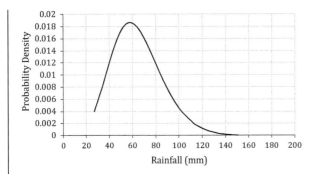

Figure 4.49 Cumulative probability distribution from the GEV model of the data given in Table 4.22.

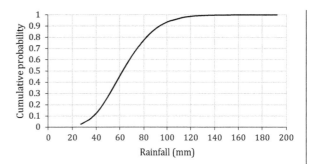

Figure 4.50 Cumulative probability distribution of the data given in Table 4.22.

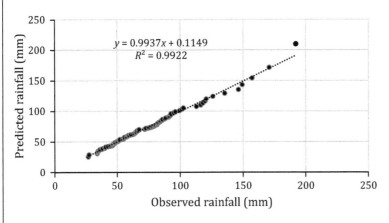

Figure 4.51 Plots of observed versus predicted rainfall according to the GEV distribution.

Example 4.9: Discuss the current practice of PMP studies in the United States with an example of an application of HMR 51 and HMR 52 documents.

The watershed used for this example contains Sayers Dam, located on Bald Eagle Creek in Centre County, Pennsylvania. Sayers Dam is about 14 miles upstream

---

Background information to Example 4.9

The National Weather Service of the NOAA developed PMP studies since the late 1940s and published a series hydrometeorological report (HMRs). Several HMRs were developed to cover different regions of the United States. These HMRs essentially contained the DAD data that can be applicable to a region. For example, HMR 51 is applicable to the region east of longitude (meridian) 105° W. To the west of this, other HMRs are applicable (e.g., HMR 57 and 58). However, many local, regional, or state governmental agencies have conducted their own PMP studies in recent years and these new studies usually take precedence over the HMRs developed by the NWS.

In probability-based point rainfall estimation, three variables, namely depth (or intensity), duration, and frequency are used. For PMP-based modeling, the frequency of occurrence is replaced by storm area as the third variable. Development of DAD relationships in a river basin or region is more of a meteorological exercise than a hydrologic one. However, for hydrologic modeling of a specific watershed or river basin that involves PMP, the regional DAD relationships need to be transformed to a spatial and temporal distribution pattern.

HMR 52 (Hansen et al., 1982) suggested a stepwise procedure to derive the spatial distribution of point PMP estimates for areas in the United States east of 105° W meridian. Subsequently, USACE (Ely and Peters, 1984) developed a computer program to implement the procedure of HMR 52. The procedure involves five elements: (1)

development of DAD curves, which specify the PMP for a specified storm area and duration; (2) development of a standard isohyetal pattern, distributing the precipitation spatially in the form of an ellipse; (3) an orientation adjustment to place the long axis of the ellipse in relation to the direction of normal atmospheric moisture flow in the region; (4) determining the critical storm area that generates the largest PMP over the watershed or basin; and (5) applying an isohyetal adjustment, which specifies the percentage of the PMP depth that applies on each contour (designated by letters A to N) of the standard isohyetal pattern.

Instead of the HMR 52 program, a better platform that is available at present to follow the procedures of HMR 52 is the HEC-MetVue software, also developed by HEC. The HMR 52 storm characteristics from HEC-MetVue can be defined in the HEC-HMS meteorological model. It should be noted that the procedure for developing the PMP in the western United States is different than what is found in HMR 51 and 52. The biggest differences include only one PMP index map for Western US HMR documents and the lack of a predefined spatial PMP storm pattern due to the complex terrain. Similarly, other countries may have different procedures for defining the spatial and temporal distribution of PMP.

of the confluence of Bald Eagle Creek and the West Branch Susquehanna River at Lock Haven, Pennsylvania. The drainage area upstream of Sayers Dam is approximately 339 square miles, while the drainage area at the confluence with the West Branch Susquehanna is approximately 781 square miles. The primary purpose of Sayers Dam was to provide flood control for the downstream reach of Bald Eagle Creek and West Branch Susquehanna River. The dam was completed in 1969 and consists of a rolled earth-fill embankment about 6835 feet long with a top width of 25 feet. The maximum height of the dam is 100 feet above the streambed at elevation 683 feet.

Solution: The shape of the watershed is input to HEC-MetVue. After providing other necessary data, HEC-MetVue created the spatial and temporal distribution of the PMP. Figure 4.52 shows the storm ellipse (the watershed is shown in the middle).

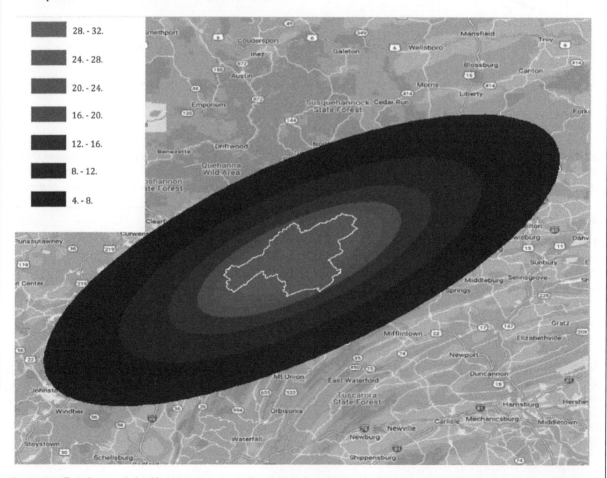

Figure 4.52 Elliptical storm calculated by HEC MetVue program from the DAD data. The watershed boundary is shown along the central region of the storm ellipse. The shading represents rainfall depths in inches. (A black and white version of this figure will appear in some formats. For the color version, please refer to the plate section.)

Table 4.24 *DAD data*

Area (mi²)	Duration				
	6 hours	12 hours	24 hours	48 hours	72 hours
10	26.14	29.96	32.53	35.81	37.35
200	17.81	21.19	23.94	26.96	28.12
1000	12.87	15.98	18.71	21.27	22.58
5000	7.79	10.92	13.24	15.95	17.22
10 000	6.03	9.1	10.98	13.56	14.99
20 000	4.25	7.17	9.03	11.69	12.74

Table 4.25 *PMP distribution*

	Isohyet area (mi²)	Depths (inches) for 6-hour increments of the PMS											
		1	2	3	4	5	6	7	8	9	10	11	12
A	10	20.34	3.64	1.77	1.08	0.85	0.74	0.64	0.54	0.43	0.33	0.25	0.21
B	25	19.11	3.51	1.75	1.08	0.85	0.74	0.64	0.54	0.43	0.33	0.25	0.21
C	50	17.87	3.38	1.73	1.08	0.85	0.74	0.64	0.54	0.43	0.33	0.25	0.21
D	100	16.64	3.29	1.71	1.08	0.85	0.74	0.64	0.54	0.43	0.33	0.25	0.21
E	175	15.56	3.21	1.7	1.08	0.85	0.74	0.64	0.54	0.43	0.33	0.25	0.21
F	300	14.33	3.13	1.7	1.08	0.85	0.74	0.64	0.54	0.43	0.33	0.25	0.21
G	450	13.25	3.06	1.69	1.08	0.85	0.74	0.64	0.54	0.43	0.33	0.25	0.21
H	700	9.71	2.5	1.43	0.9	0.72	0.62	0.54	0.46	0.36	0.27	0.21	0.18
I	1000	7.7	2.13	1.21	0.76	0.61	0.52	0.46	0.39	0.3	0.23	0.18	0.15
J	1500	5.85	1.76	1.02	0.65	0.51	0.44	0.38	0.33	0.26	0.2	0.15	0.13
K	2150	4.62	1.43	0.85	0.54	0.43	0.37	0.32	0.27	0.21	0.16	0.12	0.1
L	3000	3.54	1.18	0.67	0.43	0.34	0.29	0.25	0.21	0.17	0.13	0.1	0.08
M	4500	2.31	0.82	0.51	0.32	0.26	0.22	0.19	0.16	0.13	0.1	0.07	0.06
N	6500	1.23	0.45	0.32	0.2	0.16	0.14	0.12	0.1	0.08	0.06	0.05	0.04
O	10 000	0.46	0.15	0.12	0.08	0.06	0.05	0.04	0.04	0.03	0.02	0.02	0.01
P	15 000	0	0	0	0	0	0	0	0	0	0	0	0
Q	25 000	0	0	0	0	0	0	0	0	0	0	0	0
R	40 000	0	0	0	0	0	0	0	0	0	0	0	0
S	60 000	0	0	0	0	0	0	0	0	0	0	0	0

Once the watershed shapefile is input, the program extracts the DAD data for the centroid as given in Table 4.24.

The PMP values distributed in each of the areas enclosed by the isohyets (Figure 4.53) are given in Table 4.25.

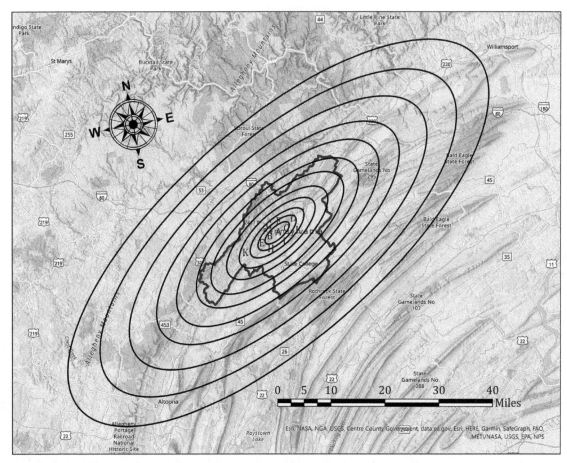

Figure 4.53 Map of PMP distribution over the watershed (A black and white version of this figure will appear in some formats. For the color version, please refer to the plate section.)

Exercises: A selection of exercises on this topic is available at www.cambridge.org/appliedhydrology.

5 Streamflow Measurements and Statistics

5.1 CONCEPTS AND PURPOSE

In Chapter 4 we dealt with rainfall, the most important *input* to a hydrological system. In this chapter we consider streamflow, that is, river or stream discharge, the major *output* from the system. Measurements of streamflow provide the data that are fundamentally important for the calibration of hydrologic models for the design of hydraulic structures and flood risk management as well as for water resources management, which requires water availability models and long-term hydrologic simulations to aid planning for water supply, irrigation, and flood control. Streamflow data are obtained from measurements using certain techniques and installation of stream-gauging stations at various points along a stream. Datasets of streamflow records at a gauging station, operating for a long period, contain long records of stage heights and discharge values. Meaningful and important information from these datasets is derived using certain statistical procedures. Statistical models of streamflow data are established by mainly analyzing a sample of discharge values. The sample means a set of measurements collected over a time period. Such a sample is also known as **time-series** data. This chapter describes the techniques used to measure discharge and the statistical procedures employed to extract valuable information from discharge time-series data.

5.2 MEASUREMENT OF STREAMFLOW

5.2.1 Sequence of Streamflow Measurements

The streamflow measurements discussed here refer to measurements of free open channel flows as opposed to ad hoc flow measurements controlled by in-line weirs and flumes. Uncontrolled streamflow measurements are done in a sequence of procedures: (1) measurement of stage heights; (2) measurement of flow velocity; (3) measurement of the cross-sectional area of the stream where stage is measured; (4) converting velocity measurements to discharge values by using the cross-sectional data; (5) developing a stage–discharge relationship called a rating curve at the gauging site from a set of measurements ranging from low to very high flow conditions, also called flood stages; and (6) continuous monitoring or measurements of stages and converting these stage heights to discharge values using the rating curve established at the gauging site at regular intervals on a daily basis.

The number of gauging stations along a river depends on the cost of installation and operation; the importance of the data for certain objectives; the accuracy needed for the data; and physical as well as hydrological characteristics of the watershed, such as its size, the extent of development within it, and the magnitude of flow. Some of these factors are interrelated. For example, large watersheds or watersheds with significant developments may entail costlier projects needing large datasets with high accuracy.

5.2.2 Measurement of Stage

Stream-gauging stations that measure **stage** usually consist of a structure in which instruments used to measure, store, and transmit the stream-stage information are housed. A variety of methods is used to measure stage, that is to say, **gauge height**. One common method uses a **stilling well** in the riverbank (Figure 5.1), or attached to a bridge pier. Water from the river enters and leaves the stilling well through underwater intake pipes allowing the water surface in the stilling well to be at the same elevation as that in the river. The stage is then measured inside the stilling well using a float or a pressure, optic, or acoustic sensor. The measured stage value is stored in an electronic data recorder on a regular interval. Several telemetry systems are used to transmit stage data from the gauging station to a central office, although satellite telemetry has become the standard. Details of the methods used by the USGS for stage measurements are given by Sauer and Turnipseed (2010). The basic stilling-well float system has traditionally been the predominant method used at gauging stations. Electronic stage sensors and water-level recorders are modern methods. Bubble gauges coupled with nonsubmersible pressure transducers eliminate the need for stilling wells. Submersible pressure transducers have become common for the measurement of stage.

At some sites, installing a stilling well is either not feasible or cost prohibitive. As an alternative, stage can be determined by measuring the pressure required to maintain a small flow of gas through a tube and bubbled out at a fixed location underwater in the stream. The measured pressure is directly related to the height of water over the tube outlet in the stream. As the depth of water above the tube outlet increases, more pressure is required to push the gas bubbles through the tube.

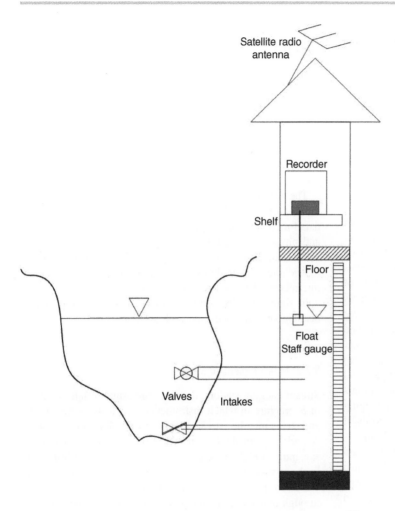

Figure 5.1 Schematic of a stilling-well configuration for measurement of river stage.

Stream gauges operated by the USGS provide stage measurements that are accurate to the nearest 0.01 foot or 0.2 percent of stage, whichever is greater. Stage at a gauging station must be measured with respect to a constant reference elevation, known as a **datum**. Sometimes, stream-gauge structures are damaged by floods or can settle over time. To maintain accuracy, and to ensure that stage is being measured above a constant reference elevation, the elevations of stream-gauge structures, and the associated stage measurement, are routinely surveyed relative to the permanent elevation benchmarks near the stream gauge.

5.2.3 Measurement of Flow Velocity

Stage provides valuable information for certain purposes, such as flood forecasting. However, discharge is what hydrologists are mostly interested in. For measurement of discharge, first the average flow velocity of water through the channel cross-sectional area at the gauging station is determined. The USGS, for example, uses numerous methods and types of equipment to measure velocity and cross-sectional area, including **current meters** and the **acoustic Doppler current profiler** (ADCP). Turnipseed and Sauer (2010) provide details of various methods to measure flow velocity. Traditionally, current meters have been the most common device for velocity measurements.

5.2.3.1 Current Meters

Current meters are the most common hydrometric instrument used for velocity measurements. However, several types of equipment and methods that are used to make current-meter measurements vary with the range of stream conditions. The most common device is the **mechanical current meter**, which basically consists of a rotating element that rotates due to the reaction of the stream current with an angular velocity proportional to the stream velocity (Figure 5.2). The device is designed such that its rotation speed varies with stream velocity, following a general linear relationship given as

$$v = a + bN_s \tag{5.1}$$

where v is the stream velocity, N_s represents the revolutions of the meter per second, and a and b are constants for the instrument.

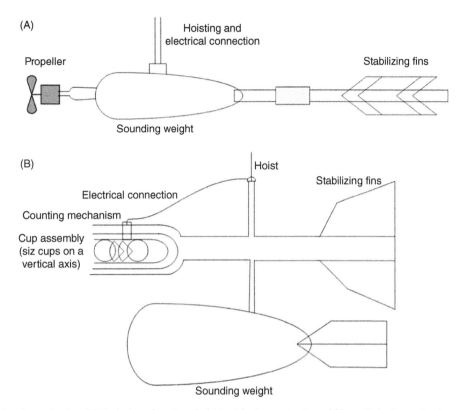

Figure 5.2 Schematics of current meters. (A) The basic configuration of a horizontal-axis current meter and (B) a vertical-axis current meter.

A common current meter is the Price AA current meter for which the values of a and b in Eq. 5.1 are 0.65 and 0.03, respectively. The Price AA current meter has a wheel of six metal cups that revolve around a vertical axis. An electronic signal is transmitted by the meter on each revolution, allowing the revolutions to be counted and timed. Because the rate at which the cups revolve is directly related to the velocity of the water, the timed revolutions are used to determine the water velocity. The Price AA meter is designed to be attached to a **wading rod** for measuring in shallow waters or to be mounted just above a weight suspended from a cable and reel system for measuring in fast or deep water. In shallow water, the **pygmy current meter,** for which the meter constants are the order of $a = 0.30$ and $b = 0.003$, is also used. It is a two-fifths scale version of the Price AA meter and is designed to be attached to a wading rod. A third mechanical current meter, also a variation of the Price AA current meter, is used for measuring water velocity beneath ice. Its dimensions allow it to fit easily through a small hole in the ice, and it has a polymer rotor wheel that hinders the adherence of ice and slush.

5.2.3.2 Acoustic Doppler Current Profiler

Advances in technology have allowed discharge measurements to be made with the ADCP. This device uses the principles of the Doppler effect, described in Chapter 4, to determine water velocity by sending a sound pulse into the water and measuring the change in frequency of that sound pulse reflected back to the ADCP by sediment or other particulates being transported in the water. The change in frequency, or Doppler shift, that is measured by the ADCP is translated into water velocity. The sound is transmitted into the water from a transducer to the bottom of the river and receives return signals throughout the entire depth (Figure 5.3). The ADCP also uses acoustics to measure water depth by measuring the travel time of a pulse of sound to reach the river bottom and back to the ADCP.

To make a discharge measurement, the ADCP is mounted onto a boat or into a small watercraft, with its acoustic beams directed into the water from the water surface. The ADCP is then guided across the surface of the river to obtain measurements of velocity and depth across the channel. The river-bottom tracking capability of the ADCP acoustic beams or a global positioning system is used to track the progress of the ADCP across the channel and provide channel-width measurements. Using the depth and width measurements for calculating the area and velocity measurements, the discharge is computed by the ADCP using discharge = area × velocity, similar to the conventional current-meter method.

The ADCP has proven to be beneficial to stream gauging in several ways: it has reduced the time it takes to make a discharge measurement; it allows discharge measurements to be made in some flooding conditions that were not previously possible; and, lastly, it provides a detailed profile of water velocity and direction for most of the cross section instead of just at point locations with a mechanical current meter, which improves the accuracy of the discharge measurements.

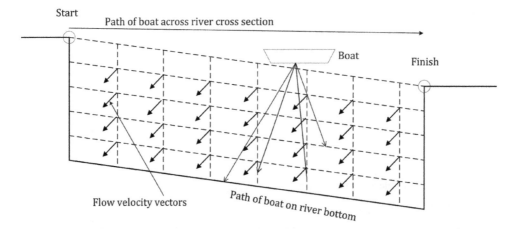

Figure 5.3 Acoustic Doppler current profiler (ADCP), mounted in a small watercraft, is used for measuring the discharge of a river. The ADCP acoustic beams are directed down into the water as it is guided across a river channel.

5.2.4 Measurement of Cross-Sectional Area

For a cross section divided into multiple imaginary vertical strips with finite width shown in Figure 5.4, the subsection width is generally measured using a cable, steel tape, or similar piece of equipment. The subsection depth is measured using a wading rod, if conditions permit, or by suspending a sounding weight from a calibrated cable and reel system off a bridge, cableway, or boat, or through a hole drilled in ice.

5.2.5 Measurement of Discharge

Discharge is the volume of water flowing through a certain section of a stream in unit time. In the United States, it is expressed in cubic feet per second (ft³/s or sometimes abbreviated as cfs [in an earlier English system it used to be called cusec]) and in SI units it is expressed in cubic meters per second (m³/s or abbreviated as cumec). Stream discharge is measured by dividing a cross section into a finite number of imaginary vertical strips (Figure 5.4). The discharge, Q, at a cross section of area A is determined from

$$Q = \iint_A \mathbf{V} \cdot d\mathbf{A} \quad (5.2)$$

in which the integral is approximated by summing the incremental discharges in each of the vertical strips using

$$Q = \sum_{i=1}^{n} \bar{v}_i(d_i w_i) = \sum_{i=1}^{n} \bar{v}_i a_i = \sum_{i=1}^{n} q_i \quad (5.3)$$

where, d_i and w_i, respectively, are the depth and width of the ith strip and $\bar{v}_i$ is the depth-integrated flow velocity through the ith strip. In each subsection shown in Figure 5.4, the area is obtained by measuring the width and depth of the subsection, and the water velocity is determined using a current meter. This method of determining stream discharge is called the **velocity–area method**.

For the selection of velocity in the vertical strip, the **midsection method** is normally used, which assumes that

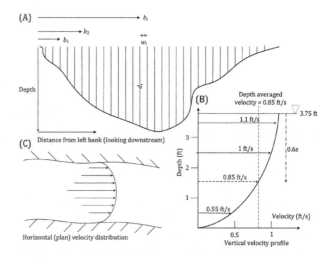

Figure 5.4 (A) Subdivision of a channel cross section into vertical strips for velocity measurements. (B) Vertical velocity profile of a natural channel (Matamek River, Quebec, Canada) adopted from Julien (2002). (C) Schematic of a horizontal velocity profile. In the vertical direction, the velocity is at a minimum near the channel bottom due to drag and increases towards the surface. The maximum velocity occurs below the exact surface due to wind drag.

the mean velocity in each vertical strip represents the mean velocity in a partial rectangular area (segment). The mean velocity in each vertical is determined by measuring the velocity at one or more selected points in that vertical. However, velocity varies with depth. This vertical variation of velocity can be expressed by a log-law as

$$v = \frac{v^*}{k} \ln\left(\frac{z}{K_s}\right) \quad (5.4)$$

where k is the von Karman constant, which is approximately 0.41, v^* is the shear velocity, K_s is the hydraulic roughness parameter describing the characteristic hydraulic friction length, and z is the upward bed-normal height above the datum. Figure 5.4 shows the vertical velocity profile of a natural channel exhibiting such a logarithmic variation, due to which certain conventions, based on empirical observations, are adopted for the representation of depth-integrated average velocity in the vertical strips. The average depth integrated velocity ($\bar{v}$) is given by

$$\bar{v} = \begin{cases} v_{0.6d} \text{ for depth of flow less than 3 m} \\ \left(\dfrac{v_{0.8d} + v_{0.2d}}{2}\right) \text{ for moderately deep streams} \\ cv_{0.5} \text{ for flood flows } (c \approx 0.85 - 0.95) \end{cases} \quad (5.5)$$

where the subscript for the velocity denotes the velocity at that depth, d, of flow measured from the water surface. The partial discharge is now computed for any partial section at location i as

$$\begin{aligned} q_i &= \bar{v}_i \left[\frac{\{b_i - b_{(i-1)}\}}{2} + \frac{\{b_{(i+1)} - b_i\}}{2} \right] d_i \\ &= \bar{v}_i \left[\frac{\{b_{(i+1)} - b_{(i-1)}\}}{2} \right] d_i \end{aligned} \quad (5.6)$$

In the midsection method, the cross sections are defined by depths at locations $i = 1, 2, 3, \ldots$, at a distance b_i where the distance from the bank into the channel is measured from left to right looking downstream (Figure 5.4). At each location, the velocities are sampled by a current meter to obtain the average of the vertical distribution of velocity according to Eq. 5.5.

The partial discharge is computed in each subsection by multiplying the subsection area by the depth-averaged velocity in that subsection. The total discharge is then computed by summing the discharge of each subsection (Eq. 5.3). Each subsection is chosen or the process is repeated such that, in each subsection, the partial discharge (q_i) is less than 5% of total discharge Q passing through the cross section.

Finally, it should be borne in mind that, in addition to the discharge measurements, the uncertainty or error associated with streamflow discharge measured at gauging stations is also important for water management applications and analysis. A study by Kiang et al. (2018) shows that uncertainty estimates vary widely when comparing recent estimation methods.

5.3 STAGE–DISCHARGE RELATIONSHIPS

5.3.1 Rating Curves

Continuous measurement of discharge at a gauging station is not practical. However, stream gauges continuously measure stage heights. A continuous record of stage heights enables the determination of a continuous record of river discharge because there is a relationship between river stage and discharge. A continuous record of stage is translated to river discharge by applying this **stage–discharge relationship** called a **rating curve**.

Stage–discharge relationships are developed for stream gauges by physically measuring the flow of the river with a mechanical current meter or ADCP at a wide range of stages. For each measurement of discharge there is a corresponding measurement of stage. Determining discharge from stage requires defining the stage–discharge relationship by measuring discharge at a wide range of river stages. In the United States, the USGS makes discharge measurements at most stream gauges every 6 to 8 weeks, ensuring that the range of stages and flows at the stream gauge are measured regularly. In addition, special measurements are made during extremely high and low stages because these occur less frequently.

When measured values of stage and discharge are plotted on arithmetic paper, the result is a parabolic curve that can be expressed as

$$Q = a(h-b)^c \quad (5.7)$$

where b is a constant representing the gauge reading for zero discharge, h is the stage height, and a and c are known as **rating-curve constants** or **station constants**. When the data are plotted on logarithmic scales, the data define a straight line and Eq. 5.7 becomes

$$\log Q = \log a + c \log(h-b) \quad (5.8)$$

An example of a stage–discharge relationship is shown in Figure 5.5A where the data are plotted on an arithmetic scale. Figure 5.5B shows the same data plotted on a log–log scale.

The values of the rating constants can be obtained using an analytical approach. From a smooth curve of Q versus h, three values of Q_1, Q_2, and Q_3 are selected such that $Q_1/Q_2 = Q_2/Q_3$. The corresponding values of stage are h_1, h_2, and h_3. Then, from Eq. 5.7,

$$\frac{h_1 - b}{h_2 - b} = \frac{h_2 - b}{h_3 - b} \quad (5.9)$$

from which the value of b is derived as

$$b = \frac{h_1 h_3 - h_2^2}{h_1 + h_3 - 2h_2} \quad (5.10)$$

From the value of b thus obtained, the constant c is determined as

$$c = \frac{\log(Q_1) - \log(Q_2)}{\log(h_1 - b) - \log(h_2 - b)} \quad (5.11)$$

Once b and c are determined, a is obtained from

$$a = 10^{[\log(Q_1) - c \log(h_1 - b)]} \quad (5.12)$$

For example, from the stage–discharge data of the Missouri River at Virgelle, Montana shown in Figure 5.5, three discharge values are selected as $Q_1 = 4540$ cfs, $Q_2 = 9310$ cfs, and $Q_3 = 19\,100$ cfs so that $Q_2/Q_1 = Q_3/Q_2 = 2.051$. The corresponding values of stages are, $h_1 = 2.97$ ft, $h_2 = 4.72$ ft, and $h_3 = 7.35$ ft. Applications of Eqs. 5.10, 5.11, and 5.12 yield, $b = -0.51011$, $c = 1.7629$, and $a = 503.8063$. Thus, for the period of record noted in Figure 5.5, the stage–discharge relationship of the Missouri River at this gauging station can be expressed as:

$$Q = 503.8063[h + 0.51011]^{1.7629}$$

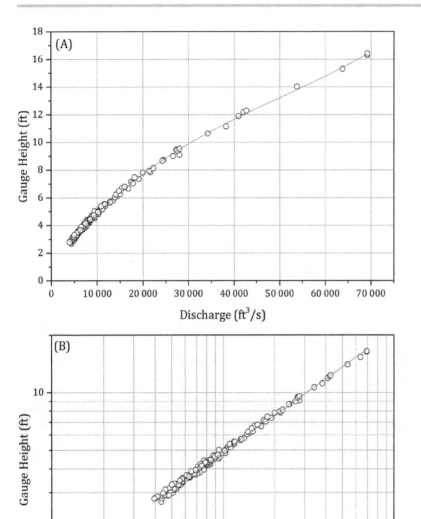

Figure 5.5 Stage–discharge relationship of the Missouri River at Virgelle, Montana (USGS Gauging Station 06109500): (A) data are plotted on (A) an arithmetic scale and (B) on a logarithmic scale. The record plotted in these figures spans a period from June 12, 1964 to March 23, 2022. Only data points with the qualification of 'good' are used in these plots. Other data points that are either 'fair' or 'poor' are discarded in creating the plots.

The determination of such an equation giving the stage–discharge relationship at a gauging station is useful for several reasons. For example, during severe flood events, it is usually very difficult to measure the discharge due to prevailing high velocities. A quantitative relationship can provide the necessary information for either very low or very high discharges and thereby cover the full range of gauge heights likely to be encountered at a gauge.

In the example of Figure 5.5, the quality of data is good and hence the plots are along a smooth curve and the simple analytical procedure described above can be used to derive the rating-curve constants. However, in many situations, the data may exhibit considerable scatter. In those cases, the values of the constants are usually obtained by the nonlinear least-squared method. Alternatively, all three constants a, b, and c can be obtained by optimization.

5.3.2 Rating-Curve Controls

A rating curve represents the integrated effect of a wide range of channel and flow parameters. The combined effect of these parameters is designated as the **control**. If the rating curve for a gauging station does not change with time, the control is said to be permanent; otherwise, it is called **shifting control**. The stage–discharge relationship must be defined for relatively stable river control sections.

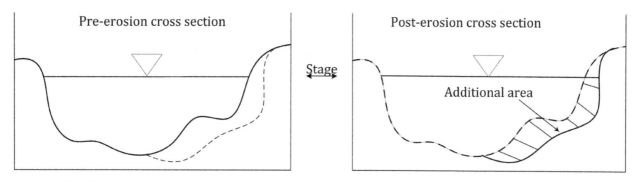

Figure 5.6 Effect of channel geometry on the stage–discharge relationship. Erosion of part of a channel results in an increased cross-sectional area, thereby increasing the capacity for conveying greater discharge at the same stage.

Shifting controls may result from (1) scour or deposition, (2) varying backwater, (3) rapidly changing flow, and (4) changes in flow caused by dredging, channel encroachment, and weed growth, etc. Thus, the stage–discharge relationship will be stable if the hydraulic characteristics of the general reach of stream are unchanging, and the bed material does not move appreciably.

The stage–discharge relationship depends upon the shape, size, slope, and roughness of the channel at the stream gauge and is different for every stream gauge. The development of an accurate relationship requires numerous discharge measurements at all ranges of stage and streamflow. In addition, these relationships must be continually checked against ongoing discharge measurements because stream channels are constantly changing.

Changes in stream channels are often caused by erosion or deposition of stream-bed materials, seasonal vegetation growth, debris, or ice. An example of how erosion in a stream channel increases a cross-sectional area for water, allowing the river to have a greater discharge with no change in stage, is shown in Figure 5.6. New discharge measurements plotted on an existing stage–discharge relationship graph will show this, and the rating can be adjusted to allow the correct discharge to be estimated for the measured stage.

Figure 5.7 shows an example of stage–discharge relationships that change with time. Such changes occur if the stream bed or hydraulic roughness, called the **rugosity coefficient**, is changing, as might occur in sand-bed streams. In such cases, frequent measurements are needed to define how the rating is changing and to define its present condition.

5.3.3 Shift-Adjusted Ratings

Stage–discharge relationships, or ratings, are usually developed from a graphical analysis of numerous current-meter discharge measurements, sometimes called calibrations. In a well-established stream-gauging program, such as that of the USGS, measurements are made at various schedules and are compiled and maintained in a data base. Each measurement is carefully made and undergoes a quality-assurance review. However, the stage–discharge relationship in a natural channel is seldom static because the physical features of the channel

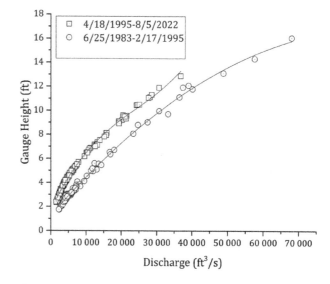

Figure 5.7 Stage–discharge relationship of the Colorado River at the Colorado–Utah state line (USGS Gauging Station 09163500) for two different periods showing distinctly different trends. Only data points with the qualification of 'good' are used in this analysis. Other data points that are either 'fair' or 'poor' are not used in the plots.

often change either gradually or abruptly. For this reason, some measurements indicate a change in the rating, often due to a change in the channel or riparian vegetation. Such changes are called **shifts** and are referred to as shifts to the base rating. The shifts may indicate a short- or long-term change in the rating for the gauge. Applying these shifts to a rating is called a **shift-adjusted rating**.

Shifts are either positive or negative, depending upon whether its value must be added to or subtracted from the recorded gauge height to adjust for the departure from the base rating. Sometimes, negative shifts are referred to as shifts to the left and positive shifts as shifts to the right. Possible causes for negative shifts include fill or deposition in the channel, temporary dams (natural or human-made), seasonal vegetative or algal growth, and debris jams. Possible causes for positive shifts include scour, gravel mining, and clearing of debris or vegetation from the channel, either by floods or humans.

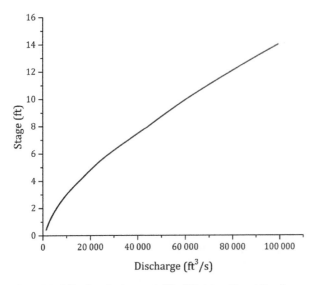

Figure 5.8 Shift-adjusted rating curve of the Yellowstone River at Forsyth, Montana (USGS gauging station 06295000). The number of data points plotted to define the rating curve is 1359.

In normal usage, the measured shifts or corrections are applied mathematically to a defined rating. The shift adjustments are applied to the individual ratings as measured data become available, resulting in an adjusted rating. Some ratings may change as often as weekly, others may not change for months. Figure 5.8 is an example of a shift-adjusted rating curve produced from the shift-adjusted stage data obtained from the USGS.

5.3.4 Backwater Effects

Backwater effects near the controls may require stage–discharge relationships that are also a function of water-surface slope. If the relationship at a gauging station is affected by backwater due to some downstream control, the slope of the water surface must be taken into account using the stage–discharge relationship from two gauges, one upstream and another downstream of the control point, called the **auxiliary gauges**,

$$\frac{Q}{Q_0} = \left(\frac{S}{S_0}\right)^k = \left(\frac{F}{F_0}\right)^m \qquad (5.13)$$

in which the difference between the stages of the two gauges indicates the fall, F, Q_0 is the normalized discharge at the stage when $F = F_0$ or $S = S_0$, where, as noted above, $S = F/L$, and L is the distance between the gauges. The exponent m usually has a value 0.5, but k is not necessarily equal to m.

5.3.5 Extrapolation of Rating Curves

The stage–discharge relationships established from observed stage–discharge data often needs extrapolation beyond the range of actual observations to arrive at the complete stage–discharge curve that is possible at a site, or for the design of hydraulic structures at design discharges outside the range of observed stage–discharge values used in the construction of the rating curve. Extrapolation of rating curves is risky and should be carefully checked. In general, there is no completely satisfactory method of extension of rating curves.

The polynomial representation (Eq. 5.7) cannot be used if there is any significant change in the hydraulic geometry of the stream at high flows. Under uniform flow conditions and gradually varying flow, the well-known Chezy and Manning equations are stage–discharge relationships, which give the relationship between discharge, slope, and stage and can be used to determine the discharge from gauge heights that are beyond the established rating curve at a station. The procedure is illustrated below with an example.

A study was undertaken by Mutreja and Bhatia (unpublished report) from the Irrigation Research Institute at Roorkee, India, to determine the stage–discharge relationship at a bridge (Tehri Girder Bridge) on the Bhagirathi River, the headwater stream of the River Ganga in Uttarakhand, India. In application of the Manning equation, it is generally assumed that, as discharge increases, $\sqrt{S}/n$ attains a stable value for higher discharges where S is the bed slope and n is the Manning roughness coefficient. To compute $\sqrt{S}/n$ from the Manning equation, the cross-sectional characteristics in the form of $AR^{2/3}$ is required (A is the flow area and R is the hydraulic radius). Since the cross section changes randomly from year to year, particularly in alluvial channels, a problem arises in deciding on the cross section to be adopted for computation of $AR^{2/3}$. Furthermore, the needed cross section during a flood is not the same as that during a low-flow period, which normally is available. To overcome the problem of selecting a cross section for analysis, curves for $AR^{2/3}$ versus gauge height were plotted from the observed cross sections of the river for a five-year period (1971–1975) and a mean curve was drawn, with lower and upper enveloping curves, to the best-fit points (Figure 5.9). These curves were used for the maximum observed gauge, corresponding to a discharge value of 1650 m³/s, to obtain the values of $AR^{2/3}$ corresponding to upper, mean, and lower envelopes. These values of $AR^{2/3}$ were used in the Manning equation to get the corresponding values of $\sqrt{S}/n$. The value of $\sqrt{S}/n$ was assumed to become constant beyond the maximum discharge of 1650 m³/s. The values of $\sqrt{S}/n$, computed in this way were used subsequently in the Manning equation to compute discharge for a given gauge height, which in turn was used to compute $AR^{2/3}$. Leonard et al. (2000) discussed development of rating curves using the Manning equation to manage instability and improve extrapolation.

While river cross sections are inherently variable, it generally takes quite some time for a change in cross section to occur. Thus, where good quality data on stage and discharge are available, a stable and representative rating curve may be established, as shown in Figure 5.5. However, all too often, the natural variability is worsened by poor data quality originating from other sources, such as human and instrumental errors. In such cases, errors lead to erroneous

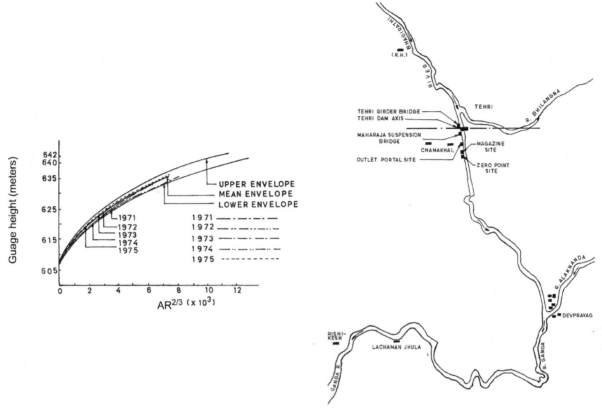

Figure 5.9 Gauge versus conveyance at Tehri Girder Bridge Site on Bhagirathi River in Uttarakhand, India (adopted from Mutreja and Bhatia, unpublished personal communication to VPS).

estimates of discharge and several of the points plotted cannot be explained by normal or natural variation inherent in the discharge process. The stage–discharge plot characteristically shows curvature, with the greatest curvature at low stage, when plotted in natural units, but often the plot can be fitted with a linear-regression equation after logarithmic transformation, thereby yielding a linear rating curve that greatly facilitates extrapolation. However, where there are erroneous data, logarithmic transformation may not remove the curvature in the stage–discharge relationship. Sefe (1996) suggested several other transformations for such cases, citing a case study from Okavango River in Botswana.

5.3.6 Daily Streamflow Data

Daily streamflow data refer to the instantaneous measurements of stage and discharge at regular intervals for every 24-hour period. In the United States, most USGS stream gauges transmit stage data, usually in 15-minute intervals, by satellite to USGS computers, where the stage data are used to estimate streamflow using the rating curve developed for each particular gauging station. The stage information is routinely reviewed and checked to ensure that the calculated discharge is accurate. The USGS maintains a quality-control process to ensure that the streamflow information across the country is of comparable quality and is obtained and analyzed using consistent methods.

Most of the stage and streamflow information produced by the USGS is available in near real time through the National Water Information System website (http://waterdata.usgs.gov/nwis). These data are extremely valuable for multiple purposes, such as derivations of unit hydrographs and other hydrologic parameters, calibration of event-based hydrologic models, and flood warning systems. In subsequent chapters, numerous examples have been developed using streamflow data recorded at 15-minute or hourly intervals. These are particularly useful to represent the hydrographs of specific storm events and hence are commonly called **flood hydrographs** or **storm hydrographs**, which typically represent a relatively short time window spanning several hours to several days. However, daily streamflow data recorded at such intervals are not always available, particularly if the dates of interest are too old or the subject stream is located in a country where such data are not openly disseminated, even if that country has an established stream-gauging program.

Daily streamflow data, in many countries also, generally refer to daily (24-hour) mean flows. For example, Hydrometric Database from the Department of Environment and Natural Resources in Canada, National River Flow Archive of the UK Centre for Ecology and Hydrology in the United Kingdom, and the Bureau of Meteorology of the Australian Government provide daily streamflow data, which are daily means of discharges in many of the rivers and streams of the respective countries. An important source of daily streamflow data for many

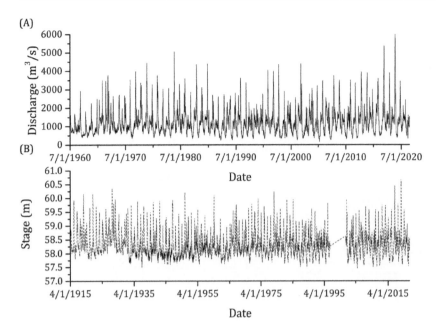

Figure 5.10 Time-series plots of (A) daily discharge and (B) stage of Ottawa River at Britannia (Station No. 02KF005), upstream of Canada's capital Ottawa. The period of record in the time series of the discharge is July 1, 1960 to December 31, 2021, and that for the stage is April 1, 1915 to December 31, 2021.

rivers of the world is the Global Runoff Data Centre of the Bundesanstalt für Gewäsrkunde (Federal Institute of Hydrology) in Germany. Figure 5.10 shows an example of time-series plots of the daily discharge and stage of Ottawa River, a major tributary of Saint Lawrence River, at Britannia, upstream of Ottawa, the capital of Canada. An **annual hydrograph** commonly refers to a plot of daily discharge for every day of a year.

Several statistical estimates derived from daily streamflow measurements at a site within a stream-gauging station network are vital for a variety of hydrologic investigations and water resources management.

5.4 STREAMFLOW STATISTICS

5.4.1 Descriptive Statistics of Discharge Time Series

Streamflow measurements provide different types of time series for river discharge as a function of time. The units of the time interval at which the discharge is given can be minutes, hours, days, months, half-years, or years. The range of time for which the discharge data are either available or given for a specific purpose is called the period of record (often abbreviated as POR). For planning, design, and management of water resources projects, certain descriptive statistics of discharge values derived from a time series carry special significance. These are listed in Table 5.1.

As an example, consider the discharge time series shown in Figure 5.10A, which contains 22 424 discharge values. The mean, minimum, and maximum of these values are 1226.81, 165, and 5980 m³/s respectively. These are the **daily** MQ, LQ, and HQ of the POR represented by the entire time series. The POR contains 42 years of daily data (1960–2021). If we determined the minimum and maximum daily discharge values of

Table 5.1 *Significant descriptive statistics of streamflow time series*

Notation	Definition
LLQ	Lowest discharge ever observed in the POR
LQ	Lowest discharge value during equal intervals in the POR
MLQ	Mean of the lowest discharge values of equal intervals in consecutive years of the POR
MQ	Mean of the discharge values of equal intervals in the POR
MHQ	Mean of the highest discharge values of equal intervals in consecutive years of the POR
HQ	Highest discharge value during equal intervals in the POR
HHQ	Highest discharge ever observed

each of these years separately and took the mean of those 42 discharge values, we would obtain daily MLQ and HLQ values. Similarly, for monthly or annual time series we would get these monthly or annual descriptive statistics. The values of LLQ or HHQ can be obtained if records of instantaneous measurements of daily discharge are available.

5.4.2 Daily Statistics

Daily statistics refers to the mean, maximum, and minimum values of every day of every month of every year in the POR, derived from the daily mean flow. Figure 5.11 is an example of daily mean values of discharge in Potomac River upstream of Washington DC. For example, a mean flow of 17 900 ft³/s for April 1 means it is the mean value of discharge that is expected to occur on April 1, based on

5.4 Streamflow Statistics | 181

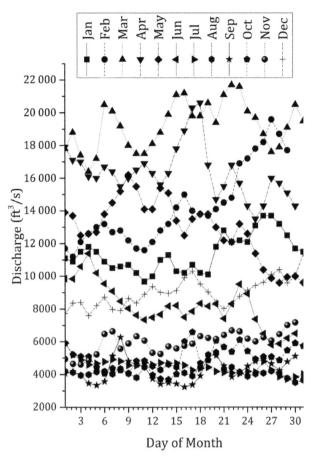

Figure 5.11 Mean of daily mean values of discharge for each day from February 1, 1895 to June 6, 2022, in Potomac River at Point of Rocks in Maryland (USGS gauge station 01638500).

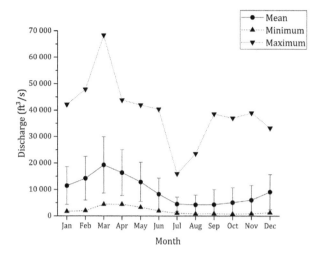

Figure 5.12 Monthly discharge characteristics based on mean monthly discharge of 126 years from January 1, 1896, to December 31, 2021 in Potomac River at Point of Rocks in Maryland (USGS gauge station 01638500). The mean, minimum, and maximum represent the respective measure of 126 discharge values of each month. The error bars associated with the means are one standard deviation of each of 126 monthly records.

the mean of 128 values (128 years in the POR), each of which is again a mean of 96 measurements, assuming that, on each day of April 1 from 1895 to 2022, measurements were taken at every 15 minutes. Such statistics are important to estimate the mean flow of every month in a stream. The daily statistics also include the maximum and minimum flow for the day. In addition to the mean, maximum, and minimum values, the USGS also provides the percentile values from 5 to 95 for daily flows in each month.

5.4.3 Monthly Statistics

Monthly statistics refers to the mean, maximum, and minimum flows for every month of each of the years within the POR. The statistics are generated from daily mean values for a given month in a given year. The data can be represented as a monthly time series just like the daily time series shown in Figure 5.10. The monthly *MQ*, *LQ*, and *HQ* are the mean, minimum, and maximum discharge values in the entire time series, whereas *MLQ* and *MHQ* are derived from the yearly minimum and maximum values. Another useful representation of monthly statistics is shown in Figure 5.12 where the mean, minimum, and maximum discharge characteristics of every month are based on the mean monthly discharge values in each of the year in the POR. Such a plot is called a **seasonal hydrograph**.

Hydrologic seasonality is reflected in the monthly statistics of discharge values, which in turn provide important information about **flow regimes** in a river. A river flow regime describes the average seasonal behaviour of flow dependent on its genetic sources and on the basin climate and physiographic setup. Typically, a flow regime of a river at a gauging station is determined from the long-term monthly mean values as shown in Figure 5.12. Mukhopadhyay and Khan (2014a), analyzed historical flow records at 11 gauging stations (Figure 5.13) within the Upper Indus Basin (Figure 4.41) with the POR from 1962 to 2010, revealing a uniform character of annual flow distributions at all gauging stations given by a Gaussian function

$$y = y_0 \frac{A e^{\frac{-4\ln(2)(x-x_c)^2}{w^2}}}{w\sqrt{\frac{\pi}{4\ln(2)}}} \tag{5.14}$$

where y is the discharge at time x, y_0 is the base discharge of the hydrograph, x_c is the coordinate of the time (x) axis where the center of the hydrograph lies (time of peak discharge), w is the width of the hydrograph between 25th and 75th percentiles of y (discharge), and A is the area under the hydrograph, which is related to the amplitude (peak discharge) of the hydrograph. Thus, the four parameters (y_0, x_c, w, A) that are used to define the distribution function each have a clear physical meaning. For all stations within the basin, except the station at Doyian, $7 < x_c < 8$ implying the time of peak discharge occurs between July and August. At Doyian, $6 < x_c < 7$ (between June and July). The uniform character of the annual hydrographs showed nonskewed unimodal flow distributions, and the peaks occurring at nearly the same at all gauging stations reflects a unique

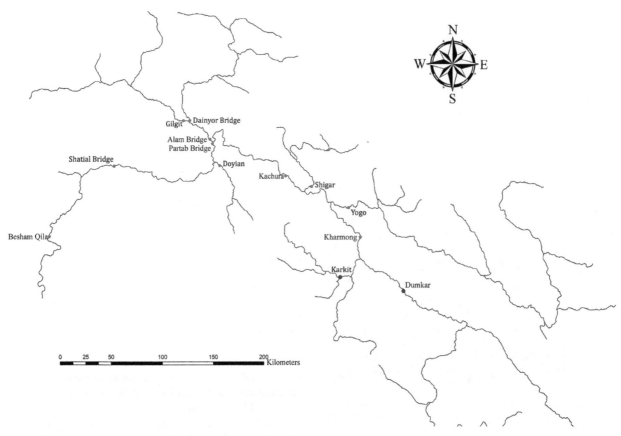

Figure 5.13 Gauging stations within the Upper Indus Basin from which monthly statistics of streamflows were analyzed. These stations all showed seasonal rhythms for flow regimes typical of Himalayan river basins (from Mukhopadhyay and Khan, 2015; Journal of Hydrology).

glacio-hydrometeorological condition controlling river flows in the entire basin. Figure 5.14 shows this flow regime based on the MQ discharge statistics from two of the gauging stations. The seasonal flow patterns throughout the basin can be generalized as shown in Figure 5.15, which is based on the concept of **water year**.[1] Streamflow starts to rise in April or May due to the melting of seasonal snow and reaches its highest snowmelt peak by the end of July. When all seasonal snow is depleted, glacial melt begins to contribute from July onward. Flow continues to rise to a higher glacial melt and monsoon peak in August. The river flows during the months of April/May to the middle of July are mostly the products of seasonal snowmelt, whereas glacial melt accounts for a dominant part of the flow during the later part of July and throughout August and September. Discharge starts declining in September and reaches a minimum in January.

Monthly statistics are also used in many practical applications. For example, gauge data are not available for every stream. So, a common practice is to define the flow characteristics at a number of gauging stations and to generalize this information in terms of drainage area and other indices describing the drainage basin. In many areas, mean monthly discharge shows a predictable correlation with the drainage area. For this reason, the **drainage-area ratio** method commonly is used to estimate streamflow for sites where no streamflow data are available. This method is based on the assumption that the streamflow for a site of interest can be estimated by multiplying the ratio of the drainage area for the site of interest and the drainage area for a nearby streamflow gauging station by the streamflow for the nearby streamflow gauging station. Thus, the drainage-area ratio method is given by

$$\overline{MQ}_{ij} = \left(\frac{A_y}{A_x}\right) MQ_{ij} \qquad (5.15)$$

where $\overline{MQ}_{ij}$ is the streamflow for month i and year j for the site of interest, A_y is the drainage area for the site of interest,

---

[1] The term water year is used by the USGS in dealing with surface-water supply in the United States and is defined as the 12-month period from October 1, for any given year, through September 30 of the following year. In the UK, the water year also runs from October of one year through September of the following year. The water year is designated by the calendar year in which it ends and includes 9 of the 12 months. This assumes that, by October, transpiration by plants will have largely ceased and soil moisture and groundwater storage will have been recharged to near their maximum levels. In the UK, water storage reaches a minimum at the end of summer and, after October, rainfall generally begins to recharge water reserves in the ground and in reservoirs once again, and the hydrological cycle moves from a time of losses to gains. Other water-year spans are appropriate for specific regions, for example, the time of disappearance of annual snowpack.

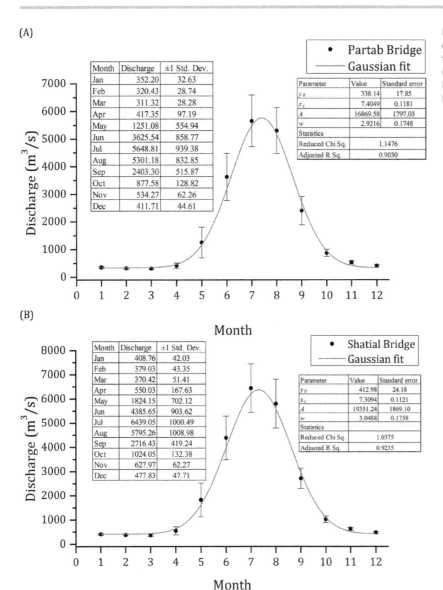

Figure 5.14 Flow regimes of the Upper Indus Basin exemplified by long-term monthly *MQ* statistics at two of the gauging stations shown in Figure 5.13: (A) Partab Bridge; (B) Shatial Bridge (from Mukhopadhyay and Khan, 2014a; Journal of Hydrology).

A_x is the drainage area for the streamflow gauging station, and MQ_{ij} is the mean streamflow for month i and year j for the streamflow gauging station.

Figure 5.16 shows the location of gauging stations along the River Rhine, which flows through four European countries, together with the variation of monthly *MQ*, *LQ*, and *HQ* values derived from the long-term monthly statistics with drainage areas.

5.4.4 Annual Statistics

Annual statistics refer to the mean flow for every year during the POR and they are generated from daily mean values for a given year. Figure 5.17 shows an example with a time-series plot of annual *MQ*. Such time-series plots are often useful in studies of the effect of climate change if there is a trend in increase or decrease of river discharge.

5.4.5 Discharge Variability

Many water management programs require streamflow data. Monthly and annual streamflow statistics are particularly important in water resources management. For planning the utilization of streamflow for water supply, the mean and variability of monthly and annual flows are obtained from continuous records of discharge at gauging stations.

The regional distribution pattern of the mean monthly and annual discharge offers valuable basic information on the availability of surface water for public, industrial, agricultural, and private water consumption as well as for

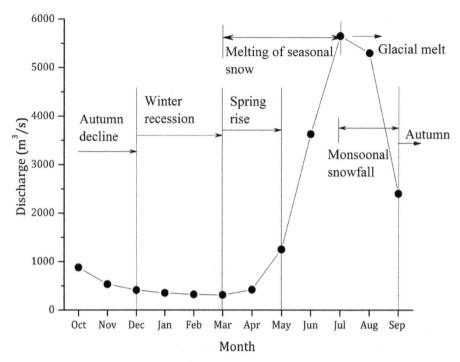

Figure 5.15 Hydrologic seasonality governing flow regime as exemplified by monthly *MQ* statistics at Partab Bridge gauging station. This plot is based on the concept of a water year.

navigation. However, the mean discharge varies considerably depending on the region, as well as the time and duration, which can be seen in a seasonal or flood hydrograph. During floods that are low-frequency, high-discharge events, large volumes of water flow through the watercourses without being able to be used for water management purposes. Furthermore, industries and providers of public water supplies, which use large amounts of water, need information on whether the water volume they require can be guaranteed or sustained without any risk of long interruptions in supply. For these reasons, it is important to quantify the variability of discharge in the major streams of a region with significant population and industries.

The characteristics of discharge variability are based on the quotients of the discharge statistics given in Table 5.1. Discharge measurements for several years, such as the example shown in Figure 5.16 for the River Rhine, are required to calculate these values for a gauging station. The ratios that are important indicators of discharge variability are usually *MHQ/MLQ*, the quotient of the mean highest and lowest discharge values, or *HHQ/LLQ*, the quotient of the highest and lowest discharges ever observed. The latter ratio is numerically higher, sometimes by more than one order of magnitude, than the ratio of mean highest and lowest discharge values. Figure 5.18 shows the spread of these two quotients for the River Rhine, one of the most important rivers in Europe. In general, higher *MHQ/MLQ* values ($\approx$ 6–8) are observed at the upstream gauges of the Alpine and High Rhine in Austria and Switzerland but, at the gauges at the Upper, Middle, and Lower Rhine flowing through Germany and entering the Netherlands, *MHQ/MLQ* values are fairly uniform ($\approx$ 3),

indicating less discharge variability. As noted above, the *HHQ/LLQ* values are an order of magnitude greater than the *MHQ/MLQ* values, but again show lesser variabilities at the Rhine at lower altitudes than at higher altitudes.

5.4.6 Peak Streamflow

The peak-flow statistic is most important for flood-frequency analysis, discussed in Section 5.6. The **peak streamflow** refers to the highest discharge recorded instantaneously on a given day in a year for the POR. This value of peak discharge is different from the daily maximum discharge for a year. Figure 5.19 gives an example where the instantaneous maximum peak discharge values at certain times on certain dates for 27 consecutive years in Ottawa River at Britannia are shown. For example, in the year 1998, the peak discharge of 4390 m^3/s was observed on April 6 at 16:00. But the mean daily discharge on April 6, 1998 was 4350 m^3/s, as shown in the 1998 annual hydrograph at this station.

5.5 FLOW-DURATION ANALYSIS

5.5.1 Concept of Flow-Duration Curves

The long-term *MQ* values discussed above represent the average discharge rates at which streamflow is potentially available for human consumption and management in a region. However, the actual flow rate varies appreciably with time. For example, Figure 5.20 shows the frequency

(A)

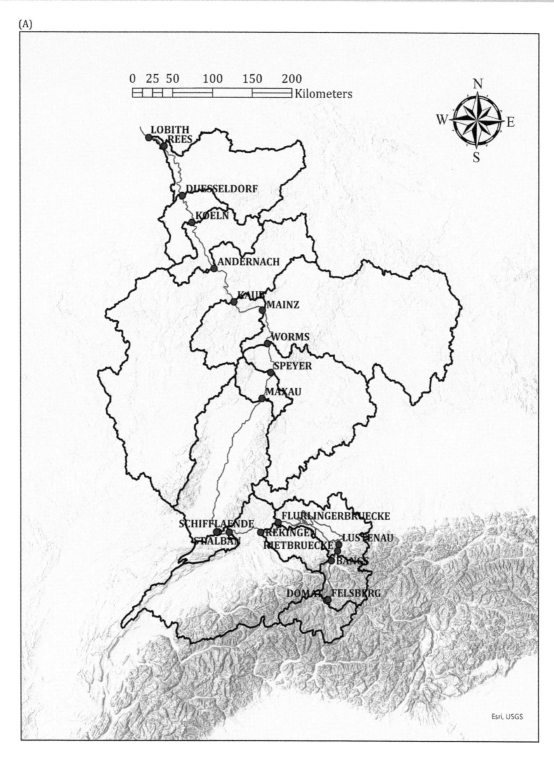

Figure 5.16 (A) Gauging stations with a long POR along the River Rhine, showing the river basin boundary. (B) Plot showing the variation of monthly *MQ*, *LQ*, and *HQ* with drainage area. A best-fit straight line through the monthly *MQ* values. (C) Bar diagram showing the POR at each of the 22 gauging stations (ISO 3166 country code: AT = Austria, CH = Switzerland, DE = Germany, NL = Netherlands). The graphs are constructed from data obtained from the Global Runoff Data Centre, Federal Institute of Hydrology in Germany.

distributions and their descriptive statistics of daily discharge in Columbia River at Dalles, Oregon, for three arbitrarily selected consecutive years separately as well as for the combined three-year data.

The **flow-duration analysis** depicts the discharge variability by showing the relation between a given flow rate, q_p, and the fraction of time, p, that the flow is equaled or exceeded. A flow-duration analysis results in a

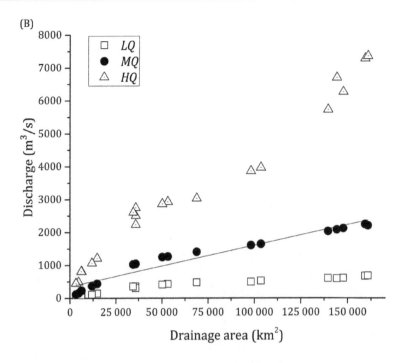

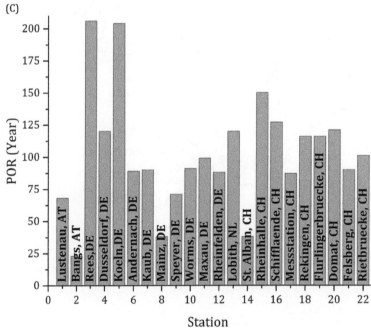

Figure 5.16 (*cont.*)

flow-duration curve, which gives a simple, yet highly informative graphical summary of the variability of a time series. In theory, the variable of the time series can be the mean discharge of daily, monthly, yearly, or of any specific duration. However, typically the variable of interest is the **mean daily streamflow**.

The flow-duration curve defines the percentage of time a particular discharge is equaled or exceeded in a given POR. The POR of the time series can be a one-year period, which can be either a calendar year or a water year. Archetypally, the time series covers several years as the POR. The fraction of time can be thought of as the frequency or probability of occurrence and, as described in Chapter 3, the cumulative fraction of time as **exceedance probability**, p, of the flow, q_p. The probability refers to the exceedance probability in a POR rather than the probability of exceedance on a specific, such as daily, time interval. Thus,

5.5 Flow-Duration Analysis

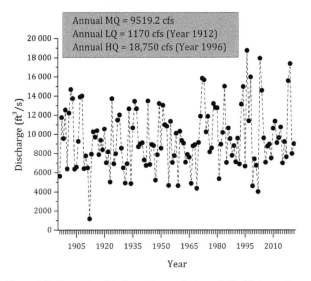

Figure 5.17 Time-series plot of annual mean discharge (*MQ*) of Potomac River from 1896 to 2021 at Point of Rocks in Maryland (USGS gauge station 01638500). The annual *LQ* and *HQ* values, respectively, are the minimum and maximum of the time series. *MLQ* and *MHQ* values can be calculated from the minimum and maximum values for each of the years in the POR.

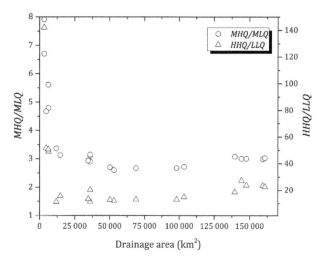

Figure 5.18 Variation of *MHQ/MLQ* and *HHQ/LLQ* in the River Rhine.

$$p = P\left[Q \geq q_p\right] = 1 - F(q) \quad (5.16)$$

where $F(q)$ is the cumulative distribution function of daily average streamflow. It follows then that the flow-duration curve can be called the **cumulative frequency** plot of a time series of the mean daily flow values, showing the percent of time during which specified discharges were equaled or exceeded in a given period. Therefore, a flow-duration curve is the relation between the magnitude of streamflow, q, at a point and the frequency with which those magnitudes are exceeded over an extended time period.

Figure 5.21 shows the flow-duration curve of the average daily streamflow of Ottawa River at Britannia for a 42-year POR shown in Figure 5.10. It combines in one curve the flow characteristics of a stream throughout the range of discharge, but the sequence of these occurrences is lost. Since it is a probabilistic description of streamflow at a given location, such a curve may be considered to represent the hydrograph of an average year with its flow arranged in order of magnitude. For example, Figure 5.21 implies that, for 25% of time, the flow in Ottawa River at Britannia can either equal or exceed 1500 m^3/s.

Even though, in statistical terms, a flow-duration curve is a cumulative frequency curve of a continuous time series, displaying the relative duration of various magnitudes, strictly speaking, the daily flow-duration curve cannot be considered a frequency curve because the daily flow on a particular day is highly correlated with the flow on the preceding day. In other words, daily discharge values of a continuous time series are not random variables in a true sense. This can be observed from autocorrelation or serial correlation, exhibited in an annual hydrograph. Figure 5.22 is an example of an annual hydrograph in which the autocorrelation with a lag of one day is close to unity implying high serial correlation.

In streams that receive major influx of runoff only during rain events or snowmelt, as can be seen in the example of Black River given in Figure 5.22, the flows may tend to be grouped near the low end, with very few large flows. In such cases, the relative frequency curve is skewed to the left. A logarithmic transformation, where the logarithms of the flows are plotted against the probability, generally reduces the skewness of the curve and produces a plot that is easily read at the extremities of data. This can be seen in Figure 5.23, where the daily mean flow-duration curve for a POR of 56 years for the Black River case is shown using plots on both arithmetic and log-probability scales. For the reason stated above, the abscissa of the flow-duration curve is labelled as the percent of time, instead of exceedance probability.

Several factors control the slope of the flow-duration curve. The time intervals of the discharge values used in the analysis is one of these. Mean daily data yield a much steeper curve than the annual data as the latter tend to group and smooth off the variation present in the shorter-interval daily data. For natural streams, the slope of the daily mean flow-duration curve for the upper end is determined by regional climate and characteristics of large precipitation events. The slope of the lower end is determined by geology, soils, and topography. The slope of the upper end is relatively flat where snowmelt is the principal cause of floods and for large streams where floods are caused by long-duration storms. Flashy watersheds and watersheds effected by short-duration storms have steep upper ends. A flat lower-end slope usually indicates that flows come from significant storage in groundwater aquifers or frequency precipitation inputs. Flow-duration curves from neighboring stations in a region yield valuable insights into hydrologic or hydrogeologic processes.

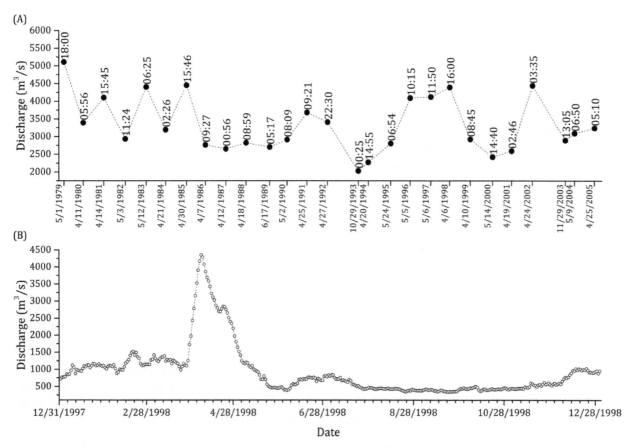

Figure 5.19 (A) Instantaneous maximum annual peak discharge of Ottawa River at Britannia for 27 consecutive years from 1979 to 2005. The dates of the discharge are labeled on the x-axis and the time of the discharge is given as data point labels on the plots. (B) Annual hydrograph for the year 1998, where the mean daily discharge for 365 days is plotted.

5.5.2 Construction of Flow-Duration Curves

5.5.2.1 Gauged Sites

An flow-duration curve from streamflow records at a gauging station can be constructed using several alternative methods described below.

Plotting Positions

This is a very simple method that involves the following steps.

1. Order the discharge values in descending magnitude: $q_1 \geq q_2 > q_3 > \ldots > q_n$; where n is the total number of observations.
2. Assign a rank, m, in ascending number to each of the ordered discharge value ($m = 1$ for the highest discharge value and subsequent rank value ascending with descending discharge values).
3. Calculate the empirical probability associated with each of the discharge values using a plotting-position formula discussed in Chapter 3. For most flow-duration curves, the Weibull plotting-position formula that requires only m and n is used. The exceedance probability, p, that is,

the probability that a given flow will be equaled or exceeded (percent of time) is

$$p = \left(\frac{m}{n+1}\right) \times 100 \quad (5.17)$$

4. Make a graph of the daily discharge values plotted as ordinates against the corresponding percent of time as abscissas.

The flow-duration curve of Ottawa River shown in Figure 5.21 is constructed using this method.

Frequency Analysis

This approach entails dividing the range of the discharge values into class intervals and calculating the relative frequency of each of the classes as shown in the histograms of Figure 5.20. The general steps in developing a flow-duration curve using this approach are as follows.

1. Determine the maximum and minimum discharge values in the time series to be analyzed. The difference between these two values is the total range of discharge.
2. Select the number of class intervals. It is recommended to have between 20 and 30 class intervals for the POR

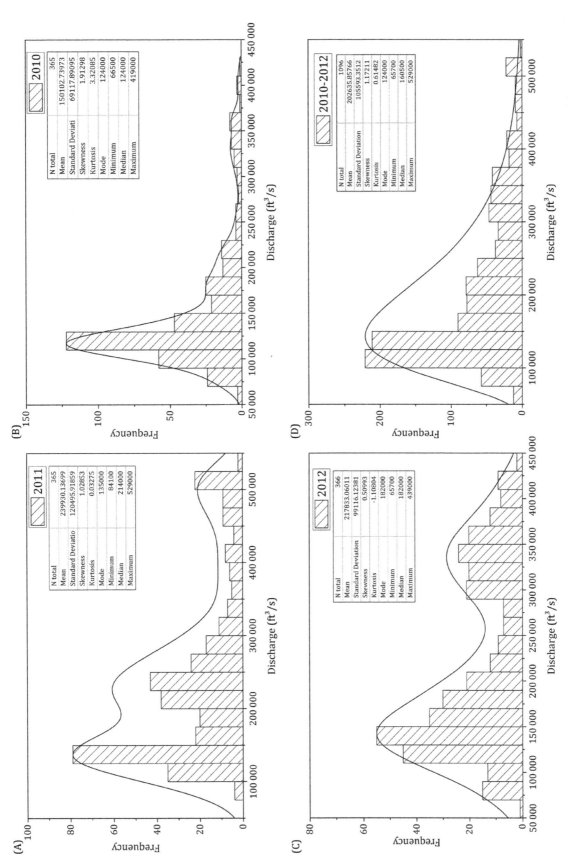

Figure 5.20 Frequency distribution and statistics of daily discharge in Columbia River at Dalles, Orgon (USGS gauge station 14105700). The curves superimposed on the histograms for (A) 2010, (B) 2011, and (C) 2012 daily discharge values are drawn by kernel smoothing (a variation of moving average technique) whereas the curve fitted to the histogram of combined three-year data (D) is a lognormal distribution that tends to capture the skewness of the frequency distribution.

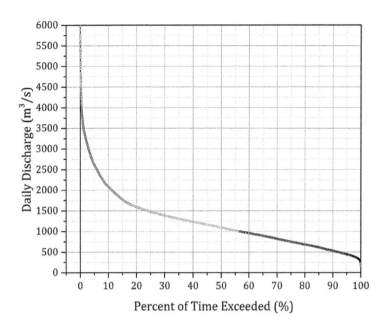

Figure 5.21 Flow-duration curve (mean daily discharge shown in Figure 5.10) of Ottawa River at Britannia, Canada.

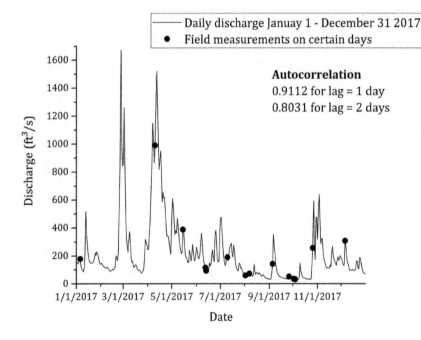

Figure 5.22 Annual hydrograph of Black River at Coventry, Vermont, with certain field measurements. The daily discharge values have high serial correlation. Autocorrelation with lag of one day and two days are shown in the figure.

(Searcy, 1963). Classes can either be of equal interval or based on log cycles. Log cycles are often used to sort data because the probability of choosing an appropriate interval spacing is higher than if the data were separated into 20 to 30 equal classes. If improper intervals are chosen, the flow-duration curve may not provide adequate information. The extreme points should be so selected as barely to include the extremes of daily discharge for the POR.

3. For the equal-interval method, determine the discharge range for each class, also called bins, by dividing the total discharge range by the desired number of class intervals. For classes based on log cycles, select classes of discharge values based on the spacing of multiples of 10^x (x is an integer including zero).

4. Count the total number of occurrences of values in each class or bin. This gives the frequency or the number of days in each class or bin. A frequency distribution can be generated by plotting the total number of occurrences in each class versus discharge.

5. Calculate the cumulative number of days for each bin/class interval. This will also give a cumulative frequency distribution plot.

5.5 Flow-Duration Analysis

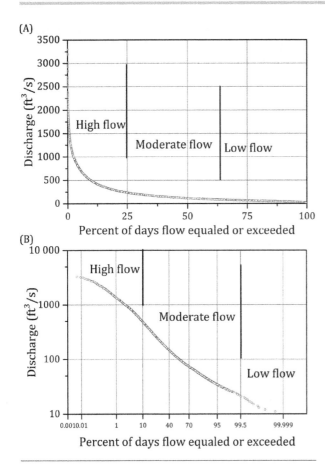

Figure 5.23 Flow-duration curve of Black River at Coventry, Vermont for a POR of 56 years (1952–2017). (A) A plot with arithmetic scales for both ordinate and abscissa. (B) A log-probability plot where flows are represented on a logarithmic scale and the percent of time is given on a probability scale.

6. Divide the cumulative number of days by the total number of observations in the entire time series. This gives the frequency with which the lower values of each class have been equaled or exceeded in the POR. Convert this value to a percentage by multiplying by 100 to get the percent of time the flow of a class interval is not exceeded.
7. Subtract the value obtained in step 6 from 100 to obtain the percent of time for which the flow in the respective class is equaled or exceeded.

Table 5.2 shows the steps of the calculations presented above for developing a flow-duration curve of daily flows of Columbia River at Dalles, Oregon, for the three-year POR shown in Figure 5.20. During this POR, the maximum and minimum values of the discharge are 529 000 and 65 700 ft^3/s, respectively, and the total number of observations was 1096. The number of classes selected is 20. Therefore, the interval of each class is 23 165 ft^3/s.

Excel Function

The two methods described above provide the insights about the concept of flow-duration curves. However, Microsoft Excel has function called PERCENTRANK.EXC that can be used to calculate the flow-duration curve in a single step. The PERCENTRANK.EXC function calculates the rank of a value in a dataset as a percentage (0 to 1, exclusive) of the dataset with numerical values that define relative standing. The value calculated by this function is multiplied by 100 and subtracted from 100 to get the percent of time a particular discharge value equals or exceeds the flow of interest. Figure 5.24 shows the comparison of the results of calculations using the frequency method and using PERCENTRANK.EXC function for the Columbia River dataset using log-probability plots.

HEC-SSP Software

Flow-duration analysis can also be performed using HEC-SSP software of USACE. Data from the USGS gauges within the United States can be directly imported via the internet. Both arithmetic and log-probability plots can be generated.

5.5.2.2 Ungauged Sites

At locations where gauging records are either unavailable or are found to be unrepresentative of the flow regime, two possible methods for synthesizing a flow-duration curve can be employed (Biedenham et al., 2000). The first method is by using records from nearby gauging stations within the same drainage basin. The second is developing a regionalized flow-duration curve. Nonetheless, these methods simply provide an approximation of the flow-duration characteristics, and that there can be considerable uncertainty in the results. The reliability of these methods is a function of the quality of the existing gauge data and the morphologic similarity between gauged and ungauged locations.

Drainage Area: Flow-Duration Curve Method

This method can be used if gauging-station data are available for a number of sites on the same river or hydrologically similar portions of the same drainage basin as the ungauged location. With this method, first, flow-duration curves for each of the gauged locations are constructed for the longest-possible common POR. This ensures comparability of the data as all the gauging stations have experienced the same or similar hydrologic events, and also makes sure that the curves represent the longest POR. Then, graphs of a specified recurrence interval of discharge versus drainage area are developed. If the data are reasonably homogeneous, power functions are fitted to the data using a regression technique (e.g., Hey, 1975). The equations generated by this method enable the synthesis of flow-duration curves at an ungauged site as a function of its upstream drainage area. It should be noted that if there is a regular downstream decrease in the discharge per unit watershed area, a graph of discharge for a given exceedance duration against upstream drainage area will produce a power function with virtually no scatter about the best-fit regression line.

Table 5.2 *Calculation of a flow-duration curve using frequency distribution, Columbia River, Dalles, Oregon (2010–2012)*

Bin (discharge, ft³/s)	Frequency	Cumulative frequency	Percent of time the flow is not exceeded	Percent of time the flow is equaled or exceeded
65 700	1	1	0.09	99.91
88 865	40	41	3.74	96.26
112 030	111	152	13.87	86.13
135 195	243	395	36.04	63.96
158 360	150	545	49.73	50.27
181 525	72	617	56.30	43.70
204 690	69	686	62.59	37.41
227 855	70	756	68.98	31.02
251 020	55	811	74.00	26.00
274 185	34	845	77.10	22.90
297 350	30	875	79.84	20.16
320 515	43	918	83.76	16.24
343 680	40	958	87.41	12.59
366 845	37	995	90.78	9.22
390 010	22	1017	92.79	7.21
413 175	22	1039	94.80	5.20
436 340	10	1049	95.71	4.29
459 505	8	1057	96.44	3.56
482 670	9	1066	97.26	2.74
505 835	11	1077	98.27	1.73
529 000	19	1096	100.00	0.00

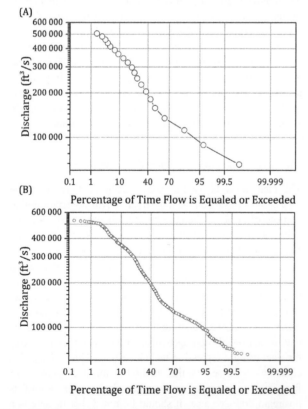

Figure 5.24 Flow-duration curves (log-probability) of Columbia River at Dalles, Oregon for the three-year period shown in Figure 5.20. (A) Curve calculated by the frequency method using the calculations presented in Table 5.2. (B) Plot of the values of each of the 1096 discharge values calculated by PERCENTRANK.EXC function of Excel (multiplied by 100 and subtracted from 100).

If there are distinct and abrupt downstream changes per unit area of the watershed, there will be scatter in the data plots and a meaningful regression equation cannot be obtained.

Regionalized Duration Curve Method

The regional scaling method is based on data from watersheds with similar characteristics. With the regionalized duration curve method, a dimensionless flow-duration curve is developed for a hydrologically similar gauged site by dividing the discharge by the bankfull discharge (Q/Q_b) or a specified recurrence-interval discharge (e.g., Q/Q_2). This ratio is used as a dimensionless index to transfer flow-duration relationships between basins with similar characteristics. If more than one gauged site is available, an average dimensionless flow-duration curve for all the sites can be developed. However, bankfull discharge does not necessarily have either a consistent return period or duration. For this reason, a specified recurrence-interval discharge for the gauged site is preferred. Then, the flow with the specified recurrence interval (e.g., Q_2) is computed for the ungauged site. For computation of flow with a specified recurrence interval at the ungauged site, a regional regression equation, such as ones developed by the USGS in the United States, can be used (USGS, 1993). Finally, the flow-duration curve for the ungauged site is derived by multiplying the dimensionless flows (Q/Q_2) from the dimensionless curve by the site Q_2.

This method is an alternative using the watershed area to generate a flow-duration curve for an ungauged site. A regional flow-duration curve can also be developed using the drainage area, which can be used to transfer flow-duration information from gauged sites to ungauged areas. However, the ratio of gauged to ungauged drainage areas should be between 0.5 and 2 for reliable results. The flow duration is dependent on the watershed conditions. For example, the effect of *urbanization* on the flow duration has been established in many watersheds. If regional flow-duration relationships are to be developed, it is recommended that a measure of watershed development is included as an independent variable. Nevertheless, it should be borne in mind that both methods simply provide an approximation to the true flow-duration curve for the site because perfect hydrologic similarity never occurs.

5.5.3 Usage of Flow-Duration Curves

Flow-duration curves are useful in assessing the general low-flow characteristics of a stream. Low-flow analysis is essential for the long-range planning of water supplies and growth of communities, especially in urban development, which can be promoted only by reliable water availability. Other uses of flow-duration curves include: (1) assessing the hydropower potential of run-of-river plants; (2) determining minimum flow release; (3) water-quality studies; (4) sediment yield studies; and (5) comparing yield potential of basins. As noted above, the chronology of the flows is lost in the assembly of data for flow-duration curves. For some studies, the low-flow sequence, or persistence, may be more important.

The area under the flow-duration curve is equal to the average for the POR. Other statistics or statistical concepts include: the median, quartiles, other percentiles, variability, and skewness. The overall slope of the flow-duration curve is an indication of the flow variability in the stream. Steeper curves are associated with increasingly variable data. The slopes and changes in the slope of the curves can be important diagnostics of streamflow conditions in a watershed. For example, if the lower end drops rapidly at the probability scale, the stream has low groundwater storage and, therefore, low or no sustained flow.

In addition to flow-duration curves, **stage-duration curves** are often used to establish vertical navigation clearances for bridges. If there have been no changes in the rating curve, stages may be used instead of flows to compute a stage-duration curve. But if there have been significant changes to the rating curve then the stage-duration curve should be derived from the flow-duration curve and the latest rating curve. Log transformation is not recommended for stages.

5.6 FLOOD-FREQUENCY ANALYSIS

5.6.1 What Is Flood-Frequency Analysis?

Flood-frequency analysis (FFA) is the method of relating the magnitude of extreme streamflow values to their frequency or probability of occurrence through the use of either graphical techniques or fitting appropriate probability density functions. Historical streamflow records provide essential information to predict the recurrence interval of discharge extremes. Information related to high-flow events is crucial for the design of flood control structures, such as reservoirs and dams, and for other engineering measures, including flood prevention, floodplain management and related economic and policy analysis, and certain aspects of water resources management.

Flood-frequency data are treated the same way as the flow-duration data, and only instantaneous peak discharges are used. The resulting flood-frequency curves are very much like flow-duration curves and are also a characteristic of a particular stream or watershed; they reflect regional similarities, vary from other stream and watershed systems because of differences in climate and geomorphology, and are affected by land use practices and channel disturbances.

The annual maximum peak discharge (Section 5.4.6) observed over an extended period of time at stream-gauging stations are the main input in FFA. These data, forming an annual maximum series (AMS), are used in most cases. However, in certain applications, a partial duration series (PDS) is also used. In some countries, such as in the UK, the PDS is also called the peak over threshold (or POT). A PDS avoids a pitfall of an AMS, which may contain peak flows that are too low, occurring in certain dry years, as well as several large values occurring in certain years with multiple high-flow events. In some cases, a time series comprising the maximum average daily discharge exceeding a certain threshold value has also been used in FFA.

In carrying out an FFA of either an AMS or PDS, an assumption is made that flood data are space and time independent, identically distributed, and uninfluenced by natural or human-made changes in the hydrological system, such as reservoir regulations. In other words, the peak streamflows are assumed to be truly random variables.

The principal objective of FFA is to estimate the probability of exceedance for a specified magnitude of streamflow or the magnitude of streamflow with a specified exceedance probability. This magnitude is referred to as a **quantile** and the probability of exceedance is usually the annual exceedance probability (AEP) or the annual recurrence interval (ARI). The analysis often makes these estimates by extrapolating the magnitudes of the past occurrences of extreme events beyond the length of the record. Obviously, the longer the POR, the more accurate the estimations are. This is important since an FFA based on a dataset covering a relatively short POR is not too reliable.

5.6.2 Flood-Frequency Analysis using the Graphical Method

The graphical method of FFA entails the assignment of exceedance probability of occurrence to each of the observed peak discharge values in a dataset and plotting the probability values against their corresponding peak discharge values. Data used in the construction of such frequency curves consist of either the maximum instantaneous flow for each year of record (AMS) or all of the independent events that exceed a selected base value (PDS). In the latter case, the base value must be smaller than any flood flow that is of importance in the analysis and should also be low enough so that the total number of floods in excess of the base value equals or exceeds the number of years of record.

The graphical method is the simplest method of probability analysis where the exceedance probability of an observed value is calculated using a chosen plotting-position formula (Chapter 3). This not only permits an inference of the parent population characteristics with a plot of observed magnitude versus the estimated exceedance probability of those data but also provides a visual summary of the data, which more quickly and completely describes the essential elements of higher-extremity streamflow at a certain location. If a best-fit line is drawn on the plot, the probability of exceeding various magnitudes can be estimated, together with any desired quantiles. Graphical representations also provide a useful check on the adequacy of a hypothesized distribution, discussed in Section 5.6.3.

5.6.2.1 Plotting-Position Formulas

The graphical method relies on plotting positions to estimate exceedance probabilities of observed events. Since we are interested in the exceedance probability (p), the data are arranged in descending order of magnitude and the variate with maximum magnitude receives rank 1. As discussed in Chapter 3, the reciprocal relationship between p and the corresponding return period (T_r) is

$$p = \frac{1}{T_r} \tag{5.18}$$

As given in Chapter 3, several plotting-position formulas are available to calculate the empirical probability associated with the discharge values in a ranked series. In the United Stated, the most used plotting-position formula in FFA is the Weibull formula according to which, if the N values are distributed uniformly between 0 and 100 percent probability, there must be $N + 1$ intervals, $N - 1$ between the data points and two at the end. Another plotting-position formula that is also widely used by the USGS and USACE is the median plotting position, also known as Chegodayev formula in Russia and eastern European countries. In the UK, the Gringorten formula is popular, and the Cunnane formula, close to the Gringorten formula, is used in Canada and in some European countries. The Hazen formula is also common. Hirsch and Stedinger (1987) proposed an alternative method of calculating plotting positions using Bayesian statistics. The HEC-SSP computer program has the capability of using this formulation, which is particularly useful in assessing both systematic and historical flood flows.

The true probability associated with the largest (and smallest) observation is a random variable with mean $1/(N+1)$ and a standard deviation of nearly $1/(N+1)$.

Hence, all plotting-position formulas give crude estimates of the unknown probabilities associated with the largest and smallest events. The plotting-position coefficients can acquire several different values. As N becomes large the choice of values for a or b becomes less important. Recall that if m was expressed in ascending rank, the plotting-position formula would give the non-exceedance probability or just the percentile.

In ordinary hydrologic frequency analysis, N is taken as the number of years of record rather than the number of events, so that the percent chance exceedance can be thought of as the number of events per hundred years. For example, to estimate the AEP of annual maximum discharge, N is the number of years of data. To express the results as percent chance of exceedance, the results obtained from a plotting-position formula are multiplied by 100.

5.6.2.2 Probability Plots

Once the exceedance probabilities are calculated using a plotting-position formula, the data are plotted with the observed values on the ordinate and the probabilities or return period on the abscissa. In earlier days, these plots were made on probability paper to linearize the distribution so that the plotted data could be represented by a straight line. Nowadays, a number of graphical software packages are available, such as Origin, which have the capability to express an axis on the probability scale as seen in various flow-duration curves presented in the preceding section. The discharge values can be represented either on an arithmetic or logarithmic scale.

A best-fit straight line can be extrapolated beyond the limits of calculated probabilities or return periods since, in many situations, the POR may not be longer than the desired return period, such as 100 years or 500 years. In certain cases, such as for undisturbed streams, the flood-frequency data tend to plot closer to a straight line when a logarithmic scale is applied to the discharge values (i.e. a log-probability plot) than do flow-duration curves and thus they lend themselves more reliably to extrapolation. However, for a log-probability plot, the data lie close to a straight line if their distribution follows a normal distribution and, as most streamflow data do not follow a normal distribution, such flood-frequency plots typically depart from linearization. If the probability plots do not exhibit a linear pattern, the deviation from the linearity can be a result of three typical conditions: asymmetry or skewness, heavy tails of the distribution, and outliers.

As an illustration of the graphical method of FFA, the AMS of the River Trent, near Nottingham in central England, are analyzed using four different plotting-position formulas and are presented in Figure 5.25. In this case, since the POR is only 63 years, significant discrepancies can be seen in the estimation of the AEP or ARI using different plotting-position formulas at the higher end of discharge values. Furthermore, the data also show deviations from a linear pattern on the probability plot.

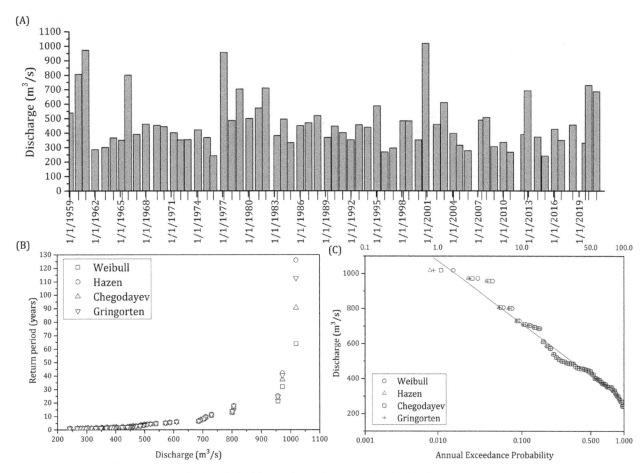

Figure 5.25 (A) AMS of the River Trent at Colwick, a suburb of the city of Nottingham in central England, for the 1959-2021 POR (the minor ticks represent the dates of occurrence of the annual peak). (B) The return period (ARI) and (C) annual exceedance probability (AEP) for the discharge values calculated using four plotting-position formulas given in Table 3.4. The AEP given in C is multiplied by 100 to get the percent probability on the upper x-axis in (C). The discharge data are from the UK National River Flow Archive.

5.6.3 Flood-Frequency Analysis using Probability Density Functions

5.6.3.1 Probability Density Functions used in FFA

In addition to the graphical method using plotting-position formulas, various pdfs are also used for FFA. With a theoretical model, the relationship between the magnitude and probability for the parent population is hypothesized. This relationship is generally represented by a cdf, which is an equation that defines the probability of exceedance as a function of specified magnitude and one or more parameters. From this, an inverse distribution, called a quantile function, is obtained, which defines the magnitude as a function of specified probability and one or more parameters.

In hydrologic engineering, a probability distribution is selected because it models well the data that are observed. The parameters for the model are selected to optimize the fit. A graphical or numerical technique is used to identify the appropriate distribution and to estimate its parameters. Subsequently, statistical tests, such as Kolmogorov–Smirnov, Anderson–Darling, and chi-squared tests determine the goodness of fit of the hypothesized model with the empirical probability distribution. These are all covered in detail in Chapter 3. The L-moment ratio diagrams are also used to assess whether a candidate distribution can simulate the site-to-site variation in statistical characteristics that is present in observed flood flows.

For FFA, the pdf used in the United States as the standard method is the log Pearson type III (LP III) distribution in accord with the recommendation originally made by the Water Resources Council in 1967. This is also a preferred model in many other countries, such as Australia and Canada. The other pdf that finds adoption in FFA is the Gumbel distribution, introduced in Chapter 4. Some studies have suggested that generalized extreme value (GEV) distribution, also presented in Chapter 4, is better suited than the LP III and Gumbel distributions, for FFA, and GEV is preferred in the UK. As a matter of fact, an important problem in hydrology is the choice of a frequency distribution function for the fitting of extreme flood series in a region or country. Investigating the relative suitability of various pdfs in this regard has received considerable attention from various corners. For example, Pekárová et al. (2019) considered a set of 50–60 pdfs to analyze the discharge maxima from different gauging stations along the River Danube and ordered the statistical distribution according to their adequacy, based on the statistical tests mentioned above, and quantified the aleatory and epistemic uncertainty (Merz and Thieken, 2009). Aleatory uncertainty is mainly due to temporal variability and the length of the time series, while the epistemic uncertainty is a consequence of the incomplete knowledge of the hydrological system. From their assessment, Pekárová et al. selected the LP III distribution and adopted the procedures used in the United States for FFA of discharge maxima from 85 gauging stations within the Danube river basin, which, with a drainage area of 817 000 km^2 straddling 19 countries, is one of the most flood-endangered regions in Europe.

Since the LP III distribution is used to estimate the extremes in many natural processes and is the most common frequency distribution used, especially in FFA, this is described in further details below.

5.6.3.2 Log Pearson Type III Distribution

Probability Density and Cumulative Distribution Functions

Karl Pearson (1857–1936) was a distinguished English mathematician and biostatistician who was credited with establishing the discipline of mathematical statistics. He developed a system for categorizing pdfs based on solutions to the differential equation

$$\frac{1}{p_X(x)} \frac{dp_X(x)}{dx} = \frac{a+x}{c_0 + c_1 x + c_2 x^2} \quad (5.19)$$

The different types of distributions in the Pearson system correspond to the relationships between the constants in Eq. 5.19. The resulting series of pdfs fit virtually any distribution, even if the data are not be theoretically justified to have a distribution like LP III, and are widely used in practical statistical work. The distributions that have $c_2 = 0$ and $c_1 \neq 0$ are called type III distributions that fit a wide range of distribution shapes with positive or negative skewness, including a good approximation to the normal distribution. The type III density function resulting from the solution to the differential equation is

$$p_X(x) = K(c_0 + c_1)^m \exp\left(-\frac{x}{c_1}\right) \quad (5.20)$$

The LP III distribution is a re-parameterized form, which in the field of statistics is more commonly known as a **three-parameter gamma distribution**, written as

$$p_X(x|\varphi, \alpha, \beta) = \frac{1}{|\beta| \Gamma(\alpha)} \left(\frac{x-\varphi}{\beta}\right)^{(\alpha-1)} \exp\left[-\frac{(x-\varphi)}{\beta}\right] \quad (5.21)$$

where α, β, and φ are *shape*, *scale*, and *location* parameters respectively, and $\Gamma(\alpha)$ is the gamma function. Note that Eq. 5.21 is just like the two-parameter (α and β) gamma distribution discussed in Chapter 3, except that the variable x has a shift by an amount given by the location parameter, φ.

If the logarithm of x follows a Pearson type III (P III) distribution, then x is said to follow the LP III distribution.

Both logarithms to base 10 and e can be used. For its application specifically in FFA, log10 is used traditionally.

Two forms of LP III distribution can be used: one in the **log space** and the other in the **real space** (Griffis and Stedinger, 2007).

In the log space, setting y as the transformed variable

$$y = \log(x) \quad (5.22)$$

$$p_Y(y|\alpha, \beta, \varphi) = \frac{1}{|\beta| \Gamma(\alpha)} \left(\frac{y-\varphi}{\beta}\right)^{(\alpha-1)} \exp\left[-\frac{(y-\varphi)}{\beta}\right] \quad (5.23)$$

is the pdf representing the LP III distribution. Equation 5.23 gives the probability density (p_Y) of the variable Y which is in log space (X is the variable discharge in real space).

In real space, the probability density of the variable X is given by

$$p_X(x|\alpha, \beta, \varphi) = \frac{1}{x|\beta| \Gamma(\alpha)} \left[\frac{\log(x)-\varphi}{\beta}\right]^{(\alpha-1)} \exp\left[-\frac{\{\log(x)-\varphi\}}{\beta}\right] \quad (5.24)$$

If we define, $z = \frac{\log(x)-\varphi}{\beta}$, the cdf of the distribution is given by

$$F_Y(y) = \frac{1}{\Gamma(\alpha)} \int_0^z z^{\beta-1} \exp(z) dz \quad (5.25)$$

The integral is evaluated to give

$$F_Y(y) = \frac{\Gamma_y(\alpha)}{\Gamma(\alpha)} \quad (5.26)$$

where, $\Gamma_y(\alpha)$ is the **incomplete gamma function**.

The pdf of the LP III distribution is nonzero for $x > 0$ (0 otherwise) and the shape parameter $\alpha > 0$. But the scale parameter, β, can be both positive and negative. When $\beta > 0$, the LP III distribution has a lower bound φ and is positively skewed. When $\beta < 0$, the distribution has an upper bound φ and is negatively skewed. The behavior of the distribution may also be described using the population skewness coefficient, γ, rather than with parameters. When $\gamma > 0$ ($\beta > 0$) the distribution has a positive skew and floods are unbounded. When $\gamma < 0$ ($\beta < 0$), the distribution of the logarithm of floods has a negative skew and an upper bound. In this situation, the log-space skew is constrained to $\gamma \geq -1.41$. Also, $\log(x) \geq \varphi > 0$, which means the distribution has a lower bound for the positive skew. The distribution is not defined for x beyond the shift, which is the lower bound for the positive skew case and the upper bound for negative skew. If $\gamma = 0$, the pdf becomes the normal distribution.

The gamma distribution is well suited for data that have a skewed distribution, meaning a nonuniform spread of data values above and below the mean value, likely seen in the case of normal distribution. In general, an AMS of peak flows is skewed, since an AMS generally tends to contain a greater number of flows whose magnitudes are higher than the rest. The lognormal distribution is assumed to reduce the skewness through logarithmic transformation of the original values. Yet, skewness can still persist in the AMS of instantaneous peak flows. The LP III distribution has been found to be capable of representing such cases and, for this very reason, we will see in the subsequent discussions the importance of the skew coefficient in its application in FFA.

The three parameters of the LP III distribution are estimated by the method of moments. Moments of the parent population can be computed numerically using sample moments: the mean ($\hat{\mu}$), which is the first moment; the standard deviation ($\hat{\sigma}$), which is the square root of the second moment; and the skewness coefficient ($\hat{\gamma}$), which is the third moment. Thus, the parameters α, β, and φ can be found from the sample data by computing the sample moments as

$$\alpha = \frac{4}{\hat{\gamma}_y^2} \tag{5.27}$$

$$\beta = \text{sgn}(\hat{\gamma}_y) \frac{\hat{\sigma}_y}{\sqrt{\hat{\alpha}}} \tag{5.28}$$

$$\hat{\varphi} = \hat{\mu} - \hat{\alpha}\hat{\beta} \tag{5.29}$$

In Eqs. 5.27–5.29, the subscript y implies that the variable is the log-transformed variable and the hat implies statistics based on a sample of the population. The value $\text{sgn}(\hat{\gamma})$ equals -1, 0, or 1, depending upon whether the value $\hat{\gamma}$ is negative, zero, or positive.[2] The equations for calculations of the sample statistics are

$$\hat{\mu}_y = \frac{\sum y}{N} \tag{5.30}$$

$$\hat{\sigma}_y = \sqrt{\frac{\sum(y - \hat{\mu}_y)^2}{N - 1}} \tag{5.31}$$

$$\hat{\gamma}_y = \frac{N \sum (y - \bar{y})^3}{(N-1)(N-2)(s_y)^3} \tag{5.32}$$

The coefficient of skewness used in fitting the LP III distribution is very sensitive to the size of the sample and, in particular, is difficult to estimate accurately from a small sample. But, the behavior of the distribution given by Eq. 5.24 depends upon the skewness coefficient to a large extent. For these reasons, the estimation of the skew coefficient in the application of the LP III distribution has a special place as discussed below.

It should be noted that if the discharge dataset has certain low values, these can have a significant effect in log transformation. This issue is discussed further in relation to the Bulletin 17C procedures presented below. Also, all datasets may not require the LP III distribution. Many datasets can be well represented by just P III distribution. For this reason, in

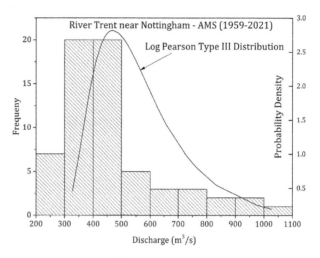

Figure 5.26 Comparison of the observed frequency distribution (histogram) and the theoretical probability distribution (LP III density function) of the AMS shown in Figure 5.25.

some countries, such as China, P III distribution instead of LP III distribution is used.

Quantile Estimation and Flood Frequency Curves

Comparison of the pdf calculated from the parameters and the observed frequency with goodness-of-fit test provides the degree of appropriateness of the theoretical model, as shown in Figure 5.26 in which the LP III distribution function is calculated from the sample statistics and plotted with the observed frequency diagram, given as a histogram. However, the quantile values of the variable being analyzed are derived from the inverse distribution of the cdf. Most, but not all, cdfs are invertible. The inverse of the cdf of the LP III distribution (Eq. 5.24) is given as

$$\hat{Y}_q = \hat{\varphi} + \hat{\beta} P^{-1}(\hat{\alpha}, q) \tag{5.33}$$

where $P^{-1}(\hat{\alpha}, q)$ is the inverse of the incomplete gamma function (given in Table 26.4 of Abramowitz and Stegun, 1970). The flood quantile is estimated as

$$\hat{Q}_q = 10^{\hat{Y}_q} \tag{5.34}$$

In Eqs. 5.33 and 5.34, q is the cumulative probability of interest, that is, $q = 1 - p$ (e.g., $q = 0.99$ for 1% probability).

Flood quantiles can also be estimated by using a frequency factor. Chow (1951) showed that with the method-of-moments estimates, many inverse distributions commonly used in hydrologic engineering could be written in the following general form:

---

[2] In mathematics, the sign function or signum function (from signum, Latin for "sign") is a function that returns the sign of a real number. In mathematical notation the sign function is often represented as sgn (x)

$$x_p = \hat{\mu}_x + K_p \hat{\sigma}_x \quad (5.35)$$

Where x_p is the quantile with a specified exceedance probability, $\hat{\mu}_x$ is the sample mean, and $\hat{\sigma}_x$ is sample standard deviation. K_p is the frequency factor (Chapter 3), which depends on the type of distribution selected. The physical sense of Eq. 5.35 is that the amount $K_p \hat{\sigma}_x$ gives the magnitude of departure from the mean value. Equation 5.35 can be written in terms of return period

$$X_T = \hat{\mu}_x + K_T \hat{\sigma}_x \quad (5.36)$$

Since in the LP III distribution the log-transformed variable is used, Eq. 5.36 is written as

$$Y_T = \hat{\mu}_y + K_T \hat{\sigma}_y \quad (5.37)$$

$$X_T = 10^{Y_T} \quad (5.38)$$

For the LP III distribution, the frequency factor, K_T, is a function of the specified probability, that is, the return period T, and the coefficient of skewness of the logarithms of the sample. For this reason, calculation of the frequency factor for the LP III distribution is more complicated than calculating the frequency factors for some other distributions (e.g., see Chapter 4 for calculation of the frequency factor for Gumbel distribution, where it is a function of return period only).

Table 5.3 shows the values of the frequency factor, K_T, as a function of skew coefficient and ARI or AEP, which can be used in Eq. 5.38

An alternative to using Table 5.3 is an analytical way of calculating K_T involving a three–step procedure as follows.

First, a variable, w, is calculated using

$$w = \begin{cases} \sqrt{\ln\left(\dfrac{1}{p^2}\right)} & \text{for } 0 < p \leq 0.5 \\ \sqrt{\ln\left(\dfrac{1}{(1-p)^2}\right)} & \text{for } p > 0.5 \end{cases} \quad (5.39)$$

Then, the standard normal variable z is calculated using an approximation:

$$z = w - \frac{2.515517 + 0.802853 w + 0.010328 w^2}{1 + 1.432788 w + 0.189269 w^2 + 0.001308 w^3} \quad (5.40)$$

When $\gamma_y = 0$, $K_T = z$. If $\gamma_y \neq 0$, K_T is calculated (from Kite, 1977)

$$K_T = z - (z^2 - 1)k + \frac{1}{3}(z^3 - 6z)k^2 - [z^2 - 1]k^3 + zk^4 + \frac{1}{3}k^5 \quad (5.41)$$

where $k = \frac{\gamma_y}{6}$.

The procedure for fitting the LP III distribution to a set of annual maximum discharge values is as follows: (1) transform the data x_i (annual maxima) to their logarithmic values (y_i); (2) compute the mean, standard deviation, and coefficient of skewness for the log-transformed values (y_i); (3) get the value of K_T for the desired return period from the tabulated values or the analytical procedure described above; (4) compute y from $y = \bar{y} + K_T s_y$; and (5) compute $x = 10^y$ for the T value considered.

The results of quantile estimations are given as a flood-frequency curve, where discharge quantiles are plotted. Usually, the magnitudes of discharge with computed AEP (or ARI) are plotted on the ordinate using a logarithmic scale and the AEP is plotted on the abscissa on a probability scale, usually from high to low AEP so that the curve has a positive slope. Figure 5.27 shows the flood-frequency curve for the River Trent near Nottingham, calculated from the AMS shown in Figure 5.25. The AEP flows are calculated from the LP III distribution, following the procedure described.

5.6.3.3 Bulletin 17B and 17C Procedures

In 1967, the US Water Resources Council recommended the use of LP III distribution to model the annual maximum streamflow series in the United States and since then all US federal and most other governmental agencies follow this procedure for FFA in the United States. The US Water Resources Council was abolished in 1981 but its duties in establishing guidelines for flood-frequency determinations were taken over by the Advisory Committee on Water Information, coordinated by the USGS. The USGS published the procedures of FFA using LP III distribution as Bulletin 17B (US Department of Interior, 1982) and subsequently revised it as Bulletin 17C (England et al., 2019).

Bulletin 17B and 17C both follow the frequency-factor method to compute flood-frequency quantiles. The frequency factors for various values of skew coefficient and AEP are tabulated in Bulletin 17B. Instead of using any equation to calculate K_T, the use of this table is recommended. The major procedure that is specific to Bulletin 17B is the computation of a weighted skew coefficient (G_W) from the **station skew coefficient** $(\hat{\gamma})$ and **generalized** or **regional skew coefficient** $(\hat{G})$. Bulletin 17C provides more advanced statistical procedures than those provided in Bulletin 17B. The key elements of the procedures of FFA according to these two valuable guidelines are summarized below.

Estimation of the Weighted Skew Coefficient

The value of the skew coefficient, $\hat{\gamma}$, calculated from the sample (Eq. 5.32) is called **station skew** coefficient. Its value is quite sensitive to extreme events and for this reason there is relatively large uncertainty in the station skew coefficient. It can be improved by weighting the station skew coefficient with the **generalized** or **regional skew** coefficient, $\hat{G}$, by pooling the information from nearby stations.

The regional skew coefficient, applicable to a particular station, is usually read from a map that gives gridded values of $\hat{G}$ as well as isolines of $\hat{G}$ values. Bulletin 17B gives such

5.6 Flood-Frequency Analysis

Table 5.3 *Values of the frequency factor K_T as a function of coefficient of skewness and ARI or AEP*

	Recurrence interval (ARI in years)									
	1.053	1.25	2	5	10	25	50	100	200	500
	Percent chance (AEP in percent)									
Skew coefficient	95	80	50	20	10	4	2	1	0.5	0.2
−3.00			−0.396	0.42	1.18	2.278	3.152	4.051	4.97	
−2.00	−1.9957	−0.6094	0.3069	0.7769	0.8946	0.9592	0.9798	0.9900	0.9950	0.9980
−1.90	−1.9891	−0.6266	0.2944	0.7882	0.9199	0.9967	1.0231	1.0370	1.0443	1.0490
−1.80	−1.9812	−0.6434	0.2815	0.7987	0.9450	1.0354	1.0686	1.0871	1.0975	1.1047
−1.70	−1.9723	−0.6596	0.2681	0.8084	0.9698	1.0751	1.1163	1.1404	1.1548	1.1653
−1.60	−1.9621	−0.6753	0.2542	0.8172	0.9942	1.1157	1.1658	1.1968	1.2162	1.2313
−1.50	−1.9508	−0.6905	0.2400	0.8252	1.0181	1.1568	1.2172	1.2561	1.2817	1.3028
−1.40	−1.9384	−0.7051	0.2254	0.8322	1.0414	1.1984	1.2700	1.3182	1.3511	1.3798
−1.30	−1.9247	−0.7192	0.2104	0.8384	1.0641	1.2403	1.3241	1.3827	1.4244	1.4623
−1.20	−1.9099	−0.7326	0.1952	0.8437	1.0861	1.2823	1.3793	1.4494	1.5011	1.5502
−1.10	−1.8940	−0.7454	0.1797	0.8481	1.1073	1.3241	1.4353	1.5181	1.5811	1.6431
−1.00	−1.8768	−0.7575	0.1640	0.8516	1.1276	1.3658	1.4919	1.5884	1.6639	1.7406
−0.90	−1.8586	−0.7690	0.1481	0.8543	1.1471	1.4072	1.5489	1.6600	1.7492	1.8424
−0.80	−1.8392	−0.7799	0.1320	0.8561	1.1657	1.4481	1.6060	1.7327	1.8366	1.9481
−0.70	−1.8186	−0.7900	0.1158	0.8570	1.1835	1.4885	1.6633	1.8062	1.9258	2.0570
−0.60	−1.7970	−0.7995	0.0995	0.8572	1.2003	1.5283	1.7203	1.8803	2.0164	2.1688
−0.50	−1.7743	−0.8083	0.0830	0.8565	1.2162	1.5674	1.7772	1.9547	2.1083	2.2831
−0.40	−1.7505	−0.8164	0.0665	0.8551	1.2311	1.6057	1.8336	2.0293	2.2009	2.3994
−0.30	−1.7256	−0.8238	0.0499	0.8529	1.2452	1.6433	1.8896	2.1039	2.2942	2.5174
−0.20	−1.6997	−0.8304	0.0333	0.8499	1.2582	1.6790	1.9450	2.1784	2.3880	2.6367
−0.10	−1.6728	−0.8364	0.0166	0.8461	1.2704	1.7158	1.9997	2.2526	2.4819	2.7571
0.00	−1.6449	−0.8416	0.0000	0.8416	1.2816	1.7507	2.0538	2.3264	2.5758	2.8782
0.10	−1.6159	−0.8461	−0.0166	0.8364	1.2918	1.7846	2.1070	2.3996	2.6697	2.9998
0.20	−1.5861	−0.8499	−0.0333	0.8304	1.3011	1.8176	2.1594	2.4723	2.7632	3.1217
0.30	−1.5553	−0.8529	−0.0499	0.8238	1.3094	1.8495	2.2108	2.5442	2.8564	3.2437
0.40	−1.5236	−0.8551	−0.0665	0.8164	1.3167	1.8804	2.2613	2.6154	2.9490	3.3657
0.50	−1.4910	−0.8565	−0.0830	0.8083	1.3231	1.9102	2.3108	2.6857	3.0410	3.4874
0.60	−1.4576	−0.8572	−0.0995	0.7995	1.3285	1.9390	2.3593	2.7551	3.1323	3.6087
0.70	−1.4235	−0.8570	−0.1158	0.7900	1.3329	1.9666	2.4067	2.8236	3.2228	3.7296
0.80	−1.3886	−0.8561	−0.1320	0.7799	1.3364	1.9931	2.4530	2.8910	3.3124	3.8498
0.90	−1.3530	−0.8543	−0.1481	0.7690	1.3389	2.0185	2.4981	2.9574	3.4011	3.9693
1.00	−1.3168	−0.8516	−0.1640	0.7575	1.3404	2.0427	2.5421	3.0226	3.4887	4.0880
1.10	−1.2802	−0.8481	−0.1797	0.7454	1.3409	2.0657	2.5848	3.0866	3.5753	4.2058
1.20	−1.2431	−0.8437	−0.1952	0.7326	1.3405	2.0876	2.6263	3.1494	3.6607	4.3226
1.30	−1.2058	−0.8384	−0.2104	0.7192	1.3390	2.1082	2.6666	3.2110	3.7450	4.4384
1.40	−1.1683	−0.8322	−0.2254	0.7051	1.3367	2.1277	2.7056	3.2713	3.8280	4.5530
1.50	−1.1308	−0.8252	−0.2400	0.6905	1.3333	2.1459	2.7433	3.3304	3.9097	4.6665
1.60	−1.0934	−0.8172	−0.2542	0.6753	1.3290	2.1629	2.7796	3.3880	3.9902	4.7788
1.70	−1.0563	−0.8084	−0.2681	0.6596	1.3238	2.1787	2.8147	3.4444	4.0693	4.8897
1.80	−1.0197	−0.7987	−0.2815	0.6434	1.3176	2.1933	2.8485	3.4994	4.1470	4.9994
1.90	−0.9838	−0.7882	−0.2944	0.6266	1.3105	2.2067	2.8809	3.5530	4.2234	5.1077
2.00	−0.9487	−0.7769	−0.3069	0.6094	1.3026	2.2189	2.9120	3.6052	4.2983	5.2146
2.10	−0.9146	−0.7648	−0.3187	0.5918	1.2938	2.2299	2.9418	3.6560	4.3719	5.3201
2.20	−0.8816	−0.7521	−0.3300	0.5738	1.2841	2.2397	2.9703	3.7054	4.4440	5.4243
2.30	−0.8498	−0.7388	−0.3406	0.5555	1.2737	2.2483	2.9974	3.7535	4.5147	5.5269
2.40	−0.8193	−0.7250	−0.3506	0.5368	1.2624	2.2558	3.0233	3.8001	4.5839	5.6282
2.50	−0.7902	−0.7107	−0.3599	0.5179	1.2504	2.2622	3.0479	3.8454	4.6518	5.7280
2.60	−0.7624	−0.6960	−0.3685	0.4987	1.2377	2.2674	3.0712	3.8893	4.7182	5.8263
2.70	−0.7361	−0.6811	−0.3764	0.4793	1.2242	2.2716	3.0932	3.9318	4.7831	5.9232
2.80	−0.7112	−0.6660	−0.3835	0.4598	1.2101	2.2747	3.1140	3.9730	4.8467	6.0186
2.90	−0.6876	−0.6509	−0.3899	0.4402	1.1954	2.2768	3.1336	4.0129	4.9088	6.1125
3.00	−0.6653	−0.6357	−0.3955	0.4204	1.1801	2.2778	3.1519	4.0514	4.9696	6.2051

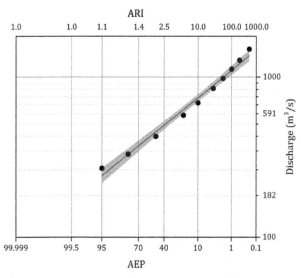

Figure 5.27 Flood-frequency curve of the River Trent near Nottingham, England, calculated using the frequency factor for LP III distribution. The shaded band represents the confidence interval between the 95% confidence limits.

a map of the United States. A map like this is produced from station skews in conjunction with applications of various statistical procedures that are given in both Bulletin 17B and 17C, although the procedures given in Bulletin 17C are more advanced and provide a more accurate picture of the regionalized character of this streamflow statistic.

Under the assumption that the $\hat{G}$ is unbiased and independent of $\hat{\gamma}$, the mean square error (designated as MSE) of the station skew, $MSE_{\hat{\gamma}}$, and that of the regional skew, $MSE_{\hat{G}}$, can be used to estimate the weighted skew coefficient, G_W. Bulletin 17B prescribes the following procedure for the estimation of G_W.

The station $MSE_{\hat{\gamma}}$ is obtained from results of bias and variance of the skew coefficients obtained from LP III distributed variables and is given by Wallis et al. (1974). An approximation to their numerical results is

$$MSE_{\hat{\gamma}} = 10^{[A - B \log(N/10)]} \quad (5.42)$$

where

$$A = \begin{cases} 0.33 + 0.08|\hat{\gamma}| & \text{if } |\hat{\gamma}| \leq 0.90 \\ -0.52 + 0.30|\hat{\gamma}| & \text{if } |\hat{\gamma}| > 0.90 \end{cases} \quad (5.43)$$

$$B = \begin{cases} 0.94 - 0.26|\hat{\gamma}| & \text{if } |\hat{\gamma}| \leq 1.5 \\ 0.55 & \text{if } |\hat{\gamma}| > 1.5 \end{cases} \quad (5.44)$$

The $MSE_{\hat{G}}$ is taken as the estimated error of the map giving the regionalized values of $\hat{G}$. This value is estimated as 0.302. Subsequently, the weight, W, is calculated as

$$W = \frac{MSE_{\hat{G}}}{MSE_{\hat{G}} + MSE_{\hat{\gamma}}} \quad (5.45)$$

from which the weighted skew is calculated as

$$G_W = W\hat{\gamma} + (1 - W)\hat{G} \quad (5.46)$$

Bulletin 17C describes the computation of the station skew by the expected moments algorithm (EMA) and discusses the advanced statistical procedures that are currently being employed to update the regional skew coefficients. An updated map covering the entire United States is in progress but updates in many separate regions are currently available from the USGS.

In an example of regional distribution of station skews from the Danube river basin, the spread of the skew coefficients is shown in a histogram instead of isoline maps, due to the paucity of gauges in the northeastern part of the basin (Figure 5.28).

Systematic Records and Historic Floods

Systematic records refer to AMS or PDS consisting of annual peak discharge data collected at regular, prescribed intervals at a gauging station. At many locations, particularly where people have occupied the floodplain for an extended period, or where past flood control projects were constructed, there can be information about major floods that occurred either before or after the period of systematic data collection. Similar information may be available at sites where the gauge has been discontinued, or where records are broken or incomplete. Data for recent floods that occurred outside the systematic data collection period are also treated as historical floods. Such historical data should be included in the FFA. Bulletin 17C provides further details on the inclusion of historical flood data in FFA.

Broken, Incomplete, and Discontinued Records

Annual peaks for certain years may be missing because of conditions not related to flood magnitude, such as gauge removal. These are considered **broken records**. The different record segments can be analyzed as a continuous record with length equal to the sum of both records if the gauge is reestablished in a nearby location, unless there is some physical change in the watershed between segments, which may make the total record nonhomogeneous.

An **incomplete record** refers to a streamflow record in which some peak flows are missing because they were too low or too high to record, or the gauge was out of operation for a short period because of flooding. Missing high and low data require different treatment. When one or more high annual peaks during the period of systematic record have not been recorded, there is usually information available from which the peak discharge can be estimated, or a flow interval estimate can be made.

Streamflow data are available at many locations where records are no longer being collected. Data from these stations are considered **discontinued records**. However, these are extremely valuable data, and should be used for FFA.

Zero flows and Potentially Influential Low Flows

In general, **potentially influential points** are **outliers** that depart significantly from the trend of the remaining data.

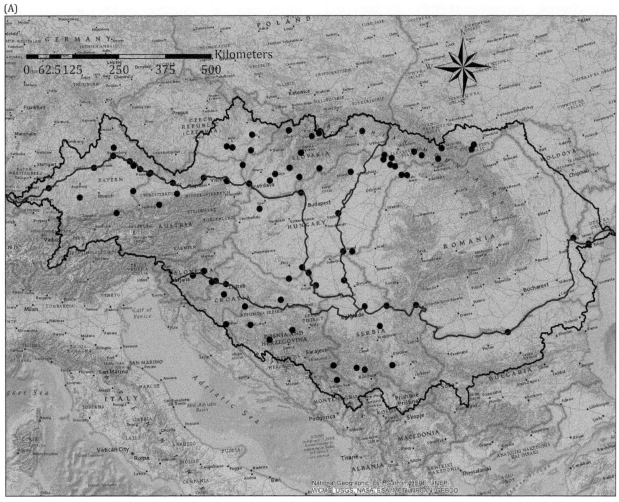

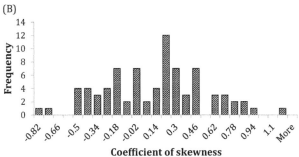

Figure 5.28 (A) Map of the Danube river basin showing the locations of the stream gauges from where annual maxima series were analyzed and station skew coefficients were determined (constructed from data given in Pekárová et al., 2019). (B) The spread of the station skew values is shown in the histogram. Note the absence of gauges in the northeastern section of the basin (mostly in Romania), limiting reliable isoline maps.

In the case of annual peak flows, low outliers may be floods caused by different processes than the larger floods in the annual peak series. For example, many rivers and streams in arid and semi-arid southwestern United States, such as in California and Arizona, have zero or very small flows for the entire year. The annual flood series for these streams typically has one or more low-magnitude or zero flood values. These data merit special attention. The logarithm of zero is negative infinity, and the logarithm of unusually small values can also be anomalous. Moreover, small flood values can have a large influence on the fitting of the flood-frequency distribution and the estimation of the magnitude of rare flood flows. These small observations are called **potentially influential low floods** (**PILFs**). Because the inclusion of these zero-flow values and outliers can significantly affect the statistical parameters computed from the data, especially for small samples, the presence of PILFs in the dataset will bias parameter estimates. The use of robust estimation procedures is recommended to focus on the largest floods – the upper tail of the flood frequency distribution – to eliminate PILFs. Robust estimation procedures are reasonably efficient when the assumed characteristics of the flood distribution are true, while not doing poorly when those assumptions are violated.

Bulletin 17C recommends and implements the multiple Grubbs–Beck test for the detection of PILFs. Statistical procedures for identifying outliers have been extensively studied, including methods for addressing the case of multiple low outliers considered here. The multiple Grubbs–Beck test was developed in Bulletin 17C as an improvement to the Grubbs–Beck test that was used in Bulletin 17B.

Regulated Flow Frequency

As mentioned at the beginning, one of the basic tenets of FFA is that the annual peak-flow series being analyzed represents natural flows of a stream. However, a large number of gauging sites in the United States have flows that are altered to various extents by regulating structures, such as dams, reservoirs, and diversions, or the flows are affected by levees. This type of situation is also prevalent in many developed or developing countries. In spite of the high significance of this issue, there is still no well-established method for estimating flood flow-frequency curves at gauge locations affected by varying degrees of regulation. Some common regulated flood-frequency methods include estimating unregulated flows using empirical relationships or synthetic floods (USACE, 1993a), graphical frequency analysis, or by applying total probability concepts (Kubik, 1990).

Expected Moments Algorithm

Several studies documented certain weaknesses and potential improvements to the methods of estimation of moments given in Bulletin 17B, including historical data, handling of low outliers, use of regional skew, and confidence intervals (Stedinger and Griffis, 2008). The EMA, developed by Cohn et al. (1997) is a generalized method-of-moments procedure to estimate the LP III distribution parameters. This method is implemented in Bulletin 17C. The EMA provides a direct fit of the LP III distribution using the entire dataset, simultaneously employing regional skew information and a wide range of historical flood and threshold-exceedance information, while adjusting for any potentially influential low floods, missing values from an incomplete record, or zero flood years. The EMA is an improved extension of the Bulletin 17B method-of-moments approach to deal in a consistent statistical framework with all of the sources of information that are likely to be available. Details of the computation of moments using the EMA, the data requirements, and preparation for EMA application are given in England et al. (2019).

Confidence Intervals for Quantiles

A streamflow record of annual peak flows at a site is only a random sample of the population of annual peaks. How well this sample really represents its parent population depends on the sample size and its representativeness. The record of the sample is used to estimate the frequency curve of that population given the fact that the estimated flood-frequency curve is an approximation of the true frequency curve of the parent population of annual flood peaks. If the same size of random sample is selected from a different POR, a different frequency curve would likely result.

To assess the accuracy of the approximation of the population frequency curve, an interval or range of hypothetical frequency curves, which with a high degree of confidence contains the population frequency curve, are constructed. Such intervals are called **confidence intervals** and their end points are called **confidence limits**. Thus, confidence limits are control curves plotted on either side of the calculated frequency curve with the property that, if the data belong to the selected probability distribution, a known percentage of the data points should lie within these two curves. Figure 5.27 shows the confidence interval of the computed flood-frequency curve.

Confidence intervals provide either a measure of the uncertainty of the estimated exceedance probability of a selected discharge or a measure of the uncertainty of discharge at a selected exceedance probability.

In the classical regression analysis of a dependent variable, y, against the independent variable, x, the confidence interval for the forecasted values $\hat{y}$ of x at a probability level of p is

$$CI_{p\%} = \hat{y} \pm t_{(\alpha, df)} S_{yx} \sqrt{\frac{1}{n} + \frac{(x_i - \bar{x})^2}{\sum_{i=1}^{n}(x_i - \bar{x})^2}} \qquad (5.47)$$

where t is the **critical t-statistic**, S_{yx} is the **standard error of the estimate** obtained from the coefficient of determination (r^2), x_i is the given value of x, $\bar{x}$ is the average of the x values, and n is the number of observations used in regression analysis. The confidence interval shown is Figure 5.27 is calculated in this procedure.

In FFA, however, the focus is on the quantiles of various distributions and computation of confidence limits differs for different distributions. Stedinger et al. (1993) give a simple formula for a confidence interval on a flood quantile, $\hat{X}_q$, with variance, $\text{Var}(\hat{X}_q)$, as:

$$\hat{X}_q \pm z_{1-\alpha/2} \sqrt{\text{Var}(\hat{X}_q)} \qquad (5.48)$$

where q is the quantile of interest (such as $q = 0.99$), $z_{1-\alpha/2}$ is the $(1-\alpha)/2$ quantile of the standard normal distribution,

α is the confidence level, and the estimated standard error of the flood quantile is given by

$$\sqrt{\text{Var}(\hat{X}_q)} = \sigma_{\hat{X}_q} \qquad (5.49)$$

Typically, the confidence level $\alpha = 0.05$, resulting in a 90% confidence interval (5% and 95% confidence limits).

The procedure for calculating confidence limits for flood quantiles using the product moments is given in Bulletin 17B whereas that for flood quantiles estimated with the EMA is given in Bulletin 17C, which is the method given by Cohn et al. (2001).

5.6.3.4 Software for FFA using LP III Distribution

The procedures of FFA using Bulletin 17B and 17C have been specifically implemented in two software packages, namely HEC-SSP developed by USACE and PeakFQ developed by the USGS. These platforms provide an estimation of the LP III distribution parameters, using both product moments and the EMA, with available historical and paleoflood data, PILFs, regional skew information, and flood quantiles for a series of AEP/ARI. Confidence intervals and plotting positions, including Hirsch–Stedinger formula are also estimated. Examples are given for their use. These programs can be used with any dataset from any country, provided the inputs are formatted appropriately.

5.6.4 Trends and Autocorrelation

As noted above, FFA is based on the assumption that the year-to-year peak discharge values are random occurrences. In other words, this implies that the time series is **stationary**, that is, there is no trend. In addition to the stationarity requirement, no **serial correlation** should exist in the time series in order for it to qualify as a set of random variables.

A trend in peak discharge can occur due to various reasons. For example, increasing urbanization of a watershed may cause a gradual increase in peak flows, whereas the construction of a dam on a river may cause a decrease in flows from pre-construction to post-construction periods, and these periods should not be lumped together in the analysis. Climate change is another important factor that can impart a long-term trend, either increasing or decreasing, in annual peak discharges.

A plot of the time-series data can often reveal the existence of trends from simple visual examination. For example, the time-series plots shown in Figure 5.19 or 5.25 do not show any apparent trend and the annual maximum flows appear to be random, and thus the data are amenable to FFA. However, if there are reasons to believe that there can be changes in the annual peak flows over time, statistical tests must be conducted to determine the existence of trends and suitability of the data for FFA.

A common test for trends in a time series is the Mann–Kendall test, which uses Kendall's τ as the test statistic to measure the strength of the monotonic relationship between annual peak streamflow and the year in which it occurred. The Mann–Kendall test is nonparametric and does not require the data to conform to any specific statistical distribution. The statistic is calculated using the difference between concordant and discordant data pairs of observed streamflow peaks with time, and not the actual magnitudes of discharge. Positive τ-values indicate that annual peak streamflows are increasing with time, whereas negative τ-values indicate the opposite for the POR.

As with other statistical tests, a p-value can be calculated for the test. The p-values will be correct only when there is no serial correlation in the annual time series. This requirement can be problematic for hydrologic time series, which can exhibit short-term or long-term persistence (Cohn and Lins, 2005).

In addition to the statistical significance of a trend, the actual magnitude of the trend should be considered. The **Theil slope** can be calculated in conjunction with Kendall's τ for this purpose (Helsel and Hirsch, 2002). It is calculated as the median of all the slopes using all the possible pairs of peak-flow values and years and is a nonparametric estimate of the slope.

An annual flood series should be examined for autocorrelation using a correlogram (Salas, 1993). In an autocorrelated time series, the value in one time step is correlated with the value in a previous (and future) time step. Autocorrelated time series can also be said to exhibit persistence. Hydrologic time series will often exhibit long-term persistence, which can affect trend analysis.

5.7 REGIONAL FLOOD-FREQUENCY ANALYSIS

5.7.1 Need for Regional Analysis

The FFA of an AMS or PDS derived from a single gauging station gives a flood frequency curve that with all likelihood is not a true representation of the probabilistic model of flood flows in that stream or the system of streams of that region. This is primarily because a flow series drawn from data from a single gauge data is just a small sample of the population. If similar FFA is conducted with a larger set of flow series drawn from multiple gauges in the same homogeneous region, and the results are pooled together, the sample size increases and the results become closer to the actual characteristics of the population. The results of such a synthesis are more reliable for practical applications to sites in the same region but without any gauging station nearby than the result taken from a single gauging station. The regional FFA refers to this process of integration of multiple at-a-station FFA.

The premise of the regional FFA is that if the gauging stations considered all belong to a region that has the same climatologic, geomorphologic, and surficial geologic characteristics, the samples drawn from those different gauging stations belong to the same population. This is called a hydrologically similar or homogeneous region.

What follows from the above is that regional FFA has two essential elements: (1) a process of synthesis of results from multiple gauge sites and (2) a process of confirming that the gauges belong to a homogeneous region.

Sauer et al. (1983), Jennings et al. (1994), and several others have performed regional flood-frequency studies for undeveloped and various levels of urbanizing watersheds. If the physical characteristics of a watershed under investigation fall within the range of data used in the regional study, the regional relationships may be used to estimate flow frequencies for existing and future land use conditions.

5.7.2 Process of Synthesis of Regional Flood-Frequency Curves

The method of integration of flood-frequency curves derived from multiple gauging stations into one regional flood-frequency curve might have been developed first by the USGS, as documented by Dalrymple, 1960). This method is known as the **index flood method**.

The index number is a widely used statistical device for comparing one group of related variables with another group. It is actually an average of certain ratios with a common denominator. A related set of index numbers, such as a series of price-index numbers, computed for a certain period, is called an **index**.

Dalrymple (1960) defined the **index flood** of a station as the **mean annual flood**. The mean annual flood at a station is considered to be the discharge value with an ARI of 2.33 years that is obtained from the flood-frequency curve for that station. Benson (1962) confirmed that the mean annual flood has a magnitude equivalent to the flood of a 2.33-year recurrence interval. The index flood has also been represented by other measures, such as the mean of the at-a-station AMS or the mean annual flow (e.g., Hosking and Wallis, 1988).

The discharge magnitudes of different ARIs are then divided by the value of the index flood for the respective site. The median value of the ratios of discharge values of a particular ARI and index flood from different sites are then computed and these median values define a **dimensionless flood-frequency curve** for the region, assuming this distribution applies to all sites in the region.

Regression equations giving correlation between the mean annual flood or index flood for the gauged sites and their corresponding basin physiographic or geomorphologic characteristics, such as drainage area, are developed. These relationships are used for a site of interest, such as an ungauged site, where the known basin physical characteristics are used to estimate the mean annual flood or index flood. This mean annual flood or index flood is then used with the dimensionless regional flood-frequency curve, also known as a **growth curve**, to get the flood quantile, Q_T. Therefore, two relationships are used in this method. A general form of the first relationship can be written as

$$Q_{IF} = f(a_i, W_j); \quad i = 1, 2, \ldots; \quad j = 1, 2, \ldots \quad (5.50)$$

where a_i represents a set of constants or parameters of the regression equation and W_j represents a set of physiographic, geomorphologic, or even climatologic characteristics of the region/basin. This relationship is established from gauged sites where Q_{IF} is known. The second relationship is given as

$$Q_T = Z_T \times \theta \quad (5.51)$$

where θ is the scaling factor, that is, the index flood estimated from Eq. 5.50, and Z_T is the dimensionless growth curve or dimensionless flood-frequency curve, which defines the frequency distribution common to all the sites in a homogenous region.

A contemporaneous procedure for regional FFA is the L-moment index-flood method (Hosking and Wallis, 1997). The steps involved in the application of the L-moment method include the following.

1. In a homogeneous region, at each site, k, the L-moment estimators, λ_1^k, λ_2^k, and λ_3^k are computed.
2. For the region, the average sample-size-weighted estimator of the normalized L-moments for a region, λ_r^R, of order $r = 2$ and 3, across all sites are computed as

$$\widehat{\lambda_r^R} = \frac{\sum_{k=1}^{K} n_k \left(\widehat{\lambda_r^k} / \widehat{\lambda_1^k} \right)}{\sum_{k=1}^{K} n_k} \quad \text{for } r = 2 \text{ and } 3 \quad (5.52)$$

Where n_k is the length of record at site k_1, and $\lambda_1^R = 1$.
3. Using the estimator of λ_r^R for $r = 2$ and 3, the values of parameters and quantiles $\widehat{x_p}$ of the regional dimensionless flood distribution are computed; often the GEV distribution is used.
4. The estimate of the 100th percentile of the flood distribution at any site k is then

$$\widehat{x_p^k} = \widehat{\lambda_1^k} \widehat{x_p} \quad (5.53)$$

where $\widehat{\lambda_1^k}$ is the sample mean for site k and $\widehat{x_p}$ is the estimated index-flood quantile for the region.

5.7.3 Homogeneous Regions

An important and challenging step in regional FFA is the delineation of homogeneous regions. The objective of this determination is to test the hypothesis that the floods at the various sites of the homogeneous region are identically distributed except for a scaling factor, or index flood. In other words, it is answering the question of whether the records differ from one another by amounts attributable to chance alone (or if the differences have any statistical significance or not), by computing these differences and setting statistically acceptable limits.

There are no universally accepted objective methods for such delineation. This is due to the complexity of factors that affect the generation of floods. Numerous procedures have been applied or developed for this purpose. These range from general knowledge of hydrological regimes in a region, variability of skewness of instantaneous flood peak data, iterative search procedures to optimally divide the basin characteristic data, simple estimations of the standard error of estimate of the reduced variate of the distribution function being used, to more advanced statistical methods such as cluster analysis, factor analysis, multivariate discriminant analysis, just to mention a few. A number of other regionalization techniques have also been developed for an objective determination of homogeneous regions. Regionalization of the basin is also based on the quartile–quartile plot method for the identification of groups of stations with a similar distribution of tail behavior, and regression equations can be developed between the mean annual maximum flow of the catchments and their corresponding physical characteristics. In general, if the plots generated in the dimensionless growth curves can be fitted with smooth curves without too much scatter, the records are most likely to come from a hydrologically homogeneous region.

With the L-moment index-flood method, the identification of homogeneous regions is done using the heterogeneity measure (H-statistic), discussed in detail in Chapter 3 in relation to regionalization of precipitation frequency estimates using L-moments.

5.7.4 Application of Regression Equations to Ungauged Sites

It is likely that sufficient information can be extracted from the **at-a-station** analysis for sites with available data. Nevertheless, it is a challenge to estimate floods for ungauged sites or sites with a short time series, which influences the reliability of estimation. In such cases, regionalization techniques assist in estimating flows of required return periods at ungauged locations of interest. The primary goal of regional frequency analysis is to investigate typical observations of the same variable at numerous measuring sites within a suitably defined region. This leads to more accurate conclusions when analyzing all data samples rather than using only a single sample. The final result of the regional analysis is a regional curve, which permits flood-frequency and quantile estimations at any location along the drainage network of a watershed.

Regression equations are developed by statistically relating the computed streamflow statistics to the physical and climatic characteristics of the watersheds for a group of gauging stations that have virtually natural streamflow conditions within a region. Regression equations enable the transfer of streamflow statistics from gauging stations to ungauged sites simply by determining the watershed and climatic characteristics needed for the ungauged site and solving the regression equations based on these input values.

Several agencies have developed regional regression relations to estimate peak discharges at ungauged sites. Regional regression equations are easy to use and provide relatively reliable and consistent findings when applied by hydraulic engineers. The regression equations are based on statistical modeling to quantify general regional relationships between flow of a specific AEP and the physiographical, hydrological, and meteorological characteristics of watersheds. A common type of regression equation is given as

$$Q_T = a + bA^c \tag{5.54}$$

where Q_T is the dependent variable flow with a recurrence interval of T and A is an independent variable such as drainage area; a, b, and c are regression coefficients derived from the dataset. A regression equation can incorporate multiple measurable physical characteristics of the watershed or river basin. For example, a form more complex than the one given in Eq. 5.54 can be

$$Q_T = b_0 A^{b_1} S^{b_2} R^{b_3} \tag{5.55}$$

where S and R, respectively, are the basin slope and the mean annual rainfall and b_0, b_1, b_2, and b_3 are the regression constants that vary with T in the subscript of Q.

In the United States, of the various regional regression equations, the ones developed by the USGS are the result of many years of detailed investigations and are quite reliable. The USGS has been involved in the development of regionalization procedures for over 50 years. These regionalization procedures are used to transfer flood statistics, such as the 100-year flood peak discharge, and other statistics, such as the mean flow and the 7-day, 10-year low flow, known as 7Q10, from gauged to ungauged sites. The USGS has traditionally used regionalization procedures that relate streamflow statistics to watershed and climatic characteristics using correlation or regression techniques. Because streamflow statistics may vary substantially between regions due to differences in climate, topography, and geology, tests of regional homogeneity form an integral part of streamflow regionalization procedures.

Based on extensive research and data collection, the USGS has developed the National Streamflow Statistics (NSS) Program. It is a computer program that is useful to engineers, hydrologists, and others for planning, management, and design applications. NSS compiles all current USGS regional regression equations for estimating streamflow statistics at ungauged sites in an easy-to-use interface. The regression equations included in NSS are used to transfer streamflow statistics from gauged to ungauged sites through the use of watershed and climatic characteristics as explanatory or predictor variables. Generally, the equations were developed on a statewide or metropolitan-area basis as part of cooperative study programs. Equations are available for estimating rural and urban flood-frequency statistics, such as the 100-year flood, for every state and US territory. Equations are available for estimating other statistics, such as the mean annual flow, monthly mean flows, flow-duration percentiles, and 7Q10.

The NSS output provides indicators of the accuracy of the estimated streamflow statistics. The indicators may include any combination of the standard error of estimate, the standard error of prediction, the equivalent years of record, or 90% prediction intervals. The program includes several other features that can be used for flood-frequency estimation. These include the ability to generate flood-frequency plots, and plots of typical flood hydrographs for selected recurrence intervals, estimates of the probable maximum flood, extrapolation of the 500-year flood when an equation for estimation it is not available, and weighting techniques to improve flood-frequency estimates for gauging stations and ungauged sites on gauged streams.

different flow levels in any given year. Computation of an n-day minimum series and non-exceedance probabilities are calculated using the LP III distribution too. The USGS and EPA have jointly developed programs to establish 7Q10 flows that are used in water-quality modeling, where mixing models are used involving wastewater effluents and 7Q10 flows to ensure that the concentration levels of the permitted pollutants are at or below the regulatory limits. Application of LP III distribution in the frequency analysis of low flows was presented by Riggs (1972). In this case, instead of the instantaneous annual maximum flows used in FFA, daily discharge data are used to conduct the FDA.

5.8 STREAMFLOWS FOR ENVIRONMENTAL HYDROLOGY

5.8.1 7Q10 Flows

An n-day frequency statistic, such as the 7Q10 – the 7-day minimum flow that occurs on average only once every 10 years – is used in the United States for permitting discharge of effluents from wastewater treatment plants to receiving streams under the National Pollution Discharge Elimination System mandated by the US Environmental Protection Agency (EPA).

When used for low flows, the n-day frequency analysis is used to estimate the probability that flow will not exceed

5.8.2 Environmental Flows

Environmental flows describe the quantity, timing, and quality of water flow required to sustain freshwater and estuarine ecosystems and the human livelihoods and well-being that depend on these ecosystems. Environmental flow programs aim to protect aquatic habitats and species while recognizing competing water demands. Often this is done at the local or watershed level, because it is relatively easier to address technical and implementation challenges at these scales. This is an important subject in ecohydrology where annual hydrograph analysis is employed.

5.9 EXAMPLES

Example 5.1: In the USGS Open-File Report 00-169 (2000), B. F. Pope presented flow-velocity and depth data collected from July 1996 through December 1998 during peak discharge events at 21 bridge crossings, adjacent to USGS stream-gauging stations in North Carolina. The data were collected during measurements of peak discharges with ARIs ranging from less than 2 years to about 100 years. The first three columns of Table 5.4 give the data from station 021413000 on Henry Fork near Henry River. Flow velocities were measured by using a current meter at several depths at each vertical section in the cross section. Velocities were measured either at a single depth, 0.6 of the total depth, or at two depths, 0.2 and 0.8 of the total depth. Velocity measured at the 0.6 depth is assumed to be the average velocity for that vertical section, and the average of the velocities measured at the 0.2 and 0.8 depths is assumed to be the average velocity for that vertical section. Calculate the total discharge that passed through this section during measurements. Did the width of the vertical sections satisfy the convention of stationing for velocity measurements? Plot the horizontal distribution of velocity and make observations.

Solution: The horizontal distribution of velocity is plotted in Figure 5.29.

Columns 1, 2, and 3 give the data measured in the field.
The values in column 4 are obtained from column 1: $A_i + 1 - A_i$.
The values in column 5 are obtained from column 2 * column 4.
The values in column 6 are obtained from column 3 * column 5.
The total discharge = 13 935.78 ft³/s (sum of all entries in colum 6).
The values in column 7 are obtained from the values in column 6, divided by 13 935.78 and multiplied by 100.

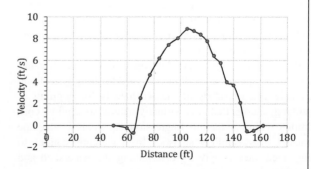

Figure 5.29 Horizontal velocity (ft/s) profile of Henry Fork, North Carolina.

Table 5.4 *Measured flow velocities at Henry Fork, North Carolina*

Distance from starting point on left bank (ft)	Depth (ft)	Velocity (ft/s)	Width (ft)	Area (ft^2)	Discharge (ft^3/s)	Percent of total discharge
50	0	0				
60	8.7	−0.24	15	130.5	−31.32	−0.22
65	11.8	−0.7	10	118	−82.60	−0.59
70	13.2	2.5	12	158.4	396.00	2.84
77	14.5	4.65	14	203	943.95	6.77
84	15.5	6.17	14	217	1338.89	9.61
91	15.4	7.43	14	215.6	1601.91	11.49
98	15.4	8.07	14	215.6	1739.89	12.49
105	15.6	8.89	12	187.2	1664.21	11.94
110	15.5	8.71	10	155	1350.05	9.69
115	14.7	8.38	10	147	1231.86	8.84
120	14.1	7.78	10	141	1096.98	7.87
125	13.4	6.41	10	134	858.94	6.16
130	12.6	5.74	10	126	723.24	5.19
135	12.9	3.98	10	129	513.42	3.68
140	12	3.72	10	120	446.40	3.20
145	10.7	2.1	10	107	224.70	1.61
150	8.8	−0.55	10	88	−48.40	−0.35
155	5.5	−0.49	12	66	−32.34	−0.23
162	0	0		−155		

Example 5.2: Table 5.5 gives the data from which the seasonal hydrograph of Potomac River at the Point of Rocks, shown in Figure 5.12, is constructed. Calculate MQ, LQ, HQ, MLQ, and MHQ.

Solution: The MQ, LQ, HQ, MLQ, and MHQ statistics are calculated by taking the averages of the columns and rows separately.

When the averages of each of the columns are taken, and the minimum and maximum values of each column we get the values shown in Table 5.6.

The lowest of the monthly minimum values is LQ. Thus, from Table 5.6, $LQ = 705.7$ ft^3/s.

The highest of the monthly minimum values is HQ, Thus, from Table 5.6, $HQ = 68\,360.0$ ft^3/s.

A process of averaging and counting of the minimum and maximum values along the rows yields the values listed in Table 5.7.

The average of the yearly LQ values gives MLQ. Thus, $MLQ = 2073.8$ ft^3/s.

The average of the yearly HQ values gives MHQ. Thus, $MHQ = 25\,953.5$ ft^3/s.

The average of the yearly MQ values gives MQ. Thus, $MQ = 9628.6$ ft^3/s, which is numerically equal to the average of all 1524 discharge values given in Table 5.5.

Table 5.5 *Mean monthly discharge (ft^3/s) through Potomac River at the Point of Rocks, Maryland*

Year	Jan	Feb	Mar	Apr	May	Jun	Jul	Aug	Sep	Oct	Nov	Dec
1896	6392	11 960	10 070	10 410	3390	6040	9281	3449	2175	12 470	6926	4721
1897	4284	42 640	20 840	10 820	22 940	5997	5315	4092	2337	1968	2096	6582
1898	14 670	8341	15 330	15 970	18 050	4178	2418	22 130	2497	13 660	8557	15 330
1899	18 670	23 710	35 230	11 750	11 600	5315	2519	2335	2345	1663	3171	4066
1900	8166	13 330	18 460	9295	4466	8395	3008	1917	1344	1333	4569	6215
1901	4948	3148	13 790	39 740	26 920	19 160	10 730	8334	7638	4302	4647	25 620
1902	17 510	25 210	54 300	28 780	5973	3186	3086	2464	1494	2767	2838	18 600
1903	17 210	22 190	26 720	28 900	6212	17 980	12 760	4826	4667	3212	2175	2926
1904	7289	14 680	11 180	7408	9367	10 160	4506	2394	1595	1165	1345	2203
1905	8628	4368	23 480	6581	4493	6980	10 200	5831	3205	2888	2267	10 650
1906	15 000	5116	15 900	22 430	5539	7007	4381	15 210	4276	16 310	6341	11 140
1907	26 950	9561	30 510	14 890	11 090	23 180	6011	3715	6430	3553	7849	17 850
1908	25 360	24 120	28 640	10 210	32 070	7250	4474	3885	2505	2778	2671	2439
1909	6951	11 290	8704	16 680	8661	11 040	2699	2109	1779	2194	1851	3132

Table 5.5 (*cont.*)

Year	Jan	Feb	Mar	Apr	May	Jun	Jul	Aug	Sep	Oct	Nov	Dec
1910	11 670	13 810	9965	10 060	5519	26 420	5503	2000	1737	1561	1441	2085
1911	12 240	11 330	9674	15 140	4330	3691	2158	2444	12 560	8508	5970	10 240
1912	7332	14 900	28 700	14 870	19 750	5282	8550	3456	6365	3094	3268	3391
1913	11 690	5890	18 200	16 100	13 600	10 440	4486	3014	1769	6850	13 490	8121
1914	19 970	15 910	19 350	20 930	9603	3193	3018	1754	1265	1404	1541	5540
1915	28 600	27 020	8225	4368	5903	20 380	2668	6762	5347	6735	3594	6655
1916	12 470	15 620	24 710	19 850	7946	13 990	7172	3596	2334	2190	1546	3249
1917	9167	7627	38 410	10 680	5594	7873	3891	2623	1379	4771	3830	2269
1918	2500	28 330	13 620	39 840	5992	3312	3356	2911	3938	1713	5610	12 960
1919	14 760	5983	12 720	8856	18 280	7127	7217	3130	1721	2419	4621	6879
1920	12 740	20 290	32 330	14 500	8704	8546	3795	7149	4965	2715	4780	9503
1921	6362	6167	12 870	4941	20 690	4145	4256	3914	3652	1754	4398	11 320
1922	8113	17 520	21 640	9120	9710	6663	4396	2079	1760	1266	990	2340
1923	6777	12 490	10 690	10 320	5415	2513	1560	2855	3768	1037	2030	6877
1924	18 680	8175	33 770	22 060	41 970	14 410	8959	3172	2678	7097	3812	4685
1925	10 400	23 590	9683	5994	11 330	2715	2417	1577	952	2762	6825	5295
1926	9213	22 970	10 800	11 480	4554	2641	2654	9600	8204	6573	18 120	12 910
1927	10 710	21 620	14 540	23 170	13 970	9420	3214	3112	1536	12 410	8399	11 830
1928	6100	10 720	14 230	21 250	21 600	16 110	9561	6236	5768	2748	1981	4595
1929	4620	8029	24 630	27 390	14 650	7071	3915	1497	1499	15 170	12 980	7432
1930	6880	8970	10 910	7528	3276	2228	1100	771	834	706	840	1420
1931	2372	2930	5400	14 200	14 230	5606	5518	3630	2063	1239	944	2335
1932	5199	10 690	14 860	14 960	24 610	3363	2743	1117	1000	5332	15 850	6644
1933	13 030	11 990	21 820	32 100	20 110	5243	4273	9122	6654	2855	2238	3413
1934	8351	2661	10 040	11 080	4498	3037	1540	3069	5353	4223	2709	16 510
1935	13 950	21 060	15 810	22 530	11 170	4472	3677	4548	8223	1823	5946	7300
1936	19 650	21 040	68 360	20 570	5613	3888	2865	2356	1629	4701	2657	10 100
1937	31 350	23 440	12 010	32 200	13 340	6183	4020	7352	5581	23 840	13 320	7332
1938	7094	8022	10 280	8019	9275	5337	4982	4616	1925	1400	1927	7700
1939	6515	31 370	18 270	16 670	7362	4948	6987	5019	1675	3299	4383	2732
1940	2848	12 130	13 970	28 010	8935	16 110	4153	7315	6201	2995	9050	11 430
1941	10 560	6494	11 530	14 150	3595	6905	6980	2346	1776	1075	1463	3312
1942	3436	5869	11 460	12 890	17 320	7094	3135	9813	3928	37 030	8232	14 980
1943	17 570	22 090	19 120	15 890	13 060	5889	4449	2147	1465	1514	2403	1439
1944	5860	7281	23 760	14 840	15 100	4159	1787	1296	2666	5499	2378	7185
1945	8567	15 430	18 060	9109	10 100	4487	3166	6071	17 600	4472	6888	12 250
1946	14 220	10 760	15 480	8198	13 510	11 870	3389	3149	1726	2824	2163	2144
1947	8969	5082	10 490	6201	8353	5697	5020	3555	1972	1305	4566	2680
1948	6085	11 380	15 780	19 930	14 980	6399	3988	4985	2624	8249	9269	23 600
1949	24 930	19 590	10 300	12 570	9489	13 050	16 000	7259	4365	2242	4988	7565
1950	7307	21 380	15 090	8337	14 410	8550	3028	1941	8818	5511	10 170	25 010
1951	14 300	26 200	18 330	22 710	11 550	15 240	4499	2388	1706	1304	2210	5939
1952	20 330	14 860	21 660	27 570	20 850	6012	4001	2871	4601	1941	11 180	10 550
1953	21 380	14 620	26 220	16 180	14 200	7761	2695	1968	1535	1210	1600	2866
1954	3315	3842	17 120	7043	7071	5578	2055	2321	1799	10 610	8036	10 870
1955	10 170	10 660	27 200	10 950	6587	10 570	2618	23 580	3943	2936	2503	1981
1956	1947	18 980	16 680	15 260	6445	5219	5528	5517	2355	3599	6490	9429
1957	8913	18 680	12 810	18 000	6709	5309	1934	1127	1251	2231	2363	8481
1958	10 150	9769	22 400	24 860	22 030	5272	5076	6196	2148	1604	1714	1884
1959	4506	5535	7302	10 830	9455	7873	1977	1987	1223	6309	3546	8036
1960	10 670	12 820	14 730	27 530	19 890	11 580	3071	2680	3219	1786	1814	1485
1961	3739	25 590	24 080	27 840	13 770	6380	2867	2294	1998	2392	2545	6662
1962	8486	10 260	38 630	19 700	8674	5829	2667	1582	1281	1486	5908	3321
1963	8704	4434	38 520	7379	4456	5869	1933	1316	1230	995	2529	3776
1964	15 420	9925	26 060	15 710	12 740	2756	2139	1371	1067	2375	2237	5340
1965	13 510	16 130	23 910	14 800	7121	2346	1470	1221	1184	1526	1164	1253

Table 5.5 (cont.)

Year	Jan	Feb	Mar	Apr	May	Jun	Jul	Aug	Sep	Oct	Nov	Dec
1966	1896	11 610	10 530	10 680	12 030	2402	1056	929	4048	5813	3698	6873
1967	7820	9044	31 410	8075	15 390	4447	4380	5276	2633	5412	3969	15 570
1968	11 290	14 580	21 690	7499	11 710	9228	2348	1793	1735	1524	4833	3507
1969	3266	5799	7252	6375	3515	1932	2557	8168	3788	2208	3407	6706
1970	10 500	17 000	14 350	29 810	7970	5869	7572	3367	1737	1809	14 750	13 950
1971	16 280	29 300	18 430	9459	13 860	13 310	2900	4640	5656	10 660	6536	13 750
1972	8296	22 210	23 210	20 960	21 910	40 400	14 740	5515	2566	11 640	18 770	32 610
1973	14 440	21 570	15 830	32 240	16 690	11 620	4830	5280	3328	5558	6380	22 410
1974	21 510	9408	9596	17 020	8487	12 760	4211	2429	2955	1809	2031	12 710
1975	12 390	18 190	28 200	13 040	17 000	8931	6743	4174	17 820	15 310	8430	6091
1976	19 880	12 760	9275	9193	4030	5383	3037	2580	1917	29 410	8969	8349
1977	3258	5324	18 650	17 690	3931	2185	1883	1402	1295	2155	11 690	16 020
1978	18 140	8670	36 410	13 250	24 910	5411	7595	10 030	3203	1784	1948	8545
1979	23 160	24 510	30 150	14 730	13 760	10 970	5134	4082	15 570	25 350	14 960	10 610
1980	14 720	6938	18 760	25 330	20 290	6627	4041	3338	1792	1634	3147	3221
1981	1703	10 290	6164	10 810	8691	10 900	4090	1845	2537	2554	2971	3980
1982	6412	23 020	24 690	10 560	6342	17 390	5421	3525	1894	2096	2595	6355
1983	3940	13 100	20 200	38 610	20 270	8756	3356	1935	1626	4134	7759	21 300
1984	8345	32 220	28 210	39 640	16 450	4687	5997	9346	3037	2915	4636	9852
1985	6228	17 040	11 260	9651	8702	7087	3513	2900	1739	3223	39 000	15 350
1986	5115	20 180	21 180	9685	6384	3204	2394	1758	1549	1478	3674	9323
1987	8874	10 540	13 160	38 790	13 010	4982	3609	1569	6114	2577	3821	9214
1988	10 380	9368	6744	9689	29 630	5322	2215	1873	2385	1641	3584	2638
1989	6016	5855	15 670	8214	29 650	10 110	10 500	6212	5045	7855	7204	3542
1990	10 980	13 190	5748	8691	9701	7475	5191	3578	2813	13 450	5503	11 370
1991	27 170	9727	20 310	12 830	5232	2493	2519	2184	1737	1444	1671	4472
1992	6377	6158	15 380	15 180	11 120	7853	5032	3601	4469	2218	5677	15 250
1993	13 790	6903	47 360	43 840	9849	4502	2836	2318	2607	2377	8467	16 830
1994	14 090	32 230	52 000	21 540	15 190	3633	3344	7804	3121	2148	3324	6534
1995	16 500	5981	8038	4717	9634	7665	9205	4133	1776	5178	8264	9569
1996	42 160	17 510	24 400	16 230	22 580	16 170	7524	17 170	38 300	12 700	17 000	27 730
1997	9338	14 930	24 680	8058	5743	9597	2854	2104	2221	2025	14 560	6082
1998	28 970	47 870	36 090	22 390	20 210	7235	4137	2738	1831	1873	1767	1738
1999	6662	4703	13 450	10 930	4868	1972	1247	1158	5041	6479	3554	7504
2000	3092	12 160	14 810	12 880	6314	7837	3626	5420	5724	3704	2332	4898
2001	4711	9131	14 560	17 230	6822	8006	3213	4940	1854	1354	1357	1784
2002	1833	1982	5460	11 860	13 600	3665	2015	1721	1424	4554	11 780	14 090
2003	17 010	13 340	36 380	23 850	25 780	30 970	8264	6108	23 120	9035	18 910	25 370
2004	9648	19 150	17 560	26 150	14 060	9636	4270	3899	18 040	8399	9888	15 920
2005	14 210	8186	19 750	17 600	8046	3991	4796	2525	1647	3300	3499	10 640
2006	15 620	11 030	4583	8822	6404	7343	6127	2033	5755	6236	16 010	6825
2007	9833	6171	25 710	18 540	5998	3357	1911	2610	1695	2058	2468	7719
2008	7071	12 200	18 600	16 710	24 620	7117	3654	2159	3117	2444	2190	10 650
2009	8765	6472	4407	14 760	21 490	11 480	3011	2699	1700	2749	5335	18 010
2010	22 070	12 320	36 980	11 130	8839	4729	1947	1463	1385	3627	2755	7135
2011	2480	8133	26 910	35 770	28 550	6728	3009	2066	9086	8313	12 790	18 860
2012	14 340	8611	15 910	6386	8703	5030	2947	3261	3933	7443	7856	5911
2013	13 740	15 590	15 990	12 940	16 550	8343	5595	3375	2672	4446	2890	16 250
2014	11 860	21 980	15 040	13 810	25 100	9484	3941	2687	1965	2958	2395	7392
2015	6120	5421	17 850	15 350	7590	7704	6997	2301	1844	6422	4891	8973
2016	10 020	29 360	11 260	5712	16 130	9786	3920	3224	1820	3815	1752	3844
2017	12 690	7561	8209	12 450	21 770	6928	4833	5388	2171	2805	5747	2549
2018	4827	16 450	10 540	19 710	27 850	33 580	9170	16 360	38 630	14 460	29 970	33 230
2019	24 260	26 100	23 870	17 740	20 300	6483	6846	3354	2132	2393	3542	5389
2020	10 340	12 500	8231	14 670	17 840	8538	3187	5273	4174	3006	5989	15 130
2021	10 960	11 300	18 570	13 870	7103	6019	2741	2385	11 240	3833	6044	2754

Table 5.6 *Average, minimum, and maximum discharge (ft³/s) of each month based on the long-term record presented in Table 5.5*

	Jan	Feb	Mar	Apr	May	Jun	Jul	Aug	Sep	Oct	Nov	Dec
Mean	11 443.0	14 226.2	19 285.9	16 384.7	12 872.7	8247.5	4488.1	4218.0	4294.6	5041.8	5981.7	9059.4
Standard deviation of mean	7112.4	8280.8	10669.9	8683.5	7451.8	6112.7	2666.2	3707.0	5716.9	5666.9	5636.0	6673.7
Minimum	1703	1982	4407	4368	3276	1932	1056	771	834.4	705.7	839.9	1253
Maximum	42 160	47 870	68 360	43 840	41 970	40 400	16 000	23 580	38 630	37 030	39 000	33 230

Table 5.7 *Yearly values of LQ, HQ, and MQ*

Year	Yearly LQ	Yearly HQ	Yearly MQ	Year	Yearly LQ	Yearly HQ	Yearly MQ	Year	Yearly LQ	Yearly HQ	Yearly MQ
1896	2175.0	12 470.0	7273.667	1938	1400.0	10 280.0	5881.4	1980	1634.0	25 330.0	9153.2
1897	1968.0	42 640.0	10 825.92	1939	1675.0	31 370.0	9102.5	1981	1703.0	10 900.0	5544.6
1898	2418.0	22 130.0	11 760.92	1940	2848.0	28 010.0	10262.3	1982	1894.0	24 690.0	9191.7
1899	1663.0	35 230.0	10 197.83	1941	1075.0	14 150.0	5848.8	1983	1626.0	38 610.0	12 082.2
1900	1333.0	18 460.0	6708.167	1942	3135.0	37 030.0	11265.6	1984	2915.0	39 640.0	13 777.9
1901	3148.0	39 740.0	14 081.42	1943	1439.0	22 090.0	8919.7	1985	1739.0	39 000.0	10 474.4
1902	1494.0	54 300.0	13 850.67	1944	1296.0	23 760.0	7650.9	1986	1478.0	21 180.0	7160.3
1903	2175.0	28 900.0	12 481.5	1945	3166.0	18 060.0	9683.3	1987	1569.0	38 790.0	9688.3
1904	1165.0	14 680.0	6107.667	1946	1726.0	15 480.0	7452.8	1988	1641.0	29 630.0	7122.4
1905	2267.0	23 480.0	7464.25	1947	1305.0	10 490.0	5324.2	1989	3542.0	29 650.0	9656.1
1906	4276.0	22 430.0	10 720.83	1948	2624.0	23 600.0	10605.8	1990	2813.0	13 450.0	8140.8
1907	3553.0	30 510.0	13 465.75	1949	2242.0	24 930.0	11029.0	1991	1444.0	27 170.0	7649.1
1908	2439.0	32 070.0	12 200.17	1950	1941.0	25 010.0	10796.0	1992	2218.0	15 380.0	8192.9
1909	1779.0	16 680.0	6424.167	1951	1304.0	26 200.0	10531.3	1993	2318.0	47 360.0	13 473.3
1910	1441.0	26 420.0	7647.583	1952	1941.0	27 570.0	12202.2	1994	2148.0	52 000.0	13 746.5
1911	2158.0	15 140.0	8190.417	1953	1210.0	26 220.0	9352.9	1995	1776.0	16 500.0	7555.0
1912	3094.0	28 700.0	9913.167	1954	1799.0	17 120.0	6638.3	1996	7524.0	42 160.0	21 622.8
1913	1769.0	18 200.0	9470.833	1955	1981.0	27 200.0	9474.8	1997	2025.0	24 680.0	8516.0
1914	1265.0	20 930.0	8623.167	1956	1947.0	18 980.0	8120.8	1998	1738.0	47 870.0	14 737.4
1915	2668.0	28 600.0	10 521.42	1957	1127.0	18 680.0	7317.3	1999	1158.0	13 450.0	5630.7
1916	1546.0	24 710.0	9556.083	1958	1604.0	24 860.0	9425.3	2000	2332.0	14 810.0	6899.8
1917	1379.0	38 410.0	8176.167	1959	1223.0	10 830.0	5714.9	2001	1354.0	17 230.0	6246.8
1918	1713.0	39 840.0	10 340.17	1960	1485.0	27 530.0	9272.9	2002	1424.0	14 090.0	6165.3
1919	1721.0	18 280.0	7809.417	1961	1998.0	27 840.0	10013.1	2003	6108.0	36 380.0	19 844.8
1920	2715.0	32 330.0	10 834.75	1962	1281.0	38 630.0	8985.3	2004	3899.0	26 150.0	13 051.7
1921	1754.0	20 690.0	7039.083	1963	994.8	38 520.0	6761.7	2005	1647.0	19 750.0	8182.5
1922	989.5	21 640.0	7133.042	1964	1067.0	26 060.0	8095.0	2006	2033.0	16 010.0	8065.7
1923	1037.0	12 490.0	5527.667	1965	1164.0	23 910.0	7136.3	2007	1695.0	25 710.0	7339.2
1924	2678.0	41 970.0	14 122.33	1966	928.9	12 030.0	5963.7	2008	2159.0	24 620.0	9211.0
1925	951.5	23 590.0	6961.625	1967	2633.0	31 410.0	9452.2	2009	1700.0	21 490.0	8406.5
1926	2641.0	22 970.0	9976.583	1968	1524.0	21 690.0	7644.8	2010	1385.0	36 980.0	9531.7
1927	1536.0	23 170.0	11 160.92	1969	1932.0	8168.0	4581.1	2011	2066.0	35 770.0	13 557.9
1928	1981.0	21 600.0	10 074.92	1970	1737.0	29 810.0	10 723.7	2012	2947.0	15 910.0	7527.6
1929	1497.0	27 390.0	10 740.25	1971	2900.0	29 300.0	12 065.1	2013	2672.0	16 550.0	9865.1
1930	705.7	10 910.0	3788.583	1972	2566.0	40 400.0	18 568.9	2014	1965.0	25 100.0	9884.3
1931	944.1	14 230.0	5038.925	1973	3328.0	32 240.0	13 348.0	2015	1844.0	17 850.0	7621.9
1932	1000.0	24 610.0	8864	1974	1809.0	21 510.0	8743.8	2016	1752.0	29 360.0	8386.9
1933	2238.0	32 100.0	11 070.67	1975	4174.0	28 200.0	13 026.6	2017	2171.0	21 770.0	7758.4
1934	1540.0	16 510.0	6089.25	1976	1917.0	29 410.0	9565.3	2018	4827.0	38 630.0	21 231.4
1935	1823.0	22 530.0	10 042.42	1977	1295.0	18 650.0	7123.6	2019	2132.0	26 100.0	11 867.4
1936	1629.0	68 360.0	13 619.08	1978	1784.0	36 410.0	11 658.0	2020	3006.0	17 840.0	9073.2
1937	4020.0	32 200.0	14 997.33	1979	4082.0	30 150.0	16 082.2	2021	2385.0	18 570.0	8068.3

Example 5.3: An AMS of Potomac River at the Point of Rocks for 127 years is given in Table 5.8. Why can this dataset be considered as an excellent example of perfect AMS for FFA? Compare the results of the FFA of this dataset using both PeakFQ and HEC-SSP programs and evaluate the relative merits of these two platforms.

Table 5.8 *Peak streamflow of Potomac River, at point of Rocks, Maryland, upstream of Washington DC, USGS Gauging Station 01638500*

Water year	Date	Gauge height (ft)	Stream-flow (ft^3/s)	Water year	Date	Gauge height (ft)	Stream-flow (ft^3/s)
1889	1889-06-02	40.2	460 000	1958	1958-05-07	14.13	72 000
1895	1895-04-10	10.7	68 500	1959	1959-06-04	11.8	55 700
1896	1896-07-26	9.4	56 000	1960	1960-05-10	20.28	124 000
1897	1896-10-01	27.2	204 000	1961	1961-02-21	17.9	102 000
1898	1898-08-12	18	127 000	1962	1962-03-23	19.45	116 000
1899	1899-03-06	18.1	128 000	1963	1963-03-21	20.47	125 000
1900	1900-03-21	9.6	57 700	1964	1964-03-06	16	87 000
1901	1901-04-22	22	161 000	1965	1965-03-06	17.35	97 800
1902	1902-03-02	29	219 000	1966	1966-02-15	14.04	71 300
1903	1903-03-01	16.6	110 000	1967	1967-03-08	22.53	144 000
1904	1904-06-01	8.6	44 500	1968	1968-03-18	14.73	76 800
1905	1905-03-11	11.9	71 400	1969	1969-03-27	7.15	27 800
1906	1906-03-29	13.1	81 300	1970	1970-04-03	16.64	92 100
1907	1907-03-15	17.6	119 000	1971	1971-02-24	15.92	86 400
1908	1908-01-13	21.6	152 000	1972	1972-06-23	37.43	347 000
1909	1909-04-16	13.3	83 000	1973	1972-10-08	18.28	106 000
1910	1910-06-18	23.5	168 000	1974	1973-12-28	21.27	132 000
1911	1911-09-01	16.1	106 000	1975	1975-03-21	26.15	181 000
1912	1912-02-28	14.8	95 400	1976	1976-01-02	18.71	109 000
1913	1913-03-28	20	139 000	1977	1976-10-11	27.25	193 000
1914	1914-03-19	12.2	73 900	1978	1978-03-16	22.01	139 000
1915	1915-06-04	20	139 000	1979	1979-02-27	25.94	178 000
1916	1916-03-29	18.3	124 000	1980	1979-10-11	13.87	69 500
1917	1917-03-13	18.1	123 000	1981	1981-04-14	9.81	41 900
1918	1918-04-16	18.6	127 000	1982	1982-06-15	16.66	92 000
1919	1919-05-11	13	80 500	1983	1983-04-26	19.21	115 000
1920	1920-03-06	16.4	109 000	1984	1984-02-16	28.14	199 000
1921	1921-05-06	14	88 800	1985	1985-02-13	15.57	84 700
1922	1922-03-17	12.8	78 800	1986	1985-11-07	36.28	309 000
1923	1923-04-16	8.8	40 700	1987	1987-04-18	23.49	153 000
1924	1924-05-13	32.2	277 000	1988	1988-05-20	17.01	96 100
1925	1925-02-13	15	89 000	1989	1989-05-17	15.11	80 900
1926	1926-02-27	11.5	60 500	1990	1990-05-31	9.3	39 800
1927	1926-11-17	15.1	89 900	1991	1991-03-25	15.63	85 000
1928	1928-05-02	21.3	145 000	1992	1992-04-23	16.25	90 000
1929	1929-04-18	24.94	180 000	1993	1993-03-06	24.94	167 000
1930	1929-10-23	17.4	110 000	1994	1993-11-29	19.25	115 000
1931	1931-05-24	8.16	36 800	1995	1995-01-17	15.9	87 200
1932	1932-05-14	23.34	158 000	1996	1996-01-21	36.34	310 000
1933	1933-04-21	19.3	123 000	1997	1996-12-03	15.65	85 200
1934	1934-01-09	8.06	36 700	1998	1998-03-22	20.24	123 000
1935	1934-12-02	19.78	128 000	1999	1999-03-20	9.42	40 600
1936	1936-03-19	41.03	480 000	2000	2000-02-20	12.54	61 600
1937	1937-04-27	33.86	310 000	2001	2001-03-23	13.23	66 600
1938	1937-10-30	24.93	175 000	2002	2002-04-24	11.21	52 300
1939	1939-02-05	19.39	124 000	2003	2003-09-21	23.12	150 000
1940	1940-04-21	15.67	93 600	2004	2003-12-12	19.58	118 000
1941	1941-04-07	12.56	69 000	2005	2005-03-30	19.37	116 000

Table 5.8 (cont.)

Water year	Date	Gauge height (ft)	Stream-flow (ft³/s)	Water year	Date	Gauge height (ft)	Stream-flow (ft³/s)
1942	1942-05-24	21.13	125 000	2006	2005-12-01	14.1	73 100
1943	1942-10-16	40.43	418 000	2007	2007-04-17	17	96 000
1944	1944-05-08	13.92	70 300	2008	2008-05-13	16.99	96 000
1945	1945-09-20	21.98	139 000	2009	2009-05-06	17.5	100 000
1946	1946-06-03	11.4	53 100	2010	2010-03-15	25.52	172 000
1947	1947-03-16	9.65	42 100	2011	2011-04-18	22.52	144 000
1948	1948-04-15	16.04	87 000	2012	2011-12-09	14.32	74 800
1949	1949-06-20	21.2	132 000	2013	2012-10-31	18.19	106 000
1950	1950-02-03	13.09	64 700	2014	2014-05-17	22.62	145 000
1951	1950-12-05	20.75	128 000	2015	2015-04-22	11.21	52 300
1952	1952-04-29	20.67	127 000	2016	2016-02-05	16	88 000
1953	1952-11-23	19.68	118 000	2017	2017-05-07	13.86	71 300
1954	1954-03-03	18.65	109 000	2018	2018-06-04	25.83	175 000
1955	1955-08-20	29.08	214 000	2019	2018-12-17	22.4	143 000
1956	1956-04-09	12.54	60 800	2020	2020-05-02	14.77	78 300
1957	1957-04-07	13.74	69 200	2021	2021-03-02	16.39	91 100

Solution:

Step 1: Evaluation of the dataset for suitability of FFA.

Two major qualifiers for an AMS to be suitable for FFA are that the data are stationary and the flows are natural, that is, unregulated. Another quality is the length of the POR, which is not necessarily a discriminant factor.

When the AMS is imported to PeakFQ program, it first conducts a trend analysis. The result of the trend analysis is as follows:

Kendall's tau, $\tau = -0.03$. *p value* $= 0.62$.

Median slope $= -51.282$

H_0: *There is no trend. Accepted.*

There is no dam or control of the streamflow upstream of the gauge. The USGS does not give any qualification code of 6, used to denote discharge affected by regulation or diversion, for any of the annual peaks. Thus, the flood flows are natural flows.

The POR is long, spanning 127 years 1895 to 2021 with a continuous record. In addition, the 1989 flow is a historic flow.

The three qualifications of this dataset make it an ideal case for conducting FFA.

Step 2: Conduct Bulletin 17B Analysis using PeakFQ.

Bulletin 17B analysis is selected in PeakFQ, Station Specification panel. The weighted skew option and the single Grubbs–Beck test for outlier detection are selected. Based on the longitude and latitude of the gauging station, the program digitally obtains the regional skew and MSE of regional skew from the generalized skew map accompanying Bulletin 17B.

The program also checks, using the USGS flow qualifier, whether the flows are regulated or unregulated. The station specification panel is shown in Box 5.1.

Step 3: Review results

The main output is given as Table 4 in the program output and is shown in Box 5.2. Column 2 gives the calculated discharge for the selected AEP (column 1) based on weighted skew values whereas column 3 gives the calculated discharge based on the station skew. The values of the skewness and associated parameters are shown in Table 5.9.

Step 4: Change the method to Bulletin 17C.

In the next step the calculation method is changed to EMA (Bulletin 17C). In this case the weighted skew option is selected but the outlier detection method is changed to multiple Grubbs–Beck test. The output (Table 4 by the program) is shown in Box 5.3.

Note that for Bulletin 17B the output in column 4 of Table 4 in Box 5.2 is blank. Previously, it used to give expected probability as described in Bulletin 17B. But that method is no longer used and its calculation has been dropped from PeakFQ.

Step 5: Run Bulletin 17B Analysis in HEC-SSP.

The input panel for HEC-SSP is shown in Box 5.4.

Step 6: Review results.

The key output from HEC-SSP is shown in Box 5.5. According to HEC-SSP, the discharge value with 0.01 AEP at this station is 438 185.0 ft³/s. According to PeakFQ, the discharge value with 0.01 AEP at this station is either 405 500.0 ft³/s (Bulletin 17B) or 407 000.0 (Bulletin 17C).

5.9 Examples

Box 5.1 PeakFQ input

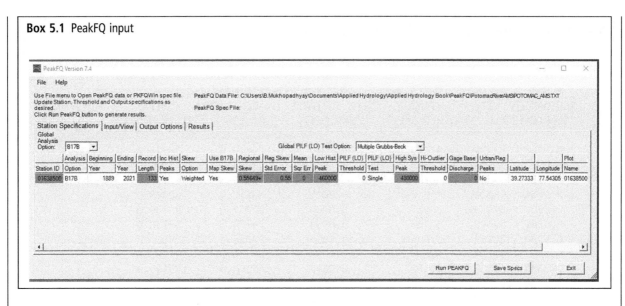

Box 5.2 Results of FFA using PeakFQ and Bulletin 17B procedure

```
    TABLE 4 - ANNUAL FREQUENCY CURVE -- DISCHARGES AT SELECTED EXCEEDANCE PROBABILITIES

        ANNUAL                             <-- FOR BULLETIN 17B ESTIMATES -->
     EXCEEDANCE   BULL.17B SYSTEMATIC LOG VARIANCE    CONFIDENCE INTERVALS
    PROBABILITY   ESTIMATE    RECORD     OF EST.    5.0% LOWER    95.0% UPPER

        0.9950     32680.     31430.      ----       27920.0       37280.0
        0.9900     35890.     34850.      ----       30950.0       40650.0
        0.9500     47190.     46750.      ----       41750.0       52420.0
        0.9000     55200.     55060.      ----       49490.0       60710.0
        0.8000     67420.     67590.      ----       61330.0       73390.0
        0.6667     82110.     82390.      ----       75520.0       88790.0
        0.5000    102100.    102100.      ----       94510.0      110200.0
        0.4292    112100.    111900.      ----      103900.0      121100.0
        0.2000    161500.    158700.      ----      148500.0      177300.0
        0.1000    209000.    202100.      ----      189500.0      233800.0
        0.0400    278900.    263800.      ----      248100.0      320200.0
        0.0200    338700.    314900.      ----      296900.0      396300.0
        0.0100    405500.    370400.      ----      350400.0      483200.0
        0.0050    480200.    431000.      ----      409200.0      582500.0
        0.0020    592900.    519600.      ----      496300.0      735300.0
1
```

Step 7: Plot the curves (probability plots).

Figure 5.30 shows the flood-frequency curve calculated by HEC-SSP. Computed as well as observed values of discharge are plotted against their exceedance probabilities. Figures 5.31 and 5.32 show the flood-frequency curves calculated by PeakFQ using Bulletin 17B and 17C, respectively. Note that Bulletin 17C gives a wider band of confidence limits. The historical flow shown by a pink triangle is within the upper 95% confidence band.

Table 5.9 *Skew calculations*

Regional skew	0.556
Standard error	0.55
MSE	0.303
Station skew (systematic record)	0.233
Bulletin 17B estimate (weighted skew)	0.356
MSE of Station Skew	0.054

Box 5.3 Results of FFA using PeakFQ and Bulletin 17C procedure

```
         TABLE 4 - ANNUAL FREQUENCY CURVE -- DISCHARGES AT SELECTED EXCEEDANCE PROBABILITIES

         ANNUAL      <- EMA ESTIMATE ->     <- FOR EMA ESTIMATE WITH REG SKEW ->
         EXCEEDANCE    WITH     WITHOUT     LOG VARIANCE    <-CONFIDENCE LIMITS->
         PROBABILITY REG SKEW   REG SKEW     OF EST.       5.0% LOWER    95.0% UPPER

           0.9950      32720.    32120.       0.0025        26420.0        38720.0
           0.9900      35920.    35400.       0.0018        29900.0        41440.0
           0.9500      47200.    46910.       0.0008        41890.0        51910.0
           0.9000      55200.    55050.       0.0005        50110.0        60030.0
           0.8000      67410.    67440.       0.0004        62060.0        73000.0
           0.6667      82100.    82300.       0.0004        75770.0        88920.0
           0.5000     102100.   102400.       0.0005        94050.0       110900.0
           0.4292     112100.   112500.       0.0005       103200.0       122100.0
           0.2000     161600.   161900.       0.0007       147300.0       179500.0
           0.1000     209300.   209000.       0.0010       187600.0       240100.0
           0.0400     279500.   277900.       0.0018       243200.0       340900.0
           0.0200     339700.   336400.       0.0027       287700.0       437600.0
           0.0100     407000.   401400.       0.0039       334800.0       556500.0
           0.0050     482300.   473700.       0.0055       384800.0       702300.0
           0.0020     595900.   582000.       0.0079       455700.0       946300.0

        *Note: If Station Skew option is selected then EMA ESTIMATE WITH REG SKEW will
               display values for and be equal to EMA ESTIMATE WITHOUT REG SKEW.
```

Box 5.4 Input panel of HEC-SSP for the AMS given in Table 5.8

5.9 Examples

Box 5.5 Results of FFA using HEC-SSP

```
<< Skew Weighting >>
---------------------------------------------------------------------
Based on 128 events, mean-square error of station skew =    0.056
Mean-square error of regional skew =                        0.303
---------------------------------------------------------------------


<< Frequency Curve >>
Potomac River AMS-POINT OF ROCKS, MD-FLOW-ANNUAL PEAK
---------------------------------------------------------------------
|   Computed      Expected     |  Percent    |  Confidence Limits        |
|    Curve      Probability    |  Chance     |    0.05         0.95      |
|      FLOW, CFS               | Exceedance  |       FLOW, CFS           |
|------------------------------|-------------|---------------------------|
|    676,653.8     714,134.6   |    0.2      |    850,390.7    560,408.9 |
|    531,068.7     552,540.5   |    0.5      |    650,367.8    449,021.6 |
|    438,184.9     451,521.0   |    1.0      |    525,923.3    376,460.4 |
|    357,960.7     365,852.8   |    2.0      |    420,865.9    312,581.3 |
|    268,335.5     271,783.6   |    5.0      |    306,803.3    239,462.3 |
|    210,854.3     212,501.0   |   10.0      |    236,016.2    191,209.5 |
|    160,411.1     161,023.6   |   20.0      |    175,983.8    147,545.5 |
|    100,413.8     100,413.8   |   50.0      |    108,353.4     92,959.3 |
|     67,311.7      67,140.0   |   80.0      |     73,268.1     61,245.6 |
|     56,040.0      55,789.9   |   90.0      |     61,570.7     50,317.4 |
|     48,785.7      48,459.6   |   95.0      |     54,057.4     43,304.1 |
|     38,741.7      38,289.9   |   99.0      |     43,620.6     33,675.0 |
|------------------------------|-------------|---------------------------|
```

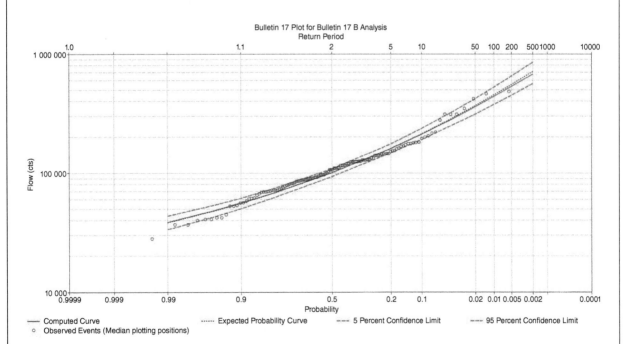

Figure 5.30 Flood-frequency curve for the AMS of Potomac River at the Point of Rocks according to the Bulletin 17B procedure done by HEC-SSP. (A black and white version of this figure will appear in some formats. For the color version, refer to the plate section.)

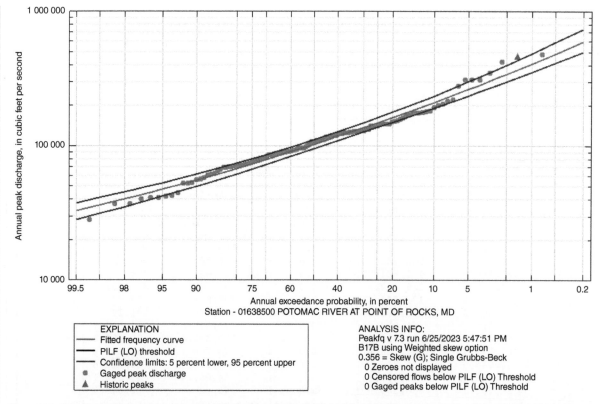

Figure 5.31 Flood-frequency curve for the AMS of Potomac River at the Point of Rocks according to Bulletin 17B procedure done by PeakFQ. (A black and white version of this figure will appear in some formats. For the color version, refer to the plate section.)

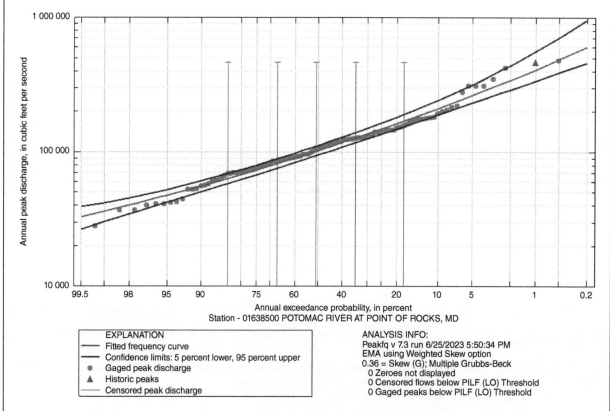

Figure 5.32 Flood-frequency curve for the AMS of Potomac River at the Point of Rocks according to Bulletin 17C procedure done by PeakFQ. (A black and white version of this figure will appear in some formats. For the color version, refer to the plate section.)

5.9 Examples

Step 8: Check goodness of fit using HEC-SSP.

The analysis presented above conducted by both the PeakFQ and HEC-SSP programs is based on the LP III distribution. However, as noted above, in many countries nowadays, the GEV distribution is also widely used in FFA. In HEC-SSP the relative merit of these two distributions, namely LP III and GEV, are evaluated. Figure 5.33 shows the pdf of the LP III compared to the observed distribution. The Kolmogorov–Smirnov test statistic is 0.056, as shown in the table in the panel to the left. Figure 5.34 shows the cdf of the LP III distribution and observed CDF. The details of the Kolmogorov–Smirnov test are shown in Box 5.6. This test statistic for GEV distribution is 0.072. Thus, the LP III distribution fits the observed data better. Nevertheless, the pdf and cdf of the GEV distribution are shown in Figures 5.35 and 5.36, respectively.

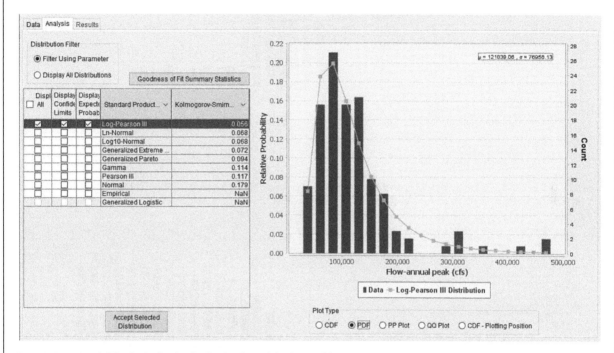

Figure 5.33 LP III probability density function fitted to the observed distribution of data given in Table 5.8 (HE-SSP analysis). (A black and white version of this figure will appear in some formats. For the color version, please refer to the plate section.)

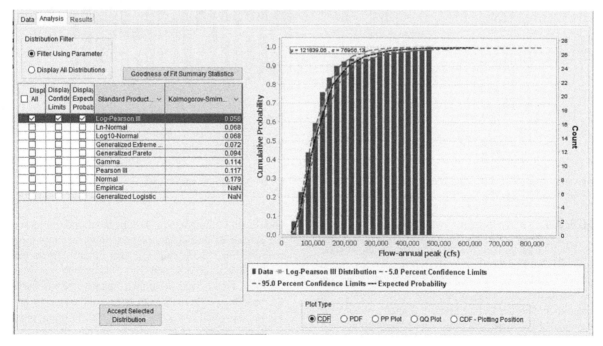

Figure 5.34 LP type III cumulative distribution function fitted to the observed cumulative distribution of data given in Table 5.8 (HE-SSP analysis). (A black and white version of this figure will appear in some formats. For the color version, please refer to the plate section.)

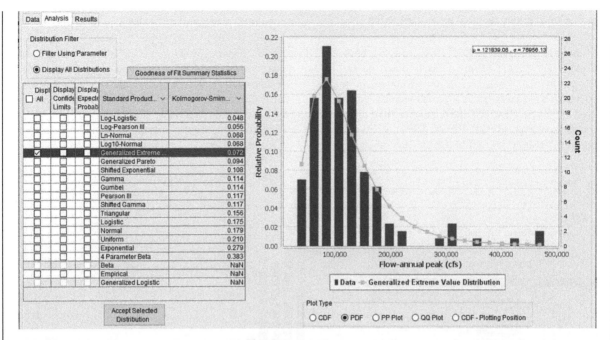

Figure 5.35 LP III probability density function fitted to the observed distribution of data given in Table 5.8 (HE-SSP analysis). (A black and white version of this figure will appear in some formats. For the color version, please refer to the plate section.)

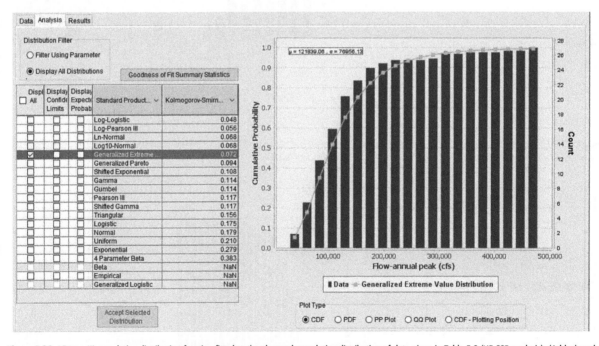

Figure 5.36 LP type III cumulative distribution function fitted to the observed cumulative distribution of data given in Table 5.8 (HE-SSP analysis). (A black and white version of this figure will appear in some formats. For the color version, please refer to the plate section.)

Step 9: Relative merits of the two platforms.

The PeakFQ program offers two immediate advantages over HEC-SSP to begin an analysis. First, it does a trend analysis and gives the value of Kendall's τ with p-value to assess the stationarity of the time series. Second, for stations within the United States it digitally obtains the regional skew coefficient based on the geographic coordinate of the gauging station. In addition, it allows the setting of thresholds for both low- and high-flow outliers. For USGS gauges it also discards flows that are regulated or affected by urbanization or other controls. Finally, the program computes the discharge for specific AEP using both the weighted skew and station skew and that gives the analyst an opportunity to assess the role of skewness in the calculations.

Box 5.6 Results of the Kolmogorov–Smirnov test using HEC-SSP to test the goodness of fit between observed distribution and LP III distribution

```
Selected Distribution: Selected Distribution: Log-Pearson III (Standard Product Moments)

|-------------------------------------------|
|                             Parameter     |
|-------------------------------------------|
|    MeanLog=5.02 ,StDvLog=0.23 ,Skew=0.33   |
|-------------------------------------------|


|-----------------------|
|  Kolmogorov-Smirnov   |
|   (Test Statistic)    |
|-----------------------|
|          0.056        |
|-----------------------|


|----------------------------------------|---------------|--------------------------------|
|   Median         Expected              |   Percent     |      Confidence Limits         |
|   Curve          Probability           |   Chance      |      0.05          0.95        |
|   Flow-annual peak, cfs                |   Exceedance  |      Flow-annual peak, cfs     |
|----------------------------------------|---------------|--------------------------------|
|     585017.9        613722.2           |    0.002      |     834549.2       414477.2    |
|     475735.5        476818.1           |    0.005      |     636687.0       356522.1    |
|     402836.1        407037.9           |    0.010      |     515551.9       315046.3    |
|     337374.5        342943.8           |    0.020      |     413094.6       275027.2    |
|     260796.7        262514.3           |    0.050      |     302823.7       223429.9    |
|     209244.8        210470.2           |    0.100      |     235046.4       185227.2    |
|     161966.2        162956.2           |    0.200      |     178153.5       146866.8    |
|     102415.9        103713.1           |    0.500      |     111024.2        94443.4    |
|      67430.0         67394.1           |    0.800      |      73127.6        62334.7    |
|      55048.4         54778.9           |    0.900      |      60202.9        50420.8    |
|      46924.5         44142.6           |    0.950      |      52279.3        42245.2    |
|      35436.7         31673.1           |    0.990      |      41858.6        30021.7    |
|----------------------------------------|---------------|--------------------------------|
```

The HEC-SSP program does not currently offer the tests noted above for specific application of FFA using the LP III distribution. Separate calculations are needed for using either the weighted skew or station skew and values of the regional skew, and its MSEs need to be obtained from outside the program and provided as inputs. HEC-SSP still calculates the expected probability, and this concept and method are described in Bulletin 17B. However, expected probability has been abandoned in Bulletin 17C. HEC-SSP offers an opportunity to test various other probability distributions and the goodness of fit, using chi-squared and Kolmogorov–Smirnov tests.

Example 5.4: S. G. Buchberger (1981) published an interesting study titled "Flood frequency analysis for regulated rivers" published in *Transportation Research Record* (vol. 832, pp. 12–21) showing how valid FFA can be conducted on regulated rivers. In this example, this study is presented in a reformatted way, and with updated calculations using Excel.

Solution:

Step 1: The annual flood series in relation to the project.

The annual peak discharge data used in the study are from Colorado River at Glenwood Springs, Colorado (Table 5.10). The flood-frequency study was conducted during the design of a segment of Interstate Highway 70 (I-70) through Glenwood Canyon by the Colorado Department of Highways. Much of the I-70 parallels and at times bridges over Colorado River.

Step 2: Test the behavior of the time-series data for its suitability for FFA.

An AMS (or PDS) is considered *well-behaved* or suitable for FFA using conventional methods if the following three conditions are satisfied:

1. The series is stationary meaning the properties of the annual flood series are time invariant
2. The series is independent, meaning the peak flows from one year are not influenced by peak flows from previous years

Table 5.10 *Peak flows of Colorado River at Glenwood Springs*

Year	Discharge (ft³/s)	Year	Discharge (ft³/s)	Year	Discharge (ft³/s)	Year	Discharge (ft³/s)
1900	20 000	1920	24 300	1940	11 100	1960	9730
1901	20 000	1921	29 000	1941	14 900	1961	7680
1902	12 000	1922	16 100	1942	16 800	1962	14 600
1903	16 500	1923	20 400	1943	13 000	1963	5470
1904	16 500	1924	24 500	1944	10 600	1964	7580
1905	22 500	1925	11 200	1945	10 600	1965	11 900
1906	22 100	1926	23 000	1946	9720	1966	4840
1907	20 400	1927	18 400	1947	14 200	1967	9200
1908	11 500	1928	27 400	1948	16 600	1968	8100
1909	27 900	1929	21 400	1949	16 300	1969	7120
1910	14 600	1930	15 500	1950	10 100	1970	13 220
1911	15 200	1931	9710	1951	14 400	1971	9970
1912	27 700	1932	17 300	1952	20 800	1972	7300
1913	12 400	1933	20 600	1953	14 000	1973	12 220
1914	28 100	1934	8140	1954	4060	1974	9620
1915	13 400	1935	21 300	1955	5400	1975	8270
1916	14 800	1936	16 900	1956	12 600	1976	4240
1917	29 400	1937	11 400	1957	18 900	1977	2340
1918	30 100	1938	20 900	1958	16 000	1978	11 180
1919	12 300	1939	13 100	1959	8480	1979	11 860

3. The series is homogeneous, that is, all the peak flows are from the same parent population

Thus, an assessment of the adequacy and applicability of flood records is the first step in determining the approach that should be taken for FFA.

Step 2A: Test of stationarity.

Instead of determining Kendall's τ, a simpler approach is taken to ascertain the stationarity of the time series. A plot of the time series, shown in Figure 5.37, indicates a linear trend line can be used to model the series. The linear regression model, $y = b_0 + b_1 t$ (t is year) through 80 data points shown by the dashed line has the equation

$$Q = -180.4t + 22037.1; t = 0, 1, 2, \ldots, 79$$

$R^2 = 0.40352$ and $r = -0.63523$

A hypothesis test is performed where the null hypothesis is that the slope of the regression (b_1) line is zero, that is, there is no trend:

$H_0 : b_1 = 0$

$H_a : b_1 \neq 0$

Test statistic : $T = r\sqrt{\left(\dfrac{n-2}{1-R^2}\right)}$

When H_0 is true, and the test statistic T is a random variable that has a t distribution with $n-2$ degrees of freedom. Note that computation of the test statistic T involves both Pearson's r (correlation coefficient) and coefficient of determination (R^2).

Let $t_c[1-(\alpha/2)]$ be the critical value of the test statistic at the α-level of significance. H_0 is accepted if

$|T| < t_c[1-(\alpha/2)]$

At a 1% level of significance for $n-2 = 78$, $t_c = 2.65$. For the value of r (-0.6352) given above and $n = 80$, $T = -7.26408$. Therefore,

$|T| > t_c$

H_0 is rejected. Thus, the AMS is not stationary. Other t-test as well as ANOVA also show that the slope of the trend line is significantly different from zero.

Step 2B: Test for independence.

A serial correlation is a measure of the degree of linear dependence among successive observations of a time series that are separated by k time units, called lag. For an AMS, k can be set as 1, for the successive year of observation. This correlation is given by the autocorrelation function

$$r_k = \dfrac{s_k}{s_0}$$

where

$$s_k = \dfrac{1}{n}\sum_{i=1}^{n-k}(x_i - \bar{x})(x_{i+k} - \bar{x}) = \dfrac{1}{n}\sum_{i=k+1}^{n}(x_i - \bar{x})(x_{i-k} - \bar{x})$$

and s_0 is the variance of the time series. For the data given in Table 5.10, $r_1 = 0.354$.

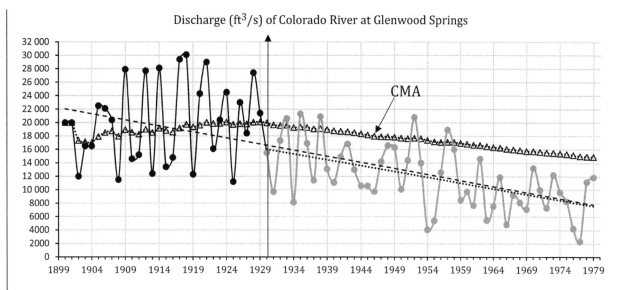

Figure 5.37 AMS and trend of flow of Colorado River at Glenwood Springs.

The autocorrelation function defined above is actually for a stationary stochastic process. The requirement for stationarity is necessary because long-term trends introduce significant positive correlation into a series. If non-stationarity is manifest as a linear trend as given above, the positive correlation expected at r_1 is also given by

$$r_1^* = \left(\frac{b_1^2}{12s_0^2}\right)[n^2 - 2n - 2]$$

For the case in hand, $r_1^* = 0.393$. Consequently, the difference $r_1 - r_1^* = -0.04$, which is insignificant. Thus, the apparent significant serial correlation results from a negative linear trend, not from dependence among successive peak flows. Thus, it can be concluded that the AMS of the Colorado River at Glenwood Springs should be considered independent.

Step 2C: Test for homogeneity.

In order to test that the data belong to the same population, the sample needs to be divided into separate groups. For this purpose, the cumulative moving average (CMA) is applied to the time series, as shown in Figure 5.37. The CMA for any given year is equal to the average value of all annual floods that have occurred from 1900 through that particular year. The CMA sequence reveals that, since 1930, there has been a progressive decrease in the mean value of the annual peak-flow time series. From this observation the dataset is divided into two groups belonging to two distinct periods:

1. Pre-regulation period: 1900 to 1929
2. Regulation period: 1930 to 1979

The **Mann–Whitney U test**, also known as the **Wilcoxon rank sum test** assesses whether two sampled groups are likely to derive from the same population, and essentially asks if these two populations have the same shape with regard to their data. In other words, we want evidence as to whether the groups are drawn from populations with different levels of a variable of interest. It follows that the hypotheses in a Mann–Whitney U test are:

I. The null hypothesis (H_0) is that the two populations are equal (there is no difference).
II. The alternative hypothesis (H_a) is that the two populations are not equal.

The first step is to assign ranks to the discharge values by ordering the data from the smallest to the largest. This is done on the combined or total sample of the two groups ($n = 80$), and assigning ranks from 1 to 80. Then, the sums of ranks in the two groups are calculated separately and designated by R_1 and R_2, respectively. Then, the test U-statistic is calculated for the two groups as:

$$U_1 = n_1 n_2 + \frac{n_1(n_1 + 1)}{2} - R_1$$

$$U_2 = n_1 n_2 + \frac{n_2(n_2 + 1)}{2} - R_2$$

In the case of the Glenwood Springs AMS, the pre-regulation period has $n_1 = 30$ and regulation period has $n_2 = 50$. The calculated values of R_1 and R_2 are 1744 and 1491, respectively. Thus, we obtain

$U_1 = 221$

$U_2 = 1284$

For the Mann–Whitney U test, the theoretical range of U is from 0 to $(n_1 \times n_2) = U_1 + U_2$. If $U = 0$, H_0 is most likely false, meaning that there is complete separation between the groups and, if $U = n_1 n_2$, there is little evidence in support of H_a. The procedure for determining exactly when to reject H_0 is as follows.

A critical value of U is determined with which to compare the smaller of the two calculated test statistics, U_1 and U_2. This can be done using a reference table of critical values of U given in many statistical texts and using the sample sizes and two-sided level of significance ($\alpha = 0.05$). However, such a table can be only used if the value of n in both groups is less than 20 (small samples). For large samples (one or both groups with $n > 20$), the value of U approaches a normal distribution and therefore the null hypothesis can be tested by a z-test.

a. Compute the standard deviation of U:

$$SD_U = \sqrt{\frac{(n_1 \times n_2)(n_1 + n_2 + 1)}{12}} = 100.62$$

b. Compute Z:

$$Z = \frac{U - \{(n_1 n_2)/2\}}{SD_U}$$

c. Compare the obtained Z value with the critical Z value to determine whether to retain or reject the null hypothesis.
   (i) If $|Z| < 1.96$, H_0 is retained
   (ii) If $|Z| > 1.96$, H_0 is rejected

In our case $|Z| = 5.26$ if U_1 is used and $|Z| = 5.31$ if U_2 is used. In either case, $|Z| > 1.96$. So H_0 is rejected.

In addition to the three conditions tested above it is assumed that the series is reliable, which implies that the flood record is free of substantial errors caused by measuring, transmitting, recording, and processing data. There is no statistical means to assess this. However, the dataset is obtained from the USGS and because of their quality-control protocol, the data are considered reliable.

Thus, the entire AMS of Colorado River at Glenwood Springs is neither stationary nor homogeneous. Therefore, the regulation period dataset should be analyzed separately for FFA.

Step 3: Investigate the regulations in the watershed.

Figure 5.38 shows the watershed drained by the Colorado River up to the gauging station at Glenwood Springs. The watershed is on the west side of the continental divide. Annual floods mostly result from snowmelt run-off during the spring.

Three areas of the watershed are shown as regulated areas. There are several major reservoirs in these areas and water from them is diverted to the east side of the

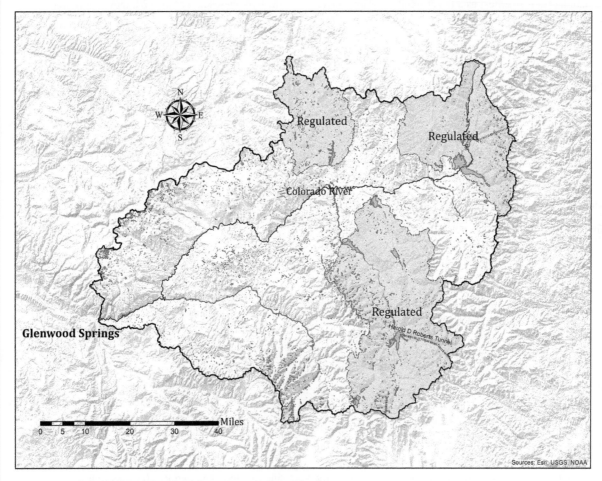

Figure 5.38 The watershed of the Colorado River up to Glenwood Springs, Colorado.

continental divide where the major population centers, such as the City of Denver and surrounding cities are located. Water there is supplied through tunnels.

Step 4: Reconstruct the flood series for the regulation period.

From the time-series plot shown in Figure 5.37, it can be said that an annual flood discharge, q, has two components, namely a deterministic component, d, which can be predicted from the linear trend, and a random component, ξ, which is the stochastic residual. So, for a year i, we can write

$$q_i = d_i + \xi_i$$

A linear regression of the data for the regulation period, shown by the dotted straight line in Figure 5.37, gives d_i as:

$$d_i = -172.23t + 16\,017; t = 0, 1, 2, \ldots, 49$$

The corresponding ξ_i is obtained by subtracting the d_i calculated using the regression model from the observed q_i, and the results are given in Table 5.11.

The mean and standard deviation of the stochastic residuals as well as those of the q_i are calculated, as given in Table 5.12.

The mean of the random component is zero but comparison of the standard deviations shows that the removal of the deterministic component causes a substantial reduction in the SD values because a measurable portion of the total variance of the regulation series is contributed by the deterministic trend.

The residual components have both positive and negative values, as given in Table 5.11 and as such cannot be interpreted as annual peak flows. Engineering judgment is applied to determine a meaningful flood series for the regulation period. The mean of the regulation period is 11 797 ft$_3$/s, as given in Table 5.12. Several other considerations led to establishing 11 500 ft$_3$/s as the mean annual flood (designated as *MAF*) for the regulation period. This value is added to the stochastic residuals to obtain the annual peak flow, $Q_i = MAF + \xi_i$.

The results are given in Table 5.13 and the resulting time-series plot is shown in Figure 5.39.

Further statistical tests show that the reconstructed AMS (Table 5.13) is stationary and homogeneous with an absence of serial correlation. Therefore, this reconstructed AMS can now be used for FFA using a conventional method.

Step 5: FFA of the regulation period reconstructed AMS.

Step 5A: Determine an appropriate probability distribution.

The mean, standard deviation, and skewness of the reconstructed AMS are:

$$\bar{x} = 1150; s_x = 3958.63; \gamma_x = 0.06$$

Because of the very small skew coefficient, the decision was to choose between the normal probability distribution and the two-parameter gamma distribution. Both fitted the data well but the normal probability

Table 5.11 *The deterministic component and stochastic residual of the flood discharge during the regulation period 1930 ($t = 0$) to 1979 ($t = 49$)*

t (years)	d_i (ft^3/s)	ξ_I (ft^3/s)	t (years)	d_i (ft^3/s)	ξ_I (ft^3/s)	t (years)	d_i (ft^3/s)	ξ_I (ft^3/s)
0	16 016.73	−516.729	17	13 088.75	1111.246	34	10 160.78	−2580.78
1	15 844.5	−6134.5	18	12 916.52	3683.48	35	9988.545	1911.455
2	15 672.26	1627.738	19	12 744.29	3555.714	36	9816.311	−4976.31
3	15 500.03	5099.972	20	12 572.05	−2472.05	37	9644.077	−444.077
4	15 327.79	−7187.79	21	12 399.82	2000.182	38	9471.843	−1371.84
5	15 155.56	6144.44	22	12 227.58	8572.415	39	9299.609	−2179.61
6	14 983.33	1916.674	23	12 055.35	1944.649	40	9127.375	4092.625
7	14 811.09	−3411.09	24	11 883.12	−7823.12	41	8955.141	1014.859
8	14 638.86	6261.141	25	11 710.88	−6310.88	42	8782.908	−1482.91
9	14 466.62	−1366.62	26	11 538.65	1061.351	43	8610.674	3609.326
10	14 294.39	−3194.39	27	11 366.42	7533.585	44	8438.44	1181.56
11	14 122.16	777.843	28	11 194.18	4805.818	45	8266.206	3.793998
12	13 949.92	2850.077	29	11 021.95	−2541.95	46	8093.972	−3853.97
13	13 777.69	−777.689	30	10 849.71	−1119.71	47	7921.738	−5581.74
14	13 605.46	−3005.46	31	10 677.48	−2997.48	48	7749.504	3430.496
15	13 433.22	−2833.22	32	10 505.25	4094.754	49	7577.271	4282.729
16	13 260.99	−3540.99	33	10 333.01	−4863.01			

Table 5.12 *Statistics of the random component and the observed discharge during the regulation period*

Mean (ξ_i)	0
SD_{qi} (regulation period)	4687.69
$SD_{\xi i}$ (stochastic residual)	3958.63
Mean (q_i)	11 797
Mean annual flood	11 500

Table 5.13 *AMS for the regulation period*

t (years)	Q_i (ft³/s)	T (years)	Q_i (ft³/s)
0	10 983.27	25	5189.117
1	5365.504	26	12 561.35
2	13 127.74	27	19 033.58
3	16 599.97	28	16 305.82
4	4312.206	29	8958.052
5	17 644.44	30	10 380.29
6	13 416.67	31	8502.52
7	8088.908	32	15 594.75
8	17 761.14	33	6636.988
9	10 133.38	34	8919.222
10	8305.609	35	13 411.46
11	12 277.84	36	6523.689
12	14 350.08	37	11 055.92
13	10 722.31	38	10 128.16
14	8494.545	39	9320.391
15	8666.778	40	15 592.62
16	7959.012	41	12 514.86
17	12 611.25	42	10 017.09
18	15 183.48	43	15 109.33
19	15 055.71	44	12 681.56
20	9027.948	45	11 503.79
21	13 500.18	46	7646.028
22	20 072.42	47	5918.262
23	13 444.65	48	14 930.5
24	3676.883	49	15 782.73

distribution was chosen since its frequency factor is given by the standard normal variate.

Figure 5.40 shows the normal probability curve and frequency histogram.

For normal distribution, the frequency factor is obtained from tables. The discharge for a return period T is thus obtained from

$$Q_T = \bar{x} + s_x K_T$$

The three design discharge values obtained from the frequency factors are given in Table 5.14.

These are the flood flows of the Colorado River needed for the design of the I-70 through Glenwood Canyon. The previous estimate of Q_{100} was 26 500 ft³/s. Although inflated estimates may be condoned or provide an extra margin of safety, overconservativeness was not warranted since the economy of the I-70 project was linked inextricably to the magnitude of the design discharge.

Step 6: Confirm the goodness of fit of the probability model.

A chi-squared test provided a slightly better fit than the two-parameter gamma distribution. The Kolmogorov–Smirnov test statistic is 0.07384 with a degree of freedom of 50 and p-value of 1. This implies that the data were significantly drawn from a normally distributed population.

Table 5.14 *Discharge for three selected AEP values*

AEP	ARI (years)	K_T	Q_T (ft³/s)
0.5	2	0	11 500
0.02	50	2.05	19 615.19
0.01	100	2.32	20 684.02

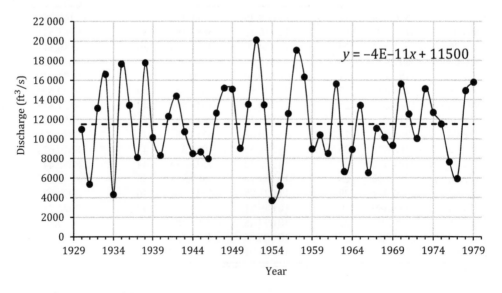

Figure 5.39 Reconstructed AMS during regulation.

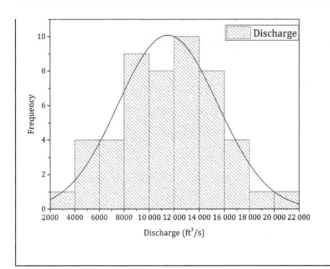

Figure 5.40 Normal probability curve and frequency histogram.

Exercises: A selection of exercises on this topic is available at www.cambridge.org/appliedhydrology.

6 Geomorphologic Concepts and Watershed Characteristics

6.1 CONCEPTS AND PURPOSE

Geomorphologic characteristics of a drainage basin govern its hydrologic response to rainfall, such as the shape and size of the direct runoff hydrograph and the timing of peak discharge. Therefore, these characteristics should be quantified during hydrologic investigations. The geomorphologic characteristics of drainage basins and streams, which are important for hydrologic modeling and design, include certain morphometric parameters and geomorphologic features. The objective of this chapter is to present the geomorphologic concepts related to hydrologic parameters and features and discuss how the parameters are quantified. Geomorphologic investigations are often required for river engineering works or other scientific purposes. The time parameters for the movement of water derived from the morphometric parameters of the watershed are also presented here. These time parameters are of fundamental importance in engineering hydrology.

6.2 HIERARCHICAL STRUCTURE OF A DRAINAGE BASIN

The land surface of the Earth possesses drainage (stream) systems whose spatial patterns are derived from the land surface topography. This can be described by a network of drainage lines indicating the principal pathways of water movement along the lowest elevations of the valleys. A drainage basin or drainage area refers to any portion of the land surface with a physical boundary, defined by the topography, that diverts all runoff to a common point having the lowest elevation within that defined boundary, called the **outlet** of the basin. Thus, it is a drainage unit defined by a chosen outlet point, also called a **pour point** or **sink**, in a drainage system.

In hydrologic literature, the terms **drainage area**, **catchment**, **watershed**, and **basin** have been used somewhat synonymously. In the United States, watershed is the standard term for a drainage unit, whereas in Europe, Australia, and in certain Asian countries, the standard term is catchment. Hence, it is necessary to establish a convention following a logic for the usage of this nomenclature. In this book, a convention is adopted and recommended to distinguish these terms by the scale of the drainage units represented by these terms. This convention is used because a common current practice which will continue in the future is to delineate drainage units with the aid of GIS using digital elevation data, also known as digital elevation models (DEMs), that are now widely available at various spatial resolutions and accuracy. Traditionally, drainage areas are delineated from topographic maps by connecting ridgelines in the topography to points on the drainage network. However, if GIS is used to delineate drainage areas, a rule has to be established to define the spatial scale of the smallest drainage unit, called the contributing area, that is used to define a stream. Because of this, the following convention is adopted from the Arc Hydro data model that describes the drainage units and drainage lines of a land surface topography in a systematic hierarchical structure (Maidment, 2002).

A *catchment* represents any landscape subdivision into drainage areas that are defined by a consistent set of rules applied consistently through the landscape. Such rules include the threshold contributing area method used in the ArcGIS software or a system where a drainage area is defined around each labeled reach in a map of a stream network. These rules can result in a very large number of small drainage areas, each connected to the unit downstream. Initially, catchments are delineated based on such a set of rules that can be interpreted by the programming code in GIS. The usual rule is to establish a minimum threshold value of the area that will define a drainage line, that is, a stream. This also produces a very large number of catchments even within a relatively small drainage unit. In most practical applications, such a large number of finer subdivisions of individual drainage units may not be necessary. The size of the minimum contributing area in a DEM analysis can vary, depending on the scale and objective of analysis. This offers the flexibility of not having a rigid upper limit of spatial extent for a drainage unit to be designated as a catchment and the choice can vary with the scope of the investigation. Therefore, certain other rules, based on geomorphologic characteristics or administrative preference or requirements can be applied to integrate those catchments to a manageable number of units. For example, a stream ordering system, which is discussed in Section 6.5, can be used to define catchments. In this case, a rule can be established to integrate all smaller catchments initially delineated, based on a threshold value of contributing area into a larger catchment that drains to a stream of a certain order. Similarly, for certain design evaluations, a rule can be established to set the lower and upper limits of the areas of the catchments in a study area.

Watersheds represent any arbitrarily defined subdivision of the land surface into drainage areas. Watersheds may drain to a point on the stream network, to a segment of a stream or river, or to a water body. There are an infinite number of ways watersheds can be defined for any land surface, and the method chosen in a particular instance entirely depends on the scope of the study as opposed to using a defined set of rules used for the delineation of catchments. Nonetheless, a convention can be adopted whereby once the catchments are delineated, those are used to define a watershed.

Basins are a set of watersheds that are selected administratively to be the principal description of drainage areas over a particular region. Basins represent the units that organizations have defined historically to partition the land surface in question into well-known and well-understood drainage areas, usually associated with the principal rivers and streams of the region.

In summary, catchments are considered to be the drainage units with the smallest spatial scale. Several catchments constitute a watershed, and several watersheds make up a basin. Thus, a basin is the entire area of the land surface drained by a major river system consisting of an extensive channel network. This convention allows a convenient description of the physical system under consideration without ambiguity. Certain examples follow for exemplification of this concept.

The USGS divided the landmass of the United States into 22 broad hydrologic units, each having a two-digit hydrologic unit code (HUC). Each of these units is further subdivided into subregions, accounting units, cataloging units, watersheds, and sub-watersheds in a hierarchical system of coding where each subdivision is designated by two digits. Thus, a two-digit HUC is further subdivided into units having four-, six-, eight-, ten-, and twelve-digit HUCs. In this hierarchical system of nesting drainage units at various spatial scales, each with an assigned HUC, a drainage unit with a six-digit HUC is generally a major river basin and the drainage units with an eight-digit HUC within such a six-digit HUC are the major watersheds of the river basin, commonly used for hydrologic data archiving and delivery.

Figure 6.1 shows the two-digit HUCs of the contiguous United States. The Platte River basin system is used here to illustrate the six- and eight-digit HUCs and a line diagram of the system is shown in Figure 6.2. Figure 6.3 shows the South Platte River basin, which has a six-digit HUC, 101900. There are 18 major watersheds within the South Platte River basin, each with an eight-digit HUC, also shown in Figure 6.3. Figure 6.4 shows catchments, each with a twleve-digit HUC, within the Middle South Platte River–Cherry Creek watershed (HUC 1019003 in Figure 6.3).

In this book, the term watershed is used to refer to a drainage area in relation to various hydrologic methods that are described in different chapters. All of the methods and principles that refer to a watershed are equally applicable to its constituent catchments. In HEC-HMS, the largest drainage unit designated as *basin* is subdivided into a number of sub-units, which are called *subbasins*. Therefore, in accordance with the convention presented here, a subbasin in an HEC-HMS model can be a catchment when the larger drainage unit is a watershed, or it can be a river basin if the larger unit is a basin.

6.3 MORPHOMETRIC PARAMETERS OF A WATERSHED

6.3.1 Watershed Area

On a topographic map, the drainage divide is traced beginning with the chosen drainage outlet on the stream, which ultimately drains the entire area of the watershed, and following the divide around the watershed and back to the outlet point. The area enclosed by this boundary line is the watershed area. The outlet point is selected based on the purpose of investigation. It can be a stream gauging station, a confluence point of two streams, or even can be the inlet or upstream point of a culvert or bridge crossing along a highway. Thus, a watershed boundary intersects a stream only at the outlet point and the watershed area depends on the selection or definition of this outlet point.

It should be also noted at this point that some drainage basins contain, within their boundaries, areas that do not contribute runoff to the drainage system. These areas are isolated from the drainage system and comprise closed drainage areas that might form lakes or swamps, or simply areas where drainage infiltrates into soil or becomes groundwater. The playa lakes are good examples of the latter case. Sometimes such areas are called **interior drainage areas** and are excluded from the total drainage area to give the effective or contributing drainage area. However, certain river basins that do not ultimately drain to the ocean are also called **interior drainage basins**. The Tarim basin in central Asia is an example.

The watershed area is the most fundamental morphometric parameter required for hydrologic computations and modeling. It is highly correlated with several other hydrologic parameters. The amount of runoff generated from a rainfall event over a watershed area is directly proportional to the watershed area. This should be intuitively obvious. What is not so obvious is that this proportionality generally follows a power function given as

$$Q = dA^k \tag{6.1}$$

where Q denotes a measure of watershed discharge, A is the watershed area, and d and k are parameters. Leopold and Miller (1956) and Hack (1957) were perhaps the earliest workers who made observations to deduce such a relationship. Subsequently, this area–discharge relationship gained a central role in regional flood frequency as discussed in Chapter 5.

It is important to understand the type of discharge that is represented by Q in Eq. 6.1. In the literature, it has been variously called the mean annual flow or maximum flood. In relation to a diagram showing log–log plot of annual

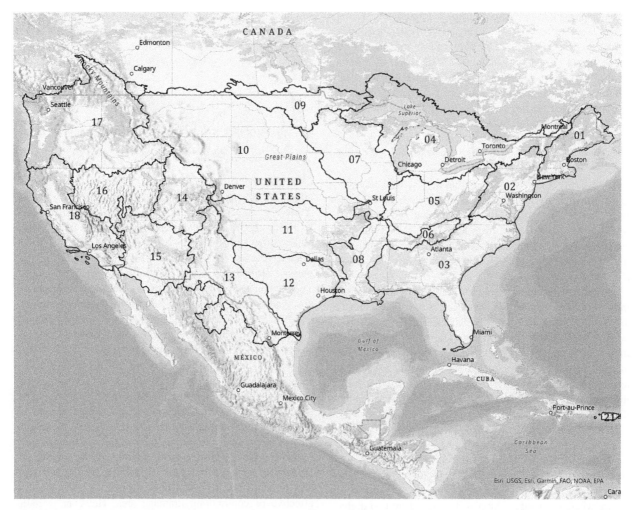

Figure 6.1 The 18 hydrologic units of the contiguous United States. Each hydrologic unit has a two-digit hydrologic unit code (HUC). Shapefile provided by USGS (courtesy: Kimberley A. Jones, Supervisory GIS Specialist, Utah Water Science Center).

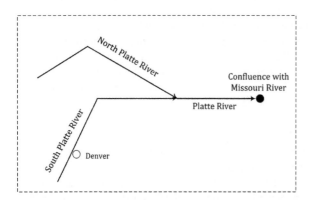

Figure 6.2 Line diagram showing the major rivers of the Platte River system.

discharge against drainage area, Hack (1957) stated that he plotted "average discharge determined at all the gauging stations in the Potomac River basin" above Washington, DC. From this plot he noted, "enlargement of drainage area is accompanied by proportional increase in discharge." Unfortunately, Hack did not explicitly explain what was meant by "annual discharge" except that he mentioned that he used the data from USGS Water Supply Paper 1111, which is no longer available from the USGS. However, several other water supply papers of the USGS published during that time are still available and show that these reports listed monthly and annual statistics of streamflow measurements at existing gauging stations of major river basins of the United States for water supply planning. In Chapter 5 we discussed these statistics and the use of plots of long-term average monthly average flows against drainage areas for water resources planning. So, it appears that Hack most likely used long-term average annual discharge at respective gauging stations and the corresponding area.

Leopold and Miller (1956) made several observations related to hydraulic (width, depth, and velocity of flow) and physiographic factors of several ephemeral streams in New Mexico during flood flows to conclude, "a whole series

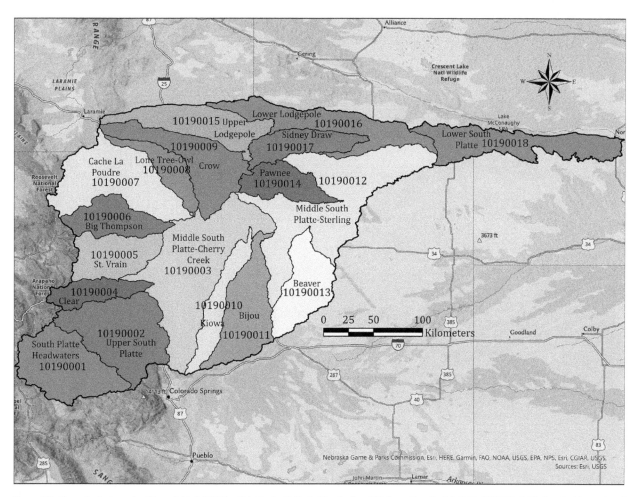

Figure 6.3 The watersheds, each with an eight-digit HUC, within the South Platte River basin, which has with a six-digit HUC of 101900. (A black and white version of this figure will appear in some formats. For the color version, please refer to the plate section.)

of hydraulic and drainage-network factors are interrelated in the form of power or exponential functions." Subsequently, Leopold et al. (1992) presented an equation:

$$Q_b \propto A_d^{0.75} \tag{6.2}$$

in which Q_b represents the "flood or bankfull discharge," and A_d is the drainage area. They discussed the variability of the exponent in Eq. 6.2 in different climatic regions and recommended 0.75 as "a general average of wide applicability." Nevertheless, they made a significant statement in relation to this type of power function relating drainage areas and other hydrologic parameters. They stated, "The values of these exponents need not be considered as wholly unexplained empiricism." In other words, they implied that these exponents could have certain physical basis.

A **bankfull discharge** is defined as the discharge that a channel can convey when reaching the floodplain level. Thus, the surface at the top of the bankfull channel is the floodplain, which is inundated whenever the river or stream experiences a **bankfull flow**. Return periods of 1.5 and 2 years are most often cited in the literature as the recurrence frequency of bankfull discharge; however, it is quite variable.

Goodrich et al. (1997) plotted 2-year and 100-year floods against drainage areas using annual peak flow data from 18 nested subbasins of Walnut Gulch basin, located in semiarid Arizona, and showed linearity on a log–log plot to present the power law function as

$$Q_p = dA^k \tag{6.3}$$

in which Q_p is the basin peak discharge associated with a specified return period.

The power function given in Eq. 6.3 has also been found to be applicable to individual event-based rainfall–runoff analysis where the peak discharge resulting from a single rainfall event is plotted against the drainage area (Furey and Gupta, 2007).

From the discussion presented above, Eq. 6.3 is called the **peak-discharge power law**. Examples of applicability of this relationship are given below for both types of analysis.

Ayalew et al. (2015) and Chen et al. (2020) used rainfall–runoff data from 52 individual rainfall events occurring between 2002 and 2013 within Iowa River basin (32 400 km^2) of the mid-western United States. The peak discharge values observed at 35 gauging stations are plotted

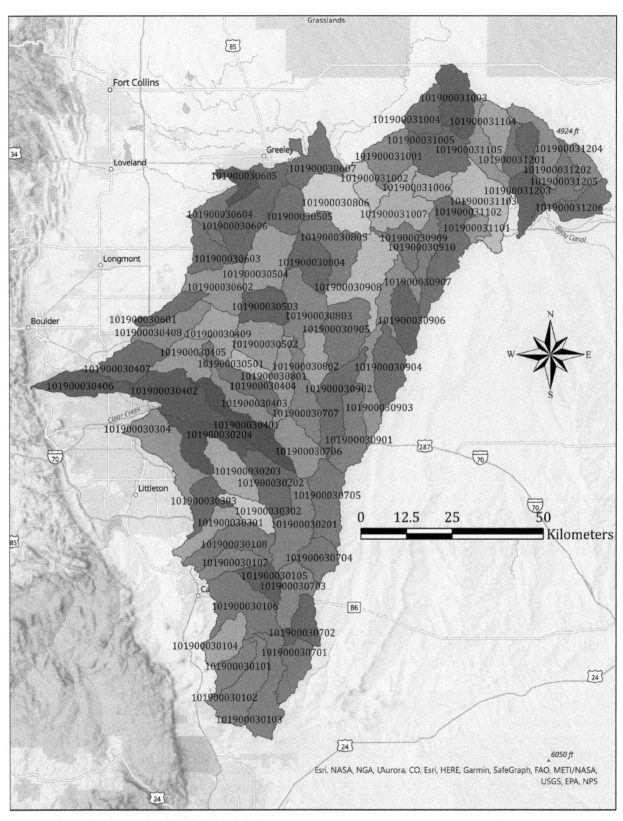

Figure 6.4 The catchments, each with a 12-digit HUC, within the Middle South Platte River–Cherry Creek Watershed (eight-digit HUC 1019003) within the South Platte River basin (six-digit HUC) shown in Figure 6.3. (A black and white version of this figure will appear in some formats. For the color version, please refer to the plate section.)

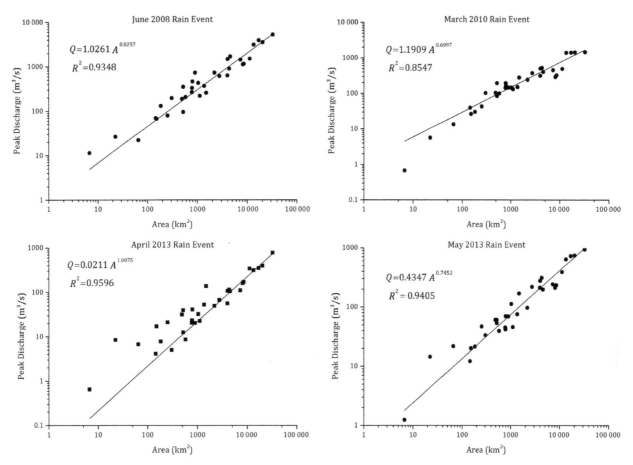

Figure 6.5 Observed peak-discharge power law or spatial scaling of peak discharges with drainage area for four rainfall–runoff events in the Iowa River basin (data courtesy of Dr. Bo Chen, China).

against the drainage areas corresponding to the gauging station locations, as shown in Figure 6.5. Note that both the coefficient and exponent change from event to event within the same river basin.

Annual peak discharge values from 24 USGS gauging stations within the headwater area (10 422 km^2) of Arkansas River basin, located upstream of Pueblo, Colorado, are analyzed below to derive 2-year and 100-year flood flows following the method of flood-frequency analysis discussed in Chapter 5. These gauging stations are located upstream of any reservoir regulation, controlling river flows, and have relatively long periods of record. Figure 6.6 shows the relationships derived from this analysis. In this case, the exponents of the power functions representing peak discharge of two different recurrence intervals as a function of drainage area are quite similar.

The drainage area is an important scale variable for the power law discussed above which is also called a scaling relationship. The coefficient d in Eq. 6.3 is the **scaling intercept** and exponent k is the **scaling slope** or **scaling exponent**.

One interesting attribute of power laws is their **scale invariance**. Given a relation $f(x) = \alpha x^\beta$, scaling the argument, x, by a constant factor, ζ, causes only a proportionate scaling of the function itself. That is, $f(\zeta x) = \alpha(\zeta x)^\beta = \zeta^\beta(\alpha x^\beta) = \zeta^\beta f(x) \propto f(x)$. Therefore, scaling by a constant ζ simply multiplies the original power function by the constant ζ^β. Thus, it follows that all power laws with a particular scaling exponent are equivalent up to constant factors since each is simply a scaled version of the others. This behavior is what produces the linear relationship when logarithms are taken of both $f(x)$ and x, and the straight line on the log–log plot is often called the **signature** of a power law. With real data, such straightness is a necessary, but not sufficient, condition for the data following a power-law relation. Accurately fitting and validating power-law models to an observed set of data is an active area of research in hydrology (e.g., Chen et al., 2020).

The physical interpretation of the mathematical description of scale invariance is that the relationship holds true irrespective of the scale of the argument. Most watershed characteristics are scale invariant, that is, the spatial scale has no effect on these relationships - they would apply to small, medium, and large watersheds.

Many attempts have been made to correlate the scaling parameters with rainfall duration and depth for event-based data and flood magnitude for flood-frequency studies as well as other physical characteristics of watersheds. Even though

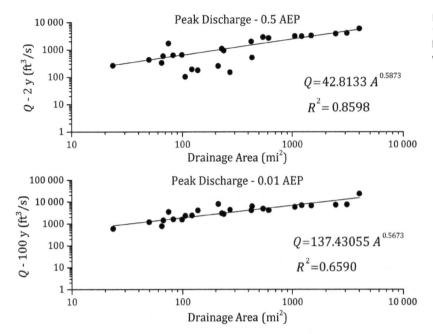

Figure 6.6 Peak discharge rates for 2-year and 100-year return periods versus drainage area in the headwater (upstream of any reservoir regulation) watersheds of Arkansas River basin within Colorado.

this has been mostly investigated within theoretical frameworks (e.g., Gupta, 2004), one impetus behind such investigations is that if the scaling parameters can be derived empirically from measurable variables such as rainfall depth and duration, a very simple relationship can be used to predict the peak discharge, which in reality is governed by a multitude of physical factors that cannot be directly measured. Precisely for this reason, we will see in subsequent chapters that hydrologic models require estimates of many parameters which by no means is an easy task.

6.3.2 Watershed Centroid

The centroid of a watershed is the center of gravity or the first moment of the watershed area. The coordinates of the centroid of a watershed are defined by the method of moments where a square grid is overlain the watershed boundary. Each grid has an equal area. Because centroid is the first moment of the area, its coordinates are given in the x-y plane as

$$\bar{x} = \frac{1}{A} \sum_{i=1}^{N} x_i a_i \tag{6.4}$$

and

$$\bar{y} = \frac{1}{A} \sum_{i=1}^{M} y_i a_i \tag{6.5}$$

where x_i is the distance between the y-axis and the center of the ith grid square, a_i is the area of the ith grid square, y_i is the distance between the x-axis and the center of the ith grid square, N is the total number of grid squares in the x-direction, and M is the total number of grid squares in the y-direction.

6.3.3 Watershed Length or Longest Flow Path

A geomorphologic definition of watershed length was given by Schumm (1956) as the longest dimension of a watershed parallel to its principal drainage channel. However, in hydrology it is slightly modified to define the flow path that extends from the watershed outlet to the hydraulically most remote point upstream. This dimension, called the **longest flow path** and designated by L, is very important in calculating the watershed time of concentration (t_c), which is discussed later in this chapter. L a key parameter in the NRCS unit hydrograph method and is a factor in determining the value of $(t_c + R)$, as it affects the storage coefficient, R, and t_c, and both are used in Clark unit hydrograph method. These are discussed in Chapter 9. In general, L measures the distance a particle of water travels from the upstream boundary to the outlet of the watershed. It is determined from topographic maps aided by areal photographs. Nowadays, GIS tools are commonly used to delineate this line from a DEM. However, sometimes manual adjustments applying engineering judgment may be required as well.

6.3.4 Centroidal Flow Path and 10–85 Flow Path

The **centroidal flow path**, L_c, is a segment of the longest flow path. It begins at the watershed outlet and extends upstream along the longest flow path up to the point on the longest flow path that is nearest to the watershed centroid. This point is

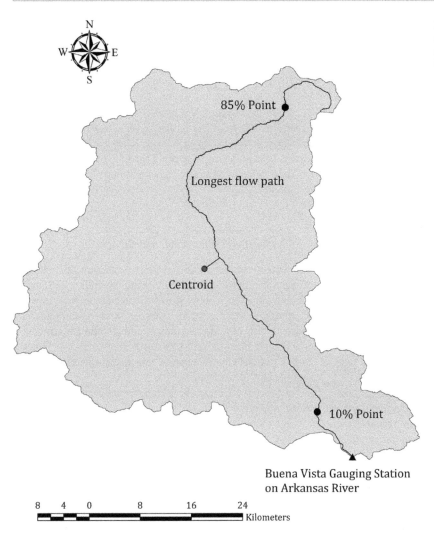

Figure 6.7 Centroidal flow path, and 10–85 points of the headwater watershed of Arkansas River basin up to the USGS gauging station located at Buena Vista in Colorado.

determined by dropping a perpendicular from the watershed centroid to the longest flow path. Both L and L_c are used in Snyder's unit hydrograph method, discussed in Chapter 9.

The 10–85 flow path is also a segment of the longest flow path. Measuring from the outlet in the upstream direction, the 10–85 flow path, $L_{10\text{-}85}$, begins at a point representing 10% of the total length of the longest flow path and ends at a point representing 85% of the total length of the longest flow path; both counted from the outlet, that is, excluding the lower 10% and upper 15% of L. These two points are also used to calculate the 10–85 slope of the flow path (slopes are discussed below).

Figure 6.7 shows the locations of centroid and 10–85 points along the longest flow path of a headwater watershed of Arkansas River basin in Colorado.

6.3.5 Watershed Shape

The watershed shape has an influence on the hydrograph shape, especially for small watersheds. For example, if a watershed is narrow and elongated along its main watercourse, water will take a long time to travel from the watershed extremities to the outlet and the resulting runoff hydrograph will be flatter, having a broad peak and long period of total runoff duration. On the other hand, if a watershed is oval in shape, where the outlet is nearly equidistant from the watershed extremities, the resulting hydrograph will have sharp peak and short time base.

Several quantitative measures have been used in hydrologic literature to describe watershed shape. Two of the commonly mentioned of these are the **shape factor** and **elongation ratio**.

The shape factor, B, is simply determined from watershed length, L, and area, A, as

$$B = \frac{L^2}{A} \tag{6.6}$$

The elongation ratio, E, is a dimensionless ratio used to categorize the general shape of a watershed (Schumm, 1956). It is a ratio of the diameter of a circle with the same area as the watershed and the watershed length. Thus,

$$E = \frac{\left(\frac{2}{\sqrt{\pi}}\right)A^{0.5}}{L} = \frac{1.1284 A^{0.5}}{L} \quad (6.7)$$

This ratio typically ranges from 0.2 to 1.0, with lower values representing elongated watersheds and values closer to 1 representing circular or oval-shaped watersheds.

The shape factor and elongation ratio generally show an inverse relationship as shown in Figure 6.8B for nine headwater watersheds of Arkansas River basin in Colorado (Figure 6.8A).

6.3.6 Relationship between Watershed Length and Watershed Area

Hack (1957) demonstrated the applicability of a power function relating length and area for streams of the Shenandoah Valley and adjacent mountains in Virginia. He observed the equation

$$L = 1.4 A^{0.6} \quad (6.8)$$

where L is the length of the longest stream in miles from the outlet to the divide and A is the corresponding area in square miles. Hack also corroborated his equation through the measurements of Langbein (1947), who had measured L and A for nearly 400 sites in the northeastern United States. Gray (1961) later refined the analysis, finding a relationship $L \propto A^{0.568}$. Many other researchers have validated Hack's original study, and although the exponent in the power law may vary slightly from region to region, it is generally accepted to be slightly below 0.6. Equation 6.8 rewritten as $L \propto A^h$ with $h > 0.5$ is usually termed **Hack's law**. Rigon et al. (1996) explored the significance of Hack's law in the internal structure of river basins. The coefficient in Eq. 6.8 becomes 1.312 if L is measured in km and A in km² (Singh, 1992).

The relationship between A and L of the nine watersheds of the same river basin, moving progressively downstream (Figure 6.8A), is shown in Figure 6.8C. In this example, the exponent of Hack's law is 0.552. Values of A and L for 21 watersheds from different parts of the United States, measured from DEMs, are given by Rigon et al. (1996). Plots of these data, shown in Figure 6.9, show a value of 0.615 for Hack's exponent.

Hack (1957) noted, "it is fairly well demonstrated that in the northeastern United States the length of a stream at any locality is, on average, proportional to the 0.6 power of its drainage area at the locality." Bras (1990) points out that the fact that watershed length is not proportional to the square root of the watershed area, implies that river basins are not fully geometrically similar, that is, the ratio of the area to the square of length is not constant for all basins. He also notes that the equation that would represent geometrically similar watersheds is

$$L = 1.73 A^{0.5} \quad (6.9)$$

or

$$\frac{A}{L^2} = \frac{1}{3} \quad (6.10)$$

There are two possible explanations for the exponent h not being equal to 0.5. One is that larger watersheds are more elongated than smaller watersheds since the value of A/L^2 falls as area increases. The other possible explanation could be that large watersheds are measured with a coarser resolution.

6.3.7 Watershed Slope, Relief, and Aspect

The **watershed slope** has a profound effect on the velocity of overland flow, watershed erosion potential, and local wind system. When coupled with slope orientation or aspect, it influences the receipt of solar or short-wave radiation, which in turn has a great control on microclimate, distribution of precipitation, and snowmelt, particularly, in mountainous watersheds where snow and glacial melts contribute significantly to streamflow.

The watershed slope usually represents the average slope of the watershed given as

$$S = \frac{H_v}{H_h} \quad (6.11)$$

where H_v is the vertical fall in feet or meters and H_h is the horizontal distance over which the fall occurs. When H_v in Eq. 6.11 represents the elevation difference between the watershed outlet and the highest point in the watershed boundary or perimeter, that is, the drainage divide, S is then the maximum watershed **relief** or **relief ratio**, R_h. Both slope and relief can be expressed either in degrees or as a percentage.

Since elevations across a watershed vary greatly, measurement of the average watershed slope can be made by a variety of procedures. Horton (1932) presented two methods for the estimation of the average watershed slope. He called these methods contour length and contour area methods. Singh (1992) has given a modified version of Horton's method.

At present, DEMs are used with GIS to estimate the average watershed slope. A DEM is actually a square grid representing the horizontal surface of the earth in east–west (x-) and north–south (y-) directions. The grid has equally spaced cells (Δx and Δy). An elevation value (vertical or z-direction) is assigned to each cell to represent the elevation of the ground at that point. These are called elevation raster values. For each elevation raster value within a watershed, the elevations of the surrounding eight cells are used to calculate the slope in eight directions. The maximum of these eight slopes is taken as the slope of the cell being evaluated. This process is performed in each of the cells of the entire grid that covers the watershed. The average of these values is considered as the average watershed slope.

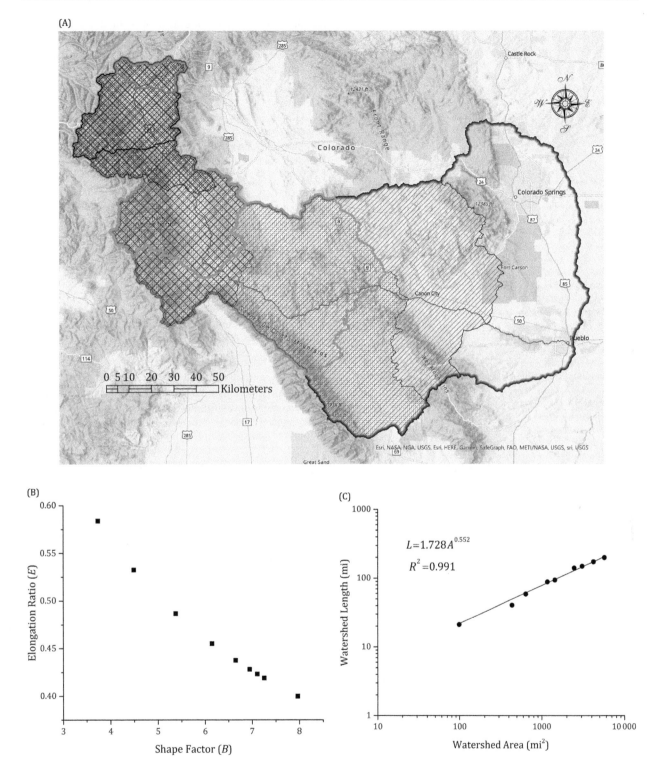

Figure 6.8 (A) Nine watersheds, each delineated from a USGS gauging station located on the headwater section of Arkansas River from its point of origin in the eastern slope of the southern Rocky Mountains up to a point in Pueblo, Colorado. (A black and white version of this figure will appear in some formats. For the color version, please refer to the plate section.) (B) Watershed shape factor and elongation ratio for each of the nine watersheds. (C) Watershed area and longest flow path in each of the watersheds to illustrate the exponent of Hack's law.

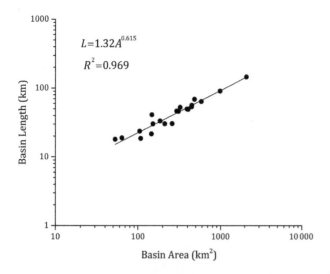

Figure 6.9 Relationship between watershed length and area for 21 different watersheds in West Virginia, Kentucky, Pennsylvania, Idaho, New York, and Alabama. Data are from Rigon et al. (1996).

Another method, frequently used in the United States, uses contours to estimate the average slope (as a percentage) of a watershed, and applies the formula

$$S = \frac{100(C_L \times C_I)}{A} \quad (6.12)$$

where C_L denotes the total length (m or ft) of all contours within the watershed boundary, C_I denotes the contour interval (m or ft), and A is the watershed area (m² or ft²).

Figure 6.10 shows the calculations of the slope of a catchment within South Platte River basin in Colorado using both a DEM and contours. In this case, the two methods yield similar results.

A property of slope, which is important for modeling snowmelt in mountainous watersheds is the **aspect**. The aspect is the orientation of the slope, measured clockwise in degrees from 0° to 360°, where 0° is north-facing, 90° is east-facing, 180° is south-facing, and 270° is west-facing (Figure 6.10D).

Aspect can have a strong influence on temperature and thus on hydrologic processes of snowmelt and evapotranspiration. This is because the angle of the Sun in the northern and southern hemispheres is less than 90°. In the northern hemisphere, the north-facing slopes are often shaded, while southern-facing slopes receive more solar radiation for a given surface area insolation, because the slope is tilted toward the Sun and is not shaded directly by the Earth itself. The further north or south, and the closer to the winter solstice, the more pronounced the effects of aspect of this are, and on steeper slopes the effect is greater, with no energy received on slopes with an angle greater than 22.5° at 40° north on December 22 (winter solstice).

The aspect of a slope can have a very significant influence on its microclimate and vegetation. For example, in the northern hemisphere, a south-facing slope that is more open to sunlight and warm winds will generally be warmer and drier due to higher levels of evapotranspiration than a north-facing slope. This is seen in the Swiss Alps, where farming is much more extensive on south-facing than on north-facing slopes. In the Himalayas, this effect is seen to an extreme degree, with south-facing slopes being warm, wet, and forested, while the north-facing slopes are cold and dry, but with significant snow cover and much more heavily glaciated areas.

Figure 6.11 shows the DEMs of three very high-altitude watersheds of the Karakoram Mountains that drain to the Upper Indus River, one of the few rivers in the world that is almost entirely fed by snow and glacial melts. These three watersheds have some of the most outstanding and large glaciers outside the polar regions of the world. Figure 6.12 shows the variation of the aspects, in the form of spider diagrams, and the distribution of accumulation and ablation zones of the glaciers with respect to aspect in each of these three watersheds.

6.4 HYPSOMETRY AND HYPSOMETRIC CURVES

Hypsometry refers to an accounting process to calculate the areas of a watershed that lie within a certain elevation range or elevation band. This is very important for watersheds with great variations in elevations such as mountainous watersheds, and in particular where snow and glacial melts are significant components of streamflow or surface runoff. This is because the extents of snow- and ice-covered areas greatly vary with elevation. Furthermore, in many mountainous regions, precipitation, vegetative cover, and evapotranspiration change with altitude.

When elevation bands are plotted against their area of occupancy, the resulting curve is called **hypsometric curve**, which gives a visual idea of the distribution of areas in different elevation ranges. Figure 6.13 shows the hypsometric curves of the three watersheds of the Karakoram mountains discussed above.

Sometimes a hypsometric curve is also represented in a dimensionless form. In this case, the curve plots the percent area (area divided by total watershed area) of the watershed occurring above a given percent elevation contour. The percent elevation is defined as the given elevation divided by the watershed maximum relief, H. A combination of percent area and actual elevation is often more insightful, since it directly conveys the percent area of the watershed that lies in a given elevation range. In addition to the practical usage of hypsometric data and curves, they have also been used for scientific interpretation of tectonic processes of Earth causing the evolution of different landforms.

Hypsometric analysis or the relation of horizontal cross-sectional drainage basin area to elevation, was developed in its modern dimensionless form by Langbein (1947). He stated, "The area–altitude distribution curve has several applications. For example, snow surveys generally show an increase in depth of cover and water equivalent with increase in altitude; the area–altitude relation provides a means for

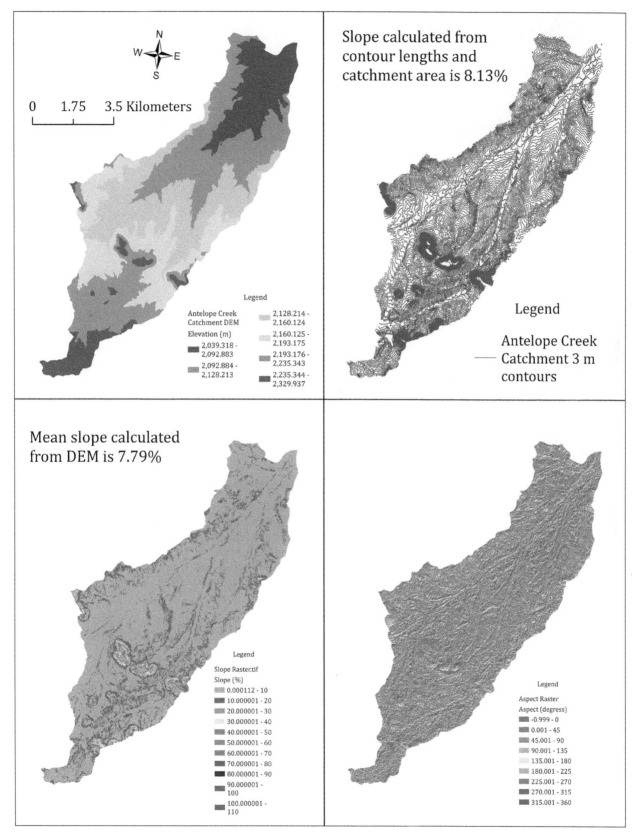

Figure 6.10 DEM, slope, contours, and aspect of Antelope Creek Catchment, one of the catchments shown in Figure 6.4. (A black and white version of this figure will appear in some formats. For the color version, please refer to the plate section.)

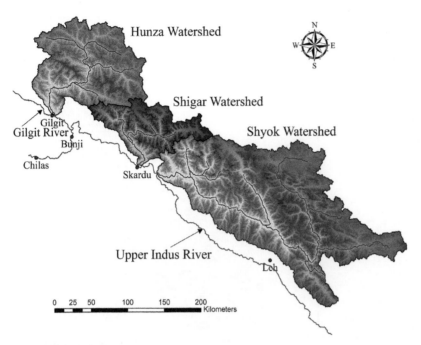

Figure 6.11 Three high-altitude, highly glaciated watersheds of Karakoram Mountains to the north of Greater Himalayas. From Mukhopadhyay, B., Khan, A., and Gautam, R. 2015. Rising and falling river flows: contrasting signals of climate change and glacier mass balance from the eastern and western Karakoram. *Hydrological Sciences Journal*, Volume 60, No. 11-12, 2062–2085. (A black and white version of this figure will appear in some formats. For the color version, please refer to the plate section.)

estimating the mean depth of snow or its water equivalent over a drainage basin. Several studies have shown a significant variation in annual precipitation and runoff with respect to altitude. The obvious variation in temperature with change in altitude is further indication of the utility of the area–altitude distribution curve." Vivoni et al. (2008) showed from modeling studies hypsometric controls on surface and subsurface runoff generation.

6.5 STREAM ORDER

If a map of a stream network, also called a **drainage net**, within a watershed is inspected, it will be immediately revealed that if flow is traced from one of the uppermost streams in the watershed toward the outlet, the uppermost stream joins another stream, which in turn joins another stream, and so on.

First-order streams are those that have no tributaries. The junction of two first-order streams forms a second-order stream. Similarly, a third-order stream is formed at the junction of two second-order streams.

The flow in a first-order stream is entirely dependent on the surface or overland flow that is called the source. A second-order stream receives flows from two first-order streams that form it and from the overland flow. It may also receive flow from another first-order stream that directly flows into it. Strahler (1957) proposed this stream ordering system following the procedures given as:

1. Streams that originate at a source are defined to be first-order streams.
2. When two streams of order w join, a stream of order $(w+1)$ is formed.
3. When two streams of different orders join, the stream segment immediately downstream of the confluence has the order of the higher-order stream at the confluence.
4. The order of the watershed is the highest-order stream with order W present in the watershed.

The main watercourse is usually considered to be the length of the stream of the highest order present in the watershed. An example of classification of streams according to the ordering system for a watershed in Texas is shown in Figure 6.14.

6.6 HORTON'S LAWS

Horton (1945) suggested several empirical laws based on a stream ordering system that was somewhat different from Strahler's ordering system. However, numerous studies later confirmed that Horton's laws hold true using Strahler's stream ordering system. These laws are given below.

6.6.1 Law of Stream Numbers

The **law of stream numbers** states that the number of streams of a given order follows an inverse geometric relationship with stream order that can be stated as

$$N_w = R_b^{W-w} \tag{6.13}$$

where N_w is the number of streams with order w, W is the order of the watershed (i.e., the order of the highest-order

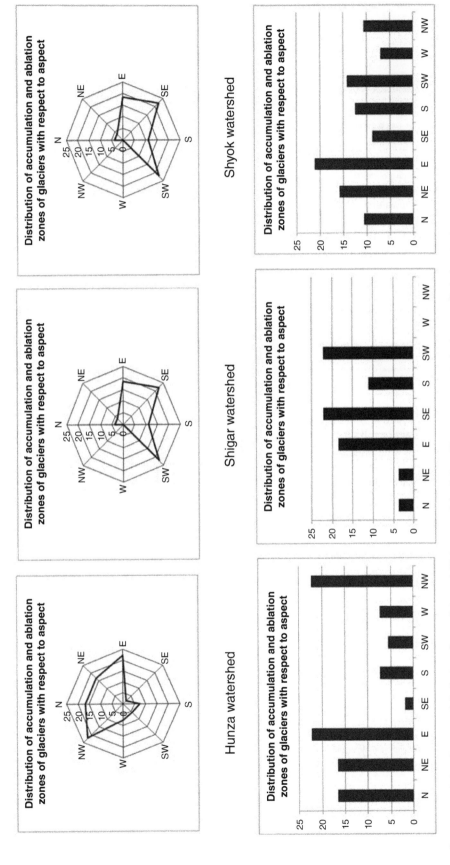

Figure 6.12 Variation of aspects and distributions of accumulation and ablation zones of the glaciers with respect to aspect in the three watersheds shown in Figure 6.11.

$$R_b = \frac{N_{w-1}}{N_w} \quad (6.14)$$

which implies that on average there are R_b streams of order $W - 1$. The bifurcation ratio varies normally between 3 and 5. This law is an expression of topological phenomena and is a measure of drainage efficiency. Also, elongated basins have higher bifurcation ratios than rounder basins.

It follows from Eq. 6.13 that the total number of streams in the drainage network can be given as

$$\sum_{w=1}^{W} N_w = 1 + R_b + R_b^2 + \ldots + R_b^{W-1} = \sum_{w=1}^{W} R_b^{W-w} = \frac{R_b^W - 1}{R_b - 1} \quad (6.15)$$

6.6.2 Law of Stream Length

This law relates the length of streams with their orders. The average length of streams of each higher order increases as a geometric sequence and can be formulated as

$$\overline{L}_w = \overline{L}_1 R_l^{w-1} \quad (6.16)$$

where $\overline{L}_w$ is the average length of stream of order w, $\overline{L}_1$ is the average length of first-order stream, and R_l is called **stream length ratio** given as

$$R_l = \frac{\overline{L}_w}{\overline{L}_{w-1}} \quad (6.17)$$

The stream length ratio generally varies between 1.5 and 3.5. The average length of a stream of order w in Eq. 6.16, is given by

$$\overline{L}_w = \frac{1}{N_w} \sum_{i=1}^{N_w} L_{wi} \quad (6.18)$$

in which L_{wi} is the length of the ith stream of order w and N_w is the number of streams of order w. Thus, the numerator of Eq. 6.18 gives the total length of all streams of order w.

Horton (1945) observed that the laws of stream number and length can be combined to produce an expression for the total length L_w, of all streams of a given order. Thus, from Eq. 6.18,

$$L_w = N_w \overline{L}_w \quad (6.19)$$

6.6.3 Law of Stream Areas

The stream area of order w, designated by A_{wi}, is the area of the watershed that contributes runoff to the stream segment of order w and its tributaries, that is, all lower-order streams. Thus, $\overline{A}_W$ is the total watershed area. The average stream area of each order, $\overline{A}_w$, is given by

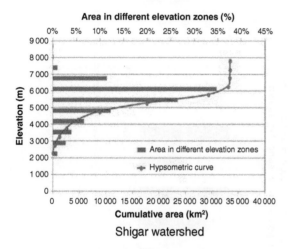

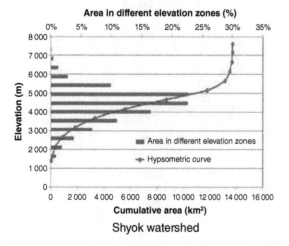

Figure 6.13 Hypsometric curves of the three watersheds shown in Figure 6.11, within the Upper Indus Basin straddling the northern slopes of the Greater Himalayas and southern slopes of the Karakoram Mountains. The elevation data used for computation of the curves are given in Mukhopadhyay and Dutta (2010).

stream in the network) and R_b is a constant for the network, called the **bifurcation ratio**. The bifurcation ratio for a given stream network is defined as

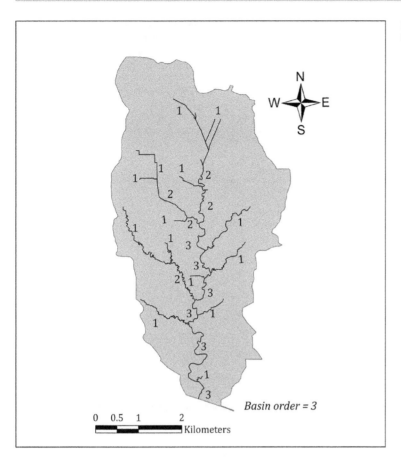

Figure 6.14 Delineation of stream orders is shown for Graham Branch Watershed near Denton, Texas.

$$\overline{A}_w = \frac{1}{N_w} \sum_{i=1}^{N_w} A_{wi} \quad (6.20)$$

The law of stream order states that the mean drainage area of progressively higher orders forms a geometric sequence given as

$$\overline{A}_w = A_1 R_a^{w-1} \quad (6.21)$$

where $\overline{A}_1$ is the mean drainage area that contributes runoff to all first-order streams and R_a is called the **stream area ratio** defined by

$$R_a = \frac{\overline{A}_w}{A_{w-1}} \quad (6.22)$$

The value of R_a has been found to vary between 3 and 6. The law of stream area was actually formulated by Schumm (1956).

6.6.4 Law of Stream Slopes

Horton (1945) also introduced the law of stream slopes, which states that the average slope of streams of each order tends to approximate an inverse geometric series that can be expressed as

$$\overline{S}_w = \overline{S}_1 R_s^{W-w} \quad (6.23)$$

where $\overline{S}_w$ is the average slope of streams of order w; $\overline{S}_1$ is the average slope of first-order streams; R_s is the **slope ratio** defined by

$$R_s = \frac{\overline{S}_w}{\overline{S}_{w-1}} \quad (6.24)$$

The value of R_s is approximately 0.55.

6.6.5 Implications of Horton Ratios

The bifurcation ratio, stream length ratio, and stream area ratio are obtained by plotting the values of N_i and corresponding values of $\overline{L}_w$, and $\overline{A}_w$ on a logarithmic scale against stream order on a linear scale and then deriving these values from the slopes of the plotted lines. Figure 6.15 shows an example of Horton's ratios of Emory River watershed in Tennessee based on the data given by Morisawa (1962).

These ratios have been used as measures of geometric similarity among the channel networks of watersheds. For example, Valdes et al. (1979) provided plots to show that the ratios R_b, R_l, and R_a of a basin (Mamon) are consistent with

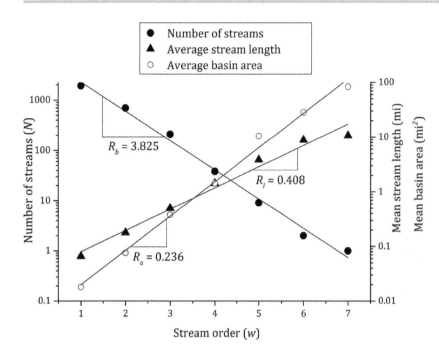

Figure 6.15 Number, mean length, and mean drainage area of streams in the drainage network of Emory River watershed, Tennessee, plotted as function of stream order, with bifurcation, length, and area ratios (R_b, R_l, and R_a) calculated from the slopes of the regression lines (data from Morisawa, 1962).

those of a subbasin (Mamon 5) within it, indicating their geometric similarity. Geometric similarities in watershed morphology provide both qualitative and quantitative bases for grouping regions. This in turn can be used for simple correlations of hydrologic and geomorphologic features to aid the basis for the choice of characteristics to use in multiple regression of peak runoff intensity.

6.7 STREAM SLOPE AND STREAM POWER

Usually, a slope of the reach of a stream is calculated by the difference in elevation between the most upstream point of a stream reach and the most downstream point of the reach, called **reach relief**, and dividing it by the length of the horizontal projection of these two points. The stream slope has a profound effect on the velocity of flow in the stream, and consequently on the flow characteristics of runoff from a watershed or its constituent catchments. It is a fundamental parameter needed in the use of the Manning equation and channel flow routing using the Muskingum–Cunge method described in Chapter 12.

In general, the slope varies in different reaches of the same stream. A stream can be divided into N number of reaches, each having nearly uniform slope S_i and the **equivalent slope**, S_e, of the stream is then calculated using

$$S_e = \left(\frac{\sum_{i=1}^{N} L_i \sqrt{S_i}}{\sum_i^N L_i} \right)^2 \quad (6.25)$$

This method from Johnstone and Cross (1949) is designed to estimate the slope that would result in the same total time of travel if the actual stream length, roughness, channel cross section, and any other pertinent factors other than slope were unchanged.

Another slope parameter that is used in hydrology is called the 10–85 slope S_{10-85}, which is the slope of the 10–85 flow path, L_{10-85}, discussed above. It is the change in elevation divided by the length between two points located respectively 10% upstream from the basin outlet and 15% downstream from the basin divide on the longest flow path.

One important parameter that is derived from stream slope is the **stream power** given as

$$\Omega = \rho g Q S_e = \gamma Q S_e \quad (6.26)$$

where ρ and g, respectively, represent density of water and acceleration due to gravity (γ is specific weight of water). Stream power has the unit of ML/T^3 and, in SI, Ω is given as watts. It represents the amount of energy the water in a river or stream is exerting on the sides and bottom of the river and thereby its erosion-causing potential.

Stream power is the rate at which a river does work per unit of length. The work done by rivers is manifested by scouring of bed and banks and deposition of sediments. Langbein and Leopold (1964) used the concept of stream power to put forth a theoretical discussion about how natural channels should follow two ends of a spectrum. On the one hand, rivers can do work such that the expenditure of energy is uniformly distributed as the river flows from upstream to downstream. In this case, the work per unit area, $\frac{\gamma Q S}{w}$ is constant along the channel length (w denotes the width of the river cross section). On the other hand, a river can follow a course where it does minimum work and the total rate of work being done along the whole length of river from headwater to distance L along the river is then $\int_0^L \gamma Q S \, dx$, where x denotes the river distance.

6.8 LONGITUDINAL PROFILE OF A STREAM

6.8.1 Importance of a Longitudinal Profile

A **longitudinal profile** of a stream, simply called the **stream profile**, is a graph of the elevation of the channel bed (ordinate) versus the distance along the channel (abscissa). The shape of a river profile results from fluvial erosion and deposition of sediments transported by rivers. Study of river profiles has both practical and theoretical importance. The practical aspect of examination of river profiles seeks to find how river profiles and sediment yields will react to engineering modifications of a river profile by projects involving dams and reservoirs, dredging, or straightening, etc. The theoretical aspect of the study of river profiles is an essential element of landscape evolution. It is useful in the reconstruction of past landscapes and rates of landform changes.

6.8.2 Concavity of Stream Profiles

Rivers increase in size downstream. Tributaries, drainage area, discharge, depth, and width increase downstream. Bed particle size, on the other hand, decreases downstream. The gradient generally flattens downstream. Thus, in general the longitudinal profile of a river is concave upward, referred to as **upconcavity**. Analysis of longitudinal profiles of many rivers has shown that nearly all profiles are concave, but in some there are convex portions, and there are of course some channels that exhibit straight profiles. The concave profile is a manifestation of downstream flattening of the stream slope.

The slope of a stream is determined by conditions imposed from upstream, but the elevation and location of each point of the profile are also determined by the downstream base level. The downstream decrease in slope is attributed, in part, to the decrease in the grain size of the bed material due to abrasion and sorting. The reduction in median particle size with the downstream distance can be expressed as

$$d = d_0 e^{-\beta x} \tag{6.27}$$

where, d_0 is the median particle size at the reference section and β is the coefficient of particle size reduction.

For a river system in equilibrium under uniform uplift rate, that is, where the erosion rate is uniform along the river, the river profile must progressively decrease in slope as the drainage area increases. For this reason, the equilibrium profile of a river system is generally concave resembling the graph of a hyperbola. Indeed, there are examples, also, where the longitudinal profiles of streams do not show smooth upconcavity (e.g., Phillips and Lutz, 2008).

The general equation used to express the erosion rate, $-(dz/dt)$, is commonly given as

$$-\frac{dz}{dt} = K A^m S^n \tag{6.28}$$

where z is the height above a datum, t is time, A is the drainage area, S is the slope and K, m, and n are constants. Since discharge can be used as a proxy to the drainage area, as we have seen above, Q has also often been used to replace A in equations like Eq. 6.28.

6.8.3 Controls of Stream Profiles

In order to explore a stream profile, it is necessary to understand the variables that influence the change of slope downstream. Major variables controlling the slope are discharge, sediment load, and caliber. Grove Karl Gilbert (1877), a pioneering American geologist who helped launch the USGS, and was its first senior geologist, explored **declivity** (downward gradient) of stream profiles, and concluded that stream gradient had an inverse relationship with discharge. He further noted that steeper channels and channels with greater drainage areas eroded faster. Langbein and Leopold (1964) discussed how the longitudinal profile can be considered to be a function of the following variables: (i) discharge, Q; (ii) load delivered to the channel, W_L; (iii) size of debris, d_s; (iv) flow resistance factor, n (v); velocity, v; (vi) width, b; (vii) depth, d; and (viii) slope, S.

6.8.4 Longitudinal Profile as a Graded Form

Langbein and Leopold (1964) illustrated that the theoretical longitudinal profile of a river doing "least work" is sharply concave. In contrast, the theoretical profile for a "uniform distribution of work" has a nearly constant slope. It has only a slightly concave upward slope because, as the rate of doing work increases downstream with discharge, the surface area of the streambed also increases (Figure 6.16). The rate of doing work per unit area of streambed is constant on a river that widens downstream but decreases its slope only gradually. The least-work profile, on the other hand, is a curve in which the greatest loss of altitude takes place where the discharge is least, near the head of the river. The profile is very steep near the head and almost horizontal near the mouth.

The two tendencies toward equilibrium oppose each other, and the resulting equilibrium is most likely to be some predictable intermediate condition (Figure 6.16). Langbein and Leopold (1964) concluded from their studies of river dynamics that both the downstream profiles and channel cross sections of rivers approach the equilibrium form predicted from the principles of least work and uniform distribution of work.

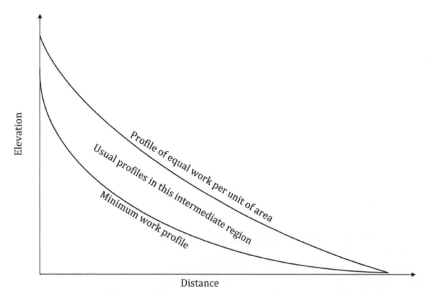

Figure 6.16 Schematic longitudinal profiles of rivers showing the tendency to balance minimum total work with maximum distribution of work (adopted from Langbein and Leopold, 1964).

6.8.5 Equations of Longitudinal Profiles

A variety of continuous mathematical functions has been used to describe the typical concave-up longitudinal profiles of alluvial rivers. Exponential, logarithmic, and power functions are the most widely accepted and all have been fitted to actual river profiles with some success. Each is supported to varying degrees by empirical, semi-empirical, rational, or heuristic arguments.

6.8.5.1 Exponential Form

An exponential decay function was fitted to the longitudinal slope profile of a stream by Shulits (1941) as

$$S = S_0 e^{-\alpha x} \tag{6.29}$$

in which S is the slope at a distance x downstream of a reference section where the slope is S_0; and α is a coefficient of slope reduction.

Since $S = \frac{dz}{dx}$, Eq. 6.29 can be written as

$$\frac{dz}{dx} = S_0 e^{-\alpha x} \tag{6.30}$$

or

$$dz = S_0 e^{-\alpha x} dx \tag{6.31}$$

Upon integration,

$$z = \frac{S_0}{\alpha}[e^{-\alpha x} - 1] \tag{6.32}$$

where x and z are the respective longitudinal and vertical coordinates of the stream profile.

The general exponential form of a stream profile is written as

$$z = a \times b^x \tag{6.33}$$

in which a and b are arbitrary constants. If a stream profile follows an exponential form, a plot of altitude, z, on a logarithmic scale with the horizontal distance, x, on an arithmetic scale should fit a straight line of the form

$$\log z = c + dx \tag{6.34}$$

where $c = \log a$ and $d = \log b$.

6.8.5.2 Power Form

Since the longitudinal profile of the graded stream is known to follow a hyperbolic curve, it can also be expressed as a power function

$$z = ax^b \tag{6.35}$$

In this case, a plot of both elevation, z, and horizontal distance, x, on logarithmic scales should yield a straight line given by

$$\log z = \log a + b \log x \tag{6.36}$$

From Eq. 6.36, b is obtained as the mean gradient (negative) of the stream and $\log a$ is the vertical height (elevation) of the reference section (intercept of the profile on the z-axis on a logarithmic plot).

6.8.5.3 Logarithmic Form

A logarithmic form of a stream profile results in a straight line when elevation, z, is plotted on an arithmetic scale on the ordinate against the horizontal distance, x, scaled logarithmically on the abscissa, yielding

$$z = a + b\log x \quad (6.37)$$

However, a problem with the logarithmic form is that, at the stream head ($x = 0$), elevation becomes undefined, and the constants cannot be evaluated. Broscoe (1959) resolved this issue by introducing a horizontal distance to the stream head from the watershed divide along the longest flow path.

6.8.5.4 Polynomial Form

The longitudinal profile of an alluvial channel is the long-term consequence of erosion and deposition of the sediment pile over which the river flows. Based on such a premise, Rice and Church (2001) offered the solution of a differential equation providing the one-dimensional relationships between bed elevation and rate of change of sediment load downstream. They showed that the most plausible solution of such an equation is a quadratic function given as

$$z = ax^2 + bx + c \quad (6.38)$$

where z denotes the bed elevation, x is the downstream distance; and a, b, and c are constant coefficients. These authors tested their model by examining the longitudinal profiles of various segments of a stream where each segment is considered to be a homogeneous reach having no significant input of water or sediment from lateral sources. In each segment or sediment link, grain size typically fines downstream in a relatively systematic manner.

6.8.6 Profile Segmentation

Broscoe (1959) examined the longitudinal profiles of streams of **low order** in six small watersheds from Pennsylvania and Colorado. He noted that for such streams the individual profiles were not simple curves but segmented, each segment corresponding to a given order. Individual segments can be best represented by linear curves even if those segments are slightly concave up, whereas a composite profile of the segments can be best represented by logarithmic curves. Profile segmentation can be attributed to changes in the relationship between sediment load and discharge at a change in stream order. Linear segments may be natural stable forms of low-order stream profiles in dry climates. The composite profiles, best fitted by logarithmic curves, contradict the derivation of the longitudinal profile as controlled by particle abrasion following Sternberg's law of abrasion (Shulits, 1941).

In most natural cases, the entire stream profile often cannot be represented by a single mathematical function and the entire profile needs to be broken down into several homogeneous segments, each of which can be represented by one of the mathematical functions discussed above.

An extreme example, perhaps unique in the world, of a river profile of distinct segmentation is shown by the longitudinal profile of Brahmaputra River (Figure 6.17). The Brahmaputra descends from the high-altitude plateau of Tibet to lower elevation near Mount Namcha Barwa in the Nyainquetanglha Mountains making a sharp turn, known as Tsangpo Grand Canyon or Big Bend of Tsangpo George where the river profile is extremely steep, as shown in Figure 6.17. Each segment of the river has a distinct gradient, also given in Figure 6.17.

6.9 HYDRAULIC GEOMETRY OF STREAMS

The **cross-sectional area of flow** and **average velocity of flow** at a river station give the discharge, also called the volumetric flux passing at that station. This is the continuity equation of open channel flow, which is widely known as

$$Q = vA \quad (6.39)$$

The average velocity at the cross section is usually given by another well-known formula, namely, the Manning equation given in US customary units and SI, respectively, as

$$v = \frac{1.486}{n}(R_h)^{2/3}\sqrt{S} \quad (6.40)$$

$$v = \frac{1}{n}(R_h)^{2/3}\sqrt{S} \quad (6.41)$$

in which R_h refers to hydraulic radius, which is the ratio of flow cross-sectional area (A_q) and wetted perimeter (P_w): $R_h = A_q/P_w$. When either Eq. 6.40 or Eq. 6.41 is used in Eq. 6.39, we obtain a factor K, called the **conveyance**, as

$$K = \frac{C}{n}A_q R_h^{2/3} \quad (6.42)$$

in which C is either 1.486 or 1. As the name implies, K is a measure of the carrying capacity of the channel section. These formulations show the importance of flow cross-sectional geometry in controlling the magnitude of volumetric flux.

The **hydraulic geometry** of a channel refers to the way in which the width (w), depth (d), and velocity (v) of flow change with changes in channel discharge (Q). These three parameters have a singular property that can be expressed as

$$Q = w \cdot v \cdot d \quad (6.43)$$

Thus, a change in Q must fully be reflected by changes in width, depth, and velocity. Since velocity at a cross section varies along both depth and width, in Eq. 6.43, it is the average velocity. Likewise, unless the channel is perfectly rectangular, d in Eq. 6.43 should represent the **hydraulic depth** given as the ratio of the flow area and top width of the flow: $d = A_q/w$.

Discharge in a stream can change in two ways:

1. the changing dimension of flow at a single gauging location as discharge changes during the passage of a flood;
2. the general increase in discharge progressively downstream of a channel as the river collects runoff from a progressively greater drainage area.

246 6 Geomorphologic Concepts and Watershed Characteristics

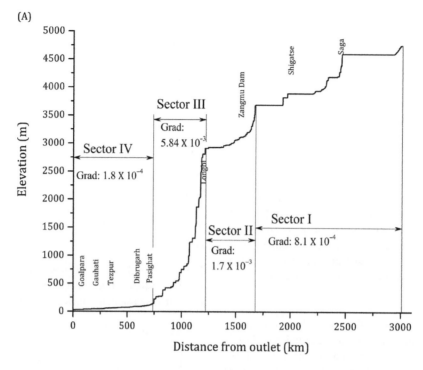

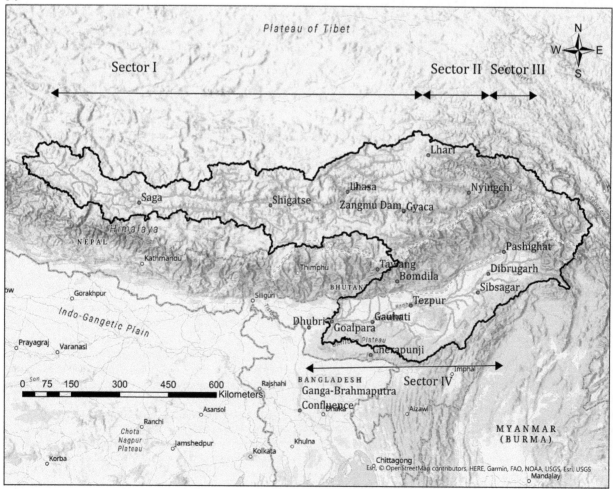

Figure 6.17 (A) Longitudinal profile of Brahmaputra River. (B) Map of the upper-middle Brahmaputra basin with locations given on the longitudinal profile, showing the four sectors of the longitudinal profile (the longitudinal profile is constructed from the Shuttle Radar Topography Mission (SRTM) elevation data discussed in Mukhopadhyay and Singh, 2011). (A black and white version of this figure will appear in some formats. For the color version, please refer to the plate section.)

6.9 Hydraulic Geometry of Streams

The relationship of the first kind is known as **at-a-station hydraulic geometry** and that of the second kind as **downstream hydraulic geometry**.

Leopold and Maddock (1953) first developed hydraulic geometry relations from different regions of the United States. They studied discharges of various flow frequencies at a point – the at-a-station change of hydraulic properties, w, d, and v – with varying discharges. Similarly, they also studied discharges of the same flow frequency at different points moving downstream in a basin, that is, variation of the same hydraulic parameters with variation of discharge in the downstream direction of flow. This is not necessarily equivalent to discharges resulting from the same event at different points. However, the results are reasonable indicators of what may occur in that situation. They expressed these relations as power functions of discharge.

$$w = aQ^b \tag{6.44}$$

$$d = cQ^f \tag{6.45}$$

$$v = kQ^m \tag{6.46}$$

Multiplication of these three equations results in

$$w \cdot d \cdot v = (a \cdot c \cdot k) Q^{(b+f+m)} \tag{6.47}$$

Equating Eq. 6.43 and Eq. 6.47 results in

$$Q = (a \cdot c \cdot k) Q^{(b+f+m)} \tag{6.48}$$

This gives

$$a \cdot c \cdot k = 1 \tag{6.49}$$

and

$$b + f + m = 1 \tag{6.50}$$

Although every stream and stream system have differences, repeated measurements of hydraulic geometry yield remarkably similar values. For example, values classically reported (data of the 1950s and 1960s) for the hydraulic geometry exponents, b, f, and m, are given in Table 6.1.

The power function relationships indicate linear relationships between width, velocity, depth, and discharge on a logarithmic plot. Thus, graphically, these exponents equal the slopes of the lines of discharge plotted against each channel parameter on the log-log plot.

Physically, these values indicate that increasing discharge at one location in a channel (at-a-station) is primarily reflected by increasing depth and velocity; relative to the magnitude of the discharge increase, the flow width changes least of the three (i.e., $b < f$ or m). In contrast, the physical dimensions of channels (depth and width) tend to increase sufficiently in the downstream direction to account for nearly all of the associated increase in discharge. It can be noted from Table 6.1 that the velocity exponent (m) is small but positive indicating an increase in average velocity in the downstream direction as well. However, since for the downstream hydraulic geometry relation, the exponent of v is 0.1, it tends to indicate that velocity remains nearly constant along the stream.

The parameters a, c, and k do not have any such universal values. These parameters depend on the specific size of the channels being measured and also on the units of measurement being used. Bieger et al. (2015) used a very large dataset comprising bankfull hydraulic geometry parameters from a large number of sites scattered throughout the contiguous United States to develop regional regression equations in the form of power functions predicting bankfull width, depth, and cross-sectional area as a function of drainage area for the physiographic division and provinces of the United States

Leopold and Maddock (1953) discovered the hydraulic geometry relations in power-law forms. Because of their great practical value in various river engineering projects, such as design of stable and natural channels, river flow control works, river improvement works, and irrigation schemes, there is a large body of literature describing the derivation of these relations using distinct types of theories. The hydraulic geometry relations of alluvial channels are useful in river engineering since they provide insights into the intra- and inter-basin variations of width, depth, and velocity with discharge. Furthermore, hydraulic geometry relationships are frequently used for stream classification and hydrologic modeling studies.

The hydraulic geometry of a stream remains constant for any stream as long as its drainage environment is not changed by factors such as urbanization, agricultural development, or any other setting that affects runoff generation. The establishment of hydraulic geometry relationships requires the development of detailed cross-sectional profiles at various locations of a stream from field surveys, using appropriate horizontal and vertical benchmarks and data and measurements of hydraulic parameters for flows of varied frequencies, which can be established from flow-duration curves (Chapter 5) using discharge records spanning several years at that location.

Table 6.2 shows an example of data required for the development of at-a-station hydraulic geometry. The data were collected by Stall and Yang (1970) in an effort to evaluate channel characteristics of 12 river basins of the United States representing widely different physiographic regions but having relatively uniform physiography and varying in sizes from 1532 to 8410 square miles. Discharge, cross-sectional area, width, depth, and velocity of a stream at a particular location in the stream system are each related to the frequency of occurrence of the discharge, in percent of days per year established from flow-duration

Table 6.1 *Empirical values of hydraulic geometry exponents*

Geometry type	b (width exponent)	f (depth exponent)	m (velocity exponent)
At-a-station	0.26	0.40	0.34
Downstream	0.5	0.4	0.1

Table 6.2 *At-a-station (2-0750) hydraulic geometry for Dan River at Danville, Virginia, in the Roanoke River basin based on a period of record from 1931 to 1967 (37 years)*

Flow frequency (%)	Discharge (ft^3/s)	Flow cross-sectional area (ft^2)	Velocity (ft/s)	Width (ft)	Mean depth (ft)
90	760	704	1.07	426	1.66
80	930	789	1.17	432	1.82
70	1100	867	1.26	438	1.98
60	1300	952	1.35	443	2.15
50	1520	1039	1.45	448	2.32
40	1790	1139	1.56	453	2.52
30	2130	1256	1.69	458	2.74
20	2700	1435	1.88	466	3.09
10	3850	1751	2.2	478	3.66

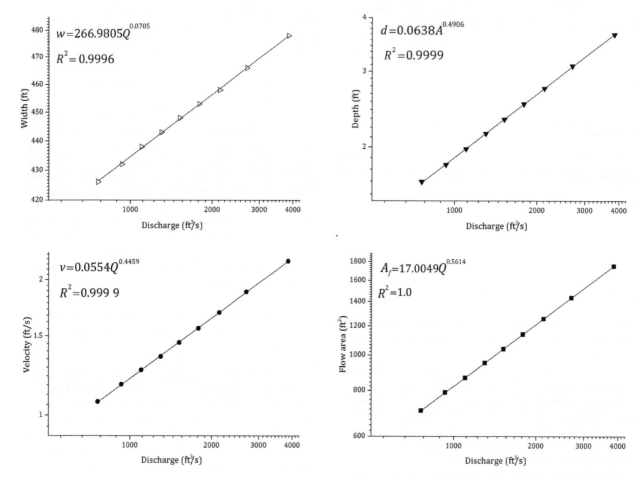

Figure 6.18 At-a-station hydraulic geometry of Dan River at Danville, Virginia for the period of record: 1931–1967, derived from the data given in Stall and Yang (1970). In this case, $ack = 0.9437$ and $b + f + m = 1.007$.

curves and the drainage area contributing flows to that particular location.

Figure 6.18 shows the hydraulic geometry relationships in the form of power-law functions based on the data given in Table 6.2.

It should be noted here that if the Manning equation is used to solve for velocity from discharge values, an appropriate numerical method, such as the Newton–Raphson iteration technique, should be used, since there is no analytical solution for determination of the flow depth given the flow rate because the flow area and hydraulic radius are complicated functions of the depth.

6.10 DRAINAGE DENSITY AND STREAM FREQUENCY

Horton (1932) defined the **drainage density** as the length of drainage line per unit area. It is then given as

$$D_d = \frac{L_T}{A_W} \quad (6.51)$$

where L_T (miles or km) is the total length of all streams of all orders in the watershed and A_W is the watershed area (square miles or square km). Equation 6.51 can also be written as

$$D_d = \sum_{w=1}^{W} \sum_{i}^{N_w} \frac{L_{wi}}{A_W} = \sum_{w=1}^{W} \frac{N_w \overline{L}_w}{A_W} \quad (6.52)$$

which can be simplified using the laws of stream lengths and stream numbers to

$$D_d = \frac{\overline{L}_1}{A_W} \frac{R_l^W - R_b^W}{R_l - R_b} = \frac{\overline{L}_1}{A_W} \frac{r^W - 1}{r - 1}; \quad r = \frac{R_l}{R_b} \quad (6.53)$$

From the definition, it follows that D_d is a measure of the closeness (density) of the stream spacing or degree of drainage development in a watershed and is a metric to describe the efficiency in which a watershed is drained by stream channels. Several studies have suggested that drainage density is a reflection of climatic variables, such as the mean annual precipitation on surface topography (Moglen et al., 1998). Melton (1957) found a strong correlation between the drainage density and the ratio of average annual precipitation and average annual evaporation.

If the streams in a watershed are fed by Hortonian overland flow from all of their contributing areas, drainage density can be used to calculate the average overland flow length, L_o, by

$$L_o = \frac{1}{2D_d} \quad (6.54)$$

The **stream frequency** is defined as the number of stream segments per unit area:

$$F = \frac{\sum_{w=1}^{W} N_w}{A_W} \quad (6.55)$$

where N_w is the number of streams of order w.

The issue of scale in river basins is fundamental to hydrology. The drainage density has been suggested as a basic length scale. Stream frequency, like drainage density, is a quantity dependent on the scale of resolution. Therefore, drainage density is a fundamental concept in hydrologic analysis.

Melton (1957) analyzed 156 watersheds covering a broad range in size, relief, cover, and climate and established a relationship between drainage density and stream frequency, which is given as

$$F = 0.694 D_d^2 \quad (6.56)$$

Equation 6.56 implies that the dimensionless ratio F/D_d^2 approaches a constant (0.694) independently of scale.

Another related and important geomorphic concept is the **drainage texture**, which means the relative spacing of drainage lines. Horton (1945) pointed out that what is commonly referred to as drainage texture really includes both drainage density and stream frequency.

6.11 DRAINAGE PATTERN

The **drainage pattern** refers to the spatial pattern or geometric arrangement of streams in a network. It refers to the particular plan or design that the individual stream courses collectively form. Within a watershed, the stream system achieves a particular drainage pattern with the passage of time to its network of stream channels and tributaries, governed by local geologic factors. Drainage patterns or *drainage nets* are classified based on their form and texture. Their shape or pattern develops in response to the local topography and subsurface geology.

Drainage channels develop where surface runoff is concentrated, and earth materials provide the least resistance to erosion. The texture is governed by the infiltration characteristics of the soil, and the volume of water available in a given period of time to enter the surface. If the soil has only a moderate infiltration capacity and a small amount of precipitation strikes the surface over a given period of time, the water will likely be absorbed rather than evaporate away. If a large amount of rainfall hits the surface, more water will evaporate, infiltrate into the surface, or pond on level ground. On sloping surfaces, this excess water will produce surface runoff. Drainage density will be low where the surface is flat, and the soil infiltration is high because the water will be absorbed into the surface. The fewer the number of channels, the coarser will be the drainage pattern.

Drainage patterns determine the locations where significant runoff hydrographs will be observed. It also controls the shape of the hydrographs. For this reason, it is important to identify the drainage pattern of a watershed. The following types of drainage patterns are commonly observed on land surfaces.

1. **Dendritic drainage pattern**: Dendritic drainage patterns are the most common form of a drainage net and look like the branching pattern of tree twigs or roots. They are characterized by irregular branching of tributary streams in many directions and at almost any angle, although usually at considerably less than a right angle and typically tributaries join larger streams at an acute angle. Dendritic patterns develop in regions underlain by homogeneous material. That is, the subsurface geology has a similar resistance to weathering so there is no apparent control over the direction the tributaries take.
2. **Trellis drainage pattern**: Trellis drainage patterns display a system of subparallel streams with tributaries joining the main streams at nearly right angles. In a trellis pattern, the main river flows along a strike valley and smaller tributaries feed into it from the steep slopes on the sides of mountains. These tributaries enter the main river at right angles, causing a trellis-like appearance of the river system. Trellis drainage develops in folded topography like that found in the Appalachian Mountains of North America.
3. **Parallel drainage pattern**: Parallel drainage patterns form where there is a pronounced slope to the surface.

Tributary streams tend to stretch out in a parallel-like fashion following the slope of the surface. A parallel pattern also develops in regions of parallel, elongate landforms like outcropping resistant rock bands. All forms of transitions can occur between parallel, dendritic, and trellis patterns.

4. **Rectangular drainage pattern**: In rectangular drainage patterns, both the main stream and its tributaries display right-angled bends. The rectangular drainage pattern is found in regions that have undergone faulting. Movements of the surface due to faulting offset the direction of the stream. As a result, the tributary streams make sharp bends and enter the main stream at high angles. Streams follow the path of least resistance and thus are concentrated in places where exposed rock is the weakest.
5. **Radial drainage pattern**: Radial drainage patterns develop around a central elevated point. In this pattern, streams diverge from this point. Radial drainage is common to such conically shaped features as volcanic cones, domes, and various other types of isolated conical or subconical hills. The tributaries from a central high point follow the slope downwards and drain down in all directions.
6. **Centripetal drainage pattern**: Centripetal drainage patterns are just the opposite of tradial pattern, as streams flow toward a central depression, meaning that the drainage lines converge into a central depression. This pattern is typical in the western and southwestern portions of the United States where basins exhibit interior drainage. During wetter portions of the year, these streams feed ephemeral lakes, which evaporate away during dry periods. Salt flats are created in these dry lake beds, as salt dissolved in the lake water precipitates out of solution and is left behind when the water evaporates away.
7. **Deranged drainage pattern**: Deranged or contorted patterns develop from the disruption of a preexisting drainage pattern. It is marked by irregular stream courses, which flow into and out of lakes or swamps and have only a few short tributaries.
8. **Reticulate drainage pattern**: Reticulate drainage patterns usually occur on floodplains and deltas where rivers often interlace with each other forming a net.
9. **Barbed drainage pattern**: These patterns usually have only local extent and are only found at or near the headwater portions of drainage systems. The tributaries join the main stream in boathook bends that point upstream.
10. **Annular drainage pattern**: These patterns are usually found around maturely dissected domes that have alternating belts of strong and weak rock encircling them. They have ring-like plans such as Red Valley, which nearly encircles the Black Hills in South Dakota, United States.

Figure 6.19 shows certain types of drainage patterns that can be observed in nature.

6.12 TIME OF CONCENTRATION AND LAG TIME

6.12.1 Definitions of the Time Parameters

Some of the morphometric parameters described above are used to estimate watershed characteristics related to the time of water movement in various segments of a watershed. Estimation of these time parameters is supremely important for the computation of runoff hydrographs, which is central to the subject of applied hydrology. The time parameters are defined first, and subsequent discussions pertain to the commonly used methods in their estimation using geomorphic concepts.

6.12.1.1 Travel Time

Travel time, denoted by t_t, is the time water takes to travel from one point on a watershed to another. It is given by

$$t_t = \frac{l}{V} \tag{6.57}$$

in which l is the distance between two points (ft or m), and V is the average velocity of flow between the two points (ft/s or m/s).

6.12.1.2 Time of Concentration

Time of concentration, t_c, denotes the time required for an entire watershed to contribute to runoff at the point of interest for a specific hydrologic study, which is typically the outlet point of a catchment or watershed. Because of the travel time to the watershed outlet, only part the watershed may be contributing to surface water flow at any time t after precipitation begins. The boundaries of these contributing areas are lines of equal time of flow to the outlet and are called **isochrones** (Figure 6.20A). As time progresses, this area grows and, if rainfall duration is sufficiently long, ultimately the entire watershed area contributes flows to the outlet (the entire watershed boundary defines an isochrone). For this reason, the time of concentration is defined as the time required for runoff to travel from hydraulically the most remote point of the watershed to the outlet. The hydraulically most distant point is the point with the longest travel time to the watershed outlet, and not necessarily the point with the longest flow path to the outlet. It is evident, then, that for the entire area of the watershed to contribute flow to the outlet, rainfall duration must be either equal to or greater than the time of concentration.

6.12.1.3 Lag Time

The lag time, t_l, is the delay between the time runoff from a rainfall event over a watershed begins until runoff reaches its maximum peak. Conceptually, it may be thought of as a weighted time of concentration where, if for a given rainfall event, the watershed is divided into bands of area, the travel

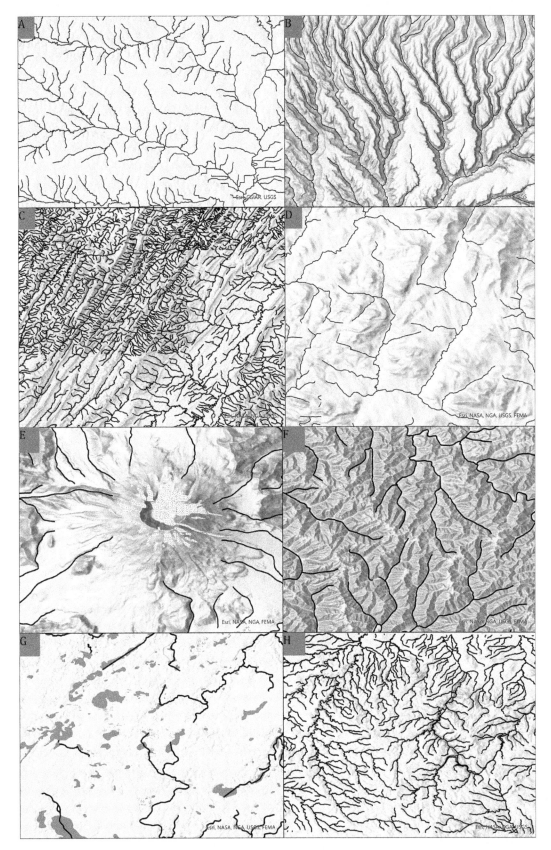

Figure 6.19 Selected examples of types of drainage patterns: (A) dendritic pattern, Missouri River basin – Nebraska/Missouri; (B) parallel pattern, Mesa Verde – Colorado; (C) trellis pattern, Monterey – Virginia; (D) rectangular patterns, Adirondacks Mountains, Elizabethtown – New York; (E) radial pattern without tributaries, Mount Saint Helen – Washington; (F) centripetal pattern, Burkes Garden – Virginia; (G) deranged pattern, Galesburg – Michigan; and (H) radial pattern with tributaries, Black Hills – South Dakota.

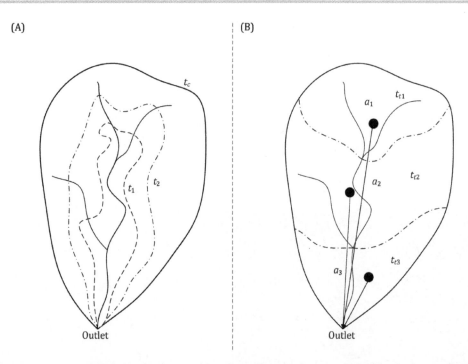

Figure 6.20 (A) Schematic representation of the concept of isochrones and time of concentration. Isochrones at t_1 and t_2 define the area contributing flow to the outlet for rainfall duration t_1 and t_2. At time of concentration, the isochrone becomes the watershed boundary. (B) Conceptual illustration of watershed lag time.

times from the centroids of the areas to the main watershed outlet can be given as

$$t_l = \frac{\sum a_i q_i t_{ti}}{\sum a_i q_i} = \frac{\sum a_i q_i t_{ti}}{Aq} \qquad (6.58)$$

in which t_l is the total watershed lag time (h), A is the total watershed area (mi^2 or km^2), a_i is the area in the ith band (mi^2 or km^2), q_i is the runoff (in or mm) from the ith band with area a_i, and t_{ti} is the travel time from the centroid of a_i to the point of reference. The concept is presented in Figure 6.20B.

The lag time is typically estimated from the time of concentration using an empirical relationship that was developed by various researchers at NRCS for average natural watershed conditions and an approximately uniform distribution of runoff. This relationship is given as

$$t_l = 0.6 t_c \qquad (6.59)$$

The lag time is also defined as the time difference between the first moment of rainfall excess and that of the instantaneous unit hydrograph. This is the definition used in unit hydrograph theory. In the Snyder unit hydrograph method, it is defined as the difference between the centroid of the rainfall excess and the peak of the unit hydrograph. Boyd (1978) observed that solution of the stream order laws produced a relation between lag time and drainage area which had the same form as the relations developed between different watersheds.

6.12.1.4 Time to Equilibrium

If an inflow of excess rainfall continues at a steady rate for an indefinite period of time, the outflow at the outlet of a watershed continues to increase and its value asymptotically approaches the value of inflow. The time elapsing before there is no substantial difference between inflow and outflow (usually less than 3%) is called the **time to equilibrium** (Bell and Kar, 1969). Even though these conditions rarely occur in nature, and it is not usually possible to determine the time of equilibrium, t_{eq}, from rainfall–runoff data, the concept has been found useful in the studies of overland flow (e.g., Singh, 1976). When kinematic wave theory is used to model overland flow over planes or different shapes of watersheds (Chapter 10), the maximum discharge equals the rainfall intensity times the plane or watershed area to indicate 100% contribution of runoff to the outlet. Therefore, the time to equilibrium is treated as synonymous with the time of concentration (Overton and Meadows, 1976). However, using kinematic wave approximation of surface and subsurface flows, Beven (2020) argued that there is a difference between the time of concentration and time of equilibrium. He further opined that the definition of time of concentration as given above is wrong when applied to watershed responses where, in terms of how surface and subsurface flows produce hydrographs, the time to equilibrium is the correct concept. From the viewpoint of kinematic wave theory, t_c is the travel time of the wave (celerity) not the average velocity of water particle, which is less than wave celerity. These concepts are discussed in Chapter 10.

6.12.2 Types of Flows

There can be four principal types of flow that water from rainfall, that is not abstracted, follows to make its way to the

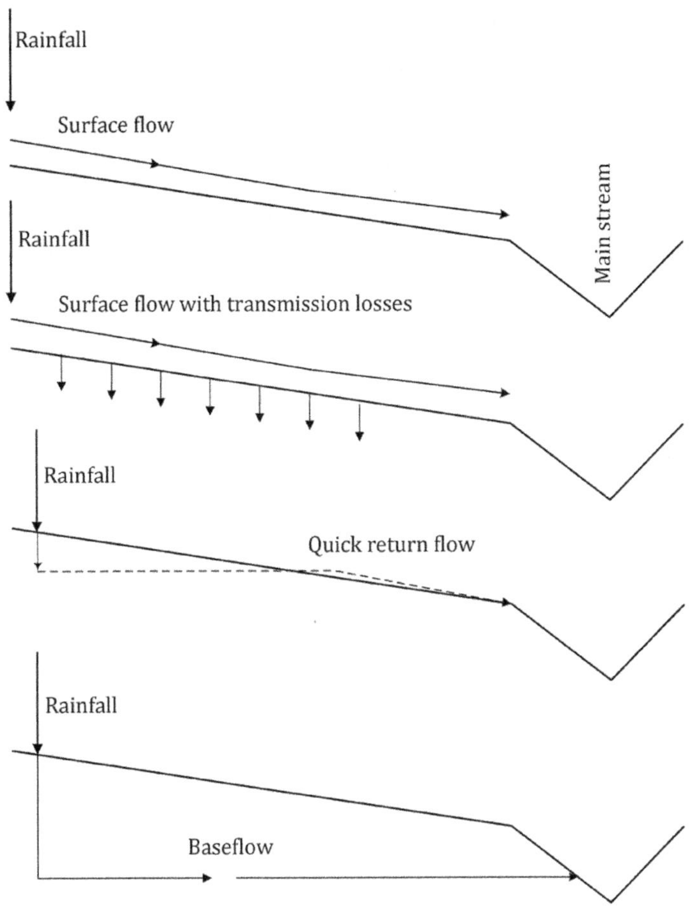

Figure 6.21 Conceptual representation of types of flow.

stream system. Distinctions are made for these flow types for realistic estimation of travel times accordingly.

1. **Surface flow**: Surface flow represents the overland flow of water. Three types of overland flow are recognized: (i) sheet flow; (ii) shallow concentrated flow; and (iii) channel flow.
2. **Surface flow with transmission losses**: This is also the overland flow component but during flow some water is lost by further infiltration, the subject matter of Chapter 7.
3. **Interflow or quick return flow**: Interflow or quick return flow is the water that first infiltrates the ground at a point, then moves horizontally to resurface at another point downslope.
4. **Baseflow**: Baseflow is the groundwater component of stream flow, discussed in Chapter 8. Water from rainfall infiltrates vertically, reaches the groundwater table, and moves very slowly toward a stream. Baseflow contributes to stream flow below the line where the water table surface intersects the stream slope.

Figure 6.21 is a conceptual representation of the flow paths described above. In the following we will be concerned with the travel time for surface flow.

6.13 METHODS FOR ESTIMATION OF TIME OF CONCENTRATION

6.13.1 Watershed Lag Method

The **watershed lag method** was developed by Mockus (1961) and this is one of the methods NRCS uses for the estimation of the time of concentration. In this method, first the watershed lag time (h) is estimated using

$$t_l = \frac{L^{0.8}\left[\left(\frac{1000}{CN} - 10\right) + 1\right]^{0.7}}{1900 S^{0.5}} \quad (6.60)$$

in which L denotes the flow length (ft), S represents the average watershed slope, given as percent, and CN is the curve number of the watershed, discussed in Chapter 7. Once the lag time is calculated, t_c is calculated using Eq. 6.59.

In Eq. 6.60, the flow length is the longest flow path discussed in Section 6.3.3. Mockus (in USDA, 1973) developed an empirical relationship between flow length (ft) and watershed area, A (in acres), which is also a power function like Hack's law and is given as

$$L = 209 A^{0.6} \quad (6.61)$$

The average watershed slope in Eq. 6.60 is estimated as discussed in Section 6.3.7.

The lag method was developed using data from 24 watersheds, ranging in size from 1.3 acres to 9.2 square miles. However, it spans a broad set of conditions ranging from heavily forested watersheds with steep channels and a high percentage of runoff resulting from subsurface flow to meadows, providing a high retardance (low CN) to surface runoff, and to smooth land surfaces and large paved areas (high CN). Thus, Eq. 6.60 was originally developed for nonurban watersheds. However, it has been adapted to small urban watersheds under 2000 acres. Nonetheless, because the lag equation was developed for rural areas, it tends to overestimate lag time for two reasons. First, the increased amount of impervious area allows water from overland flow sources and side channels to reach the main channel at a much faster rate than under natural conditions. Second is the extent to which a stream (usually the major watercourse in the watershed) has been changed over natural conditions to allow higher flow velocities. The Federal Highway Administration (1985) suggested adjustment factors to correct for channel improvement and impervious area that should be multiplied to the result obtained Eq. 6.60. They developed the channel improvement factor as a function of percent of hydraulic length of the channel that is modified and the curve number of the watershed. Similarly, the impervious factor is given as a function of percent of impervious area and curve number of the watershed.

6.13.2 Velocity Method

The **velocity method** is perhaps the most common method used by practicing engineers to estimate the watershed time of concentration. It was originally developed by NRCS and for this reason it is also known as the NRCS method of computation of time of concentration. NRCS (2010) termed this the velocity method mainly because in it the flow path is divided into consecutive flow segments, and the travel time for each of the segments is calculated by estimating the velocity of flow in those segments. Then all of the travel times are added to get the total watershed time of concentration,

$$t_c = t_{t1} + t_{t2} + t_{t3} + \ldots + t_{tm} \quad (6.62)$$

where t_{ti} denotes the travel time in flow segment i and n indicates the number of flow segments.

As noted above, surface flow can be of three types, namely sheet flow, shallow concentrated flow, and channel flow. Therefore,

$$t_c = t_{sh} + \sum t_{sc} + \sum t_{ch} \quad (6.63)$$

in which the travel times for sheet flow, shallow concentrated flow, and channel flow are denoted by t_{sh}, t_{sc}, and t_{ch}, respectively, and the summation sign is used to indicate that there can be multiple segments of both shallow concentrated and channel flows.

6.13.2.1 Sheet Flow

Sheet flow refers to flow over plane surfaces, usually occurring in the headwater area of the overland segment of a catchment, that is, near the ridgelines that define the catchment boundary. The travel time for a sheet flow segment is given by a simplified version of the kinematic wave equation using the Manning equation, developed by Welle and Woodward (1986) after studying the impact of various parameters on the estimates.

$$t_{sh} = \frac{0.007(n_{ol}L_{sh})^{0.8}}{(P_2)^{0.5} S_{sh}^{0.4}} \quad (6.64)$$

where n_{ol} is the overland roughness coefficient, L_{sh} is the length of sheet flow (ft), S_{sh} is the land surface slope of the sheet flow segment, and P_2 is the rainfall depth (in) for a frequency-based design storm with 24-h duration and a 2-year return period.

With sheet flow, the friction value is an effective roughness coefficient that includes the effect of raindrop impact, drag over the plane surface, obstacles such as litter, crop ridges, and rocks, and erosion and transportation of sediments. Table 6.3 provides the roughness coefficients for sheet flow for various surface conditions. These values are for very shallow flow depths of about 0.1 foot or so.

Because Eq. 6.64 was developed through the simplification of the kinematic wave equation for overland flows, several assumptions were made that led to set the upper limit of the length of sheet flow to be 100 feet. However, McCuen and Spiess (1995) indicated that the use of flow length as the limiting variable in Eq. 6.64 could lead to less accurate designs and proposed that the limitation should instead be based on

$$L_{sh} = \frac{100\sqrt{S_{sh}}}{n_{ol}} \quad (6.65)$$

Table 6.3 *Manning's roughness coefficients for sheet/overland flow*

Surface type	Manning's roughness coefficient, n_{ol}
Smooth surface (concrete, asphalt, gravel, or bare soil)	0.011
Fallow (no residue)	0.05
Cultivated soils: residue cover $\leq 20\%$	0.06
Cultivated soils: residue cover $> 20\%$	0.17
Short-grass prairie	0.15
Dense grasses	0.24
Bermuda grass	0.41
Range (natural)	0.13
Woods: light underbrush	0.4
Woods: dense underbrush	0.8

6.13.2.2 Shallow Concentrated Flow

After 100–300 feet, sheet flow usually becomes shallow concentrated flow collecting in swales, small rills, and gullies. Shallow concentrated flow is assumed not to have a well-defined channel, with depth of flow in the range of 0.1–0.5 foot. NRCS (2010) developed empirical coefficients, C_{sc}, for the equation

$$V_{sc} = C_{sc}\left(\sqrt{S_{sc}}\right) \tag{6.66}$$

giving the velocity of shallow concentrated flow, V_{sc}, as a function of the slope, S_{sc}, of the shallow concentrated flow path. The coefficients depend on the type of surface or flow type. Seven types of flow and the corresponding values of C_{sc} are given in Table 6.4. Plots of Eq. 6.66 show that, for each flow type, the velocity linearly increases with the slope and depends on the surface type, with forest flow having the lowest velocities and pavement surface having highest values (Figure 6.22).

Once a flow type (surface) is recognized and its velocity is computed, the travel time is calculated from Eq. 6.57.

6.13.2.3 Channel Flow

Shallow concentrated flow becomes channel flow when the flow path becomes well defined along channels with depth of flow greater than 0.5 foot. In urban areas, this may include, in addition to the overland component, pipes of different shapes in a storm drainage system. In either case, the flow velocity for channel flow is computed using the Manning equation:

$$V_{ch} = \frac{1.486}{n_{ch}}(R_h)^{2/3}\sqrt{S_{ch}} \tag{6.67}$$

Usually, bankfull discharge is assumed to calculate these two parameters.

Manning's roughness coefficients for various type of open channels, streets and gutters, and closed conduits are given in Tables 6.5–6.7, respectively.

6.13.3 Kerby Method

Kerby (1959) developed an equation for computing the time of concentration for overland flow. For small watersheds, where overland flow is an important component of overall travel time, the Kerby method can be used. The equation giving travel time for overland flow, t_{ov}, is given as

$$t_{ov} = C_1(N \times L_{ov})^{0.467} S_{ov}^{-0.235} \tag{6.68}$$

where C_1 is a coefficient (0.828 for the US unit system and 1.44 for SI), L_{ov} is the overland flow length (ft or m), S_{ov} is slope of the terrain conveying the overland flow, and N is a dimensionless retardance coefficient.

The Kerby method is restricted to overland flow distances of less than 1200 feet (366 m). Even though this length is

Table 6.4 *Manning's roughness (n) and slope coefficients (C_{SC}) in Eq. 6.66 for different surface types*

Surface type	Manning's coefficient, n	Slope coefficient, C_{sc}
Pavements and small upland gullies	0.025	20.328
Grassed waterways	0.050	16.135
Nearly bare and untilled (overland flow); and alluvial fans in western mountain regions	0.051	9.965
Cultivated straight row crops	0.058	8.762
Short grass pasture	0.073	6.962
Minimum tillage cultivation, contour or strip-cropped, and woodlands	0.101	5.032
Forest with heavy ground litter and hay meadows	0.202	2.516

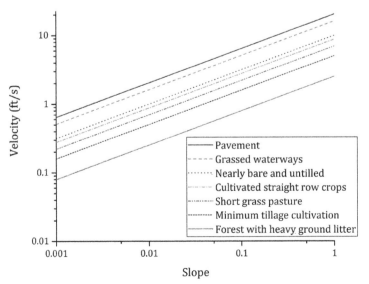

Figure 6.22 Velocity versus slope for shallow concentrated flow (velocity method of time of concentration).

Table 6.5 *Manning's roughness coefficients for open channels*

Type of channel	Manning's coefficient, n
Natural streams	
Minor streams (top width at flood stage < 100 ft)	
Clean, straight, full, no rifts or deep pools	0.025–0.033
Same as a, but more stones and weeds	0.030–0.040
Clean, winding, some pools and shoals	0.033–0.045
Same as c, but some weeds and stones	0.035–0.050
Same as d, lower stages, more ineffective	0.040–0.055
Same as d, more stones	0.045–0.060
Sluggish reaches, weedy, deep pools	0.050–0.080
Very weedy, heavy stand of timber and underbrush	0.075–0.150
Mountain streams with gravel and cobbles, few boulders on bottom	0.030–0.050
Mountain streams with cobbles and large boulders on bottom	0.040–0.070
Floodplains	
Pasture, no brush, short grass	0.025–0.035
Pasture, no brush, high grass	0.030–0.050
Cultivated areas, no crop	0.020–0.040
Cultivated areas, mature row crops	0.025–0.045
Cultivated areas, mature field crops	0.030–0.050
Scattered brush, heavy weeds	0.035–0.070
Light brush and trees in winter	0.035–0.060
Light brush and trees in summer	0.040–0.080
Medium to dense brush in winter	0.045–0.110
Medium to dense brush in summer	0.070–0.160
Trees, dense willows summer, straight	0.110–0.200
Trees, cleared land with tree stumps, no sprouts	0.030–0.050
Trees, cleared land with tree stumps, with sprouts	0.050–0.080
Trees, heavy stand of timber, few down trees, flood stage below branches	0.080–0.120
Trees, heavy stand of timber, few down trees, flood stage reaching branches	0.100–0.160
Major streams (top width at flood stage > 100 ft)	
Regular section with no boulders or brush	0.025–0.060
Irregular rough section	0.035–0.100
Excavated or dredged channels	
Earth, straight and uniform	
Clean, recently completed	0.016–0.020
Clean, after weathering	0.018–0.025
Gravel, uniform section, clean	0.022–0.030
With short grass, few weeds	0.022–0.033
Earth, winding and sluggish	
No vegetation	0.023–0.030
Grass, some weeds	0.025–0.033
Deep weeds or aquatic plants in deep channels	0.030–0.040
Earth bottom and rubble sides	0.028–0.035
Stony bottom and weedy banks	0.025–0.040
Cobble bottom and clean sides	0.030–0.050

Table 6.5 (*cont.*)

Type of channel	Manning's coefficient, n
Winding, sluggish, stony bottom, weedy banks	0.025–0.040
Dense weeds as high as flow depth	0.050–0.120
Dragline-excavated or dredged	
No vegetation	0.025–0.033
Light brush on banks	0.035–0.060
Rock cuts	
Smooth and uniform	0.025–0.040
Jagged and irregular	0.035–0.050
Unmaintained channels	
Dense weeds, high as flow depth	0.050–0.120
Clean bottom, brush on sides	0.040–0.080
Clean bottom, brush on sides, highest stage	0.045–0.110
Dense brush, high stage	0.080–0.140
Lined channels	
Asphalt	0.013–0.016
Brick (in cement mortar)	0.012–0.018
Concrete	
Trowel finish	0.011–0.015
Float finish	0.013–0.016
Unfinished	0.014–0.020
Gunite, regular	0.016–0.023
Gunite, wavy	0.018–0.025
Riprap (n-value depends on rock size)	0.020–0.035
Vegetal lining	0.030–0.500

Table 6.6 *Manning's roughness coefficients for streets and gutters*

Type of gutter or pavement	n
Concrete gutter, troweled finish	0.012
Asphalt pavement: smooth texture	0.013
Asphalt pavement: rough texture	0.016
Concrete gutter with asphalt pavement: smooth texture	0.013
Concrete gutter with asphalt pavement: rough texture	0.015
Concrete pavement: float finish	0.014
Concrete pavement: broom finish	0.016

Note: For gutters with small slope or where sediment may accumulate, n values should be increased by 0.02 (Federal Highway Administration, 1984).

considered an upper limit of the flow distances, the method is only applied for shorter distances. Table 6.8 gives the retardance coefficients that can be used in the Kerby equation.

6.13.4 Kirpich Method

Kirpich (1940) developed an equation to estimate the travel time for the channel flow as

$$t_{ch} = C_2 (L_{ch})^{0.770} S_{ch}^{-0.385} \qquad (6.69)$$

Table 6.7 *Manning's roughness coefficients for closed conduits*

Material	Manning's coefficient, n
Asbestos-cement pipe	0.011–0.015
Brick	0.013–0.017
Cast iron pipe	
Cement-lined and seal coated	0.011–0.015
Concrete (monolithic)	
Smooth forms	0.012–0.014
Rough forms	0.015–0.017
Concrete pipe	0.011–0.015
Box (smooth)	0.012–0.015
Corrugated-metal pipe (2-1/2 in × 1/2 in corrugations)	
Plain	0.022–0.026
Paved invert	0.018–0.022
Spun asphalt lined	0.011–0.015
Plastic pipe (smooth)	0.011–0.015
Corrugated-metal pipe (2-2/3 in × 1/2 in annular)	0.022–0.027
Corrugated-metal pipe (2–2/3 in × 1/2 in helical)	0.011–0.023
Corrugated-metal pipe (6 in × 1 in helical)	0.022–0.025
Corrugated-metal pipe (5 in × 1 in helical)	0.025–0.026
Corrugated-metal pipe (3 in × 1 in helical)	0.027–0.028
Corrugated-metal pipe (6 in × 2 in structural plate)	0.033–0.035
Corrugated-metal pipe (9 in × 2-1/2 in structural plate)	0.033–0.037
Corrugated polyethylene	0.010–0.013
Smooth	0.009–0.015
Corrugated	0.018–0.025
Spiral rib metal pipe (smooth)	0.012–0.013
Vitrified clay	
Pipes	0.011–0.015
Liner plates	0.013–0.017
Polyvinyl chloride (PVC) (smooth)	0.009–0.011

Note: Manning's coefficient for corrugated pipes is a function of the corrugation size, pipe size, and whether the corrugations are annular or helical (see Arcement and Schneider, 1989).

where C_2 is a coefficient (0.0078 for the US unit system and 0.0195 for SI), L_{ch} is channel flow length (ft or m), and S_{ch} is slope of the channel. Kirpich developed this equation for small drainage basins in Tennessee and Pennsylvania with basin areas from 1 to 112 acres (0.40-45.3 ha). Also, the exponent for the channel slope varied from −0.385 for Tennessee watersheds to −0.5 for the basins in Pennsylvania.

6.13.5 Kerby–Kirpich Method

Equations 6.68 and 6.69 are combined to estimate the total travel time ($t_t = t_{ov} + t_{ch}$).

Table 6.8 *Retardance coefficient (N) for the Kerby equation*

Generalized terrain description	Retardance coefficient, N
Pavement (smooth impervious surface)	0.02
Smooth, bare, packed soil	0.10
Poor grass, cultivated row crops, or moderately rough packed surfaces	0.20
Pasture, average grass	0.40
Deciduous forest	0.60
Dense grass, coniferous forest, or deciduous forest with deep litter or dense grass	0.80

$$t_t = C_1(N \times L_{ov})^{0.467} S_{ov}^{-0.235} + C_2(L_{ch})^{0.467} S_{ch}^{-0.385} \quad (6.70)$$

The Kerrby–Kirpich method for estimating t_c is applicable to watersheds ranging from 0.25 to 150 square miles, main channel lengths between 1 and 50 miles, and main channel slopes between 0.002 and 0.02 (Roussel et al., 2005).

6.13.6 Time of Concentration Equations

There is a multitude of formulations developed by various workers for estimating the time of concentration. Table 6.9 summarizes the well-known ones in the SI system. Almost all equations involve two physical characteristics of watersheds, namely, the length of the watershed, L, and the slope of the watershed, S. It would be very convenient if nature were obliging and conformed to the simplifying assumptions that are made to represent a complex natural process in order to obtain an easy solution. Several workers have attempted to relate watershed lag time and the ratio of L to the square root of the main channel slope, S, by a simple regression equation. In many cases, however, no such luck has been evident. In some other formulations, the surface roughness coefficient has also been factored into the formulations. And in the case of application of kinematic wave theory to the overland flow equation, rainfall intensity has also been incorporated in the equation for calculation of t_c. Singh (1976) showed that t_c depends on the duration of rainfall. A theoretical background for t_c can be found in Singh (1976), where an equation for t_c was derived by analyzing surface flow using kinematic wave theory. Even though the equation was derived under some simplifying assumptions, such as a rectangular and converging cross section, it provided a basic idea about t_c in a basin. The equation can be written as

$$t_c = (i)^{\frac{1-m}{m}} \left(\frac{L}{\eta}\right)^{\frac{1}{m}} \quad (6.71)$$

where L is the length of longest flow path of a basin, i is the rainfall intensity, η is the kinematic wave friction-related

258   6 Geomorphologic Concepts and Watershed Characteristics

Table 6.9 *Common empirical formulas for t_c*

Formula	Reference	Formula	Reference
$t_c = 0.0074 \dfrac{L}{S^{0.515}}$	Kraven - I (Japan Society of Civil Engineers, 1999)	$t_c = 43.75 \dfrac{L^{0.29}}{S^{0.145}}$	Espy and Winslow (1974)
$t_c = 0.0074 \dfrac{L}{V}$	Kraven - II (Japan Society of Civil Engineers, 1999)	$t_c = \dfrac{1}{3600} \sum \dfrac{L}{V}$	SCS (1975)
$t_c = 0.0139 \dfrac{L}{S^{0.6}}$	Rziha (1876)	$t_c = \dfrac{0.257 L^{0.8}\left[\left(\dfrac{1000}{CN}\right)-9\right]^{0.7}}{1900 S^{0.5}}$	SCS (1985); lag method
$t_c = 0.0663 \dfrac{L^{0.77}}{S^{0.385}}$	Kirpich (1940)	$t_c = 5.6256 \dfrac{L^{0.9417}}{S^{0.2639} A^{0.3666}}$	Ahn and Lee (1986)
$t_c = 0.543 \left(\dfrac{L}{S}\right)^{0.5}$	Johnstone and Cross (1949)	$t_c = \dfrac{0.000524(1.1-C) L^{0.5}}{S^{1/3}}$	Federal Aviation Administration (1970)
$t_c = 0.6059 \left(\dfrac{L.N}{\sqrt{S}}\right)^{1/2.14}$	Kerby (1959)	$t_c = 1.54 \dfrac{L^{0.875}}{S^{0.181}}$	USGS (2000)
$t_c = 0.83 \left(\dfrac{nL}{\sqrt{S}}\right)^{0.47}$	Hathaway (1945)	$t_c = 1.08 \dfrac{A^{0.09} L^{0.16}}{S^{0.12}}$	Yoon and Park (2002)
$t_c = 58.1 \dfrac{L^{0.6}}{S^{0.3}}$	Carter (1961)	$t_c = 0.119 \dfrac{L^{0.777}}{S^{0.212}}$	Jung (2005)
$t_c = \dfrac{1.396 \times 10^{-6} L^{0.6} n^{0.6}}{i^{0.4} S^{0.3}}$	Morgali and Linsley (1965)	$t_c = 0.0663 \left(\dfrac{L^3}{H}\right)^{0.385}$	California Department of Transportation (1955) also USBR (1973)
$t_c = \dfrac{0.00547(0.0178 i + C) L^{0.6}}{i^{0.667} S^{0.333}}$	Izzard and Hicks (1947)	$t_c = 0.089 \dfrac{A^{0.427}}{S^{0.239}}$	Kim (2015)

Note: t_c is the concentration time (h), A is the basin area (km$_2$), L is the channel length (km), S is the basin slope or channel slope, N is the retardance coefficient, V is the mean velocity (m/s), i is the rainfall intensity (in/h), and n is the Manning's roughness coefficient.

parameter varying in space (friction parameter), and m is a constant.

The lag method and the velocity method, developed in NRCS, are two commonly use methods for the computation of runoff hydrographs. Sometimes the Kerby–Kirpich method also gets preference for small watershed studies. This preference is based on the simplicity of approach and ease of acquisition of the required input data. Furthermore, it is a rather straightforward approach and produces t_c values that Roussel et al. (2005) found to mimic the time to peak from analysis of observed rainfall and runoff data from 92 watersheds in Texas. Nonetheless, an engineer must review the applicability and limitation of each of these formulations before adopting a particular one. However, various regulatory agencies have either mandated or expressed preference for a particular method. The velocity method of NRCS has received wide acceptance in the United States.

6.14 EXAMPLES

Example 6.1: Obtain estimates of the bifurcation ratio, the length ratio, and area ratio for Youghiogheny River watershed in Maryland, USA, using the data given in Table 6.10 (from Morisawa, 1962). Also determine the drainage density, stream frequency, and average overland flow length.

Solution: Horton's laws are given as

$$N_w = R_b^{W-w}$$
$$\overline{L}_w = \overline{L}_1 R_l^{w-1}$$
$$\overline{A}_w = \overline{A}_1 R_a^{w-1}$$

Taking logarithms of each of the above equations leads to:

$\log N_w = -w \log R_b + W \log R_b$
$\log \overline{L}_w = w \log R_l + \log \overline{L}_1 - \log R_l$
$\log \overline{A}_w = w \log R_a + \log \overline{A}_1 - \log R_a$

These equations are linear in w. Thus, estimates of R_b, R_l, and R_a can be obtained by linear regression of $\log N_w$ versus w, $\log \overline{L}_w$ versus w, and $\log \overline{A}_w$ versus w.

The slope (m) of the regression line shown in Figure 6.23 is -0.6097

$$R_b = \frac{1}{anti \log m} = \frac{1}{10^{-.6097}} = 4.071$$

Table 6.10 *Geomorphic parameters of Youghiogheny River watershed in Maryland, USA*

Order of stream, w	Number of streams, N	Average length, $\overline{L}$ (mi)	Average basin area, $\overline{A}$ (mi²)
1	1798	0.124	0.038
2	452	0.34	0.212
3	62	0.951	1.005
4	13	2.572	7.079
5	4	5.644	40.85
6	1	5.3	134

Table 6.11 *Calculated values of $\log N_w$*

Order of stream, w	$\log (N_w)$
1	3.2869
2	2.8439
3	2.3201
4	1.5798
5	0.9542
6	0.3010

Table 6.12 *Calculated values of $\log \overline{L}_w$*

Order of stream, w	$\log \overline{L}_w$
1	−1.1805
2	−0.7496
3	−0.3028
4	0.1544
5	0.5857
6	0.9465

The slope (m) of the regression line shown Figure 6.24 is 0.4314.

$$R_l = \frac{1}{anti \log m} = \frac{1}{10^{0.4314}} = 0.3703$$

The slope (m) of the regression line shown in Figure 6.25 is 0.6565.

$$R_a = \frac{1}{anti \log m} = \frac{1}{10^{0.6565}} = 0.2205$$

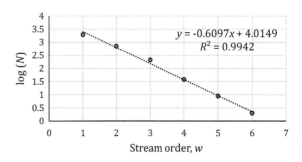

Figure 6.23 $\log (N_w)$ versus w (Youghiogheny River watershed) from values given in Table 6.11.

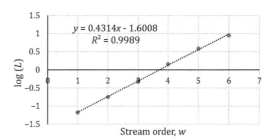

Figure 6.24 $\log \overline{L}_w$ versus w (Youghiogheny River watershed) from values given in Table 6.12.

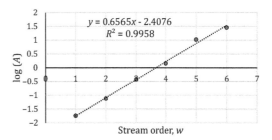

Figure 6.25 $\log \overline{A}_w$ versus w (Youghiogheny River watershed) for values given in Table 6.13.

Table 6.13 *Calculated values of* $\log \overline{A}_w$

Order of stream, w	$\log \overline{A}_w$
1	−1.7447
2	−1.1192
3	−0.4214
4	0.1553
5	1.0183
6	1.4533

For the determination of the drainage density, we need to calculate the total length of the stream using the equation:

$$L_T = \sum_{w=1}^{W} N_w \overline{L}_w$$

For our case,

$$L_T = 1798 \times 0.124 + 452 \times 0.34 + 62 \times 0.951 + 13 \\ \times 2.572 + 4 \times 5.644 + 1 \times 5.3 \\ = 496.91 \text{ mi}$$

Thus, the drainage density is

$$D_d = \frac{496.91 \text{ mi}}{134 \text{ mi}^2} = 3.71/\text{mi}$$

The stream frequency is given as

$$F = \frac{1798 + 452 + 62 + 13 + 4 + 1}{134 \text{ mi}^2} = 17.4/\text{mi}^2$$

The overland flow length is given as

$$L_o = \frac{1}{2 \times 3.71/\text{mi}} = 0.135 \text{ mi}$$

Example 6.2: For the headwater watershed of Arkansas River up to Buena Vista, shown in Figure 6.7, the longest flow path length, L, is 58.8 mile, the mean watershed slope, S, is 32%, and the curve number (CN) is 51.88, all computed by the StreamStats program of the USGS. StreamStats also gives the time of concentration for this watershed as 19.61 hours (it does not tell how this is calculated). Confirm this estimate of the time of concentration for this watershed.

Solution: Since the available data are given for L, CN, and S, the lag method developed by Mockus of NRCS is used to calculate the watershed lag time. Note the system of units used in this equation.

$L = 58.8$ mi. $L = 58.8 \text{ mi} \times 5280 \dfrac{\text{ft}}{\text{mi}} = 310464$ ft.

We first calculate the storage factor from the CN value (see Chapter 7)

$$S = \frac{1000}{51.88} - 10 = 9.2753$$

Then, the lag time is calculated as

$$t_l = \frac{(310464^{0.8})(9.2753)^{0.7}}{1900 \times 32^{0.5}} = 11.76 \text{ h}$$

Since the lag time is generally 60% of the time of concentration, we estimate:

$$t_c = \frac{11.76 \text{ h}}{0.6} = 19.61 \text{ h}$$

Example 6.3: A small catchment (area = 52.95 acres) within Graham Branch Watershed, located to the northwest of Dallas, Texas, is shown in Figure 6.26. A reach of Graham Branch flows through the catchment from north to south, shown as the main water course. The longest flow path is shown to have three segments, also shown in Figure 6.26. The topographic contours of the catchment are also shown in the figure.

The morphologic parameters of the three segments of the longest flow path are given in Table 6.14.

A channel transect along which the cross-sectional data are derived is shown in Figure 6.27. The cross-sectional data are given in Table 6.15. In this area the intensity of 24-hour, 2-year rainfall is 3.63 in/h. Determine the lag time for this catchment.

Solution:

Step 1: Calculate the travel time for the sheet flow segment.

The slope of the segment is calculated as

$$S_{sh} = \frac{(674 - 673) \text{ ft}}{100 \text{ ft}} = 0.010$$

For the short-grass prairie, $n_{ol} = 0.15$ (Table 6.3).

Table 6.14 *Morphologic parameters of the three segments of the longest flow path*

Flow path segments	Elevation upstream (ft)	Elevation downstream (ft)	Length (ft)
Sheet flow (short-grass prairie)	674	673	100
Shallow concentrated flow (grassed surface)	673	647	1104.25
Channel flow (clean, winding, some pools and shoals)	647	631	2135.06

The travel time for the sheet flow segment is calculates as

$$t_{sh} = \left[\frac{0.007 \times (0.15 \times 100)^{0.8}}{3.63^{0.5} \times 0.01^{0.4}} \right] \times 60 \frac{\min}{h} = 12.14 \min$$

Step 2: Calculate of velocity and travel time for the shallow concentrated flow segment.

The slope of the segment is calculated as

$$S_{sc} = \frac{(673 - 647) \text{ ft}}{1104.25 \text{ ft}} = 0.0235$$

For grassed waterway/surfaces, the slope coefficient $C_{SC} = 16.135$ (Table 6.4)

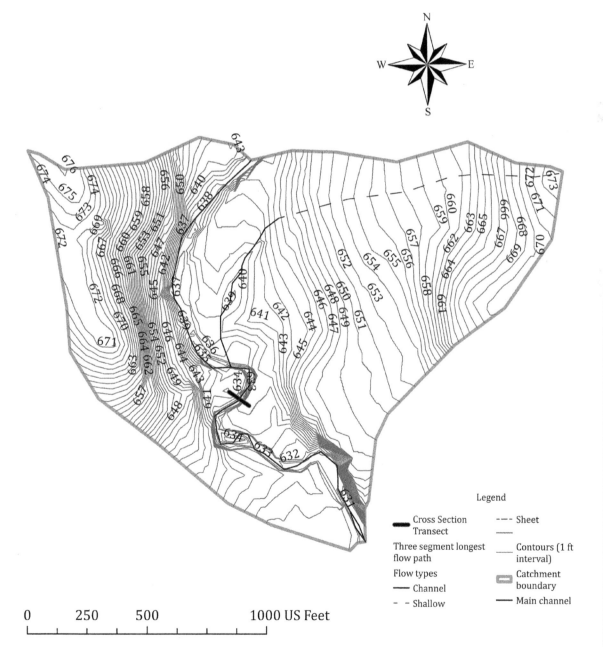

Figure 6.26 A catchment (AVA MCC01) within Graham Branch Watershed.

Thus, the flow velocity is given as

$$V_{sc} = 16.135 \times 0.0235^{0.5} = 2.48 \text{ ft/s}$$

The travel time is calculated as

$$t_{sc} = \frac{1104.25 \text{ ft}}{3600\frac{s}{h} \times 2.48\frac{ft}{s}} \times 60\frac{\min}{h} = 7.43 \text{ min}$$

Step 3: Calculate of velocity and travel time for the channel flow segment.

From the cross-sectional data given in Table 6.15, the flow area and wetted perimeter are calculated assuming bank-full discharge at 638.029 ft. The cross-sectional plot is shown in Figure 6.27. For each station elevation, the

Table 6.15 *Channel cross-sectional data*

Station (ft)	Elevation (ft)	Station (ft)	Elevation (ft)
0.000	638.029	52.042	634.260
3.253	637.704	55.294	633.947
6.505	637.516	58.547	633.335
9.758	637.468	61.800	633.815
13.010	637.436	65.052	634.981
16.263	637.417	68.305	636.263
19.516	637.371	71.557	637.987
22.768	637.329	74.810	638.299
26.021	637.232	78.063	638.295
29.273	636.987	81.315	638.268
32.526	636.704	84.568	638.248
35.779	636.364	87.821	638.232
39.031	635.960	91.073	638.203
42.284	635.661	94.326	638.185
45.537	635.238	97.578	638.164
48.789	634.729	100.831	638.140

cross section is assumed to be a rectangular strip. The mean depth of the strip is calculated from two successive depths and the area of the strip is calculated as a rectangle. Areas of all rectangles below the water surface (638.029 ft) are added to get the flow area. Similarly, for each successive station elevation below the water surface, the wetted perimeter is the hypotenuse of a right-angled triangle. The sum of all these give the total wetted perimeter. Table 6.16 shows the calculations to derive $A_q = 1956.73 \text{ ft}^2$ and $P_w = 788.34$ ft.

The slope of the segment is calculated as

$$S_{ch} = \frac{(647 - 631) \text{ ft}}{2135.06 \text{ ft}} = 0.0075$$

For a clean, winding, natural channel with pools and shoals, $n_{ch} = 0.045$ (Table 6.5).

Thus, the velocity of the channel flow is calculated as

$$V_{ch} = \frac{1.486 \times \left(\frac{1956.73}{788.34}\right)^{\frac{2}{3}} \times \sqrt{0.0075}}{0.045} = 5.24 \text{ ft/s}$$

The travel time is calculated as

$$t_{ch} = \frac{2135.06 \text{ ft}}{3600\frac{s}{h} \times 5.24\frac{ft}{s}} \times 60\frac{\min}{h} = 6.79 \text{ min}$$

Step 4: Calculate the time concentration and lag time.

The time concentration and lag time of the watershed are calculated as

$$t_c = (12.14 + 7.43 + 6.79) \text{ min} = 26.36 \text{ min}$$

$$t_l = 0.6 \times 26.36 \text{ min} = 15.82 \text{ min}$$

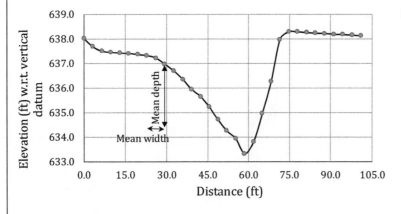

Figure 6.27 Channel cross section (station elevation)

Table 6.16 *Calculation of flow area and wetted perimeter*

Station (ft)	Elevation (ft)	Mean distance (ft)	Depth (ft)	Mean depth (ft)	Strip area (ft^2)	~ Length (ft)
0.000	638.029		0.000			
3.253	637.704	1.626	−0.325	−0.163	−0.264	1.634
6.505	637.516	4.879	−0.514	−0.419	−2.046	4.897
9.758	637.468	8.132	−0.561	−0.537	−4.367	8.149
13.010	637.436	11.384	−0.593	−0.577	−6.565	11.399
16.263	637.417	14.637	−0.612	−0.602	−8.814	14.649
19.516	637.371	17.889	−0.658	−0.635	−11.360	17.901
22.768	637.329	21.142	−0.700	−0.679	−14.355	21.153
26.021	637.232	24.395	−0.797	−0.748	−18.254	24.406
29.273	636.987	27.647	−1.042	−0.920	−25.423	27.662
32.526	636.704	30.900	−1.325	−1.184	−36.579	30.922
35.779	636.364	34.152	−1.665	−1.495	−51.066	34.185
39.031	635.960	37.405	−2.069	−1.867	−69.831	37.452
42.284	635.661	40.658	−2.368	−2.218	−90.197	40.718
45.537	635.238	43.910	−2.791	−2.580	−113.273	43.986
48.789	634.729	47.163	−3.300	−3.045	−143.632	47.261
52.042	634.260	50.415	−3.769	−3.534	−178.193	50.539
55.294	633.947	53.668	−4.082	−3.925	−210.669	53.811
58.547	633.335	56.921	−4.694	−4.388	−249.768	57.089
61.800	633.815	60.173	−4.214	−4.454	−268.021	60.338
65.052	634.981	63.426	−3.048	−3.631	−230.296	63.530
68.305	636.263	66.678	−1.766	−2.407	−160.505	66.722
71.557	637.987	69.931	−0.043	−0.904	−63.249	69.937

Example 6.4: A small culvert needs to be designed for the crossing of a proposed road over an existing small stream within a small catchment (GB 08) of Graham Branch Watershed located to the northwest of Dallas in a suburban setting. The catchment boundary (area = 1.12 km^2) with the streams and longest flow path are shown in Figure 6.28.

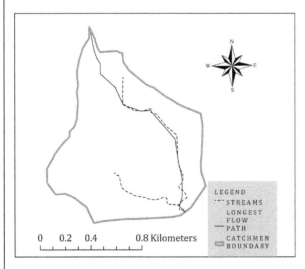

Figure 6.28 Catchment GB 08.

Because of the small size of the catchment and the nature of the project, the design engineer decided to use Kerby–Kirpich method to calculate the time of concentration for this catchment to use in the rational method for the estimation of peak discharge for the design of the culvert. The design engineer delineated the longitudinal flow path from an areal photograph and derived the longitudinal profile of this flow path as shown in the Table 6.17. The total length of the longest flow path is 1.689 km.

From a field inspection, the engineer determined that the first 220.66 m of pasture of the longest flow path is overland flow, as shown in Figure 6.29. Determine the time of concentration for this catchment.

Solution: Slope of the overland segment is calculated as

$$S_{ov} = \frac{(228.89 - 226.46) \text{ m}}{220.66 \text{ m}} = 0.01097$$

For pasture, the retardance coefficient $N = 0.4$ (Table 6.8) The time of concentration for the overland segment is calculated as

$$t_{c,ov} = \frac{1.44 \times (0.4 \times 220.66)^{0.467}}{(0.01097)^{0.235}} = 33.7 \text{ min}$$

Table 6.17

Distance (m)	Elevation (m)
0.00	228.89
66.89	228.03
126.49	227.35
189.81	227.12
220.66	226.46
570.10	222.67
1093.79	219.64
1248.20	218.18
1455.82	217.08
1688.61	212.25

For the channel section, the length of the overland flow must be subtracted from the total length. Thus,

Length of channel flow $= (1688.61 - 220.66)$ m $= 1467.95$ m

The slope of the overland segment is calculated as

$$S_{ch} = \frac{(226.46 - 212.25)\, \text{m}}{1467.95\, \text{m}} = 0.00969$$

The time of concentration for the channel segment is calculated as

$$t_{c,ch} = \frac{0.0195 \times (1467.95)^{0.77}}{(0.00969)^{0.385}} = 31.9\, \text{min}$$

The total time of concentration is calculated as

$$t_c = (33.7 + 31.9)\, \text{min} = 65.6\, \text{min}$$

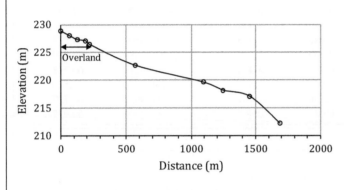

Figure 6.29 Longitudinal profile of longest flow path

Example 6.5: The DEM, HUC-12 boundaries (shapefile), and stream network (shapefile) of the catchments drained by Cherry Creek and Coal Creek within Middle South Platte–Cherry Creek Watershed of South Platte River basin (Figure 6.3) are downloaded from the USGS website. A map created from these GIS data is shown in Figure 6.30. Import these data into the latest version of HEC-HMS and use the GIS functions available in HEC-HMS to derive geomorphologic parameters of the catchments and streams. From the calculated parameters conduct the following experiments.

a. In the Snyder unit hydrograph method, the watershed lag time is estimated from watershed characteristics expressed either as $(LL_c)^{0.3}$ or $\left(\frac{LL_c}{\sqrt{S}}\right)^{0.38}$. Is there any correlation between these two derived parameters?
b. Calculate the time of concentration for the catchments using the equations that relate stream length and stream slope and make observations on which equation yields the most reasonable estimates.

Solution: In HEC-HMS the following steps are taken to derive the geomorphologic parameters of the catchments and streams.

Step 1: Create a "new project."

The DEM of the watershed is imported into the project as "terrain data." This operation is performed by choosing from the top menu Components –> Create Component –> Terrain Data. Prior to bringing the DEM into HEC-HMS as terrain data, it must be projected to a standard projection system outside HEC-HMS if the DEM is in the geographic coordinate system (GCS). In this example, the original DEM is in GCS, and it is projected to Albers equal-area conic projection system of the USGS used for conterminous United States. The vertical unit is set as meters by choosing the metric system.

Step 2: Create a new basin model.

A new basin model is created by again choosing from the top menu Components –> Create Component –> Basin Model. Now in the Component Editor panel, the terrain data is attached to the basin model and the project is saved after answering the prompts for the projection system. The terrain (DEM) appears in the desktop pane.

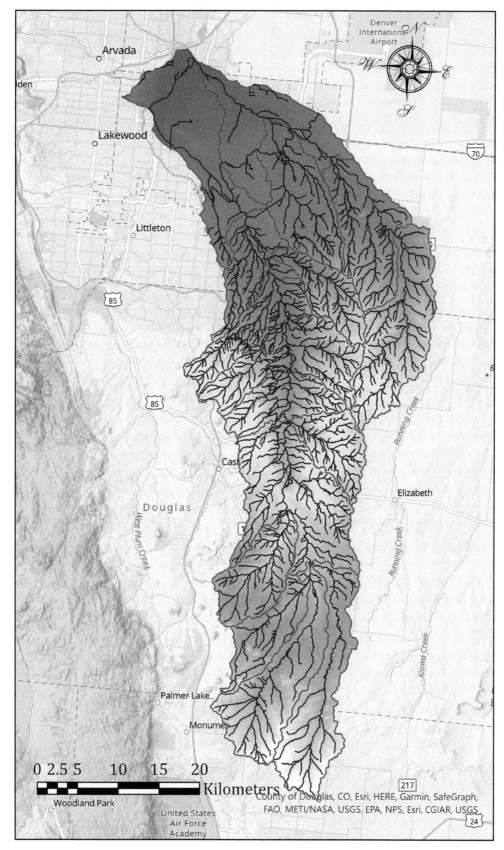

Figure 6.30 DEM, HUC 12 boundaries, and streams of the catchments of Cherry Creek and Coal Creek watersheds. (A black and white version of this figure will appear in some formats. For the color version, please refer to the plate section.)

Step 3: Run the steps from the GIS menu in sequence.

Once a basin model is created and a terrain data is attached to it, all of the steps from the GIS menu are run in sequence. Note that the "Terrain Reconditioning" step is not always necessary, and the process can begin with the "Preprocess Sink" operation or even "Preprocess Drainage" operation. However, the Terrain Reconditioning operation is necessary when the catchment delineation from the DEM needs to be matched with a known catchment boundary map and further if the streamlines delineated from the DEM need to be matched with a known stream map. In this case, we want to have the delineated catchments to match the HUC 12-digit boundaries and the streamlines to match exactly with the streamlines given in the National Hydrography Dataset (NHD) of the USGS. So, Terrain Reconditioning is run. The Terrain Reconditioning has two steps. The first step is called "build walls" and the second step is called "burn streams." The build wall step is used to delineate the catchment boundaries with the shapefile of the catchments and the burn stream step is run to get the delineated streams to match the shapefile of the streams. Either one or both steps can be skipped. In this example, both steps are run. After completion of the "Identify Stream" operation, a "break point" is placed at the watershed outlet and then the "Delineate Elements" operation is run. HEC-HMS builds the entire drainage network with appropriate downstream connections from subbasins (catchments) to junctions and reaches.

Step 4:
After completion of the GIS operations, from the parameter menu, first select "subbasin area" to calculate the area of each catchment. Then again under parameter menu select "characteristics" and then select "subbasins." This will compute all parameters of the catchments. Repeat this second step but now select "reaches" and this will compute all parameters for the streams.

The results of the catchment characteristics are given in Table 6.18.

The results of the stream characteristics are given in Table 6.19.

Comment: The parameters $(LL_c)^{0.3}$ and $\left(\frac{LL_c}{\sqrt{S}}\right)^{0.38}$ are calculated and plotted in Figure 6.31. These two derived parameters do show a strong correspondence (the correlation coefficient is shown in Figure 6.31.

Of all the formulas given in Table 6.9 relating the time of concentration with reach length and slope, the Kirpich formula yields the most reasonable results. The results of the formula given by Jung give results that are very close to the results obtained from the Kirpich formula. The results obtained using the formula given by Kraven (I) yield realistic values but consistently lower than the values obtained from the Kirpich formula. The other equations yield much higher values of t_c, which seem unrealistic for these catchments. Figure 6.32 shows the plots of t_c calculated from the Kirpich formula against channel length.

Table 6.18 *Geomorphologic parameters of the catchments*

Area (km^2)	Longest flow path length (km)	Longest flow path slope	Centroidal flow path length (km)	Centroidal flow path slope	10–85 flow path length (km)	10–85 Flow path slope	Basin slope	Basin relief (m)	Relief ratio	Elongation ratio	Drainage density (km/km^2)
90.650	24.324	0.010	14.995	0.007	18.243	0.009	0.075	263.015	0.011	0.442	0.183
17.258	12.140	0.015	6.061	0.011	9.105	0.014	0.086	185.091	0.015	0.386	0.132
24.149	11.939	0.014	5.301	0.009	8.954	0.012	0.080	186.061	0.016	0.464	0.154
18.140	16.455	0.013	7.476	0.009	12.342	0.011	0.097	240.481	0.015	0.292	0.166
20.370	11.732	0.013	4.683	0.008	8.799	0.012	0.063	151.458	0.013	0.434	0.345
17.384	11.107	0.012	4.842	0.006	8.330	0.010	0.060	132.972	0.012	0.424	0.440
123.37	38.859	0.008	23.053	0.007	29.144	0.007	0.069	325.321	0.008	0.323	0.253
3.5226	4.009	0.024	1.502	0.009	3.006	0.012	0.070	94.927	0.024	0.528	0.557
36.353	13.110	0.020	6.968	0.010	9.832	0.013	0.108	268.529	0.020	0.519	0.140
16.907	9.755	0.019	4.559	0.013	7.316	0.016	0.097	193.981	0.020	0.476	0.059
33.749	16.364	0.010	8.142	0.010	12.273	0.008	0.062	177.776	0.011	0.401	0.304
26.203	12.945	0.016	5.631	0.009	9.709	0.012	0.103	210.512	0.016	0.446	0.295
45.621	15.177	0.015	7.672	0.012	11.383	0.014	0.097	224.836	0.015	0.502	0.163

Table 6.18 (cont.)

Area (km²)	Longest flow path length (km)	Longest flow path slope	Centroidal flow path length (km)	Centroidal flow path slope	10-85 flow path length (km)	10–85 Flow path slope	Basin slope	Basin relief (m)	Relief ratio	Elongation ratio	Drainage density (km/km²)
9.2782	8.318	0.025	3.519	0.015	6.239	0.023	0.140	204.916	0.025	0.413	0.552
26.094	18.755	0.017	9.220	0.016	14.067	0.016	0.108	327.470	0.017	0.307	0.226
4.9894	4.092	0.031	1.841	0.018	3.069	0.027	0.078	129.490	0.032	0.616	0.437
28.948	14.257	0.016	6.178	0.005	10.693	0.017	0.087	226.672	0.016	0.426	0.230
16.793	11.481	0.017	6.169	0.014	8.611	0.015	0.126	200.278	0.017	0.403	0.102
57.816	17.335	0.014	8.327	0.012	13.001	0.013	0.085	247.256	0.014	0.495	0.172
0.0830	0.565	0.025	0.123	0.008	0.424	0.014	0.032	14.830	0.026	0.575	3.562
25.548	12.537	0.016	4.468	0.007	9.402	0.014	0.096	205.274	0.016	0.455	0.265
19.626	12.908	0.016	6.399	0.014	9.681	0.013	0.117	220.201	0.017	0.387	0.089
15.782	10.072	0.021	5.545	0.016	7.554	0.019	0.099	219.325	0.022	0.445	0.004
0.2937	1.569	0.035	0.599	0.021	1.177	0.033	0.073	55.428	0.035	0.390	0.923
24.746	9.786	0.016	5.167	0.012	7.340	0.014	0.092	174.727	0.018	0.574	0.134
15.723	12.415	0.018	6.355	0.015	9.311	0.016	0.085	225.245	0.018	0.360	0.006
19.638	11.409	0.015	5.776	0.006	8.557	0.013	0.067	170.729	0.015	0.438	0.327
4.1831	6.675	0.019	3.159	0.013	5.006	0.017	0.073	129.397	0.019	0.346	0.699
39.038	18.656	0.015	10.806	0.011	13.992	0.013	0.099	277.616	0.015	0.378	0.268
5.0695	4.217	0.019	1.513	0.006	3.163	0.015	0.048	78.667	0.019	0.602	0.521
44.241	22.063	0.014	12.109	0.009	16.548	0.011	0.099	310.159	0.014	0.340	0.326
30.983	15.287	0.013	5.974	0.004	11.466	0.010	0.072	195.431	0.013	0.411	0.252
57.104	22.294	0.011	12.632	0.007	16.720	0.009	0.082	239.972	0.011	0.382	0.260
12.720	8.159	0.016	3.997	0.005	6.119	0.011	0.057	130.064	0.016	0.493	0.300
36.118	18.618	0.013	8.912	0.007	13.964	0.009	0.057	235.498	0.013	0.364	0.176
2.7599	6.270	0.006	2.863	0.004	4.703	0.004	0.026	38.366	0.006	0.299	1.925
18.227	12.599	0.013	5.796	0.009	9.449	0.010	0.058	167.597	0.013	0.382	0.097
15.442	10.780	0.013	5.344	0.008	8.085	0.011	0.060	139.387	0.013	0.411	0.057
29.496	14.954	0.006	7.773	0.002	11.216	0.003	0.047	94.460	0.006	0.410	0.262
10.805	7.560	0.012	3.078	0.010	5.670	0.012	0.048	94.063	0.012	0.491	0.341
28.375	17.093	0.005	11.691	0.002	12.820	0.003	0.037	80.218	0.005	0.352	0.320
7.2805	6.051	0.011	2.596	0.005	4.538	0.010	0.048	69.505	0.011	0.503	0.533
27.211	23.087	0.009	10.674	0.006	17.315	0.008	0.050	218.186	0.009	0.255	0.268
0.9786	2.671	0.005	1.276	0.002	2.003	0.004	0.047	14.348	0.005	0.418	2.575
27.496	15.075	0.013	7.860	0.008	11.306	0.010	0.066	188.579	0.013	0.393	0.256
15.329	9.671	0.015	4.354	0.010	7.253	0.012	0.080	142.857	0.015	0.457	0.066
22.087	9.173	0.012	3.746	0.006	6.880	0.012	0.076	115.919	0.013	0.578	0.071
10.341	7.722	0.015	3.834	0.007	5.791	0.010	0.079	119.659	0.016	0.470	0.522
68.644	21.586	0.010	9.351	0.006	16.189	0.008	0.072	216.889	0.010	0.433	0.219
21.509	14.130	0.012	7.188	0.007	10.597	0.009	0.074	166.772	0.012	0.370	0.224
25.332	15.736	0.011	8.819	0.010	11.802	0.009	0.063	177.981	0.011	0.361	0.261
12.236	7.618	0.013	2.543	0.008	5.713	0.011	0.052	96.761	0.013	0.518	0.357
32.734	21.295	0.009	10.418	0.006	15.971	0.007	0.059	206.582	0.010	0.303	0.311
18.391	13.381	0.008	5.893	0.005	10.036	0.006	0.036	105.816	0.008	0.362	0.517
29.056	19.139	0.004	6.478	0.005	14.354	0.004	0.034	101.095	0.005	0.318	0.374
11.028	6.264	0.006	3.147	0.006	4.698	0.006	0.039	47.629	0.008	0.598	0.521
27.162	10.950	0.005	4.222	0.006	8.213	0.006	0.028	61.816	0.006	0.537	0.087
13.736	12.655	0.004	5.935	0.004	9.491	0.004	0.030	59.415	0.005	0.330	0.282
15.214	11.212	0.005	5.420	0.005	8.409	0.005	0.040	65.792	0.006	0.393	0.554

Table 6.19 *Stream characteristics*

Length (km)	Slope	Relief (m)	Sinuosity
7.650	0.006	44.475	1.327
7.032	0.007	47.325	1.243
1.968	0.009	16.782	1.115
10.280	0.008	80.931	1.519
7.736	0.008	65.491	1.419
5.126	0.015	77.945	1.288
2.187	0.004	9.481	1.150
6.652	0.004	27.065	1.390
0.301	0.004	1.205	1.107
6.766	0.004	27.370	1.448
0.276	0.009	2.589	1.146
6.421	0.004	27.528	1.440
2.930	0.010	29.326	1.283
2.646	0.004	11.556	1.216
7.813	0.004	32.517	1.404
3.827	0.004	14.933	1.365
5.321	0.004	19.046	1.311
7.735	0.001	11.344	1.173
3.688	0.003	11.620	1.488
3.883	0.006	22.147	1.223
2.525	0.004	9.776	1.624
5.755	0.005	28.956	1.292
5.404	0.007	38.483	1.236
15.028	0.006	89.030	1.306
4.379	0.005	22.381	1.327
9.511	0.004	41.819	1.207
10.861	0.005	53.809	1.172
3.882	0.004	14.696	1.095
8.423	0.005	40.800	1.138

Note: HEC-HMS calculates stream sinuosity as one of the reach characteristics. Sinuosity is a measure of the degree to which a channel meanders. It is commonly measured as the ratio of the channel thalweg distance and the down-valley or air distance between two points of a river reach. It is also called the sinuosity index (SI). Straight streams will have a sinuosity equaling 1, while highly meandering streams can have a value as high as 4.0. (Note that the term thalweg refers to the line extending down a channel that follows the lowest elevation of the bed. Thus, thalweg is the geomorphologic term equivalent to flow line used in engineering.)

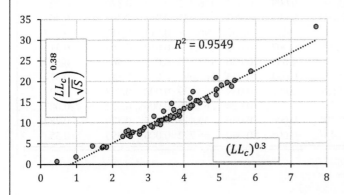

Figure 6.31 Plot of two derived watershed physical characteristics.

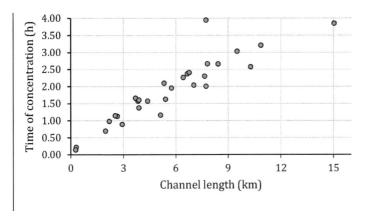

Figure 6.32 Time of concentration (Kirpich formula) as a function of channel length.

Example 6.6: The longitudinal profile of approximately six miles of the headwater of Graham Branch, a stream in suburban setting located to the northwest of Dallas is shown in Figure 6.33. The profile is derived from a DEM constructed from LiDAR-derived elevation data using GIS. The equations that best fit each of the four segments of the stream profile are also shown in Figure 6.33. What qualitative observations can be made from such a profile?

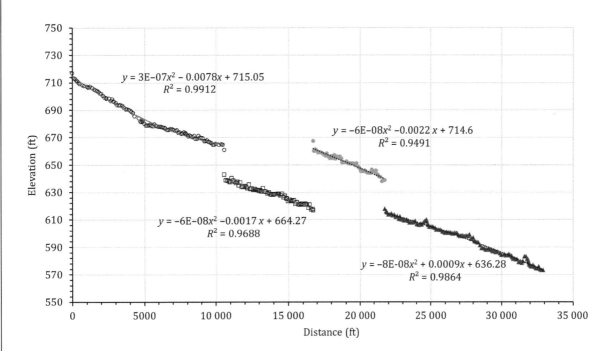

Figure 6.33 Equations that best fit each of the four segments of the stream profile.

Comment: The river profile does not show concave upward pattern. All of the four segments can be represented by a linear relationship between elevation and distance. However, fitting the profile data by quadratic polynomials gives slightly improved regression coefficients. The breaks in the profile in segment 2 are clearly a manifestation of scouring at the upstream end and deposition downstream. For this headwater stream segment, the work done by the stream appears to be uniformly distributed.

Exercises: A selection of exercises on this topic is available at www.cambridge.org/appliedhydrology.

7 Abstractions and Effective Rainfall

7.1 CONCEPTS AND PURPOSE

The total volume of rainfall that falls on any portion of the land surface throughout the duration of a rain event does not flow into the streams present in that area. A part of the rainfall is intercepted by the vegetation present on the land surface, some is stored in surface depressions, some goes into the ground by the process of infiltration, and another part of the rainfall returns to the atmosphere through the process of evapotranspiration. The process of interception by foliage, infiltration into the ground, and storage and detention in surface depressions is collectively called **abstraction**. In applied hydrology, abstraction is also loosely called losses since they do not contribute to direct runoff, even though in the hydrologic cycle no water is lost, according to the law of conservation of mass. The total volume of rainfall minus the volume of rainfall lost by abstraction is called **effective rainfall** or **rainfall excess**. This is the volume of rainfall that contributes to stream flow and is called **direct runoff**. For individual rain events, evapotranspiration is not significant; however, for long-term water balance it is quite significant, as we shall see later in Chapter 19.

This chapter deals with the collective process of abstraction, with special reference to the models that provide the quantitative estimates of the rainfall amounts that go into infiltration and surface runoff.

7.2 PHYSICAL PROCESS OF PRODUCTION OF EXCESS RAINFALL

The classical concept of direct runoff generation is known as Hortonian flow, named after Robert Elmer Horton, who propounded the theory of overland flow in a landmark paper (Horton, 1945). Accordingly, the process of overland flow results when all surface storage sites are filled, and *rainfall rate exceeds the infiltration rate.* Overland flow that does not infiltrate along the flow path to a channel results in direct runoff (Figure 7.1A). This is also known as **infiltration-excess runoff**. Such a process is applicable to urban and agricultural land surfaces where the soils reach the infiltration capacity relatively quickly or the infiltration capacity is relatively low and runoff generation is independent of the location of the surface in the landscape. However, on a landscape where the infiltration capacity of the soil cover is very high, such as on forested soils, due to well-developed surface cover and extensive tree root structure, the contribution of overland flow to direct runoff is very low. In these areas, overland flow only occurs when the soil becomes fully saturated, and any additional rainfall causes runoff. This is called **saturation-excess overland flow**. Interflow through the unsaturated soil horizon (vadose zone), groundwater (or baseflow), and overland flow are the major components of direct runoff (Figure 7.1B). As rainfall continues, the areas that contribute to these components grow in size and are termed **variable source areas**.

Figure 7.2 is a schematic representation of abstraction processes during a uniform-intensity rainfall. Depending on the land surface temperature, evaporation can initially start at a high rate, but its rate decreases rapidly and eventually reaches a low, steady-state rate. The amount of rainfall that goes to evaporation during an individual rain event causing significant runoff is negligible. Interception varies, depending upon the type of vegetation, maturity, and extent of canopy cover. Infiltration and surface storage are the major processes of abstraction.

Three periods of abstraction are shown in Figure 7.2. During the first period, there are initial losses and no rainfall excess. The rainfall loss to direct runoff during this initial period is a function of depression storage, interception, evaporation, and the high rate of infiltration capacity of the pervious ground surface. The total amount of rainfall loss during this period of no direct runoff is called the **initial abstraction** (sum of all mechanisms including infiltration). The initial period ends with the onset of ponded water on the surface. The time lapsed from the beginning of rainfall to this point is called time of ponding (T_p). The second period is marked by a declining infiltration rate and negligible role of other factors that influence rainfall loss. Effective rainfall, that is, direct runoff starts in this period. The third, and final period occurs for rainfall of sufficient duration when the infiltration rate attains a steady state or equilibrium (constant value).

An **effective rainfall hyetograph** resulting from a highly simplified representation of the process of abstraction, which in reality has much greater complexity, is shown in Figure 7.3. Surface storage is the cumulative loss

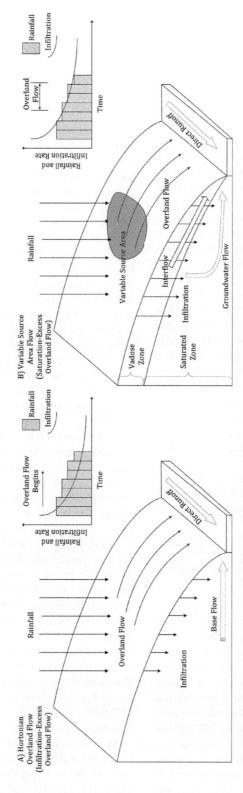

Figure 7.1 Models of generation of direct runoff. (A) Hortonian flow or infiltration-excess overland flow; (B) Variable source area or saturation-excess overland flow.

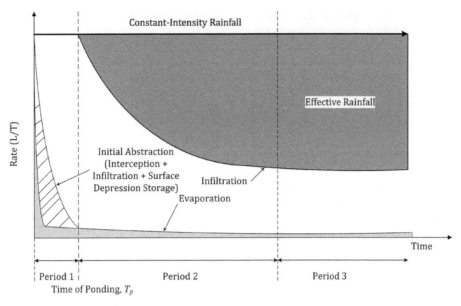

Figure 7.2 Schematic representation of variations of effective rainfall and losses with time for a rainfall occurring with uniform intensity.

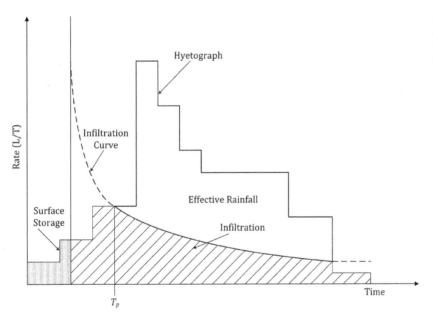

Figure 7.3 Simplified representation of an effective rainfall hyetograph produced from an observed hyetograph as well as surface storage and infiltration.

due to abstractions except infiltration and this occurs from the start of rainfall up to the time when the accumulated rainfall equals the capacity of all surface retention. From this point, infiltration begins. As long as rainfall intensity is less than the infiltration capacity of the ground, no rainfall excess occurs. Effective rainfall begins when rainfall intensity exceeds infiltration capacity. The effective rainfall hyetograph is the observed hyetograph minus all these losses.

7.3 INTERCEPTION

Interception storage occurs due to the absorption of rainfall by surface cover, mainly trees, plants, and other vegetation, such as grass and shrubs, growing on the land surface. The amount of total rainfall that is intercepted by foliage depends on vegetation characteristics, such as the type, density, and stage of growth above the ground, which in turn depends on the season of the year, rainfall intensity and duration, and wind speed.

Equations have been developed to estimate the amount of interception (e.g., Horton, 1919), however, these have little practical use because it is either not possible to quantify or estimate several of the variables in those equations. Obviously then, the estimation of the interception amount of rainfall is highly empirical and varies according to the types of vegetation present on the land surface under consideration. For example, the intercepted amount will vary from forested land to agricultural land to urbanized areas. Even within agricultural lands, it depends on the dominant type of crops present.

It should be noted that a portion of initially intercepted rainfall actually makes its way to the ground either through stemflow or from spillage from the leaves that cannot hold all of the rainfall initially falling on them. The volume of intercepted rainfall is usually small if the rain event produces a significant rainfall depth over a relatively small area. According to USACE (1994), the intercepted amount of rainfall can range from 0.03 inches to 0.5 inches (1–13 mm).

7.4 INFILTRATION

Infiltration of water into pervious land surfaces during significant rainfall events is the major form of abstraction. It is the most important hydrologic process that affects the rate and volume of direct runoff from a watershed during a rain event. The process of infiltration is extremely complex and difficult to quantify accurately due to the interplay of numerous factors that control the space–time variabilities of the amount of rainfall that infiltrates the ground and the complexities of the physical mechanisms that actually govern the infiltration process. Singh (1992) has provided an in-depth discussion of the infiltration process, the factors that affect infiltration, and the methods that are used for actual measurements of infiltration. Chief factors that control infiltration are the physical properties of soil or soil type, antecedent soil moisture conditions, vegetative cover, slope or topography of the infiltrating surface, and rainfall intensity.

7.5 MODELS OF INFILTRATION

7.5.1 Physical Properties of Soil or Ground

The principal physical properties of soil used in the mathematical formulation of the infiltration process are **porosity**, **moisture content**, and **hydraulic conductivity**.

A column of soil consists of soil or rock particles and interstitial voids. For a column of earth material, such as soil, soil–rock mixture, or fractured rock, of unit volume, the porosity is given by

$$\eta = \frac{V_v}{V_t} \quad (7.1)$$

where V_v represents the volume of void space (L^3) within a unit volume of earth material, and V_t represents the total (unit) volume (L^3) of the earth material. Actually, all of the pore spaces within a volume of earth material cannot contain water because some of the pores are either not large enough to contain water molecules or those that can are not interconnected. Therefore, the term **effective porosity**, denoted by η_e, is used to refer to the porosity available for flow of water.

A portion of the void is filled with water and the remainder by air. Therefore, another soil property called **soil moisture content**, or simply **moisture content**, denoted by θ is defined as

$$\theta = \frac{V_w}{V_t} \quad (7.2)$$

where V_w represents the volume of water within the volume of the void space.

It follows then that $0 \leq \theta \leq \eta$. Typically, for near-surface soils, $0.25 < \eta < 0.75$ and $\theta < \eta_e$, that is, the soil is unsaturated. The unsaturated soil zone is called the **vadose zone**. When, $\theta = \eta_e$, the volume of soil under consideration is saturated. Under this condition, θ is designated by θ_s.

Before any rain event, the soil column can have an initial moisture content, designated by θ_i. When rain starts and water starts to infiltrate, the ground gradually becomes saturated from the top surface and a wetting front develops, θ tends to increase toward θ_s as time progresses and the wetting front extends in depth. Figure 7.4 illustrates the

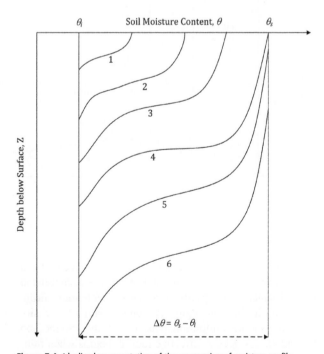

Figure 7.4 Idealized representation of the progression of moisture profile through a column of soil with time under a constant rainfall rate. Curve 1 represents the earliest time after the commencement of rainfall and curves 4, 5, and 6 signify saturated surface and initiation of direct runoff. Such a curve is also called a wetting front.

movement of moisture profile with depth and time. Initially, the soil is assumed to have a uniform water content, θ_i. At the commencement of rainfall, water is infiltrated until the rainfall exceeds the capacity of the soil to absorb water. At this point, the soil surface becomes saturated and rainfall in excess of the soil infiltration capacity is assumed to be runoff (curves 4, 5, 6 in Figure 7.4 depict this condition). At any instant, the difference between θ_s and θ_i is called **soil moisture deficit** or simply **moisture deficit** and is denoted by $\Delta\theta$. Thus, $\Delta\theta = \theta_s - \theta_i$. Soil moisture deficit is defined as the amount of water that is required to saturate the soil. The corresponding value at the commencement of a simulation or computation is called the initial deficit. Both are given as a volume ratio (Eq. 7.2).

Hydraulic conductivity, denoted by K, is the coefficient of permeability[1] of an earth material. It has the dimensions of length/time (L/T) or velocity. It signifies the velocity of flow of water through a porous media by being the volume of water that flows through a unit cross-sectional area of a soil column per unit time under a unit hydraulic gradient. When the soil is unsaturated, K is denoted by $K(\theta)$, that is, unsaturated hydraulic conductivity and, when the soil is saturated, it is denoted by K_s.

Hydraulic conductivity is a function of fluid properties as well as soil properties. It can be expressed as a product of two parts, one depending on the fluid properties and the other depending on the soil properties or the medium. Thereby, hydraulic conductivity can be expressed as

$$K = \frac{Nd^2\rho_w g}{\mu} \quad (7.3)$$

where N is a dimensionless shape factor relating to the geometry of passage, d is the mean diameter of soil grains, μ is the viscosity of the fluid (water), ρ_w is the density of the fluid (water), and g is gravitational acceleration. The hydraulic conductivity has been found to have a very large variation – over thirteen orders of magnitude.

French hydrologist Henri Darcy (1856) discovered that the velocity of groundwater was proportional to the hydraulic gradient. Darcy's law can be stated as

$$q = -K\frac{\Delta h}{L} \quad (7.4)$$

where q denotes Darcy velocity, Δh is the hydraulic gradient, that is, the difference in hydraulic heads at two points on the flow path at a distance of L. The Darcy velocity or **Darcy flux** is not the real macroscopic velocity of water, but the velocity as if the water were moving through the entire cross-sectional area normal to the flow, solid media as well as pores. The negative sign in Eq. 7.4 is due to the fact that the hydraulic head in the downstream direction is lower than that in the upstream direction caused by the head loss of flow.

During the infiltration process, as the volume of infiltrated water increases, the infiltration capacity of the soil decreases to a minimum rate equal to the soil's saturated hydraulic conductivity.

The **relative conductivity**, denoted by K_r, is the ratio of the hydraulic conductivity for a given moisture content $[K(\theta)]$ to the saturated conductivity, K_s. During an infiltration process, K_r can assume any value depending on θ but reaches its maximum limit 1 when the soil assumes θ_s.

Another important element of the flow of water in the vadose zone is the **capillary potential**. It is the hydraulic head caused by capillary forces. **Capillary suction**, denoted by ψ (dimension L) refers to the capillary potential with the opposite sign. It is a measure of the combined adhesive forces that bind the water molecules to solid walls and the cohesive forces that attract water molecules to each other. It is also called the tension head and can be measured by an instrument called a tensiometer. The capillary suction is added to the **elevation head** (z) to produce the total **hydraulic head** (h) acting on an element of water, $h = z + \psi$. Figure 7.5 is a schematic illustration of the variation of capillary suction as a function of moisture content during periods of wetting (infiltration) and drying.

The properties of soil or earth materials described above are important for modeling infiltration processes during a single rainfall event. However, for long-term or continuous simulations of rainfall–runoff processes, some additional characteristics of the soil profile are considered. These include the following soil properties.

The **field capacity** or **water holding capacity** denotes the water remaining in a soil profile after it has been thoroughly saturated and then allowed to drain freely by gravity. This can take one to several days after a wetting period and during the drying period. Usually, it is measured as water content retained in soil at -0.33 bar of hydraulic head or suction pressure. It is a dimensionless parameter since it is the ratio of the volume of water and bulk volume of soil.

The **soil matric potential** (SMP) represents the relative availability of the amount of water held in the soil profile for plant uptake or use. It is a realistic criterion for measuring soil water availability to plants as it constitutes the force with which water is held by the soil matrix, which constitutes soil particles and pore space, and is measured by a tensiometer giving its unit in force per unit area or pressure.

The **wilting point** is the water content of soil at which a plant cannot absorb any more water and it starts to wilt since the water is held so tightly by the soil particles. Usually, it is measured as water content at an SMP of -15 bar. It is also a dimensionless parameter, since it is the ratio of volume of water and bulk volume of soil.

The **available water content** is the water held in soil between its field capacity and wilting point. It is also a

---

[1] Permeability may be described in qualitative terms as the ease with which water can move through a porous media and is measured by the rate of flow in suitable units.

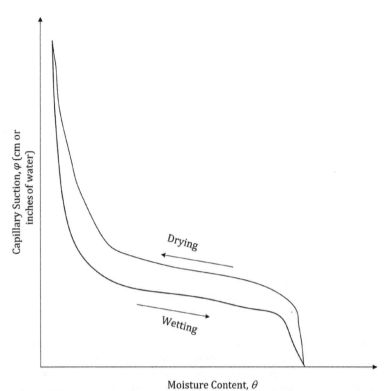

Figure 7.5 Capillary suction (cm or inches) as a function of moisture content during the periods of infiltration (wetting) and drying.

dimensionless parameter since it is that ratio of volume of water and bulk volume of soil. The available water content is usually used for irrigation and agriculture.

The **residual soil water content** or **residual moisture content**, denoted by θ_r, is the volumetric water content of a soil where a further increase in negative pore-water pressure does not produce significant changes in water content. It represents the remaining water content after a saturated soil is allowed to drain thoroughly for an extended period of time. It is dimensionless since it is the ratio of the volume of water and bulk volume of soil. Usually, the soil moisture content cannot be zero. A realistic minimum soil water content can be defined as the residual moisture content and, thus, effective porosity is porosity minus the residual moisture content: $\eta_e = \eta - \theta_r$.

The **maximum deficit** is defined as the total or maximum amount of water that is required to saturate the soil when the soil is totally dry or at its minimum moisture content. The maximum deficit can be estimated as the product of the active soil layer depth and the effective porosity. Thereby, its unit is in length (inches or cm).

7.5.2 Soil Moisture Flow Equation

For the soil–water system, the continuity equation for one-dimensional unsteady, unsaturated flow expressing the conservation of mass in terms of the Darcy flux can be written as

$$\frac{\partial \theta}{\partial t} + \frac{\partial q}{\partial z} = 0 \tag{7.5}$$

A combination of Darcy's law, applied to unsaturated flow, and the equation of continuity results in the second-order, nonlinear partial differential equation that describes the flow of water through unsaturated porous media or the vadose zone under the action of capillarity and gravity. This equation, often attributed to the American soil physicist Lorenzo Adolph Richards, who published the equation in 1931 (Richards, 1931), was actually first published by the British mathematician and physicist Lewis Fry Richardson in 1922 (Richardson, 1922). In one dimension, **Richards equation**, as it is commonly called, can be written as

$$\frac{\partial \theta}{\partial t} = -\frac{\partial}{\partial z}\left[K(\theta)\frac{\partial \psi(\theta)}{\partial z}\right] - \frac{\partial K(\theta)}{\partial z} \tag{7.6}$$

where z denotes the vertical coordinate (positive downward with dimension), t denotes time (T), $\theta(z,t)$ is the volumetric soil moisture content (dimensionless), $\psi(\theta)$ is the soil hydraulic capillary head (L), also called the **suction head** or simply capillary suction, and $K(\theta)$ is the unsaturated hydraulic conductivity L/T. Equation 7.6 mixes water content θ with capillary head ψ, but it can also be written solely as a function of water content by introducing the chain rule:

$$\frac{\partial \theta}{\partial t} = \frac{\partial}{\partial z}\left[D(\theta)\frac{\partial \theta}{\partial z}\right] - \frac{\partial K(\theta)}{\partial z} \tag{7.7}$$

where $D(\theta)$ is referred to as **soil water diffusivity** (L^2/T) and is given by

$$D(\theta) = K(\theta) \frac{\partial \psi(\theta)}{\partial \theta} \tag{7.8}$$

The first term on the right-hand side of Eq. 7.6 captures the effects of capillary action or capillarity and the second term represents the effect of gravity-driven flux.

Practical application of Eq. 7.6 or Eq. 7.7 is extremely difficult because these equations do not have closed-form analytical solutions.

7.5.3 Horton, Philip, and Holtan Equations

The infiltration equation developed by Horton and several modifications suggested by other workers subsequently, can be used to illustrate the general nature of the time-dependence of infiltration even though nowadays these equations have lost their importance in practical applications.

The term *infiltration* refers to the process of the entry of water into the ground through the land surface, whereas the term **infiltration rate**, denoted by f, refers to the rate at which water enters into the soil surface. It is expressed as volume per unit area per unit time and has the dimension of L/T (cm/h or in/h). In general, the **infiltration capacity**, f_p, is the maximum rate L/T at which soil can absorb water through its surface. It should be noted that: $0 \leq f \leq f_p$. If there is no limit of the availability of water, then $f = f_p$. As illustrated in Figure 7.6, initially $(t = 0)$, the infiltration rate is the highest but, with time, it gradually decreases and then tends to reach a constant value denoted by f_c. The term f_c is also called the ultimate or **constant** or **steady rate of infiltration capacity**. The difference between f and f_c, denoted by $f - f_c$ is called the **rate of excess infiltration**, f_e. The volume of infiltration from the beginning of rainfall $(t = 0)$ to any time t is called the **cumulative infiltration** and is denoted by F and has the dimension of L which implies that it is the accumulated depth of water infiltrated during a given time period. The amount, $F_e = F(t) - f_c t$ is called the **excess infiltration**.

It follows (τ being the dummy variable of integration) that

$$F(t) = \int_0^t f(\tau) d\tau \tag{7.9}$$

$$f(t) = \frac{dF(t)}{dt} \tag{7.10}$$

Horton (1933, 1939) developed an infiltration equation given as

$$f(t) = f_c + (f_0 - f_c) e^{-kt} \tag{7.11}$$

where k is a decay constant (T^{-1}). Horton's equation can be derived from the Richards equation (Eq. 7.6) by assuming that K and D are constant, independent of the moisture content of the soil. Under these conditions, Eq. 7.7 reduces to

$$\frac{\partial \theta}{\partial t} = D \frac{\partial^2 \theta}{\partial z^2} \tag{7.12}$$

which is the standard form of a **diffusion equation** and can be solved to get the variation of θ as a function of depth and time.

Horton's equation requires three parameters: f_0 (initial infiltration rate), f_c (final infiltration rate), and k (a measure of the rate of decrease in the infiltration rate). It is very difficult to estimate these parameters without having experimental data from an infiltrometer in the area under concern and therefore this equation cannot be readily adopted for calculation of the infiltration rate, $f(t)$, at any time.

Philip (1957) presented an analytical solution of Richards equation for homogeneous soils with a uniform initial moisture content and an excess water supply at the surface. The solution is in the form of an infinite series but, because of the rapid convergence, only the first two terms need to be considered, giving the cumulative infiltration,

$$F(t) = St^{0.5} + Kt \tag{7.13}$$

where S is a parameter called **sorptivity**. Differentiation of Eq. 7.13 gives the infiltration rate,

$$f(t) = \frac{1}{2} S t^{-0.5} + K \tag{7.14}$$

The problem of applying Philip's equation is that it requires values of two parameters S and K, which are difficult to estimate correctly without having experimental data.

Holtan (1961) advanced the Hortonian concept further by proposing an infiltration model derived from a substantial volume of field data. The Holtan model of infiltration is

$$f = f_c + a F_p^n \tag{7.15}$$

where F_p denotes the **unfilled capacity** of the soil to store water, that is, the **potential infiltration volume,** which is equal to the initial available moisture storage minus the volume of water already infiltrated, and a and n are constants. Equation 7.15 is well suited for inclusion in hydrologic modeling, because it links infiltration capacity to the soil moisture level. The exponent n has been found to be about 1.4 for many soils. However, the value of a varies from one area to another. The value of F_p ranges from a maximum of the **available water capacity** to zero. The available water capacity is a measure of the ability of a soil to store water and values for it are given for many soils by the US Department of Agriculture (USDA) (USDA, 1968). However, it should be noted that as soon as infiltration begins F_p starts decreasing and it decreases continuously as long as infiltration exceeds drainage from the soil profile. The variation of F_p as a function of time can be calculated from

$$F_p(t+1) = F_p(t) - f(t) dt \tag{7.16}$$

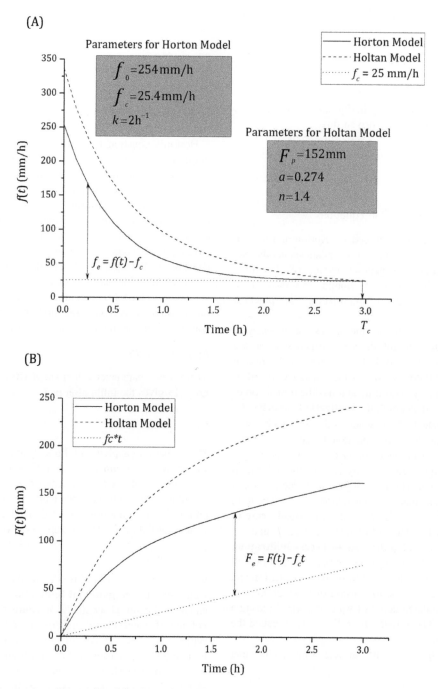

Figure 7.6 (A) Infiltration capacity and (B) cumulative infiltration curves according to Horton and Holtan models. The time required to attain the constant infiltration capacity, f_c, is denoted by T_c.

Figure 7.6A shows the infiltration rate curves according to Horton and Holtan models with the model parameters for a Bluegrass Turf on Type A soil (hydrologic soil group). The corresponding cumulative infiltration curves are shown in Figure 7.6B. Such curves have the assumption that the surface is saturated at the beginning of rainfall. However, for real soils, there is a period of constant infiltration succeeded by the period when the infiltration rate exponentially declines. Such curves are shown in Figure 7.7. As pointed out by Mein and Larson (1971), there are three general cases, which are also shown in Figure 7.7:

Case I: $i < K_s$: If the rainfall intensity (i) is less than the saturated hydraulic conductivity of a deep homogeneous soil, direct runoff will never occur; all rainfall will go into infiltration.

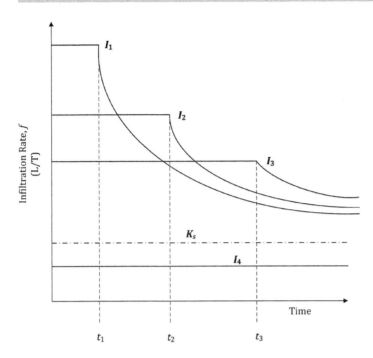

Figure 7.7 Realistic infiltration curves as a function of rainfall intensity. The rainfall intensities are in the order $I_1 > I_2 > I_3 > I_4$. For rainfall intensities $I_1, I_2, I_3,$ and I_4 the corresponding start of infiltration times are t_1, t_2, t_3. For rainfall intensity I_4, infiltration is constant from the beginning of rainfall.

Case II: $K_s < i < f_p$: When the rainfall intensity is greater than the saturated hydraulic conductivity, but less than infiltration capacity, infiltration is steady as shown by the straight-line segments of the infiltration curves shown in Figure 7.7. Direct runoff does not occur under such conditions.

Case III: $i > f_p$: When the rainfall intensity is greater than the infiltration capacity, the infiltration rate exponentially declines as shown by the curved segments of the infiltration curves shown in Figure 7.7. Direct runoff is only generated under such conditions, but the time of surface saturation depends on the rainfall intensity, also shown by the three hypothetical cases in Figure 7.7.

7.5.4 A Generalized Equation of Infiltration

From the basic continuity equation of hydrology, Singh and Yu (1990) developed a generalized equation of infiltration, which, with its analytical solution, showed that almost all of the commonly known equations or models of infiltration are special cases of that generalized equation. The relationship given by Singh and Yu (1990) is

$$f(t) = f_s(t) + \frac{a[F_p(t)]^n}{[F(t)]^m} \tag{7.17}$$

where $f_s(t)$ denotes the seepage rate or steady infiltration rate from a unit volume of soil column that has stored infiltrated water, and $a, n,$ and m are positive real constants for a given soil–vegetation–land use complex. The special cases such as the Horton, Philip, Holtan, and Green–Ampt models arise from selection of the various values of these constants.

In addition to the infiltration models described above, other such models were developed by several early workers in this field, such as Green and Ampt (1911), Kostiakov (1932), Overton (1964), and Huggins and Monke (1966). Of these, the Green–Ampt model gained greater popularity and is widely used today. This is one of the physically based models of infiltration available in HEC-HMS and is discussed at length.

7.5.5 Infiltration Model of Green and Ampt

7.5.5.1 Green–Ampt Equation

Green and Ampt (1911), in a classic paper, proposed a model, simpler than Richards equation yet based on approximate physical theory derived from Darcy's law, and it has an exact analytical solution. Furthermore, the parameters of this model can be measured or estimated from soil properties. For these reasons, it is a favored model of various governmental agencies, such as Harris County Flood Control District in Texas and Maricopa County Flood Control District in Arizona.

The idealized conditions in the development of the Green–Ampt model include: (1) an unlimited supply of ponded water at the surface with a negligible depth; (2) a homogeneous soil column; and (3) advancement of the wetting front in depth as a piston-type displacement or slug flow. Figure 7.8 depicts this idealized condition and is used in the derivation of Green–Ampt equation presented below.

Darcy's law can be written for the two points having hydraulic heads h_1 and h_2, shown in Figure 7.8. Point 1 is at the ground surface and point 2 is just at the interface of the wetting front and the dry soil beneath.

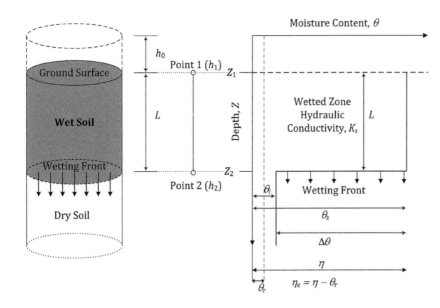

Figure 7.8 Infiltration model of Green and Ampt. The diagram on the left shows the infiltration into a soil column of unit cross-sectional area while that on the right shows the resulting wetting front and key parameters used in calculations. This infiltration model is described as a piston-type movement or slug flow. Contrast the soil moisture profile shown to the right resulting from slug flow with those shown in Figure 7.4.

$$q = -K_s \frac{h_1 - h_2}{Z_1 - Z_2} \quad (7.18)$$

In this case, the Darcy flux q is constant throughout the depth and is equal to f, because q is positive upward, while f is positive downward. As shown in Figure 7.8, $h_1 = h_0$. But the intrinsic assumption in the Green–Ampt formulation is that h_0 is negligible ($h_0 \sim 0$). Therefore, $h_1 = 0$. Furthermore, $h_2 = Z_2 + \psi$; $Z_2 = L$ and $Z_1 - Z_2 = -L$ (see Figure 7.8). From these, we can write,

$$f = -K_s \frac{0 - (-L - \psi)}{L} \quad (7.19)$$

$$f = K_s \left(\frac{L + \psi}{L} \right) \quad (7.20)$$

From the definition of F, the depth of infiltrated water (cumulative infiltration) is

$$F = (\theta_s - \theta_i)L = \Delta\theta L \quad (7.21)$$

Substitution of L from Eq. 7.21 into Eq. 7.20 yields

$$f = K_s \left[\frac{\psi \Delta\theta + F}{F} \right] = K_s \left[1 + \frac{\psi \Delta\theta}{F} \right] \quad (7.22)$$

Equation 7.22 is the original form of Green–Ampt equation. However, this equation contains F. Thus, from Eq. 7.10 we write

$$\frac{dF}{dt} = K_s \left[\frac{\psi \Delta\theta + F}{F} \right] \quad (7.23)$$

Equation 7.23 is solved for F by the method of separation of variables,

$$\left[\frac{F}{F + \psi \Delta\theta} \right] dF = K_s dt \quad (7.24)$$

Splitting the left-hand side of Eq. 7.24 into two fractions as

$$\left[\left(\frac{F + \psi \Delta\theta}{F + \psi \Delta\theta} \right) - \left(\frac{\psi \Delta\theta}{F + \psi \Delta\theta} \right) \right] dF = K_s dt \quad (7.25)$$

and integration of Eq. 7.25 written as

$$\int_0^{F(t)} \left(1 - \frac{\psi \Delta\theta}{F + \psi \Delta\theta} \right) dF = K_s \int_0^t dt \quad (7.26)$$

allows the use of integration of $\int_0^{F(t)} \frac{x}{a+x} dx = a \ln(|x + a|)$ to yield the solution for $F(t)$ as

$$F(t) = \psi \Delta\theta \ln \left(1 + \frac{F(t)}{\psi \Delta\theta} \right) + K_s t \quad (7.27)$$

Equation 7.27 is the Green–Amt equation for cumulative infiltration. The infiltration rate is obtained from Eq. 7.22, which in a time-dependent form is

$$f(t) = K_s \left[1 + \frac{\psi \Delta\theta}{F(t)} \right] \quad (7.28)$$

The advantage of Eq. 7.27 is that all of the parameters are physical properties of the soil–water system and can be measured.

7.5.5.2 Solutions of Green–Ampt Equations

The Green–Ampt model requires the computation of the cumulative infiltration first and then, from the values of $F(t)$, the infiltration rate, $f(t)$ is calculated. Equation 7.27 is an implicit nonlinear equation in $F(t)$. Therefore, its solution is not straightforward. Several methods can be used to solve this equation.

One of the methods that can be employed to solve Eq. 7.27 is the method of successive substitution or trial and error. Given K, ψ, $\Delta\theta$, and t, a trial value of F is substituted on the right-hand side of Eq. 7.27. The initial trial value is set as $F = K_s t$. With this trial value, F is calculated for the left-hand side and this value is then again plugged into the right-hand side in the second iteration. This trial-and-error process continues until the two values in a trial converge within the tolerance limit. The final value of F is used in Eq. 7.28 to obtain f.

Another numerical technique that can be used is Newton–Raphson method of iteration. This is a bit more complex a procedure than the simple trial-and-error method, but in this method the convergence is achieved fairly quickly. Microsoft Excel has a Goal Seek function that uses a numerical method to find the root of an algebraic equation and this can be used.

An easier way to solve the Green–Ampt equation, is to calculate various values of t from a series of values of $F(t)$ by rearranging Eq. 7.27 as

$$t = \frac{F(t)}{K_s} - \frac{\psi \Delta\theta}{K} \ln\left[1 + \frac{F(t)}{\psi \Delta\theta}\right] \tag{7.29}$$

These set of values of $F(t)$ and corresponding values of t can be used in Eq. 7.28 to compute $f(t)$.

Li et al. (1976) presented two methods for solving Eq. 7.27 that are quite accurate and have been adopted by HEC-HMS. The methods involve first obtaining an explicit solution of Eq. 7.27 through a power-series expansion of the logarithmic term. The maximum error in the truncation of the power series by dropping terms higher than first order was found to be 8% and was considered acceptable for all practical purposes. However, the explicit solution was further refined to get an implicit solution by using the second-order Newton method and the calculated errors were less than 0.003%.

The explicit solution of Eq. 7.27, given by Li et al. (1976), in a dimensionless form, when transformed to dimensional form, is

$$F(t) = \frac{1}{2}\left[\frac{K_s t}{\psi \Delta\theta} + \sqrt{\frac{K_s t}{\psi \Delta\theta}\left(\frac{K_s t}{\psi \Delta\theta} + 8\right)}\right] \tag{7.30}$$

However, it is more convenient to write the implicit solution with the dimensionless parameters as

$$F_i^*(t) = \left[1 + F_e^*(t)\right]$$
$$\times \sqrt{\{F_e^*(t)\}^2 + 2\left[t^* - F_e^*(t) + \ln\{1 + F_e^*(t)\}\right]}$$
$$- \{F_e^*(t)\}^2 \tag{7.31}$$

where t^* is the dimensionless quantity defined as

$$t^* = \frac{K_s t}{\Omega} \tag{7.32}$$

and,

$$\Omega = \psi \Delta\theta \tag{7.33}$$

In Eq. 7.31, $F_e^*(t)$ and $F_i^*(t)$, respectively, represent the explicit and implicit cumulative infiltrations in dimensionless form. The dimensionless form of $F_e^*(t)$ in terms of t^* is given as

$$F_e^*(t) = \frac{1}{2}\left[t^* + \sqrt{t^*(t^* + 8)}\right] \tag{7.34}$$

In both cases,

$$F^*(t) = \frac{F(t)}{\Omega} \tag{7.35}$$

A few additional points should be noted in relation to the solution of Green–Ampt equation. If there is a considerable depth of ponded water ($h_0 \neq 0$), this must be added to the capillary suction. The term $\Delta\theta$ should be multiplied by η to obtain the effective saturation of the soil. Finally, since most laboratory determinations of hydraulic conductivity give the values as K_s, it should be adjusted for unsaturated flow. According to Bouwer (1966), the effective hydraulic conductivity for an unsaturated flow, denoted by k_s, is about half the corresponding value for saturated flow. With these adjustments, t^* and Ω should be calculated as

$$\Omega = (\psi + h_0)\eta \Delta\theta \tag{7.36}$$

$$t^* = \frac{k_s t}{\Omega} \tag{7.37}$$

$$k_s \approx \frac{1}{2} K_s \tag{7.38}$$

Table 7.1 shows the results of calculations for a loam soil having the following properties:

$\psi = 250$ mm; $K_s = 10$ mm/h; $\eta = 0.50$; $\theta_i = 0.50$; and $\theta_s = 0.90$.

The second column in Table 7.1 is obtained from Eq. 7.30 (7.34) and calculating Ω and k_s according to Eqs. 7.36 and 7.38 for the values of t given in column 1. This is the explicit solution of Eq. 7.27. The third column is the infiltration rate according to this explicit solution. The fourth column gives the implicit solution using Eq. 7.31. The corresponding infiltration rates are given in the fifth column. The sixth column gives the calculated time according to Eq. 7.29 for the values of $F(t)$ obtained in the implicit solution. The values of $f(t)$ for these values are given in the last or seventh column. As can be seen from the results, all three methods yield comparable values. Nonetheless, the implicit method gives the results with the least error, as can be seen from the calculation of $f(t)$ from the calculation of t (last column).

Table 7.1 *Solutions of the Green–Ampt equations according to three methods*

t (h)	F(t)-explicit (mm)	f(t)-explicit (mm/h)	F(t)-implicit (mm)	f(t)-Implicit (mm/h)	t (h)	f(t) (mm/h)
0.08	7.28	46.19	7.35	45.81	0.08	45.81
0.25	12.89	28.28	13.09	27.91	0.25	27.91
0.5	18.62	21.12	19.03	20.77	0.50	20.77
0.75	23.17	17.95	23.78	17.61	0.75	17.61
1	27.12	16.06	27.94	15.74	1.00	15.74
1.25	30.69	14.78	31.70	14.46	1.25	14.46
1.5	33.98	13.83	35.19	13.52	1.50	13.52
1.75	37.07	13.09	38.48	12.80	1.75	12.80
2	40.00	12.50	41.60	12.21	2.00	12.21
2.25	42.80	12.01	44.59	11.73	2.25	11.73
2.5	45.48	11.60	47.47	11.32	2.50	11.32
2.75	48.07	11.24	50.26	10.97	2.75	10.97
3	50.58	10.93	52.96	10.66	3.00	10.66
3.25	53.03	10.66	55.59	10.40	3.25	10.40

Figure 7.9 shows the results of the implicit solution in graphical format. Note that according to Green–Ampt model also, the infiltration rate tends to show an exponential decline with time and as cumulative infiltration increases, the infiltration rate also decreases nearly in an exponential fashion.

7.5.5.3 Green–Ampt Parameters

The solution and application of the Green–Ampt model require reliable estimates of model parameters, $\Delta\theta$, ψ, and K_s. Brooks and Corey (1964) studied variations of the suction head (ψ) with moisture content, θ. By performing laboratory tests on many different soils, they developed a graphical and then an empirical relationship between the soil suction head and effective saturation, known as the Brooks–Corey equation. Rawls et al. (1983) used their equation to calculate Green–Ampt parameters. They analyzed 1200 soils covering 34 states of the United States and employed all available soil survey information.

There are several systems of classification of soils. These can be grouped into two broad categories, one for engineering and another for soil science. For engineering purposes, one of the most commonly used classification system is the USDA textural classification system because of its simplicity and practical applicability. In this system, soils are classified according to the percentages of clay, silt, and sand present in the soil (Figure 7.10). Clay, silt, and sand are particles with mean diameter of <2 μm, 2–50 μm, and 50–2000 μm, respectively (particles coarser than this are gravel). Rawls et al. (1983) found that the best result in the distinction of the Green–Ampt parameters in various soils was obtained by using soil classification according to the soil texture classes. The mean values and standard deviations of the parameters for 11 soil textures in the soil horizon A (top layer) are given in Table 7.2.

From an analysis of the Brooks–Corey equation, Rawls et al. (1983) showed that $\Delta\theta$ is a function of the effective porosity (η_e) presented in Table 7.2. The variation is: $0 \leq \Delta\theta \leq \eta_e$. If the soil is effectively saturated at the start of rainfall, $\Delta\theta$ equals zero; if the soil is devoid of moisture at the start of rainfall, $\Delta\theta$ equals the effective porosity of the soil. However, if the soil has an initial saturation, S_e, then $\Delta\theta = (1 - S_e)\eta_e$. The term S_e ($0 \leq S_e \leq 1.0$) is called **effective saturation** in the Brooks–Corey equation, given as

$$S_e = \left(\frac{\psi_b}{\psi}\right)^\lambda \tag{7.39}$$

in which ψ_b, known as the **bubbling pressure** or **air-entry pressure head**, and λ, called the **pore-size distribution index**, are constants obtained experimentally. The effective saturation is the ratio of the available moisture content to the maximum available moisture content, which must be equal to porosity. Thus,

$$S_e = \frac{\theta - \theta_r}{\eta - \theta_r} = \frac{\theta - \theta_r}{\eta_e} = 1 - \frac{\Delta\theta}{\eta_e} \tag{7.40}$$

in which θ_r is the **residual saturation** after the soil has been thoroughly drained (see Figure 7.8). The denominator of Eq. 7.40 follows the definition of effective porosity given before. In most practical applications, S_e is not known and is set to zero. However, for certain event-based modeling, records of weather data giving prevailing humidity, precipitation, temperature, and wind speed prior to the event of interest can be used to estimate the initial moisture content.

The hydraulic conductivity values reported in Table 7.2 should be considered as k_s, as discussed above.

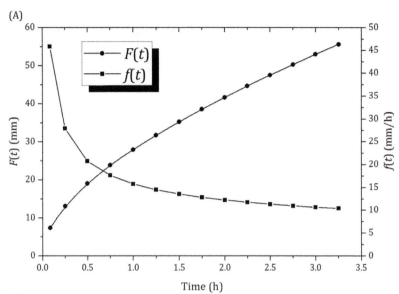
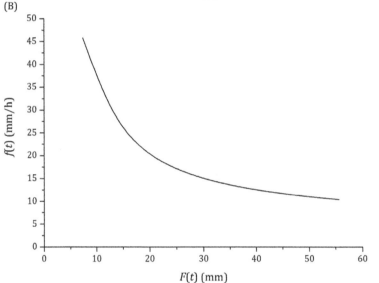

Figure 7.9 (A) Cumulative infiltration and infiltration rate and (B) relationship between the cumulative infiltration and infiltration rate according to the implicit solutions of the Green–Ampt equations (Table 7.1).

7.5.5.4 Sensitivity of Green–Ampt Model to the Model Parameters

Considering the sensitivity of the Green–Ampt model, the order of importance of the model parameters is as follows.

1. Hydraulic conductivity, K
2. Wetting-front capillary pressure, ψ
3. Effective porosity, η_e

Example 7.3 gives a detailed illustration.

7.5.5.5 Two-Layer Green–Ampt Model

The Green–Ampt model can be extended to a soil horizon having two layers with different properties, and the upper layer is more permeable than the lower layer. If the top layer has a thickness, H_1, hydraulic conductivity, K_1, capillary suction, ψ_1, and moisture deficit, $\Delta\theta_1$, and the bottom layer with the corresponding values of H_2, K_2, ψ_2, and $\Delta\theta_2$, it can be shown, following the similar derivation given for a single layer, that

$$f = \frac{K_1 K_2}{H_1 K_2 + L_2 K_1}(\psi_2 + H_1 + L_2) \tag{7.41}$$

and

$$F = H_1 \Delta\theta_1 + L_2 \Delta\theta_2 \tag{7.42}$$

Combining Eqs. 7.41 and 7.42 a differential equation is formed as

Table 7.2 *Green–Ampt infiltration parameters for various soil classes (after Rawls et al., 1983)*

Soil class	Porosity, η	Effective porosity, η_e	Wetting-front soil suction head, ψ (cm)	Hydraulic conductivity, K (cm/h)
Sand	0.437 (0.374–0.500)	0.417 (0.354–0.480)	4.95 (0.97–25.36)	11.78
Loamy sand	0.437 (0.363–0.506)	0.401 (0.329–0.473)	6.13 (1.35–27.94)	2.99
Sandy loam	0.453 (0.351–0.555)	0.412 (0.283–0.541)	11.01 (2.67–45.47)	1.09
Loam	0.463 (0.375–0.551)	0.434 (0.334–0.534)	8.89 (1.33–59.38)	0.34
Silt loam	0.501 (0.420–0.582)	0.486 (0.394–0.578)	16.68 (2.92–95.39)	0.65
Sandy clay loam	0.398 (0.332–0.464)	0.330 (0.235–0.425)	21.85 (4.42–108.0)	0.15
Clay loam	0.464 (0.409–0.519)	0.309 (0.279–0.501)	20.88 (4.79–91.10)	0.10
Silty clay loam	0.471 (0.418–0.524)	0.432 (0.347–0.517)	27.30 (5.67–131.50)	0.10
Sandy clay	0.430 (0.370–0.490)	0.321 (0.207–0.435)	23.90 (4.08–140.2)	0.06
Silty clay	0.479 (0.425–0.533)	0.423 (0.334–0.512)	29.22 (6.13–139.4)	0.05
Clay	0.475 (0.427–0.523)	0.385 (0.269–0.501)	31.63 (6.39–156.5)	0.03

Note: The numbers in parenthesis below each parameter are one standard deviation around the parameter value given.

Figure 7.10 Textural classification system of soils used by USDA.

$$t = L_2 \frac{\Delta\theta_2}{K_2} + \frac{1}{K_1 K_2}[\Delta\theta_2 H_1 K_2 - \Delta\theta_2 K_1 (\psi_2 + H_1)]$$
$$\times \ln\left[1 + \frac{L_2}{\psi_2 + H_1}\right]$$
(7.43)

In the above formulation, $K_1 > K_2$ and $L_2 < H_2$ where water has infiltrated into the lower layer at a distance of L_2.

7.5.5.6 Input Requirements for Green–Ampt Model in HEC-HMS

In HEC-HMS, the parameters of hydraulic conductivity (in mm/h or in/h) and capillary suction or wetting-front suction (in mm or in) can be entered directly from the soil texture class present in the subbasin. However, the initial soil condition can be specified in one of two ways by selecting either (i) the initial content or (ii) the initial deficit.

If the **initial content** is selected, it should be specified by providing two further parameters: (a) initial content and (b) saturated content. In terms of what described above these are θ_i and θ_s respectively. If **initial deficit** is selected, it can be input by specifying a single parameter only, which is actually $(\theta_s - \theta_i)$. In either case, this parameter is oftentimes defined with respect to the effective porosity of the soil layer. For example, if a soil layer with an effective porosity of 0.45 cm^3/cm^3 is 60% saturated, its initial content would be 0.27 cm^3/cm^3 (i.e., 0.45 cm^3/cm^3 × 0.60). Alternatively, the same soil layer could be defined using a moisture deficit of 0.18 cm^3/cm^3 (i.e., 0.45 cm^3/cm^3 – 0.27 cm^3/cm^3). The parameters must be calibrated using observed data. In order to estimate an appropriate initial moisture content, two values must be estimated for the entire study area: (1) the representative effective porosity of the soil, and (2) the soil moisture state of the soil at the beginning of the simulation. The saturated water content specifies the maximum water holding capacity in terms of volume ratio and is assumed to be the total porosity of the soil. In order to estimate the amount of moisture present within the soil throughout the study area, the antecedent conditions at the beginning of simulation must be investigated for historical events by collecting daily average humidity, daily precipitation accumulation, daily average temperature, and daily average wind speed for several weeks prior to the event being modeled.

In HEC-HMS, a single-layer Green–Ampt model is used for event-based simulations. For continuous simulations, a two-layered Green–Ampt model must be used instead. A two-layered Green–Ampt model in HEC-HMS requires additional inputs on soil properties and layer physical properties. The soil properties include specification of *field capacity* and *wilting point* of soils in both the upper and lower soil horizons. The depth of the soil layers also needs to be specified. In addition, the seepage rate (L/T) from the upper horizon or layer 1 to the lower horizon and the percolation rate (L/T) from the lower horizon or layer 2 to deep groundwater needs specification. A continuous simulation involves both periods of wetting (rain) and drying (no rain). During the drying period, infiltrated water recirculates to the atmosphere through the process of evapotranspiration and HEC-HMS models this process when the "canopy method" is used in the program in conjunction with the two-layered Green–Ampt model. The "surface method," if selected in addition to the canopy method in the HEC-HMS model, on the other hand, will allow the water in surface storage to infiltrate. The layered Green–Ampt method in HEC-HMS is intended to be used in combination with the linear reservoir baseflow method. Canopy and surface methods are discussed in Section 7.9 of this chapter and baseflow is discussed in Chapter 8.

7.5.6 Infiltration Model of Smith and Parlange

Smith and Parlange (1978) proposed an analytical model of infiltration that approximates the numerical solution of Richards equation and the Green–Ampt model during the initial times (when t is small). The model gives infiltration rate as

$$f = \frac{K_s \exp\left(\frac{FK_s}{C}\right)}{\exp\left(\frac{FK_s}{C} - 1\right)} \quad (7.44)$$

As can be seen from Eq. 7.44, the model has two parameters, saturated hydraulic conductivity (K_s) and C, which is directly proportional to initial water deficit and has a unit L^2/T. Both of these parameters are closely related to real physical soil properties, and both can be obtained by either a graphical or a regression method from infiltrometer experimental data with the relationship

$$f = K_s \left[\left(\frac{C}{K_s F}\right) + 1 \right] \quad (7.45)$$

The parameter C is related to Green–Ampt parameters according to an approximation:

$$C \simeq -\psi \Delta \theta K_s \quad (7.46)$$

Like the Green–Ampt model, the Smith–Parlange infiltration model is a conceptualization of the actual physical processes. However, it differs because it does not assume a single hydraulic conductivity as is done in the case of the Green–Ampt model. In this model the hydraulic conductivity is assumed to decrease exponentially from the saturated condition, as is often found in real soils. This means that it is less likely to overestimate the infiltration at early times during a storm event.

The parameters of the Smith–Parlange model include the initial soil water content, the saturated soil water content, and the residual water content. The first two are defined as for the Green–Ampt model. The residual water content is the water content that will remain after the saturated soil has been allowed to drain and dry for a very long time. The *bubbling pressure* is a physically measurable property of the unsaturated soil to pull water into the soil through a suction generated by capillary forces. The conductivity is the rate at which gravity alone forces water through the soil when it is effectively saturated. The pore distribution is a measure of the variation in the size of the void spaces in the soil.

The Smith–Parlange model also includes the ability to adjust the infiltration process according to the temperature. Temperature affects the viscosity and density of the water. These primary effects reduce the total gradient in the soil, which affects the conductivity and the *matric potential*. These effects were determined through theoretical analysis of the infiltration process and incorporated into the Smith–Parlange model. The temperature effects are intended to improve simulation results in desert climates where the properties of water may be significantly different from those assumed at standard temperature, as used in virtually all infiltration models.

HEC-HMS provides a single implementation of the Smith–Parlange model (Smith et al., 2002). The implementation assumes that boundary conditions and parameters are averaged over the whole subbasin. The program requires the following inputs (all defined before) from which C is evaluated: (i) initial water content (θ_i), given as a volume ratio; (ii) residual water content (θ_r), given as a volume ratio; (iii) saturated water content (θ_s), given as a volume ratio that can be approximated as total porosity of the soil (η); (iv) bubbling pressure (ψ_b) or wetting-front suction (in or mm), which is a function of soil texture; (v) pore-size distribution (λ), which determines how the total pore space is distributed in different size classes and is assumed to be a function of the soil texture; and (vi) effective saturated hydraulic conductivity (K_s), which is also a function of the soil texture.

7.6 ESTIMATION OF EFFECTIVE RAINFALL

Several methods are used to compute effective rainfall and thereby the construction of the corresponding hyetographs. These methods are described below.

7.6.1 Constant Infiltration Rate or Φ Index and Initial and Constant Loss Model

The Φ index is the mean infiltration rate occurring for the duration of the rainfall event. Obviously, then, there is a constant infiltration rate, f_c. When this constant infiltration rate or Φ index is subtracted from a rainfall hyetograph, the resulting hyetograph is called the **effective rainfall hyetograph (ERH)**, illustrated in Figure 7.11. The assumption is that, for the entire rainfall event, the rainfall intensity is greater than Φ. Otherwise the rate of infiltration will be equal to the rainfall intensity for those time intervals for which rainfall intensity is not greater than Φ. The total volume of the rainfall amount within an ERH equals the total volume of direct runoff.

In HEC-HMS, there is an option for the calculation of excess rainfall by a method called the **initial and constant loss** model. The underlying concept of this model is that the maximum potential rate of rainfall loss, f_c, is constant throughout the event. Thus, if P_e is the mean areal precipitation (MAP) depth during a time interval t to $t + \Delta t$, the excess rainfall during the interval is given by

$$P_e(t) = \begin{cases} P(t) - f_c & \text{if } P(t) > f_c \\ 0 & \text{otherwise} \end{cases} \quad (7.47)$$

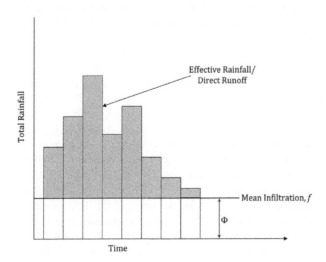

Figure 7.11 Graphical illustration of the concept of the Φ index.

Sometimes, an initial loss, I_a, is added to the model to represent interception and surface depression storage. Until the accumulated rainfall on the pervious area exceeds the initial loss volume no direct runoff occurs. Thus, the constant rate applies after the initial loss is satisfied. In this case, Eq. 7.47 is modified as

$$P_e(t) = \begin{cases} 0 & \text{if } \sum P(t) < I_a \\ P(t) - f_c & \text{if } \sum P(t) > I_a \text{ and } P(t) > f_c \\ 0 & \text{if } \sum P(t) > I_a \text{ and } P(t) < f_c \end{cases} \quad (7.48)$$

The initial and constant loss method is the simplest method but is appropriate for watersheds that lack detailed soil information. It is also suitable for certain types of flow frequency analysis. No loss is calculated for the percentage of area specified as impervious area, giving $P_e(t) = P(t)$.

The Φ index can be estimated by one of the following ways.

7.6.1.1 Derivation of Φ index from Observed Rainfall Hyetograph and Streamflow Hydrograph

Theoretically, the Φ index is that constant rate of abstraction (cm/h or in/h) that will yield an ERH with a total depth equal to the depth of direct runoff, q (in or cm) over the drainage area under consideration resulting from a given rainfall event. Interception and depression storage losses are excluded in the Φ index.

When rainfall and streamflow data are available, the amount of observed runoff, q, is determined from the observed streamflow hydrograph and knowing the drainage area. The total precipitation observed from rain gauges is determined and subtracted from the observed runoff to estimate the volume of rainfall loss. This quantity is then distributed uniformly across the storm duration. However, the value of Φ obtained in this way needs to be refined further. If the total rainfall hyetograph has M number of pulses each with duration Δt, it follows from the definition of the Φ index that

$$q = \sum_{m=1}^{M} (P_m - \Phi \Delta t) \quad (7.49)$$

where P_m is the observed rainfall depth (in or cm) in time interval m.

The number of pulses to be considered can be determined by using the value of Φ obtained in the first step. Subsequently, Eq. 7.49 can be solved algebraically to derive the value of Φ. Example 7.1 illustrates the procedure. The Φ index obtained in this way represents the constant infiltration rate of the drainage area upstream of the location of the stream gauging station. The resulting ERH represents the hyetograph of the MAP. This method of calculation of the Φ index is the most common method of estimating rainfall loss in the process of deriving the unit hydrograph of a watershed from observed rainfall and streamflow data, discussed in Chapter 9.

7.6.1.2 Estimations of Constant Infiltration Rate from Soil Type

The constant loss rate can be viewed as the ultimate infiltration capacity of soils. Skaggs and Khaleel (1982) gave estimates of infiltration rates for the four hydrologic soil groups developed by the Soil Conservation Service (SCS) – now the Natural Resources Conservation Commission (NRCS) – of the USDA. Hydrologic soil groups are discussed in Section 7.6.3.2. In the absence of better information, the values given in Table 7.3 can be used for the soil types present in the drainage area under consideration.

7.6.2 Exponential Loss Rate Method

The exponential loss rate method is an empirical method that relates loss rate to rainfall intensity and accumulated losses (Feldman, 1981). This method was originally implemented in the HEC-1 computer program and has been an option in the subsequent versions of HEC-HMS. In the original conceptualization of this method, the accumulated loss included interception, infiltration, and depression storage. It represents the incremental loss as an exponentially decreasing function of accumulated loss. The model calculates the loss rate using

$$L_R = (A_k + D_k)P^{(K_r)} \qquad (7.50)$$

where L_R is the rainfall loss rate (in/h or mm/h), A_k and D_k are two loss rate coefficients, P is the rainfall intensity (in/h or mm/h), and K_r is the exponent reflecting the influence of rainfall rate on area-average loss characteristics. It reflects the manner in which rainfall occurs within an area and may be considered a characteristic of a particular region. It varies from 0.0 up to 1.0.

The effects of soil moisture conditions accounted for by adjusting the interception and infiltration rates, resulting in

Table 7.3 *Descriptions and characteristics of four hydrologic soil groups*

HSG	Soil texture	Infiltration rate (in/h)
A	Sand, loamy sand, or sandy loam	High: 0.30–0.45
B	Silt loam or loam	Above average: 0.15–0.30
C	Sandy clay loam	Below average: 0.05–0.15
D	Clay loam, silty clay loam, sandy clay, silty clay, or clay	Low: 0.00–0.05

two loss rate factors. The two loss rate coefficients C and D are defined as follows:

$$D_k = 0.2D\left[1 - \frac{C}{D}\right]^2; C \leq D; \text{otherwise}, 0 \qquad (7.51)$$

where D, which Feldman called the "initial range," is the amount of initial accumulated loss during which the loss rate is increased; $D\,(\geq 0)$ is considered to be a function of either depression storage or primarily antecedent soil moisture deficiency which is usually storm-dependent; and C is the accumulated loss (in or mm) calculated by summing the actual losses computed for each time interval.

The other loss rate coefficient is defined as

$$A_k = \frac{S}{R^{0.1C}} \qquad (7.52)$$

where S, the "initial coefficient," specifies the starting loss rate coefficient on the exponential infiltration curve. It is assumed to be a function of infiltration characteristics and consequently may be correlated with soil type, land use, vegetation cover, and other properties of the area under consideration. In Eq. 7.52, Feldman called R the "coefficient ratio," which indicates the rate at which the exponential decrease in infiltration capability proceeds. It may be considered a function of the ability of the surface of an area to absorb rainfall and should be reasonably constant for large homogeneous areas. Figure 7.12 shows the relationship between the loss rate coefficients and accumulated loss with graphical interpretations of D, S, and R.

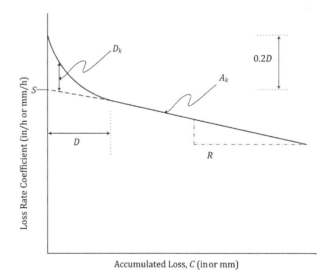

Figure 7.12 Relationship between the loss rate coefficient and accumulated loss in the exponential loss rate model given in HEC-HMS. The horizontal axis is on an arithmetic scale and the vertical axis is on a logarithmic scale. D = initial amount of loss for which the loss rate coefficient is increased to represent antecedent soil moisture conditions. S = the loss rate for average soil moisture conditions. R = rate of change of loss rate coefficient as soil moisture increases.

Due to the unknown nature of the four parameters (K_r, D, S, and R) involved in this method, it should not be used without calibration. This method includes the option for increased initial infiltration when the soil is particularly dry before the commencement of rainfall. However, the model parameters are not readily determined from measurable watershed characteristics. The method produces exponential decreases of infiltration (loss) similar to the Green–Ampt method, which has a better physical interpretation. However, if the model parameters can be calibrated from available rainfall and streamflow data and no better information is available for estimating the loss factor, this method does offer certain advantages. Example 7.2 illustrates the application of this method.

7.6.3 NRCS Curve Number Method

7.6.3.1 Origin of Curve Number Equation

Victor Mockus, a remarkable hydrologist with incredible ingenuity, from the then SCS, now NRCS, of the USDA, was the originator of a method of calculation of excess rainfall that is known as the **curve number method**. From a humble empirical beginning in the 1930s, based on limited sets of observational data of rainfall and runoff, this method has now gained almost worldwide popularity. It is often the method of choice by practicing engineers and hydrologists in the United States and in many other countries. Mockus described this method in 1965 and it was later published in a revised form by NRCS (2004a). However, most files and supporting data used in the development of this concept were never published and have been lost, destroyed, or misplaced over the years. Furthermore, this method never underwent a critical open review process. Nonetheless, owing to its simplicity, predictability, stability, and support by a major US federal agency, its widespread acceptability by the practitioners of hydrology in the United States and in many other countries, and its undoubted practicality as a design method, is understandable. It should be noted at the onset of further discussion of this method that the curve number method is an approach to compute direct runoff, that is, excess precipitation, and *not an infiltration approach*. Therefore, it is not the classical Hortonian approach discussed above.

Early observations of graphs showed that plots of direct runoff against rainfall over an area were concave-upward curves. Observations also showed that no runoff was generated for short-duration storms but, as the storm size increased, the curve tended to become asymptotic to a line parallel to a line of equality. Mockus attempted to find an equation that would fit such curves, and he discovered a proportionality relationship:

$$\frac{F}{S} = \frac{q}{P} \quad (7.53)$$

where F (in) is the actual retention of rainfall on the ground after runoff begins, S is the potential maximum retention (in) after runoff begins ($S \geq F$), q is actual runoff, that is, effective rainfall (P_e), and P is actual rainfall ($P > q$; where $q \equiv P_e$).

From the mass balance principle, $F = P - q$; therefore, Eq. 7.53 can also be written as

$$\frac{P-q}{S} = \frac{q}{P} \quad (7.54)$$

Solving for q yields,

$$q = \frac{P^2}{P+S} \quad (7.55)$$

Equation 7.55 is the rainfall–runoff equation, where there is no accounting for initial abstraction such as interception or surface depression storage. Therefore, an initial abstraction term, I_a, is introduced in Eq. 7.55 to get

$$q = \begin{cases} \dfrac{(P-I_a)^2}{P-I_a+S}; P > I_a \\ 0; P \leq I_a \end{cases} \quad (7.56)$$

Figure 7.13 shows the three components of precipitation in this approach. Equation 7.56 has two parameters, S and I_a. To remove the necessity for an independent estimation of initial abstraction, Mockus, in 1965 (in NRCS, 2004a) suggested a linear relationship between I_a and S:

$$I_a = \zeta S \quad (7.57)$$

in which ζ is called the **initial abstraction ratio**. Such a linear relationship was justified from a plot of observed measurements in experimental watersheds (NRCS, 2004a). While there was considerable scatter in the data, NRCS (2004a) reported that 50% of the data points lay within the limits $0.095 \leq \varsigma \leq 0.38$. This led NRCS to adopt a standard value of the initial abstraction ratio $\zeta = 0.2$ (the line drawn through the visual median range). However, values varying in the range $0.0 \leq \varsigma \leq 0.3$ have been documented in various

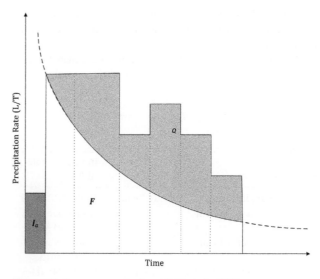

Figure 7.13 Three components of precipitation in the NRCS curve number approach.

studies. Nevertheless, if the standard relationship $I_a = 0.2S$ is adopted, Eq. 7.56 becomes,

$$q = \begin{cases} \dfrac{(P-0.2S)^2}{P+0.8S}; P > 0.2S \\ 0; P \leq 0.2S \end{cases} \quad (7.58)$$

Now, Eq. 7.58 has only one parameter, the potential maximum retention, S. Values of S can have the range $0 \leq S \leq \infty$.

For convenience in practical applications, S is mapped into a dimensionless parameter called the **curve number** and designated by CN. The chosen mapping equation is

$$S = \dfrac{1000}{CN} - 10 \quad (7.59)$$

where 1000 and 10 are arbitrarily chosen constants having the same unit as S (in). This gives a more appealing range $0 \leq CN \leq 100$. A $CN = 100$ represents a condition of zero potential retention $(S = 0)$, which represents an impermeable cover, whereas $CN = 0$ represents a theoretical upper bound to the potential retention $(S = \infty)$. Figure 7.14 is the parametric representation of Eq. 7.58. These are the types of curves, but generated from observed data, that led Mockus to develop an equation that would fit the curves. Example 7.5 illustrates this with a set of observed rainfall–runoff data.

In SI, where P, Q, S are in mm, the mapping equation is

$$S = \dfrac{25\,400}{CN} - 254 \quad (7.60)$$

7.6.3.2 Estimation of Curve Number

Hydrologic Soil Cover Complex

As noted earlier, soils can be classified for either engineering or scientific purposes. The taxonomic classification of soils developed by the USDA is used in soil science and also for mapping soil horizons in soil surveys in the United States (USDA, 1999). This is a very elaborate classification scheme based on many parameters, which arranges soil types in a systematic manner in several levels: order, suborder, group, subgroup, family, and series. The lowest level of this ordering scheme permits very specific descriptions of the soils following many criteria. There are more than 14 000 soil series recognized in the United States alone. Similarly, there are 12 orders, the most general level in the taxonomic classification of soils. Many countries in Latin America, Asia, and Africa have adopted this system of soil classification.

For hydrologic purposes, Musgrave (1955) of the then SCS, based on data for about 115 soil types, placed different types of soils into four basic groups depending on the minimum infiltration capacity (rate) obtained for bare soil after prolonged wetting and based on laboratory tests and soil texture (Figure 7.10). The four groups were A, B, C, and D, called **hydrologic soil groups** (HSGs), and their characteristics are given in Table 7.3. Subsequently, from extensive studies taking more than 4000 soil samples covering 14 000 soil series in the United States, the four HSGs were included in the National Soil Information System (NASIS) that came into being in 1995. The saturated hydraulic conductivity (K_s) was also included as one of the criteria in this very simplified, yet highly practical for applied hydrology, classification system.

In general, the soil map of any area within the United States produced by USDA gives the distribution of different soil series that are present in that area. The textural classes and the HSG that correspond to each of the map units giving soil series are also available in digital formats from USDA. The GIS method is used to derive the HSG map of a study area. Figure 7.15 shows the textural classes of soils present

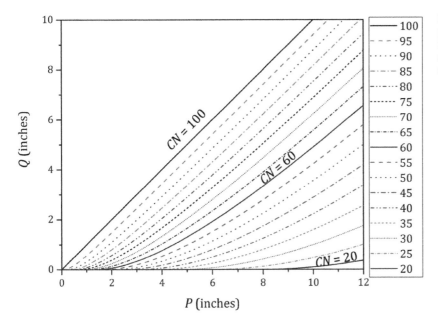

Figure 7.14 Graphical solutions of Eq. 7.58. The parameter of the curves is the curve number, *CN*. (A black and white version of this figure will appear in some formats. For the color version, please refer to the plate section.)

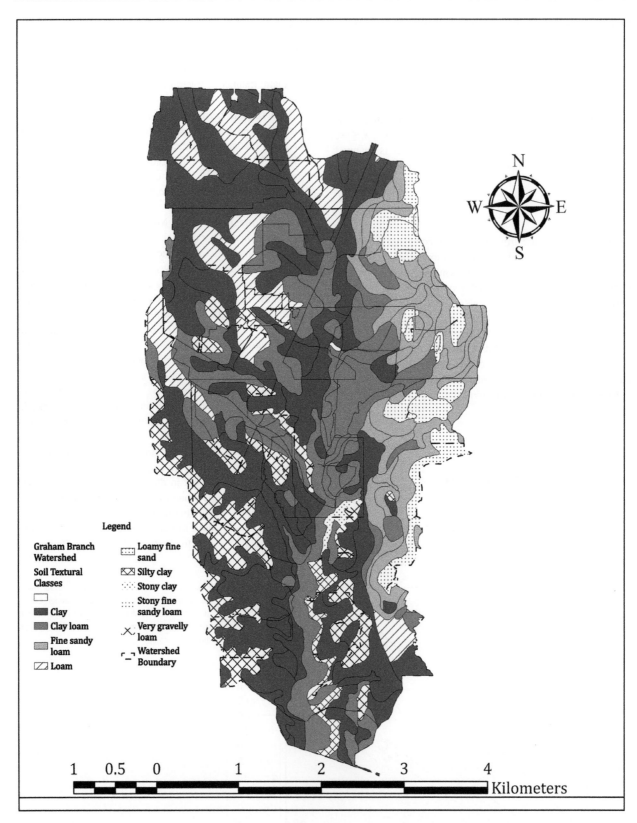

Figure 7.15 Textural classes of soils in a suburban watershed in Texas, Graham Branch Watershed, located to the northwest of Dallas.

in a suburban watershed, known as Graham Branch Watershed, located to the northwest of the City of Dallas in Texas. Figure 7.16 shows the corresponding HSG of the soil units of Graham Branch Watershed. Both of these maps were produced, using GIS technology, from the data that are freely available from the USDA's Web Soil Survey. One of the feats of the concept of placing numerous soil types into only four categories for hydrologic applications is that soils classified by another system in any country can still be categorized into the textural classes shown in Figure 7.10 based on their sand–silt–clay fractions and then one of the HSGs can be assigned to a class following the general characteristics of soils given in Table 7.3.

In many countries, the kind of information regarding soil characteristics provided by USDA for the soils of the United States are not readily available. The Food and Agriculture Organization (FAO) of the United Nations has developed the Harmonized World Soil Database (HWSD), from which soil data at coarse spatial resolutions can be obtained for the area of interest if no region-specific data are available. Nonetheless, care must be given to the system of soil classification and its correspondence to the textural classes used in the assignment of the HSG. In the HWSD, there are more than 16 000 soil mapping units. The soil classification follows a world reference base for soil resources (FAO, 1998). The primary intent of this database is to aid agricultural engineers who may be more interested in the chemical composition of soils.

Originally, the *CN* values were assigned to an area by plotting the observed runoff versus measured rainfall for 24 experimental plots from 19 states, but scattered throughout the United States. These values were then correlated with the land use. A combination of HSG (soil type) and land use and treatment class (cover type) is called a **hydrologic soil cover complex**. Curve numbers for particular combinations of soil and cover characteristics (soil cover complex) were developed by plotting the largest annual storm runoff and associated rainfall for a watershed having one soil and one cover. Laid over this plot was a graph of *CN* array constructed at the same scale. The median *CN* was selected, dividing the plotting into two equal numbers of points. When more than one site with the same soil cover complex was examined, the median *CN* values were averaged. Curve numbers were subsequently developed for many soil cover complexes and were published by NRCS (1986, 2004b). The curve number indicates the runoff potential of a complex during periods when the soil is not frozen. The higher the *CN*, the higher is the runoff potential. NRCS (2004b) provides further guidelines for their use.

Table 7.4A gives the *CN* values of hydrologic soil cover complexes in typical urban areas. The terms "good" and "poor" conditions refer to the relative runoff potential. An area in good hydrologic condition has higher infiltration rates and lower runoff rates than an area in poor condition. These reflect differences in land management. However, not all land uses have been classified in this way. There are a number of methods for determining cover type. The most common are field reconnaissance, aerial photographs, and land use maps.

Tables 7.4B and 7.4C give the suggested *CN* values for uncultivated agricultural lands or lands in natural conditions (undeveloped) and cultivated agricultural land, respectively. Agricultural land uses are often subdivided, based on *treatment* or agricultural practice, for example contoured or straight row. Therefore, treatment is a cover-type modifier (used only in Table 7.4C) to describe the management of cultivated agricultural lands. It includes mechanical practices, such as contouring and terracing, and management practices, such as crop rotations and reduced or no tillage.

The hydrologic condition indicates the effects of cover type and treatment on infiltration and runoff, and is generally estimated from the density of plants and residue cover on sample areas. For non-urban landscapes, good hydrologic condition indicates that the soil usually has a low runoff potential for that specific hydrologic soil group, cover type and treatment. Some factors to consider in estimating the effect of cover on infiltration and runoff are: (a) canopy or density of lawns, crops, or other vegetative areas; (b) amount of year-round cover; (c) amount of grass or close-seeded legumes in rotations; (d) percent of residue cover; and (e) degree of surface roughness.

It should be noted that the *CN* values suggested in Table 7.4A were developed by NRCS by considering typical urban settings in the United States. Such settings may not be applicable in their entirety in many other countries and therefore a table like this one should be developed for each country or region-specific conditions and settings. Similarly, the *CN* values for types of land use and land cover given in Tables 7.4B and 7.4C may vary. For this very reason, NRCS (2004b) has suggested separate sets of *CN* values for arid and semiarid rangelands of the southwestern parts and forest ranges in the western United States, and different sets of values for Puerto Rico and Hawaii. The *CN* values listed in Tables 7.4A–C, are for antecedent moisture condition II and with an initial abstraction of $0.2S$ as discussed above.

It is apparent from the discussion above and the examples given in Table 7.4 that, in order to estimate the curve number of an area, in addition to the soil types present in the area, the land cover type of the area must be known. A *distinction should be made between land use and land cover*. Land use denotes what people do on a plot of landscape, such as agriculture, commerce, settlement, recreation etc. Land cover defines the material on the land surface, such as crops, forest, buildings, water etc. Land cover is an excellent indicator of land use. Also, land cover changes with time as land use changes.

Land cover data in the United States at a national level have been developed by the USGS which are freely available from the National Land Cover Database (NLCD). The data are given as time series at various intervals from 2001 to the present. The data given in the NLCD follow a slightly

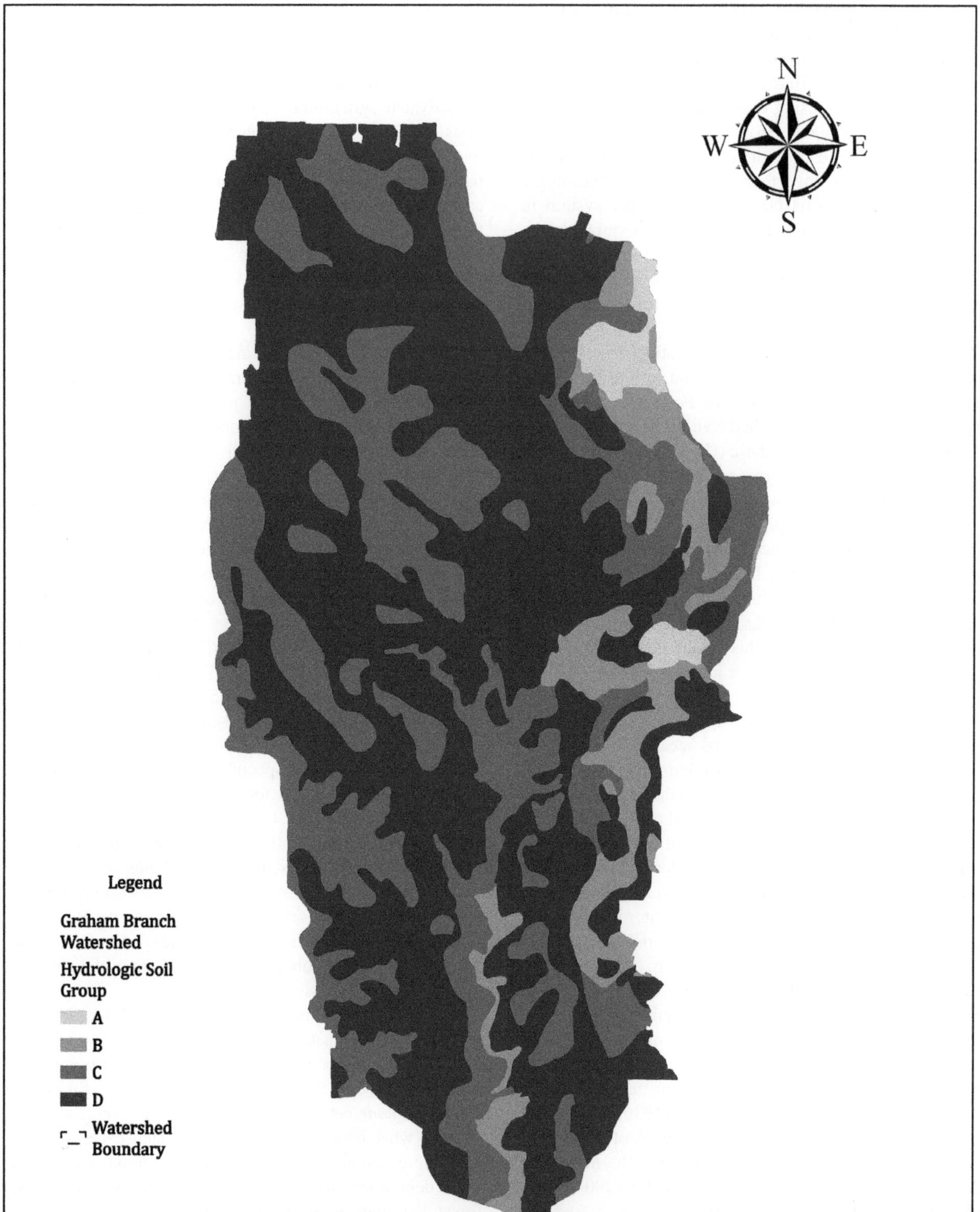

Figure 7.16 Hydrologic soil groups (HSGs) of the textural classes of soils shown in Figure 7.15.

Table 7.4A *Suggested values of* CN *for urbanized areas*

Cover description Cover type and hydrologic condition	Average percent impervious area	CN value for HSG			
		A	B	C	D
Fully developed urban areas					
Open space (lawns, parks, golf courses, cemeteries, etc.)					
Poor condition (grass cover < 50%)		68	79	86	89
Fair condition (grass cover 50% to 75%)		49	69	79	84
Good condition (grass cover > 75%)		39	61	74	80
Impervious areas					
Paved parking lots, roofs, driveways, etc. (excluding right-of-way)		98	98	98	98
Streets and roads					
Paved curbs and storm sewers (excluding right-of-way)		98	98	98	98
Paved open ditches (including right-of-way)		83	89	92	93
Gravel (including right-of-way)		76	85	89	91
Dirt (including right-of-way)		72	82	87	89
Western desert urban areas					
Natural desert landscaping (pervious areas only)		63	77	85	88
Artificial desert landscaping (impervious weed barrier, desert shrub with 1- to 2-inch sand or gravel mulch and basin borders)		96	96	96	96
Urban districts					
Commercial and business	85	89	92	94	95
Industrial	72	81	88	91	93
Residential districts by average lot size					
1/8 acre or less (town houses)	65	77	85	90	92
1/4 acre	38	61	75	83	87
1/3 acre	30	57	72	81	86
1/2 acre	25	54	70	80	85
1 acre	20	51	68	79	84
2 acre	12	46	65	77	82
Developing urban areas					
Newly graded areas (pervious areas only, no vegetation)		77	86	91	94

modified form of the land cover classification system given by Anderson et al. (1976). The land covers in the NLCD are broadly classified into the categories of developed (urbanized), barren, forest, shrubland, herbaceous, planted or cultivates, wetlands, and water. Within each category there are several subcategories, each with a numerical designation or value (e.g., 22 for low intensity developed areas, 41 for deciduous forest, 82 for cultivated crops, etc.) for the class. There are 20 classes total. Since NLCD data are given as a raster dataset (Chapter 20), a numerical code (called a grid code) is assigned to each grid representing a land cover class.

If the NLCD is used to obtain the land cover characteristics of a study area, it should be correlated with the soil cover complexes, with the suggested CN values given in the NRCS publications or with those given by the local authorities or agencies under whose jurisdiction the study area is located. This is done by finding the NRCS land cover categories and their equivalent classes in the NLCD. In Chapter 20 we discuss how such data are used to derive composite CN values of catchments using GIS technology. Table 20.1 gives recommended CN values for all combinations of NLCD classes and HSGs.

In many areas, particularly in urban municipalities in the United States, land cover maps or zoning maps at local levels have been developed based on the actual land use of the area. In such cases, these data are preferred over the NLCD because they provide greater detail at a local or regional level. Figures 7.17 and 7.18 show the land cover maps of Graham Branch Watershed, derived from the NLCD and local data, respectively.

Global land cover (GLC) data are available from the Global Environment Monitoring (GEM) unit of the Institute of Environment and Sustainability of the

Table 7.4B *Runoff CN values for uncultivated agricultural land or lands in natural conditions*

Cover type and cover description	Hydrologic condition	CN value for HSG			
		A	B	C	D
Pasture, grassland, or range – continuous forage for graving.2	Poor	68	79	86	89
	Fair	49	69	79	84
	Good	39	61	74	80
Meadow – continuous grass, protected from grazing, and generally mowed for hay.	Good	30	58	71	78
Brush – brush–weed–grass mixture with brush the major element.	Poor	48	67	77	83
	Fair	35	56	70	77
	Good	30	48	65	73
Woods – grass combination (orchard or tree farm)	Poor	57	73	82	86
	Fair	43	65	76	82
	Good	32	58	72	79
Woods	Poor	45	66	77	83
	Fair	36	60	73	79
	Good	30	55	70	77
Farmsteads – buildings, lanes, driveways, and surrounding lots.		59	74	82	86

European Commission of Joint Research Center. The GEM GLC 2000 data is considered to be the internationally standardized land cover data with 1 km × 1 km spatial resolution, producing the land cover information of the earth for the year 2000 using the land cover classification system proposed by the Food and Agriculture Organization of the United Nations Environment Programme.

Antecedent Soil Conditions

The CN values listed in a Table 7.4 (and similar tables given elsewhere) are for antecedent soil moisture condition II, which is based on median values for CN taken from sample rainfall and runoff data. However, abstractions from rainfall depend on the antecedent conditions that exist at the time a rain event occurs. Antecedent condition I is used when there has been no rain preceding the rainfall event under investigation. Antecedent condition III is used where there has been considerable rainfall prior to the rain in question. Curve numbers for antecedent conditions I and III should be adjusted from antecedent condition II using the following formulas (Chow et al, 1988):

$$CN(I) = \frac{4.2CN(II)}{10 - 0.058CN(II)} \quad (7.61)$$

$$CN(III) = \frac{23CN(II)}{10 + 0.13CN(II)} \quad (7.62)$$

where $CN(I)$, $CN(II)$, and $CN(III)$ represent curve numbers for antecedent conditions I, II, and III, respectively.

Composite Curve Numbers

For an area that consists of several soil types and land uses, a composite curve number, CN_C, of the area must be calculated using the following equation:

$$CN_C = \frac{\sum_i A_i CN_i}{\sum_i A_i} \quad (7.63)$$

where A_i is the area of the subdivision i and CN_i is the CN of the subdivision i.

However, in the present day, GIS technology is widely used to derive composite curve numbers directly from the GIS data of soil cover and land use. The procedure is described in Chapter 20.

Figure 7.19 shows the composite curve numbers of the catchments within Graham Branch Watershed. These curve numbers are derived from the soil data shown in Figure 7.16 and the land cover data shown in Figure 7.18. The GIS method was used to derive the composite CN values by intersecting the land cover characteristics and the soil types. Note that the land cover of this watershed has a significant open space under good conditions, since this is located in a currently suburban setting. Yet the CN values for the constituent catchments are quite high. This is due to the fact that the dominant soil type in this watershed is HSG D followed by C (Figure 7.16). These are clay and clay loam (Figure 7.15) with low infiltration capacity.

In relation to composite curve numbers, adjustments of curve numbers of impervious areas may be necessary. An impervious area is considered connected if runoff from it flows directly into the drainage system. The CN values for

Table 7.4C *Runoff CN values for cultivated agricultural land*

Cover type/description	Treatment	Hydrologic condition	CN value for HSG			
			A	B	C	D
Fallow	Bare soil	–	77	86	91	94
	Crop residue cover (CR)	Poor	76	85	90	93
		Good	74	83	88	90
Row crops	Straight row (SR)	Poor	72	81	88	91
		Good	67	78	85	89
	SR + CR	Poor	71	80	87	90
		Good	64	75	82	85
	Contoured (C)	Poor	70	79	84	88
		Good	65	75	82	86
	C + CR	Poor	69	78	83	87
		Good	64	74	81	85
	Contoured and terraced (C & T)	Poor	66	74	80	82
		Good	62	71	78	81
	C & T + CR	Poor	65	73	79	81
		Good	61	70	77	80
Small grain	SR	Poor	65	76	84	88
		Good	63	75	83	87
	SR + CR	Poor	64	75	83	86
		Good	60	72	80	84
	C	Poor	63	74	82	85
		Good	61	73	81	84
	C + CR	Poor	62	73	81	84
		Good	60	72	80	83
	C & T	Poor	61	72	79	82
		Good	59	70	78	81
	C & T + CR	Poor	60	71	78	81
		Good	58	69	77	80
Close-seeded or broadcast legumes or rotation meadow	SR	Poor	66	77	85	89
		Good	58	72	81	85
	C	Poor	64	75	83	85
		Good	55	69	78	83
	C & T	Poor	63	73	80	83
		Good	51	67	76	80

urban areas given in Table 7.4 were estimated considering certain percentages of directly connected impervious area and open space. So, if there is no special situation, when these values are selected, no further accounting of directly connected impervious area is required. However, if it is believed that impervious areas should be considered separately, the composite CN values should be calculated using the following equation given in NRCS (2004b):

$$CN_C = CN_P + \left(\frac{p_{imp}}{100}\right)(98 - CN_P) \quad (7.64)$$

where CN_P is the curve number for the pervious area, and p_{imp} is the percentage imperviousness. However, if a certain portion of the impervious area is not directly connected to the drainage system, the equation used to calculate the CN of the impervious area is

$$CN_C = CN_P + \left(\frac{p_{imp}}{100}\right)(98 - CN_P)(1 - 0.05R_p) \quad (7.65)$$

where R_p is the ratio of the unconnected impervious area to the total impervious area.

7.6.3.3 Temporal Variability of Effective Rainfall in the Curve Number Method

The NRCS rainfall–runoff equation (Eq. 7.55) was developed to estimate the total storm runoff from the total

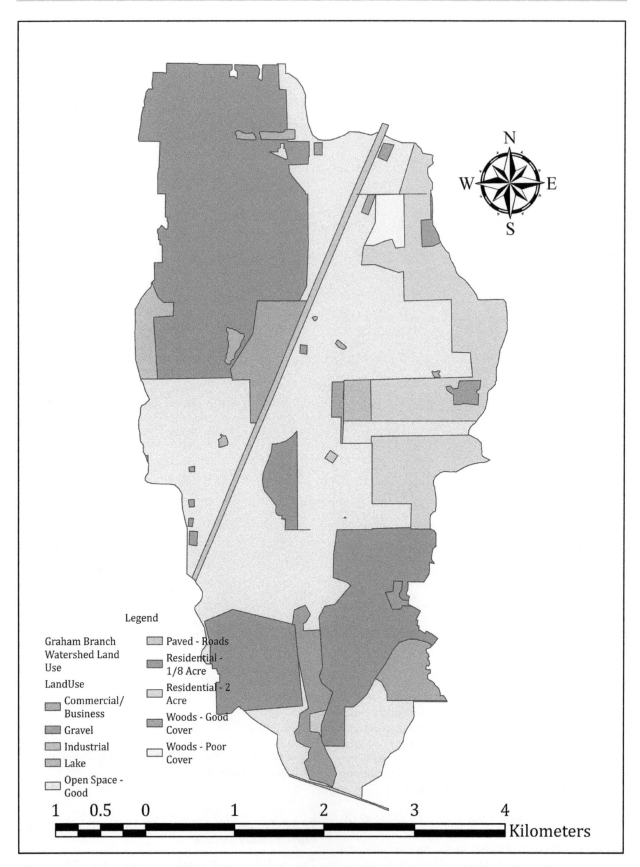

Figure 7.17 Land cover map of the watershed shown in Figure 7.15, derived from the NLCD. (A black and white version of this figure will appear in some formats. For the color version, please refer to the plate section.)

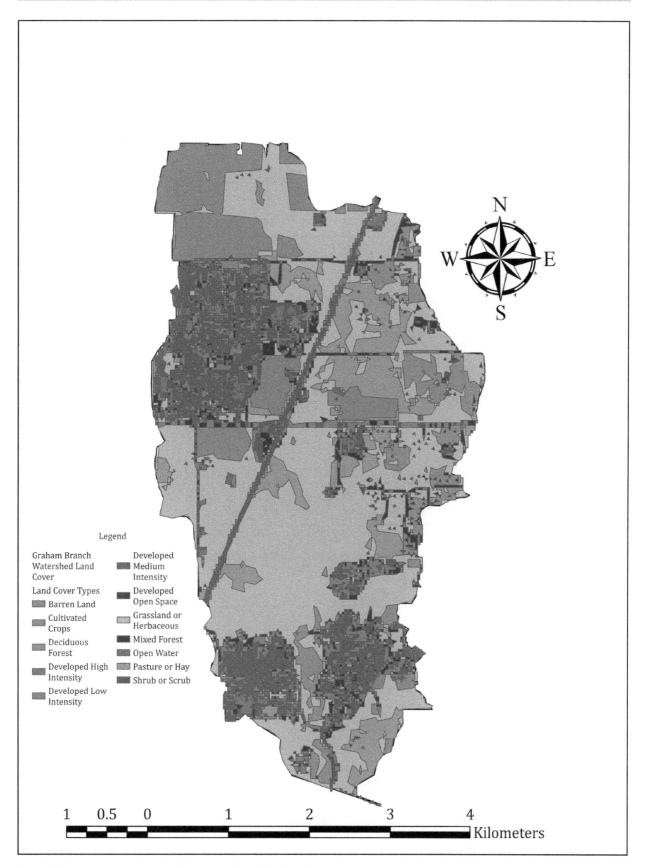

Figure 7.18 Land cover map of the watershed shown in Figure 7.15, derived from the land use and zoning maps of the local municipalities in and around the watershed. (A black and white version of this figure will appear in some formats. For the color version, please refer to the plate section.)

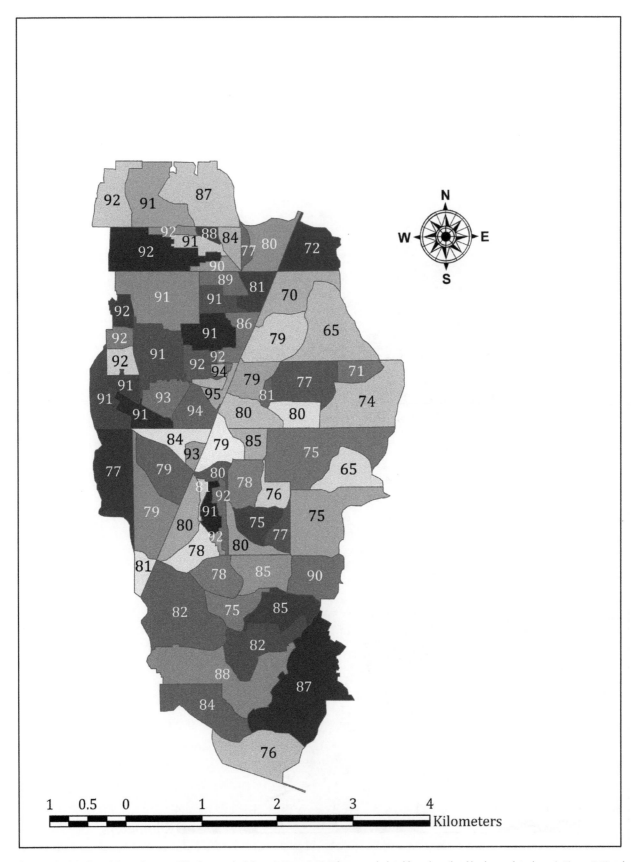

Figure 7.19 *CN* values of the catchments within the watershed shown in Figure 7.15. Values are calculated form the soil and land cover data shown in Figures 7.16 and 7.18 using the GIS method.

storm rainfall. In other words, the relationship excludes time as a variable and rainfall intensity is ignored. The time dependence of effective rainfall according to the NRCS curve number method can be seen by writing Eq. 7.53 as

$$F = \frac{S(P - I_a)}{P - I_a + S}; P > I_a \qquad (7.66)$$

and then differentiating this equation with respect to time and holding I_a and S as constant to get

$$\frac{dF}{dt} = \frac{S^2 \frac{dP}{dt}}{(P - I_a + S)^2} \qquad (7.67)$$

The presence of dP/dt (which gives rainfall intensity) in the numerator of Eq. 7.67 indicates that as rainfall intensity increases, actual retention (abstraction) increases. This is contrary to the infiltration theory with a physical basis discussed above. This problem was noted by Morel-Seytoux and Verdin (1981).

For the problem noted above, a simpler approach is usually used, as implemented in HEC-HMS to account for cumulative retention or abstraction and effective rainfall by using

$$P_e(t) = \frac{[P(t) - I_a]^2}{[P(t) - I_a + S]} \qquad (7.68)$$

where $P_e(t)$ is cumulative effective rainfall at time t, and $P(t)$ is cumulative actual rainfall. In other words, Eq. 7.58 is rewritten by substituting q and P with the corresponding cumulative values $q(t)$ and $P(t)$ at a time between the beginning and the end of the rainfall. Then, the effective rainfall depth within a time interval Δt is obtained by the difference $q(t + \Delta t) - q(t)$.

The incremental effective rainfall depth over a period $\Delta t = t_2 - t_1$ is then given by

$$\Delta P_e = P_e(t_2) - P_e(t_1) \qquad (7.69)$$

Note that the computation of cumulative effective rainfall by Eq. 7.68 is entirely dependent on the cumulative precipitation at any time. The total abstraction, therefore, just like effective rainfall, is independent of the rainfall pattern.

7.6.3.4 Shortcomings of the Curve Number Method

As noted at the very beginning of this section, the curve number method is perhaps the most popular method of computation of effective rainfall. This is not only because of its simplicity but also due to the fact that nowadays, with the availability of soil and land use maps in GIS formats combined with the widespread use of GIS technology by hydrologists, it is relatively convenient to compute *CN* values, the only parameter that is required for this method. In spite of this, there has been extensive criticism of the method, because it does not lead to an accurate reproduction of runoff hydrographs, the predicted infiltration (abstraction) rate is not in accordance with the classical theory of flow in unsaturated zones (as discussed above in relation to the problem with Eq. 7.67), the method is applied to watersheds for which it was not calibrated, the original calibration results are not available, and the method never went through a critical review process.

The results of an investigation by Rawls and Brakensiek (1986) can be used as an example to corroborate the criticisms cited by various workers for the curve number method. Rawls and Brakensiek (1986) made a comparison of the predicted runoff volumes obtained using the Green–Ampt and NRCS curve number methods. Their results showed that the Green–Ampt infiltration model gave better predictions for higher volumes of runoff. Similar observations were made by several others. Furthermore, the rainfall excess distribution provided by the Green–Ampt procedure can be used as direct input to a hydrograph model compared to the cumulative effective rainfall calculation that is used with the curve number method.

7.6.4 Use of an Infiltration Equation

An infiltration model may be used to calculate effective rainfall. The Green–Ampt model is widely used and we have chosen it in this section to illustrate the use of an infiltration model to calculate effective rainfall.

In the Green–Ampt model presented above, the assumption that there is a negligible depth of ponding water on the ground surface implies that the equations in the model are applicable if at any time the rainfall intensity is greater than the infiltration rate. This is case III described in Section 7.5.3. Mein and Larson (1973) presented a method to estimate the infiltration rate and cumulative infiltration when rainfall intensity is constant but less than the initial infiltration capacity of the soil.

The Mein–Larson model considers that, until the time of ponding, that is, when the soil becomes saturated and overland flow begins, the infiltration rate is equal to the rainfall intensity. Curves 1, 2, and 3 in Figure 7.4 represent the time less than the ponding time. Curve 4 represents the soil moisture conditions at the time of ponding, T_p. Curves 5 and 6 are at times $t > T_p$. Thus, up to $t = T_p$, $f = i$, where i denotes the rainfall intensity (L/T) and thereby $F(t) = it$. Thus, the cumulative infiltration up to the time of ponding, F_p, is given by

$$F_p = iT_p \qquad (7.70)$$

Substitution of Eq. 7.70 into Eq. 7.22 for $f(t)$ results

$$i = K\left(1 + \frac{\Omega}{iT_p}\right) \qquad (7.71)$$

Equation 7.71 is rearranged to get

$$T_p = \frac{K_s \Omega}{i^2 - iK_s} \qquad (7.72)$$

At $t = T_p$, Eq. 7.27 can be written as

$$F_p - \Omega \ln\left(\frac{\Omega + F_p}{\Omega}\right) = K_s T_p \tag{7.73}$$

Subtracting Eq. 7.73 from Eq. 7.27 and using $\ln(a/b) = \ln(a) - \ln(b)$, the cumulative infiltration after ponding is given by

$$F(t) - F_p - \Omega \ln\left[\frac{\Omega + F(t)}{\Omega + F_p}\right] = K_s(t - T_p) \tag{7.74}$$

Equations 7.70 and 7.72 are only valid if rainfall intensity is constant up to the time of ponding. In reality, however, a rainfall hyetograph shows variable rainfall intensities throughout the duration of a rainfall event. Therefore, modification to the concepts ingrained in Eqs. 7.70 and 7.74 must be made for an actual rainfall event. The procedure is relatively complex as illustrated in Example 7.4. Chow et al. (1988) presented a method that is simpler than the procedure suggested by USACE (1994). Either of these two approaches is facilitated by programming. The essential sequence is described below.

We designate a time interval by t and $t + \Delta t$, and the corresponding potential infiltration rate, cumulative infiltration, and rainfall intensity by $f_t, f_{t+\Delta t}, F_t, F_{t+\Delta t}, i_t,$ and $i_{t+\Delta t}$. Generally, when rainfall starts, its intensity is low, and the antecedent soil moisture is low. During this initial period, $f_t > i_t$. As long as $f_t > i_t$, in each interval, $F_{t+\Delta t} = F_t + i_t \Delta t$, when $F_t = i_{t-1}\Delta_{t-1}$ because $i_{t-1} < f_{t-1}$. During the period when these conditions hold true, no ponding occurs. Then comes a time when the rainfall intensity becomes greater than the potential infiltration capacity. The time when it occurs is the actual time of ponding. During the period when the condition $i_t > f_t$ holds true, Eq. 7.74 is applicable to compute the cumulative infiltration, because this is the period when ponding occurs. The corresponding infiltration rate is calculated by Eq. 7.22. Subsequently, when rainfall intensity subsides and fluctuates, there can be intervals during the rest of the duration of rainfall when either $i_t > f_t$ or $i_t < f_t$. In other words, there can be intervals when either there is no ponding or there is ponding. For this period, the condition must be evaluated at each interval and, depending on the condition in the interval, cumulative infiltration will be calculated either from only the rainfall intensity and duration (no ponding) or the infiltration equation (ponding).

7.7 SURFACE STORAGE

Surface storage or ponding is the amount of rainfall that is stored or trapped into closed depressions or other storage areas, such as small ponds and microdepressions on the surface topography, and never reaches the outlet of the basin, ultimately contributing only to evaporation and infiltration. Surface storage occurs prior to the onset of runoff. Actual measurements of surface storage and detention are extremely difficult to make. However, sometimes it is possible to identify and measure depression storage areas that are obvious from topographic and land use–land cover maps.

Linsley et al. (1982) expressed the amount of water stored at a given time by surficial depression as

$$V_d = S_d[1 - \exp(-k_d P_e)] \tag{7.75}$$

where V_d is the volume (L) of water in surface storage at time t, S_d is the maximum available surface storage capacity (L), P_e (L) is the effective rainfall (accumulated rainfall at time t minus other losses such as interception and infiltration), and k_d is a constant.

Equation 7.75 is based on the hypothesis that

$$\frac{dV_d}{dP_e} = k_d(S_d - V_d) \tag{7.76}$$

This leads to the estimation of constant k_d as follows. From Eq. 7.75, the rewritten form of Eq. 7.76 is

$$\frac{dV_d}{dP_e} = k_d S_d \exp(-k_d P_e) \tag{7.77}$$

From Eq. 7.77, it follows that when $P_e = 0$, $dV_d/dP_e = 1$, which implies that all water essentially fills all depression storage. Therefore,

$$k_d = \frac{1}{S_d} \tag{7.78}$$

The value of S_d can be estimated from topographic map or field observations.

If the value of P_e is very large, V_d approaches the value of S_d. If P_e is negligible, the value of V_d is also negligible. Thus, from Eq. 7.75, it is evident that $0 \leq V_d \leq S_d$. The rate at which the depression storage is filled declines rapidly after the initiation of rainfall. Ultimately, the amount of rain water going into surface storage goes to zero if P_e is significantly large.

Consideration for the importance of surface storage in rainfall–runoff modeling depends on several factors. The main factors are the land use and landform. If the area under consideration is a highly urbanized area with considerable impervious ground cover then surface storage is negligible. Similarly, if the land surface has steep slopes then surface storage becomes less significant, since it has been shown that surface storage decreases considerably as the slope of the ground surface increases. Table 7.5 gives some estimates of surface storage as a function of surface type.

Table 7.5 *Typical values of surface storage*

Surface type	Surface storage (in)
Turf grass, pervious area, open fields	0.20–0.50
Impervious areas, 1% slope, parking lots, roads, asphalts, flat roofs	0.06–0.12
Impervious areas, 2.5% slope, parking lots, roads, asphalts, flat roofs	0.05
Steep and smooth slopes	0.04

Another consideration when accounting for surface storage is the time frame of the rainfall–runoff event that is in question. For continuous simulation runs, surface storage can have an influence that cannot be ignored compared to a case of a relatively short-duration rainfall.

In HEC-HMS, two parameters are needed if surface storage is to be considered. The first parameter in the program is the "initial storage." This is provided as the percentage of the surface storage that is full of water at the beginning of simulation. The second is the "maximum storage," which is the maximum amount of water (in or mm) that can be held on the soil surface before surface runoff begins. The amount of storage is specified as an effective depth of water.

In most of the models, including the procedure implemented in HEC-HMS, surface runoff begins when the rainfall rate exceeds the infiltration rate, and the surface storage is filled to its maximum capacity. However, the actual rate of filling of surface storage can be quite different. As discussed by Haan et al. (1994), this is illustrated below for an ideal condition.

If interception is neglected, the surface runoff rate R can be expressed as

$$R = i - f - v_d \tag{7.79}$$

where i is the rainfall intensity (L/T), f is the infiltration rate (L/T), and v_d is the rate (L/T) at which surface depression is filled up. Thus,

$$v_d = \frac{dV_d}{dt} \tag{7.80}$$

From Eqs. 7.75 and 7.78, v_d is then given as

$$v_d = \frac{d}{dt}[S_d\{1 - \exp(-k_d P_e)\}] \tag{7.81}$$

$$v_d = S_d k_d \exp(-k_d P_e)\frac{dP_e}{dt} = \exp(-k_d P_e)\frac{dP_e}{dt} \tag{7.82}$$

Noting that $i - f = \frac{dP_e}{dt}$, v_d is given as

$$v_d = \exp(-k_d P_e)(i - f) \tag{7.83}$$

Putting this expression of v_d into Eq. 7.79 gives

$$R = i - f - \exp(-k_d P_e)[(i - f)] \tag{7.84}$$

$$R = (i - f)[1 - \exp(-k_d P_e)] \tag{7.85}$$

$$\frac{R}{i - f} = 1 - \exp(-k_d P_e) \tag{7.86}$$

Also note that $P_e = P - F$, where F is the accumulated volume of infiltration and P is the cumulative rainfall. Thus, P_e denotes overland flow and surface storage supply (volume expressed in units of L), whereas R denotes the overland flow supply or surface runoff rate. The left-hand side of Eq. 7.86 is the ratio of surface runoff supply rate to the difference in the rainfall intensity and infiltration rate.

Thus, in real situations, the case may be between these two cases, as shown in Figure 7.20, where Eq. 7.86 is plotted. A more realistic distribution can be represented by a cumulative normal probability curve with mean $= S_d$ and standard deviation $= S_d/3$ as

$$\frac{R}{i - f} = \frac{1}{S_d\sqrt{2\pi}} \int_{-\infty}^{P_e} \exp\left[-\frac{(P_e - S_d)^2}{2(S_d)^2}\right]dP_e \tag{7.87}$$

This curve is also shown in Figure 7.20.

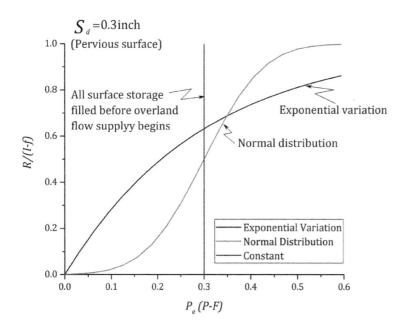

Figure 7.20 Relationship between the effective rainfall (P_e) and ratio of rate of surface runoff rate and difference between rainfall and infiltration rates [R/(I-f)]. For a similar concept, see also Tholin and Kiefer (1906).

7.8 INITIAL ABSTRACTIONS

Initial abstraction, denoted by I_a, usually includes both interception and surface depression and detention storage. In typical applications, such as in HEC-HMS, the initial abstraction is modeled as follows:

$$i_r(t) = 0 \text{ for } P(t) \leq I_a, \ t \geq 0 \tag{7.88}$$

$$i_r(t) = r_0(t) \text{ for } P(t) > I_a, \ t \geq 0 \tag{7.89}$$

where $i_r(t)$ denotes the rainfall intensity adjusted for interception and surface storage, $r_0(t)$ and I_a represent the depth of interception and surface storage, respectively, and are assumed to be uniform over the area under consideration, and $P(t)$ is the cumulative rainfall over the entire area.

In HEC-HMS, this is called initial loss. For any of the selected methods for computation of effective rainfall, the value of I_a can be provided as an option.

Initial abstraction depends on the physical properties of soil, land use, and antecedent conditions. If the watershed is in a saturated condition, I_a will approach zero. On the other hand, if dry conditions prevail prior to a rain event, I_a will increase to the maximum rainfall depth that can fall on the surface with no runoff. This will depend on the terrain topography, land use, and soil type. USACE (1994) estimates that this ranges from 10% to 20% of the total rainfall for forested areas to 0.1–0.2 inches (2.54–5.08 mm) in urban areas.

It should be noted here that in HEC-HMS, canopy interception and surface storage are treated separately in the continuous simulations discussed next, which is not the case for event modeling. For event-based simulations, interception and surface storage are lumped together into the initial loss.

7.9 ABSTRACTIONS DURING LONG-TERM RAINFALL–RUNOFF SIMULATION

Long-term rainfall–runoff simulation is also called continuous simulation. In most design applications, short-term or event-based modeling is performed. These events, either real or hypothetical, are typically short-duration rainfall events ranging from 3 to 24 hours, and in certain cases can be up to 72 hours. These events do not alternate between wetting and drying periods because they simulate the behavior of a hydrologic system during a single rainfall event only. Continuous simulations, on the other hand, model both wet and dry weather behavior extended over several months or even years. Such models are important for water resources management and planning for a watershed or river basin.

Of all the abstraction methods described above, only the layered Green–Ampt model can be used for continuous simulation and all other models are applicable to event-based rainfall–runoff modeling. For continuous simulations, the following abstraction methods, implemented in HEC-HMS, can be used. It should be noted here that, for event-based models, initial abstraction is often ignored, justifiably so. However, this can account for more than 50% of the total precipitation for a long-term hydrologic simulation.

Before becoming runoff or infiltrating into the soil, rainfall is first intercepted by plant canopies and then temporarily stored in surface depressions such as puddles. The intercepted or stored water can return to the atmosphere through evaporation. The water stored in surface depressions can continue to infiltrate into the soil, where it will be absorbed by plant roots and released to the air by plant transpiration.

7.9.1 Canopy Interception

Evapotranspiration refers to the combined process of evaporation from the land surface and transpiration by plants since it is very difficult to measure and quantify these two processes separately. Chapter 14 covers evapotranspiration in detail. Briefly, through evaporation and transpiration water is returned from the surface and soil to the atmosphere. The **potential evapotranspiration** (**PET**), meaning the theoretical evapotranspiration, serves as the upper limit for evapotranspiration rates based on atmospheric conditions. In HEC-HMS, the potential evapotranspiration is specified in "meteorological models" while the canopy method, given as a model for rainfall loss, is used to calculate the **actual evapotranspiration** (**AET**) which cannot exceed the available water capacity. The evapotranspiration method is only necessary in a continuous long-term simulation in which evapotranspiration is used to calculate losses and reset soil water deficit.

The parameters required to include canopy interception are crop coefficients and initial and maximum storage. The initial condition of the canopy should be specified as the percentage of the canopy storage that is full of water at the beginning of simulation. The canopy storage represents the maximum amount of water that can be held on leaves before throughfall to the surface begins. The amount of storage is specified as an effective depth of water that can be held in the canopy before additional precipitation falls through to the ground surface. The storage depth does not change for the duration of simulation. The crop coefficient is a ratio applied to PET to compute the amount of water actually extracted from the soil. The canopy can be set to only evaporate water from storage and extract water from the soil during dry time periods only. Alternately the canopy can be set to evaporate water from storage and extract water from the soil during both dry and wet periods. The choice for simultaneous precipitation and evapotranspiration can improve results when using a long time interval or during a snowmelt simulation.

Canopy interception can be simple or dynamic. In the dynamic mode, the parameters change with time.

As the name implies, the *simple method* is a simple representation of plant canopy. All rainfall is intercepted until the canopy storage capacity is filled. Once this happens, additional rainfall goes either to surface storage or directly to the soil, if surface storage is not included separately in the model. All PET is used to empty the canopy storage until the water in storage has been eliminated. The PET is multiplied by the crop coefficient to determine the amount of evapotranspiration from canopy storage and later the surface and soil components. Only after the canopy storage is exhausted, excess PET occurs from the surface and soil components.

There are two methods for the extraction of water from the soil. The simple method extracts water at the rate of PET and can be used with the deficit constant or soil moisture accounting methods described below. The other method, called the **tension reduction method**, can be used with the soil moisture accounting method and extracts water at the rate of PET from the gravity zone but reduces the rate when extracting from the tension zone. No water is extracted from the soil unless either the simple or tension reduction method is selected.

The dynamic canopy method includes an interception storage capacity parameter and a crop coefficient that changes with time. The storage capacity specifies the amount of water that can be held in the canopy before precipitation begins falling through to the ground surface. No throughfall occurs until the canopy storage is full. The canopy uses all the PET until the storage is exhausted. The PET is multiplied by the crop coefficient to determine the amount of evapotranspiration from canopy storage and, later, the surface and soil components. Only after the canopy storage is emptied, the remaining PET occurs from the surface and soil components.

The storage capacity (depth) does not change for the duration of simulation. However, there are two methods for specifying the crop coefficient. A time series can be used to specify how the crop coefficient changes during the plant growing season. This coefficient must be defined at a gauge Alternately, a grid may be used to specify how the crop coefficient changes at each grid cell during the growing season.

7.9.2 Surface Storage

Surface storage or depression storage is generally used for continuous simulations only. The surface storage of an impervious area such as a parking lot is generally close to zero. However, this storage for an agricultural field can be quite significant, particularly if conservation tillage practices are used. Rain throughfall from the canopy, or direct rainfall if there is no canopy, impacts the surface storage. The net rainfall accumulates in the depression storage and infiltrates as the soil has capacity to absorb water. Surface runoff begins when the rainfall intensity exceeds the infiltration rate, and the surface storage is filled. Rainfall residing in the surface storage can infiltrate after rainfall stops and is subject to PET.

Two parameters are used to model surface storage. The initial storage, specified as the percentage of the surface storage that is full of water at the beginning of the simulation, is the initial condition. The other parameter is the maximum storage which is the upper limit of the amount of water that can be held on the soil surface before surface runoff begins. This amount of storage is specified as an effective depth of water.

7.9.3 Deficit and Constant Rate Infiltration Model

The deficit and constant loss infiltration model uses a single soil layer to account for continuous changes in moisture content in the soil. This model can be used in a continuous simulation if a canopy method is used at the same time to extract water from the soil during dry periods in response to PET computed separately in the meteorological model of HEC-HMS. The soil layer dries between precipitation events as the canopy extracts soil water. No soil water extraction occurs if there is no canopy. It can also be used in combination with a surface storage model that stores water on the land surface. The water in surface storage infiltrates to the soil layer. The infiltration rate is determined by the capacity of the soil layer to absorb water. When both canopy and surface storage are incorporated in combination with the deficit and constant loss model, the system can be conceptualized as shown in Figure 7.21A. Figure 7.21B shows the corresponding system if the layered Green–Ampt method is selected as the infiltration method instead of deficit and constant loss infiltration model for a continuous simulation. The layered Green–Ampt loss method uses two layers to represent the dynamics of water movement in the soil. Both layers are functionally identical but may have separate and distinct parameters, which can be used to represent layered soil profiles, allowing for better representation of stratified soil drying between storms. Each layer is described using a bulk depth and water content values for saturation, field capacity, and wilting point.

Rainfall first fills the canopy storage. Rainfall that exceeds the canopy storage overflows onto the land surface. The new rainfall gets added to any water already in surface storage. At the beginning of a rainfall event, the soil usually is not saturated, and the precipitation (after surface/depression storage and canopy/evapotranspiration) will penetrate the soil to fill the voids in the soil layer. When the soil becomes saturated (the soil water deficit becomes zero), the infiltrated water will percolate through the bottom of the active soil layer and will get lost. If the moisture deficit is greater than zero, water infiltrates from the surface into the soil layer at a rate that is essentially infinite. This unlimited infiltration continues until the soil layer reaches saturation (the moisture deficit drops to zero) and during this period there is no percolation. The infiltration rate is defined by the constant rate while the soil layer remains at saturation. The percolation rate out of the bottom of the layer is also defined by the constant rate while the soil layer remains at saturation. Percolation stops as soon as the soil layer drops below saturation (the moisture deficit is greater than zero). The

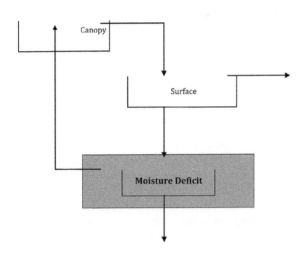
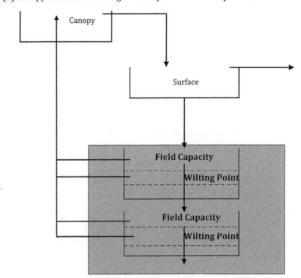

Figure 7.21 Conceptual schematic of (A) the deficit and constant loss model and (B) the layered Green–Ampt method.

moisture deficit increases in response to the canopy extracting soil water to meet the PET demand.

The deficit and constant loss infiltration model requires three parameters. The first parameter is the initial deficit (in or mm), which is the initial moisture condition of the soil. At the start of the simulation, it is the amount of water that would be required to fill the voids of soil layer and it is usually a value between zero implying fully saturated soil horizon and the maximum deficit (with residual water content). The second parameter is the maximum deficit (in or mm). This is the total amount of water the soil layer can hold, specified as an effective depth. The upper bound of this parameter is the bulk thickness of the active soil layer multiplied by effective porosity. However, in most cases such an estimate must be reduced by the permanent wilting point and for other conditions that reduce the water holding capacity. The third parameter, the constant rate (in/h or mm/h), defines infiltration and percolation rates when the soil layer is saturated. The saturated hydraulic conductivity is a good approximation.

The root zone depth can also be used to approximate the active soil layer depth. Fan et al. (2017) summarized 2200 observations of root depths and concluded that root depth varied from 0.01 m to 70 m with a distribution peak at 1.0 m. They have also provided the mean root depths of different soil textures and vegetations. Under field conditions, it is common that the lower portions of the root zone are not active during the hydrologic processes and therefore the active soil layer depth can range from half of the root zone depth to the full root zone depth. For lack of a better estimation or guidance, the active soil layer depth can be estimated as a value between 0.3 and 1.8 m. The thickness of the active soil layer is best determined through calibration.

After a rainfall event that saturates the soil layer it usually takes approximately three days for the saturated soil to drain freely by gravity to the field capacity. In such instances, the initial deficit can be estimated as the product of active soil layer depth and the difference between porosity and field capacity. Under the average antecedent moisture conditions, the wilting point (−1500 kPa water content) is a good estimate for the initial water content in the western United States, and the field capacity (−33 kPa water content) is used in the eastern states (Maidment, 1993). Correspondingly, the initial deficit can be calculated as the product of the active soil layer depth and the difference of porosity and wilting point for the western states or the product of the active soil layer depth and the difference between porosity and field capacity for the eastern United States for the average antecedent moisture condition.

7.9.4 Soil Moisture Accounting

The soil moisture accounting (SMA) model in HEC-HMS is patterned after the precipitation–runoff model originally developed by Leavesley et al. (1983) at the USGS. The present Precipitation–Runoff Modeling System developed by the USGS is used to evaluate the response of various combinations of climate and land use on streamflow and general watershed hydrology also incorporates a similar concept. Bennett and Peters (2000) described the algorithm that is implemented in the SMA model within the HEC-HMS program.

Figure 7.22 shows the storage components that illustrate the physical aspects of the watershed and are included in the SMA model to represent the dynamics of water movement in the soil. Rainfall that is not directly converted to surface runoff or stored in surface depressions and canopy storage infiltrates into the soil and is stored or conveyed into the subsurface. The subsurface is divided into three layers: (1) a

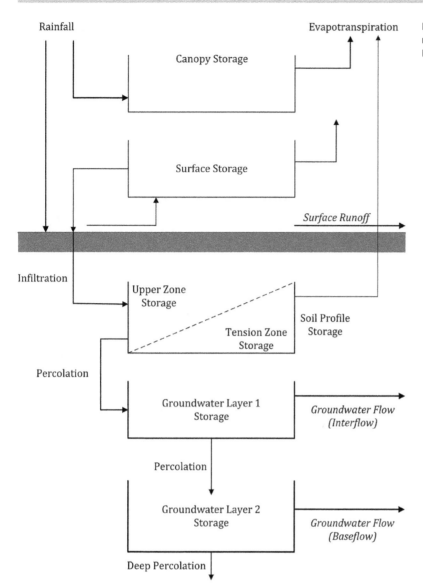

Figure 7.22 Conceptual schematic of the continuous soil moisture accounting (SMA) algorithm (after Bennett and Peters, 2000).

top soil layer, which includes an upper zone and a tension storage zone; (2) a shallow groundwater layer (GW1), which conveys interflow; and (3) a deep groundwater layer (GW2), which conveys baseflow. Groundwater layers are not designed to represent aquifer processes; they are intended to be used for representing shallow interflow processes.

7.9.4.1 SMA Model Parameters

The parameters required to model these storage layers are as follows.

1. Maximum infiltration rate (f_s^{max}, in/h or mm/h): This parameter defines the maximum rate at which water is infiltrated from the surface storage into the top soil layer. The saturated hydraulic conductivity for the upper soil layer can be used to define this parameter. This is the upper bound on infiltration; the actual infiltration in a particular time interval is a linear function of the surface and soil storage, if surface storage is included. If there is no surface storage, water will always infiltrate at the maximum rate.

2. Soil storage (S_s^{max}, in or mm): This parameter defines the volume of total (upper bound) available storage in the upper soil layer. The soil porosity is multiplied by the depth of the upper soil layer to calculate this parameter, if these values are known from the soil database or soil survey. Alternatively, if this is set to zero, the infiltrated water passes directly to groundwater.

3. Tension zone storage (in or mm): This parameter defines the volume of available storage in the tension zone or the part of the soil column adjacent to the water table and wetted by capillary forces. The volume in this zone can be computed using the field capacity or porosity of

the tension zone, multiplied by the depth of the upper soil layer. This storage volume specifies the amount of water storage in the soil that does not drain under the effect of gravity. Percolation from the soil layer to the upper groundwater layer will occur when the current soil storage exceeds the tension storage. Water in the tension storage is only removed by evapotranspiration. By definition, tension storage must be less than soil storage.

4. Soil percolation rate (p_s^{max}, in/h or mm/h): This parameter describes the maximum rate of infiltration from the upper soil layer into the upper groundwater layer (GW1). The average saturated hydraulic conductivity can be used to define this parameter. The actual percolation rate is a linear function of the current storage in the soil and the current storage in the upper groundwater.

5. Groundwater 1 (GW1) storage (S_{GW1}^{max}, in or mm): This parameter represents the maximum total storage in the upper groundwater layer. If it is set to zero, then the upper groundwater layer is eliminated, and water is percolated from the soil directly to the lower groundwater layer.

6. Groundwater 1 (GW1) percolation rate (p_{pGW1}^{max} in/h or mm/h): This parameter describes the maximum rate of infiltration from the GW1 to the GW2 layer. This value is typically used for calibration and can initially be set equal to the percolation rate. The actual percolation rate is a linear function of the current storage in the upper and lower groundwater layers.

7. Groundwater 1 (GW1) coefficient: This parameter is used as the time lag on a linear reservoir for transforming water in storage to become lateral outflow. The lateral outflow is available to become baseflow.

8. Groundwater 2 (GW2) storage (S_{GW2}^{max}, in or mm): This parameter represents the total storage in the lower groundwater layer. If it is set to zero the lower groundwater layer is eliminated, and water is percolated from the upper groundwater layer directly by deep percolation.

9. Groundwater 2 (GW2) percolation rate (p_{pGW2}^{max}, in/h or mm/h): This parameter describes deep percolation out of the GW2 layer into deep subsurface aquifer storage. This water is ultimately lost from the system being modeled. Because this parameter is difficult to estimate, it is typically set during calibration. As an initial assumption this value can be set to half of the GW1 percolation rate. The actual percolation rate is a linear function of the current storage in the lower groundwater layer.

10. Groundwater 2 (GW2) coefficient: This parameter is used as the time lag on a linear reservoir for transforming water in storage to become lateral outflow. Its value is usually larger than the GW1 coefficient. The lateral outflow is likewise available to become baseflow.

In addition to the model parameters listed above, the initial condition of the soil should be specified as the percentage of the soil that is full of water at the beginning of simulation. The initial condition of the upper and lower groundwater layers must also be specified.

7.9.4.2 SMA Process Description

The order of computations during simulation follows one of the two processes based on whether precipitation or evapotranspiration is occurring. The conceptual process are shown in Figure 7.23.

During the period of precipitation (Figure 7.23A), canopy storage is filled first (1). Precipitation in excess of canopy storage combined with water in surface storage at the beginning of time step is available for infiltration (2). If the combination of these two exceeds a calculated potential infiltration volume, the volume of water greater than the infiltration volume returns to surface storage (3). When the surface storage capacity is full, the excess is the volume of runoff (4). The volume of water available for infiltration that is equal to or less than the potential infiltration volume fills the soil storage (5). Water in soil storage then percolates to the first groundwater layer (6). Groundwater flow is routed out of the first groundwater layer (7) and the remaining water is available for percolation to the second deeper groundwater layer (8). Similarly, routing and percolation then occur from the second groundwater layer. Percolation from the second groundwater layer is to a deep aquifer and is output from the model (9).

Evapotranspiration (Figure 7.23B) is computed when precipitation is not occurring. PET is first satisfied from canopy storage (1). The surface storage capacity is then depleted once the canopy storage is empty (2). Water is next removed from soil storage (3). The evapotranspiration rate from soil storage is reduced relative to the total depth of soil storage to the tension zone storage. Infiltration can occur if water remains in the surface storage after the evapotranspiration for a computational time step has been fulfilled (4). Water infiltrates at a computed rate and fills the soil storage (5). Model simulations continue for percolation and groundwater routing as during the period of precipitation (6).

7.9.4.3 Equations Used in the SMA

Infiltration: The amount of infiltration during a time interval is a function of the volume of water available for infiltration, the state (percent full) of soil storage (ratio of the current soil storage, S_s^c, and the maximum soil storage, S_s^{max}), and the maximum soil infiltration rate (f_s^{max}). The potential infiltration rate (f_{ps}) is a function of the percentage of the filled soil storage:

$$f_{ps} = f_s^{max} - \frac{S_s^c}{S_s^{max}} f_s^{max} \qquad (7.90)$$

Once the potential infiltration rate is calculated, the actual infiltration rate is the smaller value between the water available for infiltration and the potential infiltration rate. The

7.9 Abstractions during Long-Term Rainfall–Runoff Simulation

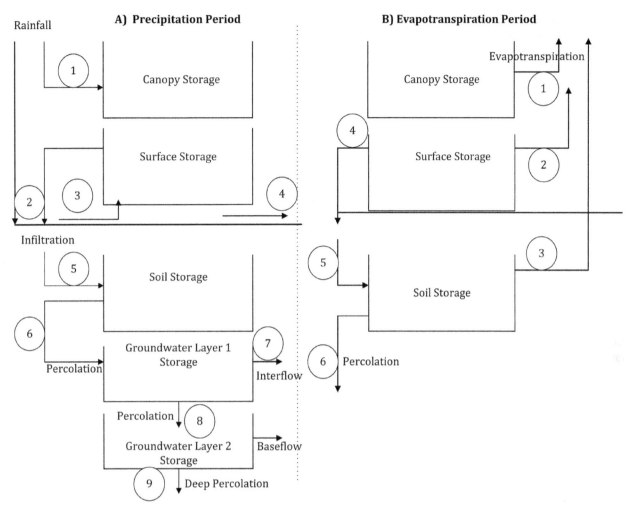

Figure 7.23 Computational orders in the SMA model during (A) wet and (B) dry periods.

water available for infiltration is precipitation that is not intercepted by the canopy during the interval, plus water in surface storage (current surface storage) at the beginning of the time step. When the soil profile storage is empty, the potential infiltration equals the maximum infiltration rate. When the soil profile is full, the potential infiltration equals zero.

Percolation: Potential percolation is a function of maximum percolation rate and the content of storage volume from which percolation is occurring. The potential soil percolation (p_{ps}) and maximum soil percolation are the potential and maximum percolation rates from the soil storage, respectively,

$$p_s = p_s^{\max}(US_f)(1 - LS_f) \tag{7.91}$$

where US_f and LS_f denote the fraction full of upper and lower storages, respectively.

The actual percolation rate linearly decreases as the full fraction of lower storage increases. As the fraction of lower storage increases so does the fraction of upper storage but not in a linear fashion. The equations for percolation between the two storage volumes are given as

$$p_{ps} = f_{ps}^{\max}\left(\frac{S_s^c}{S_s^{\max}}\right)\left(1 - \frac{S_{GW1}^c}{S_{GW1}^{\max}}\right) \tag{7.92}$$

Similarly, the percolation rate between the two groundwater layers is given as

$$p_{pGW1} = p_{pGW1}^{\max}\left(\frac{S_{GW1}^c}{S_{GW1}^{\max}}\right)\left(1 - \frac{S_{GW2}^c}{S_{GW2}^{\max}}\right) \tag{7.93}$$

In both Eqs. 7.92 and 7.93, the superscript c denotes the current condition.

When percolation is from the second groundwater layer (current GW2 storage) to the deep aquifer the resulting equation is

$$p_{pGW2} = p_{pGW2}^{\max}\left(\frac{S_{GW2}^c}{S_{GW2}^{\max}}\right) \tag{7.94}$$

Each potential percolation rate is multiplied by the length of the current time step to obtain a volume. The potential percolation is then compared against the water available for percolation, and the smaller of the two amounts is then added to the lower groundwater layer. The amount of water available for percolation is the volume of water in the upper storage at the beginning of the computational step, minus any infiltration, percolation, and/or groundwater routing from this upper storage for the computational time step.

Surface Runoff and Groundwater Flow: When the soil storage is full and the available water for infiltration exceeds the potential infiltration rate, the excess precipitation is transformed to a runoff hydrograph. This is computed separately in the transformation model. Groundwater flow is transformed to streamflow at the basin outlet using a series of linear reservoirs. The groundwater flow from each groundwater layer at the end of the time interval is calculated.

Evapotranspiration: If PET is not completely satisfied from one storage in a time interval, the unsatisfied PET volume is filled from the next available storage. The PET volume is satisfied first from the canopy storage, then from surface storage, and finally from soil storage. Within soil storage, the PET is first fulfilled from the upper zone, then the tension zone. When PET occurs from the interception storage, surface storage, or the upper zone of soil storage, the AET is equal to the rate determined in the meteorological methods of HEC-HMS. When the PET is from the tension zone, the AET is a function of this rate, the water currently in soil storage, and the maximum depth of the soil storage tension zone.

7.10 INFILTRATION MEASUREMENTS

The common devices for measurements of infiltration in the field include the double-ring infiltrometer or flooding infiltrometer, portable infiltrometer, and rainfall simulator.

The flooding or ring infiltrometer consists of a metal ring of about two feet in diameter, which is driven into the soil. Water is placed inside the ring and the level of water is recorded at regular intervals, as it falls due to infiltration. This provides the data to construct the cumulative infiltration and infiltration rate as a function of time. A second ring is added outside the first, filled with water and maintained at a constant level so that infiltration from the inner ring goes vertically down into the soil in which it is placed and there is no lateral movement of water. A variant of the double-ring infiltrometer is the portable infiltrometer, which is a single-ring infiltrometer.

Because flooding infiltrometers require ponding and thereby a hydraulic head that can drive the infiltration process, they potentially do not replicate rainfall conditions. For this reason, rainfall simulators have been used for infiltration measurements, where the water is added to a plot by a sprinkler system.

7.11 EXAMPLES

Example 7.1: Time and observed rainfall depth within Five Mile Creek Watershed in Dallas for a rain event on July 7, 1973, are given in columns 1 and 2 of Table 7.6. Times and observed discharge in Five Mile Creek near the watershed outlet are given in Table 7.7. Determine the Φ index and the duration of the effective rainfall. Show the variation of cumulative loss during rainfall. The drainage area of Five Mile Creek Watershed is 13.2 square miles.

Solution:

Step 1: The first task in the determination of Φ index is to determine the depth of total direct runoff (q) from the observed hydrograph data (Table 7.7).

The area under the hydrograph gives the total runoff volume (ft^3). This is derived by numerical integration using the trapezoidal rule in Excel. This volume is determined to be 43 089 307.2 ft^3. The drainage area is 13.2 mi^2. Thus, q is calculated as follows.

$$q = \left(\frac{43089307.2 \text{ ft}^3}{13.2 \text{ mi}^2 \times 5289^2 \text{ ft}^2/\text{mi}^2} \right) \times 12 \frac{\text{in}}{\text{ft}} = 1.41 \text{ in}$$

Step 2: The second column in Table 7.6 gives the cumulative rainfall depth. Thus, the total rainfall depth, rain ending at 20:30 is 2.57 in. The total duration of the rainfall (from 18:00 to 20:30) is 2.5 h. Total rainfall loss, due to infiltration (ignoring interception and surface storage) uniformly spread over 2.5-h duration, gives the initial estimate of Φ index as

$$\frac{(2.57 - 1.41) \text{ in}}{2.5 \text{ h}} = 0.46 \frac{\text{in}}{\text{h}} = \Phi$$

Step 3: Check the validity of the value of Φ calculated in step 2.

Column 4 of Table 7.6 gives the incremental rainfall depth at each time step calculated from the observed total rainfall depth (column 2). The Δt for each observation is given in column 3. Column 5 gives the excess rainfall at each time step assuming the value of $\Phi = 0.46$ obtained in step 2. The excess rainfall at each time step is calculated using the formula: $P_e = P - \Phi \Delta t$. Now, summing all the positive values obtained in column 5 gives the total excess precipitation = 1.55 in. This is not equal to the observed 1.41 in of direct runoff.

Step 4: Refine the value of Φ from step 4 to match the observed direct runoff.

From the calculations in step 4, it is seen that six pulses of rainfall (19:10–19:50) give positive values of excess rainfall. Taking the incremental rainfall depths at these six pulses of rainfall, Φ is derived algebraically by solving $\Phi = \frac{\sum_i^6 (P_i) - q}{\sum_i^6 \Delta t_i}$.

Table 7.6 *Data and calculations*

Time (h:min)	Rainfall total (in)	Δt (h)	Incremental rainfall (in)	Excess rainfall (in) Φ = 0.46 in/h	Excess rainfall (in) Φ = 0.64 in/h
18:00	0.03	0	0.03		
18:15	0.45	0.25	0.42	0.30	
18:30	0.49	0.25	0.04	−0.08	
18:45	0.51	0.25	0.02	−0.10	
19:00	0.53	0.25	0.02	−0.10	
19:10	0.75	0.1667	0.22	0.14	0.11
19:20	1.45	0.1666	0.7	0.62	0.59
19:30	1.93	0.1667	0.48	0.40	0.37
19:35	2.05	0.0833	0.12	0.08	0.07
19:45	2.36	0.1667	0.31	0.23	0.20
19:50	2.47	0.0833	0.11	0.07	0.06
20:00	2.47	0.1667	0	−0.08	
20:15	2.54	0.25	0.07	−0.05	
20:30	2.57	0.25	0.03	−0.09	
		Total time (h)		Total excess (in) [only positive values]	Total excess (in) [only negative values]
		2.5		1.55	1.41

Table 7.7 *Data*

Time	Flow (ft³/s)
07/07/1973@18:00:00	4
07/07/1973@18:15:00	4
07/07/1973@18:30:00	30
07/07/1973@18:45:00	1210
07/07/1973@19:00:00	1460
07/07/1973@19:10:00	2280
07/07/1973@19:20:00	3310
07/07/1973@19:30:00	5010
07/07/1973@19:35:00	7820
07/07/1973@19:45:00	8220
07/07/1973@19:50:00	8140
07/07/1973@20:00:00	6900
07/07/1973@20:15:00	6370
07/07/1973@20:30:00	4310
07/07/1973@20:45:00	3860
07/07/1973@20:50:00	1580
07/07/1973@21:00:00	1210
07/07/1973@21:30:00	880
07/07/1973@22:00:00	640
07/07/1973@22:30:00	500
07/07/1973@23:00:00	390
07/07/1973@23:30:00	300
07/08/1973@00:00:00	250

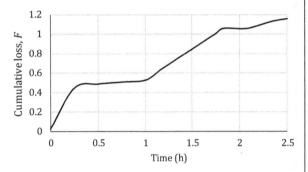

Figure 7.24 Temporal pattern of accumulated loss.

$$\Phi = \frac{(1.94 - 1.41) \text{ in}}{0.83 \text{ h}} = 0.64 \frac{\text{in}}{\text{h}}$$

Step 5: Validate the value of Φ.

The value of Φ obtained in step 4 is used again to calculate the excess precipitation for these six pulses. These values are shown in column 6 of Table 7.6. The sum of these values is 1.41 in. So, Φ = 0.64 in/h.

From column 6, it is evident that the effective rainfall (positive values of rainfall excess) took place from only 19:00 to 19:50. Thus, the duration of the effective rainfall is only 50 minutes for this storm with a total duration of 2.5 h. It should be also noted that the total rainfall loss is 1.16 in. which is significant for a rain event with a storm total precipitation of 2.57 in. Five Mile Creek Watershed is located to the southern limit of the City of Dallas and during the 1970s there was not significant development (urbanization) in this part of the city. For this reason, there were considerable soil surfaces exposed for infiltration.

The temporal variation of the cumulative loss is shown in Figure 7.24.

7 Abstractions and Effective Rainfall

Example 7.2: For the rainfall and streamflow data given for Example 7.1 (Tables 7.6 and 7.7) determine the parameters of the exponential loss rate method and discuss the significance of time variation of cumulative loss obtained in this method with that obtained from the Φ index method.

Solution: The exponential loss rate method requires two loss rate coefficients, D_k and A_k, which in turn depend on four parameters, D, S, R, and K_R. Furthermore, the coefficient D_k depends on the cumulative loss, C. The computation presented here involves both iteration and optimization carried out in three steps.

Step 1: Make initial estimates for C and all parameters for each rainfall pulse.

Initial estimates of C and the four model parameters are made to start the computation of the two loss coefficients and, using these values, loss rate (L_R) and cumulative loss at each time step can be calculated: $C_i = C_{i-1} + L_{R_i}$. The estimates of the four model parameters are made from the default values set in the vintage HEC-1 program (the current version of HEC-HMS does not have any option for any default values). These are as follows.

$D = 0.5$ in; $S = 0.2$ in/h; $R = 2.0$; and $K_R = 0.5$. An initial estimate of $C = 0.005$ was arbitrarily set to start the computation. With these values the results of the calculations are presented in Table 7.8.

Step 2: Second iteration with the values of C computed in step 1.

In the second iteration, the values of C obtained in step 1 (column 7 in Table 7.8) are obtained to recalculate the loss coefficients D_k and A_k and a new set of C. The results of the calculations are shown in Table 7.9.

Step 3: Optimization of parameters D, S, R, and K_R to match the final cumulative loss obtained from observed rainfall and streamflow data.

As shown in column 7 of Table 7.9, the final cumulative loss is 0.977 in. From the observed rainfall–runoff data in Example 7.1, the actual loss is 1.16 in. An optimization routine using the Solver program available in Excel is used to change the parameters so that the final cumulative loss equals 1.16 in.

The results of optimization give the following values of parameters:

$D = 0.504$ in; $S = 0.224$ in/h; $R = 1.979$; and $K_R = 0.452$

The final results of optimization are presented in Table 7.10.

The variation of the loss coefficients with cumulative loss, as predicted by its theoretical (hypothetical) formulation, is shown Figure 7.25. The variation of the cumulative loss with time is shown in the graph of Figure 7.26.

The cumulative loss, given by the Φ index method, shows initial high loss when the loss rate is higher than the rainfall intensity, then a nearly flat rate for some time up to one hour. After that, both methods show somewhat similar patterns. From the observations, it can be concluded that the exponential loss rate method is reliable if its parameters can be derived from optimization using observed rainfall–runoff data.

Table 7.8 *Results of calculations of step 1*

Time (h:min)	Incremental rainfall (P) (in)	C (estimated)	D_k	A_k	L_R	C (computed)
18:00	0.03	0.005	0.098	0.200	0.052	0.052
18:15	0.42	0.005	0.098	0.200	0.193	0.245
18:30	0.04	0.005	0.098	0.200	0.060	0.304
18:45	0.02	0.005	0.098	0.200	0.042	0.346
19:00	0.02	0.005	0.098	0.200	0.042	0.389
19:10	0.22	0.005	0.098	0.200	0.140	0.528
19:20	0.7	0.005	0.098	0.200	0.249	0.778
19:30	0.48	0.005	0.098	0.200	0.206	0.984
19:35	0.12	0.005	0.098	0.200	0.103	1.087
19:45	0.31	0.005	0.098	0.200	0.166	1.253
19:50	0.11	0.005	0.098	0.200	0.099	1.352
20:00	0	0.005	0.098	0.200	0.000	1.352
20:15	0.07	0.005	0.098	0.200	0.079	1.431
20:30	0.03	0.005	0.098	0.200	0.052	1.482

Table 7.9 Results of calculations of step 2

Time (h:min)	Incremental rainfall (P) (in)	C	D_k	A_k	L_R	C
18:00	0.03	0.052	0.080	0.199	0.048	0.048
18:15	0.42	0.245	0.026	0.197	0.144	0.193
18:30	0.04	0.304	0.015	0.196	0.042	0.235
18:45	0.02	0.346	0.009	0.195	0.029	0.264
19:00	0.02	0.389	0.005	0.195	0.028	0.292
19:10	0.22	0.528	0.000	0.193	0.090	0.383
19:20	0.7	0.778	0.000	0.190	0.159	0.541
19:30	0.48	0.984	0.000	0.187	0.129	0.671
19:35	0.12	1.087	0.000	0.185	0.064	0.735
19:45	0.31	1.253	0.000	0.183	0.102	0.837
19:50	0.11	1.352	0.000	0.182	0.060	0.897
20:00	0	1.352	0.000	0.182	0.000	0.897
20:15	0.07	1.431	0.000	0.181	0.048	0.945
20:30	0.03	1.482	0.000	0.180	0.031	0.977

Table 7.10 Results of calculations of step 3

Time (h:min)	Incremental rainfall (P) (in)	D_k	A_k	L_R	C	$D_k + A_k$
18:00	0.03	0.076	0.223	0.061	0.061	0.299
18:15	0.42	0.019	0.220	0.161	0.223	0.239
18:30	0.04	0.008	0.219	0.053	0.276	0.227
18:45	0.02	0.003	0.218	0.038	0.313	0.221
19:00	0.02	0.000	0.217	0.037	0.350	0.217
19:10	0.22	0.000	0.215	0.108	0.459	0.215
19:20	0.7	0.000	0.211	0.179	0.638	0.211
19:30	0.48	0.000	0.207	0.149	0.787	0.207
19:35	0.12	0.000	0.205	0.079	0.865	0.205
19:45	0.31	0.000	0.203	0.119	0.985	0.203
19:50	0.11	0.000	0.201	0.074	1.059	0.201
20:00	0	0.000	0.201	0.000	1.059	0.201
20:15	0.07	0.000	0.200	0.060	1.119	0.200
20:30	0.03	0.000	0.199	0.041	1.160	0.199

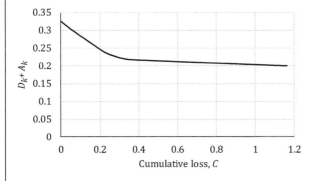

Figure 7.25 Variation of $D_k + A_k$ with C.

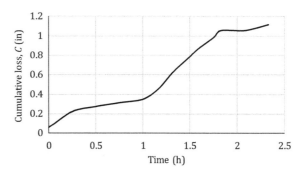

Figure 7.26 Temporal variation of cumulative loss.

Example 7.3: Richards equation is considered the best available mathematical description of moisture flow in soils. An ideal solution of this equation would be an analytical one but, except for a few special cases, no general analytical solution is available. However, several researchers have employed finite difference methods in one form or another to solve this equation numerically. Mein and Larson (1971) developed such a numerical mathematical scheme for the solution of Richards equation. They used this solution to generate infiltration data for several soil types with given physical properties and rainfall intensities. From their dataset, the infiltration data for the soil type Columbia sandy loam is presented in the first three columns of Table 7.11. The properties of the soil are given as: saturated hydraulic conductivity $K_s = 1.39 \times 10^{-3}$ cm/s; and effective porosity, $\eta_e = 0.518$. Use this information to do the following: (a) There has been claims in the literature that the Green–Ampt model can be derived from Richards equation. Does the Green–Ampt model fit the data generated from the numerical solution of Richards equation? (b) Derive the Green–Ampt model parameters that best fit this dataset. (c) Through sensitivity analysis, rank the parameters according to their degree of control on model results. (d) Make observations.

Solution:

Step 1: Fit the Green–Ampt model to the data.

The Green–Ampt equation can be written as: $f = K_s + \frac{\Omega K_s}{F}$, where K_s is the saturated hydraulic conductivity and Ω is the product of capillary suction (ψ) and effective porosity (η_e), which represents $\Delta\theta$, the soil moisture deficit. If the data presented in Table 7.11 follow the Green–Ampt model, then the Green–Ampt parameters that can be derived from these data would satisfy this equation.

For each value of F, $1/F$ is calculated as shown in column 4 of Table 7.11. If f is plotted against $1/F$ and the relationship is linear, the slope–intercept form of the equation of a straight line and the intercept gives the value of K_s and the slope gives the value of ΩK_s. The regression plot is shown in Figure 7.27.

The intercept gives the value of $K_s = 5.3272$ cm/h (1.48×10^{-3} cm/s) with lower and upper 95% confidence limits of 5.1072 cm/h (1.42×10^{-3} cm/s) and 5.5471 cm/h (1.54×10^{-3} cm/s) respectively. The slope giving the value of $\Omega = \eta_e\psi$ is 39.226 cm²/h with lower and upper 95% confidence limits of 38.3664 cm²/h and 40.0865 cm²/h, respectively. From this value of the slope and the value of K_s, the value of Ω is calculated as: $\Omega = \frac{39.226 \text{ cm}^2/\text{h}}{5.3272 \text{ cm/h}} = 0.13581$ cm.

The values of parameters thus obtained are used to solve for $F(t)$ in the Green–Ampt equation. given by: $F(t) = \Omega \ln\left(1 + \frac{F(t)}{\Omega}\right) + K_s t$. Since this is a nonlinear equation in implicit form of $F(t)$, its solution is not straightforward. The implicit method proposed by Li et al. (1976) is used to solve for $F(t)$. This method is also adopted in HEC-HMS.

The results of calculations are given in Table 7.12 and the steps involved are as follows.

First, the dimensionless parameter t^* is calculated (column 1): $t^* = \frac{K_s t}{\Omega}$.

Then, the explicit solution is calculated as a function of t^* (column 2):

$$F_e(t) = \frac{1}{2}\left[t^* + \sqrt{t^*(t^* + 8)}\right]$$

From the explicit solution, the implicit solution (dimensionless form) is obtained as (column 3):

$$F_i(t) = [1 + F_e(t)]$$
$$\times \sqrt{\{F_e(t)\}^2 + 2[t^* - F_e(t) + \ln\{1 + F_e(t)\}]}$$
$$- \{F_e(t)\}^2$$

Table 7.11 *Infiltration data (columns 1-3) for Columbia sandy loam (from Mein and Larson, 1971)*

Time (s)	F (cm)	f (cm/h)	$1/F$ (cm^{-1})
472.2	2.62	20.02	0.3817
489.7	2.72	19.54	0.3676
507.5	2.81	19.21	0.3559
531.5	2.94	18.75	0.3401
559.8	3.08	18.32	0.3247
582.3	3.19	17.6	0.3135
609.4	3.32	17	0.3012
644.1	3.48	16.7	0.2874
682.9	3.66	16.42	0.2732
714.7	3.8	15.74	0.2632
751.1	3.95	15.43	0.2532
783.9	4.09	14.93	0.2445
831.2	4.28	14.46	0.2336
883.3	4.48	14.03	0.2232
951.7	4.74	13.63	0.2110
1008.5	4.95	13.18	0.2020
1081.5	5.21	12.97	0.1919
1142.5	5.43	12.63	0.1842
1230.5	5.72	12.15	0.1748
1357.2	6.14	11.79	0.1629
1450.6	6.43	11.29	0.1555
1547.3	6.73	11.09	0.1486
1686.6	7.15	10.75	0.1399
1793.2	7.45	10.4	0.1342

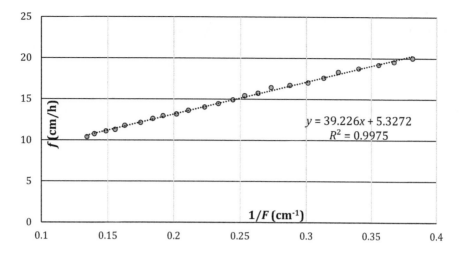

Figure 7.27 Regression of f against 1/F.

Table 7.12 *Green–Ampt model for the data given in Table 7.11*

$t*$	$F_e(t)$	$F_i(t)$	$F(t)$, cm	$f(t)$, cm/s
5.145144	6.684556	7.255948	0.985415	0.001684
5.335826	6.885662	7.472524	1.014828	0.001678
5.529776	7.089719	7.692055	1.044642	0.001672
5.791283	7.364121	7.986923	1.084687	0.001665
6.099643	7.686706	8.333078	1.131698	0.001657
6.344806	7.942492	8.607187	1.168924	0.001652
6.640090	8.249840	8.936140	1.213598	0.001645
7.018185	8.642328	9.355583	1.270562	0.001638
7.440954	9.079942	9.822451	1.333966	0.001630
7.787451	9.437731	10.203557	1.385723	0.001625
8.184069	9.846414	10.638243	1.444757	0.001619
8.541462	10.213968	11.028630	1.497775	0.001614
9.056848	10.742948	11.589600	1.573959	0.001607
9.624535	11.324332	12.205016	1.657537	0.001601
10.369829	12.085856	13.009450	1.766786	0.001594
10.988728	12.716932	13.674748	1.857139	0.001588
11.784144	13.526520	14.526595	1.972826	0.001582
12.448807	14.201923	15.235934	2.069160	0.001577
13.407665	15.174765	16.255713	2.207654	0.001571
14.788202	16.572834	17.717577	2.406187	0.001563
15.805899	17.601836	18.791045	2.551972	0.001559
16.859553	18.665998	19.899184	2.702466	0.001554
18.377381	20.197178	21.490397	2.918565	0.001549
19.538906	21.367730	22.704503	3.083451	0.001545

The dimensional form of $F(t)$ is recovered as (column 4):

$$F(t) = F_i(t) \times \Omega$$

From $F(t)$, $f(t)$ is calculated as (column 5):

$$f(t) = K_s + \frac{\Omega K_s}{F(t)}$$

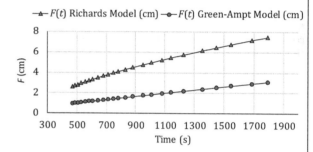

Figure 7.28 Comparison of accumulated infiltration according to the Richards and Green–Ampt models.

The results of the calculations presented in Table 7.12 and the data given in Table 7.11 are plotted to show that the Green–Ampt model does not fit the results obtained from the numerical solution of the Richards equation as presented in Mein and Larson (1971) if the Green–Ampt parameters are derived from this dataset, assuming that the data also follow the Green–Ampt equation (Figures 7.28 and 7.29).

The relative errors are calculated as: $e = \frac{Y_R - Y_{GA}}{Y_R}$ where Y_R represents a value calculated from the numerical solution of the Richards equation, and Y_{GA} represents a value calculated using the Green–Ampt equation. Table 7.13 gives the calculations with the relative errors for $F(t)$ given in column 4 and that for $f(t)$ given in column 8.

Since all errors are positive, it is suspected that these are not random errors. Plots of errors against Y_R reveal that these are systematic errors and must be related to the differences in the mathematical representations of the physical processes of infiltration in the two models (Figure 7.30 and 7.31).

Step 2: Adjust the Green–Ampt Parameters to best fit the data.

In order to adjust the Green–Ampt parameters obtained by assuming that the data given in Table 7.11 would follow the Green–Ampt model, an optimization procedure is followed. Columns 5 and 9 in Table 7.13 give $(Y_R - Y_{GA})^2$ and the sum of these quantities are at the bottom of each of these two columns. This is the objective function of the optimization method. In the optimization routine, first, the objective function giving the errors in $F(t)$ is minimized by adjusting all three model parameters: K_s, ψ, and η_e.

Ω is represented, following its definition, as: $\Omega = \psi \Delta \theta = \psi \eta = \psi \eta_e$. Since the value of $\eta_e = 0.518$, as given for Columbia sandy loam, from the regression

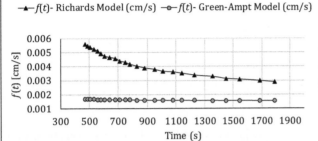

Figure 7.29 Comparison of infiltration rate according to the Richards and Green–Ampt models.

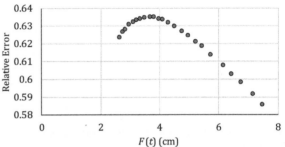

Figure 7.30 Relative error in $F(t)$ in comparisons of the Richards' and Green–Ampt equations.

Table 7.13 Results of error analysis

t (s)	$F(t)$-R	$F(t)$-GA	Relative error	Square of difference	$f(t)$-R	$f(t)$-GA	Relative error	Square of difference
472.2	2.62	0.985415	0.623887	2.671868	0.005561	0.001684	0.697234	1.5E–05
489.7	2.72	1.014828	0.626902	2.907613	0.005428	0.001678	0.690885	1.41E–05
507.5	2.81	1.044642	0.628241	3.11649	0.005336	0.001672	0.686634	1.34E–05
531.5	2.94	1.084687	0.631059	3.442186	0.005208	0.001665	0.68031	1.26E–05
559.8	3.08	1.131698	0.632566	3.795882	0.005089	0.001657	0.674319	1.18E–05
582.3	3.19	1.168924	0.633566	4.084749	0.004889	0.001652	0.662152	1.05E–05
609.4	3.32	1.213598	0.634458	4.436928	0.004722	0.001645	0.651568	9.47E–06
644.1	3.48	1.270562	0.634896	4.881617	0.004639	0.001638	0.646909	9.01E–06
682.9	3.66	1.333966	0.635528	5.410433	0.004561	0.00163	0.642537	8.59E–06
714.7	3.8	1.385723	0.635336	5.828731	0.004372	0.001625	0.62838	7.55E–06
751.1	3.95	1.444757	0.634239	6.276242	0.004286	0.001619	0.622297	7.11E–06
783.9	4.09	1.497775	0.633796	6.719631	0.004147	0.001614	0.610835	6.42E–06
831.2	4.28	1.573959	0.632253	7.322658	0.004017	0.001607	0.599803	5.8E–06
883.3	4.48	1.657537	0.630014	7.966295	0.003897	0.001601	0.589189	5.27E–06
951.7	4.74	1.766786	0.62726	8.840003	0.003786	0.001594	0.579113	4.81E–06
1008.5	4.95	1.857139	0.62482	9.565791	0.003661	0.001588	0.566255	4.3E–06
1081.5	5.21	1.972826	0.621339	10.47929	0.003603	0.001582	0.560993	4.08E–06
1142.5	5.43	2.06916	0.618939	11.29525	0.003508	0.001577	0.550527	3.73E–06
1230.5	5.72	2.207654	0.614047	12.33657	0.003375	0.001571	0.534575	3.26E–06
1357.2	6.14	2.406187	0.608113	13.94136	0.003275	0.001563	0.522657	2.93E–06
1450.6	6.43	2.551972	0.603115	15.0391	0.003136	0.001559	0.503038	2.49E–06
1547.3	6.73	2.702466	0.598445	16.22103	0.003081	0.001554	0.4955	2.33E–06
1686.6	7.15	2.918565	0.591809	17.90504	0.002986	0.001549	0.481387	2.07E–06
1793.2	7.45	3.083451	0.586114	19.06675	0.002889	0.001545	0.465208	1.81E–06
				203.552 Sum				0.00017 Sum

results, ψ is calculated as: $\psi = \frac{\Omega}{\eta_e} = \frac{0.31581 \text{ cm}}{0.518} = 0.2622$ cm.

The Solver function available in Excel under the data analysis category is used to carry on the optimization. Initially a constraint was needed to be put that $\eta_e \leq 0.5$; otherwise, initial runs showed that the process yielded $\eta_e > 1$, which is not possible.

The values of the three parameters obtained in this optimization process are as follows

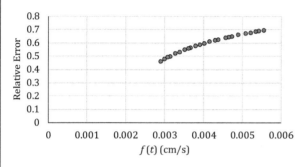

Figure 7.31 Relative error in f(t) in comparisons of Richards Equation and Green–Ampt Equation.

$K_s = 2.988 \times 10^{-3}$ cm/s

$\psi = 2.1583$ cm

$\eta_e = 0.4999$

The results of minimization of errors are given in Table 7.14.

After this optimization, both negative and positive relative errors occur, and the objective function is minimized.

Step 3: Sensitivity analysis.

A second optimization run is conducted in which the value of K_s is changed to 1.39×10^{-3} cm/s as the initial value. This is the value of K_s for Columbia sandy loam given in Mein and Larson (1971). The values of ψ and η_e remain the same as in (b) for the initial condition. The goal of optimization again is to minimize the objective function by changing the values of all three parameters simultaneously.

The values of the three parameters obtained in this second optimization process are as follows.

$K_s = 2.986 \times 10^{-3}$ cm/s

$\psi = 2.1600$ cm

$\eta_e = 0.5004$

Table 7.14 *Results of error analysis after optimization of the Green–Ampt model parameters*

t(s)	F(t)-R	F(t)-GA	Relative error	Square of difference	f(t)-R	f(t)-GA	Relative error	Square of difference
472.2	2.62	2.788339	−0.06425	0.028338	0.005561	0.004145	0.254729	2.01E−06
489.7	2.72	2.860606	−0.05169	0.01977	0.005428	0.004115	0.241803	1.72E−06
507.5	2.81	2.933604	−0.04399	0.015278	0.005336	0.004087	0.234034	1.56E−06
531.5	2.94	3.031264	−0.03104	0.008329	0.005208	0.004052	0.222041	1.34E−06
559.8	3.08	3.145373	−0.02123	0.004274	0.005089	0.004013	0.211363	1.16E−06
582.3	3.19	3.235344	−0.01421	0.002056	0.004889	0.003985	0.184932	8.17E−07
609.4	3.32	3.342887	−0.00689	0.000524	0.004722	0.003953	0.162953	5.92E−07
644.1	3.48	3.479375	0.00018	3.91E−07	0.004639	0.003915	0.156073	5.24E−07
682.9	3.66	3.630507	0.008058	0.00087	0.004561	0.003876	0.150139	4.69E−07
714.7	3.8	3.753301	0.012289	0.002181	0.004372	0.003847	0.120068	2.76E−07
751.1	3.95	3.892768	0.014489	0.003275	0.004286	0.003816	0.109571	2.21E−07
783.9	4.09	4.017518	0.017722	0.005254	0.004147	0.003791	0.085952	1.27E−07
831.2	4.28	4.195995	0.019627	0.007057	0.004017	0.003757	0.06474	6.76E−08
883.3	4.48	4.390807	0.019909	0.007955	0.003897	0.003723	0.044824	3.05E−08
951.7	4.74	4.644024	0.020248	0.009211	0.003786	0.003682	0.027367	1.07E−08
1008.5	4.95	4.852325	0.019732	0.00954	0.003661	0.003653	0.002299	7.09E−11
1081.5	5.21	5.117684	0.017719	0.008522	0.003603	0.003618	−0.00429	2.39E−10
1142.5	5.43	5.337589	0.017019	0.00854	0.003508	0.003592	−0.02393	7.05E−09
1230.5	5.72	5.652195	0.011854	0.004598	0.003375	0.003559	−0.05442	3.37E−08
1357.2	6.14	6.100344	0.006459	0.001573	0.003275	0.003517	−0.07382	5.84E−08
1450.6	6.43	6.427526	0.000385	6.12E−06	0.003136	0.00349	−0.1128	1.25E−07
1547.3	6.73	6.763764	−0.00502	0.00114	0.003081	0.003465	−0.12477	1.48E−07
1686.6	7.15	7.244169	−0.01317	0.008868	0.002986	0.003433	−0.14976	2E−07
1793.2	7.45	7.609001	−0.02134	0.025281	0.002889	0.003412	−0.18107	2.74E−07
				0.18244				1.18E−05
				Sum				Sum

Thus, the values of these parameters do not change significantly from the first optimization process. The results of minimization of errors are given in Table 7.15.

Comparative plots of $F(t)$ and $f(t)$ show that with the optimization of model parameters, the match between the results of accumulated infiltration obtained from the numerical solution of the Richards equation and Green–Ampt equation is very good (Figure 7.32). However, that for the infiltration rate is less satisfactory (Figure 7.33). This may be due to inaccuracies in approximation in the numerical experiment of Mein and Larson (1971) that yield the data given in Table 7.11.

The calculations presented above show that in order for the data presented in Table 7.11 to exhibit the infiltration characteristics of the soil in question, the hydraulic conductivity needs significant change from what is derived from this data, assuming that the data can also follow the Green–Ampt model. The adjusted value of $K_s = 2.986 \times 10^{-3}$ cm/s was outside the values of the 95% confidence limits ($K_s = 1.42 \times 10^{-3}$ cm/s [−95%] = 1.48×10^{-3} cm/s and 1.54×10^{-3} cm/s [+95%]) of the K_s initially obtained from this data with the assumption that the data can be represented by the Green–Ampt model. Thus, the adjusted K_s is more than two times than the initial value.

The value of the capillary suction changed from 0.2622 cm, obtained with the assumption that the Green–Ampt model fits the data without the adjustment of parameters, to 2.16 cm in order for the infiltration characteristics given in Table 7.11 to follow the Green–Ampt model. Thus, the adjusted ψ is more than eight times the initial value.

Table 7.15 *Results of error analysis after second optimization of the Green–Ampt model parameters*

t(s)	$F(t)$-R	$F(t)$-GA	Relative error	Square of difference	$f(t)$-R	$f(t)$-GA	Relative error	Square of difference
472.2	2.62	2.788848	−0.06445	0.02851	0.005561	0.004144	0.254808	2.01E−06
489.7	2.72	2.861107	−0.05188	0.019911	0.005428	0.004115	0.24189	1.72E−06
507.5	2.81	2.934096	−0.04416	0.0154	0.005336	0.004087	0.234128	1.56E−06
531.5	2.94	3.031744	−0.03121	0.008417	0.005208	0.004051	0.222144	1.34E−06
559.8	3.08	3.145837	−0.02138	0.004335	0.005089	0.004013	0.211478	1.16E−06
582.3	3.19	3.235795	−0.01436	0.002097	0.004889	0.003984	0.185057	8.19E−07
609.4	3.32	3.343321	−0.00702	0.000544	0.004722	0.003952	0.16309	5.93E−07
644.1	3.48	3.479785	6.17E−05	4.61E−08	0.004639	0.003914	0.156221	5.25E−07
682.9	3.66	3.630889	0.007954	0.000847	0.004561	0.003876	0.150299	4.7E−07
714.7	3.8	3.75366	0.012195	0.002147	0.004372	0.003846	0.120243	2.76E−07
751.1	3.95	3.893099	0.014405	0.003238	0.004286	0.003816	0.109756	2.21E−07
783.9	4.09	4.017822	0.017647	0.00521	0.004147	0.00379	0.086151	1.28E−07
831.2	4.28	4.196259	0.019566	0.007013	0.004017	0.003756	0.064956	6.81E−08
883.3	4.48	4.391025	0.019861	0.007917	0.003897	0.003722	0.045056	3.08E−08
951.7	4.74	4.644178	0.020216	0.009182	0.003786	0.003682	0.027618	1.09E−08
1008.5	4.95	4.852424	0.019712	0.009521	0.003661	0.003652	0.002569	8.84E−11
1081.5	5.21	5.11771	0.017714	0.008518	0.003603	0.003617	−0.00401	2.08E−10
1142.5	5.43	5.337551	0.017026	0.008547	0.003508	0.003591	−0.02363	6.87E−09
1230.5	5.72	5.652061	0.011877	0.004616	0.003375	0.003558	−0.05409	3.33E−08
1357.2	6.14	6.100067	0.006504	0.001595	0.003275	0.003516	−0.07347	5.79E−08
1450.6	6.43	6.42714	0.000445	8.18E−06	0.003136	0.003489	−0.11242	1.24E−07
1547.3	6.73	6.763262	−0.00494	0.001106	0.003081	0.003464	−0.12437	1.47E−07
1686.6	7.15	7.243494	−0.01308	0.008741	0.002986	0.003432	−0.14934	1.99E−07
1793.2	7.45	7.608189	−0.02123	0.025024	0.002889	0.003411	−0.18062	2.72E−07
				0.18244 Sum				**1.18E−05** Sum

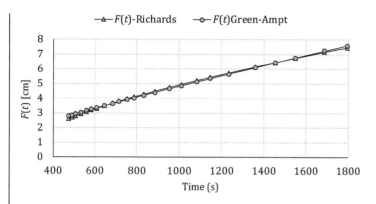

Figure 7.32 Comparison of accumulated infiltration according to the Richards and Green–Ampt models after optimization of model parameters.

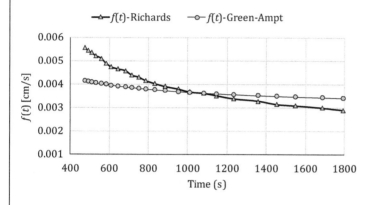

Figure 7.33 Comparison of infiltration rate according to the Richards and Green–Ampt models after optimization of model parameters.

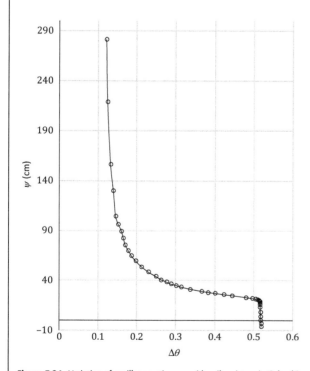

Figure 7.34 Variation of capillary suction, ψ, with soil moisture in Columbia sandy loam.

The value of η_e did not significantly change from initial 0.518 to 0.5004 with the optimization.

Thus, the Green–Ampt model is very sensitive to the model parameters K_s and ψ. In this particular example, ψ seems to have a greater influence than K_s.

Observations: The Richards equation given by

$$\frac{\partial \theta}{\partial t} = -\frac{\partial}{\partial z}\left[K(\theta)\frac{\partial \psi(\theta)}{\partial z}\right] - \frac{\partial K(\theta)}{\partial z}$$

accounts for the variation of capillary suction and hydraulic conductivity as a function of soil moisture. From the absorption data given in Mein and Larson (1971), these variations are shown in the Figures 7.34 and 7.35. Furthermore, the hydraulic conductivity in Richards equation is unsaturated hydraulic conductivity which is related to saturated hydraulic conductivity given by: $K(\theta) = K_s k_r$.

The Green–Ampt equation assumes constant $\Delta\theta$ and ψ. If the calculated hydraulic conductivity, 1.48×10^{-3} cm/s, is the unsaturated hydraulic conductivity, and $k_r = 0.5$, the saturated hydraulic conductivity is 2.986×10^{-3} cm/s.

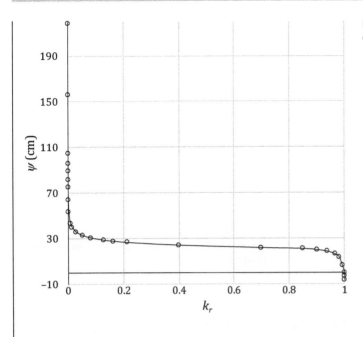

Figure 7.35 Variation of capillary suction, ψ, with relative conductivity of Columbia sandy loam.

Example 7.4: Use the Green–Ampt equation to calculate effective rainfall for the rainfall event given in Example 7.1. The dominant soil type of Five Mile Creek Watershed is clay loam.

Solution: For clay loam, from Table 7.2, $\eta_e = 0.309$, $\psi = 20.88$ cm (8.2205 in), $K_s = 0.1$ cm/h (0.0394 in/h).

From the given values of effective porosity and capillary suction, the constant Ω is calculated: $\Omega = \eta_e \psi = 2.5401$.

Columns 1, 2, 3, of Table 7.16 give the rainfall data, reproduced from Example 7.1.
The other columns in Table 7.16 are the results of computations as described step by step below.
Column 4 gives the time interval (h) between two successive rainfall measurements. This is Δt (h). Column 5 gives the incremental rainfall (in) during each of the intervals. This is obtained by subtracting the successive values of cumulative rainfall (column 3). Column 6 gives the rainfall intensity (i) in each of the

Table 7.16 *Computations for the first round*

Time (h:min)	Time (min)	Cumulative rainfall (in)	Δt(h)	Incremental rainfall (in)	Rainfal intensity (in/h)	$F'(t)$	$f'(t)$	Ponding?	$F(t)$
17:45	0	0	0		0.12	0	∞	No	
18:00	15	0.03	0.25	0.03	1.68	0.03	3.37	No	0.03
18:15	30	0.45	0.25	0.42	0.16	0.45	0.26	No	0.45
18:30	45	0.49	0.25	0.04	0.08	0.49	0.24	No	0.49
18:45	60	0.51	0.25	0.02	0.08	0.51	0.24	No	0.51
19:00	75	0.53	0.25	0.02	1.32	0.53	0.23	Yes	0.53
19:10	85	0.75	0.167	0.22	4.2	0.75	0.17	Yes	
19:20	95	1.45	0.167	0.7	2.88	1.45	0.11	Yes	
19:30	105	1.93	0.167	0.48	1.44	1.93	0.09	Yes	
19:35	110	2.05	0.083	0.12	1.86	2.05	0.09	Yes	
19:45	120	2.36	0.167	0.31	1.32	2.36	0.08	Yes	
19:50	125	2.47	0.083	0.11	0	2.47	0.08	No	
20:00	135	2.47	0.167	0	0.28	2.47	0.08	Yes	
20:15	150	2.54	0.25	0.07	0.12	2.54	0.08	Yes	
20:30	165	2.57	0.25	0.03		2.57	0.08	No	

7.11 Examples

Table 7.17 *Computations for the period of ponding*

Time (h:min)	Ponding	$K_s(t - T_p)$	F	Eq.	f	Ponding?	F
19:10	Yes	0.0065617	0.567641	0.00668735	0.215547	Yes	0.57
19:20	Yes	0.0065617	0.602755	0.00657461	0.205283	Yes	0.60
19:30	Yes	0.0065617	0.636129	0.00654282	0.196579	Yes	0.64
19:35	Yes	0.0032808	0.652297	0.00327094	0.192682	Yes	0.65
19:45	Yes	0.0065617	0.683744	0.00654772	0.185631	Yes	0.68

intervals. These values are obtained by dividing the incremental rainfall (column 5) by the corresponding time interval (column 4). Note that intensity starts from the first time interval from the cumulative rainfall and interval length of the second time interval. Basic rainfall data are thus organized up to column 6.

Now the actual computation begins.

Step 1: Calculate tentative values of cumulative infiltration and potential infiltration rate.

The cumulative infiltration during the first time interval is zero. For all subsequent time intervals, a tentative value of cumulative infiltration is obtained by using $F'_{t+\Delta t} = F_t + i_t \Delta t$. This is given in column 7. Corresponding to this tentative value of cumulative infiltration, the tentative potential infiltration rate is calculated using the infiltration rate equation of the Green–Ampt model, that is,

$$f'_{t+\Delta t} = K_s \left(\frac{\Omega}{F'_{t+\Delta t}} + 1 \right).$$

This is given in column 8. In all cases subscript t denotes the time at the beginning of interval and subscript $t + \Delta t$ denotes the end of interval. Thus, during the second time interval (18:00), i_t is 0.12 in/h and $\Delta t = 0.25$ h.

Step 2: Determine if ponding occurs or not.

For each interval, test the condition $f'_{t+\Delta t} \leq i_t$. If the condition is not true, meaning that the potential infiltration rate is greater than the rainfall rate, no ponding occurs. The result of the test is recorded in column 9.

Step 3: Calculate cumulative infiltration if no ponding occurs during an interval.

If no ponding occurs, the actual infiltration is the same as the tentative value of infiltration, that is, $F_{t+\Delta t} = F'_{t+\Delta t}$. Since $F'_{t+\Delta t}$ is calculated solely from rainfall (step 1), this is actually equal to the cumulative rainfall. This is given in column 10.

Continuing the calculations, it is found that ponding does not occur up to 18:45 or 60 minutes. But, at 75 minutes, ponding occurs and continues up to 19:45 (120 minutes). Thus, up to 60 minutes, the cumulative infiltration is 0.51 inches, which is the same as cumulative rainfall at 18:45.

Step 4: Cumulative infiltration at the time of ponding and time of ponding.

The cumulative infiltration at the time of ponding = 0.53 inches (from column 3).
The time of ponding, T_p, is 75 minutes.

Step 5: Cumulative infiltration during the time of ponding.

As shown in column 9, ponding continues up to 19:45 (120 min). During the ponding period, cumulative infiltration is calculated from

$$F_t - F_p - \Omega \ln \left(\frac{F_t + \Omega}{F_p + \Omega} \right) = K_s (t - T_p).$$

Table 7.17 shows the calculations for F during the period of ponding.

In this table, column 3 gives the calculated value of $K_s(t - T_p)$. First, a value of F is assumed and entered in column 4. Using this assumed value of Ft, the value of Ft is calculated using $F_t - F_p - \Omega \ln \left(\frac{F_t + \Omega}{F_p + \Omega} \right) = K_s(t - T_p)$ under column 5 (Eq.). This is solved by the numerical algorithm built into Excel's Goal Seek function. The right-hand side of the Eq. is calculated from the known values of K_s and T_p and the left-hand side with an assumed value of F_t is calculated and the Goal Seek was asked to find the value of F_t so that the left-hand side equals the right-hand side. After Excel finds the optimum solution, it updates the value in column 5. These are the cumulative infiltration values during the time of ponding: from 19:10 (85 min) to 19:45 (120 min).

Step 6: Recalculate the potential infiltration rate during the period of ponding.

With the values of F calculated during the period of ponding, f is calculated using the values of F obtained in step 5. These are shown in column 6 of Table 7.17. With these values of f, the condition $f'_{t+\Delta t} \leq i_t$ is tested and the test results are recorded in column 7. If the condition is satisfied, the F values calculated in step 5 are the cumulative infiltration during this interval. These values are recorded in column 8 (with only two significant figures).

Table 7.18 *Results of calculations for the remaining period of rainfall*

Time (h:min)	Ponding?	$K_s(t - T_p)$	F	Eq.	f	Ponding?	F
19:50	No	0.0032808	0.683744	0	0.185631	No	0.68
20:00	Yes	0.0098425	0.729046	0.00985662	0.176542	Yes	0.73
20:15	Yes	0.0098425	0.772247	0.00985388	0.168869	No	0.76
20:30	No	0.0098425	0.802247			No	0.76

Step 7: Period of no ponding or ponding again as rainfall intensity changes.

Table 7.16 shows that there is a period of no ponding at 19:50. This is due to no rain recorded during this interval (column 6). So, for this period, $F'_{t+\Delta t} = F_t + i_t \Delta t$. Since there is no ponding during this interval, in this case, $F_{t+\Delta t} = F'_{t+\Delta t}$.

Table 7.16 shows that during the two intervals from 20:00 to 20:15 there is ponding but at 20:30 there is no ponding. For the two intervals of ponding, F is calculated again as in step 5 and f is recalculated as in step 6 and the ponding test is again conducted. During this iteration, it is found that that ponding occurs only in the interval of 20:00. But after that no ponding occurs.

The calculations for the period 19:50–20:30 are presented in Table 7.18.

Thus, the cumulative precipitation is 0.76 inches. This is less that the loss of 1.16 inches. observed. The difference is most likely the rainfall that was in interception and surface storage.

From the calculations it appears that ponding began sometime between 60 minutes and 75 minutes. We need to find the exact time when ponding began. This is done as follows.

Since before ponding, $f > i$, and after ponding started, $f < i$, there must be a time when $f = i$. By setting $f_t = i_t$, the Green–Ampt infiltration Eq. becomes, $f_t = i_t = K_s\left(\frac{\Omega}{F_p} + 1\right)$, where F_p is the cumulative infiltration at the time of ponding. Further simplification to obtain F_p in terms of i_t, gives: $F_p = \frac{K_s \Omega}{i_t - K_s}$.

Example 7.5: Observed rainfall (P) and streamflow data (q) for several rainy days spanning a 16-year period in Buchman Branch watershed, a highly urbanized watershed, located within the City of Dallas, are given in columns 2, 3, 8, and 9 of Table 7.19. These are 24-hour total rainfall and runoff data. Derive the curve number for the watershed.

Solution: An initial value of CN is assumed, and the retention parameter, S, is calculated using, $S = \frac{1000}{CN} - 10$.

The initial abstraction is calculated as, $I_a = 0.2S$
With these two values, excess precipitation for a given rainfall value P is calculated from $q = \frac{(P-I_a)^2}{(P-I_a+S)}$ for $P > I_a$ otherwise $P = 0$. Two values of CN values

Table 7.19 *Rainfall–runoff data for Bachman Branch Watershed in Dallas, Texas*

Date	P	q	$CN = 95$	$CN = 80$	$CN = 60$	Date	P	q	$CN = 95$	$CN = 80$	$CN = 60$
09/20/64	6.13	4.24	5.54	3.90	2.01	07/12/72	2.46	0.50	1.92	0.86	0.16
09/27/64	1.15	0.35	0.69	0.13	0.01	06/07/74	3.05	0.98	2.50	1.29	0.35
11/17/64	2.85	0.99	2.30	1.14	0.28	09/16/74	3.95	1.46	3.38	2.00	0.74
05/09/65	4.84	2.55	4.26	2.75	1.21	10/30/74	3.11	1.39	2.56	1.33	0.37
02/09/66	2.07	0.59	1.55	0.61	0.07	01/31/75	3.83	2.06	3.26	1.90	0.68
04/28/66	5.17	4.59	4.59	3.04	1.40	04/07/75	2.37	1.23	1.84	0.80	0.14
04/29/66	8.23	2.80	7.63	5.84	3.51	05/26/76	1.24	0.48	0.78	0.17	0.00
06/17/66	1.85	0.52	1.34	0.47	0.04	06/18/76	1.80	0.46	1.29	0.44	0.03
04/20/67	2.12	0.50	1.60	0.64	0.08	03/26/77	5.62	3.45	5.03	3.44	1.68
05/30/67	2.48	0.56	1.94	0.88	0.17	06/12/77	3.31	0.82	2.75	1.49	0.45
03/19/68	3.04	2.24	2.49	1.28	0.35	03/23/78	1.45	0.43	0.97	0.26	0.00
04/22/68	0.93	0.35	0.50	0.06	0.03	05/28/78	3.29	1.08	2.73	1.47	0.44
05/12/68	1.18	0.61	0.72	0.15	0.00	08/04/78	2.96	0.60	2.41	1.22	0.32
08/13/68	4.44	1.38	3.87	2.41	0.99	08/21/78	0.98	0.16	0.55	0.08	0.02
10/09/68	1.88	0.39	1.37	0.49	0.04	03/30/79	3.25	2.45	2.69	1.44	0.43
05/04/69	1.88	0.70	1.37	0.49	0.04	05/03/79	3.70	2.82	3.14	1.80	0.62
05/06/69	6.22	2.54	5.63	3.98	2.07	07/17/79	1.29	0.39	0.82	0.19	0.00

Table 7.19 (cont.)

Date	P	q	CN = 95	CN = 80	CN = 60	Date	P	q	CN = 95	CN = 80	CN = 60
04/25/70	2.18	1.45	1.65	0.68	0.10						
08/31/70	3.93	1.50	3.36	1.98	0.73						
08/14/71	2.24	1.16	1.71	0.71	0.11						
10/03/71	4.45	1.64	3.88	2.42	0.99						
10/18/71	4.42	2.48	3.85	2.39	0.98						

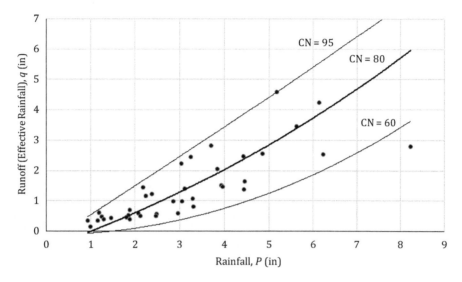

Figure 7.36 Fitting the NRCS runoff equation to observed rainfall–runoff data.

were found to bound the upper and lower limits of the data when the observed runoff, q, is plotted against the observed rainfall, P. These are $CN = 95$ and $CN = 60$, respectively. The calculated q values for these two CN values are shown in columns 4, 6, 10, and 12. A value of $CN = 80$ produces a curve that passes through the median values. These are shown in columns 5 and 11.

The plots are shown in Figure 7.36. The median value $CN = 80$ represents the CN of Bachman Branch Watershed. This is the approach that NRCS took to develop CN values for various soil cover complexes.

Bachman Branch Watershed is highly urbanized located in the central part of the City of Dallas and land cover since the 1970s have not changed significantly. A high CN value of 80 is representative of this urbanized (developed) nature of the watershed.

Example 7.6: Table 2-2a of TR-55 gives CN of 57 for a residential district of average lot size of 1/3 acre located on HSG A and an assumed impervious cover of 30%. Calculate the CN if the impervious cover increases to 50% and reduces to 20%. For pervious area with open space and fair condition, $CN = 39$.

Solution:
Case I: $CN_p = 49$; $p_{imp} = 50\%$:

$$CN_c = CN_p + \left(\frac{p_{imp}}{100}\right)(98 - CN_p)$$
$$= 39 + \left(\frac{50}{100}\right)(98 - 39) = 68.5 \approx 69$$

Case II: $CN_p = 49$; $p_{imp} = 20\%$:

$$CN_c = CN_p + \left(\frac{p_{imp}}{100}\right)(98 - CN_p)$$
$$= 39 + \left(\frac{20}{100}\right)(98 - 39) = 50.8 \approx 51$$

Example 7.7: Five Mile Creek Watershed in Dallas has two catchments with a 12-digit HUC as shown in Figure 7.37. A soil texture class map of the watershed was constructed as a vector dataset from the digital soil cover data obtained from soil survey data of USDA (Web Soil Survey, available at usda.gov). Derive spatially lumped Green–Ampt parameters of the two catchments.

Solution: The procedure involves assigning the Green–Ampt parameters to each of the soil textural classes. A CSV format of the attribute table of the soil textural class feature class is used in Excel to add these parameter values using the nested IF function of Excel. Once the attribute table of the feature class is updated, a dataset with four separate rasters is prepared by using the Feature to Raster geoprocessing tool of ArcGIS (Figure 7.38).

The results are presented in Table 7.20

Table 7.20 *Results*

HUC	Hydraulic conductivity (in/h)		Capillary suction (in)		Porosity	
	Mean	Standard deviation	Mean	Standard deviation	Mean	Standard deviation
120301050108	0.056	0.19	10.48	2.57	0.377	0.063
120301050107	0.025	0.01	10.7	1.51	0.387	0.047

Figure 7.37 Map of Five Mile Creek Watershed with HUC for two catchments.

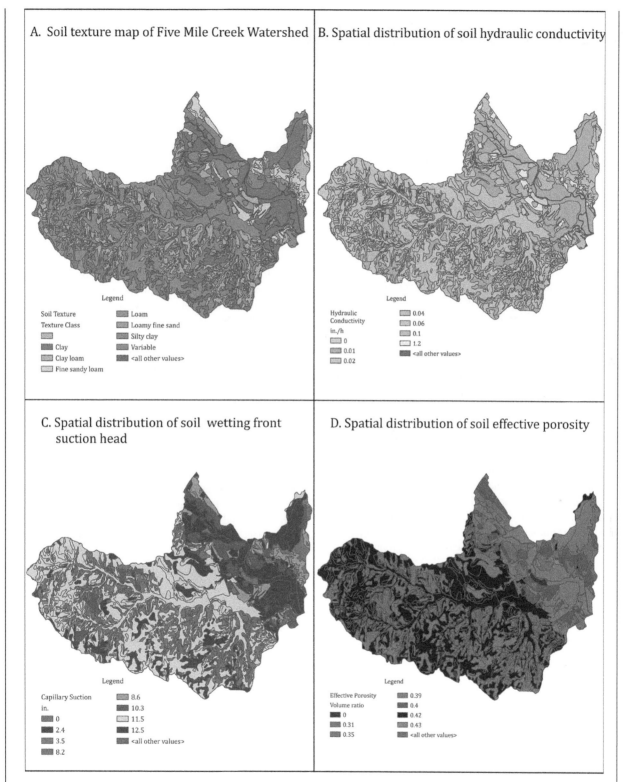

Figure 7.38 (A) Soil texture and (B)–(D) Green–Ampt parameters of Five Mile Creek Watershed, Dallas, Texas. (A black and white version of this figure will appear in some formats. For the color version, please refer to the plate section.)

Exercises: A selection of exercises on this topic is available at www.cambridge.org/appliedhydrology.

8 Groundwater and Baseflow

8.1 CONCEPTS AND PURPOSE

When a typical streamflow hydrograph constructed from discharges measured over an extended period such as weeks, months, or years at a gauging station is examined, it becomes apparent that peak streamflows are produced infrequently. Figure 8.1 shows a streamflow hydrograph constructed from discharge records collected every 15 minutes for a period of one month during the wet season at USGS gauging station 07249920 on Little Lee Creek, a second-order tributary of Arkansas River near Nicut, Oklahoma. As can be seen from the example given in Figure 8.1, two distinguishable components of a streamflow hydrograph are (1) direct runoff from an event, and (2) baseflow. Baseflow is the sustained or fair-weather runoff generated prior to an event. This latter component represents the water that was stored temporarily in the watershed plus the delayed subsurface runoff from the current event. On the other hand, peak streamflow is the result of an event of storm rainfall alone or storm rainfall and snowmelt combined.

Baseflow comes from the *groundwater* component of the hydrologic cycle. In Chapter 7 we discuss the process of infiltration by which the precipitation falling on the land surface that is porous, not compact rock or concrete or paved otherwise by impervious cover, seeps into the ground giving rise to groundwater. Groundwater also flows, albeit very slowly, from high to low elevation due to action of gravity. We also see in Chapter 7 that two types of groundwater flow, namely **interflow** and **baseflow** are distinguished based on whether groundwater is flowing through the saturated zone or unsaturated zone.

Baseflow separation, also known as **hydrograph analysis**, is the process of separating direct runoff (surface runoff and quick interflow) from baseflow. Baseflow separation is important for the derivation of unit hydrographs from observed runoff (streamflow) data. We see in Chapter 9 that the unit hydrograph concept applies only to direct or surface runoff. Therefore, direct runoff must be separated from baseflow. The converse is equally important for simulation or modeling of rainfall–runoff processes. Usually, a unit hydrograph is used to compute the direct runoff hydrograph from effective rainfall. In order to get the streamflow hydrograph from the direct runoff hydrograph by any of the unit hydrograph methods, the baseflow component must be added for accurate estimation of streamflow.

The methods of baseflow separation are somewhat arbitrary. A variety of techniques have been developed; some of which correspond to theoretical concepts of basin response. The models of baseflow estimation for streamflow

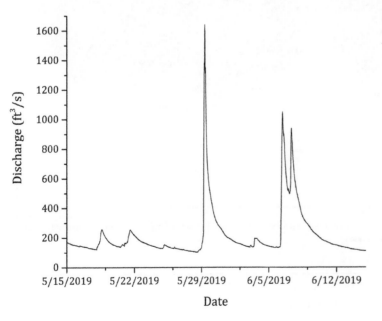

Figure 8.1 Streamflow hydrograph of Little Lee Creek at Nicut, Oklahoma, to show peaks resulting from rainstorms and low flows resulting from baseflow during the dry period.

simulation are also somewhat empirical. This chapter covers the commonly used methods of baseflow separation and baseflow models. Before presenting methods of baseflow separation, the elementary concepts of groundwater hydrology are summarized.

8.2 GROUNDWATER

8.2.1 The Groundwater Profile

During infiltration, water clings to soil particles, and this soil moisture may be drawn into the rootlets of growing plants. After the plant uses the water, it is transpired as vapor into the atmosphere. However, excess soil moisture is pulled downward by gravity. A column of soil or soil–rock admixture contains interstitial voids that are either filled with air or water. The soil horizon where all voids are not filled with water is called **unsaturated zone** or **vadose zone**. At some depth, the column is saturated with water, that is, all voids are filled up with water only. This horizon is called **saturated zone** or **phreatic zone**. The top of the saturated zone is called the **water table** (Figure 8.2). The water table also designates the bottom of the capillary zone discussed in Chapter 7.

The water table is a dynamic surface balanced by evaporation and infiltration leading to percolation. If infiltration is greater than evaporation, water percolates and the water table rises toward the land surface, whereas it becomes deeper if the rate of evaporation is greater than rate of infiltration or infiltration capacity.

8.2.2 Hydrogeologic Properties of Soil or Rock–Soil Columns

In Chapter 7 we discussed the physical properties of porous media that intrinsically govern the process of infiltration. The two most important properties that govern storage and movement of groundwater are **porosity** (η) and **hydraulic conductivity** (K). Hydraulic conductivity, K, which is the constant of proportionality in Darcy's law has also been called as coefficient of permeability.

8.2.3 Aquifers

Depending on the hydraulic conductivity, various terms are used for geologic formations containing groundwater in a saturated zone.

An **aquifer** is a geologic formation or unit that can store and transmit water at rates fast enough to supply reasonable amounts to groundwater wells. In other words, these are groundwater-bearing horizons with sufficient porosity and permeability or hydraulic conductivity to transmit and yield water in usable quantities. Unconsolidated sands, gravels, sandstones, and fractured rock formations that receive infiltration all form aquifers.

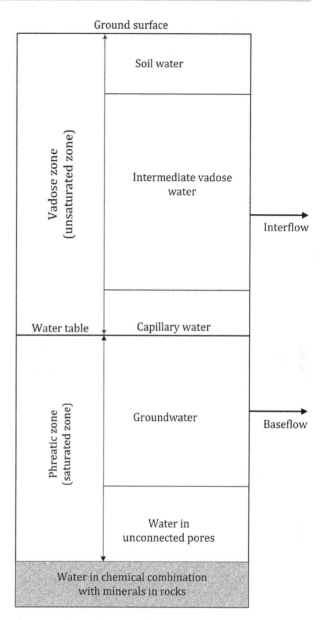

Figure 8.2 The groundwater profile. In the vadose zone, $P_{H_2O} < P_{atmosphere}$, at the water table, $P_{H_2O} = P_{atmosphere}$, and in the phreatic or saturated zone, $P_{H_2O} > P_{atmosphere}$.

A **confining layer** is a geologic unit having a little or no intrinsic permeability. Compact clay or silt horizons are examples of confining layers. Terms like aquifuge, aquitard, and aquiclude are used for confining layers, depending on the degree of permeability.

There are two types of aquifers: confined and unconfined aquifers (Figure 8.3).

An **unconfined aquifer** or **water table aquifer** is the saturated zone whose *top surface* is defined by a *water table* where $P_{H_2O} = P_{atmosphere}$. Materials with high porosity and permeability extend from the water table to the base of the aquifer. The base of an unconfined aquifer is usually a confining layer. The water table or the top of an unconfined aquifer is recharged by infiltration directly from the surface.

A. Unconfined aquifer

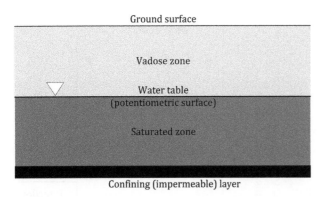

B. Confined aquifer

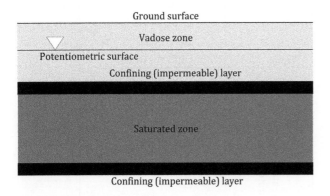

Figure 8.3 (A) Unconfined and (B) confined aquifers.

Confined aquifers are saturated horizons overlain by a confining layer. For this reason, the entire top surface of a confined aquifer does not receive infiltration. Recharge to confined aquifers occurs mostly in a recharge zone where the geologic formations forming the aquifer crop out and are exposed to the surface. However, sometimes the confining layer bounding the upper surface of the aquifer may also supply some water to it by slow leakage.

The most important distinguishing feature of unconfined and confined aquifers is the **potentiometric surface**. In the case of an unconfined aquifer, it is the real surface defined by the water table where the pressure head of the water in the aquifer is equal to the elevation of the water table with respect to any datum. In the case of a confined aquifer, it is an imaginary surface above the top of the confining layer bounding the top of the aquifer. It is the pressure head of the water in the aquifer. If a tightly cased well is placed through the confining layer, water may rise considerably above the top of the aquifer. In older literature, the term piezometric surface was often used to refer to the potentiometric surface. The potentiometric surface of an aquifer is represented by contour maps where each contour represents **equipotential lines** or the elevation up to which water will rise along that line. In the case of an unconfined aquifer, this is also called the **water table map**.

There are certain hydrologic properties of aquifers that determine flow and other hydrologic characteristics and differ from confined to unconfined aquifers. These are summarized below.

8.2.4 Aquifer Characteristics

The **specific yield**, S_y, is the ratio of the volume of water that drains from a saturated column of earth material due to gravity to the total volume of the column of material. The **specific retention**, S_r, of the column of material is the ratio of the volume of water the column can retain against gravitational drainage to the total volume of the material. The total porosity of the material is the sum of the specific yield and specific retention. The relationships are as follows:

$$\eta = S_y + S_r \qquad (8.1)$$

$$S_y = \frac{V_d}{V_t} \qquad (8.2)$$

$$S_r = \frac{V_r}{V_t} \qquad (8.3)$$

where η is the total porosity of the column of the aquifer, and $V_d, V_r,$ and V_t, respectively, denote the volume of water drained by gravity, the volume of water retained by the aquifer column, and the total volume of the column or horizon of the aquifer under consideration.

The **transmissivity**, T, or coefficient of transmissivity, is defined as the rate of flow of water at the prevailing temperature through a vertical strip of aquifer of unit width extending the full saturated thickness of the aquifer, under a unit hydraulic gradient. Thus, T is a measure of amount of water that can be transmitted horizontally by a full saturated thickness of the aquifer under a hydraulic gradient of one. It is given by the product of the hydraulic conductivity and the saturated thickness of the aquifer,

$$T = bK \qquad (8.4)$$

in which b is the saturated thickness of the aquifer and K is the hydraulic conductivity.

The **specific storage**, S_s, or elastic storage, coefficient is the amount of water per unit volume of a saturated formation that is stored or expelled from storage due to the compressibility of the mineral or soil skeleton and the pore water. It is given by

$$S_s = \rho g(\alpha + \eta \beta) \qquad (8.5)$$

where ρ is the density of water, g is the acceleration due to gravity, α is the compressibility of the aquifer skeleton, and β is the compressibility of water. The dimension of S_s is $1/L$.

The **storativity**, S, or storage coefficient is the volume of water that a permeable unit will absorb or expel from storage per unit surface area per unit change in head. For a *confined aquifer*, storativity is given as the product of the thickness of the aquifer and its specific storage. Thus,

$$S = S_s b \tag{8.6}$$

and S is a dimensionless quantity.

For an *unconfined aquifer*, storativity, specific yield, and specific storage are related by

$$S = S_y + bS_s \tag{8.7}$$

This is because in an unconfined aquifer the level of saturation rises or falls with changes in the amount of water in storage. As the water level falls, water drains from the pore spaces. This storage or release is due to the specific yield of the unit.

8.2.5 Gaining and Losing Streams

A typical stream in a humid region receives groundwater discharge or baseflow from the saturated zone. Baseflow increases downstream. This is called a **gaining stream** or **effluent stream**. The water table slopes toward the stream so that the hydraulic gradient of the aquifer is toward the stream (Figure 8.4A).

In arid regions, the water table may fall and the bottom of the stream is higher than the local water table. In this case, water may drain from the stream into the ground. This is called a **losing** or **influent stream** (Figure 8.4B).

A) Gaining stream

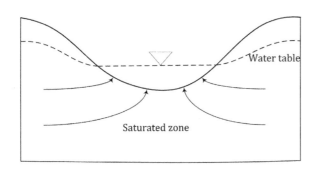

B) Losing stream

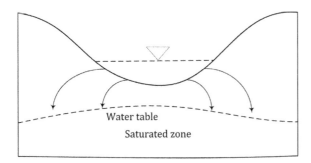

Figure 8.4 Stream–groundwater interactions: (A) gaining (A) and (B) losing streams.

A stream that is normally a gaining stream may become a losing stream during a flood, since the water surface in the stream becomes higher than the water table.

8.2.6 Equations of Groundwater Flow

8.2.6.1 Confined Aquifers

The general form of Darcy's law in three dimensions can be given as

$$\begin{aligned} q_x &= -K_x \frac{\delta h}{\delta x} \\ q_y &= -K_y \frac{\delta h}{\delta y} \\ q_z &= -K_z \frac{\delta h}{\delta z} \end{aligned} \tag{8.8}$$

where subscripts x, y, and z denote the three spatial coordinates. In general, the hydraulic conductivities in the three directions in a porous medium are not the same. For this reason, a porous medium is classified as either anisotropic or isotropic for the following two general cases: (1) steady state and (2) unsteady state or transient state.

From Eq. 8.8, it is apparent that the rate of change of hydraulic head, h, is different in different directions. The direction of the maximum change in the head is called the gradient of h and is given as

$$grad\, h = \nabla h = \frac{\delta h}{\delta x}\hat{i} + \frac{\delta h}{\delta y}\hat{j} + \frac{\delta h}{\delta z}\hat{k} \tag{8.9}$$

where $\hat{i}, \hat{j}, \hat{k}$ are unit vectors in x, y, and z directions. The symbol ∇ is known as the del operator, which itself is not a vector but when it operates on either a scalar or vector field it returns a vector.

The gradient of h or ∇h is a vector that represents the spatial rate of change of the hydraulic head. The direction of ∇h coincides with the direction in which the head changes the fastest, which is the direction of the steepest gradient.

From the continuity equation of fluid flow, it can be shown that, *under steady-state conditions*, i.e., for $\frac{\partial h}{\partial t} = 0$, the following equation holds true:

$$\nabla^2 h = 0 \tag{8.10}$$

where $\nabla^2 h$ is called the **Laplacian** operator and is simply,

$$\nabla^2 h = \frac{\partial^2 h}{\partial x^2} + \frac{\partial^2 h}{\partial y^2} + \frac{\partial^2 h}{\partial z^2} \tag{8.11}$$

Equation 8.10 is known as the **Laplace equation**. The simple interpretation of Eq. 8.10 is that under steady-state conditions, outflow equals inflow. Solutions of Eq. 8.10 with appropriate initial and boundary conditions give the steady-state potentiometric map of a confined aquifer.

If there is a steady flow of groundwater in a confined aquifer, there will be a gradient or slope to the

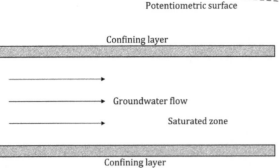

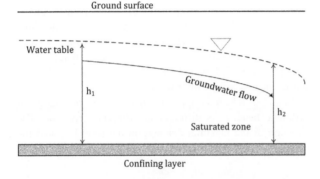

Figure 8.5 (A) Steady flow through a confined aquifer of uniform thickness. The potentiometric surface has a linear gradient. (B) Steady flow through an unconfined aquifer. The saturated thickness of the aquifer varies and therefore the potentiometric surface is nonlinear.

potentiometric surface of the aquifer. For flow of this type, Darcy's law can be used directly. In one dimension (Figure 8.5A),

$$q' = Kb\frac{dh}{dl} \qquad (8.12)$$

in which q' is the flow per unit width, also called the **specific discharge** with dimension L^2/T (e.g., m^2/d).

For *transient* flow conditions, i.e., when $\frac{\partial h}{\partial t} \neq 0$,

$$\nabla^2 h = \frac{S_s}{K}\frac{\partial h}{\partial t} = \frac{S}{T}\frac{\partial h}{\partial t} \qquad (8.13)$$

Equation 8.13 is known as the **diffusion equation**. Solutions of this equation provide potentiometric surfaces of a confined aquifer under an unsteady state.

8.2.6.2 Unconfined Aquifers

In an unconfined aquifer, the water table is also the upper boundary of the region of flow and the depth of water table decreases in the direction of flow, giving rise to nonuniform thickness of the saturated horizon of the aquifer (Figure 8.5B). Due to the nonlinear gradient of potentiometric surface, that is, the water table, Darcy's law, which gives a relationship like Eq. 8.12 cannot be applied directly in such a case. This problem was solved by Dupuit (1863) by assuming (1) the hydraulic gradient is equal to the slope of the water table and (2) for a small gradient of water table, the streamlines are horizontal and the equipotential lines are vertical. With this simplification of the situation, the **Dupuit approximation** or **Dupuit equation** of unconfined flow between two points is given by

$$q' = \frac{1}{2}K\left(\frac{h_1^2 - h_2^2}{L}\right) \qquad (8.14)$$

In the case of a confined aquifer, the saturated thickness remains relatively constant and thereby transmissivity can be assumed to be constant. But for an unconfined aquifer, due to changes in the saturated thickness, transmissivity does not remain constant.

The general flow equation for two-dimensional unconfined flow is known as the **Boussinesq equation**, named after French mathematician and physicist Joseph Valentin Boussinesq who proposed it (Boussinesq, 1904). For a homogeneous unconfined aquifer, the equation is

$$\frac{\partial}{\partial x}\left(h\frac{\partial h}{\partial x}\right) + \frac{\partial}{\partial y}\left(h\frac{\partial h}{\partial y}\right) = \frac{S_y}{K_h}\frac{\partial h}{\partial t} \qquad (8.15)$$

where S_y is the specific yield ($\approx$ porosity of the medium) and K_h is the hydraulic conductivity. Equation 8.15 is a nonlinear partial differential equation that cannot be integrated analytically. However, if the saturated thickness does not vary considerably, then the variable thickness, h (which is the hydraulic head in an unconfined aquifer), can be replaced with an average thickness b, which is assumed to be constant and thereby Eq. 8.15 can be linearized with an approximation:

$$\frac{\partial^2 h}{\partial x^2} + \frac{\partial^2 h}{\partial y^2} = \frac{S_y}{K_h b}\frac{\partial h}{\partial t} \qquad (8.16)$$

8.3 COMMON METHODS OF BASEFLOW SEPARATION

8.3.1 Subjective Methods

There are several subjective methods of baseflow separation, as shown in Figure 8.6. Three most common subjective methods are as follows.

The first method is the simplest one and consists of arbitrarily selecting the discharge marking the beginning of the rising limb as the value of the baseflow and assuming that this baseflow discharge remains constant throughout the storm duration. This is shown by line A in Figure 8.6, which is the streamflow hydrograph of Mahoning Creek at Punxsutawney, Pennsylvania, for a two-week period of

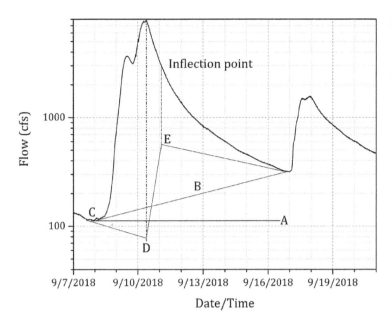

Figure 8.6 Common techniques of baseflow separation illustrated with the streamflow hydrograph of Mahoning Creek at Punxsutawney, Pennsylvania (USGS gauging station 03034000) covering a two-week period (September 7–21, 2018) with a major rainstorm occurring over the Mahoning Creek watershed.

September 2018 during which a major rainfall event took place over the Mahoning Creek watershed.

In the second method, the beginning of the groundwater recession on the falling or recession limb of the hydrograph is arbitrarily selected. It is usually assumed to occur at a theoretical inflection point and this point is connected by a straight line to the beginning of the rising limb. This is shown by line B in Figure 8.6. This method is sometimes referred to as the **straight-line method**. The first method described above is also a straight-line method but the second one is a better modification due to the slower rate of recession. Nevertheless, this method is usually applicable to a single storm event. The baseflow hydrograph has a slow recession, because the rate of movement of groundwater is very slow. Thus, the slope of the recession limb of a storm hydrograph is low, as shown in Figure 8.6, and is similar to all preceding recessions.

A third subjective method involves extending the recession prior to the storm by a line from the beginning of the rising limb to a point directly beneath the peak discharge, that is, the baseflow curve before the surface runoff began is extrapolated forward to the time of peak discharge. At this time, this point of the extrapolated baseflow line is connected to the beginning of the groundwater recession on the falling limb. This is shown by lines CD and DE in Figure 8.6. This method is referred to as the **variable-slope method**.

8.3.2 Area Method

The area method of baseflow separation is based upon a nonlinear relation between time and area (Linsley et al., 1982). The method involves determining the beginning of baseflow on the falling limb with an empirical equation relating the time in days from the peak discharge, N, marking the beginning of groundwater flow, to the basin area, A. The equation is given as

$$N = bA^{0.2} \qquad (8.17)$$

when A is in square miles, b equals 1. When A is in square kilometers, b equals 0.8. Equation 8.17 is unsuitable for smaller watersheds and should be checked for a number of hydrographs before use.

8.3.3 Master Recession Curve Method

This method represents the response of a groundwater aquifer as a **linear reservoir** whose outflow rate $Q(t)$, is proportional to the current storage, $S(t)$. Such a linear reservoir mode is represented by

$$S(t) = kQ(t) \qquad (8.18)$$

where the parameter k is an exponential decay constant having the dimension of time. This assumption leads to an equation for the groundwater recession hydrograph given as

$$Q(t) = Q_0 \exp\left[-\frac{t - t_0}{k}\right] \qquad (8.19)$$

where $Q(t)$ is the baseflow at time t; Q_0 is a reference baseflow discharge at time t_0, and k is the **recession constant** or **depletion constant** for baseflow. This method is based on a linear reservoir model of unforced basin response (that is, response from storage) and it can be used to separate the contributions to the recession flow from surface storage or surface runoff, subsurface storage or interflow, and groundwater aquifer storage or groundwater flow. For this reason, it is also known as the **three-component separation method**. It involves determining several recession constants by plotting the logarithm of $Q(t)$ against time on a linear scale.

During the rising hydrograph, the logarithm of baseflow follows a straight line with a certain slope. From the point of inflection of the falling limb of the hydrograph, the logarithm of baseflow follows another straight line with a different slope. Between the peak of the hydrograph and the point of inflection, the logarithm of baseflow is assumed to vary linearly.

8.3.4 Digital Filter Method

The digital filter method has been used in signal analysis and processing to separate high-frequency signals from low-frequency signals (Lyne and Hollick, 1979). This method has been used in baseflow separation because high-frequency waves can be associated with direct runoff, and low-frequency waves can be associated with base flow. Thus, filtering direct runoff from baseflow is similar to signal analysis and processing. The digital filter used for baseflow separation (Lyne and Hollick, 1979; Nathan and McMahon, 1990) is shown by

$$QR_{t+1} = \alpha_f Q_t + \frac{(1+\alpha_f)}{2}[(Q_{t+1} - Q_t)] \qquad (8.20)$$

where QR_{t+1} is the filtered flow that is coming from direct runoff (m³/s or ft³/s) at the time step $t+1$, Q_{t+1} is the unfiltered streamflow at the time step $t+1$ (m³/s or ft³/s), α_f is the filter parameter, and Q_t is the streamflow at the time step t (m³/s or ft³/s).

The digital filter method has no physical meaning but it removes the subjective aspect from manual separation, and it is fast, consistent, and reproducible. However, experienced hydrologists should subjectively evaluate the quality of results from the digital filter method. The value of α_f should be selected such that the computed baseflow values match the streamflow values of the hydrograph at the waning phase of the recession. When $\alpha_f = 1$, baseflow is constant throughout; the estimated contribution of baseflow to the total streamflow decreases as α_f becomes greater than unity and increases as α_f decreases further from unity.

8.4 BASEFLOW MODELS

Some conceptual models of watershed processes account explicitly for the storage of groundwater and for the subsurface movement. However, such accounting is not always necessary for rainfall–runoff modeling. The familiar models of baseflow for water resources studies are described below.

8.4.1 Constant, Monthly Varying Baseflow

This is the simplest baseflow model. It represents baseflow as a constant flow which may vary monthly. In rainfall–runoff models, a specified flow is added to direct runoff computed from rainfall for each time step of the simulation to derive the total streamflow.

8.4.2 Exponential Recession Model

The exponential recession model has often been used to explain the drainage from natural storage in a watershed (Linsley et al., 1982). This is a most widely used model of baseflow to approximate the typical behavior observed in channels when streamflow recedes exponentially after a rainfall event. It defines the relationship of Q_t, the baseflow at any time, t, to an initial value, Q_0 (at time zero) as

$$Q_t = Q_0 k^t = Q_0 \exp[-\alpha t] \qquad (8.21)$$

where k is the exponential decay constant or simply recession constant defined by a constant $\alpha = -\ln k$ or $k = \exp(-\alpha)$.

Figure 8.7 schematically represents the exponentially decreasing baseflow. Equation 8.21 will plot as a straight line using a semilogarithmic plot (with Q on the logarithmic scale). The numerical value of k depends on the time unit selected even though Eq. 8.21 does imply that it is a dimensionless parameter. Its value is always less than unity but greater than 0.9. The constant α is known as the **storage decay factor** and has the dimension of time. It denotes the time required for the flow to decrease by a factor equal to e, or one natural log cycle.

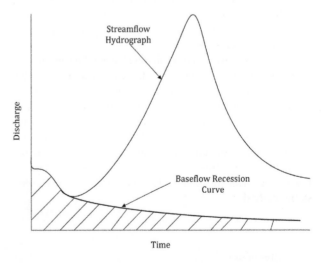

Figure 8.7 Exponential recession model of baseflow. The curve labeled baseflow recession curve represents the baseflow component of streamflow. The region representing the baseflow contribution to the total streamflow is hatched below this curve.

From Eq. 8.21 it follows that the volume of water discharged during time dt is Qdt and is equal to the decrease in storage $-dS$ during the same time interval. Therefore, the storage remaining in the basin, S_t, at time t, is obtained by integrating Eq. 8.21 and is given as

$$-dS = Q_t dt = Q_0 k^t dt \qquad (8.22)$$

Integrating Eq. 8.22 from the lower limit at t_0 when storage in the watershed $S_0 = 0$ and $Q_{t=\infty} = 0$,

$$-\int_{S_0}^{S_t} dS = \int_{t_0}^{t} Q_t dt = \int_{0}^{t} Q_0 k^t dt \qquad (8.23)$$

$$S_t = -\frac{Q_t}{\ln(k)} = -\frac{Q_t}{\alpha} \qquad (8.24)$$

As illustrated in Figure 8.7, the baseflow contribution decays exponentially from the starting flow. The total flow is the sum of baseflow and direct surface runoff. In HEC-HMS, k is defined as the ratio of baseflow at time t to baseflow one day earlier. The starting baseflow value, Q_0, is an initial condition of the model that can be specified either as flow rate (m^3/s or ft^3/s) or as flow per unit area (m^3/s/km^2 or ft^3/s/mi^2) and the recession constant, k, is specified as a dimensionless parameter.

The recession baseflow model is applied both at the start of the simulation of a storm event, and later in the event as the delayed subsurface flow reaches the watershed channels, as illustrated in Figure 8.7. Here, after the peak of the direct runoff, a user-specified threshold flow defines the time at which the recession model of Eq. 8.21 defines the total flow. That threshold may be specified as a flow rate or as a ratio to the computed peak flow. For example, if the threshold is specified as a ratio to peak of 0.10, and the computed peak is 1000 m^3/s, the threshold flow is 100 m^3/s. Subsequent total flows are computed with Eq. 8.21, with Q_0 being the specified threshold value.

At the threshold flow, baseflow is defined by the initial baseflow recession. Thereafter, baseflow is not computed directly, but is defined as the recession flow less the direct surface runoff. When the direct or surface runoff eventually reaches zero (all rainfall has run off the watershed), the total flow and baseflow are identical. After the threshold flow occurs, the streamflow hydrograph ordinates are defined by the recession model alone, unless the direct runoff plus initial baseflow recession contribution exceeds the threshold. This may be the case if subsequent precipitation causes a second rise in the hydrograph, as shown in Figure 8.1. In that case, the ordinates on the second rising limb are computed by adding direct runoff to the initial recession, as illustrated. In HEC-HMS, the exponential recession method is intended primarily for event-based simulation of rainfall–runoff process. However, the program has the ability to automatically reset after each storm event and consequently can be used for continuous simulation models.

8.4.3 Linear-Reservoir Model

The linear-reservoir baseflow model is used in conjunction with the continuous soil moisture accounting (SMA) model that is described in Chapter 7. This baseflow model simulates the storage and movement of subsurface flow as storage and movement of water through reservoirs. The reservoirs are linear: the outflow at each time step of the simulation is a linear function of the average storage during the time step. Mathematically, this is identical to the manner in which the Clark unit hydrograph model represents watershed runoff, as described in Chapter 9. The outflow from groundwater layer 1 of the SMA is inflow to one linear reservoir, and the outflow from groundwater layer 2 of the SMA is inflow to another. The outflow from the two linear reservoirs is combined to compute the total baseflow for the watershed. Infiltration or percolation computed by the loss method is connected as inflow to the linear reservoirs.

8.4.4 Nonlinear Boussinesq Baseflow

The nonlinear Boussinesq baseflow method is similar to the recession baseflow method, but assumes an unconfined groundwater layer and invokes the Boussinesq assumptions for flow in an unconfined aquifer, where the saturated thickness of the aquifer changes with time, as opposed to a confined aquifer, where the saturated thickness is assumed to remain constant. The characteristic subsurface flow length must be specified. This could be estimated as the mean distance from the subbasin boundary to the stream. The conductivity of the soil must be specified. This can be estimated from field tests or from soil texture. The drainable porosity must be specified in terms of volume ratio. The upper limit is the total porosity minus the residual porosity. The actual drainable porosity depends on local conditions.

8.5 ESTIMATING BASEFLOW MODEL PARAMETERS

8.5.1 Constant, Monthly Varying Baseflow

The parameters of this model are monthly baseflows. These are best estimated empirically, with measurements of channel flow when storm runoff is not occurring. In the absence of such records, field inspection may help establish the average flow. For large watersheds with contribution from groundwater flow and for watersheds with year-round precipitation, the contribution may be significant and should not be ignored. On the other hand, for most urban channels and for smaller streams such as those in the western and southwestern United States, the baseflow contribution may be negligible.

8.5.2 Exponential Recession Model

The parameters of this model include the initial flow, the recession ratio, and the threshold flow. As noted, the initial flow is the initial condition. For analysis of hypothetical storm runoff, the initial flow should be selected as the likely average flow that would occur at the start of the storm runoff. For frequent events, the initial flow might be the average annual flow in the channel. Field inspection may help establish this. As with constant, monthly varying baseflow, for most urban channels and for smaller streams in the western and southwestern United States, this may well be zero, as the baseflow contribution is negligible. The recession constant, k, depends upon the source of baseflow. If $k = 1.00$, the baseflow contribution will be constant, with all $Q_t = Q_0$. Otherwise, to model the exponential decay typical of natural undeveloped watersheds, k must be less than 1.00. Table 8.1 shows typical values proposed by Pilgrim and Cordery (1992) for basins ranging in size from 300 km^2 to 16 000 km^2 (120 mi^2 6500 mi^2) in the United States, eastern Australia, and several other regions. Large watersheds may have k values at the upper end of the range, while smaller watersheds will have lower values.

The recession constant can be estimated if gauged flow data are available. Flows prior to the start of direct runoff can be plotted, and an average of the ratios of ordinates spaced one day apart can be computed. This is simplified if a logarithmic axis is used for flows, as the recession model will plot as a straight line. The threshold value can be estimated also from examining the graph of the observed flows versus time. The flow at which the recession limb is approximated well by a straight line defines the threshold value.

Table 8.1 *Typical recession constant values*

Flow component	Recession constant, daily
Groundwater	0.95
Interflow	0.8–0.9
Surface runoff	0.3–0.8

8.5.3 Linear Reservoir Model

The linear reservoir model is used with the SMA model and is best calibrated using procedures consistent with those used to calibrate that model.

8.6 EXAMPLES

Example 8.1: The streamflow hydrograph of Mahoning Creek at Punxsutawney, Pennsylvania, shows a flood peak of 7810 ft^3/s on 9/10/2018 at 21:00 and then the flow recedes up to 23:45 on 9/16/2018 before the arrival of another flood peak. This portion of the hydrograph is shown in Figure 8.8. The inflection point on the recession limb is at 1:00 on 9/13/2018. The flow data are broken into three parts: (1) rising limb; (2) recession limb; and (3) baseflow and are given in Table 8.2 in two parts. Determine the baseflow recession constant and the storage of water in the watershed at the end of the event.

Solution: The discharge values of the falling limb and baseflow part of the hydrograph are plotted on a semilogarithmic graph where discharge is plotted on the logarithmic scale on the ordinate. This is shown in Figure 8.9.

The data on the baseflow part are fitted with an exponential function given as

$$Q = 2791 \times \exp(-2 \times 10^{-4} t)$$

Thus, the recession constant, $k = \exp(-2 \times 10^{-4}) = 0.9998$.

The storage delay factor, $\alpha = -\ln(0.9998) = \frac{0.0002}{\text{minute}} = 0.012 \text{ s}^{-1}$.

From the baseflow values, the value of baseflow at the end of the event is 335 ft^3/s. Therefore, the amount of water remaining in the watershed as storage is determined as

$$S_t = \frac{335 \frac{\text{ft}^3}{\text{s}}}{0.012 \frac{1}{\text{s}}} = 27916.67 \text{ ft}^3 = 0.641 \text{ acre-feet}.$$

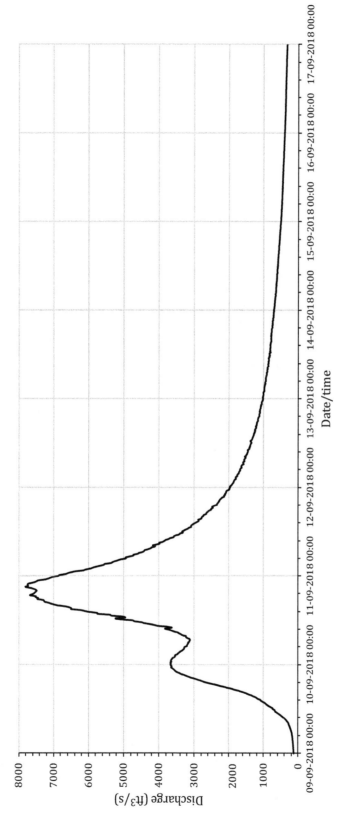

Figure 8.8 Flood hydrograph of Mahoning Creek at Punxsutawney from 9/9/2018 to 9/14/2018.

Table 8.2 *Hydrograph data*

Part A

Rising limb		Falling limb		Recession limb	
Date/Time	Flow (ft^3/s)	Date/Time	Flow (ft^3/s)	Date/Time	Flow (ft^3/s)
9/9/2018 0:00	129	9/10/2018 21:15	7740	9/13/2018 1:15	1010
9/9/2018 0:15	129	9/10/2018 21:30	7750	9/13/2018 1:30	1010
9/9/2018 0:30	131	9/10/2018 21:45	7740	9/13/2018 1:45	1010
9/9/2018 0:45	131	9/10/2018 22:00	7680	9/13/2018 2:00	988
9/9/2018 1:00	133	9/10/2018 22:15	7570	9/13/2018 2:15	988
9/9/2018 1:15	135	9/10/2018 22:30	7490	9/13/2018 2:30	982
9/9/2018 1:30	135	9/10/2018 22:45	7450	9/13/2018 2:45	982
9/9/2018 1:45	138	9/10/2018 23:00	7270	9/13/2018 3:00	982
9/9/2018 2:00	140	9/10/2018 23:15	7170	9/13/2018 3:15	976
9/9/2018 2:15	142	9/10/2018 23:30	7090	9/13/2018 3:30	971
9/9/2018 2:30	144	9/10/2018 23:45	7010	9/13/2018 3:45	971
9/9/2018 2:45	146	9/11/2018 0:00	6870	9/13/2018 4:00	954
9/9/2018 3:00	149	9/11/2018 0:15	6830	9/13/2018 4:15	959
9/9/2018 3:15	151	9/11/2018 0:30	6670	9/13/2018 4:30	954
9/9/2018 3:30	156	9/11/2018 0:45	6590	9/13/2018 4:45	942
9/9/2018 3:45	158	9/11/2018 1:00	6510	9/13/2018 5:00	942
9/9/2018 4:00	160	9/11/2018 1:15	6390	9/13/2018 5:15	937
9/9/2018 4:15	163	9/11/2018 1:30	6160	9/13/2018 5:30	942
9/9/2018 4:30	165	9/11/2018 1:45	6080	9/13/2018 5:45	937
9/9/2018 4:45	170	9/11/2018 2:00	5960	9/13/2018 6:00	931
9/9/2018 5:00	175	9/11/2018 2:15	5880	9/13/2018 6:15	920
9/9/2018 5:15	178	9/11/2018 2:30	5800	9/13/2018 6:30	914
9/9/2018 5:30	183	9/11/2018 2:45	5680	9/13/2018 6:45	909
9/9/2018 5:45	190	9/11/2018 3:00	5570	9/13/2018 7:00	914
9/9/2018 6:00	198	9/11/2018 3:15	5540	9/13/2018 7:15	898
9/9/2018 6:15	207	9/11/2018 3:30	5360	9/13/2018 7:30	898
9/9/2018 6:30	215	9/11/2018 3:45	5310	9/13/2018 7:45	898
9/9/2018 6:45	226	9/11/2018 4:00	5240	9/13/2018 8:00	887
9/9/2018 7:00	235	9/11/2018 4:15	5160	9/13/2018 8:15	892
9/9/2018 7:15	244	9/11/2018 4:30	5080	9/13/2018 8:30	887
9/9/2018 7:30	257	9/11/2018 4:45	4940	9/13/2018 8:45	871
9/9/2018 7:45	266	9/11/2018 5:00	4940	9/13/2018 9:00	871
9/9/2018 8:00	276	9/11/2018 5:15	4850	9/13/2018 9:15	876
9/9/2018 8:15	290	9/11/2018 5:30	4730	9/13/2018 9:30	865
9/9/2018 8:30	305	9/11/2018 5:45	4720	9/13/2018 9:45	860
9/9/2018 8:45	327	9/11/2018 6:00	4600	9/13/2018 10:00	865
9/9/2018 9:00	346	9/11/2018 6:15	4560	9/13/2018 10:15	854
9/9/2018 9:15	375	9/11/2018 6:30	4460	9/13/2018 10:30	854
9/9/2018 9:30	404	9/11/2018 6:45	4420	9/13/2018 10:45	854
9/9/2018 9:45	435	9/11/2018 7:00	4390	9/13/2018 11:00	860
9/9/2018 10:00	472	9/11/2018 7:15	4310	9/13/2018 11:15	839
9/9/2018 10:15	501	9/11/2018 7:30	4230	9/13/2018 11:30	844
9/9/2018 10:30	541	9/11/2018 7:45	4200	9/13/2018 11:45	828
9/9/2018 10:45	572	9/11/2018 8:00	4110	9/13/2018 12:00	833
9/9/2018 11:00	609	9/11/2018 8:15	4170	9/13/2018 12:15	839
9/9/2018 11:15	648	9/11/2018 8:30	4080	9/13/2018 12:30	807
9/9/2018 11:30	671	9/11/2018 8:45	3980	9/13/2018 12:45	823
9/9/2018 11:45	677	9/11/2018 9:00	3990	9/13/2018 13:00	812
9/9/2018 12:00	725	9/11/2018 9:15	3890	9/13/2018 13:15	817
9/9/2018 12:15	743	9/11/2018 9:30	3840	9/13/2018 13:30	817
9/9/2018 12:30	791	9/11/2018 9:45	3800	9/13/2018 13:45	812
9/9/2018 12:45	835	9/11/2018 10:00	3720	9/13/2018 14:00	807
9/9/2018 13:00	874	9/11/2018 10:15	3680	9/13/2018 14:15	807

Table 8.2 (*cont.*)

Rising limb		Falling limb		Recession limb	
Date/Time	Flow (ft^3/s)	Date/Time	Flow (ft^3/s)	Date/Time	Flow (ft^3/s)
9/9/2018 13:15	914	9/11/2018 10:30	3590	9/13/2018 14:30	812
9/9/2018 13:30	946	9/11/2018 10:45	3570	9/13/2018 14:45	807
9/9/2018 13:45	978	9/11/2018 11:00	3510	9/13/2018 15:00	802
9/9/2018 14:00	1030	9/11/2018 11:15	3430	9/13/2018 15:15	797
9/9/2018 14:15	1070	9/11/2018 11:30	3400	9/13/2018 15:30	786
9/9/2018 14:30	1100	9/11/2018 11:45	3320	9/13/2018 15:45	786
9/9/2018 14:45	1160	9/11/2018 12:00	3340	9/13/2018 16:00	786
9/9/2018 15:00	1200	9/11/2018 12:15	3300	9/13/2018 16:15	792
9/9/2018 15:15	1250	9/11/2018 12:30	3250	9/13/2018 16:30	786
9/9/2018 15:30	1320	9/11/2018 12:45	3190	9/13/2018 16:45	786
9/9/2018 15:45	1370	9/11/2018 13:00	3150	9/13/2018 17:00	792
9/9/2018 16:00	1450	9/11/2018 13:15	3090	9/13/2018 17:15	781
9/9/2018 16:15	1510	9/11/2018 13:30	3090	9/13/2018 17:30	786
9/9/2018 16:30	1590	9/11/2018 13:45	3070	9/13/2018 17:45	781
9/9/2018 16:45	1670	9/11/2018 14:00	3030	9/13/2018 18:00	776
9/9/2018 17:00	1740	9/11/2018 14:15	2990	9/13/2018 18:15	771
9/9/2018 17:15	1850	9/11/2018 14:30	2940	9/13/2018 18:30	771
9/9/2018 17:30	1930	9/11/2018 14:45	2890	9/13/2018 18:45	771
9/9/2018 17:45	2040	9/11/2018 15:00	2870	9/13/2018 19:00	766
9/9/2018 18:00	2130	9/11/2018 15:15	2880	9/13/2018 19:15	756
9/9/2018 18:15	2250	9/11/2018 15:30	2800	9/13/2018 19:30	756
9/9/2018 18:30	2350	9/11/2018 15:45	2790	9/13/2018 19:45	756
9/9/2018 18:45	2450	9/11/2018 16:00	2770	9/13/2018 20:00	746
9/9/2018 19:00	2570	9/11/2018 16:15	2740	9/13/2018 20:15	751
9/9/2018 19:15	2680	9/11/2018 16:30	2720	9/13/2018 20:30	746
9/9/2018 19:30	2770	9/11/2018 16:45	2660	9/13/2018 20:45	746
9/9/2018 19:45	2840	9/11/2018 17:00	2650	9/13/2018 21:00	746
9/9/2018 20:00	2920	9/11/2018 17:15	2620	9/13/2018 21:15	735
9/9/2018 20:15	3000	9/11/2018 17:30	2600	9/13/2018 21:30	735
9/9/2018 20:30	3100	9/11/2018 17:45	2550	9/13/2018 21:45	735
9/9/2018 20:45	3170	9/11/2018 18:00	2510	9/13/2018 22:00	724
9/9/2018 21:00	3230	9/11/2018 18:15	2500	9/13/2018 22:15	724
9/9/2018 21:15	3300	9/11/2018 18:30	2460	9/13/2018 22:30	724
9/9/2018 21:30	3380	9/11/2018 18:45	2480	9/13/2018 22:45	718
9/9/2018 21:45	3430	9/11/2018 19:00	2420	9/13/2018 23:00	718
9/9/2018 22:00	3490	9/11/2018 19:15	2350	9/13/2018 23:15	718
9/9/2018 22:15	3530	9/11/2018 19:30	2340	9/13/2018 23:30	712
9/9/2018 22:30	3560	9/11/2018 19:45	2320	9/13/2018 23:45	706
9/9/2018 22:45	3590	9/11/2018 20:00	2300	9/14/2018 0:00	712
9/9/2018 23:00	3620	9/11/2018 20:15	2280	9/14/2018 0:15	706
9/9/2018 23:15	3650	9/11/2018 20:30	2270	9/14/2018 0:30	701
9/9/2018 23:30	3640	9/11/2018 20:45	2250	9/14/2018 0:45	701
9/9/2018 23:45	3660	9/11/2018 21:00	2220	9/14/2018 1:00	695
9/10/2018 0:00	3670	9/11/2018 21:15	2190	9/14/2018 1:15	690
9/10/2018 0:15	3660	9/11/2018 21:30	2140	9/14/2018 1:30	690
9/10/2018 0:30	3670	9/11/2018 21:45	2140	9/14/2018 1:45	690
9/10/2018 0:45	3650	9/11/2018 22:00	2110	9/14/2018 2:00	684
9/10/2018 1:00	3660	9/11/2018 22:15	2110	9/14/2018 2:15	684
9/10/2018 1:15	3650	9/11/2018 22:30	2090	9/14/2018 2:30	684
9/10/2018 1:30	3630	9/11/2018 22:45	2080	9/14/2018 2:45	678
9/10/2018 1:45	3610	9/11/2018 23:00	2070	9/14/2018 3:00	678
9/10/2018 2:00	3570	9/11/2018 23:15	2020	9/14/2018 3:15	673
9/10/2018 2:15	3540	9/11/2018 23:30	1990	9/14/2018 3:30	673
9/10/2018 2:30	3540	9/11/2018 23:45	1990	9/14/2018 3:45	667

Table 8.2 (*cont.*)

Rising limb		Falling limb		Recession limb	
Date/Time	Flow (ft^3/s)	Date/Time	Flow (ft^3/s)	Date/Time	Flow (ft^3/s)
9/10/2018 2:45	3480	9/12/2018 0:00	1970	9/14/2018 4:00	662
9/10/2018 3:00	3460	9/12/2018 0:15	1970	9/14/2018 4:15	667
9/10/2018 3:15	3430	9/12/2018 0:30	1940	9/14/2018 4:30	662
9/10/2018 3:30	3390	9/12/2018 0:45	1920	9/14/2018 4:45	656
9/10/2018 3:45	3350	9/12/2018 1:00	1900	9/14/2018 5:00	656
9/10/2018 4:00	3320	9/12/2018 1:15	1880	9/14/2018 5:15	651
9/10/2018 4:15	3290	9/12/2018 1:30	1870	9/14/2018 5:30	651
9/10/2018 4:30	3280	9/12/2018 1:45	1860	9/14/2018 5:45	651
9/10/2018 4:45	3270	9/12/2018 2:00	1840	9/14/2018 6:00	651
9/10/2018 5:00	3210	9/12/2018 2:15	1810	9/14/2018 6:15	646
9/10/2018 5:15	3200	9/12/2018 2:30	1810	9/14/2018 6:30	646
9/10/2018 5:30	3190	9/12/2018 2:45	1810	9/14/2018 6:45	640
9/10/2018 5:45	3170	9/12/2018 3:00	1770	9/14/2018 7:00	646
9/10/2018 6:00	3170	9/12/2018 3:15	1770	9/14/2018 7:15	640
9/10/2018 6:15	3140	9/12/2018 3:30	1760	9/14/2018 7:30	640
9/10/2018 6:30	3130	9/12/2018 3:45	1730	9/14/2018 7:45	635
9/10/2018 6:45	3120	9/12/2018 4:00	1720	9/14/2018 8:00	630
9/10/2018 7:00	3140	9/12/2018 4:15	1710	9/14/2018 8:15	630
9/10/2018 7:15	3200	9/12/2018 4:30	1690	9/14/2018 8:30	630
9/10/2018 7:30	3230	9/12/2018 4:45	1690	9/14/2018 8:45	624
9/10/2018 7:45	3240	9/12/2018 5:00	1680	9/14/2018 9:00	624
9/10/2018 8:00	3330	9/12/2018 5:15	1650	9/14/2018 9:15	624
9/10/2018 8:15	3380	9/12/2018 5:30	1650	9/14/2018 9:30	624
9/10/2018 8:30	3430	9/12/2018 5:45	1650	9/14/2018 9:45	624
9/10/2018 8:45	3480	9/12/2018 6:00	1630	9/14/2018 10:00	614
9/10/2018 9:00	3570	9/12/2018 6:15	1590	9/14/2018 10:15	614
9/10/2018 9:15	3650	9/12/2018 6:30	1600	9/14/2018 10:30	614
9/10/2018 9:30	3720	9/12/2018 6:45	1600	9/14/2018 10:45	614
9/10/2018 9:45	3820	9/12/2018 7:00	1570	9/14/2018 11:00	614
9/10/2018 10:00	3640	9/12/2018 7:15	1570	9/14/2018 11:15	609
9/10/2018 10:15	3750	9/12/2018 7:30	1570	9/14/2018 11:30	603
9/10/2018 10:30	4010	9/12/2018 7:45	1550	9/14/2018 11:45	609
9/10/2018 10:45	4140	9/12/2018 8:00	1530	9/14/2018 12:00	598
9/10/2018 11:00	4250	9/12/2018 8:15	1550	9/14/2018 12:15	603
9/10/2018 11:15	4420	9/12/2018 8:30	1520	9/14/2018 12:30	598
9/10/2018 11:30	4500	9/12/2018 8:45	1510	9/14/2018 12:45	593
9/10/2018 11:45	4700	9/12/2018 9:00	1490	9/14/2018 13:00	598
9/10/2018 12:00	4750	9/12/2018 9:15	1480	9/14/2018 13:15	593
9/10/2018 12:15	5080	9/12/2018 9:30	1470	9/14/2018 13:30	593
9/10/2018 12:30	5230	9/12/2018 9:45	1460	9/14/2018 13:45	593
9/10/2018 12:45	4960	9/12/2018 10:00	1470	9/14/2018 14:00	588
9/10/2018 13:00	5030	9/12/2018 10:15	1440	9/14/2018 14:15	588
9/10/2018 13:15	5340	9/12/2018 10:30	1420	9/14/2018 14:30	583
9/10/2018 13:30	5400	9/12/2018 10:45	1420	9/14/2018 14:45	583
9/10/2018 13:45	5740	9/12/2018 11:00	1400	9/14/2018 15:00	583
9/10/2018 14:00	5760	9/12/2018 11:15	1400	9/14/2018 15:15	578
9/10/2018 14:15	5960	9/12/2018 11:30	1390	9/14/2018 15:30	573
9/10/2018 14:30	6110	9/12/2018 11:45	1380	9/14/2018 15:45	573
9/10/2018 14:45	6220	9/12/2018 12:00	1360	9/14/2018 16:00	573
9/10/2018 15:00	6500	9/12/2018 12:15	1370	9/14/2018 16:15	568
9/10/2018 15:15	6500	9/12/2018 12:30	1390	9/14/2018 16:30	568
9/10/2018 15:30	6560	9/12/2018 12:45	1360	9/14/2018 16:45	568
9/10/2018 15:45	6670	9/12/2018 13:00	1340	9/14/2018 17:00	568
9/10/2018 16:00	6860	9/12/2018 13:15	1330	9/14/2018 17:15	568

Table 8.2 (cont.)

Rising limb		Falling limb		Recession limb	
Date/Time	Flow (ft^3/s)	Date/Time	Flow (ft^3/s)	Date/Time	Flow (ft^3/s)
9/10/2018 16:15	6980	9/12/2018 13:30	1320	9/14/2018 17:30	563
9/10/2018 16:30	7080	9/12/2018 13:45	1310	9/14/2018 17:45	558
9/10/2018 16:45	7160	9/12/2018 14:00	1300	9/14/2018 18:00	558
9/10/2018 17:00	7260	9/12/2018 14:15	1290	9/14/2018 18:15	553
9/10/2018 17:15	7280	9/12/2018 14:30	1290	9/14/2018 18:30	553
9/10/2018 17:30	7310	9/12/2018 14:45	1270	9/14/2018 18:45	553
9/10/2018 17:45	7470	9/12/2018 15:00	1270	9/14/2018 19:00	548
9/10/2018 18:00	7450	9/12/2018 15:15	1250	9/14/2018 19:15	548
9/10/2018 18:15	7520	9/12/2018 15:30	1260	9/14/2018 19:30	543
9/10/2018 18:30	7520	9/12/2018 15:45	1250	9/14/2018 19:45	548
9/10/2018 18:45	7680	9/12/2018 16:00	1230	9/14/2018 20:00	543
9/10/2018 19:00	7620	9/12/2018 16:15	1220	9/14/2018 20:15	543
9/10/2018 19:15	7570	9/12/2018 16:30	1230	9/14/2018 20:30	538
9/10/2018 19:30	7540	9/12/2018 16:45	1210	9/14/2018 20:45	538
9/10/2018 19:45	7510	9/12/2018 17:00	1210	9/14/2018 21:00	538
9/10/2018 20:00	7490	9/12/2018 17:15	1200	9/14/2018 21:15	534
9/10/2018 20:15	7520	9/12/2018 17:30	1200	9/14/2018 21:30	534
9/10/2018 20:30	7590	9/12/2018 17:45	1190	9/14/2018 21:45	529
9/10/2018 20:45	7680	9/12/2018 18:00	1170	9/14/2018 22:00	529
9/10/2018 21:00	7810	9/12/2018 18:15	1170	9/14/2018 22:15	524
		9/12/2018 18:30	1170	9/14/2018 22:30	524
		9/12/2018 18:45	1160	9/14/2018 22:45	524
		9/12/2018 19:00	1150	9/14/2018 23:00	524
		9/12/2018 19:15	1130	9/14/2018 23:15	519
		9/12/2018 19:30	1130	9/14/2018 23:30	519
		9/12/2018 19:45	1130	9/14/2018 23:45	515
		9/12/2018 20:00	1120	9/15/2018 0:00	515
		9/12/2018 20:15	1110	9/15/2018 0:15	510
		9/12/2018 20:30	1120	9/15/2018 0:30	510
		9/12/2018 20:45	1110	9/15/2018 0:45	510
		9/12/2018 21:00	1110	9/15/2018 1:00	510
		9/12/2018 21:15	1100	9/15/2018 1:15	505
		9/12/2018 21:30	1090	9/15/2018 1:30	505
		9/12/2018 21:45	1090	9/15/2018 1:45	500
		9/12/2018 22:00	1080	9/15/2018 2:00	500
		9/12/2018 22:15	1080	9/15/2018 2:15	500
		9/12/2018 22:30	1060	9/15/2018 2:30	500
		9/12/2018 22:45	1060	9/15/2018 2:45	496
		9/12/2018 23:00	1060	9/15/2018 3:00	500
		9/12/2018 23:15	1050	9/15/2018 3:15	496
		9/12/2018 23:30	1030	9/15/2018 3:30	491
		9/12/2018 23:45	1030	9/15/2018 3:45	491
		9/13/2018 0:00	1020	9/15/2018 4:00	491
		9/13/2018 0:15	1030	9/15/2018 4:15	491
		9/13/2018 0:30	1020	9/15/2018 4:30	487
		9/13/2018 0:45	1020	9/15/2018 4:45	487
		9/13/2018 1:00	1010	9/15/2018 5:00	482

Table 8.2 (*cont.*)

Part B (Recession limb)

Date/Time	Flow (ft³/s)	Date/time	Flow (ft³/s)
9/15/2018 5:15	487	9/16/2018 0:00	409
9/15/2018 5:30	487	9/16/2018 0:15	405
9/15/2018 5:45	482	9/16/2018 0:30	405
9/15/2018 6:00	482	9/16/2018 0:45	405
9/15/2018 6:15	478	9/16/2018 1:00	400
9/15/2018 6:30	478	9/16/2018 1:15	400
9/15/2018 6:45	478	9/16/2018 1:30	400
9/15/2018 7:00	478	9/16/2018 1:45	400
9/15/2018 7:15	478	9/16/2018 2:00	400
9/15/2018 7:30	473	9/16/2018 2:15	396
9/15/2018 7:45	473	9/16/2018 2:30	396
9/15/2018 8:00	473	9/16/2018 2:45	396
9/15/2018 8:15	469	9/16/2018 3:00	396
9/15/2018 8:30	469	9/16/2018 3:15	396
9/15/2018 8:45	473	9/16/2018 3:30	392
9/15/2018 9:00	464	9/16/2018 3:45	392
9/15/2018 9:15	469	9/16/2018 4:00	392
9/15/2018 9:30	469	9/16/2018 4:15	392
9/15/2018 9:45	464	9/16/2018 4:30	392
9/15/2018 10:00	464	9/16/2018 4:45	388
9/15/2018 10:15	464	9/16/2018 5:00	388
9/15/2018 10:30	460	9/16/2018 5:15	388
9/15/2018 10:45	464	9/16/2018 5:30	388
9/15/2018 11:00	460	9/16/2018 5:45	384
9/15/2018 11:15	460	9/16/2018 6:00	384
9/15/2018 11:30	460	9/16/2018 6:15	384
9/15/2018 11:45	460	9/16/2018 6:30	384
9/15/2018 12:00	455	9/16/2018 6:45	384
9/15/2018 12:15	460	9/16/2018 7:00	384
9/15/2018 12:30	455	9/16/2018 7:15	384
9/15/2018 12:45	455	9/16/2018 7:30	380
9/15/2018 13:00	451	9/16/2018 7:45	384
9/15/2018 13:15	455	9/16/2018 8:00	380
9/15/2018 13:30	451	9/16/2018 8:15	380
9/15/2018 13:45	447	9/16/2018 8:30	380
9/15/2018 14:00	451	9/16/2018 8:45	380
9/15/2018 14:15	451	9/16/2018 9:00	380
9/15/2018 14:30	447	9/16/2018 9:15	376
9/15/2018 14:45	447	9/16/2018 9:30	376
9/15/2018 15:00	447	9/16/2018 9:45	376
9/15/2018 15:15	447	9/16/2018 10:00	376
9/15/2018 15:30	442	9/16/2018 10:15	376
9/15/2018 15:45	442	9/16/2018 10:30	376
9/15/2018 16:00	442	9/16/2018 10:45	373
9/15/2018 16:15	438	9/16/2018 11:00	373
9/15/2018 16:30	438	9/16/2018 11:15	373
9/15/2018 16:45	438	9/16/2018 11:30	373
9/15/2018 17:00	438	9/16/2018 11:45	373
9/15/2018 17:15	434	9/16/2018 12:00	369

Table 8.2 (cont.)

Date/Time	Flow (ft^3/s)	Date/time	Flow (ft^3/s)
9/15/2018 17:30	434	9/16/2018 12:15	369
9/15/2018 17:45	434	9/16/2018 12:30	369
9/15/2018 18:00	434	9/16/2018 12:45	369
9/15/2018 18:15	430	9/16/2018 13:00	369
9/15/2018 18:30	430	9/16/2018 13:15	369
9/15/2018 18:45	430	9/16/2018 13:30	365
9/15/2018 19:00	430	9/16/2018 13:45	369
9/15/2018 19:15	425	9/16/2018 14:00	369
9/15/2018 19:30	425	9/16/2018 14:15	365
9/15/2018 19:45	425	9/16/2018 14:30	365
9/15/2018 20:00	425	9/16/2018 14:45	365
9/15/2018 20:15	421	9/16/2018 15:00	365
9/15/2018 20:30	421	9/16/2018 15:15	361
9/15/2018 20:45	421	9/16/2018 15:30	361
9/15/2018 21:00	417	9/16/2018 15:45	361
9/15/2018 21:15	417	9/16/2018 16:00	361
9/15/2018 21:30	417	9/16/2018 16:15	361
9/15/2018 21:45	417	9/16/2018 16:30	357
9/15/2018 22:00	413	9/16/2018 16:45	361
9/15/2018 22:15	413	9/16/2018 17:00	357
9/15/2018 22:30	413	9/16/2018 17:15	357
9/15/2018 22:45	413	9/16/2018 17:30	357
9/15/2018 23:00	409	9/16/2018 17:45	357
9/15/2018 23:15	409	9/16/2018 18:00	353
9/15/2018 23:30	409	9/16/2018 18:15	353
9/15/2018 23:45	409	9/16/2018 18:30	353
		9/16/2018 18:45	353
		9/16/2018 19:00	350
		9/16/2018 19:15	350
		9/16/2018 19:30	350
		9/16/2018 19:45	350
		9/16/2018 20:00	350
		9/16/2018 20:15	346
		9/16/2018 20:30	346
		9/16/2018 20:45	346
		9/16/2018 21:00	346
		9/16/2018 21:15	342
		9/16/2018 21:30	342
		9/16/2018 21:45	342
		9/16/2018 22:00	342
		9/16/2018 22:15	342
		9/16/2018 22:30	338
		9/16/2018 22:45	338
		9/16/2018 23:00	338
		9/16/2018 23:15	338
		9/16/2018 23:30	338
		9/16/2018 23:45	335

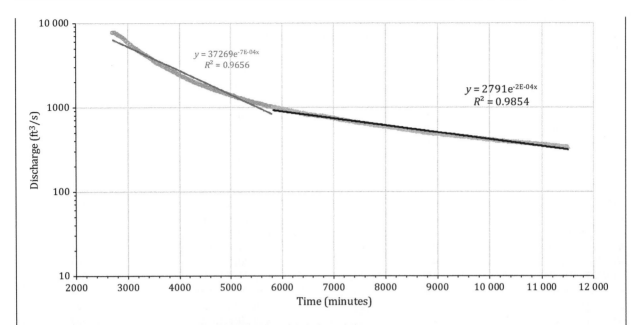

Figure 8.9 Semilogarithmic plot of the recession limb and baseflow of the hydrograph data.

Example 8.2: Use the digital filter method to determine the baseflow component for the hydrograph data given in Example 8.1.

Solution: First a trial value of the digital filter parameter a_f is selected and direct runoff is computed using

$$QR_{t+1} = \begin{cases} a_f + \left(\dfrac{1+a_f}{2}\right)[Q_{t+1} - Q_t] \\ 0 \text{ if } QR_{t+1} < 0 \\ Q_{t+1} \text{ if } QR_{t+1} > Q_{t+1} \end{cases}$$

$$QB_{t+1} = Q_{t+1} - QR_{t+1}$$

The value of a_f is adjusted so that the computed baseflow equals the end of the recession flow. The final value of a_f selected is 0.99. Figures 8.10 and 8.11 show the baseflow and streamflow components separately and Table 8.3 gives the results of calculations. The results exhibited in Figure 8.10, clearly show that baseflow is an important component of streamflow. In many hydrologic simulations, baseflow is often ignored but an attempt should be made to estimate this component of streamflow.

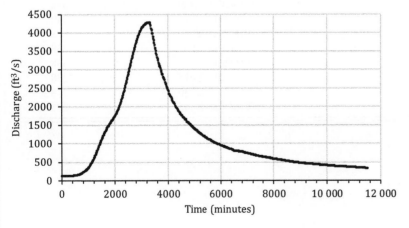

Figure 8.10 Baseflow component of flood hydrograph, QB_t.

Table 8.3 *Calculation of baseflow separation using the digital filter method*

Time (minutes)	Q_t (f³/s) (streamflow)	QR_t (f³/s) (direct runoff)	QB_t (f³/s) (baseflow)
0	129	0	129
15	129	0	129
30	131	1.99	129.01
45	131	1.9701	129.0299
60	133	3.940399	129.0596
75	135	5.890995	129.109
90	135	5.832085	129.16791
105	138	8.758764	129.24124
120	140	10.66118	129.33882
135	142	12.54456	129.45544
150	144	14.40912	129.59088
165	146	16.25503	129.74497
180	149	19.07748	129.92252
195	151	20.8767	130.1233
210	156	25.64294	130.35706
225	158	27.37651	130.62349
240	160	29.09274	130.90726
255	163	31.78681	131.21319
270	165	33.45895	131.54105
285	170	38.09936	131.90064
300	175	42.69336	132.30664
315	178	45.25143	132.74857
330	183	49.77391	133.22609
345	190	56.24118	133.75882
360	198	63.63876	134.36124
375	207	71.95738	135.04262
390	215	79.1978	135.8022
405	226	89.35082	136.64918
420	235	97.41232	137.58768
435	244	105.3932	138.60681
450	257	117.2743	139.72574
465	266	125.0565	140.94348
480	276	133.756	142.24405
495	290	146.3484	143.65161
510	305	159.8099	145.19009
525	327	180.1018	146.89819
540	346	197.2058	148.79421
555	375	224.0887	150.91127
570	404	250.7028	153.29715
585	435	279.0408	155.95918
600	472	313.0654	158.93459
615	501	338.7898	162.21024
630	541	375.2019	165.79814
645	572	402.2948	169.70516
660	609	435.0869	173.91311
675	648	469.541	178.45898
690	671	487.7306	183.26939
705	677	488.8233	188.17669
720	725	531.6951	193.30493
735	743	544.2881	198.71188
750	791	586.6052	204.39476
765	835	624.5192	210.48081
780	874	657.079	216.921
795	914	690.3082	223.69179
810	946	715.2451	230.75487
825	978	739.9327	238.06733
840	1030	784.2733	245.72665
855	1070	816.2306	253.76939
870	1100	837.9183	262.08169
885	1160	889.2391	270.76088
900	1200	920.1467	279.85327
915	1250	960.6953	289.30473
930	1320	1020.738	299.26169
945	1370	1060.281	309.71907
960	1450	1129.278	320.72188
975	1510	1177.685	332.31466
990	1590	1245.508	344.49151
1005	1670	1312.653	357.3466
1020	1740	1369.177	370.82313
1035	1850	1464.935	385.0649
1050	1930	1529.886	400.11425
1065	2040	1624.037	415.96311
1080	2130	1697.347	432.65348
1095	2250	1799.773	450.22694
1110	2350	1881.275	468.72467
1125	2450	1961.963	488.03743
1140	2570	2061.743	508.25705
1155	2680	2150.576	529.42448
1170	2770	2218.62	551.38024
1185	2840	2266.084	573.91644
1200	2920	2323.023	596.97727
1215	3000	2379.393	620.6075
1230	3100	2455.099	644.90142
1245	3170	2500.198	669.80241
1260	3230	2534.896	695.10439
1275	3300	2579.197	720.80334
1290	3380	2633.005	746.99531
1305	3430	2656.425	773.57535
1320	3490	2689.56	800.4396
1335	3530	2702.465	827.53521
1350	3560	2705.29	854.70985
1365	3590	2708.087	881.91275
1380	3620	2710.856	909.14363
1395	3650	2713.598	936.40219
1410	3640	2676.512	963.48817
1425	3660	2669.647	990.35329
1440	3670	2652.9	1017.0998
1455	3660	2616.421	1043.5788
1470	3670	2600.207	1069.793
1485	3650	2554.305	1095.695
1500	3660	2538.712	1121.2881
1515	3650	2503.375	1146.6252
1530	3630	2458.441	1171.559
1545	3610	2413.957	1196.0434
1560	3570	2350.017	1219.9829
1575	3540	2296.667	1243.3331
1590	3540	2273.7	1266.2998
1605	3480	2191.263	1288.7368
1620	3460	2149.451	1310.5494

Table 8.3 (*cont.*)

Time (minutes)	Q_t (f³/s) (streamflow)	QR_t (f³/s) (direct runoff)	QB_t (f³/s) (baseflow)
1635	3430	2098.106	1331.8939
1650	3390	2037.325	1352.675
1665	3350	1977.152	1372.8482
1680	3320	1927.53	1392.4697
1695	3290	1878.405	1411.595
1710	3280	1849.671	1430.3291
1725	3270	1821.224	1448.7758
1740	3210	1743.312	1466.688
1755	3200	1715.929	1484.0712
1770	3190	1688.82	1501.1805
1785	3170	1652.031	1517.9686
1800	3170	1635.511	1534.489
1815	3140	1589.306	1550.6941
1830	3130	1563.463	1566.5371
1845	3120	1537.878	1582.1218
1860	3140	1542.399	1597.6005
1875	3200	1586.675	1613.3245
1890	3230	1600.659	1629.3413
1905	3240	1594.602	1645.3979
1920	3330	1668.206	1661.7939
1935	3380	1701.274	1678.726
1950	3430	1734.011	1695.9887
1965	3480	1766.421	1713.5788
1980	3570	1838.307	1731.693
1995	3650	1899.524	1750.4761
2010	3720	1950.179	1769.8213
2025	3820	2030.177	1789.8231
2040	3640	1830.775	1809.2249
2055	3750	1921.917	1828.0826
2070	4010	2161.398	1848.6018
2085	4140	2269.134	1870.8658
2100	4250	2355.893	1894.1071
2115	4420	2501.484	1918.5161
2130	4500	2556.069	1943.9309
2145	4700	2729.508	1970.4916
2160	4750	2751.963	1998.0367
2175	5080	3052.794	2027.2063
2190	5230	3171.516	2058.4843
2205	4960	2871.151	2088.8494
2220	5030	2912.089	2117.9109
2235	5340	3191.418	2148.5818
2250	5400	3219.204	2180.796
2265	5740	3525.312	2214.688
2280	5760	3509.959	2250.0411
2295	5960	3673.859	2286.1407
2310	6110	3786.371	2323.6293
2325	6220	3857.957	2362.043
2340	6500	4097.977	2402.0226
2355	6500	4056.998	2443.0024
2370	6560	4076.128	2483.8724
2385	6670	4144.816	2525.1836
2400	6860	4292.418	2567.5818
2415	6980	4368.894	2611.106
2430	7080	4424.705	2655.2949
2445	7160	4460.058	2699.942
2460	7260	4514.957	2745.0425
2475	7280	4489.708	2790.2921
2490	7310	4474.661	2835.3392
2505	7470	4589.114	2880.8858
2520	7450	4523.323	2926.677
2535	7520	4547.74	2972.2602
2550	7520	4502.262	3017.7376
2565	7680	4616.44	3063.5602
2580	7620	4510.575	3109.4246
2595	7570	4415.72	3154.2804
2610	7540	4341.712	3198.2876
2625	7510	4268.445	3241.5547
2640	7490	4205.861	3284.1391
2655	7520	4193.652	3326.3477
2670	7590	4221.366	3368.6343
2685	7680	4268.702	3411.2979
2700	7810	4355.365	3454.6349
2715	7740	4242.161	3497.8386
2730	7750	4209.69	3540.3102
2745	7740	4157.643	3582.3571
2760	7680	4056.366	3623.6335
2775	7570	3906.353	3663.6472
2790	7490	3787.689	3702.3107
2805	7450	3710.012	3739.9876
2820	7270	3493.812	3776.1877
2835	7170	3359.374	3810.6259
2850	7090	3246.18	3843.8196
2865	7010	3134.119	3875.8814
2880	6870	2963.477	3906.5226
2895	6830	2894.043	3935.9574
2910	6670	2705.902	3964.0978
2925	6590	2599.243	3990.7568
2940	6510	2493.651	4016.3492
2955	6390	2349.314	4040.6858
2970	6160	2096.971	4063.0289
2985	6080	1996.401	4083.5986
3000	5960	1857.037	4102.9626
3015	5880	1758.867	4121.133
3030	5800	1661.678	4138.3217
3045	5680	1525.662	4154.3385
3060	5570	1400.955	4169.0451
3075	5540	1357.095	4182.9046
3090	5360	1164.424	4195.5756
3105	5310	1103.03	4206.9698
3120	5240	1022.35	4217.6501
3135	5160	932.5264	4227.4736
3150	5080	843.6011	4236.3989
3165	4940	695.8651	4244.1349
3180	4940	688.9065	4251.0935
3195	4850	592.4674	4257.5326
3210	4730	467.1427	4262.8573
3225	4720	452.5213	4267.4787
3240	4600	328.5961	4271.4039
3255	4560	285.5101	4274.4899
3270	4460	183.155	4276.845

Table 8.3 (*cont.*)

Time (minutes)	Q_t (f³/s) (streamflow)	QR_t (f³/s) (direct runoff)	QB_t (f³/s) (baseflow)
3285	4420	141.5235	4278.4765
3300	4390	110.2582	4279.7418
3315	4310	29.55565	4280.4443
3330	4230	0	4230
3345	4200	0	4200
3360	4110	0	4110
3375	4170	59.7	4110.3
3390	4080	0	4080
3405	3980	0	3980
3420	3990	9.95	3980.05
3435	3890	0	3890
3450	3840	0	3840
3465	3800	0	3800
3480	3720	0	3720
3495	3680	0	3680
3510	3590	0	3590
3525	3570	0	3570
3540	3510	0	3510
3555	3430	0	3430
3570	3400	0	3400
3585	3320	0	3320
3600	3340	19.9	3320.1
3615	3300	0	3300
3630	3250	0	3250
3645	3190	0	3190
3660	3150	0	3150
3675	3090	0	3090
3690	3090	0	3090
3705	3070	0	3070
3720	3030	0	3030
3735	2990	0	2990
3750	2940	0	2940
3765	2890	0	2890
3780	2870	0	2870
3795	2880	9.95	2870.05
3810	2800	0	2800
3825	2790	0	2790
3840	2770	0	2770
3855	2740	0	2740
3870	2720	0	2720
3885	2660	0	2660
3900	2650	0	2650
3915	2620	0	2620
3930	2600	0	2600
3945	2550	0	2550
3960	2510	0	2510
3975	2500	0	2500
3990	2460	0	2460
4005	2480	19.9	2460.1
4020	2420	0	2420
4035	2350	0	2350
4050	2340	0	2340
4065	2320	0	2320
4080	2300	0	2300
4095	2280	0	2280
4110	2270	0	2270

Table 8.3 (*cont.*)

Time (minutes)	Q_t (f³/s) (streamflow)	QR_t (f³/s) (direct runoff)	QB_t (f³/s) (baseflow)
4125	2250	0	2250
4140	2220	0	2220
4155	2190	0	2190
4170	2140	0	2140
4185	2140	0	2140
4200	2110	0	2110
4215	2110	0	2110
4230	2090	0	2090
4245	2080	0	2080
4260	2070	0	2070
4275	2020	0	2020
4290	1990	0	1990
4305	1990	0	1990
4320	1970	0	1970
4335	1970	0	1970
4350	1940	0	1940
4365	1920	0	1920
4380	1900	0	1900
4395	1880	0	1880
4410	1870	0	1870
4425	1860	0	1860
4440	1840	0	1840
4455	1810	0	1810
4470	1810	0	1810
4485	1810	0	1810
4500	1770	0	1770
4515	1770	0	1770
4530	1760	0	1760
4545	1730	0	1730
4560	1720	0	1720
4575	1710	0	1710
4590	1690	0	1690
4605	1690	0	1690
4620	1680	0	1680
4635	1650	0	1650
4650	1650	0	1650
4665	1650	0	1650
4680	1630	0	1630
4695	1590	0	1590
4710	1600	9.95	1590.05
4725	1600	9.8505	1590.1495
4740	1570	0	1570
4755	1570	0	1570
4770	1570	0	1570
4785	1550	0	1550
4800	1530	0	1530
4815	1550	19.9	1530.1
4830	1520	0	1520
4845	1510	0	1510
4860	1490	0	1490
4875	1480	0	1480
4890	1470	0	1470
4905	1460	0	1460
4920	1470	9.95	1460.05
4935	1440	0	1440
4950	1420	0	1420

Table 8.3 (cont.)

Time (minutes)	Q_t (f³/s) (streamflow)	QR_t (f³/s) (direct runoff)	QB_t (f³/s) (baseflow)
4965	1420	0	1420
4980	1400	0	1400
4995	1400	0	1400
5010	1390	0	1390
5025	1380	0	1380
5040	1360	0	1360
5055	1370	9.95	1360.05
5070	1390	29.7505	1360.2495
5085	1360	0	1360
5100	1340	0	1340
5115	1330	0	1330
5130	1320	0	1320
5145	1310	0	1310
5160	1300	0	1300
5175	1290	0	1290
5190	1290	0	1290
5205	1270	0	1270
5220	1270	0	1270
5235	1250	0	1250
5250	1260	9.95	1250.05
5265	1250	0	1250
5280	1230	0	1230
5295	1220	0	1220
5310	1230	9.95	1220.05
5325	1210	0	1210
5340	1210	0	1210
5355	1200	0	1200
5370	1200	0	1200
5385	1190	0	1190
5400	1170	0	1170
5415	1170	0	1170
5430	1170	0	1170
5445	1160	0	1160
5460	1150	0	1150
5475	1130	0	1130
5490	1130	0	1130
5505	1130	0	1130
5520	1120	0	1120
5535	1110	0	1110
5550	1120	9.95	1110.05
5565	1110	0	1110
5580	1110	0	1110
5595	1100	0	1100
5610	1090	0	1090
5625	1090	0	1090
5640	1080	0	1080
5655	1080	0	1080
5670	1060	0	1060
5685	1060	0	1060
5700	1060	0	1060
5715	1050	0	1050
5730	1030	0	1030
5745	1030	0	1030
5760	1020	0	1020
5775	1030	9.95	1020.05
5790	1020	0	1020
5805	1020	0	1020
5820	1010	0	1010
5835	1010	0	1010
5850	1010	0	1010
5865	1010	0	1010
5880	988	0	988
5895	988	0	988
5910	982	0	982
5925	982	0	982
5940	982	0	982
5955	976	0	976
5970	971	0	971
5985	971	0	971
6000	954	0	954
6015	959	4.975	954.025
6030	954	0	954
6045	942	0	942
6060	942	0	942
6075	937	0	937
6090	942	4.975	937.025
6105	937	0	937
6120	931	0	931
6135	920	0	920
6150	914	0	914
6165	909	0	909
6180	914	4.975	909.025
6195	898	0	898
6210	898	0	898
6225	898	0	898
6240	887	0	887
6255	892	4.975	887.025
6270	887	0	887
6285	871	0	871
6300	871	0	871
6315	876	4.975	871.025
6330	865	0	865
6345	860	0	860
6360	865	4.975	860.025
6375	854	0	854
6390	854	0	854
6405	854	0	854
6420	860	5.97	854.03
6435	839	0	839
6450	844	4.975	839.025
6465	828	0	828
6480	833	4.975	828.025
6495	839	10.89525	828.10475
6510	807	0	807
6525	823	15.92	807.08
6540	812	4.8158	807.1842
6555	817	9.742642	807.25736
6570	817	9.645216	807.35478
6585	812	4.573763	807.42624
6600	807	0	807
6615	807	0	807
6630	812	4.975	807.025

Table 8.3 (*cont.*)

Time (minutes)	Q_t (f³/s) (streamflow)	QR_t (f³/s) (direct runoff)	QB_t (f³/s) (baseflow)
6645	807	0	807
6660	802	0	802
6675	797	0	797
6690	786	0	786
6705	786	0	786
6720	786	0	786
6735	792	5.97	786.03
6750	786	0	786
6765	786	0	786
6780	792	5.97	786.03
6795	781	0	781
6810	786	4.975	781.025
6825	781	0	781
6840	776	0	776
6855	771	0	771
6870	771	0	771
6885	771	0	771
6900	766	0	766
6915	756	0	756
6930	756	0	756
6945	756	0	756
6960	746	0	746
6975	751	4.975	746.025
6990	746	0	746
7005	746	0	746
7020	746	0	746
7035	735	0	735
7050	735	0	735
7065	735	0	735
7080	724	0	724
7095	724	0	724
7110	724	0	724
7125	718	0	718
7140	718	0	718
7155	718	0	718
7170	712	0	712
7185	706	0	706
7200	712	5.97	706.03
7215	706	0	706
7230	701	0	701
7245	701	0	701
7260	695	0	695
7275	690	0	690
7290	690	0	690
7305	690	0	690
7320	684	0	684
7335	684	0	684
7350	684	0	684
7365	678	0	678
7380	678	0	678
7395	673	0	673
7410	673	0	673
7425	667	0	667
7440	662	0	662
7455	667	4.975	662.025
7470	662	0	662

Table 8.3 (*cont.*)

Time (minutes)	Q_t (f³/s) (streamflow)	QR_t (f³/s) (direct runoff)	QB_t (f³/s) (baseflow)
7485	656	0	656
7500	656	0	656
7515	651	0	651
7530	651	0	651
7545	651	0	651
7560	651	0	651
7575	646	0	646
7590	646	0	646
7605	640	0	640
7620	646	5.97	640.03
7635	640	0	640
7650	640	0	640
7665	635	0	635
7680	630	0	630
7695	630	0	630
7710	630	0	630
7725	624	0	624
7740	624	0	624
7755	624	0	624
7770	624	0	624
7785	624	0	624
7800	614	0	614
7815	614	0	614
7830	614	0	614
7845	614	0	614
7860	614	0	614
7875	609	0	609
7890	603	0	603
7905	609	5.97	603.03
7920	598	0	598
7935	603	4.975	598.025
7950	598	0	598
7965	593	0	593
7980	598	4.975	593.025
7995	593	0	593
8010	593	0	593
8025	593	0	593
8040	588	0	588
8055	588	0	588
8070	583	0	583
8085	583	0	583
8100	583	0	583
8115	578	0	578
8130	573	0	573
8145	573	0	573
8160	573	0	573
8175	568	0	568
8190	568	0	568
8205	568	0	568
8220	568	0	568
8235	568	0	568
8250	563	0	563
8265	558	0	558
8280	558	0	558
8295	553	0	553
8310	553	0	553

Table 8.3 (*cont.*)

Time (minutes)	Q_t (f^3/s) (streamflow)	QR_t (f^3/s) (direct runoff)	QB_t (f^3/s) (baseflow)
8325	553	0	553
8340	548	0	548
8355	548	0	548
8370	543	0	543
8385	548	4.975	543.025
8400	543	0	543
8415	543	0	543
8430	538	0	538
8445	538	0	538
8460	538	0	538
8475	534	0	534
8490	534	0	534
8505	529	0	529
8520	529	0	529
8535	524	0	524
8550	524	0	524
8565	524	0	524
8580	524	0	524
8595	519	0	519
8610	519	0	519
8625	515	0	515
8640	515	0	515
8655	510	0	510
8670	510	0	510
8685	510	0	510
8700	510	0	510
8715	505	0	505
8730	505	0	505
8745	500	0	500
8760	500	0	500
8775	500	0	500
8790	500	0	500
8805	496	0	496
8820	500	3.98	496.02
8835	496	0	496
8850	491	0	491
8865	491	0	491
8880	491	0	491
8895	491	0	491
8910	487	0	487
8925	487	0	487
8940	482	0	482
8955	487	4.975	482.025
8970	487	4.92525	482.07475
8985	482	0	482
9000	482	0	482
9015	478	0	478
9030	478	0	478
9045	478	0	478
9060	478	0	478
9075	478	0	478
9090	473	0	473
9105	473	0	473
9120	473	0	473
9135	469	0	469
9150	469	0	469

Table 8.3 (*cont.*)

Time (minutes)	Q_t (f^3/s) (streamflow)	QR_t (f^3/s) (direct runoff)	QB_t (f^3/s) (baseflow)
9165	473	3.98	469.02
9180	464	0	464
9195	469	4.975	464.025
9210	469	4.92525	464.07475
9225	464	0	464
9240	464	0	464
9255	464	0	464
9270	460	0	460
9285	464	3.98	460.02
9300	460	0	460
9315	460	0	460
9330	460	0	460
9345	460	0	460
9360	455	0	455
9375	460	4.975	455.025
9390	455	0	455
9405	455	0	455
9420	451	0	451
9435	455	3.98	451.02
9450	451	0	451
9465	447	0	447
9480	451	3.98	447.02
9495	451	3.9402	447.0598
9510	447	0	447
9525	447	0	447
9540	447	0	447
9555	447	0	447
9570	442	0	442
9585	442	0	442
9600	442	0	442
9615	438	0	438
9630	438	0	438
9645	438	0	438
9660	438	0	438
9675	434	0	434
9690	434	0	434
9705	434	0	434
9720	434	0	434
9735	430	0	430
9750	430	0	430
9765	430	0	430
9780	430	0	430
9795	425	0	425
9810	425	0	425
9825	425	0	425
9840	425	0	425
9855	421	0	421
9870	421	0	421
9885	421	0	421
9900	417	0	417
9915	417	0	417
9930	417	0	417
9945	417	0	417
9960	413	0	413
9975	413	0	413
9990	413	0	413

Table 8.3 (cont.)

Time (minutes)	Q_t (f³/s) (streamflow)	QR_t (f³/s) (direct runoff)	QB_t (f³/s) (baseflow)
10 005	413	0	413
10 020	409	0	409
10 035	409	0	409
10 050	409	0	409
10 065	409	0	409
10 080	409	0	409
10 095	405	0	405
10 110	405	0	405
10 125	405	0	405
10 140	400	0	400
10 155	400	0	400
10 170	400	0	400
10 185	400	0	400
10 200	400	0	400
10 215	396	0	396
10 230	396	0	396
10 245	396	0	396
10 260	396	0	396
10 275	396	0	396
10 290	392	0	392
10 305	392	0	392
10 320	392	0	392
10 335	392	0	392
10 350	392	0	392
10 365	388	0	388
10 380	388	0	388
10 395	388	0	388
10 410	388	0	388
10 425	384	0	384
10 440	384	0	384
10 455	384	0	384
10 470	384	0	384
10 485	384	0	384
10 500	384	0	384
10 515	384	0	384
10 530	380	0	380
10 545	384	3.98	380.02
10 560	380	0	380
10 575	380	0	380
10 590	380	0	380
10 605	380	0	380
10 620	380	0	380
10 635	376	0	376
10 650	376	0	376
10 665	376	0	376
10 680	376	0	376
10 695	376	0	376
10 710	376	0	376
10 725	373	0	373
10 740	373	0	373
10 755	373	0	373

Table 8.3 (cont.)

Time (minutes)	Q_t (f³/s) (streamflow)	QR_t (f³/s) (direct runoff)	QB_t (f³/s) (baseflow)
10 770	373	0	373
10 785	373	0	373
10 800	369	0	369
10 815	369	0	369
10 830	369	0	369
10 845	369	0	369
10 860	369	0	369
10 875	369	0	369
10 890	365	0	365
10 905	369	3.98	365.02
10 920	369	3.9402	365.0598
10 935	365	0	365
10 950	365	0	365
10 965	365	0	365
10 980	365	0	365
10 995	361	0	361
11 010	361	0	361
11 025	361	0	361
11 040	361	0	361
11 055	361	0	361
11 070	357	0	357
11 085	361	3.98	357.02
11 100	357	0	357
11 115	357	0	357
11 130	357	0	357
11 145	357	0	357
11 160	353	0	353
11 175	353	0	353
11 190	353	0	353
11 205	353	0	353
11 220	350	0	350
11 235	350	0	350
11 250	350	0	350
11 265	350	0	350
11 280	350	0	350
11 295	346	0	346
11 310	346	0	346
11 325	346	0	346
11 340	346	0	346
11 355	342	0	342
11 370	342	0	342
11 385	342	0	342
11 400	342	0	342
11 415	342	0	342
11 430	338	0	338
11 445	338	0	338
11 460	338	0	338
11 475	338	0	338
11 490	338	0	338
11 505	335	0	335

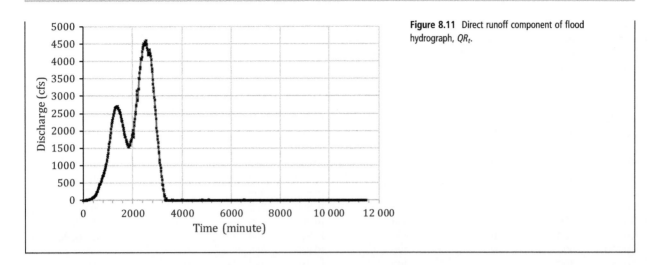

Figure 8.11 Direct runoff component of flood hydrograph, QR_t.

Exercises: A selection of exercises on this topic is available at www.cambridge.org/appliedhydrology.

9 Unit Hydrograph Models

9.1 CONCEPTS AND PURPOSE

The goal of most hydrologic engineering works is to estimate the flow rate that results from a certain rainfall event at points of interest within the drainage system or watershed under consideration. In many applications, it may be sufficient to have a good estimate of the peak flow rate that can result from a given rainfall or storm event. However, in other applications it is also necessary to know the time distribution of flow. In other words, it is necessary to develop or know the hydrograph produced by the rainfall event at the points of interest within a watershed.

Chapter 7 deals with the estimation of excess rainfall, that is, the amount of rainfall that flows over a land surface and eventually makes its way to a stream. Computations or derivations of the hydrographs that are discussed in this chapter depict the time distribution of direct or surface runoff generated by the excess rainfall over a watershed. For this reason, these are also referred to as **direct runoff hydrographs** (**DRHs**). In the literature, a DRH is also called **flood hydrograph**.

Streamflow hydrographs can be measured and, if baseflow components can be separated, they represent the DRH originating from the contributing drainage area upstream of the point of measurement. However, observed (measured) streamflow data are not always available. Even in cases where such data are available, those are not always all the data needed for design purposes. For example, DRHs may be required at points other than the point where observed streamflow data are available. Similarly, the design of hydraulic structures is typically based upon flows resulting from design storms. In that case, it is necessary to have information from DRHs resulting from the design rainfall events. For these reasons, a variety of analytical tools have been developed for the derivation of DRHs from physically based and mostly spatially aggregated (lumped) models that represent the transformation of effective rainfall to direct runoff. One such tool, the unit hydrograph method, the subject of this chapter, is the most important concept and method in applied hydrology, assuming watersheds are *linear time-invariant systems*.

9.2 RESPONSE OF A WATERSHED TO RAINFALL AS A LINEAR TIME-INVARIANT SYSTEM

The analytical procedures for the computation of DRHs that are discussed in this chapter are based on the fundamental assumption that the watersheds respond to rainfall events as linear and time-invariant systems. This assumption has been a focus of some criticisms of this approach because the relationship among watershed components is not strictly linear, as shown by several field and laboratory investigations. However, for most engineering design and applications, these procedures continue to be used and, if followed properly, they provide sufficiently accurate results for the safe design of systems or other estimation purposes. This, in turn, also suggests that the first-order features observed in the hydrographs of watersheds at the scales of engineering applications are preserved using a modeling approach adopting linear systems theory, which comprises the principle of proportionality and the principle of superposition.

The principle of superposition of a linear system, when applied to a watershed response to rainfall events, allows the superposition of responses of several different rain events to obtain the composite response of the watershed. This becomes quite handy when the unit hydrographs, discussed here, can be superimposed to derive the DRH resulting from a given pattern of rainfall excess.

The principle of proportionality of a linear system, when applied to a watershed response to rainfall events, permits the effective rainfall volumes of different magnitudes to produce watershed responses scaled up or down accordingly. In other words, the ordinates (flow rates) of the hydrographs are proportional to the volumes of rainfall excess. The principle of time invariance establishes that the hydrographs of land surface runoff resulting from a given pattern of rainfall event are constant that is, they do not change with time.

The essence of unit hydrograph theory relies upon the principles noted above, thus making the unit hydrograph an extremely flexible and powerful tool for developing **synthetic hydrographs** that do not require observed rainfall–streamflow data for their derivations.

Mathematically, the linear and time-invariant response of the watersheds to rainfall events is modeled using the theory of convolution, where the *input function* (time-distributed rainfall) is transformed through a response function (also called a transfer function) to produce the *output function* (time-distributed runoff). A mathematical function is an expression, rule, or law that defines a relationship between the independent variable or input function and the dependent variable or output function. In our present case, the transfer

9 Unit Hydrograph Models

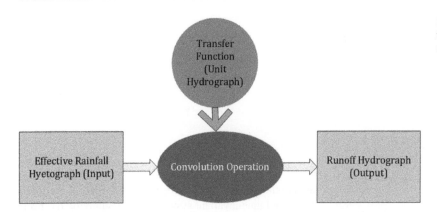

Figure 9.1 The concept of the process of transformation of an effective rainfall hyetograph (ERH) to a direct runoff hydrograph (DRH).

function is the unit hydrograph, the effective rainfall hyetograph (ERH) is the independent input function, and DRH is the dependent output function (Figure 9.1).

9.3 RESPONSE FUNCTIONS AND THE CONVOLUTION OPERATION

9.3.1 Two Views of the Time Domain

One key concept that emerges from linear systems theory is that the response function operates on the input through the mathematical operation of *convolution* to produce the output (as introduced in Chapter 2). The structure of the convolution process depends on whether the time domain of the input and output functions is *continuous* or *discrete*. In the continuous time domain, both the input and the output are continuous functions of time. This consideration facilitates the development of the basic equations from linear systems theory. These equations are then adopted in the discrete time domain, which is the time domain on which rainfall and runoff data are represented. In this case, the convolution integral becomes the **convolution summation**.

9.3.2 Response Functions in a Continuous Time Domain

To move from a continuous time domain to discrete time domain, three types of input functions and their corresponding response functions need to be considered.

The first input type, which is fundamental for the derivation of the instantaneous unit hydrograph using linear systems theory is an impulse input, which is a *Dirac delta function* (see Chapter 2) in the field of mathematics. In hydrology, a unit impulse input is a unit amount of effective rainfall that a watershed receives in an instant of time, say τ. Then, the input, I, is the **impulse function**, $I(\tau)$, and the watershed response to the impulse function is called the unit **impulse response function** (IRF) or instantaneous unit response, designated by u. Since the IRF occurs over a period of time, its value at any time t can be designated as $u(t - \tau)$, where $t - \tau$ is the time lag since the impulse was applied. Following the two principles of linear systems theory noted above, continuous input can be treated as a sum of infinitesimal impulses.

The amount of input to the system over a certain time is $I(\tau)d\tau$. Following the principle of proportionality, the direct runoff resulting from this input is the product $I(\tau)u(t - \tau)d\tau$. The response to the complete input function is obtained from the convolution integral

$$Q(t) = \int_0^t I(\tau)u(t-\tau)d\tau \qquad (9.1)$$

where $Q(t)$ is the direct runoff in a continuous time domain. The integral is due to the principle of superposition or additivity, meaning that the end result is the summation of the constituent IRFs. Note that there are three separate designations of the independent variable time, namely, τ, $t - \tau$, and t for the three dependent variables I, u, and Q, respectively, because t is a quantity at which Q is evaluated and $t - \tau$ is the lag from the time of occurrence of the impulse at time τ to t, and τ is the variable of integration in the convolution integral.

The type of input we need to consider next is called a step input, which is a continuous effective rainfall of unit intensity, starting at the origin in time $(t = 0)$. In mathematics this is known as the Heaviside function. The response function resulting from the step input, denoted by $h(t)$, is called a unit **step response function** (SRF) of the system and, from its definition, it is the convolution of 1 and $u(t)$. Thus,

$$h(t) = \int_0^t u(\tau)d\tau \qquad (9.2)$$

Therefore, the value of the unit SRF at time t equals the integral of the IRF up to that time. In other words, it is the area under the curve $u(t)$ up to t and it gives the direct runoff.

$$Q(t) = h(t) \qquad (9.3)$$

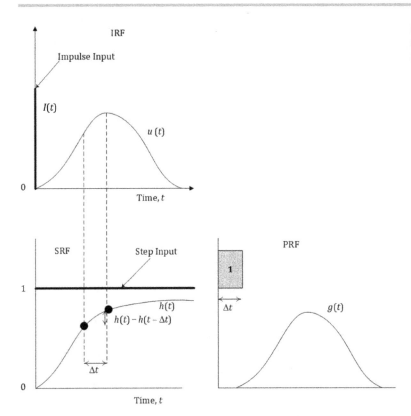

Figure 9.2 Relationships between the impulse response function (IRF), step response function (SRF), and pulse response function (PRF) in a continuous time domain.

Finally, the third type of input that we must consider is called a pulse input, which is a unit amount of rainfall occurring for a duration of Δt. Thus, the **pulse input function** is defined as

$$I(t) = \begin{cases} 1 \text{ for } 0 \leq t \leq \Delta t \\ 0, \text{otherwise} \end{cases} \quad (9.4)$$

The response function resulting from a unit pulse input, denoted by $g(t)$, is the unit **pulse response function (PRF)** of the system and it is given by

$$g(t) = \frac{1}{\Delta t}[h(t) - h(t - \Delta t)] = \frac{1}{\Delta t}\int_{t-\Delta t}^{t} u(\tau)\,d\tau \quad (9.5)$$

Thus, a PRF is the normalized difference between two SRFs, lagged by an amount Δt. This is because the response to a unit step input of rate $1/\Delta t$ beginning at time 0 is $(1/\Delta t)h(t)$. If a similar unit step input begins at time Δt, instead of 0, its response function will be lagged by time interval Δt, with a value at time t equal to $(1/\Delta t)h(t - \Delta t)$. In other words, it is the slope of the SRF curve. Figure 9.2 shows the relationship between the IRF, SRF, and PRF. The PRF is called the **unit hydrograph**.

It follows from the definitions presented above that

$$\lim_{\Delta t \to 0} g(t) = u(t) \quad (9.6)$$

The relationships between these response functions show how the concept of a unit IRF leads to a unit PRF, which leads to the formulation of a convolution sum in a discrete time domain, as presented below.

9.3.3 Response Functions in a Discrete Time Domain

For actual hydrologic applications, the convolution operation needs to be performed at discrete intervals of time. This is because rainfall data are produced as hyetographs, which represent a discrete sequence of rainfall pulses that are reduced to the same duration, Δt. On the other hand, runoff (discharge) data are represented as discrete samples at specific points in time.

If m denotes the time index for a rainfall hyetograph, the amount of rainfall for a pulse of duration Δt is given by

$$P_m = \int_{(m-1)\Delta t}^{m\Delta t} I(\tau)\,d\tau \quad m = 1, 2, \ldots, M \quad (9.7)$$

Thus, on a discrete time domain, the input function is a series of M pulses of constant amount (dimension L).

If n denotes the time index for the output, which is the direct runoff, Q, the output at the nth time interval is given by

$$Q_n = Q(n\Delta t) \quad n = 1, 2, \ldots, N \quad (9.8)$$

In other words, $Q(n\Delta t)$ is the instantaneous flow rate (dimension L^3/T, if the area is multiplied, otherwise the dimension is L/T) at the end of the nth time interval.

It should be noted at this point that the input and the output are reported with different dimensions and are with different discrete data representations. Regardless, the total depths of excess rainfall and direct runoff are equal:

$$\sum_m P_m = \sum_n Q_n \tag{9.9}$$

Now, let us define the response function, U, as

$$U_n = g(n\Delta t) \tag{9.10}$$

The time index for the response function U is $(n-m+1)$. Thus, the number of nonzero ordinates of U is: $1, 2, \ldots, (n-m+1), \ldots, (N-M+1)$.

If $(N-M+1) = N_u$, then $N = (N_u + M - 1)$ (it is worth keeping this in mind since we will see later that in applications the response function is given, and we need to find the runoff at discrete sample points).

The ordinates of the response function are given by

$$U_{n-m+1} = g[(n-m+1)\Delta t] = \frac{1}{\Delta t}\int_{(n-m)\Delta t}^{(n-m+1)\Delta t} u(\tau)\,d\tau \tag{9.11}$$

Equation 9.11 summarizes how the response function in the discrete time domain is related to the unit PRF defined in the continuous time domain.

From the definition of the PRF in the continuous time domain, the convolution sum in the discrete time domain can be derived and is given as

$$Q_n = \sum_{m=1}^{n \leq M} P_m U_{n-m+1} \tag{9.12}$$

The notation $n \leq M$ at the upper limit of the summation specifies that for each flow rate interval, n: (1) the terms are summed for $m = 1, 2, \ldots, n$ for $n \leq M$; (2) the summation is limited to $m = 1, 2, \ldots, M$ for $n > M$. Thus, the number of terms in each Q_n is n for $n \leq M$ but the number of terms in each Q_n will continue to be M for $n > M$. For example,

if $M = 4$, for $n = 3$ there are three terms for $m = 1, 2,$ and 3: $Q_3 = P_1 U_{3-1+1} + P_2 U_{3-2+1} + P_3 U_{3-3+1} = P_1 U_3 + P_2 U_2 + P_3 U_1$. If $M = 6$, there still will be only four terms in each Q_n, for $m = 1, 2, 3,$ and 4. Thus, for $n > 4$, $Q_n = P_1 U_{n-1+1} + P_2 U_{n-2+1} + P_3 U_{n-3+1} + P_4 U_{n-4+1} = P_1 U_n + P_2 U_{n-1} + P_3 U_{n-2} + P_4 U_{n-3}$. Also, note that the sum of the subscripts in each term involving the product of and U is always one greater than the subscript of Q, such as in the example given above, when subscript of Q is 3 and the sum of the subscripts of each of the PU terms is 4.

Equation 9.12 is the main **convolution equation** (or summation equation) that is used for computing direct runoff from rainfall inputs (as discrete pulses) in a watershed and the response function (as discrete sample points) of the watershed. Figure 9.3 is a graphical illustration of the discrete convolution process described by Eq. 9.12.

The dimension of the response function can be expressed in two ways. In general, the flow, Q, is represented by L^3/T (e.g., ft^3/s) and rainfall is given as L (e.g., inches). Thus, to be dimensionally correct, the ordinates of the response function need to be $L^3/T/T$ (e.g., $ft^3/s/in$, which can be read as ft^3/s flow per inch of rainfall), or, L^2/T. However, the surface runoff rate, Q, can also be expressed as the volumetric rate per unit area, $L^3/T/L^2$ (e.g., mm/h). If discharge is expressed in this fashion, the dimension of the response function is $L^3/T/L^2L$ or $1/T$.

Equation 9.12 can be represented in the form of a matrix equation. As noted above, M represents the number of effective rainfall pulses, N represents the number of direct runoff ordinates, and $N - M + 1$ is the number of nonzero ordinates so that the duration, or time base of the unit hydrograph is consistent with the time base of the effective rainfall hyetograph and the DRH. Thus, when Eq. 9.12 is expanded, we have N **linear equations** for $Q_n, n = 1, 2, \ldots, N$. These are presented in Table 9.1.

Note that once Q_M is written, further equations are also added since the number of equations will be N but the number of nonzero terms on the right-hand side will still be M. The equations presented in Table 9.1 can be represented in the form of a matrix equation given as

$$\begin{bmatrix} P_1 & 0 & 0 & & 0 & 0 & & 0 & 0 \\ P_2 & P_1 & 0 & & 0 & 0 & & 0 & 0 \\ P_3 & P_2 & P_1 & \vdots & 0 & 0 & \vdots & 0 & 0 \\ \vdots & P_3 & \vdots & & \vdots & \vdots & & & \\ P_M & P_{M-1} & P_{M-2} & \cdots & P_1 & 0 & \cdots & 0 & 0 \\ 0 & P_M & P_{M-1} & & P_2 & P_1 & & 0 & 0 \\ \vdots & \vdots & \vdots & & \vdots & \vdots & & & \\ 0 & 0 & 0 & & 0 & 0 & & P_M & P_{M-1} \\ 0 & 0 & 0 & & 0 & 0 & & 0 & P_M \end{bmatrix} \begin{bmatrix} U_1 \\ U_2 \\ U_3 \\ \vdots \\ \\ \\ \\ U_{N-M+1} \end{bmatrix} = \begin{bmatrix} Q_1 \\ Q_2 \\ Q_3 \\ \vdots \\ Q_M \\ Q_{M+1} \\ \vdots \\ Q_{N-1} \\ Q_N \end{bmatrix} \tag{9.13}$$

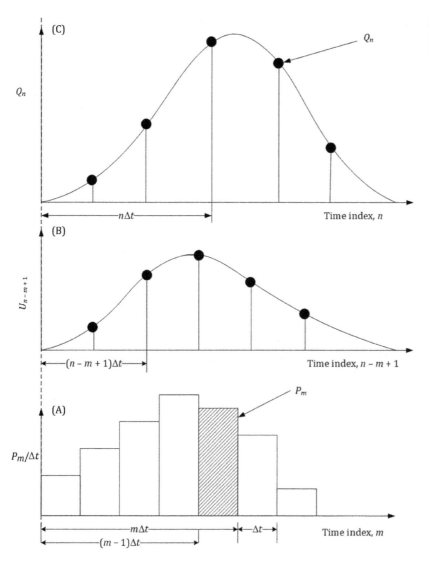

Figure 9.3 Representation of (A) excess rainfall pulses, (B) the response function, and (C) sample points of direct runoff in a discrete time domain.

Table 9.1 *Set of equations for discrete convolution of Eq. 9.12*

$Q_1 = P_1 U_1$
$Q_2 = P_2 U_1 + P_1 U_2$
$Q_3 = P_3 U_1 + P_2 U_2 + P_1 U_3$
$Q_4 = P_4 U_1 + P_3 U_2 + P_2 U_3 + P_1 U_4$
....
$Q_M = P_M U_1 + P_{M-1} U_2 + P_{M-2} U_3 + \cdots + P_1 U_M$
$Q_{M+1} = 0. U_1 + P_M U_2 + P_{M-1} U_3 + \cdots + P_2 U_M + P_1 U_{M+1}$
....
$Q_{N-1} = 0 + 0 + \cdots + 0 + 0 + \cdots + \cdots + P_M U_{N-M} + P_{M-1} U_{N-M+1}$
$Q_N = 0 + 0 + \cdots + 0 + 0 + \cdots + \cdots + 0 + P_M U_{N-M+1}$

or, in a compact matrix notation,

$$[\mathbf{P}][\mathbf{U}] = [\mathbf{Q}] \qquad (9.14)$$

where $[\mathbf{P}]$ is an $N \times (N-M+1)$ matrix, $[\mathbf{U}]$ is an $(N-M+1) \times 1$ vector, and $[\mathbf{Q}]$ is an $N \times 1$ vector. This offers a very convenient way of calculating $[\mathbf{Q}]$ from given $[\mathbf{P}]$ and $[\mathbf{U}]$.

9.4 THE UNIT HYDROGRAPH

The concept of the unit hydrograph was first proposed by Sherman (1932). The unit hydrograph is the pulse response function U described above. It is defined as a DRH resulting from 1 inch (either 1 cm or 1 mm in SI) of excess rainfall, by which we mean the rainfall that is not retained or stored in the watershed, that is distributed uniformly over the watershed at a constant rate (intensity) for an **effective duration**. The effective duration of the rainfall, also known as the **unit rainfall duration**, is the duration specified for the unit hydrograph. Thus, a unit hydrograph resulting from a 6-hour unit rainfall duration is referred to as a 6-hour unit hydrograph; so, the effective duration here is 6 hours. A unit hydrograph is the most powerful tool for the prediction of the peak flood and the flood hydrograph for gauged watersheds that have observed short-term concurrent rainfall–runoff data.

Once a unit hydrograph of a watershed is established, it can be used for any rainfall event to generate the DRH from the principles described in Section 9.3. However, there are several assumptions inherent in unit hydrograph theory. In addition to the assumptions borne in the principles of a linear and time-invariant system, the following should be kept in mind for applications of unit hydrographs.

1. Rainfall excess is treated as a lumped (basin average) quantity. Thus, the DRH is a lumped response of a watershed.
2. The duration of direct runoff, that is, the base time of the DRH resulting from an excess rainfall of given duration is constant, meaning that it is independent of the effective rainfall intensity and depends only on the effective rainfall duration.
3. The ordinates of all DRHs of a common base time are directly proportional to the total amount of direct runoff represented by each hydrograph.
4. For a given watershed, the hydrograph resulting from a given excess rainfall reflects the unchanging characteristics of the watershed.

In addition, it should also be noted that because runoff response characteristics of watersheds are not strictly linear, the unit hydrograph developed for a particular storm hyetograph should be appropriate for storms of similar characteristics. Hence, the unit hydrograph to be used with large hypothetical storms should, if possible, be derived from data for large historical events. In some cases, it is appropriate to adjust a unit hydrograph to account for anticipated shorter travel times for larger events. The duration of the precipitation excess associated with a unit hydrograph should be selected to provide an adequate definition of the DRH. Generally, a duration equal to about one-fifth to one-third of the time to peak of the unit hydrograph is appropriate (USACE, 1994). This duration should however be in resonance with design requirement.

9.5 COMPONENTS OF A DIRECT RUNOFF HYDROGRAPH

Streamflow is composed of two components, namely, direct runoff from land surfaces and baseflow from groundwater contribution. Direct runoff results from excess rainfall, which is that portion of rainfall that appears as streamflow during or shortly after a rain event. Baseflow results from subsurface runoff from prior rainfall events and delayed subsurface runoff from the current event. Thus, the total streamflow during a rainfall event includes the baseflow existing in the watershed prior to the event and the runoff due to the current rain event. For these reasons, a streamflow hydrograph during a rainfall event has two components:

1. Direct runoff, which is composed of contributions from surface runoff and quick interflow. Unit hydrograph analysis refers only to this direct runoff.
2. Baseflow, which is composed of contributions from delayed interflow and groundwater runoff.

Surface runoff includes all overland flow as well as all rain falling directly onto stream channels. It is the main contributor to peak discharge. Interflow is the portion of streamflow contributed by infiltrated water that moves laterally in the subsurface until it reaches a channel. Interflow is a slower process than surface runoff. Components of interflow are quick interflow, which contributes to direct runoff, and delayed interflow, which contributes to baseflow (Chow, 1964.) Groundwater runoff is the flow component contributed to the channel by groundwater. This process is extremely slow compared to surface runoff.

Figure 9.4 is a real streamflow hydrograph resulting from an intense rainfall event that struck the Raleigh area in North Carolina. It has (1) a rising limb representing the rising portion of the hydrograph, composed mostly of surface runoff; (2) a crest, the zone around the peak discharge; and (3) a falling or recession limb representing the portion of the hydrograph after the peak discharge, composed mostly of water released from storage in the watershed. The lower part of this recession corresponds to groundwater flow contributions.

The following time parameters describe the characteristics of a DRH typified in Figure 9.4:

Time to peak, t_p: This is the time from the beginning of the rising limb to the occurrence of peak discharge. The time to peak is largely determined by drainage characteristics, such as drainage density, slope, channel roughness, and soil infiltration characteristics. The spatial distribution pattern of rainfall also affects the time to peak.

Time of concentration, t_c: This is the time required for water to travel from the hydraulically most remote point in the watershed to the watershed outlet. For rainfall events of very long duration, the time of concentration is associated with the time required for the system to achieve the maximum or equilibrium discharge. Several empirical and physically based equations have been developed for the estimation of t_c (Chapter 6). In general, the drainage characteristics of flow path length and slope, together with the hydraulic characteristics of flow paths, determine the time of concentration. In a hydrograph, t_c is the time from the end of the rainfall to the *inflection point* on the recession limb of the hydrograph.

Lag time, t_l: This is the time interval between the center of mass of the ERH and the center of mass of the DRH. The watershed lag time is a very important concept in linear modeling of watershed response or unit hydrograph method. The lag time is a parameter that appears in theoretical and conceptual models of watershed behavior. However, in the absence of concurrent rainfall and streamflow measurements, it is often difficult to measure

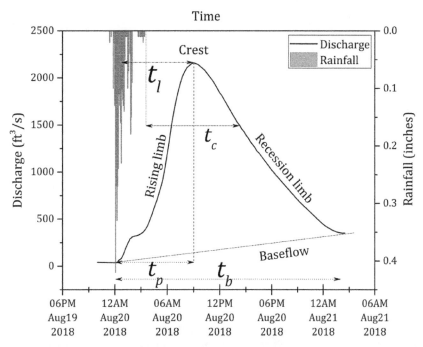

Figure 9.4 Elements of a direct runoff or flood hydrograph. The hydrograph presented in this figure is a hydrograph observed at 15-minute intervals at USGS stream gauge station 02087359 on Walnut Creek in Raleigh, North Carolina, United States. The hyetograph of the rainstorm event is constructed from rainfall data, collected at 5-minute intervals, from the USGS rain gauge station 354606078412845. Rain started at 23:25 on August 19, 2018 and stopped at 03:25 on August 20, 2018. The hydrograph started to rise at 00:15 on August 20, peaked at 09:00 on August 20, and continued to recede up to 02:15 on August 21, 2018. The rain gauge station is located near the center of the Walnut Creek watershed (also at the storm center) and the stream gauge station is approximately seven miles downstream from the rain gauge station.

in real-world situations. Many empirical equations have been proposed in the literature. The simplest of these equations is the one that gives watershed lag time as a power function of the watershed area or other measurable physical characteristics of a watershed.

Time base, t_b: This is the duration of the DRH.

9.6 FACTORS GOVERNING DIRECT RUNOFF HYDROGRAPH SHAPE

The shape of a DRH carries certain clues about the watershed characteristics. Therefore, important insights about a watershed can often be gained from an analysis of the shape of an observed DRH. The main factors that govern the shape of a DRH are as follows:

1. **Physical characteristics of a watershed**: These characteristics include the **area**, **shape** (e.g., fan or oval shape, elongated shape), and **average slope** of the watershed; the **drainage density**, **drainage network topology**, **soil type**, and **land use**, such as the **degree of urbanization** of the watershed that includes the extent of **imperviousness of the land surface**; the engineered **channelization** or **channel modifications**; and the extent of the **engineered storm drainage network system**.
2. **Rainfall characteristics**: Rainfall intensity, duration, and their spatial and temporal distribution control the shape of a DRH. In addition, storm motion also plays an important role. Storms moving in the general downstream direction tend to produce larger peak flows than storms moving in the upstream direction.

Due to the combined effect of the factors noted above, a variety of hydrograph shapes can result. Some real-world examples are cited below as illustrations.

Figure 9.5 shows two adjacent watersheds, listing of certain physical characteristics, and Figure 9.6 shows the rainfall hyetographs and stream hydrographs of a rainfall event that occurred over these two watersheds. The hydrographs show the contrasts of the characteristics of the response of these two contiguous watersheds to a rain event.

9.7 DERIVATION OF UNIT HYDROGRAPHS OF A WATERSHED FROM MEASURED RAINFALL–RUNOFF DATA: NONPARAMETRIC SYSTEM ANALYSIS

Equation 9.12 is the discrete convolution equation that can be used to compute DRH ordinates Q_n from data on rainfall excess, P_m, and unit hydrograph U_{n-m+1} for a watershed. However, the derivation of the unit hydrograph from the given data on excess rainfall (P_m) and direct runoff (Q_n) involves the reverse process called **deconvolution**. Just as integration is generally more tedious than differentiation, deconvolution is comparatively a more complex operation

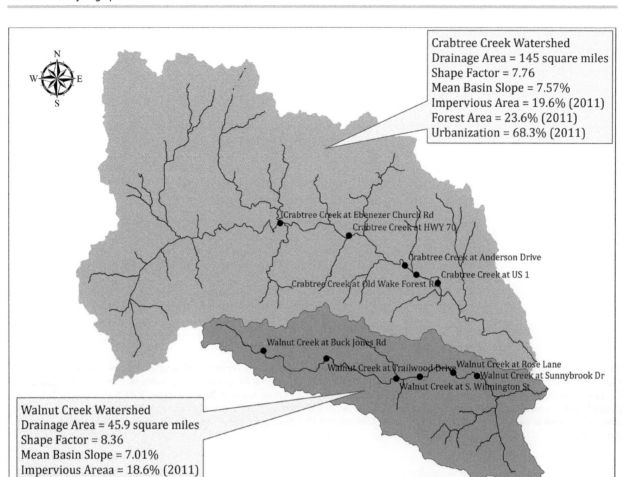

Figure 9.5 Map of Walnut Creek and Crabtree Creek watersheds that encompass the urbanized area of the cities of Raleigh and Durham in North Carolina, United States. The outlet of both these watersheds are at Neuse River. The stations shown in the map gauge both rainfall and stream runoff (USGS/City of Raleigh stations).

than convolution. Furthermore, rainfall and streamflow data come from natural systems and thereby have inherent randomness, or noise, and measurement errors. These facts further complicate a method for the derivation of the unit hydrograph of a watershed.

Several approaches have been developed to derive unit hydrographs from observed rainfall and streamflow data. These approaches can be dubbed **nonparametric system analysis**, where input and output are related in the form of a mathematical expression through the system response function. On the other hand, the next section covers the approaches that can be termed **parametric system synthesis**, where the system response is evaluated in terms of a certain number of parameters that can be estimated from the given input and output data. Both are active fields of research to strengthen the approach as a practical tool rather than to alter the tenet of unit hydrograph theory. The approaches in vogue are described in the following subsections.

9.7.1 Conventional Method

The conventional method of deriving a unit hydrograph from an observed DRH and ERH comprises four essential steps for a single, desirable, and representative rainfall event. The steps are as follows:

1. The baseflow is removed from the observed streamflow hydrograph to obtain the DRH. Typically, compared to the total direct runoff, baseflow is small. Methods for baseflow separation abound (Chapter 8).
2. The total **volume** of direct runoff is computed by calculating the area under the DRH. This area is usually calculated by a numerical integration technique such as applying the trapezoid rule. The volume is converted to **equivalent depth** of effective rainfall (in centimeters or in inches) over the entire watershed by dividing the volume by the drainage area.
3. The DRH is normalized by dividing each ordinate of the DRH by the equivalent depth (in or cm) of the total direct

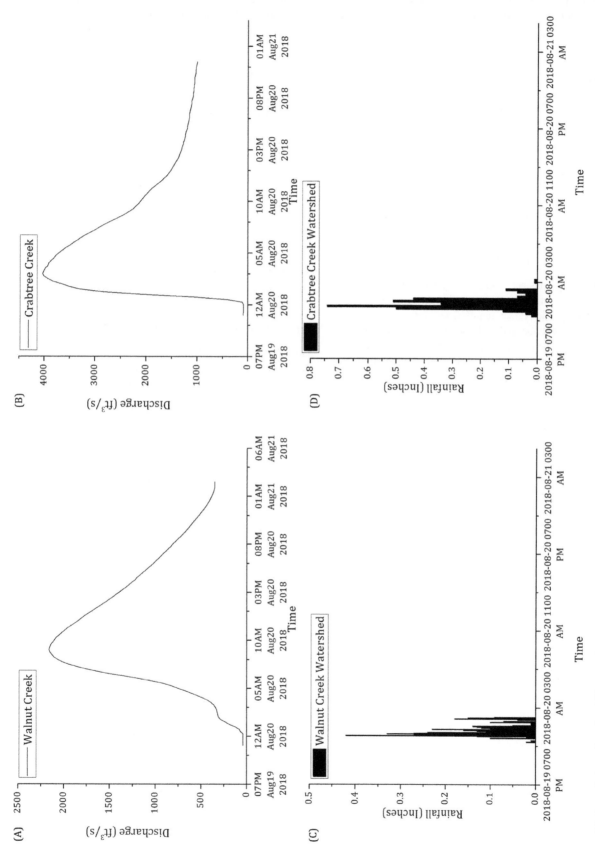

Figure 9.6 Stream hydrographs and rainfall hyetographs of a rainfall event that occurred over the entire extents of these two watersheds. (A) The stream hydrograph for Walnut Creek is from the gauging station at Sunnybrook Drive and (C) the rainfall hyetograph for Walnut Creek watershed is from Trailwood Drive (Figure 9.5). (B) The stream hydrograph for Crabtree Creek is from the gauging station at US 1 and (D) the rainfall hyetograph for Crabtree Creek watershed is from Old Wake Forest (Fig. 9.5).

runoff, that is, the effective rainfall computed in step 2. This produces a hydrograph with unit depth, that is, the unit hydrograph. The unit of the ordinates of this hydrograph is $L^3/T/L$ (e.g., $ft^3/s/in$) or $1/T$ as noted earlier.

4. The effective duration of the excess rainfall is determined from the time base of the ERH. An infiltration method, such as the Φ-index method, Green–Ampt method, or another method described in Chapter 7, is used to determine infiltration losses and associated duration. The Φ-index method usually provides a constant loss rate, which may not be the same as the actual loss rate. This loss rate is applied to the rainfall hyetograph to determine the number of m-minute pulses of excess rainfall. The actual loss rate will be the lesser quantity of the rainfall intensity or computed infiltration rate. The number of pulses of excess rainfall multiplied by m minutes is the duration of the unit hydrograph produced in step 3. This duration is the duration associated with the unit hydrograph.

Unit hydrographs are fundamentally linked to the duration of the effective rainfall event producing them. They can only be used to predict direct runoff from storms of the same duration as that associated with the unit hydrograph, or from storms that can be described as a sequence of pulses, each of the same duration as that associated with the unit hydrograph.

The method described above is used for a single storm or rainfall event. However, a problem that frequently arises in rainfall–runoff modeling is the derivation of an average unit hydrograph for a watershed or catchment from several recorded rain events. In the conventional method, several representative intense events (five to six) with approximately equal duration, producing significant amounts of direct runoff, are selected and the unit hydrograph for each event is developed. From these, an **average unit hydrograph** of that particular duration is constructed by plotting the individual hydrographs on the same graph, each beginning at the same time. The peak of the composite unit hydrograph is determined as the average of all peaks and at the average time of occurrence of all the peaks. The average unit hydrograph is then sketched to conform to the average shape of the individual hydrographs plotted on the same graph. Finally, the area under the composite curve is measured and necessary adjustments are made so that the area under this curve is satisfactorily close to one unit (in or cm). In addition to this averaging method, other methods of obtaining a watershed average unit hydrograph have also been developed and used. For example, a method of joint analysis of a group of events to determine an average unit hydrograph directly has been described by Boorman and Reed (1981). This method avoids the two-stage process of first deriving a unit hydrograph for individual events and then averaging them.

In addition, the guidelines below should be followed in the selection of the rainfall or storm events:

1. The selected rainfall events should be simple rainfall events, meaning that they should not be a sequence of events occurring together.
2. The rainfall event should have at least moderately uniform spatial distribution over the entire watershed under consideration. This in turn requires that the size of the watershed should not be too large.
3. The distribution of the rainfall amount throughout the duration of rainfall excess also should be fairly uniform.
4. The duration of the rainfall event should be approximately 10–30% of the watershed lag time.

Unit hydrographs produced from a set of selected events represent the integrated effect of events of similar magnitude and should not blindly be applied to events of very dissimilar magnitude. It should be noted that the traditional approach to the derivation of a unit hydrograph essentially relies on streamflow data and the rainfall data are only used to establish the duration of the unit hydrograph. Since in this approach the unit hydrograph is extracted from the DRH, the method can be thought of as a **backward** approach. Nevertheless, with care and computational measures taken correctly, the method produces a reliable unit hydrograph of a watershed as long as the physical conditions of the watershed remain time-invariant.

9.7.2 Method of Successive Approximation

Collins (1939) introduced a method that was used in Muskingum Watershed Conservatory District. The method employs successive approximations or trial and error to obtain a unique solution of the set of equations resulting from the expanded Eq. 9.12, given in Table 9.1. This method requires concurrent rainfall and runoff data and involves the following five steps:

1. Make a first estimate of the unit hydrograph. A constant value for unit hydrograph ordinates may be used as a first approximation.
2. Apply the first approximation of the unit hydrograph to each rainfall block, except the largest, and compute the runoff.
3. Subtract the resulting hydrograph obtained in step 2 from the actual DRH. The difference between that actual DRH and the hydrograph obtained in step 2 is assumed to be due to the omitted excess rainfall block (the largest block). From this, by proportionate adjustment, a second approximation of the unit hydrograph is obtained.
4. Compute a weighted average of the assumed unit hydrograph (step 1) and the second approximation (step 3). Use this as the revised approximation for the next trial in step 2 again. The weights are the amounts of rainfall in the largest block, omitted in step 2, and the sum of all the others, respectively.
5. Repeat the previous four steps until the method converges.

The resulting unit hydrograph may show erratic variations and even have negative values. Thus, some controls may be exercised on the method by smoothing any oscillations that may tend to occur, particularly in the later part of the unit hydrograph as the computation proceeds. Even though the iterative method has long been used for unit hydrograph derivation, in certain circumstances this method diverges and requires applications of algebraic techniques of iteration, such as the Gauss–Seidel method for convergence (Boorman and Reed, 1981).

9.7.3 Linear Algebraic Methods
9.7.3.1 Matrix Inversion

Equation 9.14 tends to indicate that, when Q_n and P_m data are available and U_{N-M+1} is required, it can be solved by finding the inverse of the matrix of $[\mathbf{P}]$, provided it is a square matrix and nonsingular. However, this is not that straightforward. Since $N > M$, it follows that $N > (N - M + 1)$. For this reason, Eq. 9.14 is overdetermined as the number of equations is greater than the number of unknowns (U-ordinates). In a very simplistic case, the equations exceeding the number $N - M + 1$ are ignored and the solution is carried forward with the number of equations equaling the number of unknowns and using the matrix multiplication given as

$$[\mathbf{U}] = [\mathbf{P}]^{-1}[\mathbf{Q}] \tag{9.15}$$

where, $[\mathbf{P}]^{-1}$ is the inverse of the matrix $[\mathbf{P}]$. However, this process usually yields some negative values of the unit hydrograph ordinates, which need to be readjusted. This method can be helpful provided the system of equations is not ill-conditioned, the data are error-free, and the number of equations is not large.

Problems in the direct solution of Eq. 9.15 may arise from the numerical problem of ill-conditioning. The coefficient matrix $[\mathbf{P}]$ is ill-conditioned when its condition number is too large. To measure it, a residual vector $[\mathbf{r}]$ is defined as

$$[\mathbf{r}] = [\mathbf{Q}] - [\mathbf{P}][\hat{\mathbf{U}}] \tag{9.16}$$

where, $[\hat{\mathbf{U}}]$ is an approximation of the exact solution of $[\mathbf{U}]$. Intuitively, when $\|\mathbf{r}\|$, any norm of $[\mathbf{r}]$ is a small number, $\|\mathbf{U} - \hat{\mathbf{U}}\|$ will also be small. However, in extreme cases this may not be true. In such cases, the following inequality can be used as an upper limit of the norm.

$$\left\|\mathbf{U} - \hat{\mathbf{U}}\right\| = \kappa(\mathbf{P}) \frac{\|\mathbf{r}\|}{\|\mathbf{P}\|} \tag{9.17}$$

where $\kappa(\mathbf{P})$ is the condition number given as $[\|\mathbf{P}\| \|\mathbf{P}^{-1}\|]$. It indicates whether $[\mathbf{P}]$ is ill-conditioned (big number) or well-conditioned (small number; condition number close to 1 is an indication of well-conditioning). In general, ill-conditioning means a small perturbation to the input will cause a great fluctuation in the output.

Because of the numerical issues described above, direct matrix inversion is not a preferred method for the derivation of unit hydrographs.

9.7.3.2 Reduction of Overdetermined Systems

The number of equations can be made equal to the number of unknowns if the system of equations is solved from two separate systems of equations. The first system consists of M equations and the matrix $[\mathbf{P}]$ is a lower triangular matrix. The second system consists of $(N - M)$ equations and the matrix $[\mathbf{P}]$ is an upper triangular matrix. This is illustrated below.

In the first system of M equations, the matrix $[\mathbf{Q}]$ starts from Q_1 to Q_M. The matrix $[\mathbf{U}]$ starts from U_1 to U_M. The matrix $[\mathbf{P}]$ is a lower triangular matrix. The main diagonal elements that equal P_1 are written and then the other parallel diagonals are written by increasing the subscript of P by 1 until the last element, P_M, is reached. Thus, the first system of equations is given by

$$\begin{bmatrix} P_1 & 0 & 0 & 0 & \cdots & 0 & 0 \\ P_2 & P_1 & 0 & 0 & & 0 & 0 \\ P_3 & P_2 & P_1 & 0 & & 0 & 0 \\ \vdots & & P_3 & & & & \\ P_{M-1} & P_{M-2} & & & & P_1 & 0 \\ P_M & P_{M-1} & & & & P_2 & P_1 \end{bmatrix} \cdot \begin{bmatrix} U_1 \\ U_2 \\ U_3 \\ \vdots \\ U_{M-1} \\ U_M \end{bmatrix} = \begin{bmatrix} P_1 \\ P_2 \\ P_3 \\ \vdots \\ P_{M-1} \\ P_M \end{bmatrix} \tag{9.18}$$

In the second system consisting of $(N - M)$ equations, the matrix $[\mathbf{Q}]$ starts from Q_{M+1} to Q_N. The matrix $[\mathbf{U}]$ starts from U_2 to U_{N-M+1}. The matrix $[\mathbf{P}]$ is an upper triangular matrix. The main diagonal elements, P_M, are written. Then the other parallel diagonals are written by decreasing the subscript of P by 1 until P_1 is reached. Zero is assigned to the other elements. Thus, the second system of equations is given as

$$\begin{bmatrix} P_M & P_{M-1} & \cdot & \cdot & \cdot & P_1 & 0 & 0 & 0 & 0 \\ 0 & P_M & P_{M-1} & & & & P_1 & 0 & 0 & 0 \\ 0 & 0 & P_M & P_{M-1} & & & & P_1 & 0 & 0 \\ 0 & 0 & 0 & P_M & P_{M-1} & & & & P_1 & 0 \\ 0 & 0 & 0 & 0 & P_M & P_{M-1} & & & & P_1 \\ 0 & 0 & 0 & 0 & 0 & P_M & P_{M-1} & & & \cdot \\ 0 & 0 & 0 & 0 & 0 & 0 & P_M & P_{M-1} & P_{M-2} & \\ 0 & 0 & 0 & 0 & 0 & 0 & 0 & P_M & P_{M-1} & \\ 0 & 0 & 0 & 0 & 0 & 0 & 0 & 0 & P_M & \end{bmatrix} \cdot \begin{bmatrix} U_2 \\ U_3 \\ \vdots \\ \\ \\ \\ \\ U_{N-M} \\ U_{N-M+1} \end{bmatrix} = \begin{bmatrix} Q_{M+1} \\ Q_{M+2} \\ \\ \\ \\ \cdot \\ \\ Q_{N-1} \\ Q_N \end{bmatrix} \tag{9.19}$$

These two systems can be solved separately. Since the matrix [U] in the second system starts with U_2, there are repeated ordinates of the derived unit hydrograph. If the repeated ordinates of the derived unit hydrograph in the two systems are different, the ordinates that ensure that the depth of the direct runoff in the unit hydrograph equals 1 inch (or 1 cm) are selected.

If the system of equations is not large, the simplification described above may produce satisfactory results. In this approach, all the available data (Q and P values) are used in the solution in contrast to the direct matrix inversion approach, where the number of equations is arbitrarily selected to be equal to the unit hydrograph ordinates and the remaining equations are neglected. This method also ensures that all the unit hydrograph ordinates are nonnegative. However, the method may not be applicable again if the matrix [P] is ill-conditioned or singular.

9.7.4 Least-Squares Method

The least-squares method, first used by Willard M. Snyder (1955), removes several of the numerical problems of the simple linear algebraic methods described above. This method obtains a set of unit hydrograph ordinates by minimizing the sum of squares of the errors. The solution of [U] is obtained from

$$[U] = \{[P]^T[P]\}^{-1}[P]^T[Q] \tag{9.20}$$

where the superscript T stands for the transpose of the matrix. The product, $[P]^T \cdot [P]$, is a square matrix, whereas the matrix [P] is typically rectangular. Equation 9.20 is obtained by multiplying both sides of Eq. 9.14 by $[P]^T$. Equation 9.20 is known as a **normal equation**.

The least-squares approach sometimes produces adequate results, particularly when the precipitation and discharge data have small errors. However, the solution by this method can be fraught with problems because of the many repeated and zero entries in [P], which create difficulties in the inversion of the matrix $[P]^T[P]$.

The application of the least-squares method to derive unit hydrograph from observed events often produces an unrealistic solution. A common problem is that the computed unit hydrograph is oscillatory or unstable and even sometimes contains negative values of the ordinates instead of a smooth curve with positive ordinate values. Various factors, such as numerical instabilities of the solution of the normal equation, can be attributed to this problem. For this reason, some modifications to the least-squares method have been proposed to ensure the stability of derived unit hydrograph. For example, the United Kingdom's Natural Environment Research Council (1975) derived unit hydrographs for more than 1400 rain events by developing a smoothing technique associated with the least-squares method.

Bruen and Dooge (1984) suggested the method of **ridge regression** for an efficient and improved estimation of physically realizable and smoothed unit hydrographs. In this procedure, the normal equation is modified as

$$[U] = \{[P]^T[P] + \lambda[I]\}^{-1}[P]^T[Q] \tag{9.21}$$

where, $\lambda(>0)$ is a parameter and [I] is a unit or identity matrix. Ridge regression is used to remove the multicollinearity that occurs when independent variables are highly correlated (Neter et al., 1990). The parameter λ is called the **biasing constant** or **regularization factor** and can be selected by a statistical method (Neter et al., 1990). This procedure may be used to derive a unit hydrograph from a single event as well as from multiple event data (Bruen and Dooge, 1984). The use of different values of λ yields a unit hydrograph without oscillations and negative ordinates.

9.7.5 Linear Programming Method

For an overdetermined system of equations, linear programming often offers good solutions. It also minimizes the sum of the absolute values of errors.

Usually, for a set of given [P] and [Q], there is no solution for [U] that will satisfy all equations given in Table 9.1. If $[\widehat{Q}]$ is an estimate, instead of Eq. 9.14, we can write

$$[P][U] = [\widehat{Q}] \tag{9.22}$$

Linear programming is a method that minimizes the absolute value of the error between [Q] and $[\widehat{Q}]$ through optimization of an **objective function** subject to simultaneously satisfying certain **linear constraints**. Two common linear constrains are

$$\Delta t \sum_{i=1}^{m} U_i = 1 \tag{9.23}$$

$$U_i \geq 0 \tag{9.24}$$

where $i = 1, 2, \ldots, m; m$ being the total number of unit hydrograph ordinates and Δt is the time interval at which the unit hydrograph ordinates are spaced.

A typical objective function, Z, can be defined as

$$Z = \min\left(\sum_{j=1}^{n} \theta 1_j + \sum_{j=1}^{n} \theta 2_j\right) \tag{9.25}$$

where $\theta 1$ and $\theta 2$ are two nonnegative slack variables that account for positive, zero, or negative differences between [Q] and $[\widehat{Q}]$, and $j = 1, 2, \ldots, n$, n being the total number of intervals or positive Q ordinates spaced at Δt. The procedure has been described by various workers (e.g., Deininger, 1969; Mays and Coles, 1980; and Zhao and Tung, 1994).

9.8 SYNTHETIC UNIT HYDROGRAPHS: PARAMETRIC SYSTEM SYNTHESIS

9.8.1 Genesis of Synthetic Unit Hydrographs

Except in occasional cases, in general, when a hydrologic investigation of a watershed is carried out, streamflow and rainfall data for that watershed are not available. Streamflow data, in particular, are unavailable for most cases. Thus, a unit hydrograph for such a watershed cannot be constructed from observed rainfall–runoff data using one of the procedures discussed in Section 9.7. Even in the rare cases where observed rainfall–runoff data are available, they may come from complex storms and thereby the data are not well suited to derive a unit hydrograph of the watershed under consideration. In addition, many hydrologic studies have a major objective to determine the effects of future conditions, such as land use change, on the way the response of the watershed to a rainfall event will change. For these determinations, there will not be any data that can be called observed data.

For the reasons listed above, several techniques have evolved to generate synthetic unit hydrographs. These are unit hydrographs that can be constructed from certain physical characteristics of a watershed and therefore do not require observed streamflow and rainfall data. The adjective *synthetic* signifies the derivation of these unit hydrographs without rainfall–runoff data. The physical characteristics of the watershed can be measured or estimated from other readily available data such as topography, land cover, and soil types. These *physical characteristics* of a watershed define certain *parameters* of a unit hydrograph that are all the pertinent properties of a unit hydrograph from which the ordinates of the unit hydrograph can be constructed.

The synthetic unit hydrographs were originally developed from observed rainfall–runoff data and, from these, the parameters of the hydrographs were related to the physical characteristics of the watersheds through empirical equations. Certain synthetic hydrographs are essentially defined by a few salient points or parameters such as peak discharge (q_p), time to peak (t_p), and time base (t_b). The remaining parts of the hydrographs are constructed from these defined points by best fitting, and thereby they are prone to some subjectivity to ensure that the area under the curve represents the unit runoff volume. The most widely used synthetic unit hydrograph procedures are discussed below.

9.8.2 Probability Density Functions

The shape of a unit hydrograph is like several probability density functions (pdfs). The unit hydrograph also has certain characteristics that are common to pdfs such as positive ordinates and unit area. Because of this, various workers attempted to derive unit hydrographs from certain types of pdfs. In this approach, the number of unknowns is generally equal to the number of parameters that are required to define a pdf and, as we have seen in Chapter 3, typically two to three parameters are required to define a pdf. The pdfs that have been used in the developments of synthetic unit hydrographs are two-parameter gamma distribution, two-parameter lognormal distribution, one-parameter chi-squared distribution, three-parameter beta distribution, and two-parameter Weibull distribution functions. Of these, the two-parameter gamma distribution function has been used widely in practical applications and the lognormal distribution function to some extent in that regard. The other pdfs mentioned have been studied mostly for academic purposes.

9.8.2.1 Two-Parameter Gamma Distribution

The use of gamma distribution with two parameters to describe a unit hydrograph has a long history in hydrologic engineering practice and can be expressed in various forms. The model developed separately by Nash (1959) and Dooge (1959) gives the following form in the context of defining an synthetic unit hydrograph with a watershed functioning as a system of cascading reservoirs of water storage:

$$q(t) = \frac{1}{\eta \Gamma(K)} \left(\frac{t}{\eta}\right)^{(K-1)} \exp\left(-\frac{t}{\eta}\right) \qquad (9.26)$$

where $q(t)$ is the depth of runoff per unit time per unit effective rainfall. The two parameters, η and K are called the scale and shape parameters, respectively, and depend on the physical as well as storage characteristics of the watershed. $\Gamma(K)$ is the complete gamma function for K, that is, $(K-1)!$. The parameter K is dimensionless, and η has the unit of time, signifying the residence time of a watershed, which can also be thought of as the storage coefficient. The area under the curve defined by Eq. 9.26 is unity.

Equations derived from the gamma distribution model of Nash (1959) and Dooge (1959), involving only the shape factor K that actually defines the shape of the hydrograph, were derived by Wu (1963) and Haan (1970). The form given by Wu (1963) is

$$\frac{q(t)}{q_p} = \left(\frac{t}{t_p}\right)^{(K-1)} [\exp\{-(K-1)\}]^{\left(\frac{t}{t_p}-1\right)} \qquad (9.27)$$

where q_p is the peak discharge from the watershed, t_p is the time to peak of the gamma distribution, and $q(t)$ is the streamflow at time t hours. Equation 9.27 can be simplified to

$$\frac{q(t)}{q_p} = \left[\frac{t}{t_p} \exp\left\{1 - \left(\frac{t}{t_p}\right)\right\}\right]^{(K-1)} \qquad (9.28)$$

Another form of the gamma distribution, also involving only the shape factor K, that can be derived from Eq. 9.26 and given by Haan (1970), is

$$\frac{q(t)}{q_p} = \left[\frac{t}{t_p} \exp\left\{1 - \left(\frac{t}{t_p}\right)\right\}\right]^{K} \qquad (9.29)$$

Equation 9.29 produces the ordinates, $q(t)$, of a gamma hydrograph. Haan (1970) defined the shape parameter K as a function of q_p and t_p, and total runoff depth, V. The parameter K defined thereby is given as

$$q_p t_p \Gamma(K) \left(\frac{e}{K}\right)^{K} = V \qquad (9.30)$$

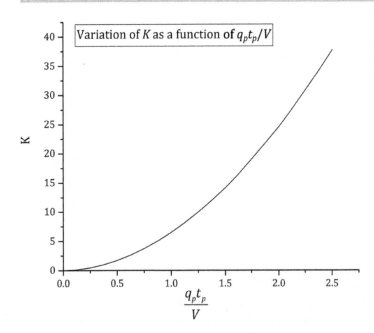

Figure 9.7 Variation of shape factor, K, of the gamma distribution function given in Eq. 9.30, as a function of the product of peak discharge depth/hour, hour (time) of peak discharge, and total runoff depth.

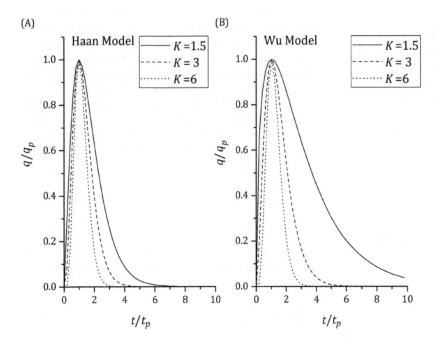

Figure 9.8 Gamma distribution curves for a set of values of shape factor K: (A) model of Haan (1970); (B) model of Wu (1963).

For a unit hydrograph, $V = 1$. Eq. 9.30 can be solved by a numerical mathematical method, such as the bisection method, to seek the root of the equation. Figure 9.7 gives an approximate solution of Eq. 9.30 as per Haan et al. (1994). It should be kept in mind that in Eq. 9.30 the unit of q_p is L/T (peak discharge depth, e.g., in/h) and that of t_p is T (e.g., h). For consistency, $q(t)$ in Eq. 9.29 should also be expressed as depth (e.g., in or mm) per hour discharge at time t.

Meadows and Blandford (1983) argued that the Wu model and the Haan model are the same. However, these two models do not produce identical distribution curves (note the difference in the exponents in these two equations). The Wu model does not work if $K = 1$ and for $1.0 \leq K < 1.5$ the recession limb of the hydrograph does not realistically represent a hydrograph. Figure 9.8 shows a comparison of the Haan and Wu models for the gamma distribution.

The equation developed by Wu (1963) was used by Meadows (2020) to develop unit hydrograph of South Carolina watersheds for South Carolina Department of Transportation. The equation developed by Haan (1970) was used by Asquith et al., (2005) to develop unit hydrographs of Texas watersheds for the Texas Department of Transportation. Meadows (2020) used frequency-based design storms, whereas Asquith et al. (2005) used observed rainfall–runoff data.

Asquith et al. (2005) developed unit hydrographs of 93 watersheds spanning an area from approximately north-northeast to south-southwest in the central part of Texas based on a large database. The database contained more than 1600 storm events with both rainfall and runoff (streamflow) data for the period 1959–1986. These data were derived from 220 historical reports of the USGS after a thorough quality control and quality assurance process and were converted into a digital database (Asquith et al., 2004). The area included the large metropolitan areas of Dallas–Fort Worth in north Texas, the City of Austin in central Texas, and the City of San Antonio in south Texas, straddling parts of four river basins: the Trinity River basin, the Brazos River basin, the Colorado River basin, and the San Antonio River basin. Thus, both urbanized (developed) and rural (undeveloped) watersheds were included in the study. The watersheds were considered small, having drainage areas less than 20 square miles. Subsequently, Asquith et al. (2011), included 24 watersheds in their study to include the other large and major metropolitan area of greater Houston with 317 databases of rainfall and runoff events from about 1965 through to about 2006.

In the study of Asquith et al. (2005, 2011), a gamma unit hydrograph was derived first by developing an analyst-directed unit hydrograph and using this with the ERH, through direct convolution, successively to arrive at the final gamma unit hydrograph that produced the observed DRH. Thus, it was a trial-and-error approach. Since in this process the unit hydrograph is extracted from the ERH, it can be thought of as a *forward* approach.

For the form of the gamma distribution given by Eq. 9.26, Singh and Chowdhury (1985) used observed hydrograph data to evaluate 12 methods, comprised of five statistical and seven mathematical methods, for the estimation of the scale parameter η and shape parameter K. From their analysis, they concluded that statistical methods, namely, the method of moments, cumulants, maximum-likelihood estimation, principle of maximum entropy, and least squares were all satisfactory for fitting a two-parameter gamma distribution to a synthetic unit hydrograph or DRH.

Bhunya et al. (2003) empirically derived the parameters K and η from 1000 sets of K, β values (β defined below) with K ranging from 1 to 40.0 and β ranging from 0.01 to 2.5. Their formulation gives

$$K = \begin{cases} 5.53\beta^{1.75} + 1.04 & \text{for } 0.01 < \beta < 0.35 \\ 6.29\beta^{1.998} + 1.157 & \text{for } \beta \geq 0.35 \end{cases} \quad (9.31)$$

where β is a dimensionless term given by

$$\beta = q_p t_p \quad (9.32)$$

where q_p is depth/hour/depth (runoff/rainfall depth) and t_p is in hours. The parameter η is directly obtained from Eq. 9.26 by setting $dq_p/dt = 0$ at $t = t_p$,

$$\eta = \frac{t_p}{(K-1)} \quad (9.33)$$

From the descriptions presented above, it should be evident that the two alternative forms of gamma distribution function (either Eq. 9.26 or Eq. 9.29) require the distribution parameters (either K and η both or only K) that can be obtained from two hydrograph parameters, namely q_p and t_p, only. Expressions and analysis of unit hydrograph in terms of only q_p and t_p are advantageous because of the critical importance of the magnitude of the peak discharge and the timing of the peak discharge for hydrologic engineering design.

The greatest practical utility of the gamma distribution function in the study of synthetic unit hydrographs comes from the fact that it can be used to define the complete hydrograph with many other synthetic unit hydrograph methods, such as the Snyder method discussed in detail later in this chapter, that provide estimates of only q_p and t_p. Methods that provide estimates of just these two parameters (sometimes also the time base, t_b) usually require manual and subjective fitting of a hydrograph through limited data points by trial and error. Such a procedure is not only tedious and cumbersome, but it also often leads to a hydrograph that does not have a unit direct runoff depth.

When an observed streamflow hydrograph is available, a gamma distribution function can be fitted to that hydrograph after the distribution parameters are evaluated from q_p and t_p, which can be obtained directly from the observed hydrograph.

9.8.2.2 Lognormal Distribution

The pdf of lognormal distribution is given by

$$q(t) = \frac{1}{t\sigma_y\sqrt{2\pi}} \exp\left[-\frac{(y-\mu_y)^2}{2\sigma_y^2}\right] \quad (9.34)$$

where $y = \log(t)$, and μ_y and σ_y, respectively, denote the mean and standard deviation of y, which are the parameters of the distribution, also known as scale and shape parameters, respectively.

9.8.3 Snyder Unit Hydrograph Model

Franklin F. Snyder (from the USGS and later from USACE) was perhaps the first person to establish a set of empirical relationships among watershed physical characteristics and the three basic parameters of a unit hydrograph. Snyder (1938) developed these relationships from a large number of gauged watersheds in the fairly mountainous Appalachian Highlands of the eastern United States. The areas of the watersheds Snyder studied ranged from 10 to 10 000 square miles (26–259 000 square kilometers).

9.8.3.1 Unit Hydrograph Parameters in the Snyder Model

In the Snyder model, there are three parameters of a *standard* unit hydrograph. These are as follows:

1. Watershed lag time, t_l: In this model, the watershed lag time (in hours) is related to the duration of the effective rainfall, t_r (in hours), through the following relationship.

$$t_l = 5.5 t_r \tag{9.35}$$

Snyder (1938) wrote in the synopsis of his paper, "The 'lag' or time from center of mass of rainfall to peak of runoff is the principal drainage-basin characteristic used in deriving the synthetic unit-graphs." Thus, for the Snyder standard unit hydrograph, if the rainfall duration is specified, the lag time can be calculated.

2. Peak discharge per unit area of the watershed, q_p: The peak rate of discharge per square mile per inch of total surface runoff is related to watershed lag time through the following relationship:

$$q_p = C_p \frac{645}{t_l} \tag{9.36}$$

where C_p is a coefficient, called the **peaking coefficient**, whose value typically ranges from 0.56 to 0.69 in Snyder's study of gauged watersheds. The factor 645 is discharge (ft^3/s) resulting from one inch of surface runoff per hour from one square mile of area (note that Snyder gave this factor as 640). As noted, the unit of q_p is $\text{ft}^3/\text{s/mi}^2$.

In SI (q_p in $\text{m}^3/\text{s/km}^2$), Eq. 9.36 becomes,

$$q_p = C_p \frac{2.78}{t_l} \tag{9.37}$$

In Eq. 9.37, the factor 2.78 is discharge (m^3/s) from 1 cm per hour of surface runoff from one square kilometer of area.

3. Duration of surface runoff or time base of the unit hydrograph, t_b: The empirical relationship Snyder used to relate the time base (in days) of a unit hydrograph to watershed lag time is given as

$$t_b = 3 + 3\left(\frac{t_l}{24}\right) \tag{9.38}$$

Physical Characteristics of the Watershed in the Snyder Model

Like three parameters of the unit hydrograph, the Snyder model requires three physical characteristics of a watershed. These are as follows.

1. Watershed area, A (mi^2 or km^2).
2. Length of the main stream of the watershed, L (mi or km); it is the distance from the outlet (or discharge station) to the upstream limit of the watershed.
3. Distance from the watershed outlet to a point on the main stream nearest to the center (centroid) of the area of the watershed, L_c (mi or km).

9.8.3.2 Relationships between Unit Hydrograph Parameters and Watershed Characteristics in the Snyder Model

The Snyder unit hydrograph parameters are derived from the watershed characteristics from the following relationships. The watershed lag time is estimated as

$$t_l = C_t (L \times L_c)^{0.3} \tag{9.39}$$

where C_t is a coefficient (dimensionless) whose value ranges from 1.8 to 2.2 in the gauged watersheds studied by Snyder. Equation 9.39 is for US customary units (L and L_c measured in miles). For SI (L and L_c measured in km), Chow et al. (1988) gave

$$t_l = 0.75 C_t (L \times L_c)^{0.3} \tag{9.40}$$

Snyder noted that larger values of C_p were associated with smaller values of C_t.

Unit hydrograph peak is calculated from

$$U_p = q_p A \tag{9.41}$$

where, U_p is the peak rate of discharge of the standard unit hydrograph in either ft^3/s or m^3/s.

The watershed lag time, calculated from either Eq. 9.39 or Eq. 9.40 is good for duration of rainfall excess that satisfies Eq. 9.35, i.e., $t_r = \frac{t_l}{5.5}$. This is the **Snyder standard unit hydrograph**. However, for other durations of rainfall excess that are significantly different from that specified by Eq. 9.35, the watershed lag time must be adjusted according to the following equation (USACE, 1959):

$$t_{lR} = t_l + \frac{t_R - t_r}{4} \tag{9.42}$$

where, t_{lR} is the *required* lag time (hours), t_R is the *required* unit hydrograph duration (hours), t_l and t_r are as defined above, respectively, denoting lag time and duration of a standard unit hydrograph. The revised (adjusted) lag time is used in Eq. 9.36 (or 9.37) to obtain the peak discharge rate per unit area of the *required* hydrograph, q_{pR} as

$$q_{pR} = C_p \frac{C}{t_{lR}} \tag{9.43}$$

where C is either 645 or 2.78 depending on the unit system in use. From Eqs. 9.36 and 9.37 we can write,

$$q_p t_l = C C_p \tag{9.44}$$

Substitution of Eq. 9.44 into Eq. 9.43 yields

$$q_{pR} = \frac{q_p t_l}{t_{lR}} \tag{9.45}$$

The peak rate of discharge of the required unit hydrograph U_{pR} (ft³/s or m³/s) is then

$$U_{pR} = C_p \frac{C}{t_{lR}} A = q_{pR} A \qquad (9.46)$$

The time at which the peak of the unit hydrograph occurs since the initiation of the rainfall excess is

$$t_{pR} = \frac{t_R}{2} + t_{lR} \qquad (9.47)$$

Assuming a triangular shape of the unit hydrograph, the time base (in hours) is approximated as

$$t_{bR} = \frac{2C}{q_{pR}} \qquad (9.48)$$

9.8.3.3 Additional Points for Defining the Ordinates of a Snyder Unit Hydrograph

From the three parameters of a unit hydrograph defined above, a number of curves passing through these three known characteristic points can be sketched satisfying the condition that the area under the curve is unity. To overcome this ambiguity associated with the Snyder model, four additional points to give the widths of a unit hydrograph at 50% and 75% of Q_p, namely, W_{50} and W_{75} respectively, were empirically defined by USACE (1959).

$$W_{50} = \frac{830}{q_p^{1.1}} \qquad (9.49)$$

$$W_{75} = \frac{470}{q_p^{1.1}} \qquad (9.50)$$

where W_{50} and W_{75} are in units of hours. The widths are distributed one-third before the occurrence of Q_p and two-thirds after the occurrence of Q_p (Chow et al., 1988).

With the seven points defined on a unit hydrograph, a smooth curve can be fitted with less ambiguity, which has an underlying area consistent with a unit depth over the watershed area.

Aron and White (1982) described a procedure to construct a complete unit hydrograph in the form of a gamma distribution function, similar to that given by Eq. 9.26, that could be derived from q_p and t_l calculated in the Snyder model. Since the gamma distribution function always guarantees a unit area under the curve, a unit hydrograph constructed using this method removes the ambiguity that is associated with using the Snyder model in which only three points of the unit hydrograph are directly obtained. In the method suggested by Aron and White (1982), the parameter of the gamma distribution function is the shape factor, K. Collins (1983) and Aron and White (1983) further provided an empirical formulation of a function ϕ, which in turn can be calculated from the watershed area and t_l, to calculate K. As a matter of fact, for the same reasons, either Eq. 9.29 or Eq. 9.30 can be used to construct the unit hydrograph using q_p and t_l obtained in the Snyder model.

9.8.3.4 HEC-HMS Procedure to Define a Snyder Unit Hydrograph

For computation of a standard Snyder unit hydrograph, HEC-HMS does not use all the relationships given above. The program requires the input of the peaking coefficient (C_p) and watershed lag time (t_l), called standard lag, as two input parameters. The program sets t_{lR} of Eq. 9.42 equal to the specified time interval and solves Eq. 9.42 to find the lag of the required unit hydrograph. Then Eq. 9.43 is solved to find the peak of the unit hydrograph. HEC-HMS then uses the computed unit hydrograph peak and time of peak to find an equivalent unit hydrograph with the Clark unit hydrograph model (discussed below). From that, it determines the time base and all ordinates of the unit hydrograph other than the unit hydrograph peak.

9.8.3.5 Estimation of Model Parameters C_p and C_t

The values of the two coefficients used in the Snyder model vary widely according to the geographic location of the watershed of interest. The coefficients should be calibrated with data from the region in which they will be applied. In general, the coefficient C_t accounts for the wide variation in topography, from flat plains to mountainous regions. The coefficient C_p accounts for flood wave conditions and watershed storage or retention capacity. C_p is a function of the lag time, duration of runoff producing rain, effective area contributing to the flow, and total drainage area. For these reasons several studies, starting from the early study conducted by the USAC (1959), have shown a wide range of values these two coefficients can assume. For example, the value of C_t can range from 0.2 to 2.2 and value of C_p, from 0.4 to 1.2.

The best means available to estimate the model parameters for a watershed is to use a unit hydrograph of the watershed that has been *derived* from observed rainfall and direct runoff. From the derived unit hydrograph of the watershed, values of its associated effective duration, t_R, its lag time, t_{lR}, and its peak discharge, q_{pR}, are obtained. Then this information, along with measured values of L and L_c from the map of the watershed can be used to estimate C_t and C_p in a reverse manner.

When the physical characteristics of an ungauged watershed are similar to a gauged watershed from where C_t and C_p values can be obtained from a derived hydrograph, these parameters can be used to obtain the required unit hydrograph.

Regionalization of Snyder Unit Hydrograph Parameters

> **Case Study 9.1** Regionalized Snyder unit hydrograph parameters from observed rainfall and streamflow data: Dallas–Fort Worth, Austin, and San Antonio areas, Texas, United States
>
> Many regional studies have been published that relate t_l to various physical characteristics of a watershed. Historically, the Snyder model has been adopted as the preferred rainfall–runoff transformation method by the various districts of USACE for hydrologic studies related to flood control engineering, flood risk management, and other water resources management projects. For regional use, different district offices of USACE developed empirical equations relating watershed lag time to other physical properties of the watershed for that region.
>
> The Los Angeles district of USACE uses the following equation, given by Linsley et al. (1982), to estimate the lag time.
>
> $$t_l = C_t \left(\frac{LL_c}{\sqrt{S}}\right)^N \tag{9.51}$$
>
> where S is the overall slope of the longest watercourse from the point of concentration to the boundary of the watershed and N is an exponent commonly taken as 0.38. If known values of the watershed lag time are plotted against $\frac{LL_c}{\sqrt{S}}$, on a log–log diagram, the resulting plots should define a straight line for watersheds of similar characteristics.
>
> In the 1970s, the USACE Fort Worth district initiated a regional study covering the metropolitan areas of Dallas–Fort Worth, Austin, and San Antonio to develop a quantitative procedure for estimating the Snyder unit hydrograph parameters by accounting for the effects of urbanization. Thomas L. Nelson (1970) and Paul K. Rodman (1977) attempted to relate t_l to $\frac{LL_c}{\sqrt{S}}$ for varying percentages of urbanization. The parameter $\frac{LL_c}{\sqrt{S}}$ is called shape factor, in which $\overline{S}$ the weighted slope of the main stream in a watershed over the distance from $0.1L$ to $0.85L$ above the discharge station.
>
> Rodman (1977) used observed rainfall–runoff data from 26 stations in the Dallas–Fort Worth area, 12 stations in the San Antonio area, and 3 stations in the Austin area. The gauging stations and the regional 8-digit HUC basins around the Dallas–Fort Worth metroplex, used in that study, are shown in Figure 9.9. He analyzed 47 storm events in the Dallas–Fort Worth area (73 rainfall–runoff models), 13 storm events in the San Antonio area (25 rainfall–runoff models), and 5 storm events in the Austin area (6 rainfall–runoff models). The number of rainfall–runoff models included the same storm events recorded at more than one observation point. Rodman computed the flood hydrographs (DRHs) for each of these events using the Snyder model and calibrated the unit hydrograph parameters with the observed flood hydrographs obtained from each of the gauging stations. In addition to the factor of urbanization, Rodman also considered the predominant soil type, namely clayey and sandy soils, which have contrastingly opposite infiltration capacity, in the watersheds from where rainfall and runoff data were collected. From this study, Rodman developed the so-called **urbanization curves** to be used to adjust t_l in this region. These curves provide guidance for estimating the lag time and peaking coefficient and are based on the length and slope of a watershed, percentage of urbanization estimated from land cover data, and percentage of sand values estimated from soil cover data. The equation, suggested by the USACE Fort Worth district, to estimate the lag time is
>
> $$\log(t_l) = 0.3833 \log\left(\frac{LL_c}{\sqrt{\overline{S}}}\right) + [Sa \times (\log 1.81 - \log 0.92) + \log 0.92] - \left(BW\frac{Ur}{100}\right) \tag{9.52}$$
>
> where $\overline{S}$ is the stream slope over the reach between 10% and 85% of L (feet/mile) as noted above; Sa is the percentage of sand factor as related to permeability of the soils (0% sand indicates low permeability, whereas 100% sand indicates high permeability), BW is the $\log(t_l)$ bandwidth between 0% and 100% urbanization [$= 0.266$ (log(hours))], and Ur is the percentage urbanization factor.
>
> Halff et al. (1979) conducted an independent investigation to determine the effect of urbanization on the relationship between the lag time and shape factor. They used observed rainfall–runoff data from 11 gauging stations in the Dallas–Fort Worth area and 50 storm events. In total, they analyzed 84 rainfall–runoff records, since some of the same storm events affected more than one gauging station. Of the 11 gauging stations used by Halff et al., only 3 were not included in Rodman's study. Figure 9.9, showing the locations of gauging stations used in the studies of Rodman and Halff et al. in relation to the regional watersheds, major rivers, and the streams on which the gauging stations were located, provides an idea of the regional extent of the Dallas–Fort Worth area that was intended to be covered by these workers.
>
> Halff et al. (1979) derived the Snyder unit hydrograph parameters from observed hydrographs through an optimization process using HEC-1. Subsequently, they conducted a multiple regression analysis to correlate t_l with the shape factor,

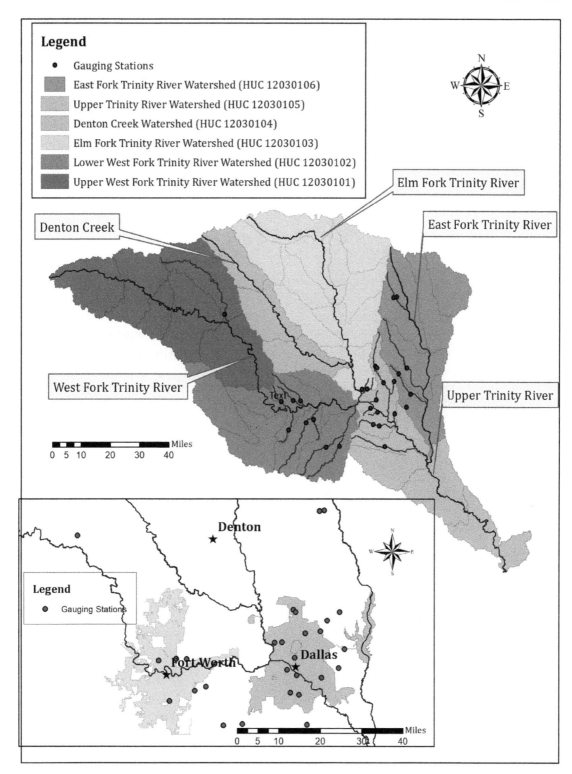

Figure 9.9 River basins and gauging stations used in the derivation of Snyder unit hydrograph parameters for Dallas–Fort Worth area.

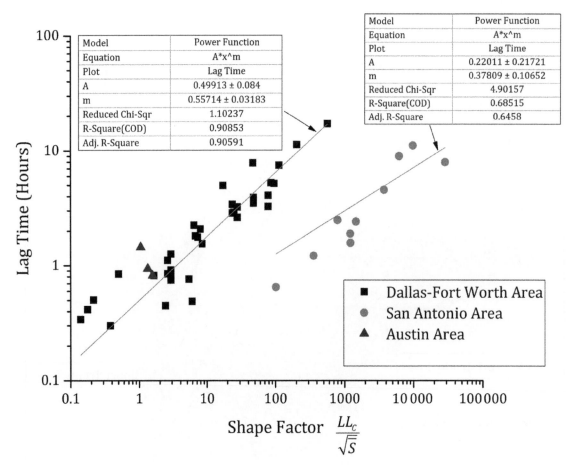

Figure 9.10 Regional relationships between watershed lag time and shape factor in parts of Texas.

percentage of urbanization of the watershed, and the percentage of channelization in the watershed. From this analysis they concluded that a strong relationship existed between t_l and shape factor, but the percentage of urbanization and the percentage of channelization did not significantly affect t_l.

Even though the study of Nelson and Rodman was perhaps the first of its kind, and detailed, the number of observations does not seem to be adequate for the large region straddling six 8-digit HUC watersheds, for the study to be conclusive. The study of Halff et al., on the other hand, contradicted the hypothesis of the effect of urbanization and channelization on t_l. Nevertheless, these studies do provide an important observation that a regional relationship between t_l and shape factor $\left(\frac{LL_c}{\sqrt{S}}\right)$ exists. Figure 9.10 shows such a relationship, expressed in the form of Eq. 9.51, for the Dallas–Fort Worth and San Antonio regions of Texas derived from the data given by both Rodman and Halff et al. In the Dallas–Fort Worth area, the value of C_t given in Figure 9.10 is 0.5 and the value of N is 0.56. The values of C_t and N derived for the San Antonio region, also shown in Figure 9.10, are 0.22 and 0.38, respectively. However, the goodness of fit for the Dallas–Fort Worth area data is better ($R^2 = 0.91$) than that for the San Antonio area data ($R^2 = 0.68$).

Another observation can be made from the results presented by Rodman. The C_p values in the Dallas–Fort Worth region varied from 0.41 to 1.01 with an average of 0.67±0.17. In the San Antonio area, the C_p values ranged from 0.63 to 1.02 with an average of 0.82±0.12. However, the C_p values do not correlate with the drainage area and lag time in either area (Figure 9.11).

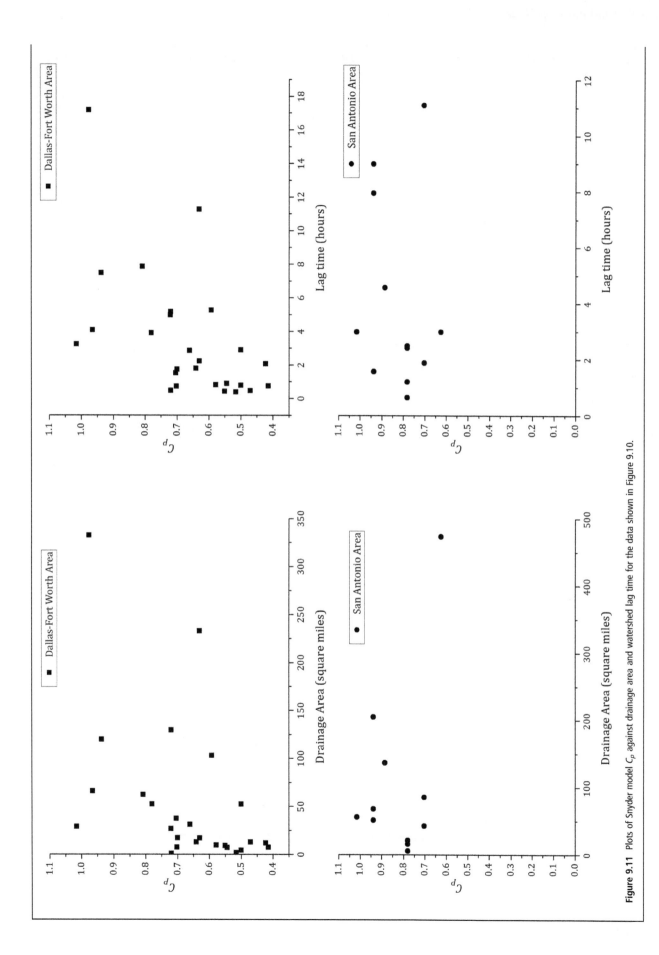

Figure 9.11 Plots of Snyder model C_p against drainage area and watershed lag time for the data shown in Figure 9.10.

Case Study 9.2 Regionalized unit hydrograph parameters following the Snyder model from observed rainfall and streamflow data: Hydrometeorological subzones of India

The Central Water Commission (CWC) has divided the entire area of India into seven major hydrologic zones, mainly based on the major river systems of the country. Each zone is further subdivided into several hydro-meteorologically homogeneous subzones, producing 26 subzones, shown in Figure 9.12. Based on rainfall–streamflow records collected

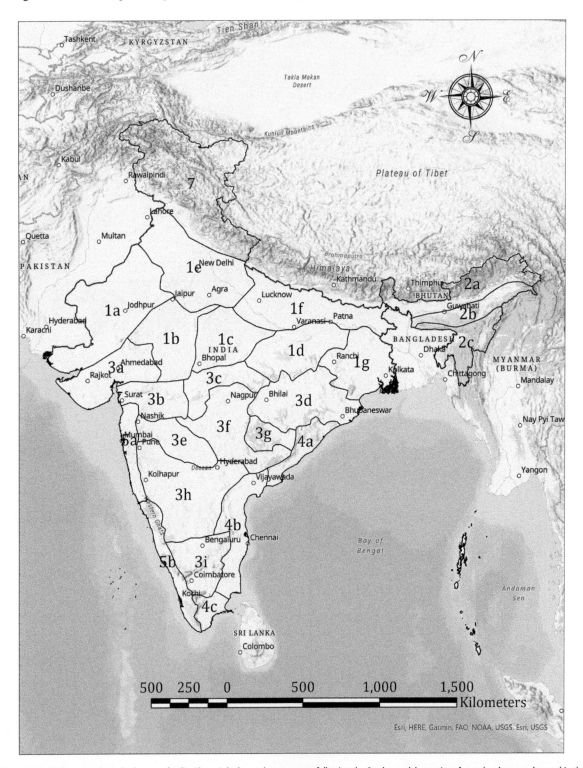

Figure 9.12 Hydrometeorological subzones of India. The unit hydrograph parameters, following the Snyder model, are given for each subzone to be used in the equations in Table 9.2. The geographic boundary of India shown in this map is reproduced from the map given in CWC (1983) and is used solely for the scientific purpose described in the text.

Table 9.2 *Equations to obtain unit hydrograph parameters of a hydrometeorological subzone of India (Figure 9.12)*

$$t_l = a_1 \left(\frac{LL_c}{\sqrt{S}}\right)^{b_1}$$

$$q_p = a_2 (t_l)^{b_2}$$

$$W_{50} = a_3 \left(q_p\right)^{b_3}$$

$$W_{75} = a_4 \left(q_p\right)^{b_4}$$

$$WR_{50} = a_5 \left(q_p\right)^{b_5}$$

$$WR_{75} = a_6 \left(q_p\right)^{b_6}$$

$$t_b = a_7 (t_l)^{b_7}$$

WR_{50} and WR_{75} are the widths of the rising limb of the unit hydrograph in hours of ordinate equal to 50% and 75% of unit hydrograph peak respectively; a_i and b_i ($i = 1 \ldots, 7$) are the coefficients specific for a subzone, and all other variables are as defined in the text.

from each of the subzones over many years, CWC derived regional unit hydrograph parameters for these different subzones. These parameters were derived following the Snyder model and were related to the physiographic characteristics of a watershed within a subzone. The general forms of the relationships are given in Table 9.2. Exercise 9.9 in the supplementary material gives an application of these equations for a flood control project in subzone 7. The exercise uses the values of the parameters a_i and b_i given by CWC.

9.8.4 The NRCS Dimensionless Unit Hydrograph

The NRCS introduced a synthetic unit hydrograph in the 1970s that was developed by Victor Mockus (1957) from analysis of small watersheds where the rainfall and streamflow were gauged (NRCS, 2007). Nowadays, it is one of the most common synthetic unit hydrograph methods widely used by engineers in both private practice and in many governmental agencies worldwide, such as the NWS and the National Operational Hydrologic and Remote Sensing Center at the NOAA.

The NRCS synthetic unit hydrograph is a special type of unit hydrograph that has certain unique characteristics, which are summarized below.

The NRCS synthetic unit hydrograph is a dimensionless curvilinear hydrograph. Dimensionless hydrographs have already been discussed in relation to the discussion of gamma distribution function. It is defined as a graph of $q(t)/q_p$ and t/t_p as ordinate and abscissa, respectively (Figure 9.13). This is also called an *index* hydrograph. NRCS (2007) noted that it was derived from many natural unit hydrographs from watersheds varying widely in size and geographic locations. The unit hydrographs were averaged, and the final product was made dimensionless. However, the derivation or justification for the shape of the curvilinear unit hydrograph has not been published by NRCS in any of the versions of their *National Engineering Handbook*. This issue was also noted by Meadows and Blandford (1983). Nevertheless, one interesting feature about this dimensionless unit hydrograph is that a gamma distribution function fits the shape of the hydrograph, even though it was developed using graphical techniques and not an equation (NRCS, 2007). NRCS (2007) gives the following form of the gamma pdf that fits the shape of the dimensionless unit hydrograph.

$$\frac{q(t)}{q_p} = \exp(K)\left[\left(\frac{t}{t_p}\right)^K\right]\left[\exp\left\{-K\left(\frac{t}{t_p}\right)\right\}\right] \quad (9.53)$$

All the terms in Eq. 9.53 are same as in Eq. 9.29 and it also has only one parameter K, the gamma equation shape factor. Equation 9.53 can be used with observed rainfall data and assuming a value of the shape factor, K, to match the observed hydrograph. Equation 9.53 produces identical curves to Eq. 9.29 even though the two forms look different. The significance of K for NRCS dimensionless unit hydrographs is discussed later in this section.

9.8.4.1 Unit Hydrograph Parameters in NRCS Model

Three parameters are required to define the NRCS dimensionless unit hydrograph.

1. Peak discharge of the unit hydrograph, U_p: The equation that is used to compute the peak discharge in this model is derived by representing the dimensionless curvilinear hydrograph by a triangular hydrograph that has the same units of time and discharge. One characteristic of this dimensionless curvilinear hydrograph is that 37.5% of the total volume of direct runoff is on the rising side. This can be seen from the mass rainfall curve in Figure 9.13. In the dimensionless form, the time to peak, t_p, represents one unit of time. Thus, if

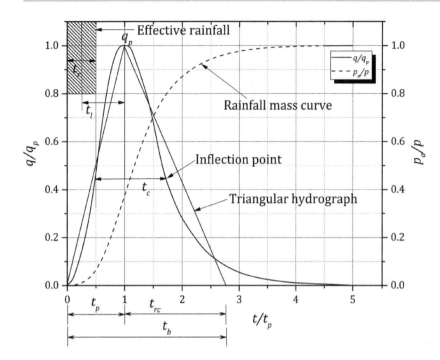

Figure 9.13 NRCS dimensionless unit hydrograph and mass curve with a triangular unit hydrograph superimposed.

one unit of time t_p represents 37.5% of the volume, the time base t_b is $1/0.375 = 2.67$ units of time. Thus, the base of the triangular hydrograph extends up to 2.67 unit of time ($t_b = 2.67 t_p$). The recession time, t_{rc} is the time from the peak to the end time. Thus, $t_b = t_p + t_{rc}$ and therefore, $t_{rc} = t_b - t_p = 2.67 - 1 = 1.67$. Also note that the ratio of the areas of the falling side of the triangle and the rising side of the triangle is $0.625/0.375 = 1.67$.

In the triangular hydrograph, the total volume is 1 inch accumulated over 1 hour. So, from the area of the triangle, the following relationships are obtained.

$$q_p \left(\frac{\text{inch}}{\text{hour}}\right) \cdot \frac{t_p + t_{rc}}{2} = 1 \text{ inch}$$

$$q_p = \frac{2}{t_p + t_{rc}}$$

To express q_p as a function of t_p, let

$$F = \frac{2 t_p}{t_p + t_{rc}}$$

Thus,

$$q_p = \frac{F}{t_p}$$

Since t_p represents one unit of time, and t_{rc} represents 1.67 unit of time,

$$F = \frac{2 \times 1}{1 + 1.67} = \frac{2}{2.67} = 0.75$$

From these, we obtain an important relationship for the dimensionless unit hydrograph which has 37.5% of volume on the rising side. That is

$$q_p = \frac{0.75}{t_p} \quad (9.54)$$

Now, to convert in/h to ft³/s/mi² we need to multiply the right-hand side of Eq. 9.54 by the factor 645.33. Thus, q_p (ft³/s/mi²) is given as

$$q_p = \frac{0.75 \times 645.33}{t_p} = \frac{484}{t_p} \quad (9.55)$$

The peak discharge rate (ft³/s) of the unit hydrograph is, therefore,

$$U_p = \frac{484 A}{t_p} \quad (9.56)$$

where the area, A, is in mi².

In SI units,

$$U_p = \frac{2.085 A}{t_p} \quad (9.57)$$

where U_p is in m³/s and area A is in km².

2. Time to peak, t_p: The general relationship involving time to peak, effective rainfall duration (t_r), and lag time of the watershed is

$$t_p = \frac{t_r}{2} + t_l \quad (9.58)$$

In the NRCS methodology, the lag time is considered as 60% of the time of concentration:

$$t_l = 0.6 t_c \quad (9.59)$$

Equation 9.59 expresses an empirical relationship, discussed in Chapter 6, in relation to the estimation of various time parameters.

Combining Eqs. 9.58 and 9.59, t_p is obtained from t_c as

$$t_p = \frac{t_r}{2} + 0.6 t_c \quad (9.60)$$

3. Duration of unit rainfall, t_r: The duration of the unit rainfall, that is, the duration of the unit hydrograph in the NRCS model is tied to the other time parameters as noted below.

In addition to the definition of the time of concentration given earlier in this chapter, in the direct runoff hydrograph, it is the time from the end of excess rain to the inflection on the recession limb of the hydrograph. Another characteristic of the dimensionless unit hydrograph, shown in Figure 9.13, is that it has a point of inflection approximately 1.7 times the time to peak, t_p. Thus,

$$t_r + t_c = 1.7 t_p \quad (9.61)$$

Substituting $t_c = 1.7 t_p - t_r$ from Eq. 9.61 into Eq. 9.58, the relationship between t_r and t_p is obtained as

$$t_r = 0.2 t_p \quad (9.62)$$

Now, substituting t_p from Eq. 9.60 into Eq. 9.62, we get the relationship between t_r and t_c as

$$t_r = 0.133 t_c \quad (9.63)$$

Finally, substituting t_c from Eq. 9.59 we get the relationship between t_r and t_l as

$$t_r = 0.222 t_l \quad (9.64)$$

The three equations presented above show that, in the NRCS dimensionless unit hydrograph, t_r, the duration of the excess rainfall or the **unit hydrograph duration**, is defined in terms of the other three time parameters, namely t_p, t_c, and t_l.

9.8.4.2 Construction of the Required Unit Hydrograph

The parameters described above are for the **standard dimensionless unit hydrograph**. The required unit hydrograph for a case in hand is calculated as follows.

The lag time of the watershed is derived by using one of the methods presented in Chapter 7 to obtain either the lag time directly or by calculating the time of concentration. From either t_l or t_c, the duration of the required hydrograph is obtained from either Eq. 9.62 or Eq. 9.63. Then either Eq. 9.58 or Eq. 9.60 is used to derive time to peak, t_p. In most modeling cases, both t_r and t_l are specified for use in Eq. 9.60. From calculated t_p, the peak of the required unit hydrograph is obtained from Eq. 9.56 (or Eq. 9.57, depending on the unit system).

The abscissa of the required unit hydrograph is obtained by multiplying the t/t_p values of the standard unit hydrograph (Figure 9.13) by the calculated t_p and the ordinates of the required unit hydrograph are obtained by multiplying the corresponding q/q_p values of the standard unit hydrograph by the calculated value of U_p. The equivalent triangular unit hydrograph is obtained by calculating the base time, t_b from the known constraint, $t_b = 2.67 t_p$.

9.8.4.3 Peak Rate Factor

The peak rate factor, designated as *PRF*, of the dimensionless unit hydrograph is the numerator of Eq. 9.55. Thus,

$$PRF = F \times C \quad (9.65)$$

where $C = 645.33$ in the US customary unit or 2.78 in SI.

For the standard dimensionless unit hydrograph, $F = 0.75$ (as shown above). Thus, for the standard case,

$$PRF = 484 \quad (9.66)$$

The *PRF* value determines the magnitude of the peak discharge of the unit hydrograph. The higher the *PRF*, the higher the peak discharge is from a watershed. As can be seen from the derivation of the value of $F(=0.75)$, for a standard dimensionless unit hydrograph, the peak rate factor essentially controls the volume of water under the rising and recession limbs of the unit hydrograph. The chief factors that control the value of *PRF* are shape and thereby the length of a watershed, watershed slope, watershed size, drainage density, stream frequency, and land use. Certain other geomorphologic and geologic characteristics of a watershed also contribute to the value of *PRF*. Several studies have shown that *PRF* can vary from less than 100 to more than 600. For example, for the watersheds in the southeastern United States covering a large area extending from Mississippi to South Carolina, including Florida where the topographic relief is low, *PRF* ranges from 174 to 476 (Sheridan et al., 2002). From this study, Sheridan et al. (2002) also concluded that of several physiographic or geomorphologic characteristics of a watershed, watershed size and main channel slope are the two major factors that influence the magnitude of

PRF. Wanielista et al. (1997) suggested *PRF* values ranging from 100 in rural very flat watersheds to 575 in urban areas with steep slopes.

Since the peak rate factor depends on a variety of physical characteristics of a watershed, it cannot be arbitrarily changed or selected without a detailed study from a gauged watershed. Since this approach is not always possible, an alternative approach can be taken based on a discussion presented by NRCS (2007). The shape factor, K, in the form of the gamma equation presented as Eq. 9.53 determines the shape of the dimensionless unit hydrographs. Depending on the different values of K chosen, Equation 9.53 produces different dimensionless unit hydrographs which have different values of the sum of the coordinates of the ordinates. From this observation, NRCS (2007) defined the peak rate factor as follows.

$$PRF = \frac{C}{\sum DUH_Y \times \Delta t} \quad (9.67)$$

where $\sum DHU_Y$ denotes the sum of the coordinates of ordinates of the dimensionless unit hydrograph, and Δt is the nondimensional time interval of the dimensionless unit hydrograph. Eq. 9.53 can be used to develop a dimensionless unit hydrograph with any desired peak rate factor. Subsequently, the observed unit hydrograph can be used to calibrate the value of the peak rate factor in the calculated dimensionless unit hydrograph.

9.8.5 Clark Unit Hydrograph Model

C. O. Clark, a hydraulic engineer from US Army Engineers Office in Winchester, Virginia, proposed a unit hydrograph method that was a significant departure from an empirical approach. Clark (1943) presented a concept that, perhaps for the first time, included the critical factor of **storage** of water in the watershed that plays a very important role in the transformation of excess precipitation to runoff. Even though storage includes both short-term storage in the soil, on the surface, and in the channels, Clark used the term **valley storage**, implying storage mostly in channels throughout a watershed. In addition, the Clark model also considered temporal distribution of runoff resulting from the spatial distribution of areas contributing to the runoff.

Clark's presentation received considerable attention from the leading hydrologists and hydraulic engineers of that time. His paper included a response to the detailed comments from 13 others, including Sherman, the proponent of the unit hydrograph theory, and Snyder, perhaps the first to propose a synthetic unit hydrograph. The form of the routing equation presented below can be traced to the discussion presented by Snyder in Clark (1945).

The Clark model derives a synthetic unit hydrograph of a watershed by conceptualizing the process into three steps: (1) developing a hydrograph that represents pure **translation** or movement of the excess rainfall from its origin throughout the watershed to the watershed outlet; (2) modifying the translated hydrograph to account for storage by **attenuation** or reduction of the runoff through calculation of an **instantaneous unit hydrograph**; and (3) converting the instantaneous unit hydrograph to a unit hydrograph *with specific duration*. Figure 9.14 illustrates the components of the Clark model.

9.8.5.1 Development of a Translation Hydrograph by the Time–Area Method

Contributions of runoff originating from different parts of a watershed arrive at the watershed outlet at different times, due to the differences in travel time of water, controlled by various factors of which the length (distance) of the flow path from the point of origin to the outlet is a main one (see Chapter 6). This fact gives rise to the **time–area method**, in which the entire area of a watershed is divided into various subareas. The travel time from each subarea is assumed to be the same and the line that separates two subareas is called an **isochrone** (Figure 9.14 and also recall the discussion related to Figure 6.20A). Thus, an isochrone represents a line of equal travel time and equal runoff at a certain time. In other words, each of the subareas between isochrones responds in the time associated with that area. The isochrones are drawn with the assumption that the travel time is directly proportional to the distance from the outlet to the isochrones.

The **time–area relationship** is defined as the fraction of the watershed area that is contributing to the discharge at the outlet as a function of time since the beginning of effective rainfall. The total time is the time of concentration of the watershed with 100% of the area being accounted for. Such a relationship for a specific watershed can only be developed from a detailed calculation of the travel time from different parts of the watershed and measurements of certain physical parameters such as slope, length, etc. This process can nowadays be greatly facilitated by using GIS technology. However, the HEC-HMS Technical Reference Manual (Feldman, 2000) provides the following relationship that is generally applicable:

$$\frac{A_c}{A_T} = \begin{cases} 1.414 \left(\frac{t}{t_c}\right)^{1.5} & \text{for } 0 \leq \frac{t}{t_c} \leq 0.5 \\ 1 - 1.414 \left(1 - \frac{t}{t_c}\right)^{1.5} & \text{for } 0.5 < \frac{t}{t_c} \leq 1.0 \end{cases} \quad (9.68)$$

where A_c is the contributing area at time t, A_T is the total area of the watershed, and t_c is the time of concentration of the

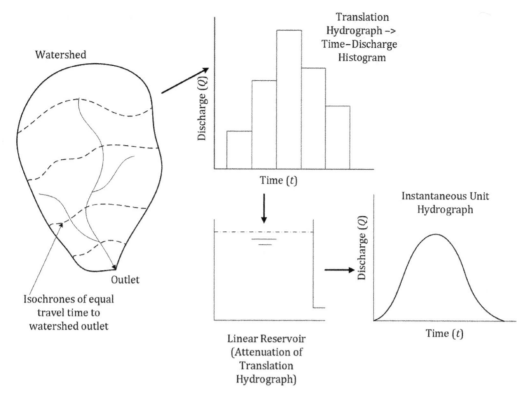

Figure 9.14 Clark model of a unit hydrograph.

watershed, which is the time when 100% of the watershed area contributes flow to the outlet. However, the time–area relationship of a watershed depends on its physical characteristics and can vary from one geographic location to another. For example, the Maricopa County Flood Control District of Arizona uses a relationship that can be expressed as

$$\frac{A_c}{A_T} = \begin{cases} 0.56\left(\dfrac{t}{t_c}\right)^2 + 0.12\left(\dfrac{t}{t_c}\right) & \text{for } 0 \le \dfrac{t}{t_c} \le 0.5 \\ -3.6\left(\dfrac{t}{t_c}\right)^2 + 7\left(\dfrac{t}{t_c}\right) & \text{for } 0.5 < \dfrac{t}{t_c} \le 1.0 \end{cases} \quad (9.69)$$

From this, the cumulative area, A_c, contributing flow to the outlet of a watershed as a function of time can be calculated. Figure 9.15 shows how the cumulative area differs as time progresses for the two different relationships given above. In the case of the relationship given in Eq. 9.68, the cumulative area will increase almost linearly in a smooth fashion with time. On the other hand, the relationship given by Eq. 9.69 indicates that for the first half of the duration of a rainfall event, the percentage of area that contributes flow at the outlet is low but after that time, it rapidly increases. Similarly, the variation of the time–area relationship that is expected to vary from an urban to natural setting of a watershed is also illustrated in Figure 9.15.

Laurenson (1964) offered a method of calculation of time–area curves from a sufficient number of points on topographic contours. This method can be implemented nowadays in GIS. Essentially, a set of points on the drainage lines defining a dense network of streams are identified. From the Manning equation it can be shown that travel time for each of these points is related to $L_i/\sqrt{S_i}$, where L_i and S_i are the length and slope of the stream segment upstream of the point. Once a travel time is assigned to each of these points, contours of equal travel time (isochrones) are constructed using one of the mathematical models of interpolation and the areas between the successive contours are calculated.

For the calculation of flow originating from each of the incremental area, the area, A_i, between two isochrones is calculated from

$$A_i = A_{c,t} - A_{c,t-1} \quad (9.70)$$

where $A_{c,t}$ is the cumulative area at time t and $A_{c,t-1}$ is the cumulative area at time $t-1$.

The incremental flow originating from each of the incremental areas between time t and $t-1$, designated as $\bar{I}$, is given by

$$\bar{I} = \frac{CA_i}{\Delta t} \quad (9.71)$$

where C is the conversion factor 645.33 for ft³/s or 2.78 for m³/s and Δt is the time increment, which is selected from

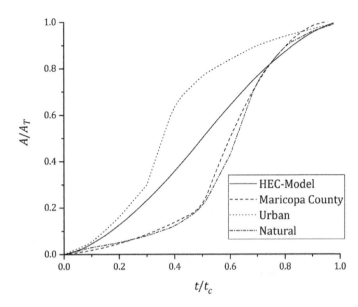

Figure 9.15 Time–area relationships showing a general relationship proposed by US Army Corps of Engineers; a relationship used by the Flood Control District of Maricopa County, Arizona, USA; and generalized representations of time–area relationship in typical urban and natural settings of a watershed.

a known or given value of t_c. From a selected Δt, the number of subareas or isochrones (N_i) for the watershed is obtained as

$$N_i = \frac{t_c}{\Delta t} \qquad (9.72)$$

Note that N_i is also the number of computational time intervals.

The incremental flow as a function of time is usually given as a bar diagram, which is the hydrograph resulting from pure translation at the outlet of the watershed.

9.8.5.2 Development of an Instantaneous Unit Hydrograph by Routing through a Linear Reservoir

Clark (1943) discussed at length the fundamental role of storage in the formation of a runoff hydrograph. He accounted for the effect of watershed storage on the translation hydrograph by routing it through a single linear reservoir, conceptually located at the watershed outlet and representing an aggregated impact of all watershed storage. A linear reservoir is a hypothetical system in which outflow from the system is directly proportional to the storage in the system. This is mathematically expressed as

$$S_t = RQ_t \qquad (9.73)$$

where S_t is the storage at time t, Q_t is the outflow at time t, and R is the constant of proportionality. This proportionality constant is called the **retardance coefficient** or, more commonly, the **storage coefficient** and it has unit of time. Now, from the principle of continuity, the rate of change of storage of the system is given as

$$\frac{dS}{dt} = I_t - Q_t \qquad (9.74)$$

where I_t is the inflow to the system at time t.

From Eq. 9.73 we can also write

$$\frac{dS}{dt} = R\frac{dQ}{dt} \qquad (9.75)$$

Therefore,

$$I_t - Q_t = R\frac{dQ}{dt} \qquad (9.76)$$

The finite difference approximation of Eq. 9.76 and simplification of the resulting algebraic equation yields

$$Q_t = C_0 \bar{I} + (1 - C_0) Q_{t-1} \qquad (9.77)$$

where C_0 is called the **routing coefficient** given by,

$$C_0 = \frac{\Delta t}{R + 0.5\Delta t} \qquad (9.78)$$

and $\bar{I}$ is the average inflow between time t and $t-1$ as defined in Eq. 9.71 and $\Delta t = t - (t-1)$ as defined above.

The ordinates of the complete **attenuated hydrograph** can then be calculated from Eq. 9.77 at incremental time Δt. The hydrograph thus calculated is an *instantaneous unit hydrograph*, which has a duration of zero hours; thereby it is abstract and needs to be converted to a unit hydrograph of finite duration.

9.8.5.3 Conversion of an Instantaneous Unit Hydrograph to a Unit Hydrograph of Required Duration

The arithmetic mean of the instantaneous discharges Q_t and Q_{t+D} yields the unit hydrograph at time t_D

9.8 Synthetic Unit Hydrographs: Parametric Synthesis

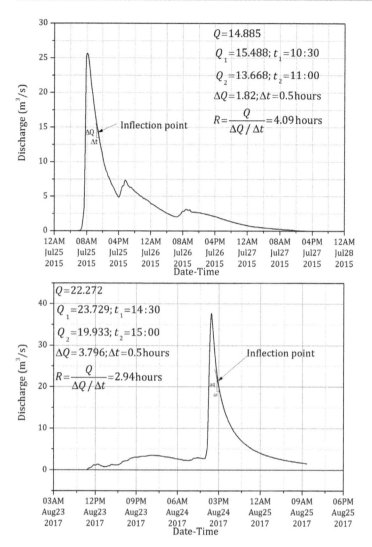

Figure 9.16 Estimation of R in the Clark model from a direct runoff hydrograph. The two hydrographs shown are from Seolmacheon Basin in Korea. The hydrographs are the result of two rain events described by Yoo et al. (2021). The average value of R estimated from five events using the method described by Yoo et al. (2013, 2019) is 3.85 hours (data courtesy of Dr. Chulsang Yoo and Dr. Wooyoung Na). Discharge at the inflection point is Q, discharge at a time before the occurrence of inflection point is Q1 and discharge at a time after the point of inflection is Q2.

$$U_D(t) = \frac{1}{2}(Q_t + Q_{t+D}) \quad (9.79)$$

where D in the subscripts denotes the desired/required unit hydrograph duration and $U_D(t)$ is the ordinate of the unit hydrograph of required duration. Equation 9.79 holds true if D is an integer multiple of the computational time interval Δt used above. The general formulation for a unit hydrograph of any duration is given as

$$U_i = \frac{1}{N}(0.5Q_{i-N} + Q_{i-N+1} + \cdots + Q_{i-1} + 0.5Q_i) \quad (9.80)$$

where U_i is the ordinate of a unit hydrograph of desired duration $D = N\Delta t$.

9.8.5.4 Estimation of Model Parameters

The development of the unit hydrograph of a watershed with the Clark model requires an estimation of two model parameters, namely t_c, the time of concentration of the watershed and R, the storage coefficient of the watershed. This coefficient is an index of the temporary storage of excess rainfall in the watershed as it drains to the outlet point and is expressed in terms of time. Clark indicated that these two parameters are interrelated.

The time of concentration, as already noted in the discussion of the NRCS dimensionless unit hydrograph, is the time interval between the end of excess rainfall and the inflection point on the recession limb of an observed (measured) direct surface runoff hydrograph. Clark (1945) noted that at this point the ratio of the discharge and the rate of change of discharge, which is the slope of the curve at the inflection point, gives the value of R. The rationale behind this estimation is that, at the point of inflection, inflow to storage has ceased and, from that time on, storage is being emptied. At the point of inflection, the continuity equation can be given as:

$$Q_{ip} = -\left(\frac{dS_{ip}}{dt}\right) \qquad (9.81)$$

where the subscript *ip* denotes the inflection point. From Eq. 9.75 we can also write

$$Q_{ip} = -R\left(\frac{dQ_{ip}}{dt}\right) \qquad (9.82)$$

From Eq. 9.82 we get

$$R = \frac{Q_{ip}}{\left(\frac{dQ_{ip}}{dt}\right)} \qquad (9.83)$$

Thus, R can be assigned the average value of $Q/(dQ/dt)$ at the point of inflection, computed from the observed flood hydrographs (Figure 9.16). Equation 9.83 can also be applied if observed short-interval flood discharges at a site are available. In such a case, if Q_1 corresponds to the discharge at the point of inflection on the recession limb of the flood hydrograph during the first event and Q_2 corresponds to the same during the second event, giving $\Delta Q = Q_2 - Q_1$, and $\Delta t = $ time interval between the two occurrences, then $R = \frac{Q_1}{\left(\frac{\Delta Q}{\Delta t}\right)}$. It should be noted that Q should be estimated after separating the baseflow. Another method for estimating R is by computing the volume remaining under the recession limb of the direct runoff hydrograph following the point of inflection and dividing this volume by the flow at the point of inflection.

Variability in rainfall events and uncertainties in the timing of recorded data may preclude, in practice, the reliable determination of t_c and R by direct measurement of the inflection point and the slope of the hydrograph. For this reason, estimates obtained using Eq. 9.83 should not be relied on too rigorously because the conceptual models are rough approximations of a natural occurrence. Yoo et al. (2013) proposed an advanced mathematical method for the derivation of both t_c and R from an observed hydrograph using the instantaneous unit hydrograph model of Nash (1957).

The interrelationship between t_c and R can be implicitly accounted for by estimating the parameters $(t_c + R)$ and $R/(t_c + R)$, rather than seeking to estimate t_c and R directly. When the best estimates of $(t_c + R)$ and $R/(t_c + R)$ are determined, the values of t_c and R can be found by a simple algebraic operation. The dimensionless ratio $R/(t_c + R)$, also called the **attenuation ratio**, has been found in several studies to be approximately constant over a region. This is possibly because R has a connotation of lag time, t_l, which is also related to t_c. Clark (1943) stated, "…, time of travel is more closely related to the variable storage capacity of the open channel; dimensionally, time is the ratio of storage to discharge". Usually, t_c is 1.6 times t_l but it can vary from 1.3 to 1.7 times t_l. For this reason, a rule of thumb often applied is that R is approximately 0.75 times t_c.

The attenuation ratio controls the magnitude and sharpness of the peak of a unit hydrograph and therefore is an indication of the nature of the watershed. Lower ratios give hydrographs with sharp and high peaks that can result from urbanized watersheds, whereas higher values of this ratio produce substantially attenuated broad-crested hydrograph peaks that typically originate from watersheds under natural conditions. Nonetheless, with these two parameters there is substantial flexibility for fitting a wide variety of runoff responses.

Observed hydrographs, derived from stream gauging stations, are not always available. For such cases, several approaches can be taken to derive these two parameters. These are discussed below with examples drawn from specific cases where such approaches were taken from detailed and careful studies.

Regionalization of Clark Unit Hydrograph Parameters

The purpose of a regional study is to develop unit hydrograph parameters, such as Snyder's t_l and C_p, Clark's t_c and R and loss rates for watersheds, where no gauged hydrograph data are available. Such regional studies may include cases where the regional parameters are derived entirely from physiographic characteristics of the watersheds within a large regional river basin or from a combination of available gauge data from certain parts of the basin and physiographic characteristics. Of the two parameters, t_c and R, t_c can be more readily derived from physiographic parameters of a watershed.

There are some standard practices that are discussed in detail in Chapter 6 to estimate the time of concentration of a watershed. In general, t_c can be calculated from certain physiographic or geomorphologic parameters of a watershed that govern the travel time of water on the land surface. The watershed parameters considered most important in this regard are the drainage area, equivalent stream slope of the longest watercourse, channel length along the longest watercourse from the outlet point of the designated watershed or subbasin to the upper limit to the watershed boundary, and the percentage of impervious area. In addition, the watercourse length for the outlet point to a point on the stream nearest to the centroid of the watershed is important. However, due to the variety of factors that have controls over the travel time in a natural system, physically based mathematical equations cannot be developed to calculate t_c. Consequently, empirical relationships giving t_c as a function of measurable physical parameters of a watershed have been developed. More than a dozen such relationships are available to use. Table 6.9 (Chapter 6) lists the common empirical formulas for t_c.

Similar attempts have also been made to estimate R from measurable physical characteristics of a watershed. Table 9.3 provides the empirical relationships that are commonly known.

Table 9.3 *Empirical formulas for R*

Formula	Reference	Formula	Reference
$R = C \dfrac{L}{\sqrt{S}}$	Clark (1945)	$R = 16.4 \dfrac{A^{0.342}}{L^{0.790}}$	USGS (2000)
$R = \dfrac{bL\sqrt{A}}{S}$	Linsley (1945)	$R = 1.625 \dfrac{A^{0.0710} L^{0.118}}{S^{0.1085}}$	Yoon and Park (2002)
$R = 1.03 A^{0.27}$	Laurenson (1962)	$R = 1.521 \dfrac{L^{0.263}}{S^{0.120}}$	Jung (2005)
$R = a t_c$	Russell et al. (1979)	$R = 0.0336 \dfrac{L^{1.253}}{A^{0.077} S^{0.126}}$	Go (2014)
$R = \dfrac{t_c}{1.46 - 0.867 \dfrac{L^2}{A}}$	Sabol (1988)	$R = 0.093 \dfrac{A^{0.238}}{S^{0.387}}$	Kim (2015)
$R = 15.282 \dfrac{A^{0.47942}}{S^{0.08572}}$	Yoon et al. (1994)		

Case Study 9.3 Derivation of Clark unit hydrograph parameters from regionalized values without streamflow data: Morel Catchment of Chambal Basin in western India

The catchment area of Morel River is approximately 3320 km² (Figure 9.17). It is located in the eastern part of the state of Rajasthan in western India and is a relatively small subbasin of the Chambal basin drained by Chambal River with a drainage area of approximately 146 630 km². Morel River is a tributary of Banas River, which is the major left bank tributary of Chambal River. Even though climatologically the area is semi-arid with mean annual rainfall of about 600 mm, several severe rainstorms have affected the area in the past 100 years causing considerable damage to the existing dams and occasional breaches. The Morel Dam, located at the outlet of Morel catchment, breached during a severe flood event in 1981. Subsequently, the National Institute of Hydrology (NIH) developed a design flood hydrograph of Morel Catchment

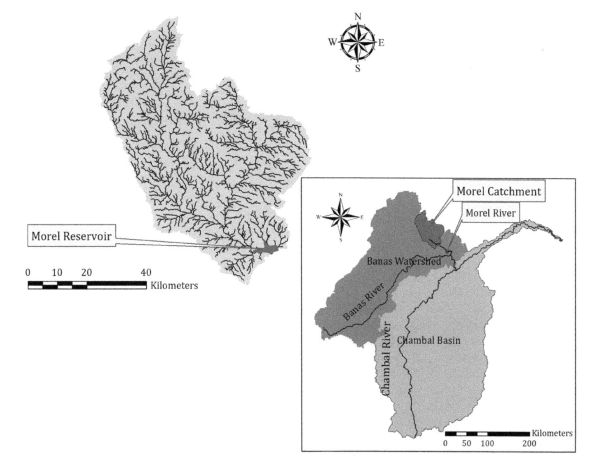

Figure 9.17 Morel catchment (up to Morel Dam) in relation to Banas River watershed within Chambal River basin, western India.

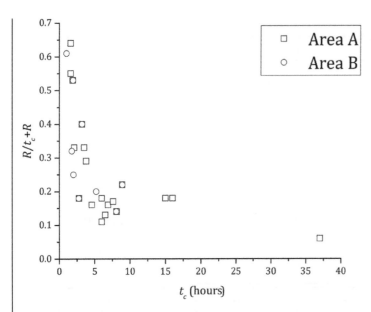

Figure 9.18 Regional values of t_c and $R/(t_c + R)$ for the derivation of t_c and R in the Morel catchment in western India. Area A denotes 19 catchments in subzone 1b and area B (four catchments within subzone 1b and five catchments from two adjacent subzones 1a and 1e). Plots are made from the data given in NIH (1998).

using the Clark unit hydrograph model (NIH, 1998). In this study, the NIH estimated the Clark unit hydrograph parameters, t_c and R, for the Morel catchment from a regionalized approach.

The Chambal basin is in the hydrologic subzone designated as subzone 1b (Figure 9.12). The Morel catchment lies at the very northern part of subzone 1b adjacent to subzones 1a and 1e.

In the regionalized approach, the NIH considered two areas. In the first case, 19 catchments all within subzone 1b, adjacent to the Morel catchment, were considered. We call this area A. Since the Morel catchment is one of the boundary catchments of subzone 1b, in the second case considered by the NIH, four catchments surrounding the Morel catchment, along with five catchments from the two other subzones, namely 1a and 1e and adjacent to Morel catchment, were considered. We call this area B. The total area of A is 7506 km² and the total area of B is 1728 km².

With the background information provided above, the steps in the derivation of t_c and R for the Clark unit hydrograph by taking a regionalized approach are as follows:

Table 9.4 *Parameters for the model (Eq. 9.84) relating the time of concentration and physiographic characteristics of the Morel catchment, western India*

Model No.	Y	X	a	b	R^2
Case I: area A data					
1	$\frac{L}{\sqrt{S_e}}$	A	0.6040	−0.3386	0.88
2	t_c	$\frac{L}{\sqrt{S_e}}$	0.920	−0.8943	0.89
3	t_c	A	0.5628	−1.363	0.81
Case II: area B data					
1	$\frac{L}{\sqrt{S_e}}$	A	0.6440	−0.66	0.94
2	t_c	$\frac{L}{\sqrt{S_e}}$	0.825	−0.835	0.79
3	t_c	A	0.5630	−1.525	0.83

- **Step 1: Computation of regional values of t_c and R.** Initial estimates of t_c and R were made, and Clark unit hydrographs were developed from a time–area relationship for each of the catchments in areas A and B. An optimization procedure, given by Rosenbrock (1960), was used to optimize the values of t_c and R such that the resulting unit hydrograph had values of W_{50}, W_{75}, WR_{50}, WR_{75}, t_l, q_p, and t_b (see Equations in Table 9.2) close to those given values according to the CWC guideline for a catchment in this subzone. Figure 9.18 shows the results. In Figure 9.18, $R/(t_c + R)$ is plotted against t_c since the former is somewhat constant for a region.
- **Step 2: Regional relationship between t_c and physiographic parameters.** The t_c values calculated in step 1 were then correlated with two easily measurable physiographic parameters to develop their interrelationships. The physiographic parameters used in this process were area of the catchment (A) and the ratio $\frac{L}{\sqrt{S_e}}$ where L is the length of the longest channel and S_e is the equivalent stream slope. Figure 9.19 shows the plots of data for areas A and B. From these plots, a general relationship can be expressed as:

$$Y = \exp[a \ln(X) + b] \tag{9.84}$$

where a is the y-axis intercept and b is a constant

(Table 9.4).

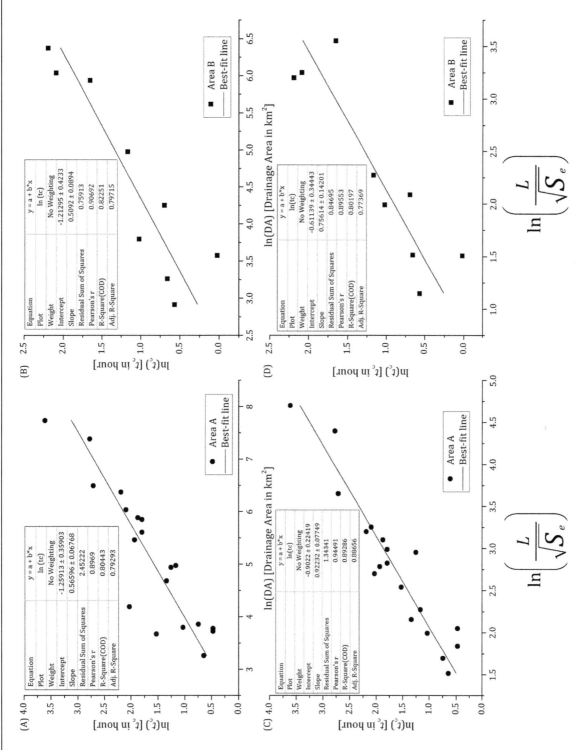

Figure 9.19 Interrelationships of t_c and two physiographic parameters: (A), (B) catchment area (A) and (C), (D) $L/\sqrt{S_e}$ for areas A and B in the Morel catchment study. Plots are made from the data given in NIH (1998).

Step 3: Regional value of R. Unlike t_c, no interrelationship between among R and other physiographic parameters could be established. Therefore, for further analysis a median value of the ratio $R/(t_c + R)$, which should be constant for a catchment and should not vary much for a region, was considered. For area A the median value of $R/(t_c + R)$ was 0.18 and that for area B was 0.5.

Step 4: Use of regional relationship for computation of t_c and R values for the Morel catchment. For the Morel catchment, $A = 3320$ km², $L = 111.6$ km, and $S_e = 1.30$ m/km, all derived from the topographic map, and $\frac{L}{\sqrt{S_e}} = 98.67$ as calculated. From these values, t_c of the Morel catchment was calculated using Eq. 9.84 and the coefficients given in Table 9.4 for both cases of areas A and B. These values of t_c were used to calculate R from the $R/(t_c + R)$ values given in step 3. The results are given in Table 9.5. From the calculated values of t_c and R it can be seen that consideration of either area A or area B yielded similar results. Similarly, either A or $\frac{L}{\sqrt{S_e}}$ can be used as an independent variable in the application of the model given by Eq. 9.84.

Table 9.5 *Computed values of t_c (hours) and R (hours) for the Morel catchment, western India*

Case	Value of t_c using			Value of R using	
	A	$\frac{L}{\sqrt{S_e}}$	$\frac{R}{(t_c+R)}$	A	$\frac{L}{\sqrt{S_e}}$
Case I: area A	24.63	24.03	0.18	5.41	5.28
Case II: area B	20.9	19.16	0.25	7.10	6.51

A derivation of the flood hydrograph using the Clark unit hydrograph where the values of t_c and R have been derived is given in Example 9.7.

Case Study 9.4 Derivation of Clark unit hydrograph parameters from regionalized values with rainfall and streamflow data: Fort Bend County, Gulf Coast of Texas, United States

Fort Bend County is located adjacent to the south of Harris County near the Gulf Coast of Texas. The mean annual rainfall in these two counties ranges from approximately 48 inches to 53 inches ($\approx$ 1200–1300 mm), which is higher than both US national and Texas averages. The urban expanse and population centers of the City of Houston, within Harris County, extends to much of Fort Bend County also. Flooding is a major problem in this area. Both the Harris County Flood Control District (HCFCD) and Fort Bend County Drainage District (FBCDD) have adopted the Clark unit hydrograph method for determinating flood hydrographs. For the estimation of the Clark unit hydrograph parameters, t_c and R, both HCFCD and FBCDD employed a regional approach using observed streamflow and rainfall records of historic storm events. The approach and procedure for the estimation of t_c and R used by FBCDD are relatively less complex than those taken by HCFCD.

For the derivation of t_c and R from observed streamflow and rainfall records, the Fort Bend County used data from five 8-digit HUCs (Figure 9.20). The procedure used is as follows:

Step 1: Selection of rainfall–runoff events. To determine t_c and R as a function of watershed characteristics, these unit hydrograph parameters were first determined through an optimization process where direct runoff hydrographs were computed using the Clark unit hydrograph method from the observed rainfall data and then compared with the observed direct runoff hydrographs until a satisfactory match between observed and computed runoff hydrographs was achieved. The observed rainfall and runoff data were collected for 34 storm events that occurred in this region between 1972 and 1984. As shown in Figure 9.20, observed runoff hydrographs were obtained from 29 gauging stations distributed regionally.

Step 2: Optimization of parameters. Since the optimization process involved event-based simulation, in addition to the two unit hydrograph parameters, t_c and R, three parameters to model rainfall **loss rate** were also optimized concurrently in a stepwise order. The term *losses* was used to indicate the amount of rainfall that did not contribute to direct runoff due to a collective effect of interception, infiltration, storage, evaporation, and transpiration. The model that was used to account for rainfall losses was the **exponential loss rate method**, in which three parameters, namely E_R, S_{TR}, and R_{TI}, were optimized.

Step 3: Stepwise optimization process. Three optimization runs were conducted for each of the 34 storm events from the 29 observation stations. In the first run, all five parameters were concurrently optimized. The results were considered to be generally of high quality with the average error in the computation of peak discharges to be 6% and half of the computed peaks higher and half lower than the observed peaks.

In the second optimization run, values of E_R and R_{TI}, two parameters of the loss functions, obtained from the first run were used and held constant. Three other parameters that were optimized in the second run were S_{TR} (one of the

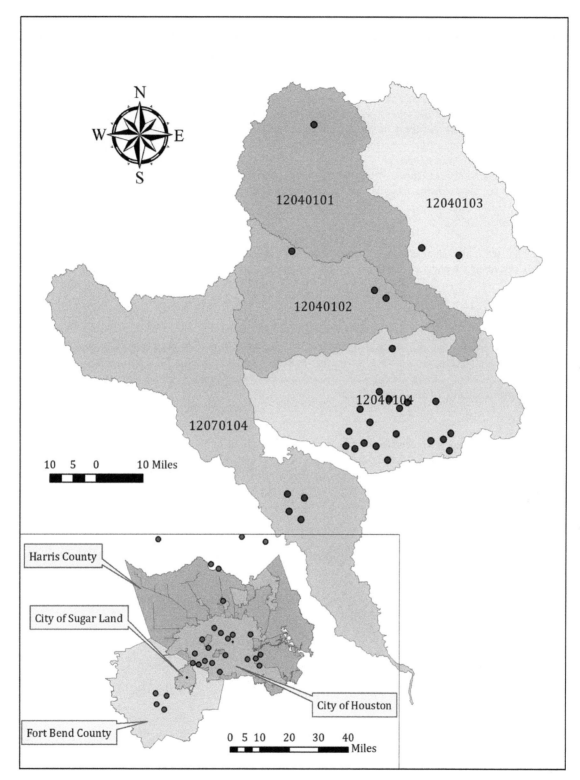

Figure 9.20 Locations of gauging stations within five hydrologic units straddling Fort Bend County, Harris County, and Montgomery County in Gulf Coast Texas. Rainfall and runoff data from these gauging stations were used to derive Clark unit hydrograph parameters to be used in Fort Bend County. (A black and white version of this figure will appear in some formats. For the color version, please refer to the plate section.)

parameters) and t_c and R. However, the parameters that were optimized for t_c and R, were (t_c+R) and $R/(t_c+R)$. From the optimized values of these latter parameters, values of $t_c/(t_c+R)$ were also computed. The values of $R/(t_c+R)$ and $t_c/(t_c+R)$ were then correlated with watershed characteristics. The physical characteristics of the watersheds that were considered are as follows:

L = length (mi) of the longest watercourse within a subbasin
S = average slope (ft/mi) of the longest watercourse in its middle 75%
N = Manning's roughness coefficient for the longest watercourse weighted in proportion to distance from upstream end
S_0 = average slope (ft/mi) of the watershed land draining into the longest watercourse
D = percentage of urban development
I = effective imperviousness ratio

From this analysis the following relationship was established and was used in the third run of optimization:

$$\frac{t_c}{(t_c+R)} = 0.38\log(S_0) \tag{9.85}$$

In the third run, only values of (t_c+R) and S_{TR} were optimized. The values of (t_c+R) were correlated with the parameters describing the physical characteristics of watersheds and the following equation was derived (Figure 9.21):

$$(t_c+R) = 128 \frac{\left(\frac{L}{\sqrt{S}}\right)^{0.57} N^{0.8}}{(S_0)^{0.11}(10)^I} \tag{9.86}$$

Step 4: Equations adopted for calculations of t_c and R. The two relationships, derived above were used in Fort Bend County for the computation of t_c and R. First, Eq. 9.86 was used to compute (t_c+R) and then this value was substituted in Eq. 9.85 to obtain the value of t_c. Subsequently, R was simply calculated as: $R = (t_c+R) - t_c$. The effective impervious ratio, I, was calculated as:

$$I = CD \times 10^{-4} \tag{9.87}$$

where C denotes average percentage of impervious cover of the developed area and D is the percentage of the subarea that is developed. Thus, in addition to the geomorphologic characteristics of the watershed, the effects of urbanization in the storage coefficient are also accounted for.

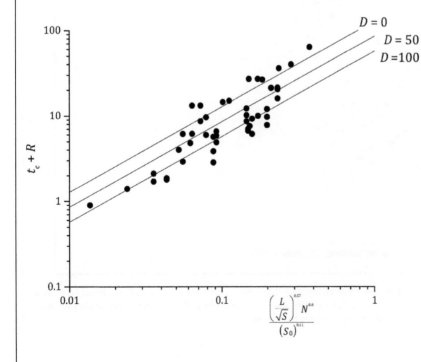

Figure 9.21 Regional relationship of $(t_c + R)$ with watershed characteristics established in Fort Bend County. Plots are generated from the data given in Appendix A of Fort Bend County Drainage Design Manual.

9.8.6 Modified Clark Method

The spatial (geographic) variation of runoff generation in the Clark unit hydrograph model originates from the time–area concept. However, Clark's original formulation still produces a lumped or spatially aggregated model. To better model the spatial variation of the characteristics of a watershed, the Clark model has been modified by the HEC to represent a watershed by a grid system consisting of square-shaped grid cells. The modified Clark unit hydrograph method is called the **ModClark** method, a distributed (or quasi-distributed more accurately) rainfall–runoff transform method.

As in the Clark unit hydrograph method, the ModClark method also explicitly accounts for translation and attenuation or storage. For the storage effect, the ModClark method employs the same parameter of the storage coefficient, R, as the Clark method does; however, the ModClark method does not need a time–area curve for translation. Instead, each cell represents a small sub-watershed, and the travel time is calculated separately based on its distance to the watershed outlet and the watershed time of concentration (Figure 9.22). This is given by

$$t_{cell} = t_c \frac{d_{cell}}{d_{max}} \tag{9.88}$$

where t_{cell} denotes the time of travel for a cell, t_c is the time of concentration of the watershed, d_{cell} is the travel distance from a cell to the outlet, and d_{max} denotes the travel distance for the cell that is most distant from the outlet.

In the ModClark method, each cell's individual runoff hydrograph is translated to the watershed outlet and at that location it is routed by a linear reservoir to account for the storage effect. Finally, the overall watershed runoff hydrograph is created by combining all the individual cell hydrographs (Figure 9.23).

The ModClark transform method requires subbasins to be discretized using structured discretization method which usually uses the standard hydrologic grid (SHG) grid system in HEC-HMS.

9.8.7 Other Synthetic Unit Hydrograph Methods

In addition to the four methods of derivation of synthetic unit hydrographs presented above several other methods for the derivation of synthetic unit hydrographs have also been developed, such as those proposed by Taylor and Schwarz (1952), Gray (1961), Croley (1980), etc. However, these are not prevalently used. Brief mentions of the ones that have seen occasional interest and application are presented below.

9.8.7.1 Gray's Method

Gray (1961) developed a dimensionless unit hydrograph procedure based on the two-parameter gamma distribution function and watershed characteristics. The geometry of this unit hydrograph is given by

$$\frac{q_t}{t_r} = \frac{25\gamma'\lambda'}{\Gamma(\lambda')}\left(\frac{t}{t_r}\right)^{\lambda'-1} \exp\left[-\frac{\gamma' t}{t_r}\right] \tag{9.89}$$

where t_r is the time from beginning of surface runoff to the occurrence of peak discharge (minutes), q_t/t_r is the percentage of flow in $0.25 t_r$ at any given (t/t_r) value, γ' is a dimensionless parameter $= (\gamma t_r)$, λ' is the shape parameter $= 1 + \gamma'$; γ is a scale parameter; and Γ is the gamma function. Even though Gray's formulation was empirical in nature, the two-parameter gamma distribution was used in the early 1960s and served as a way for the application of pdfs in the derivation of synthetic unit hydrographs.

Gray defined the ratio $1/\gamma = t_r/\gamma'$ as the storage factor, a measure of the storage property of a watershed or the travel time required for water to pass through a given reach, and related it to the watershed characteristics in the form of a power function

$$\frac{t_r}{\gamma'} = a\left(\frac{L}{S_M}\right)^b \tag{9.90}$$

where a and b are the coefficient and exponent of the power function, S_M is the slope of the main watercourse (%); and L is the length of the main watercourse (mi) in a watershed.

Equation 9.90 was applied to 33 watersheds in the midwestern United States (Nebraska, Iowa, Missouri, Illinois, Wisconsin, and Ohio), comprising three regional groups, to

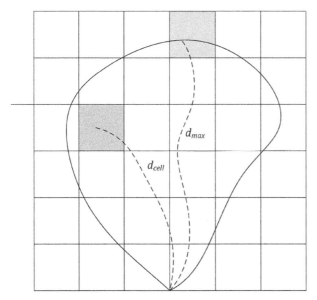

Figure 9.22 Travel time for a cell in the ModClark method.

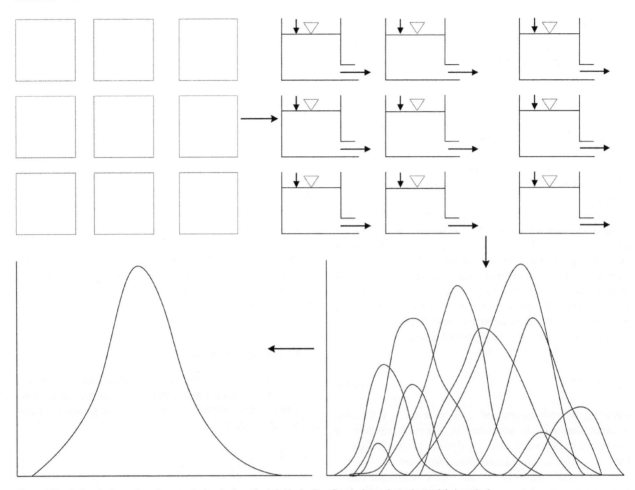

Figure 9.23 Outflow hydrograph at the watershed outlet from the individual cell outflow hydrographs in the ModClark method.

estimate the constants a and b and a regression equation representing Eq. 9.90.

9.8.7.2 Geomorphologic Unit Hydrograph

Valdes and Rodríguez-Iturbe (1979) first introduced the concept of a geomorphologic instantaneous unit hydrograph (GIUH). There followed much subsequent research, leading to various improvements and related ideas. One such area has been the derivation of the Clark unit hydrograph parameters using the GIUH approach.

Briefly, the idea presented by Valdes and Rodríguez-Iturbe (1979) relies on the topology of the channel network and some probability concepts. The premise of the probability concepts includes the following simplified assumptions:

1. Water travels through the basin, making transitions from lower to higher stream order.
2. Travel times and transition probabilities can be approximated using the Strahler stream ordering scheme.
3. A pdf analogous to an instantaneous unit hydrograph can be obtained.

The description of the topology of the channel network is derived from the applications of three Horton ratios relating the stream numbers (N), stream lengths (L), and watershed area (A) discussed in Chapter 6.

The expression derived by Valdés and Rodríguez-Iturbe (1979) yields full analytical, but complicated, expressions for the GIUH. Therefore, they suggested that it is adequate to assume a triangular instantaneous unit hydrograph and only specify the expressions for the time to peak, t_p, peak value, q_p, and time base, t_b of the instantaneous unit hydrograph. These expressions were obtained by regression of t_p as well as q_p of the instantaneous unit hydrograph derived from the analytical solutions for a wide range of parameters with those of the geomorphologic characteristics and flow velocities. The model was parameterized in terms of Horton's order laws of drainage network composition and Strahler's stream ordering scheme. The expressions for peak flow, time to peak, and time base of the instantaneous unit hydrograph are given as:

$$q_p = \frac{1.31}{L_W} R_l^{0.43} V \qquad (9.91)$$

$$t_p = \frac{0.44 L_W}{V} \left(\frac{R_b}{R_a}\right)^{0.55} R_l^{-0.38} \qquad (9.92)$$

$$t_b = \frac{2}{q_p} \qquad (9.93)$$

where, L_W is the length (km) of the main stream or highest-order stream, V is the average peak flow velocity (m/s) or characteristic velocity, q_p and t_p are in units of 1/h and h respectively and R_b, R_l, and R_a are the **bifurcation ratio**, **length ratio**, and **area ratio** as defined in Chapter 6.

On multiplying Eqs. 9.91 and 9.92, a dimensionless parameter, β is obtained as

$$\beta = q_p t_p = 0.5764 \left(\frac{R_b}{R_a}\right)^{0.55} R_l^{0.05} \qquad (9.94)$$

In Eq. 9.94, β is independent of V and L_W and thereby of the storm characteristics, and hence is a function of only the watershed characteristics. If one of the instantaneous unit hydrograph parameters q_p or t_p is known, say from observed records or some regional analysis, the terms V/L_W and L_W/V in the right-hand side of Eqs. 9.91 and 9.92 respectively, can be computed from the geomorphologic data of the watershed. And, on substituting the values of V/L_W and L_W/V the other parameters (q_p and t_p) can be obtained. Thus, with q_p and t_p known, a suitable two-parameter pdf can be used to describe the complete shape of the unit hydrograph. The GIUH provided a scientific basis for the hydrograph fitting and yielded a smooth and single-valued shape corresponding to unit runoff volume.

For the dynamic parameter, V, Valdes and Rodríguez-Iturbe (1979) in their studies assumed that the flow velocity at any given moment during the storm can be taken as constant throughout the watershed. The characteristic velocity for the watershed as a whole changes throughout as the storm progresses. For the derivation of GIUH, this can be taken as the velocity at the peak discharge time for a given rainfall–runoff event in a watershed.

Without involving the concept of probability and a trapping state, a simplified formulation of the GIUH was presented by Gupta et al. (1980) and a more practical description was given by Singh (1996) and applied by Bérod et al. (1995, 1999), which can be explained in the following steps:

1. A given watershed is ordered using the Horton–Strahler ordering scheme and the watershed order is determined. The watershed order is designated as W, which is the order of the stream of highest order as illustrated in Figure 9.24 for a watershed with $W=3$. For such a watershed, the channels are denoted as $C_i, i=1, 2$, and 3.

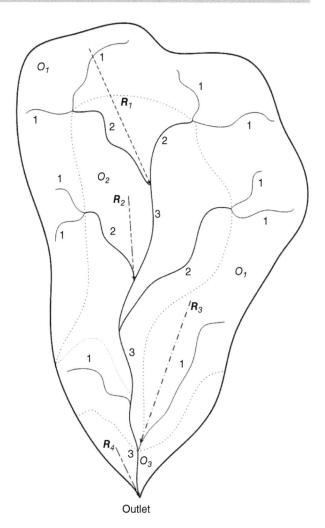

Figure 9.24 A third-order watershed showing four possible flow paths or cascades for the formulation of a GIUH.

2. The overland areas within the watershed are ordered according to the orders of the channels to which they contribute flows. For example, if a particular part of the watershed contributes overland flow to the channel of order 2, this part will be designated to be having order 2. In this manner, all overland areas are ordered. For the example presented in Figure 9.24, the overland areas are designated as $O_i, i=1, 2$, and 3.

3. The watershed is represented by a network of **cascades**. *A cascade defines the flow path*. The number of cascades, which is the number of possible flow paths in the network designated by N, is determined as $N = (2)^{W-1}$. Thus, for the example in Figure 9.24, $N = (2)^{3-1} = 4$.

4. A cascade is composed of an overland area and channels of various order. To illustrate the composition of cascades, consider the watershed shown in Figure 9.24. In this case, the number of cascades will be four as follows:

Cascade 1 (R_1): It has four elements – overland areas of order 1 draining into channels of order 1, which drain into channels of order 2, which in turn drain into channels of order 3. This can be symbolically stated as:

$$R_1 : O_1 \to C_1 \to C_2 \to C_3 \to Outlet$$

Note that all overland areas whose drainage follows this path are accumulated. Likewise, all first-order channels that follow this path are considered and their characteristics are averaged in computations.

Cascade 2: It has three elements – overland areas of order 2 draining into channels of order 2, which drain into channels of order 3. Thus, this flow path is represented as:

$$R_2 : O_2 \to C_2 \to C_3 \to Outlet$$

Likewise, all second-order channels that follow this path are considered and their characteristics are averaged in computations.

Cascade 3: It has three elements – overland areas of order 1 draining to channels of order 1, which drains directly into channel of order 3 which drain into outlet. Note that all overland areas that follow this path are accumulated. Symbolically,

$$R_3 : O_1 \to C_1 \to C_3 \to Outlet$$

Cascade 4: It has two elements-overland areas of order 3 draining into channels of order 3 which drain into the outlet. Symbolically,

$$R_4 : O_3 \to C_3 \to Outlet$$

Note that all overland areas that follow this path are accumulated.

This cascade network is shown in Figure 9.24.

The concept of representation of a watershed as series (cascade) of linear reservoirs is detailed in Section 9.9 where the theory of the instantaneous unit hydrograph is presented. All overland areas are represented by linear reservoirs and all channels are considered linear. Thus, the instantaneous unit hydrograph of the above network of cascades can be expressed as below.

Let the linear reservoir in cascades be represented by the storage–discharge relation $(S = k_{ij}Q)$ with parameter k_{ij} where i represents the cascade number ($i = 1, 2, 3, \ldots$) and j denotes the channel order ($j = 1, 2, 3, \ldots$). Linear channels be represented by travel times T_{ij}, where i again denotes the cascade number and j, the channel order. Then, the instantaneous unit hydrograph of each cascade will be:

Cascade 1 (or flow path R_1): $U_1(t) = \frac{1}{k_{11}} \exp\left(-\frac{t - T_{11} - T_{12} - T_{13}}{k_{11}}\right)$

Cascade 2 (flow path R_2): $U_2(t) = \frac{1}{k_{22}} \exp\left(-\frac{t - T_{22} - T_{23}}{k_{22}}\right)$

Cascade 3 (flow path R_3): $U_3(t) = \frac{1}{k_{31}} \exp\left(-\frac{t - T_{31} - T_{33}}{k_{31}}\right)$

Cascade 4 (flow path R_4): $U_4(t) = \frac{1}{k_{43}} \exp\left(-\frac{t - T_{43}}{k_{43}}\right)$

Thus, the instantaneous unit hydrograph of this cascade will be:

$$U(t) = \frac{1}{A}\left[\frac{A_1}{k_{11}} \exp\left(-\frac{t - T_{11} - T_{12} - T_{13}}{k_{11}}\right) \right.$$
$$+ \frac{A_2}{k_{22}} \exp\left(-\frac{t - T_{22} - T_{23}}{k_{22}}\right)$$
$$+ \frac{A_3}{k_{31}} \exp\left(-\frac{t - T_{31} - T_{33}}{k_{31}}\right)$$
$$\left. + \frac{A_4}{k_{43}} \exp\left(-\frac{t - T_{43}}{k_{43}}\right)\right]$$

For simplification, if the parameter k_{ij} for one overland area is known, for another overland area it can be proportioned according to the area. Likewise, if the travel time for a channel of given order is known then it can be proportioned for another channel according to its average length. In this manner, the instantaneous unit hydrograph will have only two parameters. Thus, the corresponding hydrograph of any cascade with n elements in such an arrangement can be written as

$$U(t) = \sum_{j=1}^{n} \frac{k_j^{n-2} \exp\left[-\dfrac{t}{k_j}\right]}{\prod_{i=1, i \neq j}^{n}(k_j - k_i)} \quad (9.95)$$

The advantage of this formulation is that it can account for the spatial distribution of rainfall as well as the spatial heterogeneity of land use and spatial watershed variability, to some extent.

9.8.8 Selection and Applicability of a Synthetic Hydrograph Method

Three methods for deriving synthetic unit hydrographs, namely the Snyder, NRCS, and Clark methods have been detailed in the preceding section. These are the most ubiquitous methods where unit hydrograph theory is applied for the derivation of a DRH. Several studies have been conducted to compare the results obtained from these methods in determining their efficacies in reproducing the observed flood hydrographs. All these three methods have been proved to be efficient in describing the runoff response of most watersheds from all around the world. In some cases, one method might seem to perform relatively better than the other but, overall, any of these methods, in principle, can be adopted for a study aimed at a hydraulic design of a certain type. Nonetheless, a few notes about the selection of the method are noteworthy.

The Snyder method requires two parameters, t_l and C_p. The lag time, t_l, also depends on another parameter, C_t,

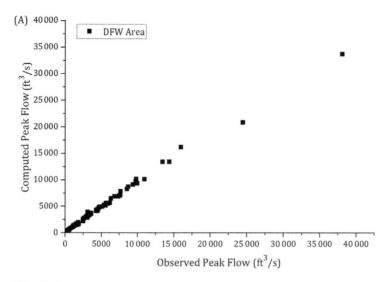

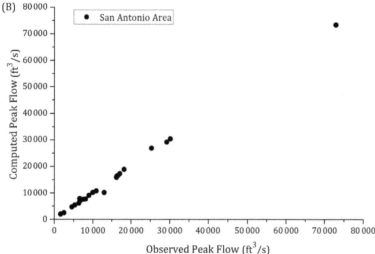

Figure 9.25 Observed and computed peak discharges in the Snyder model for the rain events at (A) Dallas–Fort Worth (DWF) and (B) San Antonio areas discussed in Halff et al. (1979) and Rodman (1977).

unless an established quantitative relationship between t_l and observed physical characteristics of a watershed is used. These parameters cannot be arbitrarily guessed. The estimation of these parameters requires either the availability of rainfall–runoff data from the watershed under investigation to derive these parameters from an optimization process or the application of regionalized equations to derive these parameters based on the physiographic (geomorphologic) characteristics of a watershed.

Figure 9.25 shows the correspondence between observed and computed peak flows from the studies of Halff et al. (1979) and Rodman (1977) described in case study 9.1. The Snyder unit hydrograph parameters in these studies were derived from calibration or optimization. This implies that the Snyder model can be used with certain confidence in areas where the estimation of the parameters can be made from available calibrated models.

In addition to what is stated above, Snyder identified several limitations of his model. These include a potential for large discrepancies between actual and synthetic unit hydrographs when the watershed shape varies greatly from a fan shape, when the predicted lag values for small floods tend to be too large, and when the application was for flat areas. In these cases, the coefficients needed further adjustments.

The NRCS method, in principle, only requires one parameter, namely t_c, from which t_l can be derived from the empirical observation that $t_l = 0.6 t_c$. The parameter t_c can be estimated from the physical characteristics of a watershed. However, the biggest challenge in adopting this method is to use the standard *PRF* value of 484 (2.08 in SI). Several studies have indicated that the value of *PRF* varies from about 600 for basins with steep slopes to 300 for flat swampy basins. The standard dimensionless unit hydrograph with a *PRF* value of 484 is based on a well-defined relationship between t_c and t_p.

To change *PRF* from 484 to a different value requires developing a completely new dimensionless unit hydrograph, which can only be done in a gauged basin where rainfall–runoff data are available. The study undertaken by Welle and Woodward (1989) exemplifies the issue.

Welle and Woodward (1989) examined the applicability of the standard NRCS dimensionless unit hydrograph in watersheds with very low relief and considerable surface storage potential. They obtained rainfall–runoff data for five historical storm events from five watersheds in a flat terrain (relief of less than 3 m or 10 feet) of the Delmarva Peninsula, located on the Atlantic Ocean coast of the northeastern United States. They observed that the standard NRCS hydrograph considerably overestimated the flows. To identify the correct *PRF* value, they created NRCS dimensionless unit hydrographs for each of the watersheds from the hydrographs derived from the Snyder method after optimization of Snyder's C_p using the observed rainfall–runoff data. The average of these hydrographs, called the **average Delmarva hydrograph**, had a *PRF* value of 284 (1.22 in SI) and was significantly different from the standard NRCS hydrograph (Figure 9.26A). The results from the application of the average Delmarva hydrograph matched relatively better with the observed peak discharges than the results from applying standard NRCS hydrograph (Figure 9.26B). This Delmarva NRCS unit hydrograph is recommended by various governmental agencies in the coastal states such as New Jersey for their flat coastal watersheds.

The NRCS unit hydrograph method is very useful but is usually limited to watersheds with areas less than 20 square miles. Furthermore, the watershed shape should be uniform with a homogeneous drainage pattern. Also, as discussed above, the accuracy of the peak flow estimate is highly dependent on the peak rate factor, which cannot be accurately estimated with ease.

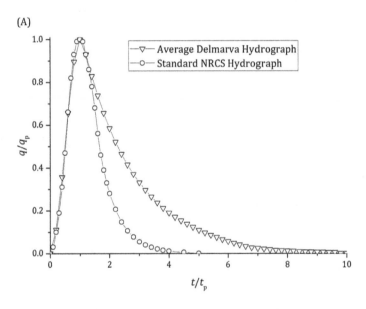

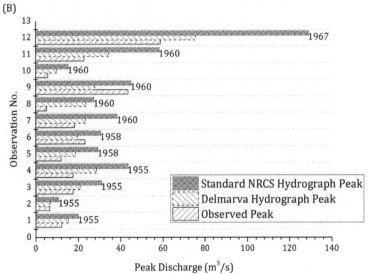

Figure 9.26 (A) The difference in geometry of the standard NRCS hydrograph and average Delmarva hydrograph. (B) Comparison of calculated peak discharge using the standard NRCS hydrograph, average Delmarva hydrograph and observed peak discharge for 12 examination points (data are from Welle and Woodward, 1989).

While the Snyder and NRCS methods are essentially empirical, the Clark method is more scientific in having a theoretical basis. Perhaps for this reason, this method has been adopted by various authorities in jurisdictions where flooding is a serious problem and accurate estimations of flood hydrographs are very important for design and construction. Another great advantage of the Clark model is that it eliminates the effect of duration of rainfall by the introduction of the instantaneous unit hydrograph. Nevertheless, this method requires two parameters of which the storage coefficient, R, cannot be easily estimated without having either observed rainfall–runoff data or regional studies that have established relationships between R and physical characteristics of a watershed. In addition, the Clark model requires a time–area relation that depends on the nature of the watershed, rainfall characteristics, and certain other controls. Nevertheless, the generalized time–area relationship, given in HEC-HMS (and previously in HEC-1), is often a good approximation. The following example illustrates the point.

Observed rainfall–runoff data were used by NIH (2001) to optimize the Clark unit hydrograph parameters for three watersheds in northeast India. The watersheds studied were the Myntdu-Leska catchment (350 km^2), which is a northern highland catchment of the Upper Meghna River basin, Dudhnai (477 km^2), and Krishnai (954 km^2) watersheds belonging to the Brahmaputra River basin. For each watershed, the mean areal precipitation was estimated from rainfall data recorded at multiple rain gauge stations, excess rainfall was estimated after accounting for rainfall loss, and selected events were used for the optimization of t_c and R with observed runoff data. The optimized or calibrated Clark unit hydrograph parameters were used for another set of events, using HEC-1, for validation. Figure 9.27A shows the correspondence between the calculated and observed

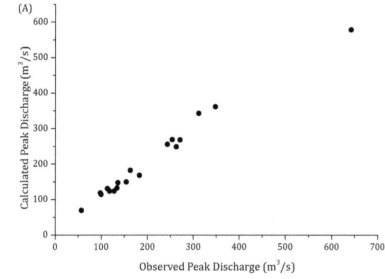
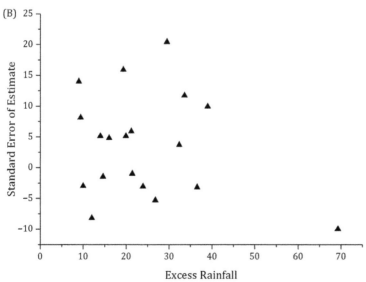

Figure 9.27 (A) Observed versus estimated peak discharge using the Clark unit hydrograph method through a calibration and validation process from 19 recorded (observed) rainfall–runoff events in three watersheds in northeast India. (B) Standard error of estimates plotted against mean areal excess rainfall for each of the events. Data used to produce the plots are from NIH (2001).

peak discharges. Figure 9.27B, in which the excess rainfall for each event was used as an index for the spatial variability of runoff, shows that there is no systematic error in the estimation of peak discharges using the Clark method.

Some of the other limitations of the Clark model include the variability in the estimate of K and the size of the watershed the method is applied to. For large watersheds, the hydrograph may rise too slowly and fall too rapidly. Although the model exhibits the effect of watershed shape to produce high peaks, sometimes the predicted values can be exaggerated.

In summary, the choice of the method depends on factors such as the availability of regional relations for parameters and thereby ease of use. Other aspects of the methods should also be considered. For example, the Snyder method requires explicit curve fitting, and the Clark method permits incorporation of basin shape and timing factors through the use of a time–area relation. Familiarity with the method is also important. As Loague and Freeze (1985) point out: "Predictive hydrologic modeling is normally carried out on a given catchment using a specific model under the supervision of an individual hydrologist. The usefulness of the results depends in large measure on the talents and experience of the hydrologist and ... understanding of the mathematical nuances of the particular model and the hydrologic nuances of the particular catchment. It is unlikely that the results of an objective analysis of modeling methods ... can ever be substituted for the subjective talents of an experienced modeler."

9.9 INSTANTANEOUS UNIT HYDROGRAPHS

9.9.1 Definition and Properties of an Instantaneous Unit Hydrograph

An instantaneous unit hydrograph is an abstract or theoretical concept but can be developed for practical applications and for theoretical investigations on the rainfall–runoff relationship in a watershed.

If the duration of the effective unit rainfall is infinitesimally small, the resulting hydrograph is an impulse response function called an instantaneous unit hydrograph. It is expressed as $u(t - \tau)$ as given in Eq. 9.1. It is also called the **kernel function** in the convolution integral given by Eq. 9.1, also known as Duhamel integral. The kernel function has the following properties:

$0 \leq u(t - \tau) \leq$ a positive peak for $(t - \tau) > 0$

$u(t - \tau) = 0$ for $(t - \tau) \leq 0$

$u(t - \tau) \to 0$ as $(t - \tau) \to \infty$

$$\int_0^\infty u(t)dt = 1 \qquad (9.96)$$

$$\int_0^\infty u(t)t\,dt = t_l$$

Compared to other unit hydrographs, the major advantage of the instantaneous unit hydrograph is that it is independent of the duration of the effective rainfall. It can be converted to a unit hydrograph of finite duration, Δt_0, by superposing several instantaneous unit hydrographs initiated at equal subintervals of an interval equal to Δt_0 and dividing the aggregated direct runoff by the number of instantaneous unit hydrographs. If Δt_0 is sufficiently small, as is typically the case to provide adequate definition to a DRH, the finite duration unit hydrograph can be developed by simply averaging the ordinates of two instantaneous unit hydrographs that are separated in time by Δt_0.

9.9.2 Model of an Instantaneous Unit Hydrograph

Several models for instantaneous unit hydrographs have been proposed by various workers. The model that is widely used both for unit hydrograph development and for streamflow routing is the one that was developed by Nash (1957). Even though this model ignores the variation in the translation time over the watershed, it approximates the real system reasonably well. The Nash model is presented below.

In the Nash model, a watershed is conceptualized as a cascade of n linear reservoirs, each with an equal storage coefficient, k, which has the dimension of time. The outflow from one reservoir becomes inflow to the next immediately downstream reservoir. In this way, the original inflow is routed through a series of linear reservoirs to produce the final outflow, which is the instantaneous unit hydrograph of the watershed (Figure 9.28). The derivation of the instantaneous unit hydrograph equation from a series of cascaded n linear reservoirs follows.

The continuity equation (Eq. 9.74) is rewritten here:

$$\frac{dS}{dt} = I(t) - Q(t) \qquad (9.97)$$

The storage–discharge relation for a linear reservoir as given in Eq. 9.73 is rewritten as

$$S = kQ \qquad (9.98)$$

Combining the two equations, we get a first-order linear differential equation

$$\frac{dQ}{dt} + \frac{Q}{k} = \frac{1}{k} \qquad (9.99)$$

For an instantaneous unit hydrograph, inflow I reduces to a delta function. Therefore, $Q(t)$ reduces to $u(t)$, and

$$\frac{du}{dt} + \frac{u}{k} = \frac{\delta(t)}{k} \qquad (9.100)$$

Note $\delta(t) = 0$ if t is other than zero; and is infinite at $t = 0$. Therefore, the solution to Eq. 9.100 is

$$u_1(t) = \frac{1}{k} \exp\left(-\frac{t}{k}\right) \qquad (9.101)$$

This is the instantaneous unit hydrograph of a single linear reservoir, that is, u_1 is the discharge from the first reservoir.

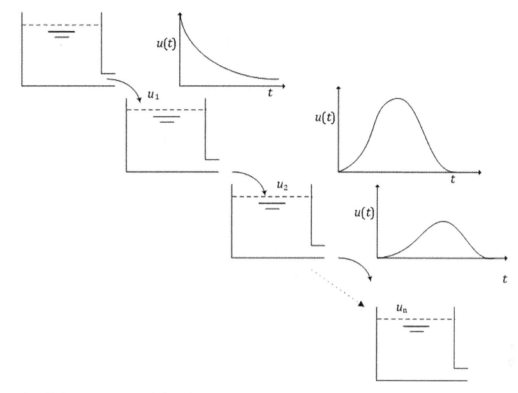

Figure 9.28 Nash model of an instantaneous unit hydrograph.

If there are n reservoirs connected in series, the instantaneous unit hydrograph of this cascade can be derived as follows. For the second reservoir, the governing equation becomes

$$\frac{du}{dt} + \frac{u}{k} = \frac{1}{k^2}\exp\left(-\frac{t}{k}\right) \tag{9.102}$$

This is also a first-order linear differential equation so it can be solved using an integrating factor, which is

$$\exp\left(\int \frac{1}{k}dt\right) = \exp\left(\frac{t}{k}\right) \tag{9.103}$$

Multiplying Eq. 9.102 by the integrating factor,

$$\exp\left(\frac{t}{k}\right)\frac{du}{dt} + \exp\left(\frac{t}{k}\right)\frac{u}{k} = \frac{1}{k^2}\exp\left(\frac{t}{k}\right)\exp\left(-\frac{t}{k}\right) \tag{9.104}$$

The left-hand side of this equation is the derivative of the product of the integrating factor and $u(t)$. Thus, Eq. 9.104 can be written as

$$\frac{d}{dt}\left[u(t)\exp\left(\frac{t}{k}\right)\right] = \frac{1}{k^2}\exp\left(\frac{t}{k}\right)\exp\left(-\frac{t}{k}\right) \tag{9.105}$$

Integration of Eq. 9.105 yields the solution as

$$u_2(t) = \frac{t}{k^2}\exp\left(-\frac{t}{k}\right) \tag{9.106}$$

It is worth recalling at this point that Eq. 9.106 is also the result of the convolution (in line with Eq. 9.1)

$$u_2(t) = \int_0^t \frac{1}{k}\exp\left(-\frac{t-\tau}{k}\right)\frac{1}{k}\exp\left(-\frac{\tau}{k}\right) \tag{9.107}$$

For the third reservoir, the governing equation becomes

$$\frac{du}{dt} + \frac{u}{k} = \frac{t}{k^2}\exp\left(-\frac{t}{k}\right) \tag{9.108}$$

To solve this equation, we again use the integrating factor, which in this is the same as before. Therefore, on multiplying Eq. 9.108 by the integrating factor, we get

$$\exp\left(\frac{t}{k}\right)\frac{du}{dt} + \exp\left(\frac{t}{k}\right)\frac{u}{k} = \frac{t}{k^2}\exp\left(\frac{t}{k}\right)\exp\left(-\frac{t}{k}\right) \tag{9.109}$$

This can be written as

$$\frac{d}{dt}\left[u(t)\exp\left(\frac{t}{k}\right)\right] = \frac{1}{k}\frac{1}{k^2} \tag{9.110}$$

Solution of the differential equation, Eq. 9.110, yields

$$u_3(t) = \frac{1}{2k}\frac{1}{k^2}\exp\left(-\frac{t}{k}\right) = \frac{t^2}{2k^3}\exp\left(-\frac{t}{k}\right) \tag{9.111}$$

Note, the instantaneous unit hydrograph of the third reservoir contains the third power of k, second power of t, and 2 in the denominator. Similarly, the instantaneous unit hydrograph of the fourth reservoir will contain the fourth power of k, third power of t, and 2 times 3 in the denominator. Likewise, the instantaneous unit hydrograph of the nth reservoir will be

$$u_n(t) = \frac{1}{(n-1)!}\frac{t}{k^n}\exp\left(-\frac{t}{k}\right) = \frac{t^2}{(n-1)!k^n}\exp\left(-\frac{t}{k}\right) \tag{9.112}$$

From the definition of the complete gamma function, $\Gamma(n) = (n-1)!$. Therefore,

$$u(t) = \frac{1}{k\Gamma(n)} \left(\frac{t}{k}\right)^{n-1} \exp\left(-\frac{t}{k}\right) \quad (9.113)$$

which is the general form of the equation for the instantaneous unit hydrograph according to the Nash model.

Dooge (1959) presented a general theory of the unit hydrograph, which includes three parts: (1) construction of isochrones for a time–area diagram, (2) representation of a watershed by a network of linear channels and linear reservoirs, and (3) routing of effective rainfall through the network. The watershed, therefore, is represented by a more realistic drainage system of a network of linear reservoirs and linear channels, as shown in Figure 9.29. Thus, the Dooge model includes the Nash model as a special case with no time–area diagram and no channels, and the Collins model as special case with no channels and only one reservoir. The general equation of the instantaneous unit hydrograph can be written as

$$u(t) = \frac{1}{T} \int_0^{t \leq T} \frac{\delta(t-\tau)}{\prod_{j=1}^{n(\tau)} (1+k_j D)} I(A) \frac{dA}{d\tau} d\tau \quad (9.114)$$

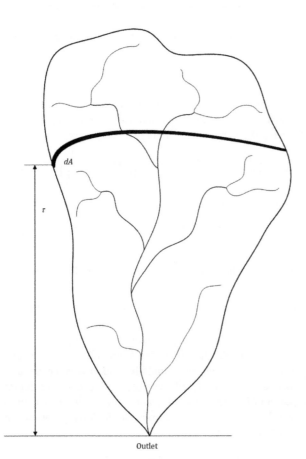

Figure 9.29 Representation of a watershed drained by a network of linear reservoirs and channels in the general unit hydrograph model of Dooge (1959).

where $dA/d\tau$ is the length of the isochrones, that is, the ordinate of the time–area diagram, T is the maximum translation time in the watershed, D is the operator, d/dt, A is the drainage area, and k is the lag time of the jth reservoir. The concept is presented in Figure 9.29 where all points on an isochrone are assumed to be drained by a single chain of reservoirs and channels. The inflow at any point is proportional to the length of the isochrones $dA/d\tau$, cutting the main stream at that point. In spite of its generality, the Dooge model has not yet found its place in practical applications because it is a little difficult to follow.

9.9.3 Estimation of Parameters of an Instantaneous Unit Hydrograph

From Eq. 9.112, it can be seen that the Nash model has two parameters, k and n. These two parameters can be estimated by the method of moments using observed rainfall–runoff data. It can be shown that the first and second moments of an instantaneous unit hydrograph about the origin $t = 0$ are respectively

$$M_1 = nk \quad (9.115)$$

$$M_2 = n(n+1)k^2 \quad (9.116)$$

The first moment, M_1, gives the time difference between the centroids of the ERH and DRH. It can be shown that the following two relationships also hold true:

$$M1_{DRH} - M1_{ERH} = nk \quad (9.117)$$

$$M2_{DRH} - M2_{ERH} = n(n+1)k^2 \quad (9.118)$$

where, $M1_{DRH}$ and $M2_{DRH}$ respectively represent the first and second moment arms of the DRH and $M1_{ERH}$ and $M2_{ERH}$ represent the first and second moment arms of the ERH, respectively. The first moment of the instantaneous unit hydrograph is the same as the lag time. Equations 9.115 and 9.116 are a direct consequence of linear theory.

Since the moment measures of the observed rainfall–runoff data can be computed, n and k can be estimated between Eqs. 9.117 and 9.118. Nash (1959, 1960) also correlated the moments of the instantaneous unit hydrograph to geomorphologic characteristics of the watershed. Other investigators have also attempted to relate the gamma instantaneous unit hydrograph parameters to watershed and storm characteristics using regression analysis (e.g., Rao et al., 1972).

9.10 S-HYDROGRAPHS

The effective rainfall of 1 inch (or 1 cm) of depth producing a unit hydrograph with duration of Δt_0 hours, has an intensity of $(1/\Delta t_0)$ inches per hour (or cm/h). If this rainfall continues indefinitely, the DRH resulting from it is called an S-hydrograph or summation hydrograph. Thus, an S-hydrograph is a hydrograph produced by a continuous excess rainfall rate for an infinite duration. This is the unit SRF as given by Eq. 9.2 and shown in Figure 9.2. The response at time t to a unit pulse of duration Δt_0 is given by Eq. 9.5. Thus, the response at time t to a unit pulse beginning at $t = \Delta t_0$ equals to $g(t - \Delta t_0)$, that is, $g(t)$ lagged by Δt_0 time units (second pulse) which is

$$g(t - \Delta t_0) = \frac{1}{\Delta t_0}[h(t - \Delta t_0) - h(t - \Delta t_0 - \Delta t_0)] \quad (9.119)$$

In this way, for a pulse beginning at $t = n\Delta t_0$, that is. the nth pulse, the response function is

$$g(t - n\Delta t_0) = \frac{1}{\Delta t_0}[h\{t - (n-1)\Delta t_0\} - h(t - n\Delta t_0)] \quad (9.120)$$

If this process continues indefinitely, summing the results produces the ordinates of the S-hydrograph given by

$$H(t) = \Delta t_0[g(t) + g(t - \Delta t_0) + g(t - 2\Delta t_0) + \ldots \infty] \quad (9.121)$$

Equation 9.121 tells that the S-hydrograph is generated by repeating an *infinite* number of unit hydrographs, each lagging the other by the duration, Δt_0, and then summing the ordinates of all the hydrographs at the corresponding times. However, note that the summation is multiplied by Δt_0 so that $H(t)$ will correspond to an input of 1, rather than $1/\Delta t_0$ as used for each of the unit pulses. Thus, the unit hydrograph and S-hydrograph are related by

$$u(\Delta t_0, t) = \frac{1}{\Delta t_0}[H(t) - H(t - \Delta t_0)] \quad (9.122)$$

where Δt_0 is the duration of the unit hydrograph and $H(t)$ is the ordinate of the S-hydrograph. The utility of S-hydrographs is that they can be used to construct unit hydrographs of a desired duration from those of different durations. The S-hydrograph is an integral curve of the instantaneous unit hydrograph. For this reason, an instantaneous unit hydrograph ordinate at time t is equal to the slope at time t of an S-hydrograph constructed for an excess rainfall intensity of unit depth per unit time.

The S-hydrograph can be obtained in two different ways: (1) in tabular form as the superposition of an infinite sequence of unit hydrographs of duration Δt_0; and (2) using Eq. 9.122, which gives the cumulative volume of the unit hydrograph per unit time.

To construct a unit hydrograph with a different duration, D, first an S-hydrograph is constructed and plotted by adding a series of (known) unit hydrographs of duration Δt_0, each lagged by a time interval Δt_0 (Figure 9.30A). The resulting superposition represents the runoff resulting from a continuous rainfall excess of intensity $1/\Delta t_0$. At the time base of the first unit hydrograph, the sum of the ordinates reaches a maximum value, and this will remain constant if the summation proceeds indefinitely. In the second step, the same S-hydrograph is plotted again, but this time lagging the first by a time unit of D (Figure 9.30B). The maximum discharge of the shifted or offset S-hydrograph occurs at a time equal to D hours less than the time base of the of the initial unit hydrograph. In the third step (Figure 9.30C) the ordinates of a new hydrograph are obtained by determining the difference between the two S-hydrographs ($S1$ and $S2$), i.e., by subtracting the ordinates of the shifted (offset) S-hydrograph from the original (unshifted) one. The ordinates of the subtracted S-hydrograph are then multiplied by $\Delta t_0/D$ yielding the unit hydrograph of duration D. Lagging the S-curve in time by an amount D and subtracting its ordinates from the original unmodified S-curve yields a hydrograph corresponding to a rainfall event of intensity $1/\Delta t_0$ and of duration D. Consequently, to convert this hydrograph with a volume of $D/\Delta t_0$ into a unit hydrograph of duration D, its ordinates must be normalized by multiplying them by $\Delta t_0/D$. The resulting ordinates represent a unit hydrograph associated with an effective rainfall of duration D.

It may be didactic to ascribe the unit hydrograph, instantaneous unit hydrograph, and synthetic unit hydrograph to three distinct characteristics of effective rainfall – volume, duration, and intensity (Singh, 1992). Table 9.6 summarizes the characteristics of effective rainfall for these three types of hydrographs.

Due to the nonlinearity present in certain hydrologic systems, fluctuations in the derived unit hydrograph ordinates are usually produced and can sometimes be very considerable. Furthermore, inaccuracies in the rainfall and runoff data, and the duration, may complicate the amplitude of the fluctuations.

9.11 APPLICATIONS OF UNIT HYDROGRAPHS

A unit hydrograph is applied to derived a DRH with the aid of the following equation:

$$Q(t) = \sum_{i=1}^{t} u[\Delta t_0, t - (i-1)\Delta t]I_i \Delta t \quad (9.123)$$

where $Q(t)$ is the ordinate of the DRH at time t after the beginning of effective rainfall, $u(\Delta t_0, t)$ is the ordinate of the unit hydrograph of duration Δt_0, I_i is the intensity of rainfall excess during block i of rainfall with duration Δt_0, and n is the total number of blocks of rainfall excess. Equation 9.115 is simply another form of the discrete convolution equation (Eq. 9.12). By adding an estimated baseflow to the DRH, the streamflow hydrograph is obtained.

In addition to its use in deriving DRHs, the unit hydrograph has many other applications in water resources engineering (Singh, 1992).

Table 9.6 *Effective rainfall characteristics for instantaneous unit hydrographs, unit hydrographs, and S-hydrographs*

Hydrograph	Effective rainfall characteristics		
	Volume	Duration	Intensity
Instantaneous unit hydrograph	Unity	Zero	Indefinite
Unit hydrograph	Unity	Finite	1/Duration
S-hydrograph	Indefinite	Indefinite	Unity

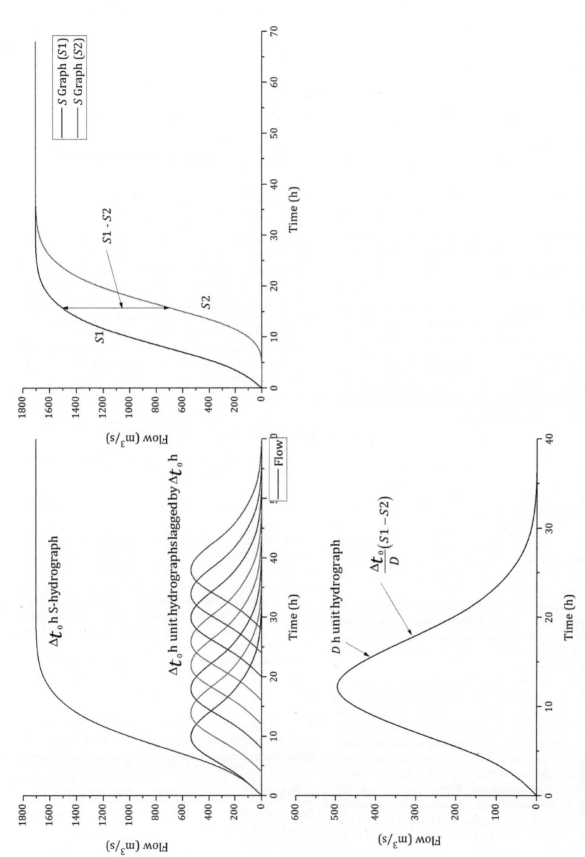

Figure 9.30 (A) Production of an S-hydrograph from summation of an infinite sequence of unit hydrograph of a certain duration (Δt_0). (B) Two S-hydrographs (S1 and S2) are generated by shifting the second by D-hour duration (in this figure, $D = 3$ h). (C) The ordinates of the unit hydrograph of D-hour duration are obtained by subtracting the ordinates of the two S-hydrographs, S1 and S2. (A black and white version of this figure will appear in some formats. For the color version, please refer to the plate section.)

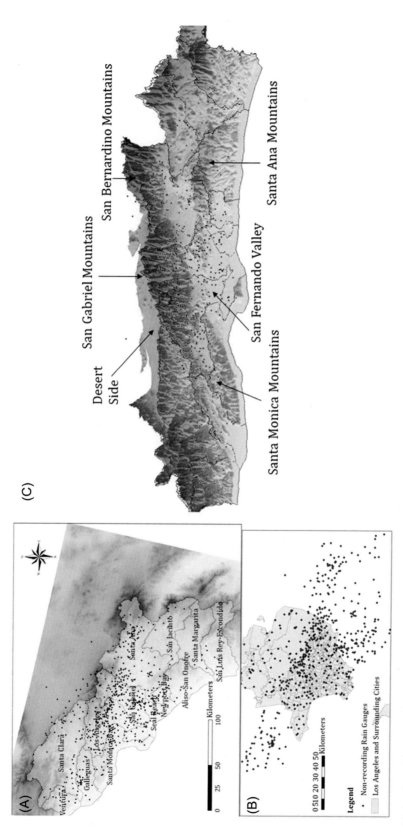

Figure 4.2 Locations of the nonrecording gauging stations that recorded the rainfalls during the rain event of February 27–March 4, 1938 in southern California. (A) Locations of the rain gauges in relation to the watersheds on the western side of the mountains that separate the coastal area from Mojave Desert. (B) Locations of the rain gauges in relation to the City of Los Angeles and surrounding areas. (C) Locations of the rain gauges in relation to the topographic features that controlled the rainfall. The digital elevation model shown in C is created from the National Elevation Dataset provided by USGS. The watershed boundaries shown in A are from the National Hydrography Dataset provided by USGS.

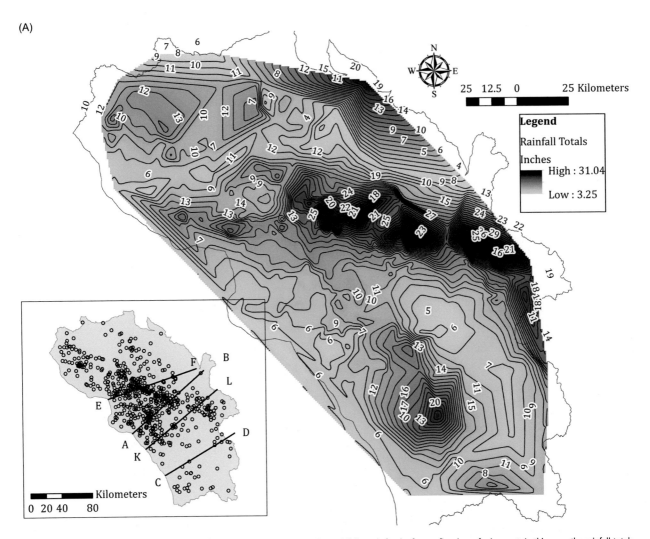

Figure 4.3 (A) Isopluvial map of the area shown in Figure 4.2B. These are based on rainfall totals for the four to five days of rain event. In this map, the rainfall totals are given in inches taken from the original source. The inset shows the rainfall and elevation profiles along four lines. (B) Rainfall profiles are shown by lines with filled circles (dots) and elevation profiles are shown by lines without any symbol.

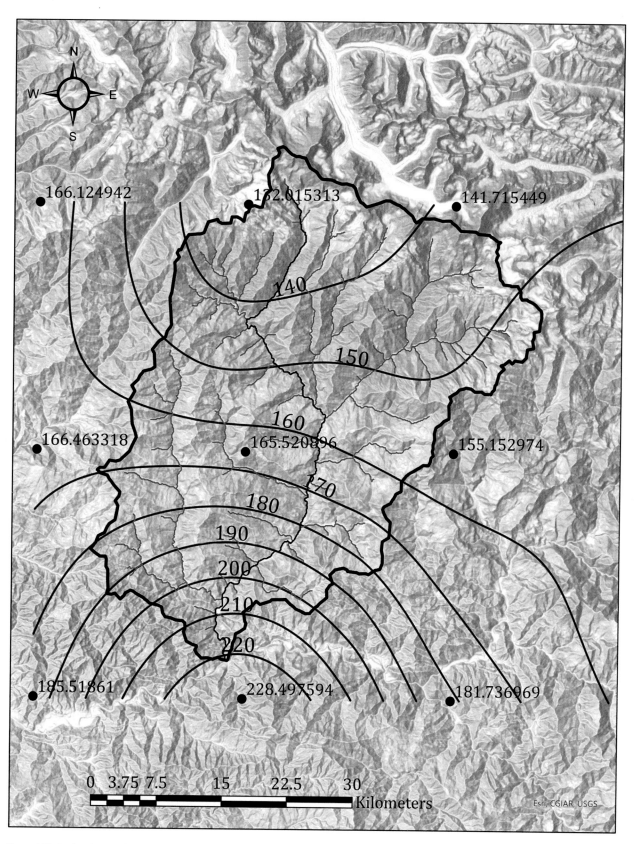

Figure 4.20 Isopluvial maps of (A) 100-year and (B) 500-year, 24-hour rainfall (mm) constructed from AMS data at the 12 grid points shown in Figure 4.16. The contours are drawn using the spline method.

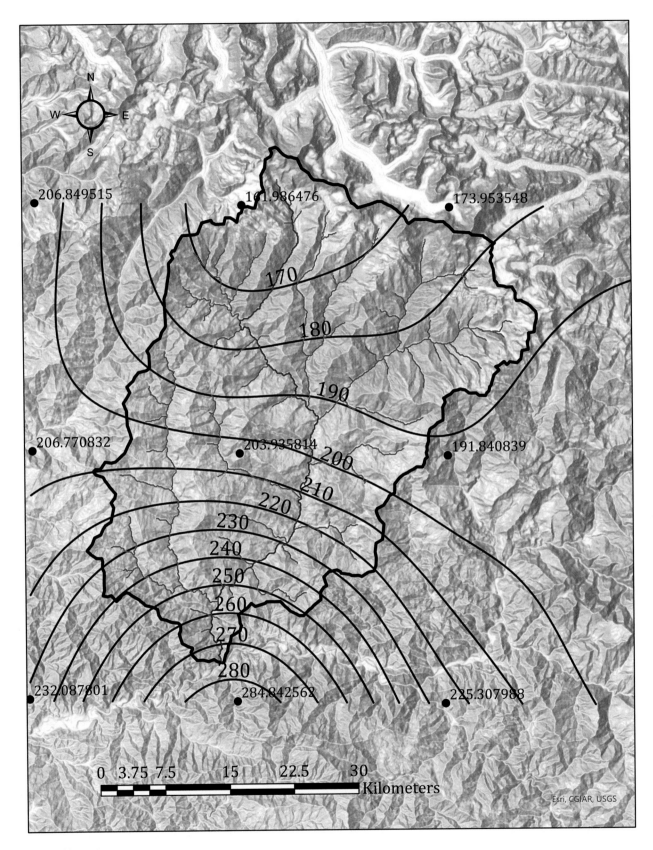

Figure 4.20 (cont.)

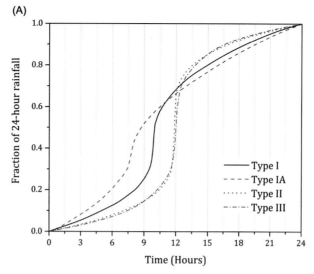
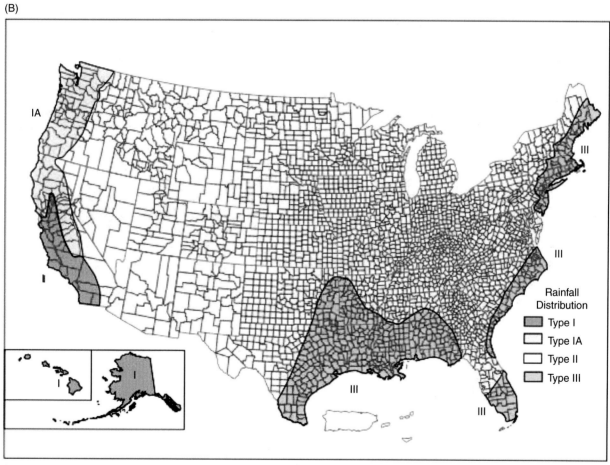

Figure 4.23 (A) Four types of time distributions of rainfall defined by NRCS. These are the mass curves shown on left. Each type is for 24-hour duration rain event. Note that each of these four rainfall distributions has an intense rainfall period somewhere near the middle and lesser rainfall intensities at the beginning and end of the storm. (B) Map of the United States showing the regions where these type curves are applicable (from NRCS, 1986).

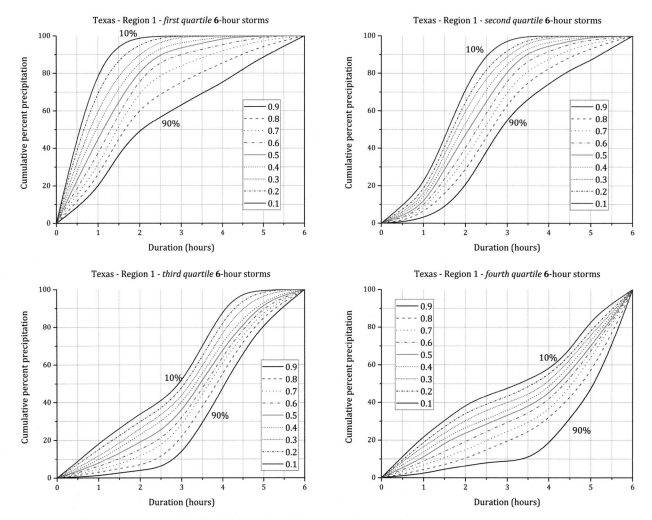

Figure 4.27 Huff curves showing temporal distributional characteristics of 6-hour storms in region 1 of Texas.

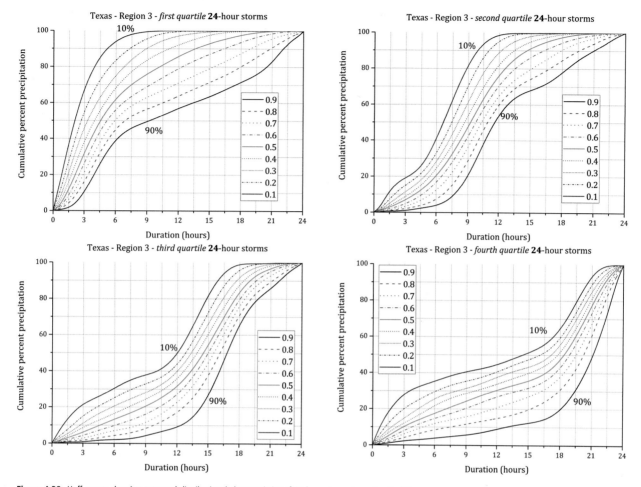

Figure 4.28 Huff curves showing temporal distributional characteristics of 24-hour storms in region 3 of Texas.

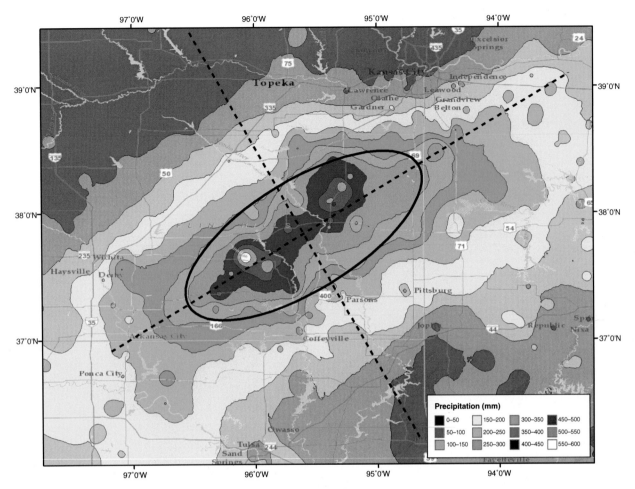

Figure 4.37 Spatial analysis of the distribution of precipitation depths of a specific storm to determine the shape and orientation of the storm (June 28–July 2, 2007 storm over Fall River, eastern Kansas). Such analyses can be done to compare the suggested idealized distributions that can be derived from application of the procedures given in HMR 52 for design PMP in regions without orographic effects.

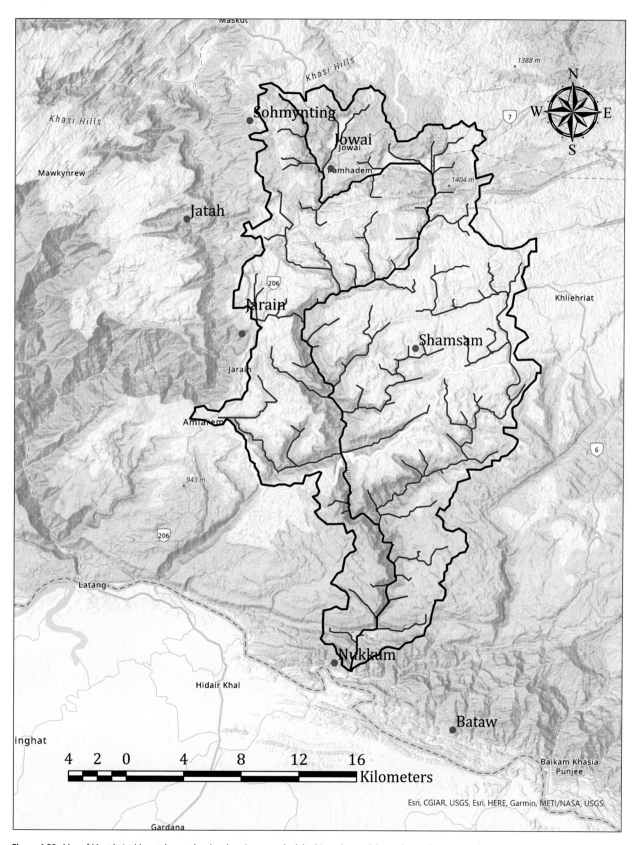

Figure 4.39 Map of Myntdu-Leshka catchment showing the rain gauges (red dots) in and around the catchment that were used to propose rain gauge network design. The main Myntdu-Leshka river course is shown by a bold dark blue line whereas other streams have thinner blue lines.

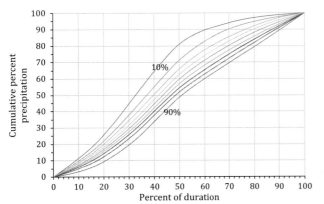

Figure 4.45 Huff curves for 6-hour second-quartile storms in the semiarid southwest United States.

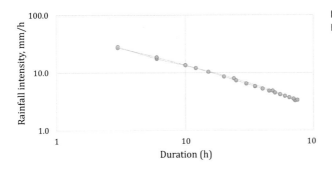

Figure 4.48 IDF curve ($F = 100$ years) in the Mandakini catchment, central Himalayas (Uttarakhand, India).

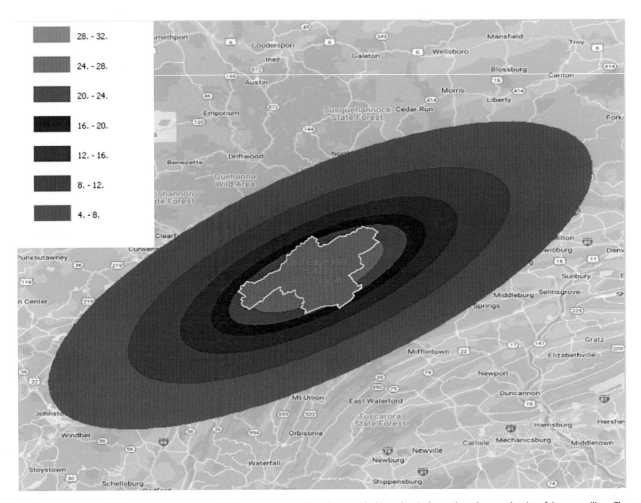

Figure 4.52 Elliptical storm calculated by HEC MetVue program from the DAD data. The watershed boundary is shown along the central region of the storm ellipse. The shading represents rainfall depths in inches.

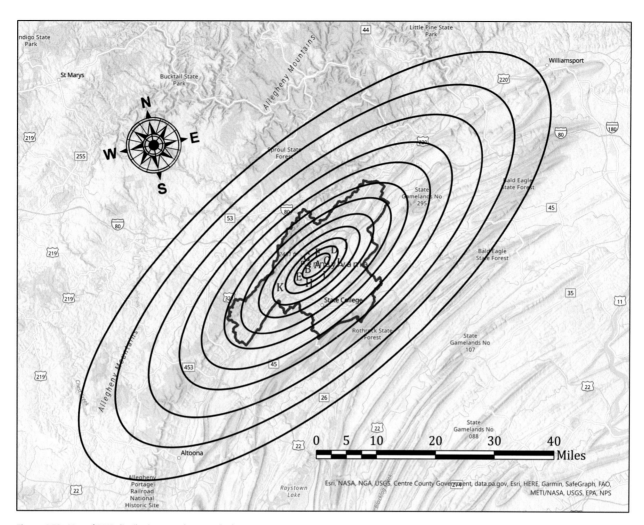

Figure 4.53 Map of PMP distribution over the watershed.

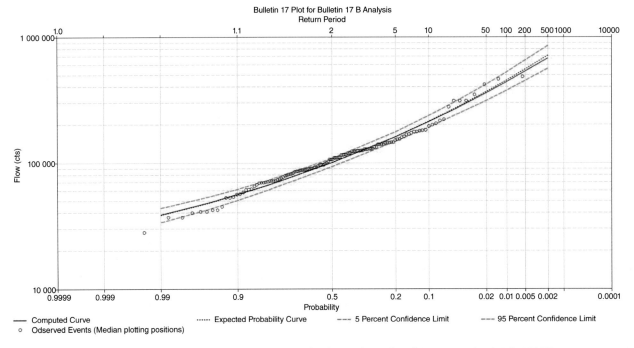

Figure 5.30 Flood-frequency curve for the AMS of Potomac River at the Point of Rocks according to the Bulletin 17B procedure done by HEC-SSP.

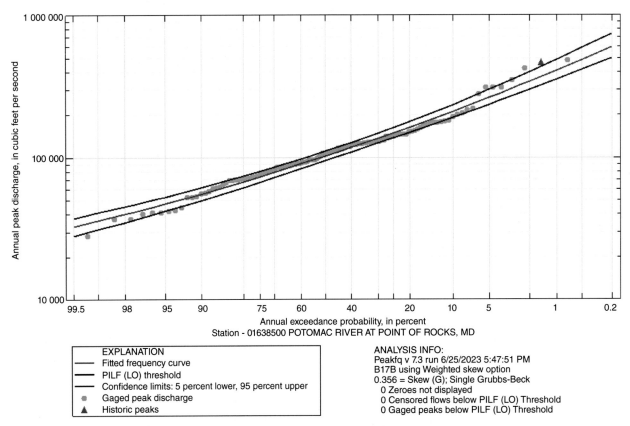

Figure 5.31 Flood-frequency curve for the AMS of Potomac River at the Point of Rocks according to Bulletin 17B procedure done by PeakFQ.

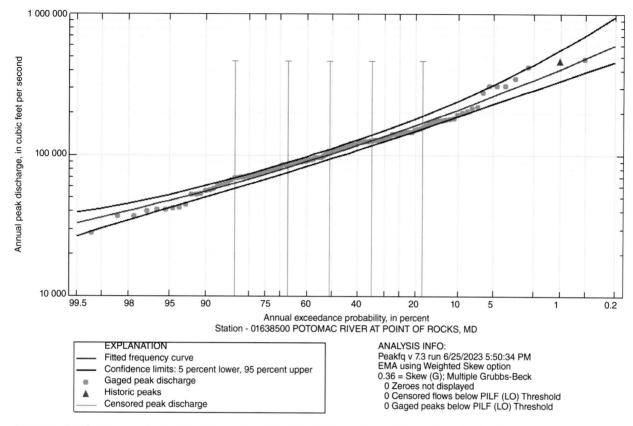

Figure 5.32 Flood-frequency curve for the AMS of Potomac River at the Point of Rocks according to Bulletin 17C procedure done by PeakFQ.

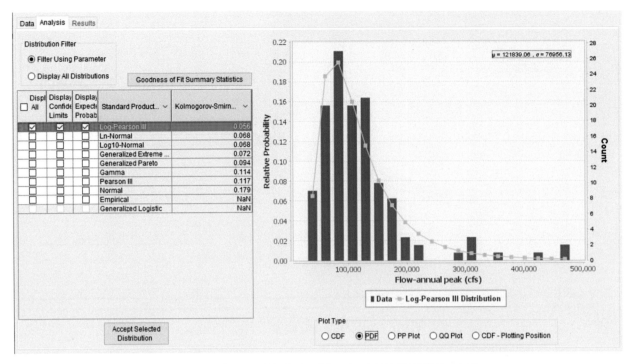

Figure 5.33 LP III probability density function fitted to the observed distribution of data given in Table 5.8 (HE-SSP analysis).

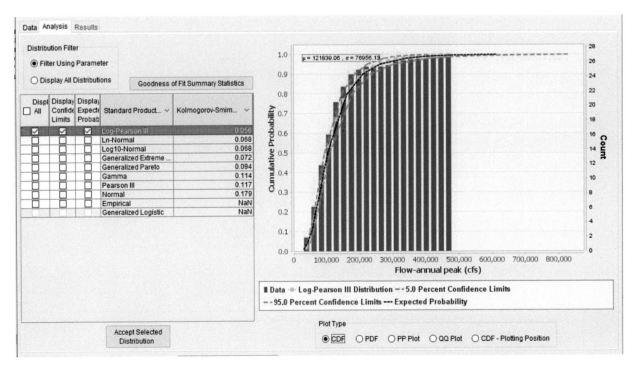

Figure 5.34 LP type III cumulative distribution function fitted to the observed cumulative distribution of data given in Table 5.8 (HE-SSP analysis).

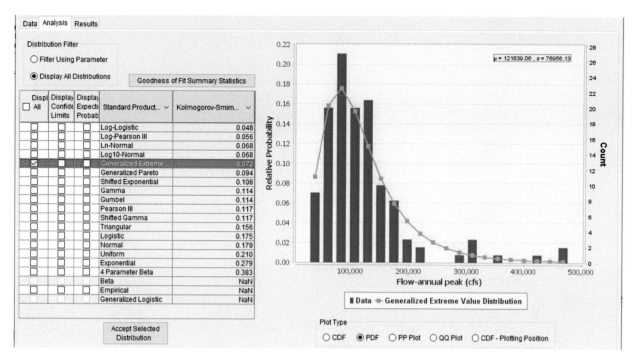

Figure 5.35 LP III probability density function fitted to the observed distribution of data given in Table 5.8 (HE-SSP analysis).

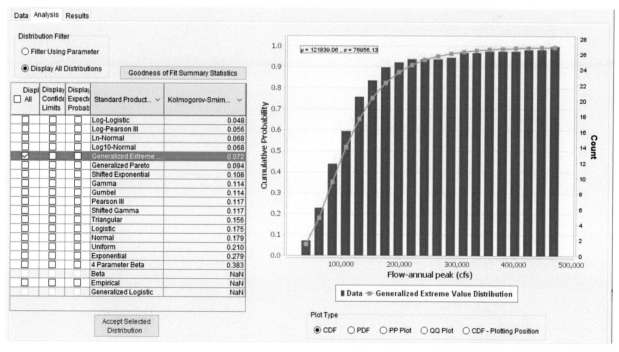

Figure 5.36 LP type III cumulative distribution function fitted to the observed cumulative distribution of data given in Table 5.8 (HE-SSP analysis).

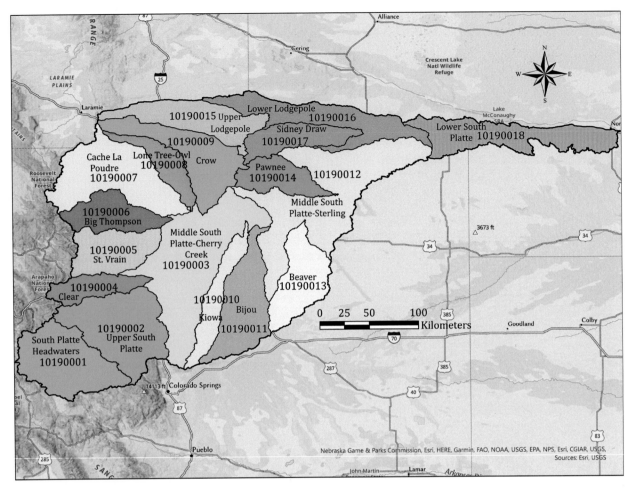

Figure 6.3 The watersheds, each with an eight-digit HUC, within the South Platte River basin, which has with a six-digit HUC of 101900.

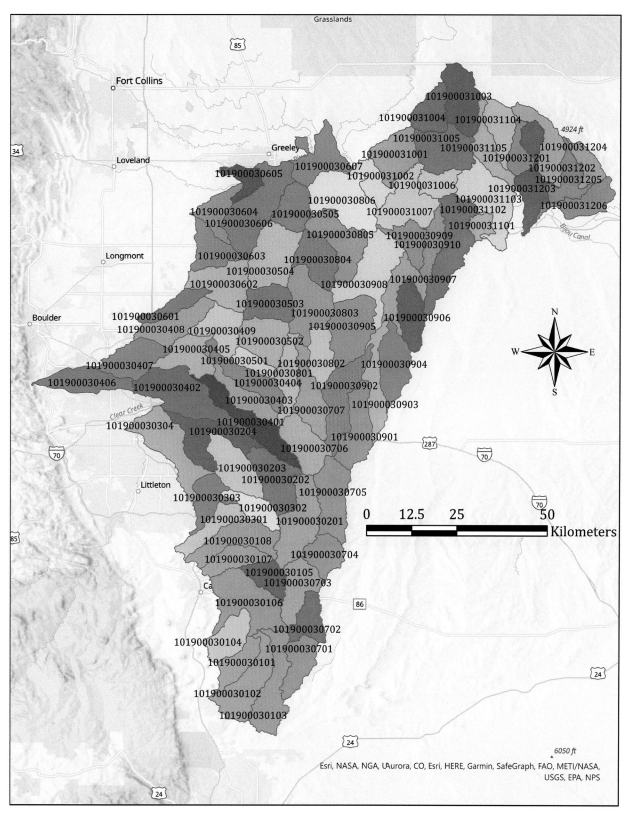

Figure 6.4 The catchments, each with a 12-digit HUC, within the Middle Platte River–Cherry Creek Watershed (eight-digit HUC 1019003) within the South Platte River basin (six-digit HUC) shown in Figure 6.3.

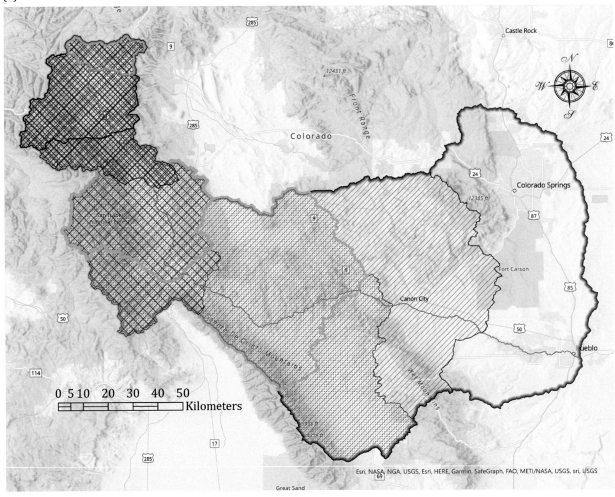

Figure 6.8 (A) Nine watersheds, each delineated from a USGS gauging station located on the headwater section of Arkansas River from its point of origin in the eastern slope of the southern Rocky Mountains up to a point in Pueblo, Colorado. (B) Watershed shape factor and elongation ratio for each of the nine watersheds. (C) Watershed area and longest flow path in each of the watersheds to illustrate the exponent of Hack's law.

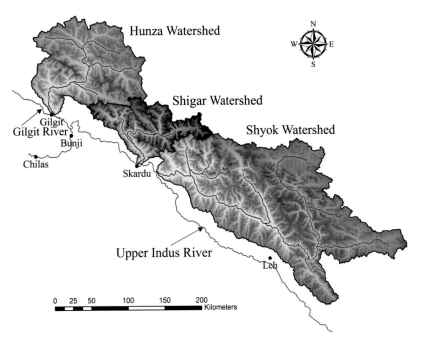

Figure 6.11 Three high-altitude, highly glaciated watersheds of Karakoram Mountains to the north of Greater Himalayas. From Mukhopadhyay, B., Khan, A., and Gautam, R. 2015. Rising and falling river flows: contrasting signals of climate change and glacier mass balance from the eastern and western Karakoram. *Hydrological Sciences Journal*, Volume 60, No. 11-12, 2062–2085.

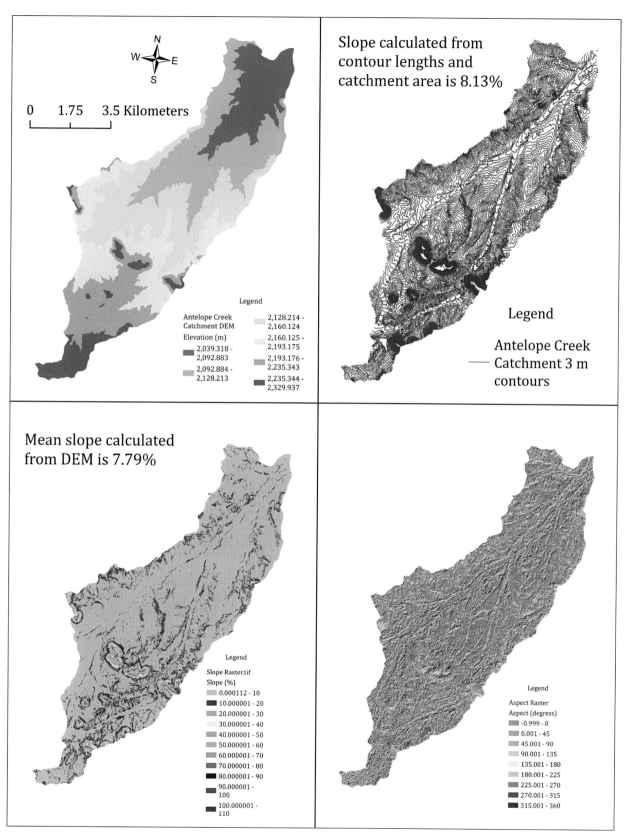

Figure 6.10 DEM, slope, contours, and aspect of Antelope Creek Catchment, one of the catchments shown in Figure 6.4.

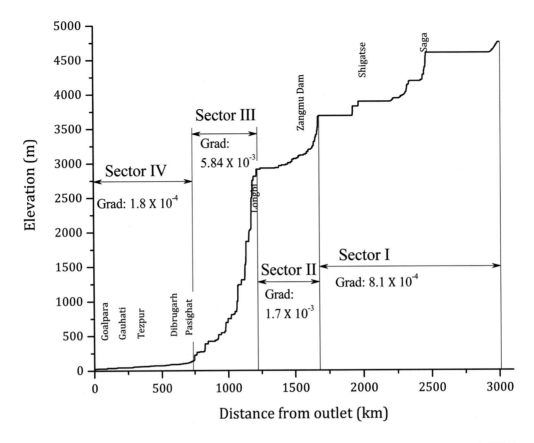
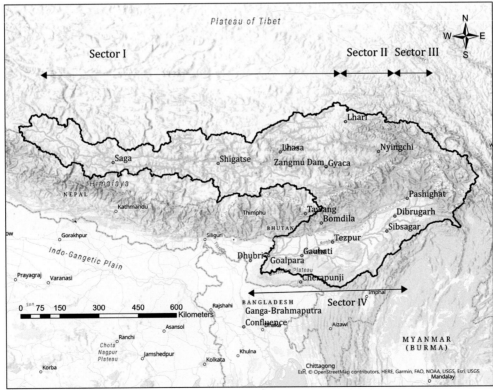

Figure 6.17 (A) Longitudinal profile of Brahmaputra River. (B) Map of the upper-middle Brahmaputra basin with locations given on the longitudinal profile, showing the four sectors of the longitudinal profile (the longitudinal profile is constructed from the Shuttle Radar Topography Mission (SRTM) elevation data discussed in Mukhopadhyay and Singh, 2011).

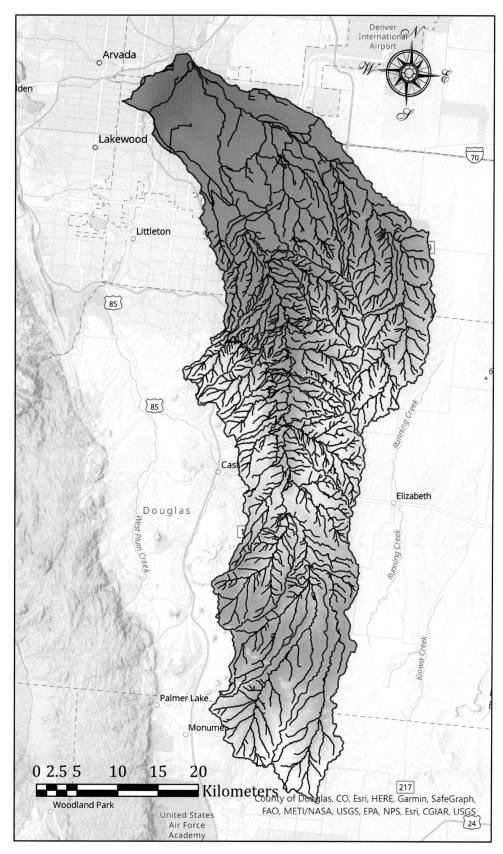

Figure 6.30 DEM, HUC 12 boundaries, and streams of the catchments of Cherry Creek and Coal Creek watersheds.

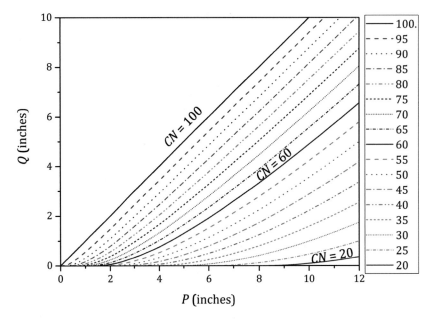

Figure 7.14 Graphical solutions of Eq. 7.58. The parameter of the curves is the curve number, *CN*.

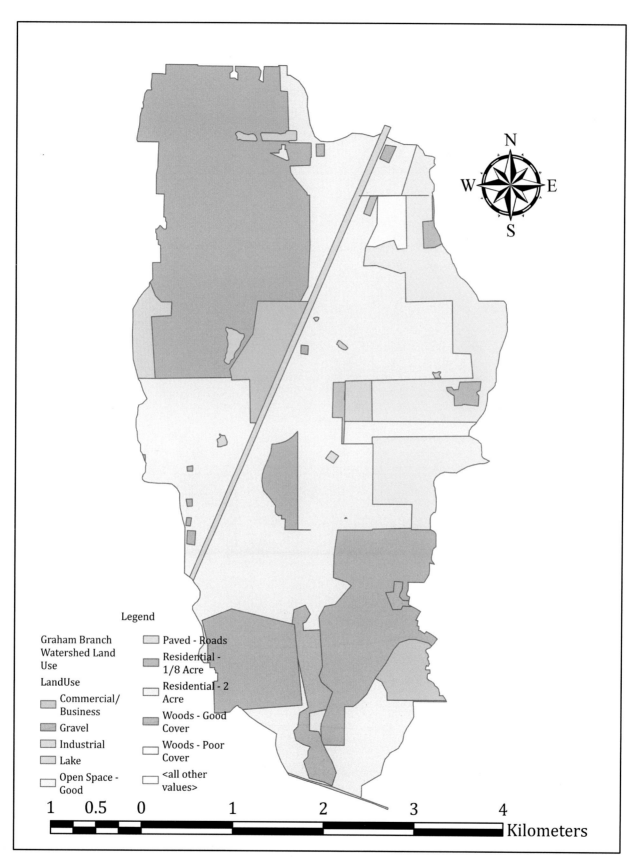

Figure 7.17 Land cover map of the watershed shown in Figure 7.15, derived from the NLCD.

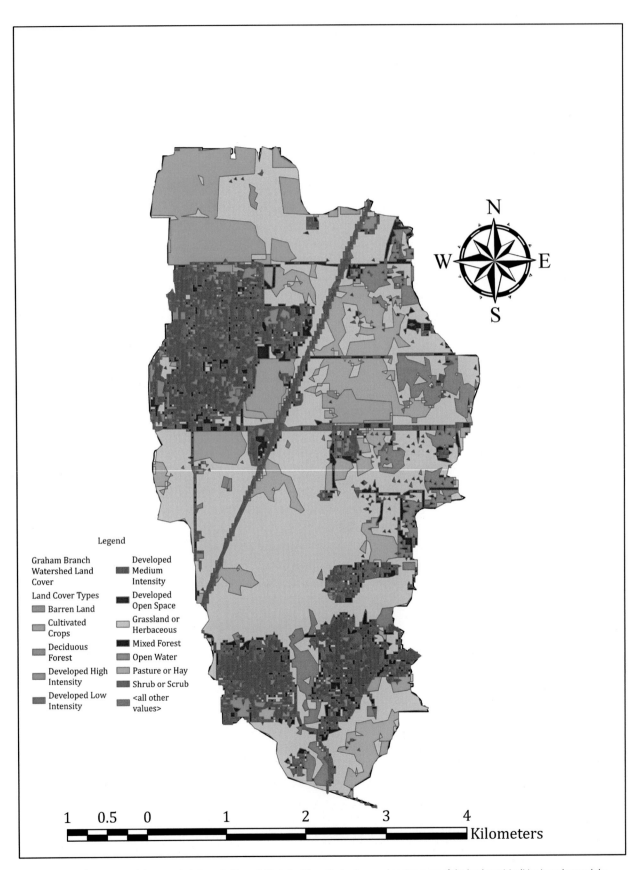

Figure 7.18 Land cover map of the watershed shown in Figure 7.15, derived from the land use and zoning maps of the local municipalities in and around the watershed.

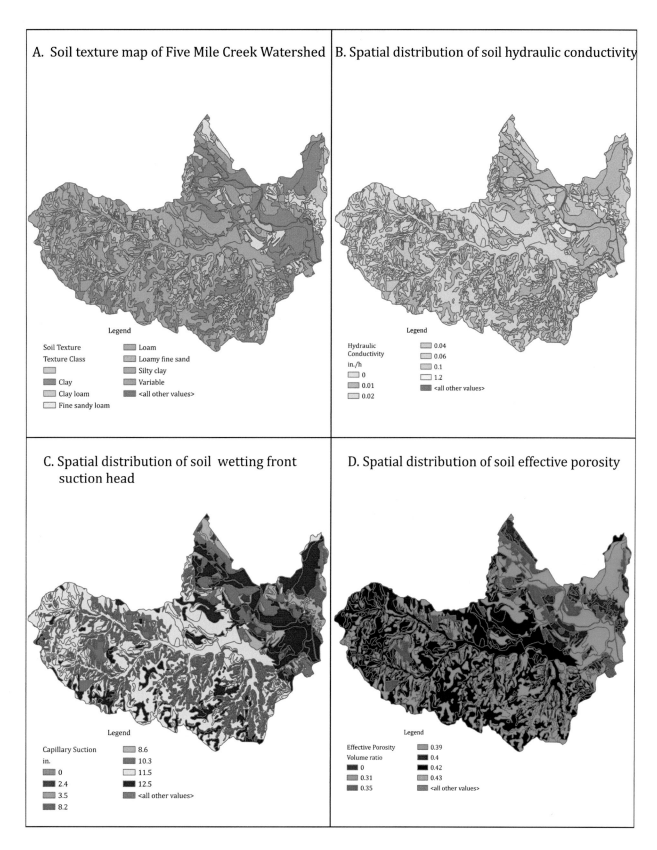

Figure 7.38 (A) Soil texture and (B)–(D) Green–Ampt parameters of Five Mile Creek Watershed, Dallas, Texas.

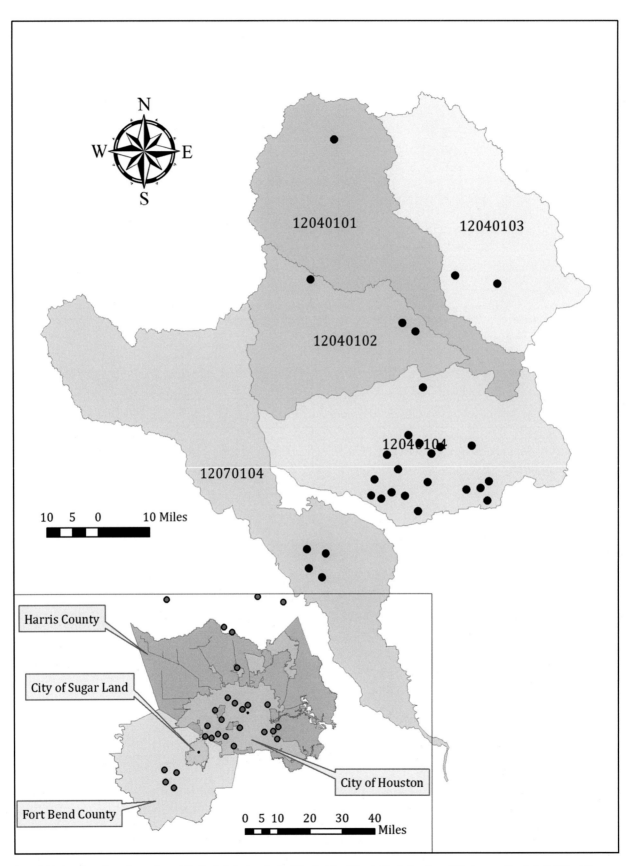

Figure 9.20 Locations of gauging stations within five hydrologic units straddling Fort Bend County, Harris County, and Montgomery County in Gulf Coast Texas. Rainfall and runoff data from these gauging stations were used to derive Clark unit hydrograph parameters to be used in Fort Bend County.

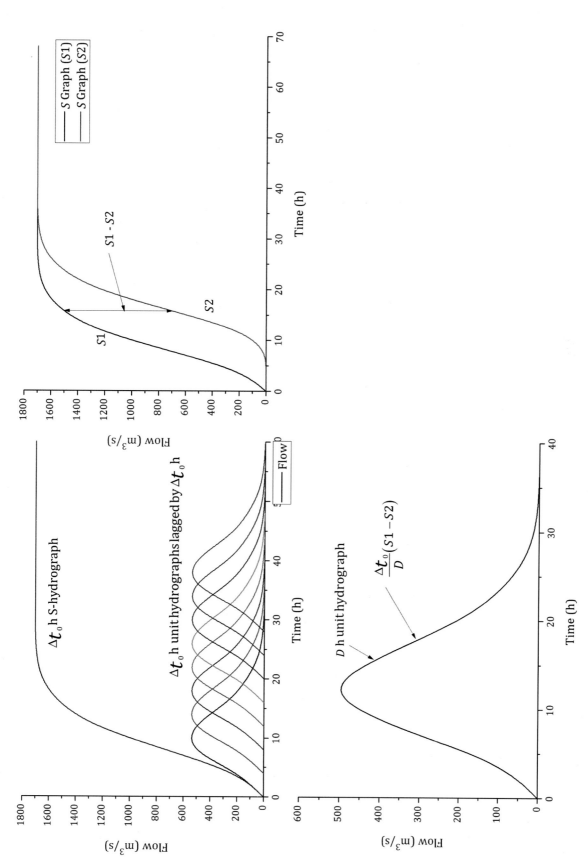

Figure 9.30 (A) Production of an S-hydrograph from summation of an infinite sequence of unit hydrograph of a certain duration (Δt_0). (B) Two S-hydrographs ($S1$ and $S2$) are generated by shifting the second by D-hour duration (in this figure, $D = 3$ h). (C) The ordinates of the unit hydrograph of D-hour duration are obtained by subtracting the ordinates of the two S-hydrographs, $S1$ and $S2$.

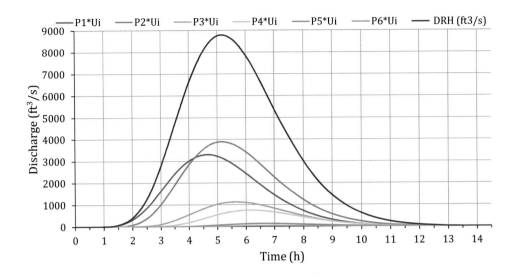

Figure 9.45 Graphical representation of the operation of superposition.

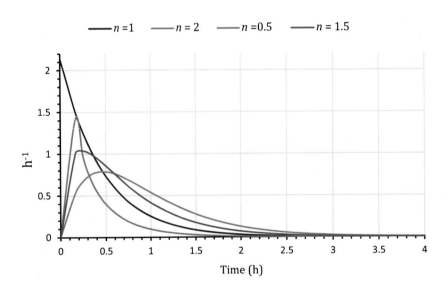

Figure 9.49 Effect of variation of n on the instantaneous unit hydrograph for constant k.

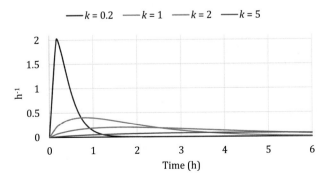

Figure 9.50 Effect of variation of k on the instantaneous unit hydrograph for constant n.

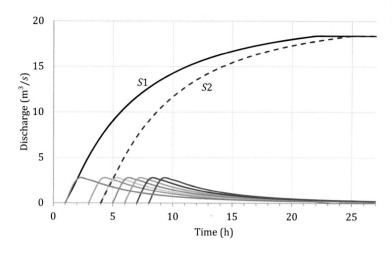

Figure 9.51 Generation of S-hydrographs.

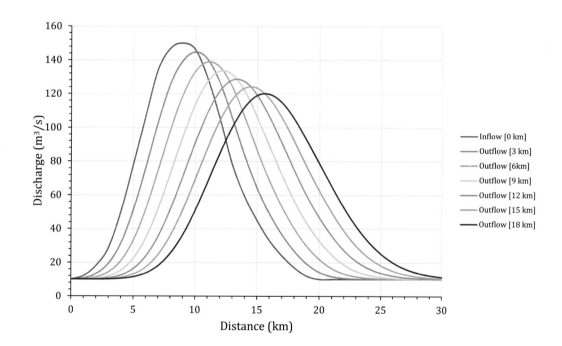

Figure 12.36 Hydrographs at different distances using the Muskingum–Cunge method.

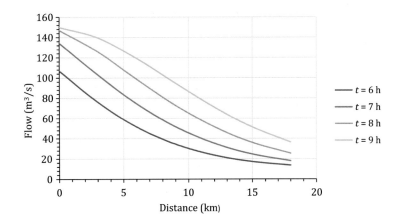

Figure 12.37 Decreasing flow downstream during the advancing flood.

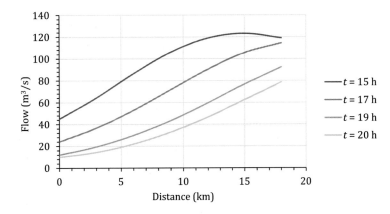

Figure 12.38 Increasing flow downstream during the receding flood.

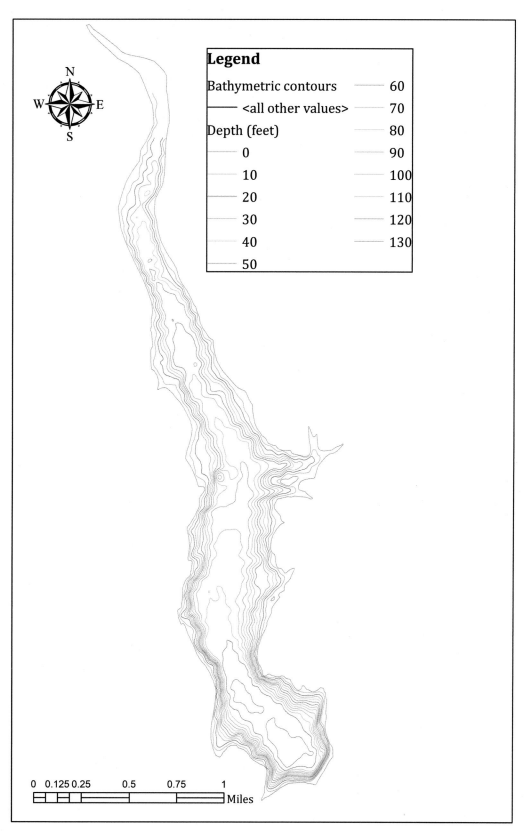

Figure 13.2 Bathymetric contours of Colebrook River Lake (reservoir). The bathymetric data are from a 2002 survey and provided by the Connecticut Department of Energy and Environmental Protection.

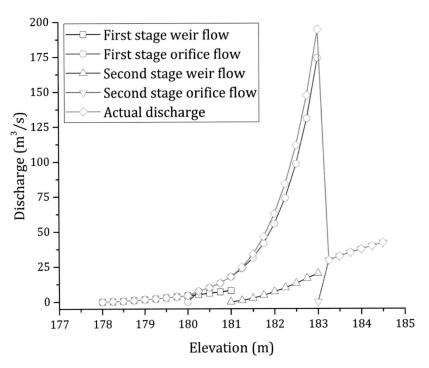

Figure 13.12 Plot of the example of elevation–discharge relationship given in Table 13.1 for a two-stage primary spillway shown in Figure 13.11.

Figure 13.19 Areal view of (A) Sirkit reservoir in Thailand and (B) the spillway of Sirkit Dam. (C) Close-up view of the spillway of Sirkit Dam. (Photographs courtesy of Rangsarit Vanijjirattikha, the Electricity Generating Authority of Thailand.) (A black and white version of this figure will appear in some formats. For the color version, please refer to the plate section.)

(B)

(C)

Figure 13.19 (*cont.*)

Figure 14.5 Global net radiation for the month of (A) January and (B) July of 2007. The source of NASA data for production of such maps for any month from 1983 onward has been described in detail by Mukhopadhyay and Singh (2011). It can be clearly seen that the oceans reflect back very considerably less radiation than the land surface. The seasonal contrast between the northern and southern hemisphere can also be seen from the reversals of radiation patterns from January to July.

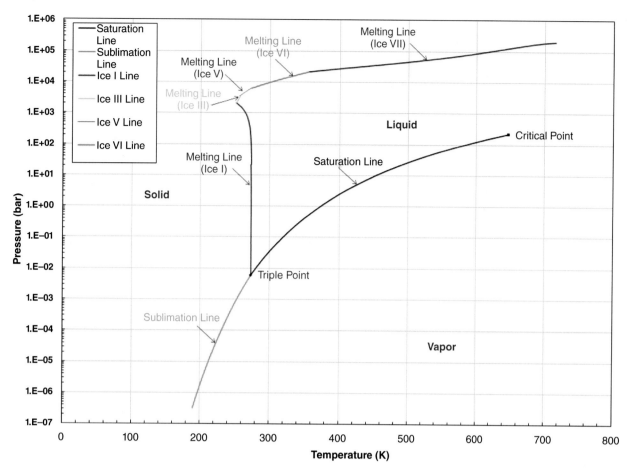

Figure 15.1 Phase diagram of water.

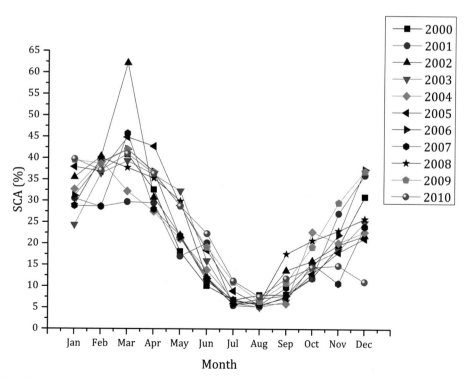

Figure 15.3 Monthly distribution of the snow-covered area (SCA; % of total basin area) for a 11-year period in Upper Indus Basin straddling the Karakoram Mountains and Higher Himalayas, derived from MODIS data. From Mukhopadhyay (2012). *Journal of Hydrology*, 412–413, 14–33.

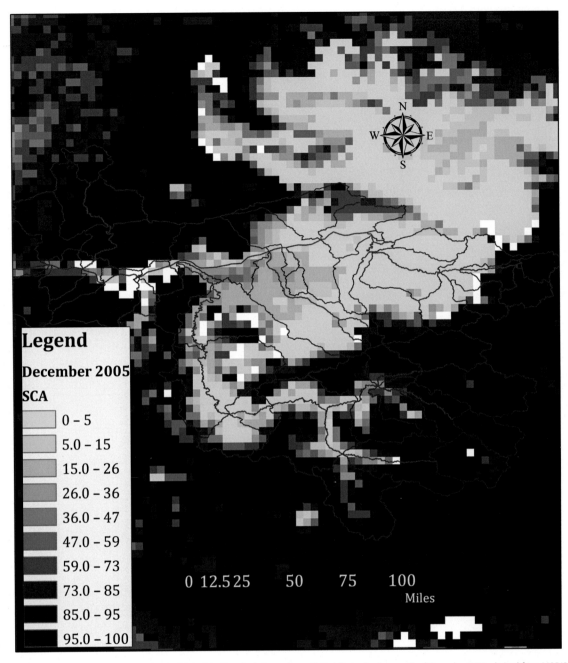

Figure 15.2 Percentage snow-covered area (SCA) of Columbia River Basin in northwest United States for the month of December 2005, derived from MODIS data.

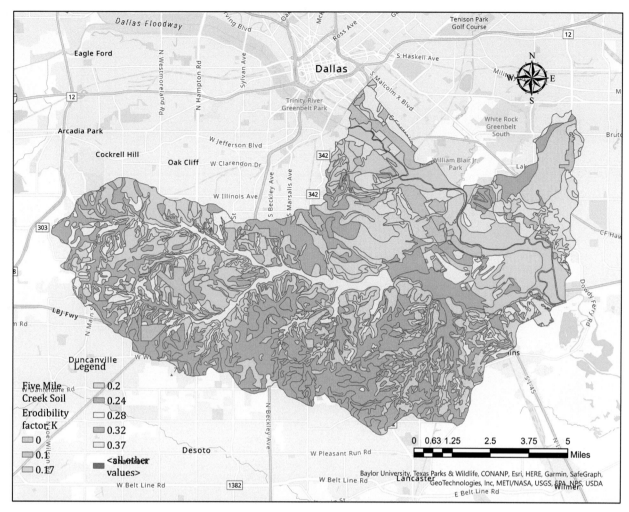

Figure 16.7 Distribution of the soil erodibility factor in Five Mile Creek Watershed in Dallas, Texas.

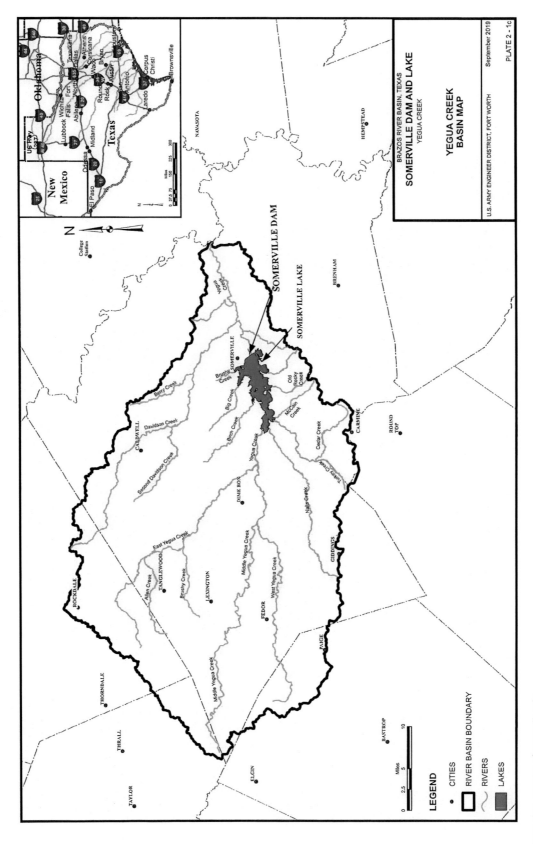

Figure 17.4 Location of Somerville Lake.

Figure 18.1 Retreat of Gangotri Glacier (1780–2001). Image source: https://earthobservatory.nasa.gov/.

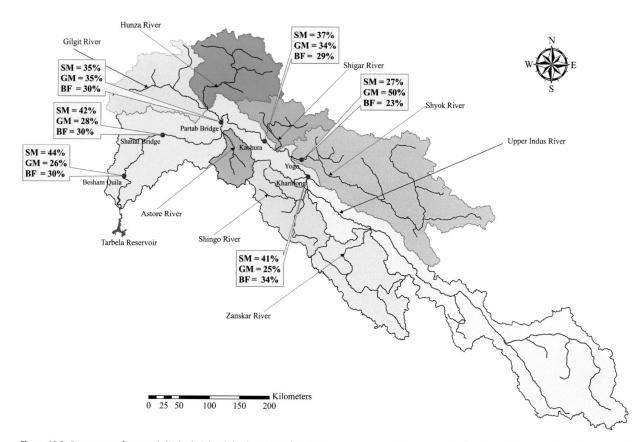

Figure 18.2 Percentages of snowmelt (SM), glacial melt (GM), and base flow (BF) at certain gauging stations along the main stem of Upper Indus River (the method of these estimations are given in Mukhopadhyay and Khan, 2015a). Revised from Mukhopadhyay, B. and Khan, A. (2015b). Boltzmann–Shannon entropy and river flow stability within Upper Indus Basin in a changing climate. *International Journal of River Basin Management*, vol. 13, no. 1, 87–95.

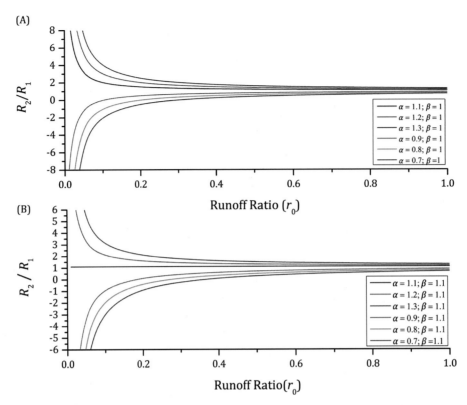

Figure 18.4 Solutions to Eq. 18.2 for a set of values for α and β. (A) The calculated curves represent changes in precipitation for a 30% decrease to a 30% increase in precipitation ($\alpha = 0.7$–1.3) with no change in evapotranspiration ($\beta = 1.0$). (B) The calculated curves represent changes in precipitation for a 30% decrease to a 30% increase in precipitation ($\alpha = 0.7$–1.3) with a 10% increase in evapotranspiration ($\beta = 1.1$). Note that if $\alpha = \beta$ then $R = \alpha$. Revised from Mukhopadhyay, B. (2013). Signature and hydrologic consequences of climate change within Upper-Middle Brahmaputra Basin. *Hydrological Processes*, vol. 27, 2126–2143.

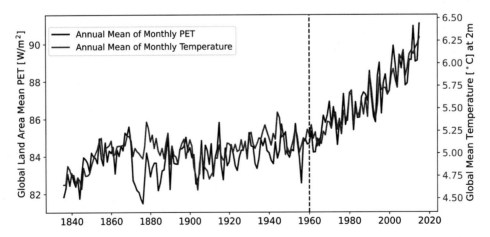

Figure 18.5 Global annual mean of monthly temperature (red solid line) and global land area annual mean of monthly PET (blue solid line).

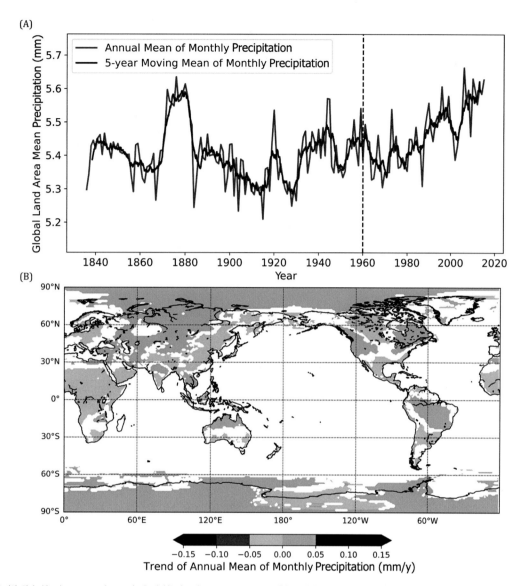

Figure 18.6 (A) Global land area annual mean (red solid line) and 5-year moving mean (blue solid line) of monthly precipitation and (B) the trend of the annual mean of monthly precipitation. The significant trends at a 5% significant level are colored.

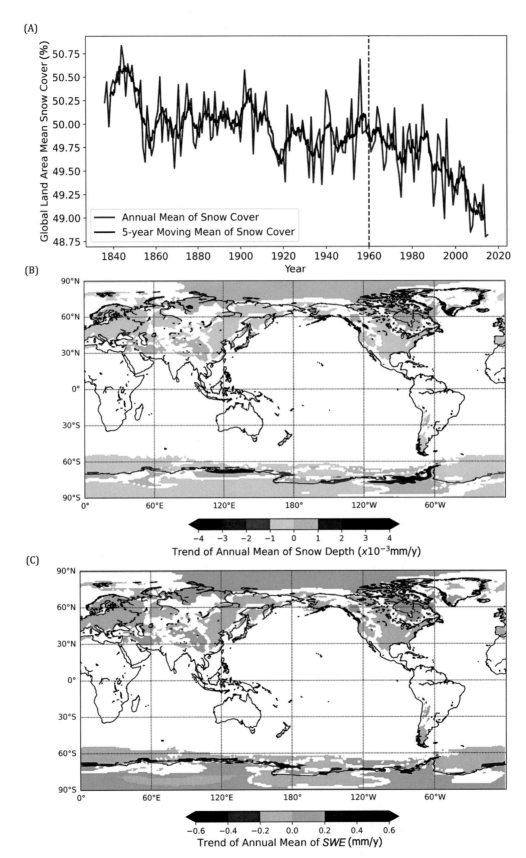

Figure 18.7 (A) Global land area annual mean (red solid line) and 5-year moving mean (blue solid line) of snow cover, (B) the trend of annual mean of snow depth, and (C) the trend of annual mean of the *SWE*. The significant trends at a 5% significant level are colored.

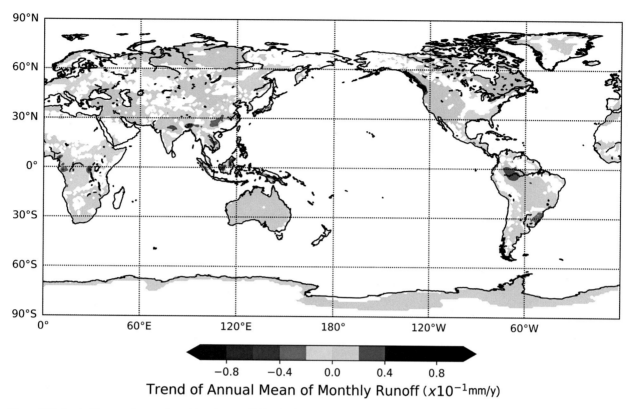

Figure 18.9 Trend of the annual mean of monthly surface runoff. The significant trends at a 5% significant level are colored.

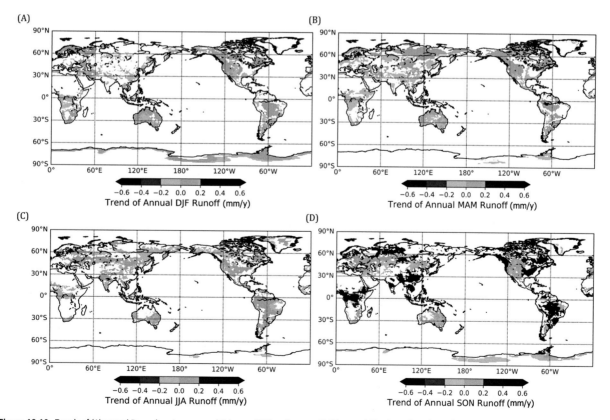

Figure 18.10 Trends of (A) annual December, January, and February (DJF) surface runoff, (B) annual March, April, and May (MAM) surface runoff, (C) annual June, July, and August (JJA) surface runoff, and (D) annual September, October, and November (SON) surface runoff. The significant trends at a 5% significant level are colored.

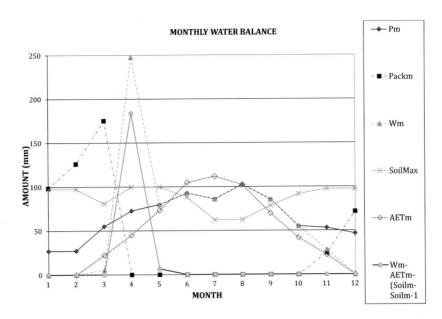

Figure 19.4 Results of water balance calculations.

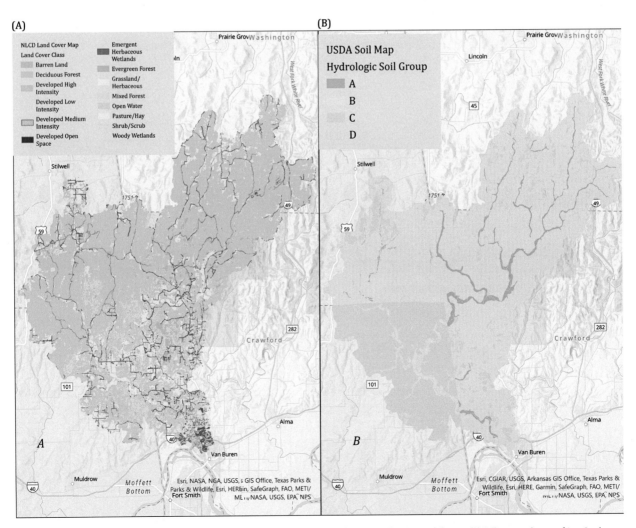

Figure 20.23 (A) Land cover polygons of Lee Creek Watershed in vector format derived from NLCD data in raster format. (B) Soil cover polygons of Lee Creek Watershed in vector format derived from the gSSURGO database.

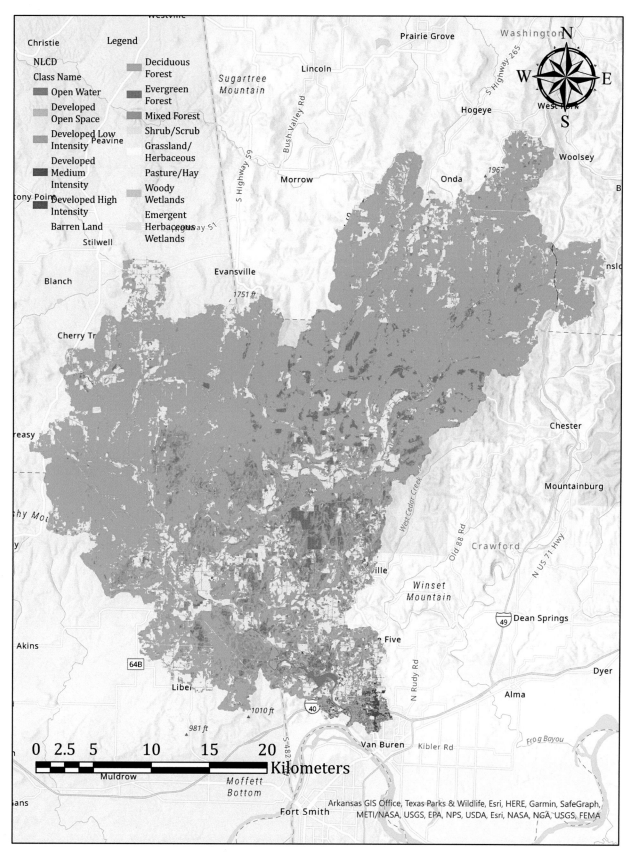

Figure 21.8 Land cover map of Lee Creek Watershed.

Figure 21.13 El Paso Hills Basin overlain on the areal imagery.

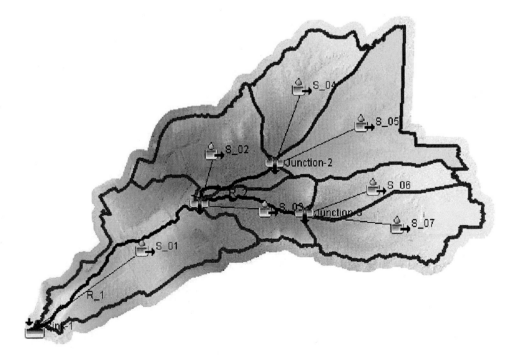

Box 21.20 HEC-HMS basin model of El Paso Hills Basin

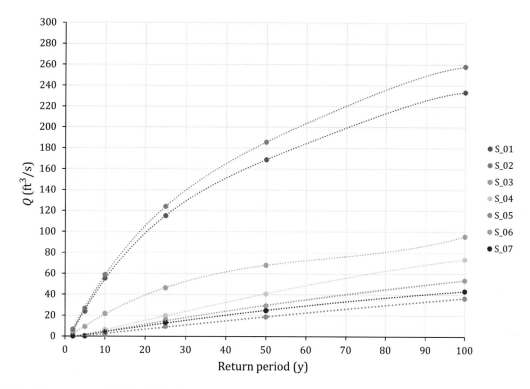

Figure 21.17 Flood-frequency curves for El Paso Hills Basin.

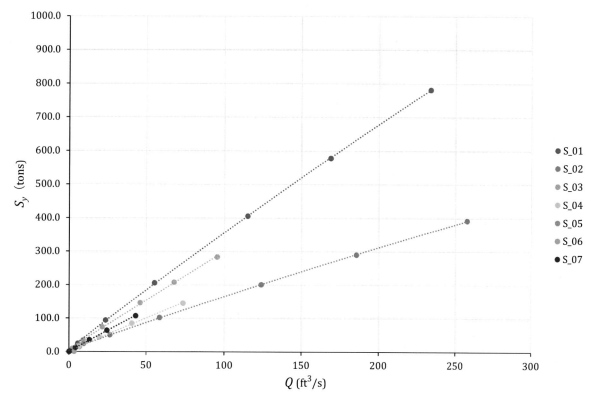

Figure 21.18 Sediment yield graphs for El Paso Hills Basin.

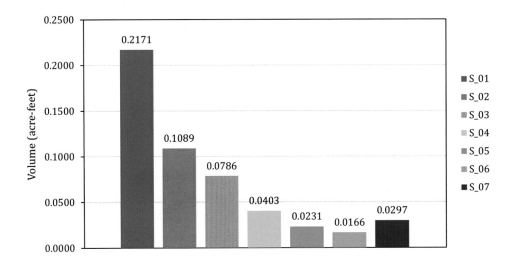

Figure 21.19 Sediment volumes for 100-year discharge from each of the catchments.

9.12 EXAMPLES

Example 9.1: The IRF of a system is given as: $u(t) = a + b\left[1 - \cos\left(\frac{\pi t}{T}\right)\right]$. Let $a = 0.1$ unit, $b = 0.2$ unit, and $T = 12$ time units. Calculate the SRF and PRF. Plot the IRF, SRF, and PRF.

Solution: The SRF is obtained by integrating the IRF. Thus, the SRF is given by

$$SRF = \int_0^t \left\{a + b\left[1 - \cos\left(\frac{\pi t}{T}\right)\right]\right\} dt$$

$$= \int_0^t a\,dt + \int_0^t b\,dt - \int_0^t \cos\left(\frac{\pi t}{T}\right) dt$$

$$= \left[at + bt - \frac{bT}{\pi}\sin\left(\frac{\pi t}{T}\right)\right]_0^t$$

For $0 \leq t \leq \Delta t$, PRF = SRF/Δt since $g(t - \Delta t) = 0$. For $t > \Delta t$, the PRF is given by

$$PRF = \frac{1}{\Delta t}\left[\left\{at + bt - \frac{bT}{\pi}\sin\left(\frac{\pi t}{T}\right)\right\} \right.$$

$$\left. - \left\{a(t - \Delta t) + b(t - \Delta t) - \frac{bT}{\pi}\sin\left(\frac{\pi(t - \Delta t)}{T}\right)\right\}\right]$$

Table 9.7 shows the results of the calculations. Figure 9.31 shows the SRF and Figure 9.32 show the plots of PRF compared with the IRF.

Table 9.7 *Calculations of IRF, SRF, and PRF*

Time, t	IRF	SRF	PRF
0	0.10	0.00	0.00
1	0.11	0.10	0.03
2	0.14	0.23	0.06
3	0.19	0.39	0.10
4	0.25	0.61	0.15
5	0.32	0.89	0.20
6	0.40	1.25	0.26
7	0.48	1.69	0.33
8	0.55	2.21	0.40
9	0.61	2.79	0.47
10	0.66	3.43	0.54
11	0.69	4.10	0.60
12	0.70	4.80	0.65
13	0.69	5.50	0.68
14	0.66	6.17	0.69
15	0.61	6.81	0.68
16	0.55	7.39	0.65
17	0.48	7.91	0.60
18	0.40	8.35	0.54
19	0.32	8.71	0.47
20	0.25	8.99	0.40
21	0.19	9.21	0.33
22	0.14	9.37	0.26
23	0.11	9.50	0.20
24	0.10	9.60	0.15

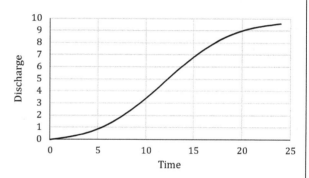

Figure 9.31 Plot of SRF.

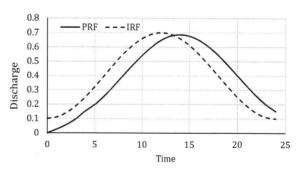

Figure 9.32 Plot of PRF and IRF.

Example 9.2: The rainfall and runoff data for a rain event that occurred on June 19, 1973, over Duck Creek Watershed in Richardson-Garland, Texas, are given in Tables 9.8. The area of the watershed is 8.05 square miles. Determine the duration and ordinates of the unit hydrograph that can be derived from these data.

Solution: The rainfall hyetograph and runoff hydrograph constructed from the given data are shown in Figures 9.33 and 9.34, respectively.

Table 9.8 *Rainfall and runoff data (Duck Creek Watershed, Richardson, Texas)*

Rainfall data		Discharge data	
Date/Time	Incremental rain (inches)	Date/Time	Runoff (ft³/s)
6/19/73 18:45	0	6/19/73 0:00	2
6/19/73 18:55	0.33	6/19/73 18:00	2
6/19/73 19:00	0.13	6/19/73 18:45	2
6/19/73 19:15	0.27	6/19/73 18:55	2
6/19/73 19:30	0.05	6/19/73 19:00	2
6/19/73 19:45	0.38	6/19/73 19:15	2
6/19/73 20:00	0.33	6/19/73 19:30	7.5
6/19/73 20:15	0.53	6/19/73 19:45	25
6/19/73 20:30	0.19	6/19/73 20:00	80
6/19/73 20:45	0.19	6/19/73 20:15	295
6/19/73 21:00	0.02	6/19/73 20:30	968
6/19/73 21:15	0.01	6/19/73 20:45	1820
6/19/73 21:30	0.01	6/19/73 21:00	2200
6/19/73 21:45	0.01	6/19/73 21:15	2320
6/19/73 22:00	0	6/19/73 21:30	2220
6/19/73 22:15	0.05	6/19/73 21:45	1900
		6/19/73 22:00	1630
		6/19/73 22:15	1390
		6/19/73 22:30	1170
		6/19/73 22:45	968
		6/19/73 23:00	825
		6/19/73 23:15	722
		6/19/73 23:30	631
		6/19/73 23:45	569
		6/20/73 0:00	515
		6/20/73 0:30	436
		6/20/73 0:45	375
		6/20/73 1:00	355
		6/20/73 1:15	305
		6/20/73 2:00	295
		6/20/73 3:00	170
		6/20/73 6:00	78
		6/20/73 9:00	39
		6/20/73 12:00	21
		6/20/73 15:00	12
		6/20/73 18:00	6.8
		6/20/73 21:00	3.8
		6/21/73 0:00	2.5

Step 1: Determine the volume under Direct Runoff Hydrograph: V_{DRH}.

From the discharge data, it is seen that the streamflow at the start of the rising limb of the observed hydrograph is only 2 ft³/s. This is determined to be the baseflow. This is subtracted from the observed discharge. The volume under the resultant DRH, V_{DRH}, is computed using the trapezoidal rule of numerical integration. Using the trapezoidal method, the total volume of discharge is calculated. The formula for a single trapezoidal approximation is

$$V_{trapezoid} = (t_{n+1} - t_n) \times \frac{U_n + U_{n+1}}{2} \times 3600 \text{ s/h}$$

where n is the number of the earliest ordinate of the pair being included in the trapezoidal approximation. By taking the sum of all trapezoidal approximations, the total volume of discharge from the storm-specific hydrograph is derived.

$$\sum_1^N V_{trapezoid,i} = 2.4 \times 10^7 \text{ft}^3$$

where N is the total number of trapezoids calculated. This value represents the total volume of discharge over the entire area. Thus,

$$V_{runoff} = 2.4 \times 10^7 \text{ft}^3$$

Step 2: Convert V_{DRH} to equivalent depth of runoff.

In order to calculate the excess rainfall produced by the storm, the volume of discharge should be divided by the total area. Thus, V_{runoff} is expressed in equivalent units of depth of runoff, V_{DRH}, by dividing this number by the area of the watershed, $Area = 8.05 \text{ mi}^2 \times 5280^2 \text{ ft}^2/\text{mi}^2$:

$$Area = 2.2 \times 10^8 \text{ft}^2$$

$$V_{DRH} = V_{runoff}/Area$$

$$V_{DRH} = \frac{2.4 \times 10^7 \text{ft}^3}{2.2 \times 10^8 \text{ft}^2} = 0.10673 \text{ ft} = 1.2807 \text{ inches}$$

Excess rainfall $= 1.2807$ in

Step 3: Establish the duration of the unit hydrograph.

The duration of the unit hydrograph is established by determining the duration of effective rainfall. This is done using the Φ-index method described in Chapter 7. The results of the calculations are presented in Table 9.9.

The steps involved in the calculation are as follows.

1. Determine the volume (expressed as depth) of rainfall loss (V_{loss}) due to infiltration etc. V_{loss} is obtained by subtracting the volume under the DRH from the total volume of rainfall, called the gross rainfall V_{gross},

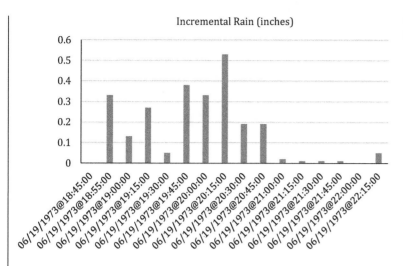

Figure 9.33 Incremental rainfall (inches) in Duck Creek Watershed near Dallas, Texas.

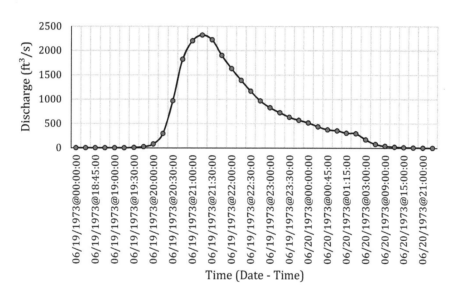

Figure 9.34 Runoff resulting from the rainfall shown in Figure 9.33 in Duck Creek.

which is obtained by calculating the cumulative rainfall. This is given in column 3 of Table 9.9.

The cumulative depth or $V_{gross} = 2.5$ inches.

$$\therefore V_{loss} = V_{gross} - V_{DRH} = (2.5 - 1.2807) \text{ inches}$$
$$= 1.219 \text{ inches}$$

2. Determine the rainfall intensity from the incremental rainfall depth and hours lapsed between two successive records of rainfall depth. The incremental depth of rainfall is converted to rainfall intensity (in/h) in column 4 of Table 9.9.
3. An initial estimate of the Φ-index is obtained:

$$\Phi - \text{index} = \frac{V_{loss}}{T_r} = \frac{1.2193 \text{ in}}{3.5 \text{ h}} 0.3484 \text{ in/h}$$

where, T_r is the observed rainfall duration (3.5 h as shown in the first column of Table 9.9)

4. When the Φ-index value obtained above is subtracted from the observed intensity (column 4) the trial effective intensity values thus obtained and given in column 5 of Table 9.9 yields a cumulative effective rainfall total of 1.94 inches (by taking only the positive values given in column 5), which is greater than the observed effective rainfall total of 1.281 inches.

Table 9.9 *Calculation of effective rainfall duration*

Hours passed since start of rainfall	Incremental depth of rainfall (in)	Cumulative depth of rainfall (in)	Intensity of rainfall (in/h)	Trial effective rainfall (in/h)	Effective rainfall intensity (in/h)	Effective rainfall depth (in)
0	0					
0.1667	0.33	0.33	1.98	1.63	1.37	0.23
0.25	0.13	0.46	1.56	1.21	0.95	0.08
0.5	0.27	0.73	1.08	0.73	0.47	0.12
0.75	0.05	0.78	0.20	−0.15	−0.41	0.00
1	0.38	1.16	1.52	1.17	0.91	0.23
1.25	0.33	1.49	1.32	0.97	0.71	0.18
1.5	0.53	2.02	2.12	1.77	1.51	0.38
1.75	0.19	2.21	0.76	0.41	0.15	0.04
2	0.19	2.4	0.76	0.41	0.15	0.04
2.25	0.02	2.42	0.08	−0.27	−0.53	
2.5	0.01	2.43	0.04	−0.31	−0.57	
2.75	0.01	2.44	0.04	−0.31	−0.57	
3	0.01	2.45	0.04	−0.31	−0.57	
3.25	0	2.45	0.00	−0.35	−0.61	
3.5	0.05	2.5	0.20	−0.15	−0.41	
					Effective rainfall total	**1.281**

Table 9.10 *The unit hydrograph ordinates of the 2-hour unit hydrograph for Duck Creek Watershed*

Date/Time	Unit hydrograph ordinate (ft³/s)
06/19/1973@00:00:00	0.00
06/19/1973@18:00:00	0.00
06/19/1973@18:45:00	0.00
06/19/1973@18:55:00	0.00
06/19/1973@19:00:00	0.00
06/19/1973@19:15:00	0.00
06/19/1973@19:30:00	4.29
06/19/1973@19:45:00	17.96
06/19/1973@20:00:00	60.90
06/19/1973@20:15:00	228.78
06/19/1973@20:30:00	754.27
06/19/1973@20:45:00	1419.53
06/19/1973@21:00:00	1716.24
06/19/1973@21:15:00	1809.94
06/19/1973@21:30:00	1731.86
06/19/1973@21:45:00	1482.00
06/19/1973@22:00:00	1271.18
06/19/1973@22:15:00	1083.78
06/19/1973@22:30:00	912.00
06/19/1973@22:45:00	754.27
06/19/1973@23:00:00	642.62
06/19/1973@23:15:00	562.19
06/19/1973@23:30:00	491.14
06/19/1973@23:45:00	442.73
06/20/1973@00:00:00	400.56
06/20/1973@00:30:00	338.88
06/20/1973@00:45:00	291.25
06/20/1973@01:00:00	275.63
06/20/1973@01:15:00	236.59
06/20/1973@02:00:00	228.78

Table 9.10 (*cont.*)

Date/Time	Unit hydrograph ordinate (ft³/s)
06/20/1973@03:00:00	131.18
06/20/1973@06:00:00	59.34
06/20/1973@09:00:00	28.89
06/20/1973@12:00:00	14.84
06/20/1973@15:00:00	7.81
06/20/1973@18:00:00	3.75
06/20/1973@21:00:00	1.41
06/21/1973@00:00:00	0.39

5. Using the procedures described in Chapter 7, the correct value of the Φ-index is determined to be 0.611 in/h. This value is used to get the effective incremental rainfall values given in column 6 of Table 9.9. The effective rainfall total thus obtained is 1.281 inches, which is the equivalent depth of runoff obtained from the DRH.
6. The effective rainfall given in column 7 show that the duration of effective rainfall is almost 2 hours.

Therefore, the duration of the unit hydrograph determined above is 2 hours.

Step 4: Determine the unit hydrograph ordinates.

The ordinates of the DRH are normalized by dividing the ordinates by 1.281 to get the ordinates of the unit hydrograph. These are given in Table 9.10 and the plots of the unit hydrograph and DRH are given in Figure 9.35.

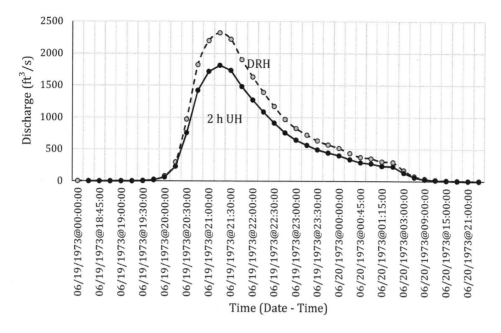

Figure 9.35 Observed DRH and derived 2-hour unit hydrograph (UH) for the rainfall–runoff event shown in Figures 9.33 and 9.34.

Example 9.3: For the rainfall–runoff event and data given in Example 9.2, do the following:
1. Determine the unit hydrograph from the NRCS dimensionless unit hydrograph parameters.
2. Derive a unit hydrograph using the gamma probability distribution method, with the form of the gamma distribution function that was given by Haan (1970). Develop also the unit hydrograph using the shape factor derived from the peak rate factor that can be established in Part 1 above.

Solution to Part 1:

Step 1: Determine the duration of the unit hydrograph from the time of concentration.

When an observed hydrograph is available, the time of concentration can be estimated as the time from the end of effective rainfall and the inflection point of the recession limb of the observed hydrograph. The inflection point on the recession limb of the hydrograph can be found by calculating the second derivative of discharge with respect to time and finding the point where the sign of the second derivative changes. In other words, it is found by calculating the slopes of two successive points of the hydrograph. This can be easily accomplished by invoking the "Slope" function in Excel. Following this procedure, the inflection point is found to be 5.25 hours and as determined in Example 9.2, the end time of the effective rainfall is 2.25 hours. Thus, $t_c = 3$ hours. From this the following calculations are performed:

$t_r = 0.133 t_c$

$t_r = 0.133 \times 3 \text{ h} = 0.399 \text{ h}$

Thus, the *duration* for the NRCS unit hydrograph is 0.399 hours or 24 minutes.

Step 2: Calculate the lag time.

With the calculated duration of the unit hydrograph, the lag time is determined using $t_l = 0.6 t_c$:

$t_l = 0.6 \times 3 \text{ h} = 1.8 \text{ h}$

Because NRCS considers the lag time as 60% of the time of concentration, empirically, a unit hydrograph with a 3-hour time of concentration yields a lag time of 1.8 hours.

Step 3: Determine the time to the peak of the unit hydrograph.

$$t_p = \frac{t_r}{2} + t_l = \frac{0.399}{2} + 1.8 = 1.9995 \text{ h}$$

Step 4: Determine the peak rate factor.

First, a dimensionless unit hydrograph is created from the hydrograph derived in Example 9.2 by dividing the coordinates of the *derived* unit hydrograph by the peak flow, q_p, and time of peak, t_p, respectively. The values of t_p and q_p are 2 h and 1809.941 ft³/s, respectively.

Table 9.11 *Coordinates of the dimensionless unit hydrographs*

Coordinates of NRCS dimensionless unit hydrograph		Dimensionless unit hydrograph (observed)	
t/t_p	q/q_p	t/t_p	q/q_p
0	0	0	0.000
0.1	0.03	0.125	0.002
0.2	0.1	0.25	0.010
0.3	0.19	0.375	0.034
0.4	0.31	0.5	0.126
0.5	0.47	0.625	0.417
0.6	0.66	0.75	0.784
0.7	0.82	0.875	0.948
0.8	0.93	1	1.000
0.9	0.99	1.125	0.957
1	1	1.25	0.819
1.1	0.99	1.375	0.702
1.2	0.93	1.5	0.599
1.3	0.86	1.625	0.504
1.4	0.78	1.75	0.417
1.5	0.68	1.875	0.355
1.6	0.56	2	0.311
1.7	0.46	2.125	0.271
1.8	0.39	2.25	0.245
1.9	0.33	2.375	0.221
2	0.28	2.625	0.187
2.2	0.207	2.75	0.161
2.4	0.147	2.875	0.152
2.6	0.107	3	0.131
2.8	0.077	3.375	0.126
3	0.055	3.875	0.072
3.2	0.04	5.375	0.033
3.4	0.029	6.875	0.016
3.6	0.021	8.375	0.008
3.8	0.015	9.875	0.004
4	0.011	11.375	0.002
4.5	0.005	12.875	0.001
5	0	14.375	0.000

Columns 1 and 2 of Table 9.11 give the coordinates of the NRCS dimensionless unit hydrograph and columns 3 and 4 give the coordinates of the dimensionless unit hydrograph derived from observed runoff presented in Example 9.2.

Figure 9.36 shows the two dimensionless unit hydrographs.

The peak rate factor is determined from the sum of the coordinates of the calculated dimensionless unit hydrograph. However, in this case, since Δt of the calculated hydrograph changed at $t/tp > 3.375$, an average Δt (0.155 h) is used in the calculation and was optimized during step 6.

$$PRF = \frac{645.33}{\Sigma DUH_Y \times \Delta t} = \frac{645.33}{9.6163 \times 0.155} = 433$$

If Δt is decreased to 0.14 h, the *PRF* value increases to 477. A PRF value of 450 is selected to obtain the optimal results in step 6.

Step 5: Determine the peak discharge of the unit hydrograph.

From the time to peak, the peak discharge of the unit hydrograph is determined, in ft^3/s:

$$U_p = \frac{450 * A}{t_p}$$

$$U_p = \frac{450 \times (8.05 \text{ mi}^2)}{1.9995 \text{ h}} \times \frac{1 \text{ h}}{3600 \text{ s}} * \frac{(5280 \text{ ft})^2}{\text{mi}^2}$$

$$U_p = 1811.703 \text{ ft}^3/\text{s}$$

Step 6: Compute the ordinates and abscissa of the dimensioned unit hydrograph.

The ordinates and abscissa of the dimensioned unit hydrograph are computed using the dimensionless unit hydrograph, time to peak, and peak flow. The coordinates of the dimensionless NRCS unit hydrograph are multiplied by t_p (2 h) and q_p (1811.703 cfs). The calculations are presented in Table 9.12 in which column 3 is obtained

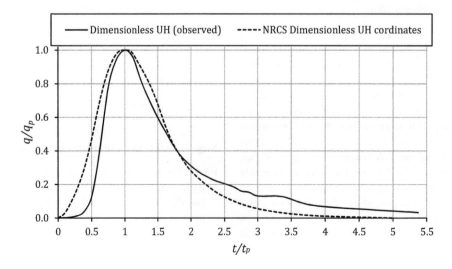

Figure 9.36 Comparisons of the NRCS dimensionless unit hydrograph (UH) with the dimensionless form of the UH derived for the rainfall–runoff event shown in Figures 9.33 and 9.34.

Table 9.12 *Comparisons of unit hydrographs calculated from NRCS method and conventional method*

Coordinates of NRCS dimensionless unit hydrograph		Unit hydrograph from t_c (NRCS)		Unit hydrograph from observed runoff	
t/t_p	q/q_p	t (h)	U (ft³/s)	t (h)	U (ft³/s)
0	0	0	0	0.00	0
0.1	0.03	0.2000	54.3511	0.25	4.294511
0.2	0.1	0.3999	181.1703	0.50	17.95886
0.3	0.19	0.5999	344.2236	0.75	60.90397
0.4	0.31	0.7998	561.6279	1.00	228.7803
0.5	0.47	0.9998	851.5004	1.25	754.2723
0.6	0.66	1.1997	1195.7239	1.50	1419.531
0.7	0.82	1.3997	1485.5964	1.75	1716.243
0.8	0.93	1.5996	1684.8837	2.00	1809.941
0.9	0.99	1.7996	1793.5859	2.25	1731.859
1	1	1.9995	1811.7029	2.50	1481.997
1.1	0.99	2.1995	1793.5859	2.75	1271.175
1.2	0.93	2.3994	1684.8837	3.00	1083.778
1.3	0.86	2.5994	1558.0645	3.25	911.998
1.4	0.78	2.7993	1413.1283	3.50	754.2723
1.5	0.68	2.9993	1231.9580	3.75	642.615
1.6	0.56	3.1992	1014.5536	4.00	562.1905
1.7	0.46	3.3992	833.3833	4.25	491.1359
1.8	0.39	3.5991	706.5641	4.50	442.725
1.9	0.33	3.7991	597.8620	4.75	400.5608
2	0.28	3.9990	507.2768	5.25	338.876
2.2	0.207	4.3989	375.0225	5.5	291.2459
2.4	0.147	4.7988	266.3203	5.75	275.6295
2.6	0.107	5.1987	193.8522	6	236.5885
2.8	0.077	5.5986	139.5011	6.75	228.7803
3	0.055	5.9985	99.6437	7.75	131.1778
3.2	0.04	6.3984	72.4681	10.75	59.34233
3.4	0.029	6.7983	52.5394	13.75	28.89035
3.6	0.021	7.1982	38.0458	16.75	14.83558
3.8	0.015	7.5981	27.1755	19.75	7.808202
4	0.011	7.9980	19.9287	22.75	3.747937
4.5	0.005	8.9978	9.0585	25.75	1.405476
5	0	9.9975	0.0000	28.75	0.39041

by multiplying column 1 by t_p and column 4 is obtained by multiplying column 2 by U_p. Figure 9.37 shows the unit hydrographs obtained from the two different methods.

Solution to Part 2:

Step 1: Compute the parameter of the gamma distribution.

The pdf (Haan model) requires an estimation of the shape factor, K.

From the observed runoff data given in columns 1 and 2 of Table 9.13, $q_p = 2320$ ft³/s and $t_p = 3.25$ h. The depth under the hydrograph, estimated in Example 9.2, $D_{DRH} = 1.2807$ in. The area of the watershed is 224 421 120 ft². From these data, q_p expressed as in/h/in, is calculated as:

$$q_p = \frac{2320 \text{ ft}^3/\text{s}}{224\,421\,120 \text{ ft}^2} \times \frac{3600 \text{ s}}{\text{h}} \times \frac{12 \text{ in}}{\text{ft}} \times \frac{1}{1.2807 \text{ in}}$$
$$= 0.34871 \frac{\text{in}}{\text{h}}/\text{in}$$

With this value of q_p, $t_p = 3.5$ h, and setting $V = 1$ in (for a unit hydrograph), the equation given as

$$q_p t_p \Gamma(K) \left(\frac{e}{K}\right)^K = V$$

is solved by iteration (using Excel's Goal Seek function) and value of K is obtained as

$$K = 8.23931$$

Step 2: Calculate q/q_p as a function of t/t_p.

From given values of t and t_p, $q(t)/q_p$ is calculated using the probability distribution function

$$\frac{q(t)}{q_p} = \left[\frac{t}{t_p} \exp\left\{1 - \left(\frac{t}{t_p}\right)\right\}\right]^K$$

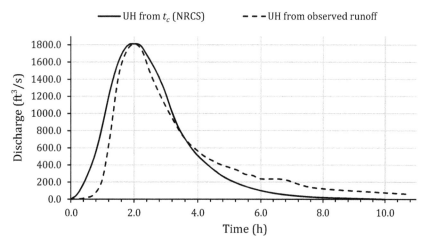

Figure 9.37 Unit hydrographs obtained from two approaches discussed.

Table 9.13 *Results of calculations for Part 2 of Example 9.3*

Observed runoff		Dimensionless gamma distribution		Dimensional gamma distribution		NRCS
		Haan model				$K = 2.2723$
t (h)	Discharge (ft^3/s)	t/t_p	q/q_p	q (in/h)	U (ft^3/s)	q/q_p
0	2	0	0	0	0	0
0.75	2	0.2308	0.0032	0.0014	7.4309	0.2051
0.9167	2	0.2821	0.0110	0.0049	25.4501	0.2881
1	2	0.3077	0.0182	0.0081	42.1888	0.3312
1.25	2	0.3846	0.0607	0.0271	140.7396	0.4617
1.5	7.5	0.4615	0.1446	0.0646	335.3995	0.5866
1.75	25	0.5385	0.2732	0.1220	633.7310	0.6992
2	80	0.6154	0.4355	0.1945	1010.3360	0.7951
2.25	295	0.6923	0.6098	0.2723	1414.7331	0.8725
2.5	968	0.7692	0.7708	0.3442	1788.2649	0.9307
2.75	1820	0.8462	0.8969	0.4005	2080.7802	0.9704
3	2200	0.9231	0.9746	0.4353	2261.1275	0.9929
3.25	2320	1.0000	1.0000	0.4466	2320.0000	1.0000
3.5	2220	1.0769	0.9771	0.4363	2266.8080	0.9936
3.75	1900	1.1538	0.9153	0.4088	2123.4311	0.9759
4	1630	1.2308	0.8265	0.3691	1917.4439	0.9488
4.25	1390	1.3077	0.7226	0.3227	1676.4944	0.9143
4.5	1170	1.3846	0.6140	0.2742	1424.5525	0.8741
4.75	968	1.4615	0.5086	0.2272	1180.0316	0.8299
5	825	1.5385	0.4118	0.1839	955.3948	0.7830
5.25	722	1.6154	0.3266	0.1459	757.7313	0.7345
5.5	631	1.6923	0.2542	0.1135	589.8266	0.6854
5.75	569	1.7692	0.1946	0.0869	451.3699	0.6367
6	515	1.8462	0.1466	0.0655	340.0715	0.5889
6.5	436	2.0000	0.0798	0.0356	185.1315	0.4979
6.75	375	2.0769	0.0578	0.0258	134.0519	0.4555
7	355	2.1538	0.0414	0.0185	95.9741	0.4154
7.25	305	2.2308	0.0293	0.0131	67.9936	0.3778
8	295	2.4615	0.0099	0.0044	22.8537	0.2797
9	170	2.7692	0.0021	0.0009	4.7798	0.1816
12	78	3.6923	0.0000	0.0000	0.0255	0.0429
15	39	4.6154	0.0000	0.0000	0.0001	0.0087
18	21	5.5385	0.0000	0.0000	0.0000	0.0016
21	12	6.4615	0.0000	0.0000	0.0000	0.0003
24	6.8	7.3846	0.0000	0.0000	0.0000	0.0000
27	3.8	8.3077	0.0000	0.0000	0.0000	0.0000
30	2.5	9.2308	0.0000	0.0000	0.0000	0.0000

Columns 3 and 4 of Table 9.13 list the results of the calculation of the complete gamma distribution.

Step 3: Convert dimensionless values to dimensional values.

The value of q_p, expressed as in/h, is

$$q_p = \frac{2320 \text{ ft}^3/\text{s}}{224\,421\,120 \text{ ft}^2} \times \frac{3600 \text{ s}}{\text{h}} \times \frac{12 \text{ in}}{\text{ft}} = 0.4466 \frac{\text{in}}{\text{h}}.$$

This value is used to obtain $q(t)$ listed in columns 5, and column 6 gives the corresponding values in ft^3/s.

Step 4: Reevaluate the shape factor from the peak rate factor.

The data provided by NRCS (2007) is used to construct Figure 9.38, which gives the variation of the shape factor K with the peak rate factor, *PRF*. The *PRF* value determined above is 450. Using the equation given in Figure 9.38, the corresponding value of K is calculated as

$$K = 10^{-5} \times (450)^2 + 4 \times 10^{-4} \times 450 + 6.73 \times 10^{-2}$$
$$= 2.2723$$

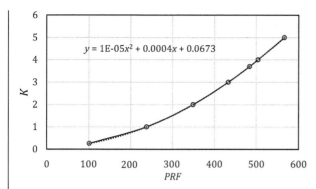

Figure 9.38 Shape factor (K) as a function of peak rate factor (PRF) in the NRCS synthetic unit hydrograph.

As noted in Section 9.8.4, the form of the gamma distribution function given by NRCS (2007) produces a curve that is identical to the one calculated by the form of the gamma distribution function given by Haan (1970). Thus, either of these two forms of the gamma distribution function can be used with the value of K derived above. The results of the calculations are given in the final column of Table 9.13.

Figure 9.39 shows the dimensionless unit hydrograph calculated using two different values of K. The calculated distribution with $K = 8.2393$ fits the observed distribution (Figure 9.37) better than the distribution calculated with $K = 2.2723$.

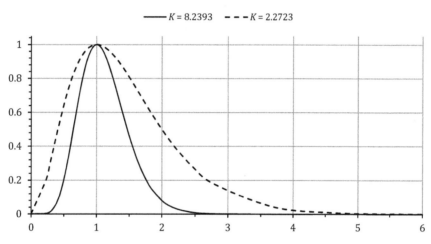

Figure 9.39 Dimensionless unit hydrograph resulting from two different values of the shape factor, K.

Example 9.4: The Bloody Run catchment, located to the northwest section of Cincinnati, Ohio has a drainage area of 1.423 square miles and the length of the main channel is 1.8363 miles. The length from the centroid of the channel to the outlet is 0.8755 miles. Calculate the peak discharge and lag time for a Snyder standard unit hydrograph for this catchment. Determine the peak discharge and time to peak for the 1-hour unit hydrograph. What is the relationship between the unit hydrograph peak and lag time? What control does the peaking coefficient has on this relationship?

Solution:

Step 1: The peaking coefficient, C_p, and the lag time coefficient, C_t, are required for the Snyder method of calculating unit hydrographs. When not given, as is the case in this example, it is necessary to estimate these parameters. Because C_t ranges from 1.8 to 2.2, an initial estimate of 2 is chosen. C_p, ranging from 0.4 to 1.2, is initially estimated at 0.6, but should be checked through model calibration.

Step 2: Using C_p and C_t, the lag time, t_l, and peak discharge per unit area for the Snyder standard unit hydrograph is evaluated:

$$t_l = C_t(L * L_c)^{0.3}$$

$$t_l = 2(1.8363 * 0.8755)^{0.3} = 2.306 \text{ h}.$$

With the assumed value of C_p, the peak discharge per unit area of the standard unit hydrograph is calculated as

$$q_p = C_p \frac{645}{t_l}$$

$$q_p = 0.6\left(\frac{645}{2.306}\right) = 167.8 \text{ ft}^3/\text{s}$$

Step 3: Based on the evaluated t_l, the t_r of the standard unit hydrograph is calculated as

$$t_l = 5.5 t_r$$

$$t_r = \frac{2.306}{5.5} = 0.42 \text{ h}$$

Step 4: Because the duration of standard unit hydrograph differs from the desired duration, the lag time must be converted to the required lag time, t_{lR}.

$$t_{lR} = t_l + \frac{t_R - t_r}{4}$$

$$t_{lR} = 2.306 + \frac{1 - 0.419}{4} = 2.451 \text{ h}$$

Where t_{lR} is the required lag time and t_R is the desired duration.

Step 5: Using the required lag time, the required peak discharge per unit area is derived:

$$q_{pR} = C_p * \frac{645}{t_{lR}}$$

$$q_{pR} = 0.6 * \frac{645}{2.451} = 157.873 \text{ ft}^3/\text{s}/\text{mi}^2$$

The peak of the 1-hour unit hydrograph for this catchment is

$$U_{pR} = q_{pR} A = 157.873 \frac{\text{ft}^3/\text{s}}{\text{mi}^2} \times 1.423 \text{ mi}^2 = 224.653 \text{ ft}^3/\text{s}.$$

Relationship between q_p and t_l:

Step 6: For three values of C_p, q_p is calculated as a function of t_l and plotted in Figure 9.40. The relationship is that of a rectangular or equilateral hyperbola, also called an orthogonal hyperbola. On a log–log the result is a 45° line, as shown in Figure 9.41.

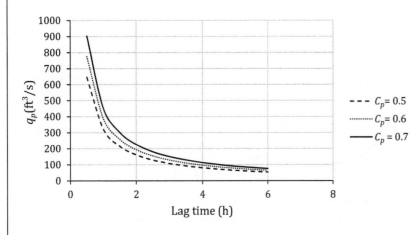

Figure 9.40 Relationship between q_p and t_l plotted on arithmetic scales.

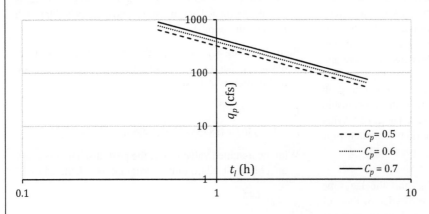

Figure 9.41 Relationship between q_p and t_l on a log–log plot.

Example 9.5: A 6-hour unit hydrograph of a watershed with an area of 1290 square miles within Saluda River basin in South Carolina was produced by USACE (1959). From the map, the length of the watershed was determined to be 92 miles and the distance from the centroid of the main stream to the outlet was found to be 47 miles. The unit hydrograph yielded the peak discharge, lag time, and time base to be 14 100 ft³/s, 34 hours, and 144 hours respectively. Based on the given information, determine the C_p and C_t.

Solution: Given the peak discharge of the derived unit hydrograph and the area of the watershed, it is necessary to obtain the peak discharge per unit area of the standard unit hydrograph:

$$U_p = 14\,100 \text{ ft}^3/\text{s}$$

$$q_p = \frac{U_p}{\text{Area}}$$

$$q_p = \frac{14\,100 \text{ ft}^3/\text{s}}{1290 \text{ mi}^2}$$

$$q_p = 10.93 \text{ ft}^3/\text{s}/\text{mi}^2$$

Following the derivation of q_p, the duration of the unit hydrograph must be checked to ensure that it agrees with the duration of the rainfall of the required hydrograph.

$$t_{lR} = 34 \text{ h}$$

$$t_r = \frac{t_l}{5.5} = \frac{34}{5.5} = 6.18 \text{ h}$$

Because t_r, the duration of the unit hydrograph, is approximately equal to t_R, the required rainfall duration (6 hours), the lag time of the derived unit hydrograph can be used with the 6-hour unit hydrograph:

$$t_r \approx t_R = 6 \text{ h}$$

$$\therefore t_l = t_{lR} = 34 \text{ h}$$

Once t_{lR} and q_p are derived, C_p and C_t. can be obtained as follows:

$$t_l = C_t(L * L_c)^{0.3}$$

$$C_t = \frac{t_l}{(L * L_c)^{0.3}}$$

$$C_t = \frac{34 \text{ h}}{(92 * 47)^{0.3}}$$

$$C_t = 2.7587$$

It should be noted that C_t is a dimensionless coefficient. In a similar manner, C_p can be derived by using the q_p equation: $q_p = C_p \frac{645}{t_l}$:

$$C_p = \frac{q_p \times t_l}{645}$$

$$C_p = \frac{10.93 \times 34}{645}$$

$$C_p = 0.576 \text{ h}$$

Example 9.6: The rainfall–runoff data for an intense rain event that occurred over the Olmos Creek watershed within the City of San Antonio on September 13, 1978, are given in Table 9.14. The drainage area above the stream gauging station (Figure 9.42) is 21.2 square miles. Rodman (1977) gave the following Snyder unit hydrograph parameters for the watershed above the gauging station: $L = 11.59$ mi; $L_c = 5.53$ mi; $S = 0.0066$ ft/ft; $C_t = 0.72$, and $640 C_p = 500$. The rainfall data represent the mean areal precipitation derived from four rain gauges located above the stream gauge. Following the procedures suggested by Aron and White (1982), construct a unit hydrograph of this watershed for a duration appropriate for the rainfall event given in Table 9.14. Subsequently calculate the DRH using the Snyder unit hydrograph constructed and the ERH.

Solution:

Step 1: Determine the ERH.

In order to determine the duration of the unit hydrograph, the duration of the effective rainfall must be determined first. The volume of rainfall lost due to abstraction (V_{loss}) is obtained by subtracting the volume of observed runoff (V_{DRH}) from the volume of gross observed precipitation (V_{GRH}) given in Table 9.11.

$$V_{loss} = V_{GRH} - V_{DRH} = (4.03 - 1.391) \text{ in} = 2.639 \text{ in}$$

In this example, a qualitative analysis is presented to determine the effective rainfall duration and the ERH.

Rainfall started at 03:30 and the streamflow hydrograph started to rise at 06:45. So, for the first 3.5 hours of rainfall, there was no direct surface runoff. Therefore, it can be assumed that 1.05 inches was initial loss and the remainder 1.589 inches (2.639 − 1.05) was lost during the rest of the rainfall duration. Rainfall essentially ceased at 11:00. Initially, a 4-hour effective duration was assumed (07:00–11:00) and a constant loss rate of 0.3973 in/h (1.589 in/4 h) was applied. But this resulted in negative intensity after 09:00. So the effective rainfall duration can be assumed to be 3 hours with a constant loss rate of 0.5297 in/h. This loss rate applied to derive

Table 9.14 *Data*

Date/Time	Cumulative rainfall (in)	Incremental rainfall (in)	Date/Time	Discharge (ft³/s)	Cumulative runoff (in)
09/13/1978@03:25:00	0	0	09/13/1978@04:30:00	0	0
09/13/1978@03:30:00	0.03	0.03	09/13/1978@05:00:00	16	0.0004
09/13/1978@03:45:00	0.07	0.04	09/13/1978@05:15:00	14	0.0007
09/13/1978@04:00:00	0.1	0.03	09/13/1978@05:30:00	13	0.0009
09/13/1978@04:30:00	0.14	0.04	09/13/1978@05:45:00	12	0.0012
09/13/1978@05:00:00	0.24	0.1	09/13/1978@06:00:00	10	0.0015
09/13/1978@05:15:00	0.34	0.1	09/13/1978@06:45:00	90	0.0048
09/13/1978@05:30:00	0.42	0.08	09/13/1978@07:00:00	243	0.0078
09/13/1978@05:45:00	0.51	0.09	09/13/1978@07:05:00	396	0.0102
09/13/1978@06:00:00	0.53	0.02	09/13/1978@07:10:00	549	0.0135
09/13/1978@06:45:00	0.65	0.12	09/13/1978@07:15:00	702	0.0178
09/13/1978@07:00:00	1.05	0.4	09/13/1978@07:20:00	855	0.023
09/13/1978@07:05:00	1.25	0.2	09/13/1978@07:25:00	1010	0.0291
09/13/1978@07:10:00	1.56	0.31	09/13/1978@07:30:00	1160	0.0362
09/13/1978@07:15:00	1.79	0.23	09/13/1978@07:35:00	1310	0.0442
09/13/1978@07:20:00	2.03	0.24	09/13/1978@07:40:00	1470	0.0531
09/13/1978@07:25:00	2.22	0.19	09/13/1978@07:45:00	1620	0.063
09/13/1978@07:30:00	2.42	0.2	09/13/1978@07:50:00	1770	0.0737
09/13/1978@07:35:00	2.67	0.25	09/13/1978@07:55:00	1930	0.0855
09/13/1978@07:40:00	2.95	0.28	09/13/1978@08:00:00	2080	0.1677
09/13/1978@07:45:00	3.24	0.29	09/13/1978@09:00:00	7450	0.7117
09/13/1978@07:50:00	3.48	0.24	09/13/1978@10:00:00	4250	1.0221
09/13/1978@07:55:00	3.69	0.21	09/13/1978@11:00:00	1780	1.152
09/13/1978@08:00:00	3.83	0.14	09/13/1978@12:00:00	1060	1.2294
09/13/1978@09:00:00	3.92	0.09	09/13/1978@13:00:00	590	1.2725
09/13/1978@10:00:00	3.97	0.05	09/13/1978@14:00:00	382	1.3144
09/13/1978@11:00:00	4.01	0.04	09/13/1978@16:00:00	184	1.3412
09/13/1978@12:00:00	4.01	0	09/13/1978@18:00:00	130	1.3602
09/13/1978@13:00:00	4.01	0	09/13/1978@20:00:00	103	1.3753
09/13/1978@14:00:00	4.03	0.02	09/13/1978@22:00:00	78	1.3866
			09/14/1978@00:00:00	60	1.391

the cumulative effective rainfall as shown in column 5 of Table 9.15.

In order to calculate a DRH from an ERH using convolution, the rainfall pulses should be given at the discrete intervals at which the unit hydrograph ordinates are calculated, here at 0.5-hour intervals. Therefore, an equivalent ERH is constructed with six pulses of rainfall as given in Table 9.16.

Step 2: Construct a 3-hour unit hydrograph.

The Snyder unit hydrograph parameters are calculated first.

$$t_l = 0.72 \times (11.59 \times 5.53)^{0.3} = 2.5083 \text{ h}$$

$$C_p = \frac{500}{640} = 0.781$$

The peak discharge of the standard unit hydrograph per unit area is:

$$q_p = 0.781 \frac{645}{2.5083} = 200.898 \frac{\text{ft}^3/\text{s}}{\text{mi}^2} / \text{in}$$

The rainfall duration of the standard unit hydrograph is

$$t_r = \frac{2.5083 \text{ h}}{5.5} = 0.4561 \text{ h}$$

But $t_R = 3$ h. Therefore,

$$t_{lR} = 2.5083 + \frac{3 - 0.4561}{4} = 3.1443 \text{ h}$$

$$q_{pR} = \frac{200.898 \times 2.5083}{3.1443} = 160.2622 \frac{\text{ft}^3/\text{s}}{\text{mi}^2} / \text{in}$$

$$U_{pR} = 160.2622 \frac{\text{ft}^3/\text{s}}{\text{mi}^2} / \text{in} \times 21.2 \text{ mi}^2 = 3397.558 \frac{\text{ft}^3/\text{s}}{\text{in}}$$

$$t_{pR} = \frac{3 \text{ h}}{2} + 3.1443 \text{ h} = 4.6443 \text{ h}$$

Step 3: Construct a unit hydrograph.

The two unit hydrograph parameters, U_{pR} and t_{pR}, namely the unit hydrograph peak flow rate and time to peak, are used in the procedure suggested by Aron and White (1982) using the gamma distribution function.

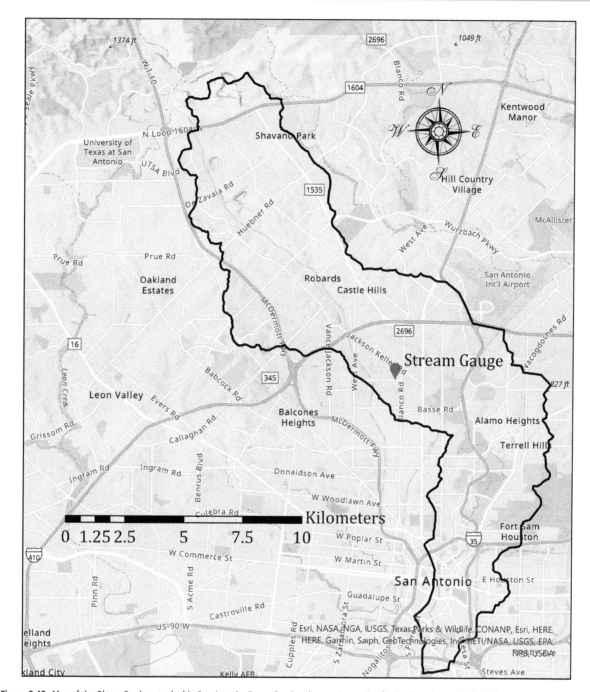

Figure 9.42 Map of the Olmos Creek watershed in San Antonio, Texas showing the gauging station for the data given in Table 9.14.

The form of the gamma distribution function suggested by Aron and White is given as $u(t) = \frac{t^\alpha \exp\left(-\frac{t}{\beta}\right)}{\beta^{\alpha+1}\Gamma(\alpha+1)}$, where α is a dimensionless parameter and β is a parameter with the dimension of time t and α and β are related to the time to peak: $t_p = \alpha\beta$. The unit runoff volume, V, and watershed area, A, are related by a factor C_V: $V = C_V A$, where $C_V = 1.008$ for volumes measured in (ft³/s)·h, areas in acres, and rainfall is 1 inch; and $C_V = 2.778$ for volumes measured in (m³/s)·h areas in square kilometers, and rainfall depth as 1 cm. At time of peak, $u(t_p) = U_p$ and $\beta = t_p/\alpha$, therefore it can be shown that

$$U_p = \frac{C_V A}{t_p} \cdot \frac{\alpha^{\alpha+1}}{\exp(\alpha)\Gamma(\alpha+1)} = \frac{C_V A}{t_p} \phi(\alpha)$$

The function $\phi(\alpha)$ can be evaluated for a given value of α. Table 9.17 gives a list of values of $\phi(\alpha)$ and Figure 9.43 shows the relationship graphically.

Table 9.15 Results of calculations to derive ERH

Incremental rainfall (in)	Intensity (in/h)	Effective intensity (in/h)	Incremental rainfall (in)	Cumulative rainfall (in)	Time (hh:mm)
0.4	0.5333	0.0037	0.0009	0.0009	07:00
0.2	0.8000	0.2703	0.0225	0.0235	
0.31	3.7170	3.1874	0.2655	0.2890	
0.23	2.7611	2.2314	0.1859	0.4748	
0.24	2.8812	2.3515	0.1961	0.6710	
0.19	2.2782	1.7485	0.1457	0.8166	
0.2	2.4010	1.8713	0.1559	0.9725	07:30
0.25	3.0012	2.4715	0.2061	1.1786	
0.28	3.3573	2.8276	0.2355	1.4142	
0.29	3.4814	2.9517	0.2459	1.6600	
0.24	2.8812	2.3515	0.1961	1.8562	
0.21	2.5180	1.9883	0.1656	2.0218	
0.14	1.6807	1.1510	0.0959	2.1177	08:00
0.09	1.0804	0.5508	0.5508	2.6684	09:00
0.05	0.0500	−0.4797	0.0500	2.7184	10:00

Table 9.16 The ERH to be used in the convolution operation to calculate the DRH.

Rainfall pulse, M	Interval (h)	Rainfall (in)	Time
1	0.5	0.97	07:00–07:30
2	1	1.15	07:30–08:00
3	1.5	0.33	08:00–08:30
4	2	0.22	08:30–9:00
5	2.5	0.04	09:00–9:30
6	3	0.01	09:30–10:00
Accumulated rainfall		**2.72 in**	

Table 9.17 Relationship between $\phi(\alpha)$ and α

α	$\phi(\alpha)$
0	0.0000
0.5	0.2420
1	0.3679
1.5	0.4625
2	0.5413
2.5	0.6102
3	0.6721
3.5	0.7288
4	0.7815
4.5	0.8308
5	0.8773
5.5	0.9215
6	0.9637
6.5	1.0042
7	1.0430
7.5	1.0805
8	1.1167
8.5	1.1518
9	1.1858
9.5	1.2189
10	1.2511
10.5	1.2825

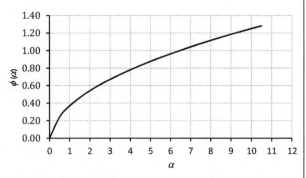

Figure 9.43 Graphical representation of the function $\phi(\alpha)$.

Thus, for a given set of U_p and t_p, $\phi(\alpha)$ can be calculated. For our example,

$$\phi(\alpha) = \frac{U_p t_p}{C_V A} = \frac{3397.558 \times 4.6443}{1.008 \times 21.2} = 1.154$$

From this value of $\phi(\alpha)$, the value of α is calculated by the method of iteration using Excel's Goal Seek function in the definition equation:

$$\phi(\alpha) = \frac{\alpha^{\alpha+1}}{\exp(\alpha)\Gamma(\alpha+1)}.$$

From this,

$$\alpha = 8.5367$$

If time is expressed in terms of t_p, the gamma distribution function can be written as

$$U(zt_p) = U_p z^\alpha \exp[(1-z)\alpha]$$

Thus, a set of zt_p values is used to represent 0.5-hour time intervals and the corresponding values of z are calculated from the value of t_p. The unit hydrograph ordinates

Table 9.18 *Unit hydrograph*

Time (h)	U (ft^3/s/in)
0.0	0.00
0.5	0.04
1.0	5.70
1.5	72.29
2.0	335.14
2.5	895.60
3.0	1689.11
3.5	2504.72
4.0	3114.70
4.5	3385.99
5.0	3310.55
5.5	2970.66
6.0	2483.36
6.5	1956.09
7.0	1464.66
7.5	1049.83
8.0	724.42
8.5	483.45
9.0	313.22
9.5	197.65
10.0	121.81
10.5	73.48
11.0	43.47
11.5	25.27
12.0	14.45

are then calculated. The unit hydrograph is given in Table 9.18 and is plotted in Figure 9.44.

Step 4: Derive the DRH.

The DRH is obtained by a direct convolution operation as presented in Table 9.19.

The principle of superposition illustrated in Figure 9.45 results in the DRH from the unit hydrograph, as shown in Figure 9.46.

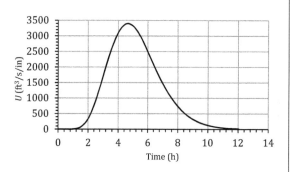

Figure 9.44 Unit hydrograph, $U(t)$.

Table 9.19 *Derivation of DRH using convolution*

Time (h)		U (ft^3/s/in)	P_1 0.972 P_1*U_i	P_2 1.145 P_2*U_i	P_3 0.330 P_3*U_i	P_4 0.220 P_4*U_i	P_5 0.040 P_5*U_i	P_6 0.010 P_6*U_i	DRH (ft^3/s)
0.0	U_1	0.00	0.00						0.00
0.5	U_2	0.04	0.04	0.00					0.04
1.0	U_3	5.70	5.55	0.04	0.00				5.59
1.5	U_4	72.29	70.30	6.53	0.01	0.00			76.85
2.0	U_5	335.14	325.92	82.78	1.89	0.01	0.00		410.60
2.5	U_6	895.60	870.97	383.79	23.89	1.26	0.00	0.00	1279.91
3.0	U_7	1689.11	1642.65	1025.62	110.75	15.93	0.23	0.00	2795.17
3.5	U_8	2504.72	2435.82	1934.32	295.96	73.83	2.89	0.06	4742.88
4.0	U_9	3114.70	3029.02	2868.33	558.18	197.31	13.41	0.72	6666.96
4.5	U_{10}	3385.99	3292.85	3566.85	827.71	372.12	35.82	3.35	8098.71
5.0	U_{11}	3310.55	3219.48	3877.53	1029.28	551.81	67.56	8.96	8754.62
5.5	U_{12}	2970.66	2888.95	3791.13	1118.93	686.19	100.19	16.89	8602.28
6.0	U_{13}	2483.36	2415.05	3401.91	1094.00	745.95	124.59	25.05	7806.54
6.5	U_{14}	1956.09	1902.28	2843.86	981.68	729.33	135.44	31.15	6623.75
7.0	U_{15}	1464.66	1424.37	2240.05	820.65	654.46	132.42	33.86	5305.81
7.5	U_{16}	1049.83	1020.95	1677.28	646.41	547.10	118.83	33.11	4043.67
8.0	U_{17}	724.42	704.49	1202.23	484.01	430.94	99.33	29.71	2950.72
8.5	U_{18}	483.45	470.15	829.58	346.93	322.67	78.24	24.83	2072.41
9.0	U_{19}	313.22	304.61	553.63	239.39	231.28	58.59	19.56	1407.06
9.5	U_{20}	197.65	192.22	358.69	159.76	159.59	41.99	14.65	926.90
10.0	U_{21}	121.81	118.45	226.35	103.51	106.51	28.98	10.50	594.29
10.5	U_{22}	73.48	71.46	139.49	65.32	69.01	19.34	7.24	371.85

Table 9.19 (cont.)

Time (h)		U (ft³/s/in)	P_1 0.972 P_1*U_i	P_2 1.145 P_2*U_i	P_3 0.330 P_3*U_i	P_4 0.220 P_4*U_i	P_5 0.040 P_5*U_i	P_6 0.010 P_6*U_i	DRH (ft³/s)
11.0	U_{23}	43.47	42.28	84.14	40.25	43.54	12.53	4.83	227.58
11.5	U_{24}	25.27	24.58	49.78	24.28	26.83	7.91	3.13	136.51
12.0	U_{25}	14.45	14.06	28.94	14.37	16.19	4.87	1.98	80.40
12.5				16.55	8.35	9.58	2.94	1.22	38.64
13.0					4.78	5.57	1.74	0.73	12.82
13.5						3.18	1.01	0.43	4.63
14.0							0.58	0.25	0.83
14.5								0.14	0.14

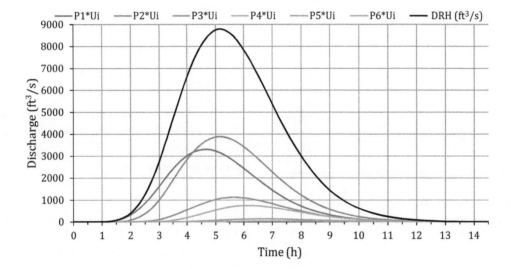

Figure 9.45 Graphical representation of the operation of superposition. (A black and white version of this figure will appear in some formats. For the color version, please refer to the plate section.)

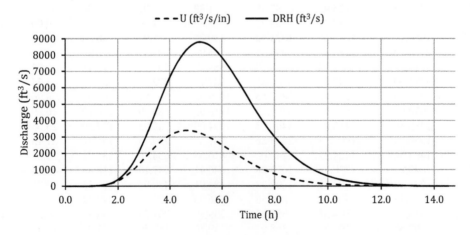

Figure 9.46 Plots of the unit hydrograph (U) and the DRH for the rainfall–runoff event in the Olmos Creek watershed, San Antonio, Texas.

9.12 Examples

Example 9.7: A watershed has an area of 810 km² and a time of concentration of 6 hours. Given a storage coefficient of 5.4, create a unit hydrograph of the required duration using the Clark unit hydrograph method.

Solution:

Step 1: The first step in the Clark unit hydrograph method is determining the time interval to be utilized throughout the Clark calculations. Because the time of concentration is 6 hours, the catchment area is divided into six isochrones, each representing 1 hour of equal travel time.

$$N_i = \frac{t_c}{\Delta t}$$

where N_i is defined as the number of computational time intervals. The HEC equation is used to calculate the cumulative area for successive isochrones by first calculating the A/A_T ratio for each area bounded by isochrones.

$$\frac{A_c}{A_T} = \begin{cases} 1.414(t/t_c)^{1.5} & \text{for } 0 \leq t/t_c \leq 0.5 \\ 1 - 1.414\left(1 - \frac{t}{t_c}\right)^{1.5} & \text{for } 0.5 \leq t/t_c \leq 1.0 \end{cases}$$

$$A_{cumulative,n} = A_{ratio} * A_{total}$$

where A_{total} for this problem is 810 km². Using the HEC equation and summing each incremental contributing area to find the cumulative area yields Table 9.20.

Table 9.20 *Area ratio and cumulative area for time intervals of 1 hour*

Time (h)	A_{ratio}	$A_{cumulative}$ (km²)
1	0.096211	77.930516
2	0.272124	220.42079
3	0.499924	404.93884
4	0.727876	589.57921
5	0.903789	732.06948
6	1	810

Using the cumulative area and the time distribution based on the time interval, the incremental area between each isochrone is derived, as shown in Table 9.21:

$$A_i = A_{c,t} - A_{c,t-1}$$

where A_i is the incremental area, $A_{c,t}$ is the cumulative area at time t and $A_{c,t-1}$ is the cumulative area at time $t-1$. This equation is used in column 4 of Table 9.21.

Following the derivation of the incremental area, the incremental flow is calculated to create a translation hydrograph for the watershed. According to the Clark unit hydrograph method, the incremental flow graph is used at the translation hydrograph prior to modification with respect to storage. The calculation of incremental flow is accomplished by using the following equation:

$$\bar{I} = \frac{CA_i}{\Delta t}$$

where $C = 2.78$ for SI calculations and $\bar{I}$ is the incremental flow for the given incremental area over the time interval. The incremental flow calculations are displayed in the last column of Table 9.21.

Step 2: The next step in the Clark unit hydrograph method involves routing the translation hydrograph through the linear reservoir to obtain the instantaneous unit hydrograph. In order to calculate the routing coefficient, C, the storage coefficient, R, must be estimated. The problem statement specifies that a storage coefficient of 5.4 should be used.

$$C_0 = \frac{\Delta t}{R + 0.5\Delta t}$$

$$C_0 = \frac{1}{5.4 + 0.5(1)}$$

$$C_0 = 0.169$$

The derived routing coefficient will be used in the calculation of the discharge at each time interval. The equation for the discharge calculation is shown below and is used in column 3 of Table 9.22:

$$Q_t = C_0\bar{I} + (1 - C_0)Q_{t-1}$$

Table 9.21 *Calculation of incremental area between isochrones*

Isochrones	Travel time (h)	Cumulative area (km²)	Incremental area (km²)	Incremental flow (m³/s)
1	1	78	78	216.4737
2	2	220	142	395.8063
3	3	405	185	512.5502
4	4	590	185	512.8899
5	5	732	142	395.8063
6	6	810	78	216.4737

Table 9.22 *Unit hydrograph derivation from the instantaneous unit hydrograph*

Time interval, Δt (1 h)	$I_{average}$	Q_t	Q_{lag}	Q_{UH}
0	0	0	0	0
1	216.4737	36.6904502	0	18.34523
2	395.8063	97.5575441	36.69045	67.124
3	512.5502	167.895274	97.55754	132.7264
4	512.8899	226.368944	167.8953	197.1321
5	395.8063	255.087141	226.3689	240.728
6	216.4737	248.542482	255.0871	251.8148
7	0	206.416638	248.5425	227.4796
8	0	171.430767	206.4166	188.9237
9	0	142.374705	171.4308	156.9027
10	0	118.243399	142.3747	130.3091
11	0	98.2021449	118.2434	108.2228
12	0	81.5577136	98.20214	89.87993
13	0	67.7343723	81.55771	74.64604
14	0	56.2539702	67.73437	61.99417
15	0	46.719399	56.25397	51.48668
16	0	38.8008568	46.7194	42.76013
17	0	32.2244404	38.80086	35.51265
18	0	26.7626708	32.22444	29.49356
19	0	22.2266249	26.76267	24.49465
20	0	18.4594004	22.22662	20.34301
21	0	15.3306884	18.4594	16.89504
22	0	12.7322667	15.33069	14.03148
23	0	10.5742554	12.73227	11.65326
24	0	8.78200869	10.57426	9.678132
25	0	7.29353264	8.782009	8.037771
26	0	6.05734067	7.293533	6.675437
27	0	5.03067276	6.057341	5.544007
28	0	4.17801636	5.030673	4.604345
29	0	3.46987799	4.178016	3.823947
30	0	2.88176308	3.469878	3.175821
			2.881763	1.440882
				0

From the instantaneous unit hydrograph shown in column 3 of Table 9.22, the unit hydrograph, Q_{UH}, is obtained by lagging the instantaneous unit hydrograph by the same time interval, column 4, and averaging both these hydrographs, as shown in column 5.

Example 9.8: A short-duration (25 minutes) rainfall with high intensity (2.76 in/h) occurred in a catchment (7.65 square miles) of Five Mile Creek Watershed in Dallas. The storm produced significant runoff in the creek. The rainfall data are given in columns 1–3 in Table 9.23. The streamflow data collected near the outlet of the catchment are given in columns 7 and 8 of Table 9.24. (1) Determine an instantaneous unit hydrograph using the Nash model for this catchment from the given data. (2) Calculate the DRH from the instantaneous unit hydrograph and compare it with the observed DRH. (3) Discuss the effect of variation of n for the constant k and the variation of k for the constant n on the instantaneous unit hydrograph. (4) Show that the parameter k of a linear reservoir model of a watershed is the lag time of the watershed.

Solution to Part 1:

The observed rainfall total for the 25 minutes event was 1.15 inches, as shown in column 3 of Table 9.23 and the observed runoff total (effective rainfall) was 0.6615 inch, as shown in column 7 of Table 9.23. Thus, the rainfall loss due to abstraction was 0.4888 inch. The soil type of the catchment under consideration is HSG D and

Table 9.23 *Rainfall data*

Date/Time	Time (h)	Observed cumulative rainfall (in)	Observed incremental rainfall (in)	Cumulative infiltration (in)	Incremental infiltration (in)	Effective cumulative rainfall	Effective incremental rainfall (in)
05/26/1976@18:15:00	0	0	0	0	0	0	0
05/26/1976@18:20:00	0.0833	0.2	0.2	0.1289	0.1289	0.0711	0.0711
05/26/1976@18:25:00	0.1667	0.58	0.38	0.2494	0.1206	0.3306	0.2594
05/26/1976@18:30:00	0.25	0.93	0.35	0.3624	0.1130	0.5676	0.2370
05/26/1976@18:35:00	0.3333	1.13	0.2	0.4685	0.1060	0.6615	0.0940
05/26/1976@18:40:00	0.4167	1.15	0.02	0.4885	0.0200	0.0000	0.0000

Table 9.24 *Calculation of moments of ERH*

Time (h)	Moment arm (h), x_i	From start to center of block (min)	Effective incremental rainfall, f_i (in)	$x_i * f_i$	$(x_i^2) * f_i$
0			0		
0.0833	0.04165	2.50	0.0711	0.002963	0.000123
0.1667	0.08335	5.00	0.2594	0.021622	0.001802
0.25	0.125	7.50	0.2370	0.029625	0.003703
0.3333	0.16665	10.00	0.0940	0.01566	0.00261
0.4167	0.20835	12.50	0.0000	0	0

Table 9.25 *Rainfall data and analysis*

Date/Time	Time (h)	Moment arm (h), x_i	From start to center of block (min)	Runoff (ft³/s)	Accumulated runoff (in)	Incremental runoff, f_i (in)	$x_i * f_i$	$(x_i^2) * f_i$
05/26/1976@18:00:00	0			2	0.0073			
05/26/1976@18:10:00	0.1667	0.0833	5.00	2	0.0073	0	0.00000	0.00000
05/26/1976@18:15:00	0.25	0.1250	7.50	2	0.0074	0.0001	0.00001	0.00000
05/26/1976@18:20:00	0.3333	0.1667	10.00	2	0.0074	0	0.00000	0.00000
05/26/1976@18:25:00	0.4167	0.2083	12.50	2	0.0074	0	0.00000	0.00000
05/26/1976@18:30:00	0.5	0.2500	15.00	323	0.0129	0.0055	0.00138	0.00034
05/26/1976@18:35:00	0.5833	0.2917	17.50	500	0.0213	0.0084	0.00245	0.00071
05/26/1976@18:40:00	0.6667	0.3333	20.00	931	0.037	0.0157	0.00523	0.00174
05/26/1976@18:45:00	0.75	0.3750	22.50	1610	0.0642	0.0272	0.01020	0.00383
05/26/1976@18:50:00	0.8333	0.4167	25.00	2070	0.0991	0.0349	0.01454	0.00606
05/26/1976@18:55:00	0.9167	0.4583	27.50	2120	0.1348	0.0357	0.01636	0.00750
05/26/1976@19:00:00	1	0.5000	30.00	2370	0.1748	0.04	0.02000	0.01000

Table 9.25 (cont.)

Date/Time	Time (h)	Moment arm (h), x_i	From start to center of block (min)	Runoff (ft³/s)	Accumulated runoff (in)	Incremental runoff, f_i (in)	x_i*f_i	$(x_i^2)*f_i$
05/26/1976@19:05:00	1.0833	0.5417	32.50	2240	0.2125	0.0377	0.02042	0.01106
05/26/1976@19:10:00	1.1667	0.5833	35.00	2220	0.25	0.0375	0.02188	0.01276
05/26/1976@19:15:00	1.25	0.6250	37.50	2100	0.2854	0.0354	0.02213	0.01383
05/26/1976@19:20:00	1.3333	0.6667	40.00	2000	0.3191	0.0337	0.02247	0.01498
05/26/1976@19:25:00	1.4167	0.7083	42.50	1630	0.3466	0.0275	0.01948	0.01380
05/26/1976@19:30:00	1.5	0.7500	45.00	1400	0.3702	0.0236	0.01770	0.01328
05/26/1976@19:35:00	1.5833	0.7917	47.50	1120	0.3891	0.0189	0.01496	0.01184
05/26/1976@19:40:00	1.6667	0.8333	50.00	858	0.4036	0.0145	0.01208	0.01007
05/26/1976@19:45:00	1.75	0.8750	52.50	756	0.4163	0.0127	0.01111	0.00972
05/26/1976@19:50:00	1.8333	0.9167	55.00	736	0.4287	0.0124	0.01137	0.01042
05/26/1976@19:55:00	1.9167	0.9583	57.50	641	0.4395	0.0108	0.01035	0.00992
05/26/1976@20:00:00	2	1.0000	60.00	603	0.4497	0.0102	0.01020	0.01020
05/26/1976@20:05:00	2.0833	1.0417	62.50	551	0.459	0.0093	0.00969	0.01009
05/26/1976@20:10:00	2.1667	1.0834	65.00	560	0.4684	0.0094	0.01018	0.01103
05/26/1976@20:15:00	2.25	1.1250	67.50	584	0.4783	0.0099	0.01114	0.01253
05/26/1976@20:20:00	2.3333	1.1667	70.00	572	0.4879	0.0096	0.01120	0.01307
05/26/1976@20:25:00	2.4167	1.2084	72.50	503	0.4964	0.0085	0.01027	0.01241
05/26/1976@20:30:00	2.5	1.2500	75.00	482	0.5045	0.0081	0.01012	0.01266
05/26/1976@20:35:00	2.5833	1.2917	77.50	494	0.5129	0.0084	0.01085	0.01401
05/26/1976@20:40:00	2.6667	1.3334	80.00	449	0.5204	0.0075	0.01000	0.01333
05/26/1976@20:45:00	2.75	1.3750	82.50	407	0.5273	0.0069	0.00949	0.01305
05/26/1976@20:50:00	2.8333	1.4167	85.00	374	0.5336	0.0063	0.00892	0.01264
05/26/1976@20:55:00	2.9167	1.4584	87.50	362	0.5397	0.0061	0.00890	0.01297
05/26/1976@21:00:00	3	1.5000	90.00	336	0.5482	0.0085	0.01275	0.01913
05/26/1976@21:10:00	3.1667	1.5834	95.00	318	0.5589	0.0107	0.01694	0.02682
05/26/1976@21:20:00	3.3333	1.6667	100.00	312	0.5695	0.0106	0.01767	0.02944
05/26/1976@21:30:00	3.5	1.7500	105.00	277	0.5788	0.0093	0.01628	0.02848
05/26/1976@21:40:00	3.6667	1.8334	110.00	264	0.5877	0.0089	0.01632	0.02991
05/26/1976@21:50:00	3.8333	1.9167	115.00	242	0.5938	0.0061	0.01169	0.02241
05/26/1976@21:55:00	3.9167	1.9584	117.50	235	0.5978	0.004	0.00783	0.01534
05/26/1976@22:00:00	4	2.0000	120.00	225	0.6073	0.0095	0.01900	0.03800
05/26/1976@22:20:00	4.3333	2.1667	130.00	195	0.6171	0.0098	0.02123	0.04600
05/26/1976@22:30:00	4.5	2.2500	135.00	180	0.6232	0.0061	0.01373	0.03088
05/26/1976@22:40:00	4.6667	2.3334	140.00	170	0.6289	0.0057	0.01330	0.03103
05/26/1976@22:50:00	4.8333	2.4167	145.00	160	0.6343	0.0054	0.01305	0.03154
05/26/1976@23:00:00	5	2.5000	150.00	145	0.6417	0.0074	0.01850	0.04625
05/26/1976@23:20:00	5.3333	2.6667	160.00	130	0.6504	0.0087	0.02320	0.06187
05/26/1976@23:40:00	5.6667	2.8334	170.00	110	0.6579	0.0075	0.02125	0.06021
05/27/1976@00:00:00	6	3.0000	180.00	100	0.6612	0.0033	0.00990	0.02970

for this soil type the parameters of the Holtan model are: $f_c = 0.1$ in/h; $a = 1$; $F_p = 1.406$ in; and $k = 2$ h^{-1}. With these parameters, the Holtan model was applied to account for the infiltration. The cumulative infiltration calculated accordingly is given in column 5 of Table 9.23. From these, the effective incremental rainfall for each of the observed time intervals is given in column 8. The calculations of each step can be followed in the results given in other columns of this table. Note that the calculated total infiltration is 0.4885 inch (column 5) and the calculated total accumulated rainfall is 0.6615 inch (column 7) which match the observed values.

In order to calculate the parameters of the instantaneous unit hydrograph model of Nash, the first and second moments of the ERH must be evaluated. This is done in Table 9.24.

The first moment of the ERH is calculated from the results given in column 5 (which are products of column 2 and 4):

$$M1_{ERH} = \frac{\sum x_i f_i}{M0_{ERH}} = \frac{0.06987 \text{ in} \times \text{h}}{0.6615 \text{ in}} = 0.11 \text{ h}$$

The second moment is calculated from results given in column 6:

$$M2_{ERH} = \frac{\sum x_i^2 f_i}{M0_{ERH}} = \frac{0.008238 \text{ in} \times \text{h}^2}{0.6615 \text{ in}} = 0.012 \text{ h}^2$$

The first moment of the DRH is calculated from the results given in column 8 (which are products of column 3 and 7)

$$M1_{DRH} = \frac{\sum x_i f_i}{M0_{DRH}} = \frac{0.62771 \text{ in} \times \text{h}}{0.6612 \text{ in}} = 0.9494 \text{ h}$$

The second moment of the DRH is calculated from results given in column 9:

$$M2_{DRH} = \frac{\sum x_i^2 f_i}{M0_{DRH}} = \frac{0.85688 \text{ in} \times \text{h}^2}{0.6612 \text{ in}} = 1.2959 \text{ h}^2$$

The instantaneous unit hydrograph parameters n and k are related to the moment measures of the ERH and DRH as:

$$nk = M1_{DRH} - M1_{ERH} = (0.9494 - 0.1056) \text{ h} = 0.8437 \text{ h}$$

Also,

$$M2_{DRH} - M2_{ERH} - 2nkM1_{ERH} = n(n+1)k^2 = n^2k^2 + nk \cdot k$$

which can be written to solve for k with the known values as

$$k = \frac{M2_{DRH} - M2_{ERH} - 2nkM1_{ERH} - n^2k^2}{nk}$$

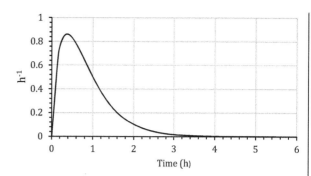

Figure 9.47 Instantaneous unit hydrograph.

Thus,

$$n = \frac{0.8437 \text{ h}}{0.4662 \text{ h}} = 1.81$$

With these values of n and k determined from the observed rainfall and runoff data, the instantaneous unit hydrograph is calculated using the equation

$$u(t) = \frac{1}{k\Gamma(n)} \left(\frac{t}{k}\right)^{(n-1)} \exp\left(-\frac{t}{k}\right)$$

Figure 9.47 shows the calculated instantaneous unit hydrograph.

Solution to Part 2: Calculation of the DRH from the instantaneous unit hydrograph.

The DRH is calculated from the instantaneous unit hydrograph using convolution. The convolution operation is presented in Table 9.26. Note that the convolution gives discharge values in in/h, which are converted to ft³/s using the area of the watershed as 7.65 square miles. The calculated DRH and observed DRH are shown in Figure 9.48.

Solution of Part 3:

The instantaneous unit hydrograph for four different values of n are calculated by keeping the value of

$$k = \frac{1.2959 \text{ h}^2 - 0.0124 \text{ h}^2 - 2 \times 0.8437 \text{ h} \times 0.1056 \text{ h} - (0.8437 \text{ h})^2}{0.8437 \text{ h}} = 0.4663 \text{ h}$$

Table 9.26 *Convolution operation for calculation of the DRH from the instantaneous unit hydrograph (IUH)*

Time (h)	Time (h) IUH	0.1667 P_1 (in) 0.330556	0.25 P_2 (in) 0.2370	0.3333 P_3 (in) 0.0940	DRH from IUH (in/h)	DRH from IUH (ft³/s)
0	0.0000	0.0000			0.0000	0.0000
0.1667	0.6984	0.2309	0.0000		0.2309	1139.77
0.25	0.8110	0.2681	0.1655	0.0000	0.4336	2140.68
0.3333	0.8562	0.2830	0.1922	0.0656	0.5409	2670.10
0.4167	0.8578	0.2836	0.2029	0.0762	0.5627	2777.87
0.5	0.8316	0.2749	0.2033	0.0805	0.5586	2757.88
0.5833	0.7879	0.2605	0.1971	0.0806	0.5381	2656.69
0.6667	0.7342	0.2427	0.1867	0.0781	0.5076	2505.71
0.75	0.6755	0.2233	0.1740	0.0740	0.4713	2326.76
0.8333	0.6152	0.2034	0.1601	0.0690	0.4324	2134.85
0.9167	0.5558	0.1837	0.1458	0.0635	0.3930	1940.11
1	0.4987	0.1649	0.1317	0.0578	0.3544	1749.56
1.0833	0.4451	0.1471	0.1182	0.0522	0.3175	1567.65
1.1667	0.3952	0.1306	0.1055	0.0469	0.2830	1397.02
1.25	0.3495	0.1155	0.0937	0.0418	0.2510	1239.23
1.3333	0.3080	0.1018	0.0828	0.0371	0.2218	1094.91
1.4167	0.2705	0.0894	0.0730	0.0328	0.1953	963.99
1.5	0.2370	0.0783	0.0641	0.0289	0.1714	846.13
1.5833	0.2071	0.0684	0.0562	0.0254	0.1500	740.69
1.6667	0.1805	0.0597	0.0491	0.0223	0.1310	646.77
1.75	0.1571	0.0519	0.0428	0.0195	0.1142	563.54
1.8333	0.1364	0.0451	0.0372	0.0170	0.0993	490.07
1.9167	0.1182	0.0391	0.0323	0.0148	0.0862	425.39
2	0.1024	0.0338	0.0280	0.0128	0.0747	368.65
2.0833	0.0885	0.0293	0.0243	0.0111	0.0646	319.01
2.1667	0.0764	0.0252	0.0210	0.0096	0.0558	275.66
2.25	0.0659	0.0218	0.0181	0.0083	0.0482	237.9063
2.3333	0.0567	0.0188	0.0156	0.0072	0.0415	205.09
2.4167	0.0488	0.0161	0.0134	0.0062	0.0358	176.59
2.5	0.0420	0.0139	0.0116	0.0053	0.0308	151.91
2.5833	0.0360	0.0119	0.0099	0.0046	0.0264	130.55
2.6667	0.0309	0.0102	0.0085	0.0039	0.0227	112.09
2.75	0.0265	0.0088	0.0073	0.0034	0.0195	96.17
2.8333	0.0227	0.0075	0.0063	0.0029	0.0167	82.44
2.9167	0.0194	0.0064	0.0054	0.0025	0.0143	70.62
3	0.0166	0.0055	0.0046	0.0021	0.0122	60.45
3.1667	0.0122	0.0040	0.0039	0.0018	0.0098	48.34
3.3333	0.0089	0.0029	0.0029	0.0016	0.0074	36.42
3.5	0.0065	0.0021	0.0021	0.0011	0.0054	26.54
3.6667	0.0047	0.0015	0.0015	0.0008	0.0039	19.31
3.8333	0.0034	0.0011	0.0011	0.0006	0.0028	14.02
3.9167	0.0029	0.0010	0.0008	0.0004	0.0022	10.87
4	0.0025	0.0008	0.0007	0.0003	0.0018	8.97

Note, the first row gives time in hours for three rainfall pulses P_1, P_2, P_3, calculated in Table 9.24 (column 1). Effective rainfall stopped at 0.3333 hour.

$k = 0.4663$ h constant (as derived above). The results are shown in Figure 9.49.

The instantaneous unit hydrograph for four different values of k are calculated by keeping the value of $n = 1.81$ constant (as derived above). The resulting IUH are shown in Figure 9.50.

Solution to Part 4:

The linear reservoir model is given as

$$u(t) = \frac{1}{k\Gamma(n)} \left(\frac{t}{k}\right)^{(n-1)} \exp\left(-\frac{t}{k}\right)$$

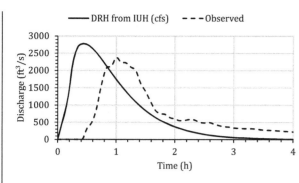

Figure 9.48 Calculated and observed DRH.

Take the first moment about the time (t) axis

$$M_1 = \int_0^\infty \frac{t}{k} \exp\left(-\frac{t}{k}\right) dt$$

Integration by parts yields

$$M_1 = -t \exp\left(\frac{t}{k}\right) - k \exp\left(-\frac{t}{k}\right)$$

By taking the limits from 0 to ∞

$$M_1 = k$$

Now recall the watershed lag time is the first moment of the instantaneous unit hydrograph since

$$t_l = M_{1DRH} - M_{1ERH} = M_{1IUH}$$

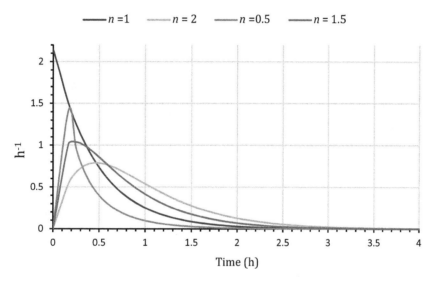

Figure 9.49 Effect of variation of n on the instantaneous unit hydrograph for constant k. (A black and white version of this figure will appear in some formats. For the color version, please refer to the plate section.)

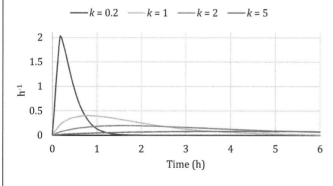

Figure 9.50 Effect of variation of k on the instantaneous unit hydrograph for constant n. (A black and white version of this figure will appear in some formats. For the color version, please refer to the plate section.)

Therefore,

$t_l = k$

Now consider an ERH input as a rectangular pulse with an intensity of duration D and intensity of any value, say $1/D$. Then its first moment will be $D/2$.

For this ERH, one can compute the DRH which can be obtained by solving the differential equation. Then, the DRH will be

$$Q(t) = \frac{1}{D}\left[1 - \exp\left(\frac{t}{k}\right)\right]; 0 \leq t \leq D$$

$$Q(t) = Q(D)\left[\exp\left(\frac{t}{k}\right)\right]; t \geq D$$

Now by taking the first moment of the DRH, the lag time is obtained.

Example 9.9: Rainfall–runoff data from a storm on May 23, 2003, from Lighvan River basin, located in the northwest of Iran, are given in Table 9.27. The 1-hour unit hydrograph derived from the observed rainfall–runoff data is also given in this table. The rainfall given is effective rainfall for a storm duration of 148 minutes. Compute the unit hydrograph using the least-squares method and compare it with the observed unit hydrograph.

Solution: Number of rainfall pulses, $M = 1$

Number of runoff hydrograph ordinates, $N = 8$

Therefore, the number of unit hydrograph ordinates is given by $N - M + 1 = 8 - 1 + 1 = 8$

So [P] will be an 8×8 matrix, [UP] will be a 8×1 vector, and [Q] is an 8×1 vector.

Step 1: Form the [P] matrix.

$$[\mathbf{P}] = \begin{vmatrix} 0.04 & 0 & 0 & 0 & 0 & 0 & 0 & 0 \\ 0 & 0.04 & 0 & 0 & 0 & 0 & 0 & 0 \\ 0 & 0 & 0.04 & 0 & 0 & 0 & 0 & 0 \\ 0 & 0 & 0 & 0.04 & 0 & 0 & 0 & 0 \\ 0 & 0 & 0 & 0 & 0.04 & 0 & 0 & 0 \\ 0 & 0 & 0 & 0 & 0 & 0.04 & 0 & 0 \\ 0 & 0 & 0 & 0 & 0 & 0 & 0.04 & 0 \\ 0 & 0 & 0 & 0 & 0 & 0 & 0 & 0.04 \end{vmatrix}$$

Table 9.27 *Data*

Time (h)	P (mm)	Q (m³/s)	Unit hydrograph (m³/s)
1	0.04	0.0000	0.0000
2		0.0810	2.0256
3		0.1854	4.6341
4		0.2496	6.2400
5		0.2496	6.2400
6		0.1224	3.0588
7		0.0201	0.5013
8		0.0000	0.0000

Step 2: Transpose the [P] matrix to get $[\mathbf{P}]^T$.

$$[\mathbf{P}]^T = \begin{vmatrix} 0.04 & 0 & 0 & 0 & 0 & 0 & 0 & 0 \\ 0 & 0.04 & 0 & 0 & 0 & 0 & 0 & 0 \\ 0 & 0 & 0.04 & 0 & 0 & 0 & 0 & 0 \\ 0 & 0 & 0 & 0.04 & 0 & 0 & 0 & 0 \\ 0 & 0 & 0 & 0 & 0.04 & 0 & 0 & 0 \\ 0 & 0 & 0 & 0 & 0 & 0.04 & 0 & 0 \\ 0 & 0 & 0 & 0 & 0 & 0 & 0.04 & 0 \\ 0 & 0 & 0 & 0 & 0 & 0 & 0 & 0.04 \end{vmatrix}$$

Step 3: Multiply [P] by $[\mathbf{P}]^T$ to get a matrix, [Z].

$$[\mathbf{Z}] = \begin{vmatrix} 0.0016 & 0 & 0 & 0 & 0 & 0 & 0 & 0 \\ 0 & 0.0016 & 0 & 0 & 0 & 0 & 0 & 0 \\ 0 & 0 & 0.0016 & 0 & 0 & 0 & 0 & 0 \\ 0 & 0 & 0 & 0.0016 & 0 & 0 & 0 & 0 \\ 0 & 0 & 0 & 0 & 0.0016 & 0 & 0 & 0 \\ 0 & 0 & 0 & 0 & 0 & 0.0016 & 0 & 0 \\ 0 & 0 & 0 & 0 & 0 & 0 & 0.0016 & 0 \\ 0 & 0 & 0 & 0 & 0 & 0 & 0 & 0.0016 \end{vmatrix}$$

Step 4: Invert [Z] to get $[\mathbf{Z}]^{-1}$.

$$[\mathbf{Z}]^{-1} = \begin{vmatrix} 625 & 0 & 0 & 0 & 0 & 0 & 0 & 0 \\ 0 & 625 & 0 & 0 & 0 & 0 & 0 & 0 \\ 0 & 0 & 625 & 0 & 0 & 0 & 0 & 0 \\ 0 & 0 & 0 & 625 & 0 & 0 & 0 & 0 \\ 0 & 0 & 0 & 0 & 625 & 0 & 0 & 0 \\ 0 & 0 & 0 & 0 & 0 & 625 & 0 & 0 \\ 0 & 0 & 0 & 0 & 0 & 0 & 625 & 0 \\ 0 & 0 & 0 & 0 & 0 & 0 & 0 & 625 \end{vmatrix}$$

Step 5: Multiply $[Z]^{-1}$ by $[P]^T$ to obtain a matrix, $[T]$.

$$[T] = \begin{vmatrix} 25 & 0 & 0 & 0 & 0 & 0 & 0 & 0 \\ 0 & 25 & 0 & 0 & 0 & 0 & 0 & 0 \\ 0 & 0 & 25 & 0 & 0 & 0 & 0 & 0 \\ 0 & 0 & 0 & 25 & 0 & 0 & 0 & 0 \\ 0 & 0 & 0 & 0 & 25 & 0 & 0 & 0 \\ 0 & 0 & 0 & 0 & 0 & 25 & 0 & 0 \\ 0 & 0 & 0 & 0 & 0 & 0 & 25 & 0 \\ 0 & 0 & 0 & 0 & 0 & 0 & 0 & 25 \end{vmatrix}$$

Step 6: Form the vector [Q] from the observed flow data.

$$[Q] = \begin{vmatrix} 0 \\ 0.081024 \\ 0.185364 \\ 0.2496 \\ 0.2496 \\ 0.122352 \\ 0.020052 \\ 0 \end{vmatrix}$$

Step 7: Multiply [T] by [Q] to get the ordinates of the unit hydrograph as a vector [UH].

$$[UH] = \begin{vmatrix} 0 \\ 2.0256 \\ 4.6341 \\ 6.24 \\ 6.24 \\ 3.0588 \\ 0.5013 \\ 0 \end{vmatrix}$$

The unit hydrograph derived above is the same as one derived from the observed flow and effective rainfall data.

Example 9.10: A 1-hour unit hydrograph derived from a storm event on June 15, 2003 over Lighvan River basin in northwest Iran is given in Table 9.28. Derive a 3-hour unit hydrograph for this river basin using the S-hydrograph method.

Table 9.28 *One-hour unit hydrograph data for Lighvan River basin*

Time (h)	Q (m^3/s)
1	0
2	2.598663
3	2.549699
4	2.140653
5	1.757242
6	1.445937
7	1.199935
8	1.005543
9	0.850703
10	0.726097
11	0.62476
12	0.541509
13	0.472469
14	0.414715
15	0.366017
16	0.324655
17	0.289286
18	0.258858
19	0.23253
20	0.209631
21	0.189619
22	0.172049
23	0

Solution:

Step 1: Develop the S-hydrograph by lagging the unit hydrograph by 1 hour and sum up the ordinates when the ordinates tend to stabilize to a constant value. This is shown in Table 9.29 where the lagging ordinates are shown only up to 8 hours but the S-hydrograph is actually taken over up to 27 hours. The value of the summation started to stabilize from 21 hours, which is very close to the base of the original unit hydrograph given in Table 9.27.

Step 2: The S-hydrograph derived in step 1 is now further lagged by 3 hours (column 3, Table 9.30) and the ordinate values of the lagged hydrograph are subtracted from those of the original S-hydrograph (column 2, Table 9.30). Since the duration of the original S-hydrograph is 1 hour and we want to derive a unit hydrograph with duration D hours, we first get the ratio:

$$\frac{\Delta t_0}{D} = \frac{1}{3} = 0.33333$$

The subtracted value of the ordinates (column 4, Table 9.30) is multiplied by 0.33333 to get the ordinates of the 3-hour unit hydrograph.

Figure 9.51 shows the S-hydrographs and Figure 9.52 shows the two unit hydrographs.

Table 9.29 *Creation of S-hydrograph by lagging the 1-hour unit hydrograph*

Time (h)	1 h	2 h	3 h	4 h	5 h	6 h	7 h	8 h	....	S-hydrograph
1	0									0
2	2.5987	0.0000								2.5987
3	2.5497	2.5987	0.0000							5.1484
4	2.1407	2.5497	2.5987	0.0000						7.2890
5	1.7572	2.1407	2.5497	2.5987	0.0000					9.0463
6	1.4459	1.7572	2.1407	2.5497	2.5987	0.0000				10.4922
7	1.1999	1.4459	1.7572	2.1407	2.5497	2.5987	0.0000			11.6921
8	1.0055	1.1999	1.4459	1.7572	2.1407	2.5497	2.5987	0.0000		12.6977
9	0.8507	1.0055	1.1999	1.4459	1.7572	2.1407	2.5497	2.5987		13.5484
10	0.7261	0.8507	1.0055	1.1999	1.4459	1.7572	2.1407	2.5497		14.2745
11	0.6248	0.7261	0.8507	1.0055	1.1999	1.4459	1.7572	2.1407		14.8992
12	0.5415	0.6248	0.7261	0.8507	1.0055	1.1999	1.4459	1.7572		15.4407
13	0.4725	0.5415	0.6248	0.7261	0.8507	1.0055	1.1999	1.4459		15.9132
14	0.4147	0.4725	0.5415	0.6248	0.7261	0.8507	1.0055	1.1999		16.3279
15	0.3660	0.4147	0.4725	0.5415	0.6248	0.7261	0.8507	1.0055		16.6939
16	0.3247	0.3660	0.4147	0.4725	0.5415	0.6248	0.7261	0.8507		17.0186
17	0.2893	0.3247	0.3660	0.4147	0.4725	0.5415	0.6248	0.7261		17.3079
18	0.2589	0.2893	0.3247	0.3660	0.4147	0.4725	0.5415	0.6248		17.5667
19	0.2325	0.2589	0.2893	0.3247	0.3660	0.4147	0.4725	0.5415		17.7993
20	0.2096	0.2325	0.2589	0.2893	0.3247	0.3660	0.4147	0.4725		18.0089
21	0.1896	0.2096	0.2325	0.2589	0.2893	0.3247	0.3660	0.4147		18.1985
22	0.1720	0.1896	0.2096	0.2325	0.2589	0.2893	0.3247	0.3660		18.3706
23	0.0000	0.1720	0.1896	0.2096	0.2325	0.2589	0.2893	0.3247		18.3706
24	0.0000	0.0000	0.1720	0.1896	0.2096	0.2325	0.2589	0.2893		18.3706
25	0.0000	0.0000	0.0000	0.1720	0.1896	0.2096	0.2325	0.2589		18.3706
26	0.0000	0.0000	0.0000	0.0000	0.1720	0.1896	0.2096	0.2325		18.3706
27	0.0000	0.0000	0.0000	0.0000	0.0000	0.1720	0.1896	0.2096		18.3706

Table 9.30 *Derivation of 3-hour unit hydrograph from lagged S-hydrographs of 1-hour duration*

Time (h)	S-hydrograph ($S1$)	Lagged S-hydrograph ($S2$)	($S1-S2$)	3-Hour unit hydrograph ordinates
1	0.0000		0.0000	0.0000
2	2.5987		2.5987	0.8662
3	5.1484		5.1484	1.7161
4	7.2890	0.0000	7.2890	2.4297
5	9.0463	2.5987	6.4476	2.1492
6	10.4922	5.1484	5.3438	1.7813
7	11.6921	7.2890	4.4031	1.4677
8	12.6977	9.0463	3.6514	1.2171
9	13.5484	10.4922	3.0562	1.0187
10	14.2745	11.6921	2.5823	0.8608
11	14.8992	12.6977	2.2016	0.7339
12	15.4407	13.5484	1.8924	0.6308
13	15.9132	14.2745	1.6387	0.5462
14	16.3279	14.8992	1.4287	0.4762
15	16.6939	15.4407	1.2532	0.4177
16	17.0186	15.9132	1.1054	0.3685
17	17.3079	16.3279	0.9800	0.3267
18	17.5667	16.6939	0.8728	0.2909
19	17.7993	17.0186	0.7807	0.2602
20	18.0089	17.3079	0.7010	0.2337
21	18.1985	17.5667	0.6318	0.2106
22	18.3706	17.7993	0.5713	0.1904
23	18.3706	18.0089	0.3617	0.1206

Table 9.30 (*cont.*)

Time (h)	S-hydrograph (S1)	Lagged S-hydrograph (S2)	(S1−S2)	3-Hour unit hydrograph ordinates
24	18.3706	18.1985	0.1720	0.0573
25	18.3706	18.3706	0.0000	0.0000
26	18.3706	18.3706	0.0000	0.0000
27	18.3706	18.3706	0.0000	0.0000

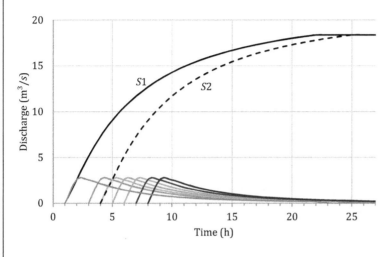

Figure 9.51 Generation of S-hydrographs. (A black and white version of this figure will appear in some formats. For the color version, please refer to the plate section.)

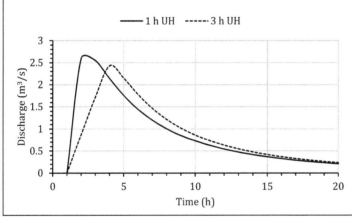

Figure 9.52 1-hour and 3-hour unit hydrographs.

Exercises: A selection of exercises on this topic is available at www.cambridge.org/appliedhydrology.

10 Kinematic Wave Model of Overland Flow

10.1 CONCEPTS AND PURPOSE

In Chapter 9 we used the unit hydrograph theory to derive direct runoff hydrographs (DRHs). Another theory that is also used in certain cases to derive DRHs is the kinematic wave theory. This theory or model can be regarded a descent of the general equation of unsteady flow in open channels, known as the dynamic wave model, which is presented in Chapter 12. In this chapter, we focus on applications of the kinematic wave model that can be applied to overland flow to obtain DRHs from rainfall data. This is a conceptual model of watershed response to rainfall, as if the watershed is acting as a very wide, open channel. There are certain advantages in applying the kinematic wave model over the unit hydrograph model to derive DRHs, which will be also discussed in this chapter.

10.2 KINEMATIC WAVE EQUATION FOR CHANNEL FLOW

In Chapter 12, we see that the general form of the equation of flow of water in an open channel is the equation that describes gradually varied unsteady flow known as the Saint Venant equation. An approximation to this equation was first named as the kinematic wave approximation by Lighthill and Whitham (1955). Kinematic is usually defined as the description of motion without considering the forces giving rise to the motion. In the Saint Venant equation, there are terms involving gravitational, frictional, inertial, and pressure forces. In the kinematic wave approximation, the inertial and pressure forces are completely dropped off from the equation of motion, these being considered as insignificant. This assumption creates a balance between gravitational and frictional forces, which leads to the condition of flow where the **energy slope** (also called the **friction slope**) of the water surface, designated by S_f, is equal to the bed slope, designated by S_0, of the channel, that is, $S_f = S_0$. This also implies that, if this condition holds true, the discharge, Q, through a channel cross section can be represented by a uniform flow formula such as the Chézy or Manning equations, which can be given by a general power relationship of the form

$$Q = \alpha A^m \tag{10.1}$$

where α and m are parameters called **kinematic wave parameters** and A is the cross-sectional area of flow. The parameters α and m depend on the geometry of the channel cross-sectional area of the flow and the roughness or friction coefficient of the channel. Table 10.1 gives α and m for the common cross-sectional geometries of channel flow.

Equation 10.1 does not give Q as a direct function of time (t), which is necessary for the computation of flow hydrographs. For this reason, Eq. 10.1 is substituted in the equation of continuity. For unsteady flow through a channel, the continuity equation can be stated as the difference between inflow and outflow in a control volume, in a given time interval that equals the change in volume of water in the control volume. This can be represented as

$$(Q^{j+1} - Q^j)\Delta t + (A_{i+1} - A_i)\Delta x = q(\Delta x \Delta t) \tag{10.2}$$

where x denotes the direction of flow or the space coordinate, t denotes time or time coordinate, the superscript j represents the time index, subscript i represents the space index, Δt and Δx, respectively, represent the time interval and the distance between an upstream and downstream cross section, and q is called the **lateral inflow**; the right-hand side of Eq. 10.2 is the inflow rate. The lateral inflow is the flow that is coming into the control volume. Dividing by $\Delta t \Delta x$, Eq. 10.2 can be written in the differential form as

$$\frac{\partial A}{\partial t} + \frac{\partial Q}{\partial x} = q \tag{10.3}$$

Combining Eqs. 10.1 and 10.3 yields the differential equation for $A(x,t)$ as

$$\frac{\partial A}{\partial t} + \alpha m A^{(m-1)} \frac{\partial A}{\partial x} = q \tag{10.4}$$

Equation 10.4 is the **kinematic wave approximation** of the equation of motion of water in a channel. In Eq. 10.3, q is in

Table 10.1 *Kinematic wave parameters for various cross-sectional geometries of channel flow*

Cross-sectional geometry	α	m
Rectangular with bottom width, W	$\dfrac{1.486}{n}\sqrt{S}W^{(-2/3)}$	5/3
Square with bottom width = water depth	$\dfrac{0.72}{n}\sqrt{S}$	4/3
Triangular with side slope 1V:ZH	$\dfrac{0.94}{n}\sqrt{S}\left(\dfrac{Z}{1+Z^2}\right)^{1/3}$	4/3
Circular with diameter D	$\dfrac{0.804}{n}\sqrt{S}D^{(1/6)}$	5/4

volume per unit time and unit width. Solutions of Eq.10.4 give the variation of the cross-sectional flow area as a function of space (x) and time (t), from which the volumetric flow rate through the channel cross section can be calculated from Eq. 10.1.

10.3 KINEMATIC WAVE EQUATION FOR OVERLAND FLOW

The kinematic wave equation of unsteady flow in a channel, given by Eq. 10.4, is applied to overland flow by assuming the overland flow surface as a very wide, rectangular open channel. Figure 10.1 shows a conceptual representation of a watershed by two planar surfaces over which water flows to a channel that runs along the line of intersection of the two planes. The water then flows down the channel to the outlet. The kinematic wave overland flow model represents the behavior of overland flow on the plane surfaces and the routing of the composite flow through a series of channel elements. Figure 10.2 shows a conceptual representation of overland and channel flow from a catchment for the kinematic wave model.

For a rectangular channel with width B, in which the depth of the water is h, the hydraulic radius, R is

$$R = \frac{\text{Flow area}}{\text{Wetted perimeter}} = \frac{Bh}{B+2h}$$

For a very wide rectangular channel, $B \gg h$. For such a channel, $R \approx \frac{Bh}{B} = h$.

The Manning equation is then written as

$$Q = \frac{C}{N} h^{2/3} \sqrt{S_o}(Bh) = \frac{C}{N} h^{5/3} \sqrt{S_o} B \qquad (10.5)$$

where $C = 1$ in SI units and $C = 1.486$ in US customary units, and S_o denotes the slope of the overland surface. Note that in Eq. 10.5 the friction coefficient is denoted by N. This surface roughness, N, is a resistance factor that depends on the cover of the planes and this is not Manning's n for a channel. Usually, values of N are higher than typical n values. Table 10.2 lists suggested values of N for certain types of land surface. Also, the subscript o is used to denote overland.

For a strip of overland plane with length L_o and unit width, $B = 1$, as shown in Figure 10.2, Eq. 10.5 becomes

$$q = \left(\frac{C}{N}\sqrt{S_o}\right) h^{5/3} \qquad (10.6)$$

In Eq. 10.6, discharge is denoted by q to imply that it is flow per unit width. Equation 10.6 can be cast in the form of Eq.10.1 as

$$q = \alpha_o h^{m_o} \qquad (10.7)$$

where

$$\alpha_o = \frac{C}{N}\sqrt{S_o} \qquad (10.8)$$

$$m_o = \frac{5}{3} \qquad (10.9)$$

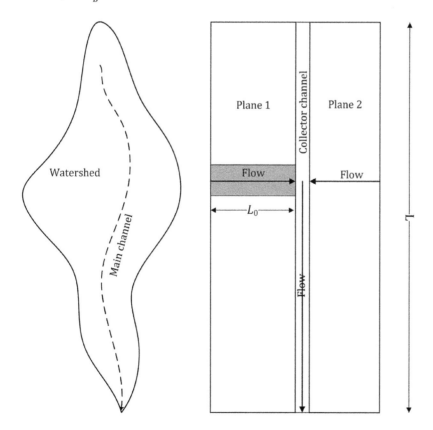

Figure 10.1 A conceptual representation of a simple watershed by two planar surfaces and a collector channel. Flows from overland planes are uniformly distributed along the collector or main channel.

Table 10.2 *Roughness coefficients for overland flow (Hjelmfelt, 1986)*

Surface	N Value
Asphalt and concrete	0.05–0.15
Bare packed soil free of stone	0.10
Fallow – no residue	0.008–0.012
Conventional tillage – no residue	0.06–0.12
Conventional tillage – with residue	0.16–0.22
Chisel plow – no residue	0.06–0.12
Chisel plow – with residue	0.10–0.16
Fall disking – with residue	0.30–0.50
No till – no residue	0.04–0.10
No till (20–40% residue cover)	0.07–0.17
No till (60–100% residue cover)	0.17–0.47
Sparse rangeland with debris	
0% cover	0.09–0.34
20% cover	0.05–0.25
Sparse vegetation	0.053–0.13
Short grass prairie	0.10–0.20
Poor grass cover on moderately rough	0.30
Bare surface	
Light turf	0.20
Average grass cover	0.40
Dense turf	0.17–0.80
Dense grass	0.17–0.30
Bermuda grass	0.30–0.48
Dense shrubbery and forest litter	0.40

The continuity equation for channel flow given by Eq. 10.3 can be written for overland flow as

$$\frac{\partial h}{\partial t} + \frac{\partial q}{\partial x} = p_e \qquad (10.10)$$

In the case of overland flow, the lateral inflow is the effective rainfall, $p_e(x,t)$. Combining Eqs. 10.7 and 10.10, the kinematic wave equation for overland flow is given as

$$\frac{\partial h}{\partial t} + \alpha_o m_o h^{(m_o-1)} \frac{\partial h}{\partial x} = p_e(x,t) \qquad (10.11)$$

Equation 10.11 can be solved to give a relationship for h in terms of x, t, and the effective rainfall. Once this found, it can be substituted back into Eq. 10.7 to obtain a value of q.

In the kinematic wave model, overland flow goes to a channel. In Figure 10.2, a channel network is represented by subcollector channels, a collector channel, and a main channel. In a natural watershed the main channel in this case becomes a third-order stream. In an urbanized watershed, the subcollector channels can be gutters along streets, the collectors can be storm drains, and the main channel can be a receiving stream. In whatever configuration the system is represented, if the length of the channel that receives overland flow, q, has a length L_c, the total discharge from the overland is computed as

$$Q = qL_c \qquad (10.12)$$

At this point, it should be noted that in the discussion above the overland flow is considered as a rectangular plane.

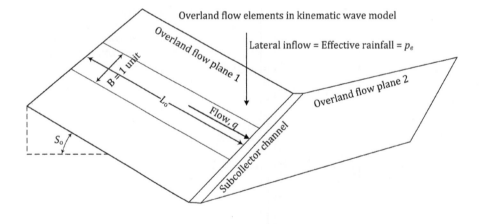

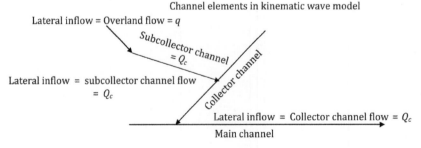

Figure 10.2 Representation of direct runoff from a catchment by planar and linear elements in the kinematic wave model. The planar elements represent overland and the linear elements represent the channel network in a catchment.

However, in certain instances, the overland flow can be better represented as either convergent or divergent planes. Singh (1996) provided the kinematic wave equations for convergent and divergent planes. Mukhopadhyay et al. (2003) modeled overland flood flows using kinematic wave equations over alluvial fans and piedmont plains, a common geomorphologic feature of semi-arid regions, as a combination of divergent and rectangular planes. However, subsequent discussions in this chapter are restricted to rectangular planes only.

10.4 ANALYTICAL SOLUTIONS OF THE KINEMATIC WAVE EQUATION

10.4.1 Limiting Characteristic and Solution Domains

Equation 10.11 is a hyperbolic partial differential equation. Hence it can be solved by the method of characteristics. If $h = f(x, t)$, the projection of the solution surface representing $h(x, t)$ on the $x - t$ plane is called **characteristic projection** and the curves given as dx/ds and dt/ds for some parameter s are called **characteristics** (Figure 10.3).

Singh (1996) presented details of derivations of the solutions of Eq. 10.11 including for cases where the kinematic wave parameters and rainfall vary in both space and time. In the following, only the key results for the cases where α, m, and p_e in Eq. 10.11 remain constant in space and time are presented to illustrate the basic nature of the space–time variability of depth and discharge of water overland during a rain event.

For the solutions of Eq. 10.11, the *initial conditions* can be given as

$$h(x, 0) = 0 \qquad 0 \leq x \leq L_o \tag{10.13}$$

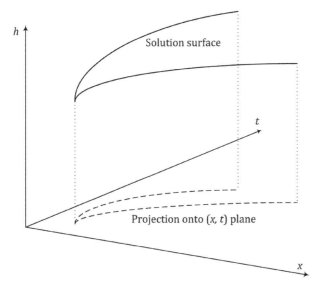

Figure 10.3 Characteristic curves resulting from the projection of the $h(x, t)$ surface onto the $x - t$ plane.

The upstream boundary conditions can be given as

$$h(0, t) = 0 \qquad 0 \leq t \leq T \tag{10.14}$$

Equation 10.13 implies an initially dry surface and Eq. 10.14 implies that, throughout the time, the water depth at the upstream of the plane is always zero. In Eq. 10.14, T denotes the rainfall duration.

The solution of Eq. 10.11 is the surface formed by all characteristic curves passing through the segment $t = 0, 0 \leq x \leq L_o$ and the segment $x = 0, 0 \leq t \leq T$. Figure 10.4 shows the projections of these characteristic curves onto the (x, t) plane. The characteristic passing through the origin (0,0) is called the **bounding** or **limiting characteristic**. By the method of characteristics, the partial differential equation (Eq. 10.11) is reduced to a system of ordinary differential equations given by

$$\frac{dt}{dx} = \frac{1}{m_o \alpha_o h^{(m_o - 1)}} \tag{10.15}$$

$$\frac{dh}{dx} = \frac{p_e(x, t)}{m_o \alpha_o h^{(m_o - 1)}} - \frac{h^{m_o}}{m_o \alpha_o h^{(m_o - 1)}} \tag{10.16}$$

The initial conditions for these equations are expressed as

$$\begin{aligned} t(0) &= t_0 \ h(0) = 0 \\ t(x_0) &= 0 \ h(x_0) = 0 \end{aligned} \tag{10.17}$$

where t_0 and x_0 are the points on the t- and x-axis, respectively, where the characteristics issue. The solution of Eqs. 10.15 and 10.16 provide the solution for Eq. 10.11.

Equation 10.15 is the differential equation of the characteristic curve. However, it is dependent on h. Thus, solutions are carried out simultaneously for the characteristics and h. For deriving solutions, two cases can be identified:

1. an equilibrium case or equilibrium hydrograph
2. a partial equilibrium or partial equilibrium hydrograph

Each of these cases is divided into three principal domains: D1, D2, D3. For the partial equilibrium case (Figure 10.4B), there are two subdomains in D3: D31 and D32. This partitioning describes the varying characteristic of surface runoff in space and time.

The equilibrium or partial equilibrium cases are distinguished according to the intersection of the bounding characteristic curve with $t = T$ line. For the equilibrium case, the bounding characteristic line defined as $t = t(x, 0)$ intersects the line $x = L_o$ before the line $t = T$ (Figure 10.4A). The notation $t = t(x, 0)$ means t is a function of x. In this case, T can be infinite and there is sufficient time to achieve equilibrium, which is defined below. For the partial equilibrium case, the limiting characteristic interests the $t = T$ line before it intersects $x = L_o$ line (Figure 10.4B). The solution domain is divided into domains D1, D2, and D3 by regions formed by the upper boundary ($x = 0$), lower boundary ($x = L_o$), and initial condition ($t = 0$) and the limiting characteristic, $t = t(x, 0)$. Note that both α_o and p_e are assumed to be constant.

10.4.2 Equilibrium Hydrographs

In this case T can be either finite or infinite. The domains D1, D2, D3 for these two situations are shown in Figure 10.5.

10.4.2.1 Solutions for Domain D1

This is the domain of the **rising hydrograph** for **unsteady uniform flow** where h is independent of x and only varies with t. The domain is bounded by the bounding characteristic, $t = (x, 0)$, $t = 0$, and $x = L_o$. Any other characteristic in this domain intersects the initial condition $t = 0$. Let x_0 be any point within the length L_o $(0 \leq x_0 \leq L_o)$. Then all characteristic curves, denoted by $t = t(x, x_0)$ in this domain are functions of x and x_0.

The boundary conditions for solutions of Eqs. 10.15 and 10.16 are

$$t(x_0) = 0$$
$$h(x_0) = 0$$

where x_0 $(0 \leq x_0 \leq L_o)$ is a parameter representing the intersection of a characteristic curve with the x-axis. Thus, the solutions begin with the initial conditions on the x-axis and are given as

$$h(x, x_0) = \left[\frac{p_e}{\alpha_o}(x - x_0)\right]^{\frac{1}{m_o}} \tag{10.18}$$

$$t(x, x_0) = p_e^{(1-m_0)/m_o}\left(\frac{x - x_0}{\alpha_o}\right)^{\frac{1}{m_o}} \tag{10.19}$$

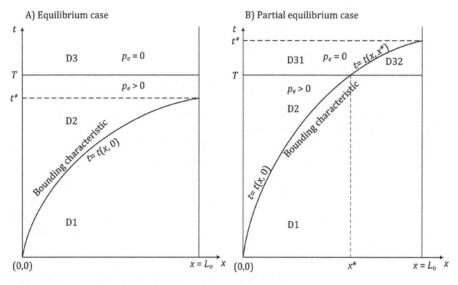

Figure 10.4 Domains of (A) equilibrium and (B) partial equilibrium cases on the x–t plane. In both cases t^* is the time on the x–t characteristic where $x = L_o$. T is the duration of rainfall. In A: $t^* \leq T$ and t^* is independent of T. In B: $t^* > T$ and t^* is dependent on T.

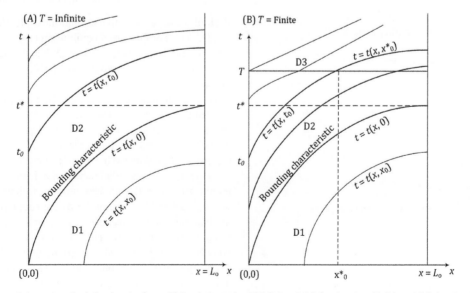

Figure 10.5 Characteristic curves and solution domains for equilibrium hydrographs: (A) infinite rainfall duration time (B) finite rainfall duration time.

Equation 10.19 gives the equation of the characteristic in domain D1. In this equation, x is a variable, x_0 is the characteristic parameter denoting the starting point of the characteristic on the x-axis: $x = x_0 \Rightarrow t = 0$. When $x = x_0, t = t^*$, that is, when the x–t bounding curve intersects the $x = L_o$ line. This is the **time of equilibrium**.

The parameter x_0 can be eliminated to give h as an explicit function of x and t as

$$h(x,t) = p_e t \qquad (10.20)$$

Therefore,

$$q = a_o (p_e t)^{m_o} \qquad (10.21)$$

Equations 10.20 and 10.21 imply that h and q are independent of x but vary with t. Thus, the flow is uniform but unsteady. Equation 10.21 gives flow off the plane per unit width. The total outflow at the end of the plane is

$$Q = qB \qquad (10.22)$$

10.4.2.2 Solutions for Domain D2

This is the domain of **equilibrium hydrograph** for **steady nonuniform flow** where h is dependent on x but does not vary with t. The domain is bounded by the bounding characteristic, $t = (x, 0)$, $x = 0$, $x = L_o$, and the line $t = T$. Any other characteristic in this domain intersects the upstream boundary condition $x = 0$. Let t_0 be any point within the time T ($0 \le t_0 \le T$). Then, all characteristic curves in this domain are functions of x and t_0. This is denoted by $t = t(x, t_0)$.

The initial conditions for solutions of Eqs. 10.15 and 10.16 are

$$t(0) = 0$$
$$h(0) = 0$$

The solutions are

$$h(x, t_0) = \left[\frac{p_e x}{a_o} \right]^{\frac{1}{m_o}} \qquad (10.23)$$

$$t(x, t_0) = t_0 + \left(\frac{1}{p_e} \right) \left(\frac{x}{a_o} \right)^{\frac{1}{m_o}} \qquad (10.24)$$

Equation 10.24 gives the equations of the characteristics in domain D2.

The parameter t_0 can be eliminated to give h as an explicit function of x and t as

$$h(x, t) = \left[\frac{p_e x}{a_o} \right]^{\frac{1}{m_o}} \qquad (10.25)$$

Therefore,

$$q = p_e x \qquad (10.26)$$

Equations 10.25 and 10.26 imply that h and q are independent of t but vary with x. Thus, the flow is nonuniform but steady.

Note that the equilibrium discharge given by Eq. 10.26 is distinct from Eq. 10.21 giving discharge during the rising phase of the hydrograph.

The total discharge off the plane is again,

$$Q = p_e x B \qquad (10.27)$$

Equation 10.27, giving the maximum discharge off the plane, is like the rational formula that is presented in Chapter 11 except that in this case no mention is made on how p_e is chosen. In the rational formula, p_e is the intensity of the rainfall for the duration equal to the time of concentration, t^*. This is discussed further below.

10.4.2.3 Solutions for Domain D3

This is the domain of **falling hydrograph** for **unsteady nonuniform flow** where h varies with both x and t. The domain is bounded by the line $t = T$, $x = 0$, and $x = L_o$. Any other characteristic in this domain intersects the line $t = T$. Note that if T is infinite, there is no domain D3. Let x_0^* be any point within the length L_o ($0 \le x_0^* \le L_o$). It is the point of intersection of the characteristic $t = t(x, x_0^*)$. Then, all characteristic curves, issuing from $t = T$, in this domain are functions of x and x_0^*. This is denoted by $t = t(x, x_0^*); t = t(x, x_0^*)$ is the elongation of $t = t(x, t_0)$ beyond $t = T$ line. Thus, D3 always starts with solutions from D2 starting on the $t = T$ line.

The boundary conditions for solutions of Eqs. 10.15 and 10.16 are

$$t(x_0^*) = T$$
$$h(x_0^*) = h(x_0^*, T) = h_0$$

where x_0^* is a parameter representing the intersection of a characteristic curve $t = t(x, x_0^*)$ with the $t = T$ line.

The solutions are

$$h_0 = \left[\frac{p_e x_0^*}{a_o} \right]^{\frac{1}{m_o}} \qquad (10.28)$$

$$h(x, x_0^*) = \left[\frac{p_e x_0^*}{a_o} \right]^{\frac{1}{m_o}} \qquad (10.29)$$

$$t(x, x_0^*) = T + \left(\frac{x - x_0^*}{a_o m_o} \right) \left(\frac{p_e x_0^*}{a_o} \right)^{\left(\frac{1}{m_o} - 1 \right)} \qquad (10.30)$$

Equation 10.30 gives the equations of the characteristics in domain D3.

The parameter x_0^* can be eliminated to give h as an explicit function of x and t as

$$t - T = \frac{x}{a_o m_o} h^{(1 - m_o)} - \frac{h}{p_e m_o} \qquad (10.31)$$

Equation 10.31 implies that h and q are dependent on both x and t. Thus, the flow is nonuniform and unsteady.

10.4.2.4 Equations for Equilibrium Hydrographs

The maximum depth of water for any value of x is obtained as

$$h(x) = \left(\frac{p_e x}{\alpha_o}\right)^{\left(\frac{1}{m_o}\right)} \tag{10.32}$$

The corresponding time is given by

$$t(x) = \left[\left(\frac{1}{p_e}\right)^{(m_o-1)/m_o}\right]\left(\frac{x}{\alpha_o}\right)^{1/m_o} \tag{10.33}$$

The corresponding discharge is given by

$$q(x) = p_e x \tag{10.34}$$

Figure 10.6 shows the equilibrium hydrographs as a function of space and time for both infinite and finite rainfall duration.

The time $t(x)$ given by Eq. 10.33 is the *time of concentration* for the length x, where $0 < x < L_o$. This is the time taken by water to travel from the upper end of the plane to the location x.

The equilibrium time and equilibrium depth refer to the particular condition when the characteristic curve $t(x, 0)$ passing through the origin (0, 0) intersects the downstream boundary $x = L_o$. This is the time taken by water to reach the end of the plane. For the equilibrium case, this is same as t^*. In this case, the equilibrium depth and time are given by Eqs. 10.32 and 10.33 with $x = L_o$.

Equations 10.32 and 10.33 can also be used to construct graphs showing depth and discharge as a function of space, that is, along the length of the plane. For $0 \leq t \leq T$, the depth and discharge evolve over the length of the plane along a curve as shown in Figure 10.7. For a certain time, h and q increase up to a certain length of the plane and then remain constant throughout the plane. This means that for a given time less than the time of equilibrium, the maximum height is achieved up to certain distance, and beyond that time the height is yet to rise with time. With a further increase in time, the length up to which the depth and discharge increase moves toward the lower boundary. Once time crosses t^*, i.e., for $t^* \leq t \leq T$, depth and discharge further increase and the maximum is reached at the lower boundary. At time after T, the evolution curve is shifted further toward the downstream end indicating that the distance at which a certain h or q can be found during rainfall has shifted further toward the downstream boundary.

10.4.3 Partial Equilibrium Hydrographs

Partial equilibrium occurs when rainfall duration, T, is so short that $t^* > T$. The characteristic curves and the solution domains D1, D2, D31, and D32 are shown in Figure 10.8. The bounding characteristic $t = t(x, 0)$ intersects $t = T$ before it intersects $x = L_o$. Domains D1 and D2 are as defined above. The solution domain D3 is further divided into D31 and D32. The subdomain D31 is bounded by the curve $t = t(x, x^*)$, which is the elongation of the curve $t = t(x, 0)$ above the line $t = T$. The subdomain D32 is bounded by $t = T$, $x = L_o$, and the limiting characteristic $t = t(x, x^*)$.

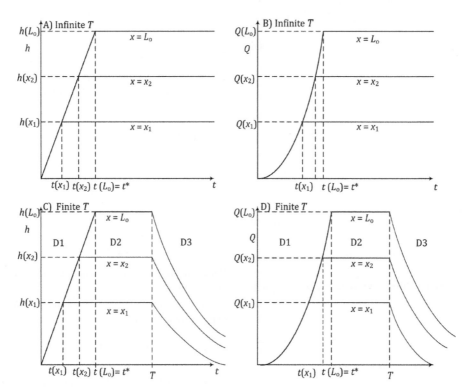

Figure 10.6 Equilibrium hydrographs for (A), (B) infinite and (C), (D) finite rainfall duration.

10.4 Analytical Solutions of Kinematic Wave Equation

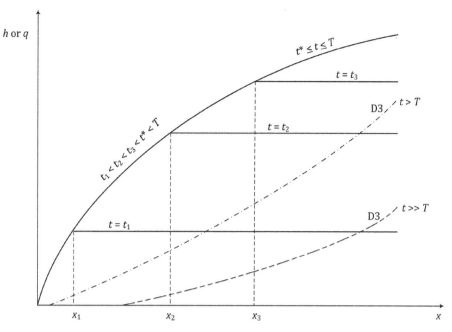

Figure 10.7 Depth or discharge as a function of space at various times.

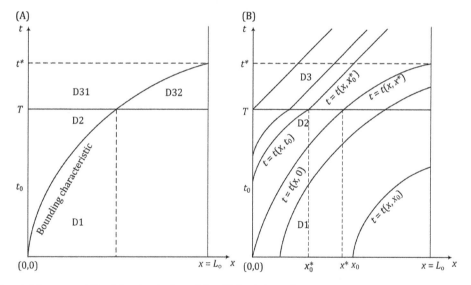

Figure 10.8 (A) Characteristic curves and (B) solution domains for partial equilibrium hydrographs.

Solutions for h and t for both domain D1 and D2 are the same as those obtained for the corresponding domains of the equilibrium case. For the solutions in domain D31 let $0 < x_0^* \leq x^*$. Once such a point is marked on the line $t = T$, the solutions are the same as those obtained for domain D3 of the equilibrium case.

For the solutions in domain D32, the initial conditions are given by

$$t(x_0^*) = T$$
$$h(x_0^*) = h(x_0^*, x_0)$$

with the solutions given by

$$h(x_0^*, x_0) = \left[\frac{p_e}{\alpha_o}(x_0^* - x_0)\right]^{\frac{1}{m_o}} \tag{10.35}$$

$$h(x; x_0^*, x_0) = h(x_0^*, x_0) \tag{10.36}$$

$$t(x; x, x_0^*) = T + \frac{1}{m_0}\left(\frac{1}{\alpha_o}\right)^{\frac{1}{m_o}}[p_e(x - x_0^*)]^{\left(\frac{1}{m_o} - 1\right)}(x - x_0^*) \tag{10.37}$$

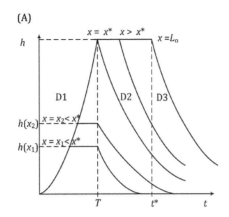 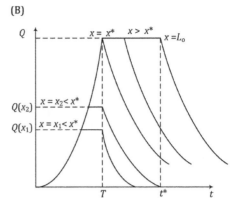

Figure 10.9 (A) Depth and (B) discharge hydrographs for partial equilibrium.

Therefore, the solutions are obtained as

$$h(x, x_0^*) = p_e T \tag{10.38}$$

$$t(x, x_0^*) = T + \left(\frac{x - x_0^*}{\alpha_o m_o}\right)(p_e T)^{(1-m_o)} \tag{10.39}$$

In domain D32, the characteristic curves are straight lines and the water depth is independent of both x and x_0^* and therefore also independent of t. Thus, in D32,

$$h(x, t) = p_e T \tag{10.40}$$

and,

$$q = \alpha_o (p_e T)^{m_o} \tag{10.41}$$

The determination of x^* is necessary for the computation of the partial equilibrium hydrograph. This quantity is determined by solving for the limiting characteristic curve where it intersects the segment $t = T$. This is given by

$$x^* = \frac{\alpha_o T^{m_o}}{p_e}(p_e)^{m_o} \tag{10.42}$$

Unlike the equilibrium case, t^* in the partial equilibrium case depends on T. In this case t^* is defined as the intersection of $t = t(x, x^*)$ with $x = L_o$ and is given by

$$t^* = T + \left(\frac{\alpha_o}{p_e x^*}\right)^{\left(\frac{m_o-1}{m_o}\right)}\left(\frac{L_o - x^*}{\alpha_o m_o}\right) \tag{10.43}$$

This is the maximum time of travel to the end of the plane.

The maximum depth and discharge at any x are obtained by denoting the distance up to which equilibrium has been achieved by $\bar{x}$, which is

$$\bar{x} = \frac{\alpha_o T^{m_o}}{p_e}(p_e)^{m_o} \tag{10.44}$$

For $x \leq \bar{x}$

$$h(x) = \left(\frac{p_e x}{\alpha_o}\right)^{\frac{1}{m_o}} \tag{10.45}$$

The corresponding time is given as in the case of equilibrium hydrograph by

$$t(x) = \left(\frac{1}{p_e}\right)^{(m_o-1)/m_o}\left(\frac{x}{\alpha_o}\right)^{1/m_o} \tag{10.46}$$

and maximum discharge is given, also as in the case of equilibrium hydrograph by:

$$q(x) = p_e x \tag{10.47}$$

For $x \geq \bar{x}$

$$h(x) = p_e T \tag{10.48}$$

$$t(x) = T\left(1 - \frac{1}{m_o}\right) + \left(\frac{x}{\alpha_o m_o}\right)(p_e T)^{(1-m_o)} x \tag{10.49}$$

$$q(x) = \alpha_o (p_e T)^{m_o} \tag{10.50}$$

Typical depth and discharge hydrographs for partial equilibrium are shown in Figure 10.9.

10.5 NUMERICAL SOLUTIONS OF THE KINEMATIC WAVE EQUATION

The analytical solutions presented above aid in understanding the behavior of the propagation of kinematic waves in space and time. However, for all practical purposes, both Eq. 10.4, representing flow through a channel element, or Eq. 10.11, representing flow over an overland element, are usually solved by finite-difference approximation of the partial derivatives and setting appropriate initial and boundary conditions. For example, the first-order partial derivative $\partial A/\partial x$ is approximated using a backward finite-difference method

$$\frac{\partial A}{\partial x} \approx \frac{\Delta A}{\Delta x} = \frac{A_i^j - A_{i-1}^j}{\Delta x} \tag{10.51}$$

A finite-difference method presents a point-wise approximation of the governing partial differential equation. The indices of the approximation refer to positions on a

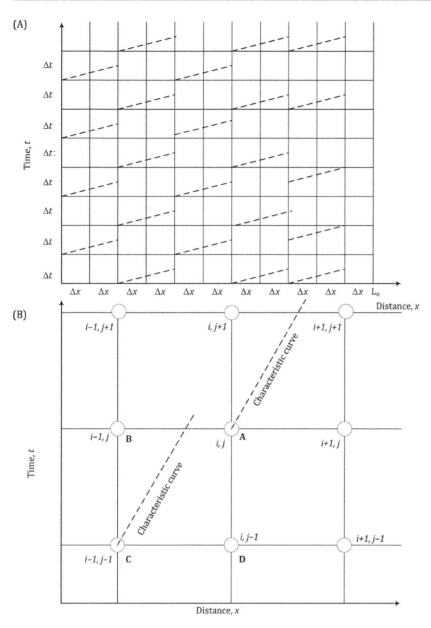

Figure 10.10 Discretization of space and time into grids for the numerical solutions of the kinematic wave equation. (A) Characteristic curves (dashed lines) on a fixed Δx–Δt grid. (B) Space–time grid used for the finite-difference method.

space–time grid, as shown in Figure 10.10A. Discretization of the space–time grid provides a convenient way to visualize the manner in which the solution scheme solves for unknown values of A at various locations and times. The index i in the subscript indicates the current location at which A is to be found along the length of the channel or overland flow plane. The index j in the superscript indicates the current time step of the solution scheme. Indices $(i-1)$, and $(j-1)$ indicate, respectively, the positions and times removed a value Δx and Δt and from the current location and time in the solution scheme. Similarly, future times and space locations that are advanced by one Δt and Δx are indicted as $(j+1)$ and $(i+1)$, respectively (Figure 10.10B). In Figure 10.10A, the characteristic solution curves are also shown. While the finite-difference solutions are always at nodes evenly spaced in the x–t space, the characteristic solution curves do not always intersect at such nodes. The characteristic curves represent locations in the x–t plane where specific flow properties, such as grid wave celerity, c, remain constant for each time step Δt. Its importance will be evident from the standard and conservation forms of the numerical solutions presented below.

In modeling a rainfall–runoff event over a watershed, computations start with the determination of overland flows, which are then input as uniformly distributed lateral inflows into the collector elements. These, in turn, modify the flows and distribute these collector flows uniformly and laterally along the main channel, as shown in Figure 10.2. The main channel routes the final wave through the catchment. A combination of several catchments thus allows for the complete description of an entire watershed during a rainfall event.

There are many numerical schemes for solutions of the governing equations. Singh (1996) described the Lax–Wendorff scheme, which uses a Taylor series expansion of the dependent variable $A(x, t + \Delta t)$, a modified Lax–Wendorff scheme, and a nonlinear finite-difference method. The scheme proposed by Leclerc and Schaake (1973) is quite simple and is implemented in HEC-HMS. This is an explicit finite-difference scheme that approximates $\partial A/\partial t$ as $\Delta A/\Delta t$, a difference in flow area in successive times, and it approximates $\partial A/\partial x$ as $\Delta A/\Delta x$, a difference in flow area at adjacent locations (USACE, 1993b). It is assumed that the kinematic wave parameters (α for channel, α_o for overland, m for channel and m_o for overland) are constant for any given system of channel and overland elements.

The governing equations are expressed in terms of finite differences using the indexing scheme described above and solved for the value of A (for channels) or h (for overland flow) at the current time as shown by point A in Figure 10.10B.

$$\alpha m A^{m-1} \frac{\partial A}{\partial x} \approx \alpha m A^{m-1} \frac{\Delta A}{\Delta x}$$
$$= \alpha m \left[\frac{A_i^{j-1} + A_{i-1}^{j-1}}{2}\right]^{m-1} \left[\frac{A_i^{j-1} - A_{i-1}^{j-1}}{\Delta x}\right] \quad (10.52)$$

where the differential of the area A in the x-direction is taken as the difference between the values known at points C and D in Figure 10.10B. The area term which is raised to the $(m - 1)$ power in Eq. 10.52 is the average area between points C and D.

The time derivative term in Eq.10.4 is evaluated between points A and D

$$\frac{\partial A}{\partial t} \approx \frac{\Delta A}{\Delta t} = \frac{A_i^j - A_i^{j-1}}{2} \quad (10.53)$$

The lateral inflow term (q for channel flow, p_e for overland flow) is handled as an average lateral inflow that occurs within a time step Δt

$$\frac{q_i^j + q_i^{j-1}}{2} = \bar{q} \quad (10.54)$$

Combining Eqs. 10.52, 10.53, and 10.54, the resulting algebraic equation for a finite-difference approximation is

$$\frac{A_i^j - A_i^{j-1}}{\Delta t} + \alpha m \left[\frac{A_i^{j-1} + A_{i-1}^{j-1}}{2}\right]^{m-1} \left[\frac{A_i^{j-1} - A_{i-1}^{j-1}}{\Delta x}\right] = \bar{q} \quad (10.55)$$

With the solution scheme proposed, the only unknown value in Eq. 10.55 is the current value at a given location, A_i^j. All other values of A are known from either a solution of the equation at a previous location and time, or from an initial or boundary condition. The solution for the unknown is given as

$$A_i^j = \bar{q} \Delta t + A_i^{j-1} - \alpha m \left[\frac{\Delta t}{\Delta x}\right] \left[\frac{A_i^{j-1} - A_{i-1}^{j-1}}{2}\right]^{m-1} \left[A_i^{j-1} - A_{i-1}^{j-1}\right] \quad (10.56)$$

Equation 10.56 is called the **standard form** of the finite-difference approximation. This standard form of the finite-difference equation is applied when the stability factor, R as defined below, is less than unity:

$$R = \begin{cases} \dfrac{\alpha}{\bar{q}\Delta x}\left[\left(\bar{q}\Delta t + A_{i-1}^{j-1}\right)^m - \left(A_{i-1}^{j-1}\right)^m\right] & \text{for } \bar{q} > 0 \\ \alpha m \left(A_{i-1}^{j-1}\right)^{(m-1)} \left(\dfrac{\Delta t}{\Delta x}\right) & \text{for } \bar{q} = 0 \end{cases} \quad (10.57)$$

The standard form of the equation applies where the average *grid wave celerity*, c is less than the ratio of the computational space interval to the time step:

$$c < \frac{\Delta x}{\Delta t} \quad (10.58)$$

This corresponds to a solution below a diagonal connecting points C and A in Figure 10.10B. However, if c is greater than $\Delta x/\Delta t$, it is possible for flood wave characteristics to propagate more rapidly through space and time than the numerical approximation method can account for. This would correspond to a solution found above a diagonal connecting A and C in Figure 10.10B.

If R is greater than unity or $c > \Delta x/\Delta t$, the **conservation form** of the finite-difference equation applies. In this form, the temporal derivatives are evaluated between points B and C in Figure 10.10B, rather than A and D, while the rapidly advancing flood wave characteristics can be handled more accurately.

The conservation form of the spatial derivative of the discharge is evaluated between points B and A as

$$\frac{\partial Q}{\partial x} = \frac{\Delta Q}{\Delta x} = \frac{Q_i^j - Q_{i-1}^j}{\Delta x} \quad (10.59)$$

and the temporal derivative of area is evaluated between points B and C

$$\frac{\partial A}{\partial t} \approx \frac{\Delta A}{\Delta t} = \frac{A_{i-1}^j - A_{i-1}^{j-1}}{2} \quad (10.60)$$

Substitution of these conservation form derivatives into the continuity equation produces

$$\frac{Q_i^j - Q_{i-1}^j}{\Delta x} + \frac{A_{i-1}^j - A_{i-1}^{j-1}}{\Delta t} = \bar{q} \quad (10.61)$$

In Eq. 10.61 the only unknown is Q_i^j which is given by

$$Q_i^j = Q_{i-1}^j + \bar{q}\Delta x - \frac{\Delta x}{\Delta t}\left[A_{i-1}^j - A_{i-1}^{j-1}\right] \quad (10.62)$$

From the value obtained for Q_i^j, the corresponding flow area is computed as

$$A_i^j = \left[\frac{Q_i^j}{\alpha}\right]^{\frac{1}{m}} \quad (10.63)$$

whereas, in the case of solution from Eq. 10.56, Q_i^j is derived from the flow area as

$$Q_i^j = \alpha \left(A_i^j\right)^m \tag{10.64}$$

The accuracy and stability of the finite-difference scheme depends on approximately maintaining the condition similar to the Courant condition for grid velocity given by $c\Delta t \approx \Delta x$, where c is the average kinematic wave speed in an element. The kinematic wave speed is a function of flow depth and, consequently, varies during the routing of the hydrograph through and element. Since Δx is a fixed value, the finite-difference scheme utilizes a variable Δt internally to maintain the desired relationship between x, t, and c.

The accuracy of the finite-difference scheme depends on the selection of the distance increment, Δx. The distance increment is initially chosen by the formula $\Delta x = c\Delta t_m$, where c in this instance is an estimated maximum wave speed depending on the lateral and upstream inflows and t_m is the time step equal to the minimum of (1) one-third of the plane or reach length divided by the wave speed, that is, one-third the travel time through the reach, the travel time being the element length divided by the wave speed; (2) one-sixth of the upstream hydrograph rise time for a channel; and (3) the specified computation interval. Usually, Δx is chosen as the minimum of this computed length and the reach, or plane, length divided by the number of distance steps (segments) specified in the input. During the solution, Δt is varied to satisfy the Courant condition. Consequently, the accuracy of the finite-difference solution depends on both the selection of Δx and the interpolation of the kinematic wave hydrograph to the user-specified computational time interval.

10.6 DISTINGUISHING FEATURES OF THE KINEMATIC WAVE MODEL

The application of the kinematic wave model differs from the unit hydrograph method in several ways. Moreover, in certain respects, it offers advantages over the traditional unit hydrograph method. For example, even though the unit hydrograph method is routinely used and has been successful in predicting DRHs with reasonably good accuracy, it is often difficult to associate the physical characteristics of a watershed to the parameters necessary to develop a unit hydrograph. It is even more difficult to define some of the parameters such as the t_c and R for the Clark unit hydrograph or C_t and C_p for the Snyder unit hydrograph for watersheds that have no recorded data. Because it is important to develop the best representation of the actual runoff situation, particularly for the design of stormwater management systems in urbanized watersheds, it is desirable to relate runoff processes directly to measurable physical features of the watershed.

First, the kinematic wave model is a **distributed model**, as we have seen that runoff is computed as a function of both space (x) and time (t) rather than a lumped approach like the unit hydrograph method. The *distributed approach* allows the model to capture different responses from both pervious and impervious areas in a single subbasin. This is especially advantageous in small urban watersheds.

Second, the kinematic wave model produces a **nonlinear response** to rainfall excess as opposed to the linear response of the unit hydrograph. The kinematic wave theory offers the benefits of a *nonlinear response* without needing an unduly complicated solution procedure. In many situations, particularly for the simulation of urban stormwater runoff processes, it is desirable for the modeling technique to be able to reproduce nonlinear runoff characteristics rather than being limited to linear responses.

For modeling unsteady overland flow, any model will require considerable parameter adjustment to account for the complexities of the watershed and the specific flows that occur within the watershed. The kinematic wave method relates basin and flow characteristics directly to the two routing parameters, α and m. These two parameters are directly related to the shape of the channel, the boundary roughness, and the slope of the channel or overland flow surfaces. Several studies have developed sets of appropriate values for these parameters for a large range of flow and boundary conditions.

When applying the kinematic wave approach to modeling subbasin runoff, the various physical processes of water movement over the basin surface, infiltration, flow into stream channels, and flow through channel networks are considered. Parameters such as roughness, slope, area, overland lengths, and channel dimensions are used to define the process. The various features of the irregular surface geometry of the basin are generally approximated by either of the two types of basic flow elements: an overland flow element, or a stream or channel flow element. In the modeling process, overland flow elements are combined with channel flow elements to represent a subbasin. The entire basin is modeled by linking the various subbasins together. Although the geometry of natural watersheds is far more complex than this idealized representation, several studies beginning from an early study by Wooding (1965, 1966) have shown that the agreement between observed hydrographs in natural drainage areas and computed hydrographs using the kinematic wave theory is quite good.

The kinematic wave approximation has been proven to be an accurate and efficient method of simulating stormwater runoff from small watersheds for both overland flow and stream channel routing (Overton and Meadows, 1976). The kinematic wave method is designed principally for representing *urban watersheds* although it can be used for undeveloped watersheds as well.

10.7 IMPLEMENTATION OF KINEMATIC WAVE MODEL IN HEC-HMS

In HEC-HMS, the distributed outflow from a catchment or subbasin can be obtained by utilizing combinations of three conceptual elements: overland flow planes, subcollector as well as collector channels, and a main channel, as shown in Figure 10.2.

In the kinematic wave model, the overland surface is conceptualized as one or two planes. Typically, one plane is used for impervious surfaces and one for pervious surfaces. The kinematic wave equations are used to calculate the runoff for each plane and then a weighted composite runoff is calculated using the representative percentage of the entire watershed or catchment for each plane.

The composite runoff is directed to a subcollector, which is used to represent the primary collection in the stormwater management system. The most common collection system is gutters along streets. In areas without gutters, unlined ditches next to the street may perform a similar stormwater collection function. The composite runoff is applied to the subcollector as a uniform lateral boundary condition, with scaling based on the representative area of a typical subcollector. Water is routed through a representative subcollector using the kinematic wave equations. There is no seepage from the subcollector so it may be difficult to use it for representing vegetated swales and other types of best management practices.

The outflow from the representative subcollector is directed to a collector, which is used to represent the next step up in the stormwater management system. Conceptually, the collector receives inflow from multiple gutters or ditches. The outflow from the subcollector is scaled up using the representative area of a typical collector and applied to the collector as a uniform lateral boundary condition. Water is routed through a representative collector using the kinematic wave equations. Again, there is no seepage from the collector.

The outflow from the representative collector is directed to a main channel. The channel may be used to represent the final step in the stormwater management system, or optionally may be used to represent the river passing through the subbasin. Conceptually, the main channel receives inflow from multiple collectors. The outflow from the representative collector is scaled up using the area of a typical collector and the area of the subbasin. The scaled flow is applied to the channel as a uniform lateral boundary condition. Water is routed through the channel to become the outflow from the subbasin. The channel routing may be performed using either the kinematic wave equations or the Muskingum–Cunge equations discussed in Chapter 12.

Figure 10.11 shows an idealized representation of an urban catchment in kinematic wave modeling. Overland flows are calculated separately from channel flows that include subcollector, collector, and main channels. After the overland runoff is routed down the length of the overland flow strips, it is then distributed uniformly along the collector system, which represents rivulets, channels, gutters, sewers, and storm drains. Once the runoff flows enter the collector system, they move through it, picking up additional uniformly distributed lateral inflow from adjacent runoff strips. These collector flows eventually reach a main channel where they are routed as open channel stream flows through the system using a channel routing method. The individually calculated flows are combined properly to preserve continuity and accuracy.

In HEC-HMS, the input required for each of the planes includes *length, slope, surface roughness,* and *the percentage of total area* represented by the plane. For impervious areas, the length of the plane should be the average flow length from the point where precipitation falls to where runoff first enters a collection gutter or channel. For pervious areas, this should likewise be the average flow length. When using the two planes to represent pervious and impervious areas, it is helpful to adopt a convention for

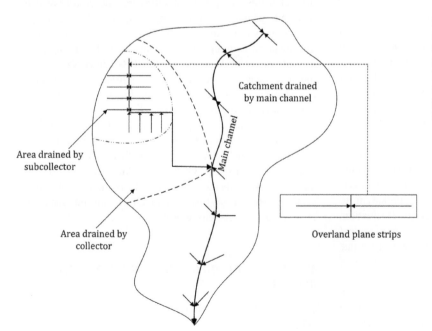

Figure 10.11 Kinematic wave representation of an urban catchment. The catchment has subareas drained by subcollector channels. Overland flow is uniformly distributed along the subcollector system. The subcollector system flow is uniformly distributed along the collector channel. Collector channel flow is uniformly distributed along the main channel, which drains the entire catchment.

which the plane number (one or two) represents the pervious area. The slope should represent the average slope along the flow line from the point where precipitation falls to where runoff first enters a gutter or channel. Roughness is the principal difference between planes representing impervious or pervious areas. Roughness coefficients for natural areas are much higher than for developed areas. However, in both cases the surface roughness coefficients are higher than the typical Manning's roughness coefficients used for open channel flow.

The input for a subcollector or collector includes *length, slope, Manning's n, shape* and *associated parameters,* and *area*. The typical area of each subcollector is used to determine how to apply the composite outflow from the subcollector to the collector channel. It essentially determines the number of subcollector channels in the subbasin.

The outflow from the subcollector enters a collector as lateral inflow. The collector channel is intended to represent small ditches or open channels that are part of an engineered stormwater management system. The parameters for the collector are exactly the same as for the subcollector. The area served by a typical collector must be entered; this is used to apply lateral inflow from the collector to the main channel. Options for the cross-sectional shape are exactly the same as for the subcollector.

The surface runoff on the two planes is always routed using the kinematic wave method. However, in HEC-HMS there is an option for the choice of using either kinematic wave or Muskingum–Cunge routing in the subcollector, collector, and main channel. All three channels must use the same method or, in other words, the method of channel routing should not be intermixed.

10.8 EXAMPLES

Example 10.1: A small partially urbanized watershed is shown in Figure 10.12. The watershed is divided into two catchments.

The upper catchment has an area of 1.8 square miles and is in a natural state, which is primarily rolling pastureland with a few wooded areas. The amount of impervious area is negligible and some tributary streams flow into a main creek that drains the entire watershed. The typical flow length of the overland surface (L_o) to tributary streams is 1500 feet. The overland surface roughness (N) is 0.4 and representative ground slope (S_o) is 0.04. The tributary channels have a typical slope (S) of 0.025 and Manning's n of 0.10. Typical length (L) of the tributary channel is 2500 feet and typical sections of the channels are triangular with 1:1 side slope. The area contributing to a typical tributary channel is 0.8 square mile. The creek in this catchment has a length (L) of 4000 feet and is approximately triangular with side slopes of 4 (H) to 1 (V). The mean slope of the creek is 0.01 and Manning's n is 0.05.

The lower catchment has an area of 2.2 square miles and is mostly urbanized and 30% of the area is completely impervious. The storm drainage system serving the area is grass-lined triangular roadside ditches with a length of 2500 feet, 1:1 side slope, 0.008 of longitudinal slope and Manning's n of 0.020. The typical flow length of the impervious overland surface, (L_{o1}) to collector system is 50 feet. This overland surface roughness (N_1) is 0.30 and representative slope (S_{o1}) is 0.06. The typical flow length of the pervious overland surface (L_{o2}) to the collector system is 180 feet. This overland surface roughness (N_2) is 0.40 and representative slope (S_{o2}) is 0.01. The area contributing to the collector system is 0.35 square mile. The creek in this catchment has a length (L) of 4000 feet and is trapezoidal in cross section with side slopes of 2 (H) to 1 (V). The mean slope of the creek is 0.003 and Manning's n is 0.025.

On May 1, 1979 a 1-hour storm starting at 12:00 produced 2 inches of rainfall with the temporal distributions as given in Table 10.3. The infiltration rate in the upper

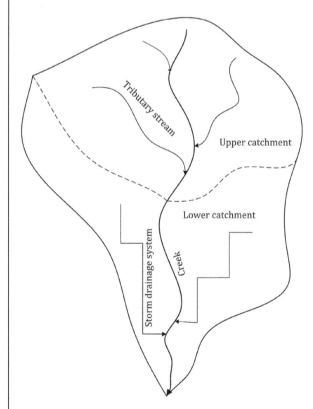

Figure 10.12 The watershed of Example 10.1.

Table 10.3 *Temporal distribution of rainfall*

Time	Percentage of precipitation
12:00	0
12:05	6.00
12:10	12.00
12:15	15.00
12:20	25.00
12:25	20.00
12:30	10.00
12:35	8.00
12:40	4.00
13:00	0.00

catchment is 0.5 in/h. The infiltration rates in the impervious surface of lower catchment is 0.02 in/h and that for the pervious portion of the catchment is 0.20 in/h. Use the kinematic wave runoff and routing options of HEC-HMS to compute the direct runoff hydrographs from both catchments.

Solution: The basin model of HEC-HMS is shown in Box 10.1.

The upper catchment can be modeled as a single overland plane. The inputs for the land component of the upper catchment are shown in Boxes 10.2 and 10.3.

The channel flows in the upper catchment can be modeled as collectors representing the tributary streams and the creek as the main channel. The inputs for the channel components of the upper catchment are shown in Boxes 10.4 and 10.5.

The lower catchment is modeled as two overland planes. The inputs for the land component of the upper catchment are shown in Boxes 10.6–10.8.

The storm drainage system is modeled as the collector channel and the creek is modeled as the main channel. Boxes 10.9 and 10.10 show the inputs for the channel component of the catchment.

Boxes 10.11, and 10.12 show the computed hydrographs for the upper and lower catchments. The DRH at the outlet is given in Box 10.13.

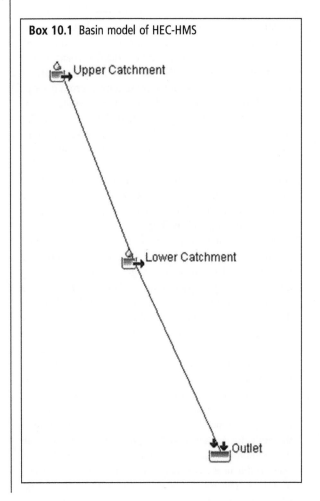

Box 10.1 Basin model of HEC-HMS

Box 10.2 Subbasin input for the upper catchment

Box 10.3 Inputs for plane 1 in upper catchment

Basin Name:	Basin 1
Element Name:	Upper Catchment
*Length (FT)	1500
*Slope (FT/FT)	0.04
*Roughness:	0.40
*Area (%)	100
Routing Steps:	5

Tabs: Subbasin, Loss 1, Loss 2, Channel, Collector, Subcollector, **Plane 1**, Plane 2, Options

Box 10.4 Inputs for the collector in upper catchment

Basin Name:	Basin 1
Element Name:	Upper Catchment
*Length (FT)	2500
*Slope (FT/FT)	0.025
*Manning's n:	0.10
Subreaches:	5
*Area (MI2)	0.80
Shape:	Triangle
*Side Slope (xH:1V)	1

Tabs: Subbasin, Loss 1, Loss 2, Channel, **Collector**, Subcollector, Plane 1, Plane 2, Options

Box 10.5 Inputs of the channel in the upper catchment

Basin Name:	Basin 1
Element Name:	Upper Catchment
Route Upstream:	No
Routing Method:	Kinematic Wave
*Length (FT)	4000
*Slope (FT/FT)	0.010
Subreaches:	5
Shape:	Triangle
*Manning's n:	0.05
*Side Slope (xH:1V)	4

Tabs: Collector, Subcollector, Plane 1, Plane 2, Options, Subbasin, Loss 1, Loss 2, **Channel**

Box 10.6 Subbasin input for the lower catchment

Basin Name:	Basin 1
Element Name:	Lower Catchment
Description:	Urbanized
Downstream:	Outlet
*Area (MI2)	2.2
Latitude Degrees:	
Longitude Degrees:	
Discretization Method:	---None---
Canopy Method:	---None---
Snow Method:	---None---
Surface Method:	---None---
Loss Method:	Initial and Constant
Transform Method:	Kinematic Wave
Baseflow Method:	---None---

Tabs: Collector, Subcollector, Plane 1, Plane 2, Options, **Subbasin**, Loss 1, Loss 2, Channel

Box 10.7 Inputs for plane 1 in the lower catchment

Basin Name:	Basin 1
Element Name:	Lower Catchment
*Length (FT)	180
*Slope (FT/FT)	0.01
*Roughness:	0.40
*Area (%)	70
Routing Steps:	5

Tabs: Subbasin, Loss 1, Loss 2, Channel, Collector, Subcollector, Plane 1, **Plane 2**, Options

Box 10.8 Inputs for plane 2 in the lower catchment

Basin Name:	Basin 1
Element Name:	Lower Catchment
*Length (FT)	50
*Slope (FT/FT)	0.06
*Roughness:	0.30
*Area (%)	30
Routing Steps:	5

Tabs: Subbasin, Loss 1, Loss 2, Channel, Collector, Subcollector, **Plane 1**, Plane 2, Options

Box 10.9 Inputs for the collectors in the lower catchment

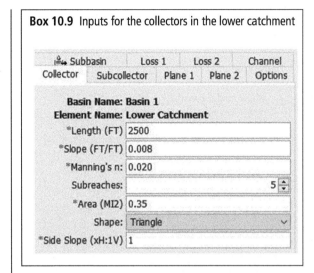

Box 10.10 Inputs for the main channel in the lower catchment

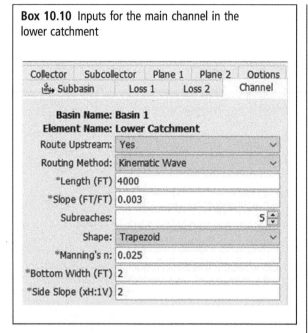

Box 10.11 DRH from upper catchment

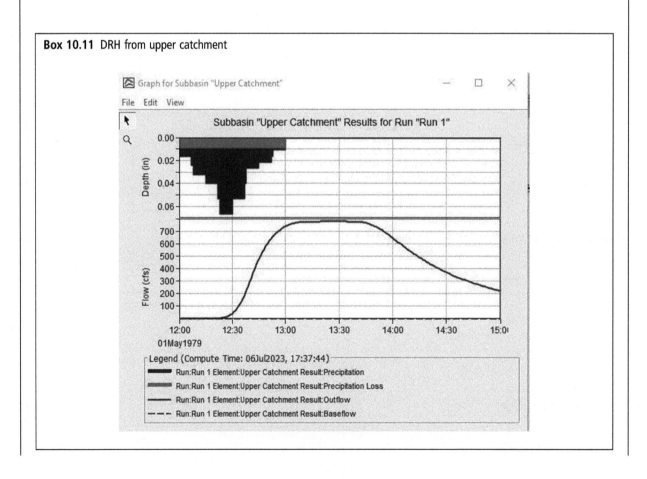

10.8 Examples 441

Box 10.12 DRH from lower catchment

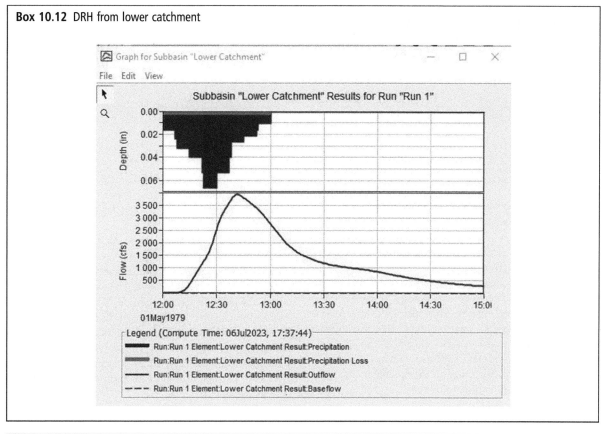

Box 10.13 DRH at the outlet

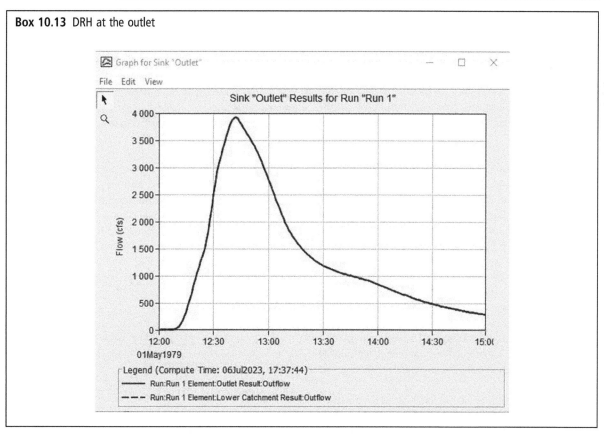

Exercises: A selection of exercises on this topic is available at www.cambridge.org/appliedhydrology.

11 The Rational Method

11.1 CONCEPTS AND PURPOSE

In certain cases of rainfall–runoff analysis, it may be sufficient to have a reasonable estimate of the peak discharge only and the derivation of a direct runoff hydrograph (DRH) may not be necessary. This is mostly true for estimating peak discharge from relatively small drainage areas, not exceeding a few hundred acres (1 acre $\approx$ 4000 m^2). The **rational method** has been traditionally used for such cases, since it was developed first by Mulvaney (1850); a more comprehensive treatment was presented by Kuichling (1889). The rational method is also sometimes modified in a way to produce a simplistic runoff hydrograph. This chapter covers the concepts pertaining to the rational method as well as to the **modified rational method**.

11.2 THE RATIONAL METHOD EQUATION

The rational method relates the peak discharge to the contributing drainage area, the average rainfall intensity for a duration equal to a watershed response time, and a coefficient that represents hydrologic abstractions and hydrograph attenuation. The coefficient is generally termed the **runoff coefficient**. Equation 11.1 represents the rational formula:

$$Q_p = CIA \qquad (11.1)$$

in which C is the runoff coefficient, Q_p is the peak discharge resulting from a rainfall intensity, I, uniformly falling over an area, A, for a period equal to the time of concentration.

The runoff coefficient, C, relates the ratio of the input volume rate, given by the product IA, to the output volume rate, Q_p. The runoff coefficient is dimensionless, if all other terms in Eq. 11.1 are in compatible units, such as I in in/h and A in acre or I in m/s and A in m^2. The value of C can range from 0 to 1. A value of zero implies no peak discharge or runoff produced for a given rainfall intensity whereas the value 1 gives a perfect conversion of rainfall intensity to peak discharge. The runoff coefficient can vary with rainfall intensity, I, which is, by originating arguments, the rainfall rate when the duration of rainfall equals the time of concentration.

The modified rational method extends the idea presented in Eq. 11.1 to parameterize simple runoff hydrographs, where the peak of the hydrograph is the peak discharge estimated by the application of the rational method, and the time base of the hydrograph from start to end is typically twice the time of concentration. This results in a triangular hydrograph that can be used to model systems that represent unsteady flow phenomena.

Several issues are associated with the use of rational method for estimating peak discharge for the design of drainage structures, and when the modified rational method is used to estimate inflow hydrographs for the design of stormwater management systems, such as detention basins discussed in Chapter 13. The use of these methods depends on estimates of the time of concentration and the runoff coefficient, both of which can vary substantially depending on a number of factors. In addition, design rainfall intensity also has a substantial impact on peak discharge estimates using the rational method.

The rational method conceptualizes the runoff process if the following holds true. If the product of rainfall intensity and area is constant over the time interval required to completely drain a watershed or for any longer duration, the runoff rate will be equal to the product of rainfall intensity and area, which means an *equilibrium condition* where inflow equals outflow. A mass balance relating the input and output under such an equilibrium is the rational runoff formula given by Eq. 11.1.

Key assumptions associated with the rational formula for practical applications are: (1) rainfall intensity is constant for a time period for at least as long as the characteristic time of the watershed; (2) the maximum runoff occurs when the rainfall intensity lasts at least as long as the characteristic time; (3) the contributing area is constant during the rainfall event; (4) the runoff coefficient is constant during the rainfall event. In practice, the runoff coefficient is typically estimated from tabular data or from runoff studies that relate the volume of runoff to the volume of precipitation (Wanielista et al., 1997). The latter approach is the conceptual source of the concept of the **volumetric runoff coefficient**.

11.3 THE RUNOFF COEFFICIENT

11.3.1 Selection of the Runoff Coefficient

The determination of C is based on the character of the drainage area, including the type of soil, type and amount of vegetation or land cover, and land use, such as the degree

Table 11.1 *Runoff coefficients for urban watersheds*

Type of area	Runoff coefficient
Business	
Downtown areas	0.70–0.95
Neighborhood areas	0.30–0.70
Residential	
Single-family areas	0.30–0.50
Multi-units, detached	0.40–0.60
Multi-units, attached	0.60–0.75
Suburban	0.35–0.40
Apartment dwelling areas	0.30–0.70
Playgrounds	0.30–0.40
Parks, cemeteries	0.10–0.25
Industrial	
Light areas	0.30–0.80
Heavy areas	0.60–0.90
Railroad yards	0.30–0.40
Unimproved areas	
Sand or sandy loam soil, 0–3%	0.15–0.20
Sand or sandy loam soil, 3–5%	0.20–0.25
Black or loessial soil, 0–3%	0.18–0.25
Black or loessial soil, 3–5%	0.25–0.30
Black or loessial soil, > 5%	0.70–0.80
Deep sand area	0.05–0.15
Steep grassed slopes	0.70
Lawns	
Sandy soil, flat 2%	0.05–0.10
Sandy soil, average 2–7%	0.10–0.15
Sandy soil, steep 7%	0.15–0.20
Heavy soil, flat 2%	0.13–0.17
Heavy soil, average 2–7%	0.18–0.22
Heavy soil, steep 7%	0.25–0.35
Streets	
Asphaltic	0.85–0.95
Concrete	0.90–0.95
Brick	0.70–0.85
Drives and walks	0.75–0.95
Roofs	0.75–0.95

of urbanization or development. Table 11.1 lists suggested ranges of C values for urban watersheds for various combinations of land use and soil or surface type.

Table 11.2 shows an alternate, systematic approach for developing the runoff coefficient. This table applies to rural watersheds only, considering the watershed as a series of four aspects. For each of these, a component of the resultant runoff coefficient is assigned. The four assigned components are added to form an overall runoff coefficient for the specific watershed. Thus, the runoff coefficient for rural watersheds is given by

$$C = C_r + C_i + C_v + C_s \quad (11.2)$$

where C is the overall runoff coefficient for a rural watershed, C_r, C_i, C_v, and C_s are the components of the coefficient accounting for *watershed relief*, *soil infiltration*, *vegetal cover*, and *surface type*, respectively. The most appropriate values for C_r, C_i, C_v, and C_s can be selected from Table 11.2. While this approach was developed for application to rural watersheds, it can be used as a check against mixed-use runoff coefficients computed using other methods. In so doing, the designer would use judgment, primarily in specifying C_s, to account for partially developed conditions within the watershed.

For areas with a mixture of land uses, a **composite runoff coefficient** should be used. The composite runoff coefficient is weighted, based on the area of each respective land use and can be calculated as

$$C_w = \frac{\sum_{i}^{n} C_i A_i}{\sum_{i=1}^{n} A_i} \quad (11.3)$$

where C_w is the composite or **weighted** runoff coefficient, C_i is the runoff coefficient for area i, A_i is the area for land cover i, and n is the number of distinct land uses.

Sometimes C is multiplied by a factor C_a to account for antecedent moisture conditions in accordance with the ARI of the rainfall used in the calculation. The generally recommended values of C_a are 1.0, 1.1, 1.2, and 1.25 for rainfall with ARIs of 1 or 2 years, 25 years, 50 years, and 100 years respectively.

11.3.2 Determination of the Runoff Coefficient

There are two approaches for the determination of a runoff coefficient from the analysis of rainfall and runoff data. The first approach is to invert the rational method equation and solve for the coefficient value. The second approach is to determine the coefficient value as the ratio of runoff volume to precipitation volume. The latter procedure gives a volumetric runoff coefficient.

The inverted runoff coefficient, C^k, is uniquely a function of peak discharge, Q_p^k, and rainfall intensity, I^k. C^k is defined for the kth event as

$$C^k = \frac{Q_p^k}{I^k(t_d)A} \quad (11.4)$$

where $I^k(t_d)$ is the maximum average rainfall intensity (in/h) for a rainfall duration of t_d. The selection of t_d is not trivial. Cleveland et al. (2011) discussed this in detail.

The volumetric coefficient, C_v^k, is defined for the kth event as

$$C_v^k = \frac{P_k}{R_k} \quad (11.5)$$

where P_k is the total runoff for the event, and R_k is the total rainfall for the event, k.

Wanielista et al. (1997) and several other workers alluded to C_v^k, although it is not usually qualified as a volumetric coefficient. Other sources where a volumetric concept is implied and logically could serve as a basis for runoff

Table 11.2 *Runoff coefficients for rural watersheds*

Watershed characteristic	Extreme	High	Normal	Low
Relief – C_r	0.28–0.35 (steep, rugged terrain with average slopes above 30%)	0.20–0.28 (hilly, with average slopes of 10–30%)	0.14–0.20 (rolling, with average slopes of 5–10%)	0.08–0.14 (relatively flat land, with average slopes of 0–5%)
Soil infiltration – C_i	0.12–0.16 (no effective soil cover; either rock or thin soil mantle of negligible infiltration capacity)	0.08–0.12 (slow to take up water; clay or shallow loam soils of low infiltration capacity or poorly drained)	0.06–0.08 (normal; well-drained light or medium textured soils, sandy loams)	0.04–0.06 (deep sand or other soil that takes up water readily; very light, well-drained soils)
Vegetal cover – C_v	0.12–0.16 (no effective plant cover, bare or very sparse cover)	0.08–0.12 (poor to fair; clean cultivation, crops or poor natural cover, less than 20% of drainage area has good cover)	0.06–0.08 (fair to good; about 50% of area in good grassland or woodland, not more than 50% of area in cultivated crops)	0.04–0.06 (good to excellent; about 90% of drainage area in good grassland, woodland, or equivalent cover)
Surface storage – C_s	0.10–0.12 (negligible; surface depressions few and shallow, drainageways steep and small, no marshes)	0.08–0.10 (well-defined system of small drainageways, no ponds or marshes)	0.06–0.08 (normal; considerable surface depression, such as storage lakes and ponds and marshes)	0.04–0.06 (much surface storage, drainage system not sharply defined; large floodplain storage, large number of ponds or marshes)

Source: Texas Department of Transportation (2019).

coefficient estimation, are maps at the basin scale of annual precipitation depths and runoff depths (Moody et al., 1986). The maps of the ratio of runoff depth to rainfall depth for a location will give volumetric runoff coefficients.

11.3.3 Cautionary Notes

For typical practical applications, a runoff coefficient is selected from tabulated values, such as in Table 11.1 and 11.2, that relate this hydrologic parameter to land use and soil type. However, it should be noted here that there have been observations that contradict such a relationship. In one study, Merz et al. (2006) back-calculated runoff coefficients from hourly runoff data, hourly precipitation data, and estimates of snowmelt. They analyzed a total of about 50 000 events in 337 Austrian catchments, with catchment areas ranging from 80 km² to 10 000 km² over the period 1981–2000. Their results indicated that the spatial distribution of runoff coefficients was highly correlated with the mean annual precipitation but little correlated with soil type and land use. Furthermore, an analysis of the runoff coefficients by flood types indicated that runoff coefficients were the smallest for flash floods, and they increased in the order: short rain floods, long rain floods, rain-on-snow floods, and snowmelt floods. The conclusion that came out of this study is that, in the type of climate and at the scale of catchments examined in that study, the main controls for event runoff coefficients were climate and runoff regime through the seasonal catchment water balance and hence antecedent soil moisture, in addition to event characteristics. Catchment characteristics, such as soils and land use, affect runoff coefficients to a lesser degree.

Another assumption that is at the heart of the design rainfall–runoff calculations such as the rational method is that there is a direct correspondence of the rainfall return period, T_P, and flood return period, T_Q, even though their relationship is still poorly understood. The results of a study conducted by Viglione et al. (2009) indicate that T_Q can be much higher than T_P of the associated rainfall event. The ratio T_Q/T_P depends on the average wetness of the watershed. In a dry system, T_Q can be of the order of hundreds of times of T_P. In contrast, in a wet system, the maximum flood return period is never more than a few times that of the corresponding storm. This is because a wet system cannot become much wetter than it normally is. The presence of a threshold effect in runoff generation related to the storm volume reduces the maximum ratio of T_Q/T_P, since it decreases the randomness of the runoff coefficient and increases the probability of a wet situation.

11.4 DRAINAGE AREA

The rational method lacks a firm physical model of initial abstraction, which increases with watershed size (Asquith and Roussel, 2007). The general phenomenological argument is that the depression storage and channel storage, both increase as the watershed drainage area increases. The implication for the rational method is that there is a watershed size limit beyond which the initial abstraction of the watershed can no longer be ignored. In other words, there is some upper bound of the size of a watershed for the applicability of the rational method. Most agencies in the United States have empirically set this upper limit as 200 acres.

11.5 CHARACTERISTIC TIME

The rational method requires an estimate of the time of concentration, which establishes the duration of a rainfall input required for the system to reach equilibrium. This is the watershed characteristic time, which is also used in computing the rainfall intensity from the IDF information discussed in Chapter 4. The physical characteristics of a watershed, such as slope, affect the characteristic time, as discussed in Chapter 6. For example, steeper slopes yield smaller times of concentration and a proportionate increase in rainfall intensity for a given rainfall frequency.

11.6 IMPLICATIONS FOR THE RATIONAL METHOD

The application of the rational method requires the evaluation of IDF information for the watershed characteristic response time. A given probability of occurrence, such as an AEP, is specified for the IDF information used in the rational method. An underlying implication is that the probability given as a specified AEP of runoff is the same as the probability, that is, the AEP of rainfall used in the rational method. The structure of the intensity equation is given by Eq. 4.47 or Eq. 4.48 (Chapter 4). The event duration or average time is in fact a watershed characteristic and not a rainfall characteristic; yet t_c is required to determine an intensity. This is the double appearance of time that Kuichling (1889) alluded to in his extensive treatise on the method.

11.7 THE MODIFIED RATIONAL METHOD

The modified rational method (MRM) is a method to parameterize simple runoff hydrographs. The MRM produces a runoff hydrograph and volume while the original rational method produces only the peak discharge. The rational method was originally developed for estimating peak discharge for sizing drainage structures, such as storm drains and culverts. The MRM, which has found widespread use in engineering practices since the 1970s, is typically used to size detention or retention facilities for a specified recurrence interval and allowable outflow rate.

The MRM is based largely on the same assumptions used in the conventional rational method and is a conceptual extension of the rational method for the development of runoff hydrographs. For the MRM, a stormwater runoff hydrograph from a design storm intensity is approximated as either a triangular or trapezoidal hydrograph, depending on the relation between the storm duration and time of concentration, with the peak or plateau discharge less than or equal to CIA.

The MRM was developed with the intent of using the rational method for hydraulic structures involving storage on small watersheds (Poertner, 1974). Modified rational method analysis refers to a procedure for manipulating the basic rational method to reflect the fact that storms with durations greater than the normal time of concentration for a basin will result in a larger volume of runoff even though the peak discharge is reduced (Poertner, 1974). Three types of hydrographs can be produced with the MRM. These are similar to the *equilibrium* and *partial equilibrium* hydrographs discussed in Chapter 10 in relation to analytical solutions of kinematic wave equations for overland flow.

If the rainfall duration, t_d, equals the time of concentration, t_c, the resulting hydrograph is triangular in shape with a base equal to $2t_c$ and with a peak discharge given by Eq. 11.1 as shown in Figure 11.1. If the storm duration is greater than the watershed time of concentration, the resulting hydrograph is trapezoidal in shape with a uniform maximum discharge that forms a plateau determined from Eq. 11.1 for the difference between the time of concentration and rainfall duration, as shown in Figure 11.2. In both cases of triangular and trapezoidal hydrographs, the rising and falling limbs are linear and each has a duration of t_c.

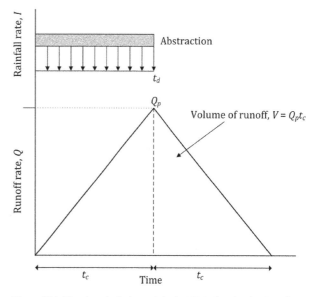

Figure 11.1 The triangular hydrograph in the MRM when the duration of rainfall is equal to the time of concentration.

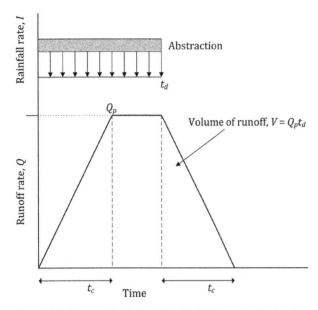

Figure 11.2 The trapezoidal hydrograph in the MRM when the duration of rainfall is greater than the time of concentration.

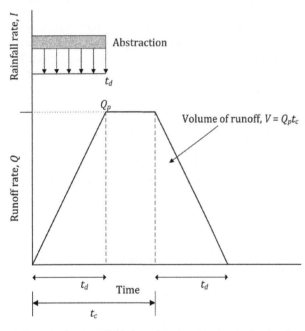

Figure 11.3 The trapezoidal hydrograph in the MRM when the duration of rainfall is less than the time of concentration.

If the rainfall duration is less than the watershed time of concentration, the resulting hydrograph is trapezoidal in shape with a maximum uniform plateau discharge being some fraction of $Q_p = CIA$ from the end of rainfall to the time of concentration. This value is denoted as Q'_p. The linear rising and falling portions of the hydrograph each have a duration of $t_d < t_c$, as shown in Figure 11.3. Smith and Lee (1984) and Walesh (1989) reported the MRM hydrograph for the case when the storm duration is less than the time of concentration of the drainage area and stated that Q'_p can be calculated using

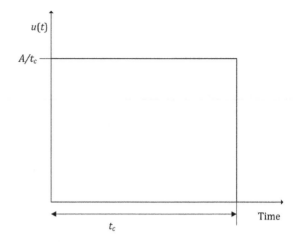

Figure 11.4 The instantaneous unit hydrograph of the rational method.

$$Q'_p = CIA \frac{t_d}{t_c} \qquad (11.6)$$

Smith and Lee (1984) examined the rational method as a unit hydrograph. They noted that if the rate of change of the contributing area is constant, so that the accumulated tributary area increases and decreases linearly and symmetrically with the time, the instantaneous unit hydrograph response function, $u(t)$, is of rectangular shape (Figure 11.4),

$$u(t) = \frac{dA}{dt} = \frac{A}{t_c} \qquad (11.7)$$

Using the rectangular response function presented as Eq. 11.7 in conjunction with uniform rainfall intensity, Smith and Lee (1984) derived the resulting direct runoff hydrographs, $Q(t)$ in watershed depth per time, by convolution as

$$Q(t) = \int_0^t i_e(\tau) u(t-\tau) d\tau \qquad (11.8)$$

where τ is the time used for integration and $i_e(\tau) = CI$ is the excess rainfall intensity, averaged as in the rational method. The decoupling of rainfall into excess rainfall by use of $i_e(\tau) = CI$ is fundamental in the interpretation of the MRM as a special case of unit hydrograph theory. Of importance, because mass must be preserved, C in this context is a volumetric runoff coefficient. Two types of outflow hydrographs, either triangular or trapezoidal in shape, were obtained by Smith and Lee from Eq. 11.8, depending upon the duration of rainfall.

Wanielista et al. (1997) discussed the rational hydrograph in the context of the contributing area and assumed that the latter varies linearly with time. They derived the triangular hydrograph when $t_d = t_c$ and the trapezoidal hydrograph when $t_d > t_c$ from the rational method, similar to the results presented by Walesh (1975).

Singh and Cruise (1992) presented a systems-based approach for the analysis of the rational formula. Like

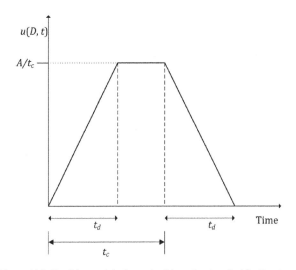

Figure 11.5 The D-hour unit hydrograph of the rational method for $D < t_c$.

Smith and Lee (1984), they assumed the watershed is represented as a linear, time-invariant system whose instantaneous unit hydrograph is a uniform rectangular distribution of base time equal to the time of concentration of the basin (Figure 11.4). They used convolution to derive the S-hydrograph and D-hour unit hydrograph for the rational method as shown in Figure 11.5. The hydrograph shown in Figure 11.5 represents the unit hydrograph that results from the MRM. This unit hydrograph has a symmetric trapezoidal shape when the duration is less than the time of concentration ($t_d < t_c$). Extending the original rational method for computing the peak discharge to the MRM for developing runoff hydrographs only assumes an ideal watershed with an equal time–area histogram and no watershed storage or hydrograph attenuation.

Many authors suggest that the rational method, and especially the MRM, requires a uniform rainfall distribution for application. This requirement is incorrect based on the interpretation of the MRM as a unit hydrograph method. The MRM can be used to generate a DRH for any nonuniform rainfall event using unit hydrograph convolution. The conclusion that the MRM is a unit hydrograph method is fundamentally important and this interrelationship establishes a continuity of methodology from very small watersheds to quite large watersheds (Cleveland et al., 2011).

11.7.1 Runoff Coefficient in the Modified Rational Method

The precise definition of the runoff coefficient in the MRM varies. The value can be defined either as the ratio of the total depth of runoff to the total depth of rainfall or as the ratio of the peak rate of runoff to rainfall intensity for the time of concentration. However, because the MRM generates a hydrograph and should preserve volumes, the method implies that the runoff coefficient is conceptually a volumetric runoff coefficient. Cleveland et al. (2011) examined in detail the volumetric runoff coefficients for the application of the MRM.

11.7.2 Characteristic Time in the Modified Rational Method

As in the rational method, the watershed characteristic time in the MRM is an exceedingly important value, exercising great influence on the convolution integral. Because the MRM is designed to produce a hydrograph from either a uniform or non-uniform rainfall distribution, the characteristic time plays a less important role in the specification of rainfall intensity.

11.8 IMPLICATIONS FOR THE MODIFIED RATIONAL METHOD

The application of the MRM in practice requires specification of the watershed characteristic time, a feature shared by the rational method and by all the synthetic unit hydrograph methods discussed in Chapter 9. The conversion of rainfall into excess rainfall in the MRM is through the application of the runoff coefficient. That is, effective rainfall is a simple fraction of incoming rainfall. Therefore, the runoff coefficient used in the MRM is a volumetric runoff coefficient and represents a conceptual process substantially different from the rate-based runoff coefficient used for estimating only peak discharge in the conventional rational method. Cleveland et al. (2011) discussed the issues associated with runoff coefficient specification, timing, and the implied probability matching of runoff and rainfall-generating runoff.

11.9 APPLICATIONS OF THE RATIONAL METHOD

11.9.1 Time–Area Considerations in Storm Drainage Design

The rational method is typically used for designing urban stormwater management systems, which mostly contain a network of storm drains and in certain cases are accompanied by detention basins. Storm sewers, for which the more appropriate term is storm drains, are designed to remove the excess runoff from urbanized areas as fast as possible so that flooding is prevented and any negative influence on transportation is minimized. An urban storm drainage system may constitute close conduits in which overland flows enter through inlets, open channels, and a combination of both.

Traditionally and typically, urban storm drains, meant to convey excess runoff from impervious covers efficiently, are designed by the application of a rational formula. Although the application of the rational formula in the design of an urban storm drainage or sewer system is a very simplistic approach, it is still widely applied even in some urbanized areas where drainage and flooding issues are serious and complex.

The method of using a rational formula in the design of a storm drainage system is a steady-state method. It is a spatially lumped model and it does not take into account variations of the following with time:

1. the rate of change of rainfall intensity;
2. flow velocity and discharge;
3. temporary storage in the drain systems; and
4. the rate of increase in the area contributing to the drain system.

Most of the design methods for storm sewers, involving a rational formula, have attempted to address item number (4) above through a concept called a time–area diagram, which is *not* the same as the **time–area diagram** discussed in Chapter 9 in relation to the Clark unit hydrograph method. For the application of a rational formula, the catchment is divided into subareas that are small enough for the contributing area to be directly proportional to time during the time of concentration. This concept is illustrated below.

Figure 11.6A shows two subareas served by a storm drain pipe. The time of concentration for subarea A_1 is T_1 and this area contributes to the storm drain through inlet, I_1. Subarea A_2 has a time of concentration T_2 and the runoff from this area is captured through inlet, I_2. The travel time from I_1 to I_2 is T_{12}. Let $T_2 > T_1 > T_{12}$. Until time T_{12}, only area A_1 contributes to the flow. By time T_1, area A_1 contributes fully. Area A_2 begins to contribute from T_{12} Up to time T_1, the total area that contributes to the flow is given by $A_1 + A_2(T_1 - T_{12})/T_2$. At time $T_2 + T_{12}$, which is the time of concentration of the composite area $A_1 + A_2$, the whole catchment contributes to the flow. This illustration shows how the contributing area increases with time according to rational method. It can be seen that when the rational formula is applied to the design of urban drainage systems, the time of concentration is estimated from the sum of the time of flow in the sewer pipe and the time of entry. The time of flow is computed with the assumption that, for the given discharge, the pipe flows full and the velocity remains constant throughout the run of the pipe. The time of entry is an allowance for the time taken for water from the most remote point in the subarea to reach the inlet serving that subarea.

In the time–area concept enumerated above, when applied to the design of a storm sewer system, the design of the pipes is carried out starting from the most upstream subarea. In Figure 11.6, pipe P_1 will carry flow Q_1, which is given by $Q_1 = C_1 I_1 A_1$. In this case I_1 is a function of T_1. But the pipe downstream of I_2 must have the capacity to convey flow from both subareas. This flow, designated by Q_2, will be calculated on the basis of the total area $(A_1 + A_2)$ and the time of concentration, which is the sum of $T_2 + T_{12}$. This procedure is illustrated with a storm sewer network in Example 11.1.

The rational method may yield erroneous results if, for a drainage system, the contributing area does not increase uniformly with time, or if the peak runoff rate is produced by a design storm whose duration is less than the time of concentration. Consider the drainage area A_1, which is a subarea of a catchment with total area A. Let the time of concentration for the area A_1 be TC_1 and that of the total area A be TC. The runoff generated from these two areas are Q_1 and Q. Since $A_1 < A$, $Q_1 < Q$. If the runoff coefficient for both these areas is assumed to be the same, then

$$\frac{Q_1}{Q} = \frac{i_1 A_1}{iA} = \frac{(A_1/A)}{(i/i_1)} \tag{11.9}$$

Since $TC > TC_1$, $i < i_1$. Thus, if the ratio (A_1/A) exceeds the ratio (i/i_1), we will get an anomaly where Q_1 exceeds Q. Thus, for the rational formula to be correctly applied to the design of a storm sewer system, the contributing area must increase uniformly with time. In other words, in practice, the *subareas should be kept to similar times of concentration and size.*

In cases where rainfall duration is either equal to or greater than the longest time of concentration of a catchment or subarea within a drainage area, the peak discharge occurs when the entire drainage area under consideration contributes flow at the farthest downstream point. In such cases, the appropriate t_c for the calculation of rainfall intensity and thereby the peak discharge at any point in the drainage area is the longest time of flow to that point. However, in some situations, the peak flow may occur when only part of the upstream catchment is contributing, due to a higher intensity of rainfall resulting from lower t_c, producing a greater discharge than that if the entire

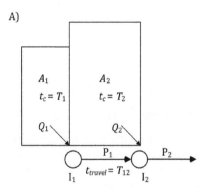

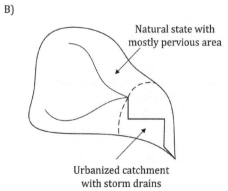

Figure 11.6 Time–area concept in the rational method.

upstream area is considered. This is known as the **partial area effect**.

The partial area effect usually occurs where the upstream catchments with mostly pervious areas and with long times of concentration come together with flows from the downstream catchments, which are mostly impervious, that is, urbanized areas with shorter times of concentration (Figure 11.6B). In the case of the partial area effect, the peak flows calculated for the subareas with larger times of concentration, with the rainfall intensity corresponding to the time of concentration of the smaller areas, are reduced by a factor of t_{cs}/t_{cl}. Here, t_{cs} is the time of concentration of the smaller subarea and t_{cl} is the time of concentration of the larger subarea.

11.9.2 General Design Procedure

In the rational method, each sewer in a network is designed individually and independently, except the computation of sewer flow time for the purpose of rainfall duration to calculate rainfall intensity for the next sewer. The time of concentration for each of the flow paths is the sum of the inlet time plus the upstream sewer flow time. For each sewer, the design peak discharge is calculated using the rational formula

$$Q_P = I_k \sum_{j=1}^{n} C_j A_j \qquad (11.10)$$

where I_k is the design rainfall intensity of the kth sewer in a network, C is the runoff coefficient, and A is the area. The subscript j represents the jth subarea upstream to be served by the kth sewer line. The summation sign implies inclusion of all subareas upstream of the sewer being designed. Each sewer has its own design j because each sewer has its own flow time of concentration and design storm. The only information needed from upstream sewers for the design of a current sewer is the upstream flow time for the determination of the time of concentration.

Once the peak discharge is computed for a sewer, its size can be computed from the application of the Manning equation with the assumption that the pipe flows full for the design flow. For a circular pipe, the minimum required diameter is given by

$$D_r = \left[3.208 \frac{n}{k_n} \frac{Q_P}{\sqrt{S_o}} \right]^{\frac{3}{8}} \qquad (11.11)$$

where $k_n = 1$ for SI units and 1.486 for US customary units.

The time of concentration for any catchment or subarea can be calculated using any of the methods described in Chapter 6. However, for storm sewers receiving flows from overland flows, the kinematic wave time of concentration is often preferred. However, this equation requires rainfall intensity a priori, which poses a problem by requiring a complicated iterative procedure. An equation closely related to the kinematic wave time of concentration equation, proposed by Yen and Chow (1983) for computation of overland flow time of concentration, is quite handy in this regard. This equation is given as

$$t_c = K \left(\frac{NL_o}{\sqrt{S_o}} \right)^{0.6} \qquad (11.12)$$

where K is a constant and all other terms in Eq. 11.12 are as defined in Chapter 10. The values of K are given in Table 11.3.

Table 11.3 *Values of K for Yen and Chow formula*

		Light rain	Moderate rain	Heavy rain
Rain intensity	(in/h)	< 0.8	0.8–1.2	> 1.2
	(mm/h)	< 20	20–30	> 30
For L_o in feet with	t_o in h	0.025	0.018	0.012
	t_o in min	1.5	1.1	0.7
For L_o in meters with	t_o in h	0.050	0.036	0.024
	t_o in min	3.0	2.2	1.4

Design starts at the upper end of a storm sewer system and proceeds downstream, following flow direction and pathway, in the order of pipe sections between two manholes or intercepts to laterals, submains, and the main. The design of the pipe section (size D and slope S) alternates with the computation of design flow (Q). The general steps are as follows:

Step 1. Determine the total drainage area or sum of subareas contributing to the intercept points such as the inlets.

Step 2. Compute the time of concentration, t_c, for the intercept point. This is the longest of all possible routes, including sewer flow time in the upstream pipe section that has already been designed.

Step 3. Compute the peak flow, resulting from a design storm, at the intercept point, including flow in the upstream pipe section.

Step 4. Design the pipe section immediately downstream, that is, between the current intercept point and the next one:
  a. Determine the pipe slope by selecting the ground slope first.
  b. Compute the pipe size, such as the diameter for a circular pipe, necessary to just flow full under uniform flow conditions and to carry the design flow capacity using the Manning equation.
  c. Check the flow velocity, using the continuity equation.
  d. Adjust the pipe size and slope, if necessary.

Step 5. Compute the time of concentration and peak flow at the next intercept point down slope:
  a. Design the next pipe section by repeating steps 1 to 4.

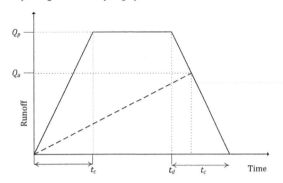

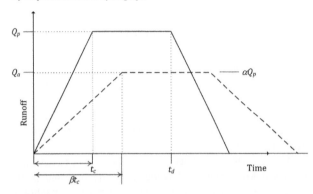

Figure 11.7 Approximation of detention basin volumes in the MRM.

b. Design all pipe sections in a sewer lateral.
c. Design all upstream laterals of a storm sewer submain.
d. Design all upstream submains of a storm sewer main.
e. Design the main of the storm sewer system.

Step 6. Check the ground cover above the pipe (minimum 2–3 ft).

11.9.3 Detention Basins

A detention basin reduces the peak inflow by temporarily storing the excess stormwater and then releasing the water volume at specified rates over an extended period. The main parameter required for the detention basin is the storage volume needed to control the outflow rate. The MRM is a simplified means for the determination of volume of the detention required for the given inflow and outflow peaks. This is the simplest but least accurate method for the design of a detention basin. A more accurate method is described in Chapter 13.

The basic approach assumes that the inflow hydrograph to the detention basin resulting from the stormwater runoff hydrograph for the design storm is trapezoidal in shape. The peak runoff rate is calculated using Eq. 11.1. It is assumed that the peak of the outflow hydrograph falls on the recession limb of the inflow hydrograph, and the rising limb of the outflow hydrograph can be approximated by a straight line (Figure 11.7A). The storage volume is determined by the critical duration of inflow, and using a constant outfall release rate. With these assumptions, the detention volume, V_d, is approximated as

$$V_d = Q_p t_d - \frac{Q_a(t_d + t_c)}{2} \quad (11.13)$$

where Q_a is the allowable or specified peak outflow rate.

If the outflow hydrograph is also trapezoidal in shape (Figure 11.7B) then also from the geometric relationships of the two trapezoids, the detention volume is given as

$$V_d = Q_p t_d - Q_a t_d \left[1 + \frac{t_c}{t_d}\left(1 - \frac{\alpha + \beta}{2}\right)\right] \quad (11.14)$$

where α and β are the proportions by which Q_p decreases and t_c increases, respectively.

In practical applications, $Q_p, Q_a, t_c,$ and t_d are used for the proposed and existing conditions. Thus, t_d is the t_c of the proposed conditions.

11.10 EXAMPLES

Example 11.1: Figure 11.8 shows an existing residential neighborhood block served by a storm drainage system. Table 11.4 gives the characteristics of the catchments for the determination of overland flow. The storm drains are circular concrete pipes with Manning's $n = 0.015$. The length (L) and slope (S) of the pipes are given in Table 11.6. The IDF relationships in this location are given by $I(\text{in/h}) = \frac{100 F^{0.2}}{t_d + 25}$, where t_d is the rainfall duration in minutes and F is ARI in year. Assuming equilibrium conditions, determine the sizes of the four storm sewer lines for a rainfall frequency of 25 years using the rational method.

Solution: The overland flow time to the inlets, t_o, is calculated using Eq. 11.12 with $K = 0.7$ for heavy rain. These values are given in Table 11.4. For equilibrium conditions, $t_o \equiv t_c = t_d$.

Calculations of rainfall intensity and pipe sizes alternate as noted in the design procedures for storm drains. These are given in Tables 11.5 and 11.6, respectively.

The determination of design rainfall intensity (I) for the storm drain pipes is given in Table 11.5. The entries in this table are as follows:

Table 11.4 *Characteristics of catchments*

Catchment	Area (acres)	L_o (ft)	S_o	N	t_o (min)	C
I	3	300	0.010	0.015	6.87	0.8
II	2	400	0.0081	0.016	9.04	0.7
III	4	450	0.012	0.030	12.58	0.4
IV	6	650	0.010	0.020	12.99	0.6
V	8	700	0.010	0.021	13.98	0.5

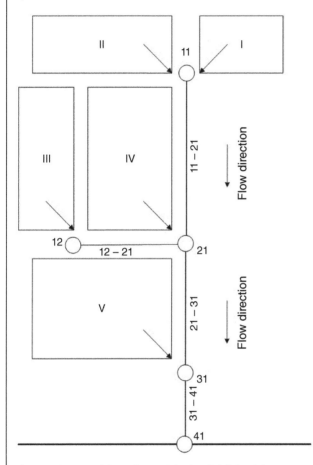

Figure 11.8 Layout of the catchments, inlets (small circles) and the storm drainpipes.

Column 1. Pipe number identified by the inlet numbers at its two ends.

Column 2. The catchment or the pipe number immediately upstream that drains directly through the manhole or junction into the sewer pipe being considered.

Column 3. The area, A, of the directly drained catchment (Table 11.4).

Column 4. The value of the runoff coefficient, C, for each catchment (Table 11.4).

Column 5. The product of C and the corresponding catchment area, A.

Column 6. Summation of $C_i A_i$ for all the areas drained by the pipe. It is equal to the sum of contributing values in column 5.

Column 7. The values of the time to get to the inlet of the sewer system for the catchments drained. These are overland flow times (t_o), that is, the time of concentration of each catchment calculated in Table 11.4.

Column 8. The travel time (t_t) for the water in the pipes. These are calculated in Table 11.6 (column 10).

Column 9. The time of concentration for each of the possible critical flow paths given by $t_o + t_t$ (column 7 + column 8).

Column 10. The design rainfall duration (t_d) is assumed to be equal to the longest of the different times of concentration of different flow paths to arrive at the entrance of the pipe being considered.

Column 11. The rainfall intensity for the rainfall duration is calculated using the IDF relationship with $F = 25$.

Table 11.6 shows the calculations of the pipe size (column 8) and travel time (column 10). The length (column 6) and slope (column 5) of each of the pipes are given. The peak flows (Q_p) that pass through each pipe is given in column 4 in the same row where the pipe number (column 1) is given. From these, the pipe diameter is calculated in column 7 using

$$D_{calc} = \left(0.0324 \frac{Q_p}{\sqrt{S}}\right)^{3/8}$$

Column 8 gives the value of the diameter of the pipe, rounded to the nearest integer for commercially available pipes.

Column 9 gives the flow velocity from which travel time (t_t) in column 10 is calculated using

$$t_t = \frac{L}{V}.$$

The example demonstrates that, in the rational method, each pipe in the storm drain system is designed individually, except the computation of the pipe flow time for the purpose of rainfall duration determination for the next downstream pipe.

Table 11.5 Calculation of design rainfall intensity

Sewer	Contributing element	Area (A_i)	Runoff coefficient (C_i)	A_iC_i	SA_iC_i	Inlet time, t_o (min)	Upstream sewer flow time, t_t (min)	Time of concentration, t_c (min)	Rainfall duration, t_d (min)	Rainfall Intensity (in/h)
11–21	I	3	0.8	2.4		6.87		6.87		
	II	2	0.7	1.4		9.04		9.04	9.04	5.59
12–21	III	4	0.4	1.6	3.8	12.58		12.58	12.58	5.07
					1.6					
21–31	IV	6	0.6	3.6		12.99		12.99		
	11–21			3.8		9.04	1.42	10.46		
	12–21			1.6		12.58	0.65	13.22	13.22	4.98
					9					
31–41	V	8	0.5	4		13.98		13.98		
	21–31			9		12.99	0.98	13.96	13.98	4.88
					13					

Table 11.6 *Calculation of travel time and pipe diameter*

Sewer line	Contributing element	Rainfall Intensity (in/h)	$Q = CIA$	S	L (ft)	D_{calc} (ft)	D_{used} (ft)	$V = (Q/A)$ (ft/s)	Travel time, t_t (min)
	I	5.97	14.34						
	II	5.59	7.83						
11–21			22.16	0.0085	600	2.16	2	7.06	1.42
12–21	III	5.07	8.11	0.0300	400	1.17	1	10.32	0.65
	IV	5.01	18.04						
	11–21		22.16						
	12–21		8.11						
21–31			48.31	0.0100	400	2.81	3	6.83	0.98
	V	4.88	19.53						
	21–31		48.31						
31–41			67.85	0.015	300	2.95	3	9.60	0.52

Exercises: A selection of exercises on this topic is available at www.cambridge.org/appliedhydrology.

12 Channel Routing

12.1 CONCEPTS AND PURPOSE

In hydrologic modeling of a watershed or river basin, the drainage unit is subdivided into several subbasins based on various considerations. Runoff from the subbasins flows into channel networks and eventually makes its way to the watershed or basin outlet. Direct runoff hydrographs (DRHs) from one or more subbasins contribute to a point on a stream, called the hydrologic junction. Flows from one hydrologic junction move to the next hydrologic junction downstream where additional flows from the lower subbasins join the flows coming from the upstream hydrologic junction through the channel. If there is no tributary inflow between two successive hydrologic junctions then, as the hydrograph travels along a channel from an upstream hydrologic junction, it attenuates in magnitude and gets delayed in time when it reaches the next downstream hydrologic junction. This change in the hydrograph principally happens due to the storage, S, of water in the channel. It is necessary then to know how the hydrograph from the upper hydrologic junction changes when it reaches the next downstream hydrologic junction. Similarly, if there is a stream gauging station at a certain location on a channel, it may be necessary to compute the hydrograph at a location further downstream without any stream gauge, from the known hydrograph constructed from measured flows at the gauging station. In general, channel routing is the procedure to determine the **outflow hydrograph** at a point on a channel from a known hydrograph, called the **inflow hydrograph**, upstream. Figure 12.1 shows this basic concept of channel routing.

There are many operational procedures that use channel routing. For example, if a gauging station at a point on a river records flows resulting from a severe storm event, the known hydrograph at this gauging station can be used to issue flood warnings, giving the timings and stage of the arrival of the peak flood discharge at downstream points, by calculating the hydrographs at these downstream points from the measured hydrograph by a channel routing procedure. Channel routing procedures also generate river and water information that is used for flood control, navigation, recreation, reservoir operation, irrigation for agriculture, and water supply plans.

At the outset, it should be noted that channel routing is distinctly different from reservoir routing, which is discussed in Chapter 13. In the case of a reservoir, the storage,

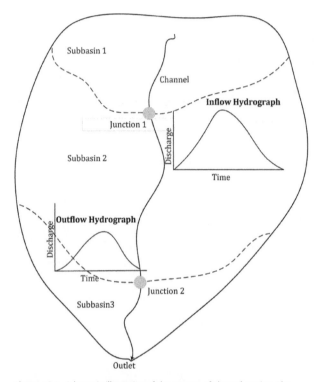

Figure 12.1 Schematic illustration of the concept of channel routing. The hydrograph at junction 1, called the inflow hydrograph is known. The hydrograph at junction 2, called the outflow hydrograph, is calculated using a method of channel routing.

S, is assumed to be a unique function of only outflow, Q, that is, $S = f(Q)$. This is known as an **invariable relationship**. On the contrary, in the case of a channel, S is assumed to be a function of both inflow, I, and outflow, Q, that is, $S = f(I, Q)$. This is known as a **variable relationship**. For this reason, channel routing is more complex than reservoir routing.

12.2 GOVERNING EQUATIONS

12.2.1 Equations of Unsteady Flow

The propagation of a flood wave in a channel reach is a gradually varied unsteady flow. Adhémar Jean Claude Barré de Saint-Venant was a French mechanician and mathematician who developed the equations called shallow water equations that describe unsteady open channel flow (Saint-Venant, 1871). These are one-dimensional forms of conservation of mass and momentum equations, known as the **Saint-Venant equations**. These equations describe the time-dependent movement of flood waves in an open channel by considering the four principal forces, namely *inertia*, *pressure*, *gravity*, and *friction*, acting upon the flood wave. These are called shallow water equations mainly because the velocity is depth integrated for the horizontal length scale, which is much greater than the vertical length scale.

For some methods of solving Saint-Venant equations, the **nonconservation** form of the continuity equation is used, in which the average flow velocity, V, is the dependent variable, instead of the discharge, Q. The nonconservation form of the continuity equation is written as

$$A\frac{\partial V}{\partial x} + VB\frac{\partial h}{\partial x} + B\frac{\partial h}{\partial t} = q_L \quad (12.1)$$

where A is the cross-sectional flow area, B is the top width of the water surface, x is the distance along the channel, t is the time, and q_L is the lateral inflow per unit length. If the lateral inflow is zero, then

$$A\frac{\partial V}{\partial x} + VB\frac{\partial h}{\partial x} + B\frac{\partial h}{\partial t} = 0 \quad (12.2)$$

In addition to the lateral inflow, rain, infiltration, etc. are also ignored. The continuity equation accounts for the volume of water in a reach of an open channel, including that flowing into the reach, that flowing out of the reach, and that stored in the reach.

The nonconservation form of the momentum equation is given as

$$S_f = S_0 - \frac{\partial h}{\partial x} - \frac{V}{g}\frac{\partial V}{\partial x} - \frac{1}{g}\frac{\partial V}{\partial t} \quad (12.3)$$

where g is the acceleration due to gravity, S_0 is the channel bed slope, and S_f is the friction slope. The momentum equation accounts for forces that act on a body of water in the open channel. It equates the sum of the gravitational, pressure, and friction forces to the product of fluid mass and acceleration. Various simplifications of Eq. 12.3 result, depending on the significance or importance of each of the terms for a flow condition. These are as follows:

1. If all terms, except S_f and S_0 are neglected, then

$$S_f = S_0 \quad (12.4)$$

The friction slope, S_f, represents frictional effects and S_0 signifies gravitational effects. In other words, here the only two forces that are considered in the momentum balance are **frictional** and **gravitational forces** and these two forces balance each other. Equation 12.4 corresponds to what is represented by the **kinematic wave** approximation of the unsteady flow equation.

2. If, in addition to S_f and S_0, the second term on the right-hand side of Eq. 12.3 is also considered (and the other two terms on the right-hand side are neglected),

$$S_f = S_0 - \frac{\partial h}{\partial x} \quad (12.5)$$

The term, $\left(\frac{\partial h}{\partial x}\right)$, is the pressure differential or pressure head gradient, that is, the change in the water depth, h, along the channel in the x-direction due to the variation in the **pressure force**. Equation 12.5 represents the **diffusion wave** approximation of the unsteady flow equation. This approximation is the basis of the Muskingum–Cunge routing model that is discussed in this chapter.

3. If all terms, except the last term on the right-hand side of Eq. 12.3 are considered, then

$$S_f = S_0 - \frac{\partial h}{\partial x} - \frac{V}{g}\frac{\partial V}{\partial x} \quad (12.6)$$

The term, $\frac{V}{g}\left(\frac{\partial V}{\partial x}\right)$ represents the **convective acceleration** or velocity head gradient, which is the change in velocity along the channel in the x-direction. Equation 12.6 represents **the quasi-steady dynamic wave** approximation of the unsteady flow equation.

4. When all terms in Eq. 12.2 are considered, it is the full dynamic wave equation and is the **dynamic wave** model, which is also known as the **gravity wave**. The term, $\frac{1}{g}\left(\frac{\partial V}{\partial t}\right)$ represents the **local acceleration** or rate of change in velocity over time. The local and convective acceleration terms represent the effect of **inertial forces** on the flow.

From these considerations, it can be seen that kinematic waves do not attenuate, while most flood waves attenuate to some extent. On the other hand, dynamic waves attenuate too much and therefore do not represent flood waves in typical cases. The diffusion wave (diffuses or attenuates) lies in the middle range of attenuation. For this reason, it is probably the most applicable wave form for practical applications. Also, an order-of-magnitude analysis shows that local acceleration and convective acceleration are of the same order of magnitude but opposite in sign, so they tend to counterbalance each other, further explaining the reason for the adequacy of the diffusion wave approximation.

The kinematic wave approximation (Eq. 12.4) represents **steady uniform** flow. The diffusion wave approximation (Eq. 12.5) represents **steady nonuniform** flow or **gradually varied** flow. The quasi-steady dynamic wave approximation (Eq. 12.6) also represents steady nonuniform flow. The full dynamic or gravity wave approximation (Eq. 12.2) represents **unsteady nonuniform** flow.

12.2.2 Spatially Lumped form of the Continuity Equation

The **conservation** form of the continuity equation in space and time is given as

$$\frac{\partial Q}{\partial x} + \frac{\partial A}{\partial t} = q_L \quad (12.7)$$

where Q is the discharge and q_L is the lateral inflow per unit of channel length. In the mass balance equation given by

Eq. 12.7, the term $\frac{\partial A}{\partial t}$ indicates the rate of change in storage, whereas the term $\frac{\partial Q}{\partial x}$ indicates the difference between inflow and outflow or simply the net outflow. When lateral inflow is ignored,

$$\frac{\partial Q}{\partial x} + \frac{\partial A}{\partial t} = 0 \qquad (12.8)$$

If we are only interested in determining the outflow at the downstream end due to inflow at the upstream end of a channel reach, and ignore the variability of flow along the channel, Eq. 12.8 can be integrated in space and written in a spatially lumped form as an ordinary differential equation

$$\frac{dS}{dt} = I(t) - Q(t) \qquad (12.9)$$

where $\frac{dS}{dt}$ is the rate of change of storage in the channel reach under consideration, and $I(t)$ and $Q(t)$ are the inflow and outflow rates, respectively.

The finite-difference (backward-difference) approximation of Eq. 12.9, used in most channel routing procedures, is

$$S^j = S^{j-1} + 0.5\Delta t\left[\left(I^j + I^{j-1}\right) - \left(Q^j + Q^{j-1}\right)\right] \qquad (12.10)$$

where the superscript j is the temporal index (Figure 12.2).

The form of the continuity equation given in Eq. 12.10 is combined with a storage–discharge relation in channel routing methods that are classified as *hydrologic methods* whereas, in the methods classified as *hydraulic methods*, the differential form of the continuity equation is combined with the momentum equation, which expresses the conservation of momentum during flow. It can also be noted here that Eq. 12.10 is used for reservoir routing (Chapter 13), which is the simplest form of flow routing. The known storage–discharge relationship is sometimes given in the form

$$S(t) = KQ(t) \qquad (12.11)$$

where K is a coefficient with the unit of time (e.g., hours), known as the **storage–time constant** or simply the **storage constant**.

12.3 CHARACTERISTICS OF FLOOD WAVE MOVEMENT THROUGH A CHANNEL AND FLOOD HYDROGRAPHS

12.3.1 Characteristics of Flood Wave Movement

Flood waves have two principal kinds of movements. These are

1. Translation action: This is the downstream movement of a flood wave without a change in shape. This is nearly possible in a prismatic channel in which the stage and discharge are uniquely defined at all points. This is referred to as **uniformly progressive flow** and is a special case of unsteady flow.
2. Reservoir action: This refers to the modification of a flood wave by pondage, storage, or stagnation of water within the channel, as happens in a reservoir.

12.3.2 Relationship between Inflow and Outflow Hydrographs and Storage: Time Graphs due to Pure Translation of Uniformly Progressive Flow

The storage generated in a reach of a channel by a uniformly progressive wave or translational movement is completely different from that generated in a reservoir. The characteristics of this type of flow and the resulting hydrographs are as follows:

1. The storage in the channel varies in shape and volume as the flood wave travels downstream. (This is tantamount to prism storage plus wedge storage discussed in Section 12.4).
2. The storage increases rapidly in relation to outflow but also decreases rapidly when the flood wave recedes.
3. Storage increases when peak inflow occurs, and it continues to increase until outflow rises to equal the receding inflow. At this point the storage reaches a maximum and the rising outflow equals the receding inflow. Thus, the maximum storage occurs at a time that is after the peak inflow time but before the peak outflow time. After passing this point, the outflow continues to rise until the peak outflow rate reaches a magnitude that is equal to the peak inflow rate.
4. There is no attenuation in the peak discharge; in other words, translation involves maintaining the same hydrograph shape as the flood wave moves downstream.

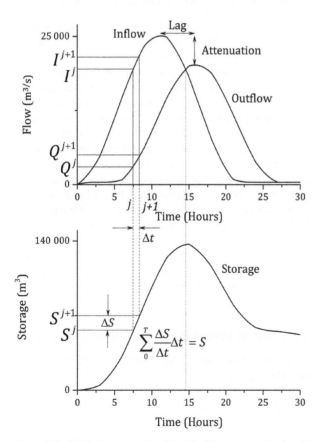

Figure 12.2 Graphical representation of the finite-difference approximation of the spatially lumped continuity equation (Eq. 12.10).

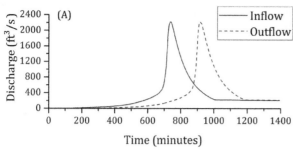
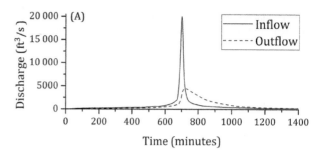
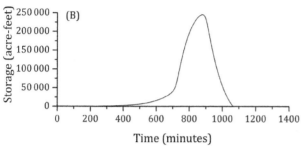
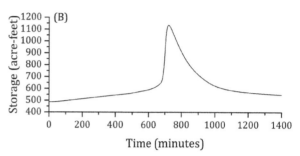
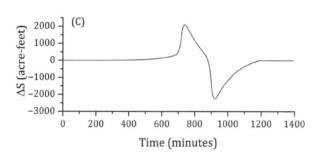
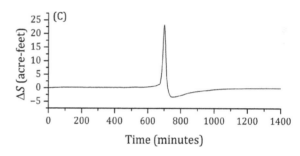

Figure 12.3 (A) Calculated outflow and (B) storage hydrographs for a given inflow hydrograph in the case of a uniformly progressive wave. (C) The variation of change in storage with time (ΔS–t graph) shows that at the instant of peak storage when inflow equals outflow, the net change in storage equals zero. At this time, the flood crest is at a central position in the reach of the channel, the upstream stages are falling, and the downstream stages are rising, and the storage in the reach at this instant is constant. The time of peak storage is 885 minutes, which is after the time of peak inflow (at 745 minutes) and before peak outflow (925 minutes). Both inflow and outflow peaks are 2217 ft^3/s. The time lag between the two peaks is 180 minutes.

Figure 12.3 shows the calculated outflow, storage, and ΔS–t graphs for a given inflow hydrograph in a channel reach where flow is assumed to be uniform.

12.3.3 Relationship between Inflow and Outflow Hydrographs and Storage–Time Graphs due to Pure Reservoir-Type Storage

1. Storage in the reservoir increases as long as inflow exceeds outflow, and continues to increase until peak inflow occurs.
2. Both storage and outflow reach a maximum when inflow equals outflow. This equality occurs only after inflow has receded from its peak. Thus, in the case of a reservoir pondage, peak storage and peak outflow occur at the same time and this time is after the peak inflow time.

Figure 12.4 (A) Calculated outflow and (B) storage hydrographs for a given inflow hydrograph in the case of a reservoir. The maximum inflow is 20 019 ft^3/s and maximum outflow is 4464 ft^3/s amounting to slightly over 78% attenuation of peak discharge. The peak inflow occurs at 723 minutes and peak outflow and peak storage occur 20 minutes later at the same time of 743 minutes, where peak outflow equals receding inflow. (C) The variation of storage differential concurrent with inflow and outflow.

In other words, pure reservoir action causes the peak of the outflow hydrograph to fall on the receding limb of the inflow hydrograph. This is the fundamental difference between channel storage and reservoir storage.
3. Because the peak outflow occurs at a time of receding inflow, reservoir or pondage storage always attenuates peak discharge.

Figure 12.4 shows the calculated outflow, storage, and ΔS–t graphs for a given inflow hydrograph in the case of a reservoir-type storage. Note that the peak outflow occurs at a point on the receding limb of the inflow hydrograph.

12.3.4 Inflow and Outflow Hydrographs due to Combined Actions of Reservoir Storage and Translational Movement of Flood Waves

Flood wave movement in natural channels is intermediate between the two ideal conditions described above; therefore, the attenuation of the peak in the outflow hydrograph of a channel reach is not as pronounced as is seen in the outflow hydrograph of a reservoir. The relative influence of channel flow and reservoir pondage determines the degree of the attenuation of the peak outflow. In this case also the following holds true.

1. The outflow peak does not occur at the point of the intersection of the inflow and outflow hydrographs but as in the uniformly progressive flow case it occurs after inflow equals outflow.
2. Like the two ideal cases noted above, the maximum storage occurs when receding inflow equals rising outflow and at that instant the net change in storage equals zero ($\Delta S/\Delta t = 0$).

Figure 12.5 shows the calculated outflow, storage, and ΔS–t graphs for a given inflow hydrograph in the case of a natural

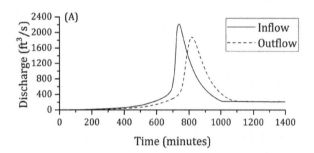

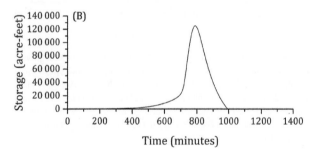

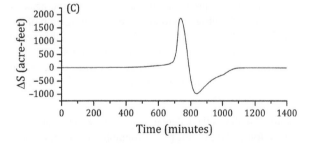

Figure 12.5 (A) Calculated outflow and (B) storage hydrographs from a given inflow hydrograph in the case of a channel exhibiting characteristics of both translational and reservoir actions. The maximum inflow is 2217 ft^3/s and maximum outflow is 1883 ft^3/s amounting to about 15% attenuation of peak discharge (contrast this with the degree of attenuation in the case of a reservoir shown in Figure 12.4). The peak inflow occurs at 745 minutes and peak outflow occurs 80 minutes later at 825 minutes. The peak storage occurs at 795 minutes, in between the peak inflow and peak outflow time but at a time when inflow equals outflow. (C) The ΔS–time graph shows that the net change in storage equals zero when inflow equals outflow and storage is at the maximum value (remember that this is not the point of maximum outflow, this occurs after this time; see Figure 12.3).

channel. It should be noted, however, that in the case of observed hydrographs, obtained from gauging stations, the actual behavior of the wave is sometimes obscured by the effects of local tributary inflow.

Another important characteristic of channel flow is that in the case of movement of a flood wave through a channel, the storage–discharge curve is generally looped (Figure 12.6). This is the **variable relationship** mentioned above. The storage for a given discharge on the rising or receding part of the flood wave will be correspondingly greater or less than the storage corresponding to steady flow.

12.4 PRISM STORAGE AND WEDGE STORAGE

As noted above, the flow in an open channel during a flood is gradually varied unsteady flow. The water surface is not parallel to the channel bottom, and it varies with time. During the advance of a flood wave, inflow always exceeds outflow, thus producing a wedge of storage. On the other hand, during recession, the outflow exceeds inflow, creating a negative wedge of storage. Thus, the total volume of water in storage in a channel during flood flow has two components: (1) prism storage and (2) wedge storage (Figure 12.7).

12.4.1 Prism Storage

Prism storage is the volume of water formed by an imaginary plane parallel to the channel bottom. This is the volume that would exist if the flow were uniform. The prism storage is actually the reservoir storage, and it is directly proportional to outflow discharge given as an invariable relationship:

$$S_p = KQ \qquad (12.12)$$

If storage can be expressed as a linear function of discharge, such as that given by Eq. 12.12, it is known as **linear storage** or a **linear reservoir**.

12.4.2 Wedge Storage

Wedge storage is the wedge-like volume of water formed between the actual water surface and the top surface of the prism storage. This is the volume that is formed due to the passage of the flood wave. This volume changes with time. It can be positive during an advancing flood and negative during a receding flood. It is a function of both inflow and outflow and can be expressed as

$$S_w = KX(I - Q) \qquad (12.13)$$

where K, as noted above (Eq. 12.11), is the storage constant having dimensions of time, essentially implying the travel time of the flood wave through the routing reach and X is a dimensionless parameter, known as the **weighting factor**,

12.4 Prism Storage and Wedge Storage

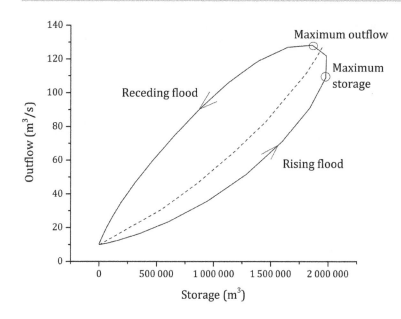

Figure 12.6 Looped storage–outflow relationship (variable relationship) for a river reach during the passage of a flood wave. Note that the maximum storage and maximum outflow do not occur at the same time. The dashed curve represents a reservoir-type storage (invariable relationship).

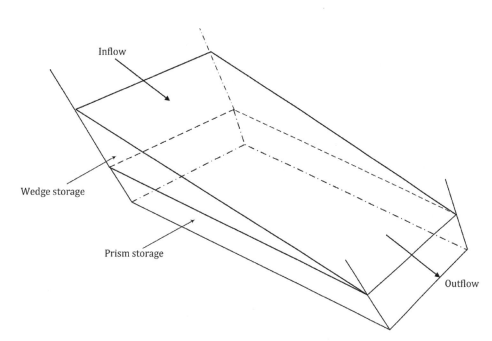

Figure 12.7 The concept of prism and wedge storage in a channel reach during the passage of a flood wave.

expressing the relative influence or weight of inflow and outflow on storage levels.

The total storage is therefore

$$S = S_p + S_w = KQ + KX(I - Q) = K[XI + (1-X)Q] \quad (12.14)$$

where K and X depend on the channel characteristics. Equation 12.14 is known as the **Muskingum equation**.

In general, the relationships between I, Q, and S are *nonlinear* given as (Chow, 1959; Singh and Scarlatos, 1987)

$$S = KQ^m + KX(I^m - Q^m) = K[XI^m + (1-X)Q^m] \quad (12.15)$$

where m is a constant exponent. The value of m varies from 0.6 for rectangular channels to a value of about 1.0 for natural channels. In the Muskingum equation, $m = 1.0$ and hence it is a *linear* model. Sometimes the relationships between I, Q, and S are also expressed as

$$S = K[XI + (1-X)Q]^m \quad (12.16)$$

in which m is not equal to 1 and is not the same as in Eq. 12.15.

In Eq. 12.14, the value of X varies between 0.0 (for full reservoir-type storage) and 0.5 (for a full wedge). Obviously, when $X = 0$, Eq. 12.14 reduces to Eq. 12.12. When $X = 0.5$

both the inflow and outflow are equally important in determining storage. In natural streams, X is between 0.0 and 0.3, with a mean value near 0.2. For a given reach, the values of X and K are assumed to be constant, but their values may change for another reach of the same channel. Since X is the weighting factor, by definition it should be between 0 and 1.00 but the reason it is between 0 and 0.50 is seen from Eq. 12.13 in which X weights the difference of I and Q, but not individually I and Q, that is, X is a partial weight.

It should be also noted here that in Eq. 12.2, the first term $\left(A\frac{\partial V}{\partial x}\right)$ gives the prism storage, the second term $\left(VB\frac{\partial h}{\partial x}\right)$ gives the wedge storage, and the third term $\left(B\frac{\partial h}{\partial t}\right)$ gives the rate of rise of the water surface.

12.5 FLOOD WAVE SPEED OR CELERITY

Flood movement in a river is the phenomenon of wave propagation. Flood routing is essentially the determination of the wave propagation. A clear conception of water waves is important for the study of flow routing. A wave is a disturbance that can be characterized by a variation in a flow, such as a change in flow rate or water surface elevation. The wave **celerity** is the velocity with which such a variation travels along the channel. Celerity, in general, is different from the average velocity of water flow in the channel. The celerity depends on the type of wave under consideration. For example, the celerity of small gravity waves that occur in shallow water in channels as a result of any momentary change in the local depth of water is equal to the critical flow velocity. Since channel routing deals with change in discharge downstream, celerity is a basic notion used in subsequent sections of this chapter. The equation frequently used to calculate celerity in relation to channel routing is derived from considering a particular type of *uniformly progressive flow*.

As a special case of unsteady flow, a uniformly progressive flow has a stable wave profile that does not change in shape as it moves downstream. One common type of uniformly progressive flow, which approximates most flood waves in natural channels, is the **monoclinal rising wave** or simply **monoclinal wave**, which is a translational wave of stable form travelling downstream at a constant velocity (Chow, 1959). This is shown in Figure 12.8 as a simple step increase in discharge or water surface elevation advancing downstream. Dynamically, the monoclinal flood wave is very different from simple surges (Henderson, 1966). It is substantially influenced by bed slope and friction slope.

Considering the flow into and out of the control volume ABCD in Figure 12.8, it can be shown that, for steady flow $(dQ/dt = 0)$,

$$Q_d \Delta t - Q_u \Delta t = (A_d - A_u)\Delta x \qquad (12.17)$$

where Q_d, Q_u, A_d, and A_u represent discharge and flow areas downstream and upstream, respectively, over a distance Δx and time interval Δt. Equation 12.17 can be rewritten as

$$c = \frac{\Delta x}{\Delta t} = \frac{Q_d - Q_u}{A_d - A_u} \qquad (12.18)$$

where c is the speed of the flood wave, or celerity. In other words,

$$c = \frac{dQ}{dA} \qquad (12.19)$$

Seddon (1900), in a study of gauge heights on the Mississippi and Missouri rivers, derived this relationship by showing that the mathematical principle developed by Kleitz (1877) for uniformly progressive waves is applicable to observations made in natural rivers. For this reason, it is also known as **Kleitz–Seddon law**.

While a monoclinal wave is a special type of gravity wave, it can be also shown that c, given by Eq. 12.19, is the celerity of the kinematic wave where Q is a function of h

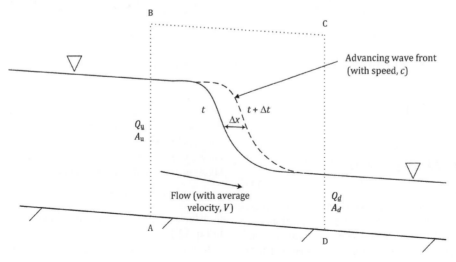

Figure 12.8 Uniformly progressive flow and the relation of wave speed to change in flow area with discharge. The celerity of the wave that is shown as a step increase in the water height is the speed with which this step advances.

alone. In that case, it is designated by as c_k and is called **kinematic wave celerity** given by dx/dt. The key idea here is that there is a unique relation between discharge and flow cross-sectional area. If Q is plotted against A, the plot is a monotonically increasing curve and c at any point on the curve is the slope of the tangent at that point. Equation 12.19 also implies that the speed of a flood wave is controlled by the change in the storage volume.

The relationship between celerity and average velocity of flow can be derived from the Manning equation. For simple cross-sectional geometries, such as trapezoidal, rectangular, and triangular channels, the relationship between discharge and flow area can be expressed as a power function given as

$$Q = \alpha A^m \tag{12.20}$$

where α is a coefficient and m is the exponent of the power function.

Differentiating Eq. 12.20 with respect to area one gets

$$\frac{dQ}{dA} = \alpha m A^{(m-1)} = \frac{m(\alpha A^m)}{A} = m\frac{Q}{A} = mV \tag{12.21}$$

where V is the average velocity at this cross section.

As noted in Eq. 12.19, dQ/dA is also the celerity. Therefore,

$$c = mV \tag{12.22}$$

Thus, in theory, the celerity of a flood wave is equal to a factor, m, multiplied by the average velocity of flow. Values of m can be estimated using the Manning equation, which can be written in either of the following forms:

$$Q = \frac{1.486}{nP^{2/3}} S_f^{1/2} A^{5/3} \quad \text{(in US customary units)} \tag{12.23}$$

$$Q = \frac{1}{nP^{2/3}} S_f^{1/2} A^{5/3} \quad \text{(in metric units)} \tag{12.24}$$

where P is the wetted perimeter, and n is Manning's roughness coefficient.

For cross sections where the hydraulic radius (A/P) can be approximated by the depth as with a wide rectangular channel with top width, B, Eq. 12.23 takes on the form

$$Q = \alpha A^m \tag{12.25}$$

where

$$\alpha = \frac{U\sqrt{S_f}}{nB^{2/3}} \tag{12.26}$$

and U is either 1.486 or 1 depending on the unit system, and

$$m = \frac{5}{3} = 1.67 \tag{12.27}$$

Thus, for wide rectangular channels,

$$c = 1.67V \tag{12.28}$$

Similarly, for a triangular channel, $m = 1.33$ and for a wide parabolic channel, $m = 1.44$.

Thus, the celerity can be estimated from the average velocity (Eq. 12.22) if the value of m is known.

The concept of hydraulic geometry also leads to a similar relationship where celerity can be estimated from discharge. The Manning equation, written as (in metric units)

$$Q = \frac{A(A/P)^{2/3}}{n}\sqrt{S_f} \tag{12.29}$$

can be applied to natural channels, where in most cases the top width of flow (B) is much greater than depth and hence $P \approx B$. In these cases, the Manning equation can be viewed as

$$A = eQ^f \tag{12.30}$$

Equation 12.30 is also known as hydraulic geometry (Chapter 6) where e and f are constants called hydraulic geometry coefficients (Leopold and Maddock, 1953). From the hydraulic geometry relationship c can be also expressed as a function of discharge. Differentiating Eq. 12.30 with respect to Q, one gets

$$\frac{dA}{dQ} = efQ^{(f-1)} \tag{12.31}$$

Inverting Eq. 12.31, c is given as a function of discharge:

$$c = \frac{dQ}{dA} = \frac{Q^{(1-f)}}{ef} \tag{12.32}$$

Even though the description concerning celerity and the Manning equation generally deals with simple cross sections, it is also applicable to most natural channel cross sections where the value of m typically assumes 1.5 (which will be the case if the Chezy equation is used). However, this concept does not always apply directly to cross sections with both channel and floodplain portions, discussed below.

The kinematic wave celerity can also be expressed in terms of depth of water and top width of the water surface as

$$c_k = \frac{1}{B}\frac{dQ}{dh} = \frac{dx}{dt} \tag{12.33}$$

For a dynamic wave,

$$\frac{dx}{dt} = V \pm c_d \tag{12.34}$$

where c_d is the **dynamic wave celerity**, given by

$$c_d = \sqrt{g\frac{A}{B}} \tag{12.35}$$

Thus, dynamic waves can propagate both upstream and downstream. This means that dynamic waves attenuate in both upstream and downstream directions. If an observer stands on the bank of a river and remains stationary, they will see a kinematic wave at two points upstream and downstream rising at the same rate as the flood wave propagates. But, in the case of dynamic wave, they will see that the water surface elevations at the upstream and downstream points are rising at different rates.

12.6 CHANNEL ROUTING METHODS

Routing is a process of prediction of the temporal and spatial variations of a flood hydrograph as it moves through a channel reach. The effects of storage and flow resistance within a channel reach are reflected by changes in hydrograph shape and timing as the flood wave moves from upstream to downstream. The methods of routing are generally classified into two categories: **hydrologic routing** and **hydraulic routing**. Flood forecasting, reservoir and channel design, watershed hydrology, and floodplain studies generally involve some form of routing.

Hydrologic routing employs the spatially lumped continuity equation and an analytical or an empirical relationship between storage within the reach and discharge at the outlet. Hydraulic routing techniques are based on the solution of the Saint-Venant equations or the dynamic wave equations.

Typically, in watershed simulation studies, hydrologic routing is utilized on a reach-by-reach basis from upstream to downstream. For example, it is often necessary to obtain a discharge hydrograph at a point downstream from a location where a hydrograph has been observed or computed. For such purposes, the upstream hydrograph is routed through the reach with a hydrologic routing technique that predicts changes in hydrograph shape and timing. Local flows are then added at the downstream location to obtain the total flow hydrograph. This approach is adequate if the flow in the channel is not significantly controlled by the water surface elevation at some point downstream, which is commonly called the **backwater effect**, or there are no discontinuities in the water surface because of jumps or bores. When there are downstream controls that will influence the routing process through an upstream reach, the channel configuration should be treated as one continuous system. This can only be accomplished with a hydraulic routing technique that can incorporate backwater effects as well as internal boundary conditions, such as those associated with culverts, bridges, and in-line weirs or small dams.

The main differences between the two types of channel routing methods can be summarized as follows.

1. **Hydrologic or lumped routing method**: In this method, the inflow upstream and the relationship of storage in the channel and the outflow downstream are known. Then, (1) outflow is calculated as a function of time alone at a particular location and (2) calculations are governed by the continuity equation and a known storage–discharge relationship, usually given in the form of Eq. 12.11.
2. **Hydraulic or distributed routing method**: Under this method, (1) outflow is calculated as a function of space and time throughout the system, and (2) calculations are governed by the continuity and momentum equations, which are partial differential equations. No prior knowledge of the storage–discharge relationship is required in this approach, but a friction relation is needed. However, computationally it is an elaborate approach since partial differential equations of open channel flow are solved by a numerical method that must ensure numerical stability and other conditions.

There is a gamut of hydrologic routing procedures. not all of which are commonly used. The methods that are widely used in the present days for standard engineering applications, include:

1. The **modified Puls method**: This is a purely hydrologic method of routing, and it is a **lumped model**.
2. The **Muskingum method**: This is also a lumped hydrologic method.
3. The **Muskingum–Cunge method**: Even though this method is often classified as a hydrologic method, it has a hydraulic basis; it can be described as a diffusive wave model, and can also be cast as a **distributed model**.

12.7 MODIFIED PULS METHOD

12.7.1 Routing Equation

The modified Puls method is also known as *storage routing* or *storage indication*, or as the *level-pool* routing method. It is based on the use of the lumped continuity equation. Originally, it was developed by L. G. Puls (1928) of USACE for reservoir routing. The so-called *modified Puls method* was later developed by the US Bureau of Reclamation (1949). In channel routing, where the river or the stream is treated like a linear reservoir, the method is adopted with some further modification (USACE, 1994).

The modified Puls method is often used and preferred for channel routing mainly because the only input required for the application of this method, in typical cases, is derived from a hydraulic model of the channel. For this reason, it is also considered to provide more reliable results, particularly in cases where the projects involve channel or floodplain developments causing changes in the storage capacity of the channel.

The lumped form of the continuity equation essentially states that for a given time interval, Δt, the difference between inflow, I, and outflow, Q, is equal to the change in storage, ΔS, and can be written as

$$I(t) - Q(t) = \frac{\Delta S}{\Delta t} \qquad (12.36)$$

which can be cast as

$$\bar{I}_t - \bar{Q}_t = \frac{\Delta S_t}{\Delta t} \qquad (12.37)$$

where $\bar{I}_t$ is the average upstream flow (inflow) during period Δt, $\bar{Q}_t$ is the average downstream flow (outflow) during period Δt, and ΔS_t is the change in storage in the reach during the time interval Δt.

A numerical or finite-difference form of this mass balance equation can be written as

$$\frac{1}{2}(I_1 + I_2) - \frac{1}{2}(Q_1 + Q_2) = \frac{1}{\Delta t}(S_2 - S_1) \qquad (12.38)$$

In general, using j as the time index in the superscript (we will later use a space index in the subscript),

$$\frac{I^{j+1}+I^j}{2} - \frac{Q^{j+1}+Q^j}{2} = \frac{S^{j+1}-S^j}{\Delta t} \quad (12.39)$$

We always know the inflows, I, for all times, and the initial storage, S_1, and initial outflow, Q_1, must be given or known. Thus, Eq. 12.39 contains *two unknowns*, S^{j+1} and Q^{j+1}. By putting all knowns on the left-hand side, Eq. 12.39 can be rearranged as

$$\frac{1}{2}(I^j + I^{j+1})\Delta t + \left(S^j - \frac{Q^j \Delta t}{2}\right) = \left[S^{j+1} + \frac{(Q^{j+1})\Delta t}{2}\right] \quad (12.40)$$

Equation 12.40 is the **modified Puls equation**, also known as the **storage-indication routing equation**. It is one equation in two unknowns (Q^{j+1} and S^{j+1}) for a given time. Therefore, another equation must be provided for its solution. This is the storage–discharge equation, a relationship obtained independently. This required storage–discharge relationship is typically derived from a separate hydraulic model of the channel under consideration.

12.7.2 Solution Procedure

12.7.2.1 Single Reach Calculation

1. Develop the relationship for storage, S, versus discharge, Q, from the given storage–discharge data for the channel reach. A plot of this relationship is called an S-curve. This must be a monotonically increasing function.
2. Select the routing interval, Δt, in such a manner that the discharge varies *linearly* within the time interval.
3. Develop the $\left(S + \frac{Q\Delta t}{2}\right)$ versus Q relationship using the S versus Q relationship as developed in step 1. This curve or table is constructed for a given Q, knowing the storage S and selected value of Δt. This is called the $S + \frac{Q\Delta t}{2}$ curve or the **storage-indication curve**. Note that no subscript is used here since this relationship holds for all time. Thus, the modified Puls method requires the construction of two curves as shown in Figure 12.9, which also gives two "look-up tables."
4. For an initial outflow, Q_1 (known), the storage S_1 is computed or read from the S-curve. From this value of S_1, the quantity, $S_1 - \frac{Q_1 \Delta t}{2}$ is computed for the corresponding discharge value of Q_1.
5. Calculate the average inflow, $\frac{I_1+I_2}{2}$, for the time interval, Δt. Compute the total inflow volume by multiplying $\frac{I_1+I_2}{2}$ by Δt
6. From steps 4 and 5, calculate $\left(\frac{I_1+I_2}{2}\right)\Delta t + \left(S_1 - \frac{Q_1 \Delta t}{2}\right)$. In accordance with Eq. 12.40, this quantity is equal to $S_2 + \frac{Q_2 \Delta t}{2}$.
7. Determine the outflow Q_2 from using the value of $S_2 + \frac{Q_2 \Delta t}{2}$ computed in step 6 and the $S + \frac{\Delta t}{2}$ versus Q curve developed in step 3.
8. With the above computation steps, the value of discharge Q_2 corresponding to the known inflow I_2 is computed. Estimate $S_2 - \frac{Q_2 \Delta t}{2}$ by using $\left(S_2 + \frac{Q_2 \Delta t}{2}\right) - Q_2 \Delta t$. (These two quantities are calculated in steps 6 and 7.)

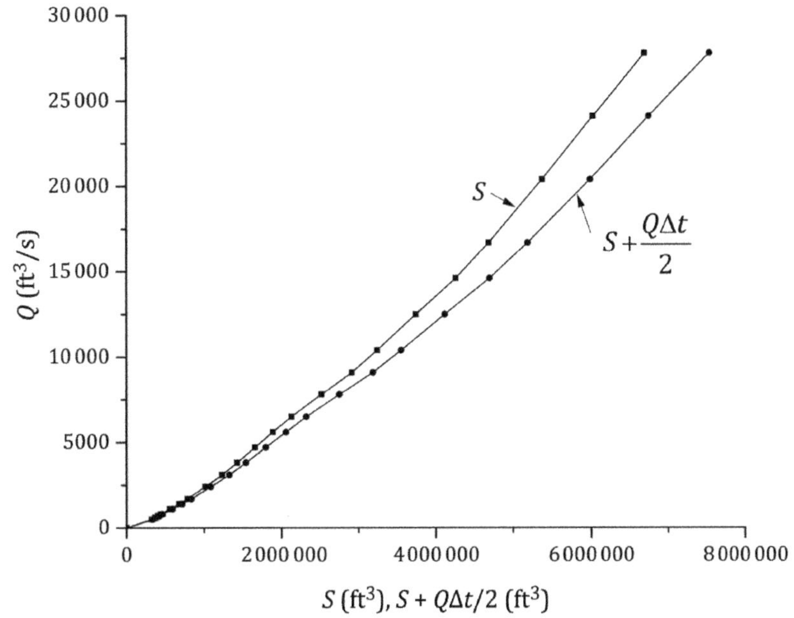

Storage (acre-feet)	Storage (ft³)	Discharge (ft³/second)	$\left(S + \frac{Q\Delta t}{2}\right)$ (ft³)
0	0	0	0
7.38	321472	500	336472
8.33	362854	600	380854
9.22	401622	700	422622
10.08	439084	800	463084
12.79	557131	1100	590131
15.55	677356	1400	719356
18	784078	1700	835078
23.27	1013639	2400	1085639
28.28	1231874	3100	1324874
32.72	1425280	3800	1539280
38.01	1655712	4700	1796712
43.37	1889193	5600	2057193
48.89	2129644	6500	2324644
57.82	2518633	7800	2752633
66.87	2912851	9100	3185851
74.44	3242599	10400	3554599
85.91	3742231	12500	4117231
97.77	4258851	14600	4696851
107.57	4685738	16700	5186738
123.36	5373549	20400	5985549
138.26	6022592	24100	6745592
153.58	6689929	27800	7523929

Figure 12.9 S-curve and $S + \frac{Q\Delta t}{2}$ curve for use in the modified Puls method. The data used in the construction of these curves are shown in the table.

9. In the next routing time step, Q_2 becomes Q_1 and the estimate of $S_2 - \frac{Q_2 \Delta t}{2}$ becomes $S_1 - \frac{Q_1 \Delta t}{2}$.
10. The recursive process from steps 4 to 9 continues for the entire inflow hydrograph to generate the entire routed outflow hydrograph.

12.7.2.2 Assumptions and Limitations

According to Lawler (in Chow, 1964), this method is used quite satisfactorily for reservoir routing; it gives poorer approximation for open channel routing. The procedure described above has the following assumptions and limitations:

1. At the heart of the modified Puls method lies the assumption that storage in a river reach depends primarily, if not entirely, on outflow, which is known as the invariable storage–discharge relationship. In the case of a reservoir, for which this method was originally developed, a unique storage–outflow relationship is likely to exist. However, as noted at the very beginning of this chapter, routing in open channels is complicated by the fact that storage in a river reach is not a function of outflow alone. Therefore, some routing error will inevitably be introduced, when the modified Puls routing method is applied to a channel.
2. The method neglects the variable slope of the water surface with respect to time that occurs during the passage of a flood wave down a channel. In other words, as in the case of a level pool, the storage considered in the model is the prism storage only and neglects the wedge storage. As discussed above, during the passage of a flood wave, the water surface in a channel is not uniform due to the wedge storage. The storage and water surface slope within a river reach, for a given outflow, are greater during the rising stages (positive wedge) of a flood wave than during the falling stages (negative wedge) as shown in Figure 12.10.
3. Although this method is used quite satisfactorily for reservoir routing, it gives poorer approximations for open channel routing.

12.7.2.3 Overcoming the Limitations of the Modified Puls Method for Channel Routing

Because of the limitations of the modified Puls method in the case of channel routing, arising from the inherent assumptions underlying the method, it has been adopted by the Hydrologic Engineering Center (HEC) of the USACE with the introduction of a concept of **sub-reaches**. This is implemented in HEC-HMS.

The concept of sub-reaches in the modified Puls method was perhaps first used by Strelkoff (1980a). For HEC, he evaluated channel routing methods using the kinematic wave model, diffusion wave model, and modified Puls method alongside a dynamic wave model, the latter being the standard against which the efficacies of the other three approaches were evaluated to identify the method that is computationally simpler than solving the full Saint-Venant equations but can still approximate its results.

For modified Puls method, Strelkoff (1980a) conceptualized a channel reach consisting of a number of sub-reaches, each with equal length and then apportioning the storage of the entire reach equally in each of the sub-reaches. Figure 12.11 shows the concept for a reach subdivided into three sub-reaches. He argued that the method of apportioning the total reach storage amongst sub-reaches gives a relationship between the storage in a sub-reach and discharge, *not at the end of that sub-reach, but at the end of the entire reach.*

Computationally, this means, dividing the storage values in the storage–discharge relationship given for the entire

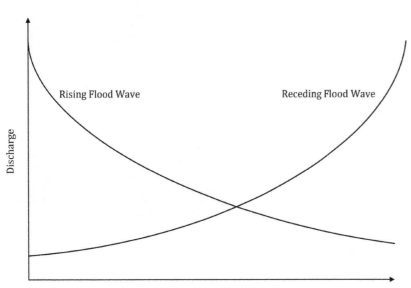

Figure 12.10 Variation of discharge with distance downstream during a rising and falling flood wave.

12.7 Modified Puls Method

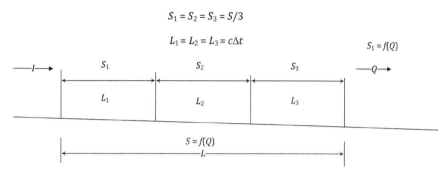

Figure 12.11 The concept of sub-reaches in apportioning reach storage for adaptation of the modified Puls method in channel routing.

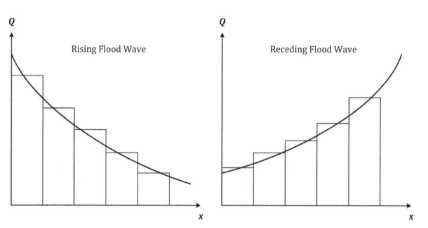

Figure 12.12 Discharge apportioning along the downstream direction of a channel during rising and falling flood waves.

reach by the number of sub-reaches without changing the discharge values and this *reduced* storage–unchanged discharge relationship is used in the modified Puls method. Evidently then, as the number of sub-reaches increases, attenuation decreases due to greater reduction of storage. Strelkoff (1980a), from his numerical experiments showed that when the number of sub-reaches increases to a large number, the results of the modified Puls method approach the results of the kinematic wave model, in which without formation of kinematic shock there is no attenuation of an outflow hydrograph. The modified Puls equation agrees with the kinematic wave equations where a stage–discharge relationship is introduced into the continuity equation.

A few points in relation to the concept presented above is noteworthy. In USACE (1994), it has been stated that to apply the modified Puls method to a channel routing problem, the storage within the river reach is approximated with a series of cascading reservoirs, similar to the model of Nash shown in Figure 9.28 of Chapter 9 where the outflow hydrograph from the upstream reservoir becomes the inflow hydrograph to the next downstream reservoir. If that were the case, with an increasing number of sub-reaches (reservoirs), the attenuation would increase. This is precisely what has been graphically depicted in Figure 9.8 in USACE (1994). However, the concept presented by Strelkoff (1980a, b) is not a concept of such cascading reservoirs.

Due to the formation of positive and negative wedges during the passage of a flood wave in a channel (Figure 12.10), the relationship between storage and discharge at the outlet of a channel is not a unique relationship, rather it is a looped relationship as shown by a storage–discharge function for a river in Figure 12.6. USACE (1994) reasoned that the cascading reservoir approach can approximate the looped storage–outflow effect when evaluating the river reach as a whole. Each reservoir is assumed to have a level pool and, therefore, a unique storage–discharge relationship. The rising and falling flood wave is simulated with different storage levels in the cascade of reservoirs, thus producing a looped storage–outflow function for the total river reach, as depicted graphically in Figure 12.12. But Strelkoff (1980a) argued that the looped storage–outflow occurs by summation of the increments: $\delta S = (I - Q)\delta t$.

12.7.3 Input Data for the Modified Puls Method

12.7.3.1 Storage–Discharge (Outflow) Relationship

A storage–discharge function defines the amount of *outflow* for a specific amount of storage in the reach. Storage is the independent variable and the values used must cover the entire range of storage that may be encountered during simulation. Usually, the first storage will be zero and the

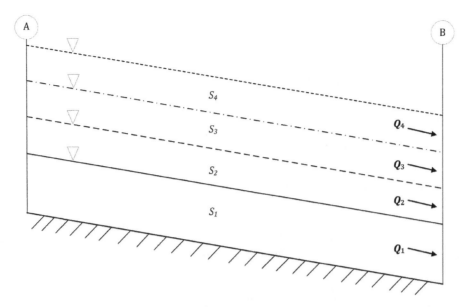

Figure 12.13 Steady-state water-surface profiles between cross sections A and B (reach AB) of a channel corresponding to a series of flows (Q_1, Q_2, etc.).

maximum storage should be slightly more than the volume in the stream reach when it is at maximum flow. The function must be monotonically increasing. The storage–outflow relationship required for the modified Puls routing method can be obtained in several ways such as:

1. steady-flow profile computations using a hydraulic model such as HEC-RAS;
2. historical records of observed high-water marks and corresponding discharge;
3. normal-depth calculations;
4. observed inflow and outflow hydrographs; and
5. optimization techniques applied to observed inflow and outflow hydrographs.

Water-Surface Profiles (thus Storages) Computed with a Hydraulic Model

Steady-state water-surface profiles, computed for a range of discharges with HEC-RAS (USACE, N.D.), or a similar program, define a relationship of storage to outflow between two channel cross sections (Figure 12.13). Strelkoff (1980b) noted that for the uniform flows developed for each steady-state discharge and normal-depth rating curve postulated at the end apportioning of reach storage into sub-reaches is precisely the correct distribution of storage. He also noted that the water volumes computed from the steady-state profiles are not simple prism storage but include wedge storage also.

After a successful steady-state HEC-RAS multi-profile run, the paired data, starting with 0:0, including $S_1 : Q_1; S_2 : Q_2; S_3 : Q_3 \ldots$, are entered into HEC-HMS to define a storage–discharge function for a reach. The storage–discharge function must be defined in the paired data manager of HEC-HMS before it can be used in the reach element.

Some additional notes about using the data from HEC-RAS runs are given below.

Cycling between HEC-HMS and HEC-RAS

Cycling is the process of balancing the peak flows computed using HEC-HMS with the flows used within the HEC-RAS storage–outflow model. Each time a new HEC-RAS storage–outflow model is developed, the flow rate is usually unknown, so that flow rate must be initially estimated, then adjusted to match the HEC-HMS results. Each time the HEC-RAS model is run with a different flow for a routing reach, the storage–outflow values for that reach change. These values are entered into HEC-HMS to calculate a new peak flow from the routing reach. The HEC-RAS storage–outflow model needs to be run again with the new flow. This cycling process must be repeated until the flow rates used in the various routing reaches are consistent with those computed using HEC-HMS. Typically, the cycling between HEC-HMS and HEC-RAS models continues until the flows used in the HEC-HMS and HEC-RAS models for each reach match within 5%.

Effective Storage Areas in HEC-RAS

Effective storage areas are those portions of a channel, floodplain, or reservoir that may reasonably be expected to contain water during a storm event. When the effective storage areas are determined, considerations should be given to the following:

- areas of ineffective flow in a cross section that may or may not contribute to the attenuation of the routed hydrograph

(an **ineffective flow area** is an area where, during a high-flow event, water enters and gets stored but does not flow);
- the consistency of discharges between routing reaches;
- the impact of altering the shape of storage outflow curve; and
- some overbank areas may be ineffective in terms of flow conveyance when determining the water-surface profile but are effective storage areas; these areas will influence hydrograph attenuation; therefore, consider these areas during storage–outflow calculations.

Examples of ineffective flow areas include: (1) a densely developed subdivision made up of numerous structures, fences, and mature trees that significantly hinder flow but allow storage of stormwater during flooding conditions; (2) areas adjacent to bridge or culvert crossings; (3) constructed stormwater detention facilities; (4) constructed sand pits; (5) levees (stop banks); and (6) berms.

Use of the ineffective flow option in HEC-RAS allows the program to account for storage. Use of a blocked obstruction, on the other hand, eliminates potential storage areas. Caution should be exercised when selecting the method used to reflect obstructions and ineffective flow areas to ensure that storage is not improperly excluded from a routing reach.

Cross sections should be reviewed to ensure that all storage is accounted for. Extending cross sections may be necessary to account for the entire storage within the cross section.

A typical procedure for determining the effective storage areas usually involves the following. A topographic map overlaid with an aerial photograph and cross-section alignments is used to determine the areas of effective and ineffective storage. These maps should be visually inspected, and the floodplain and water flow paths should be distinguished. The areas that may or may not be effective storage areas should be determined. Adjustments should be made to ineffective flow areas as needed to properly account for storage volumes in those areas.

It should be also checked that the water-surface profile is not significantly altered when the ineffective flow areas are used in the storage–outflow function. Care should be exercised to ensure that results obtained from models used to compute water-surface profiles and those used to compute storage–outflow data are consistent. The geometric data used for both purposes should be identical.

Historical Data

Records of observed high-water marks and corresponding discharge – that is, observed stage–discharge data – can be used to compute storage–outflow relationships. However, a range of elevation–discharge data are required for this type of calculation and in most cases such data are not available. If limited sets of data are available, these can be used to calibrate the hydraulic model of the reach and the results from that calibrated model can be used in the manner described above.

Normal-Depth Calculations

If cross-sectional data with estimated Manning's friction coefficients and estimated slopes of the energy grade line are available, these can be used in the Manning equation to solve for normal depths for a range of discharges. However, this procedure assumes uniform flow conditions, which never occur in nature (Chow, 1959). Nevertheless, this concept can be used if such flow conditions can be reasonably assumed. With this assumption of uniform flow conditions, the slope of the energy grade line is equal to the channel bed slope. Therefore, this approach should not be used in backwater areas.

Observed Inflow and Outflow Hydrographs

If observed inflow and outflow data are available from gauging stations for an event with significant flood flow, these data can be used to calculate storage using Eq. 12.10. If there is any tributary inflow between two gauging stations, that should be also accounted for. The total storage should be computed from a base level storage at the beginning or end of the routing sequence.

Optimization Techniques Applied to Observed Inflow and Outflow Hydrographs

Observed inflow and outflow hydrographs can be used to determine the storage–outflow function by optimization or trial and error. The process involves defining a candidate function and using that function to route the inflow hydrograph. The calculated outflow hydrograph is then compared with the observed hydrograph. If the match is not satisfactory, the function is adjusted, and the process is repeated until an adequate match between the observed and calculated hydrograph is found. This method can also be used for the calibration of a model.

12.7.3.2 Model Parameter: Number of Sub-reaches

As noted above, the modified Puls method, which was originally intended for reservoir routing, has been adopted by HEC for channel routing with the introduction of the concept of sub-reaches. For this reason, the modified Puls method of routing in HEC-HMS requires the specification of the **number of sub-reaches** as an input parameter. This parameter was called "number of routing steps" in the predecessor HEC-1 program.

As discussed above, the number of sub-reaches is a critical parameter as it has a significant impact on the attenuation of an inflow hydrograph through the routing reach. The selection of a single routing reach provides the maximum attenuation. As the number of sub-reaches increases, the amount of attenuation decreases. The maximum attenuation, corresponding to one sub-reach, can be used for routing through ponds, lakes, wide floodplains, channels with in-line weirs or small dams, and in general channels in which the flow is heavily controlled by downstream conditions in the limit of a large reservoir. A *single sub-reach* essentially implies a *level pool*, as in the case of a true reservoir. Increasing the number of sub-reaches may approach zero attenuation in the limit of the kinematic wave model without any shock formation.

As illustrated in Figure 12.11, the number of sub-reaches, N_S, can be approximated from

$$N_s = \frac{L}{c \Delta t} \quad (12.41)$$

where L is the actual reach length, and Δt is the simulation time step. Note that the denominator of Eq. 12.41 gives the length the flood wave travels for the given time interval. Thus, the program tries to divide the reach into a specific number of individual sub-reaches that will have travel time that is approximately equal to the modeling time step, Δt. In other words, the travel time through a sub-reach should be approximately equal to the simulation time step for an idealized channel. It should be noted here that Strelkoff (1980a) opined that there is no formal way of correctly determining N_s.

Equation 12.41 implies that the number of sub-reaches or routing steps is the ratio of the average travel time of the flood waves ($t_{FW} \approx c$) from the upstream to the downstream end of a reach ($t_{FW} = L/c$) and the computational time step or time interval Δt_{Model} that is chosen or specified in the control specification of the HEC-HMS model. Therefore,

$$N_s = \frac{t_{FW}}{\Delta t_{Model}} \quad (12.42)$$

The travel time of the flood wave, t_{FW}, is estimated from the calculated flow velocities in the HEC-RAS model. However, the velocity implied in the calculation of the number of sub-reaches is the velocity (actually speed) of the flood wave that is, the celerity in the channel. As discussed in Section 12.5, this is different from the channel flow velocity and is generally 1.3 to 1.7 times faster than the average flow velocity. Generally speaking, in typical natural channels, the celerity is typically 1.5 times the average water velocity; this is also the case if the Chezy equation is used. Thus, flood wave travel time is less than the water travel time. HEC-RAS provides the travel time of water (t_W) from an upstream cross section to a downstream cross section. Therefore, for use in Eq. 12.42, a general relationship is

$$t_{FW} = \frac{t_W}{1.5} \quad (12.43)$$

The following guidelines should be used for determining the number of sub-reaches.

1. In subdividing a routing reach, the sum of the lengths of the sub-reaches must be equal to the length of the original reach.
2. If the HEC-HMS computational time interval is modified, the number of sub-reaches used for all routing reaches should also be adjusted.
3. In certain instances, the calculated number of sub-reaches may produce results that can seem counterintuitive. For example, reaches with significant channel and overbank storage tend to exhibit lower channel velocities (higher travel time), which consequently result in an increased sub-reach parameter and decreased attenuation. In such cases, proper judgment is required to reduce the number of sub-reaches, to correctly capture the attenuation and thereby valley storage.

12.7.3.3 Initial Conditions

As noted, above, the initial outflow must be specified for the computation to start. Two alternatives are available in HEC-HMS: (1) user-specified discharge and (2) inflow = outflow.

For the first option, a discharge value must be entered. The initial storage in the reach is calculated from the specified discharge and the storage–discharge function. In the second option, it is assumed that the initial outflow is the same as the initial inflow to the reach from upstream elements. This is essentially the same as assuming a steady-state initial condition. The initial storage is computed from the first inflow to the reach and storage–discharge function.

12.8 MUSKINGUM METHOD

12.8.1 Historical Background

The classical Muskingum method of channel flow routing, based on Eq. 12.14, was developed by G. T. McCarthy and others of USACE in connection with studies of the Muskingum Conservancy District Flood Control Project in eastern Ohio. McCarthy (1938), in an unpublished manuscript used the term Muskingum method because he tested this newly developed hydrologic method of flood routing using data from the Muskingum River. Chow (1959) referenced this method as the Muskingum method and it was described in detail by both Carter and Godfrey (1960) and USACE (1960). Since then, this method has gained popularity worldwide and been historically used widely.

12.8.2 Routing Equation

The Muskingum equation (Eq.12.14) for change in storage, ΔS, for a given time interval, Δt, can be written as

$$S^{j+1} - S^j = K[\{XI^{j+1} + (1-X)Q^{j+1}\} - \{XI^j + (1-X)Q^j\}] \quad (12.44)$$

The finite-difference approximation of the change in storage in terms of inflow and outflow, that is, the continuity equation, is also given as

$$S^{j+1} - S^j = \bar{I} - \bar{Q} = \left(\frac{I^{j+1} + I^j}{2}\right)\Delta t - \left(\frac{Q^{j+1} + Q^j}{2}\right)\Delta t$$
(12.45)

Equating the right-hand sides of Eqs. 12.44 and 12.45, we obtain

$$\frac{K}{\Delta t}\left[XI^{j+1} + (1-X)Q^{j+1} - XI^j - (1-X)Q^j\right]$$
$$= \frac{1}{2}\left(I^{j+1} + I^j - Q^{j+1} - Q^j\right)$$
(12.46)

Further simplification of Eq. 12.46, writing the only unknown Q^{j+1} on the left-hand side, yields

$$Q^{j+1} = C_1 I^{j+1} + C_2 I^j + C_3 Q^j$$
(12.47)

where

$$C_1 = \frac{\frac{\Delta t}{K} - 2X}{2(1-X) + \frac{\Delta t}{K}}$$
(12.48)

$$C_2 = \frac{\frac{\Delta t}{K} + 2X}{2(1-X) + \frac{\Delta t}{K}}$$
(12.49)

$$C_3 = \frac{2(1-X) - \frac{\Delta t}{K}}{2(1-X) + \frac{\Delta t}{K}}$$
(12.50)

Equation 12.47 is the routing equation for the original Muskingum method of channel routing with $C_1, C_2,$ and C_3 called the **routing coefficients**. Note that this equation is a lumped model, since it solves for outflow as a function of time only. At a time step after the initial conditions or the first time step (where $Q^j = Q^1$ is a known quantity), the calculated Q^{j+1} at the previous time step becomes Q^j. Also note that $C_1 + C_2 + C_3 = 1$, and $2KX < \Delta t \leq K$ or $2KX < \Delta t \leq 2K(1-X)$ for numerical accuracy and theoretical stability (K and Δt must be in the same units). It may be noted that C_1 and/or C_2 can be negative depending on the values of the parameters and Δt, which sometimes may cause outflow to become negative at the beginning.

12.8.3 Estimation of Routing Parameters in the Muskingum Method

12.8.3.1 Estimation of K

The Muskingum K parameter is equivalent to the travel time of the flood wave from the upstream end to the downstream end of the river reach. Therefore, K accounts for translation in routing and is a function of flow and channel characteristics. This parameter can be estimated in multiple ways including:

1. using known hydrograph data;
2. comparing flow length to a flood wave speed derived from rating curves; and
3. Using regional regression equations, developed from observed data in a similar region.

Using Known Hydrograph Data

The travel time of a flood wave moving through a reach can be estimated by taking the difference between similar points on known inflow and outflow hydrographs. These points can be the peaks of either hydrograph, the centroid of the area underneath each hydrograph, or between some reference flow on the rising limb of either hydrograph.

An example is given from the Santa Fe River basin in northcentral Florida. Sant Fe River is an east tributary of Suwannee River, which flows from Georgia through Florida and discharges into Gulf of Mexico on Florida's west coast. The National Weather Service has implemented a flood forecasting system for Santa Fe River. Figure 12.14 shows the USGS gauging stations along this river and Figure 12.15 shows the hydrographs at three gauging stations in the central part of the basin during Tropical Storm Elsa, which affected much of north Florida at different times from June 30 to July 10, 2021. The peak discharge at the Worthington Springs gauge was 8340 ft^3/s and occurred at 05:15 on July 9. The peak at the gauge near (north) High Springs was 7040 ft^3/s and occurred on July 11 at 10:45. So, the travel time for the flood wave from Worthington Springs to High Springs was 53.5 hours and the attenuation of the peak discharge was 1300 ft^3/s. The peak discharge at Fort White gauge was 7070 ft^3/s and occurred on July 12 at 04:45. In this case the travel time of the flood wave was 18 hours with negligible attenuation. It should be pointed out that Santa Fe River is typical of many rivers in karst geomorphology. It drops into a large sink hole in O'Leno State Park north of High Springs and reappears in the adjacent River Rise Preserve State Park.

Comparing Flow Length to a Flood Wave Speed derived from a Rating Curve

The travel time, t_{FW}, of a flood wave through a reach can be estimated by dividing the length of the reach, L, by the flood wave velocity, V_{FW}, which is the celerity, c:

$$t_{FW} = \frac{L}{V_{FW}} = \frac{L}{c} = K$$
(12.51)

Using the Kleitz–Seddon law, a flood wave velocity can be approximated from the discharge rating curve at a station whose cross section is representative of the routing reach. The slope of the discharge rating curve is equal to dQ/dh. The flood wave velocity, and therefore the travel time, K, can be estimated from

$$V_{FW} \equiv c = \frac{1}{B}\frac{dQ}{dh}$$
(12.52)

where B is the top width of water surface at the station where the rating curve is established, h is the stage height, and therefore dQ/dh is the slope of the flow-stage rating curve;

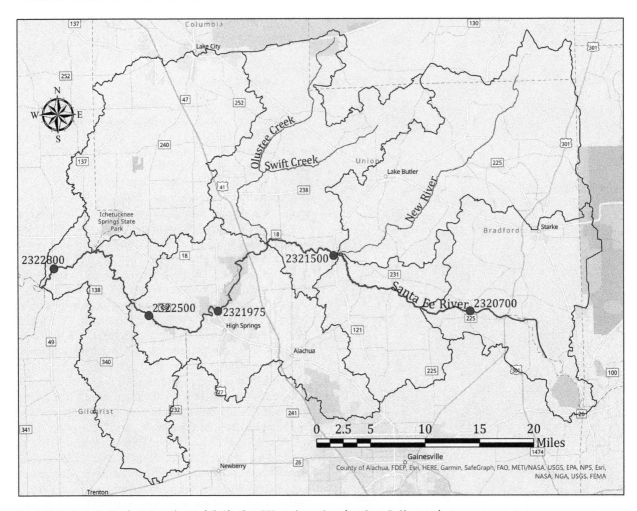

Figure 12.14 Santa Fe River basin in north central Florida. The USGS gauging stations along Santa Fe River are shown.

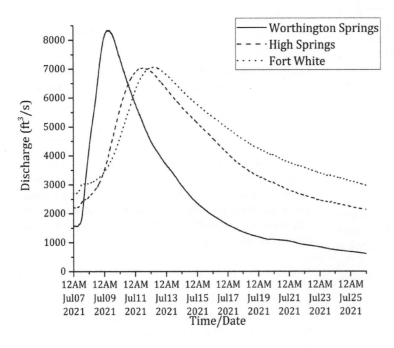

Figure 12.15 Hydrographs at USGS gauging stations 2321500 (Worthington Springs), 2321975 (near High Springs) and 2322500 (Fort White) on Santa Fe River in Florida during Tropical Storm Elsa.

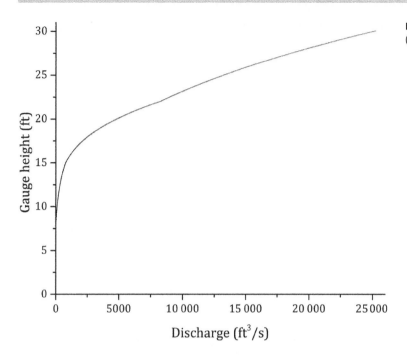

Figure 12.16 Rating curve at gauging station 2321500 (Worthington Springs).

both variables must be estimated for the flow rate in question and at a cross section that is representative of the routing reach. Figure 12.16 shows the rating curve at gauging station 2321500 (Worthington Springs) of Santa Fe River (Figure 12.14). The slope of the curve for a flow rate between 5000 and 10 000 ft³/s is 1640 ft³/s/ft and, at a flow rate of approximately 10 000 ft³/s, the top width of the river is 2000 ft. The celerity is then about 0.82 ft/s conforming to the slow-flowing nature of this river in a flat coastal region. The distance between the Worthington Springs gauging station and the gauging station to the north of High Springs is 20 miles. Thus, the travel time is approximately 36 hours.

Using Regional Regression Equations

In a watershed or river basin, if the parameter K can be determined by one or both of the two methods mentioned above for several streams, it may be possible to develop regression equations correlating K with certain physical characteristics of the streams, such as reach length, bed slope, roughness, etc. Such a regression equation can then be used for streams in other parts of the same watershed or river basin where hydrograph data are not available.

12.8.3.2 Estimation of X

The Muskingum X parameter is a dimensionless coefficient that lacks a strong physical meaning. The parameter X is used to account for the storage component of the routing. For a given flood hydrograph there is a unique value of X for which the storage in the calculated outflow hydrograph should nearly be the same as an observed outflow hydrograph. This parameter is considered as a weighting factor that must range between 0.0, giving maximum attenuation or pure reservoir storage, and 0.5, with no attenuation and giving pure translational movement of the flood wave. When this parameter is set to a value of 0, the storage within the reach is computed solely as a function of the outflow. This is equivalent to level-pool routing and results in the maximum possible amount of attenuation. When this parameter is set to a value of 0.5, equal weight is given to both inflow and outflow when determining storage within the reach. This results in no attenuation to the inflow hydrograph as it progresses through the reach. For most applications, an initial estimate of 0.25 or 0.30 is further refined through model calibration.

12.8.3.3 Simultaneous Determination of K and X

When measured inflow and hydrograph data are available, the parameter X can be determined by a graphical procedure outlined by Linsley et al. (1982) in which the instantaneous storage (S) values are plotted on the x-axis against the corresponding weighted discharge values $[XI + (1 - X)Q]$ on the y-axis for various values of X. The value of X that produces the narrowest loop between S and $[XI + (1 - X)Q]$, that is, the data plot most nearly as a single-valued curve, is considered the best value of X. In accordance with Eq. 12.16 with $m = 1$, the slope of the curve which is a straight line gives the value of $1/K$.

The values of K and X determined by the graphical method described above are mostly valid for the river reach whose data are used and, more importantly, for the flood event for which the data are available. Use of these parameter values to other reaches of the same river or to other flood events of significantly different magnitudes in the same river reach should be carried out with caution. A better approach is to use the data from multiple flood events with different values of magnitude and duration, to determine the average or representative parameters for the river reach.

12.8.3.4 Number of Sub-reaches

This parameter is required for using Muskingum method in HEC-HMS. The number of sub-reaches (N_s) parameter affects attenuation. One sub-reach results in the maximum amount of attenuation and increasing the number of sub-reaches approaches zero attenuation. An initial estimate of this parameter can be obtained by dividing the Muskingum K parameter by the simulation time step, Δt:

$$N_s = \frac{K}{\Delta t} \qquad (12.53)$$

For natural channels that vary in cross-sectional dimension, slope, and storage, the number of sub-reaches can be treated as a calibration parameter. It may be used to introduce numerical attenuation, which can better represent the movement of flood waves through the natural system. The sub-reach parameter appropriately adjusts Δt in the program.

12.8.4 Limitations of the Muskingum Method

Some known limitations of Muskingum method are as follows.

1. It may produce negative flows in the initial portion of the hydrograph (Perumal, 1992).
2. It is generally limited to moderate to slow rising hydrographs being routed through channels with mild to steep slopes.
3. It is not applicable to steeply rising hydrographs such as in the case of a dam break.
4. It neglects backwater effects caused by structures like downstream dams, constrictions, bridges, culverts, and tidal influences.

12.8.5 A Special Case of the Muskingum Method: Lag and K method

The lag and K routing method is based on a routing technique that was originally proposed by Wilson (1941) as a graphical method of solution of the Muskingum equation. The method is still extensively used by the NWS in the United States. The method is a special case of the Muskingum method where channel storage is represented by the prism component alone with no wedge storage (i.e., Muskingum $X = 0$). The lack of wedge storage means that the method should only be used for slowly varying flood waves. For example, the flood forecasting system for Santa Fe River in Florida, mentioned above, is based on this method.

Operationally, the lag and K method has been and continues to be a practical and widely used method of storage routing between flow points. It is also a very flexible method of routing since both the lag and K elements can be either constant or variable. The lag term accounts for the travel time of the flood wave as it moves downstream. The K term accounts for attenuation of the flood wave. Examination of historical flood hydrographs of varying magnitude provides a basis for establishing lag and K relationships within a reach. The first process in a normal operation is to lag or delay in hours an inflow hydrograph in order to create what is called a lagged inflow graph. The K part of the operation is then used to attenuate the lagged inflow graph in order to create an outflow hydrograph at the downstream flow point. Though normally used together, lag and K can also be used separately to account for lag with no attenuation or attenuation with negligible lag.

Lag and K parameters can be constants or variable functions. The lag versus inflow relationship, known as a variable lag curve, is defined at points, and placed in tabular form and then plotted as shown in Figure 12.17. Significant changes in the slope of the curve should always be represented with a point in the table. There is no limit within reason to the number of points that can be used. However, the lag versus inflow curve represents an average relationship. It is not exact and from four to ten points should be adequate to describe even the most complex situation. Instead of being related to inflow as lag is, K is related to outflow at the downstream point. K versus outflow relationships at different discharges are determined through investigation of historical flow records. If K is constant throughout the entire range of outflow values, only one value of K is needed for all flows. However, if K varies with outflow, it is necessary to describe a K versus outflow relationship in tabular form. Figure 12.17 shows an example. Values in the variable K table are taken from points along the K versus outflow curve. In HEC-HMS, when using variable lag and K functions, lag is entered as an inflow-lag curve and K is entered as an outflow-attenuation curve as shown in Figure 12.17.

Evaluation of historical flood hydrographs provides the basis for lag and K relationships within a reach. Care must be exercised when using lag functions with multiple intercepts (i.e., lag is the same for more than one flow rate). This may result in numerically attenuated peak flow rates.

12.9 MUSKINGUM–CUNGE METHOD

12.9.1 Historical Background

Application of the Muskingum method of channel routing requires stream gauge data from multiple locations for

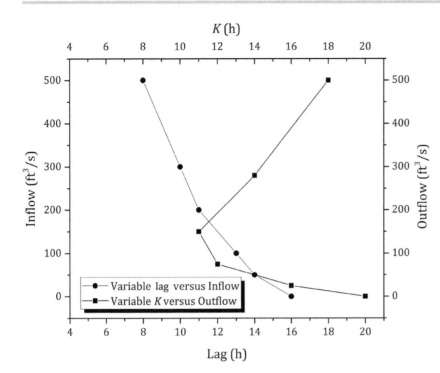

Figure 12.17 An example of variable lag versus inflow and variable K versus outflow relationships used in the lag and K method of channel routing.

estimation of the storage constant, K, and the weighting parameter, X. In typical engineering applications of ungauged streams, such data are not available and hence it is difficult to estimate these two parameters that are essential for this method to be applied. Furthermore, even if such data are available for a reach of a river, they cannot be readily applied to other reaches of the same river. Cunge (1969), a Polish–French engineer, published the equations used in the Muskingum–Cunge method. It is based on the physical characteristics of a channel, such as rating curves, cross-sectional data, and channel slope, and hence offered an attractive alternative to the Muskingum method. This method is widely used now a days because, unlike the Muskingum method where the parameters are estimated using streamflow data, the Muskingum–Cunge method uses parameters that can be derived from only channel characteristics.

It should be also noted that while the classical Muskingum method is a purely hydrologic method, the Muskingum–Cunge method has a distinct hydraulic basis. Another important difference between these two methods should be borne in mind. The Muskingum method, based on the storage concept, is a lumped method, where the parameters K and X are reach averages. The Muskingum–Cunge method, on the other hand, is a *distributed model*, where the model parameters (K and X) are based on values evaluated at channel cross sections. It is necessary, therefore, when using the Muskingum–Cunge method that the routing parameters are evaluated at channel cross sections that are representative of the channel reach under consideration.

12.9.2 Routing Equation

In the Muskingum–Cunge method, variation of outflow as a function of space is introduced by setting, $I^j = Q_i^j$, $I^{j+1} = Q_i^{j+1}$, and $Q_{i+1}^j = Q^j$ in Eq. 12.46 where subscript i is the space index in a space (x)–time (t) grid. This yields

$$\frac{K}{\Delta t}\left[XQ_i^{j+1} + (1-X)Q_{i+1}^{j+1} - XQ_i^j - (1-X)Q_{i+1}^j\right]$$
$$= \frac{1}{2}\left(Q_i^{j+1} + Q_i^j - Q_{i+1}^{j+1} - Q_{i+1}^j\right)$$

(12.54)

Now simplifying Eq. 12.54 in the same manner to bring the unknown quantity Q_{i+1}^{j+1} to the left-hand side of the equation, we get

$$Q_{i+1}^{j+1} = C_1 Q_i^{j+1} + C_2 Q_i^j + C_3 Q_{i+1}^j \quad (12.55)$$

Here, the coefficients, C_1, C_2, and C_3 are also as given in Eqs. 12.48–12.50. Equation 12.55 is the routing equation for the Muskingum–Cunge method. The finite-difference grid, representing the solution of Eq. 12.55, is shown in Figure 12.18.

It should be pointed out at this point that Cunge (1969) showed that, if K and Δt in Eq. 12.54 are constants, it is also an approximate solution of the kinematic wave equation, which is essentially the combination of the continuity equation with the friction slope equaling the bed slope.

In the Muskingum–Cunge method, not only the continuity equation is discretized in space and time, but also K and X are evaluated differently, giving it an approximate solution of the modified diffusion wave equation as discussed below.

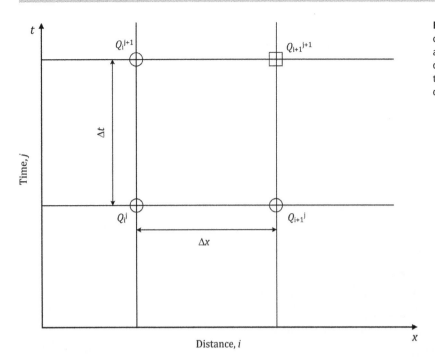

Figure 12.18 Finite-difference grid or discretization of Muskingum–Cunge routing equation in space (x) and time (t) with *i* denoting the space index and *j* denoting the time index. The known quantities at a time are shown by the circles and the only unknown quantity is shown by the square.

12.9.3 Input Data for the Muskingum–Cunge Method

The Muskingum–Cunge method requires the following data, which are all physical characteristics of a channel.

1. representative channel cross section;
2. reach length, L;
3. Manning's roughness coefficient, n (for main channel and overbanks); and
4. friction slope (S_f) or channel bed slope (S_0).

12.9.4 Evaluation of K and X in the Muskingum–Cunge Method

12.9.4.1 Evaluation of K

The value of K is given as

$$K = \frac{\Delta x}{c} \tag{12.56}$$

Where c is the celerity of the flood wave in the x-direction or the speed at which the flood wave travels downstream. The right-hand side of Eq. 12.56 represents the time of propagation of a given discharge along a reach of length Δt.

12.9.4.2 Evaluation of X

The value of X is given as

$$X = \frac{1}{2}\left(1 - \frac{Q_0}{BS_0 \, c\Delta x}\right) \tag{12.57}$$

where Q_0 is called the **reference flow** or **index flow**. For numerical stability of the finite-difference solution, it is required that $0 \leq X \leq \frac{1}{2}$. (Chow et al., 1988). In the variable coefficient method of solution, discussed below, the discharge in Eq. 12.57 is not constant as the reference discharge and hence only Q is used instead of Q_0.

It should be noted that the choice of which flow rate to use as the reference flow is not completely clear (USACE, 1994). Most of the workers take it as the peak flow from the inflow hydrograph: $Q_0 = Q_{peak}^{inflow}$ However, according to Ponce (1983), reference flows based on peak flow values tend to accelerate the wave much more than would happen in nature, while the converse is true if base flow is used as the reference flow. According to USACE (1994), a reference flow based on the average of the base flow (Q_b) and peak inflow is the most suitable choice. Accordingly,

$$Q_0 = Q_b + \frac{1}{2}\left(Q_{peak}^{inflow} - Q_b\right) \tag{12.58}$$

On the other hand, Merkel (2002) notes that at least for the linear or constant coefficient method (discussed below), having the reference flow equal to the peak flow is far superior to having it equal to the average of peak flow and base flow.

In the same vein, the reference celerity, c_0, is calculated as

$$c_0 = m\frac{Q_0}{A_0} \tag{12.59}$$

where Q_0 is again the reference flow, normally taken as the peak discharge of the inflow hydrograph, A_0 is the cross-sectional area corresponding to the reference discharge, and m is the coefficient as discussed earlier.

In summary, in the Muskingum–Cunge formulation, the values of K and X are determined, based on the channel cross section and reach hydraulic properties. These, in turn, govern the routing coefficients (Eqs. 12.48–12.50).

12.9.5 Solution Procedures

There are two formulations for the estimation of routing coefficients and the numerical method of solution of the Muskingum–Cunge routing equation. In both methods, the space–time grid shown in Figure 12.18 is used in the numerical scheme.

1. **Constant coefficient** or **linear** method: In this formulation, the routing coefficients are held constant throughout the computation. This is the linear approach where the routing parameters (K and X) are calculated based on average or representative hydraulic variables, c_0 and Q_0 as defined in Eqs. 12.58 and 12.59. Again, as discussed above, the reference discharge can be either the peak inflow or the average of the peak inflow and base flow. The values of K and X thus calculated are used to calculate the routing coefficients, C_1, C_2, and C_3, which do not vary as the solution marches through the space–time grid shown in Figure 12.18. This is a three-point direct solution technique where the unknown value of discharge is computed as a *linear combination* of the three other known flow values (Eq. 12.55). The TR-20 program, developed and used by the USDA, implements this technique for the Muskingum–Cunge method (which is the only routing method used and recommended by the USDA). One of the characteristics of this formulation is that the volume of inflow (upstream end of the reach) is always equal to the volume of outflow (downstream end of the reach).
2. **Variable coefficient** or **nonlinear** method: In this formulation, the routing coefficients, C_1, C_2, and C_3, vary, since they are calculated at each time (Δt) and space (Δx) interval of computation to reflect flow characteristics, primarily wave celerity, friction slope, and top width, of the rising and receding flood wave. This is done because hydraulic parameters Q, B, and c change according to channel properties and flow depth and, consequently, both K and X change with respect to time and space. This is an iterative four-point averaging scheme, proposed by Ponce (1986), to solve for Q, B, and c. This four-point iterative procedure bases the routing coefficients on the three known discharges and the estimate of the outflow at the second time step. Since the outflow at the second time step is not known, iterations are needed to converge to the final discharge, based on some error tolerance. Therefore, it is a *nonlinear* process. This scheme is implemented in HEC-HMS. Merkel (2002) notes that the primary disadvantage of the four-point scheme is that this method does not always conserve hydrograph volume (volume of inflow is not always equal to volume of outflow). Ponce and Yevjevich (1978) also noted that there is small but perceptible loss of mass in the nonlinear method of solution of the Muskingum–Cunge equation.

The way of calculating the parameters has a definite bearing on the overall accuracy of the routing method. Ponce and Yevjevich (1978) pointed out that the assumption of constant parameters makes the solution dependent on the reference values chosen to evaluate these parameters. A more physically realistic approach is to consider the parameters K and X to vary in space and time according to flow variability. The physical and numerical significance of the routing coefficient can be seen by substituting Eqs. 12.56 and 12.57 into Eqs. 12.48–12.50 by defining

$$C = c \frac{\Delta x}{\Delta t} \tag{12.60}$$

and

$$D = \frac{Q}{S_o c \Delta x} \tag{12.61}$$

where C is the Courant number, defined as the ratio of wave celerity, c, to grid celerity, $\Delta x/\Delta t$, and D is a type of cell Reynolds number, being the ratio of hydraulic diffusivity to grid diffusivity. With these, the routing coefficients can also be expressed in the Courant and cell Reynolds numbers as

$$C_1 = \frac{-1 + C + D}{1 + C + D} \tag{12.62}$$

$$C_2 = \frac{1 + C - D}{1 + C + D} \tag{12.63}$$

$$C_3 = \frac{1 - C + D}{1 + C + D} \tag{12.64}$$

12.9.5.1 Calculation of c

From the discussion presented above, it should be evident that for the calculation of parameter K an estimation of wave celerity is the first step in the Muskingum–Cunge method. Wave celerity is given by Eq. 12.22 and estimated from the average velocity for a given hydraulic exponent m.

Accurate estimation of the value of m is particularly important for the constant coefficient method, since in this formulation K is calculated at the very beginning from the celerity value (with a specified Δx also) and does not change for the remainder of the calculation process. In this method, the value of m used in the calculation of c has a major effect on the calculation of travel time, K, of the hydrograph through the reach. The influence of initial estimate of c (c_0) is less in the variable coefficient method due to the iterative process involved in the calculation. In this case, a rough estimate does not induce a significant error in the calculation. In any event, a discussion of the estimation of m and thereby c is noteworthy.

The Kleitz–Seddon law indicates that celerity is the slope of the discharge versus the flow–area curve at a station whose cross section is representative of the routing reach. From this relationship, the celerity can be approximated from the station rating curve dQ/dh and knowing the top width, B, of the water surface $\left(c = \frac{1}{B}\frac{dQ}{dh}\right)$. However, in typical cases of hydrologic modeling pertaining to any design of water resources project, measured station rating curves are not available or, even if available at a point, may not be used for other reaches of the same natural channel. On the other hand, a hydraulic model of the channel under consideration can be built using a program like HEC-RAS. From the results of such a hydraulic model, m can be estimated as the slope of a log–log plot of discharge versus flow–area relationship. Alternatively, data can be plotted on arithmetic scales and the exponent of a best-fit power function can be used for the direct estimation of m. Figure 12.19 shows an example of such discharge–area curves of six cross sections along seven miles of a reach of a natural channel, named Graham Branch, located in Denton County in north Texas. These data are generated from an HEC-RAS model of this channel for a range of flows. A power function in the form of Eq. 12.20 is fitted to the data to derive the exponent m. From this randomly chosen example, it can be seen that in every cross section of the same channel, the power function does not always perfectly fit the discharge–area relationship. This is mainly because of floodplain storage. Thus, if water leaves the channel and fills up the floodplain, the curve fitting procedure is not quite accurate since the data may scatter around a trend line. NRCS (2014) proposed a procedure to calculate m for each discharge value from the log–log plot of data in a discharge-rating table listing monotonically increasing pairs of discharge–area values. The procedure is as follows.

In general, the first pair in the rating table gives zero area for zero. So, this value is ignored, and the calculation involves discharge–area pairs listed as the second set of values.

First, m is calculated for all discharges less than the third value of discharge (Q_3), with $S_{2,3}$ denoting the log–log slope between the second and third data points,

for $Q_k < Q_3$,

$$m = S_{2,3}$$

For all subsequent discharge values, $Q_3 < Q < Q_k$,

$$m_k = \frac{\{Q_3 S(2,3)\} + \sum (Q_k - Q_{k-1}) S_{k-1,k}}{Q_k}$$

12.9.5.2 Selection of Δx and Δt

Calculation of the values of both K and X requires the selection of an appropriate value of Δx. Furthermore, the values of Δx and Δt should be chosen for the accuracy of calculations. In general, Δt should be evaluated by looking at the following three criteria and selecting the smallest value:

(1) an arbitrarily selected computational time interval (in HEC-HMS, it is an input in the control specification), (2) the time of rise of the inflow hydrograph divided by 20 ($t_p/20$), and (3) estimating the travel time through the channel reach. Once Δt is appropriately chosen, Δx is defined as

$$\Delta x = c \Delta t \quad (12.65)$$

USACE (1994) also refers to Ponce (1983) to state that Δx must also meet the following criterion to preserve consistency in the method:

$$\Delta x \leq \frac{1}{2}\left(c\Delta t + \frac{Q_0}{BS_0 c}\right) \quad (12.66)$$

Again, Q_0 is the reference flow that is discussed above.

12.9.5.3 Selection of Reaches or Reach Lengths

As noted above, the Muskingum–Cunge method can be used with a simple cross section, such as trapezoidal, rectangular, square, triangular, or circular (pipes). However, more detailed cross sections, including the left overbank, main channel, and right overbank, can also be used to calculate the hydraulic properties of the reach. A very important assumption in this regard is that the cross section is representative of the entire reach. If this assumption is not adequate, the routing reach must be broken into smaller sub-reaches with representative cross sections for each reach. In this case, the outflow from one reach becomes the inflow for the next reach.

12.9.6 Hydraulic Basis of the Muskingum–Cunge Formulation

The *conservation* form of the continuity equation (in space and time) is given by Eq. 12.7. The diffusion form of the momentum equation is given by Eq. 12.5. When Eqs. 12.5 and 12.7 are combined and linearized, the result is the **convective diffusion equation** given as (Miller and Cunge, 1975):

$$\frac{\partial Q}{\partial t} + c\frac{\partial Q}{\partial x} = \mu \frac{\partial^2 Q}{\partial x^2} + cq_L \quad (12.67)$$

where c is the wave celerity as given in Eq. 12.22, and μ is **hydraulic diffusivity (physical diffusion)** given as

$$\mu = \frac{Q}{2BS_0} \quad (12.68)$$

The continuity equation (Eq. 12.7), discretized on the x–t space, is given as

$$Q_{i+1}^{j+1} = C_1 Q_i^{j+1} + C_2 Q_i^j + C_3 Q_{i+1}^j + C_4 q_L \quad (12.69)$$

Equation 12.69 is the same as Eq. 12.55 with the additional term $C_4 q_L$ where the coefficient C_4 is given as

12.9 Muskingum–Cunge Method of Channel Routing

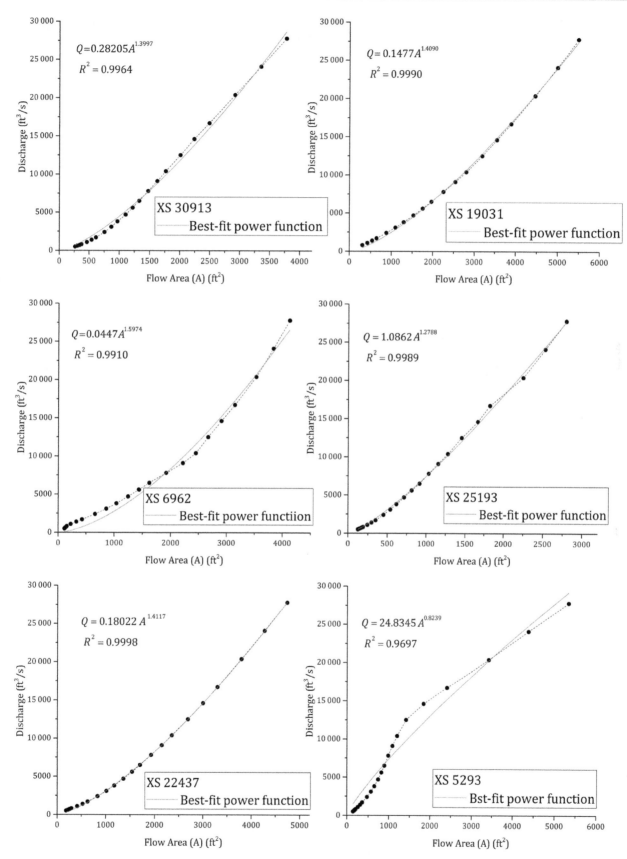

Figure 12.19 Discharge–area relationships at several cross sections of a natural channel, named Graham Branch located in Denton County in north Texas. The discharge–area data are obtained from a steady-state HEC-RAS model of the channel.

$$C_4 = \frac{2\frac{\Delta t}{K}}{2(1-X) + \frac{\Delta t}{K}} \quad (12.70)$$

The other coefficients are as given in Eqs. 12.48–12.50. These are the finite-difference approximations of the partial derivatives in Eq. 12.67. In most channel routing problems, the lateral inflow is ignored and hence the $C_4 q_L$ term is dropped when Eq.12.67 is solved.

The Muskingum–Cunge equation, with the values of K and X given by Eqs. 12.56 and 12.57, is considered an approximation of the convective diffusion equation (Eq. 12.67). Cunge (1969) evaluated the diffusion that is produced in the original Muskingum equation and analytically obtained the following diffusion coefficient:

$$\mu_n = c\Delta x \left(\frac{1}{2} - X\right) \quad (12.71)$$

In Eq. 12.71, the diffusion coefficient μ_n is called **numerical diffusion**, because the diffusion obtained with the Muskingum method is a function of how the equation is solved. It is not the *physical diffusion* given by Eq. 12.68. In the calculations, the diffusivity of the numerical model is set equal to the diffusivity of the theoretical model. Thus, equating the right-hand sides of Eqs. 12.68 and 12.71, we obtain the value of X given by Eq. 12.57.

12.9.7 HEC-HMS Inputs

12.9.7.1 Initial Conditions

There are two options for the initial condition: (1) inflow equals outflow and (2) specified discharge. In the first option, it is assumed that the initial outflow is the same as the initial inflow to the reach from upstream elements. This is essentially the same as assuming a steady-state initial condition. In the second option, a specified discharge value has to be entered.

12.9.7.2 Hydraulic Properties of the Channel

Channel hydraulic properties include: *length*, *slope*, and *Manning's n*:

The *length* is the total length of the reach element. Usually, length is measured from maps of the watershed.

The *slope* is the average bed slope for the whole reach and is assumed to represent the friction slope. Typically, it is estimated from maps if they provide sufficient vertical resolution. If field survey data are available, then those data are used to estimate this channel hydraulic property. If the slope varies significantly throughout the stream represented by the reach, it is then necessary to use multiple reaches with different slopes.

Manning's n or channel roughness coefficient usually is the average value for the whole reach. This value can be estimated from pictures of streams with known roughness coefficient or by calibration. When the tabular shape is selected (discussed below), no Manning's *n* value needs to be entered.

12.9.7.3 Space–Time Method

There are three options for the selection of space (dx) and time (Δt) intervals for computations.

The first option is "Auto Δx Auto Δt." When this method is selected, the program automatically selects space and time intervals that maintain numerical stability.

The second option is "Specified Δx Auto Δt." When this method is selected, the program uses the specified number of sub-reaches (i.e., Δx) while automatically varying the time interval to take as long a time interval as possible, while also maintaining numerical stability.

The third option is "Specified Δx Specified Δt." When this method is selected, the program uses the specified number of sub-reaches and subintervals throughout the entire simulation.

12.9.7.4 Index Method

The index method and specified parameter are used by the program to *discretize* the routing reach in both space and time. The index method is used in conjunction with the physical properties of the channel and the previously mentioned Δx and Δt interval selection to discretize the routing reach in both space and time.

Two options to specify the index method are provided within HEC-HMS. Either the *index celerity* (ft/s or m/s) or *index flow* (ft^3/s or m^3/s) must be specified. The HEC-HMS documentation does not explain what is meant by index flow and index celerity. Since the program uses the algorithms of Ponce (1986), the index flow and the index celerity most likely imply reference flow and reference celerity as discussed above.

Appropriate reference flow or reference celerity is dependent upon the physical properties of the channel as well as the event(s) in question. As noted above, a reference flow (or celerity) based upon the average values of the hydrograph in question (i.e., midway between the base flow and the peak flow) is, in general, the most suitable choice. A reference flow (or celerity) based on peak values tends to numerically accelerate the wave much more than would occur in nature, while the converse is true if a low reference flow (or celerity) is used (Ponce, 1983).

When the flow index method is used, a reference flow and cross-sectional properties are used to infer a celerity. For typical cases, using the celerity index method with a celerity of 5 ft/s works adequately.

12.9.7.5 Cross Sections

Six options are available within HEC-HMS for specifying the cross-sectional shape: circle, rectangle, trapezoid, triangle, tabular, and 8-point.

The *circle* shape is not meant to be used for pressure flow or pipe networks but is suitable for representing a free surface inside a pipe. Depending upon the chosen shape, additional information has to be entered to describe the size of the cross-sectional shape. This information may include a diameter (circle; in ft or m), bottom width (deep, rectangle, and/or trapezoid; in ft or m), or side slope (trapezoid and triangle; in ft/ft or m/m). In all cases, the cross-sectional shapes must be defined in such a way that all possible flow depths simulated will be completely confined within the defined shape.

The *tabular shape* option allows one to enter user-defined *elevation versus discharge*, *elevation versus area*, and *elevation versus width* relationships (which are all paired data objects). This option is typically used when these relationships, derived from hydraulic simulations such as the HEC-RAS model of the channel in question, are available. If the tabular shape is used, multiple curves that describe the way in which discharge, area, and top width change with elevation must be selected. These curves must be defined as elevation–discharge, elevation–area, and elevation–width functions, respectively, in the paired data manager before they can be used in the reach element. These curves must be monotonically increasing. In each of these curves, the *x*-axis defines the elevation while the *y*-axis defines the variable of interest. Elevations must be monotonically increasing. When the tabular shape is selected, no Manning's roughness coefficient values need to be entered.

When one of the standard cross-sectional shapes does not represent the channel geometry, the so-called *8-point* cross-sectional configuration is used. With the 8-point shape, a simplified cross section (which is comprised of eight pairs of distance [*x*] and elevation [*y*] values) with eight *station versus elevation* is selected. The cross section is typically configured to represent the main channel plus the left and right overbank areas. For this reason, separate Manning's *n* values are required for each overbank. Cross-sectional shapes and properties are typically estimated using GIS and field survey data, when available. Several cross sections of a reach are first cut and then are simplified to 8-point paired data to be the representative cross section of the reach (Figure 12.20). The cross section should extend from the channel invert up to the maximum water surface elevation that will be encountered during simulation. The cross section must be created in the Paired Data Manager before it can be used in the reach.

Whatever cross-sectional shape is chosen, if the channel properties or cross-sectional geometries vary significantly along the routing reach, the reach should be subdivided as a series of linked sub-reaches, with the properties of the reach defined separately.

There is an option (not required data) to enter an *invert elevation*. When used, the flow depth computed during routing is added to the invert elevation to compute the stage.

12.9.8 Advantages of the Muskingum–Cunge Method

The strength of the Muskingum–Cunge method is its theoretical basis as an analog of the diffusion wave equation. The Muskingum–Cunge method compares well with full unsteady flow equations over a wide range of flow situations (Ponce, 1983; Brunner and Gorbrecht, 1991). In addition, unlike some other hydrologic routing methods, the parameters are physically based, and therefore this method is good for ungauged channels. The solution of the routing equation is independent of the user-specified computation interval.

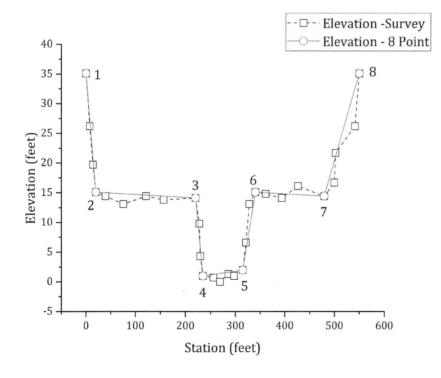

Figure 12.20 Format for describing channel geometry with eight points. Points labeled 3 and 6 represent the left and right banks of the channel at the representative cross section. Points 4 and 5 are within the channel. Points 1 and 2 represent the left overbank and points 7 and 8 represent the right overbank.

A diffusion wave applies to a wider range than competing waves, such as kinematic and dynamic waves (Ponce and Simons, 1977). The numerical properties of this model, including stability and convergence, have been extensively documented, both in theory and in practice (Ponce and Vuppalapati, 2016). The method is very stable with excellent convergence properties for values of Courant number of about 1. This property of very good stability and excellent convergence is due to its grid independence, that is, the property of a numerical scheme to reproduce the same result, regardless of grid resolution.

12.9.9 Limitations of the Muskingum–Cunge Method

The major limitations of the Muskingum–Cunge method are that it cannot account for backwater effects and begins to diverge from a full unsteady flow solution when very rapidly rising hydrographs are routed through flat channel sections (USACE, 1994). If the channel slope (S_0) is less than 0.0004, the error with the Muskingum–Cunge model is less than 5% if

$$TS_0\sqrt{\frac{g}{h_0}} \geq 30 \tag{12.72}$$

where T denotes the hydrograph duration, g is the acceleration due to gravity, and h_0 is reference flow depth.

12.10 HYDRAULIC METHODS OF CHANNEL ROUTING

12.10.1 Kinematic Wave Approximation

There can be instances of flow, such as on steep slopes, where gravitational and frictional forces approach an equilibrium and the slope of the channel and friction slope are nearly equal, even though this rarely occurs in nature. The kinematic wave equation is derived by combining the continuity equation (Eq. 12.7) and momentum equation (Eq. 12.4). In addition, by substituting A for Q through Eq. 12.20 one obtains

$$\frac{\partial A}{\partial t} + \alpha m A^{(m-1)}\frac{\partial A}{\partial x} = q_L \tag{12.73}$$

The coefficient α varies as a function of friction, cross-sectional shape, and bottom slope. The exponent m varies as a function of the type of friction relation and cross-sectional shape. The kinematic wave approximation can also be derived without approximating the Saint-Venant equation. Fundamentally, it comprises the continuity equation and a unique relation between flux and concentration.

Because of the uniform flow assumption, the kinematic wave equation does not allow for hydrograph attenuation (diffusion). It only exhibits simple translation of the hydrograph in time, provided there is no lateral inflow. The unsteadiness of the flow is preserved through the time-dependent term in the continuity equation. However, changes in depth and velocity of flow with respect to time and distance are small in magnitude when compared to the bed slope of the channel. Furthermore, this approximation cannot handle backwater effects. Since all the terms in the momentum equation that describe the propagation of the flood wave upstream (backwater) are neglected, waves or disturbances can only propagate in the downstream direction. The kinematic wave equation does not predict channel storage and any computed attenuation simply results from the approximations in the numerical solution procedures.

The kinematic wave approximation works well for steep-sloped, well-defined channels where a flood wave is gradually varied and there is no flood wave attenuation. It is sometimes applied in urban areas because the routing reaches are generally short and well defined, such as circular pipes, concrete-lined channels, etc.

The adequacy of the kinematic wave approximation is often evaluated using the **kinematic wave number**, K (Woolhiser and Liggett, 1967):

$$K = \frac{S_0 L_0}{h_0 F_0^2} \tag{12.74}$$

where L_0 is the length of reach, S_0 is the bed slope, h_0 is the normal flow depth, and F_0 is the Froude number corresponding to normal flow. The approximation is quite accurate if K is greater than 20, provided F_0 is equal to or more than 0.4. For low values of F_0, K increases as $1/F_0^2$. Thus, for small values of the Froude number, the kinematic wave approximation can be used if $F_0^2 K \geq 5$, which is the ratio of the total fall to the normal depth.

For evaluating the accuracy of the kinematic wave approximation, Singh (1974) derived a generalized kinematic wave number as

$$K_g = \frac{4g S_0 h_0}{C q_0^2} \tag{12.75}$$

in which h_0 is the initial depth, q_0 is the initial flow, such as constant rainfall intensity, and C is Chezy's coefficient. If K_g is greater than 5, the kinematic wave approximation is quite reasonable and for values greater than 10 it is excellent.

Ponce (1989) gives the following criteria to verify the applicability of the kinematic wave approximation:

$$\frac{t_0 S_0 V_0}{h_0} \geq N = 85 \tag{12.76}$$

where t_0 is the time to peak, S_0 is the bed slope, V_0 is the average velocity, and h_0 is the average depth.

12.10.2 Diffusion Wave Approximation

Because of the inclusion of the pressure differential term in the momentum equation, the diffusion equation allows attenuation (diffusion effect) of the flood wave. On mild slopes, such as flat channels with wide floodplains, there is typically a large difference between the channel and friction slope. The pressure gradient term is important for modeling wave propagation and storage effects within the channel for mild slopes and steeply rising and falling hydrographs. Most flood waves travelling in mild slope channels have some physical diffusion. The diffusion equation also allows the specification of a boundary condition at the downstream extremity of the routing reach to account for backwater effects. Due to the exclusion of inertial terms, it is limited to slow to moderately rising flood waves. Flood wave propagation through most natural channels can be described with a diffusion wave model. In most situations, not much accuracy is lost if the inertial terms are neglected and, therefore, the diffusion wave equation is typically sufficient to simulate the downstream propagation of a hydrograph.

Ponce (1989) gives the following criterion to verify the applicability of the diffusion wave approximation:

$$t_0 S_0 \sqrt{\frac{g}{h_0}} \geq M = 15 \quad (12.77)$$

In other words, a wave is termed a diffusion wave if it satisfies the inequality in Eq. 12.72. Morris and Woolhiser (1980) showed that, if $KF_0^2 \geq 5$, both kinematic wave and diffusion wave approximations are quite accurate.

12.10.3 Quasi-Steady Dynamic Wave Approximation

In general, this simplification of the dynamic wave equations is not used in channel routing.

12.10.4 Dynamic Wave Model

The full dynamic wave equation (Eq. 12.3), when combined with the continuity equation, simulates the widest range of flow situations and channel characteristics.

12.10.4.1 Solution Procedures

Numerical Solutions

To this date, there is no closed-form analytical solution of the equations described above. A semi-analytical method, called the method of characteristics, has been developed and used in the past. However, in the present time, the equations are solved by either explicit or implicit finite-difference methods. The equations are solved for incremental times (Δt) and incremental distances (Δx) along the channel.

The unsteady-state flow options in HEC-RAS can be used if hydraulic modeling is required. HEC-HMS currently does not have any option for using hydraulic methods of channel routing

12.10.4.2 Data Requirements

A full unsteady-state hydraulic model requires a series of cross-sectional data at regular intervals along the entire reach of the channel along which an inflow hydrograph from an upstream point will be routed to a downstream point. In addition, a good estimation of Manning's roughness coefficients for the channel and overbank areas is required. One of the many advantages of the hydraulic method of channel routing is that Manning's roughness coefficient is the only parameter in the model that can be used as a calibration parameter. However, in certain cases the calculations can be sensitive to the contraction and expansion coefficients of the channel cross sections.

12.11 SELECTION OF THE ROUTING METHOD

In hydrologic modeling of watersheds, a hydrologic method of channel routing is typically used. Such hydrologic methods are also used in many operational procedures. Even though the Muskingum–Cunge method has a solid hydraulic basis, it is generally classified as one of the hydrologic methods. Both of the two hydrologic methods described above, namely the modified Puls and Muskingum–Cunge methods, have advantages and limitations as noted above. Selection of one method over another depends on the consideration of several hydraulic factors that can have significant effects on flood wave propagation along a channel. These factors are briefly discussed below.

One factor that influences attenuation and lag of a flood hydrograph is the backwater effect caused by dams, bridges, culverts, channel constrictions, and significant tributary inflow at a confluence. The effect of backwater can be included in the hydraulic model to generate storage–discharge relationships to be used in the modified Puls method. However, in this case the downstream conditions giving rise to the backwater effect are time-invariant. Therefore, the modified Puls method, to some extent, is capable of accounting for backwater effect.

Another factor that quite significantly affects downstream translation and attenuation of flood waves arises when the flood flow exceeds the channel capacity and overtops the channel bank and gets stored (ponded) on the floodplain. The effect of floodplain storage can be accounted for by taking into consideration of varying conveyance between the main channel and the overbank areas. Both modified Puls and Muskingum–Cunge methods can account for the effects of floodplain storage.

The channel slope is an extremely important factor in the approximation of the full dynamic wave model by various simplifications since, as the channel slope decreases, the importance of the acceleration terms increases. The Muskingum–Cunge method is more applicable for a wide range of conditions based on channel slope.

Mixed flow regimes, where flow changes from subcritical to supercritical during the movement of a flood wave, can have a noticeable impact on the discharge hydrograph.

The Muskingum–Cunge method is well suited for ungauged streams or where no observed streamflow data are available for model calibration.

A complete hydraulic model, based on the full dynamic wave equation, can account for all of the situations listed above.

12.12 COMPARING STORAGE FROM HEC-HMS AND HEC-RAS

Since channel storage is a function of time-dependent inflow and outflow, steady-state calculations of channel and floodplain storage from HEC-RAS cannot be readily compared to the HEC-HMS outputs. From HEC-HMS outputs of inflows and outflows, storage can be calculated as a function of time using Eq. 12.10. The peak storage calculated then can be compared with HEC-RAS outputs to check the similarities in the order of magnitude of storage.

12.13 CONCLUSIONS

Since the Muskingum method of channel routing was developed in the United States in the 1930s, many other channel routing procedures have been developed. In this chapter we have dealt with three methods, namely the modified Puls method, Muskingum method, and Muskingum–Cunge method in greater detail than others. This is mainly because, of many other methods, these are the most commonly used practical methods that have received wide acceptance by the governmental agencies responsible for the approval of projects where hydrologic modeling is required. Furthermore, these methods have traditionally been used in many operational flood forecasting models even though many flood warning systems of recent years have implemented hydraulic method entailing full dynamic wave models. In addition to practical applications and methods, considerable theoretical developments in the area of channel routing have also progressed as discussed by Perumal and Price (2017), which can be found in Singh (2018).

12.14 EXAMPLES

Example 12.1: The discharge–flow area data for Hickory Creek, near the City of Denton in North Texas, are given in Table 12.1. Show that wave celerity is always greater than the average flow velocity. The discharge–flow area data are obtained from an HEC-RAS model of Hickory Creek.

Solution: The data given in Table 12.1 are plotted in Figure 12.21. A power function in the form of Eq. 12.20 is fitted through the data (in this particular example, the best fit just happens to be perfect).

Differentiating the equation $Q = 0.002 A^{1.647}$ with respect to A, we obtain

$$\frac{dQ}{dA} = (0.002).(1.647) A^{(1.647-1)} = 0.003294 A^{0.647}$$

We evaluate this derivative at a couple of flow areas to obtain the celerity directly.

Table 12.1 *Discharge–flow area data for Hickory Creek (cross section no. 3890)*

Flow area, A (ft^2)	Flow, Q (ft^3/s)
11 304.36	9301.1
20 284.55	24 379.4
24 932.75	34 145.4
28 368.06	42 304.8
31 882.08	51 290.1
40 528.2	76 225.9

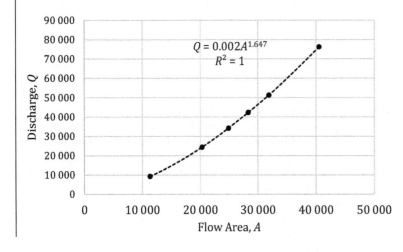

Figure 12.21 Discharge–area plot of the data given in Table 12.1.

(a) $\left.\dfrac{dQ}{dA}\right|_{A=20\,284.55} = (0.003294).(20284.55)^{0.647}$
$= 2.01 \text{ ft/s} = c$

At this point, the average velocity, V is given by:

$$V = \dfrac{Q}{A} = \dfrac{24\,379.4 \text{ ft}^3/\text{s}}{20\,284.55 \text{ ft}^2} = 1.202 \text{ ft/s}$$

Thus, $c > V$.
Also note that in this case, the exponent of Eq. 12.20, $m = 1.647$. Thus, from Eq. 12.22, we obtain

$$c = mV = 1.647 \times 1.202 = 1.98 \text{ ft/s}$$

which is very close to the value 2.01 ft/s obtained above.

(b) $\left.\dfrac{dQ}{dA}\right|_{A=31\,882.08} = (0.003294).(31882.08)^{0.647}$
$= 2.70 \text{ ft/s} = c$

At this point, the average velocity, V is given by:

$$V = \dfrac{Q}{A} = \dfrac{51\,290.1 \text{ ft}^3/\text{s}}{31\,882.08 \text{ ft}^2} = 1.609 \text{ ft/s}$$

Thus, $c > V$.
Also note that in this case, the exponent of Eq. 12.20, $m = 1.647$. Thus, from Eq. 12.22, we obtain

$$c = mV = 1.647 \times 1.609 = 2.65 \text{ ft/s}$$

which is very close to the value 2.70 ft/s obtained above. This can be shown to hold true for any pair of discharge–area values.

Example 12.2: A discharge–storage relationship for a reach of a channel is given in Table 12.2. The inflow hydrograph at the upstream end of the channel reach is given in Table 12.3. Show that, if the modified Puls method of channel routing is used by considering only one step (a single sub-reach), the channel exhibits characteristics of a pure reservoir type of storage. Confirm this observation by running this problem in HEC-HMS.

Solution: The flows in the inflow hydrograph are given in 20-minute intervals. Thus, the computational time step is set at, $\Delta t = 20$ minutes $= 1200$ seconds.

Using the Δt and given discharge(Q)–storage (S) relationship, the $(Q) - [S + (Q\Delta t/2)]$ relationship is developed (Table 12.4) and is plotted in Figure 12.22.

In the calculation steps of the modified Puls method, storage as a function of discharge can be calculated from the best-fit equation shown in Figure 12.22.

$$S = 2638Q - 0.278Q^2$$

Table 12.2 *Discharge–storage relationship*

Discharge (ft^3/s)	Storage (acre-feet)	Storage (ft^3)
0	0	0
282.3	14.55	633 796.545
564.6	25.33	1 103 372.27
1129.2	44.19	1 924 911.98
1411.5	52.09	2 269 035.19
2117.25	71.04	3 094 495.3

However, an inverse function for the $(Q) - [S + (Q\Delta t/2)]$ relationship is more convenient for direct calculations. Since the relationship shown in Figure 12.22 is monotonically increasing, it is a one-to-one relationship and, therefore, such an inverse function exists. This is shown in Figure 12.23.

A second-degree polynomial fits the data almost perfectly. However, for the sake of simplicity, the linear relationship is used:

$$Q = 0.0005 \left[S + \left(\dfrac{Q\Delta t}{2}\right)\right]$$

Calculations of the outflow are presented in Table 12.5.

The inflow–outflow hydrographs are shown in Figure 12.24.

From the plots presented in Figure 12.24, it can be seen that the peak of the outflow hydrograph intersects the recession limb of the outflow hydrograph (at 780 minutes). *This is a characteristic of reservoir-type storage.*

In addition, storage is calculated from the inflows and outflows using Eq.12.10. This is presented in Table 12.6 and Figure 12.25. The results from HEC-HMS run are also presented in Table 12.6

The peak storage (1 645 798 ft^3) occurs at 780 minutes, the time when peak outflow occurs, and inflow equals outflow. These are all characteristics of reservoir-type storage. The same observation can be made from the inflow and outflow hydrographs derived from the HEC-HMS model of this problem. Figure 12.26 shows the inflow–outflow hydrographs from HEC-HMS run of

Table 12.3 *Inflow hydrograph at the upstream end of the channel reach*

Time	Inflow (ft³/s)	Time	Inflow (ft³/s)
00:00	0	18:20	207.6
00:20	0	18:40	190.7
00:40	0	19:00	176.7
01:00	0	19:20	164.4
01:20	0	19:40	153.4
01:40	0	20:00	143.8
02:00	0	20:20	135.3
02:20	0	20:40	127.8
02:40	0	21:00	120.9
03:00	0	21:20	114.9
03:20	0	21:40	109.5
03:40	0	22:00	104.7
04:00	0	22:20	100
04:20	0	22:40	95.5
04:40	0	23:00	91.2
05:00	0	23:20	87.3
05:20	0.3	23:40	83.6
05:40	1.3	00:00	79.8
06:00	2.9	00:20	69.7
06:20	4.9	00:40	55.2
06:40	7.6	01:00	44.4
07:00	11	01:20	36.4
07:20	15	01:40	30.2
07:40	19.6	02:00	25
08:00	25	02:20	20.6
08:20	31.3	02:40	16.9
08:40	39.4	03:00	13.9
09:00	49.6	03:20	11.5
09:20	61.9	03:40	9.5
09:40	76.2	04:00	7.8
10:00	92.5	04:20	6.5
10:20	111.9	04:40	5.4
10:40	136	05:00	4.5
11:00	165.7	05:20	3.8
11:20	206.8	05:40	3.2
11:40	290.5	06:00	2.7
12:00	638.7	06:20	2.2
12:20	1179.5	06:40	1.9
12:40	1357	07:00	1.5
13:00	1178.2	07:20	1.3
13:20	998.2	07:40	1.1
13:40	873.9	08:00	0.9
14:00	784.5	08:20	0.7
14:20	707.5	08:40	0.6
14:40	633.5	09:00	0.5
15:00	566.9	09:20	0.4
15:20	509.4	09:40	0.3
15:40	458.2	10:00	0.2
16:00	411.8	10:20	0.2
16:20	369.8	10:40	0.1
16:40	331.5	11:00	0.1
17:00	298.7	11:20	0.1
17:20	272.3	11:40	0
17:40	248.8	12:00	0
18:00	227.2		

12.14 Examples

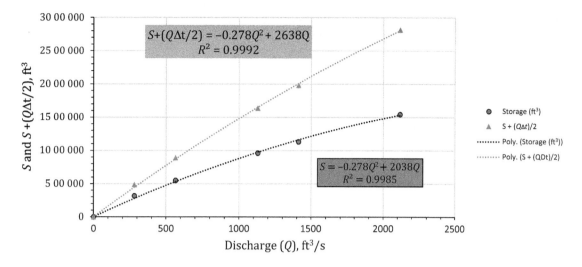

Figure 12.22 S versus Q (Table 12.2) and $(S + Q\Delta t/2)$ versus Q (Table 12.4).

Table 12.4 $(Q) - [S + (Q\Delta t/2)]$ relationship

$S + (Q\Delta t)/2$ (ft^3)	Discharge, Q (ft^3/s)
0	0
803176.55	282.3
1442132.3	564.6
2602432	1129.2
3115935.2	1411.5
4364845.3	2117.25

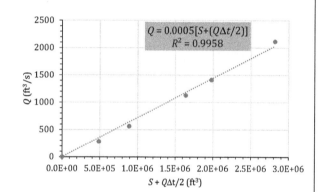

Figure 12.23 Q as a function of $(S+Q\Delta t/2)$.

Table 12.5 *Calculation of outflows from inflows using the S–Q and $(Q) - [S + (Q\Delta t/2)]$ relationships established above in the modified Puls method (number of steps/number of sub-reaches = 1)*

Time (hh:mm)	Time (cumulative min)	Inflow (ft^3/s)	S^j[$f(Q)$]	$S^j - Q\Delta t/2$	$([I^{j-1}+I^j]/2)*\Delta t$	$S^j+(Q\Delta t/2)$	Q^j	$S^j - Q\Delta t/2$
00:00	0.00	0	0.0	0.0			0.0	
00:20	20.00	0	0.0	0.0	0	0.0	0.0	0.0
00:40	40.00	0	0.0	0.0	0	0.0	0.0	0.0
01:00	60.00	0	0.0	0.0	0	0.0	0.0	0.0
01:20	80.00	0	0.0	0.0	0	0.0	0.0	0.0
01:40	100.00	0	0.0	0.0	0	0.0	0.0	0.0
02:00	120.00	0	0.0	0.0	0	0.0	0.0	0.0
02:20	140.00	0	0.0	0.0	0	0.0	0.0	0.0
02:40	160.00	0	0.0	0.0	0	0.0	0.0	0.0
03:00	180.00	0	0.0	0.0	0	0.0	0.0	0.0
03:20	200.00	0	0.0	0.0	0	0.0	0.0	0.0
03:40	220.00	0	0.0	0.0	0	0.0	0.0	0.0

Table 12.5 (cont.)

Time (hh:mm)	Time (cumulative min)	Inflow (ft³/s)	$S^j[f(Q)]$	$S^j - Q\Delta t/2$	$([I^{j-1}+I^j]/2)*\Delta t$	$S^j+(Q\Delta t/2)$	Q^j	$S^j - Q\Delta t/2$
04:00	240.00	0	0.0	0.0	0	0.0	0.0	0.0
04:20	260.00	0	0.0	0.0	0	0.0	0.0	0.0
04:40	280.00	0	0.0	0.0	0	0.0	0.0	0.0
05:00	300.00	0	0.0	0.0	0	0.0	0.0	0.0
05:20	320.00	0.3	0.0	72.0	180	180.0	0.1	72.0
05:40	340.00	1.3	183.4	412.8	960	1032.0	0.5	412.8
06:00	360.00	2.9	1051.5	1173.1	2520	2932.8	1.5	1173.1
06:20	380.00	4.9	2987.9	2341.2	4680	5853.1	2.9	2341.2
06:40	400.00	7.6	5961.9	3936.5	7500	9841.2	4.9	3936.5
07:00	420.00	11	10021.5	6038.6	11160	15096.5	7.5	6038.6
07:20	440.00	15	15367.5	8655.4	15600	21638.6	10.8	8655.4
07:40	460.00	19.6	22017.1	11766.2	20760	29415.4	14.7	11766.2
08:00	480.00	25	29914.0	15410.5	26760	38526.2	19.3	15410.5
08:20	500.00	31.3	39154.8	19676.2	33780	49190.5	24.6	19676.2
08:40	520.00	39.4	49956.5	24838.5	42420	62096.2	31.0	24838.5
09:00	540.00	49.6	63007.4	31295.4	53400	78238.5	39.1	31295.4
09:20	560.00	61.9	79298.5	39278.2	66900	98195.4	49.1	39278.2
09:40	580.00	76.2	99389.3	48855.3	82860	122138.2	61.1	48855.3
10:00	600.00	92.5	123419.4	60030.1	101220	150075.3	75.0	60030.1
10:20	620.00	111.9	151357.4	73068.0	122640	182670.1	91.3	73068.0
10:40	640.00	136	183815.9	88723.2	148740	221808.0	110.9	88723.2
11:00	660.00	165.7	222594.5	107897.3	181020	269743.2	134.9	107897.3
11:20	680.00	206.8	269798.7	132558.9	223500	331397.3	165.7	132558.9
11:40	700.00	290.5	330041.8	172375.6	298380	430938.9	215.5	172375.6
12:00	720.00	638.7	426187.5	291958.2	557520	729895.6	364.9	291958.2
12:20	740.00	1179.5	706644.4	553151.3	1090920	1382878.2	691.4	553151.3
12:40	760.00	1357	1275909.8	830020.5	1521900	2075051.3	1037.5	830020.5
13:00	780.00	1178.2	1814468.0	940456.2	1521120	2351140.5	1175.6	940456.2
13:20	800.00	998.2	2010658.4	898518.5	1305840	2246296.2	1123.1	898518.5
13:40	820.00	873.9	1937406.5	808711.4	1123260	2021778.5	1010.9	808711.4
14:00	840.00	784.5	1775389.6	721500.6	995040	1803751.4	901.9	721500.6
14:20	860.00	707.5	1611333.7	646680.2	895200	1616700.6	808.4	646680.2
14:40	880.00	633.5	1465306.9	580512.1	804600	1451280.2	725.6	580512.1
15:00	900.00	566.9	1332104.1	520300.8	720240	1300752.1	650.4	520300.8
15:20	920.00	509.4	1207579.3	466432.3	645780	1166080.8	583.0	466432.3
15:40	940.00	458.2	1093496.2	418796.9	580560	1046992.3	523.5	418796.9
16:00	960.00	411.8	990507.9	376318.8	522000	940796.9	470.4	376318.8
16:20	980.00	369.8	897002.8	338111.5	468960	845278.8	422.6	338111.5
16:40	1000.00	331.5	811556.5	303556.6	420780	758891.5	379.4	303556.6
17:00	1020.00	298.7	733183.5	272670.6	378120	681676.6	340.8	272670.6
17:20	1040.00	272.3	662251.7	246108.3	342600	615270.6	307.6	246108.3
17:40	1060.00	248.8	600584.8	223507.3	312660	558768.3	279.4	223507.3
18:00	1080.00	227.2	547630.8	203642.9	285600	509107.3	254.6	203642.9
18:20	1100.00	207.6	500721.3	185809.2	260880	464522.9	232.3	185809.2
18:40	1120.00	190.7	458314.3	169915.7	238980	424789.2	212.4	169915.7
19:00	1140.00	176.7	420287.6	156142.3	220440	390355.7	195.2	156142.3
19:20	1160.00	164.4	387155.5	144320.9	204660	360802.3	180.4	144320.9
19:40	1180.00	153.4	358587.3	134000.4	190680	335000.9	167.5	134000.4
20:00	1200.00	143.8	333546.6	124928.1	178320	312320.4	156.2	124928.1
20:20	1220.00	135.3	311458.1	116955.3	167460	292388.1	146.2	116955.3
20:40	1240.00	127.8	291986.9	109926.1	157860	274815.3	137.4	109926.1
21:00	1260.00	120.9	274774.7	103658.4	149220	259146.1	129.6	103658.4

Table 12.5 (cont.)

Time (hh:mm)	Time (cumulative min)	Inflow (ft³/s)	$S^j[f(Q)]$	$S^j-Q\Delta t/2$	$([I^{j-1}+I^j]/2)*\Delta t$	$S^j+(Q\Delta t/2)$	Q^j	$S^j-Q\Delta t/2$
21:20	1280.00	114.9	259 390.7	98 055.4	141 480	245 138.4	122.6	98 055.4
21:40	1300.00	109.5	245 609.1	93 078.2	134 640	232 695.4	116.3	93 078.2
22:00	1320.00	104.7	233 343.9	88 639.3	128 520	221 598.2	110.8	88 639.3
22:20	1340.00	100	222 387.1	84 583.7	122 820	211 459.3	105.7	84 583.7
22:40	1360.00	95.5	212 361.5	80 753.5	117 300	201 883.7	100.9	80 753.5
23:00	1380.00	91.2	202 879.7	77 109.4	112 020	192 773.5	96.4	77 109.4
23:20	1400.00	87.3	193 846.9	73 683.8	107 100	184 209.4	92.1	73 683.8
23:40	1420.00	83.6	185 345.1	70 489.5	102 540	176 223.8	88.1	70 489.5
00:00	1440.00	79.8	177 408.3	67 411.8	98 040	168 529.5	84.3	67 411.8
00:20	1460.00	69.7	169 752.6	62 844.7	89 700	157 111.8	78.6	62 844.7
00:40	1480.00	55.2	158 377.1	55 113.9	74 940	137 784.7	68.9	55 113.9
01:00	1500.00	44.4	139 079.9	45 949.6	59 760	114 873.9	57.4	45 949.6
01:20	1520.00	36.4	116 137.1	37 771.8	48 480	94 429.6	47.2	37 771.8
01:40	1540.00	30.2	95 602.4	31 092.7	39 960	77 731.8	38.9	31 092.7
02:00	1560.00	25	78 787.7	25 685.1	33 120	64 212.7	32.1	25 685.1
02:20	1580.00	20.6	65 145.5	21 218.0	27 360	53 045.1	26.5	21 218.0
02:40	1600.00	16.9	53 856.9	17 487.2	22 500	43 718.0	21.9	17 487.2
03:00	1620.00	13.9	44 415.5	14 386.9	18 480	35 967.2	18.0	14 386.9
03:20	1640.00	11.5	36 560.5	11 850.8	15 240	29 626.9	14.8	11 850.8
03:40	1660.00	9.5	30 128.6	9780.3	12 600	24 450.8	12.2	9780.3
04:00	1680.00	7.8	24 873.7	8064.1	10 380	20 160.3	10.1	8064.1
04:20	1700.00	6.5	20 515.0	6657.6	8580	16 644.1	8.3	6657.6
04:40	1720.00	5.4	16 941.1	5519.1	7140	13 797.6	6.9	5519.1
05:00	1740.00	4.5	14 046.5	4583.6	5940	11 459.1	5.7	4583.6
05:20	1760.00	3.8	11 667.6	3825.4	4980	9563.6	4.8	3825.4
05:40	1780.00	3.2	9739.0	3210.2	4200	8025.4	4.0	3210.2
06:00	1800.00	2.7	8173.4	2700.1	3540	6750.2	3.4	2700.1
06:20	1820.00	2.2	6875.3	2256.0	2940	5640.1	2.8	2256.0
06:40	1840.00	1.9	5745.0	1886.4	2460	4716.0	2.4	1886.4
07:00	1860.00	1.5	4804.1	1570.6	2040	3926.4	2.0	1570.6
07:20	1880.00	1.3	3999.9	1300.2	1680	3250.6	1.6	1300.2
07:40	1900.00	1.1	3311.6	1096.1	1440	2740.2	1.4	1096.1
08:00	1920.00	0.9	2791.8	918.4	1200	2296.1	1.1	918.4
08:20	1940.00	0.7	2339.3	751.4	960	1878.4	0.9	751.4
08:40	1960.00	0.6	1913.9	612.5	780	1531.4	0.8	612.5
09:00	1980.00	0.5	1560.3	509.0	660	1272.5	0.6	509.0
09:20	2000.00	0.4	1296.6	419.6	540	1049.0	0.5	419.6
09:40	2020.00	0.3	1068.9	335.8	420	839.6	0.4	335.8
10:00	2040.00	0.2	855.5	254.3	300	635.8	0.3	254.3
10:20	2060.00	0.2	647.9	197.7	240	494.3	0.2	197.7
10:40	2080.00	0.1	503.7	151.1	180	377.7	0.2	151.1
11:00	2100.00	0.1	384.9	108.4	120	271.1	0.1	108.4
11:20	2120.00	0.1	276.2	91.4	120	228.4	0.1	91.4
11:40	2140.00	0	232.8	60.6	60	151.4	0.1	60.6
12:00	2160.00	0	154.2	24.2	0	60.6	0.0	24.2

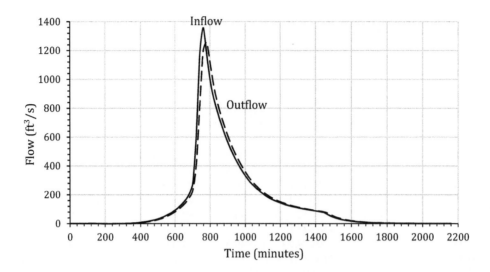

Figure 12.24 Calculated inflow–outflow hydrographs.

Table 12.6 *Storage volumes calculated from inflow and outflow*

Time (cumulative min)	Inflow (ft^3/s)	Outflow (ft^3/s)	Outflow from HEC-HMS run (ft^3/s)	Storage (ft^3)
0	0	0.0	0	0.0
20	0	0.0	0	0.0
40	0	0.0	0	0.0
60	0	0.0	0	0.0
80	0	0.0	0	0.0
100	0	0.0	0	0.0
120	0	0.0	0	0.0
140	0	0.0	0	0.0
160	0	0.0	0	0.0
180	0	0.0	0	0.0
200	0	0.0	0	0.0
220	0	0.0	0	0.0
240	0	0.0	0	0.0
260	0	0.0	0	0.0
280	0	0.0	0	0.0
300	0	0.0	0	0.0
320	0.3	0.1	0.1	126.0
340	1.3	0.5	0.4	722.4
360	2.9	1.5	1.1	2053.0
380	4.9	2.9	2.3	4097.2
400	7.6	4.9	4	6888.9
420	11	7.5	6.2	10 567.5
440	15	10.8	9.1	15 147.0
460	19.6	14.7	12.5	20 590.8
480	25	19.3	16.6	26 968.3
500	31.3	24.6	21.5	34 433.3
520	39.4	31.0	27.4	43 467.3
540	49.6	39.1	34.6	54 766.9
560	61.9	49.1	43.5	68 736.8
580	76.2	61.1	54.3	85 496.7
600	92.5	75.0	67	105 052.7
620	111.9	91.3	81.8	127 869.1
640	136	110.9	99.6	155 265.6
660	165.7	134.9	121.2	188 820.3
680	206.8	165.7	148.6	231 978.1
700	290.5	215.5	190.8	301 657.2

Table 12.6 (*cont.*)

Time (cumulative min)	Inflow (ft³/s)	Outflow (ft³/s)	Outflow from HEC-HMS run (ft³/s)	Storage (ft³)
720	638.7	364.9	312.4	510 926.9
740	1179.5	691.4	635.3	968 014.8
760	1357	1037.5	1004.9	1 452 535.9
780	1178.2	1175.6	1162.1	1 645 798.4
800	998.2	1123.1	1115.2	1 572 407.3
820	873.9	1010.9	1010.6	1 415 244.9
840	784.5	901.9	904.7	1 262 626.0
860	707.5	808.4	812	1 131 690.4
880	633.5	725.6	729.4	1 015 896.2
900	566.9	650.4	653.9	910 526.5
920	509.4	583.0	586.3	816 256.6
940	458.2	523.5	530	732 894.6
960	411.8	470.4	479.6	658 557.9
980	369.8	422.6	432.5	591 695.1
1000	331.5	379.4	389.1	531 224.1
1020	298.7	340.8	349.9	477 173.6
1040	272.3	307.6	315.8	430 689.4
1060	248.8	279.4	286.5	391 137.8
1080	227.2	254.6	265.2	356 375.1
1100	207.6	232.3	245	325 166.0
1120	190.7	212.4	225.7	297 352.4
1140	176.7	195.2	208	273 249.0
1160	164.4	180.4	192.2	252 561.6
1180	153.4	167.5	178.1	234 500.6
1200	143.8	156.2	165.7	218 624.3
1220	135.3	146.2	154.7	204 671.7
1240	127.8	137.4	144.9	192 370.7
1260	120.9	129.6	136.2	181 402.3
1280	114.9	122.6	128.5	171 596.9
1300	109.5	116.3	121.6	162 886.8
1320	104.7	110.8	115.5	155 118.7
1340	100	105.7	109.9	148 021.5
1360	95.5	100.9	104.8	141 318.6
1380	91.2	96.4	100	134 941.4
1400	87.3	92.1	95.5	128 946.6
1420	83.6	88.1	91.2	123 356.6
1440	79.8	84.3	87.2	117 970.7
1460	69.7	78.6	81.9	109 978.3
1480	55.2	68.9	73.7	96 449.3
1500	44.4	57.4	63.6	80 411.7
1520	36.4	47.2	53.8	66 100.7
1540	30.2	38.9	45.2	54 412.3
1560	25	32.1	37.7	44 948.9
1580	20.6	26.5	31.4	37 131.6
1600	16.9	21.9	26.1	30 602.6
1620	13.9	18.0	21.6	25 177.1
1640	11.5	14.8	17.8	20 738.8
1660	9.5	12.2	14.7	17 115.5
1680	7.8	10.1	12.2	14 112.2
1700	6.5	8.3	10.1	11 650.9
1720	5.4	6.9	8.3	9658.4
1740	4.5	5.7	6.9	8021.3
1760	3.8	4.8	5.7	6694.5
1780	3.2	4.0	4.8	5617.8
1800	2.7	3.4	4	4725.1
1820	2.2	2.8	3.3	3948.1
1840	1.9	2.4	2.8	3301.2
1860	1.5	2.0	2.3	2748.5

Table 12.6 (cont.)

Time (cumulative min)	Inflow (ft³/s)	Outflow (ft³/s)	Outflow from HEC-HMS run (ft³/s)	Storage (ft³)
1880	1.3	1.6	1.9	2275.4
1900	1.1	1.4	1.6	1918.2
1920	0.9	1.1	1.3	1607.3
1940	0.7	0.9	1.1	1314.9
1960	0.6	0.8	0.9	1072.0
1980	0.5	0.6	0.8	890.8
2000	0.4	0.5	0.6	734.3
2020	0.3	0.4	0.5	587.7
2040	0.2	0.3	0.4	445.1
2060	0.2	0.2	0.3	346.0
2080	0.1	0.2	0.3	264.4
2100	0.1	0.1	0.2	189.8
2120	0.1	0.1	0.2	159.9
2140	0	0.1	0.1	106.0
2160	0	0.0	0.1	42.4

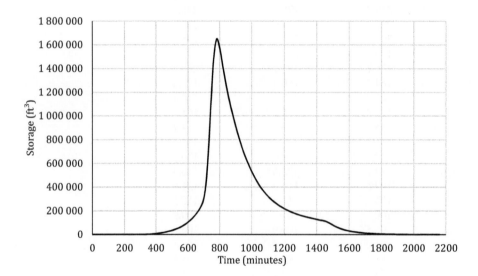

Figure 12.25 Storage as a function of time.

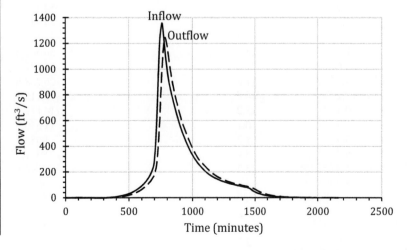

Figure 12.26 HEC-HMS results.

this problem (number of sub-reaches = 1) and here also the peak outflow occurs at a time of the recession of the inflow giving it the characteristics of a reservoir-type, not a channel-like, situation. However, as soon as the number of sub-reaches is greater than one, the peak outflow occurs after the time when inflow equals outflow, and the *attenuation decreases*.

Example 12.3: An urban drainage area had a natural stream that was a source of frequent flooding in a neighborhood that was highly developed. The local municipality evaluated several alternatives for flood management and finally adopted a measure that entailed channel improvement to provide increased conveyance. Figure 12.27 shows the drainage area, which can be divided into three catchments contributing flows at three locations of the stream that is broken into two reaches (Reach A and Reach B). Use HEC-HMS to develop the modified Puls method to route the hydrographs through the two reaches of the improved channel for a rainfall event for which the data are given in Figure 12.28. As one of the inputs to the HEC-HMS model, develop the storage–discharge relationships for the reaches from the HEC-RAS model of the channel shown in Figure 12.29. Figures 12.30 and 12.31 show the typical cross sections of the two reaches after channel improvements. Develop the procedures that are used to determine the number of steps or sub-reaches necessary in the modified Puls method in HEC-HMS from multiple HEC-RAS analysis or the so-called balancing process.

Solution: A hydrologic model of the drainage area is set up in HEC-HMS using the parameters (curve numbers and lag times) shown in Figure 12.27 for computation of effective rainfall and NRCS dimensionless unit hydrographs. The rainfall pattern given in Figure 12.28, constructed from the rainfall data given in Table 12.7, is converted to a percentage curve to input in the meteorological model of HEC-HMS.

Direct runoff from Catchment 1 goes to the upstream of Reach A, that from Catchment 2 goes to Junction 1, and that from Catchment 3 goes to the outlet. Thus, at Junction 1, direct runoff from Catchment 2 is combined with the routed flows coming from Reach A and this combined flow is routed through Reach B to the drainage area outlet where it is combined with

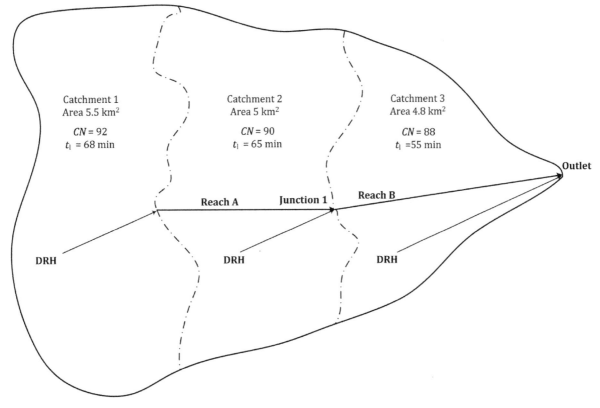

Figure 12.27 Drainage area.

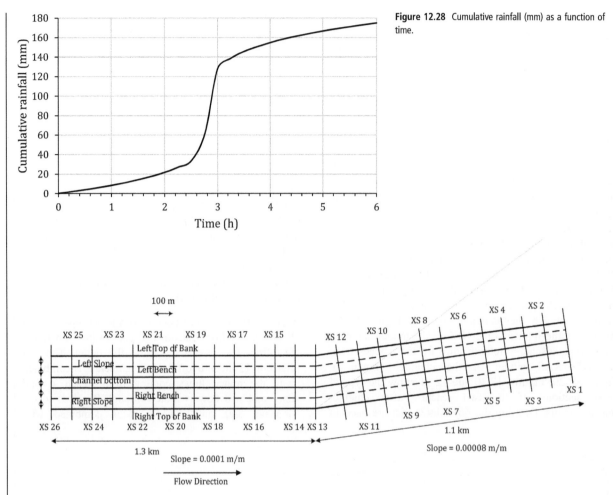

Figure 12.28 Cumulative rainfall (mm) as a function of time.

Figure 12.29 HEC-RAS cross-sectional layout.

direct runoff from Catchment 3. Since the storage–discharge relationships in the channel reaches are unknown at the beginning, both reaches are assigned to apply the lag routing method for the first HEC-HMS run. The only data required for the lag routing method in HEC-HMS is the lag time for the reach, which is provided by an educated guess to get the DRHs from the catchments. The DRH from a catchment is not affected by the choice of this lag time for a reach.

The cross-sectional layout in the HEC-RAS model is shown in Figure 12.29. The cross sections are cut at regular 100-m intervals and perpendicular to the direction of flow. Each of the cross sections in Reach A, which is 1.3 km long, has the geometry shown in Figure 12.30 and the elevations change from one cross section to another for this reach of the channel to have a slope of 0.0001 m/m. The municipality has created hike-and-bike trails on the benches as part of the linear green belt through which the channel flows. Manning's n values in the channel and the overbank areas are assigned accordingly as shown in Figure 12.30.

Each of the cross sections in Reach B, which is 1.1 km long, has the geometry shown in Figure 12.31. In this reach the channel is wider and the slope is milder than Reach A, having a uniform slope of 0.00008 m/m. The hike-and-bike trails on the benches continue in this reach also and for this reason there is no change in the Manning's n values in the channel and the overbank areas.

In the HEC-RAS model using the cross section (XS) layout shown in Figure 12.29, XS 26 represents the upstream end of Reach A and XS 12 represents the upstream end of Reach B.

A range of flows are needed to run the multi-profile steady-state HEC-RAS model and, for this reason, the event peak flow Q_1 (profile PF 1.0) is calculated from

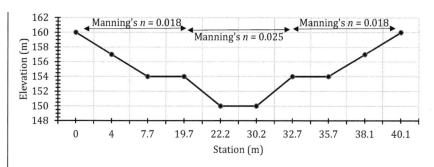

Figure 12.30 Typical section of Reach A (upstream).

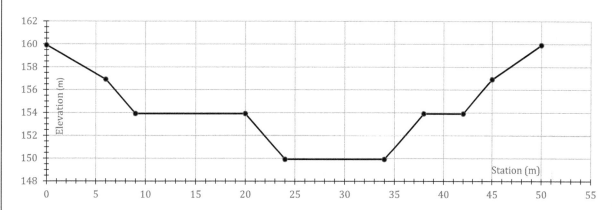

Figure 12.31 Typical section Reach B (downstream).

Table 12.7 *Rainfall data*

Time (h)	Cumulative precipitation (mm)
0.00	0.00
0.25	1.98
0.50	3.96
0.75	6.07
1.00	8.42
1.25	11.09
1.50	14.12
1.75	17.66
2.00	21.80
2.25	27.03
2.50	33.69
2.75	60.83
3.00	127.81
3.25	138.42
3.50	145.65
3.75	150.62
4.00	154.81
4.25	158.36
4.50	161.50
4.75	164.25
5.00	166.70
5.25	168.94
5.50	171.07
5.75	173.09
6.00	175.00

HEC-HMS using a computational time interval of 1 minute. Six other flows $0.25 \times Q_1$ (profile PF 0.25), $0.5 \times Q_1$ (profile PF 0.5), $0.75 \times Q_1$ peak flows (profile PF 0.75), $1.5 \times Q_1$ peak flows (profile PF 1.5), and $1.75 \times Q_1$ peak flows (profile PF 1.75), are used in the steady-state HEC-RAS model. Table 12.8 shows the initial set of flows.

After initial flows for the multi-profile HEC-RAS steady flow model are established (Table 12.8), the HEC-RAS model is run, and the Output Summary Table (Standard Table 1) is opened to view the results for each cross section. Two additional variables computed by the program but now shown in the Standard Table 1 are added by clicking Options —>Define Table. These two additional variables that are now added to the table columns are Volume (1000 m^3) and Trvl Time Ave (h). The volume variable is the cumulative volume of water under each profile from the very downstream end cross section to the current cross section and Trvl Time Avg is the travel time based on the average velocity of the entire cross section.

The outputs required for the present purpose are summarized in Excel as Table 12.9 to develop the

Table 12.8 *Initial flows (m³/s) used in the HEC-RAS after flows from HEC-HMS are derived for PF 1.0*

Source	XS	Profile number						
		1	2	3	4	5	6	7
		PF 1.0	PF 0.25	PF 0.50	PF 0.75	PF 1.25	PF 1.5	PF 1.75
Catchment 1	26	108.1	27.0	54.1	81.1	135.1	162.2	189.2
Junction 1	12	203.9	51.0	102.0	152.9	254.9	305.9	356.8
Outlet	1	292.9	73.2	146.5	219.7	366.1	439.4	512.6

Table 12.9 *Summary of HEC-RAS results for upstream and downstream cross sections of reach A and reach B, respectively, using the first trial run of discharge values obtained from HEC-HMS initial run where channel routing is done by simple lag method*

Reach A	Q (m³/s)	Volume (1000 m³)	Average travel time (h)	Storage in Reach A (1000 m³)	Travel time from upstream to downstream (h)
XS 26	108.1	260.62	0.51	129.93	0.33
Upstream end	27.0	81.92	0.64	40.42	0.41
Reach A	54.1	136.90	0.54	70.15	0.36
	81.1	201.72	0.53	101.76	0.35
	135.1	299.29	0.47	149.53	0.31
	162.2	337.17	0.44	168.84	0.29
	189.2	372.69	0.42	187.00	0.27
XS 13	108.1	130.69	0.18		
Downstream	27.0	41.50	0.23		
Reach A	54.1	66.75	0.18		
	81.1	99.96	0.18		
	135.1	149.76	0.16		
	162.2	168.33	0.15		
	189.2	185.69	0.15		

Reach B	Q (m³/s)	Volume (1000 m³)	Average travel time (h)	Storage in Reach B (m³)	Travel time from upstream to downstream (h)
XS 12	203.9	120.22	0.16	120.22	0.16
Upstream end	51.0	38.04	0.20	38.04	0.20
Reach B	102.0	61.01	0.16	61.01	0.16
	152.9	91.71	0.16	91.71	0.16
	254.9	137.82	0.15	137.82	0.15
	305.9	154.94	0.14	154.94	0.14
	356.8	170.93	0.13	170.93	0.13
XS 1	292.9	0	0		
Downstream	73.2	0	0		
Reach B	146.5	0	0		
	219.7	0	0		
	366.1	0	0		
	439.4	0	0		
	512.6	0	0		

storage–discharge relationships (paired data in HEC-HMS) for Reach A (XS 26 to XS 13) and Reach B (XS 12 to XS 1). The travel time from upstream XS to downstream XS is used to calculate the number of sub-reaches in the modified Puls routing method in the second HEC-HMS model run.

The storage–discharge data shown in Table 12.9 for Reach A and Reach B are now used in HEC-HMS as paired data sets, which must be entered starting from 0 and as monotonically increasing functions. The routing methods for Reach A and Reach B, are changed to the modified Puls method and the HEC-HMS model is re-run to get another set of peak flows for XS 26, XS 12, and XS1.

From the calculations presented in Table 12.9, the average travel time for water in Reach A (from XS 26 to XS 13) is 19.9 minutes. Thus, the approximate travel time for the flood wave is: $T_{FW} = 19.88/1.5 = 13.26$ min. If the computational time interval (Δt_{Model}) in HEC-HMS is set to 5 minutes then the number of sub-reaches required for Reach A is: $N_s = 13.26$ min/5 min $= 2.65$. Similarly, the average travel time for water in Reach B (from XS 12 to XS 1), is 9.4 minutes. Thus, the approximate travel time for the flood wave in this reach is: $T_{FW} = 9.43/1.5 = 6.29$ min. For a computational time interval (Δt_{Model}) in HEC-HMS of 5 minutes, the number of sub-reaches required for Reach B is: $N_s = 9.43$ min/5 min $= 1.26$. Thus, in the modified Puls method, the number of sub-reaches in Reach A and B are set to 2 and 1, respectively. The flows now obtained for the HEC-RAS model are given in Table 12.10 (the computational time interval used in this run is also 5 minutes).

With these revised set of flow values, a new set of storage–discharge relationships are obtained as shown in Table 12.11.

The third HEC-HMS run uses the storage–discharge values given in Table 12.11. Again, from the consideration of the travel time and computational time interval, the number of sub-reaches for Reach A and B are 2 and 1, respectively. The flows now obtained are given in Table 12.12 (the computational time interval used in this run is still 5 minutes).

Table 12.10 *Flows (m^3/s) used in the HEC-RAS after flows from HEC-HMS are derived for PF 1.0 using the first trial of the modified Puls method of channel routing*

		Profile number						
		1	2	3	4	5	6	7
Source	XS	PF 1.0	PF 0.25	PF 0.50	PF 0.75	PF 1.25	PF 1.5	PF 1.75
Catchment 1	26	106.3	26.6	53.2	79.7	132.9	159.5	186.0
Junction 1	12	192	48.0	96.0	144.0	240.0	288.0	336.0
Outlet	1	267.8	67.0	133.9	200.9	334.8	401.7	468.7

Table 12.11 *Summary of HEC-RAS results for upstream and downstream cross sections of Reach A and Reach B, respectively, using the second run of discharge values obtained from HEC-HMS run using the modified Puls method of channel routing*

Reach A	Q (m^3/s)	Volume (1000 m^3)	Average travel time (h)	Storage in reach A (1000 m^3)	Travel time from upstream to downstream (h)
XS 21	106.3	242.37	0.49	121.72	0.31
Upstream end	26.6	78.34	0.64	38.88	0.41
Reach A	53.2	127.1	0.52	64.96	0.34
	79.7	187.87	0.51	95.74	0.33
	132.9	283.78	0.46	142.21	0.29
	159.5	319.61	0.43	160.47	0.27
	186	352.99	0.41	177.56	0.26
XS 13	106.3	120.65	0.18		
Downstream Reach A	26.6	39.46	0.23		
	53.2	62.14	0.18		
	79.7	92.13	0.18		
	132.9	141.57	0.17		
	159.5	159.14	0.16		
	186	175.43	0.15		

Table 12.11 (cont.)

Reach B	Q (m³/s)	Volume (1000 m³)	Average travel time (h)	Storage in reach B (m³)	Travel time from upstream to downstream (h)
XS 12	192	110.87	0.16	110.87	0.16
Upstream end	48	36.15	0.21	36.15	0.21
Reach B	96	56.86	0.16	56.86	0.16
	144	84.39	0.16	84.39	0.16
	240	130.22	0.15	130.22	0.15
	288	146.42	0.14	146.42	0.14
	336	161.43	0.13	161.43	0.13
XS 1	267.8	0	0		
Downstream	67	0	0		
Reach B	133.9	0	0		
	200.9	0	0		
	334.8	0	0		
	401.7	0	0		
	468.7	0	0		

Table 12.12 *Flows (m³/s) from HEC-HMS are derived for PF 1.0 using the second run of the modified Puls method of channel routing*

		Profile number						
		1	2	3	4	5	6	7
Source	XS	PF 1.0	PF 0.25	PF 0.50	PF 0.75	PF 1.25	PF 1.5	PF 1.75
Catchment 1	26	106.3	26.6	53.2	79.7	132.9	159.5	186.0
Junction 1	12	193	48.3	96.5	144.8	241.3	289.5	337.8
Outlet	1	269.6	67.4	134.8	202.2	337.0	404.4	471.8

Note that in this run the flow values (Table 12.12) at Junction 1 and Outlet are not significantly different from those given in Table 12.10. So, the iterative process can be terminated after this second run. However, if these values were significantly different, the steady flow data in the HEC-RAS model would be updated with these new peak flows from HEC-HMS and the above steps would be repeated to re-run the HEC-RAS model to get a new storage–discharge relationship and travel time. The updated storage (volume) and travel time to calculate the number of sub-reaches would be used to recalculate the storage–discharge paired data and the number of sub-reaches.

The iterative process of running HEC-HMS and HEC-RAS models continues until the calculated peak flows at flow change locations converge to certain values (for example, the differences between two consecutive runs are less than 5%,). The most up-to-date storage–discharge paired data are used as the final results. This process of repeating the above steps is also called cycling or balancing.

Example 12.4: Conduct dynamic wave routing for the flows through reaches A and B given in Example 12.3 above and compare the results with the results obtained from the final modified Puls method. Why is the Muskingum–Cunge method not applicable in this case?

Solution: A routing calculation, using the dynamic wave model, is performed in the unsteady-state HEC-RAS model, which solves the full form of the Saint-Venant equations by an implicit finite-difference approximation using a space–time grid somewhat similar to that shown in Figure 12.18. The cross-sectional geometries and layout in this model are the same as in the steady-state model used in Example 12.3. The major changes in the unsteady-state model are as follows.

The DRH from Catchment 1 is input as the flow hydrograph upstream boundary condition at XS 26. The DRH from Catchment 2 is entered as an internal boundary or lateral inflow at XS 12. The DRH from

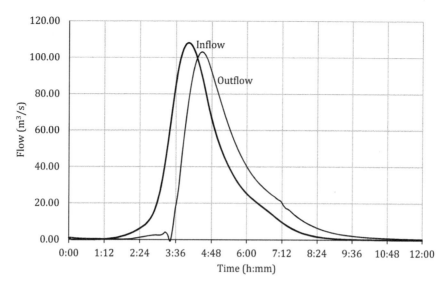

Figure 12.32 Inflow–outflow hydrographs in Reach A (dynamic wave model).

Catchment 3 is entered as another internal boundary or lateral inflow at XS 2, which is 100 m upstream of the downstream end of the channel. The downstream boundary condition at XS 1 is set as the normal depth with a friction slope equaling the channel bed slope. Note that the DRH from Catchment 3 cannot be assigned as a downstream boundary flow hydrograph at XS 1 because the time-dependent flows at this point result from combined contributions from Catchment 3 and Reach B.

In addition to specifying the upstream, downstream, and internal boundary conditions, the initial conditions should be also specified for the numerical analysis. Ideally, the initial condition at each cross section is zero flow (assuming a dry channel before the storm event). However, such a condition creates too much numerical instability at the beginning of the computed hydrographs at each cross section. A trial-and-error solution shows that an initial flow of 10 m^3/s at each of the cross sections eliminates all numerical instabilities. However, 10 m^3/s can be viewed as too high for an initially dry channel. For this reason, an initial flow of 1 m^3/s is assigned at each of the cross sections, in spite of the persistence of some numerical instabilities at times as the hydrographs start to rise; but these are neglected for the present purpose.

The computational time interval is set at 1 minute to optimize the run time and numerical stability but output at desired locations is obtained at 5-minute intervals so that the results can be directly compared with those obtained from HEC-HMS using the modified Puls method.

From the setup of the unsteady-state HEC-RAS model described above, it should be evident that the results of dynamic wave routing at Reach A can be seen from the flow hydrographs at XS 26 and XS 13. These flow hydrographs are shown in Figure 12.32.

In this case the peak of inflow hydrograph is 108.03 m^3/s, occurring at 04:00. The peak of the outflow hydrograph is 102.99 m^3/s, occurring at 04:30. Therefore, according to dynamic wave modeling, the attenuation in this reach is 5.04 m^3/s and the lag is 30 minutes.

Figure 12.33 shows the inflow and outflow hydrographs for Reach A using the modified Puls method in HEC-HMS.

In the case of routing by the modified Puls method, the peak of the inflow hydrograph is 106.32 m^3/s, occurring at 04:00. The peak of the outflow hydrograph is 102.18 m^3/s, occurring at 04:20 h. Therefore, according to the modified Puls method, the attenuation in this reach is 4.14 m^3/s and the lag is 20 minutes.

For Reach B, in the unsteady-state HEC-RAS model, since XS 12 has lateral inflow from Catchment 2, the flow at the next downstream cross section, XS 11, is compared with the flow at XS 3, which is just upstream of XS 2 that receives the lateral inflow from Catchment 3. The hydrographs at these two cross sections are shown in Figure 12.34.

Figure 12.35 shows the inflow and outflow hydrographs for Reach B using the modified Puls method in HEC-HMS.

In the case of Reach B, the modified Puls method shows peak inflow of 193.05 m^3/s at 04:10 and peak outflow of 188.40 m^3/s at 04:20, giving an attenuation of 4.65 m^3/s and lag of 10 minutes. However, the results of

498 12 Channel Routing

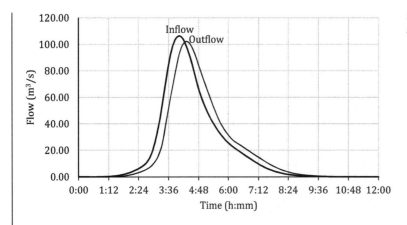

Figure 12.33 Inflow–outflow hydrographs in Reach A (modified Puls routing).

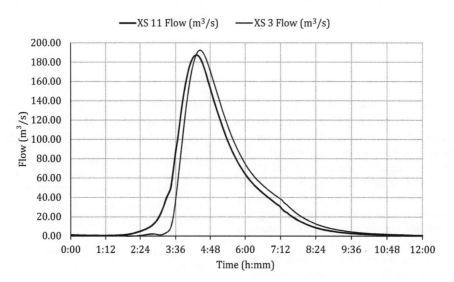

Figure 12.34 Inflow–outflow hydrographs in Reach B (dynamic wave model).

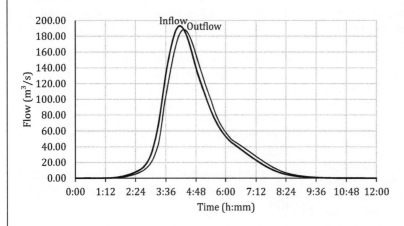

Figure 12.35 Inflow–outflow hydrographs in Reach B (modified Puls routing).

dynamic wave models at cross sections XS 11 and XS 3 show that the peak discharge at XS 11 is 187.32 m³/s at 04:20, while the peak discharge at XS 3 is 192.01 m³/s at 04:25. This clearly shows the effect of downstream boundary conditions that was not accounted for in the hydrologic method.

In this example, the channel bed slope is extremely mild (<0.0004). For this reason, the applicability of Muskingum–Cunge method is tested by taking the hydrograph duration of 10 h 35 min (38 100 seconds)

at a reference depth of 4.48 m taken from HEC-RAS results for $0.5Q_1$ (PF 0.5) noted above.

$$TS_0\sqrt{\frac{g}{h_0}} = (38100 \text{ s})(0.0001)\sqrt{\frac{9.81 \text{ m/s}^2}{4.48 \text{ m}}} = 5.64$$

Since this is less than 30, the Muskingum–Cunge method is not applicable for this channel. The hydrograph is rapidly rising.

Example 12.5: An inflow hydrograph at the upstream end of a channel is given in Table 12.13. After the passage of the flood wave, inflow from 20 hours continues to be the baseflow of 10 m³/s. The hydraulic data of the channel are given as follows: reach length = 18 km, top width of water surface at peak flow = 25.3 m, channel slope = 0.001; a celerity of 2 m/s is assumed. Use a linear solution of the Muskingum–Cunge method of channel routing to show that, during the rising flood wave, discharge decreases downstream, whereas during the receding flood wave, discharge increases downstream.

Solution:

Step 1: Either derive or estimate celerity, c. In this example, celerity is given as 2 m/s.

Step 2: Select appropriate values of Δx and Δt.
Since the river reach is 18 km, we will calculate outflow hydrographs at three points downstream.

Table 12.13 *Inflow hydrograph*

Time (h)	Inflow (m³/s)
0	10
1	12
2	18
3	28.5
4	50
5	78
6	107
7	134.5
8	147
9	150
10	146
11	129
12	105
13	78
14	59
15	45
16	33
17	24
18	17
19	12
20	10

In the first trial, set $\Delta x = 6$ km $= 6000$ m.
In the first trial set $\Delta t = \frac{\Delta x}{c} = \frac{6000 \text{ m}}{2 \text{ m/s}} = 3000$ s
With this value of Δt, check for Δx whether it satisfies Eq.12.66 (use peak inflow as the reference flow, Q_0).

$$\frac{1}{2}\left(c\Delta t + \frac{Q_0}{BS_0 c}\right)$$
$$= \frac{1}{2}\left(2 \text{ m/s} \times 3000 \text{ s} + \frac{150 \text{ m}^3/\text{s}}{25.3 \text{ m} \times 0.001 \times 2 \text{ m}^{-1}}\right)$$
$$= 4482.2 \text{ m}$$

Since this quantity is greater than the previously selected Δx, we reset $\Delta x = 3000$ m.
With this $\Delta x = 3000$ m, Δt should be $\leq \Delta x/c$ (Courant condition). We reset $\Delta t = 130$ second. With this new Δt,

$$\frac{1}{2}\left(c\Delta t + \frac{Q_0}{BS_0 c}\right)$$
$$= \frac{1}{2}\left(2 \text{ m/s} \times 1300 \text{ s} + \frac{150 \text{ m}^3/\text{s}}{25.3 \text{ m} \times 0.001 \times 2 \text{ m}^{-1}}\right)$$
$$= 2782 \text{ m} \approx 3000 \text{ m}$$

Thus, we use

$\Delta x = 3000$ m

$\Delta t = 1300$ s

Step 3: Calculate K (Eq. 12.56)

$$K = \frac{\Delta x}{c} = \frac{3000 \text{ m}}{2 \text{ ms}^{-1}} = 1500 \text{ s}$$

Step 4: Calculate X (Eq. 12.57)
Use peak discharge as the reference flow.

$$X = \frac{1}{2}\left(1 - \frac{150 \text{ m}^3/\text{s}}{25.3 \text{ m} \times 0.001 \times 2\text{m}^{-1} \times 300 \text{ m}}\right) = 0.00593$$

Step 5: Calculate routing coefficients, C_1, C_2, C_3, using Eqs. 12.48–12.50. The results are:

$$C_1 = \frac{\frac{\Delta t}{K} - 2X}{2(1-X) + \frac{\Delta t}{K}}$$

$$= \frac{\frac{1300 \text{ s}}{1500 \text{ s}} - 2 \times 0.00593}{2 \times (1 - 0.00593) + \frac{1300 \text{ s}}{1500 \text{ s}}} = 0.299428$$

$$C_2 = \frac{\frac{\Delta t}{K} + 2X}{2(1-X) + \frac{\Delta t}{K}}$$

$$= \frac{\frac{1300 \text{ s}}{1500 \text{ s}} + 2 \times 0.00593}{2 \times (1 - 0.00593) + \frac{1300 \text{ s}}{1500 \text{ s}}} = 0.307735$$

$$C_3 = \frac{2(1-X) - \frac{\Delta t}{K}}{2(1-X) + \frac{\Delta t}{K}}$$

$$= \frac{2 \times (1 - 0.00593) - \frac{1300 \text{ s}}{1500 \text{ s}}}{2 \times (1 - 0.00593) + \frac{1300 \text{ s}}{1500 \text{ s}}} = 0.392874$$

Step 6: Now calculate outflows as a function of space and time using Eq. 12.55. The results are given in Table 12.14.

The calculated hydrographs as a function of space are shown in Figure 12.36.

Table 12.14 *Results of routing calculations (flows in m^3/s) using the Muskingum–Cunge method*

Time (h)	Inflow [0 km]	Outflow [3 km]	Outflow [6 km]	Outflow [9 km]	Outflow [12 km]	Outflow [15 km]	Outflow [18 km]
0	10	10.0	10.0	10.0	10.0	10.0	10.0
1	12	10.6	10.2	10.1	10.0	10.0	10.0
2	18	13.2	11.2	10.4	10.2	10.1	10.0
3	28.5	19.3	14.3	11.8	10.7	10.3	10.1
4	50	31.3	20.9	15.3	12.4	11.1	10.5
5	78	51.0	33.1	22.4	16.3	13.1	11.4
6	107	76.1	51.5	34.4	23.6	17.2	13.7
7	134.5	103.1	74.5	51.7	35.3	24.6	18.0
8	147	125.9	98.7	72.8	51.6	36.0	25.4
9	150	139.6	119.3	94.7	71.0	51.3	36.4
10	146	144.7	133.2	113.8	91.1	69.3	50.8
11	129	140.4	138.9	127.3	108.9	87.9	67.6
12	105	126.3	135.6	133.3	121.9	104.5	84.9
13	78	105.3	123.7	131.1	128.2	116.9	100.5
14	59	83.0	105.8	121.3	127.0	123.4	112.4
15	45	64.2	86.4	106.1	119.0	123.2	119.0
16	33	49.0	68.4	88.7	105.9	116.7	119.6
17	24	36.6	52.9	71.7	90.4	105.5	114.5
18	17	26.8	40.1	56.4	74.5	91.6	104.9
19	12	19.4	29.8	43.4	59.6	76.7	92.4
20	10	14.3	21.9	32.8	46.6	62.5	78.6
21	10	11.7	16.5	24.6	35.8	49.6	64.9
22	10	10.7	13.3	18.7	27.2	38.6	52.3
23	10	10.3	11.6	14.9	20.9	29.8	41.4
24	10	10.1	10.7	12.6	16.6	23.1	32.4
25	10	10.0	10.3	11.4	13.8	18.3	25.3
26	10	10.0	10.1	10.7	12.1	15.1	20.1
27	10	10.0	10.1	10.3	11.1	13.0	16.4
28	10	10.0	10.0	10.2	10.6	11.7	13.9
29	10	10.0	10.0	10.1	10.3	10.9	12.4
30	10	10.0	10.0	10.0	10.2	10.5	11.4

Table 12.15 *Outflows (m³/s) as function of distance at various times during advancing flood*

Distance (km)	t = 6 h	t = 7 h	t = 8 h	t = 9 h
0	107.0	134.5	147.0	150.0
3	76.1	103.1	125.9	139.6
6	51.5	74.5	98.7	119.3
9	34.4	51.7	72.8	94.7
12	23.6	35.3	51.6	71.0
15	17.2	24.6	36.0	51.3
18	13.7	18.0	25.4	36.4

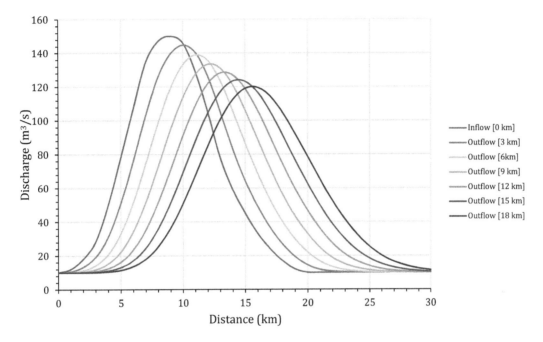

Figure 12.36 Hydrographs at different distances using the Muskingum–Cunge method. (A black and white version of this figure will appear in some formats. For the color version, please refer to the plate section.).

Step 7: Now time slices of discharges at various distances are taken for four time periods during the rising stage of the hydrographs (advancing flood). These are presented in Table 12.15 and are graphically represented in Figure 12.37.

Step 8: Now time slices of discharges at various distances are taken for four time periods during the falling stage of the hydrographs (receding flood). These are presented in Table 12.16 and are graphically represented in Figure 12.38.

Table 12.16 *Outflows (m³/s) as a function of distance at various times during receding flood*

Distance (km)	$t = 15$ h	$t = 17$ h	$t = 19$ h	$t = 20$ h
0	45.0	24.0	12.0	10.0
3	64.2	36.6	19.4	14.3
6	86.4	52.9	29.8	21.9
9	106.1	71.7	43.4	32.8
12	119.0	90.4	59.6	46.6
15	123.2	105.5	76.7	62.5
18	119.0	114.5	92.4	78.6

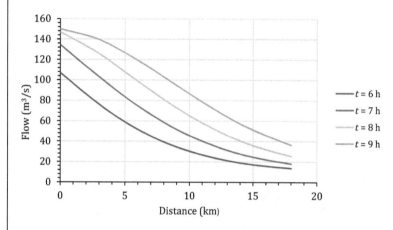

Figure 12.37 Decreasing flow downstream during the advancing flood. (A black and white version of this figure will appear in some formats. For the color version, please refer to the plate section.)

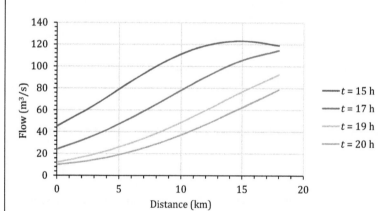

Figure 12.38 Increasing flow downstream during the receding flood. (A black and white version of this figure will appear in some formats. For the color version, please refer to the plate section.)

Example 12.6: Compute the outflow hydrograph from the data given in Example 12.5 in HEC-HMS using the Muskingum–Cunge method of routing.

Solution: A simple HEC-HMS model is set up as shown in the Box 12.1. The basin schematic is shown in Figure 12.39.

The inflow hydrograph is provided as time-series gauge data assigned to the source.

The routing method for the reach is selected as the Muskingum–Cunge method. The input data for the routing method are as follows (see Box 12.2 for HEC-HMS input):

initial condition – inflow = outflow
length = 18 000 m
slope = 0.001
Manning's $n = 0.045$
trapezoidal channel with bottom width 10 m and side slope 4H:1V
space–time method – Auto DX Auto DT
index method – celerity
index celerity = 2 m/s

The computational time interval is chosen as 20 minute (1200 seconds, which is similar to that in Example 12.5 above).

The model is run for 30 hours. After it is run, HEC-HMS provides the following messages regarding space–time discretization and calculated celerity:

$\Delta t = 1200$ s

$\Delta x = 2250$ m (note that this is similar to what is used in Example 12.5).

Consequently, the spatial steps = 8 (2250 m × 8 = 18 000 m).

Celerity ranged from 0.9 to 1.8 m/s.

HEC-HMS uses the nonlinear (four-point) method for the solution of the Muskingum–Cunge equation, as opposed to the linear method used in Example 12.5 where the routing coefficients are held constant. Since in the nonlinear method, the routing coefficients are variable in space and time, HEC-HMS does not provide any information regarding the routing coefficients.

The results of the HEC-HMS run are provided in Table 12.17 and the inflow–outflow hydrographs are shown in Figure 12.40.

Box 12.1 Basin model schematic in HEC-HMS

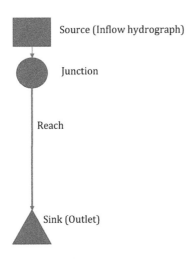

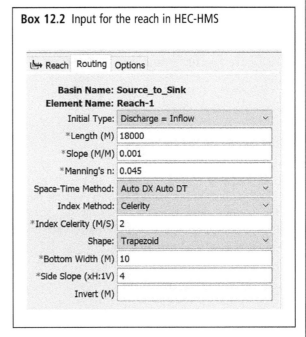

Box 12.2 Input for the reach in HEC-HMS

Figure 12.39 Model schematic

Table 12.17 *Input–output from HEC-HMS run*

Time (min)	Inflow (m^3/s)	Outflow (m^3/s)
0	10	10
20	10.7	10
40	11.3	10
60	12	10
80	14	10
100	16	10
120	18	10
140	21.5	10
160	25	10
180	28.5	10
200	35.7	10.1
220	42.8	10.2
240	50	10.5
260	59.3	10.9
280	68.7	11.6
300	78	12.5
320	87.7	13.7
340	97.3	15.3
360	107	17.4
380	116.2	20.3
400	125.3	24.9
420	134.5	31.9
440	138.7	41.3
460	142.8	52.3
480	147	64.4
500	148	76.9
520	149	89.1
540	150	100.8
560	148.7	111.4
580	147.3	120.6
600	146	128.4
620	140.3	134.5
640	134.7	139.3
660	129	142.6
680	121	144.7
700	113	145.7
720	105	145.6
740	96	144.5
760	87	142.3
780	78	139.1
800	71.7	135

Table 12.17 (*cont.*)

Time (min)	Inflow (m^3/s)	Outflow (m^3/s)
820	65.3	130.1
840	59	124.5
860	54.3	118.5
880	49.7	112.1
900	45	105.4
920	41	98.8
940	37	92.2
960	33	85.9
980	30	79.8
1000	27	74.2
1020	24	68.9
1040	21.7	63.9
1060	19.3	59.4
1080	17	55.1
1100	15.3	51.2
1120	13.7	47.5
1140	12	44
1160	11.3	40.8
1180	10.7	37.8
1200	10	34.9
1220	10	32.3
1240	10	29.8
1260	10	27.5
1280	10	25.1
1300	10	22.8
1320	10	20.5
1340	10	18.4
1360	10	16.5
1380	10	14.9
1400	10	13.5
1420	10	12.5
1440	10	11.7
1460	10	11.1
1480	10	10.7
1500	10	10.4
1520	10	10.2
1540	10	10.1
1560	10	10.1
1580	10	10
1600	10	10

The peak inflow is 150 m^3/s at 09:00 on April 1, 2001 and peak outflow is 145.7 m^3/s 11:40 on the same day. Thus, the attenuation is 4.3 m^3/s and the lag is 160 minutes. The constant coefficient calculations given in Example 12.5 show that, at 18 km, the peak outflow is 119.6 m^3/s, that is, an attenuation of 30.4 m^3/s and a lag of 7 hours.

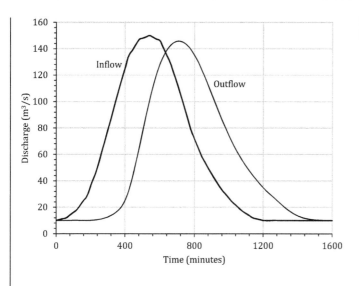

Figure 12.40 Inflow–outflow hydrographs from HEC-HMS run using Muskingum–Cunge routing.

Example 12.7: Compute the outflow hydrograph from the data given in Example 12.5 in HEC-HMS using the kinematic wave model of routing.

Solution: In this example, the routing method for the reach has been changed to kinematic wave but all other geometrical parameters for the channel are kept same.

The inputs for kinematic wave modeling in HEC-HMS are shown in Boxes 12.3 and Box 12.4. In this case either the number of sub-reaches or the time step needs to be adjusted to meet the Courant conditions.

Note that in this case the input sub-reach was adjusted to get the correct Δt for the x–t computational grid since the time step for the computation was fixed by the time intervals for the inflow hydrograph. Figure 12.41 shows the inflow and outflow hydrographs. The peak inflow is 150 m^3/s at 09:00 on April 1, 2021 and peak outflow is 149.6 m^3/s 09:20 on the same day. Thus, the attenuation is 0.4 m^3/s and the lag is 20 minutes. The attenuation decreases to zero if the computational time step is less than 20 minutes since kinematic waves do not attenuate.

Box 12.3 Specification of the kinematic wave as the channel routing method in HEC=HMS

| Reach | Routing | Options |

Basin Name: Source_to_Sink - KW
Element Name: Reach-1
Description: 18 Km Long River Reach
Downstream: Sink-1
Routing Method: Kinematic Wave
Loss/Gain Method: --None--

Box 12.4 HEC-HMS input for kinematic wave model

| Reach | Routing | Options |

Basin Name: Source_to_Sink - KW
Element Name: Reach-1
Initial Type: Discharge = Inflow
*Length (M): 1800
*Slope (M/M): 0.001
*Manning's n: 0.045
*Subreaches: 4
Index Method: Celerity
*Index Celerity (M/S): 2
Shape: Trapezoid
*Bottom Width (M): 10
*Side Slope (xH:1V): 4
Invert (M):

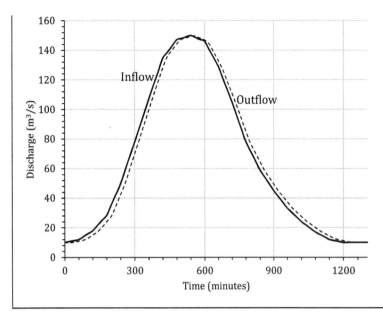

Figure 12.41 Inflow–outflow hydrographs from HEC-HMS run using the kinematic wave option.

Example 12.8: Mahoning Creek is a tributary of Alleghany River in Pennsylvania. Figure 12.42 shows the upper part of the watershed for which a hydrologic model has been developed. Observed flows at Junction 4 (confluence of Mahoning Creek and Sump Creek) and Junction 1 are given in Table 12.18. Determine the Muskingum routing parameters K and X, employing the three methods discussed in the text. Select the best

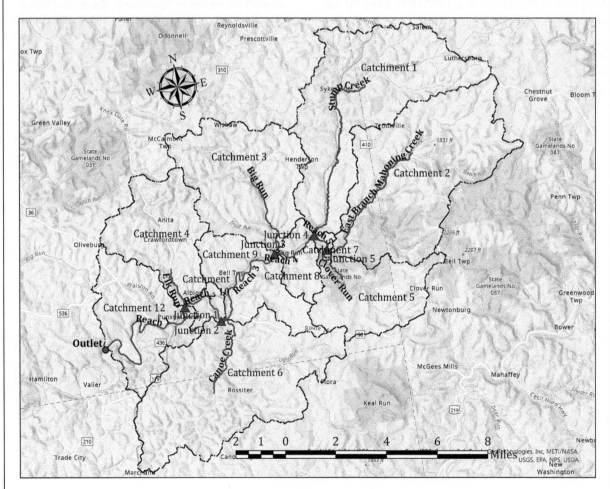

Figure 12.42 Mahoning Creek watershed, Pennsylvania.

Table 12.18 *Inflow and outflow of Mahoning Creek*

Date/Time	Inflow (ft³/s)	Outflow (ft³/s)
11Apr1994 1200	0	0
11Apr1994 1215	17	0
11Apr1994 1230	34	0
11Apr1994 1245	51	0
11Apr1994 1300	67	0
11Apr1994 1315	149	0
11Apr1994 1330	230	0
11Apr1994 1345	312	0
11Apr1994 1400	393	0
11Apr1994 1415	618	0
11Apr1994 1430	842	0
11Apr1994 1445	1066	0
11Apr1994 1500	1290	0
11Apr1994 1515	1800	0
11Apr1994 1530	2311	0
11Apr1994 1545	2821	0
11Apr1994 1600	3331	0
11Apr1994 1615	4139	0
11Apr1994 1630	4947	0
11Apr1994 1645	5755	0
11Apr1994 1700	6563	1
11Apr1994 1715	7372	1
11Apr1994 1730	8180	1
11Apr1994 1745	8989	2
11Apr1994 1800	9797	8
11Apr1994 1815	10415	1032
11Apr1994 1830	11033	2503
11Apr1994 1845	11651	3889
11Apr1994 1900	12269	4999
11Apr1994 1915	12425	5799
11Apr1994 1930	12582	6332
11Apr1994 1945	12738	6676
11Apr1994 2000	12894	7019
11Apr1994 2015	12709	7385
11Apr1994 2030	12523	7785
11Apr1994 2045	12337	8220
11Apr1994 2100	12152	8676
11Apr1994 2115	11823	9137
11Apr1994 2130	11494	9586
11Apr1994 2145	11165	10011
11Apr1994 2200	10836	10402
11Apr1994 2215	10489	10750
11Apr1994 2230	10142	11049
11Apr1994 2245	9796	11291
11Apr1994 2300	9449	11473
11Apr1994 2315	9133	11591
11Apr1994 2330	8818	11649
11Apr1994 2345	8503	11652
12Apr1994 0000	8187	11605
12Apr1994 0015	7842	11515
12Apr1994 0030	7497	11387
12Apr1994 0045	7152	11225
12Apr1994 0100	6807	11029
12Apr1994 0115	6546	10798
12Apr1994 0130	6285	10533
12Apr1994 0145	6024	10241

Table 12.18 (*cont.*)

Date/Time	Inflow (ft³/s)	Outflow (ft³/s)
12Apr1994 0200	5763	9930
12Apr1994 0215	5558	9605
12Apr1994 0230	5353	9275
12Apr1994 0245	5148	8940
12Apr1994 0300	4943	8599
12Apr1994 0315	4769	8242
12Apr1994 0330	4595	7849
12Apr1994 0345	4422	7380
12Apr1994 0400	4248	6834
12Apr1994 0415	4098	6217
12Apr1994 0430	3948	5801
12Apr1994 0445	3798	5496
12Apr1994 0500	3648	5253
12Apr1994 0515	3519	5045
12Apr1994 0530	3390	4859
12Apr1994 0545	3262	4686
12Apr1994 0600	3133	4524
12Apr1994 0615	3022	4369
12Apr1994 0630	2912	4220
12Apr1994 0645	2801	4077
12Apr1994 0700	2691	3939
12Apr1994 0715	2596	3806
12Apr1994 0730	2501	3678
12Apr1994 0745	2407	3554
12Apr1994 0800	2312	3435
12Apr1994 0815	2231	3320
12Apr1994 0830	2149	3208
12Apr1994 0845	2068	3101
12Apr1994 0900	1987	2998
12Apr1994 0915	1917	2898
12Apr1994 0930	1847	2801
12Apr1994 0945	1777	2708
12Apr1994 1000	1708	2618
12Apr1994 1015	1648	2532
12Apr1994 1030	1588	2448
12Apr1994 1045	1528	2367
12Apr1994 1100	1468	2290
12Apr1994 1115	1417	2214
12Apr1994 1130	1365	2142
12Apr1994 1145	1314	2072
12Apr1994 1200	1262	2004
12Apr1994 1215	1218	1939
12Apr1994 1230	1174	1876
12Apr1994 1245	1130	1815
12Apr1994 1300	1086	1756
12Apr1994 1315	1048	1699
12Apr1994 1330	1010	1644
12Apr1994 1345	972	1591
12Apr1994 1400	934	1540
12Apr1994 1415	901	1491
12Apr1994 1430	869	1443
12Apr1994 1445	836	1397
12Apr1994 1500	804	1352
12Apr1994 1515	775	1309
12Apr1994 1530	747	1268
12Apr1994 1545	719	1228

Table 12.18 (cont.)

Date/Time	Inflow (ft³/s)	Outflow (ft³/s)
12Apr1994 1600	691	1189
12Apr1994 1615	666	1151
12Apr1994 1630	641	1115
12Apr1994 1645	616	1080
12Apr1994 1700	592	1046
12Apr1994 1715	570	1014
12Apr1994 1730	549	982
12Apr1994 1745	528	952
12Apr1994 1800	507	922
12Apr1994 1815	489	893
12Apr1994 1830	470	866
12Apr1994 1845	452	839
12Apr1994 1900	434	813
12Apr1994 1915	418	788
12Apr1994 1930	403	764
12Apr1994 1945	387	740
12Apr1994 2000	372	718
12Apr1994 2015	358	696
12Apr1994 2030	345	674
12Apr1994 2045	332	653
12Apr1994 2100	318	633
12Apr1994 2115	307	614
12Apr1994 2130	296	595
12Apr1994 2145	284	577
12Apr1994 2200	273	559
12Apr1994 2215	263	542
12Apr1994 2230	253	526
12Apr1994 2245	244	510
12Apr1994 2300	234	494
12Apr1994 2315	225	479
12Apr1994 2330	217	465
12Apr1994 2345	209	451
13Apr1994 0000	200	437

set of parameters and compute the outflow hydrograph at Junction 1 from the inflow hydrograph at Junction 4 and compare the computed outflow hydrograph with the observed hydrograph.

Solution:

Step 1: Estimation of K from the time of peak of inflow and outflow.

The inflow and outflow hydrographs are plotted in Figure 12.43.

From the data and the plots, it is found that the inflow peak of 12 894 ft³/s occurred on April 11 at 20:00 and the outflow peak of 11 652 ft³/s occurred on the same day at 23:45. Thus, the travel time of the flood wave is estimated as 3 h 45 min, which translates to a value of $K = 3.75$ h.

Step 2: Estimation of K from the slope of the rating curve.

Figure 12.44 shows the stage–discharge relationship or the rating curve for Mahoning Creek at Punxsutawney gauge. A linear equation on the semi-log plot of the discharge-stage data is fitted to represent the rating curve. This equation is shown in Figure 12.44. From the slope of this curve,

$$\frac{dQ}{dh} = \frac{1}{0.0016 \frac{\text{ft}}{\text{ft}^3/\text{s}}} = 625 \frac{\text{ft}^2}{\text{s}}$$

The width of Mahoning Creek at this location is 200 ft. Therefore, the celerity is

$$c = \frac{1}{200 \text{ ft}} 625 \frac{\text{ft}^2}{\text{s}} = 3.125 \text{ ft/s}.$$

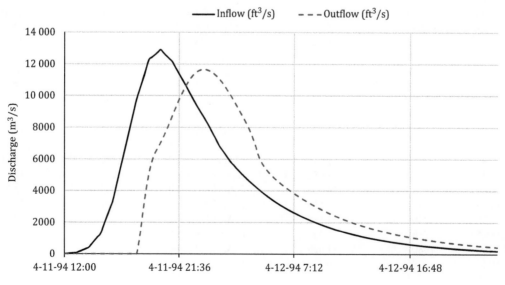

Figure 12.43 Inflow–outflow hydrographs in Mahoning Creek.

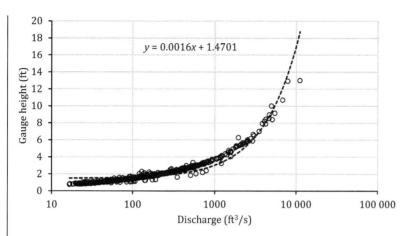

Figure 12.44 Rating curve at Punxsutawney.

The distance between Junction 1 and Junction 4 is 68 000 ft. So, the travel time for the flood wave is estimated as

$$T_{FW} = \frac{68000 \text{ ft}}{3.125 \text{ ft/s}} = 21760 \text{ s} = 6 \text{ h}$$

From this approach, the Muskingum $K = 6$ h.

Step 3: Estimation of K using a graphical method.

For each inflow and outflow, storage is calculated using Eq. 12.10. Four values of X, 0.1, 0.2, 0.3, and 0.4 are used to calculate weighted outflows using the equation: $XI + (1 - X)Q$. The calculations are presented in Table 12.19. For each set, weighted outflows are plotted against storage. The plots are shown in Figures 12.45–12.48.

From the plots presented above, it can be seen that the thinnest loop is formed for $X = 0.4$. A best-fit line through these plots has a slope of $8 \times 10^{-5} \frac{\text{ft}^3/\text{s}}{\text{ft}^3}$.

Therefore, for $X = 0.4$, $K = \frac{1}{8 \times 10^{-5}} \text{ s} = 12\ 500 \text{ s} = 3.5 \text{ h}$.

Step 4: Calculation of the outflow using K and X derived above for the Muskingum method.

$K = 3.5 \text{ h} = 12600 \text{ s}$; $X = 0.4$; $\Delta t = 15 \text{ min} = 900 \text{ s}$.
Calculate routing coefficients, C_1, C_2, C_3, using Eqs. 12.48–12.50. The results are:

$$C_1 = \frac{\frac{\Delta t}{K} - 2X}{2(1-X) + \frac{\Delta t}{K}} = \frac{\frac{900 \text{ s}}{12600 \text{ s}} - 2 \times 0.4}{2 \times (1 - 0.4) + \frac{900 \text{ s}}{12600 \text{ s}}} = -0.57303$$

$$C_2 = \frac{\frac{\Delta t}{K} + 2X}{2(1-X) + \frac{\Delta t}{K}} = \frac{\frac{900 \text{ s}}{12600 \text{ s}} + 2 \times 0.4}{2 \times (1 - 0.4) + \frac{900 \text{ s}}{12600 \text{ s}}} = 0.685393$$

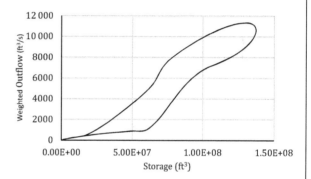

Figure 12.45 Weighted outflow versus storage $X = 0.1$.

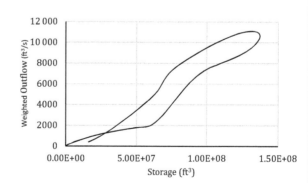

Figure 12.46 Weighted outflow versus storage $X = 0.2$

$$C_3 = \frac{2(1-X) - \frac{\Delta t}{K}}{2(1-X) + \frac{\Delta t}{K}} = \frac{2 \times (1-0.4) - \frac{900\text{ s}}{12600\text{ s}}}{2 \times (1-0.4) + \frac{900\text{ s}}{12600\text{ s}}} = 0.88764$$

Note that $C_1 + C_2 + C_3 = -0.57303 + 0.685393 + 0.88764 = 1$. Furthermore, Δt is less than K or $2K(1-X)$. Now calculate the outflows as a function of time using Eq.12.47. The results are given in Table 12.20 and the hydrograph plots are shown in Figure 12.49.

Note that as one of the limitations of Muskingum method, there are negative outflows during the initial times even though the match between the observed and calculated flows at later times is good.

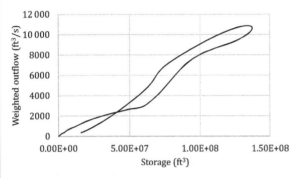

Figure 12.47 Weighted outflow versus storage $X = 0.3$.

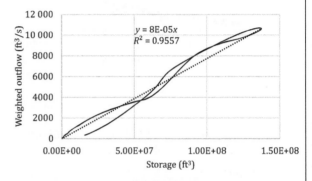

Figure 12.48 Weighted outflow versus storage $X = 0.4$.

Table 12.19 *Storage-weighted outflow calculations*

Time (min)	Inflow (ft³/s)	Outflow (ft³/s)	Storage (ft³)	Weighted outflow (ft³/s), $X = 0.1$	Weighted outflow (ft³/s), $X = 0.2$	Weighted outflow (ft³/s) $X = 0.3$	Weighted outflow (ft³/s), $X = 0.4$
0	0	0	0.00	0.49	0.49	0.49	0.49
15	17	0	7 508.71	2.16	3.83	5.50	7.16
30	34	0	30 034.85	3.83	7.16	10.50	13.84
45	51	0	67 578.40	5.50	10.50	15.51	20.51
60	67	0	120 139.38	7.16	13.84	20.51	27.19
75	149	0	216 905.85	15.32	30.15	44.98	59.81
90	230	0	387 065.88	23.47	46.46	69.44	92.43
105	312	0	630 619.48	31.63	62.77	93.91	125.05
120	393	0	947 566.63	39.78	79.08	118.37	157.67
135	618	0	1 402 113.26	62.21	123.92	185.64	247.36
150	842	0	2 058 465.29	84.63	168.77	252.91	337.05
165	1066	0	2 916 622.71	107.05	213.61	320.18	426.74
180	1290	0	3 976 585.53	129.48	258.46	387.45	516.43
195	1800	0	5 367 020.02	180.49	360.49	540.49	720.49
210	2311	0	7 216 592.44	231.51	462.52	693.54	924.55
225	2821	0	9 525 302.79	282.52	564.55	846.58	1128.61
240	3331	0	12 293 151.07	333.54	666.58	999.63	1332.68
255	4139	0	15 654 223.48	414.35	828.21	1242.07	1655.92
270	4947	0	19 742 606.18	495.16	989.83	1484.50	1979.17
285	5755	0	24 558 298.62	575.97	1151.46	1726.94	2302.42
300	6563	1	30 101 295.19	656.80	1313.09	1969.39	2625.68
315	7372	1	36 371 695.55	737.70	1474.83	2211.95	2949.08
330	8180	1	43 369 522.01	818.76	1636.71	2454.65	3272.60

Table 12.19 (cont.)

Time (min)	Inflow (ft³/s)	Outflow (ft³/s)	Storage (ft³)	Weighted outflow (ft³/s), $X = 0.1$	Weighted outflow (ft³/s), $X = 0.2$	Weighted outflow (ft³/s), $X = 0.3$	Weighted outflow (ft³/s), $X = 0.4$
345	8989	2	51 094 463.24	900.29	1798.99	2697.70	3596.41
360	9797	8	59 543 931.23	986.53	1965.48	2944.43	3923.38
375	10 415	1032	68 171 427.84	1970.59	2908.86	3847.13	4785.40
390	11 033	2503	76 232 233.07	3355.81	4208.83	5061.85	5914.87
405	11 651	3889	83 563 499.33	4665.57	5441.72	6217.87	6994.02
420	12 269	4999	90 327 440.06	5726.36	6453.31	7180.25	7907.20
435	12 425	5799	96 580 455.94	6461.69	7124.30	7786.91	8449.52
450	12 582	6332	102 374 372.26	6957.20	7582.12	8207.05	8831.97
465	12 738	6676	107 914 495.51	7281.91	7888.13	8494.34	9100.55
480	12 894	7019	113 286 471.40	7606.13	8193.69	8781.25	9368.81
495	12 709	7385	118 325 986.02	7917.53	8449.86	8982.20	9514.53
510	12 523	7785	122 853 498.97	8258.89	8732.67	9206.45	9680.23
525	12 337	8220	126 838 098.75	8632.10	9043.79	9455.48	9867.16
540	12 152	8676	130 254 639.95	9023.75	9371.30	9718.84	10 066.39
555	11 823	9137	133 027 363.75	9405.23	9673.85	9942.47	10 211.08
570	11 494	9586	135 094 644.96	9776.88	9967.66	10 158.44	10 349.22
585	11 165	10 011	136 472 279.71	10 126.77	10 242.13	10 357.50	10 472.86
600	10 836	10 402	137 186 739.35	10 445.48	10 488.89	10 532.29	10 575.70
615	10 489	10 750	137 264 762.23	10 723.95	10 697.88	10 671.81	10 645.74
630	10 142	11 049	136 739 727.12	10 957.95	10 867.34	10 776.74	10 686.13
645	9796	11 291	135 659 051.11	11 141.58	10 992.04	10 842.49	10 692.95
660	9449	11 473	134 075 389.67	11 270.27	11 067.89	10 865.51	10 663.13
675	9133	11 591	132 058 733.20	11 345.38	11 099.61	10 853.84	10 608.08
690	8818	11 649	129 678 712.03	11 366.26	11 083.13	10 800.00	10 516.87
705	8503	11 652	126 987 458.59	11 337.10	11 022.17	10 707.24	10 392.31
720	8187	11 605	124 032 356.29	11 263.21	10 921.45	10 579.69	10 237.93
735	7842	11 515	120 841 892.65	11 147.39	10 780.16	10 412.93	10 045.70
750	7497	11 387	117 439 078.10	10 997.80	10 608.85	10 219.90	9830.95
765	7152	11 225	113 856 066.92	10 817.68	10 410.40	10 003.13	9595.85
780	6807	11 029	110 123 407.46	10 606.99	10 184.78	9762.58	9340.37
795	6546	10 798	106 310 153.89	10 372.84	9947.65	9522.47	9097.28
810	6285	10 533	102 485 170.07	10 108.51	9683.69	9258.88	8834.07
825	6024	10 241	98 675 840.05	9819.58	9397.88	8976.17	8554.47
840	5763	9930	94 903 349.21	9512.88	9096.26	8679.63	8263.01
855	5558	9605	91 207 349.37	9200.49	8795.78	8391.08	7986.37
870	5353	9275	87 621 473.83	8882.36	8490.20	8098.05	7705.90
885	5148	8940	84 150 369.38	8560.67	8181.47	7802.26	7423.06
900	4943	8599	80 798 775.98	8233.06	7867.46	7501.86	7136.27
915	4769	8242	77 590 988.95	7894.27	7547.02	7199.77	6852.53
930	4595	7849	74 564 269.05	7523.62	7198.26	6872.90	6547.54
945	4422	7380	71 768 908.74	7084.25	6788.41	6492.58	6196.75
960	4248	6834	69 273 800.19	6575.81	6317.18	6058.54	5799.91
975	4098	6217	67 156 414.72	6005.08	5793.19	5581.29	5369.40
990	3948	5801	65 368 839.25	5616.06	5430.72	5245.37	5060.03
1005	3798	5496	63 770 424.81	5326.63	5156.77	4986.91	4817.05
1020	3648	5253	62 283 636.08	5092.67	4932.14	4771.60	4611.06
1035	3519	5045	60 874 383.67	4892.73	4740.10	4587.47	4434.84

Table 12.19 (*cont.*)

Time (min)	Inflow (ft³/s)	Outflow (ft³/s)	Storage (ft³)	Weighted outflow (ft³/s), $X = 0.1$	Weighted outflow (ft³/s), $X = 0.2$	Weighted outflow (ft³/s), $X = 0.3$	Weighted outflow (ft³/s), $X = 0.4$
1050	3390	4859	59 526 689.90	4712.04	4565.18	4418.32	4271.46
1065	3262	4686	58 224 672.32	4543.86	4401.38	4258.90	4116.42
1080	3133	4524	56 957 644.53	4384.54	4245.46	4106.37	3967.29
1095	3022	4369	55 725 974.47	4233.92	4099.30	3964.68	3830.05
1110	2912	4220	54 531 593.54	4089.03	3958.24	3827.44	3696.64
1125	2801	4077	53 369 086.83	3949.24	3821.70	3694.17	3566.63
1140	2691	3939	52 233 560.05	3814.15	3689.34	3564.54	3439.74
1155	2596	3806	51 127 508.70	3685.05	3564.06	3443.07	3322.08
1170	2501	3678	50 053 666.87	3560.17	3442.53	3324.88	3207.24
1185	2407	3554	49 007 899.53	3439.34	3324.59	3209.84	3095.09
1200	2312	3435	47 986 229.57	3322.42	3210.13	3097.84	2985.55
1215	2231	3320	46 990 881.10	3210.61	3101.71	2992.81	2883.91
1230	2149	3208	46 024 213.12	3102.44	2996.52	2890.61	2784.69
1245	2068	3101	45 082 657.37	2997.77	2894.45	2791.13	2687.81
1260	1987	2998	44 162 774.32	2896.47	2795.37	2694.27	2593.18
1275	1917	2898	43 266 442.29	2799.59	2701.51	2603.42	2505.34
1290	1847	2801	42 395 655.43	2705.86	2610.44	2515.02	2419.59
1305	1777	2708	41 547 325.73	2615.16	2522.06	2428.97	2335.87
1320	1708	2618	40 718 478.10	2527.38	2436.28	2345.19	2254.10
1335	1648	2532	39 910 686.68	2443.40	2354.98	2266.57	2178.15
1350	1588	2448	39 125 626.11	2362.14	2276.10	2190.06	2104.01
1365	1528	2367	38 360 626.33	2283.50	2199.54	2115.58	2031.63
1380	1468	2290	37 613 114.03	2207.39	2125.23	2043.08	1960.92
1395	1417	2214	36 884 417.64	2134.55	2054.78	1975.00	1895.22
1410	1365	2142	36 175 953.77	2064.07	1986.41	1908.75	1831.09
1425	1314	2072	35 485 414.78	1995.85	1920.06	1844.27	1768.47
1440	1262	2004	34 810 575.54	1929.82	1855.65	1781.48	1707.31
1455	1218	1939	34 152 560.03	1866.62	1794.56	1722.51	1650.45
1470	1174	1876	33 512 570.29	1805.44	1735.28	1665.11	1594.95
1485	1130	1815	32 888 614.23	1746.22	1677.73	1609.24	1540.75
1500	1086	1756	32 278 770.37	1688.89	1621.86	1554.83	1487.80
1515	1048	1699	31 683 991.00	1634.00	1568.85	1503.71	1438.56
1530	1010	1644	31 105 296.09	1580.86	1517.40	1453.95	1390.49
1545	972	1591	30 540 965.49	1529.40	1467.45	1405.50	1343.55
1560	934	1540	29 989 339.65	1479.58	1418.95	1358.32	1297.69
1575	901	1491	29 451 224.41	1431.87	1372.92	1313.97	1255.02
1590	869	1443	28 927 483.21	1385.67	1328.23	1270.79	1213.36
1605	836	1397	28 416 628.44	1340.93	1284.84	1228.75	1172.67
1620	804	1352	27 917 225.12	1297.60	1242.70	1187.81	1132.92
1635	775	1309	27 429 841.10	1256.06	1202.65	1149.23	1095.81
1650	747	1268	26 955 092.89	1215.84	1163.75	1111.67	1059.58
1665	719	1228	26 491 692.04	1176.88	1125.98	1075.09	1024.20
1680	691	1189	26 038 396.21	1139.14	1089.30	1039.46	989.62
1695	666	1151	25 595 596.37	1102.93	1054.37	1005.81	957.25
1710	641	1115	25 163 725.16	1067.86	1020.45	973.04	925.63
1725	616	1080	24 741 666.59	1033.88	987.50	941.12	894.74
1740	592	1046	24 328 344.82	1000.96	955.49	910.03	864.56

Table 12.19 (cont.)

Time (min)	Inflow (ft³/s)	Outflow (ft³/s)	Storage (ft³)	Weighted outflow (ft³/s), $X = 0.1$	Weighted outflow (ft³/s), $X = 0.2$	Weighted outflow (ft³/s), $X = 0.3$	Weighted outflow (ft³/s), $X = 0.4$
1755	570	1014	23 924 273.61	969.41	925.09	880.76	836.44
1770	549	982	23 530 002.09	938.84	895.55	852.26	808.97
1785	528	952	23 144 564.02	909.22	866.86	824.50	782.14
1800	507	922	22 767 029.71	880.51	838.98	797.45	755.91
1815	489	893	22 397 885.78	852.99	812.49	771.99	731.50
1830	470	866	22 037 655.18	826.30	786.75	747.20	707.64
1845	452	839	21 685 515.34	800.42	761.72	723.02	684.32
1860	434	813	21 340 677.98	775.31	737.38	699.45	661.52
1875	418	788	21 003 572.17	751.20	714.22	677.24	640.26
1890	403	764	20 674 657.86	727.80	691.69	655.58	619.47
1905	387	740	20 353 239.78	705.09	669.77	634.45	599.14
1920	372	718	20 038 649.50	683.02	648.43	613.84	579.24
1935	358	696	19 731 253.71	661.82	628.10	594.38	560.67
1950	345	674	19 431 440.44	641.23	608.32	575.41	542.50
1965	332	653	19 138 604.93	621.23	589.06	556.90	524.73
1980	318	633	18 852 157.64	601.82	570.33	538.84	507.35
1995	307	614	18 572 386.26	583.17	552.48	521.80	491.12
2010	296	595	18 299 591.29	565.07	535.13	505.19	475.26
2025	284	577	18 033 224.99	547.51	518.26	489.01	459.75
2040	273	559	17 772 753.27	530.49	501.86	473.23	444.60
2055	263	542	17 518 391.69	514.14	486.25	458.35	430.45
2070	253	526	17 270 369.67	498.30	471.08	443.86	416.64
2085	244	510	17 028 192.95	482.94	456.34	429.75	403.15
2100	234	494	16 791 382.99	468.05	442.03	416.00	389.97
2115	225	479	16 560 107.71	453.76	428.39	403.03	377.66
2130	217	465	16 334 549.15	439.91	415.16	390.40	365.64
2145	209	451	16 114 272.28	426.49	402.30	378.11	353.91
2160	200	437	15 898 856.84	413.49	389.81	366.13	342.46

Table 12.20 Inflow, calculated, and observed outflow

Time (min)	Inflow (ft³/s)	Calculated outflow (ft³/s)	Observed outflow (ft³/s)
0	0	0	0
15	17	−9	0
30	34	−16	0
45	51	−20	0
60	67	−21	0
75	149	−58	0
90	230	−82	0
105	312	−93	0
120	393	−94	0
135	618	−168	0
150	842	−208	0
165	1066	−219	0
180	1290	−203	0
195	1800	−328	0
210	2311	−381	0
225	2821	−371	0
240	3331	−304	0
255	4139	−359	0
270	4947	−317	0

Table 12.20 (cont.)

Time (min)	Inflow (ft³/s)	Calculated outflow (ft³/s)	Observed outflow (ft³/s)
285	5755	−188	0
300	6563	16	1
315	7372	289	1
330	8180	621	1
345	8989	1007	2
360	9797	1441	8
375	10 415	2026	1032
390	11 033	2614	2503
405	11 651	3206	3889
420	12 269	3801	4999
435	12 425	4663	5799
450	12 582	5445	6332
465	12 738	6158	6676
480	12 894	6807	7019
495	12 709	7598	7385
510	12 523	8278	7785
525	12 337	8862	8220
540	12 152	9358	8676
555	11 823	9861	9137

Table 12.20 (cont.)

Time (min)	Inflow (ft³/s)	Calculated outflow (ft³/s)	Observed outflow (ft³/s)
570	11 494	10 270	9586
585	11 165	10 596	10 011
600	10 836	10 848	10 402
615	10 489	11 046	10 750
630	10 142	11 182	11 049
645	9796	11 264	11 291
660	9449	11 298	11 473
675	9133	11 271	11 591
690	8818	11 211	11 649
705	8503	11 123	11 652
720	8187	11 009	11 605
735	7842	10 890	11 515
750	7497	10 745	11 387
765	7152	10 578	11 225
780	6807	10 391	11 029
795	6546	10 138	10 798
810	6285	9884	10 533
825	6024	9629	10 241
840	5763	9373	9930
855	5558	9085	9605
870	5353	8807	9275
885	5148	8536	8940
900	4943	8273	8599
915	4769	7998	8242
930	4595	7735	7849
945	4422	7482	7380
960	4248	7237	6834
975	4098	6988	6217
990	3948	6749	5801
1005	3798	6520	5496
1020	3648	6300	5253
1035	3519	6076	5045
1050	3390	5862	4859
1065	3262	5659	4686
1080	3133	5463	4524
1095	3022	5264	4369
1110	2912	5076	4220
1125	2801	4896	4077
1140	2691	4724	3939
1155	2596	4550	3806
1170	2501	4385	3678
1185	2407	4227	3554
1200	2312	4077	3435
1215	2231	3925	3320
1230	2149	3781	3208
1245	2068	3645	3101
1260	1987	3514	2998
1275	1917	3382	2898
1290	1847	3258	2801
1305	1777	3139	2708
1320	1708	3026	2618
1335	1648	2912	2532
1350	1588	2805	2448
1365	1528	2702	2367

Table 12.20 (cont.)

Time (min)	Inflow (ft³/s)	Calculated outflow (ft³/s)	Observed outflow (ft³/s)
1380	1468	2605	2290
1395	1417	2506	2214
1410	1365	2413	2142
1425	1314	2325	2072
1440	1262	2241	2004
1455	1218	2156	1939
1470	1174	2076	1876
1485	1130	2000	1815
1500	1086	1928	1756
1515	1048	1855	1699
1530	1010	1786	1644
1545	972	1720	1591
1560	934	1658	1540
1575	901	1595	1491
1590	869	1536	1443
1605	836	1480	1397
1620	804	1426	1352
1635	775	1372	1309
1650	747	1321	1268
1665	719	1273	1228
1680	691	1227	1189
1695	666	1181	1151
1710	641	1137	1115
1725	616	1096	1080
1740	592	1056	1046
1755	570	1016	1014
1770	549	978	982
1785	528	942	952
1800	507	908	922
1815	489	873	893
1830	470	840	866
1845	452	809	839
1860	434	779	813
1875	418	750	788
1890	403	721	764
1905	387	694	740
1920	372	669	718
1935	358	643	696
1950	345	619	674
1965	332	596	653
1980	318	574	633
1995	307	551	614
2010	296	530	595
2025	284	511	577
2040	273	492	559
2055	263	473	542
2070	253	455	526
2085	244	438	510
2100	234	421	494
2115	225	405	479
2130	217	390	465
2145	209	375	451
2160	200	361	437

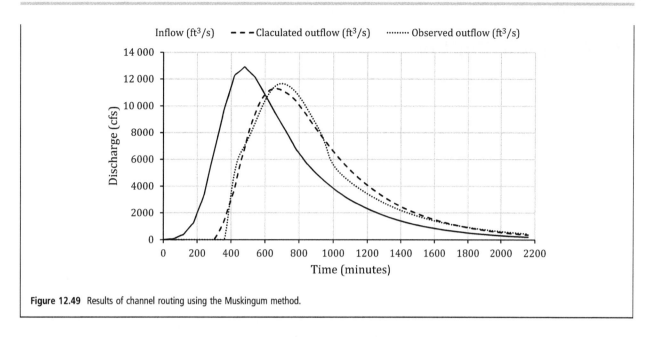

Figure 12.49 Results of channel routing using the Muskingum method.

Exercises: A selection of exercises on this topic is available at www.cambridge.org/appliedhydrology.

13 Reservoir Routing

13.1 CONCEPTS AND PURPOSE

Reservoirs behind dams on the course of major rivers are common all over the world. On lower-order tributaries of major rivers, there are flood control detention basins in communities with regulatory restrictions on the increase in surface runoff resulting from changes in land use and land cover. As discussed in Section 12.3.3 of Chapter 12, due to the storage capacity of a reservoir, a flood wave entering the reservoir at its upstream end gets attenuated and lagged when it exits the reservoir at its downstream end. The procedure used for the determination of an outflow hydrograph from an inflow hydrograph in the reservoir is **reservoir routing** and constitutes the subject matter of this chapter.

Reservoir routing has important applications in water resources engineering. These include determination of the reservoir or detention basin capacity for flood control, establishing capacities of outlet structures of reservoirs and detention basins, design and operation of reservoir spillways, determining dam height, computation of probabilities of the highest water levels, rate of change of reservoir levels, flood forecasting at locations downstream of reservoirs, and computation of dam breach outflows. Reservoir routing is used in watershed studies where one or more storage facilities exist; specifically, for watersheds where it is necessary to evaluate locations of water supply structures and regional flood control measures for watershed master plans.

13.2 THE MODIFIED PULS METHOD

Reservoir routing is based on the spatially lumped continuity equation given as Eq. 12.9 in Chapter 12. The most widely adopted procedure in the current practice for its solution is the modified Puls method, also described in detail in Chapter 12. The method is also called **level-pool routing** due to the negligible slope of the water surface in a reservoir. The solution is based on the finite-difference approximation of Eq. 12.9, given by Eq. 12.10. However, certain additional considerations necessary for reservoir routing are as follows.

Equation 12.40, used for channel routing, is rearranged a bit after multiplying this equation by $\Delta t/2$ to obtain

$$\frac{2S^{j+1}}{\Delta t} + Q^{j+1} = \left(I^j + I^{j+1}\right) + \left(\frac{2S^j}{\Delta t} - Q^j\right) \qquad (13.1)$$

In the case of channel routing, the storage–discharge relationship necessary for the solution of Eq. 12.40 can be expressed by a functional relationship, simply because in general the channel reach under consideration is assumed not to have any in-line structure to control the channel flow. However, in reservoirs, the outlet structures control the outflow from the reservoir. As will be seen in the next section, for reservoirs, expressing a storage–discharge relationship by an analytical function is not so simple due to the hydraulics of the flow control devices that can be of varied types. For this reason, the storage–indication relationship for a reservoir is typically developed in a tabular form listing $[(2S/\Delta t) + Q]$ versus Q. During the recursive process of the solution of Eq. 13.1, all of the variables on the right-hand side of the equation are known at the current computational time step to give the value of the quantity on the left-hand side of the equation. This value of $[(2S/\Delta t) + Q]$ is used to get the corresponding value of the currently unknown value of Q from the tabular form of a storage–indication table by simple linear interpolation. An illustrative example is given to show how two powerful functions of Excel can greatly aid in the interpolation process and thereby facilitate the automated computation in a spreadsheet.

The outflow from a reservoir can be either *controlled* or *uncontrolled*. These are also called regulated and unregulated reservoirs. The controlled or regulated reservoirs have spillways with gates that are operated to release water. Outlets of uncontrolled reservoirs are not controlled by gate operation. Analysis of a flood wave that passes through an unregulated reservoir is the simplest routing procedure based on Eq. 13.1.

Major dams on waterways have spillways with gates that can be raised or lowered to control the outflow from the reservoir. The dam may also have under sluices to control outflow for purposes such as agricultural irrigation and water supply. The release of water from storage through the spillway gates and under sluices depends on the state of the reservoir, magnitudes of demands and their priorities, and the operational rule or policy for the reservoir.

In gated dams, the reservoir outflow can be either (1) controlled, (2) uncontrolled, or (3) a combination of these two. In this case, both hydraulic conditions and operational rules determine the prescribed outflow. Operational rules take into account the various uses of water. In such cases, the finite-difference approximation of the continuity equation (Eq. 12.39) is modified as

$$\frac{I^{j+1}+I^j}{2} - \frac{Q^{j+1}+Q^j}{2} - \overline{Q_c} = \frac{S^{j+1}-S^j}{\Delta t} \quad (13.2)$$

where $\overline{Q_c}$ is the average controlled outflow from the reservoir during the time interval, Δt.

Rearranging the terms, Eq. 13.2 can be written in the same form as Eq. 13.1 as

$$\frac{2S^{j+1}}{\Delta t} + Q^{j+1} = (I^j + I^{j+1}) + \left(\frac{2S^j}{\Delta t} - Q^j\right) - 2\overline{Q_c} \quad (13.3)$$

When the controlled outflow, $\overline{Q_c}$, is known, the solution can be obtained by the modified Puls method. The solution of Eq. 13.3 is simple if the entire outflow is controlled. The spillway rating chart can be used to determine the outflow if the reservoir elevation and the gate opening are known.

It should be noted here that there are several other methods, such as the mass curve, Wisler–Brater, Goodrich, Steinberg, and coefficient methods, also available for reservoir routing. However, these are not used very often. Nonetheless, all of these methods, including the modified Puls method, are traditional hydrologic methods that do not involve any numerical mathematical analysis. However, reservoir routing actually involves solution of a first-order differential equation (Eq. 12.9), and therefore the procedures for its solution can be simpler and more flexible by invoking a numerical mathematical procedure, discussed in Section 13.4 of this chapter.

13.3 STORAGE–DISCHARGE RELATIONSHIPS

An essential requirement for reservoir routing computations entails establishing the relationships between reservoir elevation, storage, and discharge. These relationships depend on the geometric configuration of the reservoir and hydraulic features of the outlet structure. The relationships are typically required in tabular forms for computations. Graphical representations of these relationships aid in understanding the system being analyzed. The tables and the graphs are constructed from the data developed in steps described in the following three subsections.

13.3.1 Development of the Storage–Elevation Relationship

A storage–elevation relationship, also known as stage–storage curve, defines the relationship between the depth of water and the associated storage volume in a reservoir or detention basin. Figure 13.1 shows such a relationship for a flood control reservoir. The corresponding variation of the surface area with elevation is also depicted in this graph.

The volume of storage can be calculated by using simple geometric formulas expressed as a function of depth. The storage volume for natural basins or riverine impoundments may be developed using a contour map and the double-end area, frustum of a pyramid, prismoid, or circular conic section formulas. The double-end area formula or simply the **end-area method** is expressed as

$$V_{1,2} = \left(\frac{A_1 + A_2}{2}\right) \times \Delta h_{1,2} \quad (13.4)$$

where $V_{1,2}$ is the volume (m³ or acre-feet) between elevations (or contours) 1 and 2; A_1 and A_2 are the surface areas (m² or acre) at elevation 1 and 2, respectively, and $\Delta h_{1,2}$ is the elevation difference between elevation levels (contours) 1 and 2 (in m). Equation 13.4 is used to successively

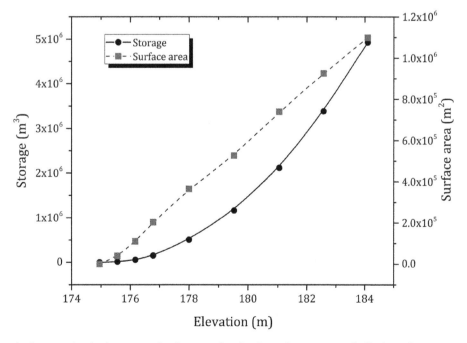

Figure 13.1 An example of storage–elevation (stage–storage) and corresponding elevation–surface area curves of a flood control reservoir.

calculate the volumes from the invert to the top of the basin and to sum these volumes. Thus, Eq. 13.4 can be generalized as

$$\Delta V = \frac{A(h) + A(h + \Delta h)}{2} \Delta h \quad (13.5)$$

$$V(h) = \sum_{i=0}^{h} \Delta V_i \quad (13.6)$$

Figure 13.2 shows the bathymetric contours of Colebrook River Lake, located on the West Branch of the Farmington River in Colebrook, straddling the border between Connecticut and Massachusetts. It is one of the three flood control dams on tributaries of the Connecticut River that were constructed as a result of major flooding from Hurricane Diane, which devastated Connecticut in August 1955. From the bathymetric contours, the elevation–storage curve for this reservoir is constructed using Eq. 13.6, shown as Figure 13.4. At present this reservoir is used for both flood control and water supply.

While Eq. 13.5 is a general formula used for reservoirs or impoundments behind dams where the reservoir geometries are generally irregular figures, as can be seen from Figure 13.2, certain reservoirs, particularly relatively smaller detention basins, have regular geometric shapes in three dimensions, for which formulas with greater accuracy are used. The **frustum of a pyramid formula** is given by

$$V_{1,2} = \left(A_1 + A_2 + \sqrt{(A_1 + A_2)}\right) \times \frac{\Delta h_{1,2}}{3} \quad (13.7)$$

Equation 13.7 is also known as the **conic method** and has been used to approximate basin volumes of irregular shapes

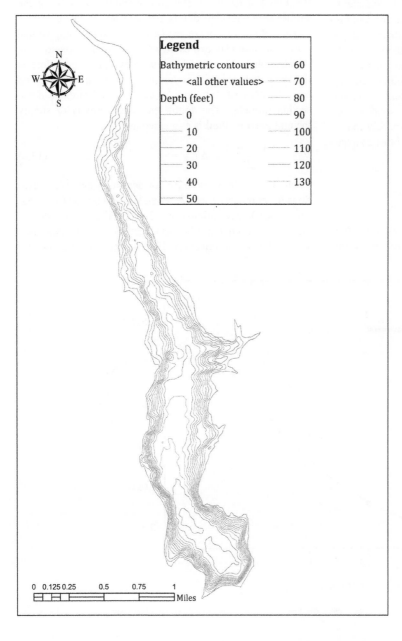

Figure 13.2 Bathymetric contours of Colebrook River Lake (reservoir). The bathymetric data are from a 2002 survey and provided by the Connecticut Department of Energy and Environmental Protection (A black and white version of this figure will appear in some formats. For the color version, please refer to the plate section.).

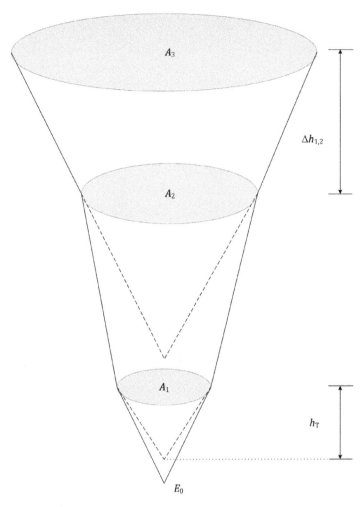

Figure 13.3 Conic method for estimation of reservoir volumes.

also. In such a case, the volume at the lowermost surface, A_1, is assumed to be zero and this surface lies at an elevation, h_T, above the apex of an inverted cone (Figure 13.3). The height of the truncated part of the cone, h_T, is given as

$$h_T = \frac{h_{1,2}}{\left(\sqrt{\frac{A_2}{A_1}} - 1\right)} \quad (13.8)$$

Note that the volume at the lowermost level is assumed to be zero even if the surface area at that level is greater than zero. Figure 13.4 shows that, for Colebrook River Lake, the storage calculated using the conic method is virtually the same as that calculated from the end-area method.

The volume of a regular quadrilateral frustum is simply

$$V = \frac{h}{3}\left[A_1 + A_2 + \sqrt{(A_1 + A_2)}\right] \quad (13.9)$$

where A_1 is the area of lower base, A_2 is the area of upper base, h is the vertical height, and V is the total volume.

One geometry sometimes used to represent a stormwater detention basin is the inverted quadrilateral frustum, one with a rectangular base ($L_0 \times W_0$) and trapezoidal side slope (Z) in both directions. As such, the length and width dimensions at any elevation, h, are given as

$$L(h) = L_0 + 2Zh \quad (13.10)$$

$$W(h) = W_0 + 2Zh \quad (13.11)$$

The horizontal area at any elevation, h, is given as the product

$$A(h) = L(h) \times W(h) = L_0 W_0 + 2Zh(L_0 + W_0) + 4Z^2 h^2 \quad (13.12)$$

The storage, that is, the volume at any elevation, h, can be found as the integral

$$V(h) = \int_0^h \left(L_0 W_0 + 2Zh(L_0 + W_0) + 4Z^2 h^2\right) dh \quad (13.13)$$

$$V(h) = L_0 W_0 h + Zh^2(L_0 + W_0) + \frac{4}{3}Z^2 h^3 \quad (13.14)$$

Thus, the general **prismoid formula** for trapezoidal basins is given by

$$V = L \times W \times D + (L + W)ZD^2 + \frac{4}{3}Z^2 D^3 \quad (13.15)$$

where V is the volume of the trapezoidal basin (ft^3 or m^3), L is the length of basin at base (ft or m), W is width of basin at

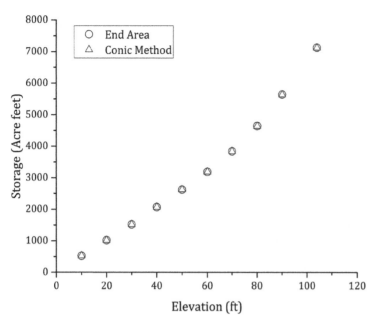

Figure 13.4 Elevation–storage curve for Colebrook River Lake.

base (ft or m), D is depth of the basin (ft or m), and Z is the side slope factor, the ratio of the horizontal to the vertical.

The circular conic section formula is

$$V = 1.047 D \left(R_1^2 + R_2^2 + R_1 R_2 \right) \quad (13.16)$$

$$V = 1.047 D \left(3 R_1^2 + 3 Z D R_1 + Z^2 D^2 \right) \quad (13.17)$$

where R_1 and R_2 are the bottom and surface radii of the conic section (ft or m), D is the depth of the basin (ft or m), and Z is the side slope factor, the ratio of the horizontal to the vertical.

For almost any type of storage basin, including ravines of complex topography, the stage–storage relationship can be expressed as a power function of the form

$$S = a h^m \quad (13.18)$$

where S is the storage (volume) in the reservoir, h is the elevation referenced to some datum, such as the reservoir bottom, and a and m are constants defining the function.

13.3.2 Elevation–Discharge Relationship

13.3.2.1 Reservoir Outlet Structures

The next set of relationship required is the elevation–discharge or stage–discharge relationship or the elevation–storage and elevation–area relationships for the reservoir or the basin. A stage–discharge curve defines the relationship between the depth of water in the reservoir and the discharge or outflow from a reservoir. This in turn depends on the elements called **flow regulators** that control the volumetric discharge exiting the reservoir through one or more structures called **outlets** or **spillways**. The elevation–discharge relationships can be derived directly from the hydraulic equations of flow through the flow regulators present in the outlet structures.

There can be various types of outlet structures. A reservoir typically has two outlets: a **principal outlet** or **principal spillway** and a **secondary outlet** or **emergency spillway**. The principal outlet is usually designed with a capacity to convey the design flows without allowing flow to enter the emergency spillway. The emergency spillway is sized to provide a bypass for floodwater during a flood that exceeds the design capacity of the principal outlet. This spillway should be designed to account for the potential threat to downstream areas if the storage facility were to fail. The principal spillway is also called the **service spillway** or **primary spillway** and the emergency spillway is also called **auxiliary spillway**. It is important to remember that the stage–discharge curve should account for the discharge characteristics of *both* the principal spillway and the emergency spillway.

The terms outlet structure and spillway in the case of reservoirs are often used interchangeably. However, for detention basins, primarily meant for the control of water quantity and sometimes also water quality, associated with land use and land cover change, the term outlet structure is more common, whereas for reservoirs intended for regional flood control or water supply, the term spillway has traditionally been used. This is perhaps because these latter types of reservoirs are impoundments with dams at their downstream ends.

The primary outlet structure of common detention basins is usually a **riser structure**, which is like a junction box or stand pipe. Flows from the basin enter the structure through one or more flow regulators and exit from the riser to a point, called the outfall, of the basin, through an over-sized pipe. A pipe culvert, weir, and orifice opening are typical flow regulators as inlets to the riser. In most cases, the inflow to

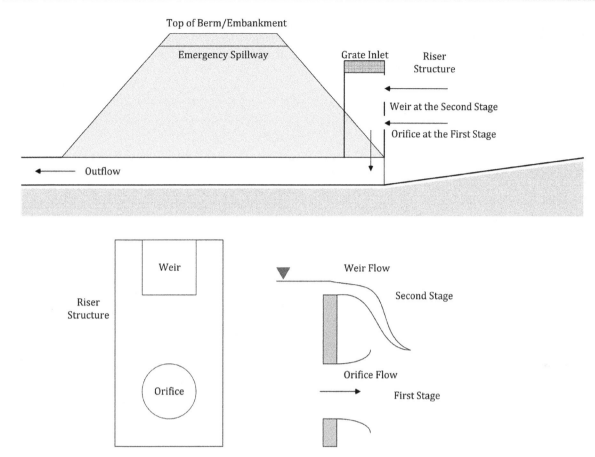

Figure 13.5 Schematic of a two-stage riser as a detention basin outlet structure.

the riser or other outlet structure can be of **orifice-type**, **weir-type**, or combinations of the two. Figure 13.5 shows schematics of a riser as a detention basin outlet structure. A riser can be a **single-stage** or **multiple-stage** riser, depending on the number of flow regulators present at different levels or elevations. For example, Figure 13.5 shows the riser with two stages, where the flow regulator at the lower elevation or first stage is an orifice and that at the higher elevation or stage is a weir. In some cases, there can be grate inlets on the top of a riser pipe to control the maximum elevation of water surface in a reservoir (Figure 13.5).

For reservoirs with dams, the primary spillways commonly fall into one of six types or are made of combinations of these types. These are (1) free overfall or straight drop; (2) overflow or ogee spillway; (3) side channel; (4) chute or trough; (5) shaft or morning glory (also called drop inlet); and (6) siphon. In addition, there are culverts, orifices, and labyrinth-type spillways. US Bureau of Reclamation (1961) provides an account of spillways of several types, their selection, and sizing. Spillways often are referred to as controlled or uncontrolled, depending on whether they are gated or ungated. Both overfall and overflow spillways can be gated or ungated. The two most common types of gates are **sluice gates** and **radial gates** (Figure 13.6). Spillways with crest gates will act as orifices under partial control of gates and as open crest weirs when full gate openings are made.

The hydraulic equations that are used to establish elevation–discharge relationships for several variants of flow regulators are given below.

13.3.2.2 Equations for Orifice Flow

The simplest flow-regulating structure is an orifice or nozzle installed into the *vertical side* of a riser structure as shown in Figure 13.5. The general equation for discharge through the orifice opening is

$$Q = C_d A \sqrt{2g\Delta h} \qquad (13.19)$$

where Q is the discharge, A is the area of the orifice or nozzle, C_d is the **discharge coefficient** (also known as **efflux coefficient**), g is the acceleration due to gravity, and Δh is the effective hydraulic head, which is the difference in the headwater and tailwater elevations measured from the centerline of the orifice (see Figure 13.7A when the outlet is unsubmerged and Figure 13.7B when the outlet is submerged). The discharge coefficient is dimensionless and therefore it is the same in both US customary and SI units.

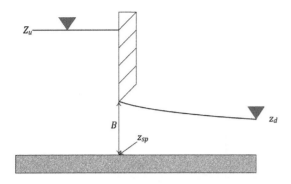

A) Sluice gate over a broad-crested spillway

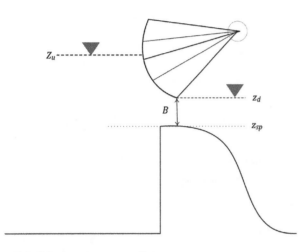

B) Radial gate over an ogee spillway crest

Figure 13.6 Two common types of gates over spillway crests: (A) sluice gate; (B) radial gate. B is the height of the opening between the spillway crest elevation, z_{sp}, and the lowest point of the gate at elevation, z_d. The water surface elevation in the reservoir is z_u.

Before the orifice opening becomes full and the inlet side is backed up as shown in Figure 13.7A, partial discharge flows through the opening unimpeded. This is shown in Figure 13.7C. A discharge equation, which can be used in such case, can be written as

$$Q = \left[0.496\left(\frac{h}{d}\right)^{1.57} - 0.04\left(\frac{h}{d}\right)^{0.5}\right]C_d\sqrt{g}(d^{2.5}) \quad (13.20)$$

Equation 13.19 can also be used for short culvert-type or pipe flows with inlets submerged (Figure 13.7A). In this case the culvert or the pipe is considered **inlet controlled**. Ideally, in that case, Δh should be greater than $1.5d$. However, if the culvert is rectangular, $A = WD$, where W is the span and D is the rise of the culvert and, in that case, Δh should be greater than $1.5D$. If the culvert or the pipe is outlet controlled, an equation that can be used for discharge computation is given as

$$Q = A\sqrt{\frac{2g\Delta h}{k}} \quad (13.21)$$

in which the factor k depends on Manning's n, length, and hydraulic radius of the pipe.

In some cases, the flow-regulating orifice has to be installed in the bottom of the storage basin similar to a bathtub drain (Figure 13.7D). These are **horizontal orifices** for which the discharge equation is generally given as

$$Q = C_d A\sqrt{2g\left(h - \frac{d}{2}\right)} \quad (13.22)$$

where h is the depth of water above the orifice opening and d is the diameter of the outlet opening (Figure 13.7D).

The discharge coefficient in Eqs. 13.19 and 13.20 can vary significantly with the shape and the type of the orifice, as shown in Figure 13.7E. However, a value of 0.6 is mostly used as a standard in the absence of detailed information. Similarly, for concrete pipes the discharge coefficient depends on the diameter and length of the pipe.

13.3.2.3 Equation for Flow-Restricting Pipes

In typical cases, instead of an actual orifice, a pipe near the bottom of reservoir is placed as an outlet and if the pipe is not too long then the orifice equation given as Eq. 13.19 can be used for flow through such a pipe when the outlet (inlet on the reservoir side) is fully submerged. However, if the pipe is long and more accurate calculations are needed, the length, slope, and the head loss through the pipe must be considered for calculation of discharge through the pipe. The equation that can be used for such a pipe, called a **flow-restricting pipe**, can be written as

$$Q = A\sqrt{2g\frac{h + S_0 L - mD}{K_L}} \quad (13.23)$$

where h is the depth of water above the inlet of the outlet pipe measured inside the reservoir, S_0 and L, respectively, are the slope and length of the pipe, m is the ratio of the water depth to the pipe diameter at the outlet end of the pipe, D is the diameter of the outlet pipe, and K_L is the sum of all loss coefficients for the outlet. The loss factor K_L is given as

$$K_L = k_e + k_f + k_b + k_o + k_t \quad (13.24)$$

where k_e is the entrance loss coefficient given by

$$k_e = \frac{1}{C_d^2} - 1 \quad (13.25)$$

k_f is the friction loss factor given by

$$k_f = f\frac{L}{D} \quad (13.26)$$

in which f is the Darcy–Weisbach friction loss coefficient, which under simplified assumptions can be expressed as a function of Manning's n as

$$f = 185\frac{n^2}{D^{1/3}} \quad (13.27)$$

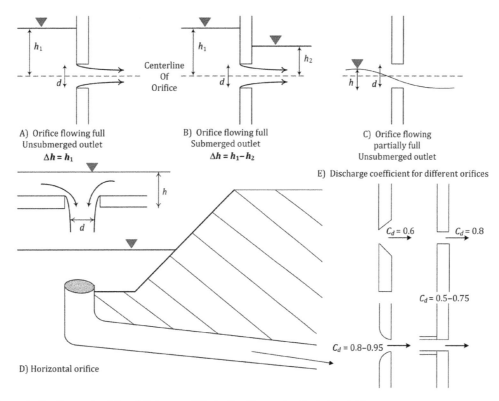

Figure 13.7 Types of orifice flows and variation of discharge coefficient with orifice opening. See text for details.

Bend losses in a closed conduit are a function of the bend radius, pipe diameter, and deflection angle of the bend. A formula that has been found to work for the calculation of bend loss is given by

$$k_b = K' k_{90} \tag{13.28}$$

where k_{90} is the loss factor for 90-degree bend, which has been found to be 0.2 for most cases, and the value of K' depends on the angle of bend.

Virtually no recovery of velocity head occurs where the pipe freely discharges into the atmosphere or is submerged under water. As a result, unless an especially shaped flared outlet is provided, k_o in Eq. 13.24 can be assumed to be 1.0.

The other loss factor included in Eq. 13.24 is the trash rack loss factor, since it is a common practice to emplace a trash rack at the inlet of the outlet pipe. The loss factor at a trash rack can be approximated using the following equation

$$k_t = 1.45 - 0.45 \left(\frac{a_n}{a_g}\right) - \left(\frac{a_n}{a_g}\right)^2 \tag{13.29}$$

in which a_n is the net open area between the rack bars and a_g is the gross area of the rack and supports.

13.3.2.4 Equations for Weir Flow

Weirs are often used as a primary overflow control device. They can provide the first level of emergency outflow, or actually be a part of the outflow regulating devices. Almost all types of spillways, when uncontrolled, are weir flows. However, depending on the crest type and cross-sectional geometry, there are variations of the equation giving discharge over weirs.

The flow over a **sharp-crested** weir having no end contractions is given by

$$Q = C_w L h^{1.5} \tag{13.30}$$

where C_w is the weir coefficient, L is the effective length of the weir crest, and h is the head above the weir crest (Figure 13.8A). In the case of a weir, the discharge coefficient (in US customary units) can be determined from

$$C_w = 3.27 - 0.4 \frac{h}{H_w} \tag{13.31}$$

in which H_w is the height of the weir crest above the invert elevation at the outlet on the reservoir side (Figure 13.8A). (In SI units the constants in Eq. 13.31 are 0.611 and −0.08, respectively.) Nevertheless, it should be noted that the weir coefficient is a function of the gravitational constant and is not dimensionless. Therefore, it has different values depending on which unit system is used. For example, a weir coefficient (C_w) of 3.00 in US customary units would be 1.66 in SI units, but both share the same discharge coefficient (C_d) of 0.56. For convenience, to convert an US customary weir coefficient to an equivalent SI weir coefficient, the US customary weir coefficient is multiplied by 0.552. For a sharp-crested weir, C_w is usually in the range of 3.2–3.3.

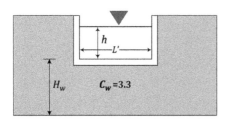
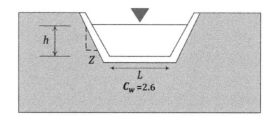

A) Sharp-crested rectangular weir (front view)

B) Broad-crested Cipolletti weir (front view)

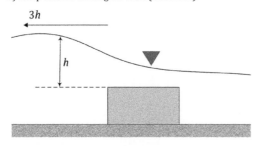
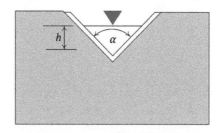

C) Broad-crested weir (side view)

D) V-notch or triangular sharp-crested weir (front view)

Figure 13.8 Type of weir flows. See text for details.

The length of the weir may need to be corrected for flow contractions at each end of a sharp-crested weir because generally the weir crest does not extend completely across the outlet structure (Figure 13.8A). In that case the effective length of the weir is given by

$$L = L' - 0.1 \times n_c \times h \tag{13.32}$$

where L' is the measured length of the weir crest and n_c is the number of end contractions.

Broad-crested or **ogee-crested** weirs are more commonly used in stormwater detention basins and reservoir spillways as overflow devices than sharp-crested weirs. The discharge over a broad-crested weir is given by

$$Q = C_w L h_t^{1.5} \tag{13.33}$$

in which h_t refers to total head above the weir crest given by $h_t = h + \frac{V^2}{2g}$ where V is the approach velocity at $3 \times h$ upstream of the crest (Figure 13.8C). However, for detention or reservoir overflow, V is usually set to zero. Eq, 13.33 is used for flow over an ungated (uncontrolled) emergency overflow spillway. The coefficient C_w for a broad-crested weir is usually in the range of 2.6–3.1. For ogee-crested weirs, the coefficient is higher, in the range of 3.2–4.1.

If there is a gate control over a broad-crested weir, the equation is

$$Q = \frac{2}{3}\sqrt{2g}\,C'_w L\left(h_1^{1.5} - h_2^{1.5}\right) \tag{13.34}$$

where h_1 is the total head to the bottom of the opening or head on the upper edge; h_2 is the total head to the top of the opening or head on the lower edge; and C'_w is the coefficient which differs with gate and crest arrangement as well as upstream and downstream flow conditions, with a typical range of 0.6–0.8.

Another broad-crested weir type is the trapezoidal or Cipolletti weir (Figure 13.8B) for which the discharge equation is

$$Q = 3.1(L + 0.8hZ)h^{1.5} \tag{13.35}$$

where Z is the side slope given as the ratio of horizontal to vertical (typically 4:1).

A **V-notch** or **triangular sharp-crested** weir should be considered, whenever the weir needs also to control low flows or for wet detention basin regular overspills from excess water entering the pond. As shown in Figure 13.8D, the water surface crest over this weir varies with depth. As a result, the weir capacity is sensitive to the water depth at low flows. The discharge over a V-notch weir is given by

$$Q = C_t h^{2.5} \tan\left(\frac{\alpha}{2}\right) \tag{13.36}$$

in which C_t is the discharge coefficient (typical value being 2.5) for a triangular weir, α is the weir notch angle in degrees, and h is the head above weir notch bottom, in feet, measured at a distance $2.5 \times h$ upstream. The coefficient varies with water depth with a typical range of 2.48–2.81.

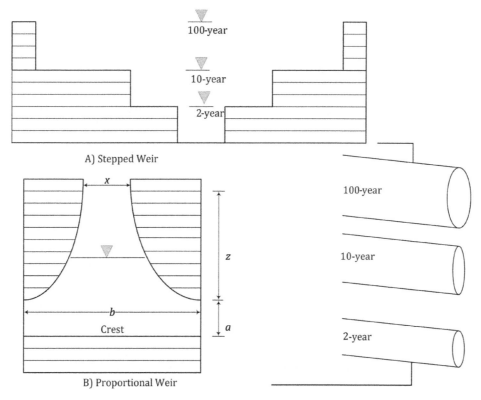

Figure 13.9 (A) Stepped weir, (B) proportional weir, and pipes of different sizes at different elevation, the number of years refer to return periods of floods.

For a V-notch weir with a notch angle of 45° the equation is

$$Q = C_t h^{2.5} \qquad (13.37)$$

and C_t is typically 2.49.

The equations for the sharp-crested weirs described above (Eqs. 13.30 and 13.36) are usually valid for a free, unsubmerged nape on the downstream side. However, when the tailwater rises above the weir crest (Figure 13.8C) the discharge equations should be corrected for submergence. Villemonte (1947) suggested the following equation with an accuracy of 5% for such a purpose:

$$\frac{Q_s}{Q} = \left[1.0 - \left(\frac{h_s}{h}\right)^n\right]^{0.385} \qquad (13.38)$$

where Q is the discharge calculated for the unsubmerged weir equation, Q_s is the discharge for the submerged weir, h is the head upstream of the weir, h_s is the tailwater depth above weir crest, and n is the exponent in the sharp-crested weir equation (i.e., 1.5 for rectangular and 2.5 for triangular).

In many detention basins, the outlets are designed to control flows arising from different frequency storms. In those situations, the outlet weir can be stepped (Figure 13.9A) for which there is also an alternative by placing pipes of different diameters at different stages or a **proportional** or **sutro weir**. Sutro or proportional weirs are designed so that the discharge is proportional to the total head. This design may be useful in some cases to meet performance requirements. The sutro weir consists of a rectangular section joined to a curved portion that provides proportionality for all heads above the elevation of the point a in Figure 13.9B. The weir may be symmetrical or nonsymmetrical. For this type of weir, the curved portion is defined by the following equation (calculated in radians)

$$\frac{x}{b} = 1 - \frac{2}{\pi}\tan^{-1}\left(\sqrt{\frac{z}{a}}\right) \qquad (13.39)$$

The head–discharge relationship is given as

$$Q = C_{pw} b \left(\sqrt{2ga}\right)\left(h_1 - \frac{a}{3}\right) \qquad (13.40)$$

The values of the discharge coefficient C_{pw} varies with a and b and for symmetric or asymmetric curved portion. A typical value is 0.6. The geometric parameters a and b are shown in Figure 13.9B.

13.3.2.5 Equation for Shaft or Morning Glory Spillway

For a **shaft** or **morning glory spillway** (Figure 13.10), the discharge equation is given as

$$Q = C_0 (2\pi R_s) h^{1.5} \qquad (13.41)$$

where C_0 is the discharge coefficient that is related to R_s, the radius of the overflow crest, and h, the total head. The shaft spillway is also called a **drop inlet.**

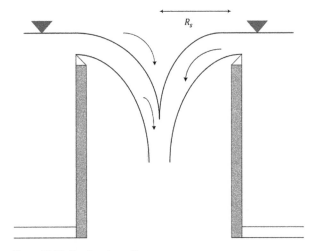

Figure 13.10 Morning glory spillway.

13.3.2.6 Elevation–Discharge Tables and Curves

As discussed above, an outlet structure or primary spillway can have a combination of two or more types of flow regulators. In instances where the primary spillway has multiple flow controls, the total discharge from all of the flow regulators as a function of stage must be calculated. In addition, the final relationship should also combine the stage–discharge relationships from both the primary spillway and emergency spillway. However, due care must be exercised in combining the stage–discharge data from individual flow regulators into one relationship. It is not merely summing up the two or more sets of relationships. The operation of the entire outlet structure must be understood in selecting the stage–discharge pairs from separate flow controls that should be added together, as illustrated with an example below.

In general, the procedure for the development of a composite stage–discharge relationship for a reservoir is to select a set of elevations, starting at the elevation of the lowest outlet and ending at the expected maximum routed water level, and to calculate discharge at each elevation through the single or multiple spillways. Usually, the elevations selected must be the same as the elevations for which the elevation–storage relationship has been developed in the previous step to eliminate interpolations needed to make elevations consistent in both tables, which will be combined in the next step to get the storage–discharge relationship for the whole reservoir. The end result will be a table of elevations with appropriate values of corresponding storage and discharge.

In cases where the outlet structure cannot be defined in terms of any of the types described above or some other standard hydraulic structure, special equations giving the stage–discharge relationship are either provided by the manufacturer of such a structure or should be developed from actual field measurements.

Figure 13.11 shows a riser outlet structure that acts as a primary spillway for the reservoir for which the stage–storage graph is shown in Figure 13.12. Table 13.1 gives the elevation–discharge values for the individual flow regulators and illustrates a useful way of keeping the calculations in order. Columns 2 to 5 show the discharges corresponding to the elevations (column 1) for the first and second stage, both weirs and orifices separately. By tabulating the data for different types of flow in separate columns, and keeping the two stages separate, the total discharges are more easily summed. Note that the totals in column 6 are not merely sums of all discharge values in a row. The operation of the outlet shown in Figure 13.11 is taken into account in recording the discharge values in column 6. At the same time, consideration should also be given as the water surface elevation inside the riser rises and eventually it gets fully submerged without any positive hydraulic head, making both stages disengaged. At elevation 180 m, the control shifts from the weir to the orifice for the first stage. Similarly, at elevation 183 m, the control shifts from weir to orifice for the second stage. Figure 13.12 shows the stage–discharge relationship for the entire structure. It should be also noted that if there were an auxiliary spillway for emergency flows, those values would also have been added in developing this relationship and in that case the stage–discharge curve would not fall back.

13.3.2.7 Tailwater Control of Outlet Hydraulics

The discharge calculated by using any of the equations given above is valid strictly when the water surface elevation in the riser box or the outfall does not control the water surface elevation in the reservoir, meaning that there is no control of the flow by the tailwater. The basic premise of hydrologic routing is that the outflow is free. If, however, the tailwater does control the headwater in the reservoir, which significantly affects the magnitude of discharge (outflow), then more detailed hydraulic computations must be carried out to develop the stage–discharge table and graph.

13.3.3 Storage–Discharge Relationship

For an uncontrolled reservoir, outflow and water in storage are both uniquely a function of the pool elevation, that is, water surface elevation in the reservoir. The elevation–storage and elevation–discharge relationships established by the procedures described above are combined to develop the storage–discharge relationship (Figure 13.13), which is the second major input required for level-pool routing.

13.4 FLOOD ROUTING BY NUMERICAL INTEGRATION OF THE CONTINUITY EQUATION

The storage equation governing the rate of change of reservoir storage volume is the continuity equation (Eq. 12.9), which can also be written as

13.4 Flood Routing by Numerical Integration of the Continuity Equation

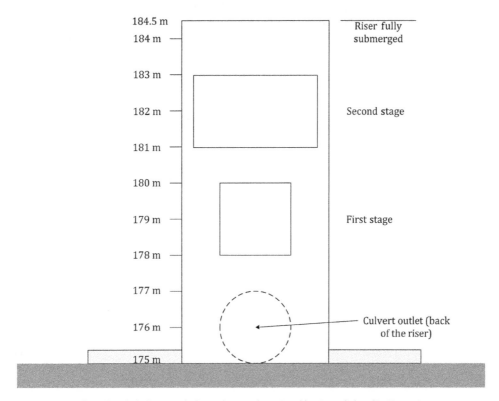

Figure 13.11 Two-stage primary spillway for which elevation–discharge data are shown in Table 13.1 and plotted in Figure 13.11.

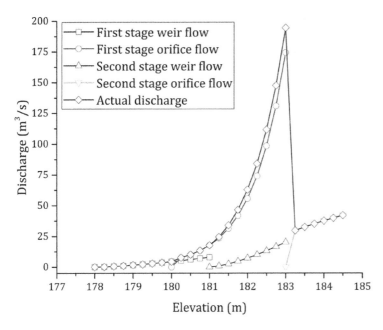

Figure 13.12 Plot of the example of elevation–discharge relationship given in Table 13. 1 for a two-stage primary spillway shown in Figure 13.11 (A black and white version of this figure will appear in some formats. For the color version, please refer to the plate section.).

$$\frac{dS}{dt} = I(t) - Q(t, S) \qquad (13.42)$$

In the modified Puls and other hydrologic methods of reservoir routing, in the storage equation the outflow Q is considered to depend only on S. However, where reservoir outflow is controlled by a spillway gate or valve, the outflow characteristics can vary. The outflow is then some known function of time, t, as well as S. Therefore, Eq. 13.42 is more general in nature. To solve Eq. 13.42, it is necessary to relate the outflow to the storage volume and time, that is, to give the form of a function $Q(t, S)$. This can be done by expressing the dependence of Q and S on the elevation of water surface, h, in the reservoir.

Table 13.1 *Elevation–discharge data for a two-stage primary outlet structure shown in Figure 13.11*

Elevation (m)	Weir flow, m³/s (first stage)	Orifice flow, m³/s (first stage)	Weir flow, m³/s (second stage)	Orifice flow, m³/s (second stage)	Total discharge (m³/s)
178.00	0.00				0.00
178.25	0.19				0.19
178.50	0.54				0.54
178.75	0.99				0.99
179.00	1.53				1.53
179.25	2.14				2.14
179.50	2.81				2.81
179.75	3.54				3.54
180.00	4.33	0.00			4.33
180.25	5.17	7.54			7.54
180.50	6.05	10.04			10.04
180.75	6.98	13.35			13.35
181.00	7.96	17.76	0.00		17.76
181.25		23.63	0.90		24.52
181.50		31.43	2.54		33.97
181.75		41.81	4.66		46.48
182.00		55.63	7.18		62.80
182.25		74.00	10.03		84.03
182.50		98.45	13.18		111.63
182.75		130.97	16.61		147.58
183.00		174.23	20.30	0.00	194.52
183.25				29.71	29.71
183.50				32.55	32.55
183.75				35.16	35.16
184.00				37.59	37.59
184.25				39.87	39.87
184.50				42.02	42.02

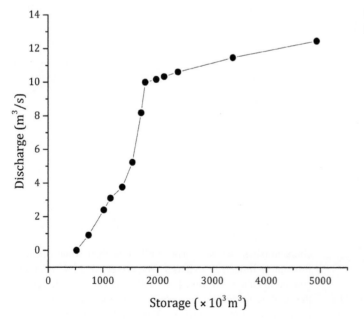

Figure 13.13 Storage–discharge relationship for the two-stage primary spillway.

Fenton (1989, 1992) pointed out that since Eq. 13.42 is a first-order ordinary differential equation for S as a function of t, it can be solved by any one of a number of methods, of varying complexity and accuracy, available for the numerical solution of the differential equation. The traditional hydrologic methods for reservoir routing are complicated, because they require the solution of a transcendental equation at each time step. But reservoir routing is actually simply the numerical integration of the differential equation, Eq. 13.42. Some of those numerical methods, such as the one presented below, are actually simpler than the hydrologic methods.

In addition to Eq. 13.42, which we will call the **S-formulation**, another form of the same differential equation expressing the rate of change of reservoir stage can also be obtained. If the reservoir surface elevation, h, changes by an amount dh, in the limit as $dh \to 0$, the change in storage, dS, is given by

$$dS = A(h)dh \tag{13.43}$$

where $A(h)$ is the plan or surface area of the water surface at elevation h. Substituting Eq. 13.43 into Eq. 13.42, writing the outflow, Q, as a function of both t (in the case of controlled discharges) and of h, an equivalent form of the storage equation is obtained as

$$\frac{dh}{dt} = \frac{I(t) - Q(t,h)}{A(h)} \tag{13.44}$$

which is a differential equation for the surface elevation itself. This equation was presented by Chow et al. (1988). It makes no use of the storage volume, S, which then does not have to be calculated for routing purposes. We will refer to it as **H-formulation**. However, it requires an elevation–area relationship.

If the form Eq. 13.42 is used, the storage volume S must be obtained as a function of h from the integral

$$S(h) = \int_0^h A(y)dy \tag{13.45}$$

which can be evaluated by a numerical approximation for various surface elevations. Then, it is necessary to express S as a function of Q, usually by creating a table of pairs of corresponding values. In those cases where the discharge is controlled, this has to be repeated every time a gate or valve is adjusted.

In the same way that the dependence of Q on S is usually represented by a table of pairs of values, in general, the H-formulation requires a similar table for A and h, to give $A(h)$, obtained from planimetric information from contour maps.

Whichever formulation is adopted, S- or H-, the problem is one of numerical integrations of a differential equation, for which there are many standard methods. A very common method is presented below subsequent to the statement of the problem in terms of numerical analysis.

The forms of storage equation can be generalized by the expression

$$\frac{dy}{dt} = F(t,y) \tag{13.46}$$

where y is the dependent variable, either S or h. In the case of Eq. 13.42, $y = S$ and $F(t,S) = I(t) - Q(t,S)$, while for Eq. 13.44, $y = h$ and $F(t,h) = [I(t) - Q(t,h)]/A(h)$. Thus, y is an unknown function of time, t, and we would like to approximate y, being told that dy/dt, the rate at which y changes, is a function of t and y itself.

The general form in Eq. 13.42 encompasses both cases where the outflow is controlled and where it is not. Numerical methods can be applied whether or not control is varying the form of the outflow function with time.

The functional forms of storage and surface area of the reservoir as a function of water surface elevation in the reservoir can be obtained from Eq. 13.18, which can be recast as

$$S(h) = ah^m \tag{13.47}$$

The derivative of Eq. 13.47 with respect to h gives

$$A(h) = amh^{m-1} \tag{13.48}$$

The discharge equation from any of the outlet works/spillway configurations can be written as

$$Q(h) = bh^m \tag{13.49}$$

where b is a constant and the exponent m is the same as for the storage function, Eq. 13.47. Hence, $Q = kS$, where $k = b/a$, a constant.

A numerical method that can be used in the solution of Eq. 13.46 is the **fourth-order Runge–Kutta method**. This is a very popular and reliable method for the solution of a first-order ordinary differential equation. The algebraic derivation of the equations used in this method, developed by the German mathematicians Carl Runge and Martin Kutta, is quite complicated. Fiorentini and Orlandini (2013) evaluated the robustness of two other computationally more intensive numerical methods along with fourth-order Runge–Kutta method. They concluded that the Runge–Kutta method is most robust and reliable but may need smaller time steps in the critical phases of simulation. The outline of the method is presented below, and Example 13.3 illustrates the computational process.

The problem in hand, in the field of differential equations, is called an initial value problem (see Section 2.14 in Chapter 2). The solution commences with the known initial conditions, either the value of S or h, at $t = 0 = t_0$, denoted by S_0 and h_0 respectively. The superscript j is used for the index of time, since the solution proceeds in time at discrete steps. The step size is an interval of time denoted by $\Delta t > 0$. Thus, $t^{j+1} = t^j + \Delta t$. The Runge–Kutta method approximates the value of y^{j+1} from the estimated value of y^j in the previous step and a weighted average of *four increments*, where each increment is a product of the size of the interval,

Δt, and an estimated slope specified by the function F on the right-hand side of Eq. 13.46. The method is called fourth order, because the local truncation error in the approximation process is in the order of $O(\Delta t^5)$, while the total accumulated error is the order of $O(\Delta t^4)$. This gives

$$y^{j+1} = y^j + \frac{\Delta t}{6}[k_1 + 2k_2 + 2k_3 + k_4] \quad (13.50)$$

for $j = 0, 1, 2, 3, \ldots, N$; and the weighting factors $k_1, k_2, k_3,$ and k_4 are defined as

$$k_1 = F(t^j, y^j) \quad (13.51)$$

$$k_2 = F\left(t^j + \frac{\Delta t}{2}, y^j + \frac{1}{2}k_1\right) \quad (13.52)$$

$$k_3 = F\left(t^j + \frac{\Delta t}{2}, y^j + \frac{1}{2}k_2\right) \quad (13.53)$$

$$k_4 = F(t^j + \Delta t, y^j + k_3) \quad (13.54)$$

The most complicated part of solving the differential equation is that of interpolating the value of the dependent variable in a table of value pairs. It is necessary to obtain $A(h)$ for arbitrary h, or $Q(t, S)$ for arbitrary S. If the time step other than the interval of the inflow hydrograph is used, it is also necessary to obtain $I(t)$ for arbitrary values of t, also from data pairs, by the method of interpolation. In any of these cases, a simple *linear interpolation* formula can be used. If there is a pair of variables, (x, y), with known pairs of values, (x_1, y_1) and (x_2, y_2), the interpolated value of y corresponding to a desired value of x in the range $x_1 \leq x \leq x_2$ is

$$y = y_1 + \left(\frac{y_2 - y_1}{x_2 - x_1}\right)(x - x_1) \quad (13.55)$$

In the case of the S-formulation, the weighting factors $k_1, k_2, k_3,$ and k_4 are first evaluated for the initial condition where the inflow I_0 and storage S_0 are known. Thus, for $j = 0$,

$$k_1 = I_0 - Q_0 \quad (13.56)$$

$$k_2 = \left(\frac{I_0 + I_1}{2}\right) - \left(Q_0 + \frac{k_1}{2}\right) \quad (13.57)$$

$$k_3 = \left(\frac{I_0 + I_1}{2}\right) - \left(Q_0 + \frac{k_2}{2}\right) \quad (13.58)$$

$$k_4 = (I_0 + I_1) - (Q_0 + k_3) \quad (13.59)$$

From the values of the weighting factors calculated above the storage for the next step, $(j + 1) = 1$, is

$$S^1 = S^0 + \left(\frac{k_1}{6} + \frac{k_2}{3} + \frac{k_3}{3} + \frac{k_4}{6}\right)\Delta t \quad (13.60)$$

Then, the recursive process continues up to the end of the inflow hydrograph time.

In the case of the H-formulation, the function in Eqs. 13.56–13.59 is replaced by $[I(t) - Q(h)]/A(h)$, and therefore the unknown h is evaluated with the initial value h_0, and A and Q are read from the tables of h versus A and h versus Q through interpolation.

It can be pointed out that Eq. 13.42 is also applicable for the modified Puls method to get Q from estimated $2S/\Delta t + Q$.

13.5 ACCURACY OF THE ROUTING METHOD

In essence, both a hydrologic method such as the modified Puls method as well as a numerical mathematical algorithm such as the Runge–Kutta method integrate the function dS/dt. Therefore, as $\Delta t \to 0$, the numerical approximation approaches the true limit of the integrand. Consequently, the accuracy of the estimation, in either method, improves as Δt gets smaller. In reality, however, for the sake of convenience, Δt is set as the time interval given for an inflow hydrograph. If the time interval in the inflow hydrograph is too large but at the same time a greater accuracy is needed, a new inflow hydrograph must be prepared with smaller time intervals by getting the values of the inflow at intermediate intervals from interpolation of the values given in the original hydrograph, using Eq. 13.55.

13.6 DETENTION BASINS

An area of modern hydrologic engineering where reservoir routing plays a crucial role is the design of detention basins. However, there is a very subtle difference between the process of reservoir routing and detention basin design.

Both channel and reservoir routing problems can be viewed as a system–load–response problem, routinely encountered in structural engineering design. The inflow hydrograph is the load, the reservoir is the system, and the outflow hydrograph is the response. In the routing procedures described above it is an analysis of the system to compute the response for a given load. In the case of detention basin design, not only the load is known but the response is given as a priori information and the task is to design the system that produces the desired response. Still, the role of reservoir routing is important.

The primary purpose of a detention basin is to reduce the runoff resulting from some actions (projects), such as an increase in the imperviousness of the land surface, which are causing the runoff being in excess of what would result without those actions, or naturally. Thus, the inflow hydrograph in this case is the one that is known to result from the project. In typical cases, these projects are developmental projects for urbanization and hence the inflow hydrograph in such situations is called a post-development condition. The outflow hydrograph is the hydrograph computed by considering the area without the project and hence is commonly called the pre-development condition. The task in the design of a detention basin is then to calculate the storage (volume) required for the detention basin and the outlet structure such that when the post-development hydrograph is routed through the detention basin, the outflow hydrograph matches the pre-development hydrograph, especially for the peak discharge.

Theoretically, the volume of a detention basin can be calculated from the given inflow and outflow hydrographs. Figure 13.14A shows the inflow and outflow hydrographs of a detention basin associated with the development of a parcel of land. The detention storage available during runoff from the design storm is shown by the hatched area. This storage volume is given by the equation

$$S^j = S^{j-1} + 0.5\Delta t \left[(I^j + I^{j-1}) - (Q^j + Q^{j-1}) \right] \quad (13.61)$$

An approximation of the detention volume is often required for initial planning. For a reservoir, the peak of the outflow hydrograph lies on the recession limb of the inflow

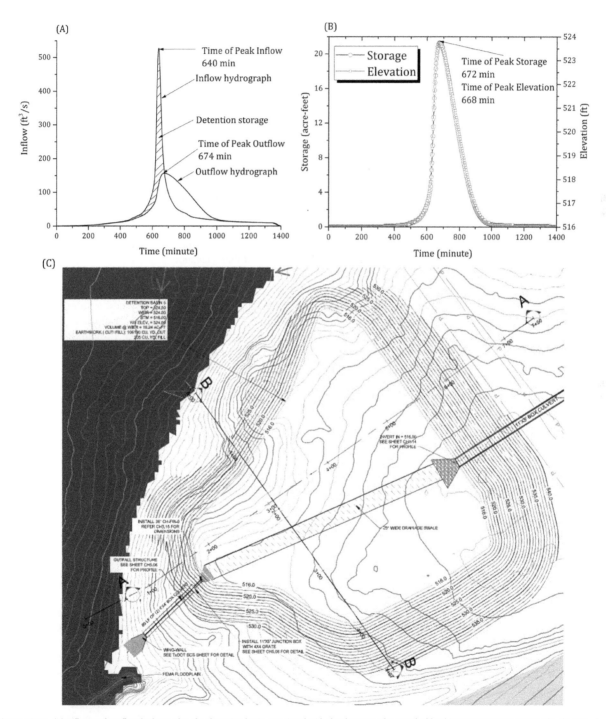

Figure 13.14 (A) Inflow and outflow hydrographs of a detention basin associated with development of a parcel of land in Tarrant County, Texas. The theoretical detention basin volume derived from the hydrographs is shown by the hatched area. (B) Variation of storage and elevation with time. Note that maximum storage occurs at about the time of peak outflow. (C) Plan view of the detention basin, which has only one culvert outlet at the bottom of the basin.

hydrograph at almost the same time when peak storage occurs (e.g., Figure 13.14B). The inflow hydrograph for post-development conditions typically exhibits a sharp peak due to the decrease in the time of concentration. Thus, if the inflow hydrograph can be assumed to be represented as an isosceles triangle, the geometric relationships give an approximation of the detention storage as

$$S_D = (I_P - Q_P)T_{IP} \qquad (13.62)$$

in which S_D denotes the required detention storage volume, I_P is the peak discharge of the inflow hydrograph, Q_P is the peak discharge of the outflow hydrograph, and T_{IP} is the time to peak of the inflow hydrograph. Empirical observation shows that this approximation works well if the outlet is principally a pipe outlet. The inflow–outflow hydrographs shown in Figure 13.14 are for the detention basin, also shown in this figure, which has only one culvert outlet at the bottom of the basin. Nevertheless, going back to the load–system–response analogy, both Eqs. 13.61 and 13.62 represent a key element of the system for the given load and desired response. For complex outlet structures and final design, Eq. 13.61 should be used for the determination of the actual volume of the detention basin.

13.7 EXISTING AND PROPOSED RESERVOIRS

Reservoir routing plays an important role in both the analysis of existing reservoirs and design of proposed reservoirs.

For a proposed reservoir, the design inflow hydrograph is developed based on a detailed hydrologic analysis and modeling of the watershed or river basin that will contribute flows to the reservoir. Typically, these are flood hydrographs of probable maximum flood (PMF) or flows that are expected to be generated from rainfall of very low frequency (high return period) such as 500- or 1000-year storms. Subsequently, the decisions on reservoir size (volume, elevation, area), spillway type, size and its stage are used in reservoir routing studies to test whether the design flood wave will pass safely and the surcharge volume of water will be retarded.

For an existing reservoir, the inflow design hydrograph is not variable once it has been constructed on the basis of the selection of the inflow design flood. The reservoir storage capacity also is not variable for a given reservoir site. However, the spillway discharge curve can be made variable. It depends not only on the type and size of the spillway but also on the manner of operating the spillway and outlet works to regulate the outflow.

The quantity of water a spillway can discharge depends on the type of control device. For a simple overflow crest, the flow will vary with the hydraulic head on the crest, and surcharge will increase with an increase in spillway discharge. For a gated spillway, however, the outflow can be varied with respect to the reservoir head by the operation of gates. The decisions for the operation of the gates give rise to **rule curves** that are used for reservoir operation. These can be done by proper routing studies.

Reservoir routing is an integral component of hydrologic modeling of watersheds containing existing reservoirs due to their controls on the attenuation and lag of hydrographs along the channels. Such models are also useful to determine the potentials of downstream flooding that can result for storms with hydrologic frequencies other than that of the design flood hydrograph. Figure 13.15 shows a medium-sized reservoir behind a small dam on Rutherford Branch, a second-order tributary of East Fork Trinity River to the northeast of Dallas. This is an example of numerous small flood control dams present throughout the United States, built by the SCS (now NRCS) of the US Department of Agriculture from the Flood Control Act of 1956. This dam (SCS Site 1b) is considered a high hazard dam as defined by Texas Administrative Code Section 299. This is an earthen structure, approximately 32 feet in height with a top of dam elevation of 712 feet above mean sea level (MSL). An approximately 140-foot earthen emergency spillway, with a crest elevation of 705 feet above MSL, is located east of the dam to convey the design storm, which is 75% of the 1-hour PMF as defined by the Texas Commission on Environmental Quality. A stand pipe is located near the center of the dam used to maintain the level-pool water surface elevation of 698 feet above MSL. The dam has a storage volume of slightly under 420.7 acre-feet at the crest elevation of the emergency spillways and 785.5 acre-feet at the top-of-dam elevation.

Figure 13.16 shows the outflow hydrographs and storage–elevation curves that can potentially result after the inflow flood hydrographs produced by two design storms are routed through this reservoir. The 100-year flood hydrograph is attenuated significantly, whereas a 12-hour PMF has the potential of overtopping the spillway.

13.8 RESERVOIR ROUTING IN HEC-HMS

There are three alternative ways by which reservoir routing can be performed in HEC-HMS. In addition to these routing methods, several options are available for the storage method. These are briefly explained below.

13.8.1 Outflow Curve Routing Method

The **outflow curve routing method** can be used when a known storage–outflow relationship can be provided for a reservoir. This method does not allow for individual representation of the components of outlet works, rather, it lumps all outflow structures into one storage–discharge relationship. The routing is always performed using only the storage–discharge curve given as an input. After the routing is performed using this curve, the program computes the elevation and surface area for each time step, depending on the selected storage method, which depends on whether tables for either elevation–area–discharge or elevation–storage–discharge are provided as inputs. The initial condition is

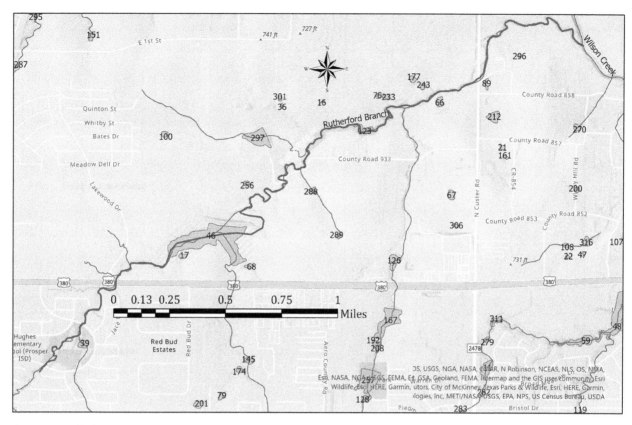

Figure 13.15 A flood control reservoir on a tributary channel on Rutherford Branch, a tributary of East Fork Trinity River, northeast Dallas, Texas.

that inflow equals outflow. The results of reservoir routing shown in Figure 13.16, done in HEC-HMS, use the outflow curve routing method.

13.8.2 Specified Release Routing Method

The **specified release routing method** can be used to model reservoirs where the total discharge is known for each time interval of simulation. Usually, this method is used when the discharge is either observed or completely specified by an external decision process, as mentioned in Section 13.6. The method can then be used to preserve the specified release and track the storage using the inflow, outflow, and conservation of mass. It uses a specified discharge and computes the storage that would result. With this method, it is possible to set a maximum capacity limit on the reservoir. The storage at the end of the previous time interval, together with the inflow volume and specified outflow volume for the current time interval, are used to calculate the storage at the end of the current time interval.

13.8.3 Outflow Structure Routing Method

The **outflow structure routing method** can be selected to model reservoirs for which data on a number of uncontrolled outlet structures can be provided as input. The outflow structure method has the option to define an outlet and a spillway with dam-top information separately. For example, a reservoir may have a spillway and several low-level outlet pipes. While there is an option to include gates on spillways, the ability to control the gates is somewhat limited. Currently, there are no gates on outlet pipes. However, a time series of releases can be provided in addition to the uncontrolled releases from various structures. This is particularly important if there is a backwater effect, the downstream control, as discussed in Section 13.3.2.7. In this case, an external analysis is used to develop the additional releases, based on an operations plan for the reservoir. Additional features in the reservoir for culverts and pumps allow the simulation of interior ponds. These types of reservoirs typically represent urban detention basins or storage areas associated with other flood control systems. The inflow–outflow hydrographs and stage–storage curves shown in Figure 13.14 for an urban detention basin were calculated in HEC-HMS using the outflow structure routing method.

13.8.4 Storage Methods

In addition to selecting a routing method, the program requires the selection of an associated **storage method**. The

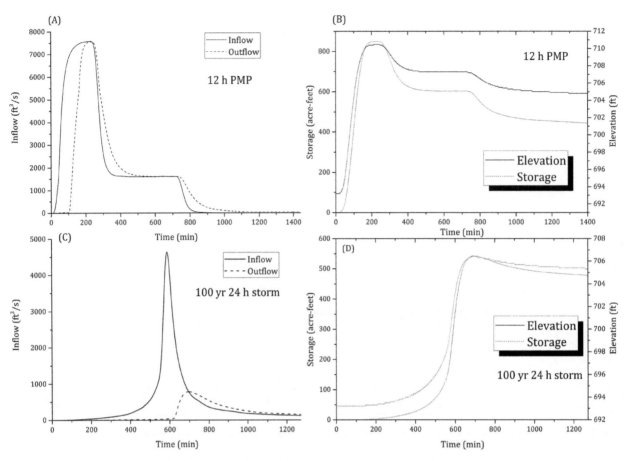

Figure 13.16 (A) Inflow–outflow hydrographs for a 12-h PMP (B) Storage–elevation curves for the 12-h PMP (C) Inflow–outflow hydrographs for a 100-year storm and (D) storage–elevation graphs for a 100-year storm. The reservoir for this example is shown in Figure 13.15.

storage method defines the relationship between detention and discharge. There are five storage methods to choose from: elevation–area–discharge, elevation–storage–discharge, storage–discharge, elevation–area, and elevation–storage. The function to be used, which is to be primary if there is more than one function in the input, must be specified.

Different sets of storage methods are available, depending on the routing method selected. For a simple reservoir, the combination of the outflow curve routing method and the storage–discharge storage method is preferred. The storage–discharge relationship can be defined with some dead pool storage associated with zero outflow. For the specified release routing method, a time series of releases is specified. Storage is post-processed using the elevation–area or the elevation–storage relationship. For the elevation–area–discharge and elevation–area storage methods, the program automatically transforms the elevation–area curve into an elevation–storage curve using the conic formula described in Section 13.3. The elevation–area–discharge storage method option is rarely used. Because HEC-HMS does not use triplet curves, if the elevation–area–discharge or elevation–storage–discharge methods are selected, ultimately the storage–discharge relationship is still used to calculate the routing. The elevation is post-processed. The user selects the primary field when entering data (i.e., which curve to key on), and interpolation is used to get the other column.

13.9 EXAMPLES

Example 13.1: The dimensions of a stormwater detention basin geometry are shown in Figure 13.17. The outlet structure of the basin consists of a 2-ft × 2-ft box culvert which has a 6-inch high weir plate at its entrance point to the riser box (see Figure 13.17B). Determine the storage–discharge relationship for this detention basin.

Solution:

Step 1: An increment of depth (D) of 0.5 ft is selected and with the increment of depth, the surface area at each level and incremental volume with the

Table 13.2 *Calculations of incremental increase in surface and volume from the bottom of the basin at elevation 100 feet*

D	L (ft)	W (ft)	Area (ft²)	Volume (ft³)	Elevation (ft)	Volume (acre-ft)	Area (acres)
0	450	120	54 000	0	100	0.00	1.24
0.5	454	124	56 296	27 572.6667	100.5	0.63	1.29
1	458	128	58 624	56 301.3333	101	1.29	1.35
1.5	462	132	60 984	86 202	101.5	1.98	1.40
2	466	136	63 376	117 290.667	102	2.69	1.45
2.5	470	140	65 800	149 583.333	102.5	3.43	1.51
3	474	144	68 256	183 096	103	4.20	1.57
3.5	478	148	70 744	217 844.667	103.5	5.00	1.62
4	482	152	73 264	253 845.333	104	5.83	1.68
4.5	486	156	75 816	291 114	104.5	6.68	1.74
5	490	160	78 400	329 666.667	105	7.57	1.80
5.5	494	164	81 016	369 519.333	105.5	8.48	1.86
6	498	168	83 664	410 688	106	9.43	1.92
6.5	502	172	86 344	453 188.667	106.5	10.41	1.98
7	506	176	89 056	497 037.333	107	11.41	2.04

(A)

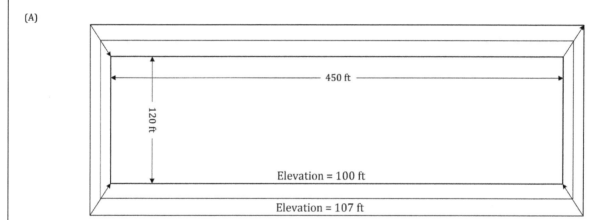

(B)

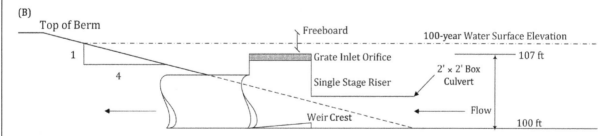

Figure 13.17 A stormwater detention basin with an outlet structure: (A) plan view, (B) elevation.

frustum of a pyramid formula are calculated as given in Table 13.2.

Step 2: From the elevation diagram it can be seen that weir flow occurs from 100.5 ft up to 102 ft and then orifice flow occurs. Since it is a weir plate, it is a sharp-crested weir and a weir coefficient of 3.3 is appropriate. The weir is across the 2-foot bottom of the concrete box. From the geometry of the box culvert, the coefficient of discharge for the orifice is taken as 0.8. Then, discharge is calculated as a function of elevation as presented in Table 13.3.

Step 3: The storage and discharge values corresponding to a common elevation from Tables 13.2 and 13.3 are combined to get the storage–discharge relationship as given in Table 13.4 and shown in Figure 13.18.

Table 13.3 *Calculations of discharge as a function of elevation*

Head (ft)	Weir flow (ft^3/s)	Orifice flow (ft^3/s)	Elevation (ft)	Total flow (ft^3/s)
0.5	0.00		100.50	0.00
1	2.33		101.00	2.33
1.5	6.60		101.50	6.60
2	12.12		102.00	12.12
2.5	18.67		102.50	18.67
3		22.24	103.00	22.24
3.5		25.68	103.50	25.68
4		28.71	104.00	28.71
4.5		31.45	104.50	31.45
5		33.97	105.00	33.97
5.5		36.32	105.50	36.32
6		38.52	106.00	38.52
6.5		40.60	106.50	40.60
7		42.59	107.00	42.59

Table 13.4 *Storage–discharge relationship*

Elevation (ft)	Storage (acre-ft)	Discharge (ft^3/s)
100.5	0.00	0.00
101	0.63	2.33
101.5	1.29	6.60
102	1.98	12.12
102.5	2.69	18.67
103	3.43	22.24
103.5	4.20	25.68
104	5.00	28.71
104.5	5.83	31.45
105	6.68	33.97
105.5	7.57	36.32
106	8.48	38.52
106.5	9.43	40.60
107	10.41	42.59

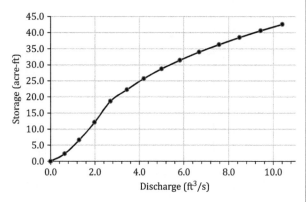

Figure 13.18 Storage–discharge relationship for the detention basin shown in Figure 13.17.

Example 13.2: The Electricity Generating Authority of Thailand (EGAT) operates 14 large and important dams with hydropower plants in Thailand. For the safe operations of spillways, EGAT implemented a monitoring system that has a reservoir routing component to simulate the reservoir water level (RWL) which becomes very crucial during the rainy season with impending flood waves entering the reservoirs, because the incorrect release of water through gated spillways can have the potential of both upstream and downstream flooding as well as inequitable distribution of water for other purposes, such as irrigation. Sirkit Dam is one of the 14 dams operated by EGAT. It is an embankment dam on the Nan River, a tributary of the Chao Phraya River, which is the major river in Thailand and flows through the capital city Bangkok. The spillway of Sirkit Dam consists of a tunnel controlled by two radial gates. The photograph in Figure 13.19A shows an areal view of the reservoir and the dam, and an areal view of the spillway is shown by in Figure 13.19B. The photograph Figure 13.19C gives a close-up view of the spillway.

The elevation-discharge-storage-surface area data for Sirkit Dam are given in Table 13.5. The PMF used in the design and specified in the construction manual of the dam is given in Table 13.6. The reservoir routing algorithm employed in the simulation routine of the monitoring system is a variation of level-pool routing. In this example, the modified Puls method is used to route the PMF.

Figure 13.19 Areal view of (A) Sirkit reservoir in Thailand and (B) the spillway of Sirkit Dam. (C) Close-up view of the spillway of Sirkit Dam. (Photographs courtesy of Rangsarit Vanijjirattikha, the Electricity Generating Authority of Thailand.) (A black and white version of this figure will appear in some formats. For the color version, please refer to the plate section.)

(c)

Figure 13.19 (cont.)

Table 13.5 *Elevation–discharge–storage–surface area data for Sirkit Dam*

Elevation from spillway crest (m above MSL)	Total discharge two gates fully open (m³/s)	Storage capacity ($\times 10^6$ m³)	Surface area (km²)
150.5	0	6759.61	210.456
157.5	1015	8381.52	238.865
158	1148	8503.55	241.030
158.5	1288	8626.41	243.223
159	1429	8750.10	245.447
159.5	1571	8874.63	247.706
160	1720	9000.00	250.000
160.5	1870	9126.21	252.334
161	2020	9253.28	254.710
161.5	2181	9381.21	257.131
162	2354	9510.00	259.600
162.5	2522	9639.67	262.121
163	2689	9770.21	264.696
163.5	2856	9901.65	267.330
164	3023	10 033.99	270.025
164.5	3201	10 167.24	272.785
165	3379	10 254.49	272.649
165.5	3557	10 381.34	275.029
166	3740	10 508.32	277.435
166.5	3853.8	10 635.40	279.867

Table 13.6 *PMF and 100-year discharge for Sirkit Dam, Thailand*

Time (h)	100-year discharge (m³/s)	PMF (m³/s)
0	500	1000
6	516	1020
12	554	1100
18	624	1300
24	730	2350
30	886	4400
36	1104	5650
42	1416	6300
48	1808	6800
54	2234	7100
60	2688	7500
66	3147	8000
72	3593	8500
78	3956	9700
84	4254	10 500
90	4384	9900
96	4417	8800
102	4306	7700
108	4111	6500
114	3824	5700
120	3523	5200
126	3193	4800
132	2854	4400
138	2548	4100
144	2289	4050
150	2054	4000

Table 13.7 *Storage–indication relationship for Sirikit Dam*

Effective storage, S (m³)	$(2S/\Delta t) + Q$ (m³/s)	Discharge, Q (m³/s)
0.00	0.00	0
1 621 917 400.38	151 192.54	1015
1 743 947 287.24	162 624.60	1148
1 866 805 964.83	174 140.40	1288
1 990 496 931.00	185 734.27	1429
2 115 024 287.31	197 406.58	1571
2 240 392 739.00	209 163.77	1720
2 366 607 594.99	221 000.33	1870
2 493 674 767.93	232 915.81	2020
2 621 600 774.12	244 921.81	2181
2 750 392 733.58	257 019.99	2354
2 880 058 370.01	269 194.07	2522
3 010 606 010.82	281 448.82	2689
3 142 044 587.09	293 786.05	2856
3 274 383 633.60	306 206.67	3023
3 407 633 288.84	318 722.60	3201
3 494 884 855.73	326 979.45	3379
3 621 732 491.25	338 902.60	3557
3 748 712 665.82	350 843.02	3740
3 875 793 426.32	362 723.56	3853.8

Solution:

Step 1: Develop the storage–indication relationship for the reservoir.

For the modified Puls method, a table giving relationship between $(2S/\Delta t + Q)$ and the corresponding Q is prepared, as shown in Table 13.7. In this case, Δt is taken as the interval of the inflow hydrograph given in Table 13.6. Thus, $\Delta t = 6\ h = 21\,600$ s.

Step 2: Set up the initial condition.

Defining the initial conditions correctly is important for the numerical approximation of the continuity equation. Different approaches can be found in various textbooks and other sources. However, the procedure outlined by Chow et al. (1988) and Singh (1992) is the correct approach and produces consistent results for both zero and non-zero initial inflow. For the modified Puls method, we use the time index $j = 1$. The calculations are presented in Table 13.8.

For $j = 1$, column 4 gives, $I^1 + I^2 = Q^1$ (inflow = outflow). Then from the $S - Q$ relationship, the initial storage is determined for the known Q^1. In this case storage values are given for $Q = 0$ and $Q = 1015$ m³/s in

Table 13.8 *Computation of outflow using the modified Puls method for routing the Sirikit Dam PMF*

Time index (j)	Time (h)	I^j (m³/s)	$I^j + I^{j+1}$	$2S^j/\Delta t - Q^j$	$2S^{j+1}/\Delta t + Q^{j+1}$	Q^{j+1}	Matched row	Interpolated Q
1	0	1000	2020	146 958.16		1000.00		
2	6	1020	2120	147 076.55	149 078.16	1000.81	1	1000.81
3	12	1100	2400	147 469.59	149 476.55	1003.48	1	1003.48
4	18	1300	3650	149 090.57	151 119.59	1014.51	1	1014.51
5	24	2350	6750	153 702.42	155 840.57	1069.07	2	1069.07
6	30	4400	10 050	161 429.00	163 752.42	1161.71	3	1161.71
7	36	5650	11 950	170 821.51	173 379.00	1278.74	3	1278.74
8	42	6300	13 100	181 107.61	183 921.51	1406.95	4	1406.95
9	48	6800	13 900	191 923.98	195 007.61	1541.82	5	1541.82
10	54	7100	14 600	203 150.88	206 523.98	1686.55	6	1686.55
11	60	7500	15 500	214 970.43	218 650.88	1840.23	7	1840.23
12	66	8000	16 500	227 466.82	231 470.43	2001.80	8	2001.80
13	72	8500	18 200	241 283.52	245 666.82	2191.65	10	2191.65
14	78	9700	20 200	256 652.32	261 483.52	2415.60	11	2415.60
15	84	10 500	20 400	271 794.15	277 052.32	2629.09	12	2629.09
16	90	9900	18 700	284 871.27	290 494.15	2811.44	13	2811.44
17	96	8800	16 500	295 455.30	301 371.27	2957.99	14	2957.99
18	102	7700	14 200	303 511.20	309 655.30	3072.05	15	3072.05
19	108	6500	12 200	309 394.86	315 711.20	3158.17	15	3158.17
20	114	5700	10 900	313 825.07	320 294.86	3234.89	16	3234.89
21	120	5200	10 000	317 203.07	323 825.07	3311.00	16	3311.00
22	126	4800	9200	319 669.93	326 403.07	3366.57	16	3366.57
23	132	4400	8500	321 376.38	328 169.93	3396.77	17	3396.77
24	138	4100	8150	322 692.33	329 526.38	3417.02	17	3417.02
25	144	4050	8050	323 871.98	330 742.33	3435.18	17	3435.18
26	150	4000	8000	324 967.90	331 871.98	3452.04	17	3452.04
27	156	4000	8100	326 128.11	333 067.90	3469.89	17	3469.89

Table 13.8 (cont.)

Time index (j)	Time (h)	I^j (m³/s)	$I^j + I^{j+1}$	$2S^j/\Delta t - Q^j$	$2S^{j+1}/\Delta t + Q^{j+1}$	Q^{j+1}	Matched row	Interpolated Q
28	162	4100	8350	327 496.22	334 478.11	3490.95	17	3490.95
29	168	4250	8850	329 308.55	336 346.22	3518.84	17	3518.84
30	174	4600	9600	331 794.37	338 908.55	3557.09	18	3557.09
31	180	5000	10 050	334 640.19	341 844.37	3602.09	18	3602.09
32	186	5050	9550	336 914.12	344 190.19	3638.04	18	3638.04
33	192	4500	8350	337 955.12	345 264.12	3654.50	18	3654.50
34	198	3850	7150	337 801.00	345 105.12	3652.06	18	3652.06
35	204	3300	6100	336 633.79	343 901.00	3633.61	18	3633.61
36	210	2800	5150	334 581.48	341 783.79	3601.16	18	3601.16
37	216	2350	4350	331 816.59	338 931.48	3557.44	18	3557.44
38	222	2000	3600	328 406.68	335 416.59	3504.96	17	3504.96
39	228	1600	3000	324 516.49	331 406.68	3445.09	17	3445.09
40	234	1400	2650	320 402.90	327 166.49	3381.79	17	3381.79
41	240	1250	2350	316 177.13	322 752.90	3287.88	16	3287.88
42	246	1100	2150	311 936.38	318 327.13	3195.38	15	3195.38
43	252	1050	2060	307 728.81	313 996.38	3133.78	15	3133.78
44	258	1010	2010	303 592.35	309 738.81	3073.23	15	3073.23
45	264	1000	1000	298 589.76	304 592.35	3001.29	14	3001.29
46	270	0	0	292 748.58	298 589.76	2920.59	14	2920.59
47	276	0	0	287 064.67	292 748.58	2841.96	13	2841.96
48	282	0	0	281 534.63	287 064.67	2765.02	13	2765.02
49	288	0	0	276 154.31	281 534.63	2690.16	13	2690.16
50	294	0	0	270 920.61	276 154.31	2616.85	12	2616.85
51	300	0	0	265 829.55	270 920.61	2545.53	12	2545.53
52	306	0	0	260 878.41	265 829.55	2475.57	11	2475.57
53	312	0	0	256 063.92	260 878.41	2407.25	11	2407.25
54	318	0	0	251 383.27	256 063.92	2340.33	10	2340.33
55	324	0	0	246 836.47	251 383.26	2273.39	10	2273.40
56	330	0	0	242 419.71	246 836.47	2208.37	10	2208.38
57	336	0	0	238 124.82	242 419.71	2147.44	9	2147.45
58	342	0	0	233 945.11	238 124.82	2089.85	9	2089.85
59	348	0	0	229 877.50	233 945.11	2033.80	9	2033.80

Table 13.7. Using the interpolation formula (Eq. 13.55), $S^1 = 15\,979\,481\,788$ m³. From these known values of S^1 and Q^1 the $(2S^1/\Delta t) - Q^1$ is calculated (column 5). Thus, only for the initial condition, the $S - Q$ relationship is used directly and column 5 is calculated first.

For $j = 2$, column 6 is calculated first using the relationship: $\frac{2S^2}{\Delta t} + Q^2 = (I^1 + I^2) + \left(\frac{2S^1}{\Delta t} - Q^1\right)$, where the right-hand quantity is obtained as the initial condition ($j = 1$). Then Q^2 is evaluated from the storage–indication relationship (Table 13.7). In this case the interpolation formula (Eq. 13.55) is used. In Excel, the interpolation task is greatly facilitated by using its MATCH and INDEX functions. The operations are shown in the last two columns of Table 13.8. The value of calculated $\frac{2S^2}{\Delta t} + Q^2$ is in Column Y (in Table 13.8 it is column 6) and the value of $2S^2/\Delta t + Q^2$ from the storage–indication relationship is in Column (define an array) SQ of the spreadsheet. Then, the value that needs to go to Column X (in Table 13.8 it is column 8) is obtained as:

$X = \text{MATCH}(Y, SQ)$.

Then the interpolated value is placed in Column Z (in Table 13.8 it is column 9) with the function

$Z = \text{INDEX}(QQ, X) + (Y - \text{INDEX}(SQ, X))$
$\quad * (\text{INDEX}(QQ, X + 1) - \text{INDEX}(QQ, X))$
$\quad / (\text{INDEX}(SQ, X + 1) - \text{INDEX}(SQ, X))$

where QQ denotes the column (define an array) where the discharge values are given in the storage–indication

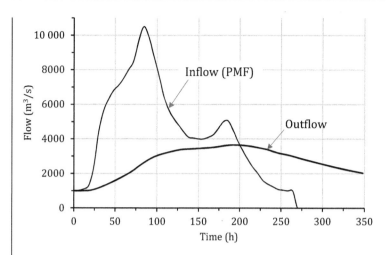

Figure 13.20 Inflow (PMF) and outflow from Sirikit Dam Spillway (two gates fully open).

table. This value of Q is then returned as the calculated value of Q^2 (column 7). From Q^2 then $\left(\frac{2S^2}{\Delta t} - Q^2\right)$ is calculated by subtracting $2*Q^2$ from $\frac{2S^2}{\Delta t} + Q^2$.

Then the rest of the calculation for $j > 2$ is just the recursive process as done for $j = 2$.
Figure 13.20 shows the inflow–outflow hydrographs.

Example 13.3: The Kurdistan region in northern Iraq is a water-stressed region. In recent years, many large or medium dams (as many as 30–35) have either been proposed or are under construction, for storing water for irrigation and for hydropower generation. One such proposed dam is Sartik Dam, on the Lesser Zab River, the second largest tributary of the Tigris River. Mustafa (2017) provided detailed data on this dam. From the data presented by Mustafa, the elevation–storage–discharge–surface area relationships of the dam site are given in Table 13.9. The proposed spillway is an ogee spillway with a radial gate, as shown in Figure 13.21, a simplified

Table 13.9 *Elevation–storage–discharge–surface area for the proposed Sarkit Dam in Kurdistan, Iraq*

Elevation (m)	Storage (m³)	Discharge (m³/s)	Effective storage (m³)	Area (m²)
320.00	332 568 424.00	0.00	0.00	26 958 134.00
320.50	346 619 608.50	66.10	14 051 184.50	28 110 242.00
321.00	360 670 793.00	186.90	28 102 369.00	29 262 350.00
321.50	375 652 420.00	343.40	43 083 996.00	29 966 008.50
322.00	390 634 047.00	528.60	58 065 623.00	30 669 667.00
322.50	406 366 036.50	738.80	73 797 612.50	31 467 349.50
323.00	422 098 026.00	971.20	89 529 602.00	32 265 032.00
323.50	438 620 547.50	1223.80	106 052 123.50	33 048 136.50
324.00	455 143 069.00	1495.20	122 574 645.00	33 831 241.00
324.50	472 146 992.50	1784.10	139 578 568.50	34 008 000.50
325.00	489 150 916.00	2089.60	156 582 492.00	34 184 760.00
325.50	506 777 223.50	2410.80	174 208 799.50	35 258 061.50
326.00	524 403 531.00	2746.90	191 835 107.00	36 331 363.00
326.50	542 987 350.50	3097.30	210 418 926.50	37 170 798.50
327.00	561 571 170.00	3461.40	229 002 746.00	38 010 234.00
327.50	580 973 166.50	3838.90	248 404 742.50	38 806 718.00
328.00	600 375 163.00	4229.10	267 806 739.00	39 603 202.00
328.50	620 550 599.00	4631.70	287 982 175.00	40 353 195.00
329.00	640 726 035.00	5046.30	308 157 611.00	41 103 188.00
329.50	661 349 227.50	5472.60	328 780 803.50	41 246 467.50
330.00	681 972 420.00	5910.30	349 403 996.00	41 389 747.00

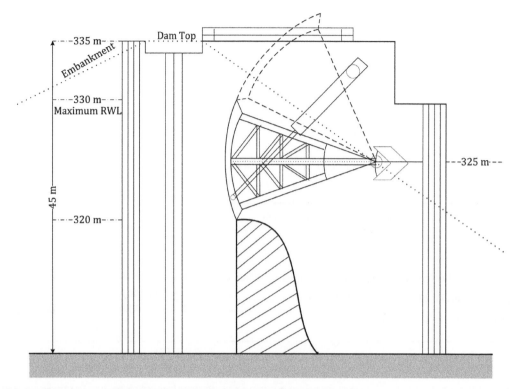

Figure 13.21 Schematic of the spillway of Sartik Dam (drawn from the concept given by Mustafa, 2017).

version drawn from the concept presented by Mustafa. Using the fourth-order Runge–Kutta method, a 100-year flood hydrograph given by Mustafa et al. is routed through the reservoir using S-formulation (Part A) and H-formulation (Part B) for the full gate opening. The orifice flows for various gate openings are given in Table 13.10 (Part C). From this table and the results obtained from Part B calculate the discharge and RWL when the gate is half open and discuss its implication for water management.

Solution:
Part A: S-formulation.

Step 1: Develop the storage-discharge relationship.

For the S-formulation of the fourth-order Runge–Kutta method, only the original storage (effective storage from spillway crest at 320 m)–discharge relationship given in Table 13.9 is necessary.

Step 2: Set up the initial conditions.

Following the customary practice in numerical mathematics, in this case we will use the time index $j=0$ for initial condition. Inflow = Outflow, so $I^0 = Q^0$. For this Q^0, the initial storage S^0 is determined using an interpolation formula from the S–Q data given in Table 13.9. In this case $Q^0 = 100$ m³/s and storage corresponding to 66 m³/s and 186.9 m³/s are given. Through interpolation, $S^0 = 17\,994\,356.31$ m³. Once the initial values are known, the weighting factors for the Runge–Kutta method are determined

$$k_1 = I^0 - Q^0 = 0 \text{ m}^3/\text{s}$$

$$k_2 = \left(\frac{I^0 + I^1}{2}\right) - \left(Q^0 + \frac{k_1}{2}\right) = 50 \text{ m}^3/\text{s}$$

$$k_3 = \left(\frac{I^0 + I^1}{2}\right) - \left(Q^0 + \frac{k_2}{2}\right) = 25 \text{ m}^3/\text{s}$$

$$k_4 = (I^0 + I^1) - (Q^0 + k_3) = 175 \text{ m}^3/\text{s}$$

Step 2: $j \geq 2$.

From the values of k_1, k_2, k_3, and k_4 calculated for $j = 0$, S^1 is calculated using Eq. 13.60. In this case $\Delta t = 6$ h $= 21\,600$ s. Now for this value of S the corresponding value of Q is found, again using interpolation in the given S–Q relationship. As described above, Excel is used to facilitate the process. Table 13.11 shows the calculations and Figure 13.22 shows the inflow–outflow hydrographs.

Part B: H-formulation for full gate opening (10 m).

Step 1: Develop the elevation–discharge and elevation–area relationships.

For the H-formulation of fourth-order Runge–Kutta method, one needs the elevation–discharge and elevation–area relationships given in Table 13.9.

Table 13.10 *Orifice flows for various gate openings. Free overflows over the spillway in the open parts are given in Table 13.9*

Reservoir water level (m) (above MSL)	Orifice flow (m³/s)									
	Gate opening, 1 m	Gate opening, 2 m	Gate opening, 3 m	Gate opening, 4 m	Gate opening, 5 m	Gate opening, 6 m	Gate opening, 7 m	Gate opening, 8 m	Gate opening, 9 m	Gate opening, 10 m
320.0										
320.5	66.1	66.1								
321.0	186.9	186.9								
321.5	228.9	343.4								
322.0	264.3	528.6								
322.5	295.5	591								
323.0	323.7	647.4	971.1							
323.5	349.6	699.3	1048.9							
324.0	373.8	747.6	1121.4	1495.1						
324.5	396.5	792.9	1189.4	1585.8						
325.0	417.9	835.8	1253.7	1671.6	2089.5					
325.5	438.3	876.6	1314.9	1753.2	2191.5					
326.0	457.8	915.6	1373.4	1831.2	2289	2746.7				
326.5	476.5	953	1429.5	1905.9	2382.4	2858.9				
327.0	494.5	988.9	1483.4	1977.9	2472.4	2966.8	3461.3			
327.5	511.8	1023.7	1535.5	2047.3	2559.1	3071.0	3582.8			
328.0	528.6	1057.2	1585.8	2114.4	2643.1	3171.7	3700.3	4228.9		
328.5	544.9	1089.8	1634.6	2179.5	2724.4	3269.3	3814.2	4359		
329.0	560.7	1121.4	1682	2242.7	2803.4	3364.1	3924.7	4485.4	5046.1	
329.5	576	1152.1	1728.1	2304.2	2880.2	3456.2	4032.3	4608.3	5184.4	
330.0	591	1182	1773	2364	2532.9	3546.0	4137	4728	5319.1	5910.1

Table 13.11 *Fourth-order Runge–Kutta method to derive the outflow hydrograph for 100-year inflow hydrograph of Sartik Dam*

Time index (j)	Time (h)	Inflow (m³/s)	k_1 (m³/s)	k_2 (m³/s)	k_3 (m³/s)	k_4 (m³/s)	S (m³)	Q (m³/s)	Row of column 4 of Table 13.9 to match column 8	Interpolated Q
0	0	100	0	50	25	175	17 994 356.31	100		
1	6	200	89.941346	124.97067	107.45601	342.48534	19 164 356.31	110.06	2	110.06
2	12	360	222.17079	481.0854	351.6281	970.5427	22 394 564.48	137.83	2	137.83
3	18	1100	865.24105	1162.6205	1013.9308	2411.3103	32 683 870.21	234.76	3	234.76
4	24	2560	2003.5416	1621.7708	1812.6562	3990.8854	60 150 624.38	556.46	5	556.46
5	30	3800	2569.5259	1434.7629	2002.1444	4667.3815	106 458 436.6	1230.47	8	1230.47
6	36	4100	1998.1079	899.05397	1448.581	4449.527	157 257 035.9	2101.89	11	2101.89
7	42	3900	1048.7108	214.35541	631.53311	3697.1777	197 371 493.2	2851.29	13	2851.29
8	48	3280	−15.73413	−337.86707	−176.8006	2781.0665	220 547 089.3	3295.73	14	3295.73
9	54	2620	−798.17804	−684.08902	−741.13353	1992.9555	226 796 678.5	3418.18	14	3418.18
10	60	2050	−1251.3999	−890.69997	−1071.05	1339.65	220 836 275	3301.40	14	3301.40
11	66	1520	−1513.3896	−979.19482	−1246.2922	807.90259	207 029 375.7	3033.39	13	3033.39
12	72	1075	−1607.6598	−956.32989	−1281.9948	444.33506	188 466 115.6	2682.66	12	2682.66
13	78	770	−1530.6144	−880.30721	−1205.4608	214.8464	168 162 208.7	2300.61	11	2300.61
14	84	540	−1402.7319	−771.36596	−1087.0489	84.31702	148 407 914	1942.73	10	1942.73
15	90	400	−1226.133	−683.06652	−954.59979	−11.533262	130 281 033.1	1626.13	9	1626.13
16	96	260	−1094.9147	−594.95735	−844.93603	−84.978677	114 034 237	1354.91	8	1354.91
17	102	165	−957.39726	−506.19863	−731.79795	−115.59932	99 419 388.39	1122.40	7	1122.40
18	108	110	−818.5582	−419.2791	−618.91865	−109.63955	86 643 025.33	928.56	6	928.56
19	114	90	−678.77148	−359.38574	−519.07861	−109.69287	75 826 489.69	768.77	6	768.77
20	120	50	−593.47	−321.73661	−457.60492	−135.87	66 663 074.66	643.47	5	643.47
Column 1	Column 2	Column 3	Column 4	Column 5	Column 6	Column 7	Column 8	Column 9	Column 10	Column 11

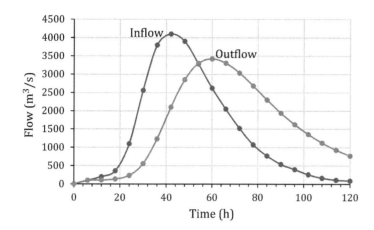

Figure 13.22 Inflow–outflow hydrographs of Sartik Dam.

Step 2: Set up the initial conditions.

As for the S-formulation, $j = 0$, inflow = outflow, so $I^0 = Q^0$. For this Q^0, the initial elevation, h^0, is determined using the interpolation formula from the h–Q relationship given in Table 13.9. In this case $Q^0 = 100$ m³/s and the elevation corresponding to 66 m³/s and 186.9 m³/s are given. Through interpolation, $h^0 = 320.64$ m. Similarly, from the h–A relationship given in Table 13.9, the initial A^0 is determined from interpolation as $28\,433\,557.08$ m². Once the initial values are known, the weighting factors for the Runge–Kutta method are determined as

$$k_1 = \frac{I^0 - Q^0}{A^0} = 0 \text{ m/s}$$

$$k_2 = \left[\frac{\left(\frac{I^0 + I^1}{2}\right)}{A^0}\right] - \left(\frac{Q^0}{A^0} + \frac{k_1}{2}\right) = 1.76 \times 10^{-6} \text{ m/s}$$

$$k_3 = \left[\frac{\left(\frac{I^0 + I^1}{2}\right)}{A^0}\right] - \left(\frac{Q^0}{A^0} + \frac{k_2}{2}\right) = 8.79 \times 10^{-7} \text{ m/s}$$

$$k_4 = \left[\frac{(I^0 + I^1)}{A^0}\right] - \left(\frac{Q^0}{A^0} + k_3\right) = 6.15 \times 10^{-6} \text{ m/s}$$

Step 2: $j \geq 2$.

From the values of k_1, k_2, k_3, and k_4 calculated for $j = 0$, h^1 is calculated using Eq. 13.50. In this case $\Delta t = 6$ h $= 21\,600$ s. Now for this value of h the corresponding values of both Q and A are found, again using interpolation in the given h–Q and h–A relationships. As described above, Excel is used to facilitate the process. Table 13.12 shows the calculations and Figure 13.23 shows the inflow–outflow hydrographs as well as the stage–time graph.

Part C: H-formulation for half gate opening (5 m).

Table 13.9 shows that when the gate is fully open (10 m), the maximum discharge is 3458.7 m³/s and the RWL is 327 m. Since the spillway crest is at 320 m, from Table 13.10 it is seen that for gate opening at 7 m, at the RWL at 327 m, the discharge is 3461 m³/s, which is within the acceptable limit of error in estimation.

It may be necessary to control the flow either for storage of the water or for downstream flood control. If the gate is lowered to 325 m from its original fully open position of 330 m, the elevation–discharge table needs to be modified from 325.5 m as per the discharge values given in Table 13.10. This is shown in Table 13.13.

Then the calculation is repeated with this elevation–discharge data and it is shown that the maximum discharge is reduced to 2588 m³/s but, at the same time, the RWL rises close to 328 m (Figure 13.24). This shows the importance of reservoir routing in spillway operation. By lowering the gate, water storage can be increased and the potential for downstream flooding is lessened. However, if the RWL rises too high, the potential of upstream flooding increases at the same time.

Table 13.12 Fourth-order Runge–Kutta method to derive the outflow hydrograph and RWL for 100-year inflow hydrograph of Sartik Dam

Time index (j)	Time (h)	Inflow (m³/s)	k_1 (m/s)	k_2 (m/s)	k_3 (m/s)	k_4 (m/s)	h (m)	A (m²)	Q (m³/s)	Row of column 4 of Table 13.9 to match column 8	Interpolated Q	Interpolated A
0	0	100	0.000E+00	1.758E−06	8.792E−07	6.155E−06	320.64	2.84E+07	100.00			2.85E+07
1	6	200	3.157E−06	4.383E−06	3.770E−06	1.201E−05	320.68	2.85E+07	109.94	2	109.9414927	2.88E+07
2	12	360	7.735E−06	1.672E−05	1.223E−05	3.372E−05	320.79	2.88E+07	137.31	2	137.3107523	2.95E+07
3	18	1100	2.936E−05	3.944E−05	3.440E−05	8.181E−05	321.15	2.95E+07	234.60	3	234.5976596	3.08E+07
4	24	2560	6.480E−05	5.253E−05	5.866E−05	1.295E−04	322.08	3.08E+07	564.03	5	564.0277355	3.32E+07
5	30	3800	7.626E−05	4.265E−05	5.946E−05	1.404E−04	323.58	3.32E+07	1269.53	8	1269.53304	3.44E+07
6	36	4100	5.659E−05	2.539E−05	4.099E−05	1.290E−04	325.10	3.44E+07	2153.43	11	2153.427699	3.67E+07
7	42	3900	2.670E−05	4.915E−06	1.581E−05	1.002E−04	326.25	3.67E+07	2918.82	13	2918.817869	3.78E+07
8	48	3280	−1.935E−06	−9.707E−06	−5.821E−06	7.327E−05	326.85	3.78E+07	3353.08	14	3353.080559	3.80E+07
9	54	2620	−2.207E−05	−1.853E−05	−2.030E−05	5.217E−05	327.00	3.80E+07	3458.67	14	3458.67225	3.77E+07
10	60	2050	−3.404E−05	−2.405E−05	−2.905E−05	3.530E−05	326.83	3.77E+07	3333.99	14	3333.989082	3.71E+07
11	66	1520	−4.154E−05	−2.677E−05	−3.415E−05	2.160E−05	326.45	3.71E+07	3060.37	13	3060.366203	3.62E+07
12	72	1075	−4.502E−05	−2.672E−05	−3.587E−05	1.213E−05	325.94	3.62E+07	2704.47	12	2704.471215	3.50E+07
13	78	770	−4.449E−05	−2.553E−05	−3.501E−05	5.962E−06	325.37	3.50E+07	2325.87	11	2325.874909	3.41E+07
14	84	540	−4.172E−05	−2.291E−05	−3.232E−05	2.321E−06	324.79	3.41E+07	1963.26	10	1963.26129	3.39E+07
15	90	400	−3.661E−05	−2.037E−05	−2.849E−05	−4.554E−07	324.25	3.39E+07	1641.79	9	1641.789762	3.35E+07
16	96	260	−3.315E−05	−1.800E−05	−2.557E−05	−2.648E−06	323.77	3.35E+07	1369.55	8	1369.547962	3.28E+07
17	102	165	−2.962E−05	−1.565E−05	−2.264E−05	−3.630E−06	323.33	3.28E+07	1135.86	7	1135.863108	3.22E+07
18	108	110	−2.578E−05	−1.320E−05	−1.949E−05	−3.490E−06	322.93	3.22E+07	938.93	6	938.9316104	3.16E+07
19	114	90	−2.185E−05	−1.156E−05	−1.670E−05	−3.563E−06	322.59	3.16E+07	780.55	6	780.5509015	3.11E+07
20	120	50	−1.935E−05	−9.675E−06	−1.451E−05	−3.232E−06	322.29	3.11E+07	652.58	5	652.5806331	
Column 1	Column 2	Column 3	Column 4	Column 5	Column 6	Column 7	Column 8	Column 9	Column 10	Column 11	Column 12	Column 13

Table 13.13 *Elevation–discharge relationship when the gate is open for 5 m (up to 325 m)*

Discharge (m³/s)	Elevation (m)
0	320
66.1	320.5
186.9	321
343.4	321.5
528.6	322
738.8	322.5
971.2	323
1223.8	323.5
1495.2	324
1784.1	324.5
2089.6	325
2191.5	325.5
2289	326
2382.4	326.5
2472.4	327
2559.1	327.5
2643.1	328
2724.4	328.5
2803.4	329
2880.2	329.5
2532.9	330

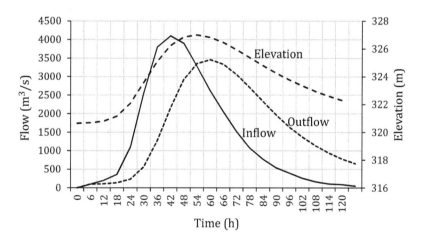

Figure 13.23 Inflow–outflow hydrographs and stage–time graph of Sartik Dam, gate fully open.

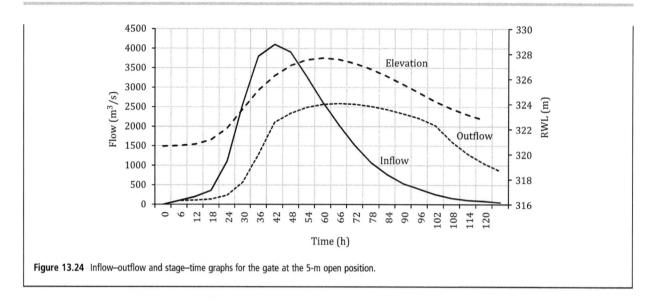

Figure 13.24 Inflow–outflow and stage–time graphs for the gate at the 5-m open position.

Exercises: A selection of exercises on this topic is available at www.cambridge.org/appliedhydrology.

14 Evaporation and Evapotranspiration

14.1 CONCEPTS AND PURPOSE

In Chapter 7 we dealt with the processes of abstraction by which rainfall amounts are lost in the ground and thereby do not contribute to direct surface runoff. In this chapter we will discuss the processes by which surface and subsurface water return to the atmosphere and thereby do not contribute directly to streamflow. These processes are **evaporation** and **evapotranspiration**. Evapotranspiration refers to the combined effects of two related processes of evaporation and **transpiration**.

Evaporation and evapotranspiration are not too significant during rainfall–runoff events resulting from convective storms that last for a few hours. Cyclonic storms have duration longer than convective storms but still the evaporation and evapotranspiration during such events are small to consider for computation of surface runoff. However, accounting for water returning to the atmosphere from the surface and subsurface becomes quite significant when long-term storage for irrigation or municipal and industrial water supply is the primary purpose of a study. In long-term continuous simulation models, evaporation from storage reservoirs can be large and evapotranspiration from soil on the land surface becomes important because the antecedent moisture deficit must be satisfied before any significant runoff can occur. Even though evaporation and transpiration are taken together, in vegetated fields transpiration is responsible for the movement of much more water than evaporation. Combined evapotranspiration is often responsible for returning 50% or even 60% of precipitation back to the atmosphere. Knowledge of evapotranspiration is needed to compute crop water requirements for irrigation, long-term simulation of rainfall–runoff and monthly water balance calculations in watersheds for water availability studies, reservoir management, low-flow management, and applications related to broad water resources planning and management.

14.2 THE PROCESSES OF EVAPORATION AND EVAPOTRANSPIRATION

In the process of evaporation water changes from its liquid state to vapor state before reaching the boiling point temperature. Transpiration is the process of vegetation extracting water from the soil through the plant root system and evaporation occurring from the leaves of plants through stomatal openings. The function of the stomata is to provide a place where carbon dioxide can dissolve into water and enter plant tissue. However, evaporation, which is unavoidable in this process, is driven by the same process as evaporation from a free water surface. The ratio of water transpired to that used to form plant matter is very large (> 800). Evapotranspiration is the combination of evaporation from the ground surface and transpiration by vegetation. It is the primary process of water transfer from land to atmosphere in the hydrologic cycle.

The term **potential evaporation** refers to climatically controlled evaporation from a surface when the supply of water to the surface, which can be ground or free water, is unlimited. **Potential transpiration** is the transpiration that would occur if water supply to plant roots and through the vascular system to the stomata was unlimited. This is controlled by climate and plant physiology. The evaporation plus transpiration from a vegetated surface with unlimited water supply is known as **potential evapotranspiration** (**PET,** designated in equations as ET_P), and it constitutes the maximum possible rate due to the prevailing meteorological conditions.

The **actual evapotranspiration** (**AET,** designated in equations as ET_A) is the amount of water that is evaporated in a normal day. In other words, if, for instance, the soil runs out of water, the AET is the amount of water that has been evaporated, and not the amount of water which could have been evaporated if the soil had an infinite amount of water supply to evaporate. Thus, the PET is the maximum value of the AET. The PET equals the AET when water supply is unlimited.

The **reference evapotranspiration** is the evapotranspiration rate from a reference surface, with unlimited water supply. The reference surface is a hypothetical grass reference crop with specific characteristics. It is commonly expressed in mm/day or cm/month. It is also called the reference crop evapotranspiration and is denoted by ET_0. In this context, another term called the **aridity index** (AI), defined as the annual precipitation (mm) divided by the annual PET (mm) is also important.

14.3 EVAPORATION THERMODYNAMICS

14.3.1 Diffusion and Vapor Pressure Deficit

Evaporation occurs by the process of **diffusion**, which is the net movement of anything like atoms, ions, molecules, energy, etc., generally from a region of higher concentration to a region of lower concentration. The mathematical theory of diffusion in an isotropic medium is based on the hypothesis that the rate of transfer of a diffusing substance through a unit area of a section is proportional to the concentration gradient measured normal to the section. This hypothesis gives the equation

$$F = -D\frac{\partial C}{\partial x} \qquad (14.1)$$

where F is rate of transfer of a substance per unit area of section, C is the concentration of the diffusing substance, x is the space coordinate measured normal to the section, and D is called the **diffusion coefficient**. If F, the amount of diffusing material, and C, the concentration, are both expressed in the same units of mass, it follows from Eq. 14.1 that D has the unit of L^2/T(e.g., cm^2/s). The fundamental differential equation of diffusion is derived from Eq. 14.1 and is given as

$$\frac{\partial C}{\partial t} = D\frac{\partial^2 C}{\partial x^2} \qquad (14.2)$$

Equations 14.1 and 14.2 are usually referred to as **Fick's first** and **second laws of diffusion**, since they were first formulated by Adolf Fick, a German physician and physiologist, in 1855.

In the case of evaporation, the molar flux of water vapor from the surface to the atmosphere is driven by the difference in partial pressures of gaseous water at the surface and the atmosphere. This is called the **vapor pressure gradient**. Since air is a mixture of several gases, each gas exerts its own pressure called the **partial pressure** of that gas and the sum of the partial pressures of all the gases present in the atmosphere is the total atmospheric pressure. This is known as Dalton's law, named after John Dalton, an English chemist and physicist, who observed this in 1801 and published his findings in 1802.

The partial pressure of the vapor resulting from evaporation of a liquid above a sample of the liquid is called **vapor pressure**. The vapor pressure of water, denoted by e, is given by Dalton's law of partial pressure:

$$e = \rho_v R_v T \qquad (14.3)$$

where ρ_v is the density of water vapor, R_v is the **gas constant** of the water vapor, and T is the absolute temperature in kelvins (K). If the total pressure exerted by moist air is p_t, the partial pressure of the dry air, p_{da}, is

$$p_{da} = p_t - e = \rho_{da} R_{da} T \qquad (14.4)$$

where R_{da} is the gas constant for dry air, which is 287 J/kg K and ρ_{da} is the density of the dry air. The density of moist air, ρ_a, is

$$\rho_a = \rho_{da} + \rho_v \qquad (14.5)$$

The gas constant for water vapor is given by

$$R_v = \frac{R_{da}}{0.622} \qquad (14.6)$$

where 0.622 is the ratio of the molecular weight of water vapor to the average molecular weight of dry air.

The ratio of the vapor pressure of water and the total pressure of air is related by the **specific humidity**, q_v

$$q_v = 0.622\frac{e}{p_t - e} = 0.622\frac{e}{p_{da}} \qquad (14.7)$$

The total atmospheric pressure in the case of moist air is

$$p_t = \rho_a R_a T \qquad (14.8)$$

in which the gas constant for the moist air, R_a, is related to that of the dry air by

$$R_a = R_{da}(1 + 0.608 q_v) = 287(1 + 0.608 q_v) \left[\frac{J}{kg\ K}\right] \qquad (14.9)$$

The density of moist air can be calculated from the ideal gas law. An adequate estimation can be made using

$$\rho_a = 3.486\left(\frac{p_a}{275 + T}\right) \qquad (14.10)$$

The **saturation vapor pressure** or **equilibrium vapor pressure** is the pressure exerted by vapor in thermodynamic equilibrium with its condensed phases, which can be solid or liquid at a given temperature in a closed system. The saturation vapor pressure is attained by an atmosphere when it reaches its maximum capacity to hold water vapor. It is an indication of the evaporation rate of a liquid. The saturation vapor pressure of any material is solely dependent on the temperature of that material. As temperature rises the saturation vapor pressure rises nonlinearly (Figure 14.1)

There are many published approximations for calculating the saturated vapor pressure over water to better approximate the relationship between temperature and saturation vapor pressure, as shown in Figure 14.1. An equation, known as the **Tetens equation**, is fairly accurate and is commonly used in engineering hydrology. It is given as

$$e_s(T) = 0.61078 \exp\left(\frac{17.27 T}{237.3 + T}\right) \qquad (14.11)$$

where e_s is the saturation vapor pressure in kPa and T is in °C (in meteorology, millibar is used for unit of pressure where 1 mbar = 0.1 kPa). Murray (1967) gave a form that is also a good approximation from 0 °C to 50 °C and has been used by some workers:

$$e_s(T) = \exp\left(\frac{16.78T - 116.9}{237.3 + T}\right) \qquad (14.12)$$

The prevailing vapor pressure relative to the saturation vapor pressure, is the **relative humidity**, R_H:

$$R_H = \frac{e_a}{e_s(T)} \qquad (14.13)$$

Thus, the relative humidity, usually expressed as %, is defined as the ratio of the actual vapor pressure of air (e_a), which is the partial pressure of water vapor in the mixture of air and water vapor, to the saturation vapor pressure at the

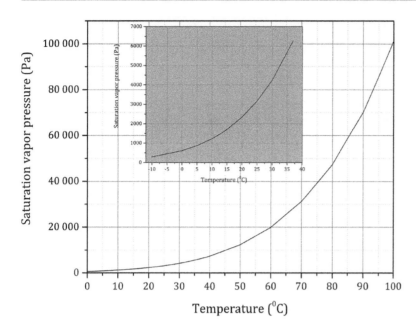

Figure 14.1 Variation of saturated vapor pressure of water with temperature.

same pressure and temperature. It is determined through the use of **psychrometric charts** if both the dry bulb temperature and the wet bulb temperature of the mixture are known. These quantities are readily estimated using a **psychrometer** or **hygrometer**.

The actual vapor pressure, e_a, is the vapor pressure of the air. Unlike the saturation vapor pressure, the actual vapor pressure cannot be determined simply by knowing the temperature of the air. To determine e_a, the air temperature and either the relative humidity or the **dew point** temperature of the air must be known. From Eq. 14.13, we can write

$$e_a = e_s \frac{R_H}{100} \qquad (14.14)$$

The dew point, T_d, is the temperature to which a parcel of air has to be cooled at constant vapor pressure to reach saturation. Thus, it is the temperature at which air becomes just saturated.

From Fick's first law, Eq. 14.1, the rate of evaporation is proportional to the difference between the saturation vapor pressure at the water temperature and the actual vapor pressure in the air, known as Dalton's law of evaporation after Dalton for first recognizing it. This law can be formulated as

$$E = K_e v_a (e_s - e_a) \qquad (14.15)$$

where E is the rate of evaporation (i.e., L/T, e.g., mm/day) and K_e is a coefficient that reflects the efficiency of vertical transport of water vapor by the turbulent eddies of the air, and v_a is the wind speed. Evaporation continues till $e_a = e_s$. If $e_s > e_a$, *condensation* takes place.

A fundamentally important quantity in the thermodynamics of evaporation is the **vapor pressure deficit**, δe, which is the difference between the saturation vapor pressure and the actual vapor pressure:

$$\delta e = e_s - e_a \qquad (14.16)$$

14.3.2 Altitudinal Variation of Temperature, Pressure, and Wind Speed

Altitudinal variation of temperature is complex in the stratified atmosphere of the Earth. In the troposphere, which extends up to about 10 000 m, the temperature decreases with increase in elevation. But then in the stratosphere, which extends from the top of troposphere to about 50 000 m, it increases with elevation. Then again, in the mesosphere, extending to about 85 000 m, the temperature decreases with elevation. In the top-most stratum of the atmosphere, called the thermosphere, nearly up to 120 000 m, the temperature steadily increases with elevation. Nevertheless, near the Earth's surface, that is, in the troposphere, the altitudinal variation of temperature is linear and can be written as

$$\frac{dT}{dz} = -\gamma_a \qquad (14.17)$$

where z denotes elevation and γ_a is called the **lapse rate**. If a parcel of air is lifted and the temperature change occurs only by compression or expansion without any exchange of heat, the process is **adiabatic** and the lapse rate is called the **dry adiabatic lapse rate**. This is nearly constant with a value of 9.8 °C/km (5.5 °F/100 ft). The lapse rate for moist air depends on several factors and is approximately 6 °C/km. The prevailing lapse rate is called the **ambient** or **environmental lapse rate**.

The variation of pressure with elevation can be expressed analytically by combining the ideal gas law of thermodynamics and hydrostatic pressure law of fluid mechanics. If the pressure variation is all hydrostatic, then

$$\frac{dP}{dz} = -\rho_a g \qquad (14.18)$$

The ideal gas law can be written either in terms of universal gas constant, R (8.31446 J/mol K), or the gas constant of the specific air, R_a. Depending on this choice, the resulting analytical expression giving the relationship between pressure and elevation will be different. Here, the derivation from the ideal gas law expressed in terms of the universal gas constant R is used, since R_a varies with specific humidity (Eq. 14.9). The ideal gas law can be written as

$$p = \frac{\rho_a}{M} RT \tag{14.19}$$

where M is the molar mass of Earth's air, ≈ 0.02896 kg/mol.

Combining these two equations, one can obtain

$$\frac{dp}{p} = -\frac{Mg}{R}\frac{dz}{T} \tag{14.20}$$

Integrating Eq. 14.20 from the surface of the Earth ($z = 0$) to an altitude z, we get

$$p = p_0 \exp\left[-\int_0^z \frac{Mg}{R}\frac{1}{T} dz\right] \tag{14.21}$$

In Eq. 14.21, M, g, and R can be held constant. The variation of T with z is given from Eq. 14.17 as

$$T = T_0 - \gamma_a z \tag{14.22}$$

Combining Eqs. 14.21 and 14.22 we get

$$p = p_0 \left(\frac{T}{T_0}\right)^{\frac{Mg}{R\gamma_a}} \tag{14.23}$$

Equation 14.23 is known as the **barometric equation**. The density variation with elevation can be derived in the same way. Figure 14.2 shows the altitudinal variation of the three intrinsic properties of air.

The standard atmospheric pressure at sea level is 101.325 kPa. Several approximations of the pressure equation can be found. For example, an equation, assuming 20 °C for a standard atmosphere, can be derived from the ideal gas law yielding P in kPa as a function of z (m)

$$P = 101.3 \left(\frac{293 - 0.0065z}{293}\right)^{5.26} \tag{14.24}$$

A polynomial approximation of pressure variation with elevation, z (m), can be given as

$$P = 101.3 - 0.01152z + 0.544 \times 10^{-6} \times z^2 \tag{14.25}$$

In addition to thermodynamics, aerodynamics also plays a vital role in the evaporation process. The ability of the air

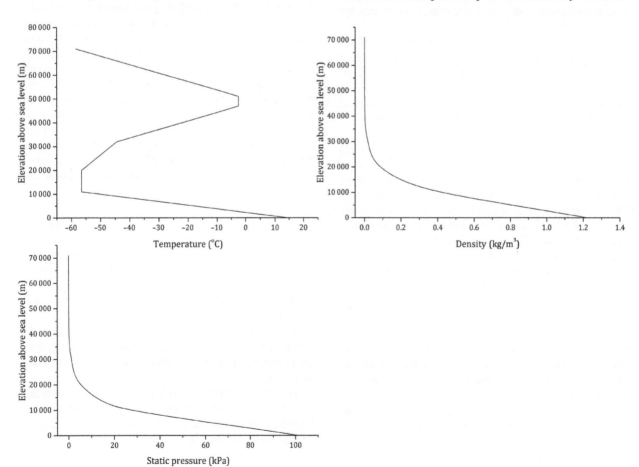

Figure 14.2 Altitudinal variation of atmospheric pressure, temperature, and air density.

Table 14.1 *Values of roughness height*

Terrain description	z_0 (m)
Open sea, fetch at least 5 km	0.0002
Mud flats, snow; no vegetation, no obstacles	0.005
Open flat terrain; grass, tundra, few isolated obstacles	0.03
Low crops; occasional large obstacles	0.10
High crops; scattered obstacles	0.25
Parkland, bushes; numerous obstacles	0.5
Regular large obstacle coverage (suburb, forest)	1.0
City center with high- and low-rise buildings	≥ 2

to transport water vapor away from the surface is a significant control for the magnitude of evaporation. Wind speed changes considerably with height. It increases more rapidly near the surface of the Earth. There are several models of altitudinal variation of wind speed. The two most widely used models in practice of various engineering disciplines are the logarithmic and power law models. The logarithmic model is presented here, since it has been adopted in the standard evaporation models commonly used.

The logarithmic law of wind speed variation with height is expressed as

$$u = \frac{1}{k} u_* \ln\left(\frac{z}{z_0}\right) \tag{14.26}$$

where u is the wind speed at height z above ground surface (m/s), k is the von Karman constant, which approximately equals 0.4, and z_0 is called the **ground roughness** (m). The term u_* is the shear velocity defined as

$$u_* = \sqrt{\frac{\tau_0}{\rho_a}} \tag{14.27}$$

where τ_0 is the shear stress of wind at ground surface.

The values of z_0 can be found in many publications. Table 14.1 lists some values. The surface roughness increases if the surface has lot of protrusions. For example forest covers with tall trees have higher roughness than grass prairies.

Another concept that is often introduced in Eq. 14.26 is the **zero-plane displacement** or **displacement height**. It is the height of a vegetated surface, usually indicated with z_d, at which the wind speed would go to zero if the logarithmic wind profile was maintained from the outer flow all the way down to the surface, that is, in the absence of the vegetation. Thus, another form of the logarithmic law is given as

$$u = \frac{1}{k} u_* \ln\left(\frac{z - z_d}{z_0}\right) \tag{14.28}$$

Both z_d and z_0 are best calculated with a wind profile measured at several levels, but theoretically they can also be calculated from two levels of wind data.

14.3.3 Variation of Saturated Vapor Pressure

The gradient of saturation vapor pressure curve, de_s/dT, from the air temperature at the surface (T_s) with elevation, z, is given as

$$\Delta = \frac{e_s - e_{sz}}{T_s - T_z} \tag{14.29}$$

where e_{sz} is the saturation vapor pressure at elevation z and the corresponding temperature at that elevation is T_z. This gradient has been tabulated and can also be approximated as

$$\Delta = \frac{4098 e_s}{(237.3 + T)^2} \tag{14.30}$$

In Eq. 14.30, T is in °C. If e_s is calculated in kPa using Eq. 14.11, the slope Δ is in kPa/°C. A polynomial approximation of Δ (kPa/°C) as a function of mean temperature can also be given as

$$\Delta = 0.20(0.00738T + 0.8072)^7 - 0.000116 \tag{14.31}$$

14.3.4 Heat Transfer

Sensible heat is the heat energy exchanged by a body or thermodynamic system in which the exchange of heat changes the temperature of the body or system, and some macroscopic variables of the body or system, but leaves unchanged certain other macroscopic variables of the body or system, such as volume or pressure. It is the heat energy that is transported by a body that has a temperature higher than its surroundings via conduction, convection, or both.

The sensible heat of a thermodynamic process may be calculated as the product of the mass, m, with its **specific heat capacity**, c, and the change in temperature (ΔT):

$$H_s = mc\Delta T \tag{14.32}$$

Latent heat, also known as **latent energy** or **heat of transformation**, is the energy released or absorbed by a body or a thermodynamic system during a constant-temperature process, such as a first-order phase transition like water going from liquid phase to vapor phase. Latent heat can be understood as energy in hidden form, which is supplied or extracted to change the state of a substance without changing its temperature. Sensible heat and latent heat are not special forms of energy. Rather, they describe exchanges of heat under conditions specified in terms of their effect on a material or a thermodynamic system.

In evaporation, energy is consumed to overcome the intermolecular forces that tightly hold the water molecules, and to move the molecules away from each other. The energy required for water to go into the vapor form is called the **latent heat of vaporization**, λ_v. Its value is about 2.45 MJ/kg. However, it actually depends upon the temperature of water and can be calculated using

$$\lambda_v = 2.501 - 0.00236T\,(°C) \qquad (14.33)$$

The heat energy used in evaporation is given by

$$H_e = E\rho_w\lambda_v \qquad (14.34)$$

where E represents the **evaporation rate** (depth/time) and ρ_w is the density of water. If E is given as m/s, ρ_w in kg/m^3, and λ_v in J/kg, the unit of H_e is W/m^2. This is the **energy flux** per unit area. The unit is typically given as MJ/m^2·d.

The **Bowen ratio** is used to describe the type of heat transfer for a surface that has moisture. Heat transfer can either occur as sensible heat, causing differences in temperature without evapotranspiration, or latent heat, the energy required during a change of state without a change in temperature. The Bowen ratio is generally used to calculate heat lost or gained in a substance. It is the ratio of energy fluxes from one state to another by sensible heat and latent heating, respectively. The ratio was named by Harald Sverdrup after Ira Sprague Bowen, an astrophysicist whose theoretical work on evaporation to air from water bodies made first use of it (Bowen, 1926).

The Bowen ratio, β, is calculated as

$$\beta = \frac{H_s}{H_e} \qquad (14.35)$$

in which both H_s and H_e are expressed as energy fluxes from one medium to another.

It can be shown that

$$\beta = \gamma\left(\frac{T_s - T_a}{e_s - e_a}\right) \qquad (14.36)$$

The factor γ is called the **psychometric constant** or **hygrometric constant**. It is the ratio of the specific heat of moist air at constant pressure to the latent heat of vaporization of water and is given as

$$\gamma = \frac{c_a p_a}{0.622\lambda_v} \qquad (14.37)$$

in which c_a is the heat capacity of air and p_a is the atmospheric pressure. Its value can be taken as $c_a = 1.013$ kJ/kg K.

The psychometric constant γ is approximately

$$\gamma \approx 66.8\ \text{Pa/°C} \qquad (14.38)$$

which enters separately into some expressions for estimating evaporation, discussed below.

14.4 CONTROLLING FACTORS OF EVAPORATION

The factors that influence the evaporation rate are the supply of solar energy to provide the latent heat of evaporation, the wind velocity, and the vapor pressure gradient that transports evaporated water away from surface. Transpiration is also affected by these factors plus the ability of the plants to extract and transmit water from the soil to the stomata.

Therefore, the rate of evaporation is dependent on (1) the vapor pressures at the water surface and air above, in other words, the relative humidity or vapor pressure gradient; (2) the air and water temperatures, which are governed by solar radiation and sunshine duration; (3) the wind speed; (iv) atmospheric pressure; (v) the quality of water; and (vi) the depth and size of the water body.

Same factors that govern water evaporation from open water surfaces, govern evapotranspiration because, essentially, transpiration is mainly due to evaporation from the stomata. Also, plant physiology plays a role, since plants can control the size of the stomata and resistance to flow through roots and vascular systems, as well as soil moisture conditions that resist flow to roots.

14.5 ESTIMATION OF EVAPORATION

14.5.1 General Considerations

The estimation of evaporation is important mainly for assessing the loss of water from open water bodies, such as storage reservoirs and lakes. The methods for estimating evaporation from free water surfaces or completely saturated soils can be classified into three categories: (1) using empirical equations, (2) using equations from analytical methods, and (3) measurements. Here we present the common analytical methods, including the water balance, energy balance, and mass transfer or aerodynamic methods, and the combination of energy and mass transfer methods, also known as the **Penman** or **combination** method. We also briefly discuss measurements using pan evaporation data

14.5.2 Water Balance Method

The water balance method is the simplest of analytical methods but may not be too reliable. The method involves writing the hydrologic continuity equation for a water body and determining evaporation from a knowledge or estimation of other variables.

Thus, considering daily average values for a reservoir, the continuity equation can be written as

$$\Delta S = P + Q_{in} + G_{in} - Q_{out} - G_{out} - E_L - T_L \qquad (14.39)$$

where ΔS is the change in storage, P denotes precipitation, Q_{in} and Q_{out} are surface water inflow and outflow, respectively, G_{in} and G_{out} are groundwater inflow and outflow, respectively, and E_L and T_L are evaporation and transpiration losses, respectively (Figure 14.3A). All values are given in terms of either volume (m^3) or depth (m).

Rearranging terms of Eq. 14.39, the daily evaporation loss in a reservoir can be estimated as

$$E_L = P + (Q_{in} - Q_{out}) + (G_{in} - G_{out}) - \Delta S - T_L \qquad (14.40)$$

The values of P, Q_{in} and Q_{out} can be measured and T_L can be ignored in most cases. However, G_{in} and G_{out} cannot be

A) Water balance

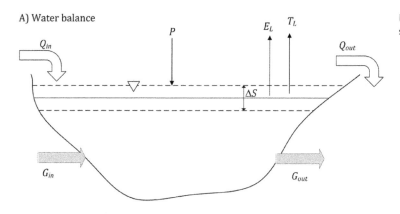

B) Energy balance

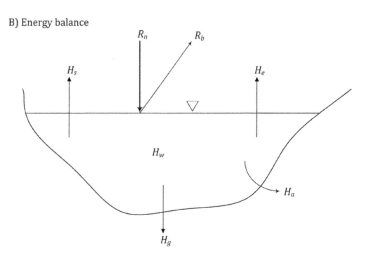

Figure 14.3 (A) Mass balance and (B) energy balance in a free surface water body.

easily measured, unless there are monitoring wells both upstream and downstream. Thus, controlled observations can provide a fairly accurate estimation of evaporation from lakes and reservoirs using the water balance approach.

14.5.3 Energy Balance Method

The energy balance method, based on the principle of conservation of energy, entails the estimation of energy available for evaporation by considering the incoming energy (which is principally shortwave radiation), outgoing energy, and energy stored in the water body in a certain interval (Figure 14.3B).

The energy balance to the evaporating surface in a one-day period is given by

$$R_n = H_s + H_e + H_a + H_g + H_w \tag{14.41}$$

where R_n is the net radiation, H_s is the sensible heat transfer to the air, H_e is the energy used in evaporation (Eq. 14.34), H_a is the advected energy, which is the net heat conducted out of the system by water flow, H_g is the heat flux into the ground, and H_w is the energy stored in the water body.

The net radiation is given by

$$R_n = R_s(1 - a_s) + R_l(1 - a_l) - R_b \tag{14.42}$$

where R_s and R_l, respectively, represent shortwave and longwave radiations with corresponding albedos of a_s and a_l, and R_b is the radiation reflected back to the atmosphere from the water surface. All units in Eq. 14.42 are in energy per unit area per unit time: R_n is given in $J/s \cdot m^2 = W/m^2$.

If the time period is short, H_w and H_a can be neglected. All other terms in Eq. 14.41 can be either measured or evaluated indirectly, except H_a, which cannot be readily measured. However, it can be estimated using the Bowen ratio

$$\beta = \frac{H_s}{\rho_w \lambda_v E_L} \approx 6.65 \times 10^{-4} \times p_a \frac{T_w - T_a}{e_s - e_a} \tag{14.43}$$

assuming constant values for heat capacity of air and latent heat of evaporation for water.

From Eqs. 14.34 and 14.41, and neglecting H_w and H_a, the evaporation is given by

$$E_L = \frac{R_n - H_g}{\rho_w \lambda_v (1 + \beta)} \tag{14.44}$$

In most situations, H_s and H_g are also ignored. In that case Eq. 14.44 is further simplified to

$$E_L = \frac{R_n}{\rho_w \lambda_v} \tag{14.45}$$

Equation 14.45 is based on the assumptions that there is (1) no water inflow or outflow to lake, (2) no change in water temperature of the lake, (3) no significant sensible heat transfer to the ground and atmosphere, and (4) no heat energy lost with water that leaves the system as vapor.

Equation 14.45 is the equation to express energy flux in terms of evaporation rate. For example, if R_n is given in MJ/m²·d, with density and latent heat of vaporization in units of kg/m³ and MJ/kg, the result obtained from Eq 14.45 multiplied by 1000 yields the **equivalent evaporation** in mm/d. A constant value of this factor, 0.408, has frequently been used in many evapotranspiration equations discussed below.

The energy balance method, where energy supply is the limiting factor, has been found to give satisfactory results, with errors of the order of 5% when applied to periods less than a week.

14.5.4 Mass Transfer Methods

Mass transfer methods are based on the concept that the rate of turbulent mass transfer of water vapor from the evaporating surface to the atmosphere is the limiting factor. Mass transfer is controlled by (1) the vapor pressure gradient and (2) the wind velocity which determines the rate at which vapor is carried away. This is simply an advancement of Dalton's law.

The evaporation rate according to mass transfer methods is given by Eq. 14.15. A generalized form of this equation can be written as

$$E_P = W(u)[e_s - e_a(z)] = W(u)\Delta e \tag{14.46}$$

where $W(u)$ is a *transport coefficient* that is a function of wind speed, u, and $e_a(z)$ is the measured vapor pressure at elevation z above the surface. Thus, evaporation is related to wind speed and is proportional to the vapor pressure deficit. Note that in Eq. 14.46 the K_e and v_a terms in Eq. 14.15 are simply combined to form a function.

Several empirical expressions for $W(u)$ have been proposed. From aerodynamic considerations, as given in Eq. 14.26, an analytical expression for $W(u)$ can be derived as

$$W(u) = \frac{0.622 k^2 \rho_a u_2}{p_t \rho_w \left[\ln\left(\frac{z_2}{z_0}\right)\right]^2} \tag{14.47}$$

In Eq. 14.47, only roughness height is considered, following Eq. 14.26. However, if the zero-plane displacement is included, as per Eq. 14.28, then

$$W(u) = \frac{0.622 k^2 \rho_a u_2}{p_t \rho_w \left[\ln\left(\frac{z_2 - z_d}{z_0}\right)\right]^2} \tag{14.48}$$

where u_2 is the wind velocity in m/s at height z_2, which is usually taken as 2 m above the surface. The unit of $W(u)$ is m/Pa·s.

Many empirical vapor transfer coefficients have been developed by fitting the aerodynamic model expressed by either Eq. 14.47 or Eq. 14.48 to local data, such as one recommended by Doorenbos and Pruitt (1977):

$$W(u) = 0.0027\left(1 + \frac{u}{100}\right) \tag{14.49}$$

in which $W(u)$ is in mm/day. Pa and u is the 24-hour wind run in km/d measured at height 2 m.

14.5.5 Combination Method or Penman Method

Evaporation can be computed by the mass transfer method when energy supply is not limiting and the energy method when vapor transport is not limiting. Typically, both factors are limiting. Hence a combination of the two methods can be used as first proposed by Penman (1948). Accordingly,

$$E_P = \frac{\Delta}{\Delta + \gamma} E_r + \frac{\gamma}{\Delta + \gamma} E_a \tag{14.50}$$

where E_r is the evaporation computed from the energy method (Eq. 14.45) and E_a is the evaporation computed from aerodynamic method (Eq. 14.46). The factor $\Delta/(\Delta+\gamma)$ is the weighting factor for E_r and the factor $\gamma/(\Delta+\gamma)$ is the weighting factor for E_a.

The combination method is the most accurate and most commonly used method if meteorological information is available. It is particularly good for small, well-monitored areas. This method needs data for net radiation, air temperature, humidity, and wind speed. If all this information is not available then an approximation known as the **Priestly–Taylor equation** can be used:

$$E = \alpha \frac{\Delta}{\Delta + \gamma} E_r \tag{14.51}$$

where $\alpha \approx 1.26$–1.3 is a coefficient that takes care of the aerodynamic effect. The approximation is based on observations that the second term in Eq. 14.50 is typically about 30% of the first term. Such an approximation is better for large areas.

All equations presented above, are suitable for daily time intervals or longer time periods.

14.5.6 Direct Measurement from Pan Evaporation

Since weather stations that are required to obtain the data to use in Eq. 14.50 are expensive to maintain, evaporative pans are often used to directly measure evaporation. An evaporation pan, a small, shallow vessel filled with water and placed in an open area and exposed to weather, is the most widely used device to measure evaporation. The water level in the pan is measured carefully to determine how much water, measured in terms of depth, is lost by evaporation. This loss is related to evaporation from a reservoir or lake.

There are many types of standardized pans for measuring evaporation. The US Class A pan is probably the most-used type. The pan, built of unpainted galvanized iron, is circular with a diameter of 1.21 m (~ 4 ft) and depth of 255 mm (~10 in), which gives it a volume of about 0.3 m³. The basin is put on a 150-mm high (~12 inches above ground) wooden frame due to air circulation around the basin. The water level is kept about 50 mm below the rim, due to the allowance for percolation in case we need to add more water. The water level is measured every day from the difference between the present and the original water level. It is recorded with a high-precision micrometer. From these measurements, the evaporation is determined based on the simple mass balance principle:

$$\Delta S = I - O$$
$$V_2 - V_1 = P - E$$
$$E_p = P - (V_2 - V_1) \tag{14.52}$$

where the input is precipitation, P, since the last pan reading, V_1 is the volume at the beginning of the time of measurement, and V_2 is the volume at the end.

Due to the Sun hitting the sides of the pan, the temperature in the pan increases, which means that the evaporation gets higher than the actual evaporation. To correct this value, the evaporation value from the pan is multiplied by a coefficient, called the **pan coefficient**, and its value depends on the climatic region where the measurements are made, and the season.

The pan coefficient should be calibrated at each site, by setting up a complete weather station to calculate E using Eq. 14.50 and compare it with the measured E_p to obtain

$$K_p = \frac{E}{E_p} \tag{14.53}$$

Once a pan coefficient for a region and season are known, it can be used to calculate

$$E = K_p E_p \tag{14.54}$$

Nevertheless, in general pans measure more evaporation than natural water bodies because of (1) less heat storage capacity of the water volume because the volume of water in a pan is smaller than in a natural water body, (2) heat transfer through pan sides, and (3) wind effects caused by pan itself.

Another kind of pan is the UK British Standard tank, which is slightly larger than the American pan and is put on the same level as the ground. The principle is about the same as for the American pan. Though the general opinion in the UK is that measurements are unreliable, and therefore calculations are more common.

Pan evaporation data have been extensively used to assess the relative efficacies of various models for the estimation of evapotranspiration from weather and climatic data, and these are presented next.

14.6 MODELS FOR ESTIMATION OF EVAPOTRANSPIRATION

14.6.1 The Penman Equation

A value of ET_A over a catchment is more often obtained by first calculating ET_P, assuming an unrestricted availability of water, and then modifying the result by accounting for the actual soil moisture content. There are several equations for calculating ET_P, based on theoretical or empirical models. Howard Penman at the Rothamsted Experimental Station, Harpenden, UK, was perhaps the first to propose a theoretically sound equation for estimating ET_P. The equation proposed by Penman directly results from the basic formula discussed in Section 14.5.5 to combine factors to account for the supply of energy and a mechanism to remove water vapor from the immediate vicinity of the evaporating surface.

Numerous variations of the equation formulated by Penman (1948) can be found in the literature mainly to express the unit of evaporation rate in a modern convention of L/T. The original form of Penman equation, using the notations adopted in this book, is

$$ET_P = \frac{\Delta R_n + \rho_a c_a (\delta e) W'(u)}{\lambda_v (\Delta + \gamma)} \tag{14.55}$$

In this form of the original Penman equation, for SI units, Δ and γ are in kPa/°C (kPa/K), R_n, is in MJ/m²·s, λ_v is in MJ/kg, ρ_a is in kg/m³, c_a in MJ/kg·K, and $W'(u)$ is called the *momentum* surface aerodynamic conductance given in m/s. This results in the units of ET_P as the mass rate of transport, that is, kg/m s, which can be multiplied by 86 400 to get kg/m² d. In order to be dimensionally homogeneous, the unit of the wind velocity function $W'(u)$ must be in m/s instead of the velocity function $W(u)$ introduced in Eq. 14.48 in which $W(u)$ has the unit of m/Pa·s.

Later, in 1933, W. Jim Shuttleworth modified the Penman equation to introduce a velocity function and rewrote it to yield the evapotranspiration rate in terms of L/T such as mm/d. With such a velocity function, and following a presentation consistent with the combination method equation presented above, an operational form of Penman equation can be presented as

$$ET_P = w \frac{R_n}{\lambda_v \rho_w} + (1-w)(\delta e) W(u) \tag{14.56}$$

where w is the weighting factor,

$$w = \frac{\Delta}{\Delta + \gamma} \tag{14.57}$$

and the velocity function, $W(u)$, is given as

$$W(u) = 6.43(1 + 0.536 u_2) \tag{14.58}$$

in which u_2 is wind velocity (m/s). The unit of $W(u)$ can be defined as discussed above. Other empirical formulations of $W(u)$ can also be found in the literature such as one given as Eq. 14.49.

A **soil heat flux factor**, denoted by G with the same unit of R_n, such as MJ/m·s has been introduced in the Penman equation to calculate reference evapotranspiration from a well-watered grass reference crop

$$ET_0 = w\frac{(R_n - G)}{\lambda_v \rho_w} + (1 - w)6.43(1.0 + 0.536u_2)(e_s - e_a) \tag{14.59}$$

14.6.2 The Penman–Monteith Equation

The Penman–Monteith equation is the most well-known formula for the estimation of evapotranspiration. The origin of the equation is the Penman equation, further improved by Monteith (1965) based on a combination of energy balance with mass transfer. The Penman–Monteith equation is given as

$$ET_P = \frac{\Delta(R_n - G) + (e_s - e_a)\left(\frac{\rho_a c_a}{r_a}\right)}{\lambda_v \rho_w \left[\Delta + \gamma\left(1 + \frac{r_s}{r_a}\right)\right]} \tag{14.60}$$

where r_a is the net aerodynamic resistance to water vapor diffusion through the air from surfaces to the height of measuring instruments (s/m), and r_s is the vegetation canopy resistance to water vapor surface (s/m). The maximum possible evapotranspiration is moderated by an aerodynamic resistance due to friction as air flows over vegetation. A bulk surface resistance is added in series with the aerodynamic resistance to account for limitations to water vapor flow at the leaf surfaces and at the soil. The parameterization is entirely dependent on the atmospheric conditions. All units of the variables must be consistent for dimensional homogeneity of Eq. 14.60 to yield ET_P in mm/d.

The stomata resistance of the whole canopy, referred to as the surface resistance, r_s, is less when more leaves are present, since there are then more stomata through which transpired water vapor can diffuse. One approximation, suggested by Allen (1986) for r_s, is

$$r_s = \frac{200}{L} \tag{14.61}$$

If h_c is the mean height of the crop, the leaf area index, L, can be estimated as (Allen et al., 1998)

$$L = \begin{cases} 24h_c \text{ clipped grass with } 0.05 < h_c < 0.15\,\text{m} \\ 5.5 + 1.5\ln(h_c) \text{ alfalfa with } 0.10 < h_c < 0.50\,\text{m} \end{cases} \tag{14.62}$$

The Penman–Monteith method implemented in HEC-HMS uses the Penman–Monteith equations for computing evapotranspiration at less than a daily time interval as detailed by Allen et al. (1998). The method requires shortwave radiation and longwave radiation to be included in the meteorological model. The algorithm of Allen et al. (1998) is followed most closely when the specified *pyranograph shortwave method* and the *FAO-56 longwave method* are selected. When shortwave data are not available, Allen et al. (1998) recommended using the Hargreaves shortwave method.

The method is of quite good accuracy and is usually used for calculations of evapotranspiration from farmlands. The accuracy is due to all the parameters of the equation but still it has deficiencies. For instance, the r_s value is a constant depending on the kind of vegetation present in the area of study. If the equation is used over a large area with different kinds of vegetation an average value for r_s needs to be estimated. The estimation becomes less accurate if the area contains spots or patches without vegetation.

14.6.3 The Food and Agriculture Organization–Penman–Monteith Equation

As described in the Irrigation and Drainage Paper 56 (Allen et al., 1998), the FAO has adopted the Penman–Monteith equation as the standard technique to compute the reference evapotranspiration. The FAO-56–PM equation is stated as

$$ET_0 = \frac{0.408\Delta(R_n - G) + \gamma\left(\frac{900}{T+273}\right)u_2(e_s - e_a)}{\Delta + \gamma(1 + 0.34u_2)} \tag{14.63}$$

where ET_0 is the reference evapotranspiration (mm/d); R_n is net radiation at the crop surface (MJ/m²·day), G is the soil heat flux density (MJ/m²·d), T is the air temperature at 2 m height (°C), e_s is the saturation vapor pressure of the air (kPa), e_a is the mean actual vapor pressure of the air (kPa), Δ is the slope of vapor pressure curve (kPa/°C), and γ is the psychometric constant (kPa/°C).

The FAO-56–PM equation determines the evapotranspiration from the hypothetical grass reference surface and provides a standard to which evapotranspiration in different periods of the year or in other regions can be compared and to which the evapotranspiration from other crops can be related.

14.6.4 The Hargreaves Equation

Hargreaves (1975) found that reference evapotranspiration can be adequately computed by using the average temperature and solar-radiation data. Hargreaves (1975) proposed the following equation to compute ET_0:

$$ET_0 = 0.0135(T + 17.78)R_s \tag{14.64}$$

where R_s is the solar radiation in terms of water evaporation (mm/d), and T is the temperature in °C. Hargreaves et al. (1985) gave the following temperature-based equation:

$$ET_0 = 0.0023R_0\sqrt{\delta T}(T + 17.8) \tag{14.65}$$

where R_0 is the extraterrestrial radiation converted to the water equivalent (mm/d), T is the air temperature (°C), and δT is the daily air temperature range (°C), usually calculated as the difference between the daily maximum and minimum temperature, which accounts for the effect of cloudiness, correlates with relative humidity and vapor pressure, and negatively with wind speed (Hargreaves and Allen, 2003)

The Hargreaves evapotranspiration method is based on an empirical relationship where reference evapotranspiration was regressed with solar radiation and temperature data. The regression was based on eight years of precision lysimeter observations for a grass reference crop in Davis, California. The method has been validated for sites around the world. It is capable of capturing diurnal variation in PET for simulation time steps of less than 24 hours. Combining the Hargreaves evapotranspiration method with the Hargreaves shortwave radiation method yields the Hargreaves evapotranspiration form equivalent to Hargreaves and Allen (2003).

14.6.5 The Priestley–Taylor Equation

The Priestley–Taylor method (Priestley and Taylor, 1972) uses a simplified energy balance approach where the soil water supply is assumed to be unlimited. Simplified forms of latent and sensible energy are used. The method is capable of capturing diurnal variation in PET through the use of a net solar radiation gage, so long as the simulation time step is less than 24 hours.

The Priestley–Taylor equation is given as

$$ET_P = \alpha \frac{\Delta}{\Delta + \gamma} \left(\frac{R_n - G}{\lambda_v \rho_w} \right) \quad (14.66)$$

where, $\alpha \approx 1.26$–1.3 is a calibration factor that accounts for the aerodynamic component. This is a purely radiation-based method

14.6.6 The Blaney–Criddle Equation

The original formulation developed by Blaney and Criddle (1950) is another empirical model, requiring only the mean daily temperature, $T(°C)$, over each month. The reference crop evapotranspiration (mm/d) as an average for a period of one month is given by

$$ET_0 = p(0.46T + 8.13) \quad (14.67)$$

where p is the mean daily percentage (for the month) of the total annual daytime hours. In other words, p is the ratio of actual daily daytime hours to annual daytime hours expressed as a percent.

The Blaney–Criddle method always refers to mean *monthly* values, both for the temperature and the ET_0. If, for example, it is found that T in April is 23 °C, it means that during the whole month of April the mean daily temperature is 23 °C. If at a local meteorological station, the daily minimum and maximum temperatures are measured, the mean daily temperature is calculated by taking the averages of all maximum and minimum temperature values.

To determine the p value, knowing the approximate latitude of the area, the number of degrees north or south of the equator, is necessary. Table 14.2 gives values of p at certain latitudes. These values can be used and interpolated for intermediate latitudes.

Blaney and Criddle (1950) calculated the evapotranspiration from a consumptive-use factor, mean monthly temperature, and percentage of total annual daylight hours occurring during the period being considered. An empirically determined consumptive-use crop coefficient, K, was then applied to establish evapotranspiration water requirements. For example, values of K for some crops with a normal growing season are: alfalfa 0.85, beans 0.65, corn 0.75, and pasture 0.75.

The simplified formula given by Eq. 14.66 was developed for the arid western portion of the United States and provides good estimates of seasonal water needs under these conditions. The main reason for the popularity of this formulation is that it does not require any weather station data like those necessary for Penman-type equations. However, Doorenbos and Pruitt (1977) noted that the effect of climate was

Table 14.2 *Daily percentage of annual daylight hours (p) for each month at different latitudes*

Latitude (°)	North: South:	Jan July	Feb Aug	Mar Sept	Apr Oct	May Nov	June Dec	July Jan	Aug Feb	Sept Mar	Oct Apr	Nov May	Dec June
60		0.15	0.20	0.26	0.32	0.38	0.41	0.40	0.34	0.28	0.22	0.17	0.13
55		0.17	0.21	0.26	0.32	0.36	0.39	0.38	0.33	0.28	0.23	0.18	0.16
50		0.19	0.23	0.27	0.31	0.34	0.36	0.35	0.32	0.28	0.24	0.20	0.18
45		0.20	0.23	0.27	0.30	0.34	0.35	0.34	0.32	0.28	0.24	0.21	0.20
40		0.22	0.24	0.27	0.30	0.32	0.34	0.33	0.31	0.28	0.25	0.22	0.21
35		0.23	0.25	0.27	0.29	0.31	0.32	0.32	0.30	0.28	0.25	0.23	0.22
30		0.24	0.25	0.27	0.29	0.31	0.32	0.31	0.30	0.28	0.26	0.24	0.23
25		0.24	0.26	0.27	0.29	0.30	0.31	0.31	0.29	0.28	0.26	0.25	0.24
20		0.25	0.26	0.27	0.28	0.29	0.30	0.30	0.29	0.28	0.26	0.25	0.25
15		0.26	0.26	0.27	0.28	0.29	0.29	0.29	0.28	0.28	0.27	0.26	0.25
10		0.26	0.27	0.27	0.28	0.28	0.29	0.29	0.28	0.28	0.27	0.26	0.26
5		0.27	0.27	0.27	0.28	0.28	0.28	0.28	0.28	0.28	0.27	0.27	0.27
0		0.27	0.27	0.27	0.27	0.27	0.27	0.27	0.27	0.27	0.27	0.27	0.27

insufficiently defined by considering only temperature and day length, because vegetative water requirements still vary widely among climates that exhibit similar temperatures and day lengths. Subsequently, this method has undergone many improvements, such as one proposed by Allen and Pruitt (1986), so much so that presently this method can hardly be considered as a merely temperature-based method. The currently preferred form of the equation is

$$ET_P = a_{BC} + b_{BC}f \tag{14.68}$$

where f is given by Eq. 14.67 and a_{BC} and b_{BC} are given as

$$a_{BC} = 0.0043 R_{H(min)} - \left(\frac{n}{N}\right) - 1.41 \tag{14.69}$$

$$a_{BC} = 0.82 - 0.0041 R_{H(min)} + 1.07\left(\frac{n}{N}\right) + 0.066(u_2)$$
$$- 0.006 R_{H(min)}\left(\frac{n}{N}\right) - 0.0006 R_{H(min)}(u_2) \tag{14.70}$$

where ET_P is the PET (mm/d), $R_{H(min)}$ is the minimum daily relative humidity in %, u_2 is the daytime wind speed at 2 m height (m/s), and the units of n and N are (h/d).

14.6.7 Thornthwaite Method

Thornthwaite (1948) derived an equation to be used for limited water conditions. This formula is based mainly on temperature with an adjustment being made for the number of daylight hours. An estimate of the monthly PET, calculated on a monthly basis, is given by:

$$ET_{P(m)} = 16 N_m \left(\frac{10\overline{T}_m}{I}\right)^a \tag{14.71}$$

where $ET_{P(m)}$ is given in mm; m stands for the months 1, 2, 3, ..., 12; N_m is the monthly adjustment factor related to hours of daylight; $\overline{T}_m$ is the monthly mean temperature (°C); and I is the **heat index** for the year, given by

$$I = \sum_{m=1}^{12} i_m = \sum_{m=1}^{12} \left(\frac{\overline{T}_m}{5}\right)^{1.5} \tag{14.72}$$

The exponent a in Eq. 14.71 is given as

$$a = 6.75 \times 10^{-7} \times I^3 - 7.71 \times 10^{-5} \times I^2 + 1.792 \times 10^{-2} \times I + 0.49239 \tag{14.73}$$

The heat index is an integral element in Thornthwaite's classification of climate. In Eq. 14.71, N_m is calculated as the maximum number of sunny hours as a function of month and latitude

Given the monthly mean temperatures from measurements at a climatologic station, an estimate of ET_P for each month of the year can be calculated. This method has been used widely throughout the world even though it was developed for the eastern United States. Therefore, its application in climates other than those similar to this area where it was developed should be made with caution.

Compared to the Penman formula, the Thornthwaite values tend to exaggerate ET_P. This is particularly marked in the summer months with the high temperatures having a dominant effect in the Thornthwaite computation, whereas the Penman estimates take into consideration other meteorological factors.

14.6.8 The Turc Method

Turc (1961) analysed data collected from 254 watersheds located in virtually all parts of the world and related evapotranspiration to rainfall and temperature as

$$ET = \frac{P}{\left[0.9 + \left(\frac{P}{I_T}\right)^2\right]^{0.5}} \tag{14.74}$$

where ET is the annual evapotranspiration (mm), P is the annual precipitation (mm) and

$$I_T = 300 + 25T + 0.05T^3 \tag{14.75}$$

in which T is the mean temperature (°C).

Toward the needs of agronomists for irrigation schemes, Turc extended his empirical method for calculating the annual AET to produce a formula for PET over a shorter period of time. The Turc short-term formula for PET over 10 days is

$$ET_P = \frac{P + a + 70}{\left[1 + \left(\frac{P+a}{L} + \frac{70}{2L}\right)^2\right]^{0.5}} \tag{14.76}$$

where P is the precipitation in a 10-day period (mm), a is the estimated evaporation in the 10-day period from the bare soil when there has been no precipitation (1 mm $\leq a \leq$ 10 mm), and L is the evaporation capacity of the air given by

$$L = \frac{(T+2)(R_{SW})^{0.5}}{16} \tag{14.77}$$

in which T is the mean air temperature over the 10-day period (°C), and R_{SW} is the mean shortwave radiation (cal/cm·d).

14.6.9 Selection of a Method

The methods of estimation of ET_P described above are the most common and widely used methods. In addition to these, many other methods of computation of ET_P have been developed by several workers. For example, Hamon (1963) developed a simple equation to estimate the ET_P given mean air temperature and day length. By this method, ET_P does not become zero when the mean air temperature is less than 0 °C but provides essentially the same annual total as that of the Thornthwaite method.

There have been numerous studies where the multiple methods were employed to compute PET and compare the results with pan evaporation data to determine which method most closely approximates the actual evapotranspiration. There has not been a universal consensus from these attempts. In some areas one method may appear better than the others but the same method may not be found to be the best method when tested in another region.

14.6.10 Estimation of Net Radiation

14.6.10.1 Extraterrestrial Radiation, R_a

All matter at a temperature above absolute zero radiates energy in the form of electromagnetic waves that travel at the speed of light. The rate at which the energy is emitted is given by the **Stefan–Boltzmann law**

$$F = \varepsilon \sigma T^4 \tag{14.78}$$

where F is the rate of energy emission per unit surface area per unit time $(E/L^2/T)$, T is the absolute temperature (K) of the surface, σ is a universal constant called the **Stefan–Boltzmann constant** $(E/L^2/T/K^4)$, and ε is a dimensionless quantity called **emissivity**. The value of $\sigma = 4.90 \times 10^{-9}$ MJ/m²·day·K Emissivity can be calculated using the **Idso–Jackson equation** given as

$$\varepsilon = -0.02 + 0.261 \exp\left[-7.77 \times 10^{-4}(273 - T)^2\right] \tag{14.79}$$

in which T is in K.

The radiation striking a surface perpendicular to the Sun's rays at the top of the Earth's atmosphere, called the **solar constant**, is about 0.082 MJ/m²·min. The local intensity of radiation is, however, determined by the angle between the direction of the Sun's rays and the normal to the surface of the atmosphere. This angle will change during the day and will be different at different latitudes and in different seasons. The solar radiation received at the top of the Earth's atmosphere on a horizontal surface is called the extraterrestrial solar radiation, R_a.

If the Sun is directly overhead, the angle of incidence is zero and the extraterrestrial radiation is 0.0820 MJ/m²·min. As seasons change, the position of the Sun, the length of the day and, hence, R_a changes as well. Extraterrestrial radiation is thus a function of latitude, date, and time of day.

14.6.10.2 Solar or Shortwave Radiation, R_s

As the radiation penetrates the atmosphere, some of the radiation is scattered, reflected or absorbed by the atmospheric gases, clouds, and dust. The amount of radiation reaching a horizontal plane is known as the solar radiation, R_s. Because the Sun emits energy by means of electromagnetic waves characterized by short wavelengths, solar radiation is also referred to as shortwave radiation. Figure 14.4 shows the five components of radiation that supply energy for the evaporation process.

For a cloudless day, R_s is roughly 75% of extraterrestrial radiation. On a cloudy day, the radiation is scattered in the atmosphere but, even with extremely dense cloud cover, about 25% of the extraterrestrial radiation may still reach the Earth's surface mainly as diffuse sky radiation. Solar

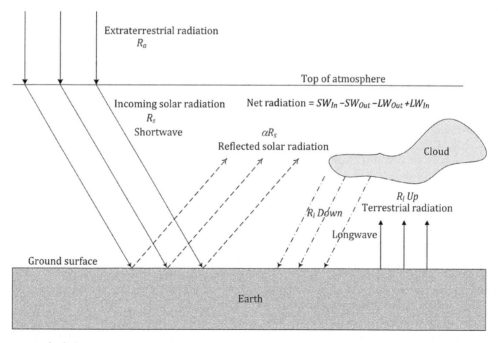

Figure 14.4 Components of radiation.

radiation is also known as global radiation, meaning that it is the sum of direct shortwave radiation from the Sun and diffuse sky radiation from all upward angles.

14.6.10.3 Relative Shortwave Radiation, R_s/R_{so}

The relative shortwave radiation is the ratio of the solar radiation, R_s, to the clear-sky solar radiation, R_{so}. R_s is the solar radiation that actually reaches the Earth's surface in a given period, while R_{so} is the solar radiation that would reach the same surface during the same period but under cloudless conditions. The relative shortwave radiation is a way to express the cloudiness of the atmosphere; the cloudier the sky the smaller the ratio. The ratio varies between about 0.33 (dense cloud cover) and 1 (clear sky). In the absence of a direct measurement of the net radiation, R_n, the relative shortwave radiation is used in the computation of the net longwave radiation.

14.6.10.4 Relative Sunshine Duration, n/N

The relative sunshine duration is another ratio that expresses the cloudiness of the atmosphere. It is the ratio of the actual duration of sunshine, n, to the maximum possible duration of sunshine or daylight hours, N. In the absence of any clouds, the actual duration of sunshine is equal to the daylight hours ($n=N$) and the ratio is one, while on cloudy days n, and consequently the ratio, may be zero. In the absence of a direct measurement of R_s, the relative sunshine duration, n/N, is often used to derive the solar radiation from extraterrestrial radiation.

14.6.10.5 Albedo and Net Solar Radiation, R_{ns}

A considerable amount of solar radiation reaching the Earth's surface is reflected. The fraction, α, of the solar radiation reflected by the surface is known as the **albedo**, which is highly variable for different surfaces and for the angle of incidence or the slope of the ground surface. It may be as large as 0.95 for freshly fallen snow and as small as 0.05 for a wet bare soil. A green vegetation cover has an albedo of about 0.20–0.25. For the green grass reference crop, α is assumed to have a value of 0.23 and for an open water surface it is 0.08.

The net solar radiation, R_{ns}, is the fraction of the solar radiation, R_s, that is not reflected from the surface. Its value is $(1-\alpha)R_s$.

14.6.10.6 Net Longwave Radiation, R_{nl}

The solar radiation absorbed by the Earth is converted to heat energy. By several processes, including emission of radiation, our planet loses this energy. The Earth, which is at a much lower temperature than the Sun, emits radiative energy with wavelengths longer than those from the Sun.

Therefore, the *terrestrial radiation* is referred to as longwave radiation. The emitted longwave radiation $(R_{l,up})$ is absorbed by the atmosphere or is lost into space. The longwave radiation received by the atmosphere $(R_{l,down})$ increases its temperature and, as a consequence, the atmosphere radiates energy of its own. Part of the radiation finds its way back to the Earth's surface, which, consequently, both emits and receives longwave radiation. The difference between outgoing and incoming longwave radiation is called the net longwave radiation, R_{nl}. As the outgoing longwave radiation is almost always greater than the incoming longwave radiation, R_{nl} represents an energy loss.

14.6.10.7 Net Radiation, R_n

The net radiation, R_n, is the difference between incoming and outgoing radiation of both short and long wavelengths. It is the balance between the energy absorbed, reflected, and emitted by the Earth's surface or the difference between the incoming net shortwave (R_{ns}) and the net outgoing longwave (R_{nl}) radiation. R_n, is normally positive during the daytime and negative during the nighttime. The total daily value for R_n, is almost always positive over a period of 24 hours, except in extreme conditions at high latitudes.

14.6.10.8 Soil Heat Flux, G

All fluxes of energy should be considered when deriving the energy balance equation. The equation for an evaporating surface can be written as

$$R_n - G - H_e - H_s = 0 \qquad (14.80)$$

in which G is the soil heat flux.

In making estimates of evapotranspiration, all terms of this energy balance should be considered. The soil heat flux, G, is the energy that is utilized in heating the soil. G is positive when the soil is warming and negative when it is cooling. Although the soil heat flux is small compared to R_n and may often be ignored, the amount of energy gained or lost by the soil in this process should theoretically be subtracted or added to R_n when estimating evapotranspiration.

14.6.10.9 Units

The standard unit used in this book to express energy received on a unit surface per unit time is megajoules per square meter per day (MJ/m^2·d). In some places, J/m^2·s, which is equivalent to W/m^2, is also used. In meteorological literature or in other places, certain other units might be used or radiation might even be expressed in units no longer accepted as standard SI units, such as cal/cm^2·d.

14.6.10.10 Calculations of Radiation

Extraterrestrial Radiation for Daily Periods

The extraterrestrial radiation, R_a, for each day of the year and for different latitudes can be estimated from the solar constant, the solar declination, and the time of the year using the equation

$$R_a = \frac{24(60)}{\pi} G_{sc} d_r [\omega_s \sin(\varphi) \sin(\delta) + \cos(\varphi) \cos(\delta) \sin(\omega_s)] \quad (14.81)$$

where R_a is the extraterrestrial radiation given in MJ/m²·d, G_{sc} is the solar constant (0.0820 MJ/m²·min), d_r is the inverse relative distance between the Earth and Sun, ω_s is the sunset hour angle (rad), φ is latitude (rad), and δ is the solar declination (rad).

The variables of Eq. 14.811 are calculated as follows.

The latitude, φ, expressed in radians is positive for the northern hemisphere and negative for the southern hemisphere. The conversion from decimal degrees to radians is given by

$$\varphi = \left(\varphi_{DD} \frac{\pi}{180}\right) \quad (14.82)$$

where φ_{DD} is latitude in decimal degree.

The inverse relative distance between the Earth and the Sun, d_r, and the solar declination, δ, are given by

$$d_r = 1 + 0.033 \cos\left(\frac{2\pi}{365} J\right) \quad (14.83)$$

$$\delta = 0.409 \sin\left(\frac{2\pi}{365} J - 1.39\right) \quad (14.84)$$

where J is the number of the day in the year between 1 (1 January) and 365 or 366 (31 December).

The sunset hour angle, ω_s, is given by

$$\omega_s = \arccos[-\tan(\varphi) \tan(\delta)] \quad (14.85)$$

Extraterrestrial Radiation for Hourly or Shorter Periods

For hourly or shorter periods, the solar time angle at the beginning and end of the period should be considered when calculating R_a. In that case the equation for calculating R_a is

$$R_a = \frac{24(60)}{\pi} G_{sc} d_r \left[(\omega_2 - \omega_1) \sin(\varphi) \sin(\delta) + \cos(\varphi) \cos(\delta)(\sin(\omega_2) - \sin(\omega_1))\right] \quad (14.86)$$

where ω_1 is solar time angle at beginning of period (rad) and ω_2 is solar time angle at the end of period (rad).

The solar time angles at the beginning and end of the period are given by

$$\omega_1 = \omega - \frac{\pi t_1}{24} \quad (14.87)$$

$$\omega_2 = \omega + \frac{\pi t_1}{24} \quad (14.88)$$

where ω is the solar time angle at the midpoint of hourly or shorter period (rad), and t_1 is the length of the calculation period (h): 1 for an hourly period or 0.5 for a 30-minute period.

The solar time angle at the midpoint of the period is

$$\omega = \frac{\pi}{12}[(t + 0.06667(L_z - L_m) + S_c) - 12] \quad (14.89)$$

where t is the standard clock time at the midpoint of the period (h). For example, for a period between 15.00 and 16.00, $t = 15.5$. L_z is the longitude of the center of the local time zone (degrees west of Greenwich). For example, $L_z = 75, 90, 105$ and $120°$ for the Eastern, Central, Rocky Mountain and Pacific time zones in the United States, and $L_z = 0°$ for Greenwich, 330° for Cairo, Egypt, 255° for Bangkok, Thailand, 267° for Kolkata, India. L_m is the longitude of the measurement site (degrees west of Greenwich) and S_c is the seasonal correction for solar time (h). When $\omega < -\omega_s$ or $\omega > \omega_s$ Eq. 14.89 indicates that the Sun is below the horizon so that, by definition, R_a is zero.

The seasonal correction for solar time is

$$S_c = 0.1645 \sin(2b) - 0.1255 \cos(b) - 0.025 \quad (14.90)$$

where

$$b = \frac{2\pi(J - 81)}{364} \quad (14.91)$$

Daylight Hours

The daylight hours, N, are given by

$$N = \frac{24}{\pi} \omega_s \quad (14.92)$$

Solar Radiation

If the solar radiation, R_s, is not measured, it can be calculated with the **Angstrom formula** that relates solar radiation to extraterrestrial radiation and relative sunshine duration:

$$R_s = \left(a_s + b_s \frac{n}{N}\right) R_a \quad (14.93)$$

in which a_s is the regression constant, expressing the fraction of extraterrestrial radiation reaching the Earth on overcast days ($n = 0$), and $a_s + b_s$ is the fraction of extraterrestrial radiation reaching the Earth on clear days ($n = N$). Depending on atmospheric conditions, including humidity, dust, etc., and solar declination, which depends on latitude and month, the Angstrom values a_s and b_s will vary. Where no actual solar radiation data are available and no calibration has been carried out for improved a_s and b_s parameters, the values $a_s = 0.25$ and $b_s = 0.50$ are recommended (Allen et al., 1998).

Clear-Sky Solar Radiation

The calculation of the clear-sky radiation, R_{so}, when $n = N$, is required for computing the net longwave radiation.

For near sea level or when calibrated values for a_s and b_s are available:

$$R_{so} = (a_s + b_s)R_a \qquad (14.94)$$

When calibrated values for a_s and b_s are not available:

$$R_{so} = (0.75 + 2 \times 10^{-5} z)R_a \qquad (14.95)$$

where z is the station elevation above sea level (m).

Net Solar or Net Shortwave Radiation

The net shortwave radiation resulting from the balance between incoming and reflected solar radiation is given by

$$R_{ns} = (1 - \alpha)R_s \qquad (14.96)$$

where the albedo, or canopy reflection coefficient, α, is 0.23 for the hypothetical grass reference crop.

Net Longwave Radiation

The rate of longwave energy emission is proportional to the absolute temperature of the surface raised to the fourth power. This relation is expressed quantitatively by the Stefan–Boltzmann law. The net energy flux leaving the Earth's surface is, however, less than that emitted and given by the Stefan-Boltzmann law due to the absorption and downward radiation from the sky. Water vapor, clouds, carbon dioxide, and dust are absorbers and emitters of longwave radiation. Their concentrations should be known when assessing the net outgoing flux. As humidity and cloudiness play an important role, the Stefan–Boltzmann law is corrected by these two factors when estimating the net outgoing flux of longwave radiation. It is thereby assumed that the concentrations of the other absorbers are constant. The net longwave radiation is calculated using

$$R_{nl} = \sigma \left[\frac{T_{max}^4 + T_{min}^4}{2} \right] (0.34 - 0.14\sqrt{e_a}) \left(1.35 \frac{R_s}{R_{so}} - 0.35 \right) \qquad (14.97)$$

in which σ is the Stefan–Boltzmann constant ($4.903 \; 10^{-9}$ MJ/K^4·m^2·d], T_{max} is the maximum absolute temperature (in K = °C + 273.16], and T_{min} is the minimum absolute temperature, during the 24-hour period.

An average of the maximum air temperature to the fourth power and the minimum air temperature to the fourth power is commonly used in the Stefan–Boltzmann equation for 24-hour time steps. The term $(0.34 - 0.14\sqrt{e_a})$ expresses the correction for air humidity, and will be smaller if the humidity increases. The effect of cloudiness is expressed by $(1.35 R_s/R_{so} - 0.35)$. The term becomes smaller if cloudiness increases and hence R_s decreases. The smaller the correction terms, the smaller the net outgoing flux of longwave radiation. Note that the R_s/R_{so} term in Eq. 14.97 must be limited so that $R_s/R_{so} \leq 1.0$.

Penman (1948) used the following equation, which can also be applied to calculate R_{nl}

$$R_{nl} = \sigma T^4 (0.56 - 0.092\sqrt{e_a}) \left[0.1 + 0.9 \left(\frac{n}{N} \right) \right] \qquad (14.98)$$

In Eq. 14.98, T is in °C.

Net Radiation

The net radiation, R_n, is the difference between the incoming net shortwave radiation, R_{ns}, and the outgoing net longwave radiation, R_{nl}:

$$R_n = R_{ns} - R_{nl} \qquad (14.99)$$

Soil Heat Flux

Complex models are available to describe the soil heat flux. Because this is small compared to R_n, particularly when the surface is covered by vegetation and calculation time steps are 24 hours or longer, a simple calculation procedure is presented here for long time steps, based on the idea that the soil temperature follows air temperature:

$$G = c_s \frac{T_i - T_{i-1}}{\Delta t} \Delta z \qquad (14.100)$$

where G is the soil heat flux (MJ/m^2·d), c_s is the soil heat capacity (MJ/m^3·C), T_i is the air temperature at time i (°C), T_{i-1} is the air temperature at time $i - 1$ (°C), Δt is the length of time interval (d), and Δz is the effective soil depth (m).

As the soil temperature lags the air temperature, the average temperature for a period should be considered when assessing the daily soil heat flux, meaning that Δt should exceed one day. The depth of penetration of the temperature wave is determined by the length of the time interval. The effective soil depth, Δz, is only 0.10–0.20 m for a time interval of one or a few days but might be 2 m or more for monthly periods. The soil heat capacity is related to its mineral composition and water content.

14.6.11 Measurements

Solar radiation can be measured with **pyranometers**, **radiometers**, or **solarimeters**. The instruments contain a sensor installed on a horizontal surface that measures the intensity of total solar radiation, that is, both direct and diffuse radiation from cloudy conditions. The sensor is often protected and kept in a dry atmosphere by a glass dome that should be regularly wiped clean.

The net longwave and net shortwave radiation can be measured by recording the difference in output between sensors facing upward and downward. In a net radiometer, the glass domes are replaced by polyethylene domes that have a transmission range for both shortwave and longwave radiation.

Where pyranometers are not available, solar radiation is usually estimated from the duration of bright sunshine. The actual duration of sunshine, n, is measured with a **Campbell–Stokes sunshine recorder**. This instrument records periods of bright sunshine by using a glass globe that acts as a lens. The Sun's rays are concentrated at a focal point that burns a hole in a specially treated card mounted concentrically with the sphere. The movement of the Sun changes the focal point throughout the day and a trace is drawn on the card. If the sun is obscured, the trace is interrupted. The hours of bright sunshine are indicated by the lengths of line segments.

Global surface radiation data are also available from different programs of NASA as described in Mukhopadhyay and Singh (2011). These data are obtained by the Clouds and the Earth's Radiant Energy System (CERES) sensors on NASA's Aqua and Terra satellites. These data are extremely useful for basin-scale monthly water balance studies in any part of the world as illustrated by Mukhopadhyay and Singh (2011). As an illustration, Figure 14.5 shows the global net radiation grids for the months of January and July of 2007.

The quantity of heat conducted into the soil, G, can be measured with systems of soil heat flux plates and thermocouples or thermistors.

14.7 MEASUREMENT OF POTENTIAL EVAPORATION

There are many ways of measuring potential evaporation. One of the most common methods is to use the irrigated **lysimeter**. Others use an **atmometer** and the standardized US Class A pan discussed above.

14.7.1 Irrigated Lysimeters

An installed irrigated lysimeter closely resembles a percolation gauge. The principal difference is in the operation of the

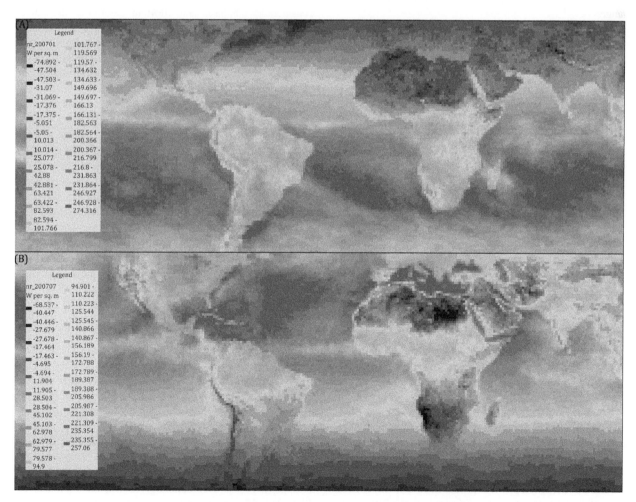

Figure 14.5 Global net radiation for the month of (A) January and (B) July of 2007. The source of NASA data for production of such maps for any month from 1983 onward has been described in detail by Mukhopadhyay and Singh (2011). It can be clearly seen that the oceans reflect back very considerably less radiation than the land surface. The seasonal contrast between the northern and southern hemisphere can also be seen from the reversals of radiation patterns from January to July. (A black and white version of this figure will appear in some formats. For the color version, please refer to the plate section.)

apparatus, with the contained soil being kept at field capacity by sprinkling a known quantity of water on the tank when rainfall is deficient. The field capacity is assured by maintaining continuous percolation from the bottom of the tank. Thus, the vegetation cover is allowed to transpire freely, and the total evaporation loss is dependent entirely on the ability of the air to absorb the water vapor. The potential evaporation is the sum of rainfall and irrigation minus percolation. One of the disadvantages of these gauges is that the soil sample is disturbed but, with careful filling of the tank and after the establishment of the vegetation cover, the gauge gradually becomes representative of the surrounding ground. In winter, with snow cover and freezing temperatures, certain difficulties in operating the gauges are encountered, but discrepancies are not of great importance, since evaporation losses are low and often negligible under such conditions. Measured values of potential evaporation using these irrigated gauges can be exaggerated in very dry periods and hot climates. Surrounding parched ground, heating and drying the air above, tends to cause increased evaporation from the continuously watered and transpiring vegetation of the gauge.

14.7.2 Atmometers

Atmometers are devices that can give direct measurement of evaporation. They basically consist of a wet, porous ceramic cup mounted on the top of a cylindrical water reservoir. The ceramic cup is covered with a green fabric that simulates the canopy of a crop. The reservoir is filled with distilled water, which evaporates out of the ceramic cup and is pulled through a suction tube that extends to the bottom of the reservoir. Underneath the fabric, the ceramic cup is covered by a special membrane that keeps rainwater from seeping into it. A rigid wire extending from the top keeps birds from perching on top of the gauge.

14.8 REFERENCE EVAPOTRANSPIRATION

Estimates of the reference evapotranspiration, ET_0, are widely used in irrigation engineering to define crop water requirements. The relationship between actual and reference evapotranspiration can be given as

$$ET_A = K_s K_c (ET_P) \qquad (14.101)$$

in which K_s is a soil coefficient with values ranging for 0 to 1, depending on the soil type, and K_c is the corm coefficient, with values ranging from 0.2 to 1.3, depending on the stage of the growth and type of crops.

These estimates are used in the planning for irrigation schemes to be developed and to manage water distribution in existing schemes. Accurate estimates of crop evapotranspiration are required in hydrologic studies and water resources modeling of stationary and changing climatic conditions.

Estimating evapotranspiration is difficult due to several factors, which include the accuracy of data measurement and the reliability and sensitivity of measuring facilities. Several models have been developed with various degrees of acceptability for determining reliable sets of data. Numerous studies have been performed using lysimeter data and have shown that, in most cases, the Penman–Monteith method is the best method for estimating the reference evapotranspiration. From several existing ET_0 equations, the FAO-56 application of the Penman–Monteith equation is currently the most widely used and can be considered as standard.

The Penman–Monteith method has two advantages over other methods. First, it is a predominately physically based approach, indicating that the method can be used globally without any need for additional parameter estimation. Second, it has been tested using a variety of lysimeters. A major downside of the application of this method is that it requires multiple physical variables like air temperature, wind speed, relative humidity, and solar radiation data. The number of meteorological stations where all these parameters are observed is limited in many areas. Allen et al. (1998) placed considerable emphasis and effort in describing alternative ways to estimate the solar radiation and humidity data required for this method using simpler or fewer measurements.

The limitation of reliable data motivated Hargreaves et al. (1985) to develop an alternative approach where only mean maximum and mean minimum air temperatures and extraterrestrial radiation are required. Because extraterrestrial radiation can be calculated for a certain day and location, only minimum and maximum temperatures are the parameters that require observation. The Hargreaves method has been tested using some high-quality lysimeter data representing a broad range in climatologic conditions (Hargreaves, 1994). The results have indicated that this equation was nearly as accurate as the Penman–Monteith equation in estimating ET_0 on a weekly or longer time step and was therefore recommended in cases where reliable data were lacking. The accuracy of this equation can be improved by adjusting the parameters to local conditions as discussed in Allen et al. (1998).

14.9 POTENTIAL EVAPOTRANSPIRATION

PET is a critical parameter in the water balance. The concept of PET has been widely accepted, and it is generally defined as the evaporation from an extended surface of short green vegetation that fully shades the ground, exerts little or negligible resistance to the flow of water, and is always well supplied with water. PET cannot exceed evaporation from a free water surface under the same weather conditions.

In humid regions where advection of sensible heat is unimportant, pan evaporation gives realistic estimates of PET. In arid regions and in regions where advection is considerable, pan evaporation may give unrealistic values and the difference between pan evaporation and PET may be quite pronounced.

Doorenbos and Pruitt (1977) presented and recommended four techniques for estimating the PET rate: the Blaney–Criddle method with the modifications that made it not just a temperature-based method, the Penman method with a correction factor, the radiation method, and the pan evaporation method.

14.10 ACTUAL EVAPOTRANSPIRATION

AET differs from PET under most circumstances. The differences are mainly due to (1) the influence of surfaces that are not extended, (2) varying heights in vegetation, (3) partial vegetative cover, (4) internal resistance in vegetation to water flow, (5) periodic water deficits such as dry seasons during which vegetation is not well supplied with water, and (6) vegetation using more water in arid and dry regions than that suggested by pan evaporation, with PET exceeding free water evaporation.

14.10.1 Evaporation from Soils

The water that evaporates from a soil horizon is stored in the small pores between the particles in the soil. From the point of field capacity, that is, the amount of water that the soil horizon can hold after free drainage, to the point of wilting point, the water in the soil is available for evaporation. During this time a pressure drop can occur from about -10 cm to -1000 m. If the water content in the soil gets larger than the field capacity the process becomes evaporation from free water surface.

14.10.2 Transpiration

Transpiration takes place through leaf pores, which are called stomata. The amount of vapor released from the pore depends on the temperature of the leaf, the light, and the amount of water in the leaf. All together these factors are called stomata resistance, r_s. The value of r_s is raised at dry weather or if the content of water in the vegetation is low.

14.10.3 Methods for Measuring Actual Evapotranspiration

Measuring AET is probably not as common as measuring PET. A method that is used for this purpose is the percolation gauge, which is generally regarded more of a research tool than a standard instrument for measuring evaporation and transpiration such as lysimeter. There are many different designs of the gauge. For example, a one-meter-deep hole is made and filled up with soil, rock, and gravel, together with a pipe from the bottom to the collection pit. The top of the hole is made indistinguishable from the surrounding vegetation. When evaporation is measured with a percolation gauge, no consideration is given to the changes in the soil water storage. This means that the measurements should be made over a time period when the gauge is saturated. This is in contrast to a lysimeter, which takes into consideration how much water is stored in the soil, since it weights the soil and gives a value on how much water is stored.

14.10.4 Estimating Actual from Potential Evapotranspiration

The calculation of PET from readily available meteorological data is much simpler than the computation or measurement of AET from a vegetated surface. However, water loss from a watershed area does not always proceed at the potential rate, since this is dependent on a continuous water supply. When the vegetation is unable to abstract water from the soil, the AET becomes less than the PET. Thus, the relationship between the AET and PET depends upon the soil moisture content.

14.11 EXAMPLES

Example 14.1: Xu and Singh (2000) evaluated eight radiation-based methods of estimating evapotranspiration for the Changins climatologic station in Switzerland (46°24′ N latitude, 6°14′ E longitude). The observed data of mean monthly temperature, relative humidity, vapor pressure deficit, wind speed, solar radiation, and pan evaporation values were for the period 1990–1994.

1. From the data given in Table 14.3, compute the evaporation using the Penman equation and the FAO-56 Penman–Monteith equation. Compare the results with observed pan evaporation data. Discuss the sources of uncertainties.
2. From the latitude information given above calculate the net radiation following the procedure given in Section 14.6.10.10.
3. Compute ET_0 using the Blaney–Criddle and Priestley–Taylor methods and compare the results. Is the radiation-based method better than the temperature-based method?

Solution:

Part 1: Table 14.4. Lists the input data after converting into the standard SI radiation and pressure units that are used for evapotranspiration calculations

Table 14.3 *Climatologic data from Changins station, Switzerland as reported in Xu and Singh (2000)*

Month	Temperature (°C)	Wind speed (m/s)	Vapor pressure deficit (mbar)	Humidity (%)	Radiation (cal/cm^2·d)	Pan evaporation, E_{pan} (mm/day)
January	2.19	2.44	1.52	80.5	88.16	0.94
February	2.77	2.45	1.79	78.25	159.33	1.15
March	7.47	2.61	3.5	69.22	266.14	2.18
April	8.98	2.71	4.29	67.76	346.34	2.78
May	14.3	2.6	5.89	68.19	457.92	3.44
June	16.7	2.54	6.37	70.74	462.61	3.47
July	20.14	2.42	9.08	66.08	513.29	4.72
August	20.53	2.4	9.9	63.84	445.13	4.9
September	15.28	2.18	4.91	75.05	295.72	2.75
October	10.03	2.35	2.45	81.16	160.04	1.41
November	6.04	2.21	1.69	82.65	84.7	0.95
December	2.75	3.08	1.52	80.37	71.98	1.02

(1 cal = 4.184 J). The table is transposed to show step-by-step calculations in the three parts.

Table 14.5 gives the step-by-step calculations, with the equations used for each one. The constants used in the calculation of this part are as follows: $c_a = 0.001013$ MJ/kg·K, $\rho_w = 1000$ kg/m^3, $\alpha = 0.23$, $\sigma = 4.9 \times 10^{-9}$ MJ/K^4·m^2·d, $n/N = 0$ (assumed), $z = 400$ m.

Notes on the calculations in Table 14.5:
  The average elevation of the area is 500 m, found from a Google search.
  The slope of the saturation vapor pressure curve, Δ, is calculated using both Eqs. 14.30 and 14.31 for comparison and illustration. Results from Eq. 14.30 are used, since this equation is more commonly used.
  The pressure is calculated using both Eqs. 14.24 and 14.25 for comparison and illustration. The result from Eq. 14.25 is used, since it does not assume a constant temperature of 20°C as in Eq. 14.24.
  For the calculation of R_{ns}, an albedo value of 0.23 is assumed. This is one of the sources of uncertainty.
  For the calculation of R_{nl}, it is necessary to know the fraction n/N. Since there are no observed data, an assumption needs to be made. This value varies from day to day and hence the average for each month is different. However, for the sake of simplicity, a constant value is used to make several trial calculations resulting in a value of R_n that would give the best fit of the calculated evapotranspiration with the observed pan evapotranspiration data. This is the greatest source of uncertainty in the calculations.

Figure 14.6 shows the comparison of results for evapotranspiration obtained from the Penman equation and the FAO-56 Penman–Monteith equation with the observed pan evaporation data.

Part 2: Table 14.6 gives step-by-step calculations used. The constants used in the calculation of this part are: $G_{sc} = 0.082$ MJ/m^2·min, $a_s = 0.25$, $b_s = 0.5$, $n/N = 0.5$ (assumed), $z = 400$ m, $\alpha = 0.23$, $\sigma = 4.9 \times 10^{-9}$ MJ/K^4·m^2·d

Notes on calculations:
  The latitude of the station given in degrees and minute is first converted to decimal degrees (60 minutes = 1 degree).
  Since no calibrated values of a_s and b_s are available, the default values recommended by Allen et al. (1998) are used.
  Like Part 1, the sources of uncertainties lie in the assumption of selecting appropriate values of albedo, α, and n/N These are the greatest sources of uncertainties. A value of n/N applicable to all months of a year is not appropriate. However, for the sake of simplicity, one value is selected by several trial calculations, whereby the calculated values of R_s match closely with observed values. The comparison can be seen in rows 10 and 11 of Table 14.6.

Part 3: For this part the only parameter for which an assumption needs to be made is the coefficient α in the Priestley–Taylor equation. A value of 1.3 is used. The calculations are given in Table 14.7. The p values in the Blaney–Criddle equation are used for latitude 45° N.

Figure 14.7 shows the how the results from the Blaney–Criddle and Priestley–Taylor methods compare with the measured pan evaporation data. The radiation-based method seems to perform better than the temperature-based method.

Table 14.4 Input data in standard SI units

Month	January	February	March	April	May	June	July	August	September	October	November	December
Temperature (°C)	2.19	2.77	7.47	8.98	14.3	16.7	20.14	20.53	15.28	10.03	6.04	2.75
Wind speed (m/s)	2.44	2.45	2.61	2.71	2.6	2.54	2.42	2.4	2.18	2.35	2.21	3.08
Vapor pressure deficit (kPa)	0.152	0.179	0.35	0.429	0.589	0.637	0.908	0.99	0.491	0.245	0.169	0.152
Humidity (%)	80.5	78.25	69.22	67.76	68.19	70.74	66.08	63.84	75.05	81.16	82.65	80.37
Radiation (MJ/ m^2·d)	3.689	6.666	11.135	14.491	19.159	19.356	21.476	18.624	12.373	6.696	3.544	3.012
Pan evaoration, E {pan} (mm/d)	0.94	1.15	2.18	2.78	3.44	3.47	4.72	4.9	2.75	1.41	0.95	1.02

Table 14.5 *Calculations for Part 1*

					Equation used						
e_s (kPa/°C)	0.7153	0.7455	1.0346	1.1465	1.6299	1.9011	2.3585	2.4160	1.7363	1.2304	0.9377	0.7444	Eq. 14.11
Δ (kPa/°C)	0.0511	0.0530	0.0708	0.0775	0.1055	0.1208	0.1458	0.1489	0.1115	0.0824	0.0649	0.0529	Eq. 14.30
Δ (kPa/°C)	0.0512	0.0531	0.0708	0.0775	0.1054	0.1206	0.1457	0.1488	0.1114	0.0824	0.0649	0.0530	Eq. 14.31
λ_v (MJ/kg)	2.4958	2.4945	2.4834	2.4798	2.4673	2.4616	2.4535	2.4525	2.4649	2.4773	2.4867	2.4945	Eq. 14.33
p_a (kPa)	96.1710	96.1710	96.1710	96.1710	96.1710	96.1710	96.1710	96.1710	96.1710	96.1710	96.1710	96.1710	Eq. 14.24
p_a (kPa)	96.6603	96.6603	96.6603	96.6603	96.6603	96.6603	96.6603	96.6603	96.6603	96.6603	96.6603	96.6603	Eq. 14.25
γ (kPa/°C)	0.0631	0.0631	0.0634	0.0635	0.0638	0.0640	0.0642	0.0642	0.0639	0.0635	0.0633	0.0631	Eq. 14.37
w	0.4476	0.4565	0.5275	0.5496	0.6232	0.6538	0.6945	0.6988	0.6359	0.5647	0.5062	0.4562	Eq. 14.57
$W(u)$	14.8394	14.8739	15.4253	15.7700	15.3908	15.1841	14.7705	14.7016	13.9433	14.5292	14.0467	17.0452	Eq. 14.58
R_{ns} (MJ/m²·d)	2.8402	5.1331	8.5742	11.1580	14.7527	14.9038	16.5366	14.3407	9.5272	5.1560	2.7288	2.3190	Eq. 14.96
e_a (kPa)	0.57579	0.58332	0.71616	0.77685	1.11146	1.34486	1.55853	1.54239	1.30307	0.99859	0.77499	0.59828	Eq. 14.14
R_{nl} (MJ/m²·d)	1.3783	1.3887	1.4628	1.4845	1.5467	1.5655	1.6116	1.6224	1.5407	1.4726	1.4239	1.3858	Eq. 14.98
R_n (MJ/m²·d)	1.4619	3.7444	7.1114	9.6734	13.2061	13.3383	14.9249	12.7183	7.9865	3.6833	1.3049	0.9332	Eq. 14.99
E_r (mm/d)	0.5857	1.5011	2.8636	3.9009	5.3525	5.4186	6.0832	5.1857	3.2400	1.4868	0.5247	0.3741	Eq. 14.56 (first part)
E_a (mm/day)	2.2556	2.6624	5.3989	6.7653	9.0652	9.6722	13.4116	14.5545	6.8462	3.5597	2.3739	2.5909	Eq. 14.56 (second part)
E_P (mm/day)	1.5082	2.1323	4.0615	5.1911	6.7516	6.8913	8.3223	8.0073	4.5531	2.3892	1.4379	1.5796	Eq. 14.56
ET_0 (mm/day)	0.6672	1.0542	2.1282	2.8111	3.9921	4.1922	5.1481	4.7831	2.6562	1.2592	0.6512	0.6686	Eq. 14.63

Table 14.6 *Calculations for Part 2*

	January	February	March	April	May	June	July	August	September	October	November	December	30-day
Day, J interval	15	45	75	105	135	165	195	225	255	285	315	345	
Latitude, φ (rad)	0.8098	0.8098	0.8098	0.8098	0.8098	0.8098	0.8098	0.8098	0.8098	0.8098	0.8098	0.8098	Eq. 14.82
Inverse Sun–Earth distance, d_r	1.0319	1.0236	1.0091	0.9923	0.9774	0.9685	0.9678	0.9754	0.9895	1.0064	1.0215	1.0311	Eq. 14.83
Declination, δ	−0.3702	−0.2361	−0.0404	0.1658	0.3288	0.4060	0.3774	0.2502	0.0579	−0.1496	−0.3180	−0.4035	Eq. 14.84
Sunset hour angle, ω_s	1.1510	1.3154	1.5283	1.7475	1.9372	2.0392	2.0001	1.8425	1.6317	1.4119	1.2178	1.1059	Eq. 14.85
Extraterrestrial radiation, R_a	11.0729	16.3863	24.4159	32.7628	39.0291	41.8099	40.6158	35.7118	28.1002	19.7044	13.0105	9.7826	Eq. 14.81
Daylight hours, N	8.7931	10.0488	11.6756	13.3497	14.7994	15.5784	15.2794	14.0760	12.4651	10.7858	9.3032	8.4481	Eq. 14.92
Solar radiation, R_s	5.5365	8.1932	12.2080	16.3814	19.5145	20.9050	20.3079	17.8559	14.0501	9.8522	6.5053	4.8913	Eq. 14.93
Clear-sky solar radition, R_{so}	8.3933	12.4208	18.5073	24.8342	29.5840	31.6919	30.7868	27.0696	21.3000	14.9359	9.8620	7.4152	Eq. 14.95
Net solar radiation, R_{ns}	4.2631	6.3087	9.4001	12.6137	15.0262	16.0968	15.6371	13.7491	10.8186	7.5862	5.0090	3.7663	Eq. 14.96
Net solar radiation, $R_{ns\,(obs)}$	3.6886	6.6664	11.1353	14.4909	19.1594	19.3556	21.4761	18.6242	12.3729	6.6961	3.5438	3.0116	Given
Saturated vapor pressure, e_a	0.5758	0.5833	0.7162	0.7768	1.1115	1.3449	1.5585	1.5424	1.3031	0.9986	0.7750	0.5983	Part 1
Temperature, T	2.19	2.77	7.47	8.98	14.3	16.7	20.14	20.53	15.28	10.03	6.04	2.75	Given
Net longwave radiation, R_{nl}	3.55	3.57	3.63	3.63	3.47	3.32	3.23	3.27	3.30	3.40	3.48	3.55	Eq. 14.97
Net radiation, R_n	0.71	2.74	5.77	8.98	11.55	12.78	12.40	10.48	7.52	4.18	1.53	0.22	Eq. 14.99
R_n from Part 1	1.46	3.74	7.11	9.67	13.21	13.34	14.92	12.72	7.99	3.68	1.30	0.93	Part 1

Table 14.7 Calculations for Part 3

Month	January	February	March	April	May	June	July	August	September	October	November	December	Equation/source used
Temperature, T (°C)	2.19	2.77	7.47	8.98	14.3	16.7	20.14	20.53	15.28	10.03	6.04	2.75	Part 2
Net Radiation, R_n	0.710	2.736	5.767	8.985	11.552	12.781	12.404	10.481	7.521	4.183	1.527	0.216	Table 14.2
p factor	0.200	0.230	0.270	0.300	0.340	0.350	0.340	0.320	0.280	0.240	0.210	0.200	Eq. 14.67
ET_P, Blaney–Criddle	1.827	2.163	3.123	3.678	5.001	5.534	5.914	5.624	4.244	3.059	2.291	1.879	Part 1
w	0.448	0.456	0.527	0.550	0.623	0.654	0.694	0.699	0.636	0.565	0.506	0.456	Part 1
λ_v (MJ/kg)	2.496	2.494	2.483	2.480	2.467	2.462	2.453	2.453	2.465	2.477	2.487	2.495	Eq. 14.66
ET_P, Priestley–Taylor	0.166	0.651	1.593	2.589	3.793	4.413	4.564	3.882	2.522	1.240	0.404	0.051	
Pan evapotranspiration, $E_{\{pan\}}$ (mm/d)	0.94	1.15	2.18	2.78	3.44	3.47	4.72	4.9	2.75	1.41	0.95	1.02	Given

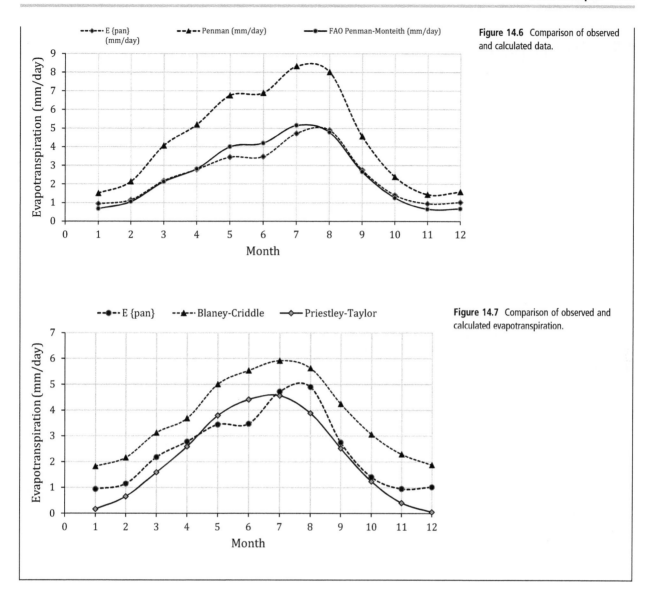

Figure 14.6 Comparison of observed and calculated data.

Figure 14.7 Comparison of observed and calculated evapotranspiration.

Exercises: A selection of exercises on this topic is available at www.cambridge.org/appliedhydrology.

15 Snowmelt

15.1 CONCEPTS AND PURPOSE

In our treatment of the various components of a hydrologic system, we have so far considered rainfall as the only input to produce surface runoff. However, there are vast regions over the globe where meltwater from snowpacks and glaciers is a significant, if not the major, contributor to streamflow. For this and other related reasons, such as the effect of climate change on the cryosphere, the estimation of both the rate and volume of meltwater generated from snow and ice has a great significance in many hydrologic investigations. Snow does not run down into the rivers until it melts. However, it is difficult to measure this important quantity of the hydrologic cycle. Several techniques, based on the principles of thermodynamics and physics, are used to calculate the rate of snowmelt generation. The subject matter of this chapter encompasses the means of quantifying meltwater from snow and ice.

15.2 ENERGY FLUX

The first law of thermodynamics is the law of conservation of energy, which leads to the fundamental equation of heat transfer. This equation can be stated in various ways. A simplified form is

$$\Delta E = dH + dW \tag{15.1}$$

where ΔE denotes the change in internal energy of a system under consideration due to changes in the amount of heat in the system and the work done on the system, denoted by dH and dW, respectively. The amount of heat can be given as

$$dH = c_p m dT + \lambda_h dm \tag{15.2}$$

where c_p is the specific heat of the material, m is its mass, λ_h is its latent heat, and T is the temperature. The temperature difference between two systems is the driving force for heat flow. For an infinitesimal transfer of heat (dh), the increase in temperature (dT) is proportional to the amount of heat supplied. Thus,

$$dh = CdT \tag{15.3}$$

The coefficient of proportionality, C, is called the **heat capacity**. In other words, the relation between the amount of heat transferred to a body and the ensuing change in temperature depends on its heat capacity:

$$C = \frac{dh}{dT} \tag{15.4}$$

If the heat capacity is expressed in terms of mass, it is called **specific heat**, which can be measured for a material either at constant pressure or constant volume and designated by c_p and c_v, respectively.

Equation 15.2 can be divided by the time interval, dt, to give rates of heating and melting.

$$\frac{dH}{dt} = c_p m \frac{dT}{dt} + \lambda_h \frac{dm}{dt} \tag{15.5}$$

The unit of the quantity is m/s, which is converted to mm per day by multiplying by 86 400 s/day and 1000 mm/m. In a system containing a mass of snowpack or glacial ice, meltwater occurs due to an influx of energy in the form of heat over snow or ice. There is virtually no work done on the system and therefore dW of Eq. 15.1 can be dropped. If ΔH denotes the change in energy input per unit area, A, of a mass of snow or ice with depth, z, using the fact that $m = \rho A dz$, Eq. 15.5 can be given as

$$\frac{1}{A}\frac{dH}{dt} = c_p \rho dz \frac{dT}{dt} + \lambda_h \rho \frac{dz}{dt} \tag{15.6}$$

The **energy flux** is defined as the rate of change of energy per unit area or **rate of energy input**, as given by the left-hand side of Eq. 15.6:

$$\overline{F} = \frac{\Delta H}{A \Delta t} \tag{15.7}$$

where $\overline{F}$ is the energy flux with unit of $E/L^2/T$ (e.g., $J/s \cdot m^2 = W/m^2$). The unit of $\Delta H/A$ is E/L^2 (e.g., J/m^2), where ΔH denotes the change in energy input per unit area of snow.

Snowmelt calculations are based on the principle of the conservation of energy, which simply states that the change in the stored energy of a system is the difference between the energy that enters the system (E_{in}) and the energy that leaves the system (E_{out}). In the case of snowmelt, the form of energy playing this role is heat energy, H. Thus, from Eq. 15.7, the **energy balance** equation for a snowpack can be written as

$$dH = \overline{F}dt \tag{15.8}$$

The flux of energy is composed of all the energy exchange mechanisms that are discussed in the subsequent sections.

15.3 PHYSICAL PROPERTIES OF SNOW

15.3.1 Snow Water Equivalent

The amount of water contained in a snowpack is the amount that ultimately enters the land phase of the hydrologic cycle. This amount is called the **snow water equivalent** (designated as SWE) and this is perhaps the most important property of a snowpack that is of interest to a hydrologist.

The snow water equivalent is used to convert the depth of snow to an equivalent depth of water. For example, a glass full of snow will not yield a full glass of water when all the snow is melted. The snow water equivalent or, simply, the **water equivalent** of a snowpack, D_m, is given as

$$SWE \equiv D_m = \frac{\rho_s}{\rho_w} D_s \tag{15.9}$$

where ρ_s is the density of snow, ρ_w is the density of liquid water, and D_s is the height of a column of snowpack. The density of snow is given as

$$\rho_s = (1 - \phi)\rho_i + \theta\rho_w \tag{15.10}$$

where ρ_i is the density of ice, and ϕ is the porosity of the snowpack. The **liquid water content**, θ, is defined as the ratio of the volume of liquid water, V_w, in the snowpack to the total volume of snow, V_s. Thus,

$$\theta = \frac{V_w}{V_s} \tag{15.11}$$

As used above, in all subsequent formulations, the subscripts s, i, and w are used for snow, ice, and liquid water, respectively.

The porosity of a snowpack is given as

$$\phi = \frac{V_w + V_a}{V_s} \tag{15.12}$$

in which V_a is the volume of air in the column of a snowpack. For example, fresh snow is almost 90 percent air, thus only 10 percent of the snow depth is actually water. Table 15.1 lists some representative values.

15.3.2 Thermal Quality

The amount of snowmelt that can be produced from a given quantity of heat energy depends on the condition of the snowpack, which is called the **thermal quality** and designated by B. The thermal quality is given by the ratio

$$B = \frac{H_s}{H_i} \tag{15.13}$$

where H_s is the heat energy required to melt a unit mass (e.g., 1 g) of snow and H_i is the heat energy required to melt a unit mass (e.g., 1 g) of pure ice at $0\,°C$.

Thus, B is expressed as a fraction. If the snowpack contains no interstitial water, $B = 1.0$. After melting begins, some interstitial water develops and $B < 1.0$. For a melting snowpack, $B = 0.95 - 0.97$. For ripe snow, the thermal quality may be lower. In other words, a thermal quality less than unity indicates that the snowpack is at $0\,°C$ and contains liquid water. If $B > 1.0$ the snowpack is at temperature less than $0\,°C$ and contains no liquid water.

Thermal quality of partially melted snow is given as

$$B = \frac{D_m}{D_m + D'_m} = \frac{D_m}{D_t} \tag{15.14}$$

where D_m and D'_m, respectively, denote the snow water equivalent that has not been melted and already melted and hence D_t denotes the total potential snow water equivalent of a snowpack.

15.3.3 Cold Content

While the snow water equivalent is useful for predicting the potential runoff volume, it does not directly provide a measure of when snow will melt. Because it is also important to predict the timing of snow melting, an index called the **cold content** is used to predict how close to melting the snowpack is. The cold content, H_{cc}, is the energy required to raise snow temperature to $0\,°C$.

The cold content or the thermal energy required to bring a snowpack at temperature T_s ($< 0\,°C$) to freezing temperature ($0\,°C$) is given as

$$H_{cc} = -c_i \rho_i D_m (T_s - T_m) = -c_i \rho_i D_m T_s \tag{15.15}$$

where c_i is the heat capacity (specific heat) of ice (2102 J/kg·K), T_s is the snow temperature and the melting temperature, T_m, is $0\,°C$. Note that H_{cc} has units of energy/area, e.g., J/m². It is essentially a storage term. On the other hand, $\overline{F}$ is used for flux or rate of energy flow. H_{cc} can be considered an energy deficit. As energy flows into the snowpack, it first has to fill the bucket represented by H_{cc} before any melt can occur.

The thermal energy required to bring the snowpack at $0\,°C$ to meltwater at $0\,°C$ is given as

$$H_m = \rho_i \lambda_f D_m \tag{15.16}$$

where λ_f is the **latent heat of fusion of ice** (0.334 MJ/kg). If D_m is in m, ρ_i is in kg/m³, and H_m is in MJ/m².

The total energy required for the completion of melting of a cold snowpack is

Table 15.1 *Types of snow and their properties*

Type of snow	Density (%)	Albedo
Fresh snow	5–15	0.9
Old snow	40–70	0.5
Glacier ice	90–98	0.4

Table 15.2 *Physical properties of different phases of water and certain other materials*

Substance	Density (kg/m³)	Specific heat (J/kg·K)	Latent heat (kJ/kg)	Thermal conductivity (W/m·K)
Water	1000.0	4186	Evaporation, λ_v: 2470	0.598 at 20°C
Snow	50–900	2100		0.04–1.0
Ice	917.0	2102	Fusion, λ_f: 334.9	2.10–2.76
			Sublimation, λ_s: 2804	
Air*	1.29	1005		0.024 at 0°C
Granitic rock	2650	760-850		2.12–3.62
Basaltic rock	2900	600-700		4.0–5.1 at 20°C

*At the pressure of air at standard sea level (101.3 kPa)

$$H_t = -c_i\rho_i D_m T_s + \rho_i \lambda_f D_m \quad (15.17)$$

Thus, $H_t = H_s$.

If the whole snowpack was entirely frozen ice, the thermal energy required to melt the pack is given by

$$H_i = H_m = \rho_i \lambda_f D_m \quad (15.18)$$

From the definition of B given in Eq. 15.13,

$$B = \frac{H_t}{H_i} = \frac{-c_i\rho_i D_m T_s + \rho_i \lambda_f D_m}{\rho_i \lambda_f D_m} \quad (15.19)$$

Equation 15.19 can also be written as

$$B = \left(1 - \frac{c_i}{\lambda_f} T_s\right) \frac{D_m}{D_t} \quad (15.20)$$

For a completely frozen ice pack, $D'_m = 0$; $D_t = D_m$. This results in

$$B = 1 - \frac{T_s}{160} \quad (15.21)$$

If the snowpack is below freezing temperature, $T_s < 0°C$, then the thermal quality of the snowpack is given by Eq. 15.21.

The physical properties of various phases of water as well as some other materials are given in Table 15.2.

Thermodynamic relationships among the three water phases determine the grain growth and **metamorphism of snow**. Whether the compound water exists as ice, water, or water vapor depends on its temperature, T, and pressure p, as shown in the phase diagram in Figure 15.1. The curves for evaporation, sublimation, and melting trace the points where two phases of water coexist in thermodynamic equilibrium. All three phases coexist at the **triple point**, where the temperature is 0.01 °C (273.16 K) and the vapor pressure is 0.611657 kPa (Guildner et al., 1976). This is not the same as the ordinary melting point of ice, which occurs at a temperature of 0 °C (273.15 K) and at 1 atmosphere of pressure (101.325 kPa). Vapor is in metastable equilibrium with respect to supercooled water along the line that extends the evaporation curve below 0.01 °C. At the **critical point**, the properties of the phases are indistinguishable.

15.4 METAMORPHISM OF SNOWPACK

15.4.1 Phases of Snowmelt

Typically, snow-covered areas have an **accumulation period** followed by a **melt** or **ablation period**. During an accumulation period the water equivalent increases. The melt period can be divided into three phases. To melt snow, energy is needed to be added to the snowpack – first to warm the pack and then to melt the pack. The melt period begins *at the end of the accumulation period* and occurs in three phases: warming, ripening, and output.

15.4.2 Warming Phase

During the warming phase, the temperature of the snowpack increases until the whole pack is at 0 °C and is under isothermal conditions. The pack must overcome the cold content, which is given by Eq. 15.15. Note that this equation does not work once the snowpack is at 0 °C.

During the warming phase, ΔH results in an average increase in temperature of the snow:

$$\Delta H = -c_i \rho_w D_m \Delta T_s \quad (15.22)$$

Thus, the temperature change due to the net rate of energy inputs during the warming phase is given as

$$\Delta T_s = \frac{\Delta H}{c_i \rho_w D_m} = \frac{\overline{F}\Delta t}{c_i \rho_w D_m} \quad (15.23)$$

15.4.3 Ripening Phase

Melting of snow occurs during the ripening phase. However, the meltwater stays in the pack. The snow is under isothermal conditions at 0 °C and cannot hold any more water. The snow has a water holding capacity given by

$$D_{wret} = \theta_{ret} D_s \quad (15.24)$$

where θ_{ret} is the maximum volumetric water content that snow can retain against gravity. In the ripening phase, ice

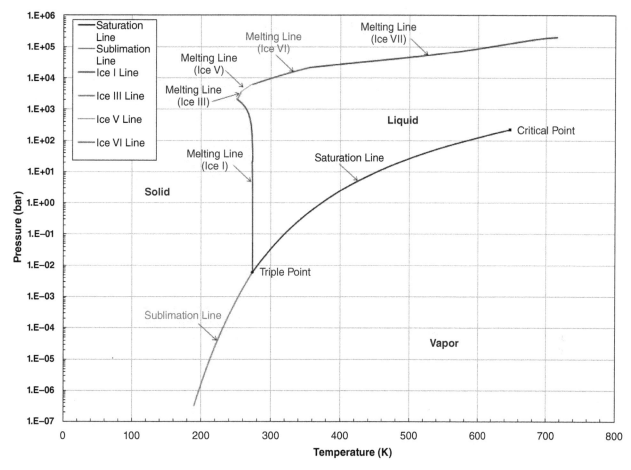

Figure 15.1 Phase diagram of water. (A black and white version of this figure will appear in some formats. For the color version, please refer to the plate section.)

melts until D_{wret}, the **liquid water holding capacity**, is satisfied.

For the completion of the ripening phase, the net energy input required is given by

$$H_{m2} = \rho_w \lambda_f D_{wret} = \theta_{ret} D_s \rho_w \lambda_f \qquad (15.25)$$

15.4.4 Output Phase

Meltwater flows out of the snowpack once it is ripe, with further input of energy. This is the output phase. The net input energy required for output phase is given by

$$H_{m3} = \rho_w \lambda_f (D_m - D_{wret}) \qquad (15.26)$$

15.4.5 Net Heat Requirement for Snowmelt Generation

From the presentation above, it is clear that the total energy required to melt a snowpack is

$$H_t = H_{cc} + H_{m2} + H_{m3} \qquad (15.27)$$

15.5 RATE OF SNOWMELT FROM ENERGY FLUX

The objective in this section is to develop formulations for the calculation of the amount and rate of water loss from snow and ice packs from given climate information.

During the ripening and output phases, ΔH results in melting. Melting during the ripening phase causes an increase in liquid water held in the snowpack. The general equation for the conversion of the rate of energy input to rate of snowmelt production follows from Eq. 15.25.

$$\Delta H = \rho_w \lambda_f \Delta D_w \qquad (15.28)$$

where ΔD_w is the increment of water output from the snowpack. This can be derived from Eq. 15.28

$$\Delta D_w = \frac{\Delta H}{\rho_w \lambda_f} \qquad (15.29)$$

The rate of production is then

$$\frac{\Delta D_w}{\Delta t} = \left(\frac{\Delta H}{\Delta t}\right)\left(\frac{1}{\rho_w \lambda_f}\right) \qquad (15.30)$$

From Eq. 15.7 we can write

$$\frac{\Delta D_w}{\Delta t} = \left(\frac{\overline{F}}{\rho_w \lambda_f}\right) \qquad (15.31)$$

Writing the left-hand side of Eq. 15.31 as flow rate $Q\,(L/T)$, the rate of snowmelt is given as

$$Q = \left(\frac{\overline{F}}{\rho_w \lambda_f}\right) \qquad (15.32)$$

In SI units, $\overline{F}$ is in W/m², $\rho_w = 1000$ kg/m³, and $\lambda_f = 334.9$ kJ/kg. Then,

$$Q\,[\text{mm/d}] = 0.258\,\overline{F}\,[\text{W/m}^2] \qquad (15.33)$$

Equation 15.33 does not include the thermal quality term. When the factor of thermal quality is included, the rate of snowmelt is given as

$$Q = \left(\frac{\overline{F}}{B\rho_w \lambda_f}\right) \qquad (15.34)$$

Equation 15.34 is the *fundamental equation used in snow hydrology* to calculate the rate of generation of snowmelt from the estimated energy flux.

15.6 ENERGY INPUTS IN SNOW MELTING

The most obvious sources of energy required to satisfy H_t in Eq. 15.27 are latent and sensible heats from the atmosphere, shortwave (i.e., solar) radiation, longwave or terrestrial radiation, rain, and terrestrial heat flow. Accordingly, several processes can cause snow to melt or sublimate or water to evaporate. Snow is affected by at least six processes, listed below:

1. radiation absorbed from sunlight and from the atmosphere,
2. heat transferred from the atmosphere by both conduction and convection,
3. heat transferred by rain falling on snow causing melting,
4. latent heating due to condensation of water on snow,
5. heat conducted from the ground, and
6. sublimation.

The total energy available for snowmelt is given as

$$H_T = H_{SW} + H_{LW} + H_H + H_E + H_P + H_G - H_I \qquad (15.35)$$

where

H_{SW} = energy input due to shortwave, i.e., solar radiation,
H_{LW} = energy input due to longwave, i.e., terrestrial radiation,
H_H = turbulent exchange of sensible heat energy with the atmosphere,
H_E = turbulent exchange of latent heat energy with the atmosphere.
H_P = energy input due to convective heat transfer with rain falling on snow,
H_G = energy input due to conductive heat transfer with the ground, and
H_I = output of internal energy stored in the snow; this energy melts the ice of the snowpack, refreezes the liquid water in the snow, and changes the temperature of snow.

The unit of all H terms in Eq. 15.35 is E/L² (e.g., J/m²).

In terms of flux, which has units of E/L²/T (e.g., W/m² or J/m²·s),

$$\frac{H_T}{dt} = \overline{F}_{SW} + \overline{F}_{LW} + \overline{F}_H + \overline{F}_E + \overline{F}_P + \overline{F}_G - \overline{F}_O \qquad (15.36)$$

15.7 ENERGY EXCHANGE MECHANISMS

15.7.1 Radiation

15.7.1.1 Physics of Radiant Energy

All bodies with a temperature above 0 K radiate energy in the form of electromagnetic waves. As discussed in Chapter 14, the transport of heat due to radiation is based on the Stefan–Boltzmann law given as

$$\frac{1}{A}\frac{dH}{dt} = \sigma T^4 \qquad (15.37)$$

where σ is the Stefan–Boltzmann constant (5.67037×10^{-8} J/m²·K⁴·s = 5.67×10^{-8} W/m²·K⁴).

This is the maximum or so-called **black body** radiation rate. Most solids and thick clouds radiate almost as well as black bodies but only water vapor and carbon dioxide radiate appreciably in the clear atmosphere. One simple approximate formula for the net rate of radiation leaving the ground under clear-sky conditions is

$$\frac{1}{A}\frac{dH}{dt} = \sigma T^4 \left[1 - 0.6 - 4.8(10)^{-3}\sqrt{e}\right] \qquad (15.38)$$

The rate of energy emission per unit surface area per unit time $(E/L^2/T)$ or Eq. 15.37 can also be written as

$$\overline{F} = \varepsilon \sigma T^4 \qquad (15.39)$$

where ε is a dimensionless quantity called emissivity. The value of ε ranges from 0 to 1, depending on the material and texture. For example, ice at 0 °C has an emissivity of 0.97 (for clean snow, ε can be assumed to be 0.97). It is the ratio of re-emitted over absorbed energy. A surface with $\varepsilon = 1$ is called a black body. Equation 15.39 shows that the rate of radiation flux is related to temperature. The wavelength, λ, of radiated energy also depends on temperature. This is given by **Wien's displacement law**:

$$\lambda_{max} = \frac{2897}{T} \qquad (15.40)$$

Thus, the maximum wavelength decreases as the temperature of the radiating body increases.

When energy strikes an object there are three ways by which the radiant energy can interact with the matter.

1. Reflection: By this interaction the fraction of incident energy is reflected back in a new direction without being diminished. The property of a matter that causes this phenomenon is called **reflectance**. The reflectance integrated over the visible wavelengths (0.4–0.7 μm) is called **albedo** or "whiteness."
2. Absorption: By this interaction a fraction of incident energy is absorbed by a surface. This energy changes the temperature of the surface and thereby can cause phase transformation, such as melting and evaporation. The property of matter that results in the absorption of energy is given by its *emissivity*.
3. Transmission: By this interaction the fraction of incident energy is transmitted through the matter without being diminished or a change in direction of propagation, and without causing any change in the matter.

The sum of the fractions of incident energy that are reflected, absorbed, and transmitted is always equal to unity.

15.7.1.2 Solar Heating

The Sun is the main source of heat and, therefore, of water in the air. Shortwave radiation is the most important source of energy to the snowpack. Sunlight varies according to latitude, time of day, and season of the year.

The Sun radiates energy approximately as a black body with a temperature of 6000 K. Its radiation spectrum extends from ultraviolet to infrared, with a maximum in the visible range. If $T_{sun} = 6000$ K, $\overline{F}_{sun} = (1)(5.67 \times 10^{-8})(6000^4) = 7.35 \times 10^7$ W/m², the amount of solar energy that the Earth receives, can be approximated. The Sun's energy arrives at the outer edge of the atmosphere at an average rate of 1.74×10^{17} W. This quantity of energy, divided by the area of the planar projection of the Earth (approximately a circle with an area of about 1.28×10^{14} m²) is the solar constant, G_{sc}, which is thus 1367 W/m². This is the global average rate of energy input at the top of the Earth's atmosphere.

The amount of shortwave radiation that any point on Earth receives is not 1367 W/m². All of the Sun's energy arriving at the top of atmosphere does not reach the surface mainly because the atmospheric gases absorb the energy at specific wavelengths.

15.7.1.3 Shortwave or Solar Radiation

As noted above, the atmospheric gases, in particular oxygen (O_2) and ozone (O_3), absorb solar radiation at specific wavelengths. Water vapor also absorbs some of the Sun's energy in the near infrared range. This absorption process in the lower stratosphere shields the biota from the ultraviolet range of the spectra. Thus, virtually all of the Sun's energy arriving at the surface is at wavelengths less than 4 μm, and concentrated between 0.4 μm and 0.7 μm. This energy is referred to as solar or shortwave radiation.

The net flux of solar radiation (energy) $\overline{F}_{SW}$ entering a snowpack is given as

$$\overline{F}_{SW} = \overline{F}_{SW}^{In} - \overline{F}_{SW}^{Out} = \overline{F}_{SW}^{In}(1-\alpha) \qquad (15.41)$$

where $\overline{F}_{SW}^{In}$ is the incoming solar radiation incident on a snowpack, $\overline{F}_{SW}^{Out}$ is the radiation reflected from the snow surface, and α is the albedo. This ranges from 0.45 for old snow to about 0.85 for a fresh snow surface. The net flux, $\overline{F}_{SW}$, is always positive and is usually measured with instruments called **pyranometers**, one facing upward to measure the incoming radiation and one facing downward to measure the outgoing or reflected flux. However, such data are only available at certain permanent locations.

Clear-sky solar radiation can be calculated for a given latitude, φ, and Julian day, J, of the year, which governs the declination angle, δ, of the Sun, and the slope, azimuth angle, and aspect. The procedures to calculate this quantity have been presented in Chapter 14. Also, as discussed in Chapter 14, various factors, the chief ones being cloud cover, forest canopy, and albedo, determine the actual amount of flux that is received by a surface, such as a snow-covered area. Empirical functions are used to modify the calculated clear sky shortwave radiation to include the influence of slope, aspect, canopy, cloudiness, and other variables.

15.7.1.4 Longwave or Terrestrial Radiation

The average temperature of the Earth's surface is about 290 K (17 °C). So, the Earth's surface radiates approximately as a black body at this temperature and emits 20 units[1] of energy in the infrared range between wavelengths of 4 μm and 50 μm. Energy in this wavelength band is referred to as terrestrial or longwave radiation.

Many naturally occurring and human-introduced gases in the atmosphere strongly absorb longwave radiation so that 14 of the 20 units radiated by the surface are absorbed to heat the atmosphere. The absorption of this energy is called the **greenhouse effect**. The most important **greenhouse gases** are water vapor, which accounts for 65% of the absorption, carbon dioxide (33%), methane, nitrous oxide, O_3, and chlorinated fluorocarbons (2% combined).

The net input of longwave energy is the difference between the incident flux emitted by the atmosphere, clouds, and overlying forest canopy and the outgoing radiation from materials at or near the Earth's surface, such as snowpack, bare soils and rocks, paved surfaces, etc.:

$$\overline{F}_{LW} = \overline{F}_{LW}^{In} - \overline{F}_{LW}^{Out} \qquad (15.42)$$

---

[1] We take 100 units of radiant energy input and trace out its fate in the Earth–atmosphere system prior to its ultimate reflection or re-radiation back to space.

$\overline{F}_{LW}$ is typically negative (the energy out of the snowpack has a net cooling effect). Incoming longwave radiation ($\overline{F}_{LW}^{In}$) can come from clouds, trees, atmosphere, etc. $\overline{F}_{LW}^{Out}$ occurs because the Earth's surface, including snowpack, reflects some of $\overline{F}_{LW}^{In}$, and the materials close to the Earth's surface, including snowpack itself, emit radiation as a black body.

Longwave radiation can also be measured directly by an instrument called **pyrgeometer**. Alternatively, it can also be estimated by taking the difference of the measurements of all wave radiation by radiometers and that of shortwave radiation by pyranometers.

$\overline{F}_{LW}^{In}$ is usually estimated from meteorological conditions using the Stephan–Boltzmann equation given as Eq. 15.39, which can also be written as

$$\overline{F}_{LW}^{In} = \varepsilon_{at}\sigma(T_{at} + 273.15)^4 \tag{15.43}$$

where ε_{at} is the integrated effective emissivity of the atmosphere and canopy and T_{at} is the effective radiating temperature of the atmosphere and canopy (°C).

Since the longwave reflectivity of a surface equals one minus its longwave emissivity, the following equation can provide an estimation of F_{LW}^{Out}:

$$\overline{F}_{LW}^{Out} = \varepsilon_s\sigma(T_s + 273.15)^4 + (1 - \varepsilon_s)F_{LW}^{In} \tag{15.44}$$

where subscript s designates the values of emissivity and temperature of the surface (e.g., snow surface). We usually ignore the longwave reflectance (e.g., for snow surface $\varepsilon_{ss} \approx 1.0$) and hence Eqs. 15.42–15.44 can be combined and simplified to

$$\overline{F}_{LW} = \varepsilon_{at}\sigma(T_{at} + 273.15)^4 - \sigma(T_s + 273.15)^4 \tag{15.45}$$

To apply Eq. 15.45, estimates of ε_{at} and T_{at} must be available. This can be a difficult problem because the effects of cloud and canopy covers must be accounted for. In practice, $T_{at} = T_a$ (air temperature), and when $T_a > 0°C$, $T_s = 0°C$ and Eq. 15.45 becomes

$$\begin{aligned}\overline{F}_{LW} &\approx \varepsilon_{at}\sigma(T_{at} + 273.15)^4 - \sigma(273.15)^4 \\ &= \varepsilon_{at}\sigma(T_{at} + 273.15)^4 - 27.3 \text{ MJ/m}^2 \cdot \text{d}\end{aligned} \tag{15.46}$$

where the energy flux is given in MJ/m²·d.

Several formulations are available to estimate the emissivity of atmosphere, ε_{at}, under clear-sky, cloudy-sky, and forest-canopy covers (Brutsaert, 1975; Kustas et al., 1994).

For clear sky and no forest canopy,

$$\varepsilon_{at} = 1.72\left(\frac{e_a}{T_a + 273.15}\right)^{\frac{1}{7}} \tag{15.47}$$

where e_a is near-surface (2-m high) atmospheric vapor pressure (kPa) and T_a is 2-m air temperature (°C).

For cloudy sky and no forest canopy,

$$\varepsilon_{at} = 1.72\left(\frac{e_a}{T_a + 273.15}\right)^{\frac{1}{7}} + (1 + 0.22C^2) \tag{15.48}$$

where C denotes the fraction of sky covered with cloud.

For cloudy sky with forest canopy,

$$\varepsilon_{at} = 1.72(1 - f)\left(\frac{e_a}{T_a + 273.15}\right)^{\frac{1}{7}} + (1 + 0.22C^2) + f \tag{15.49}$$

where f denotes the fraction of ground covered with canopy. In other words, it is the ratio of the horizontally projected area of forest canopy to the total area of interest. It can be determined from the examination of aerial photographs.

15.7.1.5 Snowmelt from Net Radiation

Net radiation is the algebraic sum of shortwave and longwave radiation. Thus,

$$\overline{F}_N = \overline{F}_{SW} + \overline{F}_{LW} \tag{15.50}$$

If an estimate of net radiation flux, commonly denoted by R_n, is known at a point, from Eq. 15.34, rate of snowmelt generation, Q_{SM}, can be directly calculated as

$$Q_{SM} = \frac{0.258}{B}R_n \tag{15.51}$$

If R_n is given in W/m², Q_{SM} is in mm/d.

15.7.2 Turbulent Convection

15.7.2.1 General Equation

Heat and vapor move through the air from concentrated regions to sparse regions by a turbulent process analogous to diffusion. The main difference is that turbulence requires moving blobs of fluid rather than individual molecules. Therefore, turbulent transport is proportional to the wind speed. Turbulent transport is usually determined by measuring wind at one height and temperature (or the quantity to be transported) at two different heights. Often, wind speed is measured at 10 m above the surface, and also at the surface (subscript 0) and 2 m above the surface (subscript 2). The equation for turbulent heat transport up from the ground is then

$$\frac{1}{A}\frac{dH}{dt} = C_H\rho c_p(T_0 - T_2)v \tag{15.52}$$

where C_H is the coefficient of turbulent heat transfer and v is the wind speed. Over lakes, the coefficient $C_H \approx 1.9 \times 10^{-3}$. The rougher the surface and the more unstably stratified the atmosphere, the larger is C_H, which can vary from less than 10^{-3} to more than 3×10^{-3}.

Turbulent exchange involves the transfer of sensible heat from warm air advected over the snowfield (convection), and also the latent heat of condensation of water vapor from the atmosphere condensed on the snow surface.

15.7.2.2 Sensible Heat Transfer

Heat energy that can be directly sensed by the measurement of temperature is called sensible heat. A non-radiant mode of heat transfer occurs in the form of sensible heat, in other word, due to transfer of this heat energy, the temperatures of the exchanging media change.

Recall that the specific heat, c_p, is defined as the heat capacity, C, per unit mass, whereas the heat capacity is the ratio of the amount of heat energy, Q_H, supplied to the body to its corresponding temperature rise.

$$c_p = \frac{dH_H}{mdT} \tag{15.53}$$

where H_H is the sensible heat energy, which can be positive or negative.

Turbulent exchange of sensible heat energy with atmosphere occurs due to the combination of **conduction** and **convection**. Both of these processes occur when there is a temperature difference between two adjacent masses. The rate of sensible heat transfer ($E/L^2/T$) is given by the following proportionality:

$$\overline{F}_H \propto (T_s - T_a) \tag{15.54}$$

where T_s and T_a denote the temperatures of the surface (e.g., snowpack) and overlying air mass, respectively. Thus, the rate of sensible rate transfer is proportional to the temperature difference between the surface and the atmosphere. Heat energy flows from high temperature to low temperature (if $T_s < T_a$ the flux is negative and sensible heat moves from the air to the surface). In the formulation, a flux toward the surface is negative. For snow, the convention is to denote a flux from the atmosphere to snow as positive.

It should be recognized that the proportionality in Eq. 15.54 represents a gradient-driven process. Heat energy flows in the direction of a potential gradient, where temperature is a measure of the potential. Most gradient-driven processes are represented by a diffusion equation discussed in Chapter 14.

Diffusion can be thought of in this way. In order for a material to travel along a potential gradient it often must pass through another material. The actual rate of travel is therefore not simply related to the potential gradient, but to the physical properties of both materials. For example, groundwater flows along a head gradient and must pass through soil, electricity flows along a voltage gradient and must pass through a conductor. The rate of diffusion of a substance is governed by a concentration gradient and a diffusion coefficient that incorporates all the physical properties that allow transmission to occur.

Diffusion proportionalities become equations by assigning diffusion coefficients that take into account the physical properties of both materials. The form of the diffusion equation in the case of heat transfer is

$$F_z = -D_H \frac{dh}{dz} \tag{15.55}$$

where F_z is the rate of transfer of sensible heat in the z-direction (in the case of air–snowpack interaction it is in the vertical direction), per unit area per unit time, called the **heat flux**; D_H is the diffusivity of sensible heat in turbulent air; and h is the concentration of sensible heat (E/L^3). The concentration of sensible heat can be given by

$$h = \rho_a c_a (T_a - T_b) \tag{15.56}$$

where ρ_a is the density of air, c_a is the specific heat of air, and T_b is an arbitrary base temperature, usually taken as $0\,^\circ\mathrm{C}$. Thus, the diffusion equation for sensible heat can be written as

$$F_z(h) = -D_H \frac{d(\rho_a c_a T_a)}{dz} \tag{15.57}$$

Assuming that the density and specific heat of air remain essentially constant, Eq. 15.57 is rewritten as

$$F_z(h) = -D_H \rho_a c_{pa} \frac{dT_a}{dz} \tag{15.58}$$

Equation 15.58 is Fick's first law of diffusion, introduced in Chapter 14.

Sensible heat must diffuse across the surface–atmosphere interface through the following two modes of heat transfer.

1. Conduction: In this mode, the heat of one mass heats up the other by vibrating molecules.
2. Convection: In this mode, air moves, so some of the heat that is conducted from the snow to the atmosphere is carried away by convection.

Turbulence results in an upward transfer of mass, and that mass carries heat. The velocity of air mass determines the rate of heat transfer. Mass times velocity is momentum. We can therefore estimate convective heat transfer by investigating momentum transfer. The diffusion equation for momentum is given as

$$F_z(v) = -D_M \frac{d(\rho_a v_a)}{dz} \tag{15.59}$$

where v denotes the momentum, v_a is the velocity of air, and D_M is the diffusivity of momentum in turbulent air. In the lowest level of atmosphere, ρ_a can be considered constant at the prevailing air temperature. Thus, Eq. 15.59 becomes

$$F_z(v) = -D_M \rho_a \frac{d(v_a)}{dz} \tag{15.60}$$

Considering the physics of turbulent flow, it can be shown that D_M can be given as

$$D_M = k^2 (z - z_d)^2 \frac{dv_a}{dz} \tag{15.61}$$

where k is a dimensionless constant called **von Karman's constant** (0.41), v_a is the time-averaged velocity of air at a height z above the ground, and the height z_d is the zero-plane displacement. Substituting D_M for D_H in Eq. 15.57 we obtain

$$F_z(h) = \overline{F}_H = k^2 \rho_a c_{pa}(z - z_d)^2 \frac{dv_a}{dz} \frac{dT_a}{dz} \qquad (15.62)$$

Integrating between levels z_1 and z_2 yields

$$\overline{F}_H = -\rho_a c_a \frac{k^2}{\left[\ln\left(\frac{z_2 - z_d}{z_1 - z_d}\right)\right]^2}(v_2 - v_1)(T_2 - T_1) \qquad (15.63)$$

If level 1 is taken as the nominal surface, Eq. 15.63 reduces to

$$\overline{F}_H = -\rho_a c_a \frac{k^2}{\left[\ln\left(\frac{z_2 - z_d}{z_0}\right)\right]^2}(v_2)(T_a - T_s) \qquad (15.64)$$

where z_a is the height above the snow surface at which v_a and T_a are measured and z_0 is the **surface roughness height**.[2]

The sensible heat flux of Eq. 15.54 can be given with **bulk transfer coefficient of sensible heat**, K_H:

$$\overline{F}_H = K_H v_a (T_s - T_a) \qquad (15.65)$$

The negative sign is eliminated by reversing the surface and air temperatures. The bulk sensible heat transfer coefficient combines thermal conductance and momentum transfer. Its unit is usually given as $J/m^3 \cdot K$. If v_a is given as m/s, the sensible heat flux is in $J/m^2 \cdot s$ or W/m^2.

The zero-plane displacement, z_d, for snow is negligibly small and can be set to zero. A typical value of z_a is 2 m and a value between 0.0005 m and 0.005 m can be selected for z_0. Using these values and taking the standard values of density and specific heat of air (Table 15.2) and $k = 0.41$, the value of K_H is derived as

$$K_H = 1.29 \frac{kg}{m^3} \times 1005 \frac{J}{kg \cdot K} \times \frac{(0.41)^2}{[\ln(\frac{2}{0.005})]^2} = 6.085 \frac{J}{m^3 \cdot K}$$

The specific heat of air changes with temperature. Thus, if this is taken as 1001 J/kg·K (at 250 K) and $z_0 = 0.0015$, $K_H = 4.1925$ J/m^3·K.

Brubaker et al. (1996) showed that the snow melting rate, Q_{SM}, (cm/d), from turbulent heat exchange of sensible and latent heat involving convection and condensation can be calculated from the **restricted degree day** method as

$$Q_{SM} = \kappa T_a \qquad (15.66)$$

where κ is called the **restricted degree day factor**. Note that unlike a **degree day factor**, which is discussed below, the restricted degree day factor does not vary considerably with seasons and typically assumes a value of 0.20–0.25 cm/°C per day (Martinec, 1989). This factor was obtained from an analysis of turbulent exchange of sensible heat between snow and air.

15.7.2.3 Latent Heat Transfer

Latent heat is the energy exchange that is associated with the phase change of water (Figure 15.1). It can be positive or negative. When vapor condenses on a snowpack, energy is transferred from air to snow (the snowpack heats up). When snow sublimates, energy is lost from the snowpack (the pack cools). Latent heat is transferred by a diffusive process much like sensible heat. However, latent heat transfer is governed by a vapor pressure gradient as opposed to sensible heat transfer which is governed by a temperature gradient. Thus, the rate of latent heat transfer ($E/L^2/T$) is given by the following proportionality:

$$\overline{F}_E \propto (e_a - e_s) \qquad (15.67)$$

where e_s and e_a denote the vapor pressure at the surface of the snowpack and of the overlying air mass, respectively.

If $e_a < e_s$, the vapor pressure gradient is directed upward ($\overline{F}_E$ is negative), meaning water particles will move from the snow to the air, which results in **sublimation**. If $e_a > e_s$, the vapor pressure gradient is directed downward ($\overline{P}_E$ is positive), meaning water particles will move from the air to the snow, which results in **condensation**.

An analysis similar to the one carried out for sensible heat transfer can be conducted to combine the concepts of heat capacity and momentum transfer to lead to the following equation:

$$\overline{F}_E = \lambda_v \frac{0.622 \rho_a}{p} \frac{k^2}{\left[\ln\left(\frac{z_a - z_d}{z_0}\right)\right]^2}(v_a)(e_a - e_s) \qquad (15.68)$$

where λ_v is the latent heat of vaporization, λ_s is the latent heat of sublimation, and p is the atmospheric pressure at the surface (≈ 101.3 kPa). When all constants and elevation variables are lumped together, the general form for the flux of latent heat is obtained as

$$\overline{F}_E = K_E v_a (e_a - e_s) \qquad (15.69)$$

where K_E is called the **bulk transfer coefficient for latent heat**. It is the diffusivity term for latent energy passing across the snow–atmosphere interface and is usually expressed in units of J/m^3·kPa.

Evaluation of K_E in Cases of Condensation and Sublimation

The phase diagram of water (Figure 15.1) shows that, for a cold snowpack ($T_s < 0°C$ where T_s denotes the temperature of the snowpack) at pressures well below the atmospheric pressure at sea level, the solid to vapor phase change undergoes a process of sublimation. In this case, λ_s must be used in Eq. 15.68 instead of λ_v.

---

[2] The zero-plane displacement (z_d) and surface roughness (z_0) heights are two elevations above the ground surface that depend on the average height of vegetation (z_{veg}). Typically, z_d is about $0.7 z_{veg}$ and z_0 is about $0.1 z_{veg}$. Wind velocity increases with elevation due to the lowering of frictional resistance by the surface. $V_a = 0$ when $z = z_d + z_0$.

Typical values of z_a and z_0 are 2.0 m and 0.005 m, respectively, and z_d is negligibly small (~0 m). Using the values of ρ_a, and λ_s, as given in Table 15.2, and taking p as the atmospheric pressure, the value of the bulk transfer coefficient for latent heat, K_E, for the process of *sublimation*, is obtained as:

$$K_E = 1.29 \frac{kg}{m^3} \times \frac{0.622}{101.3 \text{ kPa}} \times 2804 \frac{kJ}{kg} \times \frac{(0.41)^2}{\left[\ln\left(\frac{2}{0.005}\right)\right]^2}$$

$$\times 1000 \frac{J}{kJ} = 104.0041 \frac{J}{m^3 \cdot kPa}$$

If the value of z_0 is taken as 0.0015 m, $K_E = 72.1111$ J/m³·kPa. When v_a is given in m/s, and e_s and e_a are in kPa, $\overline{F}_E$ is the flux (W/m²).

Similarly, using the values of ρ_a, λ_v, as given in Table 15.2, and p (101.3 kPa) the value of K_E for the process of *condensation* is obtained as

$$K_E = 1.29 \frac{kg}{m^3} \times \frac{0.622}{101.3 \text{ kPa}} \times 2470 \frac{kJ}{kg} \times \frac{(0.41)^2}{\left[\ln\left(\frac{2}{0.005}\right)\right]^2}$$

$$\times 1000 \frac{J}{kJ} = 91.6156 \frac{J}{m^3 \cdot kPa}$$

If the value of z_0 is taken as 0.0015 m, $K_E = 63.5215$ J/m³·kPa. When v_a is given in m/s, and e_s and e_a are in kPa, $\overline{F}_E$ is the flux (W/m²).

The vapor pressure at the surface of snow, e_s, is calculated from the snow surface temperature, assuming it to be saturated. If the temperature of the overlying air, T_a, is below freezing, the snow surface temperature, T_s, is usually 2.5–3.0 °C (T_x) lower. In that case, calculations of e_s and e_a are carried out as follows:

$T_a < 0°C. T_s = T_a - T_x$ (cold snow − sublimation):

$$e_s = 0.611 \exp\left[\frac{17.3 T_s}{237.3 + T_s}\right] \text{kPa}$$

The saturated vapor pressure at the air temperature is calculated as:

$$e_a^* = 0.611 \exp\left[\frac{17.3 T_a}{237.3 + T_a}\right] \text{kPa}$$

The actual vapor pressure of air is then calculated from the relative humidity (expressed as a fraction):

$$e_a = R_H e_a^* \text{ kPa}$$

If T_a is above freezing, T_s is usually 0 °C (T_x). In this case, calculations of e_s and e_a are carried out as follows:

$T_a > 0°C.\ T_s = 0°C$ (melting snow − condensation)

$e_s = 0.611$ kPa

The saturated vapor pressure and the actual vapor pressure are calculated as given above. The rates of sensible and latent heat transfer can also be calculated with *stability factors*, C_S and C_L, resulting in

$$\overline{F}_H = \rho_a c_a C_S v_z (T_a - T_s) \tag{15.70}$$

$$\overline{F}_E = \rho_a \lambda_v C_L v_z \left(e_a - e_s|_{T_s}\right) \tag{15.71}$$

The stability factors are often presented as functions of the dimensionless bulk **Richardson number, Ri**, which is calculated as

$$Ri = \frac{g(T_a - T_s)z}{0.5(T_a + T_s)v_z^2} \tag{15.72}$$

in which g denotes acceleration due to gravity. When $Ri > 0$ ($T_a > T_s$), the conditions are stable and the stability factor tends to be small, indicating that sensible and latent heat transfer rates are small. If $Ri \gg 0$, any convection is effectively damped and the transfer rates drop to near zero. When $Ri = 0$ ($T_a = T_s$), the conditions are unstable, and the stability factor tends to be large, indicating that the sensible and latent heat transfer rates are greater due to augmentation of forced convection by natural convection.

15.7.3 Heat Input due to Rain on Snow

The heat input by rain is simply related to the heat content of the rain and the type of work that the rain must perform. If rain falls on a snowpack that is at the freezing point ($T_s = T_m$; where T_s is the snowpack temperature and T_m is the melting temperature) then the rain water is cooled to snow temperature and the heat released by the rain causes the melting of snow. For this case ($T_s = T_m$), the heat flux from rain is given as

$$\overline{F}_P = \rho_w c_w P(T_r - T_m) \tag{15.73}$$

where c_w is the specific heat of water (4190 J/kg·K), P is the rainfall rate (L/T), and T_r is the temperature of rain.

Typically, $T_m = 0°C$. Similarly, T_r is usually the dew-point temperature. However, during rain, the relative humidity is close to 1, hence the rain temperature is assumed to be equal to air temperature, i.e., $T_r = T_a$. Also, the heat flux from rain in causing snowmelt is usually small and hence these assumptions introduce an insignificant error in the estimation of this heat flux. Thus, Eq. 25.73 simplifies to

$$\overline{F}_P = \rho_w c_w P T_a \tag{15.74}$$

If P is given in mm/d and T_a is given in °C, the substitution of values for ρ_w and c_w results in

$$\overline{F}_P = 0.04845 P T_a \text{ (W/m}^2\text{)} \tag{15.75}$$

The heat flux is converted to the melting rate by the use of Eq. 15.34:

$$Q_{SM} = \frac{0.258 \times 0.04845}{B} P T_a = \frac{0.0125}{B} P T_a \tag{15.76}$$

where both Q_{SM} and P are in mm/d and T_a is in °C.

Equation 15.76 is the most common equation used in typical snowmelt calculations. However, if the snow temperature, T_s, is below freezing, the rain will be first cooled to the freezing point temperature by giving up sensible heat and then will freeze through the liberation of latent heat. In this case,

$$\overline{F}_P = \rho_w c_w P(T_r - T_m) + \rho_w \lambda_f P \tag{15.77}$$

Again, assuming $T_r = T_a$, P is in mm/d and T_a is in °C, and substituting the values of the physical properties of rain water,

$$\overline{F}_P = 0.04850 PT_a (W/m^2) + 3.8762 P(W/m^2) \tag{15.78}$$

Thus, in this case, the melting rate, Q_{SM} (in mm/d), when P is in mm/d, is given as

$$Q_{SM} = \frac{(0.0125)PT_a + 1.0P}{B} \tag{15.79}$$

15.7.4 Conductive Heat Transfer from the Ground

Convection is negligible at the ground–snow interface. Sensible heat transfer is, therefore, limited to conduction. Heat moves through solids from warmer regions to colder regions by the diffusion process of thermal conduction. The heat transport rate is proportional to the temperature gradient and the conduction coefficient of the substance. As a result, the equation of heat conduction is

$$\frac{1}{A}\frac{dQ}{dt} = -\kappa \frac{dT_g}{dz} \tag{15.80}$$

$$\overline{F}_G = \kappa \frac{dT_g}{dz} \tag{15.81}$$

where κ is the **thermal conductivity** of soil or ground and dT_g/dz denotes the temperature gradient in the soil or ground. The heat transport rate can only be determined if the temperature gradient is known.

If the equation for the heat transport rate due to conduction is properly combined with the first law of thermodynamics, the classical heat conduction equation results in

$$\frac{\partial T}{\partial t} = \frac{\kappa}{c\rho}\frac{\partial^2 T_g}{\partial z^2} \tag{15.82}$$

This equation, a form of Fick's second law of diffusion introduced in Chapter 14, is self-contained and can be solved as long as the *initial* temperature structure is known throughout, and the subsequent heat flow or temperature is known *at the boundary*. This equation can be used to determine if the snow that falls will melt when it hits the ground.

Typically, $\overline{F}_G$ is negligible compared to the other energy fluxes given in Eq. 15.36. A typical value of terrestrial heat flow over the continents is 0.061 W/m². However, it can be significant during the accumulation period to cause **ground melt** (Male and Gray, 1981). Ground melt is produced continually at the base of the snowpack during the accumulation period due to terrestrial heat flow. In general, $\overline{F}_G$ can be assumed to be a constant value of 173 kJ/m²·d or 2.0023 W/m² (USACE, 1960). Using Eq. 15.34, the contribution of ground heating to snowmelt is about 0.577 mm/d. USACE (1998) notes that most conceptual snowmelt models use the values in the range of 0–5 W/m².

15.7.5 Internal Energy

The cold content or heat deficit of a snowpack is the amount of energy required to raise its average temperature to the melting point. If it is positive, the temperature of the snowpack is below freezing. When the meltwater within a snowpack freezes, it releases heat and thereby reduces the cold content of the snowpack. Refreezing of meltwater occurs during diurnal temperature cycles. It is prominent in the night due to radiational cooling. Meltwater will continue to freeze within the snow cover until the total heat deficit is zero, that is, when the snow cover reaches an isothermal condition at 0 °C. This internal energy (E/L², e.g., J/m²) is calculated as (Gray and Prowse, 1992):

$$H_I = (\rho_i c_i + \rho_w c_w + \rho_v c_v) D_s (T_s - T_m) \tag{15.83}$$

The contribution of the vapor phase is negligible and hence the terms involving density (ρ_v) and specific heat (c_v) of the vapor phase can be dropped from Eq. 15.83.

Substitution of the values for $\rho_i, c_i, \rho_w,$ and c_w with the assumption $T_m = 0°C$ results in the value of the first term of Eq. 15.83 as 6113.534 (kJ/m³·K). Thus, H_I is given as

$$H_I = (6113534) D_m T_s \, [J/m^2] \tag{15.84}$$

where D_s is in m and T_s is in °C. Thus, calculation of H_I requires knowledge of both the depth and temperature of the snow cover. However, Eq. 15.83 can be modified to include the depth of the refreezing front instead of the total snow depth because this is the depth that actually contributes to the energy transfer process when temperature reversal occurs, and the snowpack becomes warmer than the overlying atmosphere.

Snow melting occurs when the temperature of snowpack is at 0 °C. When the temperature of the overlying air is below 0 °C, heat is transported from the snowpack to the atmosphere and refreezing occurs from the surface of snowpack. The depth of the refreezing front penetrates further into the snowpack with time as long as the temperature of the atmosphere remains below freezing. Bengtsson (1982) analyzed this process mathematically and showed that the depth of freezing is a function of overlying air temperature. The exact value of the depth of refreezing depends on certain physical characteristics of the snowpack, such as its thermal diffusivity, liquid water holding capacity, and density. When these properties of the snowpack are not exactly known, the depth of freezing can be used as one of the calibration parameters for a snowmelt model. For a set of given values of the properties of a snowpack, Bengtsson (1982) showed that the depth of refreezing varies from approximately 50 mm at −2 °C to 150 mm at −20 °C.

Estimation of the snow surface temperature can be made from the suggestion of Brubaker et al. (1996), who showed that the average daily snow surface temperature is well approximated by

$$T_s = \min[(T_a - 2.5), 0] \tag{15.85}$$

where temperatures are in °C and T_a indicates the air temperature. In practice, when $T_a > 0$°C, T_s is assumed to be 0 °C.

15.8 RUNOFF GENERATION FROM SNOWMELT

The soil cover and the groundwater level underneath the snowpack control the nature of snowmelt runoff. It is important to understand how snowmelt originating from a snowpack reaches streams. However, as Bengtsson (1986) pointed out, snow surface melt, percolation of meltwater through a snowpack, and melt flux to the ground surface can be described in many ways.

Snowmelt is not homogenous throughout the snowpack depth. Most of the melting occurs at the interfaces with the atmosphere and the ground (Male and Gray, 1981). Furthermore, water originating at the upper horizon near the surface of the snowpack percolates through it, arriving at the bottom. Thus, there is usually a lower horizon within the snowpack that is saturated with meltwater.

The meltwater accumulating at the bottom of a snowpack can reach a stream following one of the three common paths. It can infiltrate the ground and join a deeper unconfined groundwater aquifer, thereby raising the water table. The groundwater then can contribute to the surface water as baseflow. Alternatively, if the soil underlying the snowpack is fully saturated or impermeable either due to frozen ground or the nature of the bedrock and soil, meltwater directly drains to a stream as surface flow. A third possibility is a combination of the two paths, where meltwater can reach the stream as both surface flows as well as via groundwater. In the case of infiltration, some interflow through the ground can also join the surface flow. In any event, it is important to recognize that the snow melting starts at the snow surface and then first tends to reach the bottom of the snow cover through percolation and from there meltwater finds its course to the stream either through groundwater, or as surface flow or a combination thereof. During this percolation process, refreezing as well as storage of the meltwater can take place within the snow cover.

If infiltration of snowmelt takes place, then there is no direct surface runoff until the soil storage is filled to its *field capacity*, which is the maximum amount of water that can be held by a soil column against gravity. After reaching the field capacity, excess water can drain through the soil under gravity and may appear later as subsurface or base flow components of streamflow or as an interflow component of surface runoff. Thus, a portion of the infiltrated snowmelt returns to streamflow and the other portion is stored in the ground and groundwater.

During the melting season (in the northern hemisphere this is usually from the middle of May to the middle of September), the meltwater generated at the surface of a snowpack can also be lost due to evaporation. However, during snowmelt the ground may be frozen, at least to some extent. The hydraulic conductivity of soil is then reduced by several orders of magnitude. The frozen ground causes a lesser amount of loss of snowmelt and thus increases meltwater runoff.

From the processes described above, the snowmelt runoff rate is given by

$$Q_{SM} = Q_R + Q_H + Q_P + Q_G - Q_I - Q_S - Q_E \tag{15.86}$$

where Q_{SM} is the total snowmelt runoff; Q_R, Q_H, Q_P, and Q_G, respectively, denote snowmelt due to radiation, turbulent exchange of sensible and latent heat with the atmosphere, convective heat exchange with rain falling on snow, and heat conducted from the ground. The snowmelt lost due to refreezing, storage and infiltration in the ground, and evaporation are denoted by Q_I, Q_S, and Q_E, respectively.

15.9 SNOWMELT MODELING

A large variety of snowmelt models of different complexity and scope have been developed since computer applications in hydrologic simulation came into place. Snowmelt models can be classified into three broad categories as described below.

The simplest type of snowmelt model is known as **temperature index** method or **degree day method**. The general form of this model can be given as

$$Q = \begin{cases} f_m(T_a - T_m) & T_a > T_m \\ 0 & T_a < T_m \end{cases} \tag{15.87}$$

where Q is the snowmelt rate (depth/time, e.g., mm/d), f_m is a coefficient, T_a is the air temperature (e.g., daily mean temperature), and T_m is the threshold temperature at which snow melting begins. The coefficient f_m is known as either the **degree day factor** or **melt factor** with a unit of mm/°C per day. In spite of its superb simplicity, this approach often produces satisfactory results. This is due to the fact that many studies have shown a strong correlation between air temperature and snowmelt production. However, f_m varies with latitude, slope inclination and aspect, forest cover, and time of the year. Hock (2003) listed values of degree day factors for snow and bare ice, such as glacier surfaces from many cold regions of the Earth. For snow, the values range from 2.5 to 11.6 mm/°C per day and for ice it varies from 5.5 to 20.0 mm/°C per day.

The second type of snowmelt model can be called a **hybrid model** in which a radiation term has been added to Eq. 15.87 (Kustas et al., 1994; Brubaker et al., 1996). The general form of such a model can be given as

$$Q = f'_m T + aR_n \tag{15.88}$$

in which f'_m is the *restricted* degree day factor, R_n is the net radiation, and a is a coefficient that is typically assumed to be equal to $1/(\rho_w \lambda_f)$. As noted above, unlike the degree day factor, the restricted degree day factor varies within a small range of 2.0–2.5 mm/°C per day. Brubaker et al. (1996) provided a method to estimate this factor from representative meteorological characteristics of the basin, which requires wind speed, relative humidity, and air temperature, and is based on a simplified energy balance equation for snowmelt. In contrast to the degree day method, in which f_m can vary from month to month in the same basin, the restricted degree day factor, f'_m, for a site is held constant throughout the snowmelt season.

The third type of snowmelt model is the most comprehensive type involving the complete energy balance equation such as Eq. 15.35.

A model based on the degree day method, originating with Jaroslav Martinec, attached at the time to the Swiss Snow and Avalanche Research Institute at Davos, is a snowmelt runoff model which has come to be referred to as SRM (Martinec, 1975). The original model has undergone substantial development since 1975, by Martinec himself in collaboration with Al Rango (US. NASA, later of the US Agricultural Research Service) and Michael Baumgartner (University of Bern). The model has been amended and extended several times in the light of operational experience. An updated version was described by Martinec et al. (2008). SRM has been widely used in many parts of the world. Various enhancements have been made to the original model. Nonetheless, the basic structure of SRM is as follows.

A basin or watershed is subdivided into different elevation zones, usually in 500-m bands. Temperature from the base station at a lower elevation is usually known and then the temperature at the mean hypsometric elevation in each band is calculated using a temperature lapse rate, discussed in Chapter 14. In SRM, the lapse rate is usually taken as 6.5 °C/km. However, as shown by Mukhopadhyay and Khan (2017), monthly average adiabatic lapse rates can be calculated from observed daily temperatures at two stations at different elevations, using differences in daily maximum and minimum temperatures at two different stations for a given month of a year, and the elevation difference between these two stations, to calculate maximum and minimum lapse rates. The monthly average values are then calculated from the daily values. Selecting conjugate stations, the value of the adiabatic lapse rate is given as

$$\gamma_a = \frac{T_{s2} - T_{s1}}{Z_{s1} - Z_{s2}} \quad (15.89)$$

where T_{s2} and T_{s1} represent the monthly average temperatures at station 1 and 2, respectively; and Z_{s1} and Z_{s2}, respectively, represent the elevations of station 1 and station 2.

Typically, values of γ_a decrease from the dry adiabatic lapse rate of 0.0098 °C/m depending on the humidity and are close to the global mean lapse rate of 0.00649 °C/m, also known as the environmental lapse rate. However, in mountainous terrains, γ_a can show great diurnal and seasonal variations and can depart significantly from the standard environmental lapse rate.

Once the temperature at different hypsometric bands is calculated, a daily time step is used to calculate snowmelt for each zone and then the runoff from each zone is added together before routing. Runoff is computed using

$$Q_{n+1} = (1 - k_{n+1}) \sum (C_{Sn} a_i T_i [SCA]_i + C_{Rn} P_i) A + k_{n+1} Q_n \quad (15.90)$$

where n is the day number, i indexes the elevation zones of the basin, Q is the discharge from the basin, T is the air temperature extrapolated to the hypsometric mean height of each zone, P is the rain falling in the zone when $T >°$ C, SCA_i is the current fraction of the total basin area, A, that is snow-covered, k is a recession coefficient, a is the degree day factor, C_{Sn} is the runoff coefficient for snowmelt, and C_{Rn} is the runoff coefficient for rainfall. The recession coefficient k indicates the decline of discharge in a period without snowmelt and rainfall, and is assumed to vary inversely with discharge as

$$k_{n+1} = xQ^{-y} \quad (15.91)$$

where x and y are the constants determined for a given basin by analysis of the recession curves defined by historically observed snowmelt discharge.

The current version of HEC-HMS (version 4.11) can be used to calibrate the parameters for temperature index method from the observed snow water equivalent at snow monitoring sites. The NRCS owns and maintains a network of snow observation sites, snow telemetry. The NRCS also regularly conducts manual snow surveys throughout the mountainous regions of the western United States (www.wcc.nrcs.usda.gov/snow). Each snow telemetry site is an automated, near real-time data collection station that collects hydroclimatic data, including the snow water equivalent, snow depth, precipitation, and temperature observations. Snow telemetry data collection began in 1978 at some stations and extends through to the present. The Hydrologic Engineering Center is in the process of adding the options of hybrid radiation-derived temperature index method and the full energy balance method. The energy balance method that will be implemented in HEC-HMS is based upon the Utah Energy Balance Snow Accumulation and Melt Model, originally developed by Tarboton and Luce (1996).

15.10 SNOW-COVERED AREAS

The term *snowpack* is used when referring to the physical and mechanical properties of snow on the ground. It is defined as a laterally extensive accumulation of snow, both new snow and previous snow, on the ground that persists through winter and melts in the spring and summer. On the other hand, the term snow cover or snow-covered area is used to refer to the areal extent of the snow-covered ground.

There are a couple of ways by which the snow-covered area of a basin or watershed can be estimated, such as using areal photographs or remote sensing data. Perhaps, satellite observations provide the best estimates of spatial distribution of snow-covered areas at a regular temporal resolution, such as daily or monthly. The Earth Science Division of the Science Mission Directorate of NASA launched a coordinated series of polar-orbiting and low-inclination satellites in 1999 for long-term global observations of the land surface, biosphere, solid earth, atmosphere, and oceans as part of the Earth Observing System (EOS) umbrella program. The Moderate Resolution Imaging Spectroradiometer (MODIS) and NASA Visible Infrared Imaging Radiometer Suite (VIIRS) snow cover products are snow-covered area maps that provide swath-based and daily maps of the Earth's land areas. MODIS refers to two instruments, which are 36-channel visible to thermal-infrared sensors. They collect data as part of NASA's EOS program. The first MODIS instrument launched on board was the Terra satellite, on December 18, 1999, and the second launched on board was the Aqua satellite, on May 4, 2002.

The MODIS instrument provides high radiometric sensitivity (12 bit) in 36 spectral bands ranging in wavelengths from 0.4 μm to 14.4 μm. MODIS obtains measurements with spatial resolutions of 250 m (bands 1 and 2), 500 m (bands 3–7), and 1000 m (bands 8–36) using a continuously rotating double-sided scan mirror. In short, the MODIS instruments provide calibrated, geo-referenced radiance data from individual bands, and a series of geophysical products from land, ocean, and atmosphere disciplines, which can be used for studies of processes and trends on local to global scales.

The National Snow and Ice Data Center in Boulder, Colorado, archives and distributes snow cover data derived from MODIS instruments from both Terra and Aqua satellites, with overlapping dates for certain time periods (http://nsidc.org). A variety of snow and ice products is produced from the MODIS sensors, and products are available at a variety of spatial and temporal resolutions. Hall and Riggs (2007) provide a detailed account of the MODIS products. Mukhopadhyay (2012, 2013) discussed in detail how MODIS-derived data can be used to estimate the snow-covered area in a river basin. Figure 15.2 shows the snow-covered area of Columbia River Basin of northwest

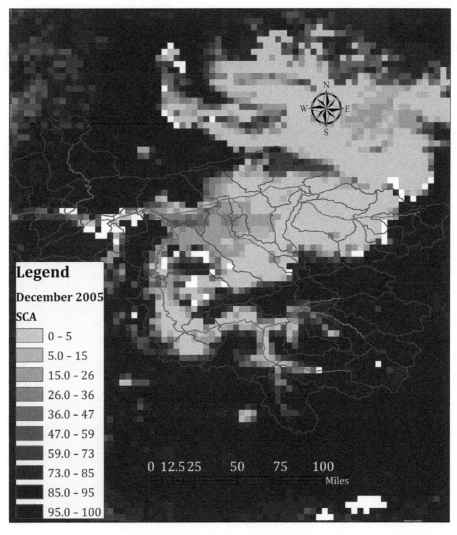

Figure 15.2 Percentage snow-covered area (SCA) of Columbia River Basin in northwest United States for the month of December 2005, derived from MODIS data. (A black and white version of this figure will appear in some formats. For the color version, please refer to the plate section.)

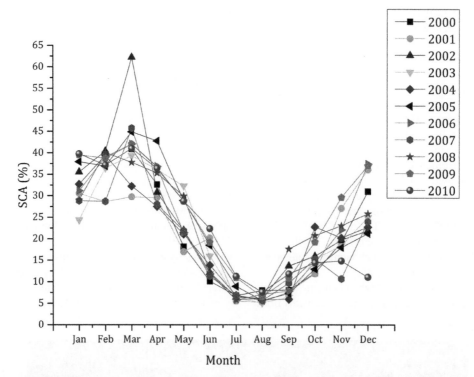

Figure 15.3 Monthly distribution of the snow-covered area (SCA; % of total basin area) for a 11-year period in Upper Indus Basin straddling the Karakoram Mountains and Higher Himalayas, derived from MODIS data. From Mukhopadhyay (2012). *Journal of Hydrology*, 412–413, 14–33. (A black and white version of this figure will appear in some formats. For the color version, please refer to the plate section.)

USA for the month of December 2005, derived from MODIS data. Figure 15.3 shows the monthly distribution of snow-covered areas in Upper Indus basin for a 12-year period. As can be seen from this figure, within this river basin, the accumulation period starts from September and peaks in March, then the ablation period starts from April and ends in August. Such curves also used to be called **recession curves**.

15.11 EXAMPLES

Example 15.1: Express an energy flux of 5.11 MJ/m²·d in W/m².

Solution:

$$\overline{F} = 5.11 \text{ MJ/m}^2\cdot\text{d} = \frac{5.11 \times 10^6 \text{ J/d}}{86\,400 \text{ s/d}} = 59.1435 \text{ W/m}^2$$

Example 15.2. For the energy flux of $\overline{F} = 59.1435 \text{ W/m}^2$ at a point, calculate the snowmelt.

Solution:

$$Q = \frac{\overline{F}}{\lambda_f \rho_w} = \frac{59.1435 \text{ J/s}\cdot\text{m}^2}{334.9 \text{ kJ/kg} \, 1000 \text{ kg/m}^3} = \frac{59.1435 \times 10^{-3} \text{ kJ/s}\cdot\text{m}^2 \times 86\,400 \text{ s/d} \times 10^3 \text{ mm/m}}{334.9 \text{ kJ} \times 10^3 \text{ m}^{-3}} = 15.26 \text{ mm/day}$$

15.11 Examples

Example 15.3: For $\bar{F} = 59.1435 \, \text{W/m}^2$, calculate the snowmelt at a point on a snowpack with thermal quality, $B = 0.95$.

Solution:
$$Q = \frac{\bar{P}}{B\lambda_f \rho_w} = \frac{59.1435 \, \text{J/s·m}^2}{(0.95)334.9\text{kJ/kg} \cdot 1000 \, \text{kg/m}^3}$$
$$= (1.0526) \times 15.26 \, \text{cm/d} = 16.06 \, \text{mm/d}$$

Example 15.4: Calculate the heat flux on a snowpack at 0 °C due to a rainfall of 35 mm/d when the air temperature is 4 °C.

Solution: Assume the temperature of rain is equal to the air temperature:

$$\bar{F} = 1000 \, \text{kg/m}^3 \times (4.186 \, \text{MJ/kg·K}) \times (0.035 \, \text{m/d})$$
$$\times (4°C - 0°C)$$
$$= 0.586 \, \text{MJ/m}^2 \cdot \text{d}$$

Example 15.5: The area–elevation–snow-covered area data of Gongnaisi River basin in the western Tianshan Mountains in China are given in Table 15.3. The average daily air temperature from April 5 to April 12 is given in the second column of Table 15.4. Calculate the snowmelt runoff using the SRM equation with the following data: temperature lapse rate = 4.5 °C/km, degree day factor = 0.3 cm/°C per day, snowmelt runoff coefficient = 0.3, rainfall runoff coefficient = 0.6, constants for the recession coefficient: $x = 0.9$ and $y = 0.03$.

Solution: Snowmelt runoff (m³/s) in each zone is calculated using

$$Q_{n+1} = \left[C_s f_m T_n \left(\frac{SCA}{100} \right) + C_r P_n \right] \frac{A10^4}{86400} (1 - k_{n+1}) + k_{n+1} Q_n$$

$$k_{n+1} = x Q_n^{-y}$$

where the subscript n is the index for the day. For each day, runoff from each zone is added to get the total runoff. The calculations are presented in Table 15.4.

Table 15.3 *Area–elevation–(SCA) data for Gongnaisi River basin*

Elevation zone	Mean elevation (m)	Area (km²)	SCA (%)
A	2113.6	99	10.5
B	2448.2	267	28.4
C	2899.4	190	20.3
D	3346.1	221	23.5
E	3727.4	163	17.3

Data from Hong and Guodong, *Chinese Science Bulletin*, vol. 48, no. 20, 2003.

Table 15.4 *Calculations*

Date	Temperature (°C)	Rainfall (cm)	T in Zone A (°C)	T in Zone B (°C)	T in Zone C (°C)	T in Zone D (°C)	T in Zone E (°C)	k_{n+1} (Zone A)	Q in Zone A	k_{n+1} (Zone B)	Q in Zone B	k_{n+1} (Zone C)	Q in Zone C	k_{n+1} (Zone D)	Q in Zone D	k_{n+1} (Zone E)	Q in Zone E	Total Q (m³/s)
5 Apr	10	0	0.49	−1.02	−3.05	−5.06	−6.77		0.05		0.00		0.00		0.00		0.00	0.05
6 Apr	15	2	5.49	3.98	1.95	−0.06	−1.77	0.98	0.30	0.00	40.23	0.00	198.70	0.00	0.00		0.00	239.23
7 Apr	18	5	8.49	6.98	4.95	2.94	1.23	0.93	2.63	0.81	51.51	0.77	208.17	0.00	78.33		56.96	397.59
8 Apr	17.5	8	7.99	6.48	4.45	2.44	0.73	0.87	9.32	0.80	71.93	0.77	213.89	0.79	87.96	0.80	63.81	446.91
9 Apr	16.5	4	6.99	5.48	3.45	1.44	−0.27	0.84	12.32	0.79	73.30	0.77	215.10	0.79	82.46	0.79	0.00	383.18
10 Apr	17	1	7.49	5.98	3.95	1.94	0.23	0.83	11.55	0.79	62.85	0.77	217.63	0.79	68.48	0.00	11.39	371.90
11 Apr	17.5	0	7.99	6.48	4.45	2.44	0.73	0.84	9.80	0.79	51.01	0.77	221.16	0.79	54.57	0.84	9.56	346.10
12 Apr	16	0	6.49	4.98	2.95	0.94	−0.77	0.84	8.35	0.80	41.59	0.77	219.08	0.80	43.66	0.84	0.00	312.68

Exercises: A selection of exercises on this topic is available at www.cambridge.org/appliedhydrology.

16 Erosion and Sedimentation

16.1 CONCEPTS AND PURPOSE

Erosion and sedimentation are natural processes caused by water and wind. Water is a universal agent of erosion and sedimentation, with far greater impact than wind, although wind action in arid regions can also be quite significant. Since our focus in this book is hydrology, our chief concerns centering on erosion and sedimentation are those carried out by water. Both rainwater and surface water play significant roles in these processes.

The topics that are covered in this chapter include the common methods of estimating the amount of soil that can potentially be eroded from the land surface during rainfall events and the amounts of sediments that can be transported and deposited by streamflow at points downstream of upland watershed. These estimations are fundamentally important for the design of dams, reservoirs, stable channels, debris basins or sediment traps, soil conservation practices, and watershed management.

16.2 EROSION PROCESSES

Soil erosion is the process of dislodgement and transport of soil particles from the land surface by water and wind. The dislodgement of soil particles at the soil surface by the energy imparted to the surface by falling raindrops is the primary agent of erosion. The energy is imparted to the soil surface by forces resulting from impulses produced by the momentum of falling raindrops and then transported by runoff by the force of gravity. Thus, water erosion on the land surface occurs by the combined action of rainfall and runoff.

In general, there are three erosion processes on upland watersheds: **surface erosion**, **gully erosion**, and **soil mass movement**. Surface erosion involves the detachment and subsequent removal of soil particles and small aggregates from the land surface by water or wind. Gully erosion is the detachment and movement of either individual soil particles of large aggregates in a well-defined channel. Soil mass movement includes erosion in which cohesive masses of soil are displaced by processes such as soil creep or soil slumps and landslides.

In addition to erosion on upland watersheds, erosion also occurs in channels by the shearing action of flowing water. Even though gully erosion can be called channel erosion, the principal types of channel erosion include stream-bank erosion, valley trenching, degradation, and floodplain scour.

The major factors that control soil erosion are: (1) precipitation, (2) topography of the land surface, (3) land cover and use, (4) wind, (5) soil texture, and (6) soil structure.

16.3 TYPES OF LAND SURFACE EROSION

16.3.1 Rill Erosion

Surface runoff combined with the beating action of raindrops causes rills to be formed on the soil surface. Rills may develop due to topographic variations, tillage operations, or irregularities on the surface. **Rill erosion** involves soil detachment, principally by the concentrated flow of water on the land surface. Rills carry runoff from inter-rill areas as well as rain falling directly on them. Rill erosion is a type of surface erosion.

16.3.2 Sheet Erosion

Another type of surface erosion is **sheet erosion**, which involves a relatively uniform removal of soil between rills and thus is also called inter-rill erosion. Sheet erosion is the movement of semi-suspended layers of soil particles over the land surface

16.3.3 Gully Erosion

Gullies often start as rills and increase in size until they become permanent topographic features. Some gullies can be like channels. **Gully erosion** is the massive removal of soil by a large concentration of runoff.

16.4 ESTIMATION OF EROSION FROM THE LAND SURFACE

16.4.1 Prediction of Soil Loss

Models have been developed for estimating the amount of erosion caused by water on the land surface component of a watershed. These include the **universal soil loss equation (USLE)**, **modified universal soil loss equation (MUSLE)**, and **revised universal soil loss equation (RUSLE)**, among several others. The USLE or MUSLE are the most widely used and accepted models for estimating soil erosion, given as mass per unit area per unit time. These methods simulate the erosion process from the previous area of a land surface. A method known as **build-up and wash-off method** has been developed by the Hydrologic Engineering Center to simulate erosion from an impervious land segment.

16.4.2 Universal Soil Loss Equation

The basic USLE, developed by Wischmeier and Smith (1965, 1978) for annual soil loss from a watershed, is given as

$$S_e = R \times K \times (L \times S) \times C \times P \qquad (16.1)$$

where S_e is the soil loss per unit area, expressed in units selected for K, for the period selected for the unit of R; R is the **rainfall–runoff factor** or **rainfall erosivity factor**, K is the **soil erodibility factor**, L is the **slope length factor**, S is the **slope steepness factor**, C is the **crop management factor**, and P is the **support practice factor**.

In the United States, where US customary units are employed, S_e is usually computed as tons per acre per year. But other units can also be selected. For example, in SI units, S_e is expressed as metric tons per hectare per year (t/ha·y), R is given in mm/ha·y, and K is in t/ha·R (where R is expressed in terms of units of EI, as described below).

For a specific area, the rainfall erosivity factor, R is usually expressed in terms of the average rainfall **erosion index**, designated as EI, units, plus a factor for runoff from snowmelt or applied water where such runoff is significant. The soil erodibility factor, K, specific for a soil horizon, is the soil loss rate per erosion index unit for a specified soil as measured on a unit plot, which is defined as 72.6 feet in length of a uniform 9% slope continuously in clean-tilled fallow. The slope length factor, L, is the ratio of soil loss from a given slope length to soil loss from a 72.6-foot length under identical conditions. The slope steepness factor, S, is the ratio of soil loss from the field slope, that is, the slope under consideration to that from the 9% slope under identical conditions. The product LS is a combined dimensionless factor for slope length and slope gradient, and is called the **topographic factor**. The crop management factor, C, is the ratio of soil loss from an area with specified cover and management, that is, the conditions of interest to that from an identical area in tilled continuous fallow, which is the condition under which K is determined. The support practice factor, also known as the **erosion control practice factor**, P, is the ratio of soil loss with a support practice like contouring, strip cropping, or terracing to that with straight row farming up and down the slope.

Procedures for the estimation of each of the factors in Eq. 16.1, are further discussed below. Good estimations of these factors enable one to predict, fairly accurately, the soil mass eroded from a land surface due to rainfall–runoff events for a given period. Benavidez et al. (2018) reviewed these factors and analyzed how different studies around the world have adapted the equations to local conditions. They compiled these studies and equations to serve as a reference for other researchers working with these models.

Equation 16.1 provides an estimate of sheet and rill erosion from rainfall events on pervious areas of upland watershed. It does not include erosion from stream banks, snowmelt, or wind and it does not include eroded sediment deposited at the base of slopes and at other reduced-flow locations before runoff reaches the streams or reservoirs.

16.4.2.1 Rainfall–Runoff Factor, R

The rainfall–runoff factor or rainfall erosivity factor, R, is an index that characterizes the effect of raindrop impact and rate of runoff associated with a rainstorm. It is determined by calculating the EI for a specified period, usually one year or one season within the year. The EI averaged over a number of these periods, n, equals R:

$$R = \frac{\sum_{i=1}^{n} EI_n}{n} \qquad (16.2)$$

The energy of a rainstorm depends on the amount of rain and all the component rainfall intensities of the storm. For any given mass in motion, the energy is proportional to the velocity squared. Therefore, rainfall energy is directly related to rainfall intensity given by (Laws and Parsons, 1943).

$$E = \begin{cases} 916 + 331(\log I_i) & \text{for } I \leq 3 \text{ in/h} \\ 1074 & \text{for } I > 3 \text{ in/h} \end{cases} \qquad (16.3)$$

where E is the kinetic energy per inch of rainfall in foot-tons/acre and I is rainfall intensity in each rainfall intensity period of the storm (in/h). In SI units,

$$E = \begin{cases} 0.119 + 0.0873 \log(I) & \text{for } I < 76 \text{ mm/h} \\ 0.283 & \text{for } I \geq 76 \text{ mm/h} \end{cases} \qquad (16.4)$$

The total kinetic energy of a storm, K_e, is obtained by multiplying E by the depth in inches of rainfall in each intensity period, i, and summing:

$$K_e = \sum_{i=1}^{n} [916 + 331(\log I_i)] \qquad (16.5)$$

The EI for an individual storm is calculated by multiplying the total kinetic energy, K_e, of the storm by the maximum amount of rain falling within 30 consecutive minutes (I_{30}) multiplied by 2 to obtain in/h and dividing the result by 100 to convert from hundreds of foot-tons/acre to foot-tons/acre

$$EI(storm) = \frac{2 K_e I_{30}}{100} \qquad (16.6)$$

In SI units, R is calculated by multiplying the total kinetic energy of the rain, E, by I_{30} and then dividing the result by

100 to obtain in units of MJ/mm·ha·h. The EI for a specific period, such as a year or a season, is the sum of the EI values of individual storms computed for all significant storms during that period. Usually, only storms with rainfall totals >0.5 inches, are selected. The R is then determined as the sum of the EI values for all such storms that occurred during a 20–25 year period, divided by the number of years, as given in Eq. 16.2.

16.4.2.2 Soil Erodibility Factor, K

The soil erodibility factor, K, indicates the susceptibility of soils to erosion and is expressed as soil loss per unit area per unit of R for a unit plot. A unit plot is 72.6 feet long (22.13 m), on a uniform 9% slope, maintained in continuous fallow, with tillage when necessary to break surface crusts and to control weeds.

The soil erodibility factor describes the resistance to soil erosion and depends on soil properties, organic matter content, bulk density, particle size, and shape. The value of the K factor ranges from 0 to 1, where 0 indicates the least vulnerability to soil erosion and 1 indicates high vulnerability.

In the United States, representative values for most of the soil types and texture classes have been complied by NRCS and, for a specific area, K can be obtained from the Web Soil Survey data, discussed in Chapter 7. Typical values of K range from 0.05 for unconsolidated loamy sand to 0.75 for silty and clayey loam soils.

The soil erodibility factor can be calculated, based on observed soil parameters such as organic matter content, texture, coarse fragments, structural class, and permeability by using the nomograph developed by Wischmeier et al. (1971). The nomograph can be described by

$$K = 2.77 \times 10^{-7}(12 - OM)M^{1.14} + 4.28 \times 10^{-3}(s - 2) + 3.29 \times 10^{-3}(p - 3)$$
(16.7)

$$M = (100 - Lc)(Ls - Armf)$$
(16.8)

where Lc is the percentage of clay (<0.002 mm); Ls is the percentage of silt (0.002 mm to 0.05 mm); $Armf$ denotes the percentage of very fine sand (0.05 mm to 0.1mm); OM is the percentage of organic matter content; s is the code of the structure size, type, and grade based on field observations; and p is the code of permeability.

Another method for calculating K as cited in Benavidez et al. (2018) is

$$K_{USLE} = f_{c\,sand} \times f_{cl-si} \times f_{orgc} \times f_{hi\,sand}$$
(16.9)

$$f_{c\,sand} = 0.2 + 0.3 \times \exp\left[-0.256 \times m_s \times \left[1 - \frac{m_{silt}}{100}\right]\right]$$
(16.10)

$$f_{cl-si} = \left(\frac{m_{silt}}{m_c + m_{silt}}\right)^{0.3}$$
(16.11)

$$f_{orgc} = 1 - \frac{0.0256 \times orgc}{orgC + \exp(3.72 - 2.95 \times orgC)}$$
(16.12)

$$f_{hi\,sand} = 1 - \frac{0.7 \times \left(1 - \frac{m_s}{100}\right)}{\left(1 - \frac{m_s}{100}\right) + \exp\left[-5.51 + 22.9 \times \left(1 - \frac{m_s}{100}\right)\right]}$$
(16.13)

where m_s is the percentage of sand content (0.05–2-mm diameter particles), m_{silt} is the percentage of silt content (0.002–0.05-mm diameter particles), m_c is the percentage of clay content (<0.002-mm diameter particles), and $orgC$ is the percentage of organic carbon content. In other words, m_s, m_{silt}, m_c, and $orgC$ are the topsoil sand fraction, silt fraction, clay fraction, and organic carbon content, respectively, expressed in percentages. For the use of K in MUSLE, K_{USLE} is converted as

$$K = 0.1317 K_{USLE}$$
(16.14)

In the United States, the soil horizon maps produced by USDA have the K factor as one of the attributes. Thus, the K factor of the soil of a watershed can be derived from the soil data obtained from the USDA Web Soil Survey (usda.gov).

16.4.2.3 Topographic Factor, LS

The topographic factors L and S indicate the effects of slope length and steepness, respectively, on erosion. Slope length refers to overland flow, flow from where it originates to where runoff reaches a defined channel or to where deposition begins. In general, slopes are treated as uniform profiles. The selection of slope lengths requires field inspection and judgement.

The slope length factor, L, is defined as

$$L = \left(\frac{\lambda}{72.6}\right)^m$$
(16.15)

where λ is the field slope length in feet, and m is an exponent, affected by the interaction of the slope length with the gradient, soil properties, type of vegetation, etc. The exponent value ranges from 0.3 for long slopes with gradient of less than 5% to 0.6 for slopes greater than 10%. The average value of 0.5 is applicable to most cases.

The slope gradient factor, S, is defined as

$$S = \frac{0.43 + 0.30s + 0.043s^2}{6.613}$$
(16.16)

where s is the slope gradient (percent).

Foster and Wischmeier (1974) adapted the LS factors for use on irregular slopes. This is especially useful on wildland sites, which rarely have uniform slopes. They described the combined factor as

$$LS = \frac{1}{\lambda_e} \sum_{j=1}^{n} \left[\left(\frac{S_j \lambda_j^{m+1}}{(72.6)^m} - \frac{S_j \lambda_{j-1}^{m+1}}{(72.6)^m} \right) \left(\frac{10\,000}{10\,000 + s_j^2} \right) \right]$$
(16.17)

where λ_e is the overall slope length, j is the sequence number of segment from top to bottom, n is the number of segments, λ_j is the length (ft) from the top to the lower end of the jth slope segment, λ_{j-1} is the slope length above segment j, S_j is the S factor for segment j (Eq. 16.16), and s_j is the slope (%) for segment j.

For uniform slopes, LS can be determined as

$$LS = \left(\frac{\lambda_e}{72.6}\right)^m S\left(\frac{10\,000}{10\,000 + s^2}\right) \quad (16.18)$$

A generalized form can be used to compute LS from

$$LS = \left(\frac{\lambda}{72.6}\right)^m (65.41 \sin^2\theta + 4.56 \sin\theta + 0.065) \quad (16.19)$$

where λ is the actual slope length in feet, θ is the angle of slope measured in radians, and the value of the exponent m ranges from 0.5 for slopes equal to or greater than 5% to 0.2 for slopes equal to or less than 1%.

In SI units, LS can be calculated as

$$LS = L \times S \quad (16.20)$$

$$L = \left(\frac{\lambda}{22.13}\right)^m \quad (16.21)$$

$$m = \frac{\beta}{1+\beta} \quad (16.22)$$

$$\beta = \frac{\sin\theta}{3\sin\theta^{0.8} + 0.56} \quad (16.23)$$

$$S = \begin{cases} 10.8\sin\theta + 0.03 & \text{for } \theta < 9\% \\ 16.8\sin\theta - 0.5 & \text{for } \theta \geq 9\% \end{cases} \quad (16.24)$$

In Eq. 16.21, λ is defined as the distance from the origin point of overland flow to deposition points given in meters and in Eq.16.20 S is expressed by θ, the slope angle. In SI units, LS can also be computed using

$$LS = \left(\frac{\lambda}{22.13}\right)^m \left[0.065 + 0.0456(S) + 0.006541(S^2)\right] \quad (16.25)$$

in which the slope steepness is given in %. Eq. 16.25 can also be used in US customary units by taking 72.6 as the denominator of λ. General values of m are 0.2, 0.3, 0.4, and 0.5 for the corresponding S of <1%, $1 \leq S < 3$, $3 \leq S < 5$, and ≥ 5, respectively. Typical values of LS range from 0.1 for short and flat slopes to 10 for long or steep slopes.

16.4.2.4 Cover Management Factor, C

The cover factor, C, represents an integration of several factors that affect erosion, such as vegetative cover, plant litter, soil surface, and land management. In most cases, the value of C is not constant over the year. Although treated as an independent variable in the USLE, the true value of this factor probably is dependent upon all other factors in Eq. 16.1. Therefore, the value of C should be established experimentally. It is the most complicated factor in USLE.

This factor is the proportion of soil losses from land cultivated under specific conditions to the compared loss under ploughed, fallow conditions. It reflects the combined effects of the cultivation of the field (dates and types), crops, seasonal erosion index distribution, cultivation history (cycle), and harvest yield level (organic material creation potential). C is a function of the vegetation and the management. It is utilized to decide the ability of the soil and crop management systems in terms of reducing the soil loss. The C factor is a ratio comparing the erosion from land under a particular crop and the management system to the corresponding losses from persistently unploughed and ploughed land. It can be controlled by choosing the crop type and culturing strategy that relates to the field and afterward multiplying these elements together.

Generally, the value of C ranges from 1 to 0.1. Values may be as small as 0.0001 for forest soils with a well-developed soil O horizon under a dense tree canopy. The value for the bare ground is 1, which is more vulnerable to erosion, whereas C for thick vegetation cover or fully mulched land is 0.1, which is significantly less vulnerable to erosion. Values of C of 0.00, 0.01, 0.03, 0.06, and 0.37, corresponding to water and wetland, urban areas, forest areas, paddy fields, and cropland, respectively, have been used by various workers.

16.4.2.5 Erosion Control Practice Factor, P

The soil loss ratios for erosion control practices vary with slope gradient. Practices, characterized by P, including strip-cropping and terraces, are not applicable to most forested and rangeland watersheds. The factor P shows the ratio of soil loss based on conservation practices to the soil loss based on certain cultivation. It reflects the impacts of *conservation practices* that will decrease the value of this factor and velocity of water runoff and consequently reduce the total soil loss. The most frequently used supporting cropland practices are cross-slope cultivation, contour cultivating, and strip cropping.

The value of P ranges from 0 to 1. Values close to 0 indicate higher erosion resistance capacity, whereas those close to 1 represent lower erosion resistance capacity. Generally, the assumption that $P = 1$ will be due to insufficient data of existing contour location and indicates no conservation practice. Table 16.1 gives values of P based on the cultivation method and slope (Kim and Julien, 2006). Table 16.2 shows values for rangeland and forestland in the western United States (Morgan, 2009).

General values of P values are: 1.00 for no support practice or up-and-down slope, 0.75 for cross slope farming, 0.50 for contour farming, 0.38 for strip-cropping on cross slopes, and 0.25 for strip-cropping on contours. Nevertheless, it is difficult to establish general ranges for the value of P, since it is usually highly specific to the effect of specific soil

16.4 Estimation of Erosion from the Land Surface

Table 16.1 *Practice factor values based on cultivation and slope*

Slope (%)	Contouring	Strip cropping	Terracing
0.0–7.0	0.55	0.27	0.10
7.0–11.3	0.6	0.30	0.12
11.3–17.6	0.8	0.40	0.16
17.6–26.8	0.9	0.45	0.18
>26.8	1	0.50	0.20

Table 16.2 *Practice factor values for the western United States*

Erosion control practice	P factor value
Contouring: 0–1° slope *	0.60*
Contouring: 2–5° slope *	0.50*
Contouring: 6–7° slope *	0.60*
Contouring: 8–9° slope *	0.70*
Contouring: 10–11° slope *	0.80*
Contouring: 12–14° slope *	0.90*
Level bench terrace	0.14
Reverse-slope bench terrace	0.05
Outward-sloping bench terrace	0.35
Level retention bench terrace	0.01
Tied ridging	0.10–0.20

* Use 50% of the value for contour bands or if contour strip-cropping is practiced.

conservation practices, sometimes called best management practices, present in a watershed.

16.4.3 Modifications of the Universal Soil Loss Equation

The USLE has been modified for use in rangeland and forest environments. The cropping management factor, C, and the erosion control practice factor, P, used in the USLE have been replaced by a vegetation management factor, VM, to form the modified soil loss equation (MSLE) as

$$S_e = R \times K \times (LS) \times VM \quad (16.26)$$

The vegetation management factor, VM, is the ratio of soil loss from land managed under specified conditions of vegetative cover to that from the fallow conditions on which the K factor is evaluated.

Williams (1975) modified the USLE by replacing the R factor with a runoff factor. This modification is based on the assumption that the total discharge and peak discharge rate resulting from a storm on the watershed depend upon the duration, amount, and intensity of the storm. The modified USLE, or MUSLE, can be expressed as

$$S_y = B(V_q Q_p)^{0.56} \times K \times (LS) \times (C \times P) \quad (16.27)$$

where S_y is the sediment yield from an individual storm, V_q is the volume of runoff in acre-feet, Q_p is the peak flow rate (ft^3/s), B is a constant, and the other terms have the same meanings as in USLE. The parameters LS, C, and P are dimensionless.

In US customary units, $B = 95$, and S_y in Eq. 16.27 is given in tons of sediment yield from an individual storm through sheet and rill erosion only on the pervious surface. In SI units, $B = 11.8$, V_q is in m^3, Q_p is in m^3/s, and K has a unit of metric tons per hectare per units of R (t/ha/R).

The quantities V_q and Q_p can be determined from rainfall–runoff modeling. USLE was developed to directly estimate the sediment yield at the outlet of a watershed rather than soil loss on a storm-by-storm basis. Therefore, the MUSLE is more suitable for estimating sediment yield for individual storms on small watersheds. The MUSLE allows the estimation of soil losses for each precipitation event throughout the year, thereby becoming an event model rather than an average annual runoff model.

16.4.4 Build-Up and Wash-Off Method

Erosion from impervious areas is fundamentally different from pervious areas, where the MUSLE approach may apply. A common approach to modeling impervious areas is the build-up and wash-off method, similar to the Storm Water Management Model (Huber and Dickinson, 1988) and the Soil and Water Assessment Tool (Neitsch et al., 2005). In this approach, sediment accumulates on the watershed between storm events. All of the accumulated sediment or possibly only a portion of it washes off during a storm event. The build-up and wash-off method tracks the time between storm events to accumulate sediment from an impervious land segment. The build-up of solids is calculated based the Michaelis–Menton formulation as (Pak et al., 2008).

$$SED = \frac{SED_{mx} \times td}{t_{half} + td} \quad (16.28)$$

where SED is the solid build up (kg/curb km) td days after the last occurrence of $SED = 0$ kg/curb km, SED_{mx} is the maximum accumulation of solids possible for the urban land type (kg/curb km), t_{half} is the length of time needed for solid build-up to increase from 0 kg/curb km to 0.5 SED_{mx} (days), and td is a day with surface runoff less than 0.1 mm. The build-up and wash-off method computes a time series of load in HEC-HMS as described later. An equivalent time series of concentration can be computed by dividing the load by the flow volume for each time interval.

Wash-off is represented by an exponential curve, which is a function of the accumulated sediment at the beginning of the storm and the peak flow rate of the storm (Huber and Dickinson, 1988). Therefore, it is necessary to know the peak flow that immediately follows the initiation of direct runoff. The exponential decrease of wash-off is given as

$$Y_{sed} = SED_0 \left[1 - \exp(-urb_{coef} \times Q_{peak} \times t)\right] \quad (16.29)$$

where Y_{sed} is the cumulative amount of solids washed off at time t (kg/curb km), SED_0 is the amount of solid build-up on the impervious area at the beginning of the precipitation event (kg/curb km), urb_{coef} is the wash-off coefficient (0.039–0.390 mm^{-1}), and Q_{peak} is the peak runoff rate (mm/h).

A street cleaning operation is performed to control the build-up of solids in urban areas. The sweep process (Huber and Dickinson, 1988) is simulated based on

$$SED = SED_0 (1 - fr_{av} \times reff) \quad (16.30)$$

where SED is the amount of solids remaining after sweeping (kg/curb km), SED_0 is the amount of solids present prior to sweeping (kg/curb km), fr_{av} is the fraction of the curb length available for sweeping, and $reff$ is the removal efficiency of the sweeping equipment.

16.5 SEDIMENT YIELD

Sediment is the product of erosion whether it occurs as surface erosion, gully erosion, soil mass movement, or channel erosion. The amount of sediment contributed to a channel depends on many factors, such as the proximity of source of erosion to the channel, characteristics of sediment particles, and the efficiency by which sediment particles are transported from one part of the landscape to another.

The **sediment yield** is the total sediment outflow from a watershed or drainage basin, measurable at a point of reference and in a specified period of time. Thus, the sediment yield from a basin is that portion of the eroded soil that leaves the basin.

Although the types of erosion occurring on a watershed can be determined, it is difficult to determine how much of the eroded material is transported by a channel and how much will be deposited into a channel over time. An erosion model, such as the MUSLE, gives an estimate of the amount of soil erosion from the upland area of a watershed. All of the eroded soil mass is not carried to the outlet of the watershed or to any point downstream. During the transport of eroded material, part of it is deposited along the course.

The **sediment delivery ratio**, designated as SDR, is defined as the fraction of gross erosion that is transported from a given catchment in a given time interval. It is a dimensionless scalar relating the erosion rate to sediment transport given by

$$SDR = \frac{Y_s}{Y_e} \quad (16.31)$$

where Y_s is the sediment yield at a point over a period such as a year and Y_e is the gross or total erosion from the watershed above the point at which the sediment yield is measured. Both Y_s and Y_e are given as mass/area/year. Note the in Eq.16.27, S_y is a **single storm sediment yield** as opposed to Y_s in Eq. 16.31, which gives the sediment yield over a period such as a year. The gross erosion is composed of rill and inter-rill, gully, and channel erosion. Conventionally, the point at which the sediment yield is measured is the outlet of a drainage area.

The sediment delivery ratio accounts for the amount of sediment that is actually transported from the eroding sources to the catchment outlet compared to the total amount of soil that is detached over the same area above that point. It often has a value between 0 and 1 due to sediment deposition caused by the change of flow regime and reservoir storage.

16.6 DETERMINATION OF SEDIMENT YIELD

16.6.1 Need for Sediment Yield Studies

The determination of sediment yield is required in studies of sedimentation for reservoir projects, local flood protection channel projects, navigation projects, alternative future land use studies, and many other projects. Each reservoir project needs a sediment yield analysis, and most yield studies calculate reservoir storage depletion resulting from the deposition of sediment during the project life, which can be different for a flood control reservoir to that of a navigation reservoir.

16.6.2 Methods for Determining the Sediment Yield

Methods to estimate the sediment delivery ratio can be roughly grouped into three categories. The first deals with specific sites where sufficient sediment yield and streamflow data are available. Methods such as sediment rating curve–flow duration, or a reservoir sediment deposition survey, are often used. These methods are based on direct measurements. USACE (1989) provided details of these methods.

As an example, Figure 16.1 shows sediment yields at various gauging stations within the Upper Indus Basin. Much of the flow of the Indus originates in the mountainous regions of the Karakoram and Himalayas, resulting in a high sediment yield, which creates a number of operational and maintenance problems for downstream water use in Pakistan. The Water and Power Development Authority of Pakistan routinely collects flow and sediment data at various gauging stations. The sediment yield data shown in Figure 16.1 were estimated from such direct measurements over a long period of record.

Even though the methods involving direct field measurements provide quite accurate estimates of sediment yield, those measurements are not available in every instance. Furthermore, limited data are not suitable for estimating the spatial distribution of sediment yield for a large watershed because the measurements required are rarely available at each of the constituent catchments.

The second category of the methods of sediment yield determination uses empirical relationships which relate the

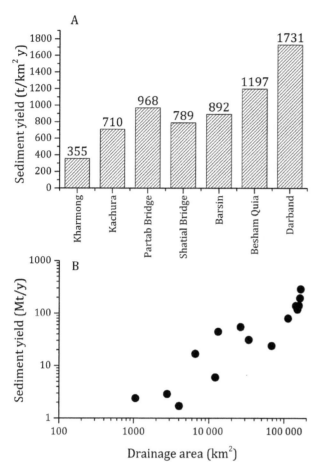

Figure 16.1 (A) Sediment yields at various stations on the main stem of the Upper Indus River from upstream to downstream. Gauging stations plotted from left to right are from upstream to downstream with increasing drainage areas. (B) Sediment yields at various stations on the Upper Indus River and its major tributaries plotted against the areas drained to the points of the gauging stations. See Figure 5.13 and Figure 18.2 for the locations of the stations and the basin map. The data are from Ali and De Boer (2007).

determination of sediment yield is known as the **sediment delivery ratio method**, which entails the determination of the sediment delivery ratio by some other means. For example, Dendy and Bolton (1976) developed the following two regression equations for sediment yield related to drainage area and mean annual runoff, from about 800 reservoirs throughout the continental United States.

For watersheds having a mean annual water runoff equal to or less than 2 inches,

$$S_e = 1280(Q^{0.46})(1.43 - 0.26\log A) \quad (16.33)$$

For watersheds having a mean annual water runoff greater than 2 inches,

$$S_e = 1958[\exp(0.055Q)](1.43 - 0.26A) \quad (16.34)$$

where S_e is unit sediment yield for the watershed, given as tons per square mile per year, Q is the mean annual water runoff for the watershed, given in inches, and A is the watershed area, given in square miles. These two empirical relationships have also been used to estimate the sediment yield.

Despite its simplicity, Eq. 16.32 provides little understanding of physical processes that underlie the sediment transport in a large watershed. Furthermore, for a given catchment area, differences in values of α and β cause orders of magnitude variation in the SDR value, suggesting the strong dependence of the sediment delivery ratio on additional properties which represent heterogeneity in factors influencing it, such as the hydrologic regime and physical characteristics of the watershed including vegetation, topography, and soil properties, and their complex interactions.

The third category of the methods of sediment yield determination includes models based on fundamental hydrologic and hydraulic processes. In the majority of these models, sediment delivery and deposition are predicted through the coupling between direct surface runoff and erosion–deposition in channels, based on their sediment transport capacities and an appropriate sediment routing, accounting for reservoir sedimentation. This a reliable method, provided the models are calibrated properly. As an example, Figure 16.2 shows yearly rainfall and sediment yields in a small catchment of the Lower Indus Basin for 35 years. Direct measurements of rainfall and runoff data were available only for 1975–1980. Hydrologic models, discussed in Section 16.11, are used to develop a calibrated model, which is subsequently used to develop a time series of sediment yields at the catchment outlet from the available rainfall data.

sediment delivery ratio to the most important morphological characteristics of a catchment, such as the catchment area. A widely used method is a sediment delivery ratio–area power function

$$SDR = \alpha A^{\beta} \quad (16.32)$$

where A is the catchment area (in km^2), α and β are empirical parameters. Statistical regression-based sediment measurements show that the exponent β is in the range of -0.01 to -0.25. Although the sediment delivery ratio experiences a wide variation for a given area, it roughly varies inversely as the 0.2 power of the drainage area, which suggests that the sediment delivery ratio decreases with drainage area. However, as exemplified in Figure 16.1, this does not hold true in every case.

Based on estimated values of gross erosion SDR–A curves have been developed for various areas. Once the sediment delivery ratio is determined from the area relationship, S_y can be determined from Eq. 16.31. This method of

16.7 TEMPORAL DISTRIBUTION OF S_Y OR E_T

The sediment yield or the gross erosion computed at the outlet of a catchment or only from the land surface at a point gives the total amount of sediment delivered at a point. A procedure is needed to translate this yield to a time

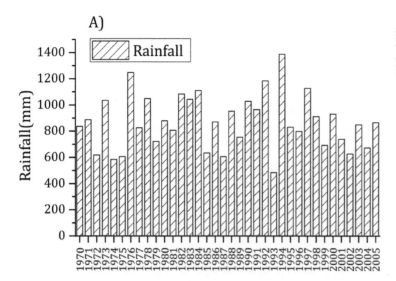
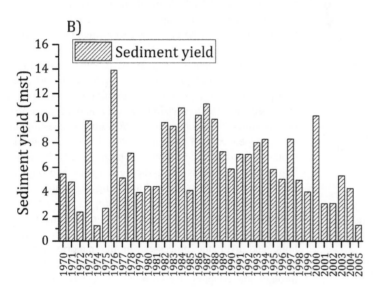

Figure 16.2 (A) Yearly rainfall (1970–2005) and (B) sediment yields in a small catchment (214 km²) drained by Papin River in the Lower Indus Basin in Pakistan (mst = monthly sediment yield in tons). Data from H. Ahmad Waseem (University of Engineering and Technology, Lahore).

distribution of sediment yield. Such a graph is also called **sedigraph**. A sedigraph is constructed by making the assumption that concentration is a power function of water discharge:

$$C_s = kQ_w^a \qquad (16.35)$$

where C_s is the concentration of sediments in the water (M/L³, e.g., mg/l), Q_w is the volumetric flow rate (L³/T, e.g., m³/s) of water, and k and a are constants. Based on a value of a, k can be calculated from the runoff and sediment yield as described below.

The mass flow rate of sediments, also called the **sediment discharge**, is given by

$$Q_s = C_s Q_w \qquad (16.36)$$

where Q_s denotes the mass transport rate of sediments (M/T, e.g., tons/day).

Substituting Eq. 16.35 in Eq. 16.36 we can write

$$Q_s = kQ_w^{a+1} \qquad (16.37)$$

The total sediment yield for a storm duration, D, is given by

$$S_y = \int_0^D kQ_w^{a+1} dt \qquad (16.38)$$

Since k is a constant, it can be determined as

$$k = \frac{S_y}{\int_0^D Q_w^{a+1} dt} = \frac{S_y}{\sum_{i=1}^n Q_{w_i}^{a+1} \Delta t_i} \qquad (16.39)$$

where the storm is divided into n increments for the purpose of evaluating the integral. Thus, given a runoff hydrograph and the total sediment yield, a sedigraph can be calculated

(Haan et al., 1994). The values of *a* are typically near 0.5–1.0. A small value flattens the sedigraph compared to the hydrograph, whereas a large value heightens the sedigraph. The coefficient *k* indicates an index of erosion severity, whereas the exponent *a* indicates the erosive power of the flow.

The time series of the sediment load is then given as

$$C_{st} = k^i Q_{wt}^a \qquad (16.40)$$

where C_{st} is the sediment concentration at time t, k^i is the proportion of the load for the current event i to the total annual load, and Q_{wt} is the watershed discharge rate at time t.

16.8 ENRICHMENT RATIO

The particle size distribution of source material on the land surface will vary from the particle size distribution within the stream channel at the basin outlet due to changing hydrodynamic forces as water flows over the land surface and concentrates within streams and channels. The degree of clay and silt enrichment in the suspended sediment is largely the result of preferential deposition of the courser fraction during the transport and delivery of sediment from its source to the basin outlet. A quantity called the **enrichment ratio** is used to convert the watershed particle size distribution to an outlet particle size distribution (Slattery and Burt, 1997). The enrichment ratio, designated as *ER*, is calculated as

$$ER = \frac{\text{percentage sediment in a given grain size class in outlet}}{\text{percentage sediment in a given size class in watershed}} \qquad (16.41)$$

Significant errors may result, if the transportability of sediment is inferred from the dispersed particle size distribution rather than actual or effective sediment sizes. Knowledge of the changes in the particle size distribution from the source material at the point of erosion to the catchment discharge point is important to understand the comprehensive sediment transport process at work in the watershed. The physical relationship between the particle size characteristics of eroded material and source material is complex but simple ratios can be used to update the erosion and sediment delivery processes operating throughout the watershed (Slattery and Burt, 1997). The enrichment ratio converts the watershed particle size distribution into an outlet particle size distribution that reflects in-stream gradations. An *ER* value greater than 1 represents an enrichment condition: a given size class forms a greater percentage of the transported load at the outlet than at the source. An *ER* value less than 1 represents a depletion condition: a given size class forms a greater percentage at the source than in the transported load at the outlet.

16.9 SEDIMENT LOADS IN CHANNELS

16.9.1 Types of Sediment Loads

Both water and sediment make up the flow in a stream. The sediments in streamflows are referred to as the **sediment load**. In other words, the term load refers to the sediment that is in motion in a stream. It is usually given as a measure of the concentration of sediments. The units are mg/l (mass/volume), which can also be expressed as parts per million (ppm), the ratio of the mass of dry sediment in a water–sediment mixture to the mass of the mixture, times 10^6. If the concentration is less than 16 000 mg/l, the concentration in ppm is essentially the same as mg/l. For concentrations greater than 16 000 mg/l, these units are related by the following equations

$$C_{ppm} = \frac{10^6}{SG\left(\dfrac{10^6}{C_{mg/l}} + \dfrac{1}{SG} - \dfrac{1}{SG_s}\right)} \qquad (16.42)$$

$$C_{mg/l} = \frac{10^6}{\left(\dfrac{10^6}{C_{ppm}}\dfrac{1}{SG_s} - \dfrac{1}{SG} + \dfrac{1}{SG_s}\right)} \qquad (16.43)$$

where SG_s and SG are specific gravities of sediments and water, respectively.

The sediment discharge of a stream is defined as the mass rate of transport through a given cross section of the channel. It is usually given as mass per unit time, such as tons/day, at a given point on the channel. Sediment load denotes the material that is being transported, whereas sediment discharge denotes the rate of transport. The sediment load is described with a variety of terminology: generally. it is defined based on the mode of transport, by its availability in the streambed, or by the method of measurement. The flowing water can transport sediments in three major ways. Accordingly, sediment loads are classified as three types, namely **bed load**, **suspended load**, and **wash load** on the basis of the mode of transport.

16.9.2 Bed Load

The bed load is the part of the sediment load composed of relatively coarse-grained particles, which move along the streambed. The movement of bed load occurs by traction, rolling, sliding, hopping, or saltation along the bed by the action of moving water at a rate related to stream discharge. Saltation includes the process of bouncing of sediment particles along the streambed or movement by the impact of bouncing particles. Particles are moved when eddies formed by turbulent flow dissipate part of their kinetic energy into mechanical work. The bed load consists of sand, gravel, or larger cobbles.

16.9.3 Suspended Load

The suspended load is the sediment load that is constituted of relatively finer particles, which are distributed throughout the flow cross section and stay in suspension by turbulence in the flowing water for an appreciable length of time without contact with the streambed. The particle suspension is the result of vertical velocity fluctuations characteristic of turbulent flow. A suspended load moves with the same velocity as the flowing water.

For most streams, there is a correlation between suspended load and stream discharge. During a rainfall–runoff event, the rising limb of the hydrograph is associated with higher rates of sediment transport and degradation. As the flood peak passes and the rate of discharge drops, the amount of sediment in suspension also diminishes rapidly and aggradation occurs. If sufficient measurements of discharge and sediment concentration are available, a relationship can be developed for use as a **sediment rating curve**. The relationship most often takes on a power function form given by

$$C = aQ^b \tag{16.44}$$

where C is the concentration (e.g., mg/l or kg/m^3) of suspended sediment load, Q is the stream discharge, and a and b are constants for a particular stream.

16.9.4 Wash Load

Part of the suspended load is called the wash load. This is made up only of silt and clay, whereas suspended sediment also includes sand-sized particles. The wash load is in near-permanent suspension and is, therefore, transported through the stream without deposition. The discharge of the wash load through a reach depends only on the rate with which these particles become available in the catchment area and not on the transport capacity of the flow.

16.9.5 Bed-Material Load

Based on its availability in the streambed, the sediment load can be divided into bed-material load and wash load. As described above, the wash load consists of the finest particles in the suspended load, which are continuously maintained in suspension by the flow turbulence and, thus, significant quantities are not found in the bed. Particles that move as a suspended load or bed load and periodically exchange with the bed are part of the bed-material load. This is the sediment load that can be calculated from the composition of the streambed. The bed-material load typically consists of all of the bed load, and the proportion of the suspended load that is represented in the bed sediments. It generally consists of grains coarser than 0.062 mm, with the principal source being the channel bed. Its importance lies in the fact that its composition is that of the bed, and the material in transport can therefore be actively interchanged with the bed. For this reason, the bed-material load exerts a control on the river channel morphology. The bed load and wash load, and the sediment that rides high in the flow and does not extract non-negligible momentum from it, together constitute the total load of sediment in a stream. With the exception of coarse materials, which are mainly transported as bed load, total bed-material load equations should be used for the determination of sediment transport capacity, discussed next, in natural channels.

16.10 SEDIMENT TRANSPORT IN THE STREAM

16.10.1 Sediment Transport Capacity

Sediment transport capacity refers to the volumetric or mass rate, such as tons/day, at which the river can transport sediment. It is calculated as

$$Q_s = C_s Q_w \tag{16.45}$$

where Q_s denotes mass transport rate of sediments (M/T), C_s is the concentration of sediments in the water (M/L^3), and Q_w is the volumetric flow rate (L^3/T) of water. The **sediment discharge rating curve** is sometimes called a suspended sediment transport graph or a suspended sediment transport relationship. It is a relationship between water discharge and sediment discharge as shown in Figure 16.3, as an example. The sediment discharge is the quantity of sediment per unit of time passing a cross section, and it is expressed as tons/day. The equation to convert from concentration to sediment discharge is

$$Q_s = kC_s Q_w \tag{16.46}$$

in which $k = 0.0027$ if C_s is given in mg/l but Q_w is given in ft^3/s.

The sediment discharge rating curve is plotted as water discharge Q_w versus sediment discharge, Q_s on a log–log grid (Figure 16.3). The relationship is often expressed as

$$Q_s = aQ_w^b \tag{16.47}$$

The sediment transport capacity can also be defined as the maximum sediment load that a particular discharge can transport at a certain slope. It is also called the **sediment transport potential** of a river. It depends on the cross-sectional geometry and hence can vary from reach to reach in a channel. Generally, the sediment transport capacity is comprised of both bed load and suspended load and, as noted above, it is the bed-material load.

Sediment processes within a reach are directly linked to the capacity of the streamflow to carry the eroded soil. The transport capacity of flow can be calculated from flow parameters and sediment properties. If the stream can transport more sediment than is contained in the inflow, additional sediment will be eroded from the streambed and entrained in the flow. However, if the flow in the reach cannot transport the sediment of the inflow, sediment will settle and be deposited on the reach bed.

There are multitudes of equations to model sediment discharge. The models can be classified according to their applicability to the types of loads, such as bed load or suspended load. These models can be also classified based on different approaches to sediment dynamics, such as the shear stress approach, unit stream power approach, etc. The most common equations are presented below.

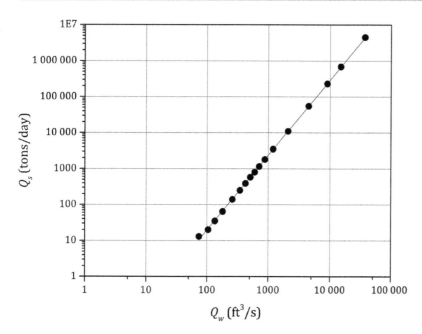

Figure 16.3 Sediment discharge rating curve of Elkhorn River at Waterloo, Nebraska, based on sampling from August 1948 to November 1950 (data from USACE (1998). The linear relationship on the log–log plot is represented by $Q_s = 0.0013 Q_w^{2.084}$. For most sediment rating curves, represented by $Q_s = a Q_w^b$, the exponent $b \approx 2$.

16.10.2 Sediment Transport Functions

16.10.2.1 The Ackers–White Function

The transport function of Ackers and White (1973) is a total load function developed under the assumption that fine sediment transport is best related to the turbulent fluctuations in the water column and coarse sediment transport is best related to the net grain shear, with the mean velocity used as the representative variable. The transport function was developed in terms of particle size, mobility, and transport.

A dimensionless size parameter is used to distinguish between fine, transitionary, and coarse sediment sizes. Under typical conditions, fine sediments are silts less than 0.04 mm, and coarse sediments are sands greater than 2.5 mm. Since the relationships developed by Ackers and White are applicable only to noncohesive sands greater than 0.04 mm, only transitionary and coarse sediments apply. Original experiments were conducted with coarse grains up to 4 mm; however, the applicability range was extended to 7 mm.

This function is based on over 1000 flume experiments, using uniform or near-uniform sediments with flume depths up to 0.4 m. A range of bed configurations was used, including plane, rippled, and dune forms; however, the equations do not apply to upper phase transport (e.g., antidunes) with Froude numbers in excess of 0.8.

Ackers and White defined a mobility number for a sediment as

$$F_{gr} = U_*^n \left[gd\left(\frac{\gamma_s}{\gamma} - 1\right) \right]^{-1/2} \left[\frac{V}{\sqrt{32} \log\left(\frac{\alpha D}{d}\right)} \right]^{(1-n)} \quad (16.48)$$

where U_* is the shear velocity, V is the average flow velocity, n is the transition exponent that depends on sediment size, α is the coefficient in rough turbulent flow equation of fluid dynamics (usually, $\alpha = 10$), d is the sediment particle size, D is the water depth, and γ_s and γ are the specific weights of sediments and water, respectively.

The general transport equation for the Ackers–White function for a single grain size is represented by a dimensionless sediment transport function:

$$G_{gr} = C \left(\frac{F_{gr}}{A} - 1 \right)^m \quad (16.49)$$

in which G_{gr} is a sediment transport parameter, C is a coefficient, A is called the sediment mobility parameter, and m is an exponent.

The dimensionless grain diameter is expressed as

$$d_{gr} = d \left[\frac{g\left(\frac{\gamma_s}{\gamma} - 1\right)}{\nu^2} \right]^{1/3} \quad (16.50)$$

in which ν is the kinematic viscosity.

The values of A, C, m, and n were determined by Ackers and White based on the best-fit curves of laboratory data with sediment size greater than 0.04 mm and Froude number less than 0.8:

For $1 < d_{gr} \leq 60$,

$$n = 1.00 - 0.56 \log d_{gr} \quad (16.51)$$

$$A = 0.23 d_{gr}^{-\frac{1}{2}} + 0.14 \quad (16.52)$$

For $d_{gr} > 60$,

$$n = 0.00 \quad (16.53)$$

$$A = 0.17 \quad (16.54)$$

$$m = 1.50 \quad (16.55)$$

$$C = 0.025 \tag{16.56}$$

For the transition zone,

$$m = \frac{9.66}{d_{gr}} + 1.34 \tag{16.57}$$

$$\log C = 2.86 \log d_{gr} - (\log d_{gr})^2 - 3.53 \tag{16.58}$$

The rate of sediment transport, X, in terms of mass flow per unit mass flow rate, which is the concentration by weight of fluid flux, can be written as

$$X = G_{gr}\left(\frac{d}{D}\right)\left(\frac{\gamma_s}{\gamma}\right)\left(\frac{V}{U_*}\right)^n \tag{16.59}$$

The concentration mass flow rate or sediment discharge can be calculated for a given flow rate.

16.10.2.2 The Engelund–Hansen Function

Engelund and Hansen (1972) obtained a sediment transport function that is a total load predictor. It gives adequate results for sandy rivers with substantial suspended load. It is based on flume data with sediment sizes between 0.19 mm and 0.93 mm. It has been extensively tested, and found to be fairly consistent with field data. The general transport equation for the Engelund–Hansen function is given by

$$g_s = 0.05 \gamma_s V^2 \sqrt{\frac{d_{50}}{g\left(\frac{\gamma_s}{\gamma} - 1\right)} \left[\frac{\tau_0}{(\gamma_s - \gamma)}\right]^{\frac{3}{2}}} \tag{16.60}$$

where g_s is the unit sediment transport, and τ_0 is the bed-level shear stress, and d_{50} is particle size of which 50% is smaller, and all other terms are as defined above.

16.10.2.3 The Laursen–Copeland Function

Laursen (1958) developed a functional relationship between the flow condition and the resulting sediment discharge. The Laursen method is a total sediment load predictor, derived from a combination of qualitative analysis, original experiments, and supplementary data. The transport of sediments is primarily defined based on the hydraulic characteristics of the mean channel velocity, depth of flow, energy gradient, and on the sediment characteristics of gradation and fall velocity. Copeland and Thomas (1989) extended the range of applicability to gravel-sized sediments. The range of applicability is 0.011–29 mm, median particle diameter.

The general transport equation for the Laursen–Copeland function for a single grain size is represented by

$$c_m = 0.01 \gamma \left(\frac{d_s}{D}\right)^{\frac{7}{6}} \left(\frac{\tau_0}{\tau_c} - 1\right) f\left(\frac{U_*}{\omega}\right) \tag{16.61}$$

in which c_m is the sediment discharge concentration, in weight/volume, τ_c is critical bed shear stress, ω is the fall velocity of particles of mean particle diameter d_s and all other terms are as defined above. The function, $f(U_*/\omega)$ as a function of the ratio of shear velocity to fall velocity was given by Laursen (1958) in graphs that can be used in Eq. 16.61.

The Laursen–Copeland function is a popular function used in sediment transport capacity calculations. The fall velocity is determined by one of the methods presented below.

16.10.2.4 Meyer-Peter–Müller Function

The Meyer-Peter–Müller (1948) bed load transport function is based primarily on experimental data and has been extensively tested and used for rivers with relatively coarse sediment. The transport rate is proportional to the difference between the mean shear stress acting on the grain and the critical shear stress. Applicable particle sizes range from 0.4 mm to 29 mm with a sediment specific gravity range of 1.25 to in excess of 4.0. This method can be used for well-graded sediments and flow conditions that produce other-than-plane bed forms. The Darcy–Weisbach friction factor is used to define the bed resistance. Results may be questionable near the threshold of incipient motion for sand bed channels.

The general transport equation for the Meyer-Peter–Müller function is

$$\gamma\left(\frac{K_s}{K_r}\right)^{\frac{3}{2}} RS = 0.047(\gamma_s - \gamma)d_m + 0.25\left(\frac{\gamma}{g}\right)^{\frac{1}{3}} g_s^{\frac{2}{3}} \tag{16.62}$$

where γ and γ_s are the specific weights of water and sediments in metric tons per cubic meter (t/m³), R is the hydraulic radius (m), S is the energy slope, K_s is a roughness coefficient, K_r is a roughness coefficient based on grains, and g_s is the unit sediment transport rate given in metric tons per second per meter (mass/time/unit width). The quantity $(K_s/K_r)S$ is the slope that is adjusted such that only a portion of the total energy loss, namely, that due to the grain resistance, S_r, is responsible for bed load motion.

The energy slope can be obtained from Strickler's formula:

$$S = \frac{V^2}{K_s^2 R^{4/3}} \tag{16.63}$$

If the energy loss due to grain resistance can also be calculated from Strickler's formula:

$$S_r = \frac{V^2}{K_s^2 R^{4/3}} \tag{16.64}$$

then

$$\frac{K_s}{K_r} = \left(\frac{S_r}{S}\right)^{\frac{1}{2}} \tag{16.65}$$

However, test results showed the relationship to be of the form

$$\left(\frac{K_s}{K_r}\right)^{\frac{3}{2}} = \frac{S_r}{S} \tag{16.66}$$

The coefficient K_r was determined by Müller as

$$K_r = \frac{26}{d_{90}^{1/6}} \tag{16.67}$$

where d_{90} is the size of the sediment for which 90% of the material is finer. From Eqs. 16.64, 16.66, and 16.67, Eq. 16.62 can be solved for the sediment transport rate, g_s.

16.10.2.5 The Toffaleti Function

Toffaleti (1968, 1969) presented a procedure to compute the unsampled load based on the concept of Einstein (1950) and Einstein and Chien (1953). Thus, the Toffaleti method uses a modified Einstein total load function that breaks the suspended load distribution into vertical zones, replicating two-dimensional sediment movement. Four zones are used to define the sediment distribution: the upper zone, middle zone, lower zone and bed zone (Figure 16.4). Sediment transport is calculated independently for each zone and is summed to arrive at the total sediment transport. The sediment discharge per unit width (tons/day/foot) for each zone is given as:

bed zone,

$$g_{sb} = M(2d_m)^{1+n_v-0.756z} \tag{16.68}$$

lower zone,

$$g_{ssL} = M \frac{\left(\frac{R}{11.24}\right)^{1+n_v-0.756z} - (2d_m)^{1+n_v-0.756z}}{1+n_v-0.756z} \tag{16.69}$$

middle zone,

$$g_{ssM} = M \frac{\left(\frac{R}{11.24}\right)^{0.244z}\left[\left(\frac{R}{2.5}\right)^{1+n_v-z} - \left(\frac{R}{11.24}\right)^{1+n_v-z}\right]}{1+n_v-z} \tag{16.70}$$

upper zone,

$$g_{ssU} = M \frac{\left(\frac{R}{11.24}\right)^{0.244z}\left(\frac{R}{2.5}\right)^{0.5z}\left[R^{1+n_v-z} - \left(\frac{R}{2.5}\right)^{1+n_v-z}\right]}{1+n_v-1.5z} \tag{16.71}$$

The total in all zones is

$$g_s = g_{sb} + g_{ssL} + g_{ssM} + g_{ssU} \tag{16.72}$$

where

$$M = 43.2 C_L (1+n_v) V R^{0.756z-n_v} \tag{16.73}$$

In the above equations, g_{sb}, g_{ssL}, g_{ssM}, and g_{ssU}, respectively, represent the sediment discharge (tons/day/foot) in the bed, lower, middle, and upper zones; M is a the sediment concentration parameter, C_L is the sediment concentration in the lower zone, R is the hydraulic radius, d_m is the median particle diameter, z is an exponent describing the relationship between the sediment and hydraulic characteristics, and n_v is a temperature exponent. The general transport equations for the Toffaleti function given above are for a single grain size. Thus, the equations should be used for all grain sizes present in proportion to their distribution in the total sediment.

This method was developed using an exhaustive collection of both flume and field data. The flume experiments used sediment particles with mean diameters ranging from 0.3 to 0.93 mm, however, successful applications of the Toffaleti method suggests that mean particle diameters as low as 0.095 mm are acceptable.

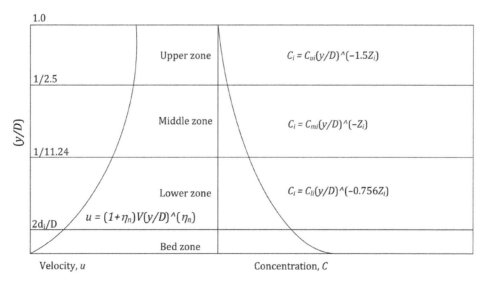

Figure 16.4 Velocity and concentration profiles in three vertical zones of a channel in the sediment transport formulation of Toffaleti (1969). The exponent in the velocity distribution function is given by the empirical relation $\eta_v = 0.1198 + 0.00048T$, where T is water temperature in °F. D denotes total depth of the water column and y denotes the depth at which velocity and concentration are given by the relationships shown in the figure.

16.10.2.6 Yang Functions

Yang (1972) reviewed the basic assumptions used in the derivation of conventional sediment transport equations. He concluded that the assumption that sediment transport rate could be determined from water discharge, average flow velocity, energy slope, or shear stress is questionable. Consequently, the generality and applicability of any equation derived from one of these assumptions is also questionable.

Yang (1973) developed the method under the premise that a unit stream power is the dominant factor in the determination of the total sediment concentration. The research is supported by data obtained in both flume experiments and field data under a wide range of conditions, found in alluvial channels. Principally, the sediment size range is between 0.062 mm and 7.0 mm, with the total sediment concentration ranging from 10 ppm to 585 000 ppm. Channel widths range from 0.44 ft to 1746 ft, depths from 0.037 ft to 49.4 ft, water temperature from 0 °C to 34.3 °C, average channel velocity from 0.75 to 6.45 ft/s, and slopes from 0.000043 to 0.029.

Yang (1984) expanded the applicability of his function to include gravel-sized sediments. The general transport equations for sand and gravel using the Yang function for a single grain size are given as follows:

For sand, $d_m < 2$ mm,

$$\log C_t = 5.435 - 0.286 \log \frac{\omega d_m}{v} - 0.457 \log \frac{U_*}{\omega}$$
$$\times \left(1.799 - 0.409 \log \frac{\omega d_m}{v} - 0.314 \log \frac{U_*}{\omega} \right)$$
$$\times \log \left(\frac{VS}{\omega} - \frac{V_c S}{\omega} \right)$$

(16.74)

For gravel, $d_m \geq 2$ mm,

$$\log C_t = 6.681 - 0.633 \log \frac{\omega d_m}{v} - 4.816 \log \frac{U_*}{\omega}$$
$$\times \left(2.784 - 0.305 \log \frac{\omega d_m}{v} - 0.282 \log \frac{U_*}{\omega} \right)$$
$$\times \log \left(\frac{VS}{\omega} - \frac{V_c S}{\omega} \right)$$

(16.75)

All the terms in Eqs. 16.74 and 16.75 are as defined earlier, with v denoting the kinematic viscosity and V_c is the critical velocity that gives a critical unit stream power $V_c S$.

16.10.3 Sediment Properties

16.10.3.1 Specific Gravity

The specific gravity of soil particles ranges from 2.60 to 2.80. Generally, the average specific gravity of natural soil is close to 2.65. The lower values of specific gravity are typical of coarser soils, while higher values are typical of the fine-grained soil types. However, the specific gravity of soil samples that are collected from the field should be determined whenever possible.

16.10.3.2 Dry Density

The dry density of soil is the ratio of the mass of dried soil to its total volume, which is sum of the pore volume and solid volume. Values range from 1100 kg/m³ to 1600 kg/m³. The typical values of dry density of clay, silt, and sand, are 481 kg/m³, 1041 kg/m³, and 1490 kg/m³, respectively. The value for gravel, cobble, and boulder sediments is also close to 1490 kg/m³. In US customary units these are given as specific weights of 30 lb/ft³, 65 lb/ft³, and 93 lb/ft³. However, whenever possible the dry density of soil particles that are collected from the field should be determined.

16.10.3.3 Sediment Gradation

The **gradation curve**, also known as the **particle size distribution curve**, determines the percentage distribution of soil particles that represents the whole sediment load. In determining the curve, soil samples must be collected from the field and analyzed in a laboratory. To determine the particle size distribution of the soil, **sieve analysis** and **sedimentation analysis** are commonly used. Sedimentation analysis is also called **hydrometer analysis**. Hydrometer analysis is used for fine soil particles which have a size less than 75 μm (micrometers or microns) and sieve analysis is used for coarse-grained soil particles. Wet and dry sieve analysis are used to find the particle size distribution of sizes greater than 75 μm. There are also other laboratory methods for the determination of the grain size distribution.

Sediment particles are classified, based on their size, into six general categories: clay, silt, sand, gravel, cobbles, and boulders. Because such classifications are essentially arbitrary, many grading systems are found in the engineering and geologic literature. Table 16.3 shows a grade scale proposed by the Subcommittee on Sediment Terminology of the American Geophysical Union (Lane, 1947). This scale, referred to as the AGU scale, is preferred for sediment transport modeling, because the sizes are arranged in a geometric series with a ratio of two. This classification is different from the Unified Soils Classification System commonly used in geotechnical engineering. Whereas particle sizes versus percent finer are the same in sedimentation studies as they are in geotechnical studies, the size classification terminology is different.

The variation in particle sizes in a sediment mixture is described with a gradation curve, which is a cumulative size–frequency distribution curve showing particle size versus the accumulated percentage finer, by weight

Table 16.3 *American Geophysical Union Sediment Classification System*

Grain class	Size in millimeters (mm)	Size in microns (μm)
Boulders		
Very large boulders	4096–2048	
Large cobbles	256–128	
Medium boulders	1024–512	
Small boulders	512–256	
Cobbles		
Large cobbles	256–128	
Small cobbles	128–64	
Gravels		
Very coarse gravel	64–32	
Coarse gravel	32–16	
Medium gravel	16–8	
Fine gravel	8–4	
Very fine gravel	4–2	
Sands		
Very coarse sand	2.0–1.0	2000–1000
Coarse sand	1.0–0.5	1000–500
Medium sand	0.5–0.25	500–250
Fine sand	0.25–0.125	250–125
Very fine sand	0.125–0.062	125–62
Silts		
Coarse silt	0.062–0.031	62–31
Medium silt	0.031–0.016	31–16
Fine silt	0.016–0.008	16–8
Very fine silt	0.008–0.004	8–4
Clays		
Coarse clay	0.004–0.002	4–2
Medium clay	0.002–0.001	2–1
Fine clay	0.0010–0.0005	1.0–0.5
Very fine clay	0.0005–0.00024	0.5–0.24

(Figure 16.5). It is common to refer to particle sizes according to their position on the gradation curve. For example, d_{50} is the geometric mean particle size; that is, 50% of the sample is finer, by weight; $d_{84.1}$ is 1 standard deviation larger than the geometric mean size but in practice it is rounded to d_{84}; and $d_{15.9}$ is 1 standard deviation smaller than the geometric mean size and is rounded to d_{16} in practice.

Natural river sediments are typically distributed log-normally. Hence, gradation curves are plotted semi-logarithmically, and the geometric mean and geometric standard deviation are used to describe the distribution. The geometric mean size is calculated as

$$d_g = \sqrt{d_{84} \times d_{16}} \qquad (16.76)$$

The geometric standard deviation is calculated as

$$\sigma_g = 0.5\left(\frac{d_{84}}{d_{50}} + \frac{d_{50}}{d_{16}}\right) \qquad (16.77)$$

It is common practice to use these definitions for the mean sediment size and standard deviation in a mixture, even if the distribution is not log-normal.

16.10.3.4 Cohesiveness

The cohesion of a sediment particle is associated with soil type and particle size. The three most common clay minerals that have electrochemical forces causing individual particles to stick together are illite, kaolinite, and montmorillonite. Cohesion increases with decreasing particle size for the same type of material. Clays are much more cohesive than silts. Cohesive sediment is characterized by the dispersed particle fall velocity, flocculated fall velocity of the suspension, the clay and nonclay mineralogy, organic content, and the cation exchange capacity.

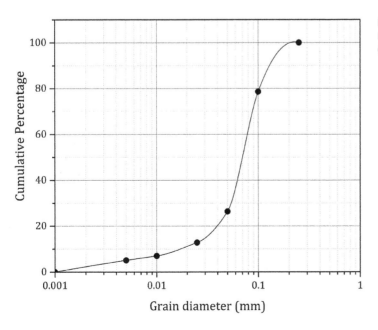

Figure 16.5 Gradation curve of a sample of bed material from Huang Ho (Yellow) River, in China. The data are given in Yang (1996).

16.10.4 The Fall Velocity

16.10.4.1 Motion and Suspension

The initiation of motion of particles in the bed depends on the hydraulic characteristics in the near-bed region. Therefore, flow characteristics in that region are of primary importance. Since determining the actual velocity at the bed level is difficult, *shear stress* is the more prevalent, though not exclusive, way of determining the point of incipient motion. Shear stress at the bed is given by

$$\tau_b = \gamma RS \tag{16.78}$$

in which γ is the specific weight of water, R is the hydraulic radius, and S is the channel slope.

Another factor that plays an important role in the initiation and continued suspension of particles is the turbulent fluctuations at the bed level. A measure of the turbulent fluctuations near the bed can be represented by the current-related bed shear velocity:

$$U_* = \sqrt{\frac{\tau_b}{\rho}} = \sqrt{gRS} \tag{16.79}$$

in which ρ is the density of water.

Additionally, the size, shape, roughness characteristics, and fall velocity of the representative particles in the stream have a significant influence on their ability to be set into motion, to remain suspended, and to be transported. The particle size is frequently represented by the median particle diameter, d_m. For convenience, the shape is typically represented as a perfect sphere, but sometimes it can be accounted for by a shape factor, and the roughness is a function of the particle size.

The **fall velocity** is a general term describing the rate of fall or settling of a particle in a fluid. The standard fall velocity of a particle is the average rate of fall that the particle would finally attain if falling alone in quiescent distilled water of infinite extent and at a temperature of 24 °C. The fall diameter of a particle is the diameter of a sphere that has a specific gravity of 2.65 and has the same standard fall velocity as the particle. Fall velocity is the most fundamental property governing the motion of a sediment particle in a fluid; it is a function of the *volume*, *shape*, and *density* of the particle and the *viscosity* and *density* of the fluid. The fall velocity of any naturally worn sediment particle may be calculated, if the characteristics of the particle and fluid are known.

The suspension of a sediment particle is initiated once the bed-level shear velocity approaches the same magnitude as the fall velocity of that particle. The particle will remain in suspension as long as the vertical components of the bed-level turbulence exceed that of the fall velocity. Therefore, the determination of suspended sediment transport relies heavily on the particle fall velocity. Several equations have been proposed by different workers for the computation of fall velocity. Some of the sediment transport functions given above require specification of fall velocity. The three equations presented below are those proposed by Rubey (1933), Toffaleti (1968), and Van Rijn (1993).

Another important parameter related to sediment suspension and movement in water is the **shape factor**, designated as SF, and defined as

$$SF = \frac{c}{\sqrt{ab}} \tag{16.80}$$

where a, b, and c are the lengths of the longest axis, the intermediate axis, and the shortest axis, respectively. These axes are the mutually perpendicular axes of the particle. The shape factor for a sphere is 1.0. Natural sediment typically has a shape factor of about 0.7. The particle shape affects the fall velocity and, hence, both the sediment diameter and fall diameter of particles. The shape factor is more important for medium sands and larger. Toffaleti used an SF value of 0.9, while Van Rijn developed his equations for $SF = 0.7$. Natural sand typically has an SF value of about 0.7.

16.10.4.2 The Rubey Velocity

Rubey (1933) developed an analytical relationship between fluid, sediment properties, and the fall velocity based on the combination of Stokes' law for fine particles, subject only to viscous resistance and an impact formula for large particles outside the Stokes' region. This equation has been shown to be adequate for silt, sand, and gravel grains. Rubey suggested that particles of the shape of crushed quartz grains, with a specific gravity of around 2.65, are the most applicable to the equation. Some of the more cubic or uniformly shaped particles tested tended to fall faster than the equation predicted. Tests were conducted in water with a temperature of 16 °C.

For quartz particles with diameter greater than 1 mm, the fall velocity can be computed by

$$\omega = F\left[gd\left(\frac{\gamma_s - \gamma}{\gamma}\right)\right]^{\frac{1}{2}} \tag{16.81}$$

where the parameter F has a value of 0.79 for particles greater than 1 mm settling in water with temperatures between 10 °C and 25 °C, and d is the particle diameter.

For smaller grain sizes,

$$F = \sqrt{\frac{2}{3} + \frac{36}{gd^3\left(\frac{\gamma_s - \gamma}{\gamma}\right)}} - \sqrt{\frac{36}{gd^3\left(\frac{\gamma_s - \gamma}{\gamma}\right)}} \tag{16.82}$$

For particle sizes greater than 2 mm, the fall velocity in 16 °C can be approximated by

$$\omega = 6.01\sqrt{d} \tag{16.83}$$

for ω in ft/s and d in ft, and

$$\omega = 3.32\sqrt{d} \tag{16.84}$$

for ω in m/s and d in m.

16.10.4.3 The Toffaleti Velocity

Toffaleti (1968) presented a table of fall velocities with a shape factor of 0.9 and specific gravity of 2.65. Different fall velocities are given for a range of temperatures and grain sizes, broken up into AGU standard grain size classes from very fine sand to medium gravel.

16.10.4.4 The Van Rijn Velocity

Van Rijn (1993) approximated the US Interagency Committee on Water Resources curves for fall velocity using non-spherical particles with a shape factor of 0.7 in water with a temperature of 20 °C. Three equations are used, depending on the particle size:

For $0.001 < d < 0.1$ mm,

$$\omega = \frac{gd\left(\frac{\gamma_s - \gamma}{\gamma}\right)}{18v} \quad (16.85)$$

For $0.1 < d < 1$ mm,

$$\omega = \frac{10v}{d}\left[\left\{1 + \frac{0.01\left(\frac{\gamma_s - \gamma}{\gamma}\right)gd}{v^2}\right\}^{0.5} - 1\right] \quad (16.86)$$

For $d \geq 1$ mm,

$$\omega = 1.1\left[\left(\frac{\gamma_s - \gamma}{\gamma}\right)gd\right]^{0.5} \quad (16.87)$$

The kinematic viscosity of water as a function of temperature, T, is given as

$$v = \frac{1.79^2 \times 10^{-6}}{1.0 + 0.0337\,T + 0.000221\,T^2} \quad (16.88)$$

16.11 SEDIMENT ROUTING

16.11.1 The Method of Sediment Routing

Sedimentation embodies the processes of erosion, entrainment, transportation, deposition, and compaction of sediment. Sediment discharge refers to the all kinds of sediment load transported by a stream.

If the sediment load entering a river reach equals the sediment load exiting the same reach, there is a sediment *equilibrium* between inflow and outflow load. However, that kind of condition rarely occurs in nature. Typically, new sediments are generated within a reach while part of the incoming sediments are deposited in the reach. *Nonequilibrium* sediment transport refers to cases where the outflowing sediment discharge from a reach does not equal the inflowing sediment discharge to that reach. All five processes of sedimentation: erosion, entrainment, transport, deposition, and consolidation are active. The nonequilibrium sediment transport condition results in an unstable streambed elevation. In such cases a numerical sedimentation model provides the computational framework for analysis.

Although sediment transport formulas are used in an analysis of nonequilibrium conditions, there are significant differences between the calculations for equilibrium sediment transport and calculations for the nonequilibrium condition. The words equilibrium and nonequilibrium refer to the exchange of sediment particles between the flow field and the bed of the cross section. Whereas the bed is the only source of sediment to a sediment transport formula, the sources for a nonequilibrium sediment condition include the bed, upstream reach, tributaries, and bank caving.

Instream sediment routing consists of calculating the transport capacity, modification of the hypothetical channel bottom to reflect deposition or erosion, and routing of the sediment load from upstream to downstream. The sediment continuity equation is solved in conjunction with sorting algorithms to solve for the actual volume of deposition or erosion.

Without physical constraints, surplus or deficit sediment computed by the sediment continuity equation would go directly into deposition or erosion within one time step. Deposition can be limited using Toffaleti's concentration relationships (Vanoni, 1975). Using the fall velocity and the expected center of mass of the material in the water column, the deposition rate, D_R, can be calculated for each grain size using

$$D_R = \frac{\omega_i \Delta t}{D_{ei}} \quad (16.89)$$

where ω_i is the settling velocity for particle size i, D_{ei} is the effective depth for sediment size i, for example, the midpoint of the depth zone in which transport is expected for the grain class, and Δt is the duration of computational time.

The temporal modifier for erosion uses the characteristic length approach, which follows the assumption that erosion takes a distance of approximately 30 times the effective depth to fully develop. Therefore, in cases where capacity exceeds supply, the amount of discrepancy is multiplied by an entrainment coefficient, EC, which limits the amount of erosion:

$$EC = 1.368 - \exp\left(-\frac{L}{30D}\right) \quad (16.90)$$

where L is the length of the control volume, and D is the effective depth (Thomas, 1994).

The other major constraint in the computation of sediment continuity is the potential supply limitation based on the amount of materials in the active layer as a result of bed mixing processes. The active layer approach divides the hypothetical sediment bed into two layers: active and inactive. Initially, sediment is removed from or added to the active layer. During each time step, the composition of this active layer is evaluated based on the deposition/erosion and updated according to d_{90} of the active layer with sediment from the inactive layer. The amount of material available to satisfy the excess capacity can be limited by the amount of material in the active layer for each time step.

There are several ways by which sediment routing calculations can be performed. Some of the common procedures are briefly mentioned below.

16.11.2 The Exner Equation

The Exner equation is a statement of conservation of mass that applies to sediment in a fluvial system such as a river to model the **aggradation–degradation process**. It was developed by the Austrian meteorologist and sedimentologist Felix Maria Exner, from whom it derives its name. The Exner equation describes the conservation of mass between sediment in the bed of a channel and sediment that is being transported. It states that bed elevation increases – aggrades – proportionally to the amount of sediment that drops out of transport and conversely decreases – degrades – proportionally to the amount of sediment that gets entrained by the flow.

The equation is given as

$$(1 - \theta_l)B\frac{\partial h}{\partial t} = -\frac{\partial Q_s}{\partial x} \tag{16.91}$$

in which θ_l is the porosity of the active layer, B is the channel width, h is the channel elevation, and x is the distance in the direction of flow.

Like most continuity equations, the Exner equation simply states that the difference between sediment entering and leaving a control volume must be stored or removed from storage. The unique feature of the Exner equation is that sediment is stored in the bed in a multiphase mixture with water, requiring porosity to translate mass change into volume change. The Exner equation translates the difference between inflowing and outflowing loads into bed change, eroding or depositing sediment.

16.11.3 The Volume Ratio Method

The **volume ratio method** links the transport of sediment to the transport of flow in the reach using a conceptual approach. For each time interval, sediment from upstream elements is added to the sediment already in the reach. The deposition or erosion of sediment is calculated for each grain size to determine the available sediment for routing. The proportion of available sediment that leaves the reach in each time interval is assumed equal to the proportion of streamflow that leaves the reach during that same interval (Neitsch et al., 2005). This means that all grain sizes are transported through the reach at the same rate, even though erosion and deposition are determined separately for each grain size. The amount of sediment transported from the reach is calculated from

$$Sed_{out} = Sed_{ch}\frac{V_{out}}{V_{ch}} \tag{16.92}$$

where Sed_{out} is the amount of sediment transported from the reach, Sed_{ch} is the amount of suspended sediment in the reach, V_{out} is the volume of outflow during the timestep, and V_{ch} is the volume of water in the reach segment.

This is the simplest of the routing methods that does not require any parameter for the model.

16.11.4 The Linear Reservoir Model

The **linear reservoir method** uses a simple linear reservoir model, as described in Chapter 12, to route each grain size through the reach. For each time interval, available sediment is calculated from the upstream sediment and local erosion or deposition. The available sediment in each grain size class is routed through a linear reservoir independently of the hydrologic routing of the flow. This allows sediment of different grain sizes to move at different speeds through the reach. The parameter that is used in this model is the **retention parameter**, R_s. The retention parameter is equivalent to the storage coefficient in the linear reservoir used to route the water through the reach.

With the linear reservoir model, the finite-difference form of sediment output at a time, t, is given as

$$Q_{s2} = 2C_1\frac{I_{s1} + I_{s2}}{2} - C_2Q_{s1} \tag{16.93}$$

in which

$$C_1 = \frac{\Delta t}{2R_s + \Delta t} \tag{16.94}$$

$$C_2 = \frac{2R_s - \Delta t}{2R_s + \Delta t} \tag{16.95}$$

16.11.5 The Advection–Diffusion Equation

Advection is the process by which particles are carried by the velocity of flow. **Diffusion** of suspended sediment occurs in a streamflow through two essential mechanisms. The first is the transport of sediment by the velocity fluctuations in turbulent flow. The second involves the mixing of sediment grains with the surrounding fluid. Diffusion of sediments in a flowing channel principally occurs in the direction of flow (x-direction) controlled by the flow velocity, v, in the x-direction and in the vertical direction (z-direction) controlled by fall velocity, ω. For gradually varied or nonuniform unsteady flow, the two-dimensional advection–diffusion equation giving sediment concentration, C_s, in space and time can be written as

$$\frac{\partial C_s}{\partial t} + v\frac{\partial C_s}{\partial x} + \omega\frac{\partial C_s}{\partial z} = \frac{\partial}{\partial x}\left(\varepsilon_x\frac{\partial C_s}{\partial x}\right) + \frac{\partial}{\partial z}\left(\varepsilon_z\frac{\partial C_s}{\partial z}\right) + C_E - C_D$$

(16.96)

where ε_x and ε_z are, respectively, diffusion coefficients in the x- and z-directions, C_E is the source from erosion in the channel and C_D is the sink due to deposition in the channel. The diffusion coefficients have dimension of L^2/T and for sake of simplicity a common assumption is that $\varepsilon_x = \varepsilon_z = \varepsilon$. Equation 16.96 can be solved by a numerical method such as the finite-difference method.

Several points regarding various terminologies found in the literature are clarified here. First of all, *a distinction should be made between diffusion and dispersion*. Technically, diffusion is more of a chemical process where mass transport occurs due to a concentration gradient at molecular or atomistic scale or due to Brownian motion. Thus, for transport of dissolved materials, that is, for solutes, diffusion is the principal process of transport in addition to advection. On the other hand, dispersion is more of a mechanical process applicable to suspended materials in a fluid. For this reason, for sediment transport, dispersion is the principal process in addition to advection for mass transport. Similarly, the term convection has also been used in place of advection in relation to various forms of Eq. 16.96. Technically, convection applies to movement of a fluid, often caused by a density gradient, whereas advection is the movement of some materials by the velocity of the fluid. From this discussion, the technically correct term for Eq. 16.96, when applied to chiefly undissolved sediment transport, is the **advection–dispersion equation**. The coefficient, ε, is the **dispersion coefficient**.

In one dimension, without any source (erosion) and sink (deposition), the advection–dispersion equation can be given as

$$\frac{\partial C_s}{\partial t} = D_x\frac{\partial^2 C_s}{\partial x^2} - v_x\frac{\partial C_s}{\partial x}$$

(16.97)

where D_x is the **hydrodynamic dispersion coefficient**, and v_x is the longitudinal velocity.

We now drop the subscript s from C and x from v. An analytical solution of Eq. 16.97 can be obtained with the following initial and boundary conditions.

$$\left.\begin{array}{l}C(x,0) = 0 \text{ initial condition}\\ C(0,t) = C_0 \text{ boundary condition}\end{array}\right\}$$

(16.98)

The solution is

$$C(x,t) = \left(\frac{C_0}{2}\right)\left[\text{erfc}\left\{\frac{(x-vt)}{2\sqrt{(Dt)}}\right\} + \exp\left(\frac{vx}{D}\right)\text{erfc}\left\{\frac{(x+vt)}{2\sqrt{(Dt)}}\right\}\right]$$

(16.99)

Sometimes, the second term within the third bracket is dropped and, if diffusion is small compared to mechanical dispersion, the dispersion coefficient, D, can be substituted as $D = \alpha_x v$, where α_x is the longitudinal dispersivity. In that case, an abbreviated solution to the one-dimensional advection–dispersion equation is given as

$$C(x,t) = \left(\frac{C_0}{2}\right)\left[\text{erfc}\left\{\frac{(x-vt)}{2\sqrt{(\alpha_x vt)}}\right\}\right]$$

(16.100)

In Eqs. 16.99 and 16.100, erfc denotes the complementary error function. Excel returns the complementary error function (ERFC) of its argument. Hence, Eqs. 16.99 and 16.100 can be solved in Excel.

For a numerical solution of Eq. 16.97, the finite-difference approximations of the spatial and temporal derivatives can be written as

$$\left.\frac{\partial C}{\partial t}\right|_i^j = \frac{C_i^{j+1} - C_i^j}{\Delta t} + O(\Delta t)$$

(16.101)

$$\left.\frac{\partial C}{\partial x}\right|_i^j = \frac{C_i^j - C_{i-1}^j}{\Delta x} + O(\Delta x)$$

(16.102)

$$\left.\frac{\partial^2 C}{\partial x^2}\right|_i^j = \frac{C_i^j - 2C_{i-1}^j - C_{i-2}^j}{\Delta x^2} + O(\Delta x)$$

(16.103)

The complete finite-difference scheme for forward in time and backward in space formulation can then be written as

$$C_i^{j+1} = C_i^j + \left[D\frac{(C_{i+1}^j - 2C_i^j + C_{i-1}^j)}{\Delta x^2} - v\frac{(C_i^j - C_{i-1}^j)}{\Delta x}\right]\Delta t$$

(16.104)

16.12 RESERVOIR SEDIMENTATION

16.12.1 Sediment Distribution within a Reservoir

When sediment-laden water reaches a reservoir, the velocity and turbulence are greatly reduced. The larger suspended particles and most of the bed load are deposited as a **delta** at the head of the reservoir (Figure 16.6). Smaller particles remain in suspension longer and are deposited further down the reservoir. The very smallest particles may remain in suspension for a long time and some may pass the dam with water discharged through sluiceways, turbines, or the spillway. Nevertheless, sedimentation occurs throughout a reservoir, not just at the delta or in the dead storage. All reservoirs, if not periodically dredged, will fill with sediments. For this reason, a resurvey of a reservoir's effective storage volume once every 5 or 10 years is needed. Bathymetric surveys of numerous reservoirs have shown decreases in reservoir volumes over time and the data have been used to estimate sedimentation rates per year.

The data obtained from surveys of existing reservoirs have been extensively used to develop empirical relationships for predicting sediment distribution patterns in reservoirs (Strand and Pemberton, 1982). Many mathematical

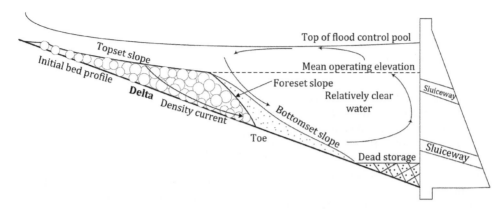

Figure 16.6 Schematic drawing of the sediment accumulation in a typical reservoir.

models involving fluid dynamics have also been proposed for sediment distributions in reservoirs.

Strand and Pemberton (1982) gave an example of a sediment distribution study for Theodore Roosevelt Dam, located on the Salt River in Arizona. The construction of the dam was completed in 1909, and a complete survey of the reservoir was made in 1981. The reservoir had an original total capacity of 1 530 500 acre-feet at an elevation of 2136 feet, the top of the active conservation pool. The purpose of this example was to: (1) compare the actual survey of 1981 with the sediment distribution procedures commonly used; (2) show all of the steps involved in a sediment distribution study; and (3) show the changes in the capacity and projected sediment depths at the dam for 100, 200, and 300 years. The total sediment accumulation in Theodore Roosevelt Lake, as determined from the 1981 survey, was 193 765 acre-feet. In the 72.4 years from the closure of the dam in May 1909 until the survey in September 1981, the average annual sediment deposited was 2676 acre-feet per year.

16.12.2 Trap Efficiency

The amount of sediment inflow to a reservoir depends on the amount of sediment yield produced by the upstream watershed. However, reservoir deposition is not synonymous with sediment yield. Some amount of inflowing sediment leaves the reservoir through the outlet and is not deposited within the pool.

The **trap efficiency** of a reservoir is the ratio of the quantity of deposited sediment to the total sediment inflow. The volume of sediment deposited in a reservoir depends on the trap efficiency and the density of deposited sediments. The trap efficiency depends mainly on the *fall velocity* of sediment particles, as well as the size, depth, and shape of the reservoir and the operational rules that govern the rate of flow through the reservoir. It generally decreases continuously after storage.

The trap efficiency is calculated by knowing the flow velocity through the reservoir and the gradation of the inflowing sediment load. It can also be estimated by comparing the settling velocity of sediment to a critical settling velocity. The rate of flow through the reservoir can be computed as the ratio of the reservoir storage capacity to the rate of flow. The greater the residence time of the reservoir, the higher the percentage of inflowing sediment settling on the bottom. The **residence time**, T_R, is given by

$$T_R = \frac{V_R}{Q_O} \qquad (16.105)$$

where V_R is the reservoir volume and Q_O is outflow rate.

The density of deposited material in terms of dry mass per unit volume is used to convert the total sediment inflow to the reservoir from mass to volume.

The ratio of the reservoir capacity to the mean annual streamflow volume can be used as an index to estimate the reservoir sediment trap efficiency. A greater relative reservoir size yields a greater potential sediment trap efficiency and reservoir sedimentation.

Churchill (1948) developed a trap efficiency curve for settling basins, small reservoirs, flood retarding structures, semi-dry reservoirs, and reservoirs that are frequently sluiced. Using data from Tennessee Valley Authority reservoirs, Churchill (1948) developed a relationship between the percentage of incoming sediment passing through a reservoir and the **sedimentation index** of the reservoir, which is the period of retention divided by velocity. Brune (1953) developed an empirical relationship for estimating the long-term reservoir trap efficiency for a large storage or normal pond reservoir based on the correlation between the relative reservoir size and the trap efficiency observed in Tennessee Valley Authority reservoirs in the southeastern United States.

The trap efficiency can be estimated by comparing the *settling velocity* of the sediment to a *critical settling velocity* (Chen, 1975). The settling velocity is computed using one of the methods described above. The critical settling velocity is computed as the discharge rate from the reservoir divided by the surface area. Chen (1975) developed a series of curves for various particle sizes, relating the trap efficiency to the ratio of the basin area to outflow rate. According to Chen,

Churchill's curves and Brune's curves for determining the trap efficiency of reservoirs are compatible in the silt range. From the studies conducted by him, Chen concluded that both Churchill's and Brune's trap efficiency curves tend to underestimate the trap efficiency for coarser materials and overestimate it for finer materials.

16.13 MODELING EROSION AND SEDIMENTATION IN HEC-HMS

16.13.1 Activation of the Sediment Component in the Basin Model

The HEC-HMS program has the capability to compute the amount of soil erosion from the land surface component of a watershed by surface runoff, transport of the eroded materials by the streams in the watershed, and sediment yield at hydrologic junctions, including reservoir sedimentation. To access the features of sediment and erosion, one needs to activate the sediment component from the basin model manager.

16.13.2 Estimation of Erosion from the Land Surface Component

There are several methods available within the program for the estimation of erosion. The method that can be applied to many regions in any country is the MUSLE. The inputs required for this method are specifications of the values of the erodibility factor, K, topographic factor, LS, cover factor, C, and practice factor, P.

In addition to the parameters described above, specification of a threshold runoff rate is required. Only some precipitation events generally cause surface erosion. The threshold value can be used to set the lower limit for runoff events that cause erosion. Events with a peak flow less than the threshold will have no erosion or sediment yield.

Each erosion method computes the total sediment load transported out of the subbasin during a storm. This calculation process is repeated for each storm during the simulation time window. The sediment load is distributed into a time series of sediment discharge from the subbasin. The distribution of sediment is based on the computed direct runoff hydrograph and the power function approach described in Section 16.7. The exponent a of Eq.16.35 is another input that is used to distribute the sediment load into a time-series sedigraph.

The grain size distribution of the surface soil characteristic of the catchment being considered is a required input. This is called the gradation curve in the program. It defines the distribution of the total sediment load into grain size classes and subclasses.

The gradation curve is defined as a diameter-percentage function in the Paired Data Manager. The grain size distribution is common to all of the surface erosion methods. All of the methods first compute the bulk sediment discharge, which includes all grain sizes. A gradation curve specifies the proportion of the total sediment discharge that should be apportioned to each grain size class or subclass. A gradation curve must be defined by the user and selected at each subbasin. A different gradation curve can be used at each subbasin to represent differences in the erosion, deposition, and resuspension processes within each subbasin. The combination of these processes is often represented by an enrichment ratio discussed in Section 16.8.

The parameters noted above are entered under the erosion tab for each of the catchments or subbasins in a watershed.

16.13.3 Sediment Transport in the Stream Component

For the entire watershed, specifications of sediment properties that are globally applicable to the watershed must be specified under the sediment tab of the basin. The sediment properties that are used for the entire watershed or basin are the following.

The transport potential method specifies how the sediment carrying capacity of the streamflow will be calculated for noncohesive sediments. One of the functions described in Section 16.10.2 can be selected if streams are present in the watershed. Each method is developed for a particular sediment grain size distribution and environmental conditions. The same transport potential method is used at all reaches within the basin model or watershed to calculate the amount of sediment that can be carried by streamflow. A cohesive transport potential method can also be selected, and the cohesive sediment is computed in addition to the noncohesive sediment.

The physical properties of sediments that must be supplied include the specific gravity of sediment grains; density of dry clay sediments; density of dry silt sediments; and density of dry sand, gravel, cobble, and boulder sediments (see Section 16.10.3). These properties are used in the sediment transport potential calculations using one of the functions selected. Some of the available transport potential methods require calculation of the fall velocity. The fall velocity is also used in several of the sediment settling methods for the reservoir sedimentation calculations. All reaches and reservoirs that require the calculation of the fall velocity will use the method selected in the sediment properties. The available methods for calculating the fall velocity are described in Section 16.10.4. The default selection is the Rubey velocity which is good for most cases.

Two different grade scale systems can be used to specify the grain size distributions. The selected system determines the number of grain sizes used for calculating erosion and sediment transport throughout the basin model. Calculations of erosion, deposition, settling, and transport are carried out separately and independently for each size in the selected grade scale. The program's clay–silt–sand–gravel system only recognizes four broad size classes. The AGU system recognizes 24 different size classifications, breaking each of the six broad size classifications into subclassifications, as given in Table 16.3.

The current version of HEC-HMS offers four models for calculations of sediment transport through streams. These are the advection–diffusion, linear reservoir, uniform equilibrium, and volume ratio methods. Of these, the advection–diffusion method is a well-known method that has been described by several authors to varying extent. HEC-HMS has followed the method that has been detailed by Fischer et al. (1979).

For all methods, the common inputs include an initial gradation curve, bed width, bed depth, active layer factor, erosion limit, deposition limit, and temperature.

The *initial gradation curve* defines the distribution of the bed sediment by grain size at the beginning of the simulation. The same curve is used in both the upper and lower layers of the two-layer bed model. The gradation curve is defined as a diameter-percentage function in the Paired Data Manager. The current functions are shown in the selection list. If there are many different functions available, a particular function applicable to the reach under consideration must be selected from the selector accessed with the paired data button next to the selection list. The selector displays the description for each function, making it easier to select the correct one.

The *width* of the sediment bed must be specified. The width should be typical of the reach and is used in computing the volume of the upper and lower layers of the bed model. The *depth* of the bed must also be specified. The depth should be typical of the total depth of the upper and lower layers of the bed, representing the maximum depth of mixing over very long time periods.

Using an *erosion limit* is optional. When the erosion limit is deactivated, erosion is limited only by the transport capacity of flow. When the erosion limit is activated, the actual erosion is reduced when the ratio of reach length to flow depth is less than 30. The erosion limit is usually encountered only in very short reaches.

Using a *deposition limit* is optional. When the deposition limit is deactivated, sediment in excess of the transport capacity is deposited completely. When the deposition limit is activated, sediment is limited by the flow depth calculated during flow routing and with the fall velocity of each grain size. The fall velocity is computed using the method selected with the basin model properties. Using a deposition limit requires the specification of the water temperature in the reach.

A fixed temperature or a temperature time-series gauge can be selected. Temperature gauges must be created in the Time-Series Data Manager before they can be used for the sediment method.

The *active layer factor* is used to calculate the depth of the upper layer of the bed model. At each time interval, the upper layer depth is computed as d_{90} of the sediment in the upper layer multiplied by the active layer factor.

One of the two simplest routing methods used in HEC-HMS is the **uniform equilibrium method**, which assumes that sediment is translated instantaneously through the reach. It is the simplest method because it does not compute any temporal lag for the sediment passing through the reach. Sediment enters the reach from upstream elements. The transport capacity for each grain size is calculated to determine the deposition or erosion state. The available sediment is computed subject to the limitations on deposition and erosion. The available sediment is then passed out of the reach regardless of the velocity determined during flow routing. No additional input other than the common inputs is required to use this method.

The other simple method of sediment routing in HEC-HMS is the volume ratio method, described in Section 16.11.3. In this method, the assumption that all grain sizes are transported through the reach at the same rate essentially limits the advection velocity of sediment to the bulk water velocity. Therefore, this method is a higher-fidelity option than the uniform equilibrium method, without requiring many more data, but it is less precise than the other available methods that require substantially more data.

The third method, the linear reservoir method, described in Section 16.11.4, requires an input of the retention parameter. The routing is performed separately for clay, silt, sand, and gravel. The value of the retention parameter is analogous to the median length of time for each sediment class to transit the reach. It may change for each sediment size class but is often close to the *travel time* for water in the reach.

The advection–diffusion method, described in Section 16.11.5, is the most detailed of the sediment routing methods and requires more data than some of the other available methods. Separate specifications of the advection and diffusion parameters for each grain size class are required. The advection parameter includes travel time and the diffusion parameter includes diffusion coefficients separately for clay, silt, sand, and gravel. This permits the large-grained sediments to move slower than the fine-grained sediments. For each time interval, the sediment from upstream elements is added to that already in the reach. After erosion or deposition is calculated, the remaining available sediment is translated in the reach by a travel time and attenuated through a diffusion process. The advection and diffusion of the sediment is linked to the velocity of water in the reach, which is calculated during flow routing.

16.13.4 Reservoir Sedimentation

The program computes the sediment balance in the reservoir over time. However, the volume of sediment accumulating in the reservoir is not linked to the storage characteristics of the reservoir used for flow routing through the reservoir. Each reservoir present in the basin model is treated separately and should use an appropriate method for the size of the reservoir compared to the inflow volume. Very large reservoirs typically trap nearly all incoming sediments. Small reservoirs often trap gravel and sand, while allowing silt and clay to pass through with some attenuation of the sediment pulses.

In the present version of HEC-HMS there are four methods by which trap efficiency is calculated whereby the amount of sediment leaving the reservoir is computed.

The *complete sediment trap* method settles all incoming sediment to the bottom of the reservoir with no sediment remaining in the outflow. This method may be used for reservoirs that have a large residence time, which is sufficient for even silt and clay particles to settle. The suitability of the method may be verified by computing the settling velocity of small sediment particles, dividing the reservoir depth by the settling velocity, and comparing to the estimated residence time. A reference such as Jiménez and Madsen (2003) may be used to compute the settling velocity. There are no parameters for the method.

The *zero sediment trap* method passes all incoming sediment to the outlet with no settling. The incoming sediment load for each grain size is totaled for the upstream elements at the end of each time interval. The total time interval sediment load for each grain size class is then transferred to the outlet of the reservoir. No settling happens for any of the grain size classes during the time interval. This may lead to large differences in the sediment concentration at the inlet and outlet of the reservoir if the inflow rate is different from the discharge rate. This method can only be used for very small reservoirs with a residency time approximately equal to the simulation time interval. There are no parameters for the method.

The *specified sediment* method uses observations of the sediment at the outlet of the reservoir to compute the sediment leaving the reservoir at each time interval. The sediment observed at the outlet may be more or less than the incoming sediment during a time interval. Settling is presumed to occur when the observed sediment is less than the incoming sediment. When the observed sediment is more than the incoming sediment, resuspension and discharge is presumed to occur. The actual physical processes are not significant; the observed sediment at the outlet of the reservoir is simply specified as a time series.

The *Chen sediment trap* method calculates trap efficiency using the method developed by Chen (1975). The settling velocity is computed using the method selected for the basin model. The critical settling velocity is computed as the discharge rate from the reservoir divided by the surface area. The computations are performed separately for each grain size class or subclass. In order to compute the surface area for each time interval of the simulation, this method can only be used if the reservoir storage characteristics are specified with the elevation–area method. The only parameter for the Chen sediment trap method is the depth-averaged temperature of the water in the reservoir. Either a fixed temperature or a temperature time-series gauge, created in the Time-Series Data Manager, can be selected.

16.14 EXAMPLES

Example 16.1: The upper North Bosque River Watershed in central Texas is a 359.8 square-mile rural watershed. The soil erodibility factor, K, ranges from 0.29 to 0.45, the topographic factor, LS, ranges from 0.78 to 5.61, and the cover management factor, C, ranges from 0.00291 to 0.02796. The practice factor, P, can be assumed to be 1. The rainfall–runoff factor, R, is 300. Estimate the possible range of sediment yield from this watershed.

Solution: Taking the mean values of soil erodibility factor, topographic factor, and cover management factor:

$K = 0.37$

$LS = 3.195$

$C = 0.015435$

$P = 1$

$R = 300$

the average sediment yield can be calculated:

$$S_e = K \times R \times (LS) \times C \times P = 0.37 \times 300 \times 3.195$$
$$\times\, 0.015435 \text{ tons/acre per year}$$
$$= 5.474 \text{ tons/acre per year}$$

The watershed area is

$$A = 359.8 \text{ mi}^2 \times \frac{640 \text{ acres}}{\text{mi}^2} = 230\,272 \text{ acres}$$

The average sediment yield is

$$S_e = 5.474 \text{ tons/acre per year} \times 230\,272 \text{ acres}$$
$$= 126\,0496 \text{ tons/year}$$

Example 16.2: The average land slope of a watershed is 2.5% and the slope length is 600 feet. Compute the topographic factor.

Solution: $S = 0.025$. Using Eq. 16.19,

$$LS = \left(\frac{600}{72.6}\right)^{0.3} \left[65.41 \times \sin(0.025)^2 + 4.56 \right.$$
$$\left. \times \sin(0.025) + 0.065\right]$$
$$= 0.4143$$

Example 16.3: In a watershed, the maximum 30-minute rainfall intensities recorded for a period are 1, 1.5, 1.75, 2, 2.25, 2.5, 2.75, and 3.0 inches per hour. Compute the rainfall–runoff factor for this watershed.

Solution: The calculations are presented in Table 16.4.

The kinetic energy E is calculated using Eq. 16.3. The erosion index, EI is calculated using Eq. 16.6. The total EI is 344.7152 foot-tons/(acre-hour). The average of eight readings is R, 43.1 foot-tons/(acre-hour).

Table 16.4 *Calculations*

I (in/h)	E	EI
1	916	18.320
1.5	974.286	29.229
1.75	996.446	34.876
2	1015.641	40.626
2.25	1032.572	46.466
2.5	1047.718	52.386
2.75	1061.419	58.378
3	1073.927	64.436

Example 16.4: Develop the soil erodibility map of Five Mile Creek Watershed in south Dallas from USDA Web Soil Survey data.

Solution: As stated earlier, soil erodibility factors for various soil textural classes in the United States have been mapped by USGS. These maps can be produced just like Green–Ampt parameters are derived from the soil database produced by USDA. Figure 16.7 shows the soil erodibility map of Five Mile Creek Watershed. An application is given in Chapter 21 (Example 21.2).

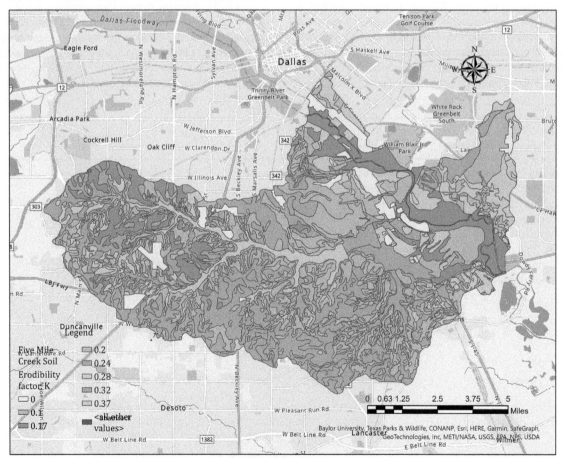

Figure 16.7 Distribution of the soil erodibility factor in Five Mile Creek Watershed in Dallas, Texas. (A black and white version of this figure will appear in some formats. For the color version, please refer to the plate section.)

Exercises: A selection of exercises on this topic is available at www.cambridge.org/appliedhydrology.

17 Reservoir Operations

17.1 CONCEPTS AND PURPOSE

A reservoir can be a single-purpose conservation reservoir or a single-purpose flood control reservoir, or a multipurpose reservoir. Multipurpose reservoirs are designed for two or more purposes. The common type of such reservoirs usually has four main objectives to serve, namely, water supply to treatment plants for drinking water, flood control, irrigation, and hydropower generation. The operation of a reservoir essentially entails creating a balance between the supply of water in the reservoir from the river flowing into it and the demand of water stored within it by the sector serviced by the reservoir. Reservoir operation is a highly complex subject. This chapter describes the common methods employed in the operations of typical reservoirs.

17.2 WHAT IS A RESERVOIR OPERATION?

Reservoirs on a river course are impoundments, primarily built to serve two main purposes: (1) flood control and (2) conservation. Thus, the active storage volume of a reservoir consists of flood control storage and conservation storage. The aim of the flood control storage is to temporarily hold some of the flood water of a river when the discharge rate reaches a stage likely to cause damage to the valley downstream, and to release the flood water gradually at a safe rate when the flood recedes. Conservation storage is meant to store the surplus water brought down by a river during periods when the natural flow exceeds current demand, and to use this stored water during periods when demand exceeds the natural flow. Conservation may be done for any one or more of the following purposes: (1) public and industrial water supply, (2) irrigation, (3) hydropower, (4) regulation of low water flow for navigation, and (5) recreation, pisciculture etc.

A **reservoir operation** refers to the rules and regulations that specify the amount of water to be released from the storage at any time depending upon the state of the reservoir, level of demands, and any information about the likely inflow in the reservoir.

The effect of watershed uses and management on reservoir operations can be illustrated by considering the storage allocation of a typical reservoir, schematically shown in Figure 17.1. This multipurpose reservoir has storage space allocated for both conservation purposes, that is, water supply, and flood control. Within the conservation storage is a buffer zone, the top of which is used as a threshold to allocate water releases from the reservoir for different demands during critical or dry periods. Once the reservoir pool elevation drops to the top of the buffer zone, water can be released only for those purposes predetermined to be most important, such as providing municipal water supplies. Other, less essential needs will not be met until the pool elevation is higher than the top of the buffer pool. Several buffer zones can be used as a mechanism to allocate water on a priority basis during periods of shortages. If the reservoir has hydroelectric power-generating capacity, the head required to drive the turbines and the corresponding storage is an added operational zone in the reservoir. Once sediment begins to encroach into the various storage spaces, operating the reservoir to meet the respective demands becomes more difficult. As discussed in Chapter 16 (Figure 16.6), sediments typically do not settle down only in the dead storage space. Coarse sediments are deposited at various inflow points in the upper reaches of the reservoir pool, where finer sediments tend to settle down near the dam. As sediment fills the active storage space, the capability of a reservoir to meet all demands becomes limited. Storage volumes in the conservation pool can be inadequate to meet demands during dry periods or droughts. Likewise, there might not be adequate storage space in the flood control pool to control major flood events. The situation of excess water in the rainy season and scarcity in the dry season poses significant challenges to effective reservoir operation. Many reservoirs in India, for example, are water-deficient in the dry season, but are threatened by dam-break disasters in the flood season. Additional uncertainty in reservoir operations is introduced due to global climate change as well as economic activities in the river basin.

Due to changes of hydro-meteorological conditions and shifting goals of water requirements from one region to the other, each reservoir has a different set of operation rules. The operation of reservoirs is a complex process that involves many decision variables and multiple objectives, as well as considerable risk and uncertainty.

17.3 RULE CURVES

Some of the problems encountered in regular reservoir operations can be overcome, at least partly, by using an

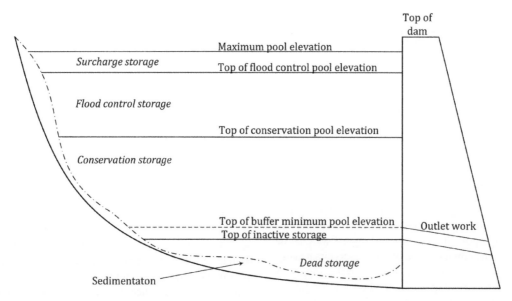

Figure 17.1 Storage allocation for a multipurpose reservoir with associated sedimentation.

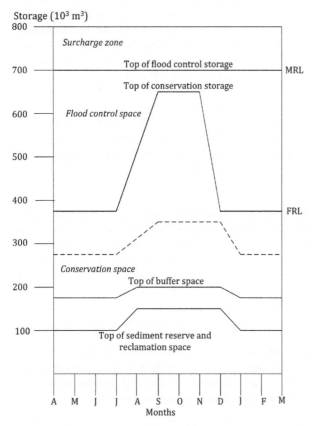

Figure 17.2 Operational rule curves illustrating seasonally varying storage requirements of a multipurpose reservoir: FRL indicates the full reservoir level, MRL, the maximum reservoir level. Names of the months are given as lettered abbreviations.

operational **rule curve** approach to reallocate storage space for different purposes seasonally, as illustrated in Figure 17.2. A rule curve may be defined as a diagram showing the reservoir storage requirements during the year. The reservoir operator is expected to maintain these levels as closely as possible while generally trying to satisfy various water needs downstream. If the reservoir storage levels are above the target or desired levels, the release rates are increased. Conversely, the release rates are lowered if the levels are below the target. These release rates may or may not be specified but will depend in part on any maximum or minimum flow requirements and on the expected inflow.

Rule curves provide target pool elevations that vary with season. Their purpose is to provide operational guidelines that allow the most efficient use of reservoir storage. Typically, the season in which irrigation supplies are needed is a period in which the threat of large floods is slight; therefore, the elevation of conservation pool can be raised into the normal flood control pool during that time. When the flood season approaches, the conservation pool is then lowered to provide flood control space. Rule curves allow demands to be met with a smaller total storage capacity in the reservoir. But even rule curve operations may not suffice when sedimentation becomes excessive.

The development of rule curves involves considerable knowledge of the variability of inflow to the reservoir and a clear understanding of seasonal water demands, priorities, and the downstream channel capacity. High- and low-streamflow sequences observed in the past are used to develop and test rule curves to ensure that water requirements and all constrains are met. When streamflow records are limited at the site, regional data can be used or streamflow records can be simulated with either deterministic or stochastic models.

Rule curves serve as guides in the operation of a reservoir. The ability to adhere to these guides is improved by providing the reservoir manager with accurate, real-time forecasts of inflow. Many operating constraints must be considered along with rule curves. For example, most reservoirs have restricted rates of change of outflow to prevent rapid surges of water downstream. In some cases, this can cause the reservoir pool elevation to deviate from the rule curve. Similarly, rule curves may occasionally be violated to accommodate the passage of migrating fish or to assist downstream efforts to remove a barge that has gone aground. However, such deviations in rule curve operations should not be allowed to affect some other parts of the system adversely.

In a multipurpose reservoir, when the demands for water are concurrently competitive, storage volumes in the reservoir must then be allocated to meet these competitive demands optimally. This partitioning process involves both the determination of required volumes and establishing rules to specify how the reservoir is to be managed. Elevations of the various zones are used as guides for operation and can vary seasonally. The five zones that are shown in Figure 17.1 are as follows:

1. Surcharge zone: This is the storage above the flood control zone associated with actual flood damage. The top of the flood control pool is used to maintain the integrity of the reservoir. Reservoir releases are usually at or near their maximum to prevent the dam from collapsing when the storage volume is within this zone.
2. Flood control zone: This is the reserve for storing large inflows during periods of abnormally high runoff. The flood control zone is evacuated of water at a time corresponding to the flood season. The reservoir is then kept at the top of conservation space, which coincides with the bottom of the flood control space to provide sufficient storage to control flooding. Once the pool elevation is in this zone, the reservoir is operated to release the maximum amount of water without causing flooding. Ideally this will coincide with the bank full conditions downstream.
3. Conservation zone: This is the zone of storage from which various water-based needs are satisfied. The ideal storage volume or level is normally located within this zone. A system of priorities may be established within this zone to ensure that vital water requirements are met.
4. Buffer zone: This is the storage beneath the conservation zone entered only in abnormally dry periods. When volumes of water are stored within this zone, releases are restricted temporarily to satisfy high-priority demands only.
5. Inactive zone: This is the dead storage beneath the buffer zone, which is, if possible, entered only under extremely dry conditions. Reservoir withdrawals are an absolute minimum. This zone contains enough space to trap and retain sediments over the life of the project.

It should be noted at this point that in recent years views have been expressed that rule curves should be made obsolete because computerized decision support systems, based on real-time modeling of watershed hydrology, can make reservoir operations easier and more efficient.

17.4 METHODS OF MATHEMATICAL PROGRAMMING

Traditionally, a reservoir operation is based on heuristic procedures, embracing rule curves and subjective judgments by the operator. This provides general operation strategies for reservoir releases according to the current reservoir level, hydrologic conditions, water demands and the time of the year. Established rule curves, however, do not allow fine-tuning and hence optimization of the operations in response to changes in the prevailing conditions. Therefore, it would be valuable to establish an analytical and more systematic approach to reservoir operation, based not only on traditional probabilistic or stochastic analysis but also on the information and prediction of extreme hydrologic events and advanced computational technology to increase the reservoir efficiency for balancing the demands from the different users.

A new reservoir operations strategy, dubbed as Forecast Informed Reservoir Operations (FIRO) is being developed at the Center for Western Weather and Water Extremes at the University of California at San Diego, which better informs decisions to retain or release water by integrating additional flexibility in operation policies and rules with enhanced monitoring and improved weather and water forecasts. The reservoir operations strategy under FIRO uses weather and water forecasts to inform decision making to selectively retain or release water from reservoirs, in order to optimize water supply reliability and environmental co-benefits and to enhance flood-risk reduction. FIRO is being developed as a collaborative effort by several governmental agencies and academic institutions in the United States and is tested in the Russian River basin (Lake Mendocino), the Santa Ana River basin (Prado Dam), and the Yuba-Feather River basin in California, and it is being expanded to other states such as Texas.

Rapid increase in computer technology since the 1980s made the development of sophisticated mathematical models for the analysis of water resource systems possible. These models are increasingly being used by system managers to determine decision alternatives that are optimal in some defined sense. The optimization of a reservoir system operation usually involves a search through large decision spaces for optimal parameter sets. Often, the decision space is too large for a complete search. This has motivated the development of various optimization procedures.

Most optimization models are based on some type of mathematical programming technique. Many successful applications of these techniques to reservoir operation studies have been reported in the literature, but no universally proven technique exists. Yeh (1985) classified the techniques that are commonly used in reservoir operations in the following six broad categories.

1. linear programming (LP);
2. nonlinear programming (NLP);
3. dynamic programming (DP);
4. discrete differential DP (DDDP), incremental dynamic programming (IDP), differential DP (DDP), successive approximation DP (SADP), and stochastic DP (SDP);
5. genetic algorithms; and
6. simulation.

The literature is replete with descriptions and discussions of various kinds of mathematical programming that can be used for optimal reservoir operations and development of rule curves by various techniques.

17.5 BASIC PRINCIPLES OF OPTIMIZATION OF RESERVOIR OPERATIONS

Figure 17.3 is a simplified line diagram of a reservoir system and various reservoir storages along the main stem of Narmada River in western India. Most of the water resources projects within a river basin are highly complex with multiple controls of flows in the river system within the basin.

The basic operation of a single reservoir in the system of reservoirs can be expressed as a continuity equation:

$$S = S_t + I_t + LI_t + P_t - E_t - Q_t - R_t \tag{17.1}$$

where S is the reservoir storage at the end of time t, S_t is the reservoir storage at the beginning of time t, I_t is the inflow into the reservoir during time t, LI_t is the local inflow to the reservoir from the surrounding area in time t, P_t is the precipitation in the reservoir in time t, E_t is the evaporation loss from the reservoir in time t, Q_t is the release to the natural channel from the reservoir in time t, and R_t is the total outflow, that is, the release from the reservoir in time t.

For a reservoir operation, the simplest operating rule is to supply all the water demanded, if available. The release is then independent of the reservoir content and season. If there is sufficient water in the reservoir to meet the required releases, the reservoir empties and this is called the *conventional rule*. The reservoir will operate under the following basic constraints. The volume of water released during any period cannot exceed the contents of the reservoir at the beginning plus the flow into the reservoir during the period:

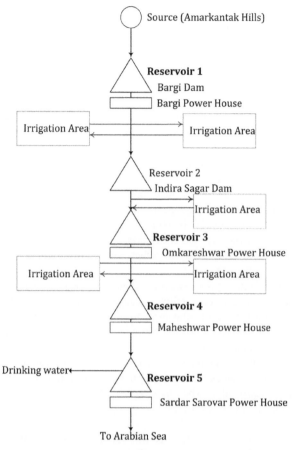

Figure 17.3 Line diagram of the reservoir system and major water allocation along the main stem of Narmada River in India.

$$R_t \leq S_{t-1} - Y_{\min} + I_t + P_t + LI_t - Q_t - E_t \tag{17.2}$$

$$R_t = R_{rt} + S_{pt} \tag{17.3}$$

where $Y_{\min}$ is the variable capacity of the reservoir up to the minimum pool of the reservoir in time t, R_{rt} is the actual irrigation release from the reservoir in time t, and S_{pt} is the reservoir spill in time t. Thus, the continuity equation for the reservoir can be defined as

$$S_t = S_{t-1} + I_t + P_t + LI_t - Q_t - E_t - R_t \quad \text{for all } t \tag{17.4}$$

The contents of the reservoir at any period cannot exceed the capacity of the reservoir, and the dead storage of the reservoir puts a lower limit on the reservoir storage, such that

$$Y_d = Y_{\min} \leq S_{t-1} \leq Y \quad \text{for all } t \tag{17.5}$$

where Y_d is the dead storage of the reservoir and Y is the total reservoir capacity at *maximum reservoir level* (MRL).

A reservoir operation according to a single rule curve such as Figure 17.2, with constraints given in Eqs. 17.2 and 17.4, and a modified constraint of Eq. 17.5, may be such that as the storage of water in the reservoir decreases, restrictions may be imposed in the uses, so that the demand falls and releases are lowered. So, the constraint is expressed as

$$Y_d \leq Y_{\min} \leq S_{t-1} \leq Y \quad \text{for all } t \tag{17.6}$$

The important question is whether the stored water in the reservoir is to be used at present or retained for use during possible drought periods.

17.6 LINEAR PROGRAMMING

In a problem where all the objective and constraint functions are linear, LP can be used in the optimization of reservoir systems. It has been one of the most widely used techniques in water resources management due to its simplicity and adaptability to a variety of systems. A typical LP model can be formulated as

$$\min_X Z = C^T X \tag{17.7}$$

subject to

$$\begin{aligned} AX &\geq b \\ X &\geq 0 \end{aligned} \tag{17.8}$$

where $X \geq 0$ is an n-dimensional vector of decision variables, C is an n-dimensional vector of objective function coefficients, b is an m-dimensional vector of the right-hand side of Eq. 17.8, A is an $m \times n$ matrix of constraint coefficients, and T represents the matrix transpose operation.

If LP is applied to a reservoir operation model, the typical planning objective might be to minimize the capacity or the cost of a reservoir that meets all of the constraints or to maximize the total net annual benefit, which is represented as a function of storage and release in each presumed operation period. It is required that the *cost function* is convex and the *benefit function* concave for the successful application of LP. Yeh (1985) described several variants of LP such as stochastic LP.

17.7 NONLINEAR PROGRAMMING

NLP techniques can be applied where either the objective function or constraints are nonlinear. NLP can effectively handle a nonseparable objective function and nonlinear constraints. A general NLP problem can be expressed in the form:

$$\text{Minimize } F = f(x_1, x_2, \cdots, x_n) \tag{17.9}$$

subject to

$$g_i(x) = 0 \quad i = 1, m \tag{17.10}$$

where

$$\underline{x}_j \leq x_j \leq \overline{x}_j \quad j = 1, n \tag{17.11}$$

in which F is to be minimized subject to m constraints expressed by function $g(x)$, n is the number of decision variables, and Eq. 7.11 is a bound constraint for the jth decision variable x_j with lower and upper bounds, respectively.

Many NLP software packages are commercially available. However, NLP has not been very popular due to the computational complexity of the approach for multireservoir system optimization. For multiple reservoir systems, the number of constraints is large because they deal with similar subsystems repeated in time or location. Therefore, NLP requires large amounts of storage and execution time when compared to other methods, limiting its applicability to large systems (Yeh, 1985). The use of NLP is further limited to problems that are smooth and continuous, because it requires the calculation of derivatives for its search procedure.

17.8 DYNAMIC PROGRAMMING

DP, a method formulated largely by Bellman (1957) is a procedure for optimizing a multistage decision process. DP is a commonly used method for the optimization of reservoir systems as these are characterized by a large number of nonlinear and stochastic features, which can be translated into a DP formulation. However, DP suffers from a problem known as the *dimensionality problem*.

DP is an enumeration procedure used to determine the combinations of decisions that optimize the overall system effectiveness as measured by a criterion function. It can treat nonconvex, nonlinear, and discontinuous objective and constraint functions. Constraints on both decision and state variables introduce no difficulties. The constraints speed up the computational procedure. The key feature of DP application is that it is usually identified as serially or progressively directed for operational or planning problems, respectively. The operation of reservoirs is a multistage decision process and DP is particularly suited to such problems. The problem is divided into stages, with a decision required at each stage. The stages usually represent different points in time and each stage should have a finite number of states associated with it. In reservoir operation studies, the state usually represents the amount of water in the reservoir at a given stage. If DP is used for the determination of reservoir releases, these form decision variables. The stage-to-stage transformation is carried out by the continuity equation subject to constraints on storages and releases.

When the returns are independent and additive, a typical recursive equation of DP can be written as

$$f_n(x_n) = \max_{d_n} \left[r_n(x_n, d_n) + f_{n-1}(x_{n-1}) \right] \tag{17.13}$$

where x_n is the *state variable*, d_n is the *decision variable*, $r_n(x_n, d_n)$ is the *objective function value* or the *return function*, n is a *stage*, $f_n(x_n)$ is the cumulative return at stage n with known $f_0(x_0)$, and $x_{n-1} = t_n(x_n, d_n)$ is the stage-to-stage *transformation function*.

If DP is applied to the determination of reservoir releases, the state variable is the storage and the decision variable is the release. The stage is represented by the time period, i. The stage-to-stage transformation is characterized by the continuity equation:

$$S_{i+1} = S_i + I_i - R_i - E_i \qquad (17.14)$$

subject to

$$S_{\min} \leq S_{i-1} \leq S_{\max} \qquad (17.15)$$

and constraints on the R release.

Several variants of DP have been developed over time to alleviate the problems of dimensionality. Notable among these are the DDDP and IDP. These are iterative techniques and start with the assumption of a trial trajectory. DDDP is specifically designed to overcome the dimensionality problem posed by DP.

17.9 SIMULATION MODELS

Simulation is a modelling technique used to approximate the behavior of a system on a computer, representing all the characteristics of the system largely by a mathematical or algebraic description. Simulation models provide the response of the system to certain inputs, which include decision rules that allow the decision makers to test the performance of either an existing system or proposed new system before building it. Reservoir operators must simultaneously meet requirements for many needs, including water supply, flood control, power generation, recreational use of the reservoir pool, environmental quality downstream of the reservoir, and the safety and structural integrity of the dam itself. Each of these needs imposes constraints on the storage and release of water from the reservoir, and the needs and constraints often conflict with one another. Setting a schedule of reservoir releases that fulfils the purpose of a reservoir, meets operating constraints, and is physically possible is not a simple task, and engineers have created reservoir simulation models to help develop those release schedules.

A typical simulation model for a water resources system is simply a model that simulates the interval-by-interval operation of the system with specified inflows at all locations during each interval, specified system characteristics, and specified operating rules. Optimization models aim to identify optimum decisions for a system operation that maximizes certain given objectives while satisfying system constraints. On the other hand, simulation models are used to explore only a finite number of decision alternatives so that the optimum solution may not necessarily be achieved. However, there are some simulation models that involve a certain degree of optimization. For given operating criteria, the performance of a reservoir system may be evaluated by analyzing the computed time sequence of levels, storage, discharge, hydropower, etc. The procedure can be repeated for several inflow sequences to arrive at a statistical measure of the system. Simulation models have been routinely applied for many years by water resources management agencies.

17.10 RESERVOIR OPERATIONS MODELING WITH HEC-RESSIM

17.10.1 Introduction to HEC-ResSim

The Hydrologic Engineering Center developed a reservoir simulation model, HEC-ResSim, designed to simulate reservoir operations for flood management as well as flow augmentation. HEC-ResSim is an effective tool for real-time decision support that can be used by agencies responsible for water resources management. HEC-ResSim uses a traditional *rule-based approach* to mimic the operational decision-making process that reservoir operators follow in setting releases to meet operating requirements for flood control, power generation, water supply, and environmental quality. Parameters that may influence flow requirements at a reservoir include the time of year, hydrologic conditions, water temperature, and simultaneous operations by other reservoirs in a system. The reservoirs designated to meet the flow requirements may have multiple and/or conflicted constraints on their operation. HEC-ResSim describes these flow requirements and constraints for the operating zones of a reservoir using a separate set of prioritized rules for each zone. Basic reservoir operating goals are defined by flexible at-site and downstream control functions and multi-reservoir system constraints. With the advanced features available within HEC-ResSim, such as outlet prioritization, scripted state variables, and conditional logic, it is possible to model more complex systems and operational requirements.

17.10.2 Representation of the Physical Reservoir System

HEC-ResSim represents a system of reservoirs as a network composed of four types of elements: *junctions*, *routing reaches*, *diversions*, and *reservoirs*. Each element is defined with enough information to be physically realistic. By combining reservoirs, reaches, junctions, and diversions, a network, capable of representing anything from a single reservoir on a single stream to a highly developed and interconnected system like that of California's Central Valley, can be built. The program's user interface allows the user to draw the network either as a line diagram, such as the one shown in Figure 17.3, or as a map drawn over georeferenced graphics.

The simplest element type is the junction. Junctions represent stream confluences or points where external flows enter the system. HEC-ResSim does not calculate runoff. All local flows must be introduced at junctions as external flows. The flow out of a junction is simply the sum of the flows into the junction. Flows at junctions can be converted to stages using rating curves.

Routing reaches represent the natural streams in the system, and the lag and attenuation of flow in a reach is computed by one of a variety of available standard hydrologic routing methods, discussed in Chapter 12, such as the Muskingum, modified Puls, coefficient, or Muskingum–Cunge methods. Losses through seepage can be specified for each routing reach.

A diversion is a more complex element. It represents a withdrawal of water from the natural stream. The quantity of the withdrawal can be specified as a constant amount or as a function of some parameter such as time or flow. Some or all of the diverted water can be routed and returned by a diversion, or it can be removed from the system entirely.

A reservoir is the most complex element of the reservoir network and is composed of a pool and a dam. The pool is assumed to be a level pool and its hydraulic behavior is completely defined by an elevation–storage–area table. The real complexity of the reservoir network of HEC-ResSim begins with the *dam*. The dam is the root of an outlet hierarchy that allows the user to describe different outlets of the reservoir in as much detail as is deemed necessary. There are two basic and two advanced outlet types. The basic outlet types are *controlled* and *uncontrolled*. An uncontrolled outlet can be used to represent an outlet of the reservoir, such as an overflow spillway, that has no control structure to regulate flow. Controlled outlets can be used to represent any outlet, such as a gate or valve, capable of regulating flow. The advanced outlet types are *power plant* and *pump*, both of which are controlled outlets with additional features to represent their special purposes. The power plant adds the ability to compute energy production to the standard controlled outlet. The pump is an even more specialized controlled outlet. Its flow direction is opposite to that of the other outlet types, and it can draw water up into the reservoir from the pool of another reservoir. The pump outlet type can model pump-back operations in hydropower systems.

17.10.3 Operation Rules

For most reservoirs, flow requirements and constraints vary depending on the state of the reservoir pool. That is, the rules change depending on the amount of water stored in the reservoir. HEC-ResSim describes this dependency by dividing the pool into elevation bands, called zones, and applying a different set of prioritized rules to each operating zone in the reservoir. An operating zone is described by a water elevation curve representing the top of the zone. When the water level in the pool exceeds the top (or bottom) of a zone, its rules no longer apply to release decisions. The top-of-zone elevation curve can be a constant or can vary seasonally.

A reservoir in HEC-ResSim must have a *target elevation*. The target elevation of a reservoir, represented as a function of time, is called its *guide curve*. It is the dividing line between the upper zones of the reservoir (typically called the flood control pool) and the lower zones (typically called the conservation pool).

The release decision logic in HEC-ResSim starts and ends with the guide curve. When the reservoir's pool elevation is above the guide curve (*in flood control*), the reservoir is operated to release more water than is entering the pool; when below the guide curve (*in conservation*), the reservoir is operated to release less water than is entering the pool. All operating rules and physical limitations act as constraints upon the reservoir's ability to meet the goal of returning the pool to its guide curve elevation. Without rules, the reservoir will be constrained only by the physical capacity of the outlets to get to and stay at the guide curve elevation.

Each reservoir operating goal is described by a flexibly defined rule that, when evaluated, specifies a minimum or maximum limit on the release from the reservoir or outlet. The rules are placed in a prioritized list in one or more reservoir zones. As each rule is evaluated, its calculated minimum and/or maximum flow is applied to an evolving *allowable range of release*. At the start of the release decision process, HEC-ResSim sets the allowable release range to the physical limits of the dam or outlet: the maximum of the range is the total maximum capacity of the outlets for the current pool elevation, the minimum of the range is the minimum release capacity of the outlets, usually zero. As a rule is applied, it may narrow the allowable release range. If a rule neither raises the minimum allowable release nor lowers the maximum, that rule will have no effect on the range. Once all rules have been evaluated and applied to the range, the allowable range is considered complete and the *desired guide curve release* is computed. The desired guide curve release is the release the reservoir would make if it were not constrained by any *limits*. The final release is the closest value to the desired guide curve release that falls inside the allowable range. Although only a small variety of rule types are available in HEC-ResSim, when combined with one another and the conditional *IF-Then-Else* rule usage logic, the user can describe very complex operation schemes.

17.11 MASS CURVES

The mass curve is a graph showing the cumulative net reservoir inflow, exclusive of upstream abstraction, as the ordinate against time as the abscissa. The ordinate may be denoted by depth (e.g., cm) or in hectare meters or in any other unit of volume. Mass curves permit a simple graphical inspection of the entire record or any portion of it for the evaluation of the **reservoir yield**.

The yield is the amount of water that can be supplied from the reservoir in a specified time interval. The time interval may vary from a day for a small distribution reservoir to a year or more for a large storage reservoir. The yield is dependent on inflow and varies from year to year. The **safe** or **firm yield** is the maximum quantity of water that can be guaranteed during a critical dry period. The critical period is often taken as the period of lowest natural flow on record of the stream. Hence, there is a finite probability that a drier period may occur with a yield even less than the safe yield.

17.12 RESERVOIR SILTATION

The important factors that affect the rate of silting in a reservoir are:

1. the quality, quantity, and concentration of sediment brought down by the river,
2. the size of the reservoir,
3. the length of the reservoir,
4. the steepness of the thalweg,
5. the ratio of the reservoir capacity to the annual runoff,
6. the method of reservoir operation,
7. the nature of the spillway,
8. the exposure of depleted material,
9. the depth and age of the sediment deposited, and
10. the depth and age of the head of the reservoir.

17.13 EXAMPLES

Example 17.1: Figure 17.4 shows the location of Somerville Lake, in east central Texas, between Austin and Houston. It is one of the many reservoirs within the Brazos River basin. The US Congress authorized the construction of Somerville Lake for flood control, water conservation, and other multipurpose uses with the passage of the Flood Control Act of September 3, 1954. The construction of Somerville Dam began on June 4, 1962 and completed on October 27, 1967. The Sommerville Dam and Somerville Lake are owned by the US Government and operated by USACE, Fort Worth District.

Figure 17.5 shows the area–capacity curve of Somerville Lake. The outlet works consist of a gated conduit of 10-ft diameter. The controls are two tractor-type gates, each 5×10 ft. The invert elevation of the conduit is 206.0 ft above mean sea level (MSL). Figure 17.6 shows the rating curves for the outlet works. The spillway is an ogee spillway with no control. The length of the crest is 1250 ft, and the crest elevation is 258.0 ft above MSL. Figure 17.7 shows the rating curve for the spillway. The full reservoir level (FRL) is 274.5 ft. The top of the dam is 280.0 ft above MSL.

The evaporation curves for Somerville Lake are shown in Figure 17.8. The bathymetric map of Somerville Lake, based on a survey conducted by the Texas Water Development Board in 2012, is shown in Figure 17.9. Table 17.1 gives the surface-area and total capacity of the reservoir based on surveys conducted in five different years.

What reservoir operation rules should be recommended?

Solution: The following reservoir operation rules were observed:

1. No flood control releases are made when the lake level is at or below the top of the conservation pool elevation of 238 feet.
2. Releases from the conservation storage are made as requested by the water management authority. The general guidelines are given in Table 17.2.
3. When pool level is between 238 and 258 feet, gated releases will not contribute to exceed the downstream control point capacities, as shown in Table 17.3.
4. When the pool elevation is between 258 and FRL release should still comply with Table 17.3.
5. All flood gates will be fully closed when the spillway flow equals or exceeds 2500 ft^3/s.

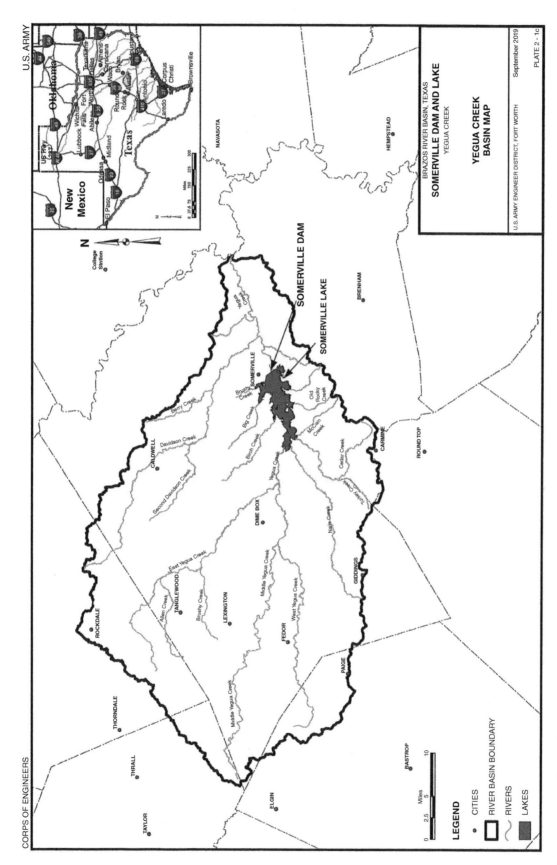

Figure 17.4 Location of Somerville Lake. (A black and white version of this figure will appear in some formats. For the color version, please refer to the plate section.)

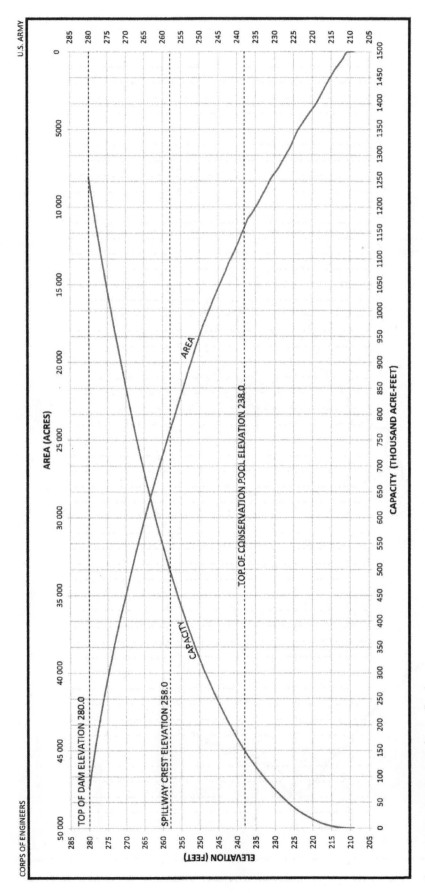

Figure 17.5 Area–capacity curve of Somerville Lake.

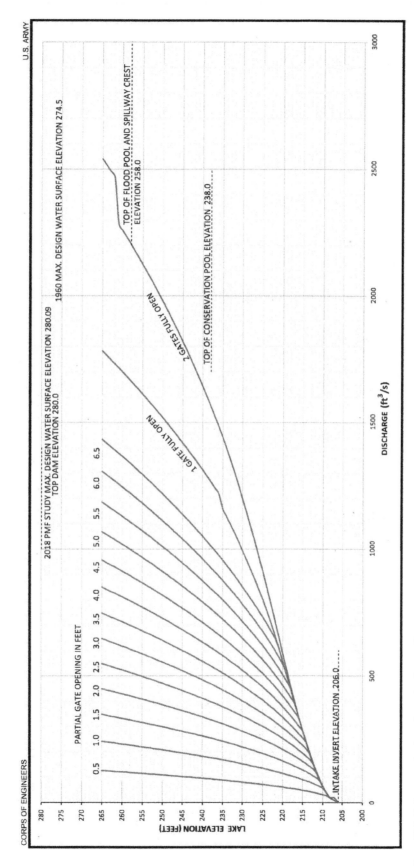

Figure 17.6 Rating curves for the outlet works of Somerville Dam.

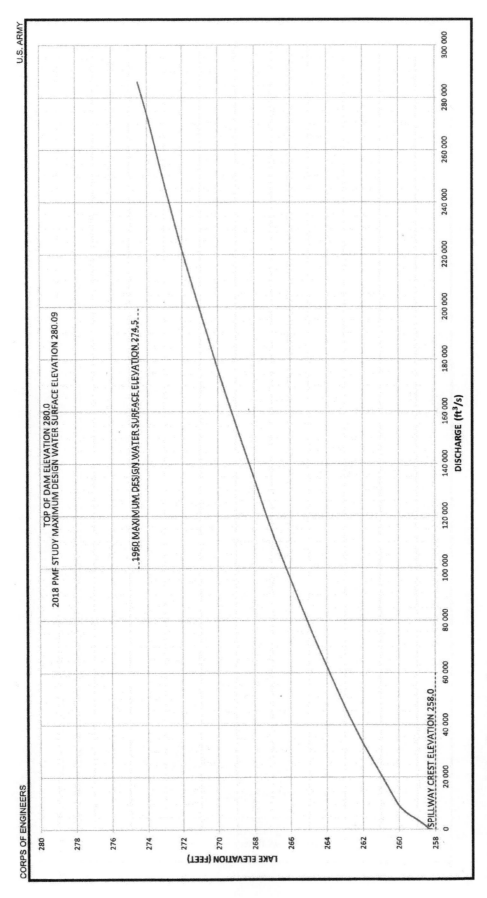

Figure 17.7 Rating curve for the spillway at Somerville Dam.

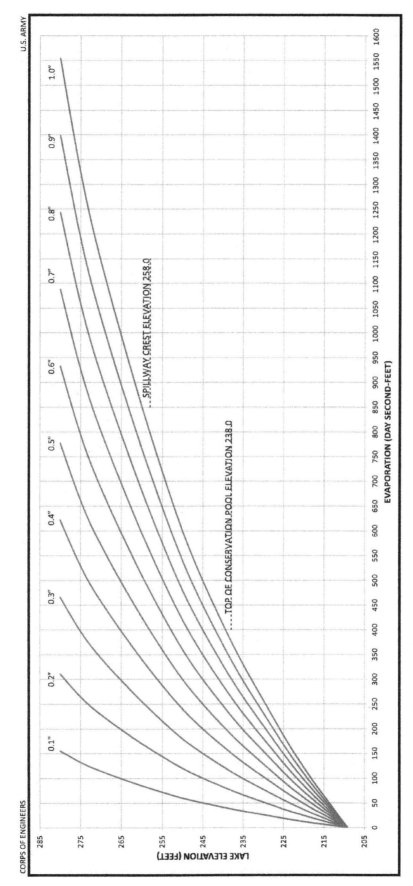

Figure 17.8 Evaporation curve for Somerville Lake (one day second-feet = 1.9835 acre-feet).

628 17 Reservoir Operations

Table 17.1 *Surface area and volume of Somerville Lake based on the survey years*

Survey year	Surface area (acres)	Total capacity (acre-feet)
1967	11 656.00	164 974.00
1992	11 591.00	159 682.00
1995	11 456.00	156 184.00
2003	11 127.00	149 072.00
2012	11 395.00	150 293.00

Table 17.2 *Low flood pool typical release rates*

Pool elevation range (ft)	Percentage of flood storage	Daily average release rates (ft^3/s)
238.0–238.5	0.0–1.7	95–185
238.5–239.0	1.7–3.4	270–505
239.0–239.5	3.4–5.1	270–505

Table 17.3 *Downstream control points*

River channel location	Control capacity (ft^3/s)
Yegua Creek near Somerville	2500
Brazos River near Hempstead	60 000
Brazos River at Richmond	60 000

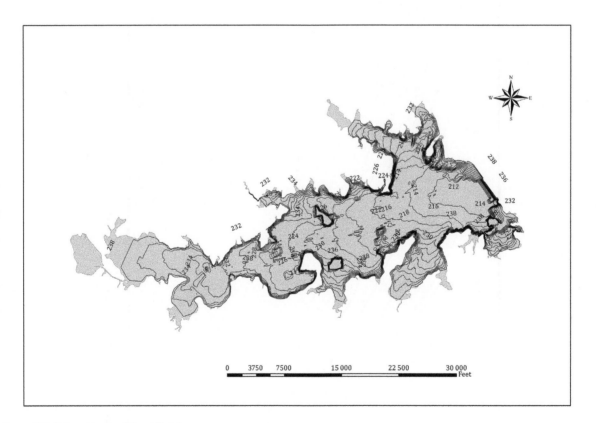

Figure 17.9 Bathymetric map of Somerville Lake.

Exercises: A selection of exercises on this topic is available at www.cambridge.org/appliedhydrology.

18 Climate Change

18.1 CONCEPTS AND PURPOSE

Climate change refers to long-term shifts in temperatures and weather patterns. These shifts may be natural, such as through variations in the solar cycle. However, since the 1800s, human activities have been the main driver of climate change, primarily due to the burning of fossil fuels like coal, oil, and gas. Climate change is synonymous with **global warming**, which refers to the rise in the temperature of the Earth's atmosphere near the surface. Even though there have been advocacies against global warming from certain corners, in the scientific literature, there is almost an unequivocal consensus that global surface temperatures have increased in recent decades and that the trend is caused by human-induced emissions of **greenhouse gases**, including water vapor, carbon dioxide, methane, nitrous oxide, and ozone. While the Earth's climate has changed throughout its history, the current warming is happening at a rate not seen in the past 10 000 years. According to the Intergovernmental Panel on Climate Change (IPCC, 2007), "Since systematic scientific assessments began in the 1970s, the influence of human activity on the warming of the climate system has evolved from theory to established fact." According to the National Centers for Environmental Information Annual 2022 Global Climate Report, the global yearly average temperature has been warming at a rate of 0.18 °C (0.32 °F) per decade since 1981. No scientific body of national or international standing disagrees with such a view. From the previous chapters, it should be obvious by now that the form of energy that runs the engine of the hydrologic cycle is heat. Therefore, if there is a dramatic change in the influx of heat energy within a system at various stages of the hydrologic cycle, there are bound to be hydrologic consequences of climate change. This chapter touches upon the plausible *hydrologic impacts of climate change*. Any design and planning of major water resources projects therefore should factor in the foreseeable effects of climate change on hydrologic processes.

18.2 CLIMATE AND HYDROLOGIC PROCESSES

Any unusual weather pattern observed at any location in recent decades is linked to the greenhouse effect and its enhancement by humans. The difficulty to establish any real link between an unusual hydrologic phenomenon and climate change lies with the fact that hydrologic systems naturally contain a huge amount of variability. The range of potential changes in climate is complex, particularly in varied landscapes with varied climatologic settings. The extreme events that have been experienced in different parts of the world in recent decades may be part of that natural variability, or they are being pushed to further extremes by climate change. It is likely that such connections cannot be verified with great certainty until it is too late to try and do anything about climate change.

Estimates of how global climate has changed over the past decades and centuries as well as predictions of how climate may change in the coming decades and centuries have been provided by several studies (e.g., IPCC, 2007). In general, these studies have employed both empirical models, based on historical climatic data, and various climate models, such as **global circulation models** (GCMs),[1] to make future projections (e.g., Thodsen, 2007). Virtually all of such currently available studies imply a warming trend of the atmosphere near the surface of the Earth and thus make global warming of the climate system almost undeniable. Historical records of global average air temperatures yield an estimate that can be best described as a linearly increasing trend in global near-surface air temperatures from 1960 to present. This trend shows a general warming in the likely range of 0.56–0.92 °C, with an average of 0.74 °C, and a more rapid warming trend over the past 50 years (Bates et al., 2008). For example, according to Lozan et al. (2001), global temperatures have increased by 0.6 ± 0.2 °C since 1990.

The IPCC (2007) predicted that the Earth may experience a global surface temperature rise of 0.2 °C per decade over the next 100 years. Even if the concentrations of all greenhouse gases and aerosols were kept constant at year 2000

---

[1] A GCM is a type of climate model. It employs a mathematical model of the general circulation of a planetary atmosphere or ocean. It uses the Navier–Stokes equations on a rotating sphere with thermodynamic terms for various energy sources (radiation, latent heat). These equations are the basis for computer programs used to simulate the Earth's atmosphere or oceans.

Table 18.1 *Predicted impacts of climate change on hydrologic processes (IPCC, 2007)*

Hydrologic events	Predicted impact
River flows	There is a high level of confidence that by the mid-twenty-first century, the annual average river runoff and water availability will increase by 10–40% at high latitudes and in some wet tropical areas, and decrease by 10–30% over some dry regions at mid-latitudes and in the dry tropics, some of which are presently water-stressed areas.
Extreme events (floods and droughts)	There is a high level of confidence that drought-affected areas will increase in extent. Heavy precipitation events, which are very likely to increase in frequency, will augment flood risk.
Snow and ice cover	There is a high level of confidence that, over the course of the twenty-first century, water supplies stored in glaciers and snow cover will decline.
Forest production	There is medium confidence that, globally, commercial timber productivity will rise modestly with climate change in the short to medium term, with large regional variability around the global trend. This may affect water yield from forest-covered catchments.
Coastal flooding	There is very high confidence that many millions more people will be flooded every year due to sea-level rise by the 2080s. The numbers affected will be largest in the mega-deltas of Asia and Africa, while small islands will be especially vulnerable.

levels, a further warming of about 0.1 °C per decade would be expected. Linked to this prediction are an increase in sea level of 15–95 cm and changes in the temporal and spatial patterns of precipitation. All these predicted changes will influence the hydrologic cycle in some way, but it is difficult to establish exactly how much. The IPCC predictions on impacts on hydrologic events are shown in Table 18.1.

At a very simple level, a temperature rise would lead to greater evaporative demand from the atmosphere, which, in turn, puts more water into the atmosphere. This may lead to higher precipitation rates, or at least changes in precipitation patterns. How this impacts the hydrology of an individual river basin is very difficult to predict. The most common method for making such predictions is to take the broad-brush predictions from a GCM, often at scale of 1° latitude and 1° longitude per grid square, and downscale it to the level of a river basin of concern. There are several methods used to downscale data. Wilby et al. (2000) showed that the choice of the method used could dramatically influence model predictions.

The hydrologic consequences of warmer temperatures also include less snowpack and a decrease in glaciated areas. In general, snowpack depth, extent, and duration are expected to decrease, particularly at lower and middle elevations, because of a combination of less precipitation falling as snow (Pierce et al., 2008) and the generation of snowmelt slightly earlier than the usual summer melting period (Luce et al., 2014). The degree of change expected because of warming varies considerably over the landscape as a function of temperature (Luce et al., 2014). Places that are warm (near the melting point of snow) are expected to be more sensitive than places where temperatures remain subfreezing throughout much of the winter despite warming (Woods, 2009).

18.3 EVAPOTRANSPIRATION

The relationship between evapotranspiration and warming climate is complicated (Roderick et al., 2014). Warmer air can hold more water, which means that even if the relative humidity stays constant, the vapor pressure deficit (Δe) increases. That difference between the actual water content of the air and that at the saturation drives a water vapor gradient between leaves and the atmosphere, which can draw more moisture out of the leaves. This is likely to cause more evapotranspiration in a warmer climate (Dai, 2013; Cook et al., 2014). However, evaporation is an energy-driven process, and there is only so much additional energy that will be available for evaporation. In addition, both the water balance and the energy balance must be considered together when accounting for future warming (Roderick et al., 2015). The observation that temperatures are warmer during drought is related more generally to the lack of water to evaporate, leading to warmer temperatures, than to warmer temperatures causing faster evaporation (Yin et al., 2014). When potential evapotranspiration (PET) models based on air temperature, including the Penman–Monteith method, are applied as postprocessing to GCM calculations, an overestimate of increased evapotranspiration is likely, because the energy balance is no longer tracked (Milly, 1992; Milly and Dunne, 2011). The reality is that most of the increased energy from increased longwave radiation will result in warming rather than increased evaporation (Roderick et al., 2015).

18.4 PRECIPITATION

Precipitation has a direct effect on hydrologic processes. However, because of the uncertainties surrounding climate change projections, it is difficult to accurately project the effects of precipitation on many hydrologic processes, such as floods, hydrologic drought, snow accumulation, groundwater recharge, etc. (Blöschl and Montanari, 2010; IPCC, 2013).

Two primary concepts are applied for precipitation change: aerodynamic, which refers to changes in wind and atmospheric circulation; and thermodynamic, which refers to how much water the air can hold (Seager et al., 2010). Aerodynamic drivers of precipitation change include changes in global circulation patterns, such as the Hadley cell extent, and changes in mid-latitude storm tracks. Changes in teleconnection patterns like the North American Monsoon System falls into this category. Thermodynamic changes reflect the fact that the atmosphere can hold more water (Held and Soden, 2006), leading to an expectation of about a 3.9% increase in precipitation per 1 °F of temperature change. There are, however, other physical limits to the disposition of energy driving the cycling of water in the atmosphere. These lead to estimates as significant as less than 1% per 1 °F of temperature change at the global scale, with individual grid cells being less or potentially negative, particularly over land (Roderick et al., 2014). Different approaches to scaling the thermodynamic contribution are a reason for differences among models, although the dynamic process modeling differences can be great as well. One outcome of thermodynamically driven changes is that, when precipitation occurs, the same total volume is expected to fall with greater intensity, leading to shorter events and longer dry periods between events. The number of consecutive dry days is projected to increase across the western United States (Muir et al., 2018). In the Pacific northwest of the United States, this projection is connected to an expected decrease in summer precipitation, but for the southwestern United States, it is more likely connected to a decrease in monsoonal moisture during the late spring (Muir et al., 2018).

Changes in orographic enhancement of precipitation over mountainous areas also have dynamic effects. Historical changes in westerly wind flows have led to a decrease in the enhancement of winter precipitation by orographic lifting over mountain ranges (Luce et al., 2013), raising the question of whether such a pattern may continue. Westerly winds are strongly correlated with precipitation in mountainous areas, but valley precipitation is not. The historical trend in westerlies was driven by pressure and temperature changes spatially consistent with those expected under a changing climate; however, the rapidity of changes in the last 60 years may have been partly enhanced by normal climatic variability.

In non-temperate regions of the world, predictions vary due to climate change. Parry (1990) suggested that rainfall in the Sahel region of Africa would stay at current levels or possibly decline by 5–10%. He also suggested a 5–10% increase in rainfall for Australia, although this may have little effect on streamflow when linked with increased evaporation from a 2 °C temperature rise. Chiew et al. (1995) highlighted large regional variations in predictions of hydrologic change in Australia. The wet tropical regions of northeast Australia are predicted to have an increase in annual runoff by up to 25%, a 10% increase in Tasmania, and a 35% decrease in South Australia by 2030. The uncertainty of this type of prediction is illustrated by southeast Australia where there are possible runoff changes of ±20% (Chiew et al., 1995). Similarly, Kaleris et al. (2001) attempted to model the impacts of future climate change on rivers in Greece but concluded that the error of the model is significantly larger than climate change impacts and therefore no firm conclusions could be made. Overall, it is difficult to make specific predictions of changes in hydrology, as the feedback mechanisms under climate change are not properly understood.

18.5 SNOWPACK

Snowpack declines are among the most widely cited changes occurring with climate change, through the effect of warmer temperatures on the fraction of precipitation falling as snow (Barnett et al., 2008). The areas of a river basin that are covered by perennial snow and ice form a vitally important component of the hydrologic cycle where contributions from snow, ice, and glacial melts to river flows are significant. Such areas include glaciers and perennial as well as seasonal snow-covered areas within the basin. Those areas have recorded the traits of global warming because they are widespread in space and are extremely sensitive to climate change. As such, the cause of a large-scale decrease of snow cover and extensive retreat of glaciers observed throughout the global land surface is attributed to global warming of the climate (Oerlemans, 2005).

About 70% of the water supply in the western United States is tied to mountain snowpacks (Service, 2004). So, changes in snowpack are highly relevant to municipal and agricultural water supplies and timing (Stewart et al., 2005). Historical trends in snowpack accumulation have been negative across most of the western United States (Mote et al., 2005; Regonda et al., 2005). Temperature sensitivity of the snowpack is highest in places that are already relatively warm (warm snowpacks), and warm snowpacks with high precipitation are likely to undergo some of the largest changes in snow storage as the climate warms (Nolin and Daly, 2006; Luce et al., 2014). The most sensitive locations in the western United States include the eastern Sierra Nevada and mid- to lower-elevation areas across Idaho, Utah, and Nevada. In contrast, many interior portions of this region are cold enough to be relatively insensitive to warming and strongly sensitive to precipitation variation (Mote, 2006; Luce et al., 2014). At the coldest and highest elevations, in the Uinta, Teton, Wind River, and some central Idaho ranges, for instance, there could be increases in the snow water equivalent (*SWE*) values if precipitation increases (Rice et al., 2017). Despite warming temperatures, a large proportion of precipitation would still fall as snow in these areas. This means that the future of snow, and consequently hydrology in these regions, will depend on one of the more uncertain parts of GCM projections, namely, precipitation.

Precipitation uncertainty can be substantial, but it does not translate into equal uncertainty in snowpack changes everywhere. Muir et al. (2018) estimated the sensitivity of April 1 *SWE* values using data from 524 SNOTEL stations across the western United States in a space-for-time model (Luce

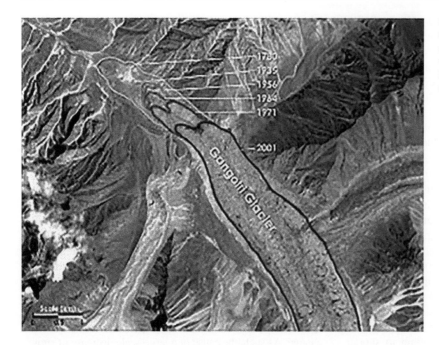

Figure 18.1 Retreat of Gangotri Glacier (1780–2001). Image source: https://earthobservatory.nasa.gov/. (A black and white version of this figure will appear in some formats. For the color version, please refer to the plate section.)

et al., 2014). This allowed them to determine where in the western United States snowpack was more sensitive to variability in precipitation or variability in temperature. They computed an index of uncertainty as the ratio, R_u, of the effects on snow, ΔS, from the likely range of precipitation values (about ±7.5% for 1 standard deviation across models) in the numerator to ΔS from the relatively certain temperature change in the denominator:

$$R_u = \frac{\Delta S \text{ across precipitation uncertainty } (\pm 7.5\%)}{\Delta S \text{ expected from warming}} \quad (18.1)$$

They found a strong certainty of large changes in April 1 *SWE* values for the Cascade Range, Sierra Nevada, and the Southwest ($R_u < 0.2$). However, they also found substantial uncertainty ($R_u > 0.6$) in outcomes for interior locations, such as the Greater Yellowstone Area and higher elevations in Idaho and Utah, where cold temperatures leave the snowpack more sensitive to precipitation than to temperature changes. The uncertainty ratio in these colder areas suggests that relatively large increases in precipitation could help counter the effects of warming on snowpack loss.

18.6 GLACIERS

Barry (1990) pointed out that the effect of global warming on the cryosphere of mountainous terrains is most visibly manifested in the shrinkage of glaciers and in reduced snow cover. Glacial retreat and decrease in perennial snow and ice cover have been observed and documented in various sections of the Himalayan range (e.g., Kulkarni et al., 2007; Raina, 2009), Tibetan Plateau (e.g., Ren et al., 2003; Prasad et al., 2009; Tandong et al., 2009; Jianchua et al., 2009), Tien Shan, Kunlun, Quilian, and other mountain ranges in central Asia (Shi et al., 2006; Nakawo, 2009; Severskiy, 2009; Haritashya et al., 2009). Similar observations have also been made in other mountainous regions, such as the Alps (Paul et al., 2007; Abermann et al., 2009; Keiler et al., 2010), Caucasus (Shahgedanova et al., 2009), and the Andes (Vuille et al., 2008). These trends have been attributed to global atmospheric warming and, in some cases, in combination with a decrease in precipitation, another probable manifestation of climate change. Gangotri Glacier in the central Himalayas, which is the feeding glacier of River Ganga, can be cited as an example of severe glacial retreat. This glacier is retreating at a fast rate owing to increasing temperature and climate change. It has been reported that the average rate of glacier retreat is 19 m per year (Naithani et al., 2001). Remote sensing images reveal that this glacier has been receding since 1780; however, the retreat has increased since 1971 (Figure 18.1). In the last 25 years of the twentieth century, it has shrunk more than 850 m (Sharma and Owen, 1996). This will have a serious impact on the flow characteristics of the Ganga (Bolch et al., 2012; Kääb et al., 2012). It has been reported that between 2003 and 2009 approximately 174 gigatons of water was lost by Himalayan glaciers, which led to severe floods in the Indus, Ganga, and Brahmaputra rivers, affecting millions of lives (Gardner et al., 2013; Laghari, 2013).

Estimating future changes in glaciers is complex (Hall and Fagre, 2003); but empirical relationships derived for glaciers in the Pacific northwest of the United States indicate a brief future for them, with many glaciers becoming fragmented or disappearing by the 2030s. Increasing temperatures yield a rising equilibrium line altitude (ELA), decreasing the effective contributing area for each glacier as warming progresses. Warming of 5.5 °F can translate to an elevation rise of 1000 to 1600 feet in snow–rain partitioning and summer temperatures. Those changes do not directly equate to a shift in the ELA, which depends on the geometry and topography of the contributing cirque. Temperate alpine glaciers are well known for being as, or more, sensitive to precipitation variations as they are to temperature variations (McCabe and Fountain, 1995), which has very likely contributed to changes in glacial dynamics across the Pacific Northwest. Westerlies and their contribution to winter precipitation have changed over the northern part of the region since the 1940s (Luce et al., 2013), and April 1 *SWE* values at these elevations and latitudes are relatively insensitive to temperature. However, summer temperature is a strong predictor of glacial behavior, and changes in summer temperatures could affect the melt rate and additional snow contributions in glaciers because these areas receive significant spring and summer precipitation (Hall and Fagre, 2003).

One particularly important region on the globe where hydrologic effects of climate change can be discernible is the mountainous terrains of the Himalayan and other ranges in central Asia. The Himalaya–Karakoram–Hindu Kush (HKH) region along with the adjoining Tibetan Plateau and central Asian mountain ranges such as Tien Shan and Kunlun ranges of western China, have the most highly glaciated areas and the largest body of ice outside the polar region. Significantly large, both in number and size, glaciers, and perennial snow- and ice-covered areas $(SCA_P)^2$ of the world outside the polar regions are present in this high-altitude terrain. Numerous rivers originate from this region and provide water for multipurpose usage, which includes agricultural irrigation, water supply, and hydropower generation to a vast population of Tibet, western China, Pakistan, Nepal, Bhutan, northern India, Myanmar, and Thailand. This region, aptly referred to as the Water Tower of Asia is the source of the 10 largest rivers in Asia, namely the Ganga, Indus, Brahmaputra, Tarim, Mekong, Irrawaddy, Amu Darya, Salween, Yangtze Kiang, and Huang Ho (Yellow River). About 1.4 billion people, constituting more than one-sixth of world's population inhabit the basins drained by these rivers and their tributaries. Consequently, water availability in these river basins has a profound influence on the livelihood of a huge human population on this planet.

The glaciers and snowfields of the HKH mountain belts and Tibetan Plateau are found to be amongst the fastest receding glacial and snow covers in the world (Dyurgerov and Meier, 2005; Prasad et al., 2009). In addition to glacial retreats, widespread fragmentation of the glaciers has also degraded the total areal coverage of perennial snow and ice in this region (Kulkarni et al., 2007; Raina, 2009). Compared to the attention paid to the problem of glacial degeneration, less detailed investigations have been directed toward the overall retreat of the ELA to higher elevations in various parts of the HKH region due to climate warming.

The most extensive glacier cover of the world outside Alaska and the Arctic is in the Himalayan and Karakoram mountains (Bolch et al., 2012). The majority of workers who examined the cryosphere of the HKH region through remote sensing techniques using satellite-derived data and imagery or conventional *in situ* measurements in the field, observed a shrinkage of glaciers and reduction of snow-covered areas (Berthier et al., 2007; Kulkarni et al., 2007; Kääb et al., 2012). The global average glacier mass balance is undeniably negative (Cogley, 2012). Most Himalayan glaciers are losing mass at rates similar to those observed elsewhere (Bolch et al., 2012; Kääb et al., 2012). However, the Karakoram Mountains, with a glaciated area of approximately 18 000 km^2, where the glaciers account for nearly 3% of the total of ice outside the ice sheets in Greenland and Antarctica, is possibly an exception (Bolch et al., 2012; Cogley, 2012), even though observations from the Karakoram are somewhat ambiguous. Field-based studies first indicated that glaciers in the Karakoram were either stable or had experienced a positive mass balance in the recent past. Hewitt (2005) found from field studies that 10 glaciers of intermediate size (9–31 km length) and the tributaries of the larger glaciers (60–68 km) like Hispar and Biafo were advancing. The only large glacier that also showed advancement was Baltoro (60 km length, with highest and lowest/terminus elevations at 8610 m and 3490 m, respectively). In general, the thickening in their ablation zone was on the order of 5–15 m. Hewitt (2007) noted that recessions were observed for almost all of the glaciers in the Karakoram, from the 1920s to the early 1990s, with the exception of some short-term advances in the 1970s. Since the late 1990s, there has been thickening and advancements in many non-surging glaciers. Rankl et al. (2013) observed that out of 1334 Karakoram glaciers, 134 showed advance, or surging behavior, with a marked increase since 2000. The *Karakoram Anomaly* refers to the observations that are in sharp contrast to those made in all other parts of the world where glaciers are shedding mass (Hewitt, 2005; Gardelle et al., 2012, 2013; Minora et al., 2013). At the same time, it should also be acknowledged that Hewitt's observations of glacial expansion were confined only to the highest and central part of the Karakoram. Increasing winter precipitation has been attributed to the possible glacier expansion in the Karakoram.

While Hewitt's observations come from the western Karakoram, observations from the eastern Karakoram are also uncertain. One of the most well-known glaciers in the eastern Karakoram is Siachen Glacier. It has the largest glacier area (~926 km^2) in the Karakoram. Satellite-derived

---

[2] A snow-covered area, designated by *SCA*, includes both seasonal and perennial snow and glacier covers whereas subscript *P* (*SCA$_P$*) refers only to that part of the *SCA* that is perennial. This part is most significant in contributions to river flows during the melting season (summer months).

images and snow cover data at different times clearly show the recession of this glacier in recent times (Rasul et al., 2008; Mukhopadhyay, 2012). Field-based glacial mass balance studies of Siachen Glacier conducted by Bhutiyani (1999) for a five-year period (1986–1991) showed a negative mass balance for all years except for one (1988–1989), which experienced comparatively heavy winter snowfall amounts and comparatively low temperatures during the ablation season. Unlike some of the high-altitude glaciers in the central part of the western Karakoram, the glaciers in the eastern Karakoram are mostly in the receding mode, particularly at lower elevations. At much higher elevations, there might have been no significant retreat in recent years (e.g., Raina and Sangewar, 2007).

Advancements of several glaciers have taken place in the Gangrigabu range to the southeast of Namcha Barwa within the Nyainqêntanglha Mountains of southeastern Tibet. Liu et al. (2006) used older aerial photos, Landsat Thematic Mapper imagery acquired in 1999 and China–Brazil Earth Resources Satellite (CBERS) imagery from 2001, geometrically corrected to topographical maps from 1980, to observe that 102 measured glaciers in the Gangrigabu range have all retreated between 1915 and 1980, with decreases of 47.9 km^2 and 6.95 km^3 in total area and volume, respectively. However, between 1980 and 2001, 41% of the glaciers in this region advanced and the rest retreated. This resulted in a total glaciated area decrease of 2 km^2, or only 0.25% of the area covered by glaciers since 1980. The net decrease in the volume of glaciers was 0.31 km^3. Liu et al. (2006) attributed this overall marginally negative mass balance of the glaciers, due to significant glacial advancements, to the substantial increase in precipitation in the Nyainqêntanglha mountain region since the mid-1980s. Shi et al. (2006) described glacial expansions in this region as one of the two *peculiar phenomena* observed in the Tibetan Plateau. The other phenomenon they describe is the decrease of approximately 0.6 °C in temperature in the northern Tibetan Plateau, as revealed from oxygen isotope studies of ice core records.

18.7 STREAMFLOW

Arora and Boer (2001) simulated the impacts of possible future climate change on the hydrology of 23 major river basins worldwide. They concluded that in warmer climates there may be a general reduction in the annual mean discharge, although as some rivers showed an increase this is not absolute. For mid- to high-latitude rivers they concluded that there may be huge changes in the timing of large runoff events that could be linked to changing seasonal times. This confirms the findings of Middelkoop et al. (2001), who predicted higher winter discharges of the River Rhine in Europe from intensified snowmelt and increased winter precipitation. In a similar vein, Wilby and Dettinger (2000) predicted higher winter flows for three river basins in the Sierra Nevada in California. These higher winter flows reflect changes in the winter snowpack due to a predicted rise in both precipitation and temperature for the region. One of the key predictions made from several investigations on the effects of climate change on snowmelt-dominated river flows is that, in a warmer climate, a combined effect of less winter precipitation (snowfall) and melting of winter accumulation occurring earlier in spring rather than in peak summer is the shifting of peak river runoff to late winter and early spring away from summer and autumn when demand is the highest (Barnett et al., 2005; Adam et al., 2009; Mukhopadhyay and Khan, 2015a). Such suggestions made from hydrologic climate model simulations are also supported by observed streamflow data from various river basins in the western United States (USGS, 2005). Regional studies in these western states suggest that hydrologic adjustments to a warming climate have been ongoing since the middle of the twentieth century.

Arnell and Reynard (1996) used models of river flow to try and predict the effects of differing climate change predictions on the river flows in twenty-one catchments in the U.K. Their results suggested a change in the seasonality of flow and considerable regional variation. Both these changes are by and large driven by differences in precipitation. The northwest of England is predicted to become wetter while the southeast becomes drier. Overall, it is predicted that winters will be wetter and summers drier. This may place a great strain on the water resources for southeast England, where the greater percentage of people live by far. In another study, Arnell and Reynard (2000) suggested that flow duration curves were likely to become steeper, reflecting a greater variability in flow. They also predicted an increase in flood magnitudes, which in the case of the Thames and Severn rivers in England had a much greater effect than realistic land use change. These changes in river flow regimes have important implications for water resource management in the future.

Barnett et al. (2005) discussed the shifting of timing of peak flows to early spring away from summer in snow-dominated regions due to global warming. Seasonal timing of river runoff in areas where snowmelt is a prime component of river flows is also highly sensitive to projected losses of snowpack associated with warming trends (Adam et al., 2008). These have already been manifested in the discharge records from the western parts of the United States and Canada (Cayan et al., 2001, USGS, 2005, Stewart et al., 2004, 2005, Barnett et al., 2008, Burn, 2008).

Streamflow changes of significance for aquatic species, water supply, and infrastructure include annual yields, summer low flows (average, extreme), peak flows (scouring floods), peak flow seasonality, and center of runoff timing. Irrigation water for crops and urban landscapes is typically needed in summer months. Annual yields, summer low flows, and the center of runoff timing are important metrics with respect to water supply, but they are most relevant to surface water supplies rather than groundwater supplies, although changes in long-term annual means could be informative for the latter. The mean summer yield (June through September) in the western United States is used for summer low flows. The center of runoff timing is the date on which 50% of annual runoff has flowed out of a basin and is an effective index for the timing of water availability in snowmelt-driven basins. Shifts to earlier runoff in the winter or spring disconnect the streamflow timing from water supply needs such as agricultural irrigation. The center of timing can be redundant with other metrics that measure the impact more directly but, with care in interpretation, it can help clarify different potential causal mechanisms, such as changing precipitation versus changing temperature. Peak flows are important to fish and infrastructure. Scouring flows can damage eggs in fish redds if they occur while the eggs are in the gravel or alevins are emerging. Winter peak flows can affect fall-spawning fish, whereas spring peak flows affect spring-spawning fish. Spring peak flows associated with the annual snowmelt pulse are typically muted in magnitude compared to winter rain-on-snow events, for two reasons. The rain-on-snow events can generate larger water input rates (rainfall precipitation plus high melt rates), and they tend to affect much larger fractions of a basin at a time, so scouring is less of a risk to spring-spawning fish. Consequently, a shift to more midwinter events can yield higher peak flow magnitudes, which can also threaten infrastructure like roads and recreation sites, as well as water management facilities such as diversions and dams.

Historical changes in some of these streamflow metrics have been examined in northern portions of the western United States, specifically earlier runoff timing (Cayan et al., 2001; Stewart et al., 2005) and declining annual streamflows (Luce and Holden, 2009; Clark, 2010). Declining low flows (7Q10, as discussed in Chapter 5) have also been observed in the western half of the northern Rockies (Kormos et al., 2016), associated more with declining precipitation than warming temperature effects for the historical period. Low-flow changes and timing changes in projections are generally associated with expected changes in the snowpack related to temperature, for example, more melt or precipitation as rain in winter, yielding a longer summer dry period.

Snow, ice, and glacial melts constitute a significant component of river flows in all the Himalayan river basins, even though reliable quantitative estimates of meltwater contributions to river flows are mostly unavailable for individual river basins. However, the percentages of meltwater from permanent snow and ice in the river discharges vary considerably from one river to other (Singh and Bengtsson, 2004; Barnett et al., 2005). For example, from employing a detailed hydrograph separation technique to seasonal hydrographs based on long periods of streamflow records, Mukhopadhyay and Khan (2014a, 2015a) calculated that glacial melt far outweighed snowmelt in the rivers draining the Karakoram and Zanskar ranges of the Greater Himalayas in the Upper Indus Basin (Figure 18.2). In the Karakoram, on an annual basis, the glacial melt proportion varies from 43% to 50%, whereas snowmelt varies from 27% to 31%. On the other hand, snowmelt dominates over glacial melt in the rivers draining the western Greater Himalayas and the Hindu Kush. Here snowmelt percentage in river discharge varies from 31% to 53%, whereas that of glacial melt ranges from 16% to 30%. In the main stem of the Upper Indus River, the snowmelt fraction in most cases is slightly greater than the glacial melt fraction. In the main stem, the snowmelt percentage ranges from 35% to 44%, whereas glacial melt percentage ranges from 25% to 36%. The Upper Indus River just upstream of Tarbela Reservoir carries annual flows consisting of 70% meltwater of which 26% is contributed by glacial melts and 44% by snowmelts. Consequently, any hydrologic response to global climate change, such as the retreat of glacial snouts and decrease in SCA_P, will have a profound effect on regional water availability and water resources management in this part of the globe (Tandong et al., 2009; Jianchu et al., 2009).

Mukhopadhyay and Khan (2015b) examined the trend of the time series of summer inflows (April–September) at Tarbela Reservoir located at the outlet of the Upper Indus Basin (Figure 18.2). Figure 18.3 shows the time series with the trend line. The linear regression, providing the trend line, has a gentle negative slope but a very low regression coefficient (Pearson's $r = -0.1497$; adjusted $r^2 = 0.00324$) indicating no trend. To corroborate this observation, they conducted a Mann–Kendall nonparametric test of the time-series data with the following results: Kendall's $\tau = -0.091$; Sen's slope -0.065 (confidence interval $[-3.068, 2.904]$), S-statistic -121.00, p-value (two-tailed) = 0.344, $\alpha = 0.05$, with null hypothesis H_0: there is no trend in the series and alternative hypothesis, H_a: there is a trend in the series. The small negative values of τ, S, and Sen's slope do indicate a negative trend. However, as the computed p-value is greater than the significance level, $\alpha = 0.05$, one cannot reject H_0. The risk of rejecting H_0 when it is true is 34.4%. From this statistical analysis, Mukhopadhyay and Khan (2015b) argued that the summer inflows at Tarbela Reservoir for the past 53 years simply show natural variations of river flows without any statistically significant declining trend. However, trend analyses of monthly rivers flows with long periods of record from the eastern, central, and western Karakoram show both increasing and decreasing trends, which are plausible manifestations of both negative and positive glacial mass balance (Mukhopadhyay and Khan, 2014b; Mukhopadhyay et al., 2015b). The net effect of these two contrasting trends can also cause the Tarbela inflows to exhibit no distinct trend, either rising or declining.

Another prediction for the effect of a warmer climate on flow regimes of rivers with dominant contributions from summer melts of seasonal and perennial snow and ice covers is that initially (in the early part of the twenty-first century) river flows will increase due to continued increase in the melting of perennial snow and ice covers but subsequently, as the depletion of high-altitude reservoirs continues, in the summer they will decline and will be dominated by a prevailing pluvial regime (rainfall–runoff) rather than a nival regime (snowmelt–runoff) (e.g., Rees and Collins, 2006).

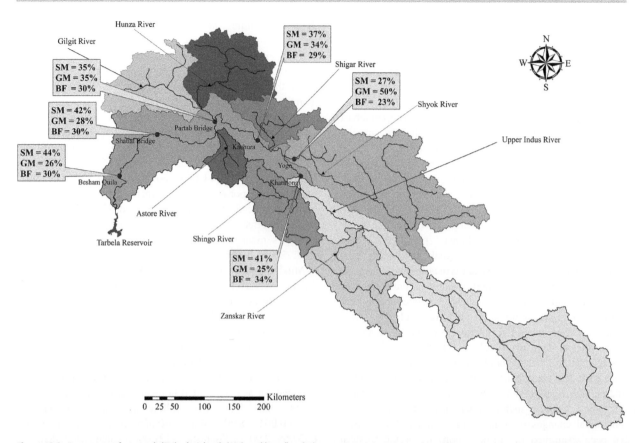

Figure 18.2 Percentages of snowmelt (SM), glacial melt (GM), and base flow (BF) at certain gauging stations along the main stem of Upper Indus River (the method of these estimations are given in Mukhopadhyay and Khan, 2015a). From Mukhopadhyay, B. and Khan, A. (2015b). Boltzmann–Shannon entropy and river flow stability within Upper Indus Basin in a changing climate. *International Journal of River Basin Management*, vol. 13, no. 1, 87–95. (A black and white version of this figure will appear in some formats. For the color version, please refer to the plate section.)

The general consensus amongst workers in the field of shrinkage of glaciers and perennial snow and ice covers is that this trend will continue in the future if climatic warming continues. The predicted hydrologic outcome of this cryospheric change is the initial increase in river discharge for a few years or decades and subsequent decline of the same with greater severity. Another hydrologic consequence of the depletion of perennial snow-covered areas in meltwater-dominated river systems that is expected to take place is nival regimes of flow progressively giving way to pluvial regimes, reflecting direct rainfall–runoff that mirrors rainfall and evapotranspiration rates (Braun and Hagg, 2009). However, Mukhopadhyay and Khan (2014b, 2015b) showed that in the central and eastern Karakoram summer flows showed rising trends, whereas in the western Karakoram the trend was declining.

Wigley and Jones (1985) analyzed the combined effects of changes in precipitation and evapotranspiration on streamflows. They expressed that effect by a very simple but quite powerful equation given as

$$R = \frac{R_2}{R_1} = \frac{\alpha - (1 - r_0)\beta}{r_0} \tag{18.2}$$

where R_2 denotes the streamflow due to changes in precipitation and evapotranspiration, R_1 denotes the same before such changes, α and β are the fractional (percentage) changes in precipitation and evapotranspiration, respectively, and r_0 is called the **runoff ratio**, which is conceptually similar to runoff coefficient of the rational method described in Chapter 11. Because

$$\frac{\partial R}{\partial \alpha} = \frac{1}{r_0} \tag{18.3}$$

and

$$\frac{\partial R}{\partial \beta} = -\frac{1 - r_0}{r_0} \tag{18.4}$$

the relative influence of changes in precipitation and evapotranspiration on changes in streamflows is given as

$$\left| \frac{\partial R}{\partial \alpha} \right| \bigg/ \left| \frac{\partial R}{\partial \beta} \right| = \frac{1}{r_0} \tag{18.5}$$

Equation 18.5 implies that the change in precipitation will have a much more amplified effect on streamflow than the change in evapotranspiration. However, as given in Mukhopadhyay (2013), the magnitude of change in R depends on the runoff ratio (Figure 18.4). For lower runoff

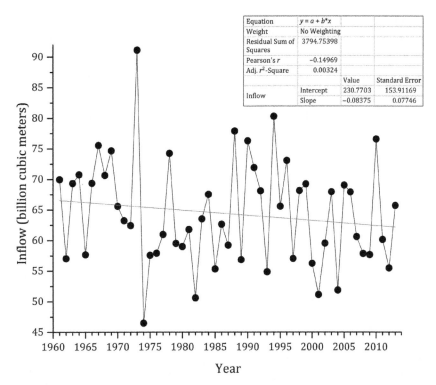

Figure 18.3 Summer inflows at Tarbela Reservoir from 1961 to 2013. The trend line shown is from a simple linear regression of the data with associated statistical parameters shown in the box. Revised from Mukhopadhyay, B. and Khan, A. (2015b). Boltzmann–Shannon entropy and river flow stability within Upper Indus Basin in a changing climate. *International Journal of River Basin Management*, vol. 13, no. 1, 87–95. Data are courtesy of Danial Hashmey (Water and Power Development Authority, Lahore, Pakistan).

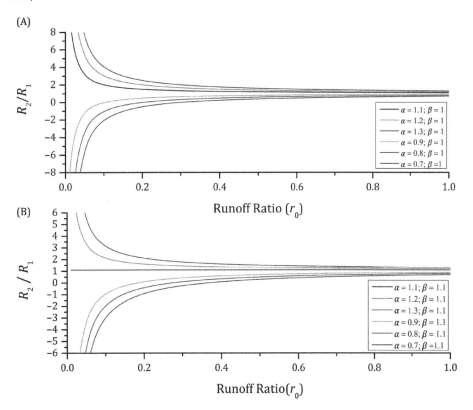

Figure 18.4 Solutions to Eq. 18.2 for a set of values for α and β. (A) The calculated curves represent changes in precipitation for a 30% decrease to a 30% increase in precipitation ($\alpha = 0.7$–1.3) with no change in evapotranspiration ($\beta = 1.0$). (B) The calculated curves represent changes in precipitation for a 30% decrease to a 30% increase in precipitation ($\alpha = 0.7$–1.3) with a 10% increase in evapotranspiration ($\beta = 1.1$). Note that if $\alpha = \beta$ then $R = \alpha$. Revised from Mukhopadhyay, B. (2013). Signature and hydrologic consequences of climate change within Upper-Middle Brahmaputra Basin. *Hydrological Processes*, vol. 27, 2126–2143. (A black and white version of this figure will appear in some formats. For the color version, please refer to the plate section.)

ratios, the changes in streamflows due to changes in precipitation are far more magnified than for higher runoff ratios. Similarly, the effects of changes in evapotranspiration on changes in streamflows are less than those caused by changes in precipitation for the entire range of runoff ratios. The runoff ratio depends on a variety of factors, such as land use, land cover, soil type, and precipitation amount and patterns. Another such factor is the **effective rainfall ratio**, p_0, which reflects that part of the rainfall that meets the evapotranspiration demand in an area. A rough approximation is

$$r_0 \approx 1 - p_0 \qquad (18.6)$$

18.8 URBAN CLIMATE CHANGE

Table 18.2 lists some of the climatic changes due to urbanization, expressed as a ratio between the urban and rural environments. This suggests that within a city there is a 15% reduction in the amount of solar radiation reaching a horizontal surface, a factor that will influence the evaporation rate. Studies have also found that precipitation levels in an urban environment are higher by as much as 10%. Atkinson (1979) detected an increase in summer thunderstorms over London, which was attributed to extra convection and condensation nuclei being available. Other factors greatly affected by urbanization are winter fog (doubled) and winter ultraviolet radiation (reduced by 30%).

Table 18.2 *Difference in climatic variables between urban and rural environments*

Climatic variable	Ratio of urban/rural environments
Solar radiation on horizontal surfaces	0.85
UV radiation: summer	0.95
UV radiation: winter	0.70
Annual mean relative humidity	0.94
Annual mean wind speed	0.75
Speed of extreme wind gusts	0.85
Frequency of calms	1.15
Frequency and amount of cloudiness	1.10
Frequency of fog: summer	1.30
Frequency of fog: winter	2.00
Annual precipitation	1.10
Days with less than 5 mm precipitation	1.10

Source: from Lowry (1967).

18.9 EXAMPLES

For the following examples, the temperature, precipitation, snow depth, snow cover, *SWE* values, PET, and surface runoff data were obtained from the 20th Century Reanalysis V3 (20CRV3) of NOAA. 20CRV3 provides monthly values for 1836 to 2015 that can cover the globe with 1 degree of spatial resolution. Glacier mass balance data were obtained from the World Glacier Monitoring Service (WGMS).

Example 18.1: Figure 18.5 shows increases in the near-surface temperature (i.e., 2 m above the surface) and PET under climate change and the accelerated increases thereof since 1960. Project the global averaged annual mean of monthly temperature (GT) using piecewise regression (PR). What will be the projected GT in 2040?

Solution: The PR is fitted to the GT for segments divided into 30-year periods as:

$$\text{GT}(t) = \begin{cases} -25.9 + 0.017(t), & 1836 \le t \le 1865 \\ 18.9 - 0.007(t - 1866), & 1866 \le t \le 1895 \\ -3.3 + 0.004(t - 1896), & 1896 \le t \le 1925 \\ -1.6 + 0.003(t - 1926), & 1926 \le t \le 1955 \\ -14.6 + 0.01(t - 1956), & 1956 \le t \le 1985 \\ -43.4 + 0.02(t - 1986), & 1986 \le t \le 2015 \end{cases} \qquad (18.7)$$

Using the fitted PR, the projected GT in 2040 is 6.81 °C.

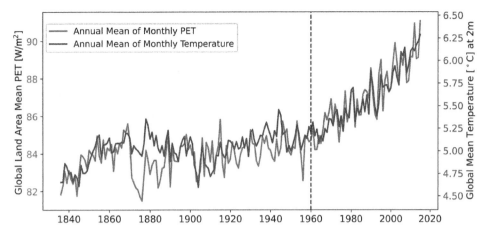

Figure 18.5 Global annual mean of monthly temperature (red solid line) and global land area annual mean of monthly PET (blue solid line). (A black and white version of this figure will appear in some formats. For the color version, please refer to the plate section.)

Example 18.2: Using the data presented in Example 18.1, project the global land area averaged annual mean of monthly PET (GPET) using PR. What will the GPET be in 2040?

Solution: PR is fitted to the GPET for segments divided into 30-year segments as

$$\text{GPET}(t) = \begin{cases} 20.0 + 0.04(t), & 1836 \leq t \leq 1865 \\ 109.3 - 0.014(t-1866), & 1866 \leq t \leq 1895 \\ 43.1 + 0.02(t-1896), & 1896 \leq t \leq 1925 \\ 63.6 + 0.01(t-1926), & 1926 \leq t \leq 1955 \\ -66.6 + 0.08(t-1956), & 1956 \leq t \leq 1985 \\ -111.7 + 0.1(t-1986), & 1986 \leq t \leq 2015 \end{cases} \tag{18.8}$$

Using the fitted PR, the projected GPET in 2040 is 92.23 W/m².

Example 18.3: Figure 18.6 shows the temporal and spatial variation of precipitation under climate change. Trends are calculated using the anomalies of annual mean of monthly precipitation as follows:

1. The annual mean of monthly precipitation is calculated for each grid.
2. The mean of annual mean of monthly precipitation is calculated for the base period (herein 1836–1900) for each grid.
3. Anomalies of the annual mean of monthly precipitation are calculated, based on the mean value for the base period for each grid.
4. The slope of linear regression for the anomalies of annual mean of monthly precipitation is calculated. When the slope is statistically different from zero at a 5% significance level, it is shown in the map.

(Note that the Mann–Kendall test for the annual mean of monthly precipitation (red solid line in Figure 18.6A) yields an increasing trend with a p-value of 0.0003.)

Project the global land area averaged annual mean of monthly precipitation (GP) using PR. Compute the projected GP in 2030.

Solution: PR is fitted to the GP for segments divided into 20-year periods as

$$\text{GP}(t) = \begin{cases} 13.3 - 0.004(t), & 1836 \leq t \leq 1855 \\ -11.7 + 0.009(t-1856), & 1856 \leq t \leq 1875 \\ 21.7 - 0.009(t-1876), & 1876 \leq t \leq 1895 \\ 10.2 - 0.003(t-1896), & 1896 \leq t \leq 1915 \\ -4.5 + 0.005(t-1916), & 1916 \leq t \leq 1935 \\ 4.2 + 0.0006(t-1936), & 1936 \leq t \leq 1955 \\ 9.8 - 0.002(t-1956), & 1956 \leq t \leq 1975 \\ -6.5 + 0.006(t-1976), & 1976 \leq t \leq 1995 \\ -3.5 + 0.005(t-1996), & 1996 \leq t \leq 2015 \end{cases} \tag{18.9}$$

Using the fitted PR, the projected GP in 2030 is 5.65 mm.

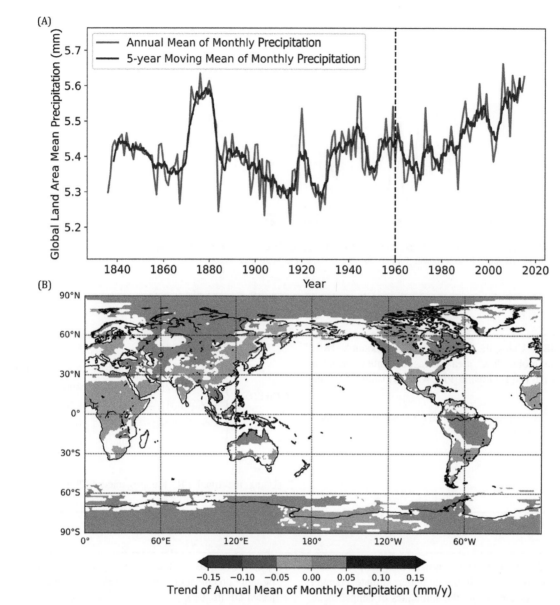

Figure 18.6 (A) Global land area annual mean (red solid line) and 5-year moving mean (blue solid line) of monthly precipitation and (B) the trend of the annual mean of monthly precipitation. The significant trends at a 5% significant level are colored. (A black and white version of this figure will appear in some formats. For the color version, please refer to the plate section.)

Example 18.4: Using the data presented in Example 18.3, project the annual mean of area mean monthly precipitation (AAP) for the Sahel region of Africa using PR. What will be the projected AAP in 2030?

Solution: PR is fitted to the AAP_Sahel for segments divided into 20-year periods as

$$\text{AAP\_Sahel}(t) = \begin{cases} -73.8 + 0.043(t), & 1836 \leq t \leq 1855 \\ -63.3 + 0.037(t - 1856), & 1856 \leq t \leq 1875 \\ 109.4 - 0.055(t - 1876), & 1876 \leq t \leq 1895 \\ -39.1 - 0.024(t - 1896), & 1896 \leq t \leq 1915 \\ 37.3 - 0.016(t - 1916), & 1916 \leq t \leq 1935 \\ 60.9 - 0.028(t - 1936), & 1936 \leq t \leq 1955 \\ 83.2 - 0039(t - 1956), & 1956 \leq t \leq 1975 \\ -2.7 + 0.004(t - 1976), & 1976 \leq t \leq 1995 \\ -19.6 - 0.007(t - 1996), & 1996 \leq t \leq 2015 \end{cases} \quad (18.10)$$

Using the fitted PR, the projected AAP for the Sahel region in 2030 is 4.4 mm, which is 0.8 mm lower than 1960.

Example 18.5: Figure 18.7 shows the temporal variation of global land area mean of snow cover, spatial variation of the trend of annual mean of snow depth, and spatial variation of the trend of annual mean of SWE. The Mann–Kendall test for the annual mean of snow cover (red solid line in Figure 18.7A) yields a decreasing trend with a p-value of ~0.0. The trends of annual mean of snow depth and *SWE* values are calculated following the same procedure as the trend of the annual mean of monthly precipitation.

Project the global land area averaged annual mean of snow cover (GSC) using PR. Compute the projected GSC in 2040.

Solution: PR is fitted to the GSC for segments divided into 30-year periods as

$$\text{GSC}(t) = \begin{cases} 85.6 - 0.019(t), & 1836 \leq t \leq 1865 \\ 37.8 + 0.007(t - 1866), & 1866 \leq t \leq 1895 \\ 73.0 - 0.012(t - 1896), & 1896 \leq t \leq 1925 \\ 41.0 + 0.005(t - 1926), & 1926 \leq t \leq 1955 \\ 62.6 - 0.006(t - 1956), & 1956 \leq t \leq 1985 \\ 101.9 - 0.026(t - 1986), & 1986 \leq t \leq 2015 \end{cases}$$

(18.11)

Using the fitted PR, the projected GSC in 2040 is 48.3%.

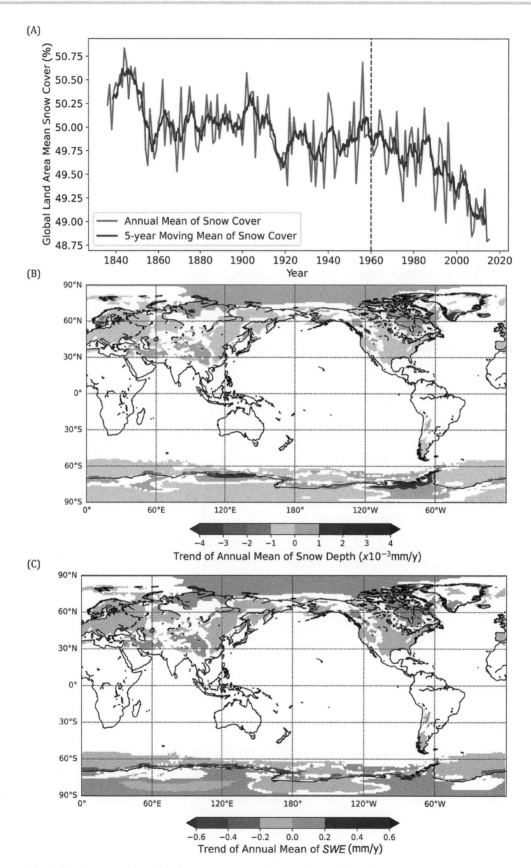

Figure 18.7 (A) Global land area annual mean (red solid line) and 5-year moving mean (blue solid line) of snow cover, (B) the trend of annual mean of snow depth, and (C) the trend of annual mean of the *SWE*. The significant trends at a 5% significant level are colored. (A black and white version of this figure will appear in some formats. For the color version, please refer to the plate section.)

Example 18.6: Using the data presented in Example 18.5, project the annual mean of area mean snow depth (GSD) for Hindu Kush Himalayan region using PR. Compute the projected GSD in 2040.

Solution: PR is fitted to the GSD for segments divided into 30-year long periods as

$$\text{GSD}(t) = \begin{cases} -486.8 + 0.28(t), & 1836 \leq t \leq 1865 \\ 180.1 - 0.08(t - 1866), & 1866 \leq t \leq 1895 \\ -99.5 + 0.07(t - 1896), & 1896 \leq t \leq 1925 \\ -32.7 + 0.04(t - 1926), & 1926 \leq t \leq 1955 \\ -153.9 + 0.09(t - 1956), & 1956 \leq t \leq 1985 \\ 899.1 - 0.43(t - 1986), & 1986 \leq t \leq 2015 \end{cases} \quad (18.12)$$

Using the fitted PR, the projected GSD in 2040 is 20.5×10^{-3}m, which is 20.7×10^{-3}m lower than in 1960.

Example 18.7: Figure 18.8 shows the average cumulative mass balance of glaciers worldwide from 1945 to 2014. The negative values represent a net loss of ice and snow compared with the base year of 1945.

Project the global annual mean of glacier mass balance (GGM) using PR. Compute the projected GGM in 2035.

Solution: PR is fitted to the GGM for segments divided into 20-year long periods as

$$\text{GGM} = \begin{cases} 772.0 - 0.39(t), & 1945 \leq t \leq 1964 \\ 77.34 - 0.03(t - 1965), & 1965 \leq t \leq 1984 \\ 1071.4 - 0.54(t - 1985), & 1985 \leq t \leq 2014 \end{cases} \quad (18.13)$$

Using the fitted PR, the projected GGM in 2035 is -38.3.

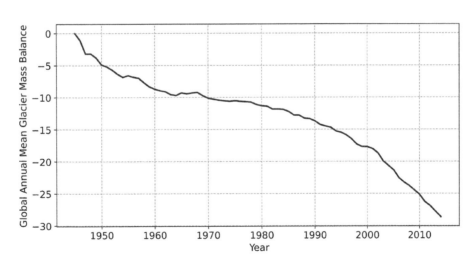

Figure 18.8 Global annual mean glacier mass balance.

Example 18.8: Figure 18.9 shows the trend of the annual mean of monthly surface runoff. Figure 18.10 shows the trend of annual DJF, MAM, JJA, and SON surface runoff.

Project the DJF area mean runoff for Nevada (DJF_Nevada) using PR. What will be the projected DJF in 2040?

Solution: DJF_Nevada is subjected to a 24-year moving average. Then PR is fitted to the moving averaged runoff for segments divided into 30-year long periods as

$$\text{DJF\_Nevada}(t) = \begin{cases} -7.6 - 0.004(t), & 1860 \leq t \leq 1889 \\ 1.9 - 0.0009(t - 1890), & 1890 \leq t \leq 1919 \\ -2.7 + 0.001(t - 1920), & 1920 \leq t \leq 1949 \\ -9.3 + 0.005(t - 1950), & 1950 \leq t \leq 1979 \\ -4.6 + 0.002(t - 1980), & 1980 \leq t \leq 2015 \end{cases} \quad (18.14)$$

Using the fitted PR, the projected DJF_Nevada in 2040 is 0.47 mm, which is 0.25 mm greater than in 1980.

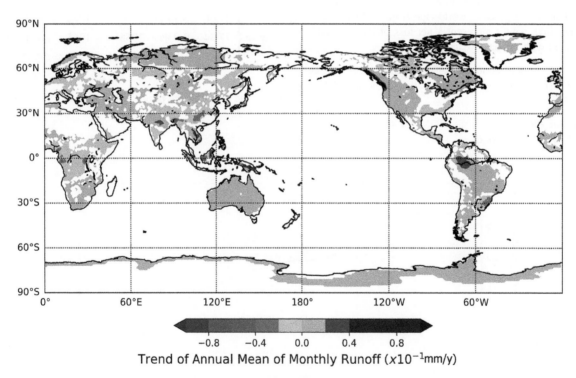

Trend of Annual Mean of Monthly Runoff ($\times 10^{-1}$ mm/y)

Figure 18.9 Trend of the annual mean of monthly surface runoff. The significant trends at a 5% significant level are colored. (A black and white version of this figure will appear in some formats. For the color version, please refer to the plate section.)

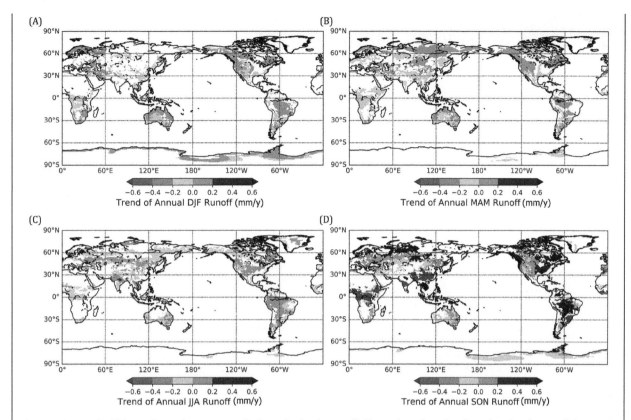

Figure 18.10 Trends of (A) annual December, January, and February (DJF) surface runoff, (B) annual March, April, and May (MAM) surface runoff, (C) annual June, July, and August (JJA) surface runoff, and (D) annual September, October, and November (SON) surface runoff. The significant trends at a 5% significant level are colored. (A black and white version of this figure will appear in some formats. For the color version, please refer to the plate section.)

Example 18.9: Using the data presented in Example 18.8, project the MMA area mean runoff for Nevada (MMA_Nevada) using PR. What will be the projected MMA in 2040?

Solution: MMA Nevada is subjected to a 24-year moving average. Then PR is fitted to the moving averaged runoff for segments divided into 30-year long periods as

$$\text{MMA\_Nevada}(t) = \begin{cases} -23.2 - 0.01t, & 1860 \leq t \leq 1889 \\ 5.6 - 0.003t - 1890, & 1890 \leq t \leq 1919 \\ -14.4 + 0.008t - 1920, & 1920 \leq t \leq 1949 \\ -4.1 + 0.003t - 1950, & 1950 \leq t \leq 1979 \\ -17.1 + 0.01t - 1980, & 1980 \leq t \leq 2015 \end{cases} \tag{18.15}$$

Using the fitted PR, the projected MMA_Nevada in 2040 is 1.37 mm, which is 0.6 mm greater than in 1960.

Exercises: A selection of exercises on this topic is available at www.cambridge.org/appliedhydrology.

19 Water Resources Management

19.1 CONCEPTS AND PURPOSE

It is often difficult to define the topic of water resources management (Davie, 2002). Is it concerned with all aspects of the hydrologic cycle or only with those of direct concern to humans, particularly water consumption and flood risk management? The term resource has a human dimension. Water is a resource because survival of humanity depends on it, and there are ways humankind can manipulate its provision. The word management implies the process of dealing with or controlling certain things or people. Therefore, water resources management is a very real topic for hydrologists to be cognizant of. Management of the water environment is not purely for its consumption. There are other uses of water that also need protection and management. In addition, while water is an invaluable resource, it also poses risks to society from flooding and pollution. So flood risk management is also an integral part of water resources management.

In this chapter certain hydrologic issues, by no means exhaustive, are explored with respect to the management of water resources and the changes that might be expected in the future arising from a compound effect of climate change, population growth, and continued sprawling of urbanization in major population centers. Broadly speaking, water resources management is the process by which decisions, intended to affect the future availability of water for beneficial usage or to manage risks of water related hazards, are implemented.

19.2 WHAT IS WATER RESOURCES MANAGEMENT?

Water resources management is the holistic approach to managing water supply and water risks to ensure sufficient quantity and quality to meet many competing demands, including drinking water and sanitation as well as energy production, food production, water transportation and navigation, recreation, and maintaining sustainable ecosystems and natural beauty. This includes planning, forecasting, and identifying the most effective and environmentally conscious methods of managing resources and maintaining equitable and cost-effective access to water. Management of the water environment also needs to be designed to maintain and enhance the amenity values of water.

Equally, managers of water resources have an obligation to protect the water environment for future generations and for other species that co-exist with the water. Therefore, water resources management needs to embrace sustainable development in its good practices. It is clear that water resources management has to embrace all of these issues and at the same time adapt to changing views on what is required of water management.

Human intervention, in various ways, can have a significant impact on almost all processes of the hydrologic cycle. For example, irrigation and changes in land use and land cover, such as vegetation, can impact rates of evapotranspiration. Land use change has a huge importance for water resources management, so that any decisions on land use need to include consultation with water resources managers.

A key part of water resources management involves water allocation: the amount of water made available to users, including both out of stream use, such as agricultural irrigation and municipal water supply, and instream environmental use, such as amenity values or recreation and supporting aquatic populations. Water allocation in a resource management context is concerned with how to ensure fair and equitable distribution of water resources between groups of stakeholders. For example, in Texas, the Texas Commission on Environmental Quality (TCEQ) uses the water availability models in evaluating water rights applications to help determine if water would be available for a newly requested water right or amendment, or if an amendment might affect other water rights. In South Africa, legislation introduced in 1998 designated that water for minimum human and ecological needs constitutes an untouchable reserve (Jaspers, 2001). This promotes human usage and instream ecology above other usages, for example, agricultural and industrial uses. In the United States, the way in which **water rights** are associated with land property rights means that there are many examples where farms have been bought specifically for the associated water right rather than the agricultural value of the land. This is particularly true in western states like Colorado, where water is a scarce resource. The city of Boulder, Colorado, steadily acquired agricultural water rights, which it has then used for municipal supply. In the 1990s, Boulder forfeited $12 million of water rights to ensure continuous flows in Boulder Creek (Howe, 1996). This was essentially a reallocation of Boulder Creek water in recognition of aesthetic and environmental needs ahead of human usage.

A key part of water resources management is the involvement of many different sectors of the community in decision making. This has led to a different approach to water management, which stresses integration between different sectors.

19.3 WATER AVAILABILITY

The distribution of water resources over the land mass of the Earth is uneven and is not related to population spread or economic development. This is revealed by analyzing and comparing the **specific water availability** for a single period of time for different regions and countries. Specific water availability represents the value of actual per capita renewable water resources and, for every design level, is determined by dividing the **gross water resources** by **population number**. In this context, water resources are assumed to be river runoff originating within a given region plus half the river flow that comes from outside. Thus, what is meant by specific water availability is the residual (after use) per capita quantity of fresh water. To asses water availability, long-term water balance calculations of river basins of a region are very important. Water availability planning requires a deep understanding of water resources, water rights, water conservation, and public policy solutions to navigate often contentious negotiations to assure future water supplies. Some specialized and sophisticated water availability and water management models, such as the Water Rights Analysis Package (WRAP) and Water Availability Modeling (WAM) System developed at Texas A&M University (Wurbs, 2020) and MODSIM developed at Colorado State University (Labadie, 2004), have been developed for simulation and predicting the amount of water that would be in a river or stream under a specified set of conditions, including future climatic conditions. Both of these models are in public domain. The WRAP/WAM system is the official model that is used by TCEQ. Another software is RiverWare, marketed by the Center for Advanced Decision Support for Water and Environmental Systems, for a licensing fee.

Water availability is also related the concept of a **safe yield**, which is often used in the determination of water supply. A safe yield is the maximum sustainable rate at which water can be withdrawn from existing sources without causing adverse impacts on ecology, environment, human health, economy, legal issues, or any other consequences.

A relatively simpler approach involving water balance calculations, which can be used for the estimation of gross water resources available for human usage in a river basin, is discussed below.

19.4 WATER BALANCE

Water balance is one of the fundamental concepts in hydrology that is highly relevant to water resources management. On the basis of the water balance approach, it is possible to make a quantitative evaluation of water resources in a river basin and how they change under the influence of human activities.

Water balance is the application of the principle of conservation of mass in hydrology. This is often referred to as the continuity equation, which states that, for any arbitrary volume and during any period of time, the difference between the total input and output will be balanced by the change of water storage within the volume. From this definition, it should be apparent that the spatial and temporal scale selected for the application of this principle is fundamentally important.

For water resources management, the water balance principle is usually applied to the spatial scale of a watershed or river basin. Consider the watershed shown in Figure 19.1. For any time period of length Δt, the water balance equation can be written as

$$\Delta S = P + G_{in} - Q - ET - G_{out} \qquad (19.1)$$

where P is the total precipitation that includes both rainfall and snowmelt, G_{in} is the groundwater inflow, Q is the streamflow, G_{out} is the groundwater outflow, ET is the evapotranspiration, and ΔS is the change in all forms of storage. The change in total water storage, ΔS, may be subdivided into several parts: changes of moisture storage in the soil or simply soil moisture (ΔM), in aquifers (ΔG), in lakes and reservoirs (ΔL), in river channels (ΔS_C), in glaciers (ΔS_G), and in snowpacks (ΔS_S). Thus, ΔS can be expressed as

$$\Delta S = \Delta M + \Delta G + \Delta L + \Delta S_C + \Delta S_G + \Delta S_s \qquad (19.2)$$

The water balance equation may be applied for any time interval such as daily, monthly, or yearly. The dimensions of all the quantities in Eqs. 19.1 and 19.2 are $[L^3]$ or $[L]$ if divided by the drainage area.

Various forms of water balance equations have been developed and applied to various spatial scales. One of the best-known water balance models is that developed by Thornthwaite (1948) and was further evolved by Thornthwaite and Mather (1955) and Mather (1978, 1979). This method usually runs on a monthly time scale and provides river runoff from a watershed. The reliable assessment of river flow characteristics is basic for the management and development of water resources in a basin.

The Thornthwaite-type water balance model requires only monthly average values of precipitation and temperature as inputs. Depending on the temperature, precipitation can be classified as snow or rain, and the snowmelt volume can be calculated using the temperature index method as described in Chapter 15. The model calculates the value of the potential evapotranspiration, designated in this chapter as *PET*, with temperature, using the Thornthwaite formula, as discussed in Chapter 14. However, any other method to estimate potential evapotranspiration can be used. For example, if radiation data are also available, a combination method can also be used. The actual mechanism of this type of model lies in the calculation of soil moisture storage, *ST*.

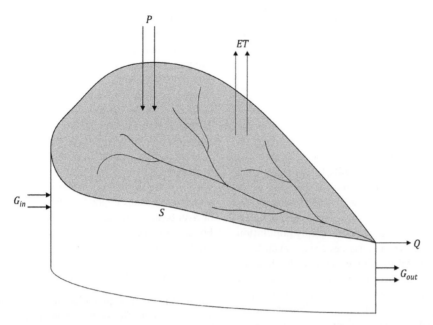

Figure 19.1 Schematic diagram of a watershed, showing the components of regional water balance: P = precipitation, ET = evapotranspiration, Q = streamflow, G_{in} = groundwater inflow, G_{out} = groundwater outflow.

In the original Thornthwaite model, the moisture status of the soil depends on the accumulated potential water loss, *APWL*. The difference in the values of potential evapotranspiration and precipitation, $(P - PET)$, is negative when there is a potential water deficit, while a positive $(P - PET)$ value represents a potential water surplus. If the $(P - PET)$ value is negative, the month is called a 'dry month' and it is subjected to the *APWL* value. When the $(P - PET)$ value is positive, the month called as a 'wet month', is subjected to a surplus value.

Available water capacity, designated as *AWC*, is a parameter used in water balance calculations. It is a simplified concept, first put forward by Veihmeyer and Hendrickson (1927), who assumed that the water readily available to plants is the difference between water content at field capacity, θ_{fc}, and permanent wilting point, θ_{wp}:

$$AWC = \theta_{fc} - \theta_{wp} \tag{19.3}$$

and is, therefore, an approximation of the range of available water that can be stored in soil and be available for growing crops. However, this concept was later to become the subject of some criticism. *AWC* is expressed as a volume fraction (0.30), as a percentage (30%), or as an amount (in mm or inches). An example of a volume fraction is water in mm (or inches) per meter or mm (or feet or inches) of soil profile. If a soil has an *AWC* value of 0.30, a 10 mm zone of the soil profile then contains 3 mm of available water.

The soil moisture term represents the amount of water held in soil storage. If the value of $(P - PET)$ is positive, the soil moisture storage value is the same as the *AWC*. On the other hand, if it is negative, the soil moisture storage is calculated as

$$ST = AWC \left(\exp\left[-\frac{APWL}{AWC} \right] \right) \tag{19.4}$$

The difference in soil moisture between months (ΔST) is calculated as

$$\Delta ST_i = ST_i - ST_{i-1} \tag{19.5}$$

A positive value of ΔST means there is enough water to add to the soil moisture storage, while negative value implies that water is removed from the storage because of evapotranspiration.

The difference between the actual evapotranspiration value (designated here as *AET*) and *PET* is in their relationship with soil moisture storage. The *PET* accounts for water removal from land surfaces only by atmospheric potential or heat, while *AET* accounts for changes in soil moisture storage in land surfaces. When the value P is higher than *PET*, soil moisture storage is saturated because of excess precipitation. Hence, the value *AET* equals *PET* because there are no changes in the soil moisture storage. When P is lower than *PET*, there are changes in the soil moisture storage. Thus, *AET* equals P subtracted by the changes in soil moisture storage:

$$AET = PET \text{ if } P > PET \tag{19.6}$$

$$AET = P - \Delta ST \text{ if } P < PET \tag{19.7}$$

The *APWL* is calculated as the cumulative sum of $(P - PET)$ values during months when $(P - PET)$ is negative. *APWL* increases during dry seasons. It is reduced during wet seasons because of soil moisture recharge. The value would be zero when soil moisture equals the *AWC* value of the soil.

The soil moisture deficit, D, is expressed as the difference between *PET* and *AET*. When soil moisture reaches the maximum soil moisture capacity, which is AWC, any excess precipitation becomes surplus (S).

$$D = PET - AET \tag{19.8}$$

$$S = P - PET \tag{19.9}$$

Thornthwaite and Mather (1955) suggested that in large watersheds for a given month, 50% of the surplus water becomes runoff and the remaining 50% is assumed to be detained and becomes runoff during the next month. Thus,

$$Q_i = 0.5 Q_i + 0.5 Q_{i-1} \tag{19.10}$$

In the Thornthwaite–Mather procedure of water balance, the $APWL$ is calculated by two different methods, depending on whether PET is greater than or less than the cumulative precipitation (P_{total}). For the months that PET is in excess of precipitation (i.e., the soil is drying out), the $APWL$ is incremented by the difference of PET and P given by

$$APWL_t = APWL_{t-\Delta t} + \left(\sum PET - \sum P\right) \tag{19.11}$$

where $APWL_t$ is the accumulated potential water loss at time t (mm), $APWL_{t-\Delta t}$ is the accumulated potential water loss at time $t - \Delta t$ (i.e., the previous month; mm), $\sum PET$ is the cumulative evapotranspiration over time period Δt (mm), and $\sum P$ is the cumulative precipitation over the time period Δt (mm).

In this method, the relationship between the $APWL$ and the amount of water stored in the root zone is expressed as

$$ST_t = ST_f \left[\exp\left(-\frac{APWL_t}{ST_f}\right)\right] \tag{19.12}$$

where ST_t is the available water stored in the root zone at time t (mm) and ST_f is the available water stored at field capacity in the root zone (mm).

For the months that PET is less than precipitation (i.e., when the moisture content increases and/or percolation occurs) the storage in the soil is incremented by the difference between PET and P,

$$ST_t = ST_{t-\Delta t} + \sum P - \sum PET \tag{19.13}$$

If the storage ST_t at time t is higher than the field capacity, percolation, $Perc_t$, is simply calculated as

$$Perc_t = ST_f - ST_{t-\Delta t} + \sum P - \sum PET \tag{19.14}$$

and the $APWL$ is set equal to zero. If, on the other hand, the moisture content in the root zone does not reach field capacity, the $APWL$ may be found by combining Eqs. 19.12 and 19.13 as

$$APWL_t = -ST_f \ln\left[ST_{t-\Delta t} + \sum P - \frac{\sum PET}{ST_f}\right] \tag{19.15}$$

and no percolation will occur.

Equations 19.11–19.15 are sufficient to describe the water balance in the root zone, unless there is an accumulation of snow. In that case, special rules apply.

McCabe and Markstrom (2007) from USGS developed a monthly water balance model referred to as the Thornthwaite water balance program. In this model, in addition to soil moisture storage, soil moisture storage withdrawal, designated as STW, is calculated:

$$STW = ST_{i-1} - \left[|P_{total} - PET| \times \left(\frac{ST_{i-1}}{STC}\right)\right] \tag{19.16}$$

where ST_{i-1} is the soil moisture storage for the previous month and STC is the soil moisture storage capacity. An STC value of 100–150 mm works for most locations. According to Eq.19.16, the value of SWT linearly decreases with decreasing ST such that as the soil becomes drier, and water becomes more difficult to remove from the soil and is less is available for actual evapotranspiration. The AET value is derived from PET, P_{total}, ST, and STW. If the sum of P_{total} and STW is less than PET, then a water deficit is calculated as $(PET - AET)$. If P_{total} exceeds PET, the AET value is equal to PET and the water in excess of PET replenishes ST. When ST is greater than STC, the excess water becomes surplus (S) and is eventually available for river runoff (Q). River runoff, Q, is generated from the surplus, S, at a specified rate (r factor). Following Thornthwaite and Mather (1955), an r factor value of 0.5 is commonly used.

Steenhuis and van der Molen (1986) extended the Thornthwaite–Mather procedure for calculating recharge from the soil moisture balance to use daily input values and to include the delay caused by percolation through the unsaturated zone. This model can be stated as

$$ST_t = ST_{t-\Delta t} + (P - AET - Q - Perc_t)\Delta t \tag{19.17}$$

where ST_t (mm) is the moisture storage of the topmost soil layer at time t, P (mm/d) is the precipitation, PET (mm/d) is the potential evapotranspiration, Q (mm/d) is the saturation excess runoff at the watershed outlet, $Perc_t$ (mm/d) is the percolation to the subsoil, and $ST_{t-\Delta t}$ (mm) is the previous time step storage of soil moisture and Δt is the time step.

During wet periods when either rainfall exceeds evapotranspiration (i.e., $P > PET$) or the moisture content is above field capacity, the moisture storage, ST_t is determined from the addition of the previous day's moisture $ST_{t-\Delta t}$ (mm), to the difference between P and PET during the time step. Conversely, when $P < PET$ and the moisture content of the soil is at or below field capacity (dry conditions), AET (mm) decreases linearly from the potential rate at field capacity to zero at the wilting point

$$AET = PET\left(\frac{ST_t}{AWC}\right) \tag{19.18}$$

The available soil water storage capacity per unit depth, AWC (mm), is defined as the difference between the soil storage at wilting point and at field capacity and varies according to the soil characteristic – porosity, bulk density, etc. Based on Eq. 19.18, the soil surface storage at time step, Δt, can be written as an exponential function:

$$ST_t = ST_{t-\Delta t}\left[\exp\left(\frac{P-PET}{AWC}\right)\right] \quad \text{when } P < PET \quad (19.19)$$

The *AWC* value can be considered uniform across the watershed when the water balance is applied in temperate climates (Steenhuis and van der Molen, 1986). Note that in Eq. 19.19 the value of *AWC* is given in unit of length (mm). A typical value of *AWC* often cited in the literature is 150 mm/m, which is equivalent to 0.15 mm. In monsoon climates, this assumption is not valid because the watershed dries out differentially, and runoff occurs only when the soil saturates, so that *AWC* is not uniform across the watershed. Thornthwaite and Mather (1955) suggested the determination of *AWC* values by considering land use, soil texture types and rooting depth by providing a water holding capacity table. For example, shallow rooted vegetation cover with a rooting depth of 0.5 m, the *AWC* value is estimated as 18.75 mm.

Easton et al. (2007) extended a water balance model for a variable-source (see Chapter 7) urbanized watershed in New York state in which, above field capacity, the excess moisture was partitioned between percolation and runoff. Percolation, $Perc_t$ (mm/d) is added to a groundwater reservoir, with groundwater storage, RS (mm/d), at each time step according to

$$RS_t = RS_{t-\Delta t} + (Perc_t - BF_{t-\Delta t})\Delta t \quad (19.20)$$

where BF is the baseflow (mm/d) to the stream and it was modeled as a linear reservoir (Chapter 8):

$$BF_t = RS_t \frac{[1-\exp(-a\Delta t)]}{\Delta t} \quad (19.21)$$

where $a\,(\text{d}^{-1})$ is the recession coefficient, a property of the aquifer, which can be calibrated from the baseflow recession curve, and Δt is the time step.

Vörösmarty et al. (1989, 1996) and Vörösmarty and Moore (1991) described a water balance model for large river basins represented by grids of climatic variables. The model simulates soil moisture variations, evapotranspiration, and runoff on single grid cells using biophysical datasets that include climatic drivers, vegetation, and soil properties. The state variables are determined by interactions among time-varying precipitation, potential evapotranspiration, and soil water content. The governing equations can be summarized as follows:

$$\frac{dW_s}{dt} = \begin{cases} -g(W_s)(ET_P - P_a) & \text{for } P_a \leq ET_P \\ P_a - ET_P & \text{for } ET_P < P_a < D_{ws} \\ D_{ws} - ET_P & \text{for } D_{ws} < P_a \end{cases} \quad (19.22)$$

$$ET_A = \begin{cases} P_a - \dfrac{dW_s}{dt} & \text{for } P_a < ET_P \\ ET_P & \text{for } ET_P < P_a \end{cases} \quad (19.23)$$

$$Q_r = \begin{cases} 0 & \text{for } P_a < D_{ws} \\ P_r - D_{ws} & \text{for } D_{ws} < P \end{cases} \quad (19.24)$$

$$Q_s = \begin{cases} 0 & \text{for } M_s < D_{ws} \\ M_s - D_{ws} & \text{for } D_{ws} < M_s \end{cases} \quad (19.25)$$

where W_s is the soil moisture (mm), $g(W_s)$ is a unitless soil drying function (see below), ET_P is the potential evapotranspiration[1] (mm/day), P_a is the precipitation (mm/d) available for soil recharge plus snowmelt, M_s, D_{ws} is the soil moisture deficit (mm/d) equal to the amount of water required within a time step to fill soil to its water holding capacity and simultaneously satisfy ET_P; ET_A is the estimated actual evapotranspiration, and Q_r and Q_s are the rainfall and snowfall excess, respectively, available for runoff and recharge of runoff detention pools. The unitless drying function is

$$g(W_s) = \frac{1-\exp(-\alpha W_s/AWC)}{1-\exp(-\alpha)} \quad (19.26)$$

where α is an empirical constant and *AWC* is the soil and vegetation-dependent available water capacity. Vörösmarty et al. (1996) set α equal to 5.0 such that the drying curve would resemble that of Pierce (1958) when $g(W_s) = ET_A/ET_P$ is plotted as a function of W_s/AWC during periods of no precipitation.

When rainfall exceeds the soil moisture deficit the excess is used to augment a rainfall-derived detention pool and to generate runoff:

$$\frac{dD_r}{dt} = (1-\gamma)Q_r - \beta D_r \quad (19.27)$$

$$R_r = \gamma Q_r + \beta D_r \quad (19.28)$$

where D_r is the rainfall runoff detention pool (mm) and R_r is the rainfall-derived runoff emerging from the grid cell. β and γ are empirical constants, set at 0.167 d^{-1} and 0.5, respectively.

Snowpack accumulates when the monthly temperature is below −1.0 °C. As detailed in Vörösmarty et al. (1989), snowmelt is a prescribed function of temperature and elevation. Snowmelt occurs at or above −1.0 °C. The expected monthly changes in the pools are calculated as the average daily change multiplied by n_d, the number of days in each month. Likewise, the associated water fluxes computed by the water balance model are initially expressed as a daily average for the duration of each month. These also are multiplied by n_d to obtain the corresponding monthly values. The time-varying changes in $W_s, D_r,$ and D_s are solved

---

[1] *PET* and ET_P (and *AET* and ET_A) are used interchangeably. But they denote the same variable: amount of either potential or actual evapotranspiration.

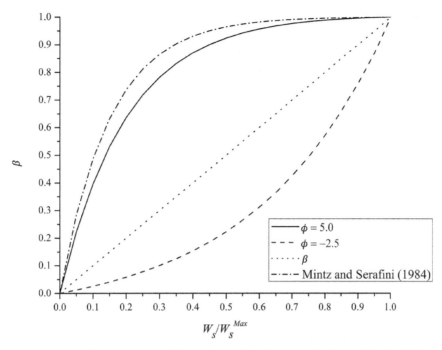

Figure 19.2 Graphs showing four types of soil moisture extraction functions. From Mukhopadhyay, B. and Singh, V. P. (2011). Hydrologic modeling at mesoscopic scales using global datasets to derive stream water availability models of river basins. In Shukla, M. K. (ed.), *Soil Hydrology, Land Use and Agriculture*, CAB International.

using a fifth- or sixth-order Runge–Kutta integration technique.

Mather (1974) presented a series of graphs to show the various types of relationships that can exist between the ratio of daily actual evapotranspiration (ET_A) to potential evapotranspiration (ET_P) and available soil moisture. Dyck (1983), on the other hand, listed several mathematical functions to describe such relationships. These functional relationships are useful for computational implementation. The following function, which can be termed as the soil moisture extraction function, can be used as a generalization:

$$\beta = \frac{1 - \exp\left[-\phi\left(\frac{W_s}{W_s^{Max}}\right)\right]}{1 - \exp(-\phi)} \quad (19.29)$$

where the value of parameter ϕ determines the nature of the daily moisture extraction. The ratio, W_s/W_s^{Max} or W_s^{Max}/AWC, provides a measure of the available soil moisture.

Figure 19.2 shows the function β for two different values of ϕ. For a value of 0.5, the relationship is almost linear, which can be simply expressed as $(ET_A/ET_P) = \phi(W_s/AWC)$. Mather (1974) showed three such linear relationships for which $\phi = 1, 1.44$, and 2.0. A linear relationship is defined by the function β as

$$\beta = \frac{W_s}{W_s^{Max}} \quad (19.30)$$

A linear relationship typifies a condition where the rate of evapotranspiration is directly proportional to the available moisture. In other words, the withdrawal of soil moisture storage linearly decreases with decreasing W_s such that, as the soil becomes drier, water becomes more difficult to remove from the soil and less available for evapotranspiration. For calculations using monthly time steps, Eq. 19.30 is a reasonable approach (McCabe and Markstrom, 2007).

The nonlinear behavior shown for a value of $\phi = +5.0$ is expected under moisture-stressed dry conditions, whereas the model with $\phi = -2.5$ depicts the soil-drying atmospheric humidity as being relatively high. Another function that has been defined by Mintz and Serafini (1984) and was later adopted by Willmott et al. (1985) in their global water balance calculations is given as

$$\beta = 1 - \exp\left[-6.68\frac{W_s}{W_s^{Max}}\right] \quad (19.31)$$

This function produces the relationship similar to that given by Eq. 19.29 with $\phi = 5.0$ (Figure 19.2).

An application of the water balance principle to arrive at water availability in a large river basin is given in Mukhopadhyay and Dutta (2010) and Mukhopadhyay and Singh (2011).

19.5 INTEGRATED WATER RESOURCE MANAGEMENT

Integrated water resources management (IWRM) is an empirical concept which was built up from the on-the-ground experience of practitioners. Many parts of the concept have been around for several decades, since the first global water conference in Mar del Plata in 1977. However, it was not until after Agenda 21 and the World Summit on

Sustainable Development in 1992, also known as the Earth Summit in Rio de Janeiro, Brazil, that the concept was made the object of extensive discussions as to what it means in practice. The Global Water Partnership (GWP) is a leading agency in promoting IWRM (www.gwpforum.org). The definition of IWRM given by the GWP is widely accepted. It states: "IWRM is a process which promotes the coordinated development and management of water, land and related resources, in order to maximize the resultant economic and social welfare in an equitable manner without compromising the sustainability of vital ecosystems." (Global Water Partnership, 2004).

IWRM helps protect the world's environment, foster economic growth and sustainable agricultural development, promote democratic participation in governance, and improve human health. Worldwide, water policy and management are beginning to reflect the fundamentally interconnected nature of hydrologic resources, and IWRM is emerging as an accepted alternative to the sector-by-sector, top-down management style that has been dominant in the past.

The basis of IWRM is that the many different uses of finite water resources are interdependent. High irrigation demands and polluted drainage flows from agriculture mean less fresh water for drinking or industrial use; contaminated municipal and industrial wastewater pollutes rivers and threatens ecosystems; if water has to be left in a river to protect fisheries and ecosystems, less can be diverted to grow crops. There are plenty more examples of the basic theme that unregulated use of scarce water resources is wasteful and inherently unsustainable.

IWRM is a cross-sectoral policy approach, designed to replace the traditional, fragmented sectoral approach to water resources and management that has led to poor services and unsustainable resource use. IWRM is based on the understanding that water resources are an integral component of the ecosystem, a natural resource, and a social and economic good.

The concepts behind IWRM lie in the so-called **Dublin Principles** (Davie, 2002). In January 1992, 500 participants, including government-designated experts from 100 countries and representatives of 80 international, intergovernmental and nongovernmental organizations attended the International Conference on Water and the Environment in Dublin, Ireland. The conference adopted what has been termed the **Dublin Statement**, which was taken forward to the Earth Summit Conference in Rio later that year. The Dublin Statement established four guiding principles for managing freshwater resources. These are as follows:

1. Fresh water is a finite and vulnerable resource, essential to sustain life, development, and the environment.
2. Water development and management should be based on a participatory approach, involving users, planners, and policy makers at all levels.
3. Women play a central part in the provision, management, and safeguarding of water.
4. Water has an economic value in all its competing uses and should be recognized as an economic good.

These four principles underlie IWRM, especially the concepts of a participatory approach and that water has an economic value (Solanes, 1998). An economic good, as used in principle four, is defined in economics as: a physical object or service that has value to people and can be sold for a non-negative price in the marketplace. A major implication from principle four is that water is not a gift or a free right to any water user, it needs to be recognized that using water restricts the usage by others and therefore there is a cost involved in the action.

The emphasis within an IWRM approach to water management is on integration between sectors involved in water resources, including local communities (a participatory approach). Although this is promoted as a new approach to resource management it is in many ways a return to traditional values, with recognition of the interconnectedness of hydrology, ecology, and land management (Davie, 2002). If there is a large amount of water from a stream allocated to agriculture, there is less water available for municipal water supply and environmental flow to sustain instream ecology. IWRM is a framework for change that recognizes this interconnection and builds structures to manage water with this in mind. It is an attempt to move away from structures that promote individual sectors competing against each other for the scarce resource of water and moves toward joint ownership of water resources management. The GWP suggested several approaches for use within IWRM as instruments for change to achieve an integrated management of water resources.

19.6 INTEGRATED RIVER BASIN MANAGEMENT

Integrated river basin management (IRBM) is essentially a subset of IWRM (Davie, 2002). It aims to promote an integrated approach to water and land management but with two subtle differences:

1. IRBM recognizes the watershed or river basin as the appropriate organizing unit for understanding and managing water-related biophysical processes in a context that includes social, economic, and political considerations.
2. There is recognition of the spatial context of different management actions and in particular the importance of cumulative effect within a river basin.

By defining a river basin as the appropriate organizing unit for managing biophysical processes there is a recognition that hydrologic pathways are important and these provide an appropriate management, as well as biophysical, boundary.

Cumulative effect refers to the way in which many small actions may individually have very little impact but when combined the impact may be large. This is true for a river basin system where individual point discharges of pollution may be small but, when accumulated within the river, they may be enough to cross an environmental threshold. Fenemor et al. (2006) defined the word 'integrated' in an IRBM context using three different connotations:

1. integration between the local community, science, and policy so that the community is linked into the planning and execution of both science and policy, and scientific research is carried out in an environment close-linked into policy requirements and vice versa;
2. integration between different scientific and technical disciplines to tackle multidimensional problems; and
3. spatial integration throughout a watershed so that the cumulative impact of different actions can be assessed.

Using this type of definition IRBM can be seen as a process that can be used to implement IWRM. One of the key principles of IRBM and IWRM is community involvement through a participatory approach: making sure that everybody can be involved in resource management, not just a few elites within a single organization. Another key principle of IRBM and IWRM is the idea of change. This ranges from extolling change in management structures to cope with modern resource management pressures to making sure the structures can cope with more inevitable changes in the future.

IRBM is promoted by UNESCO and the WMO through the Hydrology for the Environment, Life and Policy (HELP) program. Further details can be found at https://unesdoc.unesco.org/ark:/48223/pf0000214516.locale=en.

19.7 HYDROLOGY IN A CHANGING WORLD

The world is a dynamic system. This applies to both the natural Earth and the economic world. The geologic processes taking place both in the interior and on the surface of the Earth are constantly changing it. The Earth's climate is changing. Similarly, the economic and societal changes are globally true. Water is fundamental to all elements of human life and society on this planet. Then it can also be expected that hydrology is also constantly changing to keep up with the changing world. By the end of the twenty-first century people may be living in a different climate from now, their economic lives may be unlike the present population, and almost certainly their knowledge of hydrologic processes will be greater than at present. However, the hydrologic processes will still be operating in the same manner, although maybe at differing rates than those that we observe today. The previous chapters of this book have assessed these processes. Our knowledge of them will improve, and our methods of measurement and estimation will get better, but the fundamentals of the processes will still be the same.

In water resources management there is a problem concerning the statistical techniques that we use. In a frequency analysis technique, there is an inherent assumption that a storm event with similar antecedent conditions, at any time in the streamflow record, will cause the same size of storm. We assume that the hydrologic regime is stationary with time. Under conditions of land use or climate change it is quite possible that these conditions will not be met. This makes it difficult to put much faith in a technique such as frequency analysis when it is known that the hydrologic regime has changed during the period of record. These are the types of challenges facing water resources management in an ever-changing world.

19.7.1 The Role of Hydrologic Analysis in Water Resources Management

Hydrologic analysis for water resources management becomes imperative for the assessment of present and future (1) supply of water available from both surface and groundwater sources; (2) quality of water from surface and groundwater sources; (3) frequencies of flood potential to affect human activities; and (4) frequencies of drought or low stream flow.

The assessment of (1) can be done by water availability modeling or water balance studies described above; that of (2) can be done by regional flood frequency analysis described in Chapter 5; and the assessment of (3) is usually done by flow duration analysis, also detailed in Chapter 5.

When considering water supply or safe yield, water usage are classified into two types: (1) **consumptive use** and (2) **nonconsumptive use**. The former refers to the portion of withdrawn water that is either evaporated, transpired, or incorporated into a product or crop and is consequently unavailable for subsequent use. It may also include the portion of withdrawn water lost due to evaporation and leakage in transit, called **conveyance loss**. The latter category is the portion of withdrawn water that remains available for subsequent use. The portion of withdrawn water that is discharged into surface or groundwater is called the **return flow**. Thus, the conveyance loss can also be considered return flow.

19.7.2 Change in Land Use

The implications of land use change for hydrology have been an area of immense interest to researchers in hydrology over the last 50 or more years. Issues of land use change affecting hydrology include increasing urbanization, changing vegetation cover, land drainage, and changing agricultural practices leading to salination.

19.7.3 Vegetation Change

Trees have profound effects on evaporation and interception rates. Several studies estimated the amount of precipitation that is intercepted by different types of canopies. For example, Crockford and Richardson (1990) found in

Australia that the percentage of interception by a canopy cover of eucalypt forest is in the range of 5–26% and that for pine forest canopy cover it is in the range of 6–52%. The large range of values are from different size storms. The high interception losses were experienced during small rainfall events and vice versa. In contrast, the amount of annual interception in the Amazonian rain forest is only 9%, which is remarkably low, reflecting a high rainfall intensity and high humidity levels (Lloyd et al., 1988).

Hydrologic impacts of vegetation cover change were reviewed by Bosch and Hewlett (1982) in considerable depth. In general, they concluded that the greater the amount of deforestation the larger the subsequent streamflows will be, but the actual amount is dependent on the vegetation type and precipitation amount. Fahey and Jackson (1997) concluded that with the loss of forest cover both low flows and peak flows increased. The low flow response is altered primarily through the increase in water infiltrating to groundwater without interception by a forest canopy. The peak flow response is a result of a generally wetter soil and a low interception loss during a storm when there is no forest canopy cover. The time to peak flow may also be affected, with a more sluggish response in a catchment with trees. In a modeling study, Davie (1996) suggested that any changes in peak flow that result from afforestation are not gradual but highly dependent on the timing of canopy closure. There has been considerable debate in the hydrologic research literature as to how detectable the effects of deforestation are in large river catchments.

19.7.4 Urbanization

The continuing rise in urban population around the world makes it an important issue to consider under the title of change. There is no question that urban expansion has a significant effect on the hydrology of any river draining the area. Initially, this may be due to climate alterations affecting parts of the hydrologic cycle. The most obvious hydrologic impact is on the runoff hydrology, but other areas where urbanization may have an impact are point source and diffuse pollution affecting water quality, river channelization to control flooding, increased snowmelt from urban areas, and river flow changes from sewage treatment.

19.7.5 Runoff Change due to Urbanization

The changes in climate are relatively minor compared to the impact that impermeable surfaces in the urban environment have on runoff hydrology. Roofs, pavement, roads, parking lots, and other impermeable surfaces have extremely low infiltration characteristics, consequently Hortonian overland flow readily occurs. These surfaces are frequently linked to gutters and stormwater drains to remove the runoff rapidly. The result of this is far greater runoff and the time to peak discharge being reduced. Cherkauer (1975) compared two small catchments in Wisconsin, USA. The rural catchment had 94% undeveloped land, while the urban catchment had 65% urban coverage. During a large storm in October 1974 (22 mm of rain in five hours) the peak discharge from the urban catchment area was over 250 times that of the rural catchment (Cherkauer, 1975). The storm hydrograph from this event was considerably flashier for the urban catchment (i.e., it had a shorter, sharper peak on the hydrograph).

Rose and Peters (2001) analyzed a long period of streamflow data (1958–1996) to detect differences between urbanized and rural catchments near Atlanta, Georgia, USA. The stormflow peaks for large storms were between 30% and 100% larger in the urbanized catchment, with a considerably shorter recession limb of the hydrograph. In contrast to stormflows, low flows were 25–35 less in the urban catchment, suggesting a lower rainfall infiltration rate. Overall, there was no detectable difference in the annual runoff coefficient between urban and rural catchments.

There is no straightforward method for evaluating changes in the form of flood frequency distribution of an area caused by urbanization. Furthermore, typically urbanization occurs gradually over a period and, for this reason, it is difficult to separate pre-development and post-development records as two distinct datasets. Thus, unless urbanization takes place within a relatively short period, the establishment of the nature of a flood frequency distribution for a watershed under its standard state requires certain manipulation of the data. Thus, various workers developed different techniques for the estimation of mean annual flood and changes in the magnitude of higher return period floods for urbanizing watersheds. These works attempted to show that, with urbanization, the flood frequency distribution of a watershed changed.

Design flood hydrographs are usually calculated from the unit hydrographs of a watershed. Unit hydrographs of a watershed are assumed to be time invariant. However, urbanization may very well cause the shape of the unit hydrograph of a watershed to change. Consequently, the shape of the runoff hydrographs of the design flood will change. Two parameters that can describe the change in the shape of the unit hydrographs or, for that matter, the direct runoff hydrographs, are dispersion and peakedness. Certain dimensions define the shape of a unit hydrograph of a watershed. One of the essential dimensions in this regard is the lag time, the time difference between the centroid of the total rainfall hyetograph and that of the runoff hydrograph, which in turn controls the dispersivity of a hydrograph and changes its peakedness. This time parameter is mostly used as a measure of the nature of watershed response to rainfall. In addition to lag time, the time to peak or the time from the start of the rise of hydrograph to the time of peak discharge can be shortened by urbanization. This gives rise to what is dubbed by many engineers as **flashy hydrographs**. Figure 19.3 shows an example. Nonetheless, in almost all methods of applications of a unit hydrograph to derive a design hydrograph, the lag time is used as the chief scaling parameter. The effect is an increase in risk of urban flooding from even short-duration rainfall.

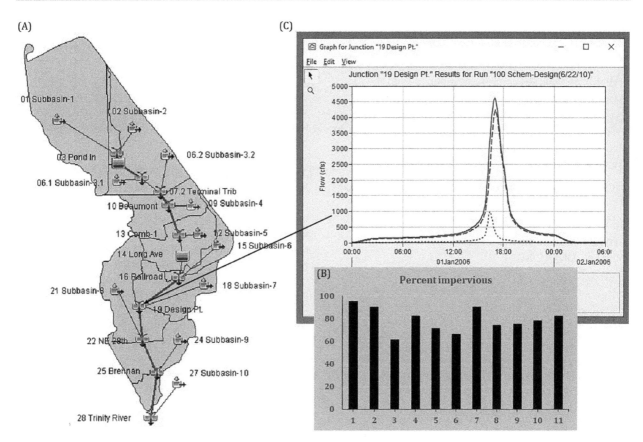

Figure 19.3 (A) An urbanized watershed located north of downtown Fort Worth, Texas. The watershed has an area of 3. 5 square miles and is drained by a channel named Lebow Creek with a stream length of 6.5 miles. (B) The percentage of impervious cover in the constituent catchments ranges from 61% to 95%. (C) The synthetic hydrograph near the downstream section, calculated in the HEC-HMS model using rainfall depths for a 100-year frequency-based design storm, shows a rapid rise and sharp peak (flashy hydrograph).

19.8 EXAMPLES

Example 19.1: A simplified water balance calculation procedure is given in Dingman (2002). Using this procedure and by programming in Excel, create a continuous simulation using the data given in Table 19.1.

Solution: A parameter that must be specified is the soil water storage capacity of the soil of the area under investigation. This is denoted by $Soil_{Max}$:

$$Soil_{Max} = \theta_{fc} \times Z_{rz} \tag{19.32}$$

where θ_{fc} is the field capacity and Z_{rz} is the vertical extent of the root zone. This is like parameter STC used in the Thornthwaite water balance program of USGS (McCabe and Markstrom, 2007). Typically, like STC, $Soil_{Max} = 100$–150 mm/m.

The monthly average values of the input parameters are used and are denoted by subscript m where $m = 1, 2, \cdots, 12$; for example, monthly average temperature is denoted by T_m.

Step 1: Calculate F_m.

Monthly precipitation is divided into rain and snow, depending on a parameter called the melt factor, F_m. This parameter is calculated as follows:

$$F_m = \begin{cases} 0 & \text{if } T_m \leq 0°C \\ 0.167 & \text{if } 0°C < T_m < 6°C \\ 1 & T_m > 6°C \end{cases} \tag{19.33}$$

The parameter melt factor here is like the degree day factor discussed in Chapter 15.

Step 2: Classify P_m, the monthly average precipitation, into snow and rain.

$$Snow_m = (1 - F_m) P_m \tag{19.34}$$
$$Rain_m = F_m P_m \tag{19.35}$$

Step 3: Calculate the snowmelt, $Melt_m$, by calculating the snow water equivalent of the snowpack, $Pack_m$, at the end of month, m.

$$Pack_m = (1 - F_m)^2 P_m + (1 - F_m) Pack_{m-1} \quad (19.36)$$

$$Melt_m = F_m(Pack_{m-1} + Snow_m) \quad (19.37)$$

Step 4: Calculate the total water input, W_m, to the system for a month.

$$W_m = Rain_m + Melt_m \quad (19.38)$$

Step 5: Calculate the potential evapotranspiration for the month, PET_m.

In the absence of radiation data, use any temperature-based method, such as the equation proposed by Hamon (1961).

Step 6: Calculate the soil moisture according to the model of Alley (1984).

1. If $W_m \geq PET_m$

$$AET_m = PET_m \quad (19.39)$$

2. If $W_m < PET_m$

AET_m is the sum of water input and an increment removed from soil storage. For this soil storage, $Soil_m$ is calculated as follows.

$$Soil_m = \min\{[(W_m - PET_m) + Soil_{m-1}], Soil_{Max}\} \quad (19.40)$$

The function min indicates the smaller of the two quantitates within the braces.

The decrease in soil storage is modeled according to the following conceptualization:

$$Soil_{m-1} - Soil_m = Soil_m \left[1 - \exp\left\{-\frac{(PET_m - W_m)}{Soil_m}\right\}\right] \quad (19.41)$$

From soil moisture calculation,

$$AET_m = W_m + Soil_{m-1} - Soil_m \quad (19.42)$$

From this calculation, the output can be produced as a table giving an annual cycle of all the inputs, soil and snowpack storage, evapotranspiration, and water available for groundwater recharge, and streamflow for the location under consideration.

Note that the computations are circular and must be iterated until all the monthly quantities converge to constant values. Such an iteration can be carried out in Excel by turning on the iteration option. To enable a circular reference, File -> Options -> Formulas -> Enable iterative calculations -> OK.

CLIMWAT is a climatic database developed by the FAO based on observed agroclimatic data of over 5000 stations worldwide. CLIMWAT provides long-term monthly mean values of nine climatic parameters, namely: (1) mean daily maximum temperature in °C; (2) mean daily minimum temperature in °C; (3) mean relative humidity in %; (4) mean wind speed in km/d; (5) mean sunshine hours per day; (6) mean solar radiation in MJ/m²·day; (7) monthly rainfall in mm/month; (8) monthly effective rainfall in mm/month; (9) reference evapotranspiration (ET_0), calculated with the Penman–Monteith method, in mm/d.

Table 19.1 gives climatic parameters for Madison, Wisconsin, from the Dane County Regional Airport as given in CLIMWAT database (longitude, latitude: −89°, 43.1°)

Table 19.2 shows the water balance calculation using the procedures given above and Figure 19.4 shows the annual cycles of all the climatic variables and water input and output in the system.

Notes:

The Hammon equation for evapotranspiration is

$$PET = 13.97 \times d \times D^2 \times W_t \quad (19.43)$$

where d is number of days in a month, D is the mean monthly hours of daylight in units of 12 hours, and W_t is a saturated water vapor density,

$$W_t = \frac{4.07 \exp(0.006T)}{100} \quad (19.44)$$

Table 19.1 *Long-term average climatic data at Dane County Regional Airport, Madison, Wisconsin*

	Month											
	J	F	M	A	M	J	J	A	S	O	N	D
Temp (max), °C	−4	−1.1	5.3	13.7	20.5	25.7	28	26.4	21.9	15.5	6.7	−1.2
Temp (min), °C	−13.8	−11.6	−5	1.2	6.8	12.3	15.3	13.8	9	3.2	−2.9	−10.3
P (mm/month)	27.2	27.4	55.1	72.6	79.8	93	86.1	102.6	85.6	55.1	53.1	46.7
ET_0 (mm/d)	0.47	0.71	1.34	2.57	3.65	4.62	4.9	4.13	2.86	2.04	1.03	0.53

Table 19.2 Results of the water balance calculation for the data given in Table 19.1

	J	F	M	A	M	J	J	A	S	O	N	D	Year
P_m (mm)	27	27	55	73	80	93	86	103	86	55	53	47	784
T (°C)	−8.9	−6.4	0.2	7.5	13.7	19.0	21.7	20.1	15.5	9.4	1.9	−5.8	
F_m (mm)	0.00	0.00	0.03	1.00	1.00	1.00	1.00	1.00	1.00	1.00	0.32	0.00	
$Rain_m$ (mm)	0	0	1	73	80	93	86	103	86	55	17	0	593
$Snow_m$ (mm)	27	27	54	0	0	0	0	0	0	0	36	47	191
$Pack_m$ (mm)	99	126	175	0	0	0	0	0	0	0	25	71	
$Melt_m$ (mm)	0	0	4	175	0	0	0	0	0	0	11	0	191
W_m (mm)	0	0	6	248	80	93	86	103	86	55	28	0	784
PET_m (mm)	0	0	25	45	73	106	121	102	70	42	22	0	606
$W_m - PET_m$ (mm)	0	0	−19	203	7	−13	−35	0	16	13	6	0	
$Soil_m$ (mm)	98	98	81	100	100	88	62	63	78	92	98	98	
$Soil_m - Soil_{m-1}$	0	0	−17	19	0	−12	−26	0	13	13	6	0	
AET_m (mm)	0	0	22	45	73	105	112	102	70	42	22	0	594
$W - AET - (Soil_m - Soil_{m-1})$	0	0	0	184	7	0	0	0	0	0	0	0	190

658 19 Water Resources Management

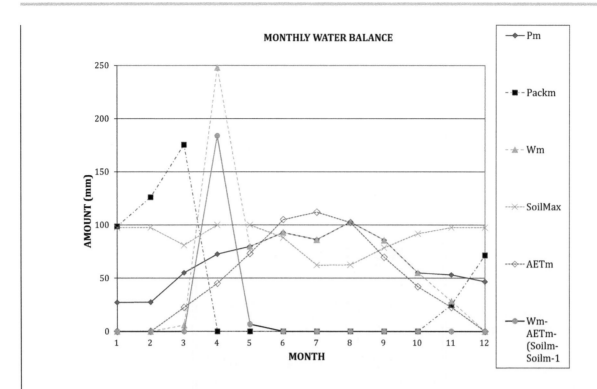

Figure 19.4 Results of water balance calculations. (A black and white version of this figure will appear in some formats. For the color version, please refer to the plate section.)

Exercises: A selection of exercises on this topic is available at www.cambridge.org/appliedhydrology.

20 Geographic Information Systems

20.1 CONCEPTS AND PURPOSE

Geographic information system (GIS) technology has become an indispensable tool in modern hydrologic studies and analysis. The unique feature of a GIS is that it associates certain information called attributes with geographic elements called features. Thus, a GIS is not just a computerized mapping tool but primarily an information management system and thereby it is an analytical or modeling platform too. There are numerous ways by which GIS technology is being used in modern hydrology. Certain essential concepts are presented in this chapter so that it can be used by students and practitioners of hydrology with an understanding of the underlying concepts of this software technology. There are several available GIS software options. However, ArcGIS, developed by the Environmental Systems Research Institute, Inc. (ESRI) is the most widely used GIS platform worldwide and is exceptionally powerful. However, it is a proprietary commercial product. Certain other GIS software products are also available in the public domain, such as QGIS, which has equally superior cartography, editing, and analysis tools.

20.2 DATABASE MANAGEMENT SYSTEMS

A GIS is, first and foremost, an information management system. It helps integrate data collected by different entities about features on the ground as well as underground. For this reason, it is important to have some basic understanding of certain concepts of storing and management of data or information with the aid of a software.

A **database** can be defined as one or more structured sets of persistent data, managed and stored as a unit and generally associated with software to update and query the data. Examples of popular databases include Microsoft Access, Oracle, and MySQL. A database is typically a collection of related files organized for efficient retrieval of information.

An **attribute** is a GIS terminology for a characteristic of a map feature. Attributes of a river might include its name, length, average depth, and so on.

In a database, a **field** refers to a column in a table. Each field contains the values for a single attribute. An **alias** is another name for a field in a table. A **record** in an attribute table refers to a single row of thematic descriptors. In SQL terms, a record is analogous to a tuple. It is a logical unit of data in a file. A table containing records and fields is called an **attribute table**. New fields that are functions of values of certain other fields can be calculated using mathematical formulas giving those relationships and can be added to an attribute table.

A **database management system** is a set of computer programs that organizes the information in a database and provides tools for data input, verification, and storage. It contains one or more attribute tables or other tabular file containing rows and columns.

A **relational database** contains data stored in tables that are associated by shared attributes. Any data element can be found in the database through the name of the table, the attribute name given as the column header, and the value of the primary key. In contrast to hierarchical and network database structures, the data can be arranged in different combinations.

A **data dictionary** refers to a catalog containing information about the datasets stored in a database. In a GIS, a data dictionary might contain the full names of attributes, meanings of codes, scale of source data, accuracy of locations, and map projections (discussed below) used.

A **schema** refers to the structure or design of a database or database object, such as a table. In a relational database, the schema defines the tables, the fields in each table, the relationships between fields and tables, and the grouping of objects within the database. Schemas are generally documented in a data dictionary. A database schema provides a logical classification of database objects.

20.3 GEODATABASES

A GIS database includes data about the spatial locations and shapes of geographic features recorded as points, lines, areas, pixels, grid cells of certain attribute called **rasters**, or a **triangulated irregular network** (TIN) representation of a continuous surface, as well as their attributes. Such a database is called a **geodatabase**. It is a geographic database that represents geographic features and attributes as objects and is hosted inside a relational database management system that provides services for managing geographic data. In a word, it is a collection of geospatial data stored in a relational database format. The services that a geodatabase provides include validation rules, relationships, and topological associations, discussed below. A geodatabase model

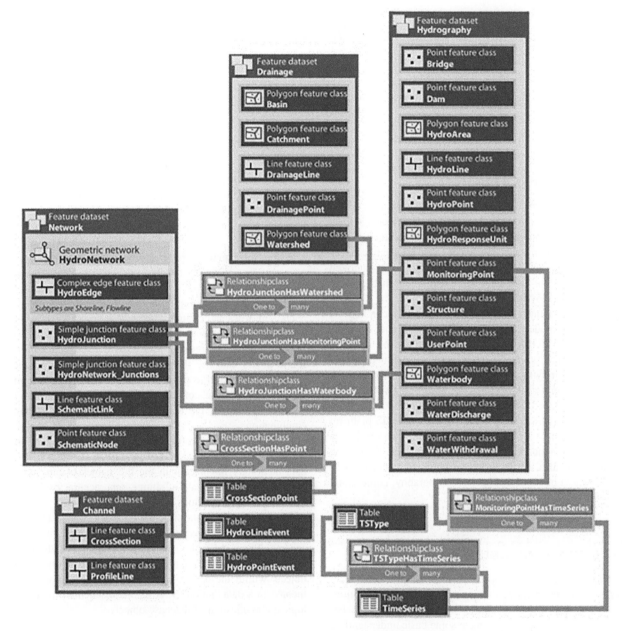

Figure 20.1 Example of schema of a geodatabase. This is the schema representing the structure of ArcHydro geodatabase developed and provided by ESRI.

is described by a schema, a linked set of relational data tables capable of storing geospatial coordinate information. Figure 20.1 shows the schema of the ArcHydro geodatabase, developed by ESRI.

A geodatabase contains **feature data sets** or **features classes**, or both, as well as grids or rasters. These are defined below.

In a geodatabase, attribute tables are associated with a class of geographic features, such as gauging stations, rivers, roads, catchments, etc. Each row represents a geographic feature. Each column represents an attribute of a feature, with the same column representing the same attribute in each row. A column is also called an **item**, such as a field with a name, and has a **data type** applied to all values in the column, which can be **text** or **numeric**.

An **object** is the representation of a real-world entity stored in a geodatabase. An object has **properties** and **behavior**. Behavior collectively represents the properties of an object in a geodatabase that describe how it can be edited and drawn. Behavior includes, but is not limited to, validation rules, subtypes, default values, and relationships.

An **attribute domain** refers to a named constraint in the database. An attribute constraint can be applied to a field of a subtype of a feature class or project class to make an attribute rule. Types of attribute domains include range and coded value domains.

20.4 DATA STRUCTURE OF GEOGRAPHIC FEATURES

Any geographic feature can be represented on a map by either a **point**, a **line**, or a **polygon**. Points are zero-dimensional objects on a map, which represent a single location on the Earth such as a gauging station. The location is recorded as an x–y coordinate. Depending upon the scale and accuracy desired, a point can also represent the location of a multidimensional feature, such as a bridge. Lines are one-dimensional objects on a map, which represent a linear feature having a beginning point and an ending point, such as a stream. The beginning point (and end point) of a line is called a **node**. Lines can be subdivided into smaller units called arcs and the point that separates two adjacent arcs is called a **vertex**. Polygons are two-dimensional objects on a map, which represent shapes which have area, such as a watershed. When the lines that form the boundaries of a polygon are defined as polygon boundaries, the polygons become distinct objects, which can be manipulated and displayed as single entities.

The points, lines, and polygons in a GIS are called **vectors**. Vector data are structured such that they are made up of x–y coordinate pairs that can be arranged to represent points, lines, and polygons. Simply put, a point is a pair of x and y coordinates, a line is a sequence of points having a beginning node and an end node with vertices in between, and a polygon is a closed set of lines. As mentioned above, another type of data structure present in a GIS is called a raster. Raster data are described by cells in a grid with one value per cell. The center of each cell in a grid has x and y coordinates and attributes of certain physical characteristics present on the land surface such as elevation, types of land cover, soil type, etc. Figure 20.2 shows the vector and raster data structure.

The term **feature** is used for a shape in a spatial **data layer**, such as a point, line, or polygon, that represents a geographic object with tabular attributes. This object has spatial coordinates. A **shapefile** is a vector data storage format for storing the location, shape, and attributes of geographic features. A shapefile is stored in a set of related files and contains one feature class. In earlier versions of GIS software, such a vector data storage format was called a **coverage**. In a shapefile, coverage, or geodatabase, a collection of spatial data with the same shape type (point, line, or polygon), is called **feature class**. In a nutshell, it is a collection of geographic objects in tabular format that have the same behavior and the same attributes. Feature classes can be stored either directly under a geodatabase or under a **feature dataset** under a geodatabase. The term **dataset** represents (1) any feature class, table, or collection of feature classes or tables in a geodatabase and (2) a named collection of logically related data items arranged in a prescribed manner.

A **spatial domain** describes the range and precision of x- and y-coordinates and z- and m- values that can be stored in a feature dataset or feature class in a geodatabase. A **spatial**

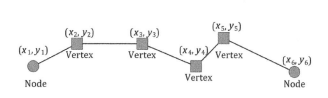

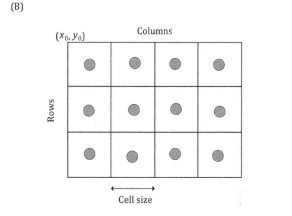

Figure 20.2 (A) Vector and (B) raster data types in a GIS.

reference describes both the projection (discussed below) and spatial domain extent for a feature dataset or feature class in a geodatabase.

In a geodatabase, a collection of feature classes that share the same spatial reference (coordinate or projection system and x–y extents) is called a feature dataset. Because the feature classes share the same spatial reference, they can participate in topological relationships with each other such as in a **geometric network**, **linear network**, or **topology**. These are explained below. Several feature classes with the same geometry may be stored in the same feature dataset. Object classes and relationship classes can also be stored in a feature dataset. A feature dataset is also a geodataset composed of graphs, networks, and feature classes and is stored under a geodatabase.

The storage format for nonspatial objects in a geodatabase is called an **object class** (compare this to feature class). It is a collection of objects stored in a table. An object class is a type of table that stores nonspatial information. It is analogous to **feature attribute table** that stores the attribute information for a specific feature class in a shapefile or coverage data. Thus, an object class is a collection of objects in tabular format that have the same behavior and the same attributes.

A **relationship** is an association or link between two objects in a database. A relationship can exist between spatial objects such as features in feature classes, nonspatial objects such as objects in object classes, or between spatial and nonspatial objects. An example can be the location of a gauging station, given as an attribute of a point feature class and a discharge value recorded at that station, which is an object class. A **relationship class** stores information about how spatial objects, such as feature classes, or nonspatial objects, such as tables, are related to other objects within a geodatabase. Annotation is an example of a relationship class.

In summary, a feature represents: (1) an object class in a geodatabase that has a field of type geometry – features are stored in feature classes; (2) a representation of a real-world object; (3) a point, line, or polygon in a coverage or a shapefile; and (4) a real-world object in a layer (discussed below) on a map. A feature class represents the following: (1) the conceptual representation of a category of geographic features – when referring to geographic features, feature classes include point, line, area, and annotation; (2) in a geodatabase, an object class that stores features and has a field of type geometry; and (3) a classification describing the format of geographic features and supporting data in a vector data format.

20.5 TOPOLOGIC DATA STRUCTURE

Many vector datasets have features that can share boundaries or corners. In a geodatabase, **topology** means the relationships between connected features in a geometric network or shared borders between features. In a vector data format such as a shapefile or feature class, topology refers to the spatial relationships between connecting or adjacent features. The topology of an arc includes it's from and to nodes and its left and right polygons. Topologic relationships are built from simple elements: points (simplest element), arcs (sets of connected points), areas (sets of connected arcs), and routes (sets of sections, which are arcs or portions of arcs). Redundant data (coordinates) are eliminated because an arc may represent a linear feature, part of the boundary of an area feature, or both.

If a topology is created in the dataset, rules defining how features share their geometry can be set up. Editing a boundary or vertex shared by two or more features updates the shape of each of those features. Topology rules can govern the relationships between features within a single feature class or between features in two different feature classes.

A feature dataset with a topology can store topologically associated features. The feature classes remain simple feature classes, and the topology manages the rules that control how features can be spatially related. Topologic editing tools can be used to maintain the topologic associations of features. The topology can be stored as a dataset within the feature dataset. It provides rules that structure how features in the feature class can be spatially related to each other and how one subtype of a feature class can be related to one subtype of another feature class.

In vector data format, topology refers to the spatial relationships between connecting or adjacent features in a geographic data layer (for example, arcs, nodes, polygons, and points). In geodatabases, topology refers to the arrangement that constrains how point, line, and polygon features share geometry. For example, street centerlines and census blocks share geometry, and adjacent soil polygons share geometry. Topology defines and enforces data integrity rules (for example, there should be no gaps between polygons). It supports topologic relationship queries and navigation (for example, navigating feature adjacency or connectivity), supports sophisticated editing tools, and allows feature construction from unstructured geometry (for example, constructing polygons from lines).

The spatial relationship between features that share geometry such as boundaries and vertices is called **topologic association**. When a boundary or vertex shared by two or more features is edited using the topology tools, the shape of each of those features is updated.

A **geometric network** is a type of topologic relationship between feature classes within a dataset. It can be thought of as a one-dimensional non-planar graph, or logical network, which is composed of features. These features are constrained to exist within the network and can, therefore, be considered network features. GIS software automatically maintains the explicit topologic relationships between network features in a geometric network. A path through a network or grid from a source to a destination is called **route**. A network is a set of lines, also called **edges**, and points called **junctions**, which are topologically connected to each other. Each edge knows which junctions are its endpoints and each junction knows which edge it connects to. Geometric networks allow modeling networks of edges and junctions, such as pipes and valves in a water system or streams and confluence points in a channel system. They allow connectivity traces and flow analysis on the features in the network to be conducted and provide some special editing functionality that is useful for networks. Figure 20.3 gives an example of geometric network, which is the physical system shown in Figure 1.5 in the online exercises.

Some vector datasets, particularly those used to model communications, material or energy flow, or transportation networks, need to support connectivity tracing and network connectivity rules. Geometric networks allow one to turn simple point and line features into network edge and junction features that can be used for network analysis. Connectivity rules of geometric networks let one control the types of network features that may be connected together when editing the network. Geometric networks, like topologies, must be created from a set of feature classes in the same feature dataset.

Topologically connected edge and junction features that represent a linear network, such as a utility or hydrologic system, form a geometric network. The connectivity of features within a geometric network is based on their geometric coincidence. A geometric network does not contain information about the connectivity of features; this information is stored within a logical network. Geometric networks are typically used to model directed flow systems such as a river network.

20.6 Geographic Data Model

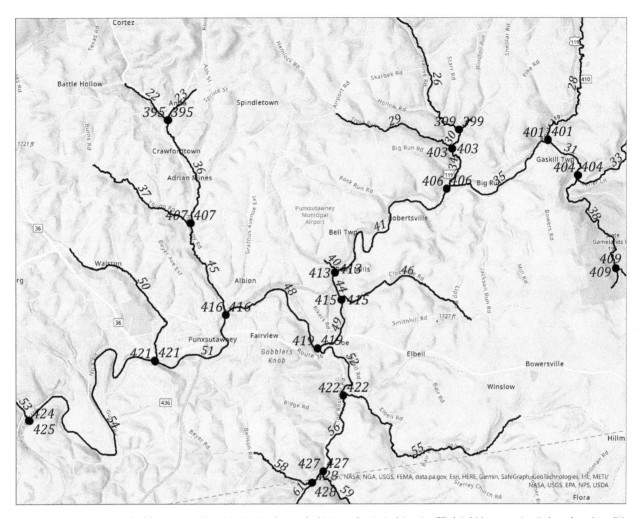

Figure 20.3 Geometric network of the streams within Mahoning Creek watershed in Pennsylvania. Each junction (filled circle) has a Junction ID shown by a three-digit number and each edge (lines) has a HydroID shown by a two-digit number. The lines represent the streams and the junctions represent the confluences. These are physical objects. In the geodatabase, the edges and junctions are topologically related.

20.6 GEOGRAPHIC DATA MODELS

All geographic information systems are built using formal models that describe how features are located in space. A formal model is an abstract and well-defined system of concepts. Models can be of four types:

1. **Conceptual model**: This is basically a set of concepts that describe a subject and allow reasoning about it. For example, a conceptual model of the hydrologic cycle is presented in Figure 1.4.
2. **Mathematical model**: This is a conceptual model expressed in symbols and equations, which describe the subject and quantify its behavior. For example, the Green–Ampt infiltration equation is a mathematical model.
3. **Data model**: This is a conceptual model expressed in a data structure, such as ascii text files and Excel tables.
4. **Geographic data model**: This is a conceptual model for describing and reasoning about the world expressed in a GIS database. Geographic data models serve as the foundation on which all types of GIS are built. They define the vocabulary for describing and reasoning about subjects located on the Earth.

Maidment (2002) elaborated the concept of a geographic data model. A data model is simply a method for describing a system using a structured set of data. Data models provide an orderly way to classify things and their relationships. A geographic data model is a representation of the real world that can be used in a GIS to produce maps, perform interactive queries, and execute analysis. Geographic data models define a vocabulary for describing and reasoning about things located on the Earth. By agreeing on a common vocabulary, practitioners can readily share data, tools, and analyses. In ArcGIS, data are stored in a geodatabase, which is a special implementation of a relational database for geospatial data. A data model is a framework for describing a subject and storing data about it (Maidment, 2002). A particular data model can be a systematic framework specially designed for creating water resources geodatabases. A geospatial data model is a framework for synthesizing data from diverse sources to represent a geographic environment. The data model can be a geodatabase model.

Data models can be defined from three angles:

1. A data model is a mathematical paradigm for representing geographic objects or surfaces as data in a GIS. The vector data model represents geography as collections of points, lines, and polygons; the raster data model represents geography as cell matrices that store numeric values; the triangulated irregular network data model represents geography as sets of contiguous, nonoverlapping triangles.
2. A data model is a set of database design specifications for objects in a GIS application. A data model describes the thematic layers used in the application, as for example, watershed boundaries, streams, and confluences. The themes are spatially represented as points, lines, or polygons, along with their attributes, their integrity rules, relationships (for example, streams cannot self-intersect, watersheds must nest within river basins), their cartographic portrayal, and their metadata requirements.
3. In information theory, a data model is a description of the rules by which data are defined, organized, queried, and updated within an information system, which usually is a database management software program.

In a general sense, a data model is an abstraction of the real world that incorporates only those properties thought to be relevant to the application at hand. It would normally define specific groups of entities, their attribute values, and the relationships between them. In a GIS, it is often used to refer to the mechanistic representation and organization of spatial data; for example, the vector data model and the raster data model. It is independent of a computer system and its associated data structure.

A hydrologic data model of a river basin describes hydrologic features of the basin, including watersheds, stream channels, water bodies, land use, soil covers, hydraulic structures, and gauging stations such as stream gauges and precipitation gauges. Any geospatial feature in the data model can be linked to a time series so that the flow of water and transport of its constituents through the landscape can also be described.

Formalization of data in a structured system becomes necessary when a project covers large areas, makes use of several different sources of GIS data, and involve running complicated hydrologic models. A greater degree of formality in the way the information is structured provides more systematic and efficient project execution, and the potential to reuse the data on subsequent projects in the same area.

20.7 TYPES OF DATA MODELS

A GIS offers the platform to develop data models for complex hydrologic studies at large spatial scales. Maidment (2002) noted that data models can be developed from two perspectives: an inventory model and a behavioral model.

A **data inventory model** answers the question: What is it and where is it? Data inventories are thematic data layers, each containing an exhaustive enumeration of all the features existing in the landscape of a given type (Figure 20.4). One source of such inventories is the water features of the landscape depicted as points, lines, and areas in map hydrography, which is the description, study, and charting of bodies of water, such as rivers, lakes, and seas on topographic maps. Other data inventories include tabular databases of stream gauges, dams, or bridges. Different layers can be terrain data in electronic (digital) format, aerial photographs, present and future planned land use, soil covers, etc.

A **behavioral data model** answers a different kind of question: How does it function? This question is answered only in a simplified way with the thought model, for instance by following the direction of flow of water from a point on a land surface through streams to seas. This requires the definition of drainage areas on the landscape, and their connection to the stream channel network. It also requires connection to a time series describing water properties and movement. In a behavioral approach, we define a system of behavior, look for pertinent features to describe the behavior, and then define how the behavior of individual features interacts with others to define the behavior of the whole system. This is the classic water resources modeling approach. Its emphasis is not on describing the environment in an exhaustive geographic detail, but rather in creating a schematic view of the landscape that highlights the features important to a particular water resources model or analysis.

20.8 THE EARTH DATUM

For almost 500 years, it has been conclusively established that the Earth is essentially a sphere, although a number of intellectuals nearly 2000 years earlier were convinced of this. In the early eighteenth century, Isaac Newton and others concluded that the Earth should be slightly flattened at the poles, but the French believed the Earth to be egg-shaped as the result of meridian measurements within France (Snyder, 1987). To settle the matter, the French Academy of Sciences, beginning in 1735, sent expeditions to Peru and Lapland to measure meridians at widely separated latitudes. This established the validity of Newton's conclusions. The Earth is not an exact ellipsoid, and deviations from this shape are continually evaluated. Actually, it is more nearly an **oblate ellipsoid** of revolution, also called an **oblate spheroid**. This is an ellipse rotated about its shorter axis. The flattening of the ellipse for the Earth is only about one part in 300; but it is sufficient to become a necessary part of calculations in plotting accurate maps at a scale of 1:100 000 or larger (map scale is discussed below). For many maps, including nearly all maps in commercial atlases, it may be assumed that the Earth is a sphere.

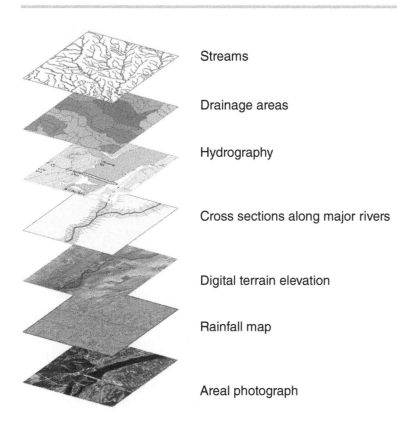

Figure 20.4 Schematic illustration of an inventory data model. The data model contains thematic layers, each of which depicts data for a certain aspect of the subject. Adapted from Maidment (2006). (Figures 20.4, 20.5, 20.6, 20.7, 20.8, 20.11–20.21 are adapted from the source cited in the Acknowledgments section of this book).

The **geoid** is the name given to the shape that the Earth would assume if it was all measured at mean sea level. This is an undulating surface that varies not more than about 100 meters above or below a well-fitting ellipsoid, a variation far less than the ellipsoid varies from the sphere. It is important to remember that elevations and contour lines on the Earth are reported relative to the geoid, not the ellipsoid. Latitude, longitude, and all plane coordinate systems, on the other hand, are determined with respect to the ellipsoid.

As noted above, the Earth's surface at sea level can be approximated as an ellipsoid called geoid. An ellipsoid that approximates the entire globe and has its center at the Earth's center of gravity is called a **global datum** or **Earth datum**. An ellipsoid that accurately represents the surface of the Earth in a specific region and has a mathematical representation allowing a point in the region to form an origin for coordinates on its surface is called a **local datum**.

Until about 1990, ellipsoids were only fitted to the Earth's shape over a particular country or continent. The polar axis of the reference ellipsoid for such a region, therefore, normally does not coincide with the axis of the actual Earth, although it is assumed to be parallel. The same applies to the two equatorial planes. The discrepancy between centers is usually a few hundred meters at most. Only satellite-determined coordinate systems are considered geocentric. Satellite data have provided geodesists with new measurements to define the best Earth-fitting ellipsoid and for relating existing coordinate systems to the Earth's center of mass.

US military efforts produced the World Geodetic System 1966 and 1972 (WGS 66 and WGS 72). The National Geodetic Survey replaced the North American 1927 Datum with a new datum, the North American Datum 1983 (NAD 83), which is Earth-centered, based on both satellite and terrestrial data.

The local datum used in North America is called the North American Datum and was first established in 1913. In 1927, a different ellipsoid was determined to be more accurate in approximating the Earth's surface in the region and a new datum was adopted: the **North American Datum of 1927 (NAD 27)**. This datum was generally accepted as the standard for almost 50 years when, through the use of satellites and other sophisticated technology, and the addition of Alaska to the region, a new datum was adopted by the federal governments of North America in 1986. This new datum, the **North American Datum of 1983 (NAD 83)**, is a global datum that fits the entire geoid.

20.9 MAP PROJECTIONS

To identify the location of points on the Earth, a graticule or network of **longitude** and **latitude** lines has been superimposed on the surface (Figure 20.5). They are commonly referred to as **meridians** and **parallels**, respectively. The concept of latitudes and longitudes was originated early in recorded history by Greek and Egyptian scientists, especially the Greek astronomer Hipparchus (second century, BCE). Claudius Ptolemy further formalized the concept (Snyder, 1987).

Given the north and south poles, which are approximately the ends of the axis about which the Earth rotates, and the equator, an imaginary line halfway between the two poles, the parallels of latitude are formed by circles surrounding the Earth and in planes parallel with that of the equator. If circles are drawn equally spaced along the surface of the sphere, with 90 spaces from the equator to each pole, each space is called a degree of latitude. The circles are numbered from 0° at the equator to 90° north and south at the respective poles. Each degree is subdivided into 60 minutes and each minute into 60 seconds of arc.

Meridians of longitude are formed with a series of imaginary lines, all intersecting at both the north and south poles, and crossing each parallel of latitude at right angles, but striking the equator at various points. If the equator is equally divided into 360 parts, and a meridian passes through each mark, 360 degrees of longitude result. These degrees are also divided into minutes and seconds. While the length of a degree of latitude is always the same on a sphere, the lengths of degrees of longitude vary with the latitude. At the equator on the sphere, they are the same length as the degree of latitude, but elsewhere they are shorter.

There is only one location for the equator and poles which serve as references for counting degrees of latitude, but there is no natural origin from which to count the degrees of longitude, since all meridians are identical in shape and size. It thus becomes necessary to choose arbitrarily one meridian as the starting point, or prime meridian. There have been many prime meridians in the course of history, swayed by national pride and international influence. In 1884, the International Meridian Conference, meeting in Washington, agreed to adopt the "meridian passing through the center of the transit instrument at the Observatory of Greenwich as the initial meridian for longitude," resolving that "from this meridian longitude shall be counted in two directions up to 180 degrees, east longitude being plus and west longitude minus" (Figure 20.6).

A **map projection** is a systematic representation of all or part of the surface of a round body, especially the Earth, on a plane. This usually includes lines delineating meridians and parallels, as required by some definitions of a map projection, but it may not, depending on the purpose of the map. Once the reference ellipsoid, that is, a datum that provides a mathematical approximation of the Earth is established for a region, a **projection system** is used to transform the three-dimensional surface of the ellipsoid to the two-dimensional surface of a map. A location or a point on the spherical Earth is given by longitude (x-coordinate denoted by λ) and latitude (y-coordinate denoted by φ). This is called geographic coordinate system (Figure 20.6). A transformation from the three-dimensional geographic coordinate system that represents a spherical coordinate system to a two-dimensional Cartesian coordinate system is a map projection (Figure 20.7). The resulting x- and y-coordinates are called a **projected coordinate system**. This is a planar coordinate system defined by a pair of orthogonal (x, y) axes drawn through an origin (Figure 20.8).

It is impossible, however, to transform a sphere to a plane surface without modifying the geometric relationships such as angles, areas distances, and directions of features on the surface of the sphere. While there are innumerable transformations that can retain one or several of these geometric relationships on a plane surface, there is no map projection system that can retain all of them. Consequently, there are over 250 different systems of map projections currently

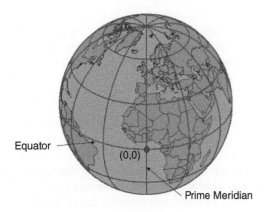

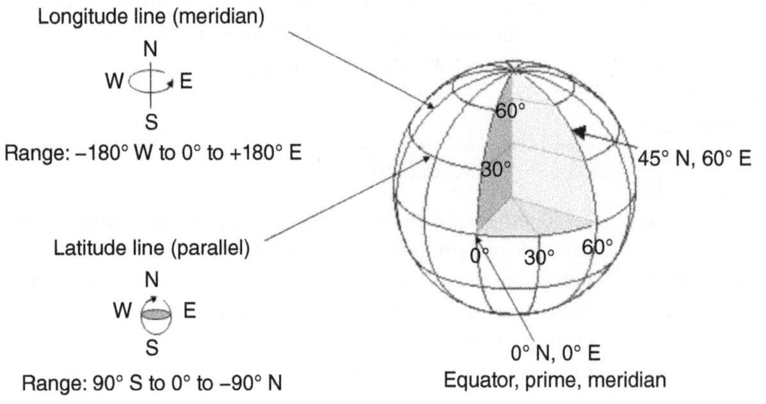

Figure 20.5 The longitude–latitude system to represent coordinates on a spherical Earth. Adapted from Maidment (2006).

Figure 20.6 The geographic coordinate system specifies the longitude and latitude of a location or point on the globe. The origin of the geographic coordinate system is the intersection of the prime meridian and equator. Adapted from Maidment (2006).

20.9 Map Projections

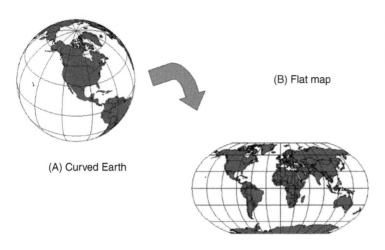

Figure 20.7 Concept of map projection. (A) curved Earth: geographic coordinates, φ and λ (latitude and longitude). (B) Flat map: Cartesian coordinates, x and y. Adapted from Maidment (2006).

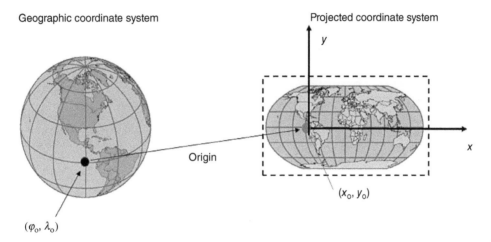

Figure 20.8 Geographic to projected coordinate system. Adapted from Maidment (2006).

devised and described in the literature of mathematical cartography, each one emphasizing a particular combination of geometric relationships.

Generally, there are three types of **developable surfaces** onto which most of the map projections are at least partially geometrically projected. Developable surfaces can be transformed to a plane without distortion and they are the cylinder, the cone, and the plane. All three are variations of the cone. A cylinder is a limiting form of a cone that has an increasingly sharp point or apex. As the cone becomes flatter, its limit is a plane. In a cylindrical projection, either a vertical transverse, oblique cylinder encloses the spherical Earth. Cylindrical projections are used primarily for complete world maps, or for maps along narrow strips of a great circular arc, such as the equator, a meridian, or an oblique great circle. To show a region for which the greatest extent is from east to west, conic projections are usually preferable to cylindrical projections.

If a cone is placed over the globe, with its peak or apex along the polar axis of the Earth and with the surface of the cone touching the globe along a **specified standard** latitude, a conic (or conical) projection can be produced, which is true to scale and without distortion (Figure 20.9). The meridians are projected onto the cone as equidistant straight lines radiating from the apex, and the parallels are marked as lines around the circumference of the cone in planes perpendicular to the Earth's axis, spaced for the desired characteristics. The parallels may not be projected geometrically for any useful conic projection. When the cone is cut along a meridian, unrolled, and laid flat, the meridians remain straight radiating lines, but the parallels are now circular arcs centered on the apex. The angles between meridians are shown smaller than the true angles.

Meridians are drawn on the cone from the apex to the points at which the corresponding meridians on the globe cross the standard parallel. Other parallels are then drawn as arcs centered on the apex in a manner depending on the projection. If the cone is cut along one meridian and unrolled, a conic projection results. A secant cone results if the cone cuts the globe at two specified parallels. Meridians and parallels can be marked on the secant cone somewhat as above, but this will not result in any of the common conic projections with two standard parallels. They are derived from various desired scale relationships instead, and the

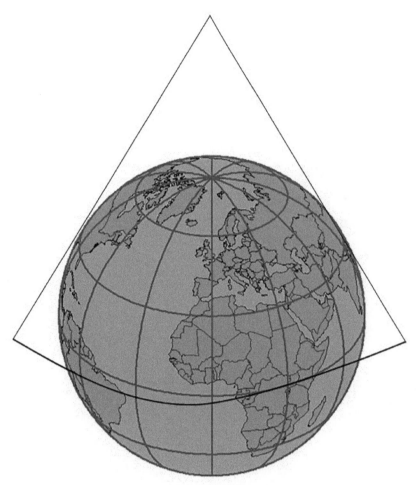

Figure 20.9 Conic projection.

spacing of meridians as well as parallels is not the same as the projection onto a secant cone.

There are three important classes of conic projections: equidistant or simple, conformal, and equal-area projections. The **equidistant conic projection**, with parallels equidistantly spaced, originated in a rudimentary form with Claudius Ptolemy. It eventually developed into commonly used present-day forms, which have one or two standard parallels selected for the area being shown. It is neither a conformal nor equal-area projection, but the north–south scale along all meridians is correct, and it can be a satisfactory compromise for errors in shape, scale, and area, especially when the map covers a small area. The **Lambert conformal conic projection** with two standard parallels is used frequently for large- and small-scale maps. The parallels are more closely spaced near the center of the map. The Lambert projection has also been used slightly in the oblique form. The **Albers equal-area conic projection** with two standard parallels is used for sectional maps of the United States and for maps of the conterminous United States. The Albers parallels are spaced more closely near the north and south edges of the map.

The most used transformations, in hydrologic mapping, are described below. These projections use the same reference ellipsoid, but they produce different reference grids and their resulting maps differ, because each one emphasizes a different aspect of accuracy, depending upon how the maps are to be used. Detailed descriptions of various types of map projections that are used by USGS have been described by Snyder (1987, 1993) and Snyder and Voxland (1989).

20.9.1 Albers Equal-Area Conic Projection

One of the most commonly used projections for maps of the conterminous United States, Europe, and Australia is the equal-area form of the conic projection that uses two standard parallels to reduce some of the distortion found in a projection with only one standard parallel. This projection was first presented by Heinrich Christian Albers, a native of Lüneburg, Germany, in a German periodical of 1805.

The Albers is a conic projection. All meridians are equally spaced straight lines converging to a common point. The parallels and both poles are represented as circular arcs centered on the point of convergence of the meridians. The spacing of the parallels decreases toward the poles. When the standard parallels are set on the northern hemisphere, the fan shape of the graticule is oriented up. When the standard parallels are on the southern hemisphere, the fan shape of the graticule is oriented down. The graticule is symmetric across the central meridian. The standard parallels can be at any latitude, except set at opposite poles. Although scale and shape are not preserved, distortion is minimal between the standard parallels.

This projection is best suited for equal-area mapping of land masses in mid-latitudes extending in an east-to-west orientation rather than those extending north to south. It is best practice to place standard parallels at one-sixth of the latitude range below the top and above the bottom of the area to be mapped. After ellipsoidal equations were developed, the projection became standard for equal-area maps of the United States.

Snyder (1987) described generating formulae for the projection, as well as the projection's characteristics. Coordinates from a **spherical datum** can be transformed into Albers equal-area conic projection coordinates with the following formulas:

$$x = \rho \sin \theta \tag{20.1}$$

$$y = \rho_0 - \rho \cos \theta \tag{20.2}$$

where

$$\rho = \frac{R}{n}\sqrt{C - 2n \sin \varphi} \tag{20.3}$$

$$\rho_0 = \frac{R}{n}\sqrt{C - 2n \sin \varphi_0} \tag{20.4}$$

$$n = \frac{1}{2}(\sin \varphi_1 + \sin \varphi_2) \tag{20.5}$$

$$\theta = n(\lambda - \lambda_0) \tag{20.6}$$

$$C = \cos^2 \varphi_1 + 2n \sin \varphi_1 \tag{20.7}$$

in which R is the radius, λ is the longitude, λ_0 is the reference longitude, φ is the latitude, φ_0 is the reference latitude, and φ_1 and φ_2 are the standard parallels. Thus, λ_0, φ_0 are the latitude and longitude for the origin of the rectangular coordinates, which are given along with $R, \varphi_1, \varphi_2, \varphi$, and, λ.

In Eqs. 20.1 and 20.2, x and y represent rectangular coordinates whereas ρ and θ represent polar coordinates.

The y-axis lies along the central meridian, λ_0, y, increasing northerly. The x-axis intersects perpendicularly at φ_0, x increasing easterly. If $(\lambda - \lambda_0)$ exceeds the range $\pm 180°$, $360°$ should be added or subtracted to place it within the range. Constants n, C, and ρ_0 apply to the entire map, and thus need to be calculated only once. If the projection is designed primarily for the northern hemisphere, n and ρ are positive. For the southern hemisphere, they are negative.

Scale distortion is most often calculated as the ratio of the scale along the meridian or along the parallel at a given point to the scale, at a standard point or along a standard line, which is made true to scale. These ratios are called **scale factors**. The factor along the meridian is called h and that along the parallel, k. The term **scale error** is frequently applied to $(h-1)$ and $(k-1)$.

For the Albers equal-area projection,

$$h = \frac{\cos \varphi}{\sqrt{C - 2n \sin \varphi}} \tag{20.8}$$

$$k = \frac{1}{h} \tag{20.9}$$

Equation 20.9 satisfies the requirement for an equal-area projection when meridians and parallels intersect at right angles.

If just one of the two standard parallels of the Albers projection is placed on a pole, the result is the **Lambert equal-area conic projection** (see the section below).

For an ellipsoid, the formulas for the parameters, ρ, θ, ρ_0, C, n are different, with two additional parameters m and q, as given in Snyder (1987):

$$\rho = \frac{a}{n}\sqrt{C - nq} \tag{20.10}$$

$$\theta = n(\lambda - \lambda_0) \tag{20.11}$$

$$\rho_0 = \frac{a}{n}\sqrt{C - nq_0} \tag{20.12}$$

$$C = m_1^2 + nq_1 \tag{20.13}$$

$$n = \frac{m_1^2 - m_2^2}{q_2 - q_1} \tag{20.14}$$

$$m = \frac{\cos \varphi}{\sqrt{1 - e^2 \sin^2 \varphi}} \tag{20.15}$$

$$q = (1 - e^2)\left\{\frac{\sin \varphi}{(1 - e^2 \sin^2 \varphi)} - \left(\frac{1}{2e}\right)\ln\left[\frac{(1 - e \sin \varphi)}{1 + e \sin \varphi}\right]\right\} \tag{20.16}$$

where e in Eqs. 20.15 and 20.16 represents the eccentricity of the ellipsoid given by $e = \sqrt{\left(1 - \frac{b^2}{a^2}\right)}$ in which a is the equatorial radius or semimajor axis of the ellipsoid of reference and b is the polar radius or semi-minor axis of the ellipsoid of reference.

For the scale factor,

$$k = \frac{\sqrt{C - nq}}{m} \tag{20.17}$$

$$h = \frac{1}{k} \tag{20.18}$$

with the same subscripts 1, 2, or none applied to m and φ in Eq. 20.15, and 0, 1, 2, or none applied to q and φ in Eq. 20.16, as required by Eqs. 20.10, 20.12, 20.14, and 20.17. As with the spherical case, p and n are negative, if the projection is centered in the southern hemisphere.

The inverse formulas to calculate longitude and latitude from a given (x, y) Cartesian coordinate are given in Snyder (1987).

20.9.2 Lambert Conformal Conic Projection

The Lambert conformal conic projection is a conic map projection used for aeronautical charts, portions of the State Plane Coordinate System (SPCS) in the United States, and many national and regional mapping systems. It is one of seven projections introduced by Johann Heinrich Lambert[1] in his 1772 publication *Anmerkungen und Zusätze zur Entwerfung der Land- und Himmelscharten* (Notes and Comments on the Composition of Terrestrial and Celestial Maps).

Conceptually, the projection seats a cone over the sphere of the Earth and projects the surface conformally onto the cone. The cone is unrolled, and the parallel that is touching the sphere is assigned a unit scale. That parallel is called the **reference parallel** or **standard parallel**. This projection is conformal in the sense that lines of latitude and longitude, which are perpendicular to one another on the Earth's surface, are also perpendicular to one another in the projected domain.

By scaling the resulting map, two parallels can be assigned a unit scale, with the scale decreasing between the two parallels and increasing outside them. This gives the map two standard parallels. In this way, the deviation from the unit scale can be minimized within a region of interest that lies largely between the two standard parallels. Unlike other conic projections, no true secant form of the projection exists because using a secant cone does not yield the same scale along both standard parallels.

Coordinates from a spherical datum can be transformed into Lambert conformal conic projection coordinates with the following formulas, where, as above, λ is the longitude, λ_0 the reference longitude, φ the latitude, φ_0 the reference latitude, R the radius of the Earth, and φ_1 and φ_2 are the two the standard parallels:

$$x = \rho \sin[n(\lambda - \lambda_0)] \qquad (20.19)$$

$$y = \rho_0 - \rho \cos[n(\lambda - \lambda_0)] \qquad (20.20)$$

where

$$\rho = RF \cot^n\left(\frac{\pi}{4} + \frac{\varphi}{2}\right) \qquad (20.21)$$

$$\rho_0 = RF \cot^n\left(\frac{\pi}{4} + \frac{\varphi_0}{2}\right) \qquad (20.22)$$

$$F = \left(\frac{1}{n}\right) \cos \varphi_1 \tan^n\left(\frac{\pi}{4} + \frac{\varphi_1}{2}\right) \qquad (20.23)$$

$$n = \frac{\ln(\cos \varphi_1 \sec \varphi_2)}{\ln\left[\tan\left(\frac{\pi}{4} + \frac{\varphi_2}{2}\right) \cot\left(\frac{\pi}{4} + \frac{\varphi_1}{2}\right)\right]} \qquad (20.24)$$

As above, the y-axis lies along the central meridian, λ_0, y, increasing northerly; and the x-axis intersects perpendicularly at φ_0, x, increasing easterly. If $(\lambda - \lambda_0)$ exceeds the range $\pm 180°$, $360°$ should be added or subtracted. Constants n, F, and ρ_0 need to be determined only once for the entire map.

If only one standard parallel, φ_1, is desired, Eq. 20.24 is indeterminate, but $n = \sin \varphi_1$. The scale along meridians or parallels is determined as

$$k = h = \frac{\cos \varphi_1 \tan^n\left(\frac{\pi}{4} + \frac{\varphi_1}{2}\right)}{\cos \varphi \tan^n\left(\frac{\pi}{4} + \frac{\varphi}{2}\right)} \qquad (20.25)$$

Formulae for ellipsoidal datums are more involved and can be found in Snyder (1987).

The Lambert equal-area conic projection, also developed by Johann Lambert, is a conic, equal-area projection that represents one pole as a point. The Albers projection is a generalization of this projection with two standard parallels. The Lambert equal-area conic projection can be viewed as an extreme case of the Albers projection, and is also known as the **Lambert azimuthal equal-area projection**.

20.9.3 Universal Transverse Mercator Projection

The Mercator projection was perhaps the first map projection to be regularly identified when atlases of over a century ago gradually began to name the projections used. While the projection was apparently used by Erhard Etzlaub (1462–1532) of Nuremberg on a small map on the cover of some sundials constructed in 1511 and 1513, the principle remained obscure until Gerardus Mercator (1512–1594) independently developed it and presented it in 1569 on a large world map of 21 sections totaling about 1.3 by 2 meters (Snyder, 1987). Mercator, born in Rupelmonde in Flanders (now in Belgium), was probably originally named Gerhard Cremer (or Kremer), but he always used the latinized form. To his contemporaries and to later scholars, he is better known for his skills in map and globe making, for being the first to use the term **atlas** to describe a collection of maps in a volume, for his calligraphy, and for first naming North America as such on a map, in 1538. To the world at large, his name is identified chiefly with his projection, which he specifically developed to aid navigation. His 1569 map is entitled: *Nova et Aucta Orbis Terrae Descriptio ad Usum Navigantium Emendate Accommodata* (A new and enlarged description of the Earth with corrections for use in navigation). He described in Latin the nature of the projection in a large panel covering much of his portrayal of North America.

There is no suitable geometric construction of the Mercator projection. For the sphere, the formulas for rectangular coordinates are as follows:

$$x = R(\lambda - \lambda_0) \qquad (20.26)$$

$$y = R \ln\left[\tan\left(\frac{\pi}{4} + \frac{\varphi}{2}\right)\right] \qquad (20.27)$$

where R is the radius of the sphere at the scale of the map as drawn, and λ and φ are given in radians. It is a cylindrical projection where the globe is enclosed in a cylinder.

---

[1] Lambert was born August 26, 1728, in the Republic of Mulhouse, Alsace, generally referred to as either Swiss or French, to a tailor's family. Although his formal schooling ended at the age of 12, he continued his quest for knowledge by teaching himself mathematics, astronomy, and physics. He soon developed a reputation as a child prodigy and mathematical genius and polymath. Before the age of 20 he was an expert in mathematics and philosophy, and fluent in five languages. In 1765, Frederick the Great appointed him Chief Counselor for Construction, where he worked on the Directing Committee of Frederick's Academy with Leonhard Euler and Joseph-Louis Lagrange, both known for their contributions to mathematics and map projections.

Since the regular Mercator projection has little error close to the equator (the scale 10° away is only 1.5% larger than the scale at the equator), it has been found very useful in the transverse form, with the equator of the projection rotated 90° to coincide with the desired central meridian. This is equivalent to wrapping the cylinder around a sphere or ellipsoid representing the Earth so that it touches the central meridian throughout its length, instead of following the equator of the Earth. The central meridian can then be made true to scale, no matter how far north and south the map extends, and regions near it are mapped with low distortion. Like the regular Mercator projection, the map is conformal.

The transverse Mercator projection in its spherical form was invented by the prolific Alsatian mathematician and cartographer Johann Heinrich Lambert (1728–77). It was the third of seven new projections that he described in 1772 in his classic publication.

While its use in the spherical form is limited, the ellipsoidal form of the transverse Mercator projection is probably used more than any other projection for geodetic mapping. In the United States, it is the projection used in the SPCS for states with predominant north–south extent. The Lambert conformal conic projection is used for the others, except for the panhandle of Alaska, which is prepared on the oblique Mercator projection. Alaska, Florida, and New York use both the transverse Mercator and the Lambert conformal conic projections for different zones. Except for narrow states, such as Delaware, New Hampshire, and New Jersey, all states using the transverse Mercator are divided into two to eight zones, each with its own central meridian, along which the scale is slightly reduced to balance the scale throughout the map.

The Universal Transverse Mercator (UTM) projection and grid were adopted by the US Army in 1947 for designating rectangular coordinates on large-scale military maps of the entire world. The UTM is the ellipsoidal transverse Mercator projection, to which specific parameters such as central meridians have been applied. The Earth, between latitudes 84° N and 80° S, is divided into 60 zones, each generally 6° wide in longitude. Bounding meridians are evenly divisible by 6°, and zones are numbered from 1 to 60, proceeding east from the 180th meridian from Greenwich, with minor exceptions. There are letter designations from south to north (Figure 20.10). Thus, Washington, DC, is in grid zone 18S, a designation covering a quadrangle from longitude 72° to 78° W and from latitude 32° to 40° N. Each of these quadrangles is further subdivided into grid squares 100 000 meters on a side with double-letter designations, including partial squares at the grid boundaries. From latitude 84° N and 80° S to the respective poles, the Universal Polar Stereographic (UPS) projection is used instead.

The NOAA website states that the system was developed by USACE, starting in the early 1940s. However, a series of aerial photos found in the Bundesarchiv–Militärarchiv (the military section of the German Federal Archives), apparently dating from 1943 to 1944, bear the inscription UTMREF followed by grid letters and digits, and projected according to the transverse Mercator. This finding would indicate that something called the UTM Reference system was developed in the 1942–1943 time frame by the Wehrmacht, the regular armed forces of Nazi Germany, from 1935 until it effectively ceased to exist in 1945, and then was formally dissolved in August 1946. It was probably carried out by the Abteilung für Luftbildwesen (Department for Aerial Photography). From 1947 onward, the US Army employed a very similar system, but with the now-standard 0.9996 scale factor at the central meridian as opposed to the German factor of 1.0. Prior to the development of the UTM coordinate system, several European nations demonstrated the utility of grid-based conformal maps by mapping their territory during the interwar period. Calculating the distance between two points on these maps could be performed more easily in the field, using the Pythagorean theorem, than was possible using the trigonometric formulas required under the graticule-based system of latitude and longitude. In the post-war years, these concepts were extended into the UTM or Universal Polar Stereographic (UPS) coordinate system, which is a global or universal system of grid-based maps.

The rectangular coordinates for the transverse Mercator applied to the sphere are

$$x = \frac{1}{2} R k_0 \ln\left[\frac{1+B}{1-B}\right] \quad (20.28)$$

$$y = R k_0 \left[\arctan\left\{\frac{\tan\varphi}{\cos(\lambda - \lambda_0)}\right\} - \varphi_0\right] \quad (20.29)$$

where

$$B = \cos\varphi \sin(\lambda - \lambda_0) \quad (20.30)$$

And k_0 is the scale factor along the central meridian, λ_0. The origin of the coordinates is at (φ_0, λ_0). The y-axis lies along the central meridian, λ_0, y increasing northerly, and the x-axis is perpendicular, through φ_0 at λ_0, x, increasing easterly.

20.10 MAP SCALES

The scale of a map is the mathematical relationship between the size of the features on a map and the actual size of the corresponding objects in the real world. All maps are modeled representations of the real world and therefore the features are reduced in size when mapped. In other words, the scale is the measurement of the amount of reduction a mapped feature has to its actual counterpart on the ground (Figure 20.11).

Scale is commonly expressed as a ratio, or representative fraction, such as 1:24 000. This scale means one unit on the map is equal to 24 000 units on the Earth. Another way of thinking about it is that the objects on the Earth are 24 000 times larger than the features on the map that represent them.

There are large-scale and small-scale map scales. Generally, a large-scale map, such as a map of city streets or a building plan, covers a small area in more detail. A typical large-scale map is 1:100, read as one to one hundred, meaning a measurement of 1 unit (inch or m) on a map represents 100 units (feet or km) on the ground.

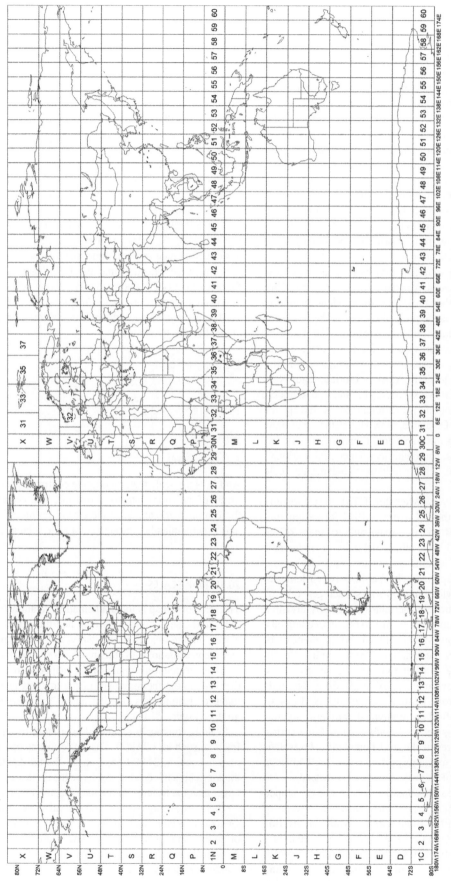

Figure 20.10 Universal Transverse Mercator (UTM) grid zone designations for the world shown on a linear equidistant projection, compiled by Dr. Alan Morton (www.dmap.co.uk, Aberystwyth). As with the SPCS, each geographic location in the UTM projection is given x- and y-coordinates, but in feet, according to the transverse Mercator projection, using the meridian halfway between the two bounding meridians as the central meridian, and reducing its scale to 0.9996 of the true scale (a 1:2500 reduction). Reproduced with permission.

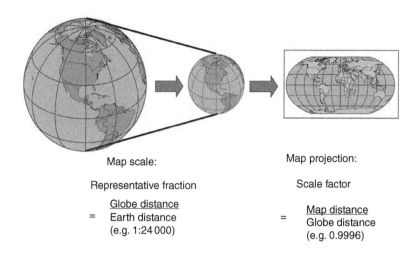

Figure 20.11 Illustration of the concept of a map scale. Adapted from Maidment (2006).

A small-scale map, such as a world map, covers a large area in less detail. It displays a large geographic area such as an entire city, state, or country. The scale of such a map is considered small because the numeric value of the ratio defined by the scale, such as 1:2000, is small compared to that of a large-scale map drawn at a larger scale, for example, 1:50. GIS maps are dynamic, meaning the scale can be changed to see more or less detail as desired.

20.11 GEOPROCESSING AND GEOVISUALIZATION

There are three principal functionalities of a GIS. These are (1) maintaining a geodatabase, (2) providing tools for geoprocessing, and (3) creating maps called geovisualizations.

Geoprocessing is the analytical capability of a GIS. A GIS is a set of information transformation tools that derive new geographic datasets from existing datasets. Geoprocessing is a GIS operation used to manipulate GIS data. A typical geoprocessing operation takes an input dataset, performs an operation on that dataset, and returns the result of the operation as an output dataset. Geoprocessing allows for definition, management, and analysis of information used to form decisions. This is done by a set of geoprocessing functions which are built-in programs in the GIS software. The geoprocessing functions take information from existing datasets, apply analytic functions, and write results into new derived datasets. There are more than 1500 geoprocessing tools available within the present version of ArcGIS Pro software. The term **georeferencing** is used to denote the operation of aligning geographic data to a known coordinate system so it can be viewed, queried, and analyzed with other geographic data. Georeferencing may involve shifting, rotating, scaling, skewing, and in some cases warping or rubbersheeting the data. The HEC-HMS also offers various geoprocessing tools including stream and subbasin delineation (Section 20.12) and the derivation of various hydrologic parameters from GIS data in both vector and raster formats.

A GIS is a set of intelligent maps and other views that show features and feature relationships on the Earth's surface. Various map views of the underlying geographic information can be constructed and used as windows into the database to support *queries*, *analysis*, and *editing* of the information. This is geovisualization. A map is produced by overlaying several *layers*. As described above, a layer is a collection of similar geographic features such as rivers, lakes, counties, or cities, of a particular area or place for display on a map. It references geographic data stored in a data source, such as a shapefile or feature class, and defines how to display it.

20.12 DELINEATION OF DRAINAGE AREAS AND STREAMS

A form of geoprocessing that is routinely used as a first step in almost all hydrologic studies is the delineation of basins, watersheds, and catchments, that is, drainage areas and streams from a digital elevation model (DEM). The introduction of DEMs and their use in GIS software to delineate drainage areas, such as catchments, watersheds, basins, and streams, was perhaps the first major application of GIS technology in hydrology. DEMs have a raster data format used to represent elevations at varying spatial resolution (Figure 20.12). In a DEM, each cell of a grid system has an elevation. However, there can be some cells with no data. The spatial resolution of the grid is given by the cell size, which can vary from less than a meter to a kilometer.

The steps used to process a DEM for delineation of catchments and streams of an area are as follows. The details of these procedures were given by Maidment (2002) in the description of the ArcHydro data model, which perhaps was the first comprehensive data model developed for a river basin by a joint collaboration between ESRI and the University of Texas at Austin:

1. DEM reconditioning (optional);
2. fill sinks or pit removal (usually optional but recommended);

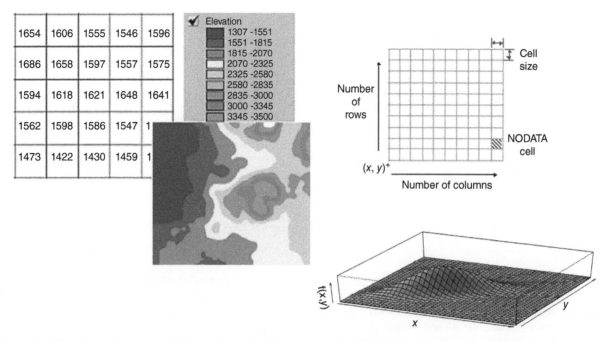

Figure 20.12 Illustration of the concept of a DEM: the representation of a continuous surface field by a grid system. Adapted from Maidment (2006).

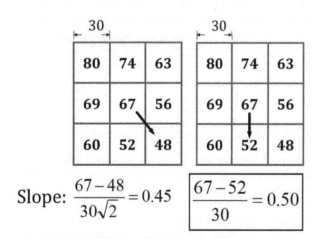

Slope: $\dfrac{67-48}{30\sqrt{2}} = 0.45$ $\boxed{\dfrac{67-52}{30} = 0.50}$

Figure 20.13 Determination of the hydrologic slope, given by the direction of steepest descent. Adapted from Maidment (2006).

3. generation of flow direction grid;
4. generation of flow accumulation grid;
5. generation of stream definition grid;
6. stream segmentation;
7. catchment grid delineation; and
8. raster to vector conversion (catchment polygon, drainage line, catchment outlet points)

The algorithm used within a GIS determines the steepest slope as the direction of flow from each cell. This is called a hydrologic slope and is determined by comparing the elevation difference between the central cell and its surrounding cells (Figure 20.13). A number or code is assigned to the cell that receives flow, or **pour point**. In the algorithm used by

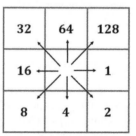

Figure 20.14 Eight-direction pour-point model. Adapted from Maidment (2006).

ESRI, these are 1, 2, 4, 8, 16, 32, 64, and 128 for east, southeast, south, southwest, west, northwest, north, and northeast (Figure 20.14). This is known as an eight-direction pour-point model, which generates the flow direction grid (Figure 20.15). The information so obtained is used to determine the number of cells that flow into each cell of the grid, yielding a grid network (Figure 20.16). This produces the flow accumulation grid in which a number corresponding to the number of cells flowing to a cell is assigned to each cell (Figure 20.17). In the next step, a threshold number is provided to the program that defines a cell as a stream. A cell in the flow accumulation grid that has a number greater than this number is considered as a stream and the resulting grid is the stream definition grid (Figure 20.18). This threshold number of grids also implies the size of the area that contributes to define a stream. A large threshold number will have fewer cells as streams than a smaller threshold number, which would produce a larger number of streams. In some algorithms, the cell with the threshold number is included in stream definition (Figure 20.19). The

20.12 Delineation of Drainage Areas and Streams

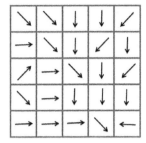

Figure 20.15 Flow direction grid. Adapted from Maidment (2006).

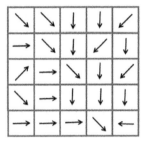 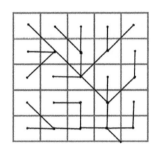

Figure 20.16 Network grid obtained from a flow direction grid. Adapted from Maidment (2006).

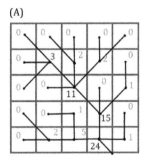 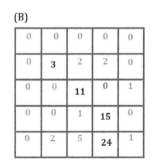

Figure 20.17 (A) Link to a grid calculator and (B) the flow accumulation grid, giving the number of cells or area draining to the grid cell. Adapted from Maidment (2006).

Figure 20.18 Stream definition grid, which selects the cells that exceed the threshold value of five (5). Adapted from Maidment (2006).

Figure 20.19 Stream network from stream definition grid. In this example the stream network is for cells that either equal or exceed the threshold of five (5). If each cell is 30 meters, five cells indicate 4500 m^2 as the contributing area. Adapted from Maidment (2006).

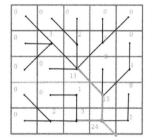

Figure 20.20 Stream segments from a stream grid. Adapted from Maidment (2006).

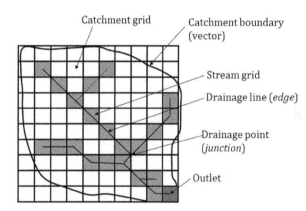

Figure 20.21 Catchment grid and stream grid from the stream threshold grid. A drainage point (junction) is formed at the intersection of two or more drainage lines (edge). Adapted from Maidment (2006).

next step is stream segmentation whereby centers of stream cells are joined (Figure 20.20). All cells contributing to a common cell form a catchment grid (Figure 20.21). The raster formats of stream grids and catchment grids are then converted to vector formats to produce a stream feature class and catchment feature class. Another feature class is formed as a drainage point as a junction of two or more drainage lines (edges). Applying this process to the entire watershed or river basin, considering all kernels which are in a 3 × 3 neighborhood, the entire drainage network and catchment boundaries are delineated.

It should be noted that the automated geoprocessing of DEMs to delineate catchments and streams may not be very accurate in flat areas. Because elevations in many DEMs are rounded off, drainage directions and the catchment boundaries may be in error. If high-resolution vector files of the stream network and catchment or watershed boundary are available, they may be used to constrain the automated delineation. These procedures are called **burning streams** and **building walls**, or DEM reconditioning (step 1 noted above). In these processes the cell elevations along the stream vector lines are lowered and the cell elevations along the catchment boundaries are raised before other geoprocessing steps. In addition, before generating a flow direction grid, another operation called sink filling is most often necessary. Because DEM creation done from digital elevation data of bare Earth obtained by remote sensing technology such as LiDAR (Light Detection and Ranging) or by satellites such as the SRTM (Shuttle Radar Topography Mission) or ASTER (Advanced Spaceborne Thermal Emission and Reflection Radiometer) data, this often results in artificial pits in the landscape. A pit is a set of one or more cells that has no downstream cells around it. Unless these pits are removed they become sinks and isolate portions of the watershed. Pit removal or sink filling is the first thing done with a DEM.

20.13 DERIVATION OF HYDROLOGIC PARAMETERS USING A GIS

Apart from the drainage area and stream delineations, the other areas where GIS technology has the greatest applications in hydrology are:

1. the derivation of physiographic or geomorphologic parameters from DEM (Chapter 6);
2. the derivation of curve numbers from land use and soil types (Chapter 7);
3. the derivation of Green–Ampt infiltration parameters from soil types (Chapter 7);
4. the derivation of snow-covered areas (Chapter 15); and
5. the derivation of regional or point net radiation (Chapter 14).

20.14 EXAMPLES

Example 20.1: Give the geoprocessing steps to derive curve numbers of catchments of a watershed from soil type and land cover data given in GIS data format.

Solution:

Step 1: Gather input data files.

For calculating curve numbers (CN values) of the catchments of a watershed, three types of input data are required in GIS format. These are the catchment boundaries, soil cover map, and land cover map. After assembling all the data from different sources, they must be projected to a common projection system.

a. Catchment boundaries: The GIS format of catchment boundaries is almost always a vector data format such as a polygon feature class or shapefile. There are various sources of these data depending on the project area and project type. Figure 20.22 shows the catchments of Lee Creek Watershed that straddles Oklahoma and Arkansas state lines in the United States. Lee Creek is a major tributary of Arkansas River. It is within the HUC eight-digit basin 11110104. The catchment boundaries shown in Figure 20.22 are at HUC twelve-digit level. Such data in the United States can be obtained from the National Hydrography Dataset developed and maintained by USGS (www.usgs.gov/national-hydrography/nhdplus-high-resolution). Various state agencies and local governments also provide such data for the river basins or watersheds within their jurisdictions. In the absence of such data or if different spatial resolutions for the areas of the catchments are required for particular investigation, the catchment boundaries can be delineated from the available DEM using the procedures described in Section 20.12.

b. **Land cover data:** The land cover map of the watershed showing the spatial variation of different types of standardized land cover is the second set of data required for the present purpose. In the United States, this data can be obtained either from the National Land Cover Database (NLCD), developed and provided by USGS, or local governmental agencies, such as the GIS department of a municipality. In the municipalities in the United States, these are often called zoning maps.

The NLCD is a raster dataset with a 30 m × 30 m cell size. The data are in the Albers equal-area conic projected coordinate system and are given as a time series at various intervals from 2001 to the present. The NLCD is a national coverage. Hence it is necessary to clip the data to the watershed extents. In most types of GIS software, there are tools to clip a raster using a polygon vector as the clipping extent.

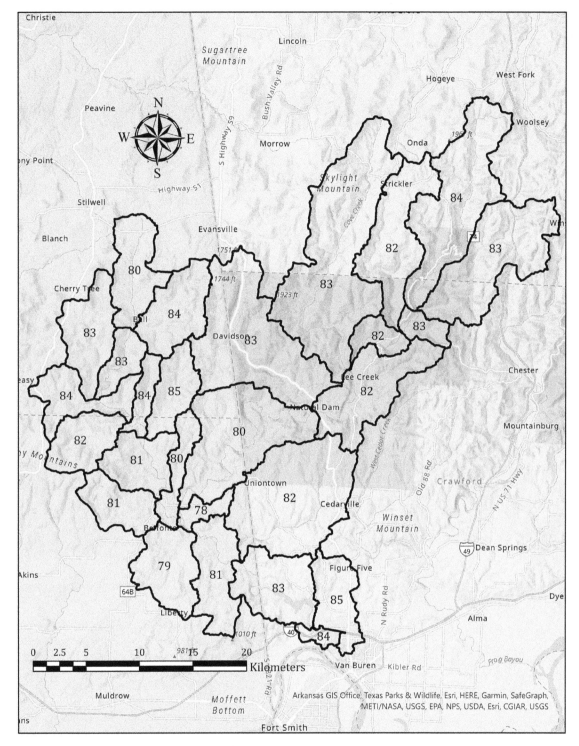

Figure 20.22 Catchment polygons within Lee Creek watershed. The composite *CN* values shown for each of the catchments are derived from the soil and land cover data of the watershed using the GIS procedures described in this example.

c. **Soil map:** The soil map of the watershed showing the spatial distribution of hydrologic soil groups (HSGs; see Chapter 7) is another required dataset. In the United States these data come in vector format from USDA. For a watershed area less than or equal to 156.25 square miles such data can be downloaded from USDA Web Soil Survey website (available at usda.gov). For larger area, the Gridded Soil Survey Geographic (gSSURGO) database, developed and maintained by the USDA of NRCS (also available

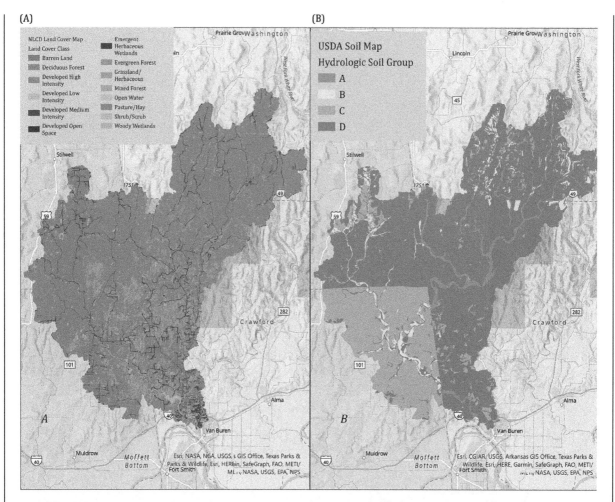

Figure 20.23 (A) Land cover polygons of Lee Creek Watershed in vector format derived from NLCD data in raster format. (B) Soil cover polygons of Lee Creek Watershed in vector format derived from the gSSURGO database. (A black and white version of this figure will appear in some formats. For the color version, please refer to the plate section.)

at usda.gov) can be used. The database contains a GIS layer of soil polygons and many attribute tables describing the properties of each soil polygon. The geodatabase file of a state or the entire United States contains vector polygons defining consistent soil types and many attribute tables with information about each soil type.

There are two alternative ways by which the geoprocessing tasks can be accomplished: either using all the data in raster format or in vector format. Raster processing requires long coding at several steps, whereas vector processing requires a very short and simple coding at the start of the processing. So here we will present the process in a vector format. Also, the commands and tools vary depending on the GIS software being used. Since ArcGIS enjoys the worldwide popularity for being the GIS platform for most hydrologic work, even though it is an expensive software, we will demonstrate geoprocessing using ArcGIS Pro V 3.1.

Step 2: Convert the land use raster data to a polygon shapefile.

1. In Geoprocessing, begin the "Raster to Polygon" operation and enter the following parameters and perform the following operation:
   a. Input Raster: land cover raster
   b. Field: the field designation describing the land coverage
   c. Output Polygon Features: the desired output file location and name
   d. Uncheck simplify polygons.
2. Run the operation to obtain a polygon file of the land cover.

Figure 20.23A shows the land cover of Lee Creek Watershed, clipped from NLCD and then converted to a polygon shapefile.

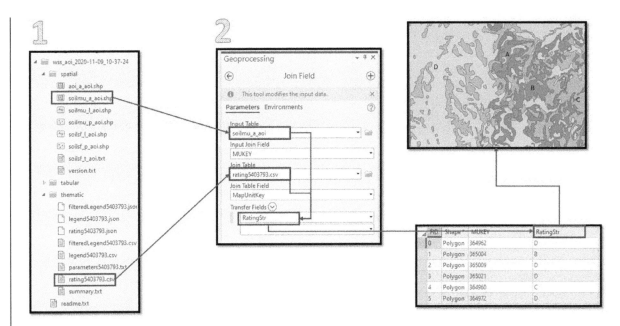

Figure 20.24 Steps to create HSG polygon shapefile from Web Soil Survey database of USDA.

Step 3: Create the shapefile for the HSGs.

The procedures for using the data from the Web Soil Survey and gSSURGO database are slightly different because of the way these two geodatabases are structured. The steps for using the data from Web Soil Survey are as follows.
1. Under the folder that was downloaded from NRCS, there is a folder called "spatial." After establishing a connection to the downloaded folder within your ArcGIS project, expand this folder.
2. There is a shapefile entitled "soilmu_a_aoi.shp." Drag it into the project and open its attribute table. There is a field called "MUKEY."
3. There are three CSV files under the folder "thematic." The file named "ratingxxxxxx.csv" contains the HSG soil groups for all of the attributes.
4. Now perform the JOIN operation under ArcGIS Pro.
5. Data Management –> Joins and Relates –> Join Field Tool. Open the tool and provide the input data as follows.
    a. Input Table: the shapefile.
    b. Input Join Field: MUKEY.
    c. Join Table: the CSV File.
    d. Join Table Field: MapUnitKey.
    e. Transfer Field: RatingStr.
    f. Run the Tool.
6. After the tool is run you will see the attribute "RatingStr" has been added to the shapefile.
7. Project the shapefile to the desired coordinate system.

Figure 20.24 shows the steps given above. Figure 20.23B shows the soil map of Lee Creek Watershed derived from the gSSURGO database for which the steps are slightly modified.

Step 4: Intersect soil type and land use.

1. In Geoprocessing, open the "Intersect" tool and enter the following parameters:
    a. Input Feature 1: soil type shapefile.
    b. Input Feature 2: land use polygon shapefile.
    c. Output Feature Class: new name.
2. Run the operation.

Step 5: Intersect the soil type–land use intersected shapefile with catchment shapefile.

1. In Geoprocessing, open the "Intersect" tool and enter the following parameters:
    a. Input Feature 1: the intersected (combined) soil type and land use shapefile obtained in step 4.
    b. Input Feature 2: catchment shapefile.
    c. Output Feature Class: new name of the resulting shapefile.
2. Run the operation.
3. Create a field for basin area
    a. Add a field to the catchment shapefile titled "basin_area."
    b. Right click the new field. Using the calculate geometry tool, calculate the area of the subbasin in this field. In the United States, the area at this step should be calculated in square feet.

Step 6: Dissolve by soil type–land use–catchment.

1. In Geoprocessing, open the "Dissolve" tool and enter the following parameters:

a. Input Features: combined soil type, land use, watershed shapefile.
b. Output Feature Class: name of the new shapefile.
c. Dissolve Fields: RatingStr, ClassName, basin_area, and FID_wbd_hydrologic_unit_12 (this may be different, but the main key is to use the field that differentiates the watersheds).
d. Check "Create Multipart features."
2. Run the operation.

Step 7: Create a curve number reference table in Excel.

Using a table of curve numbers provided by a reliable source, such as USACE or NRCS, create a table that links land usage and HSG. The engineer's best judgement should be used in trying to evaluate the NLCD land usage in terms of NRCS terminology. Alternatively, other sources may be used to evaluate the *CN* values while keeping the language used in terms of NLCD.

Format the NLCD soil type reference table in the two columns and format as given in Table 20.1. Make sure that the NLCD names match what is given in your attribute table. Insert your reference sheet into your ArcGIS project by establishing a folder connection and dragging the sheet into the map.

Step 8: Prepare the attribute table in the GIS.

1. The newly dissolved shapefile should be comprised of fields that contain the basin ID, basin area, NLCD land cover usage, and HSG soil group as shown in Box 20.1.
2. Create a new field called "Concatenated." Right click the new field and select "Calculate Field." Using the Python 3 expression type, enter the following:

```
[First Block] : conc(NLCD,HSG)
[Second Block] :
                def conc(nlcd,hsg):
                txt = nlcd + ""-"" + hsg
                return txt
```

Table 20.1 *Reference table of curve numbers for combined land use and hydrologic soil group*

NLCD_HSG	Curve_Number	NLCD_HSG	Curve_Number
Barren Land-A	77	Emergent Herbaceous Wetlands-A	80
Barren Land-B	86	Emergent Herbaceous Wetlands-B	80
Barren Land-C	91	Emergent Herbaceous Wetlands-C	80
Barren Land-D	94	Emergent Herbaceous Wetlands-D	80
Grassland/Herbaceous-A	75	Evergreen Forest-A	36
Grassland/Herbaceous-B	80	Evergreen Forest-B	60
Grassland/Herbaceous-C	87	Evergreen Forest-C	73
Grassland/Herbaceous-D	93	Evergreen Forest-D	79
Cultivated Crops-A	72	Herbaceous-A	50
Cultivated Crops-B	81	Herbaceous-B	71
Cultivated Crops-C	88	Herbaceous-C	81
Cultivated Crops-D	91	Herbaceous-D	89
Deciduous Forest-A	57	Mixed Forest-A	57
Deciduous Forest-B	73	Mixed Forest-B	73
Deciduous Forest-C	82	Mixed Forest-C	82
Deciduous Forest-D	86	Mixed Forest-D	86
Developed High Intensity-A	88	Open Water-A	98
Developed High Intensity-B	92	Open Water-B	98
Developed High Intensity-C	93	Open Water-C	98
Developed High Intensity-D	94	Open Water-D	98
Developed Low Intensity-A	81	Pasture/Hay-A	40
Developed Low Intensity-B	88	Pasture/Hay-B	61
Developed Low Intensity-C	90	Pasture/Hay-C	73
Developed Low Intensity-D	93	Pasture/Hay-D	79
Developed Medium Intensity-A	84	Shrub/Scrub-A	35
Developed Medium Intensity-B	89	Shrub/Scrub-B	56
Developed Medium Intensity-C	93	Shrub/Scrub-C	70
Developed Medium Intensity-D	94	Shrub/Scrub-D	77
Developed Open Space-A	52	Woody Wetlands-A	86
Developed Open Space-B	68	Woody Wetlands-B	86
Developed Open Space-C	78	Woody Wetlands-C	86
Developed Open Space-D	84	Woody Wetlands-D	86

Another formula that can used in calculate field box is:

`!NLCD_Land! + '-' + !RatingStr!`

One of these formulas will create a field of the combined NLCD land usages and HSG soil types formatted in the same way as your reference table.

3. Create a field called "shape_area_weight." Right click the new field and select "calculate geometry." You will now calculate the acreage of each shape in this field. This will be used to create a weighted curve number for each subbasin.

Step 9: Insert the *CN* value in the attribute table of the intersected land use and soil type shapefile.

1. Execute the "Join Field" tool in geoprocessing.
   a. Input Table: dissolved shapefile.
   b. Input Join Field: concatenated field.
   b. Join Table: *CN* reference table worksheet.
   d. Join Table Field: the "NLCD_HSG" column.
   e. Transfer Fields: the "Curve_Number" Column.
2. Create a field in the dissolved shapefile attribute table that is titled "Weighted_CN." Execute a "Calculate Field" that computes the weighted curve number, specifically:

   [Curve Number field]*[Shape_Area field]/[Basin Area field]

   This will yield a weighted curve number based on the percentage of the total subbasin that each particular shape occupies (Box 20.2).
3. Dissolve the dissolved shapefile by Basin ID. Using the "statistics field(s)" option, take the sum of the weight curve number for each subbasin. This will calculate the overall curve number for each subbasin (see Boxes 20.3 and 20.4).

Steps 2 and 3 above can be also done in an alternative way. The zonal statistics tool can be used after converting the vector data to raster and calculating the mean *CN* value for each catchment used as the zone.

Step 10: Label curve numbers.

The calculated curve numbers have large significant digits. In order to use or label *CN* values with required significant digits, the following procedure can be used. In the new curve numbers shapefile, go to "Labeling." Select the tag symbol in the "Label Class" section (Box 20.5).

Type the command shown in Box 20.6 and apply the label to round the curve number to the nearest whole number on the map.

Figure 20.22 shows the *CN* values calculated for the catchments of Lee Creek Watershed, labeled on the map.

Box 20.1 Attribute table of dissolved shapefile

Shape*	NLCD_Land_Cover_...	RatingStr	basinid	Basin Area
Polygon	Woody Wetlands	D	6	3725.46
Polygon	Woody Wetlands	C	19	175.8013
Polygon	Woody Wetlands	B	19	175.8013
Polygon	Shrub/Scrub	D	19	175.8013
Polygon	Shrub/Scrub	D	8	1117.535
Polygon	Shrub/Scrub	D	2	1630.495
Polygon	Shrub/Scrub	D	16	2392.094
Polygon	Shrub/Scrub	D	11	2614.528
Polygon	Shrub/Scrub	D	10	2775.06
Polygon	Shrub/Scrub	D	6	3725.46
Polygon	Shrub/Scrub	C	19	175.8013

Box 20.2 Addition of weighted curve number to the attribute table

	OBJECTID*	Shape*	NLCD_Land_Cover_...	RatingStr	basinid	Basin Area	Shape_Length	Shape_Area	Concatted	Shape Area	Curve Number	Weighted_CN
1	328	Polygon	Woody Wetlands	D	6	3725.46	787.4	29062.441873	Woody Wetlands-D	0.667182	86	0.015401
2	103	Polygon	Woody Wetlands	C	19	175.8013	4308.930288	260796.077731	Woody Wetlands-C	5.987054	86	2.928799
3	27	Polygon	Woody Wetlands	B	19	175.8013	140.727478	765.899155	Woody Wetlands-B	0.017583	86	0.008601
4	213	Polygon	Shrub/Scrub	D	19	175.8013	37.057157	42.559166	Shrub/Scrub-D	0.000977	77	0.000428
5	229	Polygon	Shrub/Scrub	D	8	1117.535	4724.4	174374.651251	Shrub/Scrub-D	4.003091	77	0.275819
6	247	Polygon	Shrub/Scrub	D	2	1630.495	4949.311141	176135.628391	Shrub/Scrub-D	4.043518	77	0.190955

Box 20.3 The geoprocessing steps

Box 20.4 Generation of weighted curve number

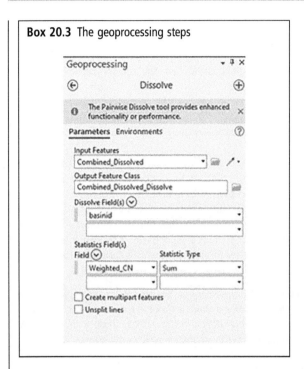

	OBJECTID *	Shape *	name	SUM_Weighted
1	1	Polygon	Sub - 1	89.383572
2	2	Polygon	Sub - 10	87.108534
3	3	Polygon	Sub - 11	84.161696
4	4	Polygon	Sub - 12	89.049356
5	5	Polygon	Sub - 13	86.814393
6	6	Polygon	Sub - 14	82.93308
7	7	Polygon	Sub - 2	85.034543
8	8	Polygon	Sub - 3	88.17481
9	9	Polygon	Sub - 4	88.834278
10	10	Polygon	Sub - 5	89.086015
11	11	Polygon	Sub - 6	90.31762
12	12	Polygon	Sub - 7	86.078085
13	13	Polygon	Sub - 8	85.301589
14	14	Polygon	Sub - 9	87.992343

Box 20.5 Selection of the label characteristics

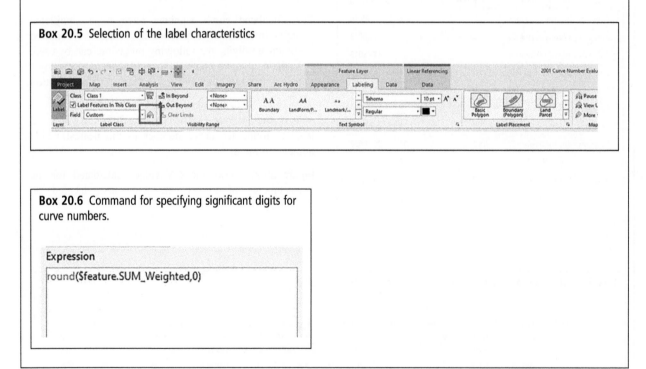

Box 20.6 Command for specifying significant digits for curve numbers.

Expression

round($feature.SUM_Weighted,0)

Exercises: A selection of exercises on this topic is available at www.cambridge.org/appliedhydrology.

21 Hydrologic Modeling

21.1 CONCEPTS AND PURPOSE

The ultimate objective of mastering the theories and procedures described in the previous chapters of this book is to build models of hydrologic systems that can be used for certain purposes, such as the design of hydraulic structures or planning for infrastructure related to water resources. If the hydrology is not done right, everything else that follows will be in error. In other words, hydrologic models are the foundations of all water resources engineering and planning. For this reason, the models must be as accurate as possible to reflect the real world. In this chapter, the underlying principles, and methods of calibration of hydrologic models are presented. Examples of major applications of hydrologic models are also given in this chapter.

21.2 TYPES OF HYDROLOGIC MODELS

Hydrologic models can be classified in various ways and accordingly there are various types of hydrologic models. Jain and Singh (2019) provided a good account, discussing how various types of hydrologic models can be classified, and their characteristics. In this book, the primary emphasis is placed on a type of models that can be called **mathematical models**. In such models, various hydrologic processes are represented by a system of equations, which can either be physically based or purely empirically derived. The primary objective of these models is to simulate the rainfall–runoff process. Such models are most useful and widely used by engineers for design, management, and planning of water resources projects. The software that is based on this principle and has global acceptance and routine use in hydrologic engineering is HEC-HMS. For this reason, throughout this book, the usage of HEC-HMS in conjunction with the discussions of hydrologic processes have been discussed in detail. The HEC-HMS technical reference manual available online (USACE, N.D.) is an excellent resource. The models developed in HEC-HMS can further be called **deterministic** models because of the absence of probabilistic elements of model outputs.

The **mathematical and deterministic** rainfall–runoff models that have received the most emphasis in this book can be used to simulate a single rainfall–runoff event lasting for a few minutes to several days. These are called **event-based models**. The mathematical models can also be used to simulate rainfall–runoff processes spanning long durations ranging from several months to several years, covering alternating dry and rainy seasons. These are called **continuous simulations**. Both event-based and continuous models built in HEC-HMS can further be either *lumped* models or *distributed* models (see Chapter 12). In a lumped model a single value of certain parameters is considered to be representative of an entire catchment. In a distributed model, the catchment is subdivided into small elemental areas, which typically are square grids, and each grid has its own value of a certain parameter. In HEC-HMS, certain parameters, such as Clark unit hydrograph parameters, Green–Ampt parameters, rainfall, etc. can be used in distributed modeling.

Once a model has been developed to address a hydrologic issue that can be of assistance in hydraulic design or planning, it is necessary to determine the right values of the parameters used in the model. Results of a model will be useful only if the assumptions are valid and the parameters are correct. For this reason, after a suitable model has been selected and built, the next task is to find the values of the model parameters such that the model correctly simulates the behavior of the watershed and channels. Model parameters can be determined either by direct measurement or through calibration. In general, it is not possible to obtain the parameters of hydrologic models by field measurements. Even if certain parameters can be measured at one or more locations, they cannot represent the entire watershed due to their spatial and possibly temporal variability. In other cases, the parameters do not correspond to a measurable feature of a watershed. The following sections describe how observed streamflow can be used to optimize model performance by automatically estimating parameters, that is, by computerized calibration.

21.3 MODEL CALIBRATION THROUGH OPTIMIZATION

As the term parametric hydrology has been introduced in Chapters 1 and 2, each hydrologic model has a set of parameters. The value of each parameter is specified to use the model for estimating a direct runoff hydrograph or flood hydrograph. Previous chapters identified the parameters associated either with certain processes such as infiltration or methods such as channel routing, and described how those could be estimated from various characteristics of the watershed and channels. For example, the kinematic wave direct runoff model described in Chapter 10 has a parameter, N, that represents the overland roughness and it can be estimated from knowing the land use of a watershed. Some parameters can be estimated directly from field measurements. For example, the watershed or catchment area can be measured directly in the field using standard surveying procedures or from maps developed through surveying and using GIS technology. Certain parameters can be estimated indirectly from field observations. In this case, a field observation does not result in a value that can be input directly into a model. However, field observations can provide a strong recommendation for a parameter in a model. For example, soil texture is correlated with parameters like hydraulic conductivity and porosity, which are used in the Green–Ampt model. However, certain parameters cannot be estimated by field observation or measurement of channel or watershed characteristics. The parameter C_p in the Snyder unit hydrograph model is an example. This parameter has no direct physical meaning. Similarly, the parameter X in the Muskingum method of channel routing cannot be measured. It is simply a weight that indicates the relative importance of upstream and downstream flows in computing the storage in a channel reach. Even for parameters that can be derived from field observations, there is often enough uncertainty in the measured parameter value to require some adjustment of the estimates in order for the model to closely follow the observed streamflow.

How then can the appropriate values for the parameters be selected? If rainfall and streamflow observations are available, **calibration** is the answer. Calibration uses observed hydrometeorological data in a systematic search for parameters that yield the best fit of the computed results to the observed runoff. This search is often referred to as **optimization**.

The quantitative measure of the goodness of fit between the computed result from the model and the observed flow is called the **objective function**. Different objective functions measure the degree of variation between computed and observed hydrographs in different ways. Some functions report a value that decreases as the agreement between the simulated and observed values increases; and some do the opposite, increasing as the goodness of fit increases. The key to automated parameter estimation is a **search method** for adjusting parameters to direct the *objective function value* toward a better goodness of fit and find optimal parameter values. An optimal value for the objective function is obtained when the parameter values best able to reproduce the observed hydrograph are found. Constraints are set to ensure that unreasonable parameter values are not used.

Optimization is an iterative parameter estimation procedure. Initial values for all parameters are required at the start of the optimization trial time window. A hydrograph is computed at a target element by computing all of the upstream elements. The target must have an observed hydrograph for the time period over which the objective function will be evaluated. *Only parameters for upstream elements can be estimated.* The value of the objective function is computed at the target element using the computed and observed hydrographs. Parameter values are adjusted by a search method and the hydrograph and objective function for the target element are recomputed. This process is repeated until the value of the objective function is sufficiently small, or the maximum number of iterations is reached.

21.4 GOODNESS-OF-FIT INDICES

21.4.1 Objective Functions

To compare a computed hydrograph to an observed hydrograph, an **index of the goodness of fit** must be calculated. This index is given by an objective function, designated by Z. Thus, an objective function is a numerical measure of the closeness between the observed output of interest and that computed by the model. A calibration algorithm tries to find the model parameters that optimize the objective function. There are several choices to define an objective function, depending upon the needs of the analysis. A few objective functions that are commonly used in hydrologic modeling are described below. The goal of a calibration scheme is to find reasonable estimates of one or more of model parameters that yield the optimum value of the objective function. The goal of some optimization procedures can be to minimize the value of the objective function, whereas for others it can be the maximization of this function. It depends on how an objective function is formulated to achieve a specific goal of a model.

21.4.2 Sum of Absolute Errors

This objective function compares each ordinate of the computed hydrograph with the observed, weighting each equally (Stephenson, 1979). The index of comparison, in this case, is the difference in the ordinates. However, as differences may be positive or negative, a simple sum would allow positive and negative differences to offset each other. In hydrologic modeling, both positive and negative differences are problematic, as overestimates and underestimates are equally undesirable. To reflect this, the function sums the absolute differences. The objective function is defined as

$$Z = \sum_{i=1}^{N} |Q_o(i) - Q_c(i)| \qquad (21.1)$$

where N denotes the number of computed hydrograph ordinates, Q_o denotes the observed flows, and Q_c, the calculated flows, computed with a selected set of model parameters.

Thus, the function given by Eq. 21.1 is an implicit measure of fit of the magnitudes of peaks, volumes, and times of peak of the two hydrographs. If the value of this function equals zero, the fit is perfect: all computed hydrograph ordinates equal exactly the observed values. Of course, this seldom happens in real-world modeling.

21.4.3 Sum of Squared Residuals

This is a commonly used objective function for model calibration (Diskin and Simon, 1977). It too compares all ordinates, but uses the squared differences as a measure of fit. Thus, a difference of 10 m³/s scores 100 times worse than a difference of 1 m³/s. Squaring the differences also treats overestimates and underestimates as undesirable. The function is defined as

$$Z = \sum_i^N [Q_o(i) - Q_c(i)]^2 \qquad (21.2)$$

The function given by Eq. 21.2 too is an implicit measure of the comparison of the magnitudes of peaks, volumes, and times of peak of the observed and computed hydrographs.

21.4.4 Percentage Error in Peaks

This function measures only the goodness of fit of the computed-hydrograph peak to the observed peak. It quantifies the fit as the absolute value of the difference, expressed as a percentage, thus treating overestimates and underestimates as equally undesirable. It is calculated as

$$Z = 100 \frac{Q_c^{peak} - Q_o^{peak}}{Q_o^{peak}} \qquad (21.3)$$

Equation 21.3 does not reflect errors in volume or peak timing. This objective function is a logical choice if the information needed for designing or planning is limited to peak flow or peak stages. This might be the case for a floodplain management study that seeks to limit development in areas subject to inundation, with flow and stage uniquely related.

21.4.5 Peak-Weighted Root-Mean-Square Error

This function compares all ordinates by squaring the differences and then weighting the squared differences (USACE, 1998). The weight assigned to each ordinate is proportional to the magnitude of the ordinate. Ordinates greater than the mean of the observed hydrograph are assigned a weight greater than 1.00, and those smaller, a weight less than 1.00. The peak observed ordinate is assigned the maximum weight. The sum of the weighted, squared differences is divided by the number of computed hydrograph ordinates, thus yielding the mean-square error. Taking the square root yields the root-mean-square error. The function is defined by

$$Z = \sqrt{\frac{1}{N} \left[\sum_{i=1}^{N} \{Q_o(i) - Q_c(i)\}^2 \left(\frac{Q_o(i) + Q_o^{mean}}{2 Q_o^{mean}} \right) \right]} \qquad (21.4)$$

where Q_o^{mean} is the average of the observed-hydrograph ordinates. Equation 21.4 is an implicit measure of comparison of the magnitudes of peaks, volumes, and times of peak of the two hydrographs. The purpose of the weighting function is to weight deviations between observed and computed ordinates more heavily for higher observed discharges. This will tend to produce a relatively good fit for high discharges compared with low discharges, which is generally desired in flood flow analysis.

Equation 21.4 is a variant of an objective function called weighted least squares, which can be generalized as

$$Z = \sum_i^N w_i [Q_o(i) - Q_c(i)]^2 \qquad (21.5)$$

where w_i is the weight at time or step i. Weights w_i specify the importance given to fitting a particular hydrograph feature such as the peak discharge. If the weights are equal to 1.0, the weighted least-squares function becomes the simple least-squares function. The minimum possible value of Z is zero, which is obtained when the model perfectly reproduces all observed values. However, in reality, a zero value is not attained, as the model results rarely, if at all, match the observed values exactly.

21.5 MEASURES OF MODEL PERFORMANCE

The model performance is typically evaluated by comparing simulated and observed variables of interest. Many statistical indices are used in hydrologic modeling to evaluate the model performance. Commonly used such indices include root-mean-square error, designated as $RMSE$, the ratio of $RMSE$ and the standard deviation, RSR, the coefficient of determination, COD, the Nash and Sutcliffe efficiency, NSE, and the percentage bias, PB. Some of these indices such as $RMSE$, COD, and NSE were introduced in Chapter 3. Here these functions are defined with respect to comparison of observed and calculated hydrograph ordinates.

The root-mean square error,

$$RMSE = \sqrt{\frac{1}{N} \sum_{i=1}^{n} [Q_o(i) - Q_c(i)]^2} \qquad (21.6)$$

indicates the error in the units of the variable. An $RMSE$ value of 0 indicates a perfect match, whereas a value less than half of the standard deviation is considered low and unacceptable.

The ratio of $RMSE$ and the standard deviation, RSR,

$$RSR = \frac{\sqrt{\sum_{i=1}^{N} [Q_o(i) - Q_c(i)]^2}}{\sqrt{\sum_{i=1}^{N} [Q_o(i) - Q_o^{mean}]^2}} \qquad (21.7)$$

varies from zero, indicating perfect model simulation, to a large positive value. A low RSR indicates better model simulation.

The coefficient of determination, COD,

$$COD = \left[\frac{\sum_{i=1}^{N}\{Q_o(i) - Q_o^{mean}\}\{Q_c(i) - Q_c^{mean}\}}{\sqrt{\sum_{i=1}^{N}[Q_o(i) - Q_o^{mean}]^2}\sqrt{\sum_{i=1}^{N}[Q_c(i) - Q_c^{mean}]^2}}\right]^2 \quad (21.8)$$

lies in the range 0–1. Higher values of COD imply better simulation. Models with systematic over- or underprediction may also give good COD values. Hence, COD should not be used alone. The coefficient of a regression line, b, obtained by plotting the $Q_o(i)$ and $Q_c(i)$ values multiplied by COD, that is, $b \times COD$, is a better measure of a model performance.

The Nash and Sutcliffe efficiency, NSE,

$$NSE = 1 - \left[\frac{\sum_{i=1}^{N}[Q_o(i) - Q_c(i)]^2}{\sum_{i=1}^{N}[Q_o(i) - Q_o^{mean}]^2}\right] \quad (21.9)$$

is a very popular index developed by Nash and Sutcliffe (1970). It lies in the range $-\infty$ to 1. NSE is 1 for a perfect fit when all $Q_o(i) = Q_c(i)$. It has the value of 0 when the numerator and denominator are equal, which shows that the model is basically a *no-knowledge* model whose prediction is the mean of observed data for all time steps. Negative values of NSE are obtained when the model performs worse than this no-knowledge model. It is easy to get comparatively high values of NSE if the data have high variance. This index is not sensitive to systematic over- or underprediction by the model. Logarithmic NSE values are calculated with logarithms of observed and predicted values. By doing this, peaks are flattened and the influence of low values increases. Logarithmic NSE values are more sensitive to systematic over- or underprediction.

The percentage bias (PB) measures the average tendency of simulated data to be larger or smaller than observed values. Its best value is 0 and it is computed as

$$PB = \frac{100 \sum_{i=1}^{N}[Q_o(i) - Q_c(i)]}{\sum_{i=1}^{N} Q_o(i)} \quad (21.10)$$

21.6 OPTIMIZATION METHODS

21.6.1 Generalities on Optimization

Optimization can be of two types: **unconstrained** and **constrained**. In its general form, an unconstrained minimization problem has the following form:

$$\min_x f(x) \quad (21.11)$$

The function f is a scalar-valued function called an objective function or the cost function or the criterion. In some cases, f can be a vector-valued function instead of a scalar-valued one. Note that the problem of maximization can easily be turned into a minimization problem by taking the negative of the objective function. A constrained minimization problem is similar to an unconstrained minimization problem except that supplemental conditions on the solution must be satisfied. Such a problem has the following form:

$$\min_x f(x) \quad (21.12)$$
$$\text{subject to } x \in C$$

where C is a set defining the constraints. These constraints can take on many forms. For instance, there can be inequality constraints such as $x_i \geq 0$, or linear equality constraints such as $Ax = B$ for some matrix A and some vector B. There can also be more complex constraints such as $h(x) \geq 0$, where h is some function.

An **optimization algorithm** is an algorithm that provides a solution to an optimization problem. A vast number of optimization algorithms have been developed in various fields of engineering, economics, operations research, and sciences.

The ultimate goal of a minimization or maximization problem is to find a global minimum or maximum of the objective function. This is called **global optimization**. When the function f is differentiable, a necessary condition for it to be a local minimizer is:

$$\nabla f(x) = 0 \quad (21.13)$$

where $\nabla f(x)$ is the gradient of function f.

A **local minimum** of the function is not necessarily a global minimum of the function. Indeed, an objective function may have several local minima. Deciding whether a local minimum is a global minimum or not is an **undecidable problem**.[1]

Convexity is a particularly interesting property for a function. If an objective function is *convex*, it has only one minimum, which is therefore a global minimum of the function. However, the converse is not true; there are non-convex functions that have only one (global) minimum. If the function f is twice differentiable, f is convex if H_f, the Hessian matrix or Hessian which is a square matrix of second-order partial derivatives of a scalar-valued function, is positive definite.

21.6.2 Search Procedures

The goal of calibration is to identify reasonable parameters that yield the best fit of computed to observed hydrographs, as measured by one of the objective functions.

---

[1] In computability theory and computational complexity theory, an undecidable problem is a decision problem for which it is proved to be impossible to construct an algorithm that always leads to a correct yes-or-no answer.

Mathematically, this corresponds to searching for the parameters that minimize the value of the objective function. When there are more than two parameters, the objective function forms a surface in the parameter space, called the **response surface**. An optimization algorithm searches the response surface subjected to constraints on the feasible values of parameters and finds the optimum value of the objective function. When there are multiple parameters, the objective function forms a complex shape. The optimization method is a trial-and-error search procedure. Trial parameters are selected, the models are exercised, and the error is computed. If the error is unacceptable, the trial parameters are changed and the process is reiterated.

Two different approaches commonly used for model optimization can be classified as **deterministic** and **stochastic**. Deterministic optimization begins with initial parameter estimates and adjusts them so that the simulated results match the observed values, such as streamflow, as closely as possible. Two broad categories of deterministic optimization techniques are **local search** and **global search**. Stochastic optimization produces a collection of equally probable parameter sets that represent a sample from the joint distribution of the parameter population. The most well-known stochastic method is the **Monte Carlo** method. A **Markov chain** is another well-known stochastic model.

Deterministic and stochastic optimization are philosophically different approaches to the optimization problem. Deterministic optimization seeks to minimize the difference between model outputs and observed data by changing model parameters to find a single optimum set. Parameters optimized in this way may be used as parameter values in an ordinary simulation run. With the same parameters, a deterministic optimization will arrive at the same optimum parameters with each trial. Stochastic optimization infers what the likely model parameter values are in light of the observed data, and can only do so by creating a number of sets of parameters. This approach treats the parameters with uncertainty and does not return a single set of optimized parameters. In order to use parameter sets generated by a stochastic optimization method, the user must use uncertainty analysis and populate tables of sampled parameters. Stochastic optimizations will result in different parameter sets with each trial.

21.6.3 Deterministic Search: Local Search Methods

Local search methods are employed to find the minimum of **unimodal functions**. Unimodal functions have only one extreme value. For unimodal functions, any algorithm that continuously proceeds downhill, a direction of improving function value, will eventually reach the minimum function value. A direct search method uses only the information on the function value in the decision-making process. Starting at the initial point, a local search algorithm selects a direction to move and a step size. Reaching the new point, the function value is evaluated. Based on the differences in function values between initial and new point(s), the best trial direction to move to improve the function and the step size are predicted. A step is taken in the trial direction, and function value at the new point is evaluated. If the function value is lower than the previous point, it replaces the old value and the procedure is repeated. If the function value at the new point is worse than the previous point, the step size is reduced and another trial is made. The search is terminated when the strategy is unable to find a direction in which improvement is possible.

A typical direct search method makes use of the *gradient* of the objective function, since it is known that the value of a function changes fastest by moving in the direction of the gradient. Since a gradient is a local property which changes with location, it is necessary to evaluate the gradient at many points. Thus, these methods are iterative. A simple gradient search algorithm is

$$x^{(k+1)} = x^{(k)} - \alpha^{(k)} \nabla f\left\{x^{(k)}\right\} \tag{21.14}$$

where $x(k)$ is the solution at the kth step, and $\alpha^{(k)}$ is the step-length parameter. At the optimum point, the gradient value will be very close to zero. Two direct search methods implemented in HEC-HMS are described below. An optimization method introduced by Rosenbrock (1960) is also a good optimization algorithm used in hydrologic model calibration.

21.6.3.1 Univariate Gradient Algorithm

The **univariate gradient algorithm** is one of the **direct search** methods. The univariate gradient search algorithm makes successive corrections to the parameter estimate. If $x^{(k)}$ represents the parameter estimate with the objective function $f\{x^{(k)}\}$ at iteration step k, the search defines a new estimate $x^{(k+1)}$ at iteration step $k+1$ as

$$x^{(k+1)} = x^{(k)} + \Delta x^{(k)} \tag{21.15}$$

in which $\Delta x^{(k)}$ is the correction to the parameter. The goal of the search is to select the value so the estimates move toward the parameter that yields the minimum value of the objective function. One correction does not, in general, reach the minimum value, so Eq. 21.15 is applied recursively.

The gradient method is often based on Newton's method which is applicable if the function f is differentiable twice. It relies on a second-order Taylor series expansion (Chapter 2) of the function f around the point x, to define $\Delta x^{(k)}$. A second-order Taylor series approximation of the objective function can be written as

$$f\left\{x^{(k+1)}\right\} = f\left\{x^{(k)}\right\} + \left\{x^{(k+1)} - x^{(k)}\right\}\frac{df\{x^{(k)}\}}{dx} + \frac{\left\{x^{(k+1)} - x^{(k)}\right\}^2}{2}\frac{d^2f\{x^{(k)}\}}{dx^2} \tag{21.16}$$

Each step of Newton's method consists of finding the minimum of the quadratic approximation of the function f around the current point x. Ideally, $x^{(k+1)}$ should be selected so that $f\{x^{(k+1)}\}$ is a minimum. That will be true if the derivative of $f\{x^{(k+1)}\}$ is zero. To find this, the derivative of Eq. 21.16 is found, as discussed below, and set to zero, ignoring the higher-order terms. That yields

$$\{x^{(k+1)} - x^{(k)}\}\frac{df\{x^{(k)}\}}{dx} + \frac{\{x^{(k+1)} - x^{(k)}\}^2}{2}\frac{d^2f\{x^{(k)}\}}{dx^2} = 0 \quad (21.17)$$

Equation 21.17 is rearranged and combined with Eq. 21.15, yielding

$$\Delta x^{(k)} = -2\frac{\dfrac{df\{x^{(k)}\}}{dx}}{\dfrac{d^2f\{x^{(k)}\}}{dx^2}} \quad (21.18)$$

The first and second derivatives of Eq. 21.18 can be evaluated by numerical approximation at each iteration k. These are computed as follows:

1. Two alternative parameters in the neighborhood of $x^{(k)}$ are defined as $x^{(k1)} = 0.99x^{(k)}$ and $x^{(k2)} = 0.98x^{(k)}$, and the objective function value is computed for each.
2. Differences are computed, yielding $\Delta_1 = f\{x^{(k1)}\} - f\{x^{(k)}\}$ and $\Delta_2 = f\{x^{(k2)}\} - f\{x^{(k1)}\}$.
3. The derivative $df\{x^{(k)}\}/dx$ is approximated as Δ_1 and $d^2f\{x^{(k)}\}/dx^2$ is approximated as $(\Delta_2 - \Delta_1)$. When these approximations are substituted in Eq. 21.18, this yields the correction in Newton's method.

If more than a single parameter is to be found in a calibration, this procedure is applied successively to each parameter, holding all others constant. For example, if Snyder's C_p and t_p are optimized, C_p is adjusted while holding t_p at the initial estimate. Then, the algorithm adjusts t_p, holding C_p at its new, adjusted value. This successive adjustment is repeated for a specified number of times such as four or five. Then, the algorithm evaluates the last adjustment for all parameters to identify the parameter for which the adjustment yielded the greatest reduction in the objective function. That parameter is adjusted, using the procedure defined above. This process continues until additional adjustments do not decrease the objective function by at least a certain percentage, such as 1%.

21.6.3.2 Downhill Simplex Method

The **downhill simplex** algorithm is a simple optimization algorithm seeking the vector of parameters corresponding to the global extreme (maximum or minimum) of any n-dimensional function, $f(x_1, x_2, \cdots, x_n)$, searching through the parameter space or search area. It is a heuristic search method that does not make any assumption on the objective function to minimize. In particular, the objective function need not satisfy any condition of differentiability. It relies on the use of simplices, that is, polytopes of dimension $n + 1$. For instance, in two dimensions, a simplex is a polytope[2] with three vertices, a triangle. In three dimensions, a simplex is a tetrahedron.

One of the many algorithms available to use the simplex method is the Nelder and Mead algorithm, which is a numerical method used to find the minimum or maximum of an objective function in a multidimensional space. It is a direct search method, based on function comparison, and is often applied to nonlinear optimization problems for which derivatives may not be known. However, the Nelder–Mead technique can converge to nonstationary points on problems that can be solved by alternative methods. The technique was proposed by Nelder and Mead (1965) as a development of the method of Spendley et al. (1962). The downhill simplex method starts from an initial simplex. Each step of the method consists in an update of the current simplex. These updates are carried out using four operations: reflection, expansion, contraction, and multiple contraction. This is illustrated in Figure 21.1.

The Nelder and Mead algorithm searches for the optimal parameter value without using derivatives of the objective function to guide the search. Instead, this algorithm relies on a simpler method of direct search. In this search, parameter estimates are selected with a strategy that uses knowledge gained in prior iterations to identify good estimates, to reject bad estimates, and to generate better estimates from the pattern established by the good estimates.

In the Nelder and Mead search method, the simplex is a set of alternative parameter values. For a model with n parameters, the simplex has $n + 1$ different sets of parameters. For example, if the model has two parameters, a set of three estimates of each of the two parameters is included in the simplex. Geometrically, the n model parameters can be visualized as dimensions in space, the simplex as a polyhedron in n-dimensional space, and each set of parameters as one of the $n + 1$ vertices of the polyhedron. In the case of the two-parameter model, the simplex is a triangle in two-dimensional space, as illustrated in Figure 21.2A.

The Nelder and Mead algorithm evolves the simplex to find a vertex at which the value of the objective function is a minimum. To do so, it uses the following operations as described in USACE (N.D.):

1. Comparison: The first step in the evolution is to find the vertex of the simplex that yields the worst (i.e., greatest) value of the objective function and the vertex that yields the best (i.e., least) value of the objective function. In Figure 21.2B, these are labeled W and B, respectively.

---

[2] In elementary geometry, a polytope is a geometric object with flat sides or faces. Polytopes are the generalization of three-dimensional polyhedral shape to any number of dimensions. Polytopes may exist in any general number of dimensions n as an n-dimensional polytope or n-polytope. For example, a two-dimensional polygon is a 2-polytope and a three-dimensional polyhedron is a 3-polytope. In this context, flat sides mean that the sides of a $(k + 1)$-polytope consist of k-polytopes that may have $(k - 1)$-polytopes in common.

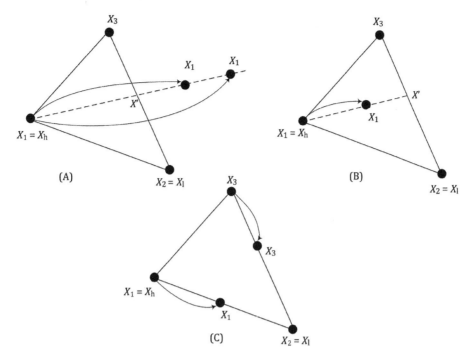

Figure 21.1 Illustration in two dimensions of the four fundamental operations applied to the current simplex by the downhill simplex method of Nelder and Mead: (A) reflection and expansion, (B) contraction, and (C) multiple contraction. The subscript h is the index of the "worst vertex."

2. Reflection: The next step is to find the centroid of all vertices, excluding vertex W; this centroid is labeled C in Figure 21.2B. The algorithm then defines a line from W through the centroid, and reflects a distance WC along the line to define a new vertex, R, as shown in Figure 21.2B. The algorithm for reflection is

$$x_i^{(reflected)} = x_i^{(centroid)} + 1.0 \left[x_i^{(centroid)} - x_i^{(worst)} \right] \quad (21.19)$$

3. Expansion: If the parameter set represented by the vertex R is better than, or as good as, the best vertex, the algorithm further expands the simplex in the same direction, as illustrated in Figure 21.2C. This defines an expanded vertex, labeled E in Figure 21.2C. If the expanded vertex is better than the best, the worst vertex of the simplex is replaced with the expanded vertex. If the expanded vertex is not better than the best, the worst vertex is replaced with the reflected vertex. The algorithm for expansion is

$$x_i^{(expanded)} = x_i + 2.0 \left[x_i^{(reflected)} - x_i^{(centroid)} \right] \quad (21.20)$$

4. Contraction: If the reflected vertex is worse than the best vertex, but better than some other vertex (excluding the worst), the simplex is contracted by replacing the worst vertex with the reflected vertex. If the reflected vertex is not better than any other, excluding the worst, the simplex is contracted. This is illustrated in Figure 21.2D. To do so, the worst vertex is shifted along the line toward the centroid. If the objective function for this contracted vertex is better, the worst vertex is replaced with this vertex. The algorithm for contraction is

$$x_i^{(contracted)} = x_i^{(centroid)} - 0.5 \left[x_i^{(centroid)} - x_i^{(worst)} \right] \quad (21.21)$$

5. Reduction: If the contracted vertex is not an improvement, the simplex is reduced by moving all vertices toward the best vertex. This yields new vertices R_1 and R_2, as shown in Figure 21.2E. The algorithm for reduction is

$$x_{i,j}^{(reduced)} = x_i^{(best)} + 0.5 \left[x_{i,j} - x_i^{(best)} \right] \quad (21.22)$$

The Nelder and Mead search terminates when either of the following criteria is satisfied:

$$\sqrt{\sum_{j=1}^{n} \frac{(z_j - z_c)^2}{n-1}} < \varepsilon \quad (21.23)$$

in which n is the number of parameters; j is the index of a vertex; c is the index of the centroid vertex; z_j and z_c are objective function values for vertices j and c, respectively; and ε is a predefined value called tolerance. The number of iterations reaches 50 times the number of parameters. The parameters represented by the best vertex when the search terminates are reported as the optimal parameter values.

The general stopping criterion used by Nelder and Mead (1965) is

$$\sqrt{\frac{1}{(n+1)} \sum_{i=0}^{n} \left[f(x_i) - \bar{f}(\bar{x}_i) \right]^2} \leq \varepsilon \quad (21.24)$$

with $\bar{f}(\bar{x}_i)$ the average of the values $\{f(x_i)\}_{i=0}^{n}$. This criterion has the advantage of linking the size of the simplex with

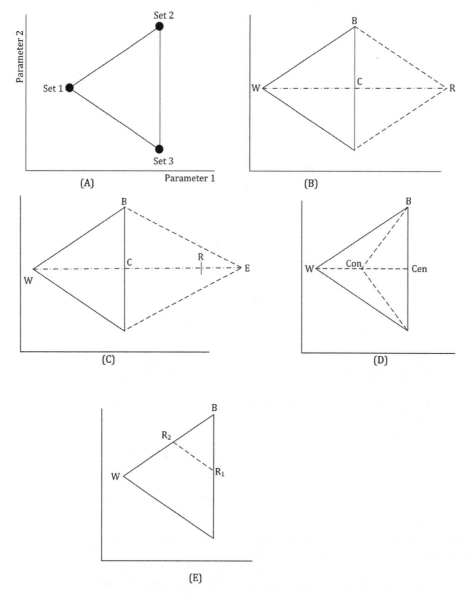

Figure 21.2 (A) Initial simplex for a two-parameter model. (B) Reflection of a simplex. (C) Expansion of a simplex. (D) Contraction of a simplex. (E) Reduction of a simplex.

an approximation of the local curvature of the objective function. A very accurate minimum is often obtained for a high curvature of the objective function. On the contrary, a minimum located in a flat valley of the objective function carries less information. Therefore, it does not make much sense to shrink an optimal simplex, which would be almost flat.

21.6.4 Deterministic Search: Global Search Methods

A response surface for more than two parameters is highly complex and can have multiple optima. It is difficult to find the global optimum in such cases, since the optimization algorithm may be trapped by one of the local optima. It is difficult to visualize the response surface if there are more than three parameters. In such cases, different starting values of parameters may give different optimal parameter values. Multi-start approaches run the local search algorithm repeatedly from different starting parameter sets in the feasible parameter space. The **shuffled complex evolution–University of Arizona (SCE–UA) global optimization method** developed by Duan et al. (1992) is a powerful multi-start algorithm that can be used to calibrate rainfall–runoff models. It combines the strengths of the simplex procedure of local search with the concepts of controlled random search competitive evolution and complex shuffling. This combination gives the SCE–UA global optimization method good global convergence properties over a broad range of problems encountered in hydrology.

21.6.5 Stochastic Search: Differential Evolution Method

The **differential evolution** search method behaves similarly to the simplex search method. However, iterations of the search vary the parameters at random instead of in a deterministic way. This means that two differential evolution searches set up in an identical way may proceed in different ways, and reach convergence in a different amount of time. However, any differential evolution search will settle on a single optimal parameter set based on the selected objective function and time series.

21.6.6 Stochastic Search: Markov Chain Monte Carlo Method

The general principle of **Markov Chain Monte Carlo (MCMC)** optimization is an algorithm that seeks to visit the plausible parameter sets in a parameter space by a random walk, and visit parameter sets that are more likely to have created the observed dataset more often. A MCMC search proceeds by generating a sequence of parameter values by making random transitions from one state of a Markov chain to another using a proposed distribution. Each state is a set of model parameter values for the parameters being optimized. If the present state, namely, the parameter values of the Markov chain, implies good agreement between the simulated and observed data, the iteration is less likely to make large jumps away from the present state. This produces repeated samples from a region of the parameter space that is associated with a higher likelihood that the parameter set produced the observed values. If the agreement is not good, larger random leaps across the parameter space are likely to occur. Multiple chains, which are initialized with different starting conditions, are used to assess whether the samples have escaped the starting conditions and the model has begun to draw samples from the highest-likelihood regions of the parameter space, which are unknown at the outset. The state of drawing from the highest-likelihood region of the parameter space is called *equilibrium*. It is desirable to have many samples from this region in order to characterize the statistical properties of the parameters associated with this high-likelihood space.

21.6.7 Constraints on the Search

Parameter optimization is essentially a mathematical procedure. It is not based on hydrologic principles. Therefore, it is quite possible that some of the parameter values obtained after automatic calibration make little sense from a hydrologic viewpoint. The mathematical problem of finding the best parameters for a selected model or models is what systems engineers refer to as a **constrained optimization problem**. That is, the range of feasible, acceptable parameters, termed **decision variables**, is limited for the systems engineer. For example, a Muskingum X parameter that is less than 0.0 or greater than 0.5 is unacceptable, no matter how good the resulting fit might be. Thus, searching outside that range is not necessary, and any value found outside that range is not accepted. For this reason, **constraints** giving the feasible range of parameters are also placed to address this problem but the user also needs to ensure that the chosen parameter set is consistent with the watershed and channel characteristics. Table 21.1 lists the constraints used by HEC-HMS for several parameters of certain hydrologic models.

During the search with either the univariant gradient or simplex algorithm, HEC-HMS checks at each iteration to ascertain that the trial values of parameters are within the feasible range. If they are not, the program increases the trial value to the minimum or decreases it to the maximum before it continues. In addition to these inviolable constraints, HEC-HMS also considers user-specified **soft constraints**. These constraints define desired limits on the parameters. For example, the default range of feasible values of the constant loss rate is 0–300 mm/h. However, for a watershed with dense clay soils, the rate is likely to be less than 15 mm/h, a much greater value would be doubtful. A desired range of 0–15 mm/h could be specified as a soft constraint. Then, if the search yields a candidate parameter outside the soft-constraint range, the objective function is multiplied by a **penalty factor**. This penalty factor is defined as

$$\text{Penalty factor} = 2 \prod_{i=1}^{n} (|x_i - c_i| + 1) \quad (21.25)$$

in which x_i is the estimate of parameter i, c_i is the maximum or minimum value for parameter i, and n denotes the number of parameters. This persuades the search algorithm to select parameters that are nearer the soft-constraint range. For example, if the search for the uniform loss rate leads to a value of 300 mm/h when a 15 mm/h soft constraint was specified, the objective function value would be multiplied by $2(300 - 15 + 1) = 572$. Even if the fit was otherwise quite good, this penalty will cause either of the search algorithms to move away from this value and toward one that is nearer 15 mm/h.

21.7 MODEL VALIDATION

After calibrating a model, it is necessary to examine the suitability of calibrated parameters. This is done by a process called **validation**, which involves running the model for the input data called **validation data** that were not used for model calibration. If the model has been calibrated properly, it will perform well on the validation data. If not, it is necessary to go back to calibration and improve it till the results of model validation are acceptable. Such a scheme is called a **split-sample test** where part of the data is used to calibrate the model and the remaining part is used for validation. If the results of validation are not acceptable, either the data contain some error or the assumptions behind the model do not hold good for the study watershed and another model may be chosen. A common practice is to use about 60–70% of the data for calibration and the remaining for validation.

Table 21.1 *Calibration parameter constraints*

Model	Parameter	Minimum	Maximum
Initial and constant-rate loss	Initial loss	0 mm	500 mm
	Constant loss rate	0 mm/h	300 mm/h
SCS loss	Initial abstraction	0 mm	500 mm
	Curve number	1	100
Green–Ampt loss	Moisture deficit	0	1
	Hydraulic conductivity	0 mm/mm	250 mm/mm
	Wetting front suction	0 mm	1000 mm
Deficit and constant-rate loss	Initial deficit	0 mm	500 mm
	Maximum deficit	0 mm	500 mm
	Deficit recovery factor	0.1	5
Clark's unit hydrograph	Time of concentration	0.1 h	500 h
	Storage coefficient	0 h	150 h
Snyder's unit hydrograph	Lag	0.1 h	500 h
	C_p	0.1	1.0
Kinematic wave	Lag	0.1 min	30 000 min
	Manning's N		
Baseflow	Initial baseflow	0 m^3/s	100 000 m^3/s
	Recession factor	0.000011	
Muskingum routing	K	0.1 h	150 h
	X	0	0.5
Kinematic wave routing	N-value factor	0.01	10

The concept of **equifinality of parameters** states that, for a given model structure and watershed, there may be more than one set of parameters that may simulate the hydrologic response of the watershed nearly equally well (Beven, 2001). In calibrating a model, even if the optimum parameters have been found, there may be some other sets of parameters that might simulate the catchment response equally well. This concept recognizes the approximations and simplifications of both the model structure and the data used for calibration.

21.8 SENSITIVITY ANALYSIS

Small changes in values of some parameters cause large changes in model results. Such parameters are termed **sensitive parameters**. Sensitivity analysis is carried out to identify such parameters before starting calibration and is often helpful to select the parameters that have the greatest influence on the model results. Another objective of sensitivity analysis is to determine the **region of indifference** for parameters, which is the region around the best value of parameters in which the function value varies insignificantly. The size and shape of this indifference region also give an idea of the amount by which the parameter can be in error without significantly impacting the results.

21.9 OPTIMIZATION OF MODELS IN HEC-HMS

HEC-HMS offers both deterministic and stochastic methods of optimization. Selecting the search method for the optimization trial is accessed from the "compute" tab of the Watershed Explorer in the optimization trial Component Editor.

Two deterministic search methods are available for optimizing the objective function and returning optimal parameter values. The **univariate method** evaluates and adjusts one parameter during the optimization simulation. If the univariate method is selected then the user can choose only one parameter for the program to adjust. The simplex method uses a downhill simplex to evaluate all parameters simultaneously and determine which parameter to adjust. The default method is the simplex method for its versatility. If the simplex method is chosen, then at least two parameters must be selected. Two controls should be specified for the iteration to stop. These are **tolerance** and **maximum number of iterations**. The tolerance determines the change in the objective function value between two successive iterations that will terminate the search. When the objective function changes less than the specified tolerance, the search terminates. The maximum number of iterations also can be used to limit the search. The search stops when the maximum number of iterations is reached regardless of changes in the objective function value or the quality of the estimated parameters.

In HEC-HMS, the differential evolution has two additional parameters beyond the two required for the simplex method. The **population size** controls the number of parameter sets and therefore the model evaluations that occur in each iteration. The default value of 30 is typically a good choice for hydrologic modeling applications. Using too few population members makes it more difficult to assess convergence, and using too many makes the search take longer. A tolerance value of 0.01 is a good value to start with; however, if one finds that there is still a notable variability in the parameter traces in a convergence trial, a smaller value may be required. Using too small of a value will increase the search time without substantially improving parameter estimates. Using too large a value will result in false convergence. The **seed value** initializes the pseudorandom number generator that creates the parameter random samples in the search. A search with the same seed will always produce the same random numbers, so two searches with the same settings and seed will always produce the same results. The seed value is initialized by the system clock when a new differential evolution search is created.

The present version of HEC-HMS offers only differential evolution as the stochastic search method. The MCMC option has been turned off in the HEC-HMS user interface and is expected to be available in the future versions. The MCMC search requires specifications of several parameterizations in order to operate. The "pool size" controls the number of independent Markov chains used in the simulation. The "initial sample" controls the number of burn-in samples to be taken before beginning to assess sample convergence. The "min iterations" and "max iterations" control the allowable range of iterations for simulation. The "convergence" allows the user to set the value for the Gelman–Rubin statistic that discriminates between pre- and post-equilibrium samples (the default value of 1.2 is generally sufficient; higher discrimination would be enforced using a lower number not lower than 1.0). The "equilibrium sample" controls the number of samples drawn after the simulation has achieved a state of equilibrium.

A start date and time and an end date and time for the optimization trial must be specified. The time control information is not specified in the wizard used to create the optimization trial. The time control information must be entered after the trial is created using the Component Editor for the optimization trial. The start date should be entered using the indicated format for numeric day, abbreviated month, and four-digit year. The end date should be entered using the same format. The start time and end time are entered using the 24-hour time format. A time interval should be chosen from the available options which range from 1 minute to 1 day. Finally, the start time and end time must each be an integer number of time intervals after the beginning of the day.

One of the principal tasks when creating an optimization trial using the wizard is the selection of components that will be used to compute optimization results. The components include the basin model and the hydrologic element in the basin model where the objective function will be computed. The components also include the meteorological model. These components are selected when creating a new optimization trial with the wizard. However, one can change the basin model and meteorological model one wishes to use at any time using the Component Editor for the optimization trial. The Component Editor from the Compute tab of the Watershed Explorer is accessed to select the components. If necessary, click on the Optimization Trials folder to expand it and view the available optimization trials. The Component Editor contains a basin model selection list that includes all of the basin models in the project where the basin model has at least one element with observed data. The Component Editor also contains a meteorological model selection list that includes all of the meteorological models in the project.

For running an optimization model, in addition to selecting the search method, an objective function must be selected. As discussed in Section 21.4.1, the objective function measures the goodness of fit between the computed results and observed data at the selected element. For the minimization goal, there are 14 different goodness-of-fit functions within HEC-HMS. These functions decrease as the agreement between simulated and observed values increases. The maximization goal can be used in two different ways: to maximize an element property, such as flow volume or peak discharge, reservoir stage, etc., or to maximize a goodness-of-fit statistic that increases in value as the goodness of fit increases.

Table 21.2 lists the objective functions available within HEC-HMS that are minimized and Table 21.3 lists those that are maximized. These objective functions are used in conjunction with a deterministic search method such as the simplex method. Only one goal and objective function are available for MCMC optimization: a minimization goal with the sum of squared residuals objective function. This combination allows the MCMC algorithm to assess the likelihood that a particular parameter set is the one that produces the observed data.

Fleming and Neary (2004) discussed the development of parameterization and calibration methodologies for the 12-parameter soil moisture accounting (SMA) algorithm available in HEC-HMS for continuous hydrologic modeling.

The maximization of an element time-series statistic, such as the flow volume, peak discharge, or especially the peak reservoir pool elevation, is used in conjunction with hazard analyses such as those required for dam safety studies. A particularly important optimization is when the trial is used in conjunction with the HMR 52 storm precipitation method (see Chapter 4) to maximize a statistic.

21.10 UNGAUGED WATERSHEDS

The rainfall and runoff data necessary to search for the calibration parameters required to develop the existing conditions of a watershed are often not available. Streamflow data may be missing, rainfall data may be sparse, or the available data may be unreliable. Furthermore, for most design work, runoff estimates are required for the forecasted future conditions of the watershed and for with-project conditions, and rainfall and runoff data are never available for these conditions. In the absence of data required for

Table 21.2 *Objective functions with minimization goal and their objectives*

Objective function	Motivation
First lag autocorrelation	Minimize systematic bias in residuals
Maximum of absolute residuals	Minimize the largest single distance between observed and simulated values
Maximum of squared residuals	Minimize the largest single distance between observed and simulated values
Mean of absolute residuals	Minimize the average distance between observed and simulated values
Mean of squared residuals	Minimize the average distance between observed and simulated values, with larger weight to larger errors
Peak-weighted *RMSE*	Minimize the average distance between observed and simulated values, with larger weight to data greater than the mean
Percentage error in discharge volume	Minimize the difference between observed and simulated volume
Percentage error in peak discharge	Minimize the difference between observed and simulated peak discharge value
Root-mean-square error	Minimize the average distance between observed and simulated values, with larger weight to larger errors; a classical choice
Sum of absolute residuals	Minimize the average distance between observed and simulated values
Sum of squared residuals	Minimize the average distance between observed and simulated values, with larger weight to larger errors
Time-weighted *RMSE*	Minimize the average distance between observed and simulated values, with larger weight to data near the end of the time window
Variance of absolute residuals	Minimize the variation in residual values
Variance of squared residuals	Minimize the variation in residual values, with larger weight to larger residual values

Table 21.3 *Objective functions with maximization goal and their objectives*

Objective function	Motivation
Coefficient of determination	Maximizes explained variance in observed data; also called R^2
Discharge volume	Maximizes the total volume discharged over the objective function time window
Index of agreement	Maximizes the dimensionless index of agreement statistic
Nash–Sutcliffe	Maximizes the dimensionless Nash–Sutcliffe efficiency statistic
Peak elevation	Reservoir element only; maximizes the single maximum reservoir elevation over the objective function time window
Relative index of agreement	Maximizes the dimensionless Index of Agreement statistic with less weight to large values
Relative Nash–Sutcliffe	Maximizes the dimensionless Nash–Sutcliffe efficiency statistic with less weight to large values

parameter estimation for either existing or future conditions, the stream and contributing catchment are declared **ungauged**.

To estimate runoff from an ungauged catchment, for existing or forecasted future conditions, the analyst can select a model that uses only parameters that can be observed or inferred from measurements, or extrapolate parameters from parameters found for gauged catchments within the same region. In practice, some combination of these solutions typically is employed, because most models include both physically based parameters and parameters that can be subjected to calibration, called the calibration parameters.

Physically based parameters are those that can be observed or estimated directly from measurements of catchment or channel characteristics. *Calibration parameters*, on the other hand, are lumped, single-valued parameters that have no direct physical significance. They must be estimated from rainfall–runoff data. If the data necessary for estimating the calibration parameters are not available, one solution is to use a flood–runoff analysis model that has only physically based parameters. For example, parameters of the

Muskingum–Cunge routing model are the channel geometry, reach length, roughness coefficient, and slope. These parameters may be estimated with topographic maps, field surveys, photographs, and site visits. Therefore, that model may be used for analysis of an ungauged catchment. Similarly, curve numbers can be estimated from land use and the hydrologic soil groups present in a catchment. Green–Ampt parameters can be estimated from the soil texture of the soil horizons. On the other hand, parameters for initial and constant loss cannot be estimated without rainfall–runoff data. Parameters of the Clark and the NRCS empirical unit hydrograph models have a strong link to the physical processes and thus can be estimated from the observation or measurement of catchment characteristics.

Another alternative for estimating parameters for an ungauged catchment is through extrapolation of gauged-catchment results. This extrapolation is accomplished by developing equations that predict the calibration parameters for the gauged catchments as a function of measurable catchment characteristics. The assumption is that the resulting predictive equations apply for catchments other than those from which data are drawn for the development of equations. The steps in developing predictive relationships for calibration parameters for a rainfall–runoff model are as follows.

1. Collect rainfall and discharge data for the gauged catchments in the region. The catchments selected should have hydrologic characteristics similar to the ungauged catchment of interest. For example, the gauged and ungauged catchments should have similar geomorphologic and topographic characteristics. They should have similar land use, vegetative cover, and agricultural practices. The catchments should be of similar size. Rainfall distribution and magnitude and factors affecting rainfall losses should be similar. If possible, data should be collected for several rainfall and resulting flood events. These rainfall and discharge data should represent, if possible, events consistent with the intended use of the model of the ungauged catchment. If the rainfall–runoff model is to be used to predict runoff from large design storms, data from large historical storms should be used to estimate the calibration parameters.
2. For each gauged catchment, use the data to estimate the calibration parameters for the selected rainfall–runoff model.
3. Select and measure or estimate physiographic characteristics of the gauged catchments to which the rainfall–runoff model parameters may be related. USACE (1994) recommended certain characteristics of catchment that can be selected for such purpose. Some of these characteristics, such as the catchment area, are directly measured. Others, such as the Horton ratios, are computed from measured characteristics.
4. Develop **predictive equations** that relate the calibration parameters found in step 2 with characteristics measured or estimated in step 3. In a simple case, the results of steps 2 and 3 may be plotted with a rainfall–runoff model parameter as the ordinate and a catchment characteristic selected in step 3 as the abscissa. Each point of the plot represents the value of the parameter and the selected characteristic for one gauged catchment. With such a plot, a relationship can be fitted by eye and sketched on the plot. Regression analysis, described in Chapter 3, is an alternative to the subjective graphical approach to define a predictive relationship. Regression procedures numerically determine the optimal predictive equation. To apply a parameter-predictive equation for an ungauged catchment, the independent variables in the regression equation are measured or estimated for the ungauged catchment. Solution of the equation with these values yields the desired flood–runoff model parameter. This parameter is used with the same model to predict runoff from the ungauged catchment.

It should be noted here that the predictive equations are subject to the same errors as rainfall–runoff models. The form and parameters of the equations are not known and must be found by trial and error. The sample size upon which the decision must be based is very small by statistical standards because data are available for relatively few gauged catchments. The reliability of a regionalized model can always be improved by incorporating a larger database into the analysis. Predictive equations are also subject to input error. Many of the catchment characteristics used in predictive equations have considerable uncertainty in their measured values. For example, the accuracy of stream length and slope estimates are a function of the map scale, discussed in Chapter 20. Furthermore, many of the characteristics are strongly correlated, thus increasing the risk of invalid and illogical relationships.

21.11 APPLICATIONS OF HYDROLOGIC MODELING

Hydrologic models are used by water resources engineers for two types of applications, namely (1) design of hydraulic structures and (2) planning of water resources projects.

21.11.1 Design of Hydraulic Structures

For the design of hydraulic structures, hydrologic models provide estimates of runoff volume, peak flow rate, timing of peak flow, sediment yield, and rate plus the amount of evapotranspiration. Examples of hydraulic structures are culverts under roadways that can range from minor arterials to multi-lane interstate highways; bridges crossing minor streams to large rivers; reservoirs that can be relatively small detention basins for stormwater management at a local scale or can be large reservoirs for regional flood control and/or water supply; flood control pump stations; dams, levees, and spillways; sediment control basins; energy dissipaters and stilling basins; and erosion control measures for stream or river banks, including river training works.

In certain design applications, it may be sufficient just to obtain the estimates of peak discharge at a point of interest.

Typical examples include storm drainage systems where the rational method, discussed in Chapter 11, is used to estimate peak flows. However, it should be noted that in many cases of modern storm drainage systems designs, particularly in highly urbanized areas where overland flooding is a major issue or concern, simple application of the rational method is not sufficient. In such areas several variants of the Storm Water Management Model, commonly called SWMM, which uses the dynamic wave equation can be used (Mukhopadhyay et al., 2009a). In such cases it is necessary to derive the complete flood hydrographs at multiple locations of the drainage system.

In many other design applications, even if estimation of only the peak discharge is required, a derivation of the flood hydrograph at the point of interest may be necessary. This may be due to the size of the drainage area that is contributing flows to the point of interest or the need for more accuracy for the estimation. Examples of such cases can be major culverts and bridge crossings along highways and major roadways. In those cases, hydrologic models of the watershed or the drainage unit must be developed.

For the design of flood control measures, such as detention basins or reservoirs, pump stations, or other diversions, it is necessary to have the inflow hydrographs at the structures to be designed for calculations of storage or diversion requirements.

21.11.2 Planning of Water Resources Projects

Planning of water resources projects is done for several types of situations. Cited below are a few examples.

21.11.2.1 Flood Risk Assessments in Urban Developments

Hydrologic studies involving modeling are typically undertaken to analyze both existing and potential flooding problems in urban watersheds that are experiencing continued developments that can either exacerbate already existing flooding problems or can give rise to potential flooding problems in the future. These are commonly known as **flood studies**. Characteristics of urban watersheds typically include (1) engineered drainage systems for stormwater management, (2) short watershed lag times giving flashy direct runoff hydrographs (see Chapter 19), and (3) the occurrence of localized damage of properties adjacent to drainage channels due to flooding.

The objectives of urban flood studies are to (1) characterize existing flood impacts, (2) predict the impact of future development, and (3) identify and plan solutions to current and future flooding, including controls on land use. To meet these objectives of an urban flood study, typically peak flow, total runoff volume, and hydrograph timing, are calculated for the watershed under current development and future development conditions by developing hydrologic models for existing and proposed conditions. These values are subsequently used in alternative analysis and planning. An often-used practice is to employ the results from hydrologic models in a hydraulic model to calculate the peak stage and delineations of floodplain limits for various frequencies of discharge. In the United States, the floodplain limit or the water surface elevations that result from a discharge with an annual exceedance probability (AEP) of 0.1 is called the **base flood elevation**, commonly known as the BFE or 100-year floodplain. In general, the procedure to develop a watershed model and calculate these values includes the following steps:

1. Appropriate hydrologic modeling methods are selected to represent the watershed.
2. Data pertinent to the watershed characteristics are collected.
3. Regional studies and equations, if available, are utilized to estimate parameter values.
4. The models are developed with various precipitation events, using either historical or hypothetical frequency-based events as needed.
5. The model of the existing conditions is run and if historical data are available, then this model is calibrated to optimize the model parameters.
6. The results are analyzed to determine the required values such as the peak flow or total runoff volume.
7. The watershed model representing the existing conditions is then modified to reflect future changes in the watershed.
8. The models for proposed or future conditions are run with the same precipitation events as used in step 5.
9. The results of the two sets of models are compared to quantify the impact of watershed changes.
10. The development and modification of the watershed hydrologic models are analyzed *to determine the impacts of future developments or existing flooding problems*. These models can also be augmented to develop one or more options or alternatives for the mitigation of either existing, potential, or future flooding problems at a planning level. This is discussed further below under flood-loss reduction studies.

21.11.2.2 Flood-Frequency Studies

Flood-frequency studies, as discussed in Chapter 5, relate the magnitude of discharge, stage height, or volume of flow to the probability of occurrence or exceedance. The resulting flood-frequency functions provide information required for (1) evaluating the economic benefits of planned flood control engineering measures that are also known as flood-damage reduction projects, (2) sizing and designing water control measures if a target exceedance level or reliability is specified, (3) establishing reservoir operation criteria and reporting performance success, (4) establishing regulations for floodplain management, and (5) developing requirements for regulating local land use.

To meet the objectives of a flood-frequency study, peak flows, stages, and volumes for specified annual exceedance probabilities, also known as quantiles (Chapter 5), are required. The flow and stage frequency curves are often used for planning flood risk management and flood-damage calculations whereas the volumes are often used for sizing flood control structures such as detention ponds. The flood control measures collectively are referred to as a *project* when a study is undertaken for a watershed of interest. Typically, the quantile functions of discharge, stage, and volume are calculated for four scenarios: (1) watershed conditions *existing* at the time of study, *without* the project; (2) watershed conditions *existing* at the time of study, *with* the project; (3) watershed conditions *at some point in the future*, *without* the project; and (4) watershed conditions *at some point in the future*, *with* the project. As an example, the with-project condition might refer to the construction of a proposed detention basin in the subject watershed, while the without-project condition refers to the absence of this detention facility. The without-project future condition, therefore, is the most likely future condition of the watershed if no action is taken to resolve whatever problem is addressed by the study.

The general procedure for developing the frequency functions for a watershed where flood-risk reduction measures are being planned and the ensuing flood-damage calculations can be summarized as follows.

A set of rainfall–runoff-routing models of the watershed is developed reflecting the characteristics of the watershed and its constituent channels for the case of interest under existing or current and future conditions, without or with the impacts of the project. The existing, without-project model should be calibrated using observed data if available or verified using regional flood-frequency equations or flow estimates.

The rainfall estimates giving the intensity–duration–frequency (IDF) or depth–duration–frequency (DDF) relationships of the watershed under investigation are obtained from reliable sources such as governmental meteorological agencies. In the United States these are easily obtained from either NOAA Atlas 14 or other publications of the NOAA (e.g., for the Pacific northwest).

In the traditional approach to design flood estimation, the intention is to estimate the flood of a selected frequency from a design rainfall of the same frequency (Pilgrim and Cordery, 1975). Therefore, for a selected frequency, the IDF or DDF functions are used to define a precipitation hyetograph, which is then used in the rainfall–runoff-routing model to compute the peak flow, stage, or volume. Subsequently, the frequency or the AEP of precipitation is assigned to the peak flow, stage, or volume, with the assumption that if the median or average values of all the other parameters are used, the frequency of the derived flood should be approximately equal to the frequency of the design rainfall.

The process is repeated for a range of frequency events. The results are assembled to yield a complete frequency function for one of the four watershed conditions. The flood-frequency functions derived from hydrologic modeling are used for various alternative analyses and flood-damage calculations discussed next.

21.11.2.3 Flood-Loss Reduction Studies

Flood-frequency functions developed following the procedures described above provide quantitative information about the risk of flooding in a watershed. If the flow-frequency functions are combined with rating and elevation-damage information, the expected annual damage can be computed. This computation is the foundation for assessment and comparison of the effectiveness of flood-loss reduction plans. In the United States, flood-loss reduction studies are conducted by USACE for flood control projects that may need funding from the federal government. The procedures are described in detail by USACE (1995). It involves economic analysis of the planned alternative solutions of flood-damage reduction projects. Figure 21.3 shows the flood-frequency curves of a 15-square-mile urban watershed in northern California where future developments are proposed. As can be seen from the future conditions curve, the proposed developments would increase flood quantiles. These curves were developed using hydrologic models of the watershed using HEC-HMS. To mitigate this adverse impact of the proposed development, several alternatives were analyzed in these models. The flood-frequency curves for three alternatives are also shown in this figure.

21.11.2.4 Flood Warning Systems

A **flood warning system** (FWS) is an integrated system of data collection and transmission equipment, forecasting models, response plans and procedures, and human resources. Together these increase the **warning** or **lead time** of an impending flood that can potentially cause damage to properties and lives. With an increased lead time, public officials and citizens can take actions to reduce damage and to protect lives. An FWS is classified as a nonstructural damage reduction alternative. It does not itself reduce flood flows or flood stages, but it can substantially reduce damage incurred due to a specified stage.

There are many robust FWSs based on real-time observations of rainfall, runoff, and hydrologic and hydraulic models. In addition to the prediction of direct runoff hydrographs by one of the methods detailed in Chapter 9, channel routing, by either a hydrologic or hydraulic method described in Chapter 12, is the main component of an FWS. For example, as noted in Chapter 12, the NWS uses the lag and K method for their national FWS. Developing a real-time FWS in a watershed or river basin is a complex process, which requires multiple entities to collect, analyze, interpret, and communicate the appropriate data and, most importantly, to disseminate alerts and react to the pending event. Figure 21.4 shows a simplified representation of a computer-based FAS where real-time observations on rainfall and runoff in some upstream sections of a channel are

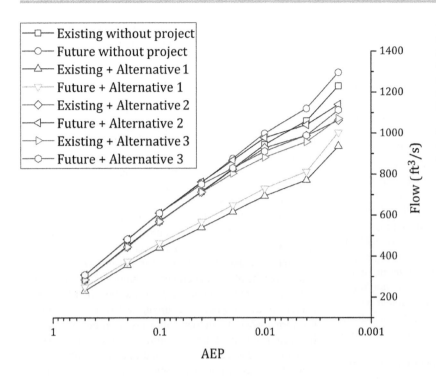

Figure 21.3 Flood-frequency functions of an urban watershed in northern California for different types of scenarios. The future curve without project refers to the watershed conditions that would potentially result from certain proposed developments. The curves with three alternatives to mitigate the impacts of development are also shown. Alternative 1 is a proposed detention basin, alternative 2 is a variant of alternative 2 with a diversion to reduce the required detention volume, and alternative 3 is a diversion without any detention.

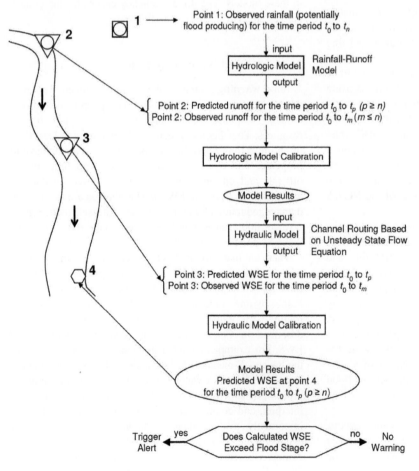

Figure 21.4 Schematic illustration of a flood alert system based on real-time coupling of rainfall observations (detection phase) with hydrologic and hydraulic modeling. In this figure t_0 denotes the initial time; t_m and t_n, the time after m or n units of time (usually hours); t_p is the time of peak discharge; and WSE = water surface elevation (after Mukhopadhyay et al., 2009b).

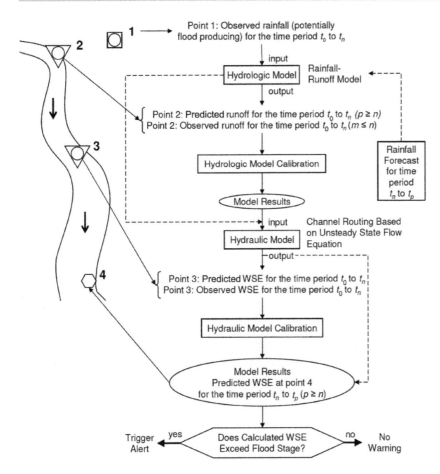

Figure 21.5 Schematic illustration of a flood alert system based on real-time coupling of rainfall observations (detection phase) and meteorological forecasts (predictive phase) with hydrologic and hydraulic modeling. Definitions are the same as in Figure 21.4 (after Mukhopadhyay et al., 2009b).

used to develop flood warnings at downstream points on the channel. This type of prediction is commonly referred to as the **detection phase** since flood-producing rainfall observations are used as an alert for potential flooding. In practice, the observation points generally form a more complex network and multiple locations receive warnings. In more advanced systems, the real-time observations of rainfall are augmented with data from weather forecast stations to run predictive models for the immediate future possibilities. Figure 21.5 is another simplified FWS where real-time observations on rainfall as well as weather forecast data from weather stations are used in conjunction. In this FWS, hydrologic and hydraulic models are driven by both real-time data gathered in the detection phase and forecasts of meteorological conditions such as rainfall and temperature. The forecasted weather data are used to generate predictive runoff and hydraulic models.

Hydrologic modeling plays a critical role in providing information for planning, designing, implementing, and operating an FWS. Studies are required to: (1) identify vulnerable areas for which a flood warning is an effective flood-damage reduction alternative; (2) establish rainfall and water-level thresholds for threat recognition; (3) link the thresholds to the vulnerability assessment, so that those who should be notified can be identified; (4) identify locations for rainfall and water-level sensors; and (5) develop and provide the tools for forecasting.

In a simple case, the hydrologic model predicts the flows for which the water level can reach a predetermined **threshold stage**. This threshold may be the elevation at which water flows out of the banks and damages property or threatens lives. If the level reaches the threshold, a warning is issued. A comparison of observations to thresholds may occur at the observation site or elsewhere. In the latter case, the data transmission components relay the observations to a central site, such as a base station at a flood warning center, for analysis.

In addition to the threshold stage, another important variable that is central to a FWS is a warning time or lead time, as noted above. The maximum potential warning times vary from storm to storm and location to location in a watershed. For example, if damageable property in the watershed is near the outlet, and if a short duration thunderstorm is centered near the outlet, the maximum potential warning time will be small. On the other hand, if the storm is centered far away from of the watershed outlet or if a forecast of precipitation is available before it occurs, the maximum potential warning time for this same location will be greater. Likewise, the antecedent soil moisture condition of the watershed plays a role in determining the maximum potential warning time.

An expected warning time for a site, can be given as (USACE, N.D.)

$$E[T_w] = \int T_w(p)\,dp \tag{21.26}$$

where $E[T_w]$ is the expected value of warning time, p is the AEP of the event being considered, and $T_w(p)$ is the warning time for an event with a specified AEP. The value of $E[T_w]$ can be calculated by a numerical approximation of Eq. 21.26:

$$E[T_w] = \sum_i T_w(p_i)\Delta p_i \tag{21.27}$$

From Eq. 21.27, it is evident that a hydrologic model with a set of frequency-based design storms can provide an estimation of the warning time. Figure 21.6 illustrates how the warning time will be computed for each event. First, the entire rainfall hyetograph is used to compute the entire runoff hydrograph. This is the runoff that would occur if the entire rainfall event occurred. The time that passes between the onset of rainfall and the exceedance of the threshold is the maximum potential warning time, T_{wp} (Figure 21.6). The question posed is: If an FWS is implemented, when will the system operators be able to detect or forecast the exceedance of the threshold flow? This time is called the recognition time (the detection phase described above) and is denoted by T_r (Figure 21.6). In other words, this is the time that passes before the threshold exceedance can be detected. Without the FWS, T_r will approach T_{wp}, and little or no time will remain for notification and action. The maximum mitigation time or the time available for taking precautionary measures, T_w, is then the difference between T_{wp} and T_r, that is,

$T_w = T_{wp} - T_r$. This difference can be estimated by hydrologic modeling. The procedure involves the following.

The threshold flow at a station where the warning would be given is established either from empirical observations of past events or from the hydraulic model of the channel of interest giving the flow that results in the overbank inundation. Typically, the data from the observation points are continually collected and transmitted to the central or base station of the FWS. However, a time interval, ΔT, is selected to represent the likely interval between successive forecasts or examination of data for identifying a threat with the FWS. A hydrologic model such as the HEC-HMS model of the watershed is run for each of the selected frequency-based storms. For each storm, the model is run recursively, using progressively more rainfall data in each run. For example, in the first run with the 0.01 AEP rainfall event, only 60 minutes of the storm can be used. With these data, a hydrograph is computed and examined to determine if the exceedance of the threshold is predicted. If not, 30 minutes of additional data can be added, and the computations repeated. When the exceedance of the threshold is detected, this defines the earliest detection time for each frequency storm. The time of the computation in each case is referred to as the **time of forecast**. The difference in time from the start of the precipitation event to the time of forecast where an exceedance is detected is T_r.

The computation scheme, presented above, simulates the gradual formation of a storm, observation of the data over time, and attempted detection of the flood event. The hydrologic model is run for each of the selected frequency-based storms using the entire precipitation event. The difference in time from the start of the precipitation event to the exceedance of the threshold is T_{wp}. After finding the detection time and maximum potential warning time for each frequency-based event, the expected values are computed

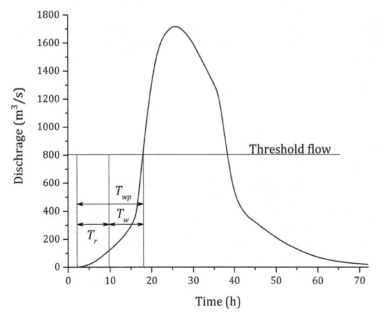

Figure 21.6 Discharge hydrograph for a 0.01 AEP event at Turning Basin, the outlet of Buffalo Bayou Watershed shown in Figure 21.7, illustrating the critical time parameters of an FWS.

using Eq. 21.27. Table 21.4 shows the warning stages and the available warning time for downtown Houston at the gauging stations along Buffalo Bayou shown in Figure 21.7.

21.11.2.5 Flood Forecasting

Flood forecasts or, more accurately called, streamflow forecasts are fundamental to the effective operation of flood control reservoirs and levee systems. Flood forecasting is discussed in Chapter 12, dealing with channel routing. Streamflow forecasts provide hydrographs (time of flood flows and volume) and peak flow estimates to inform reservoir operation. Forecasts may also support emergency operations by providing estimates of the timing and extent of expected hazardous or damaging flood conditions in the form of inundation mapping. Streamflow forecasts can be developed using a precipitation–runoff model of the watershed. The forecast simulations are based on recent meteorological and hydrologic conditions in the watershed including snowpack and soil moisture conditions. The forecast simulations may also incorporate predicted future meteorological conditions. The objective of precipitation–runoff modeling within a flood forecast is to provide reliable hydrograph estimates in a time-sensitive manner. The modeling effort must produce accurate results, which are used to provide information about the flood magnitude and to inform the future operation of reservoirs and other flood risk management projects.

Table 21.4 *Calculated warning stage and available warning time at various stations upstream of Texas Avenue along Buffalo Bayou, rainfall* AEP = 0.02 (Mukhopadhyay et al., 2009b)

Upstream gauging station	Warning stage (m)	Maximum available warning time (h:min)
Dairy Ashford	18.85	14:22
West Belt Drive	16.60	12:50
Gessner	15.92	12:05
Piney Point	14.97	11:28
San Felipe	14.01	10:36
Voss	12.98	10:07
Farther Point	11.32	8:58
Loop 610 Frontage	9.80	7:23
Shepherd	8.16	4:33

21.11.2.6 Reservoir Spillway Capacity Studies

Reservoir design demands special care because of the potential risk to human life. Chapter 13 describes how reservoir routing can optimize the reservoir spillway sizing. The location and capacity of a reservoir must be selected such that the net benefit is maximized. However, the capacity thus found may well be exceeded by rare meteorological events with inflow volumes or inflow rates greater than the reservoir's design capacity. This capacity exceedance may present a significant risk to the public downstream of the reservoir. Unless the reservoir has been designed to release the

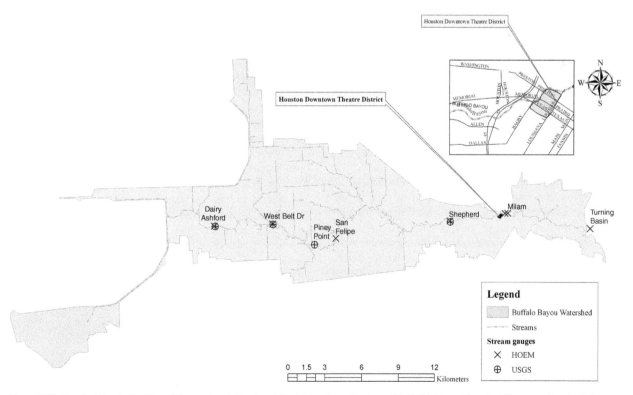

Figure 21.7 Map depicting the locations of the gauging stations in relation to downtown Houston within Buffalo Bayou Watershed. Downtown Houston is between Shepherd and Turning Basin.

excessive water in a controlled manner, the reservoir may fill and overtop. This may lead to catastrophic dam failure. Accordingly, the design a dam, and particularly the dam's spillway must be designed to pass safely a flood event caused by an occurrence of a rare event such as a PMP- one much larger than the design capacity of the reservoir. A spillway capacity study provides the information necessary for this design.

Spillway capacity studies are required for both proposed and existing spillways. For proposed spillways, the studies provide flow rates required for sizing and configuring the spillway. For existing spillways, the studies ensure that the existing configuration meets current safety requirements. These requirements may change as additional information about local meteorology becomes available, thus changing the properties of the likely extreme events. Further, as the watershed changes due to development or natural shifts, the volume of runoff into the reservoir due to an extreme event may change, thus rendering a historically safe reservoir unsafe. In that case, the spillway will be modified or an auxiliary spillway may be constructed.

21.11.2.7 Erosion and Sediment Routing studies

Erosion on overland surfaces, reservoir sedimentation, and in-stream sediment transport are important in watershed management and natural resources conservation planning studies. Processes involving sediment transport and deposition also affect evaluation and implementation of water quality *best management practices* and the evaluation of *total maximum daily loads* of regulated pollutants in the water bodies of the United States.

Specific erosion and sediment control objectives are usually specific to an area and depends on the objectives of study or project. For example, in certain areas it may be necessary to design sediment traps along with detention facilities. For designing such systems, it is necessary to estimate sediment yields from various design storms.

Surface runoff and erosion models can be used as a tool to model sediment loads from pervious and impervious areas in a watershed and then route the sediment downstream while modeling erosion and deposition within river reaches and reservoirs, as detailed in Chapter 16. Models are a useful tool for estimating flow and sedimentation in areas where observed data are unavailable, using data from watersheds with similar characteristics, or in conjunction with sensitivity and uncertainty analyses. Models can be especially useful for interpolating or extrapolating flow and sediment records in watersheds where observed data are available but not at the location of interest, or for predicting system response to projected future conditions (USACE, N.D.).

Potential soil erosion and sedimentation study objectives include (1) estimating the sediment yield either on an annualized basis or for individual frequency-based design storms for designing sediment control facilities; (2) estimating the sediment time series (sedigraphs) for each grain class either at selected locations or at selected drainage elements; (3) estimating the volume of deposition or erosion within a reach element; (4) predicting accumulated sediment at the reservoir bottom and accumulated sediment discharge from a reservoir for a given analysis period; (5) estimating changes in volume of eroded soil because of alternative land use management practices; (6) computing the quantity and gradation of sediment produced from a watershed to generate a sediment load boundary condition for a more detailed river hydraulics and sedimentation model, using an open channel hydraulics model; (7) using carefully constructed models designed to address these objectives to inform soil erosion and sediment management decisions; and (8) Estimating gross erosion for developing best management practices for conservation practices. A calibrated sediment model with accurate field data can predict regional and long-term sediment trends.

21.12 HYDROLOGIC SYSTEM SETTINGS

In Chapter 1, we introduced the concept of hydrologic system and thus far we have considered a watershed as a hydrologic system. The so-called **watershed approach** provides a framework to assess and manage water resources on a drainage basin. It focuses attention not just at one location for any issue, such as point discharge of industrial and municipal wastewater, stream instability due to erosion and sedimentation, or flooding in the stream corridors, but also on the effects of anthropogenic land uses on the waters in the entire watershed. As a result, hydrologic systems can also be classified based on the type of the setting of a watershed being considered. Thus, there are branches of hydrology called urban hydrology, agricultural hydrology, forest hydrology, coastal hydrology, wetland hydrology, lake hydrology, arid zone hydrology, karst hydrology, cryospheric hydrology, etc. In Example 21.1, we draw an example dealing with a forested watershed and another from an arid region in Example 21.2. This choice has been made mainly because we have cited many examples of watersheds in urban settings and the importance of urban hydrology for infrastructure development in the population centers. Exercise 21.2 goes back to urban hydrology.

Forest hydrology has gained increasing importance in a changing climate for forest and water resources management. In general, forest hydrology deals with the water balance of forests and natural woodlands. However, here we present rainfall–runoff modeling because this becomes important to address certain issues of forest management. For example, a question of ethics arises for storage and diversion of water for agriculture and letting the forests die of thirst. On the other hand, during flood events, it is important to delay discharge as long as possible by water retention in headwater catchments because forests and trees need water to grow and survive during drought periods. Several climate models predict that extreme weather events, such as heat, drought, and torrential rain, will continue to increase. There will be more precipitation in winter months and less in summer months.

The dominant characteristic of the climate of arid region is water deficiency or aridness. Basically, aridity involves a comparison between water supply and water need. In arid regions, evaporation is far greater than rainfall. McMahon (1979) defines arid areas as those with an average annual precipitation less than 500 mm and average annual potential evapotranspiration greater than 800 mm. Most formal definitions are in terms of the causes of aridity and are often based on comparisons between precipitation and some measure of potential evaporation. A good example is the classification published by UNESCO (1979), with its world map and explanatory notes. The degree of aridity is based on the ratio of mean annual precipitation to mean annual potential evaporation estimated by the Penman approach. The values of the ratio define three degrees of aridity: <0.03 for the hyper-arid zone, 0.03–0.20 for the arid zone, and 0.20–0.50 for the semiarid zone.

The generally low vegetation cover and high rainfall intensities experienced in arid zones contribute to runoff processes that appear to be dominated by an excess of rainfall over infiltration rates (Hortonian runoff) or through saturation excess on areas where very thin soils exist. The land surface of an arid region has sparse vegetation to protect the land surface from the erosive force of the raindrop impact. High-intensity thunderstorms are common throughout the area, and the infrequent but large precipitation excesses exert extreme erosive shear forces on the land surface. Thus, although the annual water yields from arid watersheds are generally low, the extremely high-intensity storms generate high sediment concentrations and yields per unit area. Runoff volumes differ markedly because of variability in precipitation and topography. In such environments, high-magnitude, low-frequency events are assumed to be dominant with respect to both river channel processes and soil erosion. Engineers working with sedimentation problems in an arid region must therefore understand hydrology, soils, and hydraulics. A lack of observed data provides a major problem for rainfall–runoff modeling in arid regions.

21.13 EXAMPLES

Example 21.1: The watershed used in this example is Lee Creek Watershed, which we introduced in Chapter 20 and derived curve numbers of the constituent catchments at the HUC 12-digit level. This watershed is located at the southern limit of Ozark National Forest. The land cover map of the watershed, given in Figure 21.8, shows that the watershed is mostly deciduous forest with patches of evergreen forest and pastures. Table 21.5 gives the numerical figures of the different land cover types derived from the land cover data using GIS software. The two major land uses of the watershed are forest land use (81%) and agricultural land use (13%).

The rain event that we will use resulted from a powerful weather system that impacted much of the central part of the United Sates from December 26 to 29, 2015. December is a winter month and such heavy torrential rain in a winter month is in line with the climate prediction noted above. There was major flooding, and damages at various locations. No rainfall data from any precipitation gauge located within the watershed boundary are available. However, the 24-hour storm total precipitation from a station at Van Buren is available from NOAA. As shown in Figure 21.8, Van Buren is located near the outlet of the watershed. We will use a lumped model where the rainfall data from Van Buren (Table 21.6) is assumed to be uniformly distributed over the entire watershed area.

Table 21.7 gives the rainfall depths for a certain duration frequency, at Van Buren. These precipitation frequency data are from NOAA Atlas 14 discussed in Chapter 4. Table 21.6 shows that for 24 hours on December 28, the rainfall depth was 4.89 inches, which is very close to 24-hour rainfall depth with a 10-year return period.

According to the data given in Table 21.6, the 48-hour rainfall total at Van Buren was 6.48 inches, which is again close to 10-year 48-hour rainfall total given in NOAA Atlas 14 (Table 21.7). From these observations, a 10-year, 48-hour frequency based storm is used in the HEC-HMS.

There are four USGS stream gauging stations within Lee Creek watershed. Figure 21.9 shows the streams and the gauging stations. The most downstream station, 7250085, is downstream of Leek Creek reservoir. The station upstream of the reservoir, 7250000, has flow records only up to 1992. The next station, slightly upstream of this station, is 7249985. This is located at Short, Oklahoma. We will use the streamflow records from this gauging station. The other station, 7249800, is further upstream and the station 7249920 is not on the main stem of Leek Creek.

Figure 21.10 shows the discharge hydrographs at station 7249985 (near Short, Oklahoma) and Figure 21.11 shows the discharge hydrograph at station 7250085, which is the outflow from a reservoir that is located along Leek Creek at this location. This is Lee Creek reservoir for which storage-elevation data are not readily available. For this reason, we will use only the hydrograph data from station 7249985 as

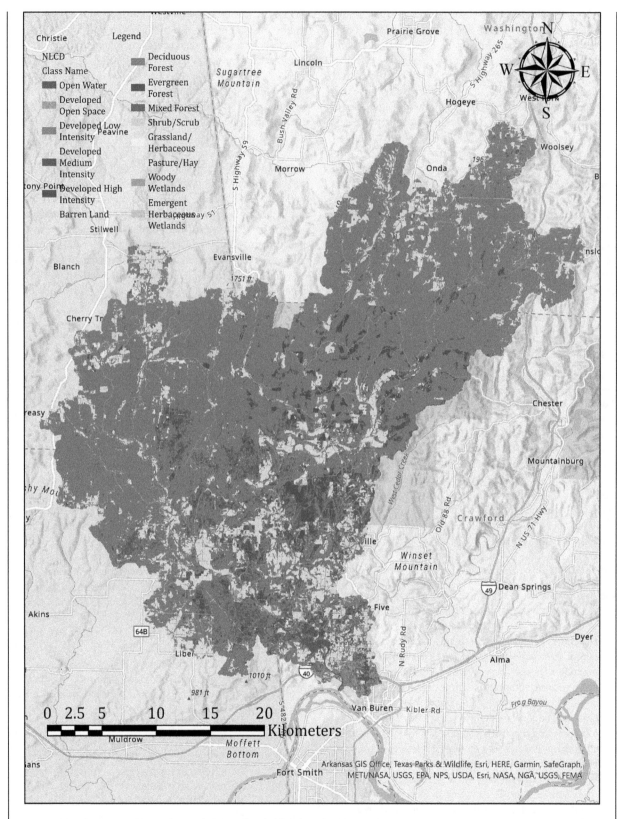

Figure 21.8 Land cover map of Lee Creek Watershed. (A black and white version of this figure will appear in some formats. For the color version, please refer to the plate section.)

Table 21.5 *Area and percentage of land cover classes within Lee Creek Watershed*

Land cover	Area (km^2)	Percentage of watershed area
Barren land	0.79	0.07
Deciduous forest	815.20	68.36
Developed high intensity	0.88	0.07
Developed low intensity	9.64	0.81
Developed medium intensity	3.41	0.29
Developed open space	35.19	2.95
Emergent herbaceous wetlands	0.92	0.08
Evergreen forest	37.70	3.16
Grassland or herbaceous	11.72	0.98
Mixed forest	110.62	9.28
Open water	5.38	0.45
Pasture or hay	147.37	12.36
Shrub or scrub	9.73	0.82
Woody wetlands	3.95	0.33

Table 21.6 *Twenty-four-hour rainfall totals at Van Buren, Arkansas during the storm event of December 27–28, 2015*

Date	24-hour rainfall (in)
12/26/2015	0
12/27/2015	1.59
12/28/2015	4.89
12/29/2015	0.1
12/30/2015	0

our calibration data for parameter optimization in HEC-HMS. Figure 21.12 shows the structure of the basin model developed in HEC-HMS for an event-based rainfall–runoff model. The gauging station 7249985 is located at Junction 2 which then will be used as the calibration point.

The NRCS curve number method and NRCS dimensionless unit hydrograph method for rainfall to runoff transformation have been selected for this exercise. The lag time is calculated by the lag method (Eq. 6.60). The parameters for abstraction and transformation are given in Table 21.8. The longest flow path and basin slope for the catchments are calculated from a DEM of the watershed using the GIS platform within HEC-HMS. The initial abstraction, I_a, is calculated using Eq. 7.57.

Use the Muskingum method for channel routing to derive the routing parameters, K and X, by optimization.

Table 21.7 *Rainfall depth (in) for certain duration and exceedance probability at Van Burn, obtained from NOAA Atlas 14*

	Exceedance probability					
Duration	1/2	1/5	1/10	1/25	1/50	1/100
5 min	0.445	0.549	0.626	0.725	0.798	0.869
10 min	0.651	0.804	0.917	1.06	1.17	1.27
15 min	0.794	0.981	1.12	1.3	1.43	1.55
30 min	1.18	1.46	1.67	1.93	2.13	2.31
60 min	1.57	1.98	2.29	2.7	3.01	3.31
2 h	1.95	2.5	2.92	3.47	3.89	4.31
3 h	2.2	2.83	3.32	3.99	4.5	5.03
6 h	2.67	3.38	3.95	4.75	5.4	6.07
12 h	3.21	3.92	4.5	5.36	6.06	6.81
24 h	3.78	4.58	5.25	6.21	7.01	7.86
2 d	4.37	5.47	6.36	7.62	8.62	9.66
3 d	4.75	6.02	7.04	8.46	9.59	10.8

Solution: The steps to conduct an optimization run in HEC-HMS are as follows:

Step 1: Run the model with an initial estimate of the parameters to be optimized.

This step is necessary. In this example, the following set up is made to make the initial run.

a. From a careful look at the observed hydrograph, the 48-hour time period is selected from midnight of December 27, 2015 to midnight of December 29, 2015. The control specification is shown in Box 21.1. Note that the time interval is selected at 15-minute interval to match the time interval of the hydrograph data obtained from USGS.

b. A meteorological model is set up as frequency-based storm using NOAA Atlas 14 data, as given in Table 21.7 and as per the discussion given above. Note that HEC-HMS offers centering of the design storm at 25%, 33%, 50%, 67%, and 75% of the time interval from the start to the end of the storm. This was discussed in detail in Chapter 4. The selection made here is based on making a few trial runs to see which selection gives a close match between the timing of the computed hydrograph peak and the observed hydrograph peak. A choice of 50% is deemed appropriate. No areal reduction is applied during the initial run. However, if the computed hydrograph gives a significantly higher peak than the observed, this can be adjusted appropriately. The setup of the meteorological model is shown in Box 21.2.

c. The observed hydrograph is entered in the Time Series data manager and selecting the Discharge Gauges as the data type. The date, time, and interval must be specified accurately as shown Box 21.3. The

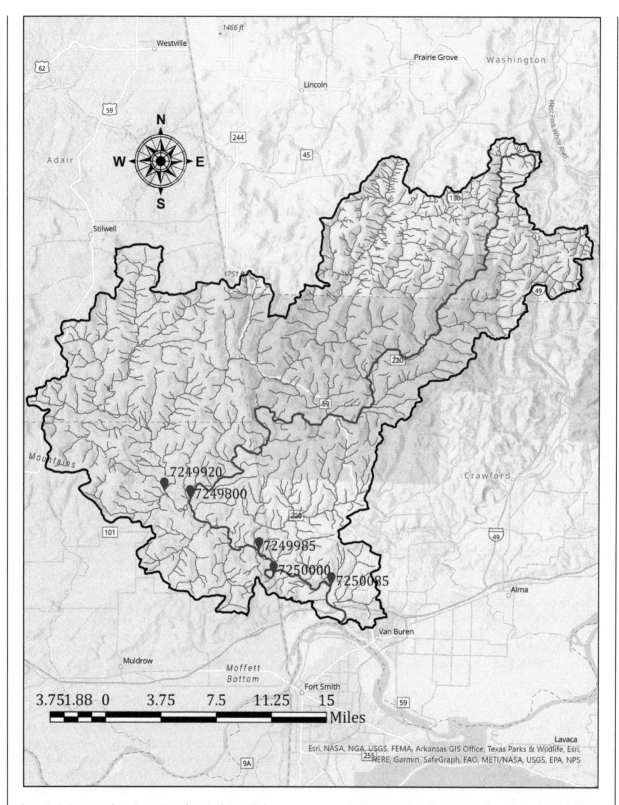

Figure 21.9 Streams and gauging stations of Lee Creek watershed.

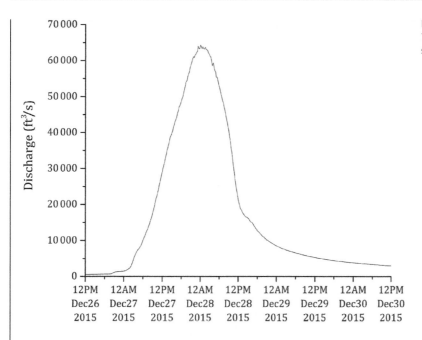

Figure 21.10 Hydrographs at gauging station 7249985 during the December 26–28, 2015 storm event.

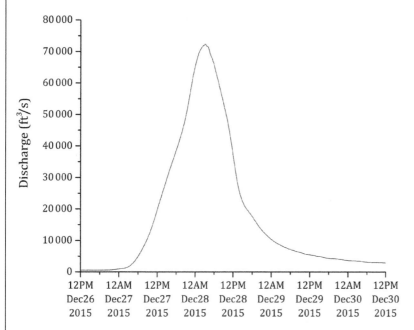

Figure 21.11 Hydrographs at gauging station 7250085 during the December 26–28, 2015 storm event.

location of this observed flow must be specified since, to optimize, the program needs to know where to compare the computed values with the observed values. This is done by selecting Junction 2 in the basin model and in the parameter edit window, selecting the option tab and in the observed flow option by providing the name of the time-series data (Box 21.4).

d. The basin model is set up with the transformation method as NRCS dimensionless unit hydrograph and loss method as the curve number method with the data given in Table 21.8. The parameters of the basin model are shown in Boxes 21.5 and 21.6. Note that because CN values were calculated by intersecting land use and soil cover data simultaneously, there was no need to enter the percentage of impervious covers in the subbasins (Box 21.6).

e. The reach routing method is set up with initial estimates of $K = 1$ h and $X = 0.25$ for all reaches (Box 21.7).

f. The model is run and the computed and observed hydrographs are compared at Junction 2 as shown in Box 21.8. Note that the computed hydrograph peak

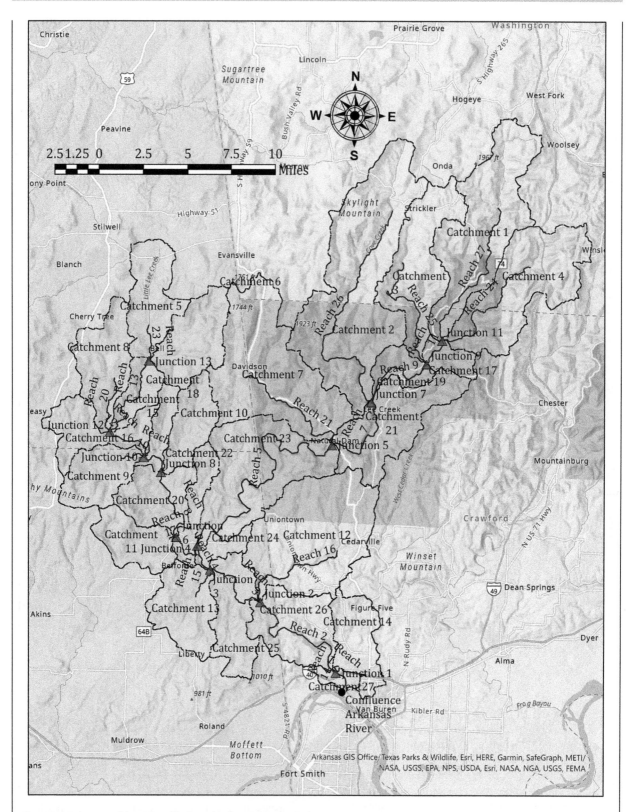

Figure 21.12 Structure of the HEC-HMS basin model of Lee Creek watershed.

Table 21.8 *Model parameters for rainfall–runoff modeling of Lee Creek watershed*

Catchment	CN	Slope (%)	Longest flow path (ft)	Lag time (h)	Lag time (min)	t_c (min)	I_a (in)
Catchment 1	83.6	21.59	98 695.71	2.40	143.76	239.60	0.3920
Catchment 2	82.9	20.19	118 932.79	2.95	176.79	294.65	0.4128
Catchment 3	82.4	18.22	91 697.13	2.56	153.73	256.22	0.4277
Catchment 4	83.2	23.26	80 648.67	1.99	119.58	199.30	0.4046
Catchment 5	80.1	11.75	46 798.28	2.00	120.10	200.17	0.4958
Catchment 6	84.5	16.57	40 573.21	1.30	78.29	130.49	0.3680
Catchment 7	83.2	18.70	86 110.62	2.34	140.53	234.22	0.4045
Catchment 8	83.3	14.43	52 032.18	1.78	106.58	177.63	0.4018
Catchment 9	81.7	14.12	50 523.63	1.85	110.94	184.90	0.4488
Catchment 10	84.7	16.39	40 304.62	1.29	77.54	129.23	0.3603
Catchment 11	80.8	15.31	46 172.54	1.70	101.80	169.67	0.4741
Catchment 12	82.3	11.41	71 677.48	2.66	159.86	266.44	0.4297
Catchment 13	79.5	10.21	39 186.36	1.90	114.20	190.33	0.5173
Catchment 14	84.7	9.81	33 804.99	1.46	87.32	145.53	0.3624
Catchment 15	83.4	16.02	36 753.18	1.27	76.28	127.13	0.3982
Catchment 16	83.9	15.83	48 076.62	1.56	93.54	155.89	0.3841
Catchment 17	83.0	20.46	22 796.56	0.78	46.67	77.79	0.4097
Catchment 18	83.9	15.63	30 913.08	1.10	66.02	110.03	0.3828
Catchment 19	81.7	19.29	36 443.35	1.22	73.00	121.66	0.4478
Catchment 20	80.5	16.03	37 145.28	1.41	84.40	140.66	0.4831
Catchment 21	81.8	17.64	61 141.34	1.92	115.16	191.93	0.4453
Catchment 22	80.0	13.12	39 766.21	1.67	100.04	166.74	0.4985
Catchment 23	80.3	15.64	107 555.87	3.36	201.73	336.21	0.4916
Catchment 24	77.8	7.14	33 584.50	2.12	126.99	211.66	0.5709
Catchment 25	80.7	13.05	42 036.09	1.71	102.84	171.39	0.4791
Catchment 26	82.8	12.74	61 781.91	2.21	132.36	220.60	0.4166
Catchment 27	84.3	10.67	18 623.35	0.88	52.53	87.56	0.3714

Box 21.1 Control specification in HEC-HMS

Control Specifications

Name: 10yr-48h
Description: Three days runoff
*Start Date (ddMMMYYYY): 27Dec2015
*Start Time (HH:mm): 00:00
*End Date (ddMMMYYYY): 29Dec2015
*End Time (HH:mm): 00:00
Time Interval: 15 Minutes

and observed hydrograph peak (black line) occur almost at the same time, confirming our choice of the storm centering at 50%, but obviously overall there is a large discrepancy between the observed and computed hydrographs that we desire to resolve through optimization.

Step 2: Set up an optimization trial run.

To perform optimization, select Compute −> Create Optimization Trial.

The time window for the observed flow, model simulation, precipitation data, and optimization run time must be all the same (Box 21.9).

Step 3: Select the search method.

In this example we have selected the simplex method, maximum number of iterations as 500, and set a tolerance limit of 0.01 (Box 21.10).

Step 4: Select the objective function from the choices listed in Table 21.2.

In this example we have we have selected peak-weighted *RMSE* as the goodness-of-fit index (Box 21.11).

Step 5: Assign the parameters that need to be optimized and specify for which element those parameters should be optimized.

As many parameters as one chooses can be selected. In this example we optimize the Muskingum K and X for all the nine reaches that are *upstream of Junction2*. Thus, 18 parameters are added to the Optimization Trials run. The number of parameters is added by

Box 21.2 Meteorological model in HEC-HMS

Frequency Storm

Met Name:	10-yr-48
Annual-Partial Conversion:	--None--
Annual-Partial Ratio:	1.00
Storm Duration:	2 Days
Intensity Duration:	5 Minutes
Intensity Position:	33 Percent
Area Reduction:	--None--
Spatial Distribution:	Uniform For All Subbasins

Duration	Depth (IN)	Are
5 Minutes	0.626	
15 Minutes	1.12	
1 Hour	2.29	
2 Hours	2.92	
3 Hours	3.32	
6 Hours	3.95	
12 Hours	4.5	
1 Day	5.25	
2 Days	6.36	
4 Days		
7 Days		
10 Days		

Box 21.3 Specification of the observed gauge data time series in HEC-HMS

Time-Series Gage

Gage Name:	Short_Gauge
Description:	Gauge at Short
Data Source:	Manual Entry
Units:	Cubic Feet Per Second
Time Interval:	15 Minutes
Station ID:	7249985
Latitude Degrees:	35.51731
Longitude Degrees:	-94.46438

Box 21.4 Specification of the junction where the gauge data are used in HEC-HMS

Junction | Options

Basin Name:	LeeCreekWatershed
Element Name:	Junction2
Observed Flow:	Short_Gauge
Observed Stage:	--None--
Stage-Discharge:	--None--
Ref Flow (CFS):	
Ref Label:	

Step 6: Run the optimization and compare the hydrographs.

Box 21.14 shows the computed and observed hydrographs at Junction 2. The main reason the rising limb of the computed hydrograph does not match well with the observed hydrograph is because the actual storm hyetograph might have been quite different from the synthetic hyetograph constructed from the frequency-based design storm as discussed in Chapter 4. Nonetheless, the match of the peak discharge and receding limbs of the hydrographs are acceptable.

Step 7: Look at the optimized parameters and see how the values changed from the initial estimates.

Box 21.15 shows the values of the parameters after optimization run.

Step 8: Look at objective function summary.

This should be examined to assess the statistics for goodness of fit and errors (Box 21.16).

Step 9: Look at the graphs of objective function and each parameter to see how those changed with each iteration.

The results of optimization of each parameter can be viewed by clicking on the parameter. An example is given in Box 21.17.

In this example more than 400 iterations are done by the program. The run time was 7 minutes and 37 seconds.

Step 10: Look at the final graph.

By clicking on the observed data, the final comparison between the observed and computed hydrographs can be examined (Box 21.18).

The example given here illustrates how optimization can be done in HEC-HMS, which offers great many

right-clicking the Optimization Trials run icon in the window explorer (Box 21.12).

By clicking on each parameter, its definition is given in the compute dialog box. Each parameter should be selected individually, and all specifications must be given (Box 21.13).

Box 21.5 Data for the hydrograph method in HEC-HMS

SCS Unit Hydrograph [LeeCreekWatershed]

Filter: --None--

Subbasin	Graph Type	Lag Time (MIN)
Catchment1	Standard (PRF 484)	143.7585269
Catchment10	Standard (PRF 484)	77.537
Catchment11	Standard (PRF 484)	101.8036819
Catchment12	Standard (PRF 484)	159.8638611
Catchment13	Standard (PRF 484)	114.1952004
Catchment14	Standard (PRF 484)	87.320
Catchment15	Standard (PRF 484)	76.279
Catchment16	Standard (PRF 484)	93.535
Catchment17	Standard (PRF 484)	46.675
Catchment18	Standard (PRF 484)	66.021
Catchment19	Standard (PRF 484)	72.996
Catchment2	Standard (PRF 484)	176.7923398
Catchment20	Standard (PRF 484)	84.398
Catchment21	Standard (PRF 484)	115.1588013
Catchment22	Standard (PRF 484)	100.0441238
Catchment23	Standard (PRF 484)	201.7269823
Catchment24	Standard (PRF 484)	126.9932789
Catchment25	Standard (PRF 484)	102.8352792
Catchment26	Standard (PRF 484)	132.3598281
Catchment27	Standard (PRF 484)	52.534
Catchment3	Standard (PRF 484)	153.7335585
Catchment4	Standard (PRF 484)	119.5789567
Catchment5	Standard (PRF 484)	120.104278
Catchment6	Standard (PRF 484)	78.294
Catchment7	Standard (PRF 484)	140.5336104
Catchment8	Standard (PRF 484)	106.5804551
Catchment9	Standard (PRF 484)	110.9404413

Compute: All Elements   Calculator...

Box 21.6 Data for the excess rainfall method in HEC-HMS

SCS Curve Number [LeeCreekWatershed]

Filter: --None--

Subbasin	Initial Abstraction (IN)	Curve Number	Impervious (%)
Catchment1	0.39202	83.7	
Catchment10	0.36029	84.7	
Catchment11	0.47411	80.9	
Catchment12	0.42966	82.3	
Catchment13	0.51727	79.7	
Catchment14	0.36238	84.7	
Catchment15	0.39818	83.4	
Catchment16	0.38408	83.9	
Catchment17	0.40968	83	
Catchment18	0.38281	83.9	
Catchment19	0.44780	81.7	
Catchment2	0.41285	82.9	
Catchment20	0.48314	80.5	
Catchment21	0.44531	81.8	
Catchment22	0.49847	80.1	
Catchment23	0.49163	80.3	
Catchment24	0.57088	77.8	
Catchment25	0.47914	80.7	
Catchment26	0.41658	82.8	
Catchment27	0.37137	84.4	
Catchment3	0.42774	82.4	
Catchment4	0.40463	83.2	
Catchment5	0.49579	80.2	
Catchment6	0.36805	84.5	
Catchment7	0.40455	83.2	
Catchment8	0.40183	83.3	
Catchment9	0.44882	81.7	

choices to arrive at the best answer. For example, the objective function can be changed to obtain better convergence. Similarly, the *CN* values and lag times can also be optimized following the same procedures by simply adding more parameters to the optimization trial. Alternatively, since the *CN* values and lag times were derived from the physical data, those are considered reasonably good estimates. So, after the Muskingum routing parameters are optimized, a separate optimization trial could be set up to further refine these two parameters. Finally, there are options in HEC-HMS to conduct uncertainty analyses with different sets of optimized parameters.

The HEC-HMS basin model of Lee Creek watershed is shown in Box 21.19.

Box 21.7 Initial estimated values of the channel routing parameters in HEC-HMS

Muskingum [LeeCreekWatershed]

Filter: --None--

Reach	Initial Type	Initial Discharge (CFS)	Muskingum K (HR)	Muskingum X	Number of Subreaches
Reach11	Discharge = Inflow		1	0.25	1
Reach9	Discharge = Inflow		1	0.25	1
Reach7	Discharge = Inflow		1	0.25	1
Reach5	Discharge = Inflow		1	0.25	1
Reach13	Discharge = Inflow		1	0.25	1
Reach12	Discharge = Inflow		1	0.25	1
Reach10	Discharge = Inflow		1	0.25	1
Reach8	Discharge = Inflow		1	0.25	1
Reach6	Discharge = Inflow		1	0.25	1
Reach4	Discharge = Inflow		1	0.25	1
Reach3	Discharge = Inflow		1	0.25	1
Reach2	Discharge = Inflow		1	0.25	1
Reach1	Discharge = Inflow		1	0.25	1

Box 21.8 Comparison of the observed and computed flood hydrograph at Junction 2 without optimization

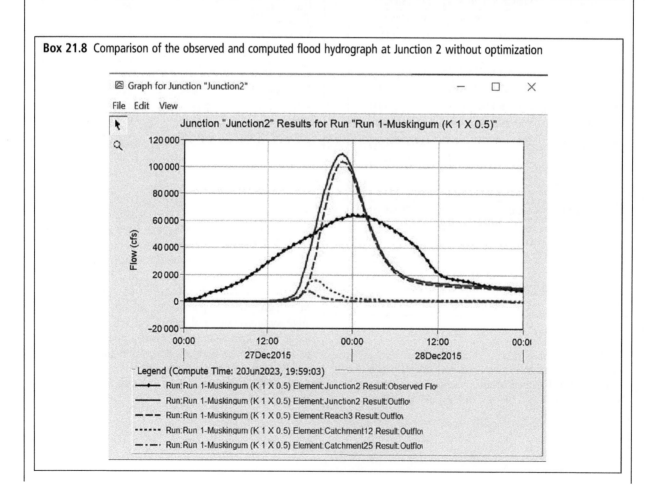

Box 21.9 Setup of the optimization trial run in HEC-HMS

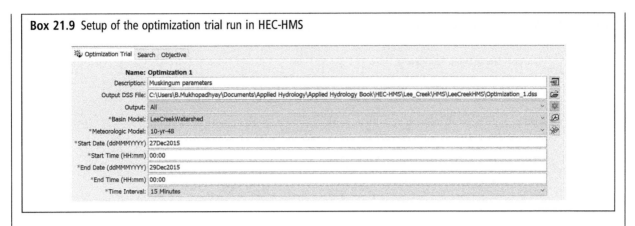

Box 21.10 Specification of the optimization method in HEC-HMS

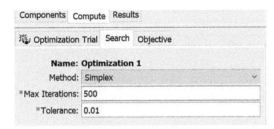

Box 21.11 Specification of the objective function to be used for optimization in HEC-HMS

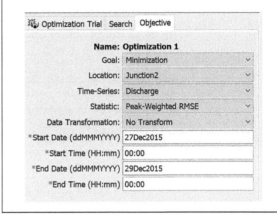

Box 21.12 Selection of the number of parameters to be optimized in HEC-HMS

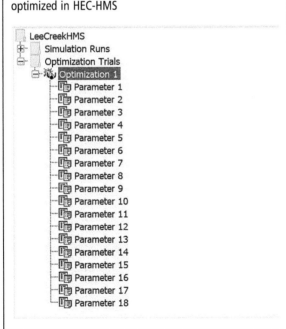

Box 21.13 Defining the parameter to be optimized in HEC-HMS

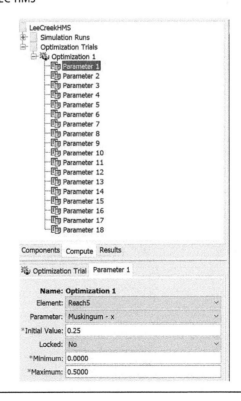

Box 21.14 Comparison of observed and computed flood hydrograph at Junction 2 after optimization in HEC-HMS

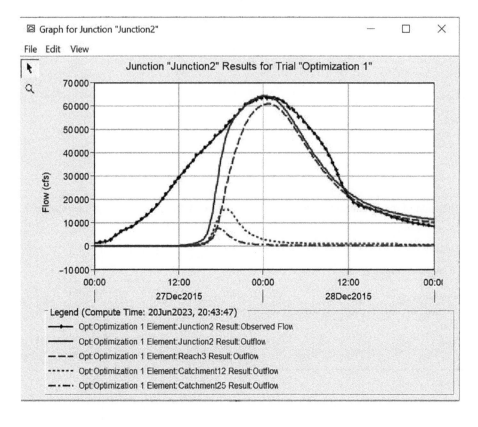

Box 21.15 Values of the optimized parameters

Element	Parameter	Units	Initial Value	Optimized Value
Reach5	Muskingum - x		0.25	0.25701
Reach5	Muskingum - K	HR	1	5.1261
Reach6	Muskingum - K	HR	1	0.0293546
Reach6	Muskingum - x		0.25	0.22970
Reach11	Muskingum - K	HR	1	1.7769
Reach11	Muskingum - x		0.25	0.23355
Reach9	Muskingum - K	HR	1	2.0743
Reach9	Muskingum - x		0.25	0.23257
Reach7	Muskingum - K	HR	1	1.8809
Reach7	Muskingum - x		0.25	0.23231
Reach13	Muskingum - K	HR	1	2.1681
Reach13	Muskingum - x		0.25	0.23207
Reach12	Muskingum - K	HR	1	1.9070
Reach12	Muskingum - x		0.25	0.23102
Reach10	Muskingum - K	HR	1	1.4754
Reach10	Muskingum - x		0.25	0.22965
Reach8	Muskingum - K	HR	1	1.0996
Reach8	Muskingum - x		0.25	0.22903

21.13 Examples

Box 21.16 Summary of the objective function after optimization in HEC-HMS

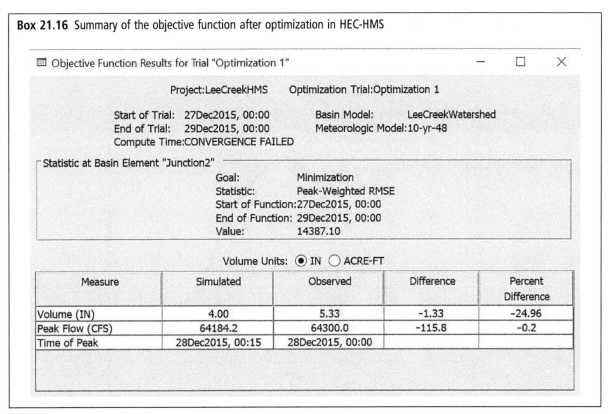

Box 21.17 Example of an optimization process in HEC-HMS

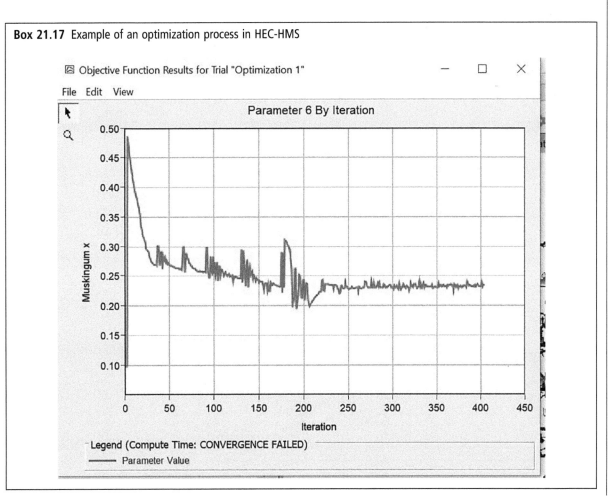

Box 21.18 Evaluation of the optimization results in HEC-HMS

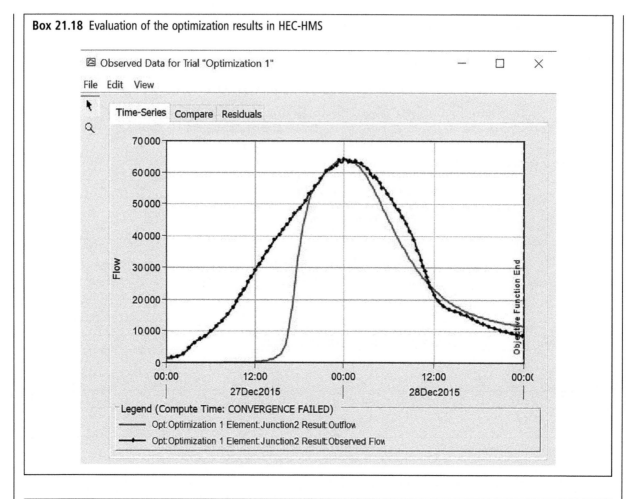

Box 21.19 HEC-HMS basin model of Lee Creek Watershed

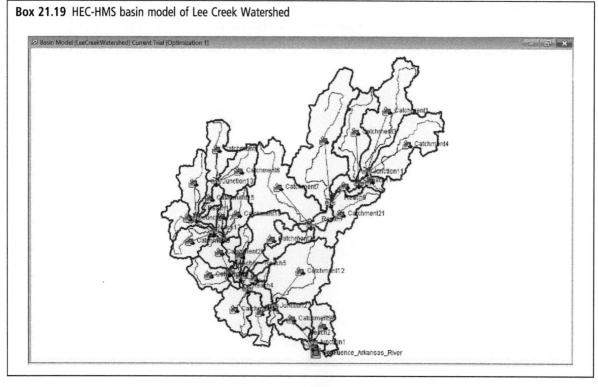

Example 21.2: In this example we use a watershed, designated as El Paso Hills Basin, from El Paso area of west Texas. According to CLIMWAT database, the average annual rainfall in this area is 254.17 mm (10 in) and average annual evapotranspiration is 1952 mm (77 in). Thus, the ratio of annual rainfall to annual evapotranspiration is 0.13, making it an arid zone, as noted in Section 21.12. Figure 21.13 shows the watershed and its constituent catchment boundaries overlain on the areal imagery. The total watershed area is 2.2 square miles. As can be seen from this imagery, most of the surface is bare ground with some residential developments in the catchments designated as S_01 and S_02. The streams shown in Figure 21.13 are ephemeral streams or arroyos. Sedimentation during high-intensity rainfall was identified as a hazard for the residential areas and for this reason a sediment control basin needed to be designed. How can design values of the eroded sediments be estimated from hydrologic modeling?

Solution: The steps to estimate sediment yield for design rainfall events in HEC-HMS are as follows:

Step 1: Construct HEC-HMS basin model.

Delineate the watershed and catchment boundaries and major streams using a DEM and following the procedures described in Chapter 6. The basin model overlying the DEM of the area is shown in Box 21.20.

Step 2: Develop a meteorologic model.

In arid zones, rainfall tends to be more variable in both space and time than in humid regions. For this reason, estimates of DDF of rainfall are derived at the centroid of each catchment from NOAA Atlas 14 and entered under the "Meteorological Model."

Step 3: Use the kinematic wave model for rainfall to runoff transformation.

Since most of the surface is bare ground and the area is in the arid region, the kinematic wave model is considered to be better suited than any unit hydrograph method for the transformation of rainfall to runoff. Each catchment

Figure 21.13 El Paso Hills Basin overlain on the areal imagery. (A black and white version of this figure will appear in some formats. For the color version, please refer to the plate section.)

Box 21.20 HEC-HMS basin model of El Paso Hills Basin

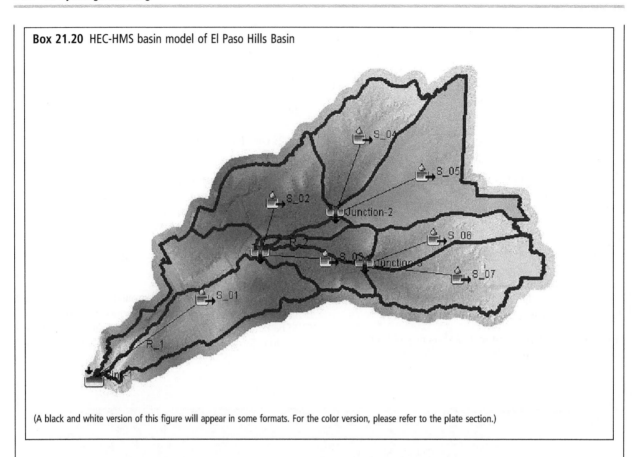

(A black and white version of this figure will appear in some formats. For the color version, please refer to the plate section.)

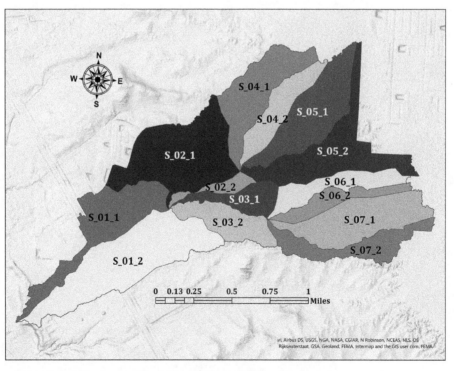

Figure 21.14 Approximation of overland surfaces by planes in kinematic wave modeling.

is modeled as two overland planes as shown in Figure 21.14. Table 21.9 shows the physical parameters of the planes in each of the catchments in the kinematic wave model. The characteristics of the main channels in each catchment are given in Table 21.10.

Each catchment has collector channels and catchment S_02 and S_07 have sub collector channels (Box 21.21).

Table 21.9 *Physical parameters of the planes in each catchment*

Subbasins	Length (ft)	Upstream elevation (ft)	Downstream elevation (ft)	Slope	Roughness[1]	Area (%)[2]
S_01_1	1177	3814.68	3774.58	0.0341	0.053	37
S_01_2	1470	3771.77	3750.53	0.0145	0.053	63
S_02_1	2167	3949.37	3864.68	0.0391	0.053	90
S_02_2	405	3889.58	3871.46	0.0447	0.053	10
S_03_1	400	3936.02	3884.13	0.1297	0.053	41
S_03_2	1200	3934.13	3881.63	0.0438	0.053	59
S_04_1	1450	3994.36	3964.47	0.0206	0.053	63
S_04_2	715	3968.00	3954.07	0.0195	0.053	37
S_05_1	1570	3980.63	3969.23	0.0073	0.053	47
S_05_2	1168	3993.00	3950.32	0.0365	0.053	53
S_06_1	500	3991.00	3966.12	0.0498	0.053	58
S_06_2	615	3983.76	3966.07	0.0288	0.053	42
S_07_1	1400	4003.93	3988.87	0.0108	0.053	63
S_07_2	450	4009.28	3987.71	0.0479	0.1	37

[1] Manning's *N* is selected for bare surface with sparse vegetation (see Table 10.2). [2] Percentage area equals the area of the plane/total area of the catchment expressed in percent.

Table 21.10 *Physical characteristics of the main channels in each subbasin*

Subbasin	Length (ft)	Upstream elevation (ft)	Downstream elevation (ft)	Slope (ft/ft)	Shape	Manning's n	Bottom width (ft)	Side Slope (xH:1V)
S_01	7789.91	3828.98	3668.30	0.0206	Triangle	0.04		15.53
S_02	3599.34	3890.33	3828.98	0.0170	Triangle	0.04		13.89
S_03	4402.47	3910.48	3828.98	0.0185	Trapezoid	0.04	300	19.09
S_04	5108.94	4000.00	3890.33	0.0215	Triangle	0.04		13.79
S_05	6827.40	4015.99	3890.33	0.0184	Triangle	0.04		25.91
S_06	4974.45	4009.69	3910.48	0.0199	Triangle	0.04		20.00
S_07	6414.43	4014.64	3910.48	0.0162	Triangle	0.04		35.45

Box 21.21 Collector and subcollector channel characteristics

Kinematic Wave [SOC1&SOC2_EX Green_Ampt]

Filter: --None--

Flow Planes | SubCollector and Collector Channels | Main Channel

Subbasin	Length (FT)	Slope (FT/FT)	Manning's n	Subreaches	Area (%)	Shape	Diameter (FT)	Width (FT)	Side Slope (xH:1V)
S_02 (SubCollector)	1136.37	0.0484	0.1	5	0.0373	Trapezoid		85	7.22
S_02 (Collector)	2226.98	0.0393	0.1	5	0.05022	Triangle			10.67
S_03 (SubCollector)				5		Trapezoid			
S_03 (Collector)	1714.30	0.0513	0.1	5	0.03143	Trapezoid		70	16
S_01 (SubCollector)				5		Trapezoid			
S_01 (Collector)	2896.00	0.0357	0.1	5	0.06458	Triangle			260
S_06 (SubCollector)				5		Trapezoid			
S_06 (Collector)	1391.89	0.0291	0.1	5	0.14751	Trapezoid		50	13.75
S_07 (SubCollector)	1478.51	0.0271	0.1	5	0.0176	Triangle			11.54
S_07 (Collector)	4640.28	0.0172	0.1	5	0.10560	Triangle			37.78
S_04 (SubCollector)				5		Trapezoid			
S_04 (Collector)	4187.58	0.0262	0.1	5	0.06844	Triangle			22
S_05 (SubCollector)				5		Trapezoid			
S_05 (Collector)	3409.75	0.0249	0.1	5	0.08933	Trapezoid		270	13.18

Step 4: Derive Green–Ampt Infiltration parameters from the soil textural class.

In arid and semiarid regions, current knowledge indicates that the predominant runoff mechanism is Hortonian overland flow where the rate of rainfall exceeds the potential rate of infiltration (Pilgrim et al., 1988). Short times to ponding result in the rapid onset of runoff after the start of rain, and the production of runoff with small depths of rain.

Given the greater evaporative demands in arid zones, the state of soil moisture is typically low. For this reason, initial infiltration in fine to medium textures soils is dominated by suction terms. Figure 21.15 shows the map of the soil texture classes present in this watershed. For these soil textural classes, the Green–Ampt parameters are derived, as shown in Table 21.11. However, for the reasons given above, the initial moisture content is assumed to be 20% of effective porosity. This assumption is made to generate more runoff and a short period of ponding. Using the Green–Ampt model requires specification of impervious areas separately. Impervious covers in terms of percentages of the total area of a plane representing the overland surface are derived from the NLCD land cover data and these values are also listed in Table 21.11.

Step 5: Calculate the MUSLE parameters.

Table 21.12 gives the parameters used in the estimation of the sediment yield for the reach of the frequency-based storm events using the MUSLE equation (see Chapter 16).

The soil erodibility factor, K, is obtained from the soil map (Figure 21.16)

In addition to MUSLE parameters, grain size distributions of the surface soil samples must also be input for sediment calculations. Laboratory analyses of the surface soil samples giving the grain size distributions, according to the AGU classifications discussed in Chapter 16, are simplified to represent the grain size with four major classes since this is one of the options available in HEC-HMS. Table 21.13 gives the gravel–sand–silt–clay fractions expressed in percentages for representative surface soils from each of the seven catchments of El Paso Hills Basin.

The following selections are made under the "Sediment" routine of HEC-HMS: (1) Sediment Transport Function – Yang; (2) Fall Velocity – Rubey (3) Specific Gravity – 2.65 (4) Sediment Routing – Volume Ratio. These parameters are discussed in Chapter 16.

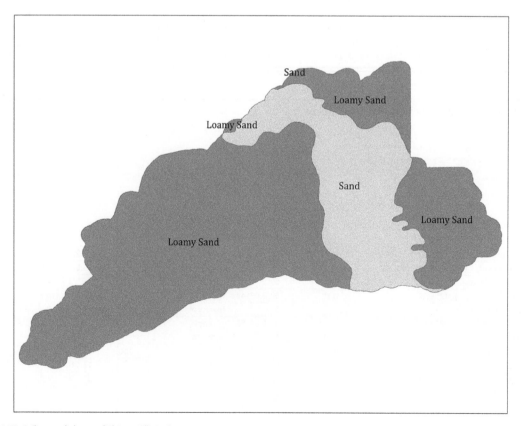

Figure 21.15 Soil textural classes of El Paso Hills Basin.

Table 21.11 *Green–Ampt parameters of the catchments of El Paso Hills Basin*

Catchment	Initial content (θ_i)	Saturated content (θ_s)	Suction (in)	Conductivity (in/h)	Percentage impervious
Plane 1-S_01	0.0808	0.4038	2.413	1.177	12
Plane 1-S_02	0.0821	0.4103	2.403	1.258	11
Plane 1-S_03	0.0820	0.4102	2.413	1.177	16
Plane 1-S_04	0.0803	0.4014	2.229	2.551	0
Plane 1-S_05	0.0802	0.4010	2.144	3.184	0
Plane 1-S_06	0.0815	0.4074	2.146	3.171	1
Plane 1-S_07	0.0802	0.4010	2.202	2.754	0
Plan2-S_01	0.0808	0.4038	2.413	1.177	0
Plane 2-S_02	0.0821	0.4103	2.403	1.258	0
Plane 2-S_03	0.0820	0.4102	2.413	1.177	1
Plan 2-S_04	0.0803	0.4014	2.229	2.551	1
Plane 2-S_05	0.0802	0.4010	2.144	3.184	0
Plane 2-S_06	0.0815	0.4074	2.146	3.171	1
Plane 2-S_07	0.0802	0.4010	2.202	2.754	2

Table 21.12 *MUSLE parameters*

Subbasins	K^1	C^2	P^3	L (ft)^5	S (decimal)^5	S (%)	θ^6(degrees)	θ^6(radians)	m^4
S_01	0.2396	0.8	1	7789.91	0.0206	2.0627	1.1816791	0.00036	0.3
S_02	0.2344	0.6	1	3599.34	0.0170	1.7043	0.9763761	0.000297	0.3
S_03	0.2399	1	1	4402.47	0.0185	1.8510	1.0604426	0.000323	0.3
S_04	0.1525	1	1	5108.94	0.0215	2.1467	1.2297711	0.000375	0.3
S_05	0.1122	1	1	6827.4	0.0184	1.8405	1.0544133	0.000321	0.3
S_06	0.1132	1	1	4974.45	0.0199	1.9945	1.1426043	0.000348	0.3
S_07	0.1396	1	1	6414.43	0.0162	1.6239	0.9303221	0.000283	0.3

Footnotes: [1] From Figure 21.16. [2] $C = 1$ for bare ground; for S_O1 and S_02 the values are lower due to presence of developments/urbanization. [3] $P = 1$ due to no conservation practice. [4] $m = 0.2$ for $S < 1\%$; 0.3 for $1 \leq S < 3$; 0.4 for $3 \leq S < 5$; 0.5 for $S \geq 5$. [5,6] $\theta = Tan^{-1}(S_{decimal})$ θ is in radians.

Table 21.13 *Percentage gravel–sand–silt–clay fractions of representative surface soils*

	Catchment						
	S_01	S_02	S_03	S_06	S_07	S_05	S_04
Gravel	0.00	5.59	0.77	4.19	3.91	1.06	1.87
Sand	84.58	88.37	92.75	77.46	83.93	80.80	81.37
Silt	12.90	3.60	4.40	15.80	8.10	14.20	14.20
Clay	2.48	2.46	2.07	2.52	4.08	3.98	1.62

Step 6: Compute the simulation runs to obtain the results.

Figure 21.17 shows the flood-frequency graphs for each of the catchments. The sediment yield as a function of discharge for each of the catchments is shown in Figure 21.18. The volumes of sediments from each of the catchments for a flood-frequency of 0.01 AEP are shown as a bar diagram in Figure 21.19. The total mass of sediment yield is 1852 tons, which is equivalent to 0.51 acre-feet, assuming a specific gravity of 2.65. However, the actual mass or volume of sediment that is expected to be transported at the outlet of the basin depends on the sediment delivery ratio.

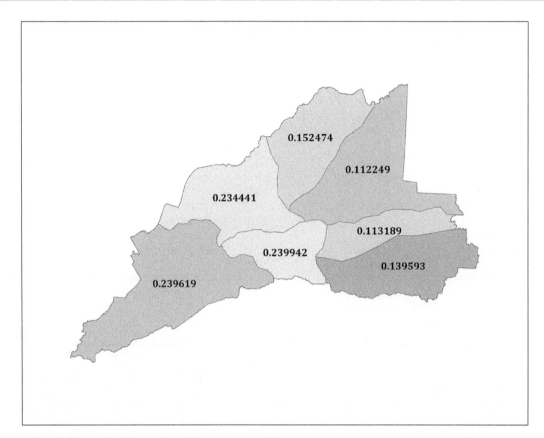

Figure 21.16 Soil erodibility factors of El Paso Hills Basin.

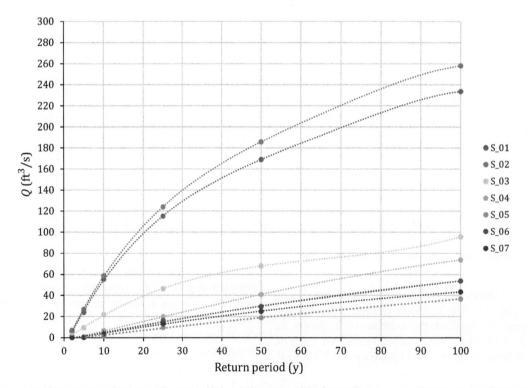

Figure 21.17 Flood-frequency curves for El Paso Hills Basin. (A black and white version of this figure will appear in some formats. For the color version, please refer to the plate section.)

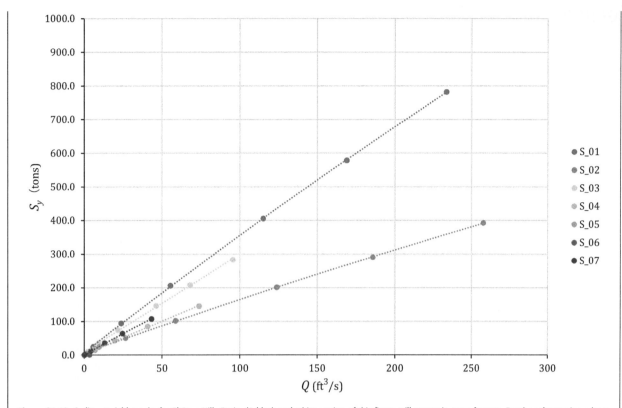

Figure 21.18 Sediment yield graphs for El Paso Hills Basin. (A black and white version of this figure will appear in some formats. For the color version, please refer to the plate section.)

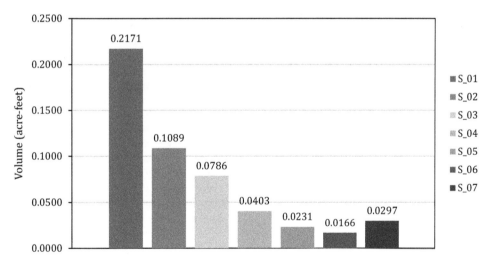

Figure 21.19 Sediment volumes for 100-year discharge from each of the catchments. (A black and white version of this figure will appear in some formats. For the color version, please refer to the plate section.)

Exercises: A selection of exercises on this topic is available at www.cambridge.org/appliedhydrology.

References

Abermann, J., Lambrecht, A., Fischer, A., and Kuhn, M. (2009). Quantifying changes and trends in glacier area and volume in the Austrian Ötztal Alps (1969–1997–2006). *Cryosphere Discussion*, 3, 415–441.

Abramowitz, M. and Stegun, I. A. (1970). *Handbook of Mathematical Functions with Formulas, Graphs, and Mathematical Tables*. National Bureau of Standards, Applied Mathematics Series 55. National Bureau of Standards, Washington, DC.

Ackers, P. and White, W. R. (1973). Sediment transport: new approach and analysis. *Journal of the Hydraulics Division, American Society of Civil Engineers*, 99(HY11), 2040–2060.

Adam, J. C., Hamlet, A. F., and Lettenmaier, D. P. (2009). Implications of global climate change for snowmelt hydrology in the twenty-first century. *Hydrological Processes*, 23, 962–972.

Ahn, S. J. and Lee, E. H. (1986). Derivation of the synthetic unit hydrograph at ungauged small watershed. *Journal of Korea Water Resources Association*, 19(2), 157–166.

Ali, K. F. and De Boer, D. H. (2007). Spatial patters and variation of suspended sediment yield in the upper Indus River basin, northern Pakistan. *Journal of Hydrology*, 334, 368–387.

Ali, M. H. and Shul, L. T. (2009). Potential evapotranspiration model for Muda Irrigation Project, Malaysia. *Water Resources Management*, 23, 57–69.

Allen, R. G. (1986). A Penman for all seasons. *Journal of Irrigation and Drainage Engineering*, 112(4), 348–368.

Allen, G. A. and Pruitt, W. O. (1986). Rational use of the FAO Blaney–Criddle formula. *Journal of Irrigation and Drainage Engineering*, 112(2), 139–155.

Allen, G. A., Pereira, L. S., Raes D., and Smith. M. (1998). *Crop Evapotranspiration: Guidelines for Computing Crop Water Requirement*. UN Food and Agricultural Organization (FAO) Irrigation and Drainage Paper 56, FAO, Rome.

Alley, W. M. (1984). On the treatment of evapotranspiration, soil moisture accounting, and aquifer recharge in monthly water balance models. *Water Resources Research*, 20, 1137–1149.

Anderson, J. R., Hardy, E. E., Roach, J. T., and Witmer, R. E. (1976). *A Land Use and Land Cover Classification System for Use with Remote Sensor Data*. Geological Survey Professional Paper 964. United States Geological Survey, Alexandria, VA.

Arkin, P. A. and Ardanuy, P. E. (1989). Estimating climatic-scale precipitation from space: a review. *Journal of Climate*, 2, 1229–1238.

Arnell, N. W. and Reynard, N. S. (1996). The effects of climate change due to global warming on river flows in Great Britain. *Journal of Hydrology*, 183, 397–424.

Arnell, N. W. and Reynard, N. S. (2000). Climate change and UK hydrology. In M. Acreman (ed.), *The Hydrology of the UK, A Study of Change*, Routledge, London, pp. 3–29.

Aron, G. and White, E. (1982). Fitting a gamma distribution over a synthetic unit hydrograph. *Water Resources Bulletin*, 18(1), 95–98.

Aron, G. and White, E. (1983). Reply to discussion: fitting a gamma distribution over a synthetic unit hydrograph. *Water Resources Bulletin*, 19(2), 303–304.

Arora, V. K. and Boer, G. J. (2001). Effects of simulated climate change on the hydrology of major river basins. *Journal of Geophysical Research – Atmospheres*, 106, 3335–3348.

Asquith, W. H. (2003). Modeling of runoff-producing rainfall hyetographs in Texas using L-moment statistics. Ph. D. dissertation, University of Texas at Austin, Austin, TX.

Asquith, W. H. and M. C. Roussel (2007). *An Initial Abstraction, Constant-Loss Model for Unit Hydrograph Modeling for Applicable Watersheds in Texas*. Scientific Investigations Report 2007-5243. US Geological Survey, Austin, Texas.

Asquith, W. H., Thompson, D. B., Cleveland, T. G., and Fang, X. (2004). *Synthesis of Rainfall and Runoff Data Used for Texas Department of Transportation Research Projects 0-4193 and 0-4194*. Open-file Report 2004-1035. Department of Transportation, Austin, TX.

Asquith, W. H., Thompson, D. B., Cleveland, T. G., and Fang, X. (2005). *Unit Hydrograph Estimation for Applicable Texas Watersheds*. Research Report 0-4194-4. Department of Transportation, Austin, TX.

Asquith, W. H., Cleveland, T. G., and Roussel, M. C. (2011). *A Method for Estimating Peak and Time of Peak Streamflow from Excess Rainfall for 10- to 640-Acre Watersheds in the Houston, Texas Metropolitan Area*. US Geological Survey Scientific Investigations Report 2011-5104. US Geological Survey, Washington, DC.

Atkinson, B. W. (1979). Urban influences on precipitation in London. In G. E. Hollis (ed.), *Man's Influence on the Hydrological Cycle in the United Kingdom*. Geobooks, Norwich, pp. 123–133.

Ayalew, T. B., Krajewski, W. F., and Mantilla, R. (2015). Analyzing the effects of excess rainfall properties on the scaling structure of peak discharges: insights from a mesoscale river basin. *Water Resources Research*, 51(6), 3900–3921.

Barnes, B. S. (1940). Discussion on analysis of runoff characteristics by O. H. Meyer. *Transactions of the American Society of Civil Engineers*, 105, 104–106.

Barnett, T. P., Adam, J. C., and Lettenmaier, D. P. (2005). Potential impacts of a warming climate on water availability in snow-dominated regions. *Nature*, 438, 303–309.

Barnett, T. P., Pierce, D. W, Hidalgo, H. G., et al. (2008). Human-induced changes in the hydrology of the western United States. *Science*, 319, 1080–1083.

Barry, R. G. (1990). Changes in mountain climate and glacio-hydrological responses. *Mountain Research and Development*, 10 (2), 161–170.

Bates, B. C., Kundzewicz, Z. W., Wu, S., Palutikof, J. P. (eds.) (2008). *Climate Change and Water. Technical Paper of the Intergovernmental Panel on Climate Change*. IPCC Secretariat, Geneva.

Bell, F. C. and Kar, S. O. (1969). Characteristic response times in design flood estimation. *Journal of Hydrology*, 8, 173–196.

Bellman, R. (1957). *Dynamic Programming*. Princeton University Press, Princeton, NJ.

Bernard, M. M. (1932). Formulas for rainfall intensities of long duration. *Transactions of the American Society of Civil Engineers*, 96(1), 592–624.

Berthier, E., Arnaud, Y., Kumar, R., et al. (2007). Remote sensing estimates of glacier mass balances in the Himachal Pradesh, Western Himalaya, India. *Remote Sensing of Environment*, 108, 327–338.

Benavidez, R., Jackson, B., Maxwell, D., and Norton, K. (2018). A review of the (Revised) Universal Soil Loss Equation ((R) USLE): with a view to increasing its global applicability and improving soil loss estimates. *Hydrology and Earth System Sciences*, 22(11), 6059–6086.

Bengtsson, L. (1982). The importance of refreezing on the diurnal snowmelt cycle with application to a northern Swedish catchment. *Nordic Hydrology*, 13, 1–12.

Bengtsson, L. (1986). Snowmelt simulation models in relation to space and time. *IAHS Publication*, 155, 115–123.

Bennett, T. H. and Peters, J. C. (2000). Continuous soil moisture accounting in the Hydrologic Engineering Center Hydrologic Modeling System (HEC-HMS). Joint Conference on Water Resources Engineering and Water Resources Planning and Management, July 30–August 2, Minneapolis, MN.

Benson, M. A. (1962). *Evolution of Methods for Evaluating the Occurrence of Floods*. US Geological Survey, Water Supply Paper 1580-A. US Geological Survey, Washington, DC.

Bérod, D. D., Singh, V. P., Devred, D., and Musy, A. (1995). A geomorphologic nonlinear cascade (GNC) model for estimation of floods from small alpine watersheds. *Journal of Hydrology*, 166, 147–170.

Bérod, D. D., Singh, V. P., and Musy, A. (1999). A geomorphologic kinematic-wave (GKW) model for estimation of floods from small alpine watersheds. *Hydrological Processes*, 13, 1391–1416.

Beven, K. J. (2001). *Rainfall–Runoff Modelling: The Primer*. Wiley, New York.

Beven, K. J. (2020). A history of the concept of time of concentration. *Hydrology and Earth System Sciences*, 24, 2655–2670.

Bhunya, P. K., Mishra, S. K., and Berndtsson, R. (2003). Simplified two-parameter gamma distribution for derivation of synthetic unit hydrograph. *Journal of Hydrologic Engineering*, 8(4), 226–230.

Bhutiyani, M. R. (1999). Mass-balance studies on Siachen Glacier in the Nubra valley, Karakoram Himalaya, India. *Journal of Glaciology*, 45(149), 112–118.

Biedenham, D. S., Copeland, R. R., Thorne, C. R., et al. (2000). *Effective Discharge Calculation: A Practical Guide*. US Army Corps of Engineers, Engineer Research and Development Center, ERDC/CHL TR-00-15. US Army Corps of Engineers, Washington, DC.

Bieger, K., Rathjens, H., Allen, P. M., and Arnold, J. G. (2015). Development and evaluation of bankfull hydraulic geometry relationships for the physiographic regions of the United States. *Journal of the American Water Resources Association*, 1–17.

Biswas, A. K. (1970). *History of Hydrology*. North Holland, Amsterdam.

Blaney, H. F. and W. D. Criddle. (1950). *Determining Water Requirements in Irrigated Areas from Climatological and Irrigation Data*. USDA Soil Conservation Service, SCS-TP-96. US Department of Agriculture, Soil Conservation Service, Washington, DC.

Blöschl, G. and Montanari, A. (2010). Climate change impacts: throwing the dice? *Hydrological Processes*, 24, 374–381.

Bolch, T., Kulkarni, A., Kääb, A., et al. (2012). The state and fate of Himalayan glaciers. *Science*, 336, 310–314.

Bonnin, G. M. (2003). *Recent Updates to NOAA/NWS Precipitation Frequency Estimates*. National Oceanic and Atmospheric Administration, National Weather Service, Silver Spring, MD

Bonnin, G. M., Martin, D., Lin, B., et al. (2004). *Precipitation-Frequency Atlas of the United States, Volume 1 Version 5.0: Semiarid Southwest (Arizona, southwest California, Nevada, New Mexico, Utah)*. US Department of Commerce, National Oceanic and Atmospheric Administration, National Weather Service, Silver Spring, MD.

Bonta J. V. and Rao, A. R. (1987). Factors affecting development of Huff curves. *Transactions of the American Society of Agricultural Engineers*, 30(6), 1689–1693.

Boorman, D. B. and Reed, D. W. (1981). *Derivation of a Catchment Average Unit Hydrograph*. Institute of Hydrology, UK, Report No. 71. Institute of Hydrology, Wallingford.

Bosch, J. M. and Hewlett, J. D. (1982). A review of catchment experiments to determine the effect of vegetation changes on water yield and evapotranspiration. *Journal of Hydrology*, 55, 3–23.

Boussinesq, J. (1904). Recherches théoriques sur l'écoulement des nappes d'eau infiltrées dan le sol et sur le debit des sources. *Journal de Mathématiques Pures et Appliquées,* 10, 5–78.

Bouwer, H. (1966). Rapid field measurement of air entry value and hydraulic conductivity of soils as significant parameters in flow system analysis. *Water Resources Research*, 2(2), 729–738.

Bowen, I. S. (1926). The ratio of heat losses by conduction and evaporation from any water surface. *Physics Review*, 27, 779–787.

Boyd, M. J. (1978). A storage-routing model relating drainage basin hydrology and geomorphology. *Water Resources Research*, 14(5), 921–928.

Bras, R. L. (1990). *Hydrology An Introduction to Hydrologic Science*. Addison-Wesley, Boston, MA.

Bras, R. L. and Colon, R. (1978). Time averaged areal mean of precipitation: Estimation and network design. *Water Resources Research*, 14(5), 8878–8888.

Bras, R. L. and Rodriguez-Iturbe, I. (1976). Network design for the estimation of areal mean of rainfall events. *Water Resources Research*, 12(6), 1185–1195.

Braun, L. N. and Hagg, W. (2009). Present and future impact of snow cover and glaciers on runoff from mountain regions: comparison between Alps and Tien Shan. In *Assessment of Snow, Glaciers and Water Resources in Asia*. International

Hydrological Programme of UNESCO and Hydrology and Water Resources Programme of WMO, Koblenz, pp. 36–43.

Brooks, R. H. and Corey, A. T. (1964). *Hydraulics Properties of Porous Media*. Hydrology Paper No. 3. Colorado State University, Fort Collins, CO.

Broscoe, A., J. (1959). *Quantitative Analysis of Longitudinal Stream Profiles of Small Watersheds*. Project NR 389-042, Technical Report No. 18. Department of Geology, Columbia University, Office of Naval Research, Geography Branch, New York.

Brubaker, K., Rango, A., and Kustas, W. (1996). Incorporating radiation inputs into the snowmelt runoff model. *Hydrological Processes*, 10, 1329–1343.

Bruen, M. and Dooge, J. C. I. (1984). An efficient and robust method for estimating unit hydrograph ordinates. *Journal of Hydrology*, 70(1–4), 1– 24.

Brune, G. M. (1953). Trap efficiency of reservoirs. *Transactions of the American Geophysical Union*, 34(3), 408-418.

Brunner, G. W. and Gorbrecht, J. (1991). *A Muskingum–Cunge Channel Flow Routing Method for Drainage Networks*. US Army Corps of Engineers, Technical Paper 135. US Army Corps of Engineers, Washington, DC.

Brutsaert, W. (1975). On a derivable formula for long-wave radiation from clear skies. *Water Resources Research*, 11, 742–744.

Burn, D. H. (2008). Climatic influences on streamflow timing in the headwaters of the Mackenzie River Basin. *Journal of Hydrology*, 352, 225–238.

California Department of Transportation (1955). *California Culvert Practice*. Department of Public Works, Division of Highways, Sacramento, CA.

Carter, R. (1961). *Magnitude and Frequency of Floods in Suburban Areas*. US Geological Survey, Reston, VA.

Carter, R. W. and Godfrey, R. G. (1960). *Storage and Flood Routing*. Geological Survey Water-Supply Paper 1543 – B. United States Department of the Interior, Washington, DC.

Cayan, D. R., Dettinger, M. D., Kammerdiener, S. A., et al. (2001). Changes in the onset of spring in the Western United States. *Bulletin of American Meteorological Society*, 82, 399–415.

Central Water Commission (1983). Flood estimation reports for different hydro-meteorological regions of India developed in collaboration with IMD, Ministry of Railways and Ministry of Surface Transport, New Delhi, India.

Centre for Ecology and Hydrology (2008). *Flood Estimation Handbook*. Centre for Ecology and Hydrology, Wallingford.

Chen, B., Ma, C., Krajewski, W. F., Wang, P., and Ren, F. (2020). Logarithmic transformation and peak-discharge power-law analysis. *Hydrology Research*, 51(1), 65–76.

Chen, C. N. (1975). Design of sediment retention basins. In *Proceedings, National Symposium on Urban Hydrology and Sediment Control*, University of Kentucky, Lexington, KY, pp. 285–298.

Cherkauer, D. S. (1975). Urbanization's impact on water quality during a flood in small watersheds. *Water Resources Bulletin*, 11, 987–998.

Chiew, F. H. S., Whetton, P. H., McMahon, T. A., and Pittock, A. B. (1995). Simulation of the impacts of climate change on runoff and soil moisture in Australian catchments. *Journal of Hydrology*, 167, 121–147.

Chow, V. T. (1951). A general formula for hydrologic frequency analysis. *Transactions of the American Geophysical Union*, 32, 231–237.

Chow, V. T. (1959). *Open Channel Hydraulics*, McGraw Hill, New York, NY.

Chow, V. T. (1964). *Handbook of Applied Hydrology*, McGraw Hill, New York, NY.

Chow, V. T., Maidment, D. R., and Mays, L. W. (1988). *Applied Hydrology*, McGraw-Hill, New York.

Churchill, M. A. (1948). Discussion of "Analysis and use of reservoir sedimentation data" by L. C. Gottschalk. Proceedings of the Federal Interagency Sedimentation Conference, 139–140, US Bureau of Reclamation, Denver, CO.

Clark, C. O. (1943). Storage and the unit hydrograph. *Proceedings of the American Society of Civil Engineers*, 9, 1333–1360.

Clark, C. O. (1945). Storage and the unit hydrograph. *Transactions of the American Society of Civil Engineers*, 110, 1419–1446.

Clark, G. M. (2010). Changes in patterns of streamflow from unregulated watersheds in Idaho, western Wyoming, and northern Nevada. *Journal of the American Water Resources Association*, 46, 486–497.

Cleveland, T. G., Thompson, D. B., and Fang, X. (2011). *Use of the Rational and Modified Rational Method for Hydraulic Design*. Report 0-6070-1. Texas Department of Transportation, Austin, TX.

Cogley, J. G. (2012). No ice lost in the Karakoram. *Nature Geoscience*, 5, 305–306.

Cohn, T. A. and Lins, H. F. (2005). Nature's style: naturally trendy. *Geophysical Research Letters*, 32(23), 1–5.

Cohn, T. A., Lane, W. L. Baier, and W. G. (1997). An algorithm for computing moments-based flood quantile estimates when historical flood information is available. *Water Resources Research*, 33(9), 2089–2096.

Cohn, T. A., Lane, W. L., and Stedinger, J. R. (2001). Confidence intervals for expected moments algorithm flood quantile estimates. *Water Resources Research*, 37(6), 1695–1706.

Collins, M. (1983). Discussion: fitting a gamma distribution over a synthetic unit hydrograph. *Water Resources Bulletin*, 19(2), 303–304.

Collins, W. T. (1939). Runoff distribution graphs from precipitation occurring in more than one time unit. *Civil Engineering*, 9(9), 559–561.

Cook, B. I., Smerdon, J. E., Seager, R., Coats, S. (2014). Global warming and 21st century drying. *Climate Dynamics*, 43, 2607–2627.

Copeland, R. R. and Thomas W. A. (1989). *Corte Madera Creek Sedimentation Study, Numerical Model Investigation*. US Army Corps of Engineers, Waterways Experiment Station, Technical Report HL-89-6. US Army Corps of Engineers, Washington, DC.

Crockford, R. H. and Richardson, D. P. (1990). Partitioning of rainfall in a eucalypt forest and pine plantation in southeastern Australia. IV. The relationship of interception and canopy storage capacity, the interception of these forests and the effect on interception of thinning the pine plantation. *Hydrological Processes*, 4, 164–188.

Croley, T. E., II (1980). Gamma synthetic hydrographs. *Journal of Hydrology*, 47, 41–52.

Cunge, J. A. (1969). On the subject of a flood propagation computation method (Muskingum method) *Journal of Hydraulic Research*, 7(2), 205–230.

Dai, A. (2013). Increasing drought under global warming in observations and models. *Nature Climate Change*, 3, 52–58.

Dai, A., Qian, T., Trenberth, K. E., and Milliman, J. D. (2009). Changes in continental freshwater discharge from 1948 to 2004. *Journal of Climate*, 22, 2773–2792.

Dalrymple, T. (1960). *Flood-Frequency Analysis*. US Geological Survey, Water Supply Paper, 1543A. United States Department of the Interior, Washington, DC.

Davie, T. (2002). *Fundamentals of Hydrology*. Routledge Taylor and Francis Group, London and New York.

Davie, T. J. A. (1996). Modelling the influence of afforestation on hillslope storm runoff. In M. G. Anderson and S. M. Brooks (eds.), *Advances in Hillslope Processes*, Wiley, Chichester, vol. 1, pp. 149–184.

Deininger, R. A. (1969). Linear programming for hydrologic analyses. *Water Resources Research*, 5(5), 1105–1109.

Dendy, F. E. and Bolton, G. C. (1976). Sediment yield–runoff drainage area relationships in the United States. *Journal of Soil and Water Consideration*, 31(6), 264–266.

Desa, M. N. and Rakhecha, P. R. (2007). Probable maximum precipitation for 24-h duration over an equatorial region: Part 2 – Johor, Malaysia. *Atmospheric Research*, 84(1), 84–90.

Dingman, S. L. (2002). *Physical Hydrology*. Hoboken, NJ.

Diskin, M. H. and Simon, E. (1977). A procedure for the selection of objective functions for hydrologic simulation models. *Journal of Hydrology*, 34, 129–149.

Dooge, J. C. I. (1959). A general theory of the unit hydrograph. *Journal of Geophysical Research*, 64(2), 241–256.

Dooge, J. C. I. (1973). *Linear Theory of Hydrologic Systems*. Technical Bulletin No. 1468, Agricultual Research Service, United States Department of Agriculture, Washington, DC.

Doorenbos, J. and Pruitt, W. O. (1977). *Crop Water Requirements*. FAO Irrigation and Drainage Paper 24. UN Food and Agricultural Organization, Rome.

Duan, Q. Y., Gupta, V. K., and Sorooshian, S. (1993). Shuffled complex evolution approach for effective and efficient global minimization. *Journal of Optimization Theory and Applications*, 76, 501–521.

Dupuit, J. (1863). *Études théoriques et pratiques sur la mouvement des eaux dans le canaux découverts et à travers les terrains perméables*, 2nd edition. Dunod, Paris.

Dyck, S. (1983). Overview on the present status of the concepts of water balance models. New Approaches in Water Balance Computations. Proceedings of the Hamburg Workshop. International Association of Hydrological Sciences, Publication No. 148, 3–19.

Dyurgerov, M. D. and Meier, M. F. (2005). *Glaciers and Changing Earth System: A 2004 Snapshot*. Institute of Arctic and Alpine Research, University of Colorado, Boulder, CO.

Easton, Z. M., Gerard-Marchant, P., Walter, M. T., et al. (2007). Hydrologic assessment of an urban variable source watershed in the northeast United States. *Water Resources Research*, 43, 1–18.

Einstein, H. A. (1950). *The Bedload Function for Sediment Transportation in Open Channel Flow*. Technical Bulletin No. 1026. US Department of Agriculture, Soil Conservation Service Washington, DC.

Einstein, H. A. and Chien, N. (1953). *Transport of Sediment Mixtures with Large Ranges of Grain Size*. Missouri River Division Sediment Series No. 2. US Army Corps of Engineers Missouri River Division, University of California Institute of Engineering Research, Berkeley, CA.

Ely, P. B. and Peters, J. C. (1984). *Probable Maximum Flood Estimation – Eastern United States*. TP-100. US Army Corps of Engineers, Hydrologic Engineering Center, Davis, CA.

Emerson, D. G., Vecchia, A. V., and Dahl, A. L. (2005). *Evaluation of Drainage-Area Ratio Method Used to Estimate Streamflow for the Red River of the North Basin, North Dakota and Minnesota*. Scientific Investigations Report 2005-5017. US Geological Survey, Reston, VA.

Engelund, F. and Hansen, E. (1972). *A Monograph on Sediment Transport in Alluvial Streams*. Danish Technical University, Hydraulics Laboratory, Teknisk Forlag, Copenhagen.

England, J. F., Cohn, T. A., Faber, B. A., et al. (2019). *Guidelines for Determining Flood Flow Frequency*. Bulletin 17C. Chapter 5 of Section B, Surface Water, Book 4, Hydrologic Analysis and Interpretation. Techniques and Methods 4-B5. US Geological Survey, Reston, VA.

Espey, W. H. and Winslow, D. E. (1974). Urban flood frequency characteristics. *Journal of the Hydraulics Division*, 100(2), 279–293.

Fan, Y., Miguez-Macho, G., Jobbágy, E. G., Jackson, R. B., and Otero-Casal, C. (2017). Hydrologic regulation of plant rooting depth. *PNAS*, 114(40), 10572-10577.

Fahey, B. and Jackson, R. (1997). Hydrological impacts of converting native forests and grasslands to pine plantations, South Island, New Zealand. *Agricultural and Forest Meteorology*, 84, 69–82.

FAO (Food and Agriculture Organization) (1998). World reference base for soil resources. In *World Soil Resources Report 84*, Food and Agriculture Organization of the United Nations, Rome.

Federal Aviation Administration (1970). *Circular on Airport Drainage*. Report A/C 150-5320-5B. Federal Aviation Administration, US Department of Transportation, Springfield, VA.

Federal Highway Administration (1984). *Guide for Selecting Manning's Roughness Coefficients for Natural Channels*. Report No. FHWA-TS-84-204. US Department of Transportation, Springfield, VA.

Federal Highway Administration (1985). *Hydrology*. Hydraulic Engineering Circular No. 19. US Department of Transportation, Springfield, VA.

Feldman, A. D. (1981). HEC models for water resources system simulation: theory and experience. In V. T. Chow (ed), *Advances in Hydroscience,* Academic Press, New York, pp. 297–423.

Feldman, A. D. (2000). *Hydrologic Modeling System HEC-HMS: Technical Reference Manual*. US Army Corps of Engineers, Hydrologic Engineering Center, Davis, CA.

Fenemor A., Phillips, C., Davie T., et al. (2006). The promise of integrated catchment management. Presented at Resource Management under Stormy Skies: Water Allocation at the Crossroads conference, Christchurch, New Zealand, November.

Fenton, J. D. (1989). A simplified approach to reservoir routing. Proceedings of the Hydrology and Water Resources Symposium, Christchurch, New Zealand.

Fenton, J. D. (1992). Reservoir routing. *Hydrological Sciences Journal*, 37(3), 233–246.

Fiorentini, M. and Orlandini, S. (2013). Robust numerical solution of the reservoir routing equation. *Advances in Water Resources*, 59, 123–132.

Fischer, H. B., List, E. J., Koh, R. C. Y., Imberger, J., and Brooks, N. H. (1979). *Mixing in Inland and Coastal Waters*. Academic Press, San Diego, CA.

Fleming, M. and Neary, V. (2004). Continuous hydrologic modeling study with hydrologic modeling system. *Journal of Hydrologic Engineering*, 9(3), 175–183.

Foster, G. R. and Wischmeier, W. H. (1974). Evaluating irregular slopes for soil erosion prediction. *Transactions of the American Society of Agricultural Engineers*, 17, 305–309.

Frederick, R. H., Myers, V. A., and Auciello, E. P. (1977). *Five-to 60-Minute Precipitation Frequency for the Eastern and Central United States*. NOAA Technical Memorandum NWS HYDRO-35. US Department of Commerce, National Oceanic and Atmospheric Administration, National Weather Service, Silver Spring, MD.

Furey, P. R. and Gupta, V. K. (2007). Diagnosing peak-discharge power laws observed in rainfall–runoff events in Goodwin Creek experimental watershed. *Advances in Water Resources*, 30, 2387–2399.

Gardelle, J., Berthier, E., and Arnaud, Y. (2012). Slight mass gain of Karakoram glaciers in the early twenty-first century. *Nature Geoscience*, 5, 322–325.

Gardelle, J., Berthier, E., Arnaud, Y., and Kääb, A. (2013). Region-wide glacier mass balances over the Pamir–Karakoram–Himalaya during 1999–2011. *The Cryosphere*, 7, 1263–1286.

Gardner, A. S., Moholdt, G., Cogley, J. G., et al. (2013). A reconciled estimate of glacier contributions to sea level rise: 2003 to 2009. *Science*, 340(6134), 852–857.

Gilbert, G. K. (1877). *Report on the Geology of the Henry Mountains*. US Geographical and Geological Survey of the Rocky Mountain Region, Washington, DC.

Global Water Partnership (2004). *Catalyzing Change: A Handbook for Developing Integrated Water Resource Management (IWRM) and Water Efficiency Strategies*. Global Water Partnership Technical Committee, Stockholm.

Go, K. P. (2014). A study on storage coefficient and concentration time of the Clark model. MS thesis, Wonkwang University, Iksan, Korea.

Goodrich, D. C., Lane, L. J., Shillito, R. M., and Miller, S. N. (1997). Linearity of basin response as a function of scale in a semiarid watershed. *Water Resources Research*, 33(12), 2951–2965.

Gray, D. M. (1961). Interrelationships of watershed characteristics. *Journal of Geophysical Research*, 66(4), 1215–1233.

Gray, D. M. and Prowse, T. D. (1992). Snow and floating ice. In D. R. Maidment (ed.), *Handbook of Hydrology*, McGraw-Hill, New York, pp. 7.1–7.58.

Green, J. K., Xuereb, E., Johnson, G., and Moore, C. (2012). The revised intensity–duration–frequency (IDF) design rainfall estimates for Australia: an overview. 34th Hydrology and Water Resources Symposium, Engineers Australia, Sydney, Australia.

Green, W. H. and Ampt, G. A. (1911). Studies on soil physics, part I: the flow of air and water through soils. *Journal of Agricultural Science*, 4(1), 1–24.

Greenwood, J. A., Landwehr, J. M., Matalas, N. C., and Wallis, J. R. (1979). Probability-weighted moments: definition and relation to parameters to several distributions expressible in inverse form. *Water Resources Research*, 15, 1049–1054.

Griffis, V. W. and Stedinger, J. R. (2007). Log-Pearson type 3 distribution and its application in flood frequency analysis. I: distribution characteristics. *Journal of Hydrologic Engineering*, 12(5), 482–491.

Guildner, L. A., Johnson, D. P., and Jones, F. E. (1976). Vapor pressure of water at its triple point. *Journal of Research, National Bureau of Standards*, 80A(3), 505–521.

Gumbel, E. J. (1941). The return period of flood flows. *Annals of Mathematical Statistics*, 12(2), 163–190.

Gumbel, E. J. (1958). *Statistics of Extremes*, Columbia University Press, New York.

Gupta, V. K. (2004). Emergence of statistical scaling in floods on channel networks from complex runoff dynamics. *Chaos, Solitons, and Fractals*, 19, 357–365.

Gupta, V. K., Waymire, E., and Wang, C. T. (1980). A representation of an instantaneous unit hydrograph from geomorphology. *Water Resources Research*, 16(5), 855–862.

Haan, C. T. (1970). *A Dimensionless Hydrograph Equation*. File Report. Agricultural Engineering Department, University of Kentucky, Lexington, KY.

Haan, C. T. (1977). *Statistical Methods in Hydrology*. Iowa State University Press, Ames, IA.

Haan, C. T., Barfield, B. J., and Hayes, J. C. (1994). *Design Hydrology and Sedimentology for Small Catchments*. Academic Press, New York.

Hack, J. T. (1957). *Studies of Longitudinal Profiles in Virginia and Maryland*. US Geological Survey Professional Paper, 294-B. United States Government Printing Office, Washington, DC.

Halff, A. H., Novoa, J. I., and Salcedo, L. M. (1979). *Effect of Urbanization and other Factors on Synthetic Unit Hydrographs*. Rice Institute Pamphlet – Rice University Studies, 65, No. 1. Rice Institute, Houston, TX.

Hall, D. K and Riggs, G. A. (2007). Accuracy assessment of the MODIS snow products. *Hydrological Processes*, 21, 1534–1547.

Hall, M. H. and Fagre, D. B. (2003). Modeled climate-induced glacier change in Glacier National Park, 1850–2100. *BioScience*, 53, 131–140.

Hallauer, W. Jr. (2022). Introduction to linear time-invariant dynamic systems for students of engineering. LibreTexts Project, sponsored by the United States Department of Education and National Science Foundation.

Hamon, W. R. (1961). Estimating potential evapotranspiration. *Journal of the Hydraulics Division, American Society of Civil Engineers*, 87, 107–120.

Hamon, W. R. (1963). Computation of direct runoff amounts from storm rainfall. *International Association of Scientific Hydrology, Publication* 63, 52–62.

Hansen, E. M., Schwarz, F. K., and Riedel, J. T. (1977). *Probable Maximum Precipitation Estimates, Colorado River and Great Basin Drainages*. Hydrometeorological Report No. 49. National Weather Service, US Department of Commerce, Silver Spring, MD.

Hansen, E. M., Schreiner, L. C., and Miller, J. F. (1982). *Application of Probable Maximum Precipitation Estimates – United States*

East of the 105th Meridian. Hydrometeorological Report No. 52. National Weather Service, US Department of Commerce, National Oceanic and Atmospheric Administration, Washington, DC.

Hargreaves, G. H. (1975). Moisture availability and crop production. *Transactions of the ASAE*, 18(5), 980–984.

Hargreaves, G. H. (1994). Defining and using reference evapotranspiration. *Journal of Irrigation and Drainage Engineering*, 120(6), 1132–1139.

Hargreaves, G. H. and Allen, R. G. (2003). History and evaluation of Hargreaves evapotranspiration equation. *Journal of Irrigation and Drainage Engineering*, 129(1), 53–63.

Hargreaves G. L., Hargreaves G. H., and Riley J. P. (1985). Agricultural benefits for Senegal River basin. *Journal of Irrigation and Drainage Engineering*, 111(2), 113–124.

Haritashya, U. K., Bishop, M. P., Shroder, J. F., Bush, A. B. G., and Bulley, H. N. N. (2009). Space-based assessment of glacier fluctuations in the Wakhan Pamir, Afghanistan. *Climate Change*, 94, 5–18.

Hathaway, G. A. (1945). Military airfields: a symposium: design of drainage facilities. *Transactions of the American Society of Civil Engineers*, 110(1), 697–729.

Hazen, A. (1914). Storage to be provided in the impounding reservoirs for municipal water supply. *Transaction American Society of Civil Engineers*, 77, 1547–1550.

Held, I. M. and Soden, B. J. (2006). Robust responses of the hydrological cycle to global warming. *Journal of Climate*, 19, 5686–5699.

Helsel, D. R. and Hirsch, R. M. (2002). *Statistical Methods in Water Resources*. US Geological Survey Techniques of Water-Resources Investigations, Book 4. US Geological Survey, Reston, VA.

Helsel, D. R., and Hirsch, R. M., Ryberg, K. R., Archfield, S. A., and Gilroy, E. J. (2020). *Statistical Methods in Water Resources*. US Geological Survey Techniques and Methods 4-A3. US Geological Survey, Reston, VA.

Henderson, F. M. (1966). *Open Channel Flow*. Macmillan Inc., New York.

Hershfield, D. M. (1961a). *Rainfall Frequency Atlas of the United States for Durations from 30 Minutes to 24 Hours and Return Periods from 1 to 100 Years*. Technical Paper No. 40. US Department of Commerce, Weather Bureau, Washington, DC.

Hershfield, D. M. (1961b). Estimating the probable maximum precipitation. *Journal Hydraulics Division, American Society of Civil Engineers*, 87(5), 99–116.

Hershfield, D. M. (1965). Method for estimating the probable maximum precipitation. *Journal American Water Works Association*, 57, 965–972.

Hershfield, D. M. (1981). The magnitude of the hydrological frequency factor in maximum rainfall estimation. *Hydrological Sciences Bulletin*, 26(2), 171–177.

Hewitt, K. (2005). The Karakoram anomaly? Glacier expansion and the 'Elevation Effect,' Karakoram Himalaya. *Mountain Research and Development*, 25, 332–340.

Hewitt, K. (2007). Tributary glacial surges: an exceptional concentration at Panmah Glacier, Karakoram Himalaya. *Journal of Glaciology*, 53, 181–188.

Hey, R. D. (1975). Design discharge for natural channels. In R. D. Hey and T. D. Davies (eds.), *Science, Technology and Environmental Management*, Saxon House, Farnborough, pp. 73–88.

Hirsch, R. M. and Stedinger, J. R. (1987). Plotting positions for historical floods. *Water Resources Research*, 23(4), 715–727.

Hjelmfelt, A. T. (1986). Estimating peak runoff from field-size watersheds. *Water Resources Bulletin, American Water Resources Association*, 22(2), 267–274.

Hock, R. (2003). Temperature index melt modeling in mountain areas. *Journal of Hydrology*, 282, 104–115.

Hollander, M. and Wolfe, D. A. (1999). *Nonparametric Statistical Methods*. John Wiley and Sons, New York.

Holtan, H. N. (1961). *A Concept for Infiltration Estimates in Watershed Engineering*. ARS, Paper 41-51. US Department of Agriculture, Washington, DC.

Holton, J. R. and Hakim, G. J. (2013). *An Introduction to Dynamic Meteorology*, 5th ed. Elsevier, Amsterdam.

Horton, R. E. (1919). Rainfall interception. *Monthly Weather Review*, 147, 603–623.

Horton, R. E. (1932). Drainage basin characteristics. *Transactions of the American Geophysical Union*, 13, 350–361.

Horton, R. E. (1933). The role of infiltration in the hydrologic cycle. *Transactions of the American Geophysical Union*, 14, 446–460.

Horton, R. E. (1939). Analysis of runoff plat experiments with varying infiltration capacity. *Transactions of the American Geophysical Union*, 20, 693–711.

Horton, R. E. (1945). Erosional development of streams and their drainage basins: hydrophysical approach to quantitative morphology. *Bulletin Geological Society of America*, 56(3), 275–370.

Hosking, J. R. M. (1986). *The Theory of Probability-Weighted Moments*. Technical Report RC 12210. IBM Research, Yorktown Heights, NY.

Hosking, J. R. M. (1989). *Some Theoretical Results Concerning L-Moments*. Research Report RC14492. IBM Research, Yorktown Heights, NY.

Hosking, J. R. M. (1990). L-moments: analysis and estimation of distributions using linear combinations of order statistics. *Journal of the Royal Statistical Society, Series B*, 52(1), 105–124.

Hosking, J. R. M. and Wallis, J. R. (1988). The effect of intersite dependence on regional flood frequency analysis. *Water Resources Research*, 29, 271–281.

Hosking, J. R. M. and Wallis, J. R. (1993). Some statistics useful in regional frequency analysis. *Water Resources Research*, 29(2), 271–281.

Hosking, J. R. M. and Wallis, J. R. (1997). *Regional Frequency Analysis: An Approach Based on L-Moments*. Cambridge University Press, Cambridge.

Howe, C. W. (1996). Sharing water fairly. Our planet 8.3 Water. October 1996 (www.ourplanet.com).

Huber, W. C. and Dickinson, R. E. (1988). *Storm Water Management Model, Version 4; User's Manual*. US Environmental Protection Agency, Athens, GA.

Huff, F. A. (1967). Time distribution of rainfall in heavy storms. *Water Resources Research*, 3(4), 1007–1019.

Huff, F. A. (1990). *Time Distributions of Heavy Rainstorms in Illinois*. Circular 173. Illinois State Water Survey, Champaign, IL.

Huff, F. A. and Angel, J. R. (1989). *Rainfall Distributions and Hydroclimatic Characteristics of Heavy Rainstorms in Illinois*. Bulletin 70. Illinois State Water Survey, Urbana, IL.

Huff, F. A. and Neill, J. C. (1957). *Rainfall Relations on Small Area*. Bulletin 44. Illinois State Water Survey, Urbana, IL.

Huggins, L. F. and Monke, E. J. (1966). *The Mathematical Simulation of the Hydrology of Small Watersheds*. Technical Report No. 1. Indiana Water Resources Research Center, Purdue University, West Lafayette, IN.

Indian Meteorological Department (1972). *Manual of Hydrometeorology*. Indian Meteorological Department, New Dehli.

IPPC (Intergovernmental Panel on Climate Change) (2007). *Climate Change 2007: Synthesis Report, Contribution of Working Groups I, II, and III to the Fourth Assessment Report of the Intergovernmental Panel on Climate Change*. IPCC, Geneva.

IPCC (2013). *Climate Change 2013: The Physical Science Basis. Contribution of Working Group I to the Fifth Assessment Report of the Intergovernmental Panel on Climate Change*. Cambridge University Press, Cambridge.

Isaaks, E. H. and Srivastava, R. M. (1989). *Applied Geostatistics*. Oxford University Press, New York.

Iwagaki, Y. (1955). Fundamental studies on the runoff analysis by characteristics. Bulletin 10. Disaster Prevention Research Institute, Kyoto University, Kyoto.

Izzard, C. F. and Hicks, W. (1947). Hydraulics of runoff from developed surfaces. *Proceedings of the Highway Research Board*, 26, 129–150.

Jacob, C. E. (1943). Correlation of groundwater levels and precipitation on Long Island, New York: 1. Theory. *Transactions of the American Geophysical Union*, 24, 564–573.

Jacob, C. E. (1944). Correlation of groundwater levels and precipitation on Long Island, New York: 2. Correlation of data. *Transactions of the American Geophysical Union*, 25, 321–386.

Jain, S. K. and Singh, V. P. (2019). *Engineering Hydrology: An Introduction to Processes, Analysis, and Modeling*. McGraw Hill, New York, NY.

Jianchu, X., Shrestha, A., Eriksson, M. (2009). Climate change and its impacts on glaciers and water resources management in the Himalayan region. In *Assessment of Snow, Glaciers and Water Resources in Asia*. International Hydrological Programme of UNESCO and Hydrology and Water Resources Programme of WMO, Koblenz, pp. 44–54.

Japan Society of Civil Engineers (1999). *The Collection of Hydraulic Formulae*. Japan Society of Civil Engineers, Tokyo.

Jaspers F. G. W. (2001). The new water legislation of Zimbabwe and South Africa: comparison of legal and institutional reform. *International Environmental Agreements*, 1, 305–325.

Jenkinson, A. F. (1955). The frequency distribution of the annual maximum (or minimum) values of meteorological elements. *Quarterly Journal of Royal Meteorological Society*, 81(348), 158–171.

Jennings, M. E., Thomas, W. O. Jr., and Riggs, H. C. (1994). *Nationwide Summary of US Geological Survey Regional Regression Equations for Estimating Magnitude and Frequency of Floods for Ungauged Sites*. USGS Water Resources Investigations Report 94-4002. US Geological Survey, Reston, VA.

Jiménez, J. A. and Madsen, O. S. (2003). A simple formula to estimate settling velocity of natural channels. *Journal of Waterway, Port, Coastal, and Ocean Engineering*, 129(2), 70–78.

Johnson, D., Smith, M., Koren, V., Finnerty, B. (1999). Comparing mean areal precipitation estimates from NXRAD and rain gauge networks. *Journal of Hydrologic Engineering*, 4, 117–124.

Johnson, F. and Sharma, A. (2017). Design rainfall. In V. P. Singh (ed.), *Handbook of Applied Hydrology*, McGraw Hill, New York, pp. 125-3–125-13.

Johnstone, D. and Cross, W. P. (1949). *Elements of Applied Hydrology*. Ronald Press, New York.

Julien, P. (2002). *River Mechanics*. Cambridge University Press, Cambridge.

Jung, S. (2005). Development of empirical formulas for the parameter estimation of Clark's Watershed flood routing model. PhD dissertation, Korea University, Seoul, Korea.

Kääb, A., Berthier, E., Nuth, C., Gardelle, J., and Arnaud, Y. (2012). Contrasting patterns of early twenty-first century glacier mass change in the Himalayas. *Nature*, 488(7412), 495–498.

Kaleris, V., Papanastasopoulos, D., and Lagas, G. (2001). Case study of atmospheric circulation changes on river basin hydrology: uncertainty aspects. *Journal of Hydrology*, 245, 137–152.

Kao, S.-C., DeNeale, S. T., Yegorva, E., Kanney, J., and Carr, M. L. (2020). Variability of precipitation areal reduction factors in the conterminous United States. *Journal of Hydrology*, https://doi.org/10.1016/j.hydroa.2020.1000064.

Keifer, C. J. and Chu, H. H. (1957). Synthetic storm pattern for drainage design. *Journal of the Hydraulics Division, American Society of Civil Engineers*, 83(4), 1–25.

Keiler, M., Knight, J., and Harrison, S. (2010). Climate change and geomorphological hazards in the eastern European Alps. *Philosophical Transactions of the Royal Society*, 368, 2461–2479.

Kendall, M. G. (1938). A new measure of rank correlation. *Biometrika*, 30(1–2), 81–93.

Kendall, M. G. (1975). *Rank Correlation Methods*. Charles Griffin, London.

Kerby, W. S. (1959). Time of concentration for overland flow. *Journal of Civil Engineering*, 26(3), 60.

Kiang, J. E., Gazoorian, C., McMillan, H., et al. (2018). A comparison of methods for streamflow uncertainty estimation. *Water Resources Research*, 54, 7149–7176.

Kim, H. S. and Julien, P. Y. (2006). Soil erosion modeling using RUSLE and GIS on the IMHA Watershed. *Water Engineering Research*, 7(1), 29–41.

Kim, Y. (2015). Development of concentration time and storage coefficient formula in urban stream watersheds. MS thesis, Sejong University, Seoul, Korea.

Kirpich, T. P. (1940). Time of concentration of small agricultural watersheds. *Journal of Civil Engineering*, 10(6), 362.

Kleitz, M. (1877). Note sur la theorie du mouvement non permanent des liquides et sur application a la propagation des crues des rivieres [Note on the theory of unsteady flow of liquids and on application to flood propagation in rivers]. *Annales des ponts et chaussees, ser. 5*, 16(2e semestre), 133–196.

Kormos, P., Luce, C., Wenger, S. J., and Berghuijs, W. R. (2016). Trends and sensitivities of low streamflow extremes to discharge timing and magnitude in Pacific Northwest Mountain streams. *Water Resources Research*, 52, 4990–5007.

Kostiakov, A. M. (1932). On the dynamics of the coefficient of water percolation in soils and of the necessity of studying it from a dynamic point of view for purposes of amelioration. *Transactions, Sixth Communication, International Soil Science Society, Part A*, 17–29 (in Russian).

Kotz, S. and Nadarajah, S. (2000). *Extreme Value Distributions: Theory and Applications*. Imperial College Press, London.

Kubik, H. E. (1990). *Computation of Regulated Frequency Curves by Application of the Total Probability Theorem*. US Army Corps of Engineers, Hydrologic Engineering Center, Davis, CA.

Kuichling, E. (1889). The relation between the rainfall and the discharge of sewers in populous districts. *Transactions of the American Society of Civil Engineers*, 20, 1–56.

Kulandaiswamy, V. C. (1964). A basic study of the rainfall excess-surface runoff relationship in a basin. Ph.D. Thesis, University of Illinois.

Kustas, W. P., Rango, A., and Uijlenhoet, R. (1994). A simple energy budget algorithm for the snowmelt runoff model. *Water Resources Research*, 30(5), 1515–1527.

Kulkarni, A. V., Bahuguna, I. M., Rathore, B. P., et al. (2007). Glacial retreat in Himalaya using Indian remote sensing satellite data. *Current Science*, 92, 69–74.

Labadie, J. W. (2004). Optimal operation of multireservoir systems: state-of-art review. *Journal of Water Resources Planning and Management*, 130, 93–111.

Laghari, J. R. (2013). Melting glaciers bring energy uncertainty. *Nature*, 502, 617–618.

Lane, E. W. (1947). Report of the Subcommittee on Sediment Terminology. *Transactions of the American Geophysical Union*, 28(6), 936–938.

Langbein, W. B. (and others) (1947). *Topographic Characteristics of Drainage Basins*. US Geological Survey, Water Supply Paper, 968-C, 125–155. United States Department of the Interior, Washington, DC.

Langbein, W. B. (1949). Annual floods and the partial duration flood series. *Transactions of the American Geophysical Union*, 30(6), 879–881.

Langbein, W. B., and Leopold, L. B. (1964). Quasi-equilibrium states in channel morphology. *American Journal of Science*, 262 (2), 782–794.

Laursen, E. M. (1958). The total sediment load of streams. *Journal of the Hydraulics Division, American Society of Civil Engineers*, 84(1), 1–36.

Laurenson, E. M. (1962). Hydrograph synthesis by runoff routing. University of New South Wales, Water Research Laboratory, Manly Vale, NSW, Australia.

Laurenson, E. M. (1964). A catchment storage model for runoff routing. *Journal of Hydrology*, 2, 141–163.

Leavesley, G. H., Lichty, R. W., Troutman, B. M., and Saindon, L. G. (1983). *Precipitation–Runoff Modeling System: User's Manual*. Water-Resources Investigations Report 83-4238. United States Department of the Interior, Geological Survey, Denver, CO.

Leclerc, G. and Schaake, J. C. (1973). *Methodology for Assessing the Potential Impact of Urban Development on Urban Runoff and the Relative Efficiency of Runoff Control Alternatives*. Ralph M. Parsons Lab. Report 167. Massachusetts Institute of Technology, Cambridge, MA.

Leonard, J., Mietton, M., Najib, H., and Gourbesville, P. (2000). Rating curve modeling with Manning's equation to manage instability and improve extrapolation. *Hydrological Sciences Journal*, 45(5), 739–750.

Leopold, L. B. and Maddock, T. (1953). *The Hydraulic Geometry of Stream Channels and some Physiographic Implications*. US Geological Survey Professional Paper 252. United States Government Printing Office, Washington, DC.

Leopold, L. B. and Miller, J. P. (1956). *Ephemeral Streams: Hydraulic Factors and Their Relation to Drainage Net*. US Geological Survey Professional Paper 282-A. United States Government Printing Office, Washington, DC.

Leopold, L. B., Wolman, M. G., and Miller, J. P. (1992). *Fluvial Processes in Geomorphology*. Dover Publications, New York.

Li, R. M., Stevens, M. A., and Simons, D. B. (1976). Solutions to Green–Ampt infiltration equations. *Journal of Irrigation and Drainage Engineering*, 102(2), 239–248.

Lighthill, M. J. and Whitman, G. B. (1955). On kinematic waves: 1. Flood movement in long rivers. *Proceedings of the Royal Society of London, Series A*, 229, 281–316.

Linsley, R. (1945). Discussion of storage and the unit hydrograph by C. O. Clark. *Transactions of the American Society of Civil Engineers*, 110, 1452–1455.

Linsley, R. K., Kohler, M. A., and Paulhus, J. L. H. (1982). *Hydrology for Engineers*. McGraw Hill, New York.

Liu S, Shangguan D, Ding Y, Han H, et al. (2006). Glacier changes during the past century in the Gangrigabu Mountains, southeast Qinghai–Xizang (Tibetan) Plateau, China. *Annals of Glaciology*, 43, 187–193.

Lloyd, C. R., Gash, J. H. C., Shuttleworth, W. J., and Marques, D de O. (1988). The measurement and modelling of rainfall interception by Amazonian rainforest. *Agricultural Forestry Meteorology*, 42, 63–73.

Loague, K. M. and Freeze, R. A. (1985). A comparison of rainfall–runoff modeling techniques on small upland catchments. *Water Resources Research*, 21(2), 229–248.

Lowry, W. P. (1967). The climate of cities. *Scientific American*, 217 (2), 20.

Lozan, J. L., Grabl, H., and Hupfer, P. (eds.) (2001). Summary: warning signals from climate. In *Climate of 21st Century: Changes and Risks*. Wissenchaftliche Auswertungen, Berlin, pp. 400–408.

Luce, C. H. and Holden, Z. A. (2009). Declining annual streamflow distributions in the Pacific Northwest United States, 1948–2006. *Geophysical Research Letters*, 36, L16401.

Luce, C. H., Abatzoglou, J. T., and Holden, Z. A. (2013). The missing mountain water: slower westerlies decrease orographic enhancement in the Pacific Northwest USA. *Science*, 342, 1360–1364.

Luce, C. H., Lopez-Burgos, V., and Holden, Z. (2014). Sensitivity of snowpack storage to precipitation and temperature using spatial and temporal analog models. *Water Resources Research*, 50, 9447–9462.

Lyne, V. and Hollick, M. (1979). Stochastic time-variable rainfall-runoff modelling. In Proceedings of the Institute of Engineers Australia, National Conference, Perth, Australia.

Maidment, D. R. (ed.) (1993). *Handbook of Hydrology*. McGraw-Hill, New York.

Maidment, D. R. (2002). *Arc Hydro*. ESRI Press, Redlands, CA.

Maidment, D. R. (2006). GIS in Water Resources. Lecture series presented at the University of Texas at Austin, TX.

Male, D. H. and Gray, D. M. (1981). Snow cover ablation and runoff. In D. M. Gray, and D. H. Male (eds.), *Handbook of Snow*, Elmsford, NY, Pergamon Press.

Marshall, J. S. and Palmer, W. M. (1948). The distribution of raindrops with size. *Journal of Meteorology*, 5, 165–166.

Martinec, J. (1975). Snowmelt–runoff model for stream flow forecasts. *Nordic Hydrology*, 6, 145–154.

Martinec, J. (1989). Hour-to-hour snowmelt rates and lysimeter outflow during an entire ablation period. *Snow Cover and Glacier Variation, Proceedings of the Baltimore Symposium, IAHS Publication*, 193, 19–28.

Martinec, J., Rango, A., and Roberts, R. (2008). *Snowmelt Runoff Model (SRM) User's Manual*. Agricultura Experiment Station, Special Report 100. College of Agriculture and Home Economics, New Mexico State University, Las Cruces, NM.

Mather, J. R. (1974). *Climatology: Fundamentals and Applications*. McGraw-Hill, New York.

Mather, J. R. (1978). *The Climatic Water Balance in Environmental Analysis*. Lexington, MA, D. C. Heath and Company.

Mather, J. R. (1979). Use of the climatic water budget to estimate streamflow. In Mather, J. R. (ed.), *Use of the Climatic Water Budget in Selected Environmental Water Problems*, C. W. Thornthwaite Associates, Laboratory of Climatology, Publications in Climatology, Elmer, NJ, vol. 32, no. 1, 1–52.

Mays, L. W. and Coles, S. L. (1980). Optimization of unit hydrograph determination. *Journal of the Hydraulics Division, American Society of Civil Engineers*, 106(1), 85–97.

McCabe, G. J. and Fountain, A. G. (1995). Relations between atmospheric circulation and mass balance of South Cascade Glacier, Washington, USA. *Arctic and Alpine Research*, 27: 226–233.

McCabe, G. J. and Markstrom, S. L. (2007). *A Monthly Water-Balance Model Driven by a Graphical User Interface*. US Geological Survey, Open File Report, 2007-1088. United States Government Printing Office, Washington, DC.

McCarthy, G. T. (1938). The unit hydrograph and flood routing. Unpublished manuscript, presented at a conference of the North Atlantic Division, US Army Corps of Engineers, June 24.

McCuen, R. H. and Spiess, J. M. (1995). Assessment of kinematic wave time of concentration. *Journal of Hydraulic Engineering*, 121(3), 256–266.

McMahon, T. A. (1979). Hydrological characteristics of arid zones. *IAHS Publications*, 128, 105–123.

Meadows, M. E. (2020). *South Carolina Unit Hydrograph Method Applications Manual: Final Report*. FHWA-SC-20-02. Federal Highway Administration, Columbia, SC.

Meadows, M. E. and Blandford, G. E. (1983). *Improved Methods and Guidelines for Modeling Stormwater Runoff from Surface Coal Mined Lands*. Research Report No. 147. Kentucky Water Resources Research Institute, Lexington, KY.

Mein, R. G. and Larson, C. L. (1971). *Modeling the Infiltration Component of Rainfall–Runoff Process*. Bulletin 43. Water Resources Research Center, University of Minnesota, Minneapolis, MN.

Mein, R. G. and Larson, C. L. (1973). Modeling infiltration during a steady rain. *Water Resources Research*, 9(2), 384–394.

Melton, M. A. (1957). *An Analysis of the Relations among Climate, Surface Properties, and Geomorphology*. Project NR 389-042. Technical Report 11, Columbia University, Department of Geology, Office of Naval Research, Geography Branch, New York.

Merkel, W. H. (2002). Muskingum–Cunge flood routing procedure in NRCS hydrologic models, Presented at the second Federal Interagency Hydrologic Modeling Conference, Las Vegas, NV.

Merz, B. and Thieken, A. H. (2009). Flood risk curves and uncertainty bounds. *Natural Hazards*, 51(3), 437–458.

Merz, R., Blöschl, G., and Parajka, J. (2006). Spatiotemporal variability of event runoff coefficients. *Journal of Hydrology*, 331, 591–604.

Meyer-Peter, E. and Müller, R. (1948). Formulas for bed-load transport. *Proceedings of the Second Meeting of the International Association for Hydraulic Research, Stockholm, Sweden*, 39–64.

Middelkoop, H., Daamen, K., Gellens, D., et al. (2001). Impact of climate change on hydrological regimes and water resources management in the Rhine basin. *Climatic Change*, 49, 105–128.

Miller, W. A. and Cunge, J. A. (1975). Simplified equations of unsteady flow. In K. Mahmood and V. Yevjevich (eds.), *Unsteady Flow in Open Channels*, Water Resources Publications, University of Michigan, Ann Arbor, MI, vol. 1, pp. 183–249.

Miller, J. F. (1964). *Two- to Ten-Day Precipitation for Return Periods of 2 to 100 Years in the Contiguous United States*. TP-49. Weather Bureau, US Department of Commerce, Washington, DC.

Miller, J. F., Frederick, R. H., and Tracey, R. J. (1973). *Precipitation-Frequency Atlas of the Western United States*. NOAA Atlas 2. US Department of Commerce, National Oceanic and Atmospheric Administration, National Weather Service, Silver Spring, MD.

Miller, J. F., Hansen, E. M., Fenn, et al. (1984). *Probable Maximum Precipitation Estimates, United States between Continental Divide and the 103rd Meridian*. Hydrometeorological Report No. 55. National Weather Service, US Department of Commerce, Silver Spring, MD.

Milly, P. C. D. (1992). Potential evaporation and soil moisture in general circulation models. *Journal of Climate*, 5, 209–226.

Milly, P. and Dunne, K. A. (2011). On the hydrologic adjustment of climate-model projections: the potential pitfall of potential evapotranspiration. *Earth Interactions*, 15, 1–14.

Minora, U., Bocchiola, D., D'Agata, C., et al. (2013). 2001–2010 Glacier changes in the Central Karakoram National Park: a contribution to evaluate the magnitude and rate of the "Karakoram anomaly." *The Cryosphere*, 7, 2891–2941.

Mintz, Y. and Serafini, Y. (1984). *Global Fields of Monthly Normal Soil Moisture as Derived from Observed Precipitation and an Estimated Evapotranspiration*. Final Scientific Report under NASA Grant No. NAS 5-26, Part V. Department of Meteorology, University of Maryland, College Park, MD.

Mirza, M. Q., Warrick, R. A., Ericksen, N. J., and Kenny, G. J. (1998). Trends and persistence in precipitation in the Ganges, Brahmaputra, and Meghna River basins. *Hydrological Sciences Journal*, 43, 845–858.

Mockus, V. (1957). *Use of Storm and Watershed Characteristics in Synthetic Hydrograph Analysis and Application*. US

Department of Agriculture, Soil Conservation Service, Washington, DC.

Mockus, V. (1961). *Watershed Lag*. ES-1015. US Department of Agriculture, Soil Conservation Service, Washington, DC.

Moglen, G. E., Eltahir, E. A. B., and Bras, R. L. (1998). On the sensitivity of drainage density to climate change. *Water Resources Research*, 34(4), 855–862.

Monteith, J. L. (1965). The state and movement of water in living organisms. In *Proceedings of the Evaporation and Environment, XIXth Symposium, Society for Experimental Biology*, Cambridge University Press, Cambridge, pp. 205–234.

Moody, D. W., E. B. Chase, and Aronson, D. A. (1986). *National Water Summary 1985: Hydrologic Events and Surface-Water Resources*. Water Supply Paper 2300. US Geological Survey, Washington, DC.

Morel-Seytoux, H. J. and Verdin, J. P. (1981). *Extension of the Soil Conservation Service Rainfall Runoff Methodology for Ungauged Watersheds*. Report No. FHWA/RD-81/060. Federal Highway Administration, Washington, DC.

Morgali, J. R. and Linsley, R. K. (1965). Computer analysis of overland flow. *Journal of the Hydraulics Division*, 91(3), 81–100.

Morgan, R. P. C. (2009). *Soil Erosion and Conservation*. Wiley, New York.

Morisawa, M. E. (1962). Quantitative geomorphology of some watersheds in the Appalachian Plateau. *Geological Society of America Bulletin*, 73, 1025–1046.

Morris, E. M. and Woolhiser, D. A. (1980). Unsteady one dimensional flow over a plan: partial equilibrium and recession hydrographs. *Water Resources Research*, 16(2), 355–360.

Mote, P. W. (2006). Climate-driven variability and trends in mountain snowpack in western North America. *Journal of Climate*, 19, 6209–6220.

Mote, P. W., Hamlet, A. F., Clark, M. P., and Lettenmaier, D. P. (2005). Declining mountain snowpack in western North America. *Bulletin of the American Meteorological Society*, 86, 39–49.

Muir, M. J., Luce, C. H., Gurrieri, J. J., et al. (2018). Effects of climate change on hydrology, water resources, and soil. In *Climate Change Vulnerability and Adaptation in the Intermountain Region*, United States Department of Agriculture, RMRS-GTR-375, US Department of Agriculture, Fort Collins, CO, ch. 4.

Mukhopadhyay, B. (2012). Detection of dual effects of degradation of perennial snow and ice covers on the hydrologic regime of a Himalayan River basin by stream water availability modeling. *Journal of Hydrology*, 412–413, 14–33.

Mukhopadhyay, B. (2013). Signature and hydrologic consequences of climate change within upper-middle Brahmaputra basin. *Hydrological Processes*, 27, 2126–2143.

Mukhopadhyay, B. and Dutta, A. (2010). A Stream water availability model of Upper Indus Basin based on a topologic model and global climatic datasets. *Water Resources Management*, 24(15), 4403–4443.

Mukhopadhyay, B. and Kappel, W. (2017). Probable maximum precipitation. In V. P. Singh (ed.), *Handbook of Applied Hydrology*, McGraw Hill, New York, pp.126-1–126-18.

Mukhopadhyay, B. and Khan, A. (2014a). A quantitative assessment of the genetic sources of the hydrologic flow regimes in Upper Indus Basin and its significance in a changing climate. *Journal of Hydrology*, 509, 249–572.

Mukhopadhyay, B. and Khan, A. (2014b). Rising river flows and glacial mass balance in central Karakoram. *Journal of Hydrology*, 513, 192–203.

Mukhopadhyay, B. and Khan, A. (2015a). A reevaluation of the snowmelt and glacial melt in river flows within Upper Indus Basin and its significance in a changing climate. *Journal of Hydrology*, 509, 549–572.

Mukhopadhyay, B. and Khan, A. (2015b). Boltzmann–Shannon entropy and river flow stability within Upper Indus Basin in a changing climate. *International Journal of River Basin Management*, 13(1), 87–95.

Mukhopadhyay, B. and Khan, A. (2017). Altitudinal variation of temperature, equilibrium line altitude, and accumulation-area ratio in Upper Indus Basin. *Hydrology Research*, 48(1), 214–230.

Mukhopadhyay, B. and Singh, V. P. (2011). Hydrological modeling at mesoscopic scales using global datasets to derive stream water availability models of river basins. In M. K. Shukla (ed.), *Soil Hydrology, Land Use and Agriculture*, CAB International, Wallingford, pp. 20–74.

Mukhopadhyay, B., Cornelius, J., and Zehner, W. (2003). Application of kinematic wave theory for predicting flash flood hazards on coupled alluvial fan–piedmont plain landforms. *Hydrological Processes*, 17, 839–868.

Mukhopadhyay, B., Dutta, A., Nouri, F., and Kaushik, C. (2009a). Modeling urban flooding from storm sewers using dynamic wave theory. In *Hydrology and Hydraulics. Proceedings of the International Conference on Water, Environment, Energy and Society. Volume 2*, Allied Publishers, New Delhi, pp. 275–289.

Mukhopadhyay, B., Khan, A., and Gautam, R. (2015). Rising and falling river flows: contrasting signals of climate change and glacier mass balance from the eastern and western Karakoram. *Hydrological Sciences Journal*, 60(11–12), 2062–2085.

Mukhopadhyay, B., Nouri, F., Penland, C. M., and Dutta, A. (2009b). Model flood alert system: development and application for the Theater District within downtown Houston. *Journal of Hydrologic Engineering*, 14(5), 475–489.

Mulvaney, T. (1850). On the use of self-registering rain and flood gauges. *Proceedings of the Institute of Civil Engineers*, 4(2), 1–8.

Murray, F. W. (1967). On the computation of saturation vapor pressure. *Journal of Applied Meteorology*, 6, 203–204.

Musgrave, G. W. (1955). How much of the rain enters the soil? In *The Yearbook of Agriculture 1955 Water*. US Department of Agriculture, Washington, DC.

Mustafa, B. Y. (2017). Hydrological study and analysis for proposed Sartik Dam. Part 2: reservoir characteristics, simulation model, and flood routing calculations. *Journal of University of Duhok*, 20(1), 776–789.

Naithani, A. K., Nainwal, H. C., Sati, K. K., and Prasad, C. (2001). Geomorphological evidences of Gangotri glacier and its characteristics. *Current Science*, 80(1), 87–94.

Nakawo, M. (2009). Shrinkage of summer accumulation: glaciers in Asia under consideration of downstream population. In *Assessment of Snow, Glaciers and Water Resources in Asia*. International Hydrological Programme of UNESCO and

Hydrology and Water Resources Programme of WMO, Koblenz, pp. 19–25.

Nash, J. E. (1957). The form of the instantaneous unit hydrograph. *International Association of Science and Hydrology*, 3, 114–121.

Nash, J. E. (1959). Synthetic determination of unit hydrograph parameters. *Journal of Geophysical Research*, 64(1), 111–115.

Nash, J. E. (1960). A unit hydrograph study, with particular reference to British catchments. *Proceedings, Institution of Civil Engineers*, 17, 249–282.

Nash, J. E. and Sutcliffe, J. V. (1970). River flow forecasting through conceptual models. 1: discussion of principles. *Journal of Hydrology*, 10(3), 282–290.

Nathan, R. J. and McMahon, T. A. (1990). Evaluation of automated techniques for base flow and recession analysis. *Water Resources Research*, 26(7), 1465–1473.

National Centers for Environmental Information, National Oceanic and Atmospheric Administration (2022). *Annual 2022 Climate Report*. National Centers for Environmental Information, Silver Spring, MD.

Neitsch, S. L., Arnold, J. G., Kiniry, J. R., and Williams, J. R. (2005). *Soil and Water Assessment Tool Theoretical Documentation, Version 2005*. USDA–ARS Grassland, Soil and Water Research Laboratory, Temple, TX.

Nelder, J. A. and Mead, R. (1965). A simplex method for function minimization. *Computer Journal*, 7(4), 308–313.

Nelson, T. L. (1970). Synthetic hydrographs relationships, Trinity River tributaries, Fort Worth–Dallas urban area. Seminar on Urban Hydrology, Davis, California.

Neter, J., Wasserman, W., and Kutner, M. H. (1990). *Applied Linear Statistical Models*. Richard D. Irwin. Inc., Homewood, IL.

NIH (National Institute of Hydrology) (1998). *Rainfall–Runoff Modeling of Morel Catchment for Design Flood Estimation*. NIH Report CS (AR)-2/97-98. National Institute of Hydrology, Roorkee, India.

NIH (2001). *Rainfall–Runoff Analysis Using Flood Analysis and Protection Systems (FLAPS) Model*. NIH Report CS (AR)-3/2000-2001. National Institute of Hydrology, Roorkee, India.

Nolin, A. W. and Daly, C. (2006). Mapping "at risk" snow in the Pacific Northwest. *Journal of Hydrometeorology*, 7, 1164–1171.

Natural Environment Research Council (1975). *Flood Studies Report*. Natural Environment Research Council, London.

NRCS (Natural Resources Conservation Commission) (1986). *Urban Hydrology for Small Watersheds*. Technical Release 55. United States Department of Agriculture, Washington, DC.

NRCS (2004a). Estimation of direct runoff from storm rainfall. In *National Engineering Handbook*, United States Department of Agriculture, Washington, DC, chapter 10.

NRCS (2004b). Hydrologic soil-cover complexes. In *National Engineering Handbook*, United States Department of Agriculture, Washington, DC, chapter 9.

NRCS (2007). *National Engineering Handbook*. United States Department of Agriculture, Washington, DC.

NRCS (2014). Flood routing. In *National Engineering Handbook*, United States Department of Agriculture, Washington, DC, Part 630, ch. 17.

NRCS (2019). Storm rainfall depth and distribution. In *National Engineering Handbook*, United States Department of Agriculture, Washington, DC, Part 630, ch. 4.

Oerlemans, J. (2005). Extracting a climate signal from 169 glacier records. *Science*, 308(5722): 675–677.

Overton, D. E. (1964). *Mathematical Refinement of an Infiltration Equation for Watershed Engineering*. ARS 41-99. Agricultural Research Service, United States Department of Agriculture, Washington, DC.

Overton, D. E. and Meadows, M. E. (1976). *Stormwater Modeling*. Academic Press, New York.

Pai, D. S., Sridhar, L, Rajeevan, M., et al. (2014). Development of a new high spatial resolution (0.25° × 0.25°) long period (1901–2010) daily gridded rainfall data sets over India and its comparison with existing data sets over the region. *Mausam*, 65, 1–18.

Pak, J., Fleming, M., Scharffenberg, W., and Ely, P. (2008). Soil erosion and sediment yield modeling with the hydrologic modeling system (HEC-HMS). World Environmental and Water Resources Congress 2008, Honolulu, Hawaii, May 12–16.

Pani, E. A. and Haragan, D. R. (1981). A comparison of Texas and Illinois temporal rainfall distributions. Preprints, 4th Conference on Hydrometeorology, American Meteorological Society, Boston, MA, pp. 76–80.

Parry, M. (1990). *Climate Change and World Agriculture*. Earthscan, London.

Paul, F., Kääb, A., and Haeberli, W. (2007). Recent glacier changes in the Alps observed by satellite: consequences for future monitoring strategies. *Global and Planetary Change*, 56, 111–122.

Pechštädt, J., Bartosch, A., Zander, F., et al. (2009). Development of a river basin information system for a sustainable development in the Upper Brahmaputra River Basin. Available at: www.researchgate.net/publication/237592171_DEVELOPMENT_OF_A_RIVER_BASIN_INFORMATION_SYSTEM_FOR_A_SUSTAINABLE_DEVELOPMENT_IN_THE_UPPER_ BRAHMAPUTRA_RIVER_BASIN.

Penman, R. L. (1948). Natural evaporation from open water, bare soil and grass. *Proceedings of the Royal Society of London, A*, 193, 120–146.

Pekárová, P., Drobot, R., Bačová Mitková, V., Mészáros, J., and Draghia, A. F. (2019). Statistical analysis of extreme discharges. In P. Pekárová, and P. Miklánek (eds.), *Flood Regime of Rivers in the Danube River Basin,* Institute of Hydrology, Slovak Academy of Sciences, Bratislava, pp. 123–150.

Perica, S., Pavlovic, S., St Laurent, M., et al. (2018). *Precipitation-Frequency Atlas of the United States, Volume 11 Version 2.0: Texas*. NOAA Atlas 14. US Department of Commerce, National Oceanic and Atmospheric Administration, National Weather Service, Silver Spring, MD.

Perumal, M. (1992). The cause of the negative initial outflow with the Muskingum method. *Hydrological Sciences Journal*, 37(4), 391–401.

Perumal, M. and Price, R. K. (2017). Reservoir and channel routing. In V. P. Singh (ed.), *Handbook of Applied Hydrology*, McGraw Hill, New York, pp. 52-1–52-16.

Peters-Lidard, C. D., Hossain, F., Leung, et al. (2019). *100 years of progress in hydrology*. In *A Century of Progress in Atmospheric and Related Sciences: Celebrating the American Meteorological Society Centennial Meteorological*

Monographs, American Meteorological Society, Boston, MA, volume 59, 25.1–25.51.

Philip, J. R. (1957). The theory of infiltration. 1. The infiltration equation and its solution. *Soil Science*, 83(5), 345–357.

Phillips, J. D. and Lutz, J. D. (2008). Profile convexities in bedrock and alluvial streams. *Geomorphology*, 102, 554–566.

Pierce, C. H. (1938). Synoptic analysis of the southern California flood of March 2, 1938. *Monthly Weather Review*, 66(5), 135.

Pierce, L. T. (1958). Estimating seasonal and short-term fluctuations in evapotranspiration from meadow crops. *Bulletin of American Meteorological Society*, 39, 73–78.

Pierce, D. W., Barnett, T. P., Hidalgo, H. G., et al. (2008). Attribution of declining western U.S. snowpack to human effects. *Journal of Climate*, 21, 6425–6444.

Pilgrim, D. H. and Cordery, I. (1975). Rainfall temporal patterns for design floods. Journal of the Hydraulics Division, *American Society of Civil Engineers*, 101(1), 81–85.

Pilgrim, D. H. and Cordery, I. (1992). Flood runoff. In D. R. Maidment (ed.), *Handbook of Hydrology*, McGraw Hill, New York.

Pilgrim D. H., Chapman, T. G., and Doran D. G. (1988). Problems of rainfall–runoff modelling in arid and semiarid regions. *Hydrological Sciences Journal*, 33(4), 379–400.

Poertner, H. (1974). *Practices in Detention of Urban Stormwater Runoff*. APWA Special Report No. 43. American Public Works Association, Washington, DC.

Ponce, V. M. (1983). Development of physically based coefficients for the diffusion method of flood routing. Final Report to the USDA, Soil Conservation Service, Lanham, Maryland.

Ponce, V. M. (1986). Diffusion wave modeling of catchment dynamics. *Journal of Hydraulic Engineering*, 112(8), 716–727.

Ponce, V. M. (1989). *Engineering Hydrology: Principles and Practices*. Prentice Hall, Hoboken, NJ.

Ponce, V. M. and Simons, D. B. (1977). Shallow wave propagation in open-channel flow. *Journal of the Hydraulics Division, American Society of Civil Engineers,* 103(12), 1461–1476.

Ponce, V. M. and Vuppalapati, B. (2016). Muskingum–Cunge amplitude and phase portraits with online commutation. Online article, available at https://ponce.sdsu.edu/muskingum_cunge_amplitude_and_phase_portraits_with_online_computation.html.

Ponce, V. M. and V. Yevjevich, V. (1978). Muskingum–Cunge method with variable parameters. *Journal of the Hydraulics Division, American Society of Civil Engineers*, 104(12), 1663–1667.

Prasad, A. K., Yang, K.-H. S., El-Askary, H. M., and Kafatos, M. (2009). Melting of major glaciers in the western Himalayas: evidence of climatic changes from long term MSU derived tropospheric temperature trend (1979–2008). *Annals of Geophysics,* 27, 4505–4519.

Priestley, C. H. B. and Taylor, R. J. (1972). On the assessment of surface heat flux and evaporation using large scale parameter. *Monthly Weather Review*, 100, 81–92.

Puls, L. G. (1928). Flood regulation of the Tennessee River. 70th Congress, 1st Session, H. D. 185, Pt. 2, Appendix B.

Raina, V. K. (2009). Himalayan glaciers: a state-of-art review of glacial studies, glacial retreat, and climate change. Ministry of Environment and Forests, Government of India. Discussion Paper.

Raina, V. K. and Sangewar, C. V. (2007). Siachen Glacier of Karakoram Mountains, Ladakh: its secular retreat. *Journal of the Geological Society of India*, 70, 11–16.

Rankl, M., Vijay, S., Kienholz, C., Braun, M. (2013). Glacier changes in the Karakoram region mapped by multi-mission satellite imagery. *The Cryosphere*, 7, 4065–4099.

Rao, R. A., Delleur, J. W., and Sarma, B. (1972). Conceptual hydrologic models for urbanizing basins. *Journal of the Hydraulics Division, American Society of Civil Engineers*, 98, 1205–1220.

Rasul, G., Dahe, Q., and Chaudhry, Q. Z. (2008). Global warming and melting of glaciers along southern slopes of HKH ranges. *Pakistan Journal of Meteorology*, 5(9), 63–75.

Rawls, W. J. and Brakensiek, D. L. (1986). Comparisons between Green–Ampt and curve number runoff predictions. *Transactions of American Society of Agricultural Engineers*, 29(6), 1597–1599.

Rawls, W. J., Brakensiek, D. L., and Miller, N. (1983). Green–Ampt infiltration parameters from soil data. *Journal of Hydraulic Engineering*, 109(1), 62–70.

Rees, H. G. and Collins, D. N. (2006). Regional differences in response of flow in glacier-fed Himalayan rivers to climatic warming. *Hydrological Processes*, 20, 2157–2169.

Regonda, S., Rajagopalan, B., Clark, M., and Pitlick, J. (2005). Seasonal cycle shifts in hydro-climatology over the western United States. *Journal of Climate*, 18, 372–384.

Ren, J. W., Qin, D. H., Kang, S. C., et al. (2003). Glacier variations and climate warming and drying in the central Himalayas. *Chinese Science Bulletin*, 48(23), 2478–2482.

Rice, S. P. and Church, M. (2001). Longitudinal profiles in simple alluvial systems. *Water Resources Research*, 37(2), 417–426.

Rice, J., Bardsley, T., Gomben, P., et al. (2017). *Assessment of Watershed Vulnerability to Climate Change for the Uinta–Wasatch–Cache and Ashley National Forests, Utah*. General Technical Report RMRS-GTR-362. US Department of Agriculture, Forest Service, Rocky Mountain Research Station, Fort Collins, CO.

Richards, L. A. (1931). Capillary conduction of liquids through porous mediums. *Physics*, 1(5), 318–333.

Richardson, L. F. (1922). *Weather Prediction by Numerical Process*. Cambridge University Press, Cambridge.

Riggs, H. C. (1972). Low-flow investigations. In *Techniques of Water Resources Investigations of the United States Geological Survey. Book 4. Hydrologic Analysis and Interpretation*, US Geological Survey, Washington, DC, ch. B1.

Rigon, R., Rodriquez-Iturbe, I., Maritan, A., et al. (1996). On Hack's law. *Water Resources Research*, 32(11), 3367–3374.

Rivera-Giboyeaux, A. M. (2020). Radar derived rainfall and rain gauge measurements at SRS. Savannah River National Laboratory, US Department of Energy.

Roderick, M. L., Greve, P., Farquhar, G. D. (2015). On the assessment of aridity with changes in atmospheric CO_2. *Water Resources Research*, 51, 5450–5463.

Roderick, M. L., Sun, F., Lim, W. H., Farquhar, G. D. (2014). A general framework for understanding the response of the water cycle to global warming over land and ocean. *Hydrology and Earth System Sciences*, 18, 1575–1589.

Rodman, P. K. (1977). Effects of urbanization on various frequency peak discharges. United States Army Corps of Engineers Meeting, Albuquerque, New Mexico.

Rodriguez-Iturbe, I. and Mejia, J. M. (1974). The design of rainfall networks in time and space. *Water Resources Research*, 10(4), 713–728.

Rosbjerg, D. (1977). Return periods of hydrological events. *Nordic Hydrology*, 8(1), 57–61.

Rose, S. and Peters, N. E. (2001). Effects of urbanization on streamflow in the Atlanta area (Georgia, USA): a comparative hydrological approach. *Hydrological Processes*, 15, 1441–1457.

Rosenbrock, K. H. (1960). An automatic method for finding the greatest or least value of a function. *The Computer Journal*, 3(3), 175–184.

Roussel, M. C., Thompson, D. B., Fang, X., Cleveland, T. G., and Garcia, A. (2005). *Time-Parameter Estimation for Applicable Texas Watersheds*. Report No. FHWA/TX-05/0-4696-2. Texas Department of Transportation, Austin, TX.

Rubey, W. W. (1933). Settling velocities of gravel, sand, and silt particles. *American Journal of Science, 5th Series*, 25(148), 325–338.

Russell, S. O., Sunnell, G. J., and Kenning, B. F. (1979). Estimating design flows for urban drainage. *Journal of the Hydraulics Division*, 105(1), 43–52.

Rziha, F. (1876). *Eisenbahn-Unter und Oberbau*. Verlag der KK Hof-und Staatsdr., Vienna, vol. 1.

Sabol, G. V. (1988). Clark unit hydrograph and *R*-parameter estimation. *Journal of Hydraulic Engineering*, 114(1), 103–111.

Saint-Venant, B. de. (1871). Theorie du mouvement non-permanent des eaux avec application aux crues des rivieres et l' introduction des varees dans leur lit [Theory of unsteady water flow, with application to river floods and propagation of tides in river channels]. *Comptes rendus hebdomadaires des Seances de l'Academie des Science*, 73(1871), 148–154.

Salas, J. D. (1993). Analysis and modeling of hydrologic time series. In D. R. Maidment (ed.), *Handbook of Hydrology*, McGraw Hill, New York, pp. 19.1–19.72.

Sauer, V. B. and Turnipseed, D. P. (2010). *Stage Measurement at Gauging Stations*. Techniques and Methods 3-A7. US Geological Survey, Reston, VA.

Sauer, V. B., Thomas, W. O., Stricker, V. A., and Wilson, K. V. (1983). *Flood Characteristics of Urban Watersheds in the United States*. US Geological Survey, Water Supply Paper 2207. United States Department of the Interior, Washington, DC.

Sauvageot, H. (1994). Rainfall measurement by radar: a review. *Atmospheric Research*, 35, 27–54.

Schreiner, L. C. and Riedel, J. T. (1978). *Probable Maximum Precipitation Estimates, United States East of the 105th Meridian*. Hydrometeorological Report No. 51. National Weather Service, US Department of Commerce, National Oceanic and Atmospheric Administration, US Department of the Army Corps of Engineers, Washington, DC.

Schumm, S. A. (1956). Evolution of drainage systems and slopes in Badlands at Perth Amboy, New Jersey. *Geological Society of America Bulletin*, 67, 597–646.

SCS (Soil Conservation Service) (1975). *Urban Hydrology for Small Watersheds*. Engineering Division, Soil Conservation Service, US Department of Agriculture, Washington, DC.

SCS (1985). *National Engineering Handbook*. US Department of Agriculture Soil Conservation Service, Washington, DC.

Searcy, J. K. (1963). *Flow-Duration Curves. Manual of Hydrology: Part 2, Low-Flow Techniques*. United States Geological Survey Water Supply Paper. United States Department of the Interior, Washington, DC.

Seager, R., Naik, N., and Vecchi, G. A. (2010). Thermodynamic and dynamic mechanisms for large-scale changes in the hydrological cycle in response to global warming. *Journal of Climate*, 23, 4651–4668.

Seddon, J. A. (1900). River hydraulics. *Transactions of the American Society of Civil Engineers*, 43, 179–229.

Sefe, F. T. K. (1996). A study of the stage–discharge relationship of the Okavango River at Mohembo, Botswana. *Hydrological Sciences Journal*, 41(1), 97–116.

Service, R. F. (2004). As the West goes dry. *Science*, 303, 1124–1127.

Severskiy, I. (2009). Current and projected changes of glaciation in central Asia and their probable impact on water resource. In *Assessment of Snow, Glaciers and Water Resources in Asia*. International Hydrological Programme of UNESCO and Hydrology and Water Resources Programme of WMO, Koblenz, pp. 99–111.

Shahgedanova, M., Hagg, W., Hassell, D., Stokes, C. R., and Popovnin, V. (2009). Climate change, glacier retreat, and water availability in the Caucasus region. In J. A. A. Jones, T. G. Vardanian, and C. Hakopian (eds.), *Threats to Water Security*, Springer, Dordrecht, pp. 131–143.

Sharma, M. C. and Owen, L. A. (1996). Quaternary glacial history of NW Garhwal, Central Himalayas. *Quaternary Science Review*, 15, 335–365.

Sheridan, J. M., Merkel, W. H., and Bosch, D. D. (2002). Peak rate factors for flatland watersheds. *Applied Engineering in Agriculture*, 18(1), 65–69.

Sherman, L. K. (1932). Streamflow from rainfall by the unit-graph method. *Engineering News Records*, 108, 501–505.

Shi, Y., Liu, S., Shangguan, D., Li, D., and Ye, B. (2006). Peculiar phenomena regarding climatic and glacial variations on the Tibetan Plateau. *Annals of Glaciology*, 43, 106–110.

Shulits, S. (1941). Rational equation of river bed profile. *Transactions of the American Geophysical Union*, 22, 622–631.

Singh, K. P. (1976). Unit hydrographs: a comparative study. *Water Resources Bulletin*, 12(2), 381–392.

Singh, V. P. (1974). A non-linear kinematic wave model of surface runoff. Unpublished Ph. D. dissertation, Colorado State University, Fort Collins, CO.

Singh, V. P. (1976). Derivation of time of concentration. *Journal of Hydrology*, 30, 147–165.

Singh, V. P. (1992). *Elementary Hydrology*. Prentice Hall, Hoboken, NJ.

Singh, V. P. (ed.) (1995). *Computer Models of Watershed Hydrology*. Water Resources Publication, Littleton, CO.

Singh, V. P. (1996). *Kinematic Wave Modeling in Water Resources: Surface Water Hydrology*. Wiley, New York.

Singh, V. P. (2018). Hydrologic modeling: progress and future directions. *Geoscience Letters*, 5(15), 5–23.

Singh, V. P. and Bengtsson, L. (2004). Hydrological sensitivity of a large Himalayan basin to climate change. *Hydrological Processes*, 18, 2363–2385.

Singh, V. P. and Chowdhury, P. K. (1985). On fitting gamma distribution to synthetic runoff hydrographs. *Nordic Hydrology*, 16, 177–192.

Singh, V. P. and Chowdhury, P. K. (1986). Comparing some methods of estimating mean areal rainfall. *Water Resources Bulletin*, 22(2), 275–282.

Singh, V. P. and Cruise, J. (1992). Analysis of the rational formula using a systems approach. In B. C. Yen (ed.), *Catchment Runoff and Rational Formula*, Water Resources Publications, Littleton, CO, pp. 39–51.

Singh, V. P. and Frevert, D. K. (eds.) (2002a). *Mathematical Models of Large Watershed Hydrology*. Water Resources Publications, Littleton, CO.

Singh, V. P. and Frevert, D. K. (eds.) (2002b). *Mathematical Models of Small Watershed Hydrology and Applications*. Water Resources Publications, Littleton, CO.

Singh, V. P. and Frevert, D. K. (eds.) (2006). *Watershed Models*. CRC Press–Taylor and Francis, Boca Raton, CA.

Singh, V. P. and Scarlatos, P. D. (1987). Analysis of the nonlinear Muskingum flood routing. *Journal of Hydraulic Engineering*, 113(1), 61–79.

Singh, V. P. and Woolhiser, D. A. (2002). Mathematical modeling of watershed hydrology. *Journal of Hydrologic Engineering*, 7(4), 270–292.

Singh, V. P. and Yu, F. X. (1990). Derivation of an infiltration equation using a systems approach. *Journal of Irrigation and Drainage Engineering*, 116(6), 837–858.

Singh, V. P. and Zhang, L. (2017). Frequency distributions. In V. P. Singh (ed.), *Handbook of Applied Hydrology*, McGraw Hill, New York, ch. 21.

Skaggs, R. W. and Khaleel, R. (1982). Infiltration. In C. T. Haan, H. P. Johnson, and D. L. Brakensiek (eds.), *Hydrologic Modeling of Small Watersheds*, American Society of Agricultural Engineers, St. Joseph, MI.

Slattery, M. C. and Burt, T. P. (1997). Particles size characteristics of suspended sediment in hillslope runoff and stream flow. *Earth Surface Processes and Landforms*, 22(8), 705–719.

Smith, A. A. and Lee, K.-B. (1984). The rational method revisited. *Canadian Journal of Civil Engineering*, 11, 854–862.

Smith, J. A., Seo, D. J., Beck, M. L., and Hudlow, M. D. (1996). An intercomparison study of NXRAD precipitation estimates. *Water Resources Research*, 32, 2035–2045.

Smith, R. E. and Parlange, J.-Y. (1978). A parameter efficient hydrologic infiltration model. *Water Resources Research*, 14(8), 533–538.

Smith, R. E., Smettem, K. R. J., Broadbridge, P., and Woolhiser, D. A. (2002). *Infiltration Theory for Hydrologic Applications*. Water Resources Monograph, 15. American Geophysical Union, Washington, DC.

Solanes, M. (1998). Integrated water management from the perspective of the Dublin Principles. *CEPAL Review*, 64, 165–184.

Sontakke, N. A., Singh, H. N., and Singh, N. (2008). *Chief Features of Physiographic Rainfall Variations across India during Instrumental Period (1813 – 2006)*. Research Report No. RR-121. Indian Institute of Tropical Meteorology, Pune, India.

Spendley, W., Hext, G. R., and Himsworth, F. R. (1962). Sequential application of simplex designs in optimization and evolutionary operation. *Technometrics*, 4(4), 441–461.

Strahler, A. N. (1957). Quantitative analysis of watershed geomorphology. *Transactions of the American Geophysical Union*, 38, 913–920.

Stall, J. B. and Yang, C. T. (1970). *Hydraulic Geometry of 12 Selected Stream Systems of the United States*. Water Resources Center Report No. 32. University of Illinois, Urbana, IL.

Stedinger, J. R., Vogel, R. M., and Foufoula-Georgiou, E. (1993). Frequency analysis of extreme events. In D. R. Maidment (ed.), *Handbook of Hydrology*, McGraw Hill, New York, pp. 18.1–18.66.

Stedinger, J. R., and Griffis, V. W. (2008). Flood frequency analysis in the United States: time to update. *Journal of Hydrologic Engineering*, 13(4), 199–204.

Steenhuis, T. S. and Van der Molen, W. H. (1986). The Thornthwaite–Mather procedure as a simple engineering method to predict recharge. *Journal of Hydrology*, 84, 221–229.

Stephenson, D. (1979). Direct optimization of Muskingum routing coefficients. *Journal of Hydrology*, 41, 161–165.

Stewart, I. T. Cayan, D. R., Dettinger, M. D. (2004). Changes in snowmelt runoff timing in Western North America under a 'business as usual' climate change scenario. *Climate Change*, 62(1–3), 217–232.

Stewart, I. T. Cayan, D. R., Dettinger, M. D. (2005). Changes toward earlier streamflow timing across western North America. *Journal of Climate*, 18, 1136–1155.

Strand, R. I. and Pemberton E. L. (1982). *Reservoir Sedimentation Technical Guidelines for Bureau of Reclamation*. US Bureau of Reclamation, Denver, CO.

Strelkoff, T. (1980a). *Comparative Analysis of Flood Routing Methods*. RD-24. US Army Corps of Engineers, Hydrologic Engineering Center, Davis, CA.

Strelkoff, T. (1980b). *Modified-Puls Routing in Chuquatonchee Creek*. RD-23. US Army Corps of Engineers, Hydrologic Engineering Center, Davis, CA.

Snyder, F. F. (1938). Synthetic unit-graphs. *Transactions of the American Geophysical Union*, 19, 447–454.

Snyder, J. P. (1987). *Map Projections: A Working Manual*. US Geological Survey Professional Paper 1395. United States Government Printing Office, Washington, DC.

Snyder, J. P. (1993). *Flattening the Earth. Two Thousand Years of Map Projections*. University of Chicago Press, Chicago, IL.

Snyder, J. P. and Voxland, P. M. (1989). *An Album of Map Projections*. US Geological Survey Professional Paper 1453. United States Government Printing Office Washington, DC.

Snyder, W. M. (1955). Hydrograph analysis by the method of least squares. *Proceedings American Society of Civil Engineers*, 81, 1–24.

Sturges, H. A. (1926). The choice of a class interval. *Journal of the American Statistical Association*, 21, 65–66.

Tandong, Y., Wang, Y., Shiying, L., et al. (2009). Recent glacial retreat in the Chinese part of High Asia and its impact on water resources of Northwest China. In *Assessment of Snow, Glaciers and Water Resources in Asia*. International Hydrological Programme of UNESCO and Hydrology and Water Resources Programme of WMO, Koblenz, pp. 26–35.

Tarboton, D. G. and Luce, C. H. (1996). *Utah Energy Balance Snow Accumulation and Melt Model (UEB)*. Utah Water Research Laboratory, Utah State University and USDA Forest Service, Intermountain Research Station.

Taylor, A. B., and Schwarz, H. E. (1952). Unit hydrograph lag and peak flow related to basin characteristics. *Transactions of the American Geophysical Union*, 33, 235–246.

Texas Department of Transportation (2019). *Hydraulic Design Manual*. Texas Department of Transportation, Austin, TX.

Thiessen, A. H. (1911). Precipitation for large areas. *Monthly Weather Review*, 39, 1082–1084.

Thodsen, H. (2007). The influence of climate change on stream flow in Danish rivers. *Journal of Hydrology*, 333, 226–238.

Tholin, A. L. and Kiefer, C. J. (1960). The hydrology of urban runoff. *Transactions American Society of Civil Engineers*, 125 (1), 1308–1379.

Thomas, W. A. (1994). *Sedimentation in Stream Networks*, HEC-6T User's Manual, Mobile Boundary Hydraulics Software, Inc., Clinton, MS.

Thompson, C. (2011). *HIRD V3. High Intensity Rainfall Design System: the Method Underpinning the Development of Regional Frequency Analysis of Extreme Rainfalls for New Zealand*. National Institute of Water and Atmospheric Research (NIWA), Auckland.

Thornthwaite, C. W. (1948). An approach toward a rational classification of climate. *Geographical Review*, 38, 55–94.

Thornthwaite, C. W. and Mather, J. R. (1955). The water balance. *Climatology*, 8, 5–86.

Toffaleti, F. B. (1968). *A Procedure for Computation of Total River Sand Discharge and Detailed Distribution, Bed to Surface*. US Army Corps of Engineers, Committee on Channel Stabilization, Technical Report No. 5. US Army Corps of Engineers, Washington, DC.

Toffaleti, F. B. (1969). Definitive computations of sand discharge in rivers. *Journal of the Hydraulics Division, American Society of Civil Engineers*, 95(1), 225–246.

Troxell, H. C. (1942). *Floods of March 1938 in Southern California*. US Geological Survey, Water Supply Paper 844. United States Department of the Interior, Washington, DC.

Turc, L. (1961). Estimation of irrigation water requirements, potential evapotranspiration: a simple climatic formula evolved up to date. *Annals of Agronomy*, 12, 13–46.

Turnipseed, D. P. and Sauer, V. B. (2010). *Discharge Measurements at Gauging Stations*. Techniques and Methods 3-A8. US Geological Survey, Reston, VA.

UNESCO (1979). *Map of the World Distribution of Arid Regions*. MAB Technical Notes 7, UNESCO, Paris.

USACE (US Army Corps of Engineers) (N.D). HEC HMS technical reference manual, online version, available at www.hec.usace.army.mil.

USACE (1959). *Engineering and Design. Flood-Hydrograph Analyses and Computations*. Engineering Manual 1110-2-1405. US Army Corps of Engineers, Washington, DC.

USACE (1960). *Routing of Floods Through River Channels*. Engineering Manual 1110-2-1408. US Army Corps of Engineers, Washington, DC.

USACE (1982). *Hydrologic Analysis of Ungaged Watersheds Using HEC-1*, Training Document 15. US Army Corps of Engineers, Hydrologic Engineering Center, Davis, CA.

USACE (1987). *HMR 52: Probable Maximum Storm (Eastern United States)*. US Army Corps of Engineers, Hydrologic Engineering Center, Davis, CA.

USACE (1989). *Sedimentation Investigations of Rivers and Reservoirs*. Engineering Manual 1110-2-4000. US Army Corps of Engineers, Washington, DC.

USACE (1993a). *Hydrologic Frequency Analysis*. US Army Corps of Engineers, Washington, DC.

USACE (1993b). *Introduction and Application of Kinematic Wave Routing Techniques Using HEC-1*. Training Document 10. US Army Corps of Engineers, Hydrologic Engineering Center, Davis, CA.

USACE (1994). *Flood–Runoff Analysis*. Engineering Manual 1110-2-1417. US Army Corps of Engineers, Washington, DC.

USACE (1995). *Hydrologic Engineering Requirements for Flood Damage Reduction Studies*. Engineering Manual 1110-2-1419. US Army Corps of Engineers, Washington, DC.

USACE (1998). *HEC-1 Flood Hydrograph Package User's Manual*. US Army Corps of Engineers, Hydrologic Engineering Center, Davis, CA.

US Bureau of Reclamation (1949). Flood routing. *Flood Hydrology*, Pt. 6 in Water Studies, volume IV. United States Department of Interior, Washington, DC., ch. 6.10.

US Bureau of Reclamation (1961). *Design of Small Dams*. United States Department of Interior, Washington, DC.

USDA (US Department of Agriculture) (1968). *Moisture-Tension Data for Selected Soils on Experimental Watersheds*. Report 41-144. Agricultural Research Service, US Government Printing Office, Washington, DC.

USDA (1973). *Method for Estimating Volume and Runoff in Small Watersheds*. Technical Paper 149A. US Department of Agriculture, Soil Conservation Service, Washington, DC.

USDA (1999). *Soil Taxonomy: A Basic System of Soil Classification for Making and Interpreting Soil Surveys*. Agriculture Handbook Number 436. US Department of Agriculture, Washington, DC.

US Department of Interior (1982). *Guidelines for Determining Flood Flow Frequency*. Bulletin No. 17B of the Hydrology Subcommittee. Office of Water Data Coordination, US Geological Survey, Reston, VA.

USGS (US Geological Survey) (1994). *Nationwide Summary of US Geological Survey Regional Regression Equations for Estimating Magnitude and Frequency of Floods for Ungagged Site, 1993*. Water Resources Investigation Report 94-4002. US Geological Survey, Reston, VA.

USGS (2005). *Changes in Streamflow Timing in the Western United States in Recent Decades*. National Streamflow Information Program, Factsheet 2005-3018. US Geological Survey, Reston, VA.

US Weather Bureau (1957). *Rainfall intensity–frequency regime, Part 1: The Ohio Valley*. Technical Paper No. 29. US Department of Commerce, Weather Bureau, Washington, DC.

US Weather Bureau (1958). *Rainfall Intensity–Frequency Regime, Part 3: The Middle Atlantic Region*. Technical Paper No. 29. US Department of Commerce, Weather Bureau, Washington, DC.

US Weather Bureau (1960). *Generalized Estimates of Probable Maximum Precipitation West of the 105th Meridian*. Technical Paper No. 38. US Department of Commerce, Washington, DC.

Valdes, J. B., Fiallo, Y., and Rodriguez-Iturbe, I. (1979). A rainfall–runoff analysis of the geomorphologic IUH. *Water Resources Research*, 15(6), 1421–1434.

Vanoni, V. A. (1975). *Sedimentation Engineering*, ASCE Manuals and Reports on Engineering Practice No. 54. American Society of Civil Engineers, Reston, VA.

Van Rijn, L. C. (1993). *Principles of Sediment Transport in Rivers, Estuaries, Coastal Seas and Oceans*. International Institute for Infrastructural, Hydraulic, and Environmental Engineering, Delft.

Veihmeyer, F. J. and Hendrickson, A. H. (1927). The relation of soil moisture to cultivation and plant growth. *Soil Science*, 3, 498–513.

Vicente, G. A., Scofield, R. A., and Menzel, W. P. (1998). The operational GOES infrared estimation technique. *Bulletin of the American Meteorological Society*, 79, 1883–1898.

Viglione, A., Merz, R., and Blöschl, G. (2009). On the role of the runoff coefficient in the mapping of rainfall to flood return periods. *Hydrology and Earth System Sciences*, 13, 577–593.

Villemonte, J. R. (1947). Submerged-weir discharge studies. *Engineering News Record*, 12, 866.

Vivoni, E. R., Di Benedetto, F., Grimaldi, S., and Eltahir, A. B. (2008). Hypsometric control on surface and subsurface runoff. *Water Resources Research*, 44, W12502.

Vorosmarty, C. J. and Moore, B. (1991). Modeling basin-scale hydrology in support of physical climate and global biogeochemical studies: an example using the Zambezi River. *Surveys in Geophysics*, 12(1–3), 271–311.

Vorosmarty, C. J., Moore, B. Gildea, M. P. et al. (1989). A continental-scale model of water balance and fluvial transport: application to South America. *Global Biogeochemical Cycles*, 3, 241–265.

Vorosmarty, C. J., Willmott, C. J., Choudhury, B. J., et al. (1996). Analyzing the discharge regime of a large tropical river through remote sensing, ground-based climatic data and modeling. *Water Resources Research*, 32, 3137–3150.

Vuille, M., Francou, B., Wagnon, P., et al. (2008). Climate change and tropical Andean glaciers: past present, and future. *Earth Science Reviews*, 89, 79–96.

Walesh, S. (1975). Discussion of Chien and Saigal (1975). *Journal of the Hydraulics Division, American Society of Civil Engineers*, 101(11), 1447–1449.

Walesh, S. (1989). *Urban Water Management*. Wiley, New York.

Wallis, J. R., Matalas, N. C., and Slack, J. R. (1974). Just a moment. *Water Resources Research*, 10(2), 211–219.

Wanielista, M., Kersten, R., and Eaglin, R. (1997). *Hydrology: Water Quantity and Quality Control*, 2nd ed., Wiley, New York.

Ward, J. H. (1963). Hierarchical grouping to optimize an objective function. *Journal of American Statistical Association*, 58, 236–244.

Welle, P. J. and Woodward, D. E. (1986). *Time of Concentration*. Hydrology Technical Note No. N4. US Department of Agriculture, Soil Conservation Service, NENTC, Chester, PA.

Welle, P. J. and Woodward, D. E. (1989). Dimensionless unit hydrograph for the Delmarva Peninsula. *Transportation Research Record*, 1224, 79–87.

Wigley, T, M. L. and Jones, P. D. (1985). Influences of precipitation changes and direct CO_2 effects on streamflow. *Nature*, 314 (14): 149–152.

Wilby, R. L. and Dettinger, M. D. (2000). Streamflow changes in the Sierra Nevada, California, simulated using a statistically downscaled general circulation model scenario of climate change. In S. J. McLaren and D. R. Kniveton (eds.), *Linking Climate Change to Land Surface Change*, Kluwer Academic Publishers, Dordrecht, pp. 99–121.

Wilby, R. L., Hay, L. E., Gutowski, W. J. Jnr., et al. (2000). Hydrological responses to dynamically and statistically downscaled climate model output. *Geophysical Research Letters*, 27, 1199–1202.

Williams, J. R. (1975). Sediment-yield prediction with universal equation using runoff energy factor. In *Present and Prospective Technology for Predicting Sediment Yield and Sources: Proceedings of the Sediment Yield Workshop*. US Department of Agriculture, Agriculture Research Service, Washington DC, 244–252.

Williams-Sether, T., Asquith, W. H., Thompson, D. B., Cleveland, T. G., and Fang, X. (2004). *Empirical, Dimensionless, Cumulative-Rainfall Hyetographs Developed from 1959–86 Storm Data for Selected Small Watersheds in Texas*. Scientific Investigations Report 2004–5075 (TxDOT Research Report 0–4194–3). US Geological Survey, Reston, VA.

Willmott, C. J., Rowe, C. M. and Mintz, Y. (1985). Climatology of the terrestrial seasonal water cycle. *Journal of Climatology*, 5, 589–606.

Wilson, W. T. (1941). A graphical flood-routing method. *Transactions of the American Geophysical Union*, 21(3), 893–898.

Wischmeier, W. H and Smith, D. D. (1965). *Predicting Rainfall-Erosion Losses from Cropland East of the Rocky Mountains*. Agriculture Handbook No. 282. US Department of Agriculture, Washington, DC.

Wischmeier, W. H. and Smith, D. D. (1978). *Predicting Rainfall-Erosion Losses: A Guide to Conservation Planning*. Agriculture Handbook No. 537. US Department of Agriculture, Washington, DC.

Wischmeier, W. H., Johnson, C. B., and Cross, B. V. (1971). A soil erodibility nomograph for farmland and construction sites. *Journal of Soil and Water Conservation*, 26, 189–193.

WMO (World Meteorological Organization) (1966). *Climate Change*. Technical Note. Secretariat of the World Meteorological Organization, Geneva.

WMO (1969). *Manual for Depth–Area–Duration Analysis of Storm Precipitation*. Technical Paper 129, WMO – No. 237. Secretariat of the World Meteorological Organization, Geneva.

WMO (1986). *Manual for Estimation of Probable Maximum Precipitation*. Operational Hydrology, Report No. 1. WMO – No. 332. Secretariat of the World Meteorological Organization, Geneva.

WMO (1994). *Guide to Hydrological Practices: Volume II. Analysis, Forecasting and Other Applications*, WMO – No. 168. Secretariat of the World Meteorological Organization, Geneva.

WMO (2009). *Manual for Estimation of Probable Maximum Precipitation*. WMO – No. 1045, Secretariat of the World Meteorological Organization, Geneva.

WMO (2020). *Guide to Hydrological Practice. Volume I: Hydrology – From Measurement to Hydrological Information*. Secretariat of the World Meteorological Organization, Geneva.

Woods, R. A. (2009). Analytical model of seasonal climate impacts on snow hydrology: continuous snowpacks. *Advances in Water Resources*, 32, 1465–1481.

Wooding, R. A. (1965). A hydraulic model for the catchment stream problem: I. Kinematic wave theory. *Journal of Hydrology*, 3(3–4), 254–267.

Wooding, R. A. (1966). A hydraulic model for the catchment stream problem: III. Comparison with runoff observations. *Journal of Hydrology*, 4(3-4), 21–37.

Woolhiser, D. A., and Liggett, J. A. (1967). Unsteady one-dimensional flow over a plane: the rising hydrograph. *Water Resources Research*, 3(3), 753–771.

Wright, D. B., Smith, J. A., and Baeck, L. (2014). Critical examination of area reduction factors. *Journal of Hydrologic Engineering*, 19(4), 769–776.

Wu, I-Pai (1963). Design hydrographs for small watersheds in Indiana. *Journal of the Hydraulics Division, American Society of Civil Engineers*, 89(6), 35–66.

Wurbs, R. A. (2020). Institutional framework for modeling water availability and allocation. *Water*, 12, 1–26.

Xu, C.-Y. and Singh, V. P. (2000). Evaluation and generalization of radiation-based methods for calculating evaporation. *Hydrological Processes*, 14, 339–349.

Yang, C. T. (1972). Unit stream power and sediment transport. *Journal of Hydraulics Division*, 98(10), 1805–1826.

Yang, C. T. (1973). Incipient motion and sediment transport. *Journal of the Hydraulics Division, American Society of Civil Engineers*, 99(10), 1679–1704.

Yang, C. T. (1984). Unit stream power equation for gravel. *Journal of the Hydraulics Division, American Society of Civil Engineers*, 110(12), 1783–1797.

Yang, C. T. (1996). *Sediment Transport: Theory and Practice*. McGraw Hill, New York, NY.

Yeh, W. W.-G. (1985). Reservoir management and operations models: a state-of-the-art review. *Water Resources Research*, 12(12), 1797–1818.

Yen, B. C. and Chow, V. T. (1983). *Local Design Storms*. Vol. I to III, Report No. FHWA-RD-82-063 to 065. US Department of Transportation, Federal Highway Administration, Washington, DC.

Yin, D., Roderick, M. L., Leech, G., et al. (2014). The contribution of reduction in evaporative cooling to higher surface air temperatures during drought. *Geophysical Research Letters*, 41, 7891–7897.

Yoo, C., Lee, J., Park, C., and Jun, C. (2013). Method for estimating concentration time and storage coefficient of the Clark model using rainfall–runoff measurements. *Journal of Hydrologic Engineering*, 19(3), 626–634.

Yoo, C., Lee, J., and Cho, E. (2019). Theoretical evaluation of concentration time and storage coefficient with their application to major dam basins in Korea. *Water Supply*, 19(2), 644–651.

Yoo, C., Doan, H. P., Jun, C., Na, W. (2021). Hillslope contribution to Clark instantaneous unit hydrograph: application to the Seolmacheon Basin, Korea. *Water*, 13, 1707.

Yoon, K. W., Wone, S. Y., and Yoon, Y. N. (1994). A sensitivity analysis of model parameters involved in Clark method on the magnitude of design flood for urban watersheds. *Water for Future*, 27(4), 85–94.

Yoon, T. H. and Park, J. W. (2002). Improvement of the storage coefficient estimating method for the Clark model. In *Proceedings of the Korea Water Resources Association Conference*. Korea Water Resources Association, Cheongju-si, pp. 1334–1339.

Zhang, L. and Singh, V. P. (2019). *Copulas and their Applications in Water Resources Engineering*. Cambridge University Press, Cambridge.

Zhao, B. and Tung, Y.-K. (1994). Determination of optimal unit hydrographs by linear programming. *Water Resources Management*, 8, 101–119.

Index

ablation period, 576
absolute frequency, 37
   cumulative absolute frequency, 37
acoustic Doppler current profiler, 172
actual evapotranspiration, 302, 549, 560–561, 650–651
adiabatic cooling, 97
advection, 608
aggradation, 608
albedo, 562, 579
alternative hypothesis, 66
analysis of variance (ANOVA), 79
annual hydrographs, 180
annual maximum peak discharge, 193
annual maximum series, 117
antecedent soil conditions, 294
aquifers, 325
   confined, 325
   unconfined, 325
area ratio, 84, 182, 258, 319, 387
areal envelope, 143
areal reduction curves, 145
areal reduction factor, 128
aspect, 234, 236
Atlas 14 (NOAA), 128
atmometers, 565
attenuation ratio, 378
autocorrelation
   autocorrelation function, 78
   autocovariance function, 78
auxiliary gauge, 178
available water capacity, 277, 648
available water content, 275

backwater effect, 462
balanced storms, 134
bankfull discharge, 276
bankfull flow, 229
barometric equation, 552
baseflow, 253, 324
baseflow separation, 324
basins, 227, 308
Bayes' theorem, 48
Bernoulli process, 55
Bernoulli random variable, 54
bifurcation ratio, 240–241, 258, 387
binomial distribution, 55
blizzards, 98
Boussinesq equation, 328
Bowen ratio, 554
Brooks–Corey equation, 282
bubbling pressure. *See* Brooks–Corey equation
buffer zone, 617

capillary potential, 250
capillary suction, 250
catchments, 226
celerity, 460
central limit theorem, 56, 58
central moments, 43
central value, 43
centroid of a watershed, 232
centroidal flow path, 232
channel flow, 255
characteristics, 427
chi-squared distribution, 71
chi-squared test, 72
class intervals, 37
class limits, 37
coefficient of determination, 75
coefficient of variation, 44, 59, 61, 146, 150
cold fronts. *See* frontal surfaces
composite curve number, 294
composite runoff coefficient, 443
condition numbers, 32, 359
conditional probability, 48, *See* Bayes' theorem
confidence intervals, 73, 202
confidence level, 73, *See* significance levels
confidence limits, 73, 202
conic method, 518
conservation zone, 617
continuity equation, 455
   conservation form, 455
continuous simulations, 683
control space. *See* control volume
control surface, 4
control volume, 4
convective acceleration, 455
convective rainfall, 97
convergence precipitation, 146
conveyance, 245
convolution, 19–20, 33, 350, 352, 418
convolution function. *See* convolution
correlation, 73
correlation coefficient, 74
covariance, 73
critical point, 576
crop management factor, 592
cumulative distribution function. *See* probability distribution
cumulative infiltration, 277
cumulative probability distribution, 50–51
current meters, 172
curve numbers, 232
curve number method, 288
cyclonic rainfall, 97

Dalton's law, 551
Darcy's law, 275
Darcy flux, 275
data dictionaries, 659
data models, 670
database management systems, 659
declivity, 243
deconvolution, 355
degradation, 608
degree day factor, 582, 589
degree day method, 585
degrees of freedom, 68
   denominator degrees of freedom, 70
   numerator degrees of freedom, 70
Delmarva hydrographs, 390
delta functions, 609
depletion constant. *See* recession constant
depth of rainfall
   duration intensity, 99
depth–area–duration data, 140
depth–duration–frequency data, 117
design storms, 113
detention volumes, 450, 531, 698
determinant function, 30
deterministic models, 683
dew point, 551
diffusion, 549, 608
diffusion coefficient, 550
diffusion equation, 240, 328, 476, 478, 480, 581, 608
Dirac delta function, 14, 22–23, 350
direct runoff, 226
direct runoff hydrographs, 349
discharge coefficient, 521, 523–525
dispersion, 609
dispersion coefficient, 609
displacement height, 553
distributed models, 683
distributed routing method, 462
Doppler effect, 95
drainage area, 226
drainage density, 248
drainage net, 238
drainage pattern
   annular, 250
   centripetal, 250
   dendritic, 249
   deranged, 250
   parallel, 249
   radial, 250
   rectangular, 250
   reticulate, 250
   trellis, 249

Dublin Principles, 652
　Dublin Statement, 652
Duhamel integrals, 21
Dupuit approximation, 328
duration of unit rainfall, 373
durational envelope, 143
dynamic waves, 455
dynamic wave celerity, 461

effective porosity, 250
effective rainfall, 288
effective rainfall hyetograph, 271
effective rainfall ratio, 638
elevation heads, 275
elongation ratio, 233
emissivity, 561, 578–579
end area method, 517
energy slope, 424
equidistant conic projection, 668
equifinality, 692
equilibrium hydrographs, 429
equilibrium vapor pressure, 550
equivalent evaporation, 556
equivalent slope, 242
erosion control practice factor, 592
erosion index, 592
event-based models, 683
exceedance, 49
exceedance probability, 48
　annual exceedance probability, 49
excess infiltration, 277
expected moments algorithm (EMA), 202
expected value, 52
exponential distribution, 62
exponential functions, 13
extratropical cyclones, 98
extreme value (EV) distributions, 62
extreme value type I (EV I), 62, 75, 118
extreme value type III (EV III), 62

F distribution, 68
feature attribute table, 661
features classes, 660
Fick's first and second laws, 550
field capacity, 275
firm yield, 622
flashy hydrographs, 654
flood control zone, 617
flood hydrographs, 179, 349
flow-duration analysis, 185
flow-duration curve, 186
flow regimes, 181
flow regulators, 520
Fourier series, 27
Fourier transform, 18, 28, 78, 85
fractiles, 45
frequency distribution, 37
frequency curves, 40
frequency factors, 1, 59–61, 128, 145, 161, 197–198, 200, 224
friction slope. See energy slope
frontal rainfall, 97
frontal surfaces, 97
frustum of a pyramid, 518–519

gamma distribution, 2, 61, 63, 196, 223–224, 361–363, 365, 371, 385, 403–405, 409–410, 420
gamma function, 18
Gangotri Glacier, 632
gas constant, 550
gauge height. See stage
Gaussian distribution, 22, 56, 181, 645
general moments, 42
generalized extreme value (GEV) distribution, 118
geoid (shape of Earth), 665
geometric distribution, 55
geometric network, 662
global optimization, 686
gradation curves, 604
gravity waves, 455
Green–Ampt equation, 279–280
ground roughness, 553
groundwater, 324
gully erosion, 591
Gumbel distribution, 1, 62, 118–119, 122, 128, 133, 161, 167, 195, 198, See also extreme value type I distribution

Hack's law, 234
heat capacity, 574
heat index, 560
heat of transformation, 553
Heaviside step function. See step function
HEC-ResSim (software program), 620
HEC-SSP (software program), 191, 203
homoscedastic variables, 74
horizontal orifices, 522
Hortonian flow, 226
Huff curves, 110
hurricanes, 97
hydraulic conductivity, 226, 241, 247, 250, 274, 276, 278, 281–282, 284–286, 289, 304–305, 312, 316, 325–326, 328, 585, 684
hydraulic depth, 245
hydraulic geometry, 245
　at-a-station, 247
　downstream, 247
hydraulic heads, 275
hydraulic routing, 462
hydrodynamic dispersion coefficient, 609
hydrologic routing, 462
hydrologic soil groups, 289
hydrologic soil cover complex, 291
hydrometers, 604
hyetographs, 105
hyetographs, 105
hygrometer, 551
hygrometric constant. See psychometric constant
hypsometric curves, 236

identity matrix, 30
Idso–Jackson equation, 561
impulse function, 21, 350
impulse response function, 20–21, 350–351, 392
inactive zone, 617
index flood. See index flood method
index flood method, 204
index hydrographs, 371
infiltration capacity, 277

infiltration equation, 277
　Holtan, 277
　Horton, 277
　Philip, 277
infiltration rate, 250
infiltration-excess runoff, 226
infiltrometers, 308
initial abstraction, 227, 288, 302
initial abstraction ratio, 288
initial condition. See initial value problem
initial value problem, 25
instantaneous unit hydrograph, 374, 376, 387
integral transform, 18
Integrated Water Resources Management (IWRM), 651
integrating factor, 24, 35, 393
intensity–duration–frequency (IDF) curves, 117
interflow, 253, 324
interior drainage basins, 227
invariable relationship, 454
　variable relationship, 454
inverse cumulative distribution function. See percentile function
inverse of matrix, 30
isochrones, 250, 374
isohyets, 102
　isohyetal maps, 102
isopluvial maps, 122

joint probabilities, 47, 54, 633

Karakoram Anomaly, 633
Kendall's τ, 76, 203
kernel function, 18, 392
kinematic wave celerity, 461
kinematic wave parameters, 424
Kleitz–Seddon law, 460
Kolmogorov–Smirnov test, 72
kriging, 102
kurtosis, 44

lag method, 253
lag time, 250
Laplace equation, 327
Laplace transform, 18–19, 32
lapse rate, 551
　dry adiabatic lapse rate, 551
　environmental lapse rate, 551
latent energy, 553
latent heat, 553
latent heat of fusion, 575
latent heat of vaporization, 553
law of stream numbers, 238
length ratio, 240–241, 258, 387
linear differential equation, 23
linear functions, 20
linear interpolations, 530
linear mapping. See linear function
linear operations, 20
linear reservoirs, 458
linear storage, 458
linear system, 20
liquid water content, 575
L-moments, 63–64
local acceleration, 455

logarithmic functions, 14
logarithmic law of wind speed, 553
lognormal distribution, 61–62, 83, 189, 197, 363
longest flow path, 289
lumped models, 683
lumped routing method, 462
lysimeters, 565

Mann–Kendall test, 203
map projections, 666
mass curves, 105
mathematical models, 683
matrix norms, 32
maximum relative error. *See* relative error
mean (of a sample), 43
mean annual floods. *See* index flood method
mean areal precipitation, 100
median (of a sample), 45, 54, 61, 111, 119, 193–194, 203–204, 243, 288, 291, 294, 321, 382, 602–603, 606, 612, 697
melt factor, 585
metamorphism of snow, 576
method of moments, 63
  product moments, 63
midsection method, 174
mode (of a sample), 45
modified Puls equation, 463
MODSIM (river basin flow model software program), 647
moisture adjustment factor, 141
moisture content, 274, *See* soil moisture content
moisture deficit. *See* soil moisture deficit
moisture maximization factor, 141
moments about the origin, *See* general moments
monoclinal waves, 460
monsoonal rainfall, 97
Muskingum equation, 459, 468
mutually exclusive events, 48

NAD 27 (North American Datum of 1927), 665
NAD 83, 665
Nash and Sutcliffe coefficient of efficiency, 76, 685
Nelder and Mead algorithm, 688
noncentral moments. *See* general moments
normal distribution, 51, 56, 58–62, 66–70, 73, 77, 81–83, 85, 194, 196, 222, 224, 361, 420
normal equations, 75, 360
normalized error statistics, 76
normalized mean-square error. *See* normalized error statistics
null hypothesis, 66

objective function, 314, 360, 619, 684–689, 691–694, 709–711, 713, 715
ogee-crested weirs, 524
optimization algorithm, 686
order statistics, 64
orographic precipitation, 146
orographic rainfall, 98
outlets. *See* pour point
outstanding storms, 138

pan coefficient, 557
parameter estimation, 56, 62

parametric hydrology, 4, 11
partial duration series, 117
partial pressure, 550; *see also* vapor pressure gradient
peak rate factor, 373, 401, 404
peak-discharge power law, 229
PeakFQ (software program), 203
peaking coefficient, 364
Pearson type III distribution, 196
Pearson's correlation coefficient, 74
percentage mass curves. *See* mass curves
percentile function, 54
percentiles, 45
periodograms, 78
persistence (in time-series analysis), 78
plotting position coefficients, 53
plotting position formulas, 53
porosity (of soils), 274
potential evaporation, 549
potential evapotranspiration, 6, 302, 549, 559–560, 638, 647, 649–650, 656, 703
potential infiltration volume, 277
potential transpiration, 549
potentiometric surfaces, 326
pour point, 226
power functions, 13
probability density, 51
probability density functions, 50
probability distribution, 50
probability distribution functions. *See* probability distribution
probability estimates, 50
probability mass functions. *See* probability distribution
probability-weighted moments, 63–64
probable maximum floods, 147
probable maximum storms, 147
product moments, 63, 74–75, 118, 203
psychometric constant, 554
psychrometers, 551
psychrometric charts, 551
pulse input functions, 351
pyranometers, 580

quantile estimations, 124
quantile functions, 53, 64, 121, 195
quantile–quantile (Q–Q) plots, 62
quantiles, 44–45

radial gates, 521
radiometers, 564, 580
rainfall and runoff factors, 592
rainfall excess, 236
random variables, 36
rate of excess infiltration. *See* excess infiltration
rating curves, 175
rating curve constants, 175
reach relief, 242
recession constants, 329
recurrence intervals, 49
  annual recurrence intervasl, 49
reduced variates, 118
reference evapotranspiration, 549
regional skew coefficient, 198
regression techniques, 74

rejection region, 66
relational database, 659
relative conductivity, 250
relative error, 16
relative frequency, 37
  cumulative relative frequency, 37
relative humidity, 550
relief ratios, 234
required lag time, 364
required unit hydrograph, 364
residuals (in regression analysis), 74–75
residual moisture content, 250
residual saturation, 282
restricted degree day factor, 582
retardance coefficient. *See* storage coefficient
return periods, 48
Richards equation, 232
Richardson number, 583
RiverWare (software program), 647
root mean square, 43
root-mean-square error, 76, 685
routing coefficients, 376, 469
rugosity coefficient. *See* rating curves
Runge–Kutta method, 529
runoff coefficients, 442
runoff ratios, 636

Saint Venant equations, 455
samples
  random samples, 36
saturated zones, 325
  phreatic zones, 325
saturation vapor pressure, 550
saturation-excess overland flow, 253
scale parameter. *See* parameter estimation
scale invariance, 231
  scaling exponent, 231
  scaling intercept, 231
schemas, 659
seasonal hydrographs, 181
sedigraphs, 598
sediment delivery ratio, 596
sediment discharge, 598
sediment discharge rating curve, 600
sediment yield, 595
serial correlation, 203, *See* autocorrelation
service spillways. *See* principal spillways
shallow concentrated flow, 255
shape parameter. *See* parameter estimation
shape factor, 233, 606
shear stress, 606
sheet flow, 254
shifting control. *See* rating curves
shuffled complex evolution, 690
Siachen Glacier, 633
sieve analysis, 604
significance levels, 66
singular matrices, 31
sinks. *See* pour points
skewness, 44
slope length factor, 592
slope ratio, 241
sluice gates, 521
snow water equivalent, 575

Snyder standard unit hydrograph, 364
soil erodibility factor, 592
soil heat flux, 558
soil mass movement, 591
soil matric potential, 275
soil moisture content, 227
soil moisture deficit, 275
soil water diffusivity, 277
solar constant, 561
solarimeters, 564
sorptivity, 277
Spearman's ρ, 76
specific discharge, 328
specific heat, 574
specific humidity, 550
specific storage, 326
specific water availability, 647
specific yield, 326
spectral density, 78
spillways, 517
   controlled, 521
   emergency, 520
   principal, 520
   uncontrolled, 521
stage, 171
stage–discharge relationship, 175
stage-duration curves, 193
standard deviation, 44
standard error of the estimate, 75, 202
standard normal variables, 58
standard normal variates, 58, 68, 82, 224
standard unit hydrographs, 364
station constants. *See* rating curve constants
station skew coefficient, 198
stationary time series, 203
statistical hypothesis, 66
   type I error, 66
   type II error, 66
Stefan–Boltzmann constant, 561, 578
Stefan–Boltzmann law, 564, 578
step functions, 22
step response function, 350
stilling wells, 171
stochastic convergence, 46
storage coefficient, 376
storage constant, 456
storage-indication curve, 463
storage-indication routing equation, 463
storativity, 326

storm hydrographs, 179
storm profiles, 110
straight-line method, 329
stream area ratio, 241
stream frequency, 276
stream length ratio, 240
stream power, 242
Strickler's formula, 602
support practice factor, 592
surcharge zone, 617
surface erosion, 591
surface flow, 253
surface roughness height, 582
surface storage, 300
synoptic scale, 98
synthetic rainfall time distribution, 129
systems, 3
   closed systems, 4
   isolated systems, 3
   open systems, 4

t critical value, 69
t distribution, 68
Tarbela Reservoir, 635
Taylor's series, 16, 18, 20
temperature index method, 586
test statistic, 66
Theil slope, 203
thermal conductivity, 584
Thiessen polygons, 101
time domain, 19
time of concentration, 250, 430
time of equilibrium, 429
time series, 171
time to equilibrium, 252
time to peak, 371
time–area method, 374
time–area relationship, 374
topographic factor, 592
topology, 662
transfer function, 20
transient flow, 328
transmissivity, 326
travel times, 250
triple point (of water), 576
tropical cyclones. *See* cyclonic rainfall
tropical depressions, 98
Tropical Rainfall Measuring Mission (TRMM) satellite, 96

tropical storms, 98
type curves, 130
   center peaking, 130
typhoons. *See* hurricanes

unfilled capacity, 277
uniformly progressive flow, 456
unit hydrographs, 351
   effective duration, 353
   lag times, 354
   time base, 355
   time of concentration, 354
   time to peak, 354
unit impulse function. *See* impulse function
urban hydrology, 702
urbanization curves, 366

vadose zone, 274, 325
valley storage, 374
vapor pressure, 550
vapor pressure deficit, 551
vapor pressure gradient, 550
variable source areas, 253
variance, 44
variates, 36
   continuous, 36
   discrete, 36–37
velocity method, 254
velocity–area method, 174
von Karman constant, 174, 553, 581

warm fronts. *See* frontal surface
Water Availability Modeling (WAM), 647
water equivalent. *See* snow water equivalent
water rights, 646
Water Rights Analysis Package (WRAP), 647
water table, 325
water years, 182
watersheds, 226–227
watershed concept, 3
watershed lag times, 364
watershed slope, 234
Weibull distribution, 53, 62, 118–119, 122, 168, 188, 194, 361
wilting point, 253

zero-plane displacement, 581